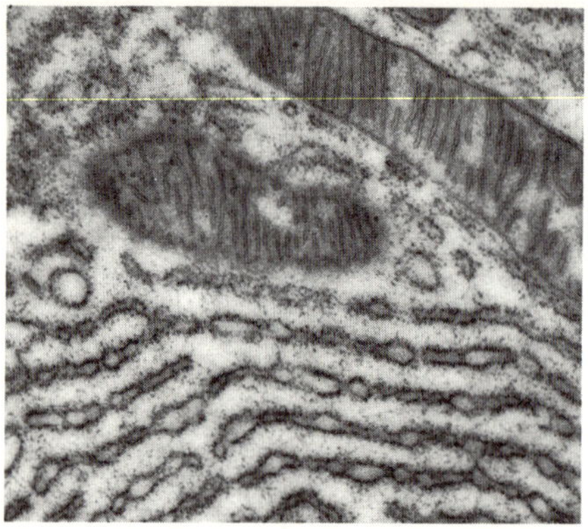

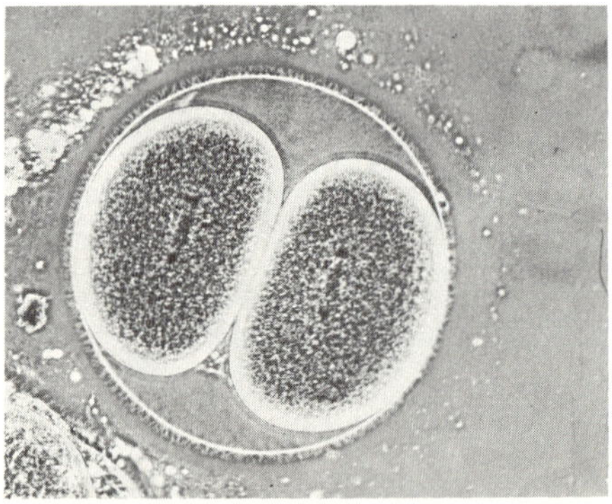

*Top left:* Mitochondria (at the top, the structures with lamellae) and the endoplasmic reticulum of an acinar cell of the pancreas (×20,000). (By courtesy of Dr. G. E. Palade, The Rockefeller University, New York.) *Top right:* The fracture faces of membranes of a freeze-etched spinach chloroplast. Most of the faces are those of internal membranes. Particles 170Å × 90Å are seen on many of the membranes (×90,000). (By courtesy of Drs. D. Branton and R. B. Park, University of California, Berkeley.) *Bottom:* A live dividing egg of *Pateria miniata,* a West Coast starfish, showing the fertilization membrane, phase optics (×185). (By courtesy of Dr. Richard Boolootian, BSCS, Boulder, Colorado.)

**W. B. SAUNDERS COMPANY**

# CELL PHYSIOLOGY

THIRD EDITION

ARTHUR C. GIESE, Ph.D.

Professor of Biology,
Stanford University

PHILADELPHIA · LONDON · TORONTO

W. B. Saunders Company: West Washington Square
Philadelphia, Pa. 19105

12 Dyott Street
London W.C. 1

1835 Yonge Street
Toronto, Ontario 295

*Spanish edition*

Editorial Interamericana, S.A.
Cedro 512, Apartado 26370
Mexico 4, D.F.

*Russian Edition*

Publisher: Izdatelstvo Inostannoy
(Publisher of Foreign Works)
Moscow, Russia

Reprinted November, 1968 and June, 1969

Cell Physiology

*To Raina*

# *Preface to the Third Edition*

*Cell Physiology* continues to be a text for the advanced undergraduate student and so to present an introductory course in the subject.

With the third edition of *Cell Physiology* I have tried to maintain the original aims of the book as outlined in the Preface to the First Edition, that is, to attempt to present a progressive development of the complex problems of the physiology of the cell in as simple terms as possible. Cell physiology is growing at an ever accelerating pace and new, important findings are appearing continually. In order to keep the book within a useful size it was necessary to eliminate some material and to select carefully for inclusion only such material as was considered most pertinent.

In a sense a textbook of such a rapidly growing field of study as that covered here can never be entirely up to date, for new progress is being made even as these pages come off the press. *Cell Physiology* is based on the belief that the basic principles and techniques, systematically introduced in connection with a selection of material, will show the relationship between the two and will appeal to the reason of the student rather than to his memory. The new facts and ideas with which he may later come in contact he should be able himself to fit into the framework presented here.

For graduate courses in the field texts of a more advanced nature and with complete documentation are available. There are also numerous books covering selected topics in cell physiology at an advanced level.

It is considered within the appropriate province of the instructor to change the order of topics, to amplify and elaborate upon a given discussion and to present the last-minute news items to his class, if he so wishes. A book which made this unnecessary would be a disservice to him, removing the joy of personal participation from his teaching.

I am indebted to Miss Pauline Schimberg for patient and understanding typing from difficult copy and to Mrs. Bonnie Pagon for a bibliographic search. Most of all, I am indebted to my wife, Raina, for her critical editing of the manuscript.

I am indebted to the many instructors who have used my book and have written criticisms, especially to Dr. R. M. Benolken of the University of Texas and to Dr. R. R. Ronkin of the University of

Delaware, and to Dr. K. Marilyn Smart of the Adelphi University. I have tried to incorporate such suggestions as were deemed within the aims of the book. Some were, however, suggestions for modifications of the book in diametrically opposite directions. Two requests adopted here are the bibliographic citation by authors and date instead of by numbers and the inclusion of complete titles in the bibliography. Inclusion of names in the body of the text may somewhat impede the flow of the material but it emphasizes that the facts under consideration were gathered and the theories presented were formulated by a large number of individuals in many laboratories. To reduce the number of citations many older references have been deleted when newer, more inclusive ones were available. A third request resulted in repositioning Chapter 13 in edition 2, Application of the Laws of Thermodynamics to the Cell, to Chapter 1 in edition 3, because the subject matter is so basic to all of cell physiology.

I am also indebted to many colleagues and to various publishers for permission to use figures and tables from other publications.

Arthur C. Giese

*Stanford, California*

# Preface to the First Edition

"Cell Physiology" as presented here is patterned after a set of lectures as they are given in my class at Stanford University. Although several excellent textbooks in cellular physiology are available and a number of good books bordering on the field have appeared, there exists no single book which gives a brief account of the subject. Students have asked repeatedly that this need be fulfilled; that a book be written which describes in simple language and in bold outline the major problems of cellular physiology, explaining their interrelationships and the current status of each of them without confusing the beginner with details or taking him into controversies upon which even the experts cannot agree.

Introducing a topic in cellular physiology by presenting controversial issues has often left students without an anchorage of fact by which to evaluate anything at all. On the other hand, I have been impressed by the tendency of many students to accept uncritically and to memorize anything written in a textbook. It is certainly not possible to avoid conflicting information on some subjects, since an open mind must be maintained on all real issues; but controversies on less important points can be minimized so that the major achievements in cellular physiology are not completely obscured. To help the student understand the scientific approach and develop an attitude of critical evaluation, I have found it extremely useful to include fundamental laboratory work and to conduct weekly discussions focusing on more specific controversial problems. And it is expected that after this primary orientation in the subject, students may become interested enough to turn to the more complete and analytical studies cited in references at the end of each chapter.

I have, moreover, confined the subject matter in this book to that dealing primarily with the cell, since this is basic to studies on multicellular organisms. By avoiding treatment of problems which relate to the organization of cells into organisms, it is possible to develop a more closely-knit body of information, such as could be presented to a class in the course of a quarter or a semester. Classes in plant physiology, comparative animal physiology, mammalian physiology and bacterial physiology are logical sequels to the course in cellular physiology.

At Stanford a one quarter course in cellular physiology is required of biology majors, most of whom take it at the junior or senior level. The student will have taken elementary botany, zoology, physics and introductory and organic chemistry. He is, therefore, unprepared for a rigorous physical-chemical treatment of the subject. However, the experience of the department over the years seems to have justified the premise that an introduction to cellular physiology at the elementary level is more beneficial than restriction of the course to a handful of graduate students with a broader background. Consequently, derivations of equations on physical-chemical principles or more extended chemical discussions, if included at all, are put in appendices because this makes the material available to those students who are really interested without impeding the flow of the subject for those who are not.

I am indebted to Miss Ruth Ogren, scientific illustrator, for preparing many of the illustrations, and to Professor Hadley Kirkman of Stanford University, Professor David Waugh of Massachusetts Institute of Technology, and Dr. Richard Boolootian for use of some photographs. The courtesy of numerous publishers who have permitted copying of figures from books or periodicals is also acknowledged with gratitude. I am also indebted to Dr. George Palade of the Rockefeller University for electron micrographs used in the frontispiece, and to Dr. John Bennett, for the figure of the metabolic mill.

I wish to express my appreciation for many helpful comments and criticisms made by my colleagues and assistants, and especially to Professor J. P. Baumberger of Stanford University, Professor Jack Myers of the University of Texas, Professor John Spikes of the University of Utah, and to my students, Dr. Ray Iverson, Dr. David Shepard and Dr. Raymond Sanders, for a critical reading of the manuscript or parts of it. However, I must assume responsibility for opinions expressed and for omissions or errors which may still remain after proofreading. My main hope is that students will find this account interesting and stimulating.

ARTHUR C. GIESE

*Stanford, California*

# Contents

# INTRODUCTION / THE SCOPE OF CELL PHYSIOLOGY

## LIFE ACTIVITIES

Cell physiology is the study of life activities of cells—plant cells, animal cells and microorganisms. The study of these activities has been a very vital one in the past few years. The present era is to the biologist what the beginning of the twentieth century was to the physicist, for never before in the history of biology has so much been achieved in so short a time toward an understanding of the fundamental problems of the living cell.

The life activities of the cell might be enumerated as follows: (1) nutrition, including in its broadest sense the uptake and preparation of food and oxygen for use in the cells, the release of energy in cells and the elimination of the wastes produced, (2) response to the environment, (3) growth and (4) reproduction or cell division, and differentiation.

The briefest comparison of plants, animals and microorganisms serves to show the fundamental similarities of their functions despite differences in structure. The body of a seed plant is composed of cells organized into a number of complex organs: root, stem, leaf, flower and, in the mature plant, the fruit with the developing seed. Through the root hairs water and nutrient salts pass into the vascular tissue and are carried through the stem to the leaves. Carbon dioxide enters through the stomata of the leaves, and during the light of day carbohydrates are synthesized from carbon dioxide and water in the chlorophyll-bearing cells. From the carbohydrates and salts the plant later synthesizes other compounds such as amino acids, fatty acids and water-soluble vitamins. These manufactured foodstuffs are then distributed in a solution (the sap) to cells in all parts of the plant, and the excess may be stored in the seed, fruit, root, stem or, less frequently, leaf. The vascular system and the surrounding woody fibers serve as a skeleton supporting the entire plant.

Although photosynthesis occurs only in some special cells of the seed plant, all cells of the plant body must be supplied with cell foods, and they must be freed of cell wastes. Plants grow, using part of the food manufactured, and after reaching maturity they reproduce, using more of it. Growing plants respond to gravity, to light and to chemicals and other factors in their environment. Thus the root grows downward in response to gravity, whereas the stem grows upward; the root grows away from the light while the stem grows toward it (for instance, everyone has seen the sunflower follow the sun). The response of plants to tactile stimuli is seen in the twining tendrils which fasten some plants to a support. Movements of plants are mainly growth movements, and although they are too slow for us to see, they can be readily demonstrated in time-lapse photography. The more rapid turgor movements can be seen in the sensitive plant *Mimosa.*

The essential architectural plan of a vertebrate, a triploblastic form (embryonically three-layered), consists of an outer tube, the body wall, enclosing an inner tube, the gut, and supported by an internal skeleton consisting of bones. The head constitutes the anterior end and contains many of the sense organs or receptors. The trunk, consisting of a body wall and the internal organs in it, is supported by the appendages. The internal organs are organized into various organ systems. For example, the organs of the digestive system prepare food for use in the body, and the digested food is absorbed into the blood and is distributed over the entire body by the organs of the circulatory system. As in the plant, regardless of the organ in which it is found, the individual cell requires the same essential cell foods. The respiratory system takes up the oxygen which is distributed to all cells by way of the circulatory system, and it voids to the air the carbon dioxide brought to it by the circulatory system. The blood flushes wastes from cells in all parts of the body, carrying these wastes to the kidneys where they are extracted from the blood and voided to the exterior. The animal may be considered to have developed a set of services of supply—digestive, respiratory, circulatory and excretory—which maintain the cellular environment. The needed supplies of oxygen and food are brought to the cells, and the wastes of cellular activity are removed from the cells and voided to the exterior. By way of receptor organs (such as the eyes and ears) sensitive to light, sound, pressure, gravity, heat or chemicals, the animal receives information about the environment. Response to the information so gathered is made possible by a nervous and muscular system supported by a bony skeletal system. After reaching sexual maturity the animal reproduces.

As can be seen, the same fundamental activities of life—nutrition, response, growth and reproduction—are therefore performed by plants and animals, though in somewhat different ways.

However, complex organs like those found in higher plants and animals are not necessary for the performance of the fundamental biological functions in all organisms. Respiration takes place in bacteria, protozoa and unicellular plants, all without a respiratory system; likewise, digestion can occur without the aid of a digestive system and excretion, without the aid of an excretory system. Microorganisms get food and oxygen directly from the environment and void their wastes directly to it. They respond to the environment, and if conditions are favorable, they grow and reproduce.

The common denominator required for the performance of these fundamental activities of plants, animals and microorganisms is clearly the cell, not the complex of organs and systems which has developed in multicellular plants and animals. These systems are concerned primarily with maintaining the internal environment in contact with the cells of the organisms. It is through the cell of the multicellular animal and plant that all the flow of matter and flux of energy occurs. The cell is a fundamental unit, not only of structure but also of function of all living things.

## VITALISTIC AND MECHANISTIC INTERPRETATION OF LIFE ACTIVITIES

The mind of man demands an explanation for the phenomena of nature. Primitive peoples developed mythologies which conceived of the earth and man as being controlled by various supernatural forces. Such forces were later suggested to explain biological phenomena. Accounts of these early conceptions may be found in various histories of biology (Nordenskjold, 1936; Singer, 1950; Dawes, 1952).*

*References to the literature cited are given in the back of each chapter. General references, often not cited, are included among them. Only a few of the general references are starred to indicate a few books or reviews covering the material of the chapter. Other titles make possible additional general reading.

As information was accumulated from observation and experimentation, it eventually became clear that the grosser functions in the animal and plant, such as digestion, respiration, circulation and excretion, are subject to the laws of physics and chemistry and some theorists, the "mechanists," think that all the phenomena of life are thus explainable. Others, the "vitalists," maintain that although digestion, respiration, excretion and circulation might be interpreted in terms of chemical and physical laws, more complex processes such as the orderly development of the embryo and the complex mental life of man can be interpreted best in terms of something apart from and in addition to the physical-chemical laws. They postulate a vital force, categorically equivalent to matter and energy, to perform this role (Sinnott, 1953).

The cell physiologist finds it profitable to operate on the working hypothesis that the physiological activities of the cell are the consequence of the operation of a physical-chemical system which constitutes the cell. If the functions of the cell are the consequences of the operation of a physical-chemical system performing work, then the cell may be likened to a machine. It is the purpose of the cell physiologist, first, to determine the nature of the machinery of the cell and, second, to determine the principles which govern its operation, with the hope of better understanding its functioning. For this reason the book begins with a consideration of thermodynamics and energy flow in biological systems. To do its work the cell must liberate chemical energy from cell foods and couple the energy-liberating reactions with energy-requiring reactions. The mechanistic hypothesis has been fruitful in stimulating experimentation and accumulating useful information (Reichenbach, 1953).

## EXPERIMENTAL APPROACH IN CELL PHYSIOLOGY

Cell physiology employs the experimental method, many examples of which will be given in later chapters. The solution to a problem in science is often achieved by slow and laborious collection of data. Pertinent observations are made and data already available are focused on the problem in question and a possible explanation, a hypothesis, is formulated. Carefully planned experiments are then prepared for testing it, and it may happen that after much work an investigator finds that the attempted explanation is inadequate and that he has only cleared the ground for further analytical work. An alternative hypothesis is then constructed and tested. Occasionally decisive experiments result in dramatic advances in knowledge and understanding (Gabriel and Vogel, 1955; Suñer, 1955). Such decisive experiments have occurred more frequently in physics and chemistry, in which the total number of variables is fewer. In the biological system, in which there are a great number of variables, similar progress in theoretical knowledge may be expected only as the physiologist gains control of these factors.

The recent analysis of the structure of proteins and nucleic acids and the deciphering of the genetic code are considered by some to be such advances in biology. These findings have created a revolution in biological thought such as had not occurred since the concept of evolution had stirred the biological world. Although it is difficult to evaluate a matter so close at hand, the birth of a whole new field called *molecular biology* resulting from these discoveries and its rapid growth attest to the significance of the findings (Maddox, 1964).

Experimentation makes possible selection of the most likely explanation for a phenomenon. As the unlikely explanations are eliminated, the remaining one becomes more probable, but still subject to revision as more decisive data become available.

By the use of the experimental method, the following cellular problems have been studied:

1. The nature of the organization of cells of plants, animals and microorganisms as a key to an understanding of cellular activities.
2. The nature of the cell environment

and the resistance and adjustment of the cell to the variations of the environment.

3. The nature of the cell membrane and the regulation of transport of materials into and out of the cell across the membrane.
4. The nature of cell foods and their interconversions and the mechanism by which the process of respiration releases energy from cell foods.
5. Use of the energy liberated in respiration for performing the various types of work done by the cell, e.g., for the maintenance of irritability or readiness to respond, for motility required for the responses, for osmotic work and for the manufacture of secretions and of new constituents leading to cell growth and cell division.

A consideration of each of these problems in turn constitutes the major portion of this book. This approach leads from the problems of organization of the cell to cellular dynamics. It also leads from the more descriptive aspects of cell physiology to the more complex studies of respiration, irritability, growth and cell division. It has the advantage that much of the descriptive information is already available when the more difficult problems are analyzed. At the end, a brief chapter on the history of cell physiology brings into perspective some of the major trends of research in the field.

## THE RELATION OF CELL PHYSIOLOGY TO VARIOUS FIELDS OF PHYSIOLOGY

Much of *animal physiology* deals with the work of the various organs of the body and with the coordination of the various organs in the integrated behavior of the animal as a whole. Since the animal physiologist is concerned with functions of organs his study might be called organ physiology. Medical men in the past have been primarily concerned with vertebrate organ physiology, especially of mammals, because the information gained can be more directly applied to the physiology of man in health and disease. Therefore, medical physiology most often meant mammalian organ physiology. However, to interpret organ physiology, research workers in medical physiology have often found it necessary to work at the cellular level as well. Therefore much overlapping occurs at the present time.

*Comparative animal physiology* is a study of organ functions in various types of animals, vertebrate and invertebrate, in an attempt to find fundamental relations not recognized when the function is studied in only one group of animals. The different ways in which each function may be performed in various animals are therefore viewed in the perspective of the whole animal kingdom.

*Plant physiology* concerns itself with the response, nutrition, growth and reproduction of various types of plants. Because many of these problems first required solution at the cell level, plant physiologists have contributed greatly to the development of cell physiology.

Since functioning of animals and plants depends ultimately upon the functioning of their component cells, an understanding of cell physiology is basic to all the other physiological studies. This has been recognized by researchers, many of whom, interested primarily in the higher organisms, have worked at the cell level as a background for solution of problems at the organ level.

## THE LITERATURE OF CELL PHYSIOLOGY

The most recent achievements in cell physiology and data on current studies are presented at scientific meetings and are recorded in current issues of such periodicals as *Science, Nature,* the *Journal of General Physiology, The Journal of Cellular and Comparative Physiology,* the *Journal of Experimental Biology, Experimental Cell Research, Physiological Zoology, Biochimica et Biophysica Acta, Proceedings of the National Academy of Sciences, U. S., Archiv für die gesampte Physi-*

*ologie, Compte Rendu de la Société de Biologie, Doklady Academii Nauk SSSR* (Biological Sciences Division), *Biofizika* and many others. Some of the Russian journals are now available in English translation.

Reviews or summaries of recent advances on subjects of interest to the cell physiologist appear in such journals as the *Quarterly Review of Biology, Physiological Reviews, Biological Reviews of the Cambridge Philosophical Society, Annual Review of Physiology,* annual reviews in a large number of specialized fields in cell physiology, symposia on special subjects (e.g., permeability, metabolism), and others. Some reviews will be cited in the subsequent chapters.

There are many good monographs on various special subjects of interest to the cell physiologist, e.g., on protoplasmic structure, permeability, oxidation-reduction and respiratory enzymes, bioluminescence, muscle chemistry and physiology. In this book reference will be made to such monographs when they are pertinent to the subject matter being discussed in a given chapter.

Compendia and textbooks cover much larger fields than monographs, and consequently their treatment of specific topics is usually less complete. Reference is given at the end of this chapter to textbooks of this kind which have been found most useful. Most of them will not be specifically cited again in this book, but they are fruitful sources of additional material, especially in the fields designated.

If a bibliography is to be compiled on a subject of interest, journals which abstract periodicals are helpful. The most useful abstract journals for the cell physiologist are *Biological Abstracts* and *Chemical Abstracts, International Abstracts of Biological Sciences, Berichte über die gesampte Physiologie, Berichte über die Wissenschaftliche Biologie, Inhaltsverzeichnisse sowjetscher Fachzeitschriften: III B, Chemie und Biologie, B.A.S.I.C., Current Contents,* and *Bulletin de l'Institut Pasteur.* These journals not only give references to recent original papers, reviews and monographs, but they also furnish abstracts of the contents.

Included in the references at the end of each chapter are general references in which the reader may find additional information covering the general field of the chapter, usually from a different point of view. Citations to other literature support views expressed on subjects of lesser scope. They are included as an aid to those who may wish to get a more complete treatment of some topic which interests him especially.

## LITERATURE CITED AND GENERAL REFERENCES

Abercrombie, M., Hickman, C. J., and Johnson, M. L., 1962: A Dictionary of Biology. Aldine Publishing Co., Chicago.

Annual Reviews, 1965: The Excitement and Fascination of Science. Annual Reviews, Palo Alto, Calif. (A collection of autobiographical and philosophical essays.)

Bayliss, L. E., 1959–1960: Principles of General Physiology. I. The Physico-Chemical Background. II. General Physiology. John Wiley & Sons, New York. (A revision of a famous classic treatise.)

Bourne, G. H. (ed.), 1964: Cytology and Cell Physiology. 3rd Ed. Academic Press, New York.

Brachet, J. and Mirsky, A. E., 1959–1961: The Cell. 5 Vol. Academic Press, New York.

Butler, J. A. V., 1964: The Life of the Cell. Its Nature, Origin and Development. Basic Books, New York.

Davson, H., 1964: A Textbook of General Physiology. 3rd Ed. Blakiston Division, McGraw-Hill Book Co., New York. (Especially useful for discussions of permeability, muscular activity and nervous action.)

Dawes, B., 1952: One Hundred Years of Biology. Macmillan Co., New York.

DeRobertis, E. L. P., Nowinski, W. W. and Saez, F. A., 1965: Cell Biology. 4th Ed. W. B. Saunders Co., Philadelphia. (Especially useful for structure of cells.)

DuPraw, E. J., 1968: Cellular and Molecular Biology. Academic Press, New York.

Fawcett, D. W., 1966: The Cell: Its Organelles and Inclusions. W. B. Saunders Co., Philadelphia.

Florkin, M., and Mason, H. (eds.), 1960–1964: Comparative Biochemistry. Vol. 1, 1960; Vol. 7, 1964. Academic Press, New York.

Gabriel, M. L., and Fogel, S., 1955: Great Experiments in Biology. Prentice-Hall, Englewood Cliffs, N.J. (History.)

Goldstein, L. (ed.), 1966: Cell Biology. A Book of Readings. W. C. Brown, Dubuque, Iowa.

Heilbrunn, L. V., 1952: An Outline of General Physiology. 3rd Ed. W. B. Saunders Co., Philadelphia. (Especially useful for the older literature.)

Heilbrunn, L. V., and Weber, F. (eds.), 1953–1959: Protoplasmatologia, Handbuch der Protoplasmaforschung. Springer-Verlag, Vienna. (Fourteen volumes in English, French and German.)

Kennedy, D., 1965: The Living Cell. Readings from Scientific American. W. H. Freeman & Co., San Francisco.

Lenhoff, E. S., 1966: Tools of Biology. Macmillan Co., New York.

Loewy, A. G., and Siekevitz, P., 1963: Cell Structure and Function. Holt, Rinehart & Winston, New York.

Maddox, J., 1964: Revolution in Biology. Macmillan Co., New York.

McElroy, W. D., 1964: Cell Physiology and Biochemistry. 2nd Ed. Prentice-Hall, Englewood Cliffs, N.J.

Nobel Prize Lectures in Physiology-Medicine, 1901–1962. 3 Vol.; in English, 1965: American Elsevier Publishing Co., New York.

Nordenskiöld, E., 1936: The History of Biology. Tudor Publishing Co., New York.

Paul, J., 1964: Cell Biology, A Current Summary. Stanford University Press, Stanford, Calif.

Picken, L., 1960: The Organization of Cells. Clarendon Press, Oxford.

Pledge, H. T., 1959: Science Since 1500. Harper & Row, New York.

Prescott, D. M. (ed.), 1964–1966: Methods in Cell Physiology. Vol. 1, 1964; Vol. 2, 1966. Academic Press, New York.

Reichenbach, H., 1953: The Rise of Scientific Philosophy. University of California Press, Berkeley, Calif.

Setlow, R. B., and Pollard, E. C., 1962: Molecular Biophysics. Addison-Wesley Publishing Co., Reading, Mass.

Singer, C., 1950: A History of Biology. Schumann, New York.

Singer, C., 1959: A Short History of Scientific Ideas. Clarendon Press, Oxford.

Sinnott, E., 1953: Two Ways to Truth. Viking Press, New York.

Stern, H., and Nanney, D. L., 1965: The Biology of Cells. John Wiley & Sons, New York.

Suñer, A. P., 1955: Classics of Biology. Pitman & Sons, London.

von Bertalanffy, L., 1952: Problems of Life. John Wiley & Sons. New York.

# CHAPTER 1 / APPLICATION OF THE LAWS OF THERMODYNAMICS TO THE CELL

## ENERGY AND MATTER

Energy is the capacity to do work. There are two main types of energy: kinetic and potential. Kinetic energy of moving bodies is dealt with in physics and is of less concern at this point than the kinetic energy of molecules. *Kinetic* energy of molecules is energy of motion. The amount of molecular kinetic energy in a system is proportional to the absolute temperature of that system, and at absolute zero (−273.18° C.) kinetic energy ceases to exist. For every 10° rise in temperature the kinetic energy increases by the ratio of 10 to the absolute temperature (in degrees). The kinetic energy of molecules may be inferred from the Brownian movement (see Fig. 1.1) of very small but microscopically visible particles which move about more actively as the temperature rises. The nature of Brownian movement is attributed to the random motion of molecules colliding with the visible particles and imparting to them sufficient energy for movement. *Potential* energy is energy of position or bound energy, which can be released to produce kinetic energy. Potential energy may be electric, chemical, photic (light), atomic and positional, such as that of a stone on the top of a hill.

That conversions between kinetic and potential energy occur should be obvious in a qualitative way to anyone who has ever lit a fire or hoisted a heavy object to a shelf. On a macroscopic scale the relations between potential and kinetic energy are described by the fundamental principles embodied in the first and second laws of thermodynamics.

*Matter* is that which occupies space and of which objects are made. Matter is converted to energy in nuclear fission and fusion, as, for example, in the atomic bomb and in the sun, but in ordinary chemical reactions this does not occur.

For convenience in the study of thermodynamics some arbitrary portion of matter known as the *system* and its own immediate environment (or surroundings) is considered. A system may absorb energy from the environment beyond the system or liberate it to the environment, unless an attempt is made to insulate it from the

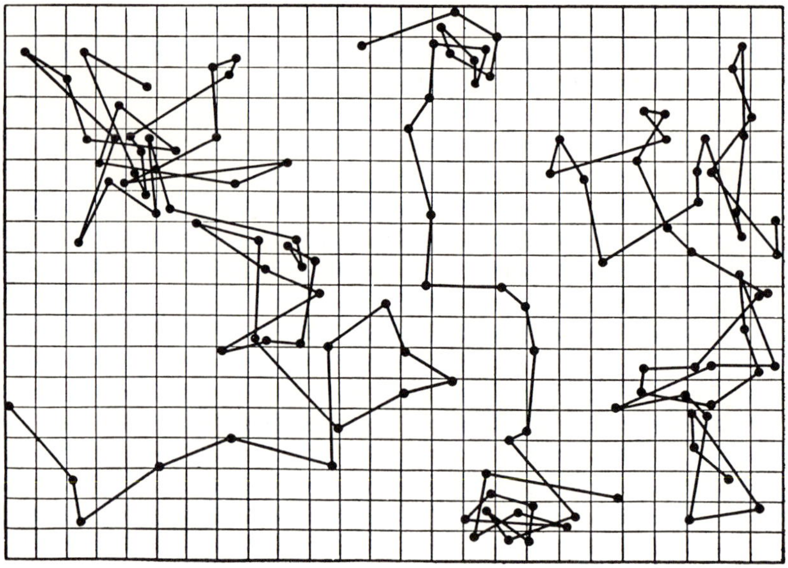

***Figure 1.1.*** Brownian movement of three particles. (From Perrin, 1913: Les Atomes. Alcan, Paris.)

environment, as in the case of a calorimeter, when it becomes an *isolated system.* The size of a system under consideration depends upon the needs of the problem at hand—it may be a single cell and its immediate environment or an organism and its environment, or it may be a larger fraction of the universe, the limit being the universe itself.

The laws of thermodynamics are thus independent of our conception of the nature and structure of matter. Classic thermodynamics deals with reactions in the equilibrium state and may therefore be called *equilibrium thermodynamics.* Equilibrium thermodynamics is not concerned with the rate of reaction. Rather, it is concerned with the change in form of the energy in a system from an *initial state* before the process has taken place to the *final state,* after the process has occurred. The number of steps taken between the initial state and the final state is also of no consequence, provided that the system arrives at the same final state. Equilibrium thermodynamics also deals with reversible reactions and indicates the direction of the reaction. For this reason the second law of thermodynamics has been called "time's arrow" by Sir Arthur Eddington. However, a new *nonequilibrium thermodynamics* dealing with steady states is being developed to deal with irreversible reactions (Prigogine, 1962). Classic thermodynamics also deals only with the macroscopic properties of matter, but a present development aims at the microscopic level (Tribus, 1966).

Because defining the energy content of a system at equilibrium in an equation of state either in the initial or final state requires specification of various properties of the system, many of which are difficult to ascertain (e.g., heat content, temperature, pressure, volume, composition), experimenters have generally concerned themselves with the *change* in energy content between the initial and final states of the system, a quantity much more amenable to measurement.

## THE FIRST LAW OF THERMODYNAMICS: THE LAW OF CONSERVATION OF ENERGY

The *law of conservation of energy* states that energy can be neither created nor destroyed but can change its form. For example, light absorbed on a black area, or electric energy in the resistance wire of a toaster, is converted to heat.

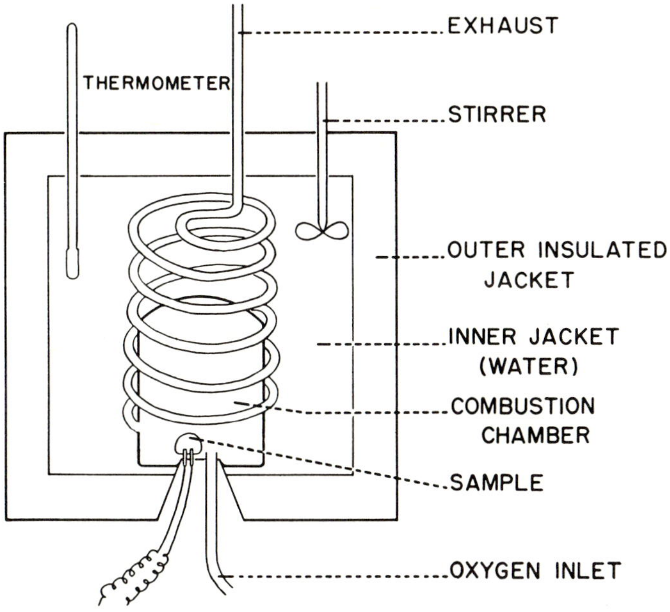


***Figure 1.2.*** A bomb calorimeter for determining heat of combustion of compounds.

If energy is neither created nor destroyed, but only changes form, it follows that the total energy in an isolated system will neither increase nor decrease. This is another way of stating the first law of thermodynamics. When a given amount of chemical, electrical or photic energy disappears in an isolated system it is transformed into an equivalent amount of heat.

Complete isolation of a system is somewhat difficult to attain because it requires a container across the boundaries of which heat neither enters nor leaves. A thermos bottle or Dewar flask, however, approximates such a container. Hot liquid placed in it will remain hot because the liquid is to some extent isolated from the surrounding environment by the vacuum between the two walls of the bottle. The greater the vacuum, the better the insulation, since heat is conducted only very slowly by the few molecules of air which remain in the vacuum. The degree of imperfection in the insulation is indicated by the time required for the liquid to cool. Heat is lost to the environment chiefly through the plug, although some is also lost by radiation.

For a study of energy exchanges in chemical systems a *calorimeter* is used. This may be nothing more than a large Dewar flask, or it may be a more elaborate piece of equipment following the same essential plan (see Fig. 1.2). The outermost jacket of the calorimeter is a good insulator and surrounds a chamber filled with water which absorbs the heat released by the sample in the combustion chamber, the innermost chamber. The water is actively stirred and its temperature is determined with a thermometer. Since perfect insulation of a calorimeter has not been attained, corrections must be made for the slight thermal leaks. Adequate corrections can be made by measuring the amount of electric energy required to produce an equal change in temperature in a comparable time lapse (Glasstone, 1960).

## THE SECOND LAW OF THERMODYNAMICS

Although the total energy in an isolated system neither increases nor decreases, the transformation of energy from one form into another proceeds in such a fashion that the capacity of the total energy to perform work decreases. That potential energy in the system capable of performing work tends to become converted to kinetic energy incapable of doing as much work is in essence a statement of the second law of thermodynamics. It tells us that spontaneous reactions have direction which is not explained by the first law of thermodynamics. Once a reaction has reached equilibrium the reactants cannot return spontaneously to the initial state.

Molecular potential energy is often spoken of as energy with a high degree of orderliness, whereas molecular kinetic energy is said to be energy of movement which may be random. As potential energy changes to kinetic energy, the randomness or disorder increases. A thermodynamic function known as *entropy* is used by the physical chemist to measure the randomness of the energy of a system that is less capable of doing work than the initial nonrandom energy. Randomness of arrangement in nature is a more probable state than a special arrangement; entropy is a measure of the randomness of arrangement in atomic mechanics. Therefore, entropy may be defined in terms of probability. Entropy tends to increase because an increase in randomness is more likely in the molecular world than the reverse. In other words, reactions proceed in such a manner as to reach maximum entropy. This is the point of equilibrium.

At absolute zero the entropy of a crystalline system is zero since the probability of randomness is nil. Entropy can therefore be measured, using absolute zero as a baseline. The second law of thermodynamics might then be restated in terms of entropy; the entropy of an isolated system may not decrease.

Stated in still other terms, the second law of thermodynamics stipulates that in any transformation of one type of potential energy into another type of potential energy, the efficiency of the process is not perfect because some of the potential energy is transformed into kinetic energy (e.g., friction). The inefficiency of potential energy conversions explains why it is exceedingly difficult to transform kinetic energy completely back into potential energy. Obviously some transformation can occur, as it does in the steam turbine or in the internal combustion engine, but in all of these cases only a small fraction of the kinetic energy is converted into useful work.* Ultimately all the potential energy available in a system may be converted to kinetic energy or heat, and, considering our universe as an isolated system, it is possible that over an almost infinite period the storehouse of potential energy may become exhausted. (This is sometimes referred to as entropy doom.)

If all the potential energy in an isolated system is converted into heat, and no work is done by the system, the total energy conversion can be measured by the amount of heat evolved, using the calorimeter. In such a system the chemical potential energy is first completely transformed to heat by combustion of the substance held in a platinum container and ignited electrically. Oxygen, usually under pressure, is supplied to the *combustion bomb* as needed. The carbon dioxide released in the reaction can be measured in the exhaust. If the volume of the water in the calorimeter is known and its rise in temperature is measured, the heat produced in the reaction can be determined. The *calorie* (often written cal.) is the amount of heat required to raise 1 gram of water 1 degree centigrade (from 14.5 to 15.5° C.). For example, complete combustion of glucose ($C_6H_{12}O_6$) releases 673,000 cal./mole. By use of the calorimetric method, the calorie content of a large number of compounds has been measured. Some of these values are given in Table 1.1.

The fundamental principle of the second law is part of the daily experiences of mankind. Everyone knows, for instance, that water runs downhill and not uphill,

---

*The efficiency of a machine receiving heat at $T_H$ which is cooled to $T_L$ is $(T_H - T_L)/T_H$. For steam at 200° C. (473° K.) cooled to 40° C. (313° K.) the calculated efficiency is 0.34. In practice it is only about 0.23. The cell, on the other hand, operates at an efficiency of about 0.4 to 0.5 (see Chapter 16).

**Table 1.1.** Heats of Combustion of Various Organic Compounds in Calories per Mole

| | |
|---|---|
| Stearic acid | 2,711,800 |
| Sucrose | 1,349,600 |
| Glucose | 673,000 |
| Glycerol | 397,000 |
| Alanine | 387,700 |
| Ethyl alcohol | 327,600 |
| Lactic acid | 326,000 |
| Acetaldehyde | 279,000 |
| Acetic acid | 209,400 |
| Pyruvic acid | 279,100 |

Data from Handbook of Chemistry and Physics, 1968. 49th Ed. Chemical Rubber Co., Cleveland; data for pyruvic acid from the International Critical Tables V: 165, 1929.

its potential energy and capacity to do useful work decreasing the while. Everyone knows that the light energy of the sun is converted to heat on absorption but that production of light from heat requires the high heat of incandescence, indicating the relatively small likelihood of heat being converted to light. No means, in fact, exists for storing light energy as such except transiently, although it is converted to, and stored as, chemical energy in green plants (Chapter 19). Electrical energy also may be stored only briefly and in small amounts in condensers. Chemical compounds, although often serving as good stores of potential chemical energy, also decompose, releasing their energy stores by such processes as oxidation. Light, electric or chemical potential energy having a high capacity for doing work, then, tends to become transformed to molecular kinetic energy capable of performing less work. Ultimately, all the potential energy in a system is converted to kinetic energy or heat.

Chemical compounds are, relatively speaking, the best stores of potential energy, provided they are kept dry. It is this fact that makes life possible. Plant cells trap the vast reserves of the sun's light energy, transforming the light into chemical potential energy in carbohydrates for later use. The carbohydrates may be converted to other foods in the plant and stored. Animals and microorganisms tap this reserve of chemicals in plants from which they obtain the energy necessary for their life activities. Man uses such energy reserves from his crops, and he exploits the energy reserves of coal and oil, plant products of ancient geological eras. An attempt is being made at present to convert chemical and photic energy directly to electrical energy for special use in space satellites. Because such direct conversions occur with higher efficiency than indirect conversions (e.g., conversion of chemical energy of wood and of coal to heat and then conversion of the heat to electrical energy in a conventional electric plant), such studies may have vast practical applications in the future (Kaye and Welsh, 1960; Sutton, 1966). Fuel cells which convert waste organic matter to electricity are also under extensive study (Yeager, 1961; Lewis, 1966). Man is also learning to use atomic energy effectively, opening an entirely new energy reserve (Egli, 1960).

## FREE ENERGY

The heat of reaction, ΔH, also called the *enthalpy* (energy content of the products minus the energy content of the reactants), is written with a negative sign, −ΔH, if heat is lost to the environment, and with a plus sign, +ΔH, if heat is gained from the environment during the reaction. For example, for the combustion of glucose in a calorimeter in which no external work is done:

$$C_6H_{12}O_6 + 6\,O_2 \rightarrow 6\,CO_2 + 6\,H_2O \qquad (1.1)$$

$$\Delta H = \text{energy content of products} - \text{energy content of reactants} = -673{,}000 \text{ cal./mole (at } 25^\circ \text{ C.)}.$$

Of greater theoretical interest than the heat of reaction is the *change in free energy* (ΔF),* which represents the maximum

*The free energy is given as ΔF° if determined at standard state, that is, 25° C., and for gases, 1 atmosphere pressure. For liquids ΔF° implies the change in free energy when one mole of reactant forms one mole of product, each being maintained at one molal concentration. For biological reactions, which in the cell occur at about pH 7.0, the designation ΔF′ is often used.

work that can be obtained from a given reaction under isothermal conditions. The change in free energy (free energy of products minus the free energy of reactants) of a reaction occurring reversibly at constant temperature (isothermally) and pressure may be defined in terms of changes in entropy ($\Delta S$) and the heat of reaction:

$$\Delta F = \text{free energy of products} - \text{free energy of reactants, or}$$

$$\Delta F = \Delta H - T\,\Delta S \qquad (1.2)$$

where T is the absolute temperature.

The free energy change ($\Delta F$) of a reaction bears no simple relation to the change in energy content ($\Delta H$). Though the two quantities are often similar in size and sign, they are sometimes very different.

The change in free energy ($\Delta F$) of a reaction may be greater than the $\Delta H$. Thus if heat is absorbed from the surroundings during the course of a reaction, more useful work may be accomplished than would be possible if only the energy liberated in the reaction were available for performance of work. For instance, in the combustion of glucose, $\Delta F$ (at 25° C.) is $-686{,}000$ calories per mole compared to a $\Delta H$ of $-673{,}000$ calories per mole. The difference between $\Delta F$ and $\Delta H$ is the entropy term:

$$\Delta F = \Delta H - T\,\Delta S$$
$$-686{,}000 = -673{,}000 - 13{,}000$$

Dividing by 298 to obtain the entropy term at standard state (25° C., or 298° absolute, 1 atmosphere pressure), gives $\Delta S = +43.6$ cal./mole. On the other hand, when the work performed during a reaction is less than the energy liberated in the reaction, the $\Delta F$ of the reaction is less than the $\Delta H$.

Free energy changes occurring during a variety of types of reactions encountered in biological processes are given in Table 1.2. When the free energy change in a reaction is small, the kinetic energy of activated molecules may be sufficient to react (Chapter 10).

It is possible to relate the change in free energy to the equilibrium constant for a reaction. Thus, in the reaction:

$$A + B \rightleftharpoons C + D \qquad (1.3)$$

at equilibrium, application of the law of mass action gives the following expression:

$$\frac{[C][D]}{[A][B]} = K \qquad (1.4)$$

where K is the equilibrium constant.

**Table 1.2.** Standard Free Energy Changes at pH 7 and 25° C. of some Chemical Reactions

| | $\Delta F'$ |
|---|---|
| *Oxidation* | |
| glucose + 6 $O_2$ → 6 $CO_2$ + 6 $H_2O$ | −686,000 |
| lactic acid + 3 $O_2$ → 3 $CO_2$ + 3 $H_2O$ | −326,000 |
| palmitic acid + 23 $O_2$ → 16 $CO_2$ + 16 $H_2O$ | −2,338,000 |
| *Hydrolysis* | |
| sucrose + $H_2O$ → glucose + fructose | −5,500 |
| glucose 6-phosphate + $H_2O$ → glucose + $H_3PO_4$ | −3,300 |
| glycylglycine + $H_2O$ → 2 glycine | −4,600 |
| *Rearrangement* | |
| glucose 1-phosphate → glucose 6-phosphate | −1,745 |
| fructose 6-phosphate → glucose 6-phosphate | −400 |
| *Ionization* | |
| $CH_3COOH$ + $H_2O$ → $H_3O^+$ + $CH_3COO^-$ | +6,310 |
| *Elimination* | |
| malate → fumarate + $H_2O$ | +750 |

From Lehninger, 1965: Bioenergetics, W. A. Benjamin, New York.

The change in free energy ($\Delta F$) for the reaction is related to its equilibrium constant (K) in the following way:

$$\Delta F = -RT \ln K \qquad (1.5)$$

where R is the gas constant (about 2.0 cal./mole/degree), T is the absolute temperature and ln is the natural logarithm (2.3 times the logarithm to the base 10). $\Delta F$ can thus be determined by the ratio of product(s) to reactant(s). For example, in the conversion of glucose-1-phosphate to glucose-phosphate (in the presence of the enzyme phosphoglucomutase) the proportion of glucose-6-phosphate to glucose-1-phosphate was found at equilibrium to be 0.019 to 0.001. The equilibrium constant $K_{eq}$ is therefore 19. Substitution of this value in equation 15.5 enables one to determine the $\Delta F$ value at standard state (1 molal concentration, 25° C.):

$$\Delta F = -RT \ln K = -1.987 \times 298 \times \log 19 = -1745 \text{ cal./mole}$$

(Lehninger, 1965).

If K is 1, no reaction will occur when equivalent concentrations of the reactants are mixed. If K is greater than 1 (and $\Delta F$ is minus) reaction 1.3 will proceed to the right with a decrease in free energy. If K is less than 1 (and $\Delta F$ is plus), reaction 1.3 will proceed to the right only if energy is supplied, as can be seen by the relation between K and $\Delta F$ in Table 1.3.

**Table 1.3.** The Numerical Relationship Between Equilibrium Constant and $\Delta F'$ at 25° C.

| K eq. | $\Delta F'$ |
|---|---|
| 0.001 | 4089 |
| 0.01 | 2726 |
| 0.1 | 1363 |
| 1.0 | 0.0 |
| 10.0 | −1363 |
| 100.0 | −2726 |
| 1000.0 | −4089 |

From Lehninger, 1965: Bioenergetics, W. A. Benjamin, New York.

## ENTROPY AND THE SECOND LAW OF THERMODYNAMICS

The total energy change in a reaction consists of two components—the utilizable energy change ($\Delta F$) and the nonutilizable energy change ($T\,\Delta S$), or change in entropy times the absolute temperature (T):

$$\Delta H = \Delta F + T\,\Delta S \qquad (1.6)$$

Equation 1.6, rearranged, is the same as equation 1.2.

For a specific reaction $A \rightleftharpoons B$ (1.7)

The $\Delta H$ can be determined if the heat of combustion of each of the reactants, A and B, is determined. This can be done in a bomb calorimeter (Fig. 1.2). If, for this hypothetical example, the heat of combustion of A is taken to be 5000 cal./mole and that of B, 2000 cal./mole, then for the reaction at the arbitrary temperature of 27° C. (300 degrees absolute):

$$(A - B) = (5000 - 2000) \text{ or}$$

3000 calories of heat would be given up; therefore $\Delta H = -3000$ cal./mole.

If we assume that the entropy change on going from A to B was 10, substituting this value into equation 1.2,

$$\Delta F = \Delta H - T\,\Delta S, \text{ or } (-3000) - (300 \times 10) = -6000 \text{ cal./mole.}$$

In this case the entropy term contributes significantly to the free energy change. The entropy term may be large or small (McElroy, 1964).

Although some reactions can be performed in such a way that they are practically reversible, all spontaneous reactions are to a considerable degree irreversible; that is, the equilibrium is shifted to favor the reactants, the reverse reaction being slight. The change in entropy in such a reaction indicates its degree of irreversibility. If the symbol S designates entropy, and $\Delta S$, the difference in entropy between the beginning ($S_b$) and the end ($S_e$) of the reaction, then:

$$\Delta S = S_b - S_e = q/T \qquad (1.8)$$

where q is the heat absorbed from the

surroundings and T is the absolute temperature.

The second law of thermodynamics may now be stated in terms of entropy: the entropy of a system may not decrease with the progression of spontaneous reactions. Part of a system may show a decrease of entropy while another part shows an increase. This is true in coupled reactions and occurs in many biological syntheses. However, the entropy of the entire system cannot decrease.

## COUPLED REACTIONS

A reaction which occurs with a decrease in free energy ($-\Delta F$) is said to be *exergonic.* One which occurs with an increase in free energy ($+\Delta F$) is said to be *endergonic.* The oxidation of glucose is an example of the first, and photosynthesis is an example of the second. In biochemical reactions in the cell the free energy is not truly liberated but is conserved in high energy bonds and is thus passed on to other molecules to be used by them to form new bonds. This is illustrated in Figure 1.3 for a whole series of biological reactions, in which cell work is accomplished at the expense of such bond energy. Each of these cases is dealt with in more detail elsewhere.

In many synthetic biological reactions the energy is supplied in the form of labile high energy phosphate bonds (see Chapter 16) from exergonic reactions coupled with endergonic reactions. By this means a complex compound may be synthesized seemingly in contradiction to the second law of thermodynamics. However, the energy of the exergonic reaction is utilized in the synthesis so that the net effect is the conversion to kinetic energy of free energy available in the system consisting of all the components entering into the coupled reactions. We shall consider one specific example of coupled exergonic and endergonic reactions. The reaction between coenzyme A and acetic acid to form the

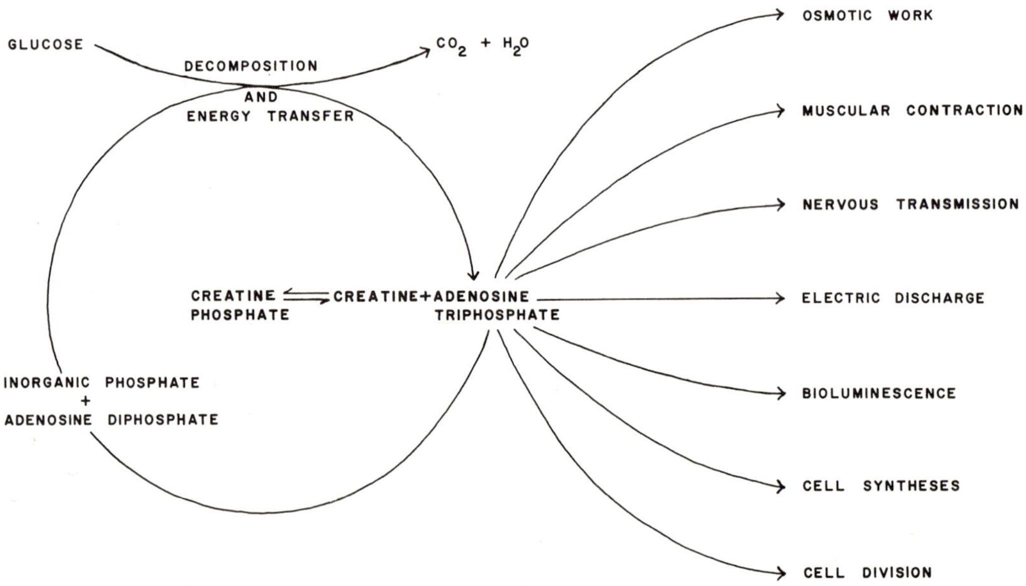


***Figure 1.3.*** High energy phosphate bonds and cell work. Energy from glucose (or other cell foods) made available during its anaerobic and aerobic decomposition in the cells is transferred to an inorganic phosphate which is thus linked by a resonant, high energy phosphate bond to adenosine diphosphate (ADP) making it into adenosine triphosphate (ATP). The ATP may transfer its high energy phosphate ($\Delta F$ = about 7000 calories per mole) to creatine (C) forming creatine phosphate (CP) as a storage depot of high energy bonds; or it may transfer the energy in coupled reactions performing all kinds of cell work. Various types of cell work are discussed in subsequent chapters. High energy phosphate bonds are probably stored in metabolizing parts of all cells. (For other high energy phosphate compounds, see Chapter 16.)

compound acetyl coenzyme A is endergonic and proceeds with an increase in free energy. However, if it is coupled to the breakdown of adenosine triphosphate (to adenosine diphosphate and inorganic phosphate), a reaction which is exergonic and proceeds with a decrease in free energy, then the acetic acid is combined with coenzyme A to form acetyl CoA because there is a net decrease in free energy of the coupled endergonic and exergonic reactions. This is possible only if the algebraic sum of the free energies of the endergonic and exergonic reactions is negative (as discussed before).

In such coupled reactions occurring in the cell, the participants normally share atoms. The major problem, however, is to determine, at the molecular level, the nature of the constraint which prevents the reaction from merely going downhill without doing work and which makes possible instead coupling and transference of the energy. In some cases the constraint is the stability of the high energy molecule, which "forbids" the uncoupled breakdown. For example, in the presence of the appropriate enzyme, creatine phosphate readily transfers its high energy phosphate to adenosine diphosphate, forming, as a result, adenosine triphosphate. But it does not do this by itself without the enzyme. Enzymes thus facilitate reactions along pathways useful to the cell (see Chapter 15) (Wilkie, 1960).

## CHEMICAL POTENTIAL

It is evident from the preceding discussion that the free energy applies to the entire system under consideration and is therefore an extensive property. On the other hand, such quantities as density, concentration, temperature and pressure are intensive properties. In biological studies it is often necessary to know not only the total free energy but also some intensive properties, especially the concentration of the reactants. The value of such information can be illustrated by an example. A solution of high concentration of a given substance is capable of doing more work than a solution of lower concentration, just as a gas under high pressure can do more work than one at low pressure. The *chemical potential* measures the capacity of a substance to do useful work by undergoing chemical or physical change.

An example of chemical potential is the phosphate transfer potential of ATP, which often drives a coupled endergonic chemical reaction (see Chapters 16, 19 and 20). An example involving a physical change is the chemical potential of a solute of high concentration which can move spontaneously to a cell or to another solution containing a lower concentration of the material (Snell *et al.,* 1965).

When the chemical potential is negative, for example, when a solute must move against a concentration gradient, movement can occur only if the reaction is coupled with an exergonic reaction of sufficiently high chemical potential. When the movement of ions must occur against an electrical gradient it will be accomplished only if coupled with a reaction of sufficiently high chemical potential (see Active Transport in Chapter 14.)

## THE CELL AND THERMODYNAMICS

Biological reactions all take place in solution—gases are dissolved before being used in the cell; therefore, with few exceptions, such reactions may be considered as systems occurring at constant pressure and at constant volume (changes in volume being negligible). Even more important, cells can function when they are at uniform temperature. Thermal gradients may occur temporarily in cells when heat is produced and thermal conductivity is limited, but the thermal gradients are incidental, not essential to cellular functions. The second law of thermodynamics, therefore, can be applied in a simplified form because, in a system at constant temperature there is no heat gradient for conversion of heat energy into potential energy or work. Comparing a cell to a heat engine is thus misleading, since a heat engine converts heat into work only because of the thermal gradients within it (Wilkie, 1960).

Some potential energy capable of doing work in the cell is continually converted into heat during cell reactions and this kinetic energy used to maintain the temperature of the cell is incapable of being transformed again into potential energy. As discussed previously, the different kinds of free energy (e.g., mechanical energy, electrical energy, chemical energy, light) can be interconverted from one form to another, but the interconversion is never complete. In the cell, that part of the free energy not converted into a new form must appear as heat (in accordance with the first law of thermodynamics). The degree to which free energy is transformed to heat is therefore a measure of the inefficiency of the conversion process. Consequently, free energy must be continually supplied to the cell from the outside if the cell is to remain alive.

It must be pointed out, however, that this kinetic energy is not useless to the cell. The living cell requires an adequate supply of thermal energy to maintain its cellular constituents in a physical state which makes metabolism possible and to supply the energy of activation for many vital reactions (Allen, 1960).

When there are no coupled reactions, no work is done and the heat produced ($\Delta H$) equals the sum of the entropy term ($T\,\Delta S$) plus the free energy term ($\Delta F$). When coupling between two reactions is perfect and no free energy is transformed to kinetic energy, as in a reversible process, the work done equals the free energy change ($\Delta F$). In the cell, only a fraction of the free energy, whose size ($\epsilon$) depends upon the effectiveness of the coupling mechanism, is made available for work:

$$\text{Work done} = -\epsilon\,\Delta F \qquad (1.9)$$

The remainder of the free energy is then given off as heat.

One simple rule which emerges from such studies is that actual absorption of heat can occur only if an entropy term is involved. The free energy term can lead to evolution of heat or to no heat change at all. It is important to realize that when heat is measured experimentally, it comes from two sources which are quite different and distinct. The first ($T\,\Delta S$) is incidental and reversible; the second ($\Delta F$) is irreversible.

It should also be obvious from a thermodynamic point of view that a living cell must obtain as food some simple molecules with considerable energy which can be built into any of the complex and improbable molecules (molecules of low entropy) characteristic of living materials. Such molecules are the proteins which are denatured with a large increase in entropy, indicating that the structure becomes random during denaturation. It is for this reason that the temperature coefficient for protein denaturation is so high (see Chapter 10).

The cell is never in equilibrium with its environment, except perhaps in death. Rather, the cell maintains a *steady state* and constitutes an *open system* in continual energy exchange with its environment. Since classic thermodynamic relationships were developed for a system in equilibrium, the question arises whether such reasoning is applicable to any particular reaction occurring in a cell.

The problem of accumulation, storage and expenditure of free energy (i.e., energy with maximum work capacity) by cells is independent of whether these cells are in a steady state condition or in equilibrium states. A given reaction occurring in the cell cannot liberate more energy or less energy than it liberates in the test tube, where it is possible to determine only the direction in which a reaction occurs and to measure its $\Delta F$. Such thermodynamic properties of the reaction in the cell cannot be determined by the classic methods because the cell is not in equilibrium with its environment and the reactions do not therefore come to equilibrium. A different approach is necessary for this purpose, and the subject of thermodynamics of irreversible processes or open systems is being intensively studied at the present time. This is one of the instances in which a biological problem has prompted study of a neglected field of physical chemistry (Eyring *et al.*, 1960; Prigogine, 1962; Katchalsky and Curran, 1965).

Classic thermodynamics is limited in another respect. Although it predicts, for a reaction of known ΔF, the direction in which the reaction will go, it can predict nothing as to the rate of the reaction. A new physical-chemical reaction-rate theory, which relates the free energy change to the rate of reaction, has been used for such studies. The relationship between the reaction rate and the free energy change is linear only when the reaction proceeds near equilibrium. From reaction-rate studies it can be shown that an open system, in which several reactions are proceeding, does not show sustained oscillations. Indeed, because each individual reaction continues to proceed at a constant rate the overall reactions settle down to a steady state in which the amounts and concentrations of the various reactants remain constant even though the reactions are proceeding. This steady state is also one in which occurs the lowest possible rate of energy dissipation, or entropy increase. When extraneous changes are introduced into the system such a system likewise alters spontaneously in such a way as to reduce to a minimum its energy dissipation, or entropy increase. The stationary state thus has somewhat the appearance of an equilibrium, and is perhaps similar to what goes on in the living cell (Wilkie, 1960).

## ENERGY EXCHANGE AND LIFE

When a gram molecular weight of glucose is burned in a calorimeter, 6 moles of oxygen are used, 6 moles of carbon dioxide are evolved and 673,000 calories of heat are liberated. If the cell obeys the laws of thermodynamics, it should also use 6 moles of oxygen, evolve 6 moles of carbon dioxide and liberate 673,000 calories of heat in a calorimeter for each gram molecular weight of glucose burned. This is, of course, on the assumption that only carbohydrate is being used by the living cells during the experiment. (This subject will be further discussed in Chapter 18).

At about the time of the American Revolution, and even before the law of conservation of energy was fully appreciated by physical scientists, the great French scientists, Lavoisier and Laplace, devised a primitive calorimeter to study the nature of energy conversions in living organisms (Fig. 1.4). Around a cage containing a guinea pig they placed an ice jacket which absorbed heat produced during the experiment, and the heat was measured by the amount of water obtained from the melted ice. They isolated this system by surrounding it with another ice jacket, which absorbed all heat entering from the environment. So long as ice was present, both ice jackets remained at a temperature of 0° C.; therefore no heat was transferred from one jacket to the other although some ice melted in each, as shown by control tests, and the inner chamber was, therefore, effectively isolated. Lavoisier and Laplace collected the water from the ice melted in the inner chamber, and from its volume (heat of fusion is 79.7 calories; that is, 79.7 calories are required to melt one gram of ice) they determined the total amount of heat produced by the guinea pig in 10 hours (enough to melt 341 grams of ice). In a parallel experiment they determined the amount of carbon dioxide produced in the same period. Comparison of the energy liberated by the oxidation of the same amount of carbon, on the one hand, in chemical combustion (enough to melt 326.7 grams of ice), and, on the other, in the animal, indicated that the values were of the same order of magnitude. The difference in the values probably stems from the fact that in the organism carbohydrate is burned rather than carbon alone. When this is taken into consideration, as in modern calorimetry, the agreement is much closer (Gabriel and Vogel, 1955).

The calorimeters in use today for biological research are considerably more elaborate, but the essential plan is still the same—to isolate a given system by good insulation and to measure the heat

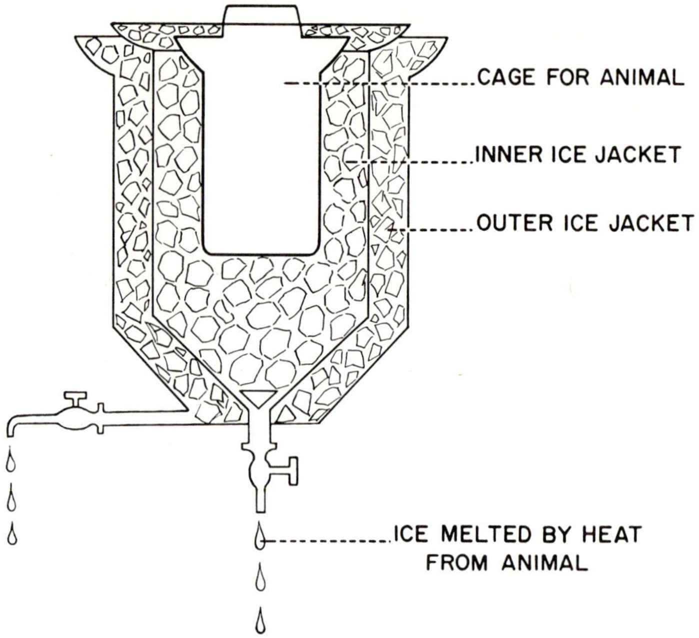


***Figure 1.4.*** Diagram of type of calorimeter used by Lavoisier and Laplace.

evolved, the oxygen consumed and the carbon dioxide produced. Figure 1.5 is a diagram of a modern calorimeter. The heat is caught in water passed through the chamber at a constant rate, and the temperature of the water is determined at the entry and exit points by thermocouples. The amount of oxygen entering is determined by flow meters, the amount of carbon dioxide is determined after absorption in soda lime (a mixture of sodium hydroxide and calcium hydroxide) and

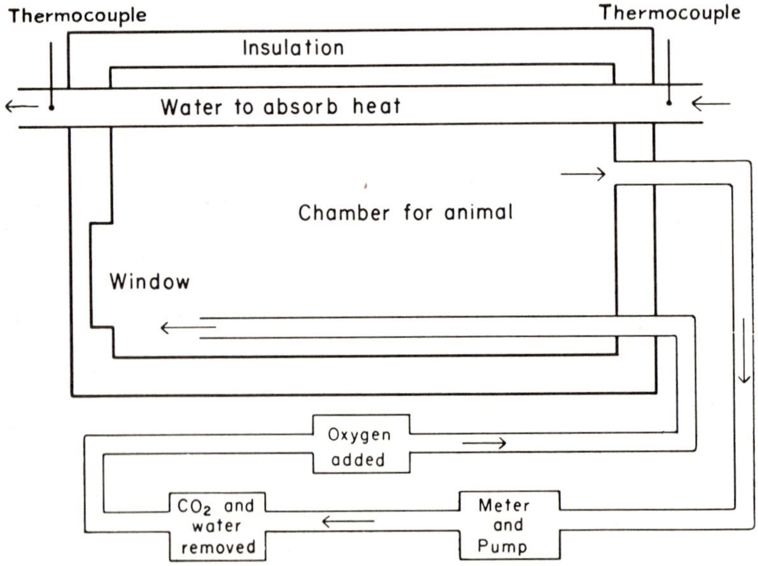


***Figure 1.5.*** Diagram of a modern type of calorimeter.

water vapor is absorbed in sulfuric acid. Undoubtedly the oxidation of a given compound in the cell occurs by pathways different from those in the bomb calorimeter. However, it has been demonstrated that in the cell or in a bomb calorimeter, regardless of the pathway by which the combustion occurs, if the compound is oxidized to the same degree (in the case of the carbohydrate cited, completely to carbon dioxide and water), the amount of energy released is the same. Since according to the first law of thermodynamics energy is neither created nor destroyed, this could have been inferred.

The findings through calorimetry are of profound significance since they indicate that in living cells, as in the inanimate world, the energy transformations are described by the same physical-chemical laws. Since energy transformations underlie all the work done by the cell, they are basic to any further studies in cell physiology.

The cell oxidizes only a fraction of the food it obtains; much chemical potential energy is stored in compounds as reserve potential energy for future use. These reserves accumulate in different places in different organisms. In unicellular organisms and cells, excess food is stored in the form of oil globules and starch or glycogen granules. Occasionally protein is also stored. Multicellular plants store reserves in bulbs, tubers, roots, seeds and fruit as starches, sugars, fats or even proteins. Multicellular animals usually store glycogen in the muscles, and fat in fat bodies. In man, glycogen is stored in the liver and muscles, and fat is stored subcutaneously and in the greater and lesser omenta that suspend the digestive organs. When the diet provides insufficient food for the organism to maintain itself, the reserves are utilized first, and when these are gone the cell substance is itself attacked.

Heat is produced by all living cells. Witness the warmth of the tissues of man, the heat of the manure pile, resulting from the metabolic action of microorganisms, or the heat of germinating seeds, accumulated from the metabolic activities of the cells, or the heat produced in a beehive by cell activities of bees. Even the cold blooded fish produce heat, but this is rapidly dissipated into the environment since water is a fair conductor of heat.

## THE LAW OF CONSERVATION OF MATTER AND LIFE

In the course of chemical reactions elements enter into various combinations with one another to form compounds, but the elements can be recovered from the compounds with neither an increase nor a decrease in the amount of each of the elements participating in the reactions. In other words, matter is neither created nor destroyed but is only transformed from one form to another. This is a statement of the law of conservation of matter. In atomic fission, matter is converted into energy, but in ordinary chemical reactions that occur in the living cell the law has been established as a fundamental principle.

Elements enter into various combinations in the course of the chemical reactions in the cells of organisms. Thus carbohydrates are produced by the cells of green plants in photosynthesis. The carbohydrates may be converted by a plant into amino acids, fatty acids and a multitude of other organic compounds. The plant may be eaten by an animal which uses the organic compounds from the plant as cell foods to convert them into compounds necessary for the activities of its own cells. The animal cells may in turn serve as food for another animal. Both animal and plant cells, at death, serve as food for microorganisms of decomposition and decay. This series of interconversions of carbon compounds is known as the *carbon cycle* and is illustrated in Figure 1.6. The details of some of these conversions are omitted here but will be discussed to some extent in Chapters 16, 18 and 19.

Plant cells build nitrogen-containing compounds such as amino acids and proteins by adding an amino group to the carbon compounds produced in photosynthesis. The nitrogen in nitrates or ammonia obtained from the soil is used for

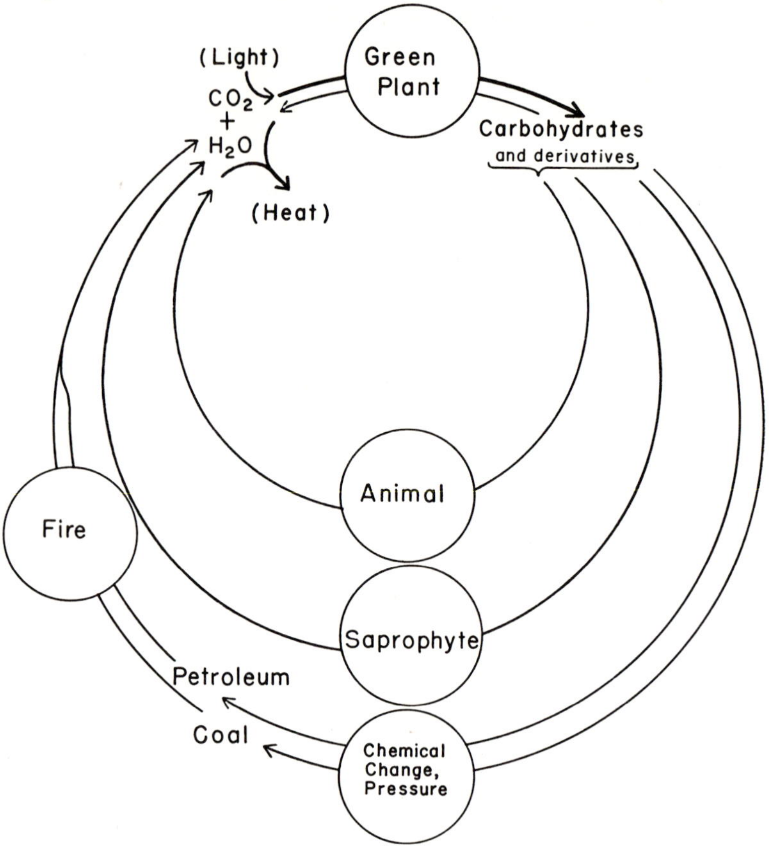


***Figure 1.6.*** The carbon cycle.

this purpose. The animal cells use plant proteins to build their own substance and void the excess nitrogen thus acquired, as well as other metabolic nitrogenous wastes, as urea, uric acid or ammonia. An animal cell may also obtain proteins by using other animal cells as food. At death, both plant and animal cells are used as food by microorganisms of decay and their nitrogen is again put into circulation. This set of relations, known as the *nitrogen cycle,* is shown in Figure 1.7. Details of some of these conversions will be discussed in Chapter 16.

Cycles similar to the carbon or nitrogen cycles could be outlined for phosphorus, sulfur and other elements used by the living cell, but the two cycles presented illustrate the major features; namely, that a given element is shuttled from compound to compound, from cell to cell and from organism to organism, all the while forming various combinations with other elements.

No reason has ever been found to doubt that the law of conservation of matter applies to the living world. Many experiments, especially recent ones using compounds marked with radioactive carbon or isotopes of nitrogen, indicate that the carbon or nitrogen in the conversions of the carbon, nitrogen or other cycles can be accounted for quantitatively. Balance sheets for experiments show that the amount of an element, supplied in nutrients, that is taken up by an organism can be traced in the organism and recovered in the various compounds synthesized or excreted by it.

Since life has existed for millions of years, it may be inferred that conservation of matter is necessary for continuance of

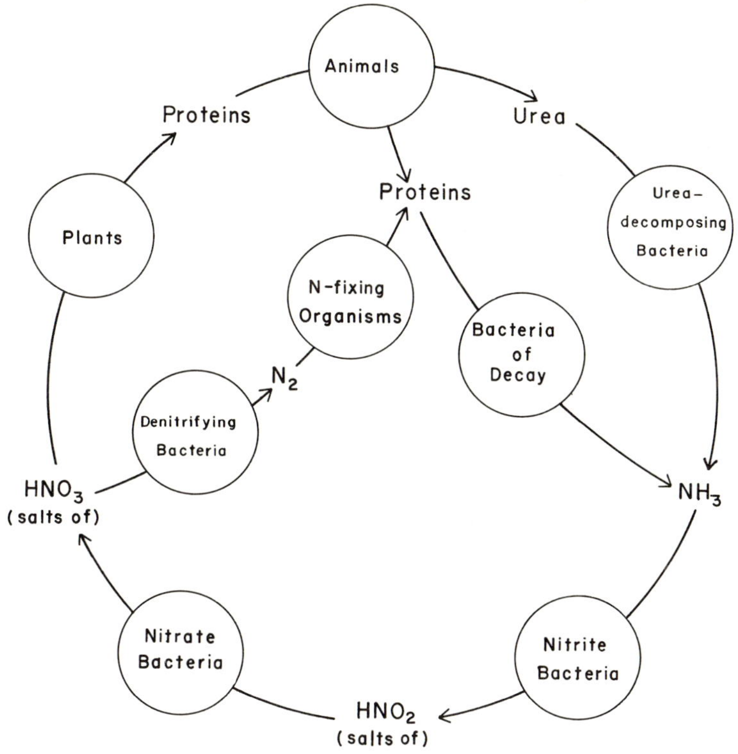


***Figure 1.7.*** The nitrogen cycle. Some animals excrete urea as the main nitrogenous waste (along with small amounts of ammonia and uric acid); others void uric acid as the main nitrogenous waste. Still others void ammonia. Uric acid, like urea, is decomposed by bacteria to ammonia and carbon dioxide.

life; the same elements must be used over and over again, and, in fact, turnover rates have been calculated for some of them (see Chapter 19 for carbon and oxygen turnover). Atoms which today form part of one organism may have formed part of another organism years ago and some day may form part of still another. Material transformations in the living cell are, therefore, governed by the same physical-chemical laws that apply to the inanimate world.

## CONVERSION OF MASS TO ENERGY

According to Einstein's theory of relativity, the equivalence of mass and energy can be represented by the equation,

$$E = mc^2, \qquad (1.10)$$

where E is the energy, m the mass converted in the atomic shrinkage and c the velocity of light. If m is given in grams and c is taken as $3 \times 10^{10}$ cm. per sec., then E will be in ergs: $E = m(9 \times 10^{20})$. To convert to calories, the value for E (in ergs) is divided by $4.18 \times 10^7$.

The total mass of a uranium-235 nucleus and a neutron is known to be greater than the mass of the immediate products resulting from atomic fission. For fission of a kilogram of uranium-235, the decrease in mass is just about 1 gram. This amounts to about $2.1 \times 10^{13}$ calories and is equivalent to 20,000 tons of TNT (trinitrotoluene), or to the daily output of the Hoover Dam ($2.3 \times 10^7$ kilowatt hours of electricity) (Los Alamos, 1950; Shilling, 1964). For the mechanism of fission the student is referred to textbooks of modern physics.

The conversion of matter to energy indicates that perhaps the laws of conservation of matter and energy should be stated in even broader terms than has been the custom in the past; e.g., matter and energy can neither be created nor destroyed although under certain conditions they can be converted from one to another. This implies that matter may also be created from energy, a view for which some evidence exists.

Whereas matter circulates between the environment and living cells, coming in to form parts of cells and then returning to the environment upon their death, and again finding its way into cells of another organism at a future time, the movement of energy is unidirectional. Potential energy comes into the green plant cell as light from the sun; some of it is stored as chemical energy in carbohydrates and other foods, and some of it is converted to kinetic energy (heat) in which form it leaves the living cell. Thus, in nature the flow of matter is cyclic, whereas, essentially, the flow of energy is not.

## LITERATURE CITED AND GENERAL REFERENCES

Alder, B. J., and Wainwright, T. E., Oct. 1959: Molecular motions. Sci. Am. *201*:113–126.

Allen, M. B., 1960: Utilization of thermal energy by living organisms. Comp. Biochem. *1*:487–514.

Bayliss, L. E., 1959–60: Principles of General Physiology. John Wiley & Sons, New York, pp. 1–75.

Bent, H. A., 1965: The Second Law: An Introduction to Classical and Statistical Thermodynamics. Oxford University Press, New York.

Blum, H. F., 1955: Time's Arrow and Evolution. 2nd Ed. Princeton University Press, Princeton, N. J.

Bray, G., and White, K., 1957. Kinetics and Thermodynamics in Biochemistry. Academic Press, New York.

Daniels, F., 1967: Direct use of the sun's energy. Am. Sci. *55*:15–47.

Edsall, J. T., and Wyman, J., 1958: Thermodynamics, electrostatics and the biological significance of matter. *In* Biophysical Chemistry. Academic Press, New York.

Egli, P. H., 1960: Direct energy conversion. Am. Sci. *48*:311–320.

Eyring, H., Boyce, R. P., and Spikes, J. D., 1960: Thermodynamics of living systems. Comp. Biochem. *1*:15–73.

Gabriel, M., and Fogel, S., 1955: Great Experiments in Biology. Prentice-Hall, New York.

Gates, D. M., 1962: Energy Exchange in the Biosphere. Row-Peterson Divison of Harper & Row, Evanston, Ill.

Glasstone, S., 1960: Elements of Physical Chemistry. 2nd Ed. D. Van Nostrand Co., New York.

Katchalsky, A., and Curran, P. F., 1965: Non-equilibrium Thermodynamics in Biophysics. Harvard University Press, Cambridge, Mass.

Kaye, J., and Welsh, J. A., 1960: Direct Conversion of Heat to Electricity. John Wiley & Sons, New York.

*Lehninger, A. L., 1965: Bioenergetics. W. A. Benjamin, New York, pp. 1–50.

Lewis, G., and Randall, M., 1961. Thermodynamics. 2nd Ed. McGraw-Hill Book Co., New York.

Lewis, K., 1966: Biochemical fuel cells. Bact. Rev. *30*:101–113.

Lindsay, R. B., 1959: Entropy consumption and valves in physical science. Am. Sci. *47*:376–385.

Los Alamos Scientific Laboratory, 1950: The Effects of Atomic Weapons. U. S. Govt. Printing Office, Washington, D. C.

McElroy, W. D., 1964: Cellular Physiology and Biochemistry. 2nd Ed. Prentice-Hall, Englewood Cliffs, N. J.

Pardee, A. B., and Ingraham, L. L., 1960: Free energy and entropy in metabolism. *In* Metabolic Pathways. Vol. 1. D. M. Greenberg, ed. Academic Press, New York, pp. 1–4.

Pasynskii, A. G., 1957: Theory of open systems and its significance to biochemistry. Uspekhi Sovremennoi Biol. *43*:263–279. Chem. Absts. *52*:1330 (1958).

Prigogine, I., 1962: Introduction to Thermodynamics of Irreversible Processes. 2nd Ed. Interscience Publishers, New York.

Shilling, C. W., 1964: Atomic Energy Encyclopedia in the Life Sciences. W. B. Saunders Co., Philadelphia.

Snell, F. M., Shulman, S., Spencer, R. P., and Moos, C., 1965: Biophysical Principles of Structure and Function. Addison-Wesley Publishing Co., Reading, Mass.

Suñer, A. P., 1955: Classics of Biology. Pitman & Sons, London.

Sutton, G. W., (ed.), 1966: Direct Energy Conversion. McGraw-Hill Book Co., New York.

Tribus, M., 1966: Micro- and macro-thermodynamics. Am. Sci. *54*:201–210.

Trincher, K. S., 1965: Biology and Information. Elements of Biological Thermodynamics. Plenum Publishing Corp., New York.

von Bertalanffy, L., 1952: Problems of Life. John Wiley & Sons, New York.

*Wilkie, D. R., 1960. Thermodynamics and the interpretation of biological heat measurements. Progr. Biophys. *10*:260–298.

Wilson, P. W., 1955: Pathways in Biological Nitrogen Fixation. Rutgers University Press, New Brunswick, N. J., pp. 110–120.

Yeager, E., 1961: Fuel cells. Science *134*:1178–1186.

## SECTION I

# Functional Organization of the Cell

Although the cell was recognized as a unit of structure and function of all living things in 1839 when Schleiden and Schwann formulated the *cell doctrine,* the nature of the cell and many details of cellular organization remained yet to be determined. For example, the origin and continuity of the nucleus were not recognized in 1839. Since then these and many other details of cellular organization have been clarified, but the functions of many of these structures have been effectively studied only recently.

A discussion of the organization of different kinds of cells and some techniques for the study of cell structure are given in Chapter 2, followed by a discussion of the biochemistry of the cell in Chapter 3 and of the colloidal behavior of chemical constituents of the cell in Chapter 4. The functional organization of cell organelles is considered in Chapter 5 and the organization and functioning of the nucleus in Chapter 6.

# CHAPTER 2 / CELL ORGANIZATION

Although the term cell was first used in biological literature by Robert Hooke in 1665 to describe a compartment seen in the structure of cork, the cell as a unit of structure and function of all living things was not recognized until about 1839 when Schleiden and Schwann developed the *cell doctrine.* In modern terms the cell may be defined as "a unit of biological activity delimited by a selectively permeable membrane and capable of self reproduction in a medium free of other living systems" (Loewy and Siekevitz, 1963).

It is common knowledge that the cell is differentiated into *nucleus* and *cytoplasm,* each of which is surrounded by a membrane, the nucleus by the *nuclear envelope* and the cytoplasm by the *cell* or *plasma membrane.* The outer portion of the cytoplasm, called the *ectoplasm,* is generally more finely granular and of a stiffer consistency than the inner portion, called the *endoplasm.*

It is familiar to all who have studied plant and animal cells that they are diverse in size and form. A brief summary of the range in size of cells is given in Table 2.1 and of the range in form in Table 2.2. It is of moment, perhaps, to point out that cells are examples of a remarkable degree of miniaturization. Thus the human sperm carries in its minute mass half of all the genetic determinants of a mature individual. The large cells

**Table 2.1.** Range in Size of Cells

| MASS IN GRAMS | ORGANISM OR CELL |
|---|---|
| $10^3$ to $10^2$ | Dinosaur eggs, ostrich eggs, cycad ovules |
| $10^1$ | *Valonia macrophysa* (mature) |
| $10^0$ | *Valonia ventricosa* (mature) |
| $10^{-1}$ | *Nitella* (large internode) |
| $10^{-2}$ to $10^{-3}$ | Frog eggs |
| $10^{-4}$ | Striated muscle of man |
| $10^{-5}$ | Human ovum |
| $10^{-6}$ | Large *Paramecium* |
| | Large sensory neuron of dog |
| $10^{-7}$ | Average *Vorticella* |
| | Human smooth muscle fiber |
| | Human liver cell |
| $10^{-8}$ | Dysentery ameba |
| $10^{-9}$ | Frog erythrocyte |
| | Malarial parasite |
| | Human sperm |
| | Smallest Protozoa (*Monas*) |
| $10^{-11}$ | Anthrax bacillus |
| $10^{-12}$ | Tubercle and pus bacteria |
| $10^{-14}$ | Smallest bacteria |
| | (Limit of microscopic vision) |
| $10^{-15}$ | Filter passing viruses, etc. |
| $10^{-16}$ | *Mycoplasma laidlawii* |

From table of comparative masses in Haldane and Huxley, 1927: Animal Biology; data on Valonia from L. R. Blinks, personal communication; on *Nitella,* original; on frog eggs, calculated from diameters; on *Mycoplasma,* from Morowitz and Tourtelotte, 1962. All values are orders of magnitude.

**Table 2.2.** Types of Cells and Their Uses in the Study of Cell Physiology

| NAME | TYPICAL SHAPE | EXAMPLES | USE IN THE STUDY OF CELL PHYSIOLOGY |
|---|---|---|---|
| | | ANIMAL | |
| Epithelial | Cuboidal or brick-like | Epidermis, glandular lining | Permeability, properties of protoplasm |
| Muscular | Spindle | Smooth, striated | Contractility, metabolism |
| Nervous | Cell body with long fibers | Sensory neuron, motor neuron, | Irritability, structure of protoplasm |
| Connective | Spheroid cells in a mass of extra-cellular material | Cartilage cells, cells in bone and connective tissue | Production of extra-cellular secretions |
| Blood | Disk-shaped or ameboid cells in fluid | Red blood cell, leucocyte | Permeability, ameboid movement, response to chemicals |
| Eggs | Usually spherical | Sea urchin egg, frog egg | Structure of protoplasm, properties of protoplasm, permeability, function of nucleus, cell division |
| Sperm | Usually flagellated cells | Sea urchin sperm | Function of nucleus, nucleoprotein |
| Protozoans | Diverse | *Amoeba*, *Paramecium* | Ameboid and ciliary movement, nuclear function, permeability, effects of environment |
| | | PLANT | |
| Epidermal | Cuboidal or brick-like | Epidermis | Permeability |
| Vascular | Elongate (some strengthened by spirals, discs, etc.) | Phloem (in xylem only cell walls remain) | Structure of protoplasm (phloem) |
| Supporting | Elongate, heavy walls | Bast fibers | |
| Undifferentiated (parenchyma) | Spheroidal | Mesophyll, interstitial cells of stem, root, etc. | Photosynthesis, structure of protoplasm |
| Unicellular algae | Diverse | *Chlorella* | Photosynthesis, growth and cell division, circadian rhythms |

**Table 2.2.** Types of Cells and Their Uses in the Study of Cell Physiology (Cont.)

| NAME | TYPICAL SHAPE | EXAMPLES | USE IN THE STUDY OF CELL PHYSIOLOGY |
|---|---|---|---|
| | | PLANT | |
| Unicellular fungi | Diverse | Yeast | Metabolism, heredity |
| Filamentous fungi | Elongate cells | *Neurospora* | Biosynthesis, heredity |
| Bacteria | Diverse | The colcn bacillus | Biosynthesis, metabolism, heredity |

listed in Table 2.1, which might appear to be exceptions to miniaturization, contain a vast store of nutrient for development of the embryo. The *Valonia* cells have a large central sap vacuole as well as a massive cell wall. In both cases the nucleus and cytoplasm represent a very small part of the total cell mass.

Although diverse in size and form, cells have a similar intracellular organization and usually possess the same formed structures or organelles (see Fig. 2.1). The presence of most of the organelles in all cells suggests that each organelle performs some fundamental role in the economy of the cell. Determination of the functions of these organelles is one of the ultimate aims of the cell physiologist.

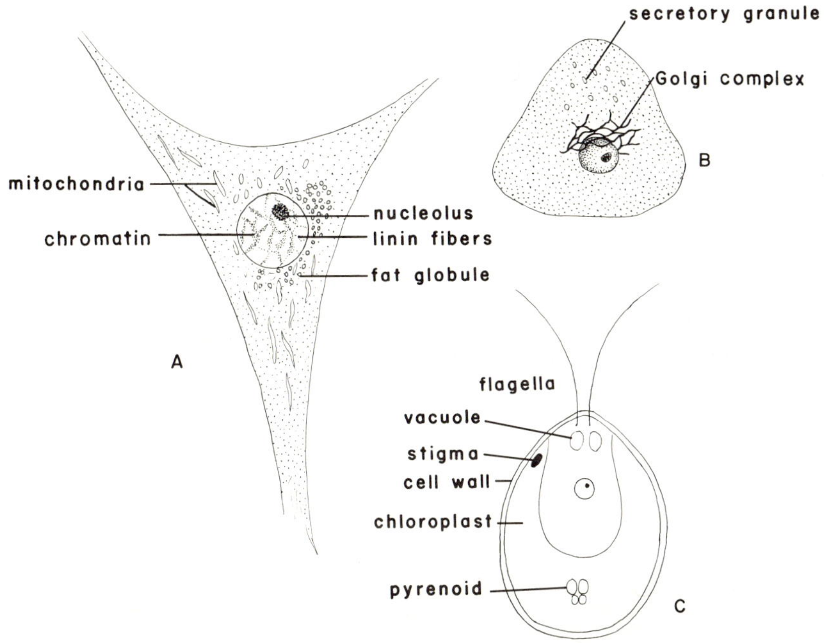


***Figure 2.1.*** Cells and cell organelles. *A* is a fibroblast, *B* is a cell of the pancreas and *C* is the flagellate *Chlamydomonas,* at the center of which is a nucleus.

## ORGANELLES SEEN IN FIXED AND STAINED CELLS

Under the microscope, a nucleus is seen in every stained* cell although in bacteria and blue-green algae it lacks a membrane and is less well organized than in true cells of higher forms of plant and animal life. Within the nucleus a *nucleolus* is commonly present and so is a *linin* network bearing the *chromatin.* Between these materials is the *nucleoplasm.* The entire nucleus is surrounded by a nuclear envelope. Just outside the nucleus of animal cells is a *centriole,* a small granule which separates into two before mitosis (cell division). At mitosis the two centrioles move apart and form the poles of the spindles and the centers of the asters.

Scattered in the cytoplasm are *mitochondria,* or chondriosomes, which have been seen in many kinds of cells examined. The appearance of mitochondria depends upon the handling of the tissues before preparation; they may appear as long rods, short rods or small granules.

The *Golgi complex,* named after its discoverer, is particularly prominent in nerve and secretory cells of animals and seems to be closely associated with secretory activities. In fixed and stained preparations (see Fig. 2.2) it appears to be a canalicular system closely associated with the mitochondria. It is absent from bacteria and blue-green algae. Its true structure and function have been the subject of long controversy which is not yet resolved.

*Lysosomes* are membrane-bound organelles containing hydrolytic enzymes. It is likely that the cell interior is protected from the enzymes by such packaging. The packaging may result from the formation of the enzymes on the surface of the endoplasmic reticulum and accumulation within it (see Chapter 5).

*Plastids* are cytoplasmic organelles which may or may not contain pigment (for example, chlorophyll is the green pigment in chloroplasts, one of the kinds of plastids in green plants). *Vacuoles,* such as the contractile vacuoles of protozoans, and the large central sap vacuoles which are characteristic of plant cells, are often seen in cells (Brown, 1960).

When a cell is cleared by centrifugation of visible granules, oil globules and various organelles, a clear matrix material remains which is thought to represent the fundamental architecture of the cell. In fixed and stained preparations under the ordinary microscope this substance is devoid of structure, but under the electron microscope, which resolves much finer detail (see Appendix 2.3), the matrix material appears to contain a network called the *endoplasmic reticulum.* Upon this reticulum can be seen fine granules called the *ribosomes.* Intercellular cytoplasmic connections called *plasmodesmata* may be found between cells in some tissues (Meeuse, 1957).

*See the appendix to this chapter for an account of methods of staining and for a description of microscopes.

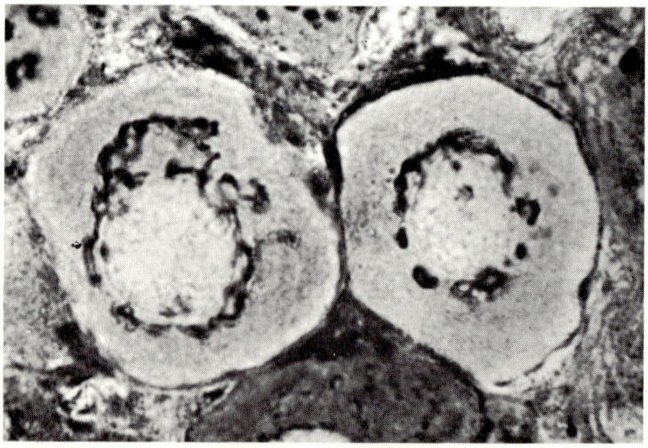

***Figure 2.2.*** The Golgi complex in cervical spinal ganglion cells of a female guinea pig. Kolatchev-Nassanov fixation (× 2000). (By courtesy of Professor Hadley Kirkman.)

It is of interest to note that although a cell contains but one nucleus it may have about 500 mitochondria, $5 \times 10^5$ ribosomes and probably $5 \times 10^8$ enzyme molecules. A green plant cell will probably have about 50 chloroplasts. This multiplicity of units is a characteristic feature of cell organization. Even the nucleus of a diploid organism has two chromosomes of a kind and two genes of a kind, one in each of the members of a homologous pair of chromosomes.

As most of the organelles just listed were first described in fixed and stained preparations (see Appendix 2.1), the possibility was suggested that the fixing process, by precipitating the proteins, produces artificial structures, or *artifacts,* which do not represent the true structures of the cell. At the turn of the century Hardy and Fisher showed that some of these organelles could be duplicated in gelatin by adding to it such fixing agents as tannic acid, chromic acid or the salts of heavy metals. Therefore, it became necessary to determine whether the structures observed in fixed cell preparations were actually present in living cells. Many new and ingenious techniques were devised to demonstrate them, and the evidence for each organelle follows.

## EXPERIMENTAL EVIDENCE FOR CELL ORGANELLES IN LIVING CELLS

When a living cell is examined under an ordinary light microscope, its main organization (nucleus, cytoplasm, plastids, granules, edges of membranes) is visible in outline because of differences in refraction of light by these structures.

When the living cell is centrifuged and observed during the process through the centrifuge microscope (see Fig. 2.3), the organelles, differing in specific gravity and size, are seen to move as discrete units. For example, in centrifuged eggs of the sea urchin *Arbacia,* the red pigment granules are thrown to the bottom of the cell; above them the yolk is deposited and then a layer of mitochondria, and finally the oil droplets are collected at the top of the egg, forming an oil cap. Between the zone of the mitochondria and the oil cap is a clear zone of fine granular protoplasm called the ground substance, within which the nucleus of the egg comes to lie (see Fig. 2.4). Clearly, the specific gravity of the nucleus is less than that of some of the other cellular constituents.

The ground substance represents the cytoplasm cleared by centrifugation of the structures suspended in it. No structure in the ground substance of a living cell is visible under the ordinary light microscope, the ultraviolet microscope or the phase microscope. However, if cells are first fragmented and then centrifuged, the fragments of higher specific gravity settle more rapidly than those of lower specific gravity, other factors being equal, and larger particles settle more rapidly than smaller ones of the same specific gravity. The various particulates within the cell can then be segregated from one another. Large structures such as nuclei

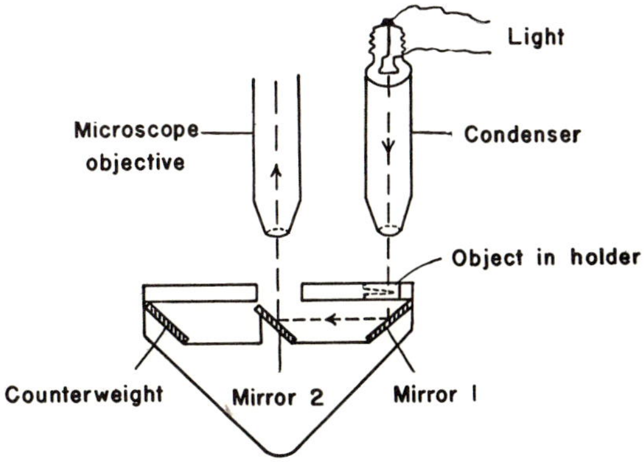


***Figure 2.3.*** Optical system for ultracentrifuge microscope. The condenser throws an image of the incandescent light on the object. The object is under the light for a fraction of a second during each revolution of the disk, but the image on mirror 1 reflected to mirror 2 is seen as a continuous moving picture because the eye, over the microscope in the axis of rotation of the centrifuge, fuses the many images coming to it per second. (From Harvey, 1938: J. Appl. Physics 9:73.)

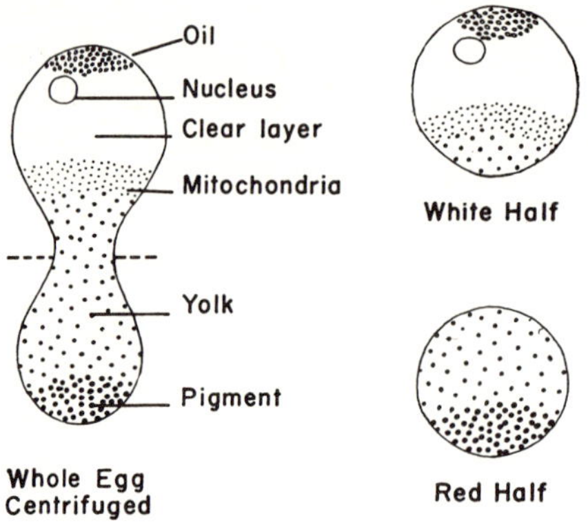


***Figure 2.4.*** Stratification in a centrifuged egg of the sea urchin *Arbacia punctulata.* (After Harvey, 1941: Biol. Bull. *81*:114.)

are readily separated from other constituents, as are also mitochondria and large granules. After all these structures have been removed, minute submicroscopic bodies remain which can be centrifuged down at very high speeds. These are called the *microsomes,* and some of them are ribosomes, the granules seen under the electron microscope in the endoplasmic reticulum. Others are fragments of the endoplasmic reticulum (see Chapter 5).

***The Nucleus and Its Constituents.*** The internal organization of the resting nucleus in a living cell is difficult to discern by ordinary microscopy, and even during mitosis and meiosis the chromosomes are seen only under rare circumstances. However, under the phase microscope (see Appendix 2.3), where slight differences in refractive index between diverse cellular constituents are exaggerated, the chromosomes of some cells may be seen quite clearly. In fact, time-lapse photography permits continual observation of a single chromosome preparation throughout a cycle of meiosis or mitosis, corroborating what was previously deduced from a random sampling of many fixed and stained preparations. Furthermore, chromosomes are seen without the distortion which usually attends fixation. Time-lapse photography makes evident not only the sequence of events, but also the dynamic nature of the process—the movements of cell organelles such as mitochondria, and the changes in shape of the entire cells (see Chapters 6 and 27).

Chromosomes are also discernible in ultraviolet photomicrographs of cells in mitosis or meiosis, because the chromosomes absorb short wavelengths of this radiation more strongly than does the cytoplasm. Much has been learned by the use of this technique (see Chapter 6) but it is not as generally useful as phase microscopy because cells are damaged by such radiation, and prolonged examination is thus impossible (Caspersson, 1950).

Chromosomes when visible may be reached by micromanipulator needles and moved about in the nucleus or in the cell (see Appendix 2.5), indicating that they are discrete units. Moreover, fragmentation and irregularities in the partition of chromosomes can be observed in cells irradiated with ionizing radiation. Chromosomes have also been released from a suspension of nuclei (obtained by fractional centrifugation) and they maintain their identity through a variety of manipulations. At present no doubt exists that chromosomes are real structural units, even though they are difficult to identify in the resting stages of cells.

***Mitochondria.*** Mitochondria are visible in living cells under the phase microscope. They are seen to undergo active movements during mitosis and meiosis. Mitochondria may also be seen in living cells stained with the vital dye Janus green B, which though not entirely harmless, permits the cell to remain alive for a long time so that one can observe how the mitochondria change in shape and size as the conditions vary.

Further evidence of the identity of mitochondria is provided by their movement as units during centrifugation of a whole cell and by their discreteness after cell fragmentation (Harvey, 1951). They may be separated from other particles in fragmented cells by washing in salt solutions followed by differential centrifugation. In suspension they maintain their identity if adequate conditions are provided. Such suspensions of mitochondria have been subjected to biochemical analyses, and many enzymes concerned with aerobic respiration, energy transfer and storage have been identified in them (see Chapter 16). Perhaps for this reason mitochondria are concentrated in regions of greatest cellular activity: at the secretory surface of a glandular cell, near the nodes of a nerve cell at which propagation is believed to occur and in active muscle fibers. In these locations mitochondria might supply most readily the high energy compounds needed for the activities of the cells.

When cells divide, the mitochondria have been seen to orient themselves relatively symmetrically on each side of the division figure, partition between the two cells being fairly equal. In some cases mitochondria have been seen to fragment after division. It is possible that the fragments then increase in size, thus reconstituting the amount of mitochondrial material needed in each daughter cell. However, it is thought that mitochondria may also develop from minute bodies in the cytoplasm. Mitochondria have been shown to contain deoxyribonucleic acid different from that found in the nucleus, which suggests that they may be self-duplicating systems (see Chapter 5).

***The Golgi Complex.*** In the living cell under favorable conditions the Golgi complex may be stained with methylene blue and observed for a limited time. It is also visible by means of phase microscopy. It seems to be closely connected to the endoplasmic reticulum but its outlines are smooth (ribosomes being lacking) and it is chemically somewhat different from the reticulum, containing much more lipid. It can be separated from other constituents of the cell by differential centrifugation (Kuff and Dalton, 1959). Some question exists whether the Golgi complex really constitutes a true organelle. Cytologists are inclined to look upon the Golgi complex as an assembly area for different kinds of secretions, especially lipids, rather than as a structural unit. Into the Golgi area flow high energy compounds from the mitochondria and various secretions, probably including nucleic acids, from the nucleus (Dalton, 1961).

***Lysosomes.*** By gently centrifuging certain kinds of cells lysosomes have been separated from other cell organelles and their enzymes studied, thus showing them to be true cell organelles (see Chapter 5).

***Plastids.*** The identity of chloroplasts as discrete structures has not been doubted because they are so readily seen in living cells of green plants. They have been separated by maceration of cells and fractional separation of the constituents (see Chapter 19). The internal structure of chloroplasts has also been studied by electron microscopy.

*Chloroplasts* have been found to contain deoxyribonucleic acid (DNA) different from that of the nucleus. Evidence has also accumulated that chloroplasts are self-replicating organelles, probably as a consequence of their extranuclear DNA (see Chapter 5).

Other plastids, however, have been less thoroughly studied, perhaps because their functions in plant cells are not as clearly evident as those of the chloroplast. Leucoplasts, within which starch grains are formed, are organized bodies in the cytoplasm and they contain enzymes which synthesize starch from the glucose formed during photosynthesis. Perhaps other synthetic functions are also localized in the plastids of plant cells.

***Microtubules and Microfilaments.*** Recently electron microscopists have discovered microtubules and microfilaments in the ground substance between the cisternae (larger membrane-bound cavities) and tubules of the endoplasmic reticulum (Porter, 1961). Because the microtubules occur in abundance in cells maintaining an odd shape, as in the case of the vertebrate red cell, they are thought to be structural elements. But they also appear to be important in cell division. Microfilaments occur in motile cells such as the plasmodium of the slime mold and may function in movement (see Chapter 5).

***Miscellaneous Cell Inclusions.*** Occasionally oil droplets and oil globules are seen in the protoplasm, particularly in cells in tissue culture, in marine eggs and in marine protozoans known as radiolarians. These oil droplets are probably stores of nutrient or in some cases devices for reducing the specific gravity of the cell, thus making floating easier. In some cells glycogen granules occur, presumably representing stores of nutrient for the cell.

At present, workers have accepted the reality of the major cell organelles and are now concerned mainly with determination of their chemical constitution, physicochemical organization and function.

## VARIATION IN ORGANIZATION OF ANIMAL CELLS

In multicellular organisms, all of the cells are not exactly alike. They have become differentiated and adapted to perform some particular function of the organism (Bloom and Fawcett, 1968).

A *tissue* is a mass of similar cells, usually continuous, held together in a supporting matrix (secreted by the cells), performing a common function and usually forming a part of an *organ.* Intercellular fluid is almost always found in tissues. This is to be expected since every cell must have water frontage from which it obtains supplies and into which it pours its wastes. But because this intercellular fluid is not easily removed or modified, both plant and animal tissues, when used in experiments cannot be considered the equivalent of suspensions of cells.

Animal tissues are usually classified as epithelial, connective, muscular, nervous and blood. Epithelial cells are generally closely packed, like bricks in a pavement, and form solid protective layers on the outside. They line external surfaces, cavities and tubules, produce secretions and proliferate cells. Epithelial cells may be isodiametric (having equal diameters in all directions), flattened, or elongated and columnar (often the columnar cells are glandular). From the epithelial cells of glands exude various secretions. Germinal cells, which give rise to gametes, spring from the epithelial cells of gonads. In epithelial cells are found the typical organelles expected in a cell. In glandular epithelial cells the mitochondria and Golgi complex are likely to be particularly well developed. Although epithelial tissue consists primarily of cells, the external epithelium may secrete a cuticle, as in the earthworm, or a skeleton, as in arthropods.

Connective tissue, in contrast to epithelial tissue, is to a large extent composed of extracellular material and may contain relatively few cells. Some examples of connective tissues are cartilage, tendon, bone and adipose or fatty tissue. During early development of the embryo the connective tissue cells lie close together, but during later development they secrete a gelatinous material (matrix) which separates the cells. In this gelatinous matrix are laid the fibers of connective tissue and the salts of bone. The chemical composition of connective tissue in tendon is diagrammatically represented in Figure 2.5.

In muscular tissue (smooth, cardiac or skeletal) the cells are considerably modified for the function of contraction. They are elongated cells and in all cases contain contractile fibrils, these being characteristic of muscle tissue and of no other tissue. The cells of the smooth or visceral musculature of vertebrates are spindle-shaped units containing a single nucleus in the center and have fine elongate fibrillae which represent the contractile elements. The skeletal musculature of vertebrates is highly differentiated for its function (see Figs. 22.7, 22.16 and 23.1). Its unit of organization is the muscle fiber, consisting of the confluent bodies of numerous cells, each of which is indicated by a spindle-shaped nucleus. Elongate striated fibrils pack the cytoplasm of the skeletal muscle fiber, and large mitochondria are present; they are especially conspicuous in the highly active flight muscles of vertebrates and insects. In the cardiac muscle of vertebrates, striated fibers are found and the unit is modified in organization from that of skeletal muscle. Muscular tissues of invertebrates are very diverse, but in all cases contractile fibrils are found.

The cells of nervous tissue are highly specialized for conduction of nervous impulses. From the nerve cell body containing the nucleus extend the long fibers called *axons* and the short fibers called

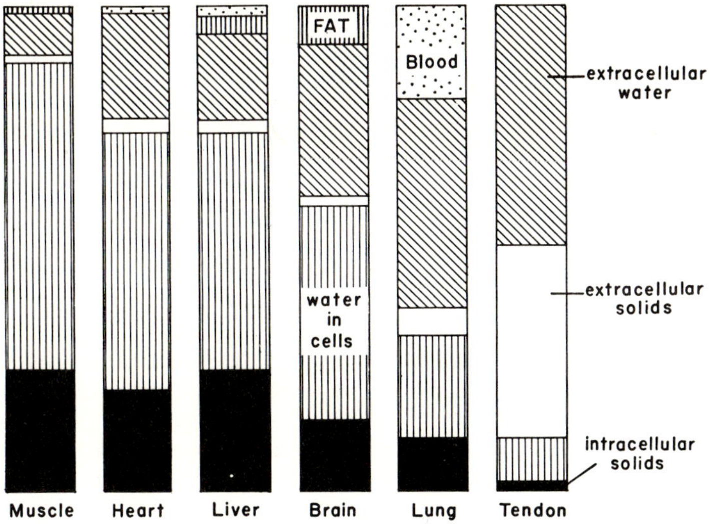


***Figure 2.5.*** Extracellular material in animal tissues. (Adapted from Lowry, 1943: Biol. Symposia *10*:239.)

*dendrites* (see Fig. 22.1). The nerve cell with all of its processes is called a *neuron.* A well developed Golgi complex appears in the cell body of a neuron, and mitochondria occur chiefly at the synapses. In the fibers are found granular deposits of nucleic acids which probably serve as nutrient in the activities of the nerve cell. The axon of a vertebrate nerve cell is called a *myelinated fiber* when it is covered with an envelope consisting of many layers of apposed cell membranes and enclosed by sheath cells (see Fig. 11.5). The myelin sheath may add much bulk to the vertebrate nervous system and, because of it, nervous tissue like the brain contains a considerable amount of extracellular material (see Fig. 2.5). However, the outgoing fibers from autonomic nerve ganglia located outside the cord are not myelinated. Invertebrate nerve fibers have little if any myelin.

Blood also has cells that are highly specialized. For instance, in vertebrate blood the red cells (erythrocytes) are mainly composed of the oxygen-carrying pigment, hemoglobin. In mammals the mature erythrocytes even lose their nuclei and have none of the characteristic cell organelles. The white cells of vertebrates and invertebrates—both granular ameboid cells called leukocytes, and the more or less spherical clear lymphocytes—are much more typical cells than the red cells since they have nuclei and the other cell organelles such as the mitochondria and the Golgi complex.

## VARIATION IN ORGANIZATION OF PLANT CELLS

Unlike animal cells, plant cells usually possess plastids and vacuoles and are covered by a cell wall. Mitochondria, ribosomes and a nucleus are present as in animal cells. Plant cells have structures called *dictyosomes* which are similar to the Golgi complex of animal cells (Jensen, 1964; Voller *et al.,* 1964; Frey-Wyssling and Mühlethaler, 1965).

Little agreement exists on classification of plant tissues, but the following might be cited as one possible classification: epidermal, undifferentiated, vascular (xylem and phloem), sclerenchyma and supporting or bast fibers (see Table 2.2).

The cells of epidermal tissues of plants line external surfaces of plant structures. Embryonic cells of plants, e.g., the *cambium,* are made up of cells similar in appearance and function to epithelial cells of animals. Undifferentiated plant cells, called *paren-*

*chyma,* exist in large numbers in the cortex of the stem, filling the spaces between conducting or supporting tissues. They also make up the mesophyll and the palisade cell layer of the leaf, in which photosynthesis occurs. Vascular tissue of plants consists of the *phloem* and the *xylem.* The phloem is constituted of elongated living cells which abut upon one another. It distributes the manufactured food throughout the entire plant. The xylem consists of tubular remains of cells; it distributes water and salts to the cells of the plant. In the adult plant the xylem ducts are entirely extracellular, the cellular material having disappeared. Cells with thickened walls, like the gritty stone cells in the pear, are representatives of *sclerenchyma* tissue. Supporting "cells" or bast fibers, consisting largely of secretions of cellulose and lignin, occur along the vascular tissues and contribute to the skeleton of the plant. Supporting fibers in the adult plant are entirely extracellular. As a result, an organ that contains an abundance of ducts and supporting fibers, like a mature stem, has much extracellular material. In contrast, a bud or a leaf is made up largely of relatively undifferentiated cells, and, except for the cell walls, consists of living cells. Unfortunately, no convenient summary of the intracellular and extracellular constituents of plant tissue, comparable to that for animal tissues shown in Figure 2.5, is available, although it would be desirable since plant tissues are extensively used in cellular research (see Table 2.2) (Esau, 1965).

## VARIATION IN ORGANIZATION OF BACTERIAL CELLS

Bacteria and blue-green algae, unlike all kinds of true cells of plants and animals, lack a distinct nucleus enclosed by a nuclear envelope, the nuclear material being scattered in the cytoplasm. The blue-green algae differ from bacteria in that they contain chlorophyll and a characteristic auxiliary pigment and differ from most bacteria in being photosynthetic (Echlin, 1966). Plant and animal cells with true nuclei are called *eukaryotic cells* and bacteria and cells of blue-green algae are called *prokaryotic cells.* Also, prokaryotic cells either lack all the other cell organelles (so characteristic of eukaryotic cells) or possess them in a primitive, undifferentiated condition. Because bacteria have been extensively investigated, they alone are considered here, but much of what is said is applicable to blue-green algae as well.

Bacterial cells are covered with a protective *cell wall* which usually consists of carbohydrates and polypeptides but may contain protein and lipids. The bacterial cell can be made to retract from this wall if it is placed in a concentrated solution of sugar or salt, so that the cell membrane (plasma membrane) now forms the outer boundary of the living material of the cell called the *protoplast* (Fig. 2.6). The cell wall may be digested with the enzyme lysozyme, exposing the protoplast which can now be made to swell or shrink like any other cell by decreasing or increasing the concentration of salts dissolved in the medium. The cell wall evidently plays some important role in addition to protection, since protoplasts divested of the cell walls do not divide* or form new cell walls (Robinow, 1960).

Some bacteria, e.g., *Bacillus megatherium,* called gram-positive bacteria, have ribonucleic acid present in the surface which takes up a stain made up of crystal violet and iodine. Bacteria which are gram-negative (e.g., *Escherichia coli,* the colon bacterium) differ in possessing cell walls heavily impregnated with lipids. Thick-walled resting stages called *spores* which resist high temperature and other unfavorable conditions are formed by some bacteria (bacilli).

Bacteria lack mitochondria and the Golgi complex although they generally are filled with bodies resembling the ribosomes of true cells. As in eukaryotic cells the ribosomes may cluster to form polyribosomes during the synthesis of proteins in bacterial cell homogenates (see Chap-

*Yeast protoplasts also fail to divide or conjugate (Holter and Ottolenghi, 1960).

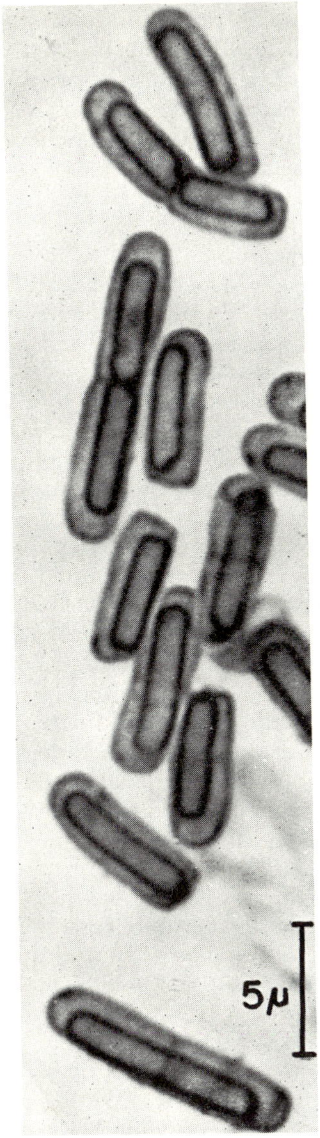


***Figure 2.6.*** Retraction of the cell contents (protoplast) from the cell wall induced by exposure to ether vapor and drying on glass before fixation. Stained with Victoria blue. (From Robinow, 1960: *In* The Cell. Vol. 4. Brachet and Mirsky, eds. Academic Press, New York.)

ters 6 and 20). The plasma membrane, which is rich in oxidative enzymes, probably performs the function usually carried out by the mitochondria of plant and animal cells. It is of interest that the plasma membrane of the large bacterium *Thiovulvum majus* sinks deep into the cytoplasm, forming a multilayered structure called the *desmosome.* In the tubercle bacillus, concentric membranes, suggesting incipient mitochondria, have been discovered in the cytoplasm with the aid of electron microscopy.

Many bacteria have been shown to possess large granules (round bodies) which grow and divide and are stained with the same dyes which affect chromosomes of true cells. These bodies appear to condense before division and become more diffuse thereafter.

Rickettsias, which cause Rocky Mountain spotted fever, endemic typhus and murine typhus, are of about the size of small bacteria and in general their organization clearly places them among the bacteria. However, to some workers they appear to be escaped mitochondria and to others, a stage between viruses and bacteria. Like bacteria, they possess a number of respiratory enzymes which are lacking in viruses. They have not yet been cultured except in cells of a host they infect (Bovarick, 1955).

## THE PPLO

The smallest bacteria, and the smallest cells known, are found in the group of organisms known as the pleuropneumonia-like organisms, also known as the PPLO (Mycoplasmataceae). *Mycoplasma laidlawii,* a free-living strain growing in sewage, is the smallest cell so far studied. It shows a definite life cycle. In culture it exists in bodies of three sizes, minute elemental bodies 0.1 $\mu$ in diameter, intermediate size bodies and full-size cells about 1 $\mu$ in diameter. Studies indicate that the small bodies develop into the large cells, which can also divide to form more of their kind (Fig. 2.7). *Mycoplasma gallisepticum,* which causes chronic respiratory disease in poultry, is somewhat larger (about 0.25 $\mu$) than *M. laidlawii* and has been especially useful in biochemical studies. Some of the chemical constituents are similar to those found in bacteria; others are like those found in animal cells. Of the nucleic acids found, some are linear and double-

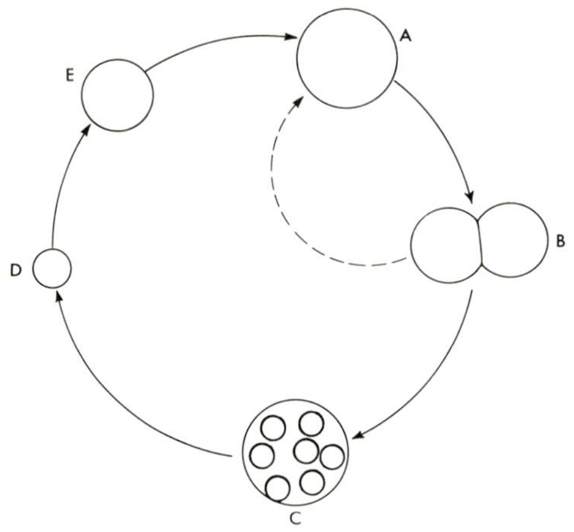


***Figure 2.7.*** Stages in the life cycle of *Mycoplasma laidlawii*. The cell at *A* is about 1 μ in diameter. It may divide (*B*) to form two cells like itself or microbodies (*C*) which are about 0.1 μ in diameter (*D*). The microbodies grow (*E*) to reform the full-sized cell (*A*). (After Morowitz and Tourtelotte.)

stranded (deoxyribonucleic acid), and others are in the form of spheroids combined with protein and resembling the ribosomes of other cells. Among the proteins are some 40 different kinds of enzymes, including those that catalyze the oxidation of pyruvic acid to carbon dioxide and water, reactions that liberate much energy and are found in a wide variety of cells, bacterial, plant and animal (see Chapter 16). The lipids found include cholesterol and cholesterol esters, both typical of animal cells. The plasma membrane is flexible and electron micrographs disclose it to be a "unit membrane" similar in dimensions to the unit membrane of cells in general.

## ORGANIZATION OF VIRUSES

Viruses are much smaller than common bacteria. For example, the large vaccinia virus measures 0.21 μ in diameter and a small virus such as that causing hoof and mouth disease is only 0.01 μ in diameter, whereas the colon bacillus, *Escherichia coli,* measures about 1 μ in diameter. However, some pleuropneumonia-like organisms, which have many of the characteristics of bacteria, have "small bodies" in the life cycle which measure only 0.125 to 0.150 μ in diameter. Viruses depend for their existence upon host cells; they lack enzymes for energy-yielding reactions. Viruses attack bacteria, plant and animal cells. Some viruses possess enzymes enabling them to penetrate cells, presumably digesting the cell wall in bacteria and mucopolysaccharides in animal cells (Weidel, 1959; Thomas, 1960).

All viruses consist of a protein coat over a nucleic acid core. Viruses contain only one kind of nucleic acid; thus ribonucleic acid (RNA) is present in plant viruses while animal and bacterial viruses contain either RNA or deoxyribonucleic acid (DNA), but never both. This distinguishes viruses from all cellular forms because bacterial, animal and plant cells always have both RNA and DNA (see Chapters 3 and 6). In general, animal viruses appear to be of a different nature from viruses of bacteria (bacteriophages or phages), but less is known of them than about phages because like plant viruses they present so many experimental difficulties. All viruses lack unit membranes characteristic of cells but they are delimited from the medium by a protein envelope called a capsid. In addition, some viruses, for example, those in mammalian cells, have an envelope formed of lipids and lipoproteins. Viruses also lack the other organelles characteristic of bacteria, including cell walls; thus viruses represent an entirely different category of organization from bacteria and other cells. (Gunsalus and Stanier, 1961; Frobisher, 1968).

Most extensively studied are the viruses (phages) which attack bacteria and which form the basis for most of the discussion that follows. It can be demonstrated by various marker techniques that, once the phage enters the bacterium, the nucleic acid of the phage takes command, altering the normal synthetic activities of the bacterium, leading to hydrolysis of its nucleic acids, and directing the synthesis of the virus type of nucleic acid and protein. The protein coat of the virus is syn-

*Figure 2.8.* Bacteriophage $T_2$ of the colon bacillus seen under the electron microscope (×37,000). Each of the particles can produce a plaque in a growth of bacteria and so become visible. (From Herriott and Barlow, 1953: J. Gen. Physiol. *36*:17.) For a diagram of the structure of a phage particle, see Figure 6.17.

thesized anew from materials taken in by the bacterial cell. In the *E. coli* phage packaging of the virus nucleic acid in protein coats begins about 10 minutes after infection and continues for another 20 minutes, when the process is complete. After a delay which varies with the virus and conditions, the bacterium is ruptured (lysed), liberating between 20 and 100 complete phage particles. In small viruses many thousands of particles may be produced (Frobisher, 1968).

Phages may affect bacteria in a number of ways; (1) They may replicate and lyse the bacteria as just described (phages that cause lysis are called virulent or *lytic phages*). (2) They may become incorporated into the genetic material of the bacterium and remain in the nonreplicating or repressed state (masked form), except that the prophage nucleic acid divides when the host cell genome replicates. That they have also altered the hereditary characteristics of the bacteria is shown by the resistance (immunity) of the bacteria so infected to attacks by homologous phages. This is achieved by the synthesis of a repressor substance which prevents multiplication of homologous phages (Frobisher, 1968). Phages that "hide" in this manner, dividing only when the bacterium divides, are spoken of as temperate or *lysogenic phages* (i.e., giving rise to lytic phages). Occasionally the masked lysogenic phage is derepressed and undergoes rapid replication, usually when its host cell has become weakened in some manner (Hogness, 1962). (3) Finally, some small phages may replicate and be released without lysing the bacterial hosts.

Damaging the bacteria infected with lysogenic phages by radiations or other unfavorable agents leads to a rapid development of the phage into an active lytic form. The particles are then released in infectious form. Lysis of an occasional bacterium in an undamaged population can spread the infection to many previously uninfected members of the population, in this way maintaining a mild infection. Lysogeny, as the phenomenon is called, indicates the intimate relationship formed by the phage and the bacterium, a conception having great influence on microbial genetics (see Chapter 6).

## APPENDIX

### 2.1 STAINING TECHNIQUE FOR LIVING AND FIXED CELLS

*Vital dyes* penetrate living cells and color certain structures without seriously injuring the cells. For example, neutral red may be used to stain the cytoplasm, although in some cells it accumulates in vacuoles instead. Janus green B stains mitochondria selectively. Methylene blue selectively stains the Golgi complex. Vital dyes are not entirely harmless, but they kill only after cells have been exposed to them for a long time, thus allowing time for study. Sometimes the poisonous constituent (e.g., the heavy metals in Janus green B) may be removed from a vital dye solution, making the dye much less toxic than the original sample. Although helpful, the vital dye technique has only limited use because many of the cell organelles are not stained by such dyes and, in all cases, scattering of light obscures boundaries of structures. However, dyes have been used to demonstrate that some cell organelles such as the mitochondria, Golgi complex and vacuoles are real structures present in the living cell (Foot, 1950).

Most of the early work on the nature of cell organelles was done with fixed and stained preparations. *Fixing agents,* such as formalin, alcohol, acids, salts of heavy metals or mixtures of these, precipitate the proteins and render them insoluble. Next, water is removed from the fixed tissues by *dehydrating agents,* such as alcohol, and the tissues are embedded in paraffin and sectioned with a microtome. The sections are affixed to slides, the paraffin removed with xylol and the slides washed in xylol-alcohol. By washing in decreasing concentrations of alcohol the sections are partially hydrated and the proteinaceous material of the cell is then differentially stained to distinguish the structures present. Natural dyes (e.g., hematoxylin) or basic anilin dyes (e.g., safranin and basic fuchsin) stain the nucleus selectively, and acid dyes (e.g., orange G, eosin and fast green) stain the cytoplasm. The sections are then dehydrated with alcohol. To reduce scattering of light, it is necessary to replace the alcoholic medium with a substance having the same refractive index as that of the protein particles. This is accomplished with a *clearing agent,* such as xylol, which infiltrates among the protein particles. The preparation is then mounted in balsam, which has a refractive index about equal to that of the cell proteins. As a result, it is possible to look at the cell and see clearly the stained structures that are within (Conn *et al.,* 1960; Davenport, 1960).

Cytochemical techniques for identification by color reactions of cell constituents and enzymes are considered in Chapter 3. Some cytochemical techniques have been developed at the electron microscope level.

### 2.2 AUTORADIOGRAPHY

If a cell is suspended in a solution containing a substance labeled by a radioactive isotope, which the cell is capable of incorporating, it is often possible to determine where in the cell incorporation occurs and to what extent. For example, $^{3}$H-uridine is presented to cells for a brief time and the cells are mounted on a slide and fixed. A layer of photographic emulsion is deposited over the preparation. The radioactive emission from the $^{3}$H-uridine will affect the silver halide in the emulsion. On development, the grains affected by the radiation (beta particles of low energy value) are reduced and appear as black grains or traces over the structure into which the $^3$H-uridine had been incorporated (Fig. 3.21). If the grains appear over the nucleus we know that the nucleus incorporates uridine into ribonucleic acid (RNA). Similar experiments can be performed with $^3$H-thymidine, which is selectively incorporated into deoxyribonucleic acid (DNA), and with $^3$H-amino acids, which are incorporated into proteins. Autoradiography has been of inestimable value in localizing the sites of a variety of syntheses in the cell. Recurrent reference to the use of this technique will be made in subsequent chapters (Prescott, 1964; Gude, 1967).

## 2.3 MICROSCOPICAL TECHNIQUES

The ordinary *light microscope* (see Fig. 2.9) makes it possible to see stained structures so long as they are larger than the limit of resolution. The smallest object, d, resolved (seen clearly and distinctly from similar sized objects separated by like small distances) by the light microscope may be determined from the following equation:

$$d = \frac{\lambda}{NA + na} \qquad (2.1)$$

In this equation λ is the wavelength* of light used, NA is the numerical aperture of the objective and na the numerical aperture of the condenser. Since for critical study the numerical aperture of the condenser is chosen equal to that of the objective, the equation is often written:

$$d = \lambda/2NA \qquad (2.2)$$

The numerical aperture is given by the equation:

$$NA = n \sin \alpha \qquad (2.3)$$

Here n is the refractive index of the light path in the formation of the microscopical image, and α is half the angular aperture of the light beam entering the condenser. When immersion oil is used between condenser and slide, and slide and objective, the numerical aperture may be as large as 1.6, although in practice it is seldom over 1.4. The factor that limits resolution with the light microscope is therefore the wave-

*Wavelength of light is given in a variety of units of length. The *angstrom unit*, Å or A, is $10^{-8}$ cm. or $10^{-10}$ meters. The *nanometer, nm.*, $10^{-9}$ meter, is coming into rather general use. However, the *millimicron*, mμ, $10^{-9}$ meter, is also in wide use. Sometimes wavelength is given in *microns*, μ, $10^{-6}$ meter.

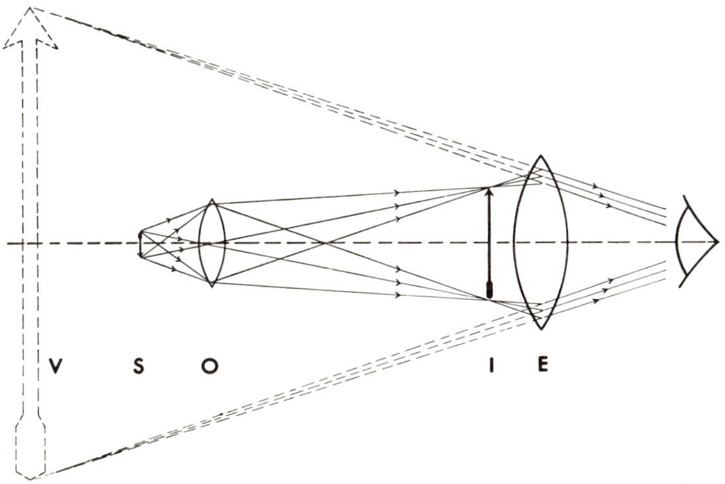


***Figure 2.9.*** Principle of the compound microscope. The intermediate image (I) of the subject (S) formed by the objective (O) is enlarged by the eyepiece (E). The virtual image (V) is seen by the eye. A real image of V is recorded with a camera. (From Photography through the Microscope, Kodak Scientific & Technical Data Book P-2, 4th Ed. Eastman Kodak Company, Rochester, New York, 1966.)

length of light. Under optimal conditions the smallest object seen is theoretically about one-third the wavelength of visible light or about 0.13 μ in blue light. Usually the limit is set at about 0.2 μ because the eye is most sensitive in the yellow-green part of the spectrum (0.55 μ) and a maximum numerical aperture of 1.6 is seldom achieved. An object of this size is visible only if the structure shows maximum contrast. In order to increase the contrast between unstained structures and to obtain greater resolution of details of cell organelles, various types of microscopes have been developed, each of which will be considered in turn (Needham, 1958; Clark, 1961).

The *dark field microscope* is an ordinary microscope with a special condenser. By use of a dark stop in the condenser the light from the center of the field is removed and the object is illuminated only by an oblique beam of light as seen in Figure 2.10. The object is therefore seen by light reflected and scattered from interfaces in the object. With dark field microscopy the outlines of the cell, the nucleus, mitochondria, oil droplets, vacuoles and various inclusions may be readily identified. Moreover, in cells undergoing division, the centrosomes, asters, spindles and chromosomes may be observed. Objects smaller than those seen with the ordinary light microscope may be detected with dark field microscopy but not resolved.

The *polarizing microscope* is like a light microscope except that the condenser and the ocular are equipped with polarizing optics, each of which transmits only plane-polarized light. The condenser is then spoken of as the *polarizer* and the ocular as the *analyzer*. If the optics of the condenser and ocular are crossed (at 90° to one another) no light is transmitted unless an interposed object rotates the plane of the light. The manner in which this occurs and the usefulness of the microscope in biology is briefly outlined in the following paragraphs.

Light, passing through a doubly refracting crystal such as calcite, is separated into two rays which vibrate at right angles to one another. Such light is said to be polarized. The *Nicol prism,* one of the best known polarizers, is made by cutting a rhomb of Iceland spar (a pure, transparent form of calcite) in halves along the line of one of the optical faces, polishing the cut faces and cementing them together with Canada balsam. As seen in Figure 2.11, unpolarized light entering one end is separated by the prism into two beams, one vibrating in the plane of the figure, the other at right angles to it. One of

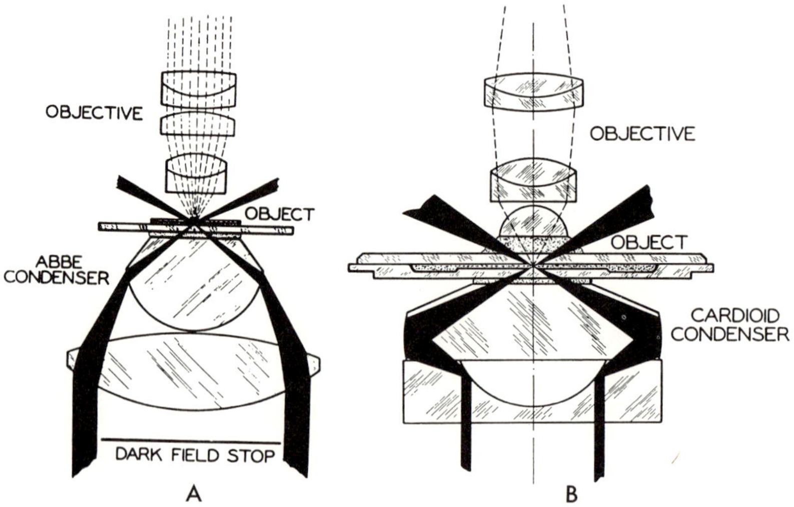


***Figure 2.10.*** Dark field microscopy. The dark field stop shown in *A* is much less effective than the dark field condenser shown in *B*. (By permission of Bausch and Lomb Inc., Rochester, N. Y.)

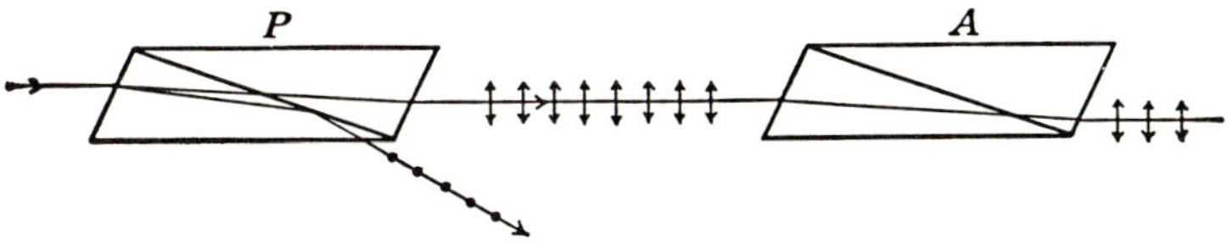


***Figure 2.11.*** Polarization of light by a Nicol prism. Only the beam in the plane of the sheet has been permitted to pass; that at right angles is deflected. The second Nicol prism, if oriented like the first, has no effect on the polarized light. However, if it is rotated through 90° the beam will be extinguished. (From Stewart and Gingrich, 1959: Physics. 6th Ed. Ginn & Co., Boston.)

these beams is totally reflected by the layer of Canada balsam and is usually absorbed by the opaque case covering part of the rhomb. The other beam is transmitted. Hence, a Nicol prism transmits plane-polarized light. If a second Nicol prism is used in series with the first, and in the same position, the polarized light is transmitted. If, however, the second prism is rotated through 90° in relation to the first prism, the light fails to pass. But if an object like quartz, which polarizes light, is placed in the beam of light between the two prisms, and if the two Nicols are at right angles to each other (crossed), instead of a plain dark field, a brilliantly lit object is seen against a dark background, indicating that the object has polarized the light passing through it. If the second prism is then rotated away from 90°, the field suddenly becomes dark. The angle required to rotate the second prism in order to extinguish the light is a measure of the degree to which the plane of polarized light has been rotated by the interposed object. The second prism is called an analyzer because by moving it one can analyze the degree to which the plane of polarized light has been rotated by an object inserted in the light path.

Various natural substances are observed to be *isotropic;* that is, light passes through them equally well in all directions. Others, which affect the passage of light unequally in different directions, are *anisotropic* or *birefringent.* The ray passing directly through a crystal is called the *ordinary ray;* the ray which is deviated is called the *extraordinary ray.* If the velocity of the extraordinary ray passing through such a material is less than the velocity of the ordinary ray, the crystal is said to be negative (e.g., quartz). If the extraordinary ray has a velocity greater than the ordinary, the crystal is said to be positive. *Crystalline* or *intrinsic birefringence* is found where the bonds between molecules (or ions) have a regular asymmetrical arrangement. The birefringence in this case is independent of the refractive index of the medium.

*Dichroism* occurs when the absorption of a given wavelength of polarized light changes with the orientation of the object. For example, the degree of absorption at wavelength 260 nm. by nucleic acids in cells depends upon whether the light is parallel to, or at right angles to, the oriented purine and pyrimidine residues. Dichroism is a sensitive indicator of orientation of molecules in a structure. However, it is seldom seen in biological objects in the visible spectrum except in some stained preparations where it results from orientation of dye molecules. It may be seen at 260 nm. in the sperm nuclei where the concentrated nucleic acids appear to be oriented.

Birefringence of inanimate objects, then, is the result of their crystalline nature. Birefringence of biological materials indicates that their molecules are oriented with respect to one another much as in a crystal. Some particles—for example, nucleoprotein molecules—line up only if the solution is flowing; what they show is therefore called *flow* or *streaming birefringence.* This indicates that the molecules are elongate and line up in the path of least resistance to a current, just as sticks line up in a brook. Organisms are more likely to show *form birefringence,* because of regular orientation of tiny particles (rods or planes) differing in refractive index from the surrounding medium. Birefringence in such cases disappears when the material is immersed in a medium of the same refractive index. Oriented rods have different refractive indices for directions parallel to and at right angles to the major axis. Therefore, by making observations with polarized light one may determine in which direction in a given structure the rods lie.

Polarizing optics have been especially useful in study of biological membranes (Chapter 11) and of the division figure in cells (Chapter 27). They are also useful in studying fibrillar structures.

The *phase contrast microscope,* which has the same resolving power as the ordinary light microscope, enables one to see structures in the living cell by taking advantage of the slight differences in refractive index between any two structures to improve their visibility (Zernike, 1955). The phase contrast microscope does this by the use of a special diaphragm in the condenser and a phase plate in the objective. The annular diaphragm used in the condenser of the phase microscope permits light to pass through the condenser as a hollow cone, the remaining light being absorbed (see Fig. 2.12). This cone is focused on the object. The phase plate placed at the back focal plane of the objective is a transparent disk containing a groove (or elevation) of such a size and shape as to coincide with the direct image of the substage annular diaphragm formed at the back focal plane when no object is viewed. If an object is placed between the condenser and the objective, in addition to the direct image a number of overlapping diffracted images of the diaphragm then appear at the back focal plane of the objective. The depth of the groove (or the elevation) in the phase plate of the objective is so made that the two sets of rays forming the direct image and the diffracted image differ in optical path by a quarter wavelength of the illuminating beam of light (Fig. 2.12*B*). Under these conditions the phase difference, which is not seen by the eye, is converted to an intensity difference which we see. In the bright contrast phase optical system the two sets of rays are added to make a brighter image as in Figure 2.12 at *D*, whereas in dark contrast phase they partially cancel one another making a more contrasting darker image, as in Figure 2.12 at *C* (Barer, 1966).

The phase contrast microscope (bright or dark contrast may be used), in which a small difference of refractive index is exaggerated, enabling one to distinguish adjacent structures, has made it possible to observe structures previously very difficult or impossible to see. The behavior of chromosomes of living cells during mitosis or meiosis can be followed with ease, and many of the other organelles may be studied. The phase contrast microscope has furnished the most convincing evidence that many of the organelles identified in fixed and stained preparations of the cell are real and not artifacts. However, the images are often surrounded by a halo and the difference in intensity between two areas of unequal refractive index diminishes with the distance from their boundary, thus causing some distortion.

The *interference microscope* avoids the residual halo which accompanies the image seen under a phase microscope because the aperture of the objective of such a microscope is not limited by a phase plate and the same area of the objective is used for both interfering beams. The relative phase of the two beams is varied continuously so that any part of an object can be given maximum or minimum contrast at will (Barer, 1966).

As observed in Figure 2.13, a horizontal glass plate, L, nearly parallel to a similar plate, U, above the object, intercepts the illuminating cone of light from the

---

***Figure 2.12.*** *A* shows the light path in a phase microscope. *B* shows the normal retardation by ¼ wavelength of light diffracted by an object and its difference in phase from the light passing through the surrounding medium. By phase optics the two waves are superimposed to reinforce each other in bright contrast phase as shown in *D* or to subtract from each other as in dark contrast phase shown in *C*. (By permission of the American Optical Company, Buffalo, N. Y.)

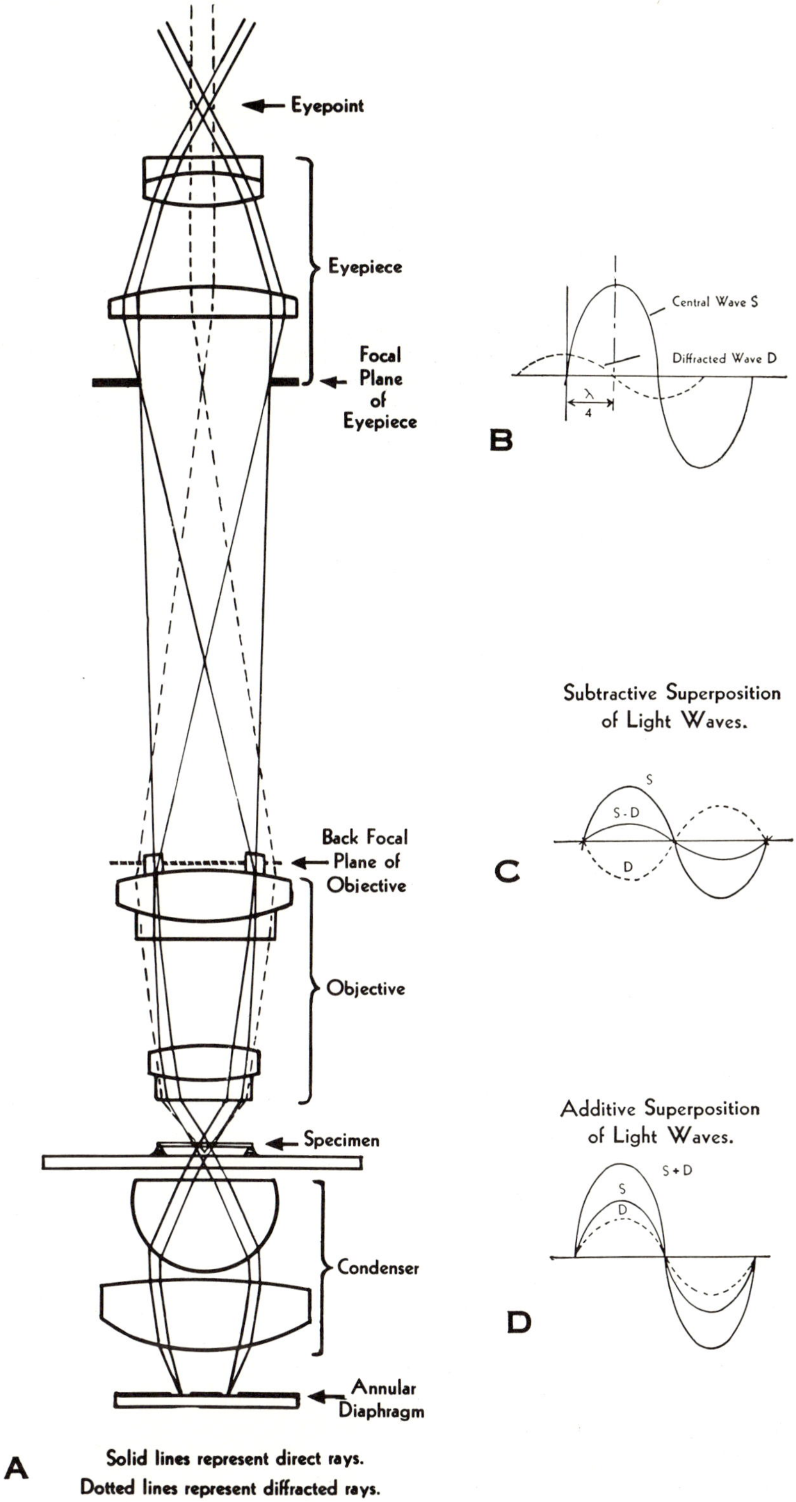


*Figure 2.12.* See opposite page for legend.

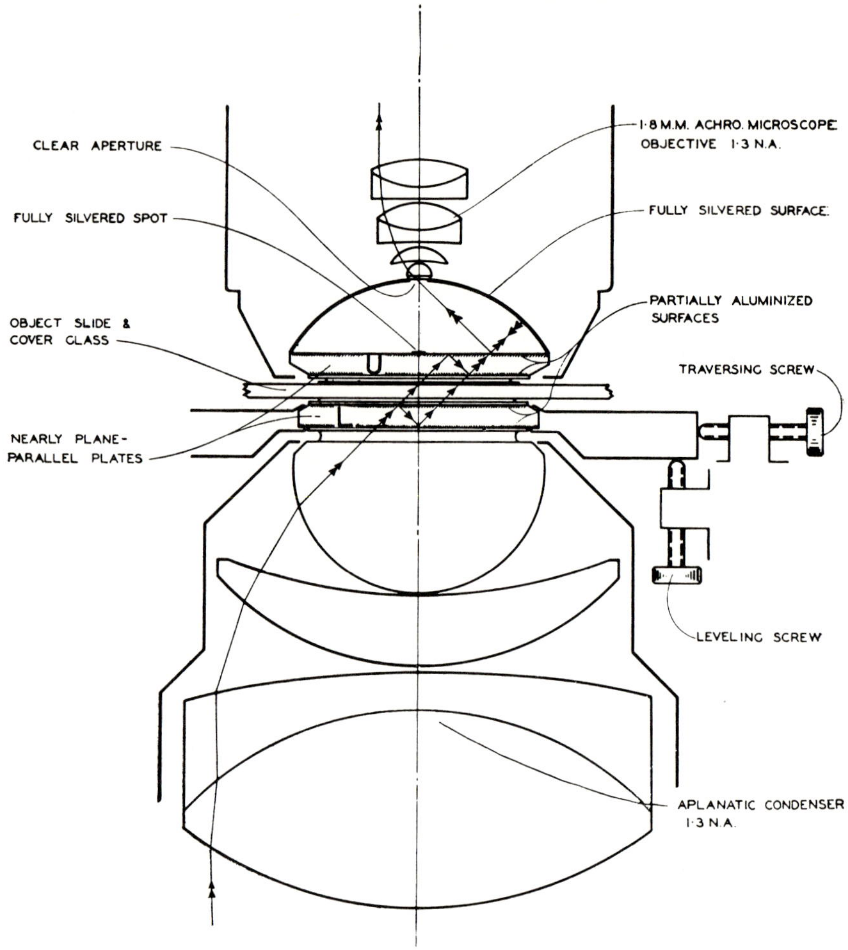


***Figure 2.13.*** Dyson interferometer microscope, light path. For explanation see accompanying text. (By permission of Cooke, Troughton and Simms, Ltd., Malden, Mass.)

condenser (both surfaces of this plate are partially aluminized). A portion of the illuminating cone passes through the plate and is focused on the object, and a portion is reflected downward at the upper surface of plate L and is then reflected upward by its lower surface. Although only one beam traverses the object, two beams traverse the object plane. The beam failing to pass through the object, called the *comparison beam,* is internally reflected twice within the plate. It will be noted that it has a diameter much larger than that of the field.

A second horizontal plate, U, identical to the first and just about parallel to it, is located above the plane of the object. Both of its surfaces are also partially aluminized, but in addition its upper surface has an axial opaque spot somewhat larger in diameter than the maximum field of the microscope. This spot prevents the direct vertical light beam from the source which has passed through the object from reaching the objective, but allows the light from the beam which has traversed the object and then enters the plate to the side of the spot to reach the objective.

Part of the illuminating beam which has traversed the object is reflected twice in plate U and part of the comparison beam traverses this plate without reflection; the two beams will emerge from the upper surface in coincidence when the plates are correctly aligned. The observer above the upper plate sees two coincident fields of view, one an image of the light source only and the other an image of the light source upon which is superimposed an image of the object. Aside from the interference resulting from passage through the object, the phase difference between the two images will be zero. The combined images may be viewed by a normal microscope objective. The interferometer plates L and U are movable so that the path difference between the two coincident fields can be varied at will over the whole area of vision. Maximum contrast cannot be obtained for objects much larger than the field of view, because the object is then traversed by a portion of the background beam.

Because an object such as a cell contains various substances, it has an optical path different from that of the surrounding water. For this reason when the background is adjusted to maximum light intensity (bright field), the cell appears darker and it is possible at a glance to note the distribution of structures in the cell much as if a stained preparation were being observed in an ordinary light microscope. More than this, it is possible to determine the average amount of dry material (dry mass) present in a given area of the cell by the use of the interference microscope, because the greater the amount of material present in the cell, the greater will be the difference between its optical path (refractive index times thickness) and that of the surrounding fluid. The difference in optical path is determined by the difference between interference bands in the surrounding fluid and in the object when the upper wedge plate is rotated so that the wedge axes are no longer parallel. The field is then crossed by parallel interference fringes. These fringes are seen as lighter and darker bands of the same color when monochromatic light is used, or bands of color on either side of white bands when white light is used. For example, one-tenth wavelength displacement for monochromatic green light (546 nm.) is equivalent to a dry mass of $10^{-13}$ gm./$\mu^2$.

The *ultraviolet microscope* looks like the ordinary light microscope but its lenses are made of quartz to permit transmission of wavelengths of invisible ultraviolet as short as 220 nm. (visible light covers the range from blue at 390 nm. to the far red at 780 nm.). The invisible ultraviolet light is then projected upon a screen which fluoresces in the visible spectrum. Usually, however, it is recorded photographically. Because chromosomes absorb more short ultraviolet radiation than cytoplasm does, they are readily seen in photographs taken with ultraviolet light at 260 nm. (Caspersson, 1950). Under the ultraviolet microscope resolution of fine structure is about twice that obtained with the ordinary light microscope, as may be judged from application of equations 2.1 and 2.2, the wavelength of short ultraviolet light being about half that of the visible. Unfortunately ultraviolet radiation damages living cells, so that prolonged observation of cells so illuminated is not possible.

An instrument which may serve as a valuable adjunct to all the microscopes considered here is the *television camera attachment.* Since the electron tube used in it "sees" at low light intensities and can be made sensitive to a restricted part of the spectrum (a narrow band in the green, blue or ultraviolet), much clearer pictures can be obtained than with ordinary microscopy (Zworykin *et al.*, 1952). Furthermore, a prolonged illumination is avoided since the image may be viewed on a phosphorescent screen which retains the image for a minute or two after exposure. Momentary illumination with light of low intensity suffices. This attachment, therefore, is especially useful for viewing living cells under ultraviolet light (Whipple *et al.*, 1954; Montgomery and Bonner, 1958).

The *electron microscope,* which was developed on the eve of World War II, has proved a powerful tool for analysis of large chemical molecules, viruses and, to some extent, cellular structures (Freundlich, 1963). In the electron microscope the beams of electrons are focused by magnets which serve as lenses (see Fig. 2.14). The object must be viewed on a fluorescent screen or photographed. The electron microscope has a resolving power 400 (theoretically 2000) times that of the light microscope. Although the wavelengths accompanying the electron beams are short (e.g., a 50,000 volt electron beam is accompanied by a wavelength of 0.05 Å), the numerical aperture of the electron microscope is only about 0.0005.

The image results from differential scattering of electrons from the molecular constituents of the cell. Scattering is greater the denser the material, regardless of its chemical composition. Even a gas scatters electrons; therefore, the biological object must be studied in a vacuum, since electron scattering from the gas molecules would obscure the differential scattering of the structures in the biological object. The preparations must be dry because the water vapor in the evacuated preparation would have the same effect as a gas. A thin slice of the object must be used; otherwise the electrons would all be scattered and the electron micrograph would reveal no detail. Since biological material varies little in density, it must be stained differentially, i.e., with some structures taking up more stain than others. *Positive staining* is obtained by fixing the cell in salts of heavy metals, e.g., osmium tetroxide or potassium permanganate. In this case the electron-dense (i.e., opaque to electrons) metal adheres differentially to structures of the cell, causing them to scatter the electrons more than the surroundings. In other cases *negative staining* is used; that is, the object is infiltrated with a heavy metal compound such as phosphotungstic acid or uranyl acetate. These materials do not adhere to the biological structures but fill the interstices between them, making these areas electron-dense in contrast to the biological structures which are then seen as relatively electron-transparent structures (Anderson, 1966; Haggis, 1967).

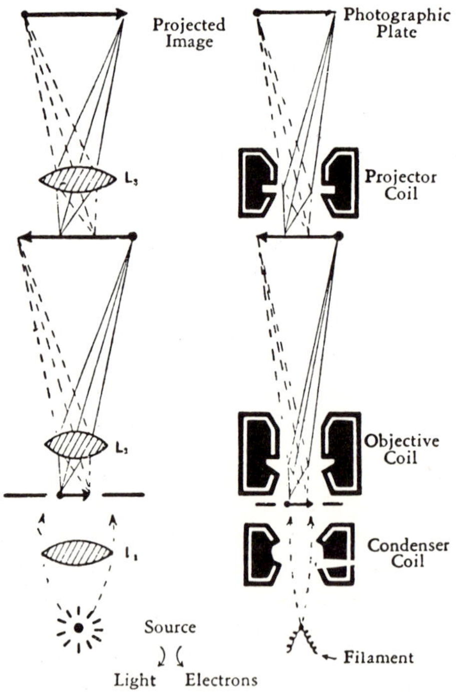


***Figure 2.14.*** Comparison of optical pathways in light and electron microscopes. (From Thompson, 1943: Endeavour *2*:125.)

The main disadvantages attendant upon use of the electron microscope are the required thinness of sections and the necessity of studying them dry (in a vacuum). The first difficulty has been overcome largely by the development of microtomes that can make sections 0.2 to 0.02 $\mu$ in thickness (Pease, 1964). However, removal of water and salts dissolved in it as well as the use of stains are likely to alter structure; therefore, membranes, cytoplasm or formed structural components seen by electron microscopy must be interpreted with some caution. Notwithstanding these facts, it will become apparent in studies presented later that much has been learned about the cell by electron microscopy, and about structural elements such as the cuticle and the cell wall.

Because of the small contrast between some biological objects and the background even after staining, *shadowcasting* with heavy metals is often used, especially to bring out three-dimensional topography. Heavy metals, especially gold, platinum and chromium, are evaporated in a vacuum at an oblique angle to the object. The metal piles up like a windrow and outlines the object. Shadowing may be performed from several directions by moving the object during the process of evaporation (Anderson, 1966).

When the object is too thick for shadowcasting a *replica* can be made. In this case the object, mounted on a grid containing holes, is subjected *in vacuo* to evaporation of platinum and carbon and becomes coated to the depth of several hundred angstroms. The preparation is then treated with hydroxide to remove the specimen and the replica is floated onto another grid and prepared for electron microscopic study.

*Surface spreading* has also been a useful technique for study of cell fine structure. The spread layer is "swept" in the trough to concentrate it, and mounted on a grid, dried and shadowed for electron microscopic study. In this manner a bacterial protoplast (bacterium from which the cell wall has been removed with the enzyme lysozyme) that is ruptured osmotically on the surface of a vessel filled with distilled water is found to liberate its chromosome as a single long fiber. The same technique has been used for a study of interphase chromosomes in cells of plants and animals and for a study of microtubules present in some cells (Wolfe, 1965).

**Freeze-etching.** The images of various cell organelles seen with the electron microscope have been criticized because they are structures seen in cells that have been fixed and stained. Are they real or are they artifacts introduced by the chemical procedures and drying? It is disquieting also that different fixatives sometimes give different images; for example, osmium tetroxide fixation shows a plasma membrane consisting of a single line whereas permanganate fixation shows the plasma membrane consisting of two outer electron-dense layers and an inner less electron-dense layer (Robertson, 1959).

It has proved possible to study the cell organelles by the use of unfixed cells for electron microscopy, in the method called *freeze-etching*. The object to be studied is placed in 20 per cent glycerol (antifreeze, see Chapter 10) and is frozen at −100° C. It is mounted on a chilled holder and splintered with a knife along natural cleavage planes, usually along surfaces of membranes. Occasionally the structure is cross-fractured, giving cross sections of organelles. The splintered preparation is freeze-dried and covered with a platinum and carbon coating in the high vacuum of a freezing ultramicrotome. When the preparation is dry the vacuum is broken and the preparation is placed in water to float the replica off the carrier. The replica is then washed in basic solution to remove the cellular material and the replica is mounted on a grid and dried for electron microscopy.

Such replicas give the outlines of various cell structures and verify the structures seen in thin fixed and stained sections of the mitochondria, smooth and rough endoplasmic reticulum, double nuclear envelope with pores and plugs, unit plasma

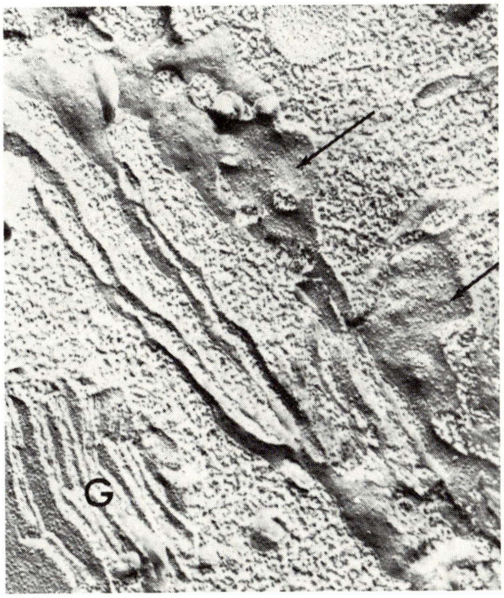


***Figure 2.15.*** A freeze-etching of an onion root tip cell. The arrows point to areas where the endoplasmic reticulum membranes have been splintered away, revealing the surface of the cytoplasm next to them. At G are seen the membranes and surfaces of the Golgi complex saccules. (By courtesy of Dr. D. Branton.)

membrane of the cell, chloroplasts and so forth (Moor and Mühlethaler, 1963; Branton, 1964). The views of structures are somewhat three-dimensional (Fig. 2.15). Since some cells (e.g., yeast) subjected to these freezing and drying procedures revive and show no evident change in properties, it is assumed that electron micrographs of cells so prepared give images of cell structures in as nearly normal condition as is at present possible. The fact that the structures compare favorably in size and appearance to those seen in fixed and stained cells is, in some measure, confirmation of the validity of the conventional method using fixatives.

## 2.4 MICROCINEMATOGRAPHY

In many cases considerable information has been obtained by time-lapse photomicrography. Details that escape the observer on a single viewing become apparent after viewing the record repeatedly. Familiar examples are mitosis and meiosis in cells (Bajer and Mole-Bajer, 1963) and irradiation of cells with a microbeam (Bessis, 1965). The components used in such studies are familiar; the details must be sought elsewhere (Rose, 1963).

## 2.5 MICROMANIPULATORS

Since the end of the last century micromanipulators of one kind or another have been in use, first for handling single cells and later for microsurgery upon them (see Fig. 2.16). The pneumatic micromanipulator of De Fonbrunne (Fig. 2.17) has had wide use for microsurgery because of its ease of operation, speed and sensitivity of response. Essentially it consists of three pumps, two horizontal and one vertical, which connect by way of rubber tubes to the pneumatic capsules of the instrument holder (for needle or pipette). The movements in the horizontal plane of a single lever or joy stick are reflected at a ratio of anywhere from 1:50 to 1:2500, according to the setting decided upon. The vertical plane level is set by the screw on the lever.

An electric micromanipulator utilizing thermal expansion wires has been described but has not had much use (Bush *et al.,* 1953). Production of microforges for the controlled manufacture of microtools has also facilitated microdissection. Important information on the function of the nucleus of cells has been gained by the use of such tools for transplantation of nuclei between cells (see Chapter 6). The micromanipulator thus makes the micro-

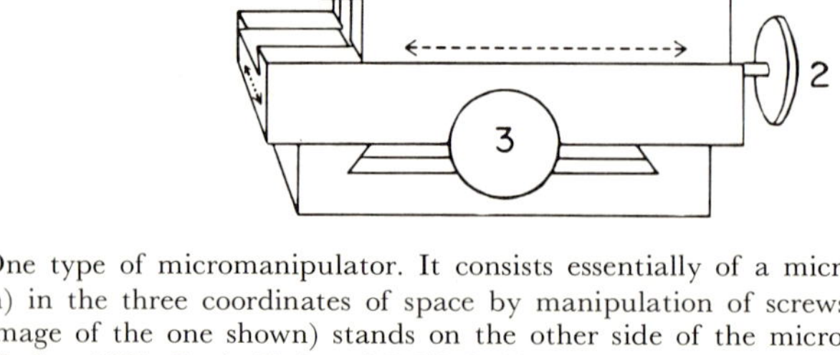


***Figure 2.16.*** One type of micromanipulator. It consists essentially of a microlathe enabling movements of the needle (n) in the three coordinates of space by manipulation of screws 1, 2, and 3. Another instrument (a mirror image of the one shown) stands on the other side of the microscope, all mounted on a heavy stand. (After Taylor, 1925: Zool., Univ. of Calif. Publ. *26*:443.)

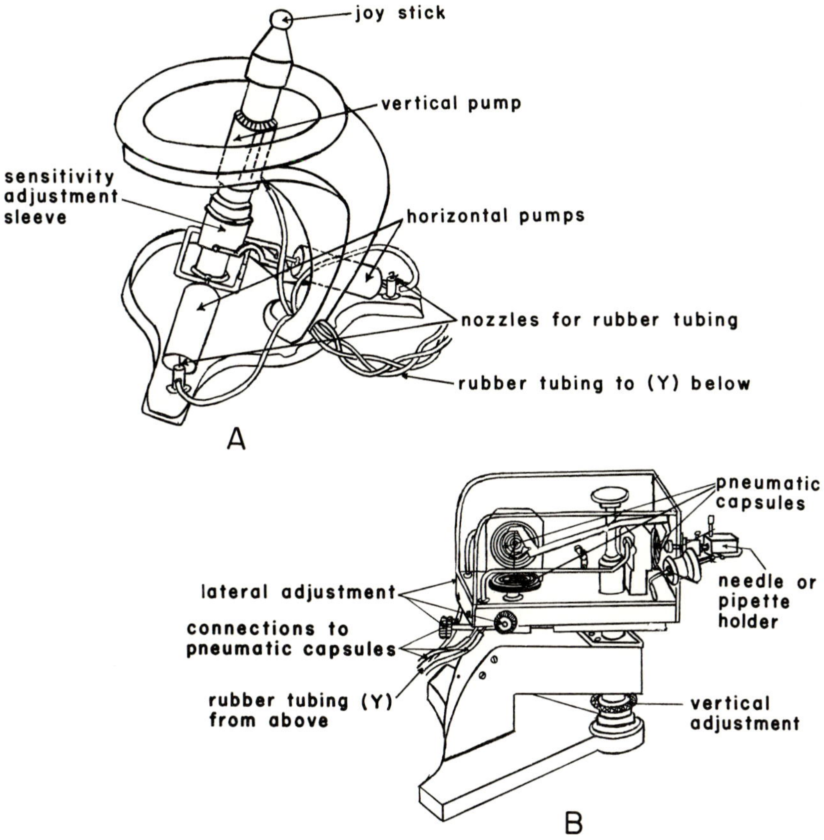


***Figure 2.17.*** The De Fonbrunne pneumatic micromanipulator. A movement of the control or joy stick is duplicated by the needle (or other instrument) with a reduction in magnitude of $1/50$ to $1/2{,}500$, set at the sensitivity adjustment sleeve. This is achieved through passage of the pressure change through three pumps, two horizontal and one vertical, attached to three pneumatic capsules with rubber tubing, the capsules in turn controlling the needle or instrument holder. Two such micromanipulators are often used together, one to hold, the other to cut or manipulate a structure viewed under the microscope. (By courtesy of A. S. Aloe Company, St. Louis, Mo.)

scope more than an observational tool, and properties of many of the cell organelles have been explored by its use. Even bacteria have been dissected under high power (El-Badry, 1964).

## LITERATURE CITED AND GENERAL REFERENCES

Ainsworth, G. C., and Sussman, A. S. (eds.), 1965: The Fungi, An Advanced Treatise. Vol. 1. The Fungal Cell. Academic Press, New York.

Anderson, T. F., 1966: Electron microscopy of microorganisms. *In* Physical Techniques in Biological Research. Pollister, A. W., ed. Academic Press, New York, Vol. 3A, pp. 319–387.

Bajer, A. and Molè-Bajer, J., 1963: Cine analysis of some aspects of mitosis in endosperm (*Haemanthus katherinae*). *In* Cinematography in Cell Biology. Rose, G. G., ed. Academic Press, New York, pp. 357–409.

Barer, R., 1966: Phase contrast and interference microscopy in cytology. *In* Physical Techniques in Biological Research. Pollister, A. W., ed. Academic Press, New York, Vol. 3A, pp. 91–158.

Bessis, M., 1965: Introductory lecture: micro-irradiation of cells. *In* Recent Progress in Photobiology. Bowen, E. J., ed. Blackwell Scientific Publications, Oxford.

*Bloom, W., and Fawcett, D. W., 1968: A Textbook of Histology. 9th Ed. W. B. Saunders Co., Philadelphia.

Blout, E. R., 1953: Ultraviolet microscopy and ultraviolet microspectroscopy. Advances Biol. Med. Phys. *3*:285–336.

Bourne, G. H., 1962: Division of Labor in Cells. Academic Press, New York.

Bourne, G. H. (ed.), 1964: Cytology and Cell Physiology. 3rd Ed. Academic Press, New York.

Bovarick, M. R., Jan. 1955: Rickettsiae. Sci. Am. *192*:74–79.

Brachet, J., and Mirsky, A. E., 1959–1961: The Cell. 5 Vol. Academic Press, New York.

Branton, D., 1964: Fine structures in freeze-etched *Allium cepa*. J. Ultrastruc. Res. *11*:401–411.

Brown, R., 1960: The plant cell and its inclusions. *In* Plant Physiology. A Treatise. Steward, F. C., ed. Academic Press, New York. Vol. 1A, pp. 3–129.

Burnet, F. (ed.), 1959: The Viruses. 3 Vol., Academic Press, New York.

Bush, V., Duryee, W. R., and Hastings, J. A., 1953: An electric micromanipulator. Rev. Sci. Instr. *24*:487–489.

Caspersson, T., 1950: Cell Growth and Cell Function. W. W. Norton & Co., New York.

Chambers, R. W., and Kopac, M. J., 1950: Micrurgical technique for the study of cellular phenomena. *In* C. E. McClung's Handbook of Microscopical Techniques. Hoeber Division, Harper & Row, New York, pp. 492–543.

Clark, G. L., 1961: The Encyclopedia of Microscopy. Reinhold Publishing Corp., New York.

Conn, H. J., Darrow, M. A., and Emmel, V. M., 1960: Staining Procedures used by the Biological Stain Commission. Williams & Wilkins, Baltimore.

Dalton, A. J., 1961: Golgi apparatus and secretion granules. *In* The Cell. Brachet, J., and Mirsky, A. E., eds. Academic Press, New York, Vol. 2, pp. 603–619.

Danielli, J. F., 1953: Cytochemistry, A Critical Approach. John Wiley & Sons, New York.

Danielli, J. F. (ed.), 1958, 1961: General Cytological Methods. Academic Press, New York, Vol. 1, 1958; Vol. 2, 1961.

Davenport, H. A., 1960: Histological and Histochemical Technics. W. B. Saunders Co., Philadelphia.

De Fonbrunne, P., 1937: Micromanipulateur-pneumatique, et micro-forge, pour la fabrication des microinstruments. Soc. Industrielle d'imprimerie, 10, rue vallier, Levallos-Perret.

DeRobertis, E. D. P., Nowinski, W. W., and Saez, F. A., 1965: Cell Biology. 4th Ed. W. B. Saunders Co., Philadelphia.

Echlin, P., June, 1966: The blue-green algae. Sci. Am. *214*:75–81.

El-Badry, H., 1964: Micromanipulators and Micromanipulation. Academic Press, New York.

*Esau, K., 1965: Plant Anatomy. 2nd Ed. John Wiley & Sons, New York.

Fernandez-Moran, H., 1956: Applications of a diamond knife for ultrathin sectioning to the study of fine structure of biological tissues and metals. J. Biophys. Biochem. Cytol. Suppl. *2*:29–30.

Foot, N. C., 1950: Vital stains. *In* McClung's Handbook of Microscopical Techniques. Hoeber Division, Harper & Row, New York, pp. 564–570.

Freundlich, M. M., 1963: Origin of the electron microscope. Science *142*:185–188.

Frey-Wyssling, A., and Mühlethaler, K., 1965: Ultrastructural Plant Cytology. American Elsevier Publishing Co., New York.

Frobisher, M., 1968: Fundamentals of Microbiology. W. B. Saunders Company, Philadelphia, Chapters 16 to 18.

Gude, W. D., 1967: Autoradiographic Techniques: Localization of Radioisotopes in Biological Material. Prentice-Hall, Englewood Cliffs, N.J.

Gunsalus, I. C., and Stanier, R. Y., 1961: The Bacteria. 5 Vol. Academic Press, New York.

Haggis, G. H., 1967: The Electron Microscope in Molecular Biology. John Wiley & Sons, New York.

Hale, A. J., 1960: The interference microscope as a cell balance. *In* New Approaches in Cell Biology, Symposium, 1958. Walker, P. M. B., ed. Imperial College, London, pp. 173–186.

Harvey, E. B., 1951: Cleavage in centrifuged eggs, and in parthenogenetic merogones. Ann. N. Y. Acad. Sci. *51*:1336–1348.

Hogness, D. S., 1962: The activity of isolated viral DNA in the genesis of proteins within bacterial cells. *In* The Molecular Control of Cellular Activity. Allen, J., ed. McGraw-Hill Book Co., New York, pp. 189–220.

Holter, H., and Ottolenghi, P., 1960: Observations on yeast protoplasts. Compt. rend. Lab. Carlsberg *31*:409–422.

Inoué, S., 1951: A method for measuring small retardations of structure in living cells. Exp. Cell Res. *2*:513–517.

Jensen, W. A., 1964: The Plant Cell. Wadsworth Publishing Co., Belmont, Calif.

Kopac, M. J., 1955: Cytochemical microsurgery. Intern. Rev. Cytol. *4*:1–29.

Kuff, C. L., and Dalton, A. J., 1959: Biochemical studies of isolated Golgi membranes. *In* Subcellular Particles. Hayashi, T., ed. Ronald Press Co., New York, pp. 114–127.

*Loewy, A. G., and Siekevitz, P., 1963: Cell Structure and Function. Holt, Rinehart & Winston, New York.

Lowry, O. H., 1943: Electrolytes in the cytoplasm. Biol. Symp. *10*:233–245.

Marshak, A., 1955: Bacterial cytology. Intern. Rev. Cytol. *4*:103–114.

Meeuse, A. D. J., 1957: Plasmodesmata (vegetable kingdom). *In* Protoplasmatologia. Vol. 2. Springer-Verlag, Vienna.

Meryman, H. T., 1960: Drying of living mammalian cells. Ann. N. Y. Acad. Sci. *85*:729–734.

Montgomery, P. P., and Bonner, W. A., May, 1958: A "flying-spot" microscope. Sci. Am. *198*:39–43.

Moor, H. and Mühlethaler, K., 1963: Fine structure in frozen-etched yeast cells. J. Cell Biol. *17*:609–620.

Morowitz, H. J., and Tourtelotte, M. E., March, 1962: The smallest living cells. Sci. Am. *206*:117–126.

Morowitz, H. J., Tourtelotte, M. E., Guild, W. R., Castro, E., and Woesse, C., 1962: The chemical composition and submicroscopic morphology of

*Mycoplasma gallisepticum* Avian. PPLO 5969. J. Molec. Biol. *4*:93–103.

Nastuk, W. L., 1955–1963. Physical Techniques in Biological Research. Vol. 1, 1955, Optical Techniques. Vol. 2, 1956, Physical Chemical Techniques. Vol. 3, 1956, Cells and Tissues. Vol. 4, 1962, Special Methods. Vol. 5, 1964, Electrophysiological Methods; Parts A and B. Academic Press, New York.

*Needham, G. H., 1958: The Practical Use of the Microscope. Charles C Thomas, Springfield, Ill.

Pease, D. C., 1964: Histological Techniques for Electron Microscopy. 2nd Ed. Academic Press, New York.

*Picken, L., 1960: The Organization of Cells. Clarendon Press, Oxford.

Pollister, A. W. (ed.), 1966: Physical Techniques in Biological Research. Vol. 3A. Cells and Tissues. Academic Press, New York.

Porter, K. L., 1961: The ground substance: observations from electron microscopy. *In* The Cell. Brachet, J., and Mirsky, A. E., eds. Academic Press, New York, Vol. 2, pp. 621–675.

Prescott, D. M. (ed.), 1964, 1966: Methods in Cell Physiology. Academic Press, New York, Vol. 1, 1964; Vol. 2, 1966.

Richards, O. W., 1954: Phase microscopy. 1950–54. Science *120*:631–639.

Robertson, J. D., 1959: The ultrastructure of cell membranes and their derivatives. Symp. Biochem. Soc. London *16*:3–43.

Robinow, G. F., 1960: Outline of the visible organization of bacteria. *In* The Cell. Brachet, J., and Mirsky, A. E., eds. Academic Press, New York, pp. 45–108.

Rose, G. G., 1963: Cinematography in Cell Biology. Academic Press, New York.

Stanier, R. Y., and Van Niel, C. B., 1962: The concept of a bacterium. Archiv. Mikrobiol. *42*:2–35.

Taylor, W. K., 1960: Flying spot microscopy. *In* Cell Biology. Walker, P. M. B., ed. Imperial College, London, pp. 187–198.

Thomas, R., 1960: Viruses. *In* The Cell. Brachet, J., and Mirsky, A. E., eds. Academic Press, New York, Vol. 4, pp. 1–44.

Voller, B., Ledbetter, M. C., and Porter, K. R., 1964: The plant cell: aspects of its form and function with electronmicrographs. *In* The Cell. Brachet, J., and Mirsky, A. E., eds. Academic Press, New York, Vol. 4, pp. 245–312.

Weidel, W., 1959: Viruses. University Michigan Press, Ann Arbor.

Whipple, A. O., Parpart, A. K., and Chang, J. J., 1954. A study of the circulation of the spleen of the living mouse. Ann. Surg. *140*:266–269.

Wolfe, S. L., 1965: Isolated microtubules. J. Cell Biol. *25*:408–413.

Zernike, F., 1955: How I discovered phase contrast. Science *121*:345–349.

Zworykin, V. K., Flory, L. E., and Shrader, R. E., 1952: Ultraviolet television microscopy. Electronics *25*:150–152.

Zworykin, V. K., Morton, G. A., Ramberg, E. G., Hillier, J., and Vance, A. W., 1945: Electron Optics and the Electron Microscope. John Wiley & Sons, New York.

# CHAPTER 3 / BIOCHEMISTRY OF THE CELL

Some early physiologists conceived of life as the result of the properties of molecules unique to protoplasm (Verworn, 1903). However, with the development of biochemistry as a science, it became evident that many different types of molecules are found in the cell and that although some of these are extremely complex, no one of them alone can be considered to confer the properties of life upon the cell. Many of these compounds have been extracted from tissues, organs or whole organisms, and analyzed and characterized. In recent years, with the development of procedures for fractionating organelles of the cell and of various micromethods, much information on the distribution of these compounds in cells and even in the parts of a single cell is now available.

## CONSTITUENTS OF THE CELL

The main classes of compounds in the cell are easy to demonstrate qualitatively. Water, proteins, lipids, salts and carbohydrates are readily identified by well known tests for each. Other classes of compounds, both organic and inorganic, are also present in small amounts (Table 3.1).

Quantitative determination of cell constituents is somewhat more difficult. It requires the exclusion of extracellular material in the analysis. For example, Table 3.2 gives data on the compounds present in per cent of wet weight in the body of an animal, a plant, a sea urchin egg and a jellyfish. At first glance one is led to believe that the relative amounts of cellular constituents are highly variable in different organisms. However, the data are not strictly comparable since in the first example they represent an analysis of the entire animal body, in the second of the entire plant body, in the third of only the eggs of the sea urchin *Paracentrotus* and in the fourth of the entire body of a jellyfish.

Only in the sea urchin egg is there an approximation of the constituents of a cell, since it contains relatively little stored material and no skeleton. The figures (Table 3.2) for the animal body include the large amount of extracellular salt present in the skeleton, the extracellular solids of connective tissues and the large amount of extracellular fatty material present in the sheaths of fibers of the nervous system. Most tissues of a multicellular animal or plant usually contain such extracellular material (Lowry, 1943). In a tendon, for example, cells make up an insignificant fraction of the tissue because most of the tissue solids and water are extracellular. On the other hand, analysis of muscular tissue, which is relatively free of extracel-

**Table 3.1.** Compounds Found in Cells of Organisms*

| CLASSES OF COMPOUNDS | SUBCLASSES | EXAMPLES | SOLUBILITY OF EXAMPLES |
|---|---|---|---|
| Protein | Albumin | Egg albumin | Dissolves in distilled water; coagulates on heating |
| | Histone | Thymus histone<br>Globin | Water soluble; relatively low molecular weight; basic; forms scum on heating |
| | Protamin | Sperm protamin | Water soluble; relatively low molecular weight; basic |
| | Globulin | Serum globulin | Soluble in salt solution |
| | Prolamine | Gliadin (wheat) | Soluble in 80% alcohol |
| | Glutelin | Glutenin (wheat) | Soluble in acid or basic solution |
| | Scleroprotein (albuminoid) | Horns, nails, hair | Insoluble in most reagents |
| Lipids | Triglycerides | Tripalmitin, tristearin | Soluble in fat solvents |
| | Sterols | Cholesterol | Soluble in fat solvents |
| | Waxes | Beeswax | Soluble in fat solvents |
| | Phospholipids | Cephalin, lecithin | Soluble in both fat solvents and water to some degree |
| Carbohydrates | Monosaccharides | | |
| | hexoses | Glucose<br>Fructose | Water soluble |
| | pentoses | Ribose<br>Deoxyribose | Water soluble |
| | Disaccharides | Sucrose<br>Maltose<br>Lactose | Water soluble |
| | Polysaccharides | | |
| | pentosans | Xylans in wood | Insoluble in water or fat solvents |
| | hexosans | Starch, glycogen | May be prepared in colloidal state in water |
| | | Cellulose | Insoluble in water or fat solvents |
| | Chitin | Insect chitin (glucosamine condensate) | Insoluble |

* This table is included here primarily for purposes of handy reference. Many of the compounds are considered in the text.

**Table 3.1.** Compounds Found in Cells of Organisms* (Cont.)

| CLASSES OF COMPOUNDS | SUBCLASSES | EXAMPLES | SOLUBILITY OF EXAMPLES |
|---|---|---|---|
| Carbohydrates | Mucopoly-saccharides | | |
| | hyaluronic acid | Intercellular cement | Insoluble (attacked by enzyme hyaluronidase) |
| | chondroitin sulfuric acid | In matrix of connective tissue | Insoluble in water or fat solvents |
| | Nitrogenous neutral hetero-polysaccharides | Blood group substances | |
| Conjugated compounds | Nucleoproteins | Nucleohistone | Soluble in strong salt solutions |
| | Chromoproteins | Hemoglobin | Soluble in water |
| | Lipoproteins | Rhodopsin (visual purple) | |
| | Phosphoproteins | Casein of milk | Water soluble |
| | Flavoproteins | Yellow enzyme | Water soluble |
| | Glycoproteins | Protein of connective tissue matrix | Water soluble |
| | Adipocellulose | On plant cell walls | Insoluble |

* This table is included here primarily for purposes of handy reference. Many of the compounds are considered in the text.

lular solid (see Fig. 2.5), gives values much closer to those for the sea urchin egg. Muscle contains somewhat less than 80 per cent water, close to 20 per cent protein, about 1 per cent carbohydrate and some lipid and salt (Mommaerts, 1950).

The plant body contains much cellulose in cell walls and skeletal tissue. The jellyfish has an unusually large amount of water and a very small amount of protein and lipid, the total solid being about 4 per cent. However, a large proportion of the water is extracellular, being present in the mesoglea, a jelly-containing layer into which a small amount of protein is secreted by the cells. If the cells lining the mesoglea were to be analyzed, their composition would probably be found to resemble that of the sea urchin egg.

The protoplasm of a plant or animal cell is likely to contain approximately 75

**Table 3.2.** Compounds in Organisms in Per Cent of Total Wet Weight

| COMPOUND | STEER* (33 MO.) | CORN PLANT* (MATURE) | SEA URCHIN EGG† (PARACENTROTUS) | JELLYFISH* |
|---|---|---|---|---|
| Proteins | 16.73 | 1.8 | 15.18 | 0.67% |
| Lipids | 24.62 | 0.5 | 4.81 | trace |
| Carbohydrates | — | 17.5 | 1.36 | trace |
| Salts (ash) | 5.24 | 1.2 | 0.34 | 3 |
| Water | 55.51 | 79 | 77.3 | 96 |

* Data from Burrell. 1947.
† Data from Ephrussi, 1933: Arch. Biol. *44*:1.

**Table 3.3.** Relative Number of Molecules of Various Types of Cellular Materials*

| SUBSTANCE | PER CENT | AVERAGE MOLECULAR WEIGHT | MOLARITY IN CELL | NUMBER OF MOLECULES RELATIVE TO PROTEIN | NUMBER OF MOLECULES RELATIVE TO DNA |
|---|---|---|---|---|---|
| Water | 85 | 18 | 47.2 | $1.7 \times 10^4$ | $1.2 \times 10^7$ |
| Protein | 10 | 36,000 | 0.0028 | 1 | $7.0 \times 10^2$ |
| DNA† | 0.4 | $10^6$ | 0.000,004 | — | 1.0 |
| RNA‡ | 0.7 | $4.0 \times 10^4$ | 0.000,175 | — | $4.4 \times 10^1$ |
| Lipid | 2 | 700 | 0.028 | $10^1$ | $7.0 \times 10^3$ |
| Other organic materials§ | 0.4 | 250 | 0.016 | 6 | $4.0 \times 10^3$ |
| Inorganic | 1.5 | 55 | 0.272 | $10^2$ | $6.8 \times 10^4$ |

* Data on all but nucleic acids from Sponsler and Bath, 1942. DNA and RNA data for rat liver cell taken from Euler and Hahn, 1948.

† Deoxyribonucleic acid; see section on nucleic acids.

‡ Ribonucleic acid.

§ Sponsler and Bath give 1.5 per cent, from which 1.1 per cent was subtracted for RNA and DNA, in making the calculations of numbers of molecules relative to DNA. The numbers of molecules relative to protein were calculated by Sponsler and Bath, using 1.5 per cent for "other organic molecules."

to 85 per cent water, 10 to 20 per cent protein, 2 to 3 per cent lipid, 1 per cent carbohydrate and about 1 per cent salt and miscellaneous substances. Although the most prominent constituent of the cell is water, the substance which gives the cell its characteristic structure is protein. Lipids are important in all membranes; carbohydrates serve as nutrient reserves.

It is instructive to compare the relative numbers of molecules of the different constituents in the cell (see Table 3.3). The relative numbers of molecules can be determined from the amount of each substance and from the average molecular weight of each class. For purposes of calculation deoxyribonucleic acid (DNA) is arbitrarily assigned a molecular weight of $10^6$, ribonucleic acid (RNA) of $5 \times 10^4$, and protein $3.6 \times 10^4$. Such a calculation shows that more than a million water molecules are present for every DNA molecule and about 18,000 for every protein molecule. From the table one may observe that in spite of the comparatively large percentage of protein content of the cell, the number of protein molecules present is relatively small. Nucleic acids (DNA and RNA together) constitute about 1 per cent of the wet weight of a cell (1.2 per cent of the liver cell, for example), yet they have a profound effect on the functioning of the cell. At present it is known that with gentle extraction procedures DNA may have a molecular weight many times $10^6$. The disparity between the number of DNA molecules and other molecules in the cell is therefore much greater than that given in Table 3.3. The data in the table are useful since they indicate the availability of the various molecules for combination with protein or for osmotic action.

## WATER, FREE AND BOUND

Cell foods reach the cell dissolved in water, and apparently all the chemical reactions of importance to life occur in aqueous solution (Dick, 1966). Water is present in large quantities in the more active tissues; for example, white matter of the brain contains about 68 per cent water, whereas the gray matter, made up of the constituents of the neurons, contains about 85 per cent (Lowry, 1943).

Water in the cell occurs in two forms, free and bound, although the partition between the two forms is quite indefinite.

*Free water* is that which is available for metabolic processes; *bound water* is that which is loosely held to a protein molecule by dipole attraction. The force of imbibition is the expression of such adsorption of water to the surface of protein molecules. Bound water is attached to protein molecules by the formation of hydrogen bonds and therefore forms part of the structure of protoplasm. Hydrogen bonds result from the attraction of two electronegative atoms for a proton, and only the most electronegative atoms such as fluorine, oxygen, nitrogen and chlorine are capable of forming them. Radicals with residual electronegativity forming hydrogen bonds in protoplasm are hydroxyl, carboxyl, keto, amino, imino, sulfhydryl and trivalent nitrogen. Each amino group of a gelatin molecule is capable of binding 2.6 molecules of water. It is estimated that in general approximately 4.5 per cent of all the water available in the cell is bound (Webb, 1965), the greater portion being free to act as a solvent in which metabolic reactants may dissolve (see also Chapters 4 and 7).

Many plant and animal cells and bacteria can withstand considerable desiccation for prolonged periods, e.g., spores of molds and bacteria and cysts of protozoans. Even mammalian red blood cells and bull sperm have been dried to within 8 per cent of their initial water content, yet revived when hydrated (Meryman, 1960).

## SALTS

Salts are present in all cells and are necessary for life. If the salts are removed by placing tissues in distilled water, death ensues. The salt requirement of cells is strikingly illustrated by the occurrence of muscle cramps following excessive perspiration, which is attended by excessive sodium chloride loss, at which time the potassium in muscle cells leaks into the extracellular fluid causing cramps. Since potassium reenters the cells after intake of sodium chloride, the cramps are quickly relieved by ingestion of a salty drink or a salted food.

The salt content in cells of a representative tissue is given in Figure 3.1. It will be noted that of the cations, potassium is present in highest concentration inside the cell, magnesium is present in lesser concentration and sodium and calcium are present in relatively small amounts (Kernan, 1965). The dominant anion of cells is phosphate, part of which is soluble in acid, and part insoluble. A second important anion is bicarbonate. In some cells

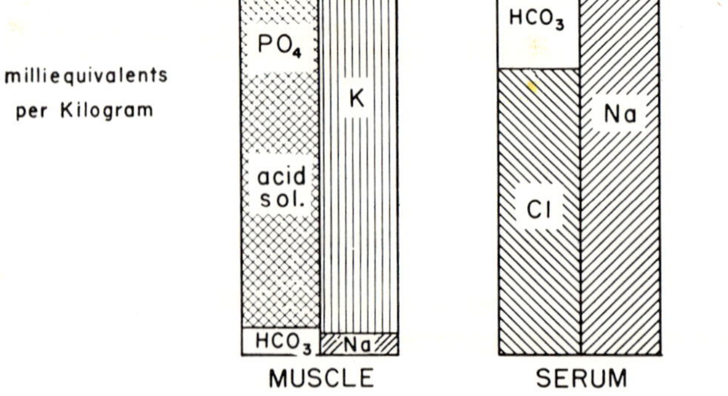


***Figure 3.1.*** Comparison of ion content of muscle tissue and blood serum. (After Lowry, 1943: Biol. Symp. *10*:241.)

**Table 3.4.** Elements Found in Cells of Various Organisms (Primary and Invariable Elements Are Found in Cells of All Organisms)

| PRIMARY | INVARIABLE | | VARIABLE | | |
|---|---|---|---|---|---|
| | SECONDARY | MICROCON-STITUENTS | SECONDARY | MICROCON-STITUENTS | |
| (1–60%) | (0.05–1%) | (0.05%) | | | |
| H | Na* | B | Ti | Li* | St |
| C | Mg* | F* | V* | Be | Mo* |
| N | S* | Si | Zn* | Al* | Ag |
| O | Cl* | Mn* | Br | Cr | Cd* |
| P* | K* | Cu* | | Co* | Sn |
| | Ca* | I* | | Ni* | Cs |
| | Fe* | | | Ge | Ba* |
| | | | | As | Pb |
| | | | | Ru | |

After Sverdrup, *et al.*, 1942.

* All elements marked with an asterisk have been found necessary for some type of enzyme action, according to Schweigart, 1951: Z. Pflanzenernähr *54*:36. Boron and silicon have been found necessary for plants. The unmarked variable elements may be only contaminants since no function for them has been found and they are found in only small quantities and not in all organisms.

(erythrocytes and liver cells) chloride is also present. Although these salts are the main ones in cells, small quantities of a large number of other elements also are found, probably as salts (Vinogradov, 1953) (see Table 3.4). For example, iron is always present in relatively sizable amounts, whereas manganese and copper are found in very small amounts (microconcentrations, less than 0.05 per cent). Vanadium, zinc, nickel, molybdenum and even tin are found in variable but always minute quantities (Sverdrup *et al.*, 1942). Some of these salts are necessary for the functioning of certain enzymes (see Chapters 15 and 16).

It is interesting that the cation present in the highest concentration in the body fluid (e.g., serum, Fig. 3.1) is sodium, and only a small amount of potassium is found. Such a difference in concentration of sodium and potassium inside and outside cells is vital to their functioning, as is most readily seen in studies on highly irritable tissues such as muscle and nerve (see Chapters 22 and 23). Sodium also appears to be necessary for syntheses occurring in the nucleus and potassium for those in the cytoplasm (Allfrey *et al.*, 1961).

## PROTEINS

Proteins are responsible for the characteristic structure of the cell. They are molecules of large molecular weight, ranging from about 6000 to many million. Their large size probably accounts for many of their peculiar properties (Kendrew, 1963).

All proteins contain carbon, hydrogen, oxygen, nitrogen and usually sulfur, and some in addition have phosphorus. Proteins are made up of chains of amino acids (see Fig. 3.2 and Table 3.5). The amino acids of a protein are united to one another by their respective amino and carboxyl groups, forming *peptide bonds* (see Fig. 3.3*A*). In the cell only phosphorylated amino acids in the presence of appropriate enzymes apparently react to form peptide bonds. The chain of peptide bonds between successive amino acid residues forms the backbone of the protein molecule and its primary structure (see Fig. 3.3*B*). However, other linkages occur as well and, in fact, are necessary to explain the native structure of a protein. These are considered later in connection with helical structure of protein molecules.

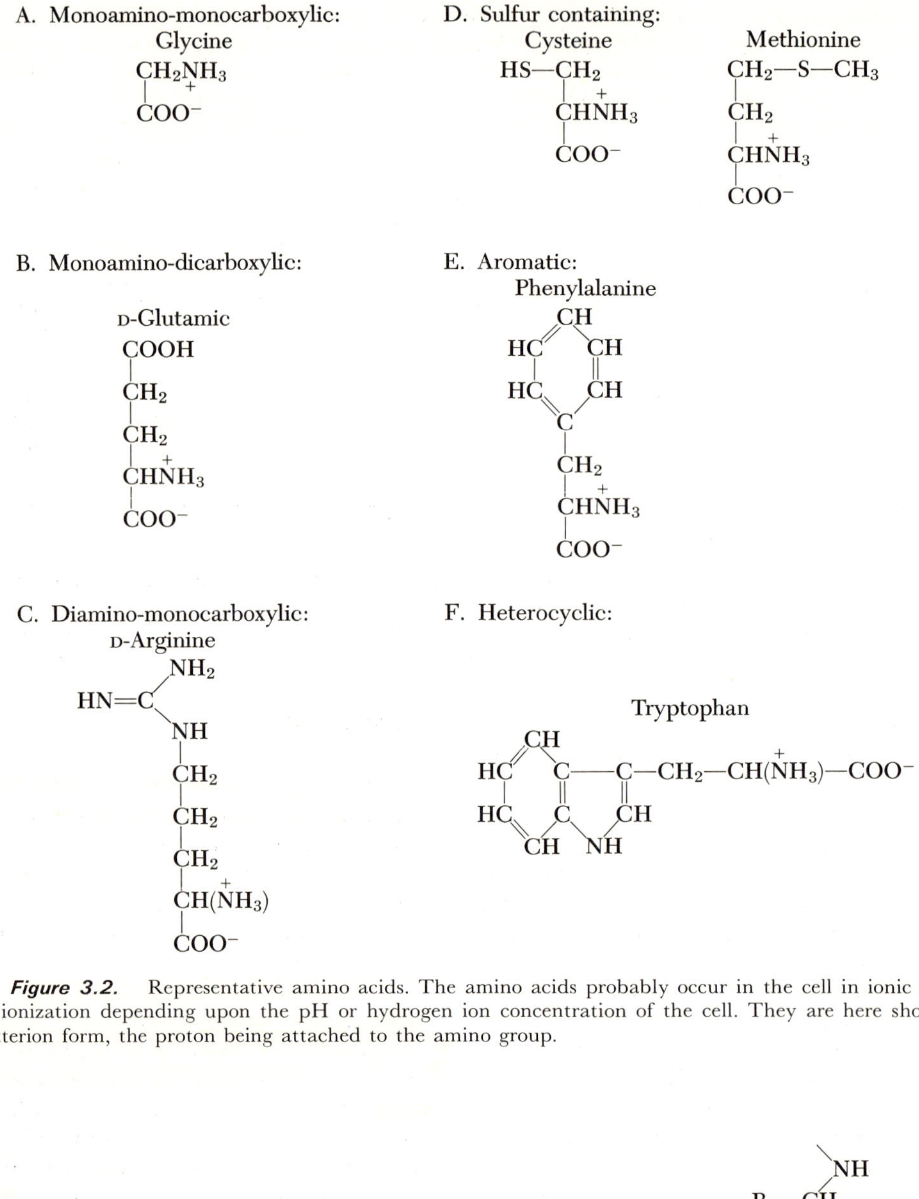


*Figure 3.2.* Representative amino acids. The amino acids probably occur in the cell in ionic forms, the ionization depending upon the pH or hydrogen ion concentration of the cell. They are here shown in zwitterion form, the proton being attached to the amino group.

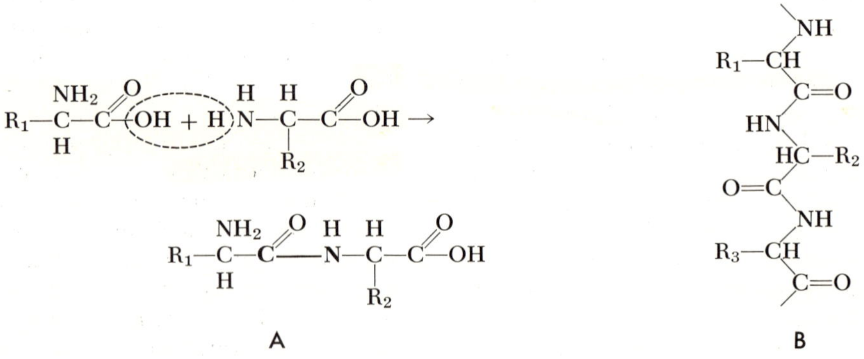


*Figure 3.3* *A,* Formation of a peptide bond between two amino acids. The N and C of the peptide bond are shown in bold type. *B,* The peptide backbone of a protein molecule.

**Table 3.5.** Types of Natural Amino Acids and Abbreviations Used for Them

| | |
|---|---|
| A. Monoamino-monocarboxylic | D. Hydroxyl-containing |
| Glycine (Gly) | Threonine (Thr) |
| Alanine (Ala) | Serine (Ser) |
| Valine (Val) | E. Sulfur-containing |
| Leucine (Leu) | Cystine (Cys or Cy-S) |
| Isoleucine (Ileu) | Methionine (Met) |
| B. Monoamino-dicarboxylic | F. Aromatic |
| Glutamic (Glu) | Phenylalanine (Phe) |
| Aspartic (Asp) | Tyrosine (Tyr) |
| C. Diamino-monocarboxylic | G. Heterocyclic |
| Arginine (Arg) | Tryptophan (Tryp) |
| Lysine (Lys) | Proline (Pro) |
| Hydroxylysine (Hlys) | Hydroxyproline (Hpro) |
| | Histidine (His) |

There are some 20 kinds of amino acids in proteins (Table 3.5), but not all the amino acids are present in all the proteins (Table 3.6), and those present occur in proportions characteristic of each protein. In Table 3.6 are given data for the amino acid composition of a few representative proteins. It is evident that some proteins, such as gelatin, contain relatively few aromatic amino acids (e.g., tyrosine and phenylalanine) per molecule, whereas most others, such as albumin, contain relatively many. In histone a larger proportion of basic amino acids is present, but in myosin a larger proportion of acidic ones is found. Sulfur-containing amino acids are completely absent from histone, and few are present per molecule of gelatin. Although only 20 types of amino acids exist, the number of proteins which can be formed from them is large. This is not difficult to comprehend if we make an analogy between the amino acids and the letters of the alphabet, since the entire English language is derived from 26 letters combined in diverse ways.

When both the molecular weight of a protein and the relative proportions of its constituent amino acids are known, an empirical formula in terms of its amino acid content may be written for it (Sanger, 1959). Thus the empirical formula for ribonuclease, an enzyme catalyzing digestion of ribonucleic acid, is:

$$Asp_4, (AspNH_2)_{11}, Glu_6, (GluNH_2)_6, Gly_3, Ala_{12}, Val_9, Leu_2, Ileu_3, Ser_{15}, Thr_{10}, Cys_8, Met_4, Pro_4, Phe_3, Tyr_6, His_4, Lys_{10}, Arg_{14}.$$

$AspNH_2$ and $GluNH_2$ refer to the amines of aspartic and glutamic acids, respectively (see Table 3.5 for meaning of the abbreviations for amino acids). Ribonuclease consists of 124 amino acids of 19 different kinds. Relatively few proteins have been so analyzed to date.

In a few cases the analysis of protein structure has been pursued to the point where not only the empirical but also the structural formula has been determined. For example, this was done for insulin (molecular weight 6000) illustrated in Figure 3.4. The procedure is long and difficult. It is necessary to determine first by end-group analysis how many polypeptide chains are present in the molecule. It is obvious that a polypeptide chain can have only one terminal α-amino group and such an amino group can be made to react with dinitrofluorobenzene to form a bright yellow dinitrophenol derivative. When a known molar concentration of the protein is then hydrolyzed by acid or enzymatically, the marked peptide can be isolated and determined

**Table 3.6.** Amino Acid Composition of Proteins*

| | FOOD PROTEINS | | | | TISSUE PROTEINS | | | |
|---|---|---|---|---|---|---|---|---|
| | GELATIN† (BOVINE BONE) | BOVINE α-CASEIN | OVALBUMIN | ZEIN | HUMAN HEMOGLOBIN | RABBIT MYOSIN | CALF THYMUS HISTONE (TOTAL) | HUMAN SERUM ALBUMIN |
| Alanine | 9.02 | 2.78 | 5.36 | 8.39 | 8.14 | 5.19 | 7.73 | — |
| Amide N | 0.24 | 1.65 | 1.02 | 2.98 | 0.90 | 1.20 | 0.71 | 0.88 |
| Arginine | 8.07 | 3.51 | 5.13 | 1.53 | 3.07 | 6.59 | 10.41 | 5.51 |
| Aspartic Acid | 5.79 | 6.96 | 8.04 | 3.98 | 9.17 | 7.69 | 4.43 | 8.99 |
| Cysteine | 0.0 | — | 1.15 | 0.70 (Cysteine and Cystine/2) | — | — | 0.0 | 0.7 |
| Cystine/2 | 0.0 | 0.37 | 0.43 | | 0.87 | 1.20 | 0.0 | 4.74 |
| Glutamic Acid | 10.18 | 18.43 | 14.48 | 23.61 | 6.32 | 19.39 | 8.52 | 15.27 |
| Glycine | 20.67 | 1.53 | 2.32 | — | 3.43 | 1.44 | 3.85 | 1.22 |
| Histidine | 0.62 | 2.58 | 2.08 | 1.17 | 7.49 | 2.13 | 2.19 | 3.09 |
| Isoleucine | 1.33 | 4.96 | 6.04 | 4.31 | 0.28 | 13.46 (Isoleucine and Leucine) | 3.96 | 1.47 |
| Leucine | 2.98 | 6.89 | 7.94 | 18.21 | 13.13 | | 7.06 | 10.27 |
| Lysine | 3.82 | 8.16 | 5.52 | 0.0 | 9.29 | 10.45 | 15.21 | 10.78 |
| Methionine | 0.55 | 2.35 | 1.57 | 2.12 | 1.41 | 2.99 | 1.00 | 1.12 |
| Phenylalanine | 2.22 | 4.15 | 6.86 | 6.50 | 7.06 | 3.86 | 2.49 | 6.95 |
| Proline | 13.07 | 6.95 | 3.04 | 8.88 | 4.22 | 1.62 | 3.90 | 4.30 |
| Serine | 3.09 | 5.30 | 6.75 | 5.84 | 4.56 | 3.59 | 4.08 | 3.06 |
| Threonine | 2.00 | 3.56 | 3.42 | 2.93 | 5.18 | 4.34 | 4.64 | 4.24 |
| Tryptophan | 0.0 | 1.82 | 1.09 | 0.14 | 1.09 | 0.73 | 0.0 | 0.17 |
| Tyrosine | 0.21 | 6.57 | 3.33 | 4.72 | 3.96 | 3.06 | 3.32 | 4.20 |
| Valine | 2.34 | 4.82 | 5.96 | 3.37 | 9.38 | 2.18 | 4.90 | 6.51 |

* In gm. anhydroamino acid per 100 gm. protein. (From Cantarow and Schepartz, 1967: Biochemistry. 4th Ed. W. B. Saunders Co., Philadelphia. Data selected from Tristram and Smith, 1963: The amino acid composition of some purified proteins, Advances Protein Chem. *18*:227.)

† Contains in addition 0.67 gm. hydroxylysine and 11.48 gm. hydroxyproline.

quantitatively. From the data the number of polypeptide chains per protein molecule can be determined. This can be checked by determination of the number of terminal carboxyl groups per protein molecule, using an enzyme specific to the terminal carboxyl groups (carboxypeptidase, see Chapter 15).

Next the disulfide bonds linking the polypeptide chains are readily broken by performic acid. The polypeptides are then separated from one another by fractional precipitation with salts. Each polypeptide must then be hydrolyzed in such ways as to cleave it in certain places only. Fortunately, selective enzymes are available and these, combined with acid or alkaline hydrolysis, made it possible to shatter the chain in many places and to note the overlapping sequences in the chromatographed products (see appendix to Chapter 19). It thus proved possible to identify all of the amino acids in the two polypeptide chains making up the insulin molecule. Furthermore, it was possible to determine the order of the various amino acids in the two polypeptide chains. Determination of the positions of the cross linkages

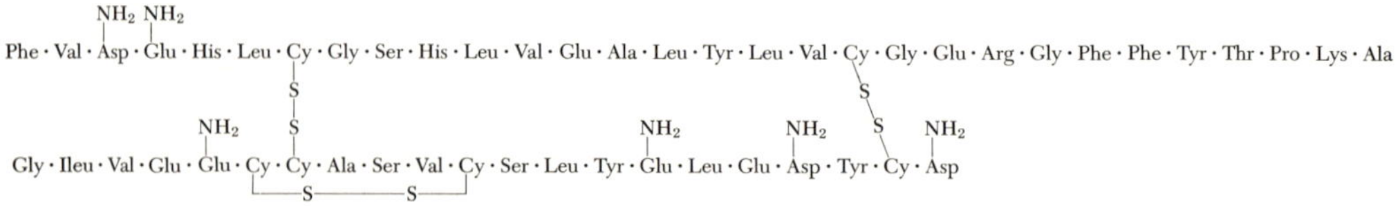


***Figure 3.4.*** Structure of the protein insulin. The abbreviations for the various amino acids are interpreted in Table 3.5. The residues indicated with $NH_2$ groups attached are amines of the corresponding amino acids. (From Sanger, 1959: Science *129*:1343.)

between the two chains by disulfide groups proved difficult, but successful methods were devised. This intensive analytical study (Sanger, 1959), continuing for 10 years, won the Nobel prize for Sanger in 1958.

At present the amino acid sequences of increasing numbers of proteins are being worked out. Among the earlier ones analyzed are: insulin (from several species of animals), adrenocorticotropic hormone, bovine glucagon, bovine pigment cell–stimulating hormone, tobacco mosaic virus and ribonuclease (Stein and Moore, 1961). The three-dimensional arrangement of ribonuclease, myoglobin and hemoglobin has also been worked out. The three-dimensional arrangement of protein molecules is being actively studied by x-ray diffraction methods (Kendrew, 1961, 1963; Perutz, 1964; Hodgkin, 1965); an example is shown in Figure 3.5). It is interesting that when hemoglobin combines with oxygen, its structure shows a change detectable by x-ray crystallography. Such an allosteric change may also accompany an enzyme on combination with its product (see Chapter 21).

***Solubility of Proteins.*** Proteins differ from one another in solubility. In the

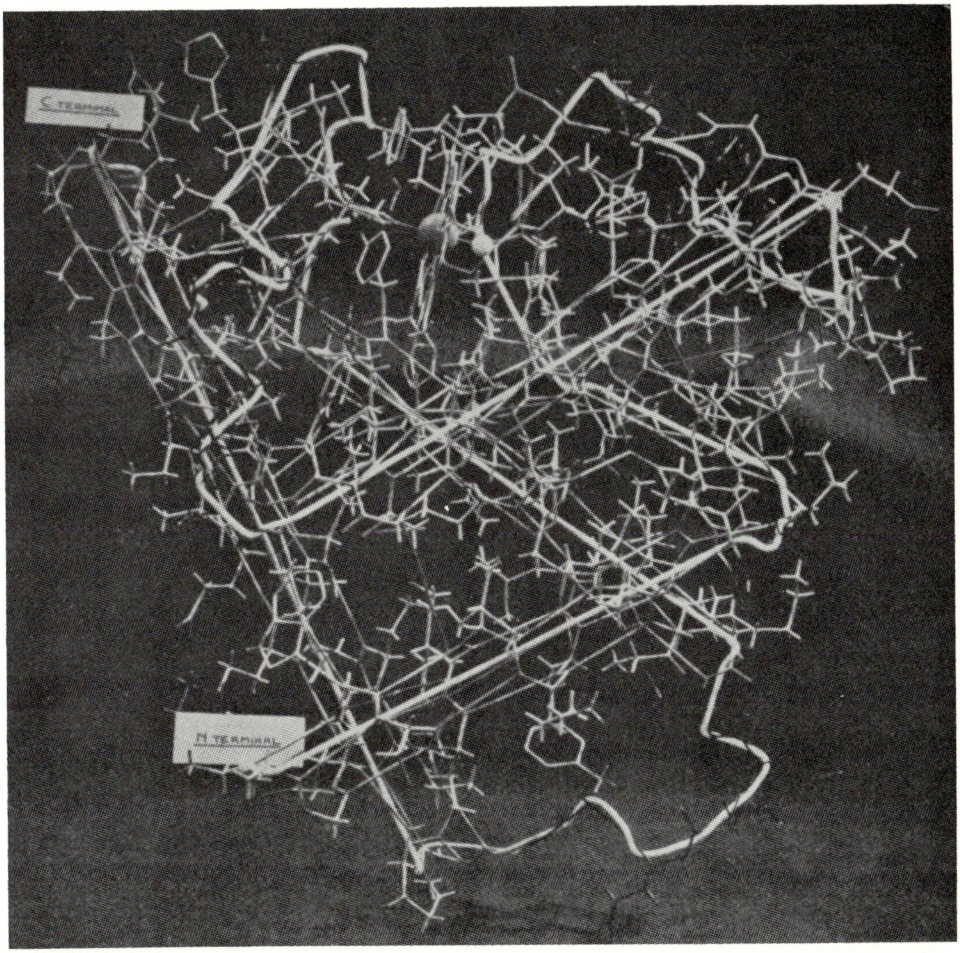

***Figure 3.5.*** Model of the myoglobin molecule, including side chains. The course of the main chain is indicated by a white cord, the N-terminal and C-terminal residues being marked. The iron atom is shown as a gray sphere; the nearby white sphere is a presumed water molecule coordinated to it. (From Kendrew, Watson, Strandberg, Dickerson, Phillips, and Shore, 1961: Nature *190*:666.)

past, before much was known of their structure, solubility was one of the main criteria for classification of proteins. At present such a classification is interesting primarily from an operational point of view. For example, the *albumins* are soluble in neutral distilled water, whereas the *globulins* are soluble only in dilute neutral salt solution. Some proteins are soluble only in acidic or basic solution (glutelins of cereals), and some are soluble in alcohol solutions (prolamines, e.g., gliadin of wheat). The *scleroproteins* or albuminoids are relatively insoluble and are hydrolyzed and degraded only under special conditions or by special reagents. They are resistant to digestive enzymes largely because disulfide linkages between respective sulfur-containing amino acids in these proteins must be reduced to sulfhydryls before the digestive enzymes can exert their influence on the peptide linkages. Reduction of disulfide occurs more readily under alkaline conditions. In the alkaline intestinal tracts of some insects (e.g., clothes moth larvae) these groups are reduced, following which the scleroproteins are digested. The dermatophyte fungi which tolerate highly alkaline conditions (see Table 8.4) are also capable of using these proteins as food.

***Proteins as Zwitterions.*** All proteins show certain characteristics in common. Like amino acids (Fig. 3.2), proteins are zwitterions; that is, they are charged either positively or negatively and at the isoelectric pH they are electrically neutral (see Chapter 8) and will not migrate toward the anode or the cathode of an electric field. As with amino acids, the isoelectric pH of a protein depends upon the relative number of basic and acidic groups present in the molecules. An excess of acid groups on a protein molecule—for example, a molecule with many polycarboxy amino acids—shifts the isoelectric point to the acid side of neutrality, as in serum albumin, which has an isoelectric pH of 4.7. An excess of basic groups on a protein—for example, one with many polyamino acids—shifts the isoelectric pH to the basic side, as in gliadin, which has an isoelectric pH of 9.0.

On the acid side of the isoelectric pH the protein dissociates as a base and is positively charged. On the alkaline side the protein dissociates as an acid and is negatively charged. The charge on proteins can be readily demonstrated by their reactions with dyes. A positively charged protein reacts only with a negatively charged dye, forming a colored precipitate. Since the protein is positively charged only on the acid side of its isoelectric pH, the reaction with the negatively charged dye occurs only at a low pH. A negatively charged protein, on the other hand, reacts only with a positively charged dye, forming a colored precipitate. The protein molecule is negatively charged only at a high pH on the alkaline side of its isoelectric pH.

As will be shown, the physical-chemical properties of a protein are at a minimum at its *isoelectric pH* (see Fig. 3.6), at which it is electrically neutral (Loeb, 1923):

1. The viscosity tends to be minimal; in other words, the protein offers least resistance to flow since there is less attraction between the molecules and less binding of water.
2. The osmotic pressure is lowest at this pH since the number of ions produced is least.
3. The least swelling occurs under these conditions since the electrically neutral protein molecules hold the least water of hydration.
4. The conductivity is at a minimum since the molecules are electrically neutral.
5. The solubility is also minimal under these conditions; and small amounts of salt precipitate them.

***Globular and Fibrous Protein Molecules.*** Protein molecules may be either *globular* or *fibrous* in shape. Some proteins may take either form, depending upon the prevailing conditions. For example, actin, an important protein of muscle, is present in either globular or fibrous form, depending upon the concentration of certain salts.

Globular proteins such as albumin and globulin consist of one or several pep-

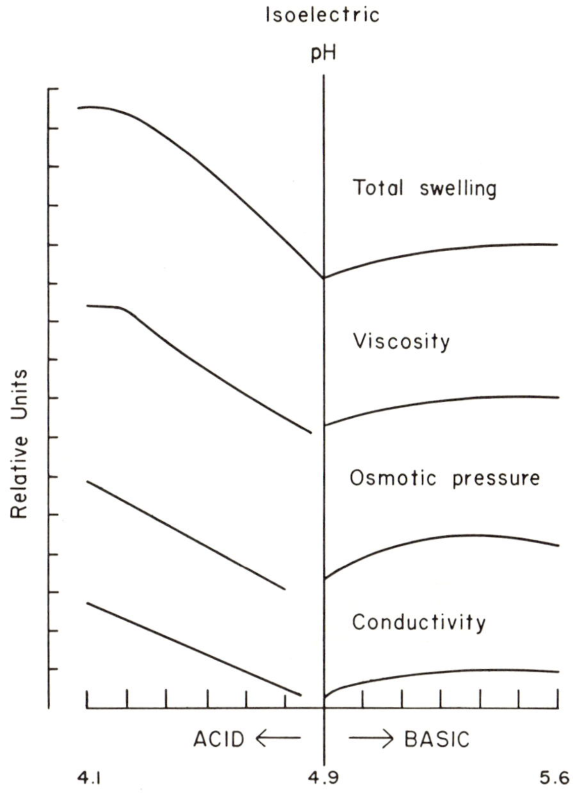


***Figure 3.6.*** Effect of pH on properties of gelatin. (After Loeb, 1923: Proteins and the Theory of Colloidal Behavior. McGraw-Hill Book Co., New York.)

tide chains in $\alpha$ helix configuration folded up in some way and held in configuration by hydrogen bonds and cohesive forces of like-to-like radicals, e.g., methyl groups of neighboring amino acids. Globular proteins may unfold when the cohesive and hydrogen bonds are broken by heat or other agents that denature the proteins (Neurath, 1963).

Two kinds of fibrous proteins have been found: (1) the keratin-myosin-elastin-fibrinogen group and (2) the collagen group. In the first are the proteins of the epidermis of mammals, amphibians and some fishes; of hair, horn and so forth, growing from mammalian epidermis; the myosin of muscle fibers; and the clotting protein, fibrinogen. All of these give x-ray diffraction patterns related to that of natural unstretched mammalian hair ($\alpha$ keratin). These fibers can be stretched in hot water or alkali to a second form called $\beta$ keratin. The latter form gives a simpler x-ray diffraction picture than the former. If the stretched fiber is released, it gradually reverts to its original size. If released immediately while still immersed, the fibers contract to two-thirds the length of the original fibers, giving a supercontracted form.

Analyses of the $\alpha$ form of the keratin-myosin-elastin-fibrinogen proteins by monochromatic x-ray bombardment indicate a fundamental crystallite structure showing regular periodic distances between refracting units (see Fig. 3.7). The only arrangement of amino acids which could be in keeping with the x-ray diffraction pictures of such fibrous proteins is thought to be a helix. From a knowledge of the stereochemical requirements of the peptide backbone (Fig. 3.3*B*), that is, of the distances between atoms and of the possible angles which they may make with one another, models have been made which summarize most of the existing data on protein structure. For example, Pauling and Corey have found that the

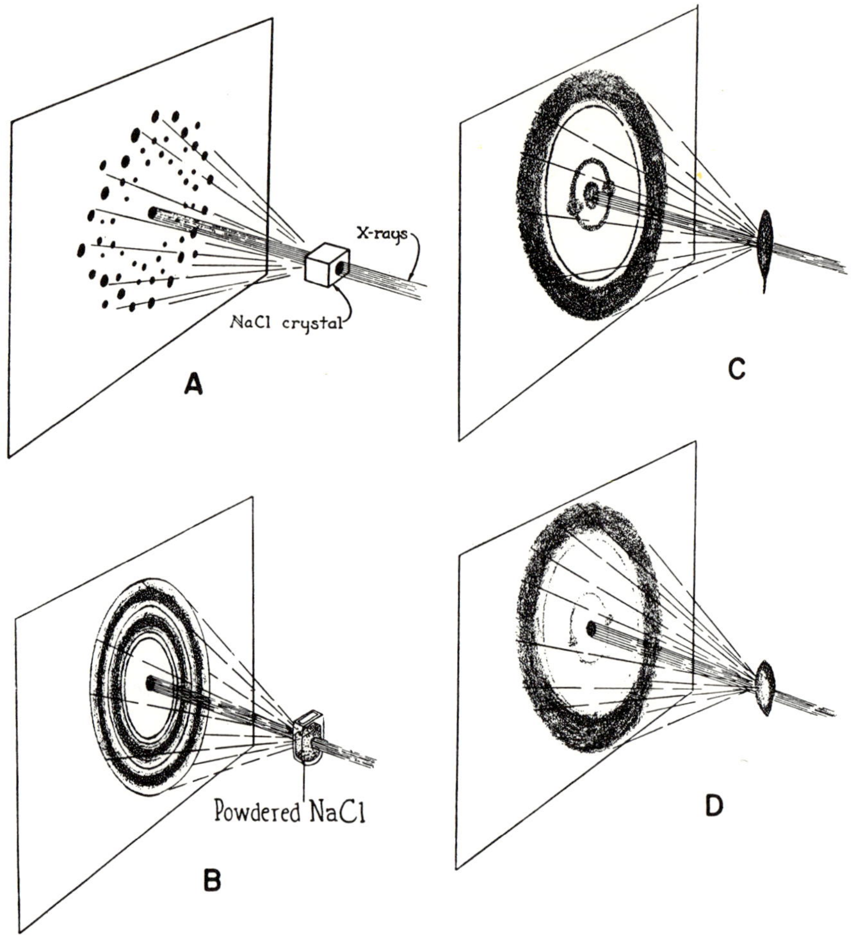


***Figure 3.7.*** X-ray diffraction diagrams for NaCl crystal (*A*), NaCl powder (*B*), resting muscle (*C*) and contracted muscle (*D*). (Modified from Amberson and Smith, 1948: Outline of Physiology. Appleton-Century-Crofts, New York.)

$\alpha$ helix, in which each turn contains 3.6 amino acid residues (see Fig. 3.8*A*), is most consistent with the facts. Agreement is still incomplete, however, unless in addition to the $\alpha$ helix there is postulated a large helix made up of $\alpha$ helices. If the radius of the large helix is 6 Å, three of these large helices would twist around one another in a model of the fiber (Fig. 3.8*B*).

X-ray analysis of the $\beta$ or stretched form of the keratin-myosin-elastin-fibrinogen proteins shows that the atomic arrangements are no longer spiral but may best be represented as a pleated ribbon polypeptide chain (Fig. 3.8*C*).

In the second or collagen group are the fibers found in tendons, cartilage and fish scales, and the partial hydrolysis product of collagen—gelatin. These fibers do not stretch on pulling, although they contract in hot water. The x-ray diffraction patterns of collagen fibers are quite distinct from those of the keratin-myosin-elastin-fibrinogen group (Hodgkin, 1965).

From x-ray analysis of the collagen group of fibrous proteins, workers postulate that here the unit is a cylindrical system parallel to the fiber axis and formed of three polypeptide chains, each coiled into a helix, all three helices having the same pitch and each chain being completely hydrogen bonded to the other two. Hydrogen bonds link glycine and hydroxyproline units in the molecule (Doty, 1957).

The banding seen in electronmicrographs is the result of regular orientation of amino acid residues in the fibers (Schmitt, 1957).

The collagen molecule is an extremely long thin rodlet—the most asymmetrical molecule yet isolated. When collagen is warmed, it breaks down and forms a gel composed of gelatin, the molecules of which are one-third of the molecular weight of the collagen. The collagen molecule has now been shown to consist of three peptide chains, each one incorporating considerable proline, hydroxyproline

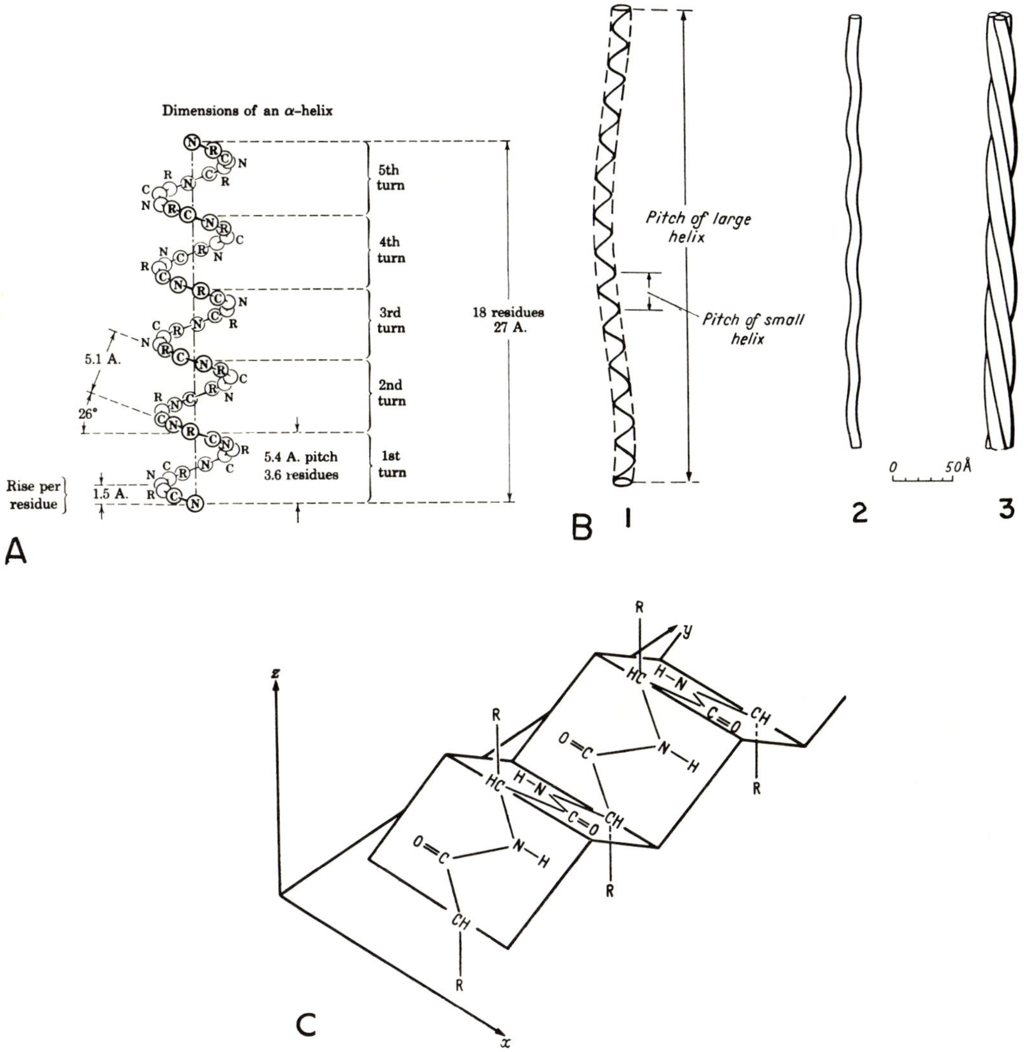


*Figure 3.8.* *A,* The Pauling-Corey 3.6 residue α helix. R refers to a radical (methyl, phenyl, etc.), C stands for the carbon atom and N for the nitrogen atom. *B,* A compound helix made up of a succession of α helices which are shown inside the large helix delimited by the dotted border. In (1) the pitch of the large helix is marked between the long arrows. The pitch of one of the many individual α helices, of which the large helix is constructed, is shown between the short arrows. The compound helix is shown in a simplified way in (2). Three such helices coil together to form a rope (3) when the radius of the large helix is 6 Å. *C,* Pleated ribbon polypeptide chain to represent the arrangement of the amino acids in the β or stretched fibers of the keratin-myosin-elastin-fibrinogen group of proteins. (*A* from Anfinsen, 1959: The Molecular Basis of Evolution. John Wiley & Sons, New York; *B* and *C* from Springall, 1954: The Structural Chemistry of Proteins. Academic Press, New York.)

and glycine (see gelatin, Table 3.6). Glycine, the smallest of the three, occurs every fourth position on each chain. This makes it possible for the bulkier units to fit into the links of the triple strand. The extra hydrogen bonds of hydroxyproline link it firmly to neighboring molecules in a fibril. As a consequence the collagen present in a tissue is more resistant to heat than the molecularly dispersed collagen in solution (Gross, 1961).

The main intent in presenting these examples of protein structure is to illustrate the possibility of explaining in terms of molecular arrangements the x-ray data and many of the facts known about the chemistry of proteins. The models at hand already help to interpret chemical and biological data with more consistency than was previously possible.

It thus appears that primary, secondary and tertiary structures occur within the protein molecule (Anfinsen, 1959; Kendrew, 1961). The *primary structure* is that expressed by the structural chemical formula and depends entirely upon chemical valence bonds: the peptide bonds which unite amino acid residues in the peptide backbone of a protein and those covalent bonds that form fixed sites of cross linkage, such as the disulfide bonds between half-cystine residues and the phosphate diester linkages of certain phosphoproteins. *Secondary structural bonds* are the result of hydrogen bond formation as, for example, the amide (CONH) linkages between C=O and NH groups on amino acid residues of the polypeptide chain. The $\alpha$ helix and pleated ribbon of Pauling shown in Figure 3.8 are results of such secondary bonds. The $\alpha$ helix has the maximum number of amide bonds for spirals of various pitch values considered and is, therefore, probably the most stable. *Tertiary structure* and disulfide linkages are considered necessary to stabilize the $\alpha$ helix in solution and it is probable that only those parts of proteins which are properly anchored by such bonds can maintain the helical configuration. The tertiary structure is the result of (1) van der Waals interactions between hydrophobic groups, agglomerated by mutual repulsion of solvent, and (2) those special hydrogen bonds which exist between hydroxyl groups of tyrosine and the $\epsilon$ amino groups of lysine and various electronegative groups along the peptide chain (Kendrew, 1963).

A protein molecule may consist of several polypeptide chains associated either end to end or laterally to form a single unit indicating *quaternary structure.* For example, hemoglobin consists of four polypeptide chains, two designated $\alpha$ and two $\beta$, each type with a characteristic amino acid sequence (Perutz, 1964). These polypeptide chains take a configuration or characteristic shape within the hemoglobin molecule.

***Denaturation.*** Denaturation of a protein is defined by Kauzmann as a "process or a sequence of processes in which the spatial arrangement of the polypeptide chains within the protein molecule is changed from that typical of the native protein to a more disordered arrangement." As defined here, denaturation may involve alteration of quaternary, tertiary and secondary structure but *not* the primary structure of the molecule. Such a modification of the unique structure of the protein molecule gives rise to definite changes in chemical, physical or biological properties. For example, denatured protein binds water less readily and becomes less soluble. Similarly, denatured enzyme is no longer active catalytically. Both types of protein, fibrous and globular, are subject to denaturation.

Native and denatured protein may exist in equilibrium with one another in solution, but various factors may affect the equilibrium between them, such as heat, acid, alkali, heavy metals, alkaloids, detergents, guanidine, urea, organic solvents, hydrostatic pressure, surface reactions and ultraviolet and ionizing radiations, all of which tend to favor denaturation. In many cases the denaturation is essentially irreversible, proceeding with a decrease in free energy.

Denaturation sometimes may not be visible although latent. For instance, when albumin is heated to 60° C. at a pH far to either side of its isoelectric pH and in the virtual absence of salts, it will remain in solution even though it is denatured. Adjustment of the solution to the isoelectric pH results in *flocculation* (a form of precipitation). Readjustment of the pH far to one side of the isoelectric pH results in dissolution of the flocculum; therefore, flocculation is reversible. However, if the flocculum is further heated, flocculation is made irreversible and the process is then spoken of as *coagulation.*

Proteins are thought to unfold during denaturation as hydrogen bonds and other weak bonds are broken by heat. Since unfolding exposes hydrophobic groups which repel water, the denatured protein is less soluble in aqueous solution and more susceptible to precipitation from solution than the native protein. Unfolding during denaturation is accompanied, in some proteins, by an increased total volume of the solution, as measured by the method of *dilatometry.* The dilatometer is a flask with a finely calibrated capillary tip. The increase in volume of a protein during denaturation may amount to 0.2 ml. per 100 grams of protein. This increase may be accounted for in the following manner: The folded, native protein has its hydrophobic radicals hidden; the hydrophilic radicals, which are exposed, attract water molecules, holding them closely. Conversely, when the molecule is denatured and unfolded, the hydrophobic radicals are exposed, repelling the water; the capacity of the protein to bind water is reduced and the total volume of the solution is thus increased (Johnson *et al.,* 1954).

If the volume of a protein solution increases during denaturation, the denaturation should be reversed by the application of pressure since the pressure would tend to decrease the volume. That this actually occurs is shown in a typical experiment (Figure 3.9). At 65° C., a temperature which denatures certain proteins at atmospheric pressure, higher pressures prevent the denaturation of the protein, and the percentage of protein remaining in solution—that is, in the native state—is higher, the higher the pressure.

When heat injury to a given biological process is reversed by pressure, it seems likely, on the basis of this reasoning, that

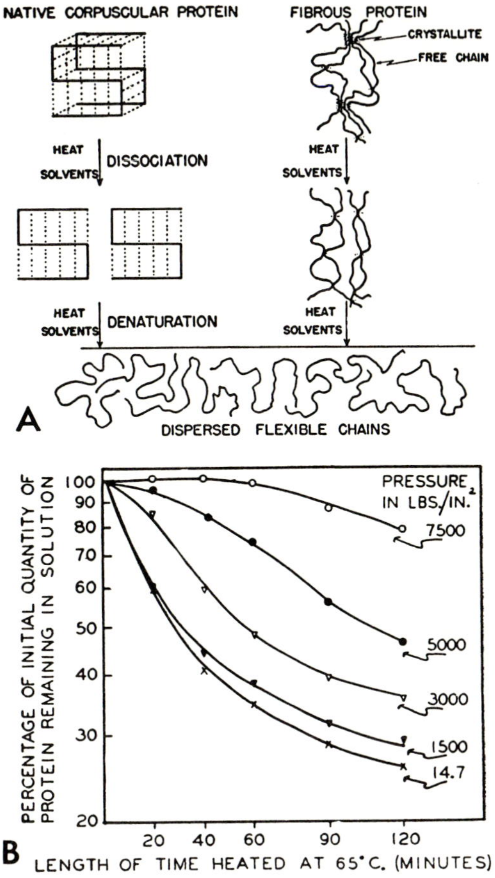


***Figure 3.9.*** *A,* Denaturation of globular and fibrous proteins. (From Cantarow and Schepartz, 1967: Biochemistry. 4th Ed. W. B. Saunders Co., Philadelphia.) *B,* Degree of precipitation of protein at 65° C. when subjected to various pressures. (From Johnson and Campbell, 1946: J. Biol. Chem. *163*: 644.)

the process is one in which the enzyme of the limiting reaction under consideration is denatured with an increase in volume. Several such cases have been effectively studied, such as the quenching of luminescence in luminous bacteria by heat, heat-stoppage of cleavage in sea urchin eggs and the heat-induced loss of infectivity of virus, all of which are reversed by moderate pressure.

***Conjugated Proteins.*** Proteins may combine with various other compounds in an organism to form so-called conjugated proteins. An example of these is hemoglobin, a chromoprotein in which

A. Purine nucleus

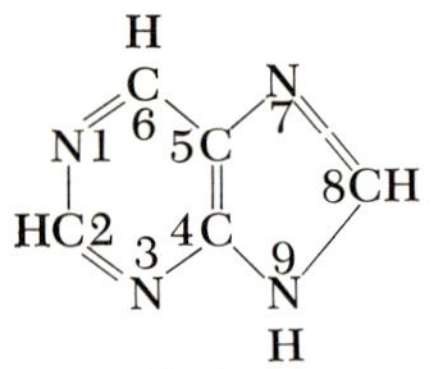

Purine

B. Pyrimidine nucleus

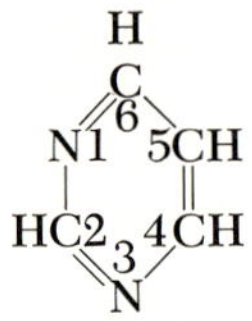

Pyrimidine

C. Purines and pyrimidines found in nucleic acids

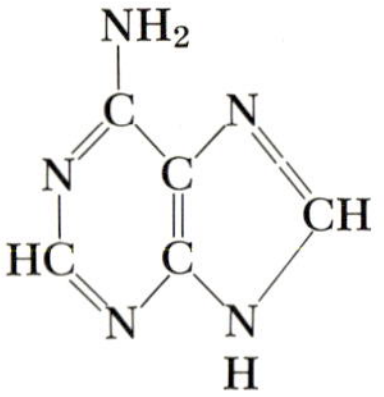

adenine
(6-aminopurine)

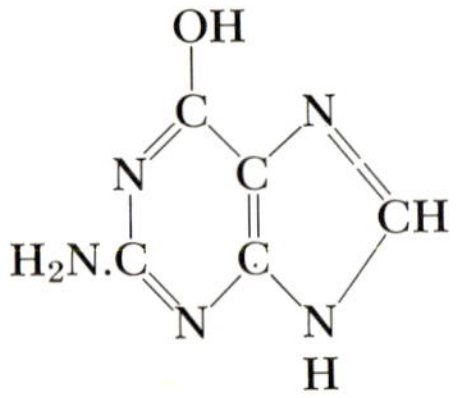

guanine
(2-amino-6-hydroxypurine)

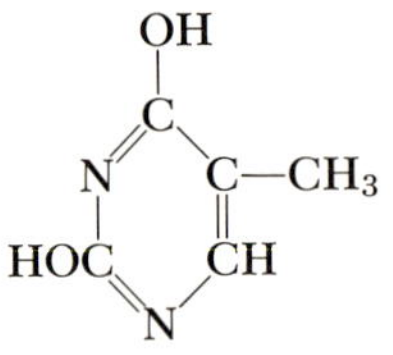

thymine
5-methyl-uracil

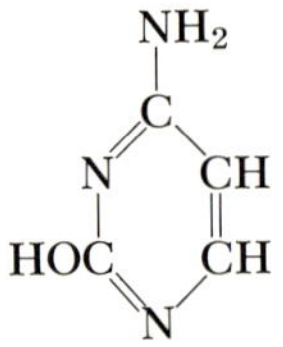

cytosine
2-hydroxy-6-amino-
pyrimidine

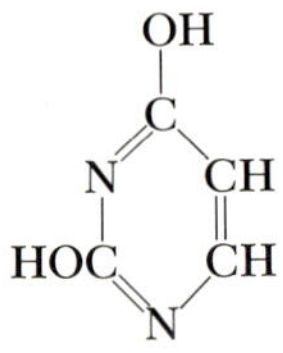

uracil
2:6-dihydroxy-
pyrimidine

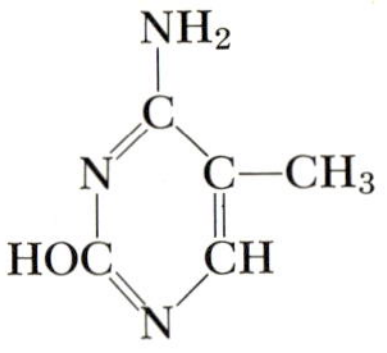

5-methyl cytosine

D. Nucleotide units

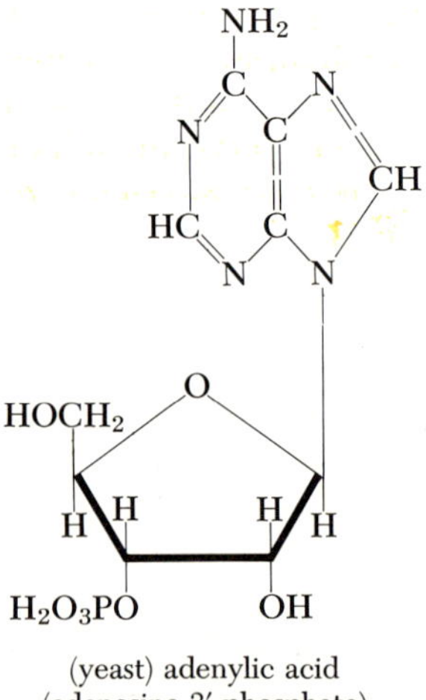

(yeast) adenylic acid
(adenosine-3′-phosphate)

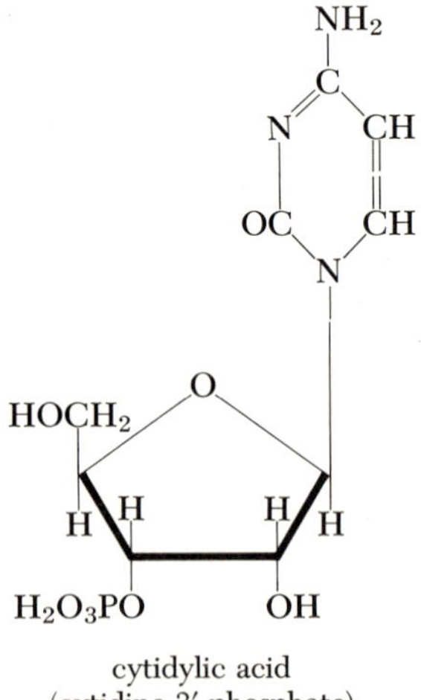

cytidylic acid
(cytidine-3′-phosphate)

***Figure 3.10.*** *A* and *B,* Purines and pyrimidines found in nucleic acids, *C. D,* Nucleotide units.

the protein *globin* is combined with the iron-containing porphyrin compound, *heme.* Hemoglobin has the unique property of loosely binding oxygen from the respiratory organ and releasing it to the tissue fluid, where the oxygen tension is low. A porphyrin chromoprotein similar to hemoglobin and serving as an electron carrier in all aerobic cells is cytochrome (see Chapter 15). Of great importance to the cell is the class of conjugated proteins known as *nucleoproteins.* They are formed by a combination of basic proteins (protamines or histones) with nucleic acid. Nucleic acid is considered to be a large molecule and is essentially a polymer of several kinds of nucleotides. Each nucleotide unit is composed of phosphoric acid, a pentose sugar and a purine or pyrimidine base. The structures of the kinds of purines and pyrimidines found in nucleic acids are given in Figure 3.10 along with that of a representative nucleotide. A polymer of nucleotides is formed by bridges between pentose and phosphoric acid residues of successive nucleotide molecules, as well as by other secondary bonds which are best represented in the Watson and Crick spiral model of nucleic acid (Fig. 3.16).

## LIPIDS

Lipids are a diverse chemical assembly which includes fats, waxes, phospholipids, carotenoids and sterols. Lipids serve not only as food reserves but also as parts of structures of the cell. They have one property in common: they are all soluble in fat solvents.

*Fat* is a triglyceride ester of glycerol and three fatty acid molecules. For instance, if three butyric acid molecules are linked to glycerol, the product is tributyrin; and if three palmitic acid molecules are linked to glycerol, the product is tripalmitin (see Fig. 3.11). Triglycerides are often spoken of as neutral fats since they have no free acid or basic groups. The three fatty acid molecules in neutral fat may be of two or three kinds. The two most prevalent fatty acids are palmitic and oleic acids, both of which are almost

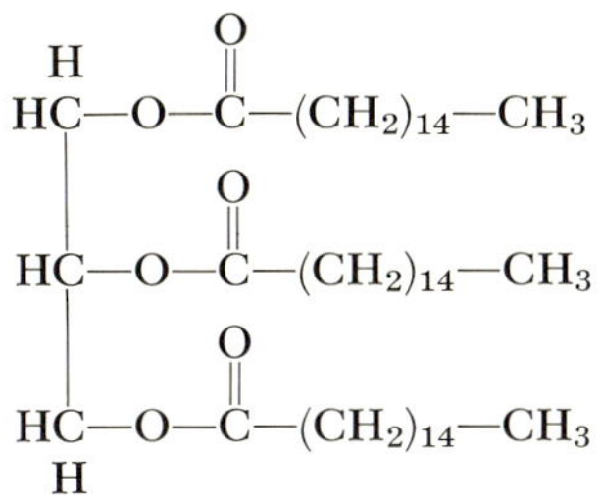

**Figure 3.11.** Tripalmitin, a triglyceride.

always present in every natural fat depot, forming the main components of the saturated and unsaturated fatty acid fractions, respectively. However, the fatty acid mixture present is often distinctive of the species of plant or animal in question. Neutral fats occur in cells primarily as food reserve (Lovern, 1957).

*Phospholipids,* containing phosphorus, are of great importance in the cell. The simplest phospholipid is a phosphatidic acid (Fig. 3.12) in which two fatty acids and one phosphoric acid have reacted with the three hydroxyl groups of glycerol. In most phospholipids the phosphoric acid, in turn, is combined with an organic base by another ester linkage, as shown in Fig. 3.13. Since phospholipids possess both acid and basic groups, they behave as zwitterions. Because of their ionic water-attracting (hydrophilic) groups and their fatty acid fat-attracting and water-repelling (lipophilic or hydrophobic) groups, phospholipids are somewhat soluble in both water and fats (or fat solvents) and therefore serve an important role in the cell in binding both types of compounds together. For example, lecithin, one of the phospholipids, is especially important as a structural material in the cell membrane

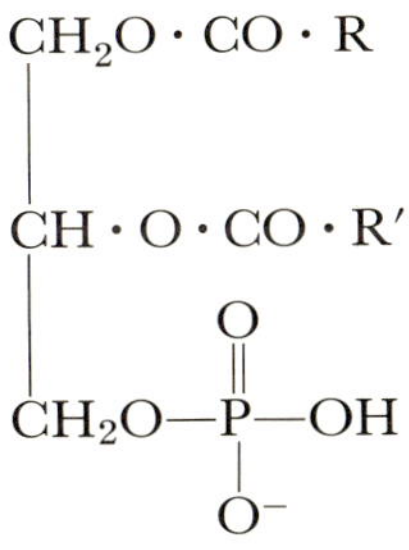

**Figure 3.12.** Phosphatidic acid.

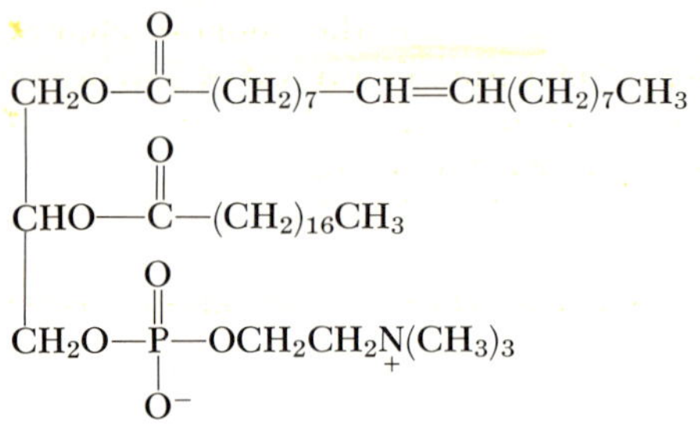


***Figure 3.13.*** Lecithin, a phospholipid.

since by its hydrophilic groups it maintains the continuity between the aqueous outside and the aqueous inside of the cell, yet fat-soluble materials dissolve in it and enter the cell because of its hydrophobic groups. Some phospholipids are characteristic of the species in which they are found. For example, phosphatidyl-inositol, in which the phosphoric acid is esterified by inositol, is unique to plant cells. Phospholipids containing large amounts of highly unsaturated $C_{20}$- and $C_{22}$-fatty acids are found in many animal cells but not in plant cells (Lovern, 1957).

*Waxes* are esters of higher aliphatic alcohols and higher fatty acids of longer chain length than the fatty acids of other lipid classes (generally $C_{24}$ to $C_{30}$ or even to $C_{36}$). For example, beeswax is a mixture of cerotic acid with myricyl palmitate. Waxes generally form protective coatings. In only a few instances do they constitute a part of the depot lipid. For example, the depot lipid of the cells in seeds of the shrub *Simmondsia californica* consists of esterified wax, and 70 per cent of the depot lipid of the sperm whale is wax ester. Depot waxes have shorter chain fatty acids than cuticular waxes (Lovern, 1957).

*Sterols* are aromatic alcohols characterized by the presence of a free hydroxyl group and high solubility in fat solvents. Sterols are found free as well as esterified with such fatty acids as palmitic or oleic. Although cholesterol (see Fig. 3.14) is by far the most prominent sterol of cells of vertebrates, a large number of sterols is known. A characteristic sterol of plant cells is $\beta$ sitosterol and one characteristic of fungi is ergosterol (one of the provitamins D). These differ from cholesterol only in the side chain attached to the top of the molecule as shown in Figure 3.14. The male and female sex hormones of vertebrates are sterols. The vitamins D also belong to this group. Sterols are found in the cell membrane and in other cell structures containing lipids (Fieser, 1955; Bloch, 1965). The sterols, being nonsaponifiable, are easily separated by saponification from most of the lipids generally found in cells. The other lipids, in turn, can be separated by chromatographic procedures.

*Carotenoids* (carotenes and xanthophylls), an important group of red and yellow plant pigments, are included among the lipids because of their solubility in fat solvents. Chemically they are tetraterpenes (a terpene has 10 carbon atoms), hydrocarbon in nature with alternate double bonds, and they may be acyclic (without a ring; for example, lycopene, the coloring matter of the tomato), or cyclic (e.g., carotene, a dicyclic tetraterpene). Carotenes consist of carbon and hydrogen only (see Fig. 19.12), whereas xanthophylls contain, in addition, oxygen (in the rings as hydroxy or carbonyl groups, or both). Lutein, the yellow xanthophyll pigment of the cells in leaves, is a dihydroxy derivative of carotene. Chromatographic analysis indicates the existence of six carotenes and a number of xanthophylls. Although carotenoids are synthesized in the cells of plants only, they may be accumulated by cells in certain tissues of animals. Thus vitamin A, which is important to growth and vision, is derived from carotene.

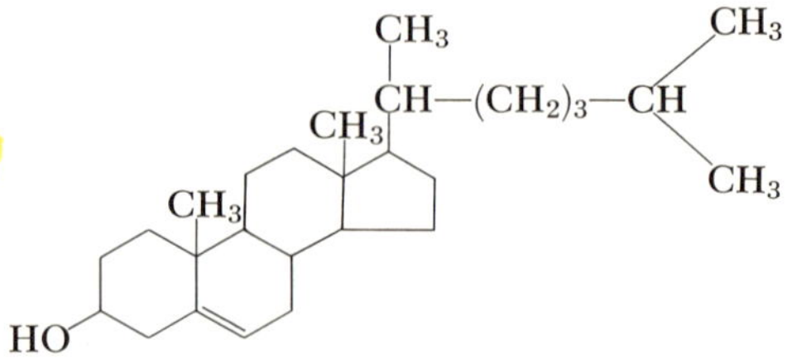


***Figure 3.14.*** Cholesterol.

## CARBOHYDRATES

*Carbohydrates* are compounds composed of carbon, hydrogen and oxygen present in the proportions of one carbon to two hydrogens to one oxygen ($CH_2O$). Representatives of the main groups of carbohydrates are the monosaccharides, oligosaccharides (composed of two or more monosaccharides), polysaccharides and mucopolysaccharides, a few of which are listed in Table 3.1 (Conn and Stumpf, 1966).

The *monosaccharides* are known as pen-

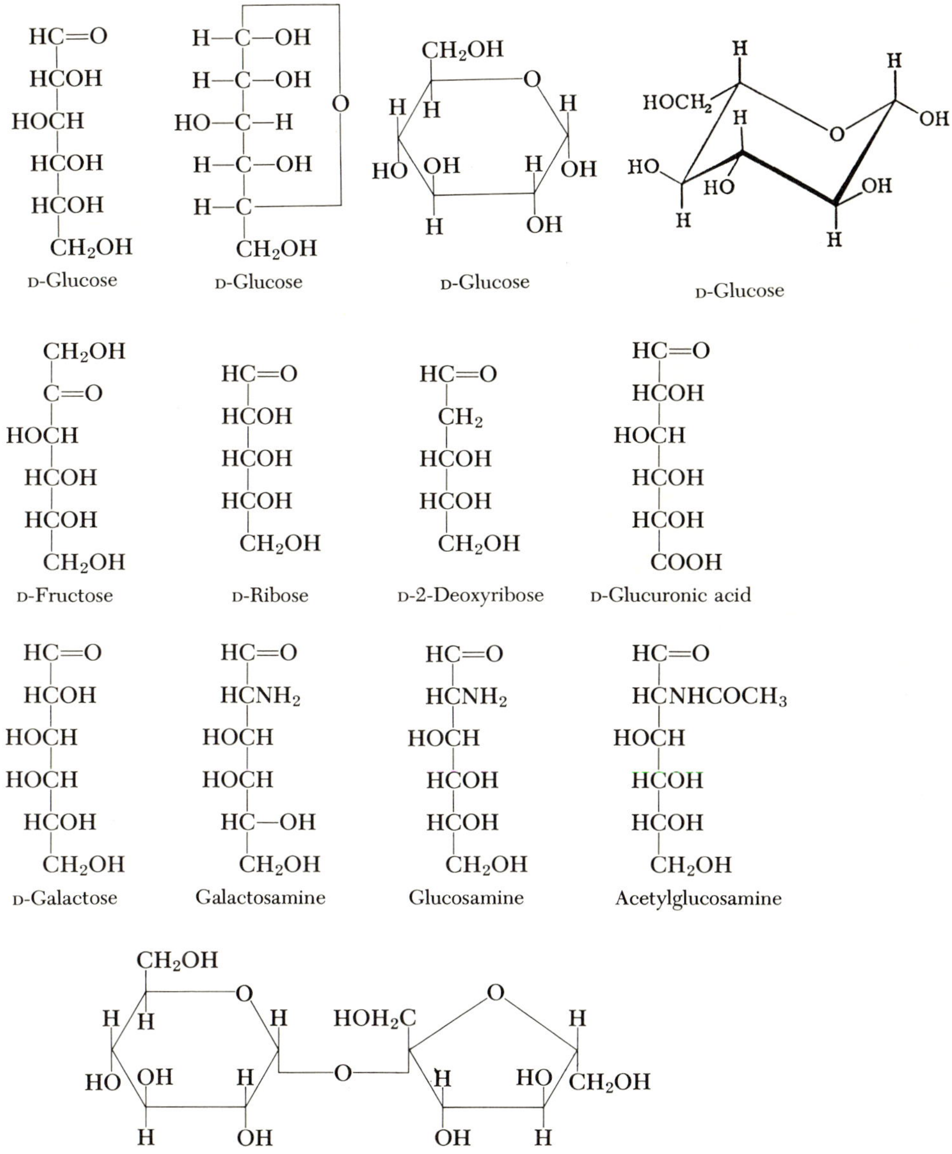


***Figure 3.15.*** Representative carbohydrates and their derivatives. Each of the substances may be represented in the four forms shown for glucose. The fourth form of glucose is one of the three-dimensional conformations.

toses if they have five carbons (see Fig. 3.15) and hexoses if they have six carbons. Ribose and deoxyribose, the sugars present in nucleic acids, are examples of pentoses. Glucose (dextrose, or grape sugar) and fructose (levulose, or fruit sugar) are examples of hexoses. Glucose is of great importance in cellular metabolism and is discussed at length in Chapters 16 and 17. Although it may be stored as glucose, especially in fruits, it is usually polymerized to an insoluble form such as glycogen (animal starch) in animal cells and starch in plant cells. There are enzymes in cells which are able to digest these insoluble food reserves, mobilizing them for use when the need arises.

The commonest *oligosaccharides* are the disaccharides of which the best known is sucrose or table sugar, which is stored in cells of such plants as the sugar cane and the sugar beet. Another important disaccharide is lactose which is present in milk secreted by cells of the mammary gland.

*Polysaccharides* are important skeletal materials in the cells of xylem, phloem and bast fibers, as well as in all cell walls of plants. Cellulose is the main polysaccharide found in cell walls. The cell walls are cemented together by the polysaccharide pectin, a D-galacturonic acid polymer (Preston, 1957; Siegel, 1962). Lignin, a complex ring polysaccharide, is also present in cell walls, especially in older cells (Nord and Schubert, 1958). D-Glucosamine acetylated on the amino group is polymerized to form part of the skeletal chitin of several invertebrate groups (Fig. 3.15). Chitin is occasionally present in some plant cell walls (e.g., yeast).

The *mucopolysaccharides* are of interest as structural materials secreted by animal cells. Intercellular cement consists of hyaluronic acid which is a polymer of glucuronic acid and acetylglucosamine. Some invading microorganisms have been shown to secrete the enzyme hyaluronidase, which enables them to penetrate host tissues more effectively because it digests the intercellular cement. The matrix of connective tissue and cartilage contains—in addition to some protein—chondroitin sulfuric acid, a mucopolysaccharide that is a polymer of galactosamine, glucuronic acid and sulfate (Conn and Stumpf, 1966).

On human red cells are found blood group substances. These are nitrogenous, neutral heteropolysaccharides containing amino acids or peptides as well as carbohydrate groups. They are therefore related to the mucopolysaccharides containing, among other substances, N-acetyl-D-glucosamine. These substances act as agglutinogens and are highly specific. Two main agglutinogens and six minor ones have been found. The blood group, or type, of an individual is determined by the nature of the hereditary blood group substance present on his red blood cells. A person having one type of blood substance has antibodies against other types but not against his own (Guyton, 1966).

## NUCLEIC ACIDS—DEOXYRIBONUCLEIC (*DNA*) AND RIBONUCLEIC (*RNA*)

About 50 years ago Miescher extracted nucleic acids and nucleoproteins from pus cells and fish sperm by the use of high concentrations of salt (1 to 2 molar NaCl). He found that about 60 per cent of the solids in sperm were nucleic acids, 35 per cent were proteins and 5 per cent were other materials, such as lipids, salts and carbohydrates. Since then extensive information has accumulated on the chemistry of the nucleic acids. Sperms have been widely used in such studies, but other types of cells have been employed as well. For example, nuclei can be easily isolated from ruptured nucleated red blood cells or from other fragmented tissues and washed free of cytoplasmic fragments by differential centrifugation (Dounce, 1954). Nucleic acid is readily extracted from such a suspension of nuclei.

Two types of nucleic acids have been extracted and identified by chemical means in all cells tested—deoxyribonucleic acid (DNA) and ribonucleic acid (RNA). Experiments show that DNA is the predominant nucleic acid in the nucleus, and RNA is predominant in the cytoplasm. A small amount of DNA is present in mitochondria (Nass *et al.,* 1965) and in chloroplasts (Gibor and Izawa, 1963). Each nucleic acid is considered to be composed of nucleotides (Fig. 3.10). Differences between the two nucleic acids include differences in pentoses, pyrimidine bases and size of molecule (Table 3.7). The pentose called deoxyribose, or D-ribodesose, and the pyrimidine thymine are present in DNA, whereas the pentose D-ribose and the pyrimidine uracil are present in RNA (see Fig. 3.15). The molecular weight of DNA is many millions whereas that of RNA is usually considered to be smaller, although the molecular weight of neither is known with accuracy since it depends partly upon the method of preparation. Thus, freshly isolated tobacco mosaic virus RNA has a molecular weight of 300,000 but decomposes spontaneously to units of molecular weight of 61,000. Alkaline hydrolysis decomposes these into units of molecular weight of 15,000. Several types of RNA are found in every cell (Davidson, 1960; Doty, 1961; Hurwitz and Furth, 1962).

***Deoxyribonucleic acid (DNA).*** The structure of DNA has been the center of attention in a number of studies, especially with x-rays. X-ray diffraction patterns indicate that dried fibers of pure DNA have a crystalline structure, although moist fibers form a less perfect pattern (paracrystalline). It has been found that the refracting units are much farther apart than would have been anticipated on the basis of a simple linear arrangement of the known constituents of the nucleic acids (phosphoric acid, pentose and base). Watson and Crick (1953; Crick, 1954) found that they could interpret the x-ray diffraction pattern by proposing a double helix as the possible structure of a nucleic acid fiber instead of a simple linear arrangement. They found that when such a fiber was coiled, refraction occurred periodically only from those units which were in line at any one time, their periodicity depending upon the pitch of the helix. From the calculated distances between atoms of the constituents of nucleic acid, they decided that the backbone of the nucleic acid consists of phosphoric acid residues and pentose sugar, and the side chains consist of the purine and pyrimidine bases. When the purine base of one helix in the model is juxtaposed to the pyrimidine base of the other helix, the fit of the two helices in the model is excellent (Fig. 3.16).

Calculation of the angles between units in the molecule predicted from x-ray diffraction data showed that this model fits the requirements for the structure of DNA best. Watson and Crick therefore propose that the two helices of DNA are linked through these bases by way of hydrogen bonds, adenine being linked to thymine, and cytosine being linked to guanine (see Fig. 3.17). In extracts of DNA the amount of adenine is always equal to that of thymine, and the amount of cytosine is equal to that of guanine (Davidson, 1960) (see Table 3.8). The model, therefore, satisfactorily accounts for the content of bases in deoxyribonucleic acid. The puzzling change in the x-ray diffraction pattern between moist and dry DNA is also interpretable with the helical model as being the result of slippage between the two helices causing

**Table 3.7.** Similarities and Differences between the Constitutions of DNA and RNA

| CONSTITUENT | DNA | RNA |
|---|---|---|
| Purine: | | |
| adenine | + | + |
| guanine | + | + |
| Pyrimidine: | | |
| cytosine | + | + |
| uracil | − | + |
| thymine | + | − |
| methyl cytosine* | + | − |
| Sugar: | | |
| ribose | − | + |
| deoxyribose | + | − |
| Phosphoric acid | + | + |

* Data from Davidson, 1960. Methyl cytosine is not always present. Methyl adenine occurs in the RNA of some microorganisms, methyl guanine in the RNA of certain plant cells and especially in the soluble RNA of liver cells along with methyl cytosine. 6-Methyl aminopurine has been found in the DNA of some strains of the colon bacillus.

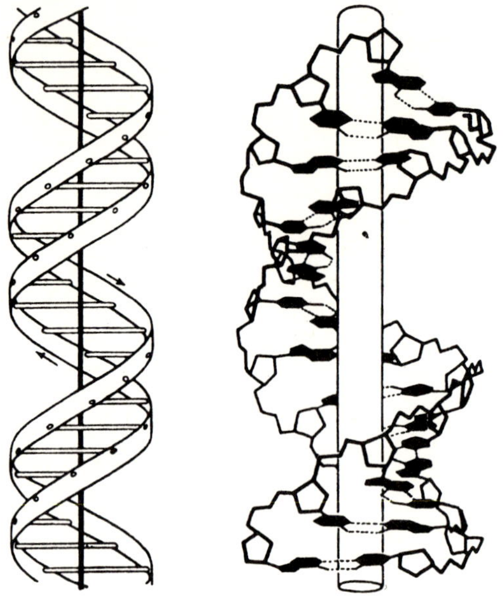

***Figure 3.16.*** *Left,* Diagrammatic representation of the DNA molecule as proposed by Watson and Crick (1953). The two phosphate-sugar chains are represented by ribbons and the pairs of bases holding the chains together are shown as horizontal rods. *Right,* Drawing of a model of the DNA helix. The dotted lines represent the hydrogen bonds binding the purine base of one chain to the pyrimidine base of the other. The rod in the center is a support in a model, not a real structure. (From Davidson, 1960: The Biochemistry of the Nucleic Acids. 4th. Ed. Methuen, London.)

**Table 3.8.** Molar Proportions of Bases (as Moles per 100 Moles Nucleotide) in Nucleic Acids from Various Sources

| RNA IN ORGAN OR ORGANISM | ADENINE | GUANINE | CYTOSINE | URACIL | 5-METHYL CYTOSINE | THYMINE |
|---|---|---|---|---|---|---|
| Ox liver | 17.1 | 27.3 | 33.9 | 21.7 | | |
| Ox kidney | 19.7 | 26.7 | 33.4 | 20.2 | | |
| Rat kidney | 19.4 | 29.5 | 30.7 | 20.4 | | |
| Sea urchin embryo | 22.6 | 29.4 | 27.2 | 20.8 | | |
| Yeast | 25.4 | 24.6 | 22.6 | 27.4 | | |
| *Serratia marcescens* | 20.3 | 31.2 | 24.3 | 24.1 | | |
| *Escherichia coli* | 25.3 | 28.8 | 24.7 | 21.2 | | |
| **DNA IN ORGAN OR ORGANISM** | | | | | | |
| Ox thymus | 28.2 | 21.5 | 21.2 | | 1.3 | 27.8 |
| Ox spleen | 27.9 | 22.7 | 20.8 | | 1.3 | 27.3 |
| Ox sperm | 28.7 | 22.2 | 20.7 | | 1.3 | 27.2 |
| Rat bone marrow | 28.6 | 21.4 | 20.4 | | 1.1 | 28.4 |
| Herring testes | 27.9 | 19.5 | 21.5 | | 2.8 | 28.2 |
| Sea urchin | 32.8 | 17.7 | 17.3 | | 1.1 | 32.1 |
| Wheat germ | 27.3 | 22.7 | 16.8 | | 6.0 | 27.1 |
| Yeast | 31.3 | 18.7 | 17.1 | | — | 32.9 |
| *Escherichia coli* | 26.0 | 24.9 | 25.2 | | — | 23.9 |
| *Mycobacterium tuberculosis* | 15.1 | 34.9 | 35.4 | | — | 14.6 |

From Davidson, 1960: The Biochemistry of the Nucleic Acids. Methuen & Co., London. For the structure of purine and pyrimidine bases, see Figure 3.10.

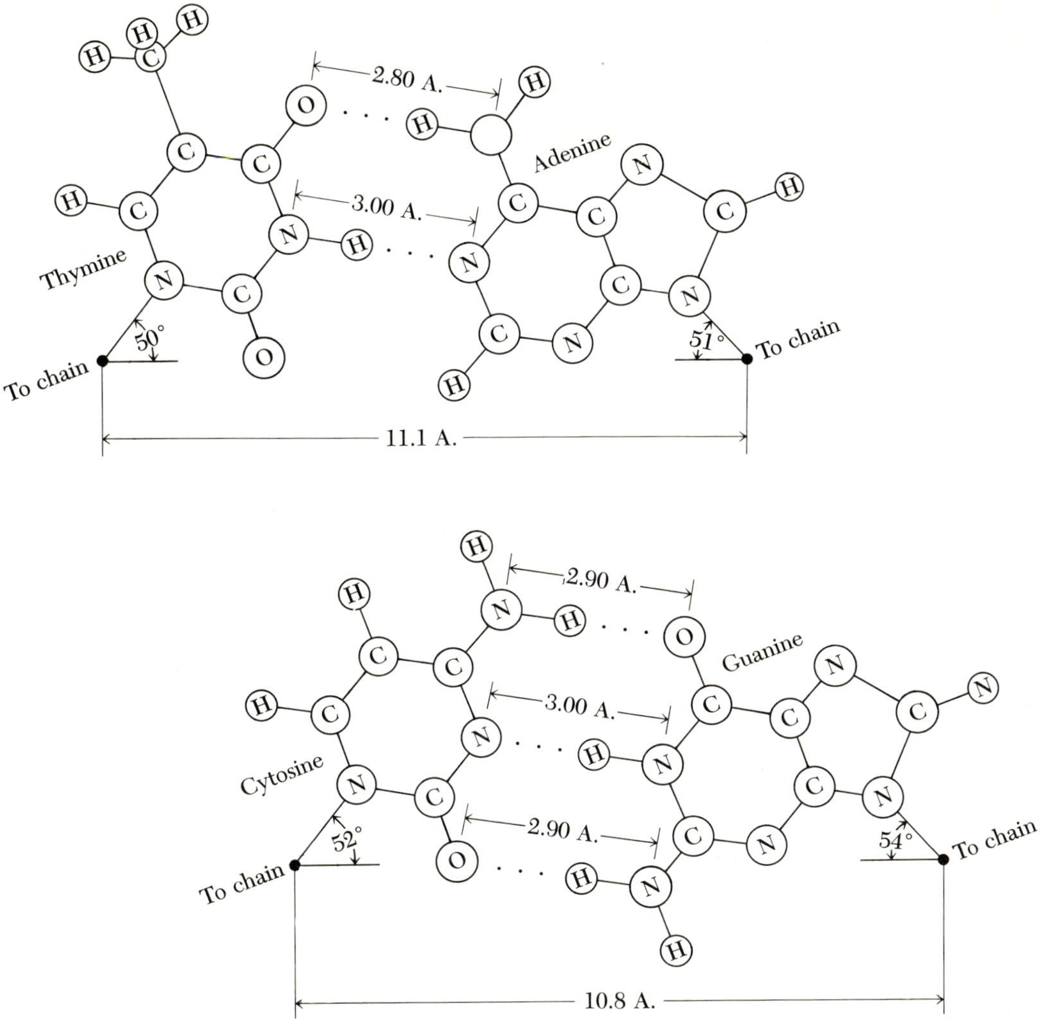


***Figure 3.17.*** The pairing of adenine and thymine (top) and cytosine and guanine (bottom) by means of hydrogen bonding. (From Anfinsen, 1958: The Molecular Basis of Evolution. John Wiley & Sons, New York.)

displacement of the perfect crystalline pattern of the dry fiber.

The DNA helix is not a unique conception for structure of fibers, since helical models of proteins (e.g., $\alpha$ helix) best explain the x-ray diffraction data on certain proteins.

Helical structure of DNA does pose some problems, however. For example, how does the new set of helices separate from the parent set when duplication occurs in the nucleus during mitosis? Some of these problems are considered in Chapters 6 and 21.

***Ribonucleic Acid (RNA).*** It is now established that each cell contains three kinds of RNA: transfer or soluble (sRNA), messenger (mRNA) and ribosomal (rRNA). Small amounts of RNA may also be present in mitochondria, plasma membrane and chloroplasts, but these may be contaminants. *Transfer RNA,* which transports the activated amino acids during protein synthesis, has a low molecular weight (about 25,000). There are probably as many (or more) kinds of transfer RNA as there are amino acids, that is, 20 (or more). *Messenger RNA,* which, as its name implies, carries information from DNA to the ribosome where protein synthesis occurs under its direction, has a variable molecular weight, of the order of 500,000. *Ribosomal RNA* is thought to have a molecular weight of several million, depending upon the method of preparation. Although transfer, messenger and ribosomal RNA are characteristic of the cytoplasm, they are all produced in the nucleus (see Chapter 6). It is not certain whether mitochondrial RNA is produced in the mitochondrion by its DNA or whether it comes from the nucleus. Very little is known about membrane RNA.

RNA is single stranded. However, the strand may twist back upon itself, so that portions of it may come in contact, at which points connections (by hydrogen bonds) may be formed, giving RNA a pseudohelical structure. Evidence from X-ray diffraction studies and from changes in ultraviolet absorption after breakage of hydrogen bonds by heat indicates that RNA has an imperfect helical structure, with hydrogen bonds between neighboring guanine and cytosine and between neighboring adenine and uracil residues. The defects in the helix in the form of unpaired bases are reflected in the somewhat unequal proportions of adenine to uracil and guanine to cytosine, as seen in the data of Table 3.8. Several possible arrangements for the sRNA molecule (alanine RNA) are shown in Figure 3.18.

It is known that ribosomal RNA is always associated with structural protein in the ribosome. The ribosome contains about 35 to 60 per cent protein, which is rich in arginine and lysine residues. The unpaired bases of the RNA helix perhaps serve as attachment sites for the amino acid residues of the protein. Pure ribosomes probably contain no lipid (Peterman, 1964).

***Denaturation of DNA and RNA.*** When DNA is heated to a temperature of 70° C. or above, the hydrogen bonds across the base pairs (adenine-thymine, guanine-cytosine), which link together the two antiparallel strands, are broken. As a consequence a number of specific properties of the DNA molecule are altered; that is, the DNA is denatured. For example, the characteristic absorption peak at 260 nm.* increases (this is called the hyperchromicity) as more hydrogen bonds are broken and the helical structure decreases, and if the absorption is plotted against the temperature, curves such as those in Figure 3.19 are obtained. As shown in Figure 3.19, the melting temperature $T_m$, at which the temperature absorbance curve rises steeply, is a function of the base composition of the DNA; it is higher the higher the guanine-cytosine to adenine-thymine ratio in the DNA molecule under consideration. Lowering the temperature permits re-forming of many of the hydrogen bonds and recovery of the physical properties dependent upon these bonds in the native DNA. From data such as these some information can be deduced about the percentage of the DNA molecule in helical form and the degree to which the physical properties depend upon the hydrogen bonds (Doty, 1961).

When a mixture of two different DNAs is melted (heated), on cooling pairing may occur so that helices containing strands of two different DNAs are formed. This will occur only if the two strands have sequences of bases in common. By so "hybridizing" DNA from different sources one can determine the degree of their similarity in sequences, using the change in physical properties as a measure. For example, the hyperchromicity will decrease in proportion to the degree of pairing and hydrogen bond formation as a mixture of two different DNA molecules is cooling. It has been shown that some organisms have sequences in common over considerable portions of the DNA mole-

---

*It has become customary to measure wavelength of light in nanometers (nm.) instead of millimicrons (mμ) in vogue until recently. 1 nm. = 1 mμ = $10^{-9}$ meter.

The angstrom unit (AU or Å or A) = $10^{-10}$ meter. It is still used to designate wavelength when citing emission lines of the elements and so forth, known with considerable precision. In practice, unfortunately, all three units are often used interchangeably. In addition, the micron (μ), $10^{-6}$ meter, may also be used.

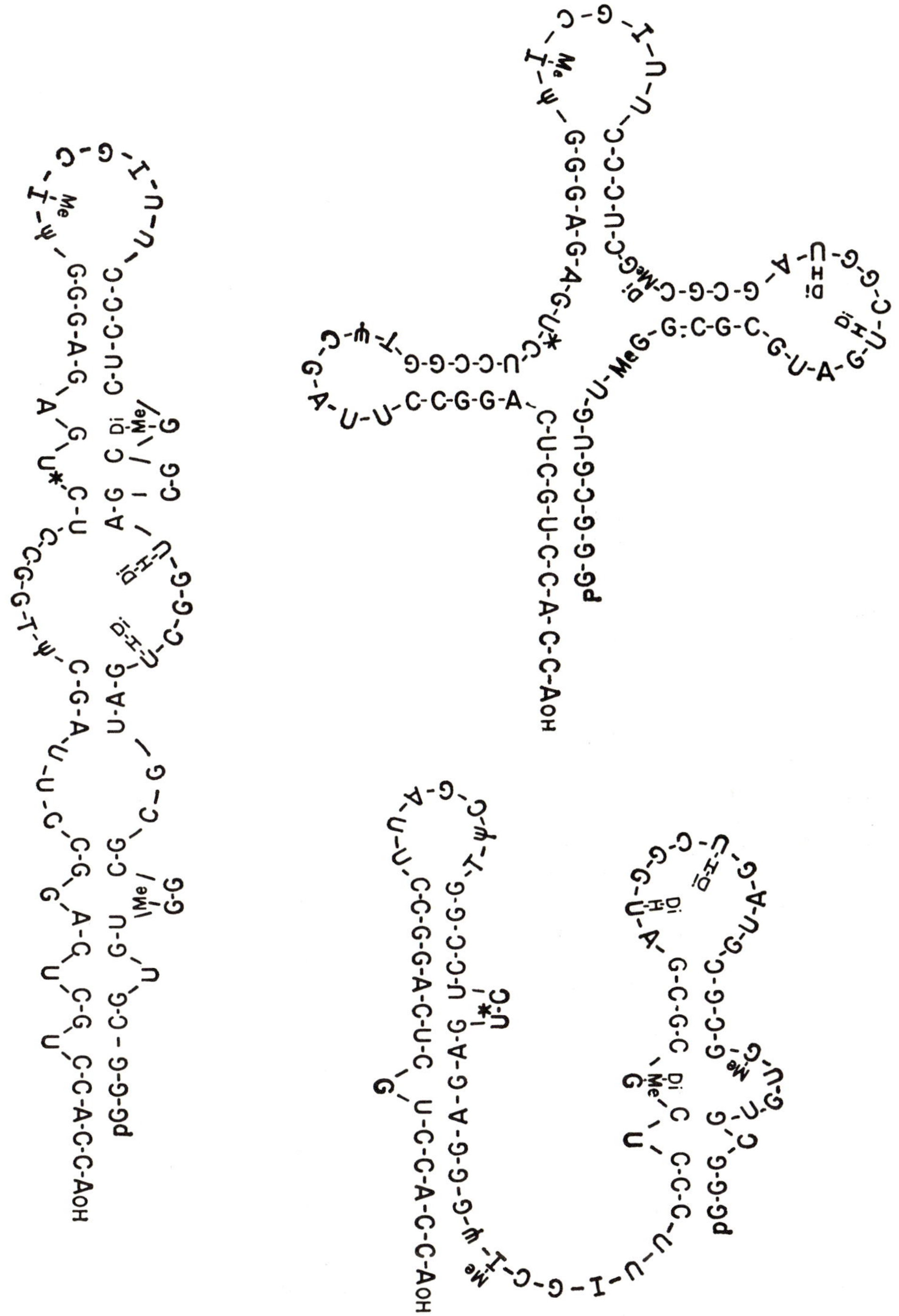

***Figure 3.18.*** Schematic representation of three possible conformations of the transfer RNA (alanine-RNA) with short, double-stranded regions. In this case the complete nucleotide sequence has been worked out. (From Holley: Science *147*:1462, 1965.)

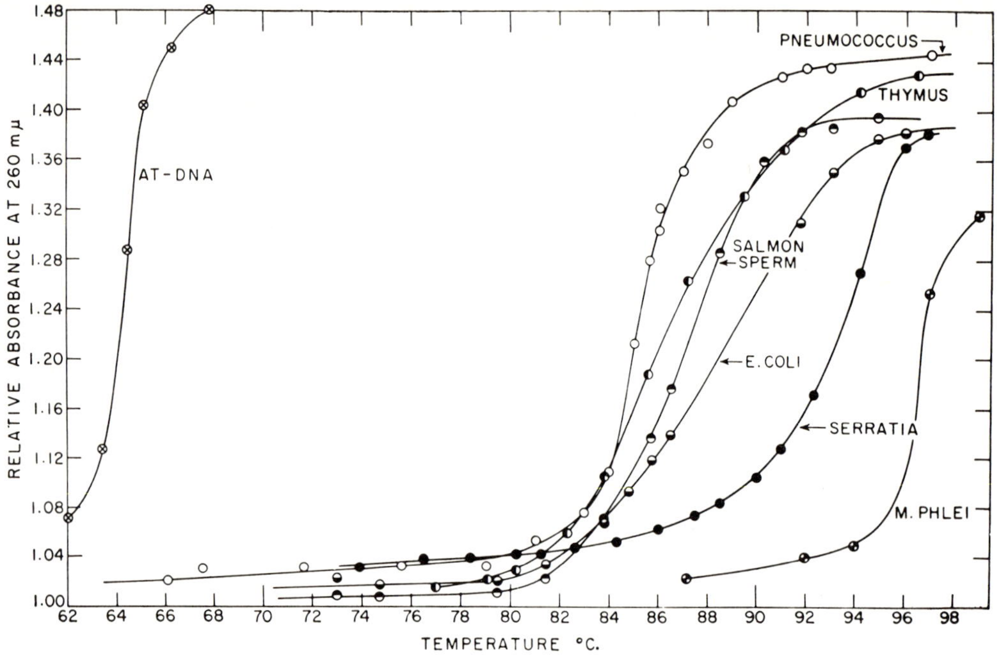


***Figure. 3.19.*** Absorbance (260 nm.)-temperature curves for various samples of DNA, showing development of hyperchromicity with rise in temperature. (From Doty: Harvey Lect. *55*:103, 1961.)

cules, whereas others have similarity over only limited stretches. In this manner relationships between organisms can be traced (Doty, 1961).

In RNA, the helical structure, to the extent that it occurs, is also dependent upon hydrogen bond formation between adjacent uracil to adenine and cytosine to guanine residues. These hydrogen bonds are progressively broken by heating (35 to 90° C.), showing characteristic melting temperatures. Also, cooling permits re-forming many of the hydrogen bonds in RNA. As discussed earlier for DNA, here also it is possible to determine the degree of breakage of hydrogen bonds by the increased absorption at 260 nm. and changes in other physical properties (Doty, 1961).

Hybridization between RNA and DNA from the same species has also been observed, as might be expected since the uracil, adenine, cytosine and guanine of RNA could, by hydrogen bond formation, pair with the adenine, thymine, guanine and cytosine, respectively, of the DNA. The hybrids are formed when DNA and RNA from the same species are heated together in appropriate salt solutions and cooled. The significance of such hybridization will be discussed in Chapter 6.

## VITAMINS AND GROWTH FACTORS

Vitamins, organic molecules of diverse chemistry, are required in minute amounts for the growth and normal functioning of cells. Sometimes the term "growth factors" is used to denote such substances required by plant cells and microbes, and "vitamins" is restricted to use as a term in animal physiology. The distinction is not valid and is not followed here.

Vitamins play an important part in cellular metabolism, acting chiefly as parts of enzymes or other biological catalysts. A table of vitamins is included here (see Table 3.9), largely to complete the survey of biologically important compounds. Of the vitamins listed, only A, D and E are oil soluble; the others are water soluble. The functions of some of these vitamins, particularly the water-soluble groups, as

**Table 3.9.** Vitamins

| CLASS | CHEMICAL NATURE | EFFECT OF LACK ON VERTEBRATES | CELLULAR FUNCTION |
|---|---|---|---|
| A | Carotenoid | Growth interference, night blindness | Part of visual purple in retina; growth |
| D | Sterol | Bone defects | Not known; mobilizes salts in gut |
| E | Tocopherol | In rat, defective implantation and testis development | Unknown |
| $B_1$ | Thiamine | Beriberi | Coenzyme for pyruvate metabolism |
| $B_2$ | Riboflavin | Cataract | Prosthetic group of flavoprotein enzyme |
| Niacin | Nicotinic acid amide | Pellagra | Part of dehydrogenase coenzyme |
| Pantothen | Pantothenic acid | Dermatitis and spectacle eye | Part of coenzyme A |
| $B_6$ | Pyridoxine | Need not demonstrated | Coenzyme for amino acid conversions |
| H | Biotin | Need not demonstrated | Coenzyme in $CO_2$ fixation in $C_4$ acids |
| Inositol | Cyclic Compound | Need not demonstrated | Unknown |
| Folic acid | Pteroylglutamic acid | Anemia | Coenzyme functioning in "one" carbon metabolism |
| $B_{12}$ | Cyanocobalamin (tetrapyrrole with cobalt in center) | Anemia | Coenzyme of an enzyme involved in methyl transfer and nucleic acid metabolism |
| Protogen | Thioctic acid | Need not demonstrated | Coenzyme in pyruvate oxidation |
| C | Ascorbic acid | Scurvy | Maintain optimal oxidation reduction potential? |
| K | 1,4-naphthoquinone acetate | Hemorrhage | Prothrombin formation in blood |
| P | A mixture of substances* | Capillary fragility | Maintains cement of capillary walls |
| Essential amino acids | Essential amino acids† | Failure in growth or normal function Wasting of tissue | Structure of cells, etc. |

* Eriodictin, hesperidin, rutin.

† Tryptophan, phenylalanine, lysine, histidine, leucine, isoleucine, threonine, methionine, valine and arginine are required by the rat. The requirement is not always the same for all types of animals.

**Table 3.9.** Vitamins (Cont.)

| CLASS | CHEMICAL NATURE | EFFECT OF LACK ON VERTEBRATES | CELLULAR FUNCTION |
|---|---|---|---|
| Essential fatty acids | Linoleic, linolenic and arachidonic | Failure in growth or functioning | Structure of cells, etc. |
| Methyl compounds | Choline, methionine | Failure in growth, etc. | Structure of cells, etc. |
| Sulfhydryl-containing compounds | Cysteine, glutathione | Failure in growth, etc. | Structure of cells, etc. |

parts of cellular enzymes are considered in Chapters 15 and 16.

## CHEMICAL COMPOSITION OF CELL ORGANELLES

Differences in the chemical nature of nucleus and cytoplasm are indicated by their different staining reactions to the same aniline dyes at the same pH. To illustrate, in the mildly acid range of pH, negatively charged (acid) aniline dyes stain the cytoplasm, not the nucleus. In a more acid solution even the nucleus (i.e., the chromatin of the diffuse stage of the chromosomes or the condensed chromosomal material) is stained. On the other hand, positively charged (basic) dyes in mildly acid solution may stain the nucleus, not the cytoplasm, but in alkaline solutions they also stain the cytoplasm. By finding the pH at which staining ceases with both positively and negatively charged dyes, it is possible to determine the approximate isoelectric pH for the cytoplasm and the chromosomes. As determined by this method, the isoelectric pH for cytoplasm appears to be in the range of 4.7 to 5.0, whereas for chromosomes it appears to be about 3.3 to 3.6.

The pH of the cytoplasm, estimated by injection of indicator solutions into the cell (Chambers, 1949), is found to be about 6.7 to 6.8. The pH of nuclei has been found to be about 7.5 to 7.8. Although these values are controversial (Caldwell, 1956) because of various possible errors, measurements for both cytoplasm and nucleus are on the alkaline side of the isoelectric pH of the proteins and nucleoproteins found in the structures. It appears that in the living cell both proteins and nucleoproteins are negatively charged. On the other hand, the nuclear envelope, with an isoelectric pH estimated to be between pH 10.5 and 11.5, has a positive charge on its outer surface. The outer surface of the cell is usually positively charged, owing to a layer of positive ions outside the cell (Bittar, 1964).

6.7-6.8
7.5-7.8

The cell membranes are made up primarily of protein and phospholipids, as well as some cholesterol. A more complete discussion of the cell membrane is given in Chapter 11, and additional information on cytoplasmic cell organelles and on the nucleus is given in Chapters 5 and 6, respectively.

The nucleus is made up, mainly, of protein and nucleic acid. On the chromosome are found ordinary protein and nucleoprotein, the latter being more prominent in the darkly staining portion. The nucleolus contains nucleic acid and the nuclear sap contains protein. The chemistry of the nucleus is of sufficient importance to be dealt with separately (see Chapter 6).

The ground substance of the cytoplasm of cells is made up mostly of proteins. The ribosomes present in the cytoplasm however, consist primarily of ribonucleic acid along with protein. A range of microsomes of different sizes and of different composition is found when cells have been fragmented and the organelles separated by centrifugation. The microsomes are fragments of the endoplasmic reticulum.

The endoplasmic reticulum, which forms a membranous network in the ground substance of the cell, consists largely of lipid (30 to 50 per cent) and protein (40 to 60 per cent), with some RNA. The lipids are mostly phospholipids (50 to 66 per cent) of which lecithin predominates, but considerable cephalin may be present (Fig. 3.13). Cholesterol (Fig. 3.14) and cholesterol esters are also found. The protein is partly structural and partly enzymatic, the endoplasmic reticulum being a region of active synthesis of various lipids as well as a region in which some drugs are detoxified, and many kinds of enzymes are found (Rothschild, 1963). It was thought that the RNA found in the endoplasmic reticulum might be an experimental contaminant, but after the ribosomes were removed the RNA still formed about 10 per cent of the total dry weight of the smooth membranes, an amount seemingly beyond contaminant quantity (Siekevitz, 1963).

The membranes of the Golgi complex are similar to those of the endoplasmic reticulum; however, there is no RNA and there is more lipid, but fewer enzymes are present than in the endoplasmic reticulum. Among the enzymes, acid phosphatase is characteristic of the Golgi complex and is found in several times the amount present in cell homogenates (DeRobertis *et al.*, 1965).

Analyses of mitochondria separated from other cell constituents and concentrated by centrifugation indicate that they are made up largely of proteins (60 to 65 per cent) and lipids (35 to 40 per cent), with a small amount of RNA. The lipids are largely phospholipids, lecithin (Fig. 3.13) being prominent, but triglycerides (Fig. 3.11), phosphatidic acid (Fig. 3.12) and cholesterol (Fig. 3.14) are present as well. Also found is a lipid characteristic of the mitochondria, called cardiolipin, which consists of three phosphatidyl and one diglyceryl radicals. The protein of mitochondria is partly structural but many enzymes have been found in the mitochondria, enzymes involved in energy liberation from nutrients (discussed in Chapters 16 and 17). Since removal of lipid lowers the activity of some of these enzymes, the structural relationships of the mitochondrial membranes and enzymes appear to be of importance to enzyme function. The structural protein of mitochondria is insoluble in water but may be solubilized by anionic detergents. Removal of detergent results once again in precipitation of the protein (Lehninger, 1965). A contractile protein has also been claimed (see Chapter 5). The DNA of mitochondria probably accounts for the self-replicating property of this organelle (see Chapter 5).

## CHEMICAL COMPOSITION OF THE SMALLEST CELLS

As was discussed in Chapter 2, the pleuropneumonia-like organisms (PPLO) range in size from 0.1 to 1.0 $\mu$. Chemical analysis of the infectious PPLO *Mycoplasma gallisepticum* indicates that the nonaqueous portion of the cell contains 4 per cent DNA and 8 per cent RNA, which is somewhat more nucleic acid in proportion to other compounds than is found in larger cells (compare Table 3.3) but in keeping with that found in bacteria. The DNA is in the form of a two-stranded filament, the RNA in the form of granules combined with protein, resembling ribosomes of larger cells. The proteins are similar to those of larger cells and over 40 different kinds of enzymes have been identified. Lipids are also present, among them cholesterol and cholesterol esters, compounds characteristic of animal cells.

It has been estimated that at least 100 kinds of enzymes are necessary for cell metabolism and that more than one molecule of each must be present because thermal agitation in the cell leads to degradation of some of them. Furthermore, these enzymes must be encoded and it presumably takes a molecular weight of about 1,000,000 of DNA to encode one of them. Even a *M. gallisepticum* cell of diameter 0.25 $\mu$ has only enough DNA to make one molecule of molecular weight

45,000,000 or 45 DNA molecules of molecular weight 1,000,000. Thus, although the biochemistry of the smallest cells is evidently not unique, these cells pose problems that are difficult to answer (Morowitz and Tourtelotte, 1962; Kleineberger-Nobel, 1962).

## LOCALIZATION OF CHEMICALS IN CELL ORGANELLES: CYTOCHEMISTRY

The aims of cytochemistry are (1) to determine the localization of various chemicals present in cells, that is, to find out in which part of the organelle they are present; (2) to measure the quantity in each locus; and (3) to establish the function of the various chemical constituents of the organelles. To accomplish these objectives the chemicals must be identified by specific microchemical reactions and quantitative determination must be made of the chemicals so identified. This is an ambitious and inclusive program (Burstone, 1963; Glick, 1963; Wied, 1966). Since the field is changing rapidly and new techniques continue to be developed and tested, it is possible here to give only a few methods as samples of the kind of attack made and of the nature of the information obtained. Other examples are given elsewhere (for nucleic acids in Chapter 6, and for the cell membrane in Chapter 11).

In one of the newer methods of choice, tissues are frozen and freed of water by evaporation in vacuum at low temperatures, a method which minimizes shrinkage as well as a change of the colloidal state. The proteins and other cell constituents are precipitated (fixed) *in situ.* A specimen dried in this manner may be directly embedded in paraffin for sectioning, and the sections may be stained or subjected to other chemical treatment for identification of cellular constituents.

If the section of tissue so prepared is to be stained or treated with reagents in aqueous solution, the proteins must be rendered insoluble in the reagents applied. A variety of fixing agents have been developed suitable for particular needs.

Of particular interest is staining specifically the chemicals in cell organelles. For example, the identification and distribution of enzymes may be determined by color reactions or the substrate film methods (Daoust, 1965). For phosphatase, this is done in the following way. The cell or section of the tissue is incubated with calcium and glycerophosphate. Calcium phosphate is deposited wherever the enzyme is present. The calcium may then be identified and evaluated by subsequent treatment of the tissue with a cobaltous salt and ammonium sulfide, upon addition of which a black precipitate of cobalt sulfide is formed. By this method phosphatase has been demonstrated to be on or in the Golgi complex, outlining its structure. Various color reactions have been developed for other enzymes and for nucleic acids, proteins, lipids and carbohydrates.

Nucleic acids can also be identified by ultraviolet spectrophotometry. This is based on the differential absorption of short wavelength ultraviolet light by nucleic acids and proteins. Nucleic acids absorb maximally at 260 nm. because of the purines and pyrimidines in their molecules, whereas proteins absorb maximally at about 280 nm. which is accounted for by the aromatic and heterocyclic amino acids present there, as shown in Figure 3.20.

In the autoradiographic method cells or sections of tissues of organisms fed nutrients containing radioactive labels (e.g., $^{32}P$) are applied directly to a photographic emulsion. After a period of exposure the film is developed, fixed and examined microscopically. The density of the developed film measures the relative concentrations of the radioactive nutrient in various cell organelles (Fig. 3.21).

Sometimes enzymes are used to remove one protoplasmic constituent, making possible the identification of remaining constituents. Thus pepsin and trypsin remove the protein from a *Drosophila* chromosome but leave nucleic acids fixed with lanthanum salts (see also Chapter 6).

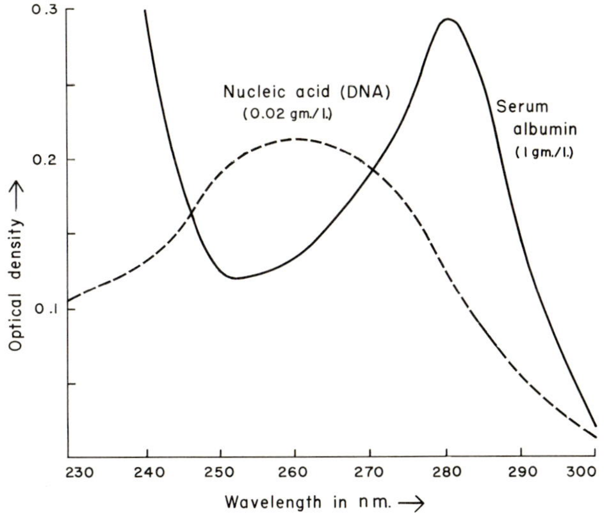


***Figure 3.20.*** Absorption spectra of nucleic acid and serum albumin. The optical density is highest where absorption is greatest; therefore the peaks in the figure indicate the regions of greatest absorption. Note that albumin has a peak at 280 nm. whereas nucleic acid has a peak at 260 nm. Also of interest is the relative absorption of ultraviolet radiation by the two compounds; for instance, a solution of only 0.02 gm. per liter of nucleic acid absorbs at its region of maximum absorption almost as much of the radiation as does a solution of 1 gm. per liter of albumin at its maximum.

Still another type of cytochemical technique, used for determining the localization of salts in the cell, is *microincineration,* in which entire cells or slices of tissues are placed in an electric furnace at a sufficiently high temperature to oxidize all the organic materials. Only the ash content of cells is then left. From the ash *in situ* it is possible to identify the kind and amount of salts present and the particular organelle in which they are found (see Fig. 3.22). Calcium is seen to be concentrated around the chromosomes, and more iron is present in the nucleus than in the cytoplasm. Iron can be recognized by the red color of its oxide. The detection of some elements is difficult. Calcium often appears in the form of crystals of calcium sulfate. Magnesium cannot be detected in the presence of calcium. Silicon can be identified because it is doubly refractile.

Not all the elements are retained in the ash, however; for example, sodium and potassium disappear because they volatilize at red heat. Distribution of potassium and sodium must be studied by first converting them into sulfates which do not volatilize at the temperatures used for incinerating the cells. Some materials may be identified by the use of x-rays (Engstrom, 1966).

It is desirable not only to identify a constituent of the cell but also to determine how much of it is present. This is difficult, but by determining how much light is transmitted through a section of an organelle stained with a specific dye and comparing it with the light transmitted by a like thickness of a standard solution of the identified substance, color tests may be made quantitative or semiquantitative. Ultraviolet spectrophotometry of nucleic acid has also been made quantitative by comparing transmissions of ultraviolet light through cell organelles and the same thickness of a solution of a known nucleic acid concentration. It is even possible by a similar calibration to make estimates of concentrations from autoradiographs.

By using the microincinerated preparation as a source of electrons in an electron microscope it has been possible to identify the distribution of some of these elements, because the electrons are given off at different temperatures by different elements (Horning, 1951).

Electron microscopy makes possible localization of enzymatic reactions within the cell organelles. For this purpose the cells are fixed in appropriate aldehydes (e.g., glutaraldehyde) which allows many enzyme systems to remain active, and from the products of their action on added sub-

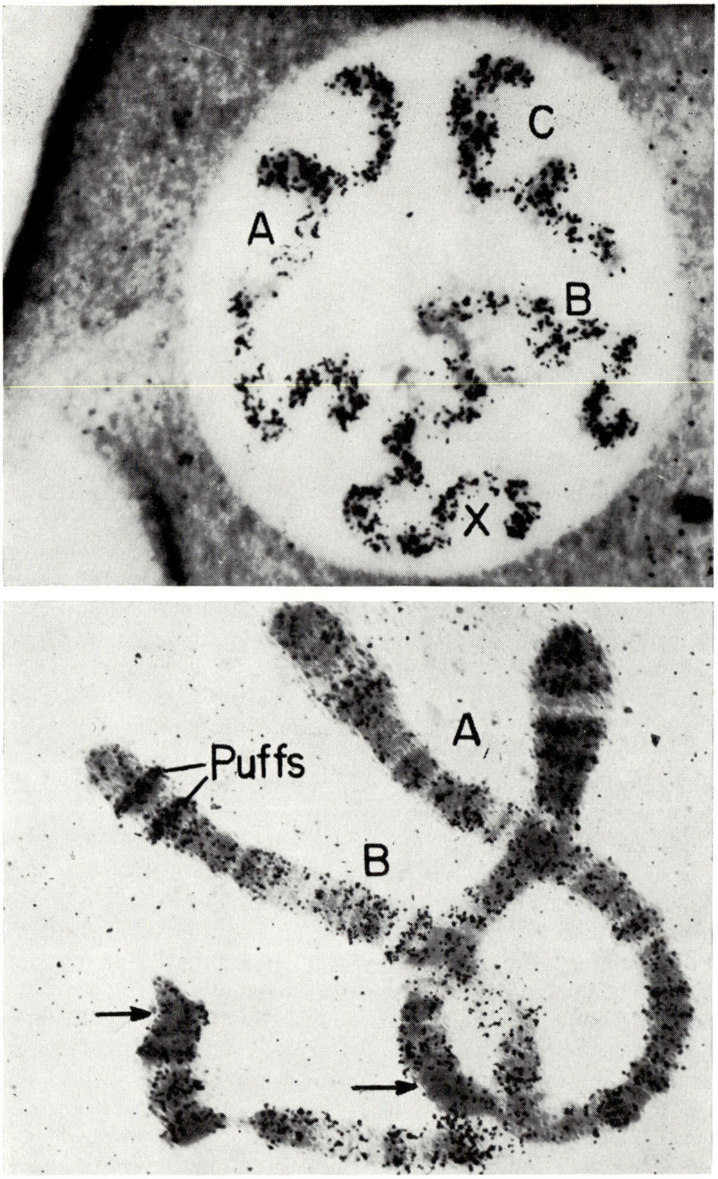


***Figure 3.21.*** The autoradiographic method. The nucleus and polytenic (multiple-stranded) chromosomes of the larvae of the fly *Rhynchosciara angelae* injected 24 hours earlier with tritiated thymidine. (Tritium emits electrons of 10,000 electron volts.) In the upper part are shown the four chromosomes of a cell. In the stained chromosomes below it can be seen that the localization of the thymidine $^{3}H$ is in the bands of the chromosomes, indicating that incorporation occurs only here. $^{32}P$ may also be used for labeling. (By courtesy of A. Ficq and C. Pavan.)

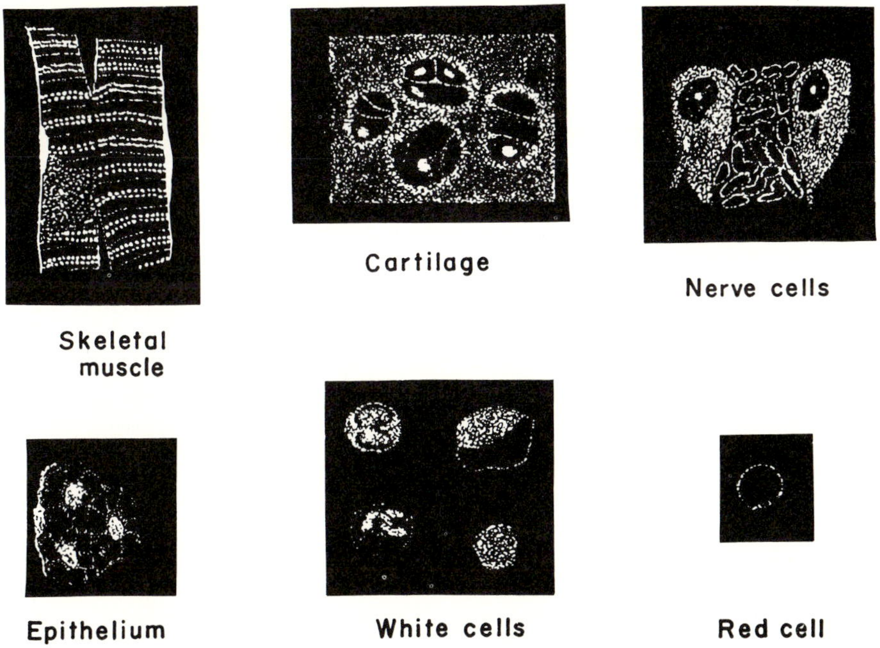


*Figure 3.22.* Microincinerated cells. Muscle from rectus abdominis of cat, cartilage from femur of cat embryo, nerve cells from cat brain, epithelium of corpus luteum of cat, blood cells from human. (Adapted from Scott, 1933: Am. J. Anat. *53*:243.)

strates, the enzymes are localized in relation to fine structure in the cells (Sabatini *et al.*, 1963).

Electron microscopy has also been combined with autoradiography for localization of labeled nutrients and salts into constituents of cell organelles (Caro, 1966). Both $^{3}H$-labeled and $^{14}C$-labeled nutrients have been used for this purpose, the first giving much better resolution than the second because of the short path traveled by the emitted radiation before absorption in the emulsion. By this means it has been possible to show, for example, that mitochondria incorporate labeled amino acids.

Cytochemistry, like many other approaches to biological problems, has its pitfalls. Purity of reagents is often of paramount importance here, since most of the work is done with minute quantities and at the cell level. Yet some biological reagents, e.g., enzyme preparations, are difficult to control. Then there is the problem of diffusion of a water-soluble constituent of a cell during the application of the very agent used for its identification. Also, the dynamic state of the cell, the continuous relocation of material in it, just prior to its fixation, must also be kept in mind (Glick, 1963).

Nonetheless, cytochemistry has added a wealth of material to our knowledge of cellular constituents and structure.

## LITERATURE CITED AND GENERAL REFERENCES

Advances in Carbohydrate Chemistry. Vol. 1. 1945: Academic Press, New York.

Advances in Protein Chemistry. Vol. 1. 1944: Academic Press, New York.

Allfrey, V. G., Meudt, R., Hopkins, J. W., and Mirsky, A. E., 1961: Proc. Nat. Acad. Sci. U. S. A. *47*:907–932.

* Anfinsen, C. B., 1959: The Molecular Basis of Evolution. John Wiley & Sons, New York.

Ansell, G. B., and Hawthorne, J. N., 1964: Phospholipids: Chemistry, Metabolism and Function. BBA Library, Vol. 3. American Elsevier Publishing Co., New York.

Arikawa, J. K., 1963: The Role of Magnesium in Biologic Processes. Charles C Thomas, Springfield, Ill.

Armington, A. E., Brooks, M. S., and Rubin, B., 1966: Purification of materials. Ann. N. Y. Acad. Sci. *137*:1–402.

Bittar, E. E., 1964: Cell pH. Butterworth, Washington, D. C.

Bloch, K., 1965: The biological synthesis of cholesterol. Science *150*:19–28. (Nobel lecture.)

Bonner, J. F., 1966: Plant Biochemistry. Academic Press, New York.

Burrell, R. C., 1947: Organic Chemistry for Students of Biological Sciences. Burgess Publishing Co., Minneapolis.

Burstone, M. S., 1963: Enzyme Histochemistry. Academic Press, New York.

Caldwell, P. C., 1956: Intracellular pH. Intern. Rev. Cytol. *5*:229–277.

*Cantarow, A., and Schepartz, B., 1967: Biochemistry. 4th Ed. W. B. Saunders Co., Philadelphia.

Caro, L., 1966: Progress in high-resolution autoradiography. Progr. Biophys. *16*:171–190.

Chambers, R., 1949: Micrurgical studies on protoplasm. Biol. Rev. *24*:246–265.

Conn, E., and Stumpf, P. K., 1966: Outlines of Biochemistry. John Wiley & Sons, New York.

Crick, F. H. C., Oct. 1954: The structure of the hereditary material. Sci. Am. *194*:54–61.

Danielli, J. F., 1953: Cytochemistry, A Critical Approach. John Wiley & Sons, New York.

Daoust, R., 1965: Histochemical localization of enzyme activities by substrate film methods: ribonucleases, deoxyribonucleases, proteases, amylase and hyaluronidase. Intern. Rev. Cytol. *18*:191–221.

*Davidson, J. N., 1960: The Biochemistry of Nucleic Acids. 4th Ed. Methuen & Co., London.

Dawson, R. M. C., 1957: The animal phospholipids. Biol. Rev. *32*:188–229.

DeRobertis, E. D. P., Nowinski, W. W., and Saez, F. A., 1965: Cell Biology. 4th Ed. W. B. Saunders Co., Philadelphia.

Dick, D. A. T., 1966: Cell Water. Butterworth, Washington, D. C.

Doty, P., Sept. 1957: Proteins. Sci. Am. *197*:173–184.

Doty, P., 1961: Inside nucleic acids. Harvey Lect. *55*:103–139.

Dounce, A. L., 1954: The use of differential centrifugation in the study of tissue. Intern. Rev. Cytol. *3*:199–275.

Engstrom, A., 1966: X-ray microscopy and X-ray absorption analysis. *In* Physical Techniques in Biological Research. Pollister, A. W., ed. Academic Press, New York, Vol. 3A, pp. 87–171.

Euler, H. V., and Hahn, L., 1948: Concentrations of RNA and DNA in animal tissues. Arch. Biochem. *17*:285–291.

Fieser, L. F., Jan. 1955: Steroids. Sci. Am. *192*:52–60.

Fieser, L. F., and Fieser, M. (eds.), 1959: Sterols. Reinhold Publishing Corp., New York.

Fox, S. W., and Foster, J. R., 1957: Introduction to Protein Chemistry. John Wiley & Sons, New York.

Freudenberg, K., 1965: Lignin: its constitution and formation from p-hydroxycinnamyl alcohols. Science *148*:595–600.

Gibor, A., and Izawa, M., 1963: The DNA content of the chloroplasts of *Acetabularia*. Proc. Nat. Acad. Sci. U. S. A. *50*:1164–1169.

Glick, D., 1961, 1963: Quantitative Chemical Techniques of Histo- and Cytochemistry. Interscience Publishers, New York, Vol. 1, 1961; Vol. 2, 1963.

Gross, J., May 1961: Collagen. Sci. Am. *204*:121–130.

Guyton, A. C., 1966: Textbook of Medical Physiology. 3rd Ed. W. B. Saunders Co., Philadelphia.

Hodgkin, D. C., 1965: The X-ray analysis of complicated molecules. Science *150*:979–988. (Nobel lecture).

Holley, R. W., 1965: Structure of a ribonucleic acid. Science *147*:1462–1465.

Horning, E. S., 1951: Micro-incineration and the inorganic constituents of cells. *In* Cytology and Cell Physiology. 2nd Ed. Clarendon Press, Oxford, pp. 287–321.

Hurwitz, J., and Furth, J. J., Feb. 1962: Messenger RNA. Sci. Am. *206*:41–49.

Jerons, F. R., 1964: The Biochemical Approach to Life. Basic Books, New York.

Johnson, F. H., Eyring, H., and Polissar, M., 1954: The Kinetic Basis of Molecular Biology. John Wiley & Sons, New York.

Kauzmann, W., 1959: Some factors in the interpretation of protein denaturation. Advances Protein Chem. *14*:1–63.

*Kendrew, J. C., Dec. 1961: The three-dimensional structure of the protein molecule. Sci. Am. *205*: 96–110.

Kendrew, J. C., 1963: Myoglobin and the structure of proteins. Science *139*:1259–1266.

Kernan, R. P., 1965: Cell K. Butterworth, Washington, D. C.

Kleineberger-Nobel, K., 1962: Pleuropneumonia-like Organisms (PPLO) Mycoplasmataceae. Academic Press, New York.

Kornberg, A., 1962: Enzymatic Synthesis of DNA. John Wiley & Sons, New York.

Kuff, C. L., and Dalton, A. J., 1959: Biochemical studies of isolated Golgi membranes. *In* Subcellular Particles. Hayashi, T., ed. Ronald Press Co., New York.

Lehninger, A. L., 1965: Bioenergetics. W. A. Benjamin, New York.

Lehninger, A., 1965: The Mitochondrion. W. A. Benjamin, New York.

Levinthal, C., June 1966: Molecular model-building by computer. Sci. Am. *214*:42–52.

Loeb, J., 1923: Proteins and the Theory of Colloidal Behavior. McGraw-Hill Book Co., New York.

Lovern, J. A., 1957: The Chemistry of Lipids of Biochemical Significance. 2nd Ed. Methuen & Co., London.

Lowry, O. H., 1943: Electrolytes in the cytoplasm. Biol. Symp. *10*:233–245.

Meryman, H. T., 1960: Drying of living mammalian cells. Ann. N. Y. Acad. Sci. *85*:729–734.

Miescher, F., 1897: Die histochemischen und physiologischen Arbeiten. 2 vol. Leipzig.

Mommaerts, W. F. H. M., 1950: Muscular Contraction. Interscience Publishers, New York.

*Morowitz, H. J., and Tourtelotte, M. E., March 1962: The smallest living cells. Sci. Am. *206*:117–126. (See also 1962: J. Molec. Biol. *4*:93–103.)

Nairn, R. C., 1962: Fluorescent Protein Tracing. Livingstone, Edinburgh.

Nass, M. M. K., Nass, S., and Afzelius, B. A., 1965: The general occurrence of mitochondrial DNA. Exp. Cell Res. *37*:516–539.

Neurath, H. (ed.), 1964: The Proteins. 2nd Ed. 4 vol., Vol. 1, 1964. Academic Press, New York.

Noller, C. R., 1965: Chemistry of Organic Compounds. 3rd Ed. W. B. Saunders Co., Philadelphia.

Nord, F. F., and Schubert, W. J., Oct. 1958. Lignin. Sci. Am. *109*:104–113.

Perutz, M. F., 1963: X-ray analysis of hemoglobin. Science *140*:863–869.

Perutz, M. F., Nov. 1964: The hemoglobin molecule. Sci. Am. *211*:64–76.

Petermann, M. I., 1964: The Physical and Chemical Properties of Ribosomes. Elsevier Publishing Co., Amsterdam.

Preston, R. D., Sept. 1957: Cellulose. Sci. Am. *197*: 157–168.

Richards, J. H., and Hendrickson, J. B., 1964: The Biosynthesis of Steroids, Terpenes and Acetogenins. W. A. Benjamin, New York.

Robbins, P. W., Wright, A., and Dankert, M., 1966: Polysaccharide biosynthesis. J. Gen. Physiol. *49*: 331–346.

Rothschild, J., 1963: The isolation of microsomal membranes. *In* The Structure and Function of the Membranes and Surfaces of Cells. Biochem. Soc. Symp. *22*:4–31.

Sabatini, D. P., Benser, K., and Barnett, R., 1963: Cytochemistry and electron microscopy. The preservation of cellular ultrastructure and enzymatic activity by aldehyde fixation. J. Cell Biol. *17*:19–58.

Sanger, F., 1959: Chemistry of insulin. Science *129*:1340–1344.

Schmitt, F. O., Sept. 1957: Giant molecules in cells and tissues. Sci. Am. *197*:205–216.

Siegel, S. M., 1962: The Plant Cell Wall. Pergamon Press, New York.

Siekevitz, P., 1963: Protoplasm: endoplasmic reticulum and microsomes and their properties. Ann. Rev. Physiol. *25*:15–40.

Smith, L., 1965: Vitamin $B_{12}$. Methuen & Co., London.

Sponsler, O. L., and Bath, J. D., 1942: Molecular structure in protoplasm. *In* A Symposium on the Structure of Protoplasm. Seifriz, W., ed. Iowa State College Press, Ames, Iowa, pp. 41–79.

Stein, W. H., and Moore, S., Feb. 1961: The chemical structure of proteins. Sci. Am. *204*:81–92.

Sverdrup, H. U., Johnson, M. W., and Fleming, R. H., 1942: The Oceans. Prentice-Hall, New York.

Verworn, M., 1903: Die Biogenhypothese. Jena.

Vinogradov, A. P., 1953: The Elementary Chemical Composition of Marine Organisms. Sears Foundation for Marine Research, #2. Yale University Press, New Haven.

Vitamins and Hormones. Vol. 1. 1943: Academic Press, New York.

Wagner, A. F., and Folkers, K., 1964: Vitamins and Coenzymes. Interscience Publishers, New York.

*Watson, J. W., and Crick, F. H. C., 1953: A structure for deoxyribose nucleic acid. Nature *171*:737–738.

Webb, S. J., 1965: Bound water in Biological Integrity. Charles C Thomas, Springfield, Ill.

Wied, G., 1966: Introduction to Quantitative Cytochemistry. Academic Press, New York.

Wolf, G., 1964: Isotopes in Biology. Academic Press, New York.

# CHAPTER 4 / COLLOIDAL PROPERTIES OF CELLS

The various constituents and organelles of the cell, as we have seen, have been studied inside the cell as well as separately in suspensions; their chemistry has been subjected to careful study, and many of the enzymes of the cell have been purified and tested; but as everyone knows, when the various enzymes and other chemicals are put together in the proportions in which they occur in the cell, the resulting mixture does not come alive. It is not just the chemicals but rather the way in which they are put together that presumably gives the spark of life. Although the specific features of organization that make cells alive have not been determined, the general principles underlying many of their properties have been shown to be those which govern the behavior of colloids. The theory of colloidal chemistry and the way in which it helps us understand the organization of cells and explains some of their activities are presented in this chapter.

## HOMOGENEOUS AND HETEROGENEOUS SYSTEMS

A *homogeneous system* is identical throughout and consists of one *phase;* for example, a solution of glucose in water, in which the sugar molecules are equally distributed among the water molecules in all parts of the solution. In a *heterogeneous system* two or more phases are present, each distinct from the other. For example, when oil is shaken with water the oil forms one phase, the water another, and each phase is distinct. A *phase* of a heterogeneous system is a component which is homogeneous throughout itself, bounded by an inter-

**Table 4.1.** Types of Heterogeneous States of Matter

| DISPERSE PHASE | DISPERSION MEDIUM | | |
|---|---|---|---|
| | GAS | LIQUID | SOLID |
| Gas | — | foam (soap bubbles) | some minerals (pumice), floating soap |
| Liquid | fog, mist, smog | emulsions (mayonnaise) | some minerals (opal, pearl), jellies |
| Solid | smoke | suspensions, colloidal solutions | metals in glass (colloidal gold in ruby glass) |

face and mechanically separable from other phases of the system. The particles of oil in water are spoken of as the *disperse phase;* the water in which they are dispersed is called the *continuous phase* or dispersing medium (McBain, 1950).

Heterogeneous systems (see Table 4.1) may consist of liquid in liquid, solid in liquid, liquid in solid, solid in gas, gas in solid, gas in liquid, liquid in gas and solid in solid. It might be added, parenthetically, that from the molecular standpoint even a homogeneous solution—for example, glucose—has heterogeneity, because a glucose molecule in its microspace differs from the water molecules which surround it. The man-made definition of a heterogeneous system is arbitrary but of practical use in differentiating the two types of systems which in reality grade one into the other. Size of particles, we shall see, is of paramount importance for entry into the cell, however.

## SURFACES, INTERFACES AND INTERFACIAL TENSION IN HETEROGENEOUS SYSTEMS

An interface always separates any two phases in a heterogeneous system. If the cohesive forces between the molecules (or atoms) in the aggregates of the dispersed particles are greater than the forces of attraction of the dispersion medium for these molecules, the aggregates of molecules will remain separate from the dispersion medium. The molecules at the surface of the aggregate are subjected to unequal forces and are pulled in toward the center of the aggregate, resulting in a tendency to contract the surface (interface) of the particle (see Fig. 4.1). An *interfacial tension* is developed as a consequence.

At liquid-gas and liquid-liquid interfaces the interfacial tension manifests itself as a *surface tension* (for details see Appendix, 4.1). The surface tension, usually designated by $\sigma$, is measured in units of force per unit of length, e.g., dynes per centimeter. The *surface energy* available per unit area of interface is given by the force acting through a given distance. For example, if in Figure 4.2 a force ($\sigma$ in dynes per centimeter $\times$ length of AB in centimeters) acts through the distance AC, the surface free energy is:

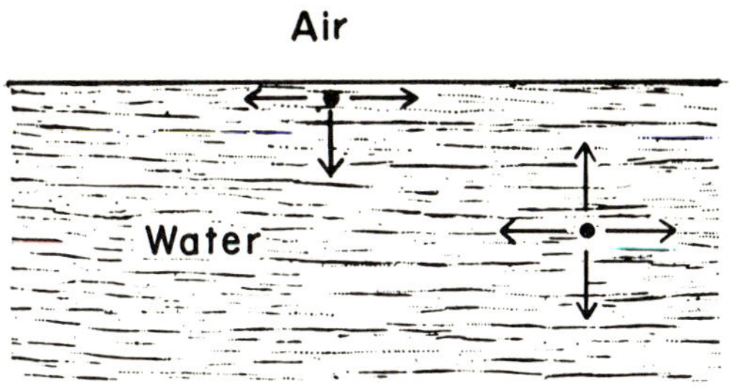


***Figure 4.1.*** Surface tension and forces in a liquid. The particle at the surface is pulled downward because of unbalanced forces. A particle deeper in the fluid is attracted by forces from all coordinates.

$$\sigma \times AB \times AC = \sigma \times \text{area} \quad (4.1)$$

In other words, increasing the surface area increases the surface free energy, this amount of energy being necessary to expand the film to this area.

According to the second law of thermodynamics, the free energy in a system tends to decrease; therefore, the surface free energy also tends to decrease and the dispersed particles in a heterogeneous system assume a shape with minimal surface area. Consequently, a drop of liquid freely suspended in another liquid or in a gas al-

***Figure. 4.2.*** Diagram illustrating the relation between surface area and free energy of the surface. When the force on AB (surface tension times the length of AB) acts through the distance AC, the surface free energy equals the area ABDC times the surface tension. (Reprinted with permission from Lloyd and Scarth, 1930: An Elementary Course in General Physiology. John Wiley & Sons, New York.)

ways assumes a spherical shape, since a sphere has a smaller area than any differently shaped object of the same volume. Because a single large globule has less surface area than many small ones, small ones coalesce unless they are *stabilized,* that is, unless they are coated with some substance which prevents their coalescence.

*Absorption* occurs *wherever there is a surface* e.g., *an interface,* because of the unbalanced valency forces. Such accumulations cause a net decrease in free energy. At interfaces of systems other than liquid-gas or liquid-liquid, the interfacial tension manifests itself mainly by the ready adsorption of substances to the interface. For example, wood charcoal, which has the extensive surface areas of the charred remains of cell walls, is used to remove colored impurities from liquors and sugar solutions by adsorbing them. As a result of adsorption, materials present in solution in great dilution may be concentrated to a considerable extent at the interface of an adsorbant (Dubinin *et al.,* 1964).

The magnitude of interfacial forces involved in adsorption may be very large. For example, water adsorbed to charcoal may be held so firmly that its volume is measurably decreased. This force will be appreciated when it is realized that water subjected to 100 atmospheres of pressure is compressed by only 0.0051. When molecules are adsorbed at an interface they lose their kinetic energy, which then appears as the *heat of adsorption.* The heat of adsorption may be very large, perhaps many thousands of calories per mole. Consequently, to free a surface of adsorbed materials, an input of energy (heat) is required, because sufficient kinetic energy must be given the molecules to enable them to escape from the forces attracting them to the adsorbing surface (McBain, 1950; Kitchener, 1963).

The attractive forces at an interface may be general, e.g., those that bind gases to a surface, or quite selective. Many substances acting at interfaces in heterogeneous systems are endowed with great and specific catalytic activity. (In homogeneous systems catalysts act by virtue of other properties.) Attraction of a molecule to the surface of a *catalyst* lowers its energy of activation for a given reaction so that less energy is required than would otherwise be the case (see Fig. 15.1). Since a catalyst is chosen specifically to attract the reactants, the reaction products are attracted to the catalyst to a lesser degree and leave its surface, freeing it for further activity.

A wide variety of catalysts (e.g., spongy platinum) have been used to selectively catalyze different chemical reactions. In biological systems, similarly, many precisely selective *biocatalysts* (*enzymes*) are known and new ones are being found continually. A substance firmly bound to the active areas on a catalyst surface, but not participating in a reaction, acts as a *catalyst poison* because it prevents the adsorption of the reactants for the desired chemical reaction. When the poison is desorbed—for example, by heat—the catalyst regains its activity.

## SUSPENSIONS AND EMULSIONS

Since all cellular activities take place in solution, the heterogeneous systems that are most important to the study of cells are suspensions, emulsions and colloids. In a *suspension* a solid is temporarily suspended in a liquid, e.g., fine sand in water. When the particles are too large to remain in suspension, the solid phase soon separates from the liquid phase. In an *emulsion* one liquid is suspended in another, e.g., oil in water. An oil in water emulsion is very unstable because the globules coalesce to form larger ones until two layers are formed—oil and water (Davies, 1964).

If a protein is shaken up with an oil in water emulsion, the emulsion is *stabilized.* A film of protein forms about each particle of oil and prevents coalescence of the oil particles. This is made possible because the oil-soluble radicals (e.g., methyl and phenyl) of the protein dissolve in oil, whereas its water-soluble radicals (e.g., carboxyl, amino and hydroxyl) dissolve in water; thus, the oil phase is linked to the water phase. The water-soluble radicals of the protein film ionize; therefore, the particles of the emulsion repel one another,

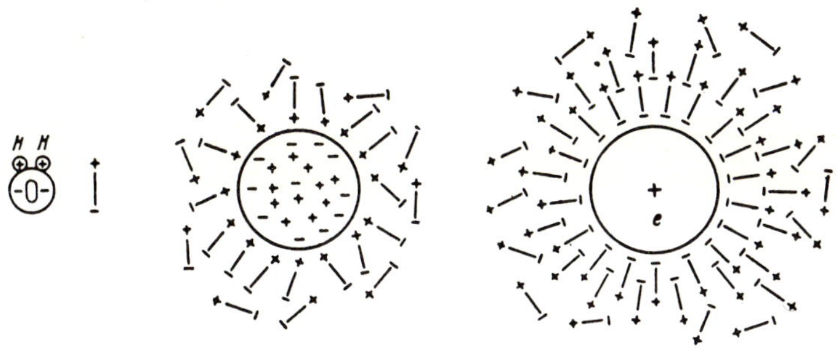


**Figure 4.3.** Hydration and stabilization. *Left,* Model and dipole diagram of a water molecule. *Middle,* Diagram of the hydration of a colloidal particle at the isoelectric point. *Right,* Hydration of a positively charged colloidal particle. (After Frey-Wyssling according to Pallmann, 1931.)

making coalescence impossible. *Hydration* (addition of water molecules) of the ionized water-soluble part of the protein adds to the stability of the coated particles (see Fig. 4.3). As expected, when the protein stabilizer is destroyed by heat or cold or chemical action, the emulsion is broken and the phases separate.

Soaps, also, may be used as stabilizers of emulsions. Sodium and potassium soaps shaken with oil and water form oil in water emulsions (see Fig. 4.4*A*), whereas calcium soaps form water in oil emulsions (see Fig. 4.4*B*). The difference in the type of emulsion formed is thought to result from the difference in relative sizes of water-soluble and oil-soluble groups in the soaps. Thus, in a divalent soap the oil-soluble "wedge" is larger than the water-soluble ionic portion, as is diagrammatically represented in Figure 4.4. Addition of a calcium salt to an emulsion stabilized by a sodium or potassium soap changes the emulsion from an oil in water type to a water in oil type when the calcium concentration reaches a critical level. This *reversal of phase* is readily observed in the laboratory.

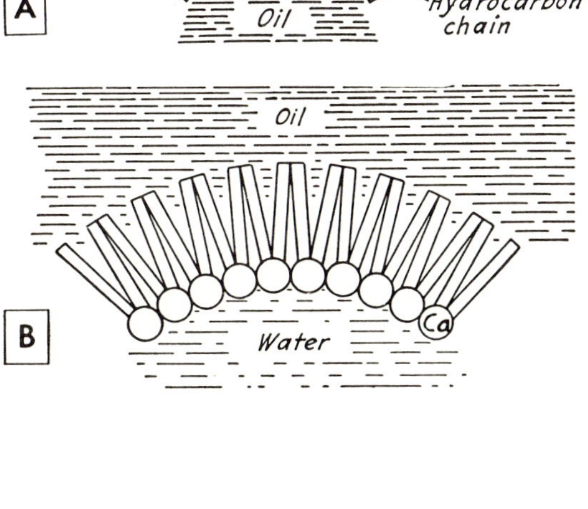


**Figure 4.4.** *A,* Oil in water emulsion. Sodium stearate (soap) molecules at the interface of oil globules in an oil in water emulsion. *B,* Water in oil emulsion. Calcium stearate (soap) molecules forming the stabilization membranes of a water in oil emulsion. (Reprinted with permission from Seifriz, 1936: Protoplasm. McGraw-Hill Book Co., New York.)

## COLLOIDS

It is difficult to define *colloid* with precision. Some have defined a colloid as a state of matter in which the chemistry and physics of surfaces are dominant. The importance of the surface in the colloidal state becomes apparent if we consider a 1 cm. cube subdivided into cubes of a colloidal dimension, for example, 1 nm. ($10^{-9}$ meter). The surface area of the 1 cm. cube is 6 cm.$^2$ whereas the total area of all the 1 nm. cubes is $6 \times 10^7$ cm.$^2$, or nearly 1.5 acres. Sometimes the colloidal state is defined as a noncrystalline state of matter in which the particles are aggregates of atoms or molecules in the range between 1 and 500 nm. This is obviously arbitrary since of two substances within the size range one may be colloidal, the other not. Nor are all colloids aggregates of molecules. It must be recognized, then, that defining colloid on the basis of size alone is rather arbitrary, especially since particles in nature grade imperceptibly from coarse to molecular (Table 4.2) and a definition of the colloidal state should include consideration of some other of its properties as well (Mysels, 1959; Sheludko, 1966).

Colloids can be distinguished from crystalloids (e.g., glucose or sodium chloride) by the following four characteristics: (1) Crystalloids in solution pass through pores in a natural membrane, but colloidal particles do not. The membrane acts as a sieve, holding the larger colloidal particles back but permitting the crystalloidal molecules to come through. (2) Scattering of light by crystalloids is slight, but scattering of light by dispersed particles of colloids is quite marked. This is best seen when viewed at right angles to the incident beam. The murky cone of scattered light is called the *Tyndall cone.* The larger the difference between the refractive indices of the dispersing medium and the colloidal particle, the greater is the scattering of light. (3) Surface-active interfaces exist between colloidal particles and the dispersing medium. As a result of these surface forces, colloids often show high resistance to flow, or they may develop structure and fail to flow, that is, they may gel. Because of the surface forces colloids are capable of adsorbing some substances from solution. (4) Colloidal particles, being large, diffuse slowly throughout a solution compared with crystalloids. If for comparative purposes sodium chloride is assigned a relative diffusion rate of 1, sucrose diffuses $\frac{1}{3}$ as fast, caramel $\frac{1}{42}$ as fast, and albumin $\frac{1}{20,000}$ as fast (McBain, 1950).

Colloids in which no attraction exists between the solvent and the colloidally dispersed particles are not stable and are spoken of as *lyophobic* colloids (literally, solvent-hating). Some examples of lyophobic colloids are a gold sol (one prepared by reducing gold chloride in water), rub-

**Table 4.2.** Size Ranges for Different States of Matter

| SIZE IN CM. | CORRESPONDING UNITS | DIVISION OF MATTER | STATE OF MATTER | VISIBLE UNDER: |
|---|---|---|---|---|
| $10^{-8}$ | Å | atoms | crystalloidal | invisible |
| $10^{-7}$ | 1 nm. | molecules | | |
| $10^{-6}$ | | | colloidal | electron microscope |
| $10^{-5}$ | | aggregates of molecules (particles) | | |
| $10^{-4}$ | 1 $\mu$ | | coarse suspensions and emulsions | ultraviolet microscope |
| $10^{-3}$ | | | | visible microscope |
| $10^{-2}$ | | | | |
| $10^{-1}$ | 1 mm | | | naked eye |

**Table 4.3.** Hydrophobic and Hydrophilic Radicals of Proteins

| HYDROPHILIC (LIPOPHOBIC) | | HYDROPHOBIC (LIPOPHILIC) | |
|---|---|---|---|
| Carboxyl | —COOH | Methyl | $—CH_3$ |
| Hydroxyl | —OH | Methylene | $—CH_2—$or$=CH_2$ |
| Aldehyde | —CHO | Ethyl | $—C_2H_5$ |
| Carbonyl | —CO | Propyl | $—C_3H_7$ |
| Amino | $—NH_2$ | Alkyl | $—C_nH_{2n+1}$ |
| Imino | =NH | Isoprene | $—C_5H_8—$ |
| Amido | $—ÇONH_2$ | Phenyl | $—C_6H_5$ |
| Imido | —CORNH | | |
| Sulfhydryl | —SH | | |

After Frey-Wyssling, 1948: Submicroscopic Morphology of Protoplasm and Its Derivatives. Elsevier, Amsterdam. The solubility of the hydrophilic groups in water decreases progressively from top to bottom, whereas the solubility of the hydrophobic groups in lipids increases from top to bottom.

ber dispersed in water and sodium chloride dispersed in alcohol. Colloids in which the solvent and dispersed particles mutually attract each other are quite stable and are designated *lyophilic* (solvent-loving) colloids. Examples are proteins in water and rubber in benzene. Lyophilic colloids are of most interest to the study of cells (see Table 4.3).

## COLLOIDAL PROPERTIES OF PROTEINS

Proteins show all the characteristics of the colloidal state, even though they may be molecularly dispersed in water (Springall, 1954). Thus, proteins do not pass through natural membranes. Solutions of proteins can therefore be freed of crystalloids (dialyzed) by being placed in cellophane sausage tubing immersed in distilled water or a buffer solution. Crystalloids from the solution in the bag pass into the distilled water or buffer, but the proteins do not.

Proteins in water show a typical Tyndall cone, although the opalescence is less marked than in a metallic colloid (such as gold sol) because the refractive index of protein is only slightly greater than that of water.

The interplay of surface-active forces at the molecular interfaces in a protein solution is manifested by the high viscosity which such a solution quite often shows. The viscosity is a measure of the resistance of the protein molecules to flow past one another when force is applied to the solution in a confined space. This resistance is thought to result from the attraction of the molecules for one another as well as from asymmetry in their structure.

A protein molecule will both attract and repel another molecule of the same kind, depending upon the distance separating them. Repulsion occurs because the overall charge (positive or negative) is similar. But if the molecules approach one another so that at the point of closest approach local residual valency forces can act, the two molecules may be attracted to each other to form a hydrogen bond between electronegative elements or radicals in each. These bonds are in a dynamic state; they are made and broken continually. Evidence of the presence of such bonds is shown by the resistance of the protein molecules to begin flowing when a force is first applied to a solution in a tube. The pressure which must be applied to start the flow is called the *yield value* and represents the force required to break these weak, temporary bonds. Even after the flow starts it may not be directly proportional to the pressure applied, but it gradually increases as more bonds are broken (see Fig. 4.5). Such a pattern of flow is called thixotropic. More generally, flow is proportional to the force applied, showing a constant increment with each

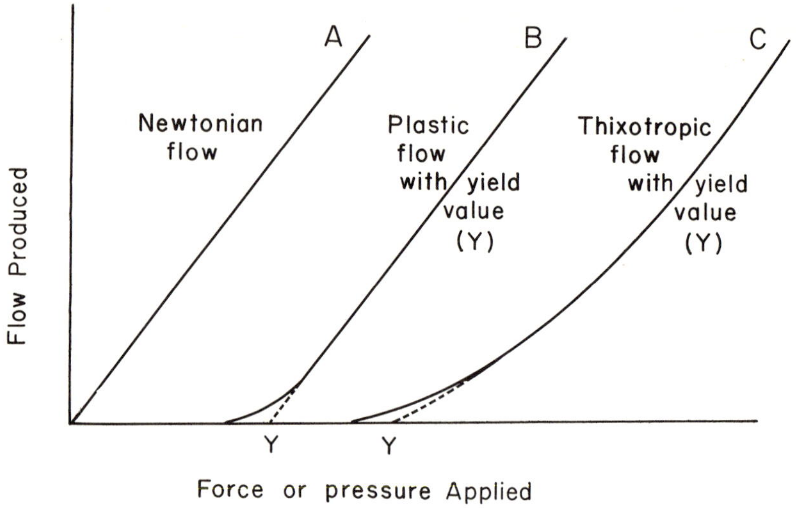


***Figure 4.5.*** Flow-force diagrams. The behavior in A is shown by gases and liquid solvents, in B by butter and clay and in C by proteins. (From McBain, 1950: Colloid Science. D. C. Heath & Co., Boston.)

increment of force, as in Newtonian flow (e.g., in ordinary fluids) where it begins as soon as force is applied, or in plastic flow (e.g., in butter), past the yield value.

Some proteins (and some other colloids) exist in either of two states—sol or gel—depending upon the conditions. The sol is fluid because the particles are not able to form more than temporary bonds with one another. Gelation of proteins probably results from the formation of numerous stronger linkages between protein molecules. Electron microscope study of a gel reveals a structure resembling a brush heap meshwork of intertwining fibers (see Fig. 4.6), verifying Nägeli's early hypothesis concerning the structure of gels.

Proteins are stable colloids; i.e., they require no stabilizing agents. No stabilization is required because, first, protein ions in solution repel one another; and second, each protein ion attracts and orients the water molecules surrounding it, thereby binding itself to the solvent. Each water molecule acts as a dipole with positive and negative ends (see Fig. 4.3).

The *hydration* of proteins deserves further notice because it is mainly upon this property that *imbibition* depends. Dry proteins adsorb (imbibe) water with great avidity—even against the osmotic pressure of saturated lithium chloride (about 1000 atmospheres). When gelatin has taken up four times its weight in water, it will no longer take up water from any solutions with a high osmotic pressure, but it will still take up water if the solution surrounding it has an osmotic pressure of only 0.2 atmosphere or less (see Fig. 4.7). Seeds also

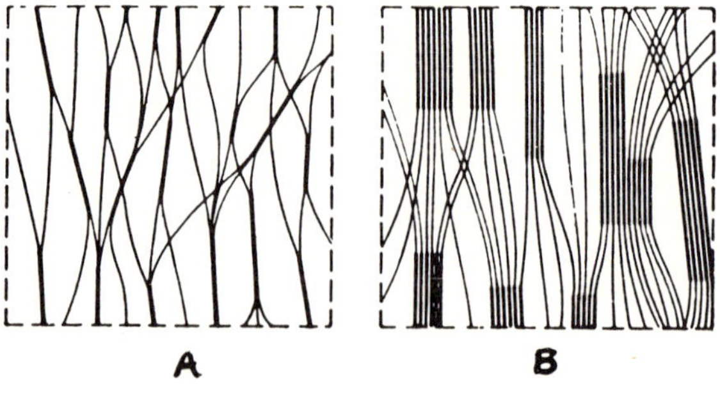


***Figure 4.6.*** Brush heap structure of gel. Diagram of gels of long polymeric particles (*A*) with some parallel orientation of the particles, and (*B*) with sites of crystalline structure. (From Frey-Wyssling, 1953: Submicroscopic Morphology of Protoplasm. 2nd Ed. Elsevier Publishing Co., Amsterdam.)

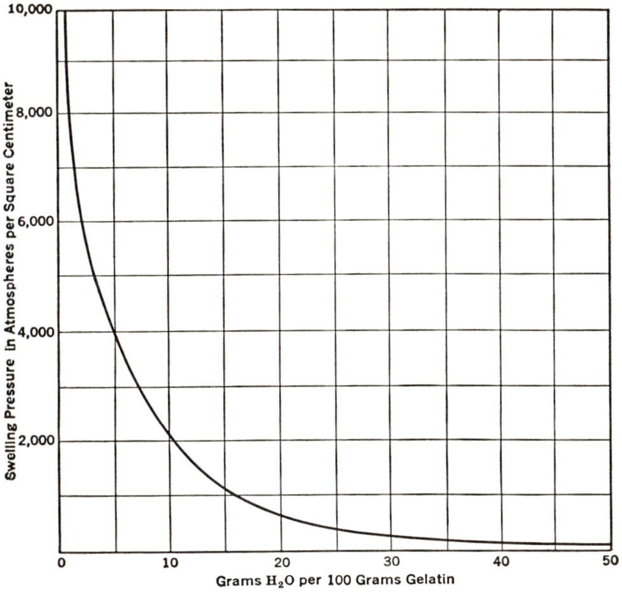


***Figure 4.7.*** Swelling pressure of isoelectric gelatin at 25° C. in atmospheres as calculated from aqueous vapor-pressure studies. (Reprinted with permission from Bull, 1943: Physical Biochemistry. John Wiley & Sons, New York.)

take up water against an osmotic pressure of more than 1000 atmospheres because of the dehydrated state of their proteins, but the force of uptake falls very rapidly as the proteins become hydrated. As previously pointed out (Chapter 3) water uptake is least at the isoelectric pH of proteins, increasing on either side.

Ions affect the uptake of water by proteins in characteristic ways. The sequence of their actions is spoken of as the *lyotropic,* or *Hofmeister, series* (Bull, 1964, p. 73). The monovalent cations of a series of chlorides, acting on a gelatin gel, for example, increase water uptake in the following order: Li > Na > K > Rb > Cs. In other words, Li increases the uptake of water more than Na, Na more than K and so on. This effect may be explained on the basis of hydration of the ions and their adsorption to the oppositely charged protein. For example, $Li^+$ has a much greater sphere of water about it than $Na^+$, since the lithium ion is smaller and the electric charge is more concentrated. Therefore, when $Li^+$ combines with a protein which is negatively charged, it adds its layer of water to the protein. The sodium ion, being larger, adds less water; the potassium ion adds still less; and so on down the series. Although adsorption of the cations to the protein weakens the attraction of water to the protein, the water of hydration added by the ions more than compensates for the loss.

The divalent calcium ion, being more strongly hydrated than the potassium ion, causes gelatin to swell to an even greater extent than the $K^+$ does and may even liquefy it.

An anionic Hofmeister series is often observed also. Trivalent citrate is a weaker swelling agent than bivalent tartrate and sulfate, and these, in turn, are weaker than the monovalent anions:

$$SCN > I > NO_3,\quad Br > Cl > \text{acetate},\quad SO_4 > \text{tartrate} > \text{citrate}$$

As might be expected, the ionic effects on the hydration (uptake of water) of proteins depend upon the pH (hydrogen ion concentration) of the medium in which the protein is dissolved, since the pH determines the sign and strength of the charge on the protein molecule (as seen in Chapter 3). For example, on the acid side of its isoelectric pH gelatin becomes positively charged and it then adsorbs anions which affect swelling in proportion to their degree of hydration.

When the charges on a protein molecule are neutralized by addition of salts (e.g., ammonium sulfate), proteins may be precipitated or *salted out.* Salting out is often used in purification since each protein is precipitated by a specific concentration of salt and can be separated in this manner from other proteins.

The salting out of protein sols by cations or anions occurs according to a lyotropic series. Here, however, the order of effectiveness of the ions is the reverse of that given before for swelling of gels.

Lyotropic series are observed in studies of: the solubility of substances in solutions of electrolytes; the catalysis of hydrolytic reactions by ions; the effect of electrolytes on surface tension; the coagulation of sols; and the permeability of artificial membranes. The order of the ions is not invariably the same, and it changes with concentration, even during an experiment, indicating that under some conditions factors other than those discussed enter into determination of ionic order. The lyotropic series appears chiefly when concentrated solutions of salts are used. Since in biological systems the salt concentration is generally rather low, it is unlikely that many of the biological salt effects are interpretable on the basis of the lyotropic series (Bull, 1964).

## COACERVATES AND TACTOIDS

Proteins and similar colloids are stabilized by electric charge and hydration. Discharge or dehydration of the dispersed particles precipitates the colloid. When partial discharge or dehydration occurs, the less fully hydrated particles of the colloid may then combine to form larger units with less solvent bound to them than the sum of what they had before. Such a system is called a *coacervate,* and such a process is called *coacervation* (Bull, 1964). For example, addition of dehydrating agents such as alcohol or acetone to a colloidal solution of protein leads to coacervation. Partial discharge of colloids by addition of salts, or by mixing two hydrophilic colloids of opposite charge, also results in coacervation. Coacervates (see Fig. 4.8) resemble some cellular inclusions and organelles, and it is possible that such structures may be a result of coacervation in protoplasm (Overbeek and Voorn, 1957; Battista, 1965).

*Tactoids* are coacervates of fibrous proteins (see Fig. 4.9). Tactoids are spindle shaped, and studies with polarized light indicate that they are made up of elongate molecules linearly arranged. Possibly the spindles and asters seen in the dividing cell are the result of tactoid formation by proteins in the protoplasm (Frey-Wyssling, 1953).

## THE CELL AS A POLYPHASIC COLLOIDAL SYSTEM

Throughout the years many theories have been proposed to explain the structure of the living material within the cell, such as the reticular theory which conceived of the material as fibrillar, the granular theory which thought of it as a system of granules and the alveolar theory which accounted for the emulsion-like appearance of some cells (see Fig. 4.10 and Chapter 28). In 1925 E. B. Wilson proposed the colloidal theory of organization of the cell substance, a theory which

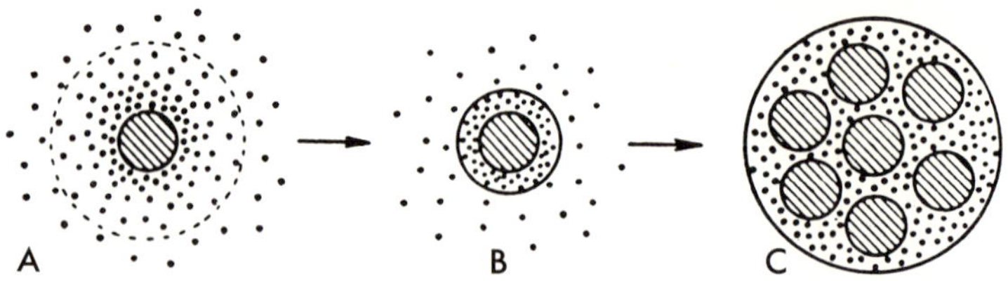


***Figure 4.8.*** Diagram of the process of coacervation. *A,* Colloidal particle with diffuse hydration layer; *B,* hydration layer reduced and delimited; *C,* beginning of the coacervation phenomenon. (From Bungenberg de Jong, 1932: Protoplasma *15*:110.)

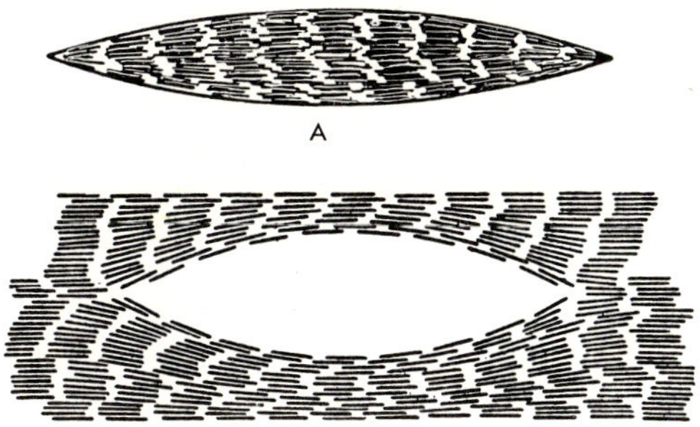


***Figure 4.9.*** Tactoids. *A,* Positive tactoid. *B,* Negative tactoid. A *positive tactoid* is formed when a fibrous colloid in solution is partially dehydrated or partially discharged. Under these circumstances the less fully hydrated particles of the colloid combine to form larger units containing less solvent than they did before. In the unit the individual fibrous protein molecules are oriented to form a spindle-shaped particle. A *negative tactoid* is formed when the fibrous particles enclose nonfibrous ones. (Adapted from Bernal and Fankuchen, 1941: J. Gen. Physiol. *25*:128.)

has been and continues to be developed by others since that time because it best explains the properties of the cell. Experimental evidence in support of the colloidal theory of cellular structure is presented in the following paragraphs (Anderson, 1957; Gross, 1957; Oster, 1957).

Any theory of the physical-chemical structure of the cell proposed must explain: (1) its appearance and rhythmic changes; (2) the variations in its viscosity from a fluidity approaching that of water to a gel approaching that of an elastic solid; and (3) the simultaneous occurrence of many different chemical reactions in a single living cell, none interfering with the others. Furthermore, such a theory must explain why some of the biochemical reactions occurring in the cell may be individually duplicated *in vitro* only at high temperatures or high pressures or both, whereas in the cell they occur at comparatively low temperatures and seemingly at atmospheric pressure.

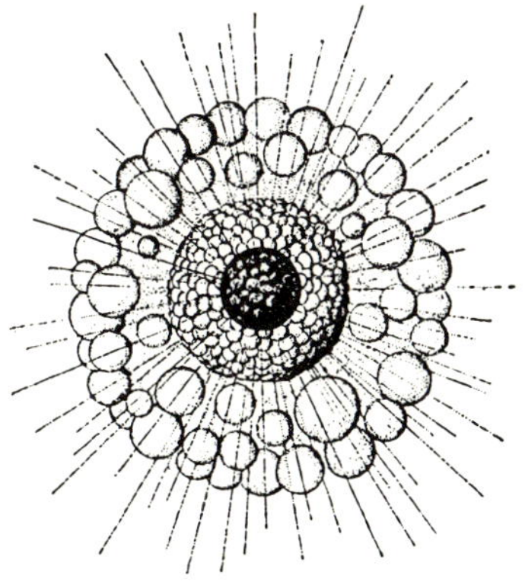

***Figure 4.10.*** A radiolarian (*Thalassicolla nucleata,* Huth) showing the frothy or alveolar appearance of the protoplasm. The rays are part of the skeleton. (From Kudo, 1946: Protozoology. 3rd Ed. Charles C Thomas, Springfield, Ill.)

Even a casual microscopical examination indicates that the cell is heterogeneous, not homogeneous. From what has been said in preceding chapters, it is evident that both cytoplasm and nucleus are composed of highly organized organelles. These organelles are not temporary aggregations but complex systems, in many cases bounded by selectively permeable membranes able to maintain their identity during and after centrifugation. Yet each of these organelles is a colloidal system, subject to the forces characteristic of this state of matter and manifesting its usual properties.

The endoplasmic reticulum, it was shown, ramifies throughout the ground substance of the cytoplasm, including the ectoplasm. It is thought that the structure of the reticulum is dynamic and that it gives way in the path of large particles moving through it under the influence of a strong centrifugal force, imposed during centrifugation, but the links which are broken are re-formed again. The redistribution of the mitochondria, granules and particles, once the centrifugal force is withdrawn, indicates that the normal random distribution of these bodies in the cell is the result of the action of kinetic energy forces acting upon them and leading to an equilibrium. This would suggest that at least part of the organization of the endoplasmic reticulum must be dependent upon the making and breaking of bonds of lesser strength than primary valency bonds (see Fig. 2.4). Most cells contain granules and many have oil droplets which on centrifugation are separated from the fundamental structure of the *ground substance* through which the particles move, indicating that they are merely suspended there (see Fig. 2.4).

The persistence of organization in the ground substance of the cell is indicated by experiments testing the effect of centrifugation on the polarity of an egg. Polarity of an egg is shown by the difference between the two ends of the egg; the end at which the polar bodies are given off designates the anterior, or head end, of the individual developing from the egg. Polarity is therefore an indication of a fundamental architectural plan of the cell. In experiments with the egg of the roundworm *Ascaris,* it was shown that regardless of the angles at which the various granules and oil droplets are layered with respect to the polar axis after centrifugation, the original polarity of the egg is maintained, provided the packed materials do not interfere with the position of the first cleavage spindle. Similar results were obtained with eggs of a number of other organisms as well (Child, 1941). In some cases, however, polarity was found to be influenced by stratification during centrifugation (e.g., the brown alga *Fucus*) (Whitaker, 1940). Polarity apparently depends upon the maintenance of some structure which resists deformation during centrifugation. This could be either the endoplasmic reticulum or the ground substance, or both (Chapter 5). It would be of interest to compare these structures before and after centrifugation in the two classes of eggs, those that show changes in polarity after centrifugation and those that do not.

By means of appropriate staining reactions it is possible to demonstrate that the ground substance is composed of protein in aqueous solution. Like a protein solution, the ground substance shows the properties of the colloidal state (Heilbrunn, 1958). Light passing through it under dark field is scattered as in the Tyndall cone of a colloid. The relative viscosity of the ground substance (see Appendix 4.2) varies from a value as low as that for water to a very high value in gelating cytoplasm such as the pseudopodium of an ameba. Chemical analyses indicate that several kinds of proteins are present in the ground substance, some fibrous, some globular. Electron micrographs made after glutaraldehyde fixation have demonstrated the presence of *microtubules* and *microfilaments* in the ground substance of a wide variety of cells (see Chapter 5). Any particle under centrifugal force would have to move around and through such structures. Recently the displacement under centrifugal force of cell organelles and their return to the original distribution subsequently after centrifugation have been studied with the electron microscope (Bouck, 1963). Even the endoplasmic reticulum was displaced; it was also the first organelle to redistribute itself after centrifugation. In addition, ribosomes, containing ribonucleic acid, may be present, adhering to the sacs and canals of the endoplasmic reticulum or free in the ground substance. Since all the particles in the disperse phase are not alike, cytoplasm is said to be a *polyphasic* colloid.

Experiments indicate that the ground substance of the cytoplasm has some of the properties of a solid. When the displacement of naturally occurring particles in the ground substance is photographed during the process of centrifugation, the particles are seen to move in a series of zigzag jumps, even in the more fluid endoplasm, as if they were tearing through a meshwork. If the ground substance were a simple viscous liquid, the movement of the particles should be smooth. Presumably the jumps are due primarily to the structure of the endoplasmic reticulum and possibly the microtubules and microfilaments, although to what extent bonding between the reticulum and the proteins of the ground substance helps account for these properties is unknown.

Further evidence of structure in the ground substance is the manner in which some plant cells plasmolyze. Long strands which attach the protoplast to the cell wall are often seen in such plasmolyzed cells (see Fig. 4.11). The reticulum and the microfilaments and microtubules demonstrated in the endoplasm as well as in the

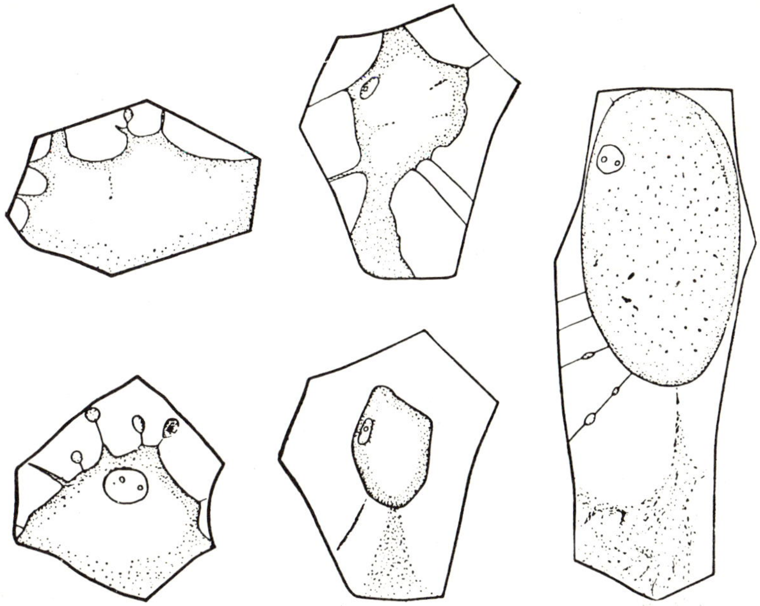

***Figure 4.11.*** Plasmolyzed plant cells showing strands of cytoplasm. (From Kuster, 1929: Pathologie der Pflanzenzelle I. Protoplasma Monographien III, Borntraeger, Berlin.)

ectoplasm of all kinds of cells (see Chapter 5) by electron microscopy may account for these structural properties.

Another indication of structure in the ground substance is provided by experiments with changes in its viscosity when sucked up into a tubule. The viscosity decreases rapidly with increased pressure, much like the thixotropic flow of protein solutions illustrated in Figure 4.5. Cytoplasm occasionally shows an unusual behavior known as *dilatancy,* i.e., an increase in viscosity with increase in applied force. (This occurs in some physical systems also, e.g., a dense suspension of quartz particles in water.) Dilatancy of the cytoplasm is most marked when it is densely packed with particles (Heilbrunn, 1958).

The fluidity of the cytoplasm under some conditions and its rigidity under others indicate that the endoplasmic reticulum and other structures should not be considered a rigid and fixed cytoskeleton, but rather a shifting and changeable structure sensitive to the environment and in a state of dynamic change.

Heilbrunn (1958) has pointed out that cytoplasm flows out of the cut end of a cell as if it were liquid in consistency and has suggested the possibility that the more rigid structure which appears in the cytoplasm may be the result of a coagulation process catalyzed by a specific enzyme, different degrees of coagulation accounting for different degrees of structural rigidity.

## NATURE OF STRUCTURAL BONDS IN THE CYTOPLASM

Frey-Wyssling (1957), in his monograph on the subject, postulates four main types of interactions between protein molecules in the cell: (1) Homopolar cohesive "bonds" (London–van der Waals forces) of the type which hold a paraffin crystal together. In proteins this type of bond is formed by the interlaced methyl groups or other hydrophobic groups of adjacent molecules (see Table 4.3). Such bonds in proteins are easily broken by heat, as they are in paraffin. (2) Heteropolar cohesive bonds, such as dipole to dipole attraction and hydrogen bonds, which result from attraction of neighboring proteins by residual valencies. These bonds are broken by mild heat, the bond strength being 5000 calories per mole or less. (3) Heteropolar valency bonds (Coulomb forces), such as those which form a salt or ester linkage. These bonds are much stronger than the first two kinds and are not likely to be broken by mild heat. (4) Homopolar valency bonds which involve formation of a bridge (as exemplified in the elimination of hydrogen between two sulfhydryl radicals). These, like those given in (3), are also relatively strong bonds. All these types of bonds are illustrated in Figure 4.12, but their postulation represents an attempt to formulate a working hypothesis for further research rather than a finite answer to the problem of cytoplasmic structure.

Booij and Bungenberg de Jong (1956) point out that the bonds between phosphatides (heteropolar valency bonds) should be added to the list since they are likely to play a role of importance in explaining such phenomena as the structure maintained in highly hydrated cytoplasm in which protein bonds of the first three classes are likely to have been broken. It should also be realized that such ions as magnesium and calcium may help in linking carboxyl groups. Water plays an important part in hydrophilic bonding as well, becoming part of the cellular structure.

The importance of weak bonds in the cell is emphasized by Watson (1965, pp. 102–103), who says: "Weak bonds are important not just in deciding which molecules lie next to each other, but also in giving shape to flexible molecules such as the polypeptides and polynucleotides. . . . Strong bonds almost never fall apart at physiological temperatures. Weak bonds are easily broken, and when they exist singly they exist fleetingly. Only when present in ordered groups do weak bonds exist for a long time. The strength of a bond is correlated with its length, so that two atoms connected by a strong bond are always closer together than the same two atoms held together by a weak bond. For example, two hydrogen atoms bound covalently to form a hydrogen molecule (H:H) are 0.074 Å apart, whereas the same two atoms, when held together by the van der Waals forces instead, are held 1.2 Å apart."

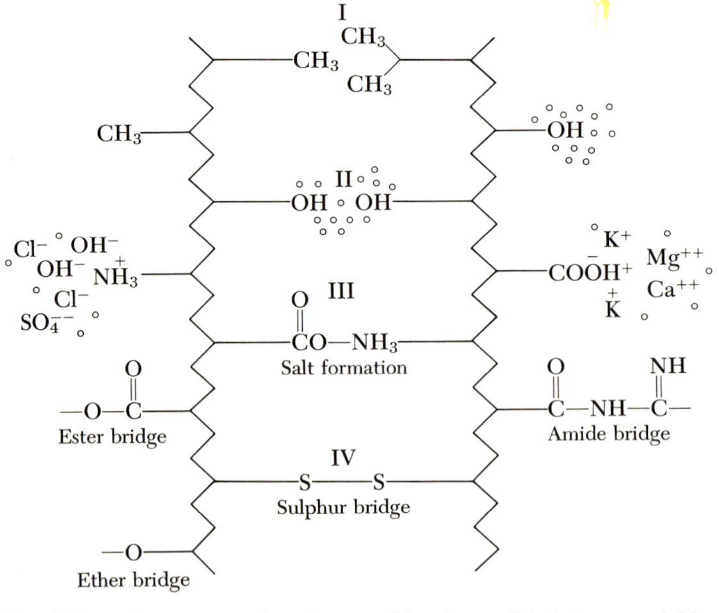


***Figure 4.12.*** Schematic representation of types of bonds possible between neighboring polypeptide chains of protein molecules in the cytoplasm. The small circles represent water molecules. (From Frey-Wyssling, 1953: Submicroscopic Morphology of Protoplasm. 2nd Ed. Elsevier Publishing Co. Amsterdam.)

The changes in free energy which accompany formation of various types of bonds are indicative of the strength of the bonds. Thus, formation of covalent bonds is accompanied by a change in free energy of between −50,000 and −100,000 cal. per mole. The weak bonds are formed with an accompanying change in free energy of between −1000 and −7000 cal. per mole. The weakest of the weak bonds in the cell are the van der Waals bonds, with free energies of formation of −1000 to −2000 cal. per mole. Hydrogen and ionic bonds are formed with free energy changes of between −3000 and −7000 cal. per mole. As might be expected, weak bonds do not require enzymes for breakage (Derjaguin, 1960; Watson, 1965).

As an inference from this conception, if the molecules in the cytoplasm are held together by weak bonds such as just described, its viscosity might be expected to fall when some of these bonds are broken, and to rise when additional bonds are formed. These bonds are susceptible to environmental conditions such as heat, pH and oxidation-reduction agents. Variation in such conditions should therefore affect the viscosity of cytoplasm. An interpretation of the manner in which the viscosity of cytoplasm is affected by the action of environmental or internal changes is considered in several examples given here.

The viscosity of an ameba varies greatly with temperature (Heilbrunn, 1958). It is high at low temperatures (see Fig. 4.13) but falls to a relatively low value with a rise in temperature from about 12.5° to 19° C. As the temperature rises further the viscosity increases to a plateau between 21° and 26° C. With a still further rise in temperature the viscosity again falls, with a low value at 30° C. When the temperature rises beyond this point, the viscosity rises abruptly as the cell is injured by heat. These changes in viscosity are interpreted by Frey-Wyssling's hypothesis as follows: Slight heat probably disrupts the homopolar cohesive bonds, lowering the viscosity. At temperatures beyond 20° C. another process sets in, namely, the dehydration of the molecular framework at those spots where hydrophilic chains come together, resulting in some sort of solidification. At the same time, however, the rupture of homopolar junctures continues, and at 25° C. it becomes of greater importance than the tendency toward solidification brought about by the simultaneous dehydration of the molecules due to heat. Further rise in

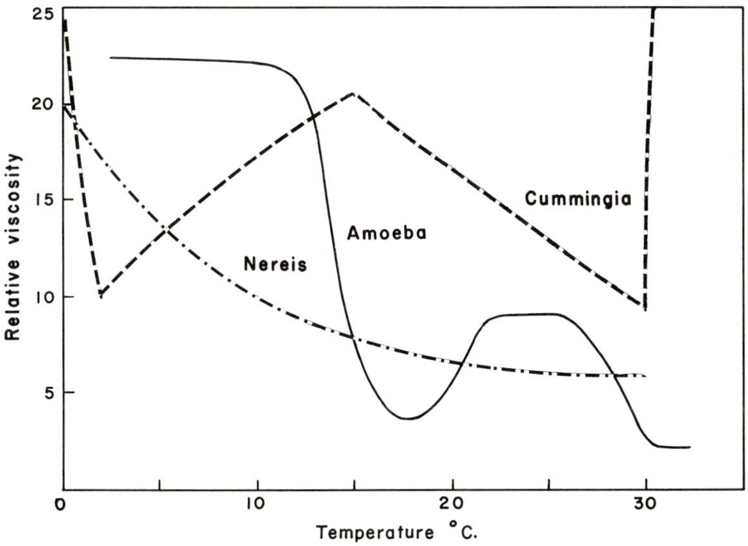


***Figure 4.13.*** Changes in viscosity of various cells with changes in temperature. (Data from Heilbrunn, 1958: The Viscosity of Protoplasm. Protoplasmatologia, *II,* C 1. Springer-Verlag Vienna.)

temperature may again raise the viscosity as a result of the predominance of the dehydration of cytoplasmic proteins. Within the physiological range of temperatures, it is doubtful that heat would affect heteropolar cohesive bonds or main valency bonds. However, it is interesting to note how differently temperature affects the viscosity of the cytoplasm of the three types of cells, *Amoeba, Nereis* eggs and *Cummingia* eggs (see Fig. 3.15). It would be rather difficult to explain all of these on the same grounds (Heilbrunn, 1958).

Experiments have shown that anaerobiosis generally causes a decrease in viscosity of cytoplasm. In the absence of oxygen as a hydrogen acceptor, metabolic hydrogen may reduce the disulfide linkage in protein to sulfhydryl, thus breaking the disulfide bonds and thereby decreasing the viscosity. A lack of oxygen may also lead to the formation of organic acids from anaerobic metabolism, and the hydrogen ions thus produced may then catalyze the breakage of linkages between amino and carboxyl groups of side chains in proteins, thereby also decreasing the viscosity.

Hypertonic solutions are found to increase the viscosity of cytoplasm. This could be caused by diminished distance between the colloidal particles following withdrawal of water, an action that would strengthen the binding between the particles. Conversely, hypotonic solutions cause the opposite effect because the particles are separated by more water.

Although the viscosity of cytoplasm varies greatly under different physiological and environmental conditions, presumably in the living state some structure always remains, even in a state of low viscosity. If all the bonds were broken, the structure which characterizes life would disappear and death would result (Frey-Wyssling, 1957).

## SUMMARY

In general, such physical properties of cytoplasm as its opalescence and its granular appearance are explainable on the basis of its being a colloid, light being scattered at each interface because of the difference in refractive indices of the dispersing medium and the suspended particles.

The cytoplasmic viscosity, its variability with changed conditions and the cytoplasmic resistance to flow upon application of force are best interpreted on the basis of the surface-active forces of a heterogeneous system and the formation of temporary weak bonds between adjacent particles. The paradox of cytoplasm behaving like a liquid while displaying the structural properties suggestive of a solid scaffolding is also best explainable on the basis of its being a colloid.

If a cell is a colloidal system it is not difficult to comprehend the occurrence of many simultaneous but different chemical reactions in a minute cell at physiological temperatures. The selectively permeable membranes surrounding certain discrete structures, e.g., the mitochondria, the Golgi apparatus and the endoplasmic reticulum, probably play an important role in separating the major reactive systems of the cell from one another. Thus the mitochondria are known to be the center of aerobic reactions liberating and storing energy; the ground substance is the center of anaerobic energy-liberating reactions; the lysosomes, for hydrolytic reactions; the Golgi complex and the endoplasmic reticulum, for the manufacture or the assembly of lipids and perhaps other materials; and the ribosomes, for the manufacture of proteins. However, each of these is also a polyphasic colloid. Because the phases are chemically different, one phase of a polyphasic colloid may attract one set of reactants, while another phase may attract others. Therefore, biochemical reactions of entirely different character might occur simultaneously at the interfaces of nearby but chemically different phases in each of the cell organelles.

It is also conceivable how a chemical present in relatively low average concentration in the entire cell might reach a very high concentration at an interface to which it is attracted and adsorbed selectively. Chemicals, coming through the cell membrane or synthesized inside, could thus be sorted out and concentrated by the different phases present in cytoplasm.

In the cell the reactants adsorbed to an interface are subjected to the distorting force of adsorption. Catalysis of chemical reactions in heterogeneous systems, whether inanimate or biological, is a result of distorting forces acting on the molecules adsorbed at interfaces. Such distorting forces lower the energy of activation required for the reactions (Chapter 1). In cells the molecules are probably adsorbed to the surfaces of enzymes (see Chapter 15). Consequently, the kinetic energy available at ordinary temperatures is sufficient to activate the molecules and to provide for continuation of the reaction. Duplication of the same reactions *in vitro* in homogeneous systems, however, may require high temperatures or high pressures, or both, because all the energy of activation and of reaction must then be supplied to the reacting molecules by the heat they obtain from the environment (Brennan, 1964).

## *APPENDIX*

### 4.1 SURFACE TENSION

The molecules in the interior of a homogeneous liquid are free to move and are attracted by surrounding molecules equally in all directions. At the surface of the liquid, where it touches air or some other liquid, however, they are attracted downward and sideways, or inward, more than upward; consequently, they are subjected to unequal stress and are held together to form a membrane. The force with which the molecules are bound is called the surface tension of the liquid. It depends upon cohesion of the molecules, or their attraction for one another. A needle placed carefully on the surface of water will float and so will smaller particles, such as powdered sulfur or talc.

Surface tension is measured in force per unit of length, generally in dynes per centimeter. Various methods are used for this purpose. A convenient one, attributed to Du Noüy, measures the weight required to just lift away a ring in contact with the solution to be tested (see Fig. 4.14). The surface tension $\sigma$ in dynes per centimeter is given by the following equation:

$$\sigma = \frac{\text{wt. in grams} \times 980}{2(2\pi \text{ r})} \qquad (4.2)$$

where r is the radius of the ring (Bull, 1964).

The surface tension of liquids that wet glass may be measured by the capillary height method. The solution rises in a capillary because it tends to contract to reduce the surface, thereby pulling the liquid upward. The surface tension is given by the following equation:

$$\sigma = \tfrac{1}{2}\,\text{hdgr} \qquad (4.3)$$

In this equation h is the height in centimeters, d is the density of the liquid, g is the acceleration due to gravity (980 cm./sec.$^2$) and r is the capillary radius in centimeters (Bull, 1964).

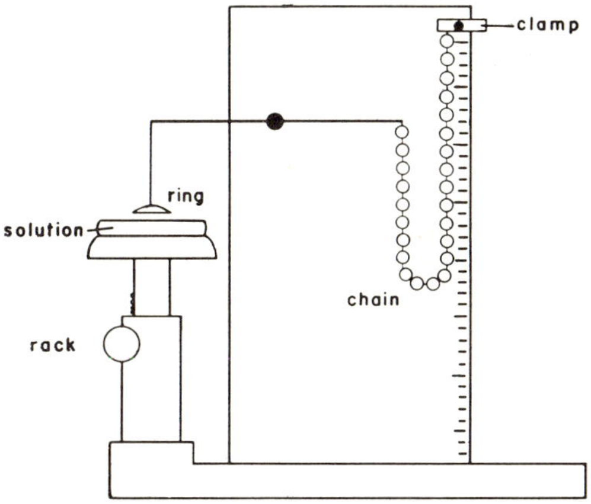


***Figure 4.14.*** Ring method of determining surface tension. A length of chain is measured on the scale on the right of the stand. Enough is measured to balance a given weight placed on the ring (100 mg., 200 mg. and 500 mg. in successive trials). The points are then connected by a line. When this is done the ring is allowed to come in contact with the test solution which is then racked up until the arm from which the ring is suspended is horizontal. The clamp is then released and moved down until the ring breaks from the surface of the solution. From the extra weight required to break the contact, the surface tension can be determined. (After Lloyd and Scarth, 1930: An Elementary Course in General Physiology. John Wiley & Sons, New York.)

**Table 4.4.** Surface Tension of Various Substances

| SUBSTANCE | TEMPERATURE °C. | SURFACE TENSION DYNES/CM. | AGAINST |
|---|---|---|---|
| Tristearin | 60 | 29.6 | air |
| Oleic acid | 20 | 32.5 | air |
| Glycerol | 20 | 63.4 | air |
| Mercury | 15 | 487 | air |
| Ethyl alcohol (100%) | 20 | 22.75 | vapor |
| Ethyl alcohol (100%) | 0 | 24.05 | air |
| Water | 18 | 73.05 | air |
| Water | 25 | 71.97 | air |
| Benzene | 20 | 40.04 | air |
| Acetic acid (1%) | 30 | 68.0 | air |
| Acetic acid (100%) | 30 | 26.6 | air |
| Sucrose 10% | 25 | 72.5 | air |
| Sucrose 55% | 25 | 75.7 | air |
| $CaCl_2$ (1M) | 25 | 3.2 + 71.97 (water) | air |
| NaCl (1M) | 20 | 1.64 + 72.75 (water) | air |
| Oil | 20 | ca. 15 | water |
| Oil with protein film | 20 | ca. 1 | water |
| *Arbacia* cell membrane | 20 | .08 – .2 | sea water |
| *Amoeba dubia* | 20 | 1.3 | fresh water |
| *Physarum* | 20 | .45 | fresh water |
| *Triturus* egg | 20 | .1 | fresh water |

Data for first 15 items from Handbook of Chemistry and Physics. 1964. 45th Ed. Chemical Rubber Co., Cleveland; the remaining six items from Davson and Danielli, 1952: Permeability of Natural Membranes. Cambridge University Press, London.

By the use of appropriate equations the surface tension may also be determined from the size of drops formed at the end of a tube. The surface tension of various solutions is given in Table 4.4. Measurement of forces at the surface of the cell are considered in Chapter 11.

## 4.2 VISCOSITY

Viscosity is a measure of the friction between molecules as they slide by one another. The unit of viscosity, the *poise,* is a viscosity such that unit force per unit area is required to cause two parallel liquid surfaces of unit area and unit distance apart to slide past one another with unit velocity. The absolute viscosity of water at 25° C. is 0.008937 poises (see Table 4.5).

For biological purposes the viscosity of various substances is usually measured relative to water. The Ostwald viscosimeter is most commonly used for this purpose (see Fig. 4.15). It consists of a **U**-shaped pipette with a capillary leg that is bent, widened to form a reservoir and continued up to the same level on the opposite side, making possible immersion of the whole in a thermostat. The time required for water to flow from a mark above the upper reservoir to one just below it is measured. The time required for the flow of a second fluid is determined in the same way. The ratio of the viscosities of the two liquids, called the relative viscosity, is directly proportional to the lengths of time required to empty the reservoir (Bull, 1964).

Viscosity may also be measured in a number of other ways: (1) The time required for a steel ball to fall a fixed distance through a solution is compared with the time required for it to fall the same distance through the solvent alone. (2) The rate of movement of an inner cylinder separated from an outer cylinder by a thin layer of the fluid whose viscosity is to be determined is also used to measure viscosity (Couette method). The outer cylinder is moved by suitable gearing, and the amount of movement induced in the inner cylinder is then determined (Bull, 1964).

For determining the viscosity within the cell several methods have been used. The rate of layering of various inclusions in a cell during centrifugation has been used for this purpose. The rate at which the layered particles return to their original locations might also serve as a measure, although it is less precise than the

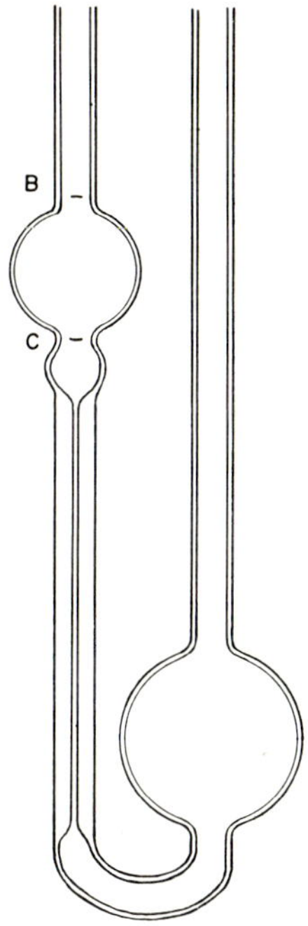


***Figure 4.15.*** Ostwald viscosimeter. To get the relative viscosity, the time required for the meniscus of a solution to move from B to C is compared with the time required for the meniscus of water to move the same distance. (After McBain, 1950: Colloid Science. D. C. Heath & Co., Boston.)

**Table 4.5.** Viscosity of Various Substances

| SUBSTANCE | TEMPERATURE °C. | VISCOSITY IN CENTIPOISES |
|---|---|---|
| Water | 25 | .8937 |
| Benzene | 10 | .758 |
| | 60 | .392 |
| Glycerin | 20 | 1490.0 |
| | 30 | 629.0 |
| Castor oil | 20 | 986.0 |
| | 30 | 451.0 |
| Pitch | 15 | $1.3 \times 10^{12}$ |
| Sucrose solution (20%) | 20 | 1.960 |
| (60%) | 20 | 56.5 |
| (60%) | 50 | 14.01 |
| Alcohol-water (50%) | 25 | 2.40 |
| Nerve fiber | 20 | 5.5 |
| *Cummingia* egg (shed) | 20 | 4.3 |
| *Cummingia* egg (during maturation) | 20 | 21 to 26 |
| *Amoeba dubia* | 18 | 2 |
| *Amoeba dubia* | varied | 2 to 14 |
| Slime molds | 20 | 9 to 18 |
| *Chara* | 20 | 10 |
| *Arbacia* egg | 20 | 7 |
| *Paramecium* | 20 | 50 |

Data for first seven items from Handbook of Chemistry and Physics. 1964. 45th Ed. Chemical Rubber Co., Cleveland; values on protoplasm from Heilbrunn, 1952: An Outline of General Physiology. 3rd Ed. W. B. Saunders Co., Philadelphia.

first. The rate of fall of starch granules in plant cells under the influence of gravity is also a measure of viscosity. In essence all these methods are akin to the falling ball method for measuring the viscosity of solutions (Heilbrunn, 1958).

## LITERATURE CITED AND GENERAL REFERENCES


Anderson, N. G., 1957: Labile colloidal complexes of cytoplasm. J. Cell. & Comp. Physiol. *49*(Suppl. I):221–242.

Battista, O. A., 1965: Colloidal macromolecular phenomena. Am. Sci. *53*:151–173.

Bayliss, L. E., 1959: Principles of General Physiology. 5th Ed. John Wiley & Sons, New York, Vol. 1, section on colloids.

Booij, H. L., and Bungenberg de Jong, H. G., 1956: Biocolloids and their interactions. *In* Protoplasmatologia. Handbuch der Protoplasmaforschung. Springer-Verlag, Vienna, Vol. 1, Chapter 2.

Bouck, G. B., 1963: Stratification and subsequent behavior of plant cell organelles. J. Cell Biol. *18*:441–457.

Brennan, D., 1964: Heterogeneous catalysis. *In* Recent Progress in Surface Science. Danielli, J. F., Pankhurst, K. G. A., and Riddiford, A. C., eds. Academic Press, New York, Vol. 2, pp. 57–110.

Bull, H. B., 1964: An Introduction to Physical Biochemistry. F. A. Davis Co., Philadelphia, Chapters 6, 9, 10, 13 and 16.

Child, C. M., 1941: Patterns and Problems of Development. University of Chicago Press, Chicago, p. 584.

Danielli, J. F., Pankhurst, K. G. A., and Riddiford, A. C. (eds.), 1964: Recent Progress in Surface Science. Vols. 1 and 2. Academic Press, New York.

Davies, J. T., 1964: Emulsions. *In* Recent Progress in Surface Science. Danielli, J. F., Pankhurst, K. G. A., and Riddiford, A. C., eds. Academic Press, New York, Vol. 2, pp. 129–167.

Derjaguin, B. V., July 1960: The force between molecules. Sci. Am. *203*:47–53.

Dubinin, M. M., Bering, B. P., and Serpinski, V. V., 1964: Physical adsorption at the gas-liquid interface. *In* Recent Progress in Surface Science. Danielli, J. F., Pankhurst, K. G. A., and Riddiford, A. C., eds. Academic Press, New York, Vol. 2, pp. 1–55.

Frey-Wyssling, A., 1957: Macromolecules in Cell Structure. Harvard University Press, Cambridge, Mass.

Gross, P. R., 1957: Labile colloidal complexes of the cell. A discussion. J. Cell & Comp. Physiol. *49*(Suppl. I):243–248.

Heilbrunn, L. V., 1958: The viscosity of protoplasm. *In* Protoplasmatologia. Handbuch der Protoplasmaforschung. Springer-Verlag, Vienna, Vol. 2, pp. 1–109.

Kerker, M., and Zettlemoyer, A. C. (eds.), 1967: Journal of Colloid and Interface Science. Vol. 23–25.

Kitchener, J. A., 1963: Surface forces in thin liquid films. Endeavour *22*:118–122.

Landau, J. V., 1959: Sol-gel transformations in amoebae. Ann. N. Y. Acad. Sci. *78*:487–500.

Loeb, J., 1922: Proteins and the Colloidal Theory of Behavior. McGraw Hill Book Co., New York.

McBain, J. W., 1950: Colloid Science. D. C. Heath & Co., Boston.

*Mysels, K. J., 1959: Introduction to Colloid Chemistry. Interscience Publishers, New York.

Oster, G., 1957: Organization in biocolloidal systems. J. Cell. & Comp. Physiol. *49*(Suppl. I):129–140.

Overbeek, J. T. G., and Voorn, M. J., 1957: Phase separation in polyelectrolyte solutions. Theory of complex coacervation. J. Cell. & Comp. Physiol. *49*(Suppl. I):7–26.

Ries, H. E., Jr., March 1961: Monomolecular films. Sci. Am. *204*:162–164.

Sheludko, A., 1966: Colloid Chemistry. Elsevier Publishing Company, Amsterdam.

Springall, H. D., 1954: The Structural Chemistry of Proteins. Academic Press, New York.

Watson, J. D., 1965: Molecular Biology of the Gene. W. A. Benjamin, New York, Chapter 4, The importance of weak chemical interactions.

Whitaker, D. M., 1940: The effects of ultra-centrifuging and of pH on the development of Fucus eggs. J. Cell & Comp. Physiol. *15*:173–188.

# CHAPTER 5 / THE STRUCTURE AND FUNCTIONS OF THE CELL ORGANELLES

Perhaps the greatest contribution to our knowledge of the structure and function of the cell organelles came from the discovery of electron microscopy on the eve of World War II, and the development of the method of fractional centrifugation for separation of the cell organelles for electron microscopic and chemical studies. The nucleus, because of its importance in cell syntheses and heredity, is treated separately in Chapter 6. The cell membrane is also considered separately, in Chapter 11, because its structure is best examined in connection with a discussion of the permeation of materials into and out of the cytoplasm.

## THE ENDOPLASMIC RETICULUM

As we have already seen, the cytoplasm of plant and animal cells is organized into a network or endoplasmic reticulum. Electron micrographic studies of ultrathin sections of a wide variety of cells cut with glass and diamond knives indicate that this network is made of thin membrane-bound cavities which vary considerably in size and shape under different physiological conditions. In some cells the network appears to consist of fine tubules of 50 to 100 nm. in diameter; in others it may be vesicular and of larger diameter (Figs. 5.1, 5.2, 5.3). In some places the enlargements form flattened sacs and vesicles called *cisternae.* Because in the initial studies the network seemed to be confined to the endoplasm of the cell, it was called the *endoplasmic reticulum.* Since that time, in thin sections of cells, the reticulum has been found to extend into the peripheral part of the cell (the ectoplasm), even connecting with the cell membrane. Study of serial sections of cells, which permits reconstruction of structure in three dimensions, has shown the endoplasmic reticulum to have connections also to the Golgi complex and the nuclear membrane.

The endoplasmic reticulum is less well developed in some cells than in others; for example, in spermatocytes few vesicles are present. In adipose tissue cells, brown fat cells and cells of the adrenal cortex, the endoplasmic reticulum is also simple. In cells engaged in protein synthesis for

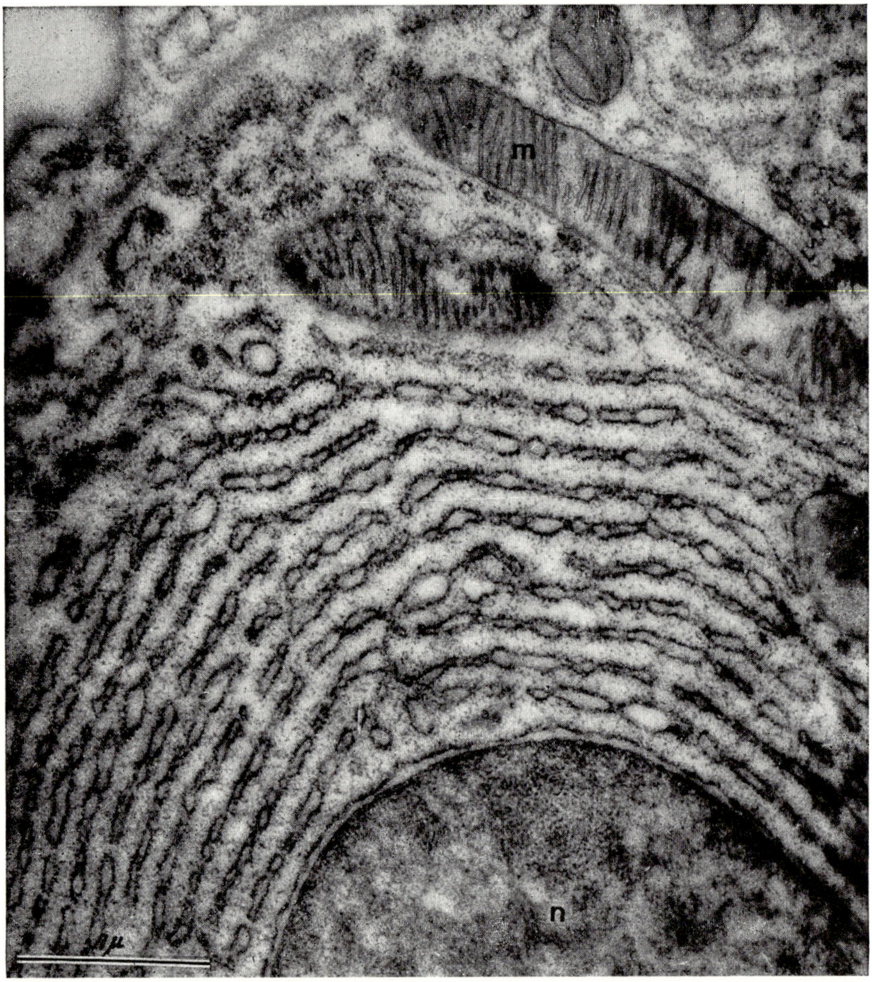


***Figure. 5.1.*** The endoplasmic reticulum as shown in an electron micrograph of a section through an acinar cell of a guinea pig pancreas. Near the nucleus (n) are paired cytoplasmic membranes of the endoplasmic reticulum to which are attached the ribosomes, the electron-dense granules on the surfaces. The mitochondria (m) show much finer lamellae. At the top of the micrograph unattached ribosomes are free in the hyaloplasm. (From Maximow and Bloom, 1957: Textbook of Histology. 7th Ed. W. B. Saunders Co., Philadelphia, by courtesy of G. E. Palade.)

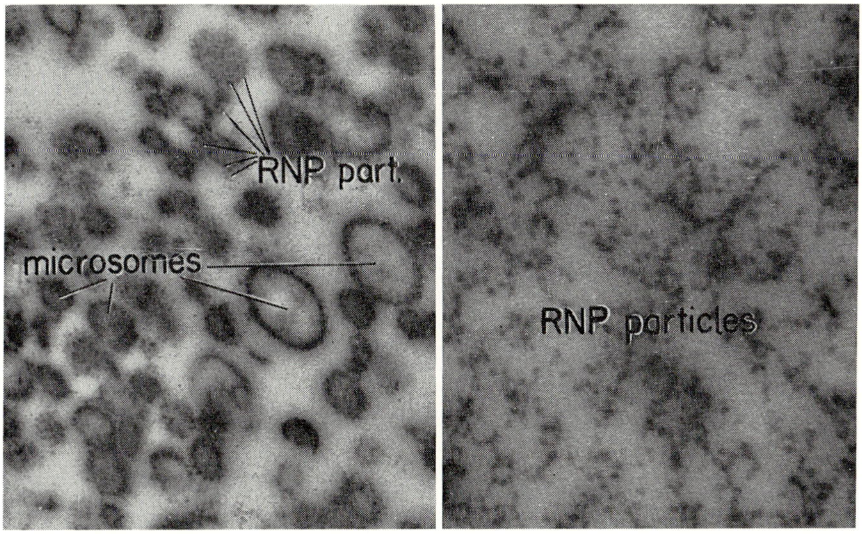


**Figure 5.2.** *Left,* Microsomal fraction isolated from homogenized pancreatic cells. The microsomes are the membranous vesicles to which are attached the ribonucleoprotein granules (RNP part) or ribosomes (×47,000). *Right,* Centrifugate of the material in the left figure after treatment with deoxycholate to solubilize the membranes (×8000). (From Palade, *In* De Robertis, Nowinski and Saez, 1960: General Cytology. 3rd Ed. W. B. Saunders Co.)

export the cytoplasm generally has a well developed rough endoplasmic reticulum. The reticulum is entirely absent from erythrocytes and bacteria (Porter *et al.,* 1945).

The membranes of the endoplasmic reticulum separate the ground substance of the cytoplasm from the contents of the lumen of the canals and vesicles. The tubules and vesicles have surfaces which may well play a role in enzymic reactions (Sjöstrand, 1964).

The membranes of the endoplasmic reticulum have been shown to possess osmotic properties, restricting entry or exit of materials into the interior (see Chapter 14). It is possible that materials are assembled inside the membrane either for synthesis or storage, and that the canalicular system allows intercommunication between the outside and the inside of the cell, especially if the membranes of the endoplasmic reticulum are truly continuous at some points with those of the cell membrane, as claimed by some workers (Robertson, 1959).

In muscle, it has been suggested that the endoplasmic reticulum is a means for conduction of the excitation from the surface of the cell to the myofibrils deep inside. The morphology of the muscle cell is consistent with such a possibility although no direct evidence for this hypothesis is available.

The membranes of the endoplasmic reticulum contain enzymes which are involved in synthesis of cholesterol, triglycerides and other lipids and therefore probably play a part in lipid metabolism (Siekevitz, 1963, 1965).

Some autoradiographic evidence from incorporation of $^3$H-glucose indicates that enzymes involved in glycogen synthesis may occur in the membranes of the endoplasmic reticulum (Coimbra and Leblond, 1966; Freeman, 1966). Injection of phenobarbitol in the rat led to proliferation of the membranes of the smooth endoplasmic reticulum, suggesting synthesis of detoxifying enzymes (Orrenius and Ericsson, 1966). The membranes of the endoplasmic reticulum also possess an electron transport system, the details of which are under study (Siekevitz, 1965).

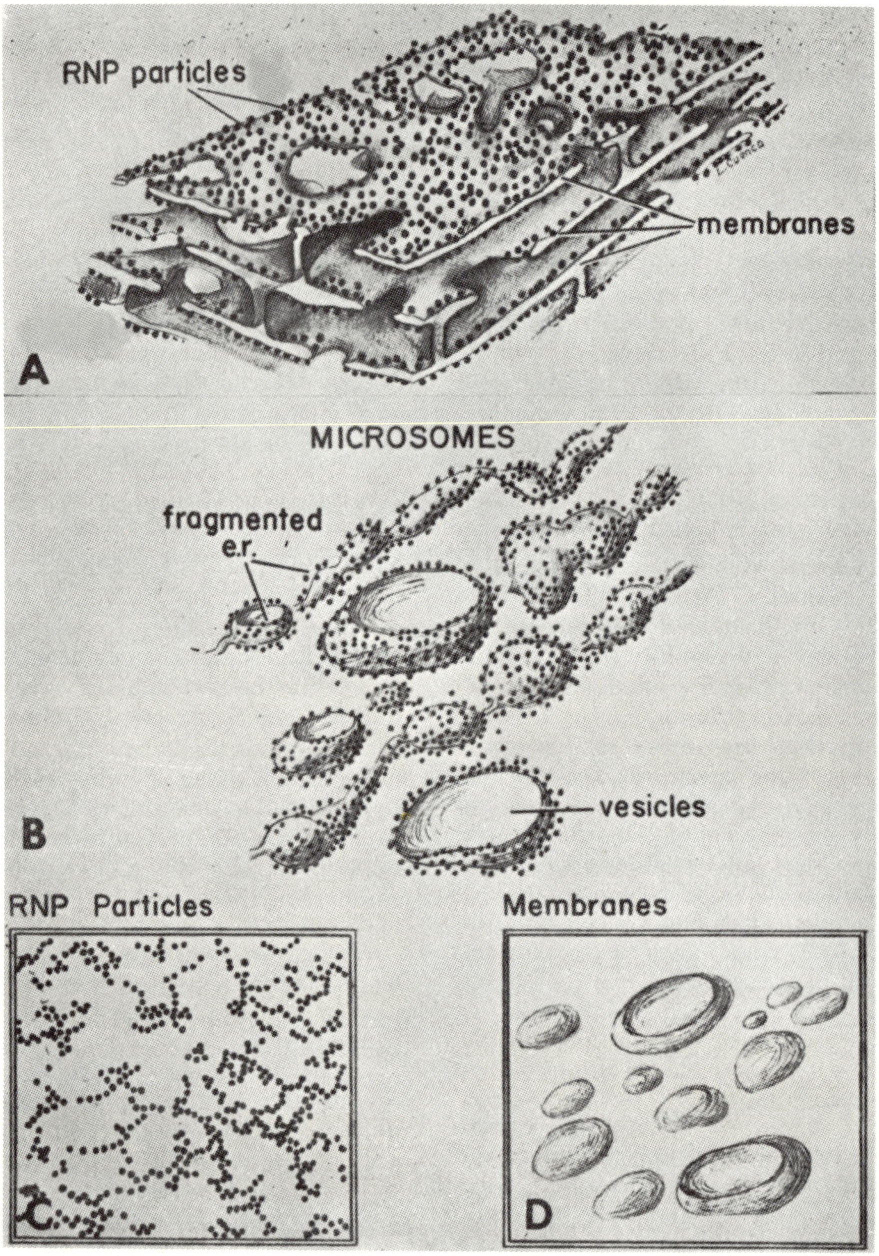


***Figure 5.3.*** Diagram of the endoplasmic reticulum of a cell and its fragmentation during disruption of the cell by homogenization. *A* shows the probable nature of the canicular system of the endoplasmic reticulum, bejewelled with the ribonucleoprotein particles (RNP). *B,* A homogenized cell with fragments of the endoplasmic reticulum forming microsomes. *C,* Ribonucleoprotein particles (RNP) removed from the membranes of the endoplasmic reticulum by action of desoxycholate. *D,* Appearance of the membranes without the RNP particles. (From DeRobertis *et al.,* 1965: Cell Biology. 4th Ed. W. B. Saunders Co., Philadelphia.)

## THE GROUND SUBSTANCE

The ground substance or hyaloplasm (Porter, 1961) lies between the canals of the endoplasmic reticulum. It appears structureless under the electron microscope except for the presence of scattered ribosomes (q.v.), and very fine microfilaments and microtubules seen in preparations fixed with glutaraldelyde (Ledbetter and Porter, 1963; Slautterback, 1963). Some soluble ribonucleic acid (RNA) occurs in the ground substance (see Chapter 21). The ground substance also contains the enzymes concerned with anaerobic metabolism. The hydrolytic (digestive) enzymes are found in the ground substance although the discoverers of *lysosomes* think that these structures are the specialized portions of the cytoplasm in which such enzymes are confined, and postulate that only when the lysosomes are ruptured are these enzymes released and so identified elsewhere.

The origin of the endoplasmic reticulum is uncertain. Invagination of the plasma membrane of the cell might give rise to a canalicular system, but there is no fully convincing evidence that such a process occurs on a large scale (Dallmer *et al.,* 1966).

## RIBOSOMES AND ERGASTOPLASM

Palade, in 1953, focused attention on some particulate units in the ground substance. These are electron-opaque particles of macromolecular (100 to 150 Å) dimensions usually attached to the outer surfaces of the endoplasmic reticulum and its vesicles. In some cells, however, e.g., those in active growth, such particles may also be free within the ground substance (Fig. 5.1 top). The number of such particles free and attached varies; it is especially large in cells carrying on active syntheses and small in less active cells or in cells which have been starved for a time. Cells from cancerous tissue may show a large number of particles.

It has long been known that basophilic granules exist in the cytoplasm and such names as the *ergastoplasm* of glandular cells or *Nissl bodies* of nerve cells had been given to the basophilic material. These granules are also highly opaque to electrons. Studies by Caspersson, Brachet and others showed that the basophilic material in the granules is ribonucleic acid (RNA). [The nucleic acid is attached to protein, forming ribonucleoprotein (RNP).] These granules containing the RNP are therefore called *ribosomes* (Haguenau, 1958).

When a suspension of cells or a tissue is homogenized, that is, subjected to mechanical disintegration, in a sucrose medium or in nonpolar solvents (Fig. 5.2), and then centrifuged after removal of the largest bodies, the remaining particles will sediment at different levels according to their size and specific gravity. The particles which settle out after prolonged centrifugation at high speeds are submicroscopic, varying in dimensions from 100 to 1500 Å. Many of these particles are larger than ribosomes. The pellet obtained by such centrifugation has 50 to 60 per cent of the ribonucleic acid of the cell and constitutes 15 to 20 per cent of the cell mass. Because of their small size the particles making up the pellet have been called *microsomes.* Electron microscope studies of ultrathin sections of the centrifugal pellet of a homogenized liver cell show that many microsomes are fragments of the vacuolar system of the endoplasmic reticulum, appearing as isolated vesicles, tubules and cisternae with attached ribosomal granules, whereas a similar preparation of homogenized pancreas cells shows round vesicles with numerous attached ribosomal granules. The range in size of microsomes obtained by centrifugation is probably the result of different degrees of fragmentation of the endoplasmic reticulum during homogenization. Thus, some microsomes consist of ribosomal granules alone; others are various-sized chunks of the reticulum or vesicles with attached ribosomal granules (Siekevitz, 1965).

The ribosomes are quite different in chemical constitution from the membranes of the endoplasmic reticulum and appear to be engaged in protein synthesis. For

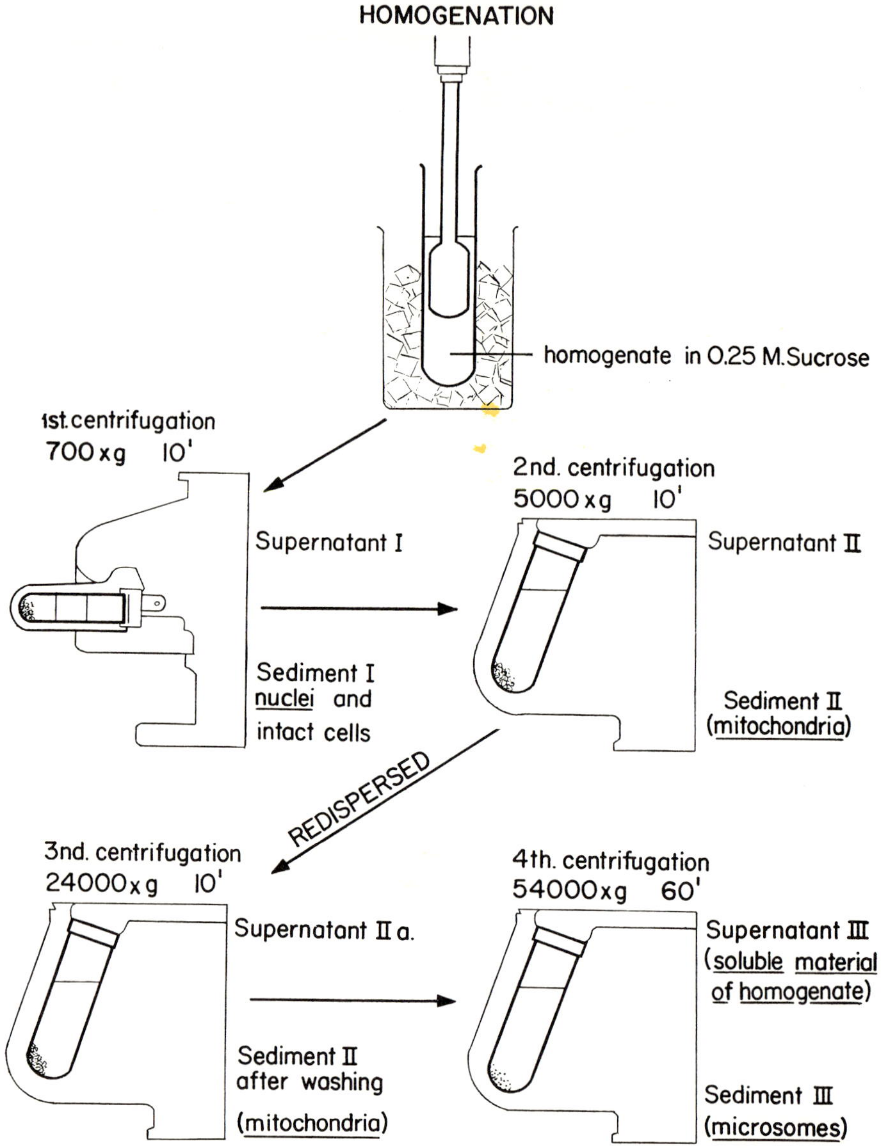


***Figure 5.4.*** Diagram of cell fractionation. (From DeRobertis *et al.,* 1965: Cell Biology. 4th Ed. W. B. Saunders Co., Philadelphia.)

example, when $^{14}C$-leucine is provided it is incorporated into protein by the microsomes with attached ribosomes. When ribonuclease is used to remove the RNA, the ribosomes disappear and synthesis of protein is reduced or ceases, depending on how long the ribonuclease had been permitted to act. The ribosomes may not all be alike. Several kinds of RNA have been identified from ribosomes by analysis of purine and pyrimidine bases (Petermann, 1964).

The synthesis of proteins presumably occurs at the surface of ribosomes, which may be attached to the outer surface of the endoplasmic reticulum. The synthesized proteins are transported through the membrane of the endoplasmic reticulum and are assembled into globules within the cisternae and canals. In cells which produce "proteins for transport" (Palade, 1956), the synthesized protein may later appear in the form of granules (250 to 350 nm. in diameter) just outside the endoplasmic reticulum, as in the case of zymogen granules produced by the cells of the pancreas.

Ribosomes are found in every type of cell, even the smallest PPLO examined. They have been studied most in bacteria but the ribosomes of animal and plant cells have many properties in common with those of bacteria. Ribosomes consist almost entirely of protein and RNA, the RNA constituting 40 to 60 per cent of the dry weight.

In electron micrographs (sections) ribosomes appear as electron-opaque bodies about 150 Å in diameter. The ribosomes may be separated from other cell constituents by fractional centrifugation (see Fig. 5.2). The fractions are collected consecutively at the bottom of the tube, and after dispersion, the ribosomal material is placed atop a sucrose density gradient (usually 5 to 20 per cent) in a centrifuge tube. Centrifugation is continued until the ribosomal particles of different sizes are banded. The rate of movement under centrifugal force is given in terms of the sedimentation constant (ratio of the rate to the acceleration), termed the Svedberg (abbreviated S) in honor of the pioneering investigator. The different bands are removed by puncturing the bottom of the tube and removing small fractions (DeDuve, 1959).

Prokaryotic cells have ribosomes mainly of 70 Svedbergs (or 70S), but these are made up of two subunits, 50S and 30S (Fig. 5.5), which appear as separate bands under certain conditions (depending upon salts, pH, and so forth). The subunits may be split into smaller units, and the 70S particles may associate into larger units (e.g., two 70S particles form a

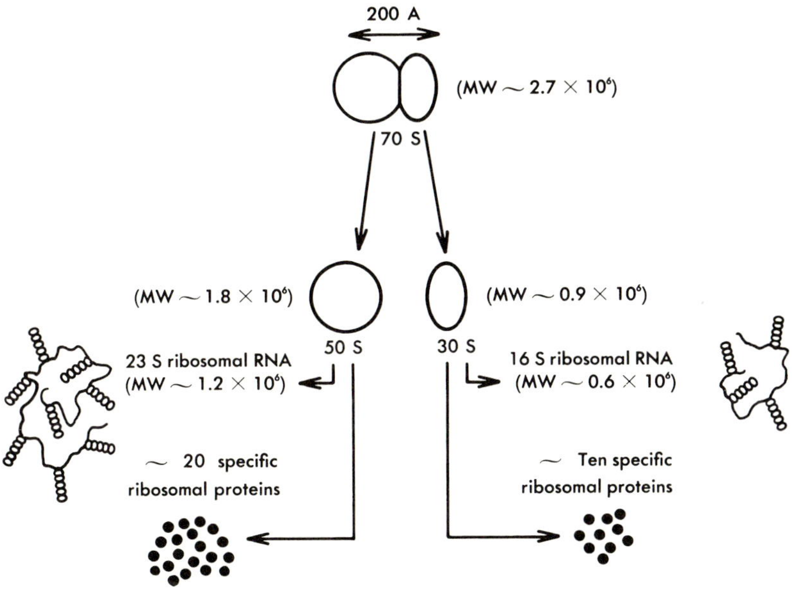


***Figure 5.5.*** The structure of the ribosome in *Escherichia coli.* It is about 200 Å long and its size is designated in terms of the sedimentation constant in Svedberg units, abbreviated S. The union of the 50S and 30S particles to form the 70S particles is dependent upon the concentration of magnesium ions. At the magnesium ion concentrations in growing cultures of the bacillus 70 to 90 per cent of the smaller units are combined to form 70S particles. Lowering the magnesium ion concentration will induce dissociation. The composition of the 50S and 30S particles is indicated below. The diagram indicates helices and unpaired portions of the RNA. (From Watson, 1965: Molecular Biology of the Gene, W. A. Benjamin, New York.)

100S unit), especially during protein synthesis. Such associations of ribosomes are called *polyribosomes* or, more simply, *polysomes* (Fig. 5.6).

In some eukaryotic cells studied the ribosomes appear to be 80S particles. Two subunits 40S and 60S are also found, the combination of the two forming the 80S particle. The 40S and 60S subunits may be further dissociated and the 80S particles may be associated to form polysomes as in the bacterial cell (Petermann, 1964).

The RNA of the ribosome is associated with structural protein of average molecular weight 25,000. The end groups appear to be the same for proteins attached to the 30S and the 50S subunits (and the 40S and 60S units in eukaryotic cells) in cells studied. It will be recalled from Chapter 3 that the RNA of the ribosome forms an imperfect helix and it is likely that the amino acid residues of the protein attach to these unpaired bases of the RNA.

Isolated ribosomes have been studied in the electron microscope by shadow casting and negative staining. In *E. coli* the 30S particle is asymmetrical and is represented by a prolate ellipsoid 140 × 170 Å. With negative staining the 50S particle is dome-shaped, and the 30S particle is flatter and fits into the 50S subunit. The ribosomes of eukaryotic cells are more spherical than those of bacteria and more highly hydrated (Petermann, 1964). The hydrated particles are much larger than the dry ones visualized in the electron micrographs.

The number of ribosomes associated into a polysome varies, perhaps with the size of the protein being synthesized. When only one protein (hemoglobin) is being produced, as in the reticulocytes (immature red blood cells which have lost the nucleus), only four to six (average five) ribosomes are attached to the messenger RNA, each presumably forming the same protein. When many proteins are being synthesized in a cell, for example, in a HeLa tumor cell, the number of ribosomes per messenger RNA varies from a few to 30 or even more (Rich, 1963).

Presumably, protein synthesis occurs on ribosomes at the outer surface of the endoplasmic reticulum to which they are attached. How the protein gets inside the canals, is not known. However, Palade (1956) has described granules of protein about 250 to 350 nm. in diameter inside the canals. It is also not known how the protein is extruded to form the larger secretion granules seen outside the reticulum, as for example, in cells of the pancreas which manufacture the enzymic proteins.

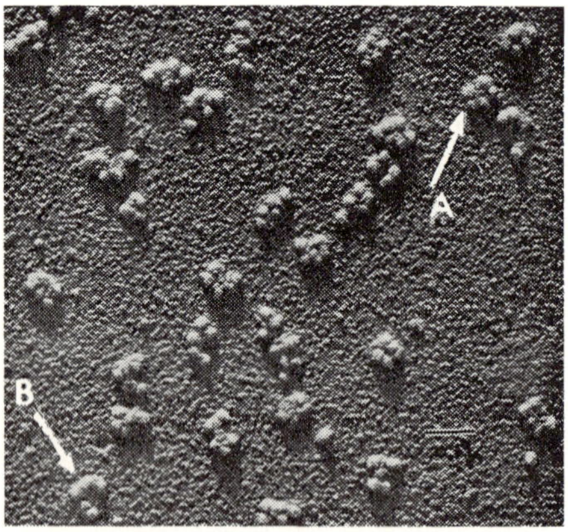


***Figure 5.6.*** Electron micrograph of polysomes in a preparation from rabbit reticulocytes synthesizing hemoglobin. Most of the ribosomes appear in clusters which are compact (A) or more open (B). The horizontal mark indicates 1000 Å. (From Warner, Rich and Hall, 1962: Science *138*:1400.)

## GOLGI COMPLEX

The Golgi complex (Fig. 5.7), like the endoplasmic reticulum, is a canalicular system with sacs, but its selective staining by osmium tetroxide and silver salt indicates its uniqueness. It was by these staining reactions (Fig. 2.2) that Golgi in 1898 first recognized the complex in nerve cells of the barn owl and the cat, long before any evidence was available for an endoplasmic reticulum. Electron micrographs of thin sections of cells indicate that the canals of the Golgi complex are of various shapes, usually consisting of stacks of flattened sacs (cisternae) associated with small vesicles and vacuoles of various sizes. The surface of the canal and sacs is always smooth, in contrast to the frequently en-

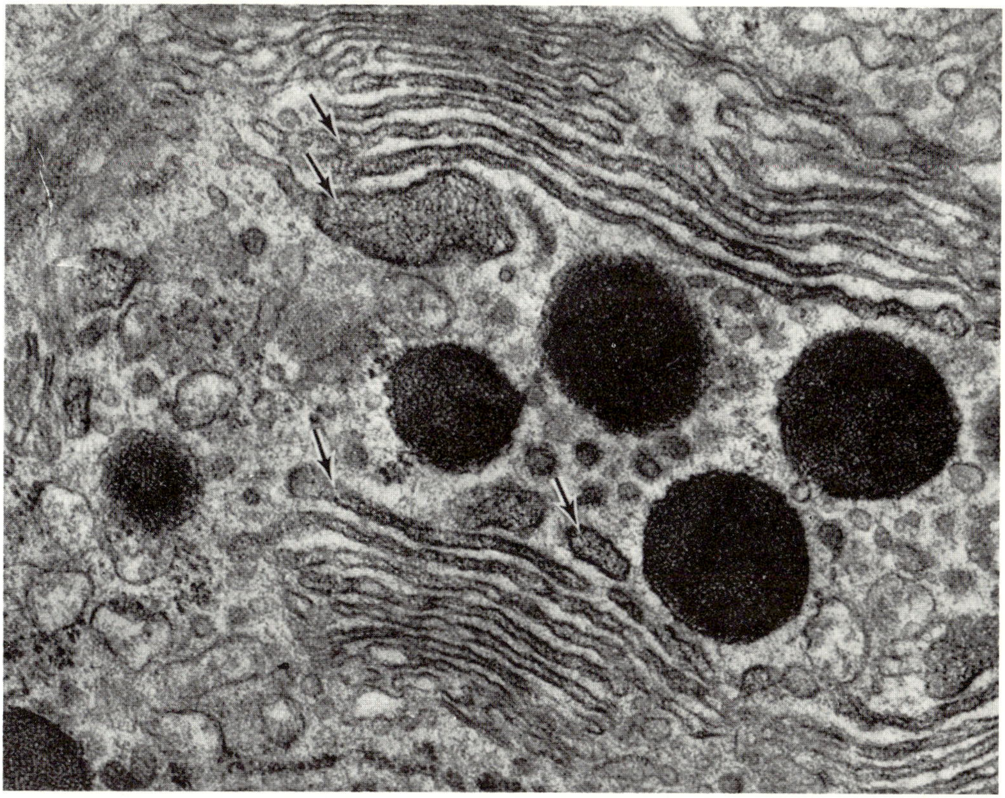

***Figure 5.7.*** The Golgi complex. Stacks of Golgi cisternae are evident and many small vesicles are associated with the thin outer cisternae. Toward the interior of the Golgi complex the cisternae are more distended and contain a flocculent material (see arrows) having a texture much like the secretion granules (large black bodies). Duodenal gland of the mouse (54,000 X). (From Fawcett, 1966: The Cell: Its Organelles and Inclusions. W. B. Saunders Co., Philadelphia, p. 127. Photograph by Dr. Daniel Friend.)

countered rough granule-encrusted outer surfaces of the endoplasmic reticulum (Dalton, 1961).

The Golgi complex is present in all eukaryotic cells, but is perhaps more prominent in vertebrate cells than in invertebrate cells or plant cells, where it has been collectively called the *dictyosome.* The Golgi complex is organized in basically different ways in different types of cells, but its organization for any one kind of cell is usually the same. Thus it is small in muscle fibers and many other types of cells, but it is large and well developed in secretory cells and in nerves, where it is clearly a reticulate structure. It occupies different positions in different kinds of cells—being polar and between the nucleus and the periphery in cells of ectodermal origin, circumnuclear in nerve cells and scattered in other parts of the cell in other cases (Kuff and Dalton, 1959).

The surface of the Golgi complex is covered by a unit membrane. The two faces of the membrane have different staining properties; only the outer part reacts with silver salts and osmic acid, indicating a difference in chemical composition of the inner and outer surfaces (Pollister and Pollister, 1957).

The inside of the Golgi complex appears to be rather fluid in consistency, since needles inserted into the canals are moved with apparent ease, and there is no indication of resistance. This might, however, be the result of injury. Nevertheless, in centrifugation experiments the entire complex moves as a unit, and is displaced toward the centripetal pole. This indicates a specific gravity of the Golgi complex less than that of the surrounding cytoplasm (Kuff and Dalton, 1959).

It has been possible to isolate for chemical study the membranes of the Golgi complex by fragmenting the cells and fractionally centrifuging them in a sucrose density gradient. The Golgi membranes (Fig. 5.8) are found in the interface between the fluids having a specific gravity of 1.09 and 1.13. The membranes appear to contain much lipid and lipoproteins and a number of enzymes, but slight amounts of nucleic acid, at least when they are carefully washed free of contaminating granules (ribosomes).

The low enzymatic content of the Golgi membranes is somewhat startling compared to the large number of enzymes found in the endoplasmic reticulum. As a consequence, it is postulated that, perhaps, the Golgi complex acts to concentrate and channel various materials rather than to actively synthesize them. However, the membranes of the Golgi complex are found to proliferate along with the membranes of the endoplasmic reticulum during the early stages in the formation of mucus in some cells (Freeman, 1966). Since uptake of monosaccharides occurs in the Golgi membranes during this process, as shown by autoradiography, these membranes may be engaging in some biochem-

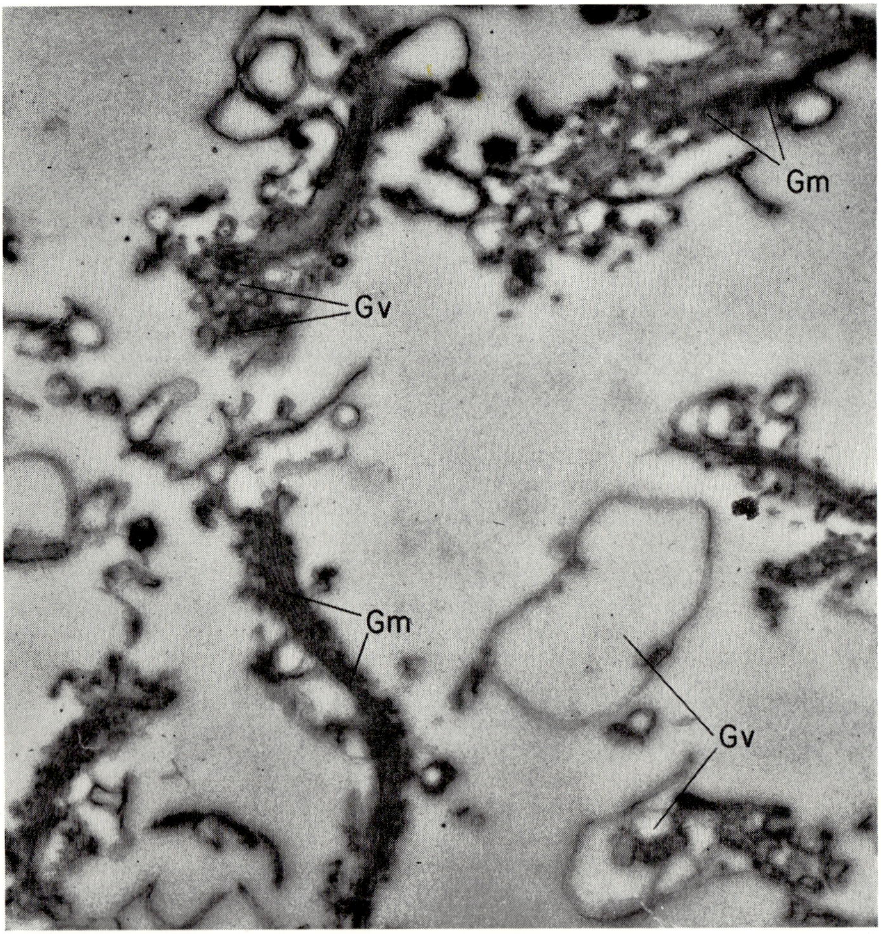


***Figure 5.8.*** Fragments of the Golgi complex after homogenization of cells of the epididymis. Gm, Golgi membranes; Gv, Golgi vacuoles. (By courtesy of A. J. Dalton.)

ical syntheses, perhaps of mucopolysaccharides (Neutra and LeBlond, 1966). Lipid globules and protein granules have been found in the Golgi channels, as, for example, in cells of the hypophysis and pancreas subsequent to their appearance in the endoplasmic reticulum (Hodge *et al.*, 1955). When labeled precursors (e.g., $^3$H-amino acids) are provided to the cell, labeling in autoradiograms is seen over the endoplasmic reticulum before it appears over the Golgi complex (Cohn and Benson, 1965).

## LYSOSOMES

The lysosomes have centrifugal properties between those of the mitochondria and the ribosomes and have been studied in only a restricted number of cell types. This class of particles ranges in size from 0.25 to 0.8 $\mu$ and under the electron microscope they appear dense and finely granular, individual granules having a diameter of about 55 to 80 Å (Fig. 5.9). They are distinguished from other types of particles in the cytoplasm mostly on the basis of their biochemical properties, namely, their content of specific enzymes which are primarily hydrolytic in function. Among the characteristic enzymes is acid phosphatase.

The lysosomes are covered by a single unit membrane which retains the enzymes. The enzymes may be released from the lysosomes by the action of a wide variety of lytic agents such as action of a blender, freezing and thawing, ultraviolet radiations, vitamin K and detergents, all of which disrupt the membranes (DeDuve and Wattiaux, 1966). These agents are called lysosome labilizers. Substances which strengthen the membranes, such as cortisone (a hormone), chloroquine (an antimalarial) and cholesterol, are spoken of as lysosome stabilizers (Weissman, 1964).

Although these enzymes have also been found in the general supernatant of all disrupted cells tested, it is quite possi-

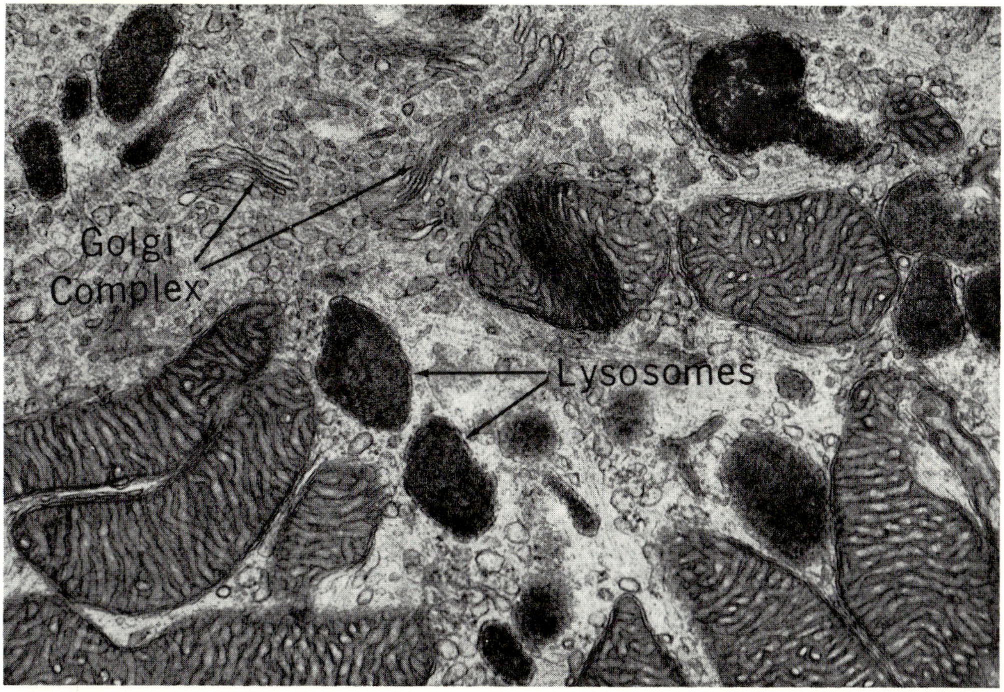


***Figure 5.9.*** Lysosomes in cells of the hamster adrenal cortex. Collidine buffered glutaraldehyde and osmium tetroxide fixation, lead citrate staining (× 25,000). Note also the Golgi complex and the numerous mitochondria. (From Fawcett, 1966: The Cell: Its Organelles and Inclusions. W. B. Saunders Co., Philadelphia, p. 197.)

ble that the lysosomes, normally present in cells, are ruptured by the treatment used to fragment cells, and their enzymes might be liberated in this manner. Lysosomes of injured or dying cells probably rupture spontaneously, in this way lysing the decrepit and useless cells of an organism. They also enable a developing organism to remove structures no longer of use, for example, cells in the tail of a metamorphosing tadpole.

Lysosomes may well play a role in turnover in normal cells. Thus, it has been noted that vacuoles (*phagosomes*) incorporating cell substance and sometimes fragments of cell organelles (e.g., mitochondria) are present in some tissue cells, and it is thought that these vacuoles unite with lysosomes which results in digestion (autodigestion) of the engulfed material in the vacuoles. That such a union between phagosome and lysosome does occur is indicated by the presence of acid phosphatase in the resulting vacuole, acid phosphatase being characteristic of lysosomes but not of phagosomes. To a lesser degree such lysosome activities may occur in other types of cells in which pinocytosis, or cell drinking (see Chapter 14), may bring in particulate material along with water in small vacuoles. After digestion of the particulate matter the membranes of the lysosomes may combine with the plasma membrane, at the same time extruding the undigestible residues (DeDuve and Wattiaux, 1966).

Lysosomes may originate either directly from the endoplasmic reticulum or from vesicles of the Golgi complex, or from both. In studies with radioactive amino acids, autoradiographic labeling in the cell occurs over the lysosomes only after labeling in the endoplasmic reticulum and the Golgi complex. Protein granules are often seen in enlargements of the endoplasmic reticulum, which, if pinched off, would produce bodies of the size and character of lysosomes (DeDuve and Wattiaux, 1966). The endoplasmic reticulum is well known for synthesis of protein for export. The lysosomes may represent another form of production for export.

By differential centrifugation it was possible to separate lysosomes from still smaller bodies (with an S value of 1.25, compared to 1.22 for lysosomes). These particles called *microbodies* are much like lysosomes in appearance under the electron microscope except for their smaller size. They contain the enzymes urate oxidase, D-amino oxidase and catalase. They do not occur in rat tissues other than the kidney and liver but they are found in the protozoan *Tetrahymena*. It is thought that they participate in metabolic oxidations involving hydrogen peroxide (peroxidosomes). (Baudhuin *et al.,* 1965; DeDuve, 1965).

## THE GOLGI–ENDOPLASMIC RETICULUM–LYSOSOME COMPLEX

Attention has been called to the similarities in the many membranes present in the cell—on its surface and bounding the endoplasmic reticulum, Golgi complex, mitochondria and nuclear surface (see Fig. 5.9). Although the membranes differ from one another to some extent, their general similarity to the plasma membrane, and the connections of the channels of the endoplasmic reticulum to the surface of the cell seen in electron micrographs, point to the plasma membrane as the possible source of all the membranes. Evidence is still scanty and circumstantial.

It is conceivable that all of the membrane systems linked in the Golgi–endoplasmic reticulum–lysosome complex are interrelated by origin. Transitions are seen from the smooth endoplasmic reticulum to the Golgi complex (vesicles and saccules) and from the endoplasmic reticulum to lysosomes, and possibly from the Golgi complex to lysosomes (Brandes, 1965). The high concentration of acid phosphatase of some portions of the Golgi complex suggests that this enzyme, so characteristic of lysosomes, is transferred from the Golgi tubules to the lysosomes, perhaps with some other hydrolytic enzymes. However, since the rough endoplasmic reticulum is the region of the membrane system in which protein synthesis occurs, it is likely that all of the enzymes in the system take origin here, pass through the canals of the smooth endoplasmic system and thence reach the Golgi complex. They may then get to the lysosomes from the Golgi complex or possibly directly from the endoplasmic reticulum, as already mentioned. Autoradiography shows that label (tracer amino acid) first appears over the ergastoplasm (rough endoplasmic reticulum), then over the Golgi region and finally within the lysosomes.

## MITOCHONDRIA

Mitochondria were first seen in 1880 by Kolliker who teased them out of muscle cells and found that they swelled in solutions containing less salt than blood and shrank in solutions containing more. They were extensively studied in various cells by many investigators thereafter, mostly in fixed and stained preparations. As already stated (Chapter 2), mitochondria were once described as the organelles which stain specifically with Janus green B. Many other names have been given to them but in the body of literature which has grown up recently the name *mitochondrion* is rather generally applied to a single unit of these organelles and *mitochondria* for several or for the mass of units, regardless of size or shape. This terminology will be followed here.

Mitochondria (Figs. 5.1, 5.9 and 5.10) are usually 0.5 to 1.0 $\mu$ in diameter and vary in length up to a maximum of 7 $\mu$. They may be filamentous or granular and may change from one form to another depending upon the physiological conditions of the cells. They may also attach to one another and then dissociate at a later time.

Mitochondria may show active movement in some cells, especially in those undergoing division. In some cells, however, they appear to be stationary (Frederic, 1958). They may aggregate in very characteristic ways about some cell structures, e.g., as rings in muscle cells. In neurons mitochondria accumulate at the internodes where the passage of the nervous impulse is supposed to occur. In secretory cells they may accumulate basally and in the retina at one end of the rod cells.

The number of mitochondria may be very large. In a liver cell as many as 2500 may be present and account for 30 to 35 per cent of the nitrogen content of the liver cells.

Because the specific gravity of mitochondria is greater than that of the ground substance (the larger fragments having first been removed), ultracentrifugation of mechanically disintegrated cells at 5000 to 24,000 times gravity will deposit the mitochondria at the bottom of a centrifuge tube. The mitochondria so separated, carefully washed free of other particles, consist of 65 to 70 per cent protein, 25 to 30 per cent lipids and 0.5 per cent RNA. The sulfur content is also relatively high because of considerable —SH present as part of the active groups of some of the enzymes (Lehninger, 1965).

Of the proteins, the largest portion probably consists of enzymes, but a structural or organizational protein is present, constituting about 30 per cent of the total protein. The organizational protein is important to the functioning of some of the mitochondrial enzymes since when it is added in the right proportion, the activity of the enzyme is increased (Woodward, 1966). An actomyosin-like protein which is an adenosine triphosphatase (splitting adenosine triphosphate) and contracts on addition of adenosine triphosphate has been reported from liver mitochondria (Onishi and Onishi, 1962), but is not known to be general. The phospholipids are also important for enzyme function since their addition to purified enzymes restores activity (Criddle *et al.,* 1962). Cell membranes are most effective in restoring enzymatic activity; this indicates the importance of the steric factor in mitochondrial enzyme function.

The membrane of the mitochondrion is capable of active transport, accumulating calcium and magnesium and phosphate against a concentration gradient (Brierley *et al.,* 1963). There is also energy-linked potassium uptake into the mitochondria (Rottenberg and Solomon,

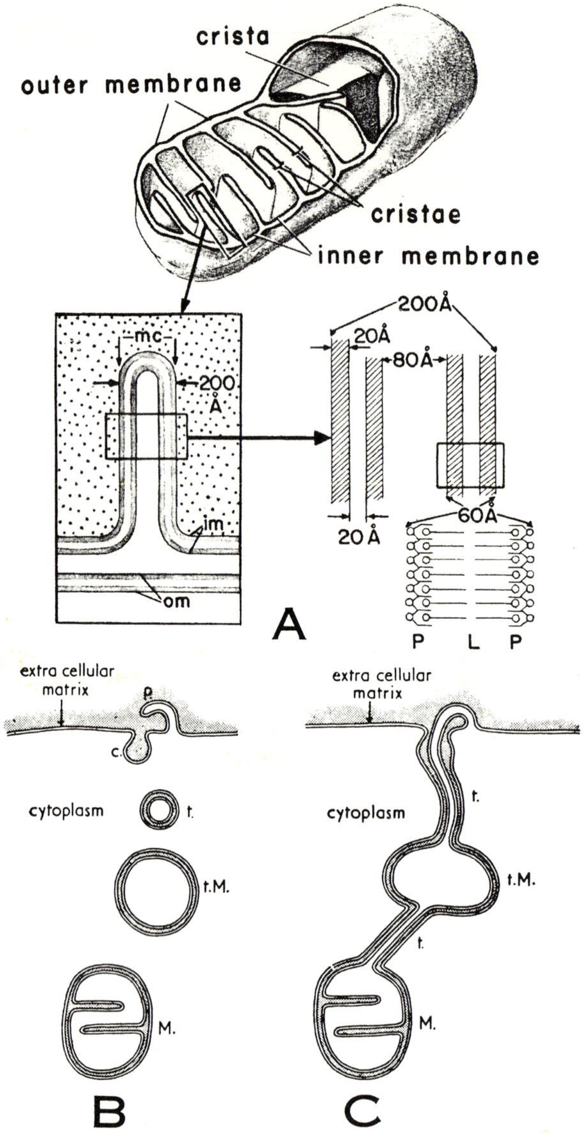


***Figure 5.10.*** *A,* Diagram of the ultrastructure of a mitochondrion. *Above,* tridimensional diagram showing outer and inner membranes and the cristae, which project into the body of the mitochondrion containing the matrix. *Below, left,* at higher magnification the structure of a single crista (mc). *Below, right* (at much higher magnification), molecular structure of a unit membrane of a crista. The outer membrane (om) and inner membrane (im) of the mitochondrion are similar. P refers to protein; L to lipid. (From DeRobertis *et al.,* 1965: Cell Biology. 4th Ed. W. B. Saunders Co., Philadelphia.)

*B* and *C,* A way in which mitochondria might originate from cell surfaces. *B* illustrates a segment through a muscle fiber, showing structures observed. In *C,* these structures are related speculatively. The protuberance (p) in *B,* passing into the invagination (c) to form a double-membrane lined tubule which enlarges into a vesicle (t.M.). Ingrowth of the inner membrane of such a vesicle to form cristae could give rise to a mitochondrion (M). (From Robertson, 1959: Biochem. Soc. Symp. *16*:38.)

1965). In this respect the mitochondrial membrane resembles the plasma membrane of the cell (see Chapters 11 and 14). Accumulation of ions in the mitochondrion affects the structure of the mitochondrion and results in its swelling (Greenawalt *et al.,* 1964).

The mitochondria have a complex organization. The overall structure is a double walled rod with rounded ends. The inner wall or membrane of the mitochondrion is 60 to 80 Å thick and in most mitochondria is extended as a sheet called the *crista* (or crest) inward almost across the internal cavity of the mitochondrion. Many such cristae may be present (Fig. 5.10). The cristae greatly extend the laminar surface exposed to the inside cavity of the mitochondrion. The cristae usually run at right angles to the long axis of the rod-shaped mitochondrion. The space in which these infoldings lie is filled with a relatively dense fluid material in which no structure is seen with the electron microscope, although fibrils have been reported in some cases. The outer membrane of the mitochondrion is about 60 Å thick and is separated from the inner membrane by a clear space.

However, many variations on the fundamental structure are found in different types of cells. In active cells such as those of the kidney the cristae may be very numerous transverse sheets, almost like a stack of coins, whereas in less active cells they may be few in number. In some neurons the cristae are longitudinal folds paralleling the long axis of the mitochondrion. In protozoan cells, in insect flight muscle cells and in adrenal cells the infolding may take the form of tubules rather than lamellar infoldings. Sometimes in mitochondria of different cells of the same species both types of infoldings are found, e.g., in insects tubules occur in flight muscle cells and lamellar cristae occur in the leg muscle cells. In *Chaos chaos,* the multinucleate ameba, the tubules have an interesting zigzag pattern and appear different from those seen in cells of other species (Pappas and Brandt, 1959). In cells of some species the mitochondria seem to branch (Rouiller, 1960).

Under the electron microscope each outer and inner unit membrane of the mitochondrion appears in turn to be made up of three layers, a middle layer (the lighter layer)—less opaque to electrons—being sandwiched between two electron-opaque (darker) layers. Robertson (1959) suggests that the lighter inner layer as seen in section is made up of two rows of lipid molecules with their nonpolar groups in the center, the two denser outer layers consisting of protein molecules as shown diagrammatically in Figure 5.10.

The membrane of the mitochondrion, like the cell membrane, is selectively permeable and the "rules" governing the relative rates of entry of various nonpolar and polar materials into the mitochondria seem to resemble those observed for the cell membrane (see Chapters 12 and 13). The mitochondria, as do entire cells, swell when placed in solutions containing a concentration of solute lower than (hypotonic to) that characteristic of the cell. Isolated

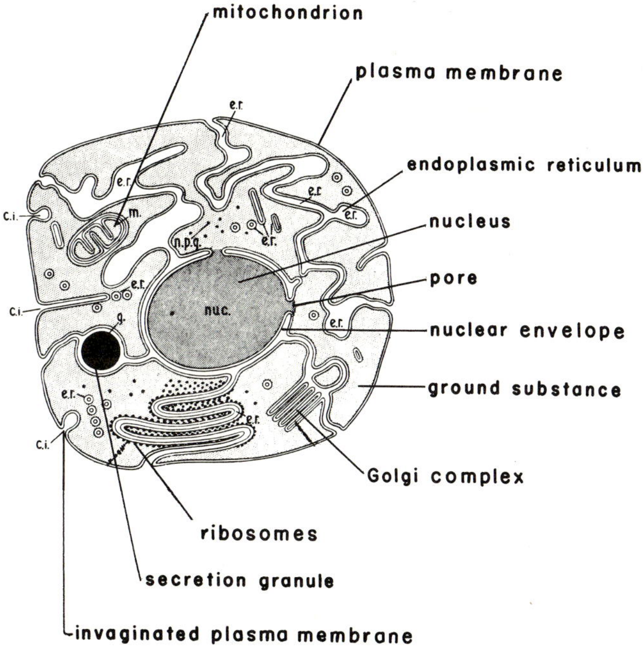


***Figure 5.11.*** Diagram of a section of a hypothetical cell illustrating possible relationships of the cell membrane to various cell organelles. The cell membrane is shown as a pair of dense lines separated by a light interzone. The invaginations of the cell surface (c.i.) are indicated in several areas. Some of these extend for a considerable distance into the cell and they may connect with the endoplasmic reticulum (e.r.). The nuclear membrane is composed of flattened sacs of the endoplasmic reticulum, and by means of the nuclear pores, nucleoplasm (nuc.) is in continuity with cytoplasm. The Golgi complex is here shown as a modified component of the endoplasmic reticulum. Secretion granules are shown as dense aggregates contained within membranes of the endoplasmic reticulum. Nucleoprotein granules are shown scattered through the cytoplasm and in some regions attached to the cytoplasmic surfaces of membranes of the endoplasmic reticulum. In some regions the endoplasmic reticulum is shown as tubules, either in longitudinal section or cross section. Little evidence has appeared to support this extreme view of the origin of membrane-bound organelles (including the mitochondria in Figure 5.10*B* and *C*). One mitochondrion is shown with its cristae formed by invagination of its inner membrane. (From Robertson, 1959: Biochem. Soc. Symp. *16*:33.)

mitochondria, placed in hypotonic solutions, may swell to five times their normal volume yet remain intact. The membrane is therefore a sturdy structure. On the basis of rate of entry of some water-soluble materials into mitochondria, it is postulated that there is unfolding of a corrugated-like lipid layer, although such foldings have not been seen under the electron microscope (Tedeschi, 1959).

Mitochondria have been fragmented by the use of the surface-active substance deoxycholate (a mild detergent), and many enzymes have been isolated from them. Some of these enzymes are firmly bound to the membranes and have resisted purification. However, it is established that all of the enzymes that have to do with the complete breakdown (aerobic) of nutrients to carbon dioxide and water in the cell occur in the mitochondria. Also in the mitochondria are found the enzymes that enable the cell to transfer the released energy used to form stable high energy compounds (high energy phosphates, such as adenosine triphosphate; see Chapter 16). Consequently, the mitochondria have been called the "powerhouses of the cell" (Siekevitz, 1957). Here is released most of the energy on the basis of which the work (osmotic, electrical, mechanical and so forth) of the cell is performed, a small amount (about 5 per cent) occurring outside the mitochondria. Therefore, their condition and activity are likely to influence practically all the functions of the cell. The involvement of mitochondria in such functions will be more effectively discussed when each of these functions is considered in turn in later chapters.

When some enzyme-catalyzed reactions occur in the test tube, many co-factors (e.g., salts, vitamin B components) must be added to obtain optimum conditions for the reactions, the required concentration of these co-factors being larger as contrasted to their action when the enzymes are inside the cell. This and the difficulty of removing some of the enzymes in an active form from the membranes make it appear that the enzymes are probably arranged in some specific steric spatial relationship which facilitates their activity inside the cell (Green, 1964). The lamellar structure of the mitochondria is ideally suited for this purpose, providing a large surface area to accommodate the enzymes and a three-dimensional matrix for juxtaposing enzymes involved in sequences of reactions (see Chapter 16). Indeed, under higher resolving power of the electron microscope the inner membrane of the mitochondrion appears to have tennis-racket-like bodies 70 Å in diameter on pedicels 35 Å long (see Fig. 16.14) which may represent electron transport bodies (Parsons, 1963).

It has already been pointed out that mitochondria make intimate contact with some portions of a cell involved in active work, being present, for example, among the fibrils in muscle cells and at internodes of nerve cells. They have also been described in contact with the ribosomes on the endoplasmic reticulum and it has been suggested that they might serve to supply energy for protein synthesis (Copeland and Dalton, 1959). If this were true, all dividing cells and secretory cells, which produce proteins, should show more frequent contacts than they actually do.

It has been postulated at various times that mitochondria may originate (1) by division, (2) anew from microbodies, or (3) from various membrane systems in the cell (plasma membrane, endoplasmic reticulum, Golgi complex and nuclear envelope as seen by attachments in electron micrographs). Some evidence is available to support each of these possibilities. Mitochondria appear to be partitioned between cells during cell division and in the aftermath of the mitotic division they multiply. Thus dumbbell-shaped mitochondria have been seen in fixed preparations, and Frederic (1958) by cinematography has observed the entire process of fission (as well as fusion in some cases). It was thought, however, that mitochondria may arise from invisible submicroscopic bodies under some circumstances, much as do chloroplasts (see Fig. 5.16). For example, mitochondria may be concentrated at one end of an *Arbacia* egg by centrifugation, and the centrifugal and centripetal halves of the egg may be pulled apart by further centrifugation, leaving one-half of the egg free of mitochondria visible with the light microscope. Mitochondria were later observed in the portion of the cells supposedly freed of them. But electron micrographs by Lansing in 1953 demonstrated that mitochondria were actually present in such eggs, accidentally caught in the oil sphere where they were unable to take up the dye Janus green used to identify them in earlier work (Lehninger, 1965).

Luck (1963) attacked the problem of the origin of the mitochondria by pulse labeling (10 minute exposure) a choline-requiring mutant of the mold *Neurospora* with $^{14}$C-choline. The choline was found incorporated into the mitochondrial phospholipid, lecithin. Thereafter the *Neurospora* was supplied with "cold" choline (nonradioactive). At various times after the "washout with nonradioactive choline" samples were removed and tested. It was found that the label, measured by autoradiography, persisted in each mitochondrion through three mass-doubling cycles of growth, the radioactivity being distributed among all the mitochondria observed. The experiments indicate that the mitochondria grow by addition of new lecithin to the already existing mitochondrial framework. This suggests that mitochondria arise by division. Had they arisen anew, there should exist newly formed, unlabeled mitochondria side by side with labeled mitochondria, the latter decreasing in proportion with each mass doubling, but with the grain count per mitochondrion remaining the same. (Fusion of mitochondria is infrequent in *Neurospora;* therefore, it will not explain the observed distribution of label.) A similar conclusion was reached from experiments with $^{14}$C-valine incorporated into proteins of ascites tumor cell mitochondria (Halder *et al.,* 1966).

On the basis of decay of label it appears that mitochondria are quite labile. Methionine labeled with sulfur-35 and acetate labeled with carbon-14 disappear at a rate to give a half-life of about 10 days for cytochrome C, soluble protein, insoluble protein and lipid in rat liver cell mitochondria (Fletcher and Sanadi, 1961). A half-life of a similar order of magnitude was found for mitochondria of cells from other organs of the rat (Wilson and Dove, 1965), as tested with tritiated water. In other words, all the protein and all the lipid in the mitochondria are replaced about every 20 days.

## MICROTUBULES AND MICROFILAMENTS

In 1963 several investigators called attention to microtubules and microfilaments in the cytoplasmic ground substance, which were resolvable in cells fixed with glutaraldehyde. The *microtubules* showed electron-opaque walls 60 to 70 Å thick and a lumen (region less electron-opaque) about 100 Å in diameter (some smaller tubules were also seen). The microtubules are often seen in the outer (cortical) cytoplasm and may be oriented with respect to the long axis of the cell. They may be quite prominent at specialized connections (desmosomes, see Chapter 11) between two cells, or connected to small "bodies" close to the centrioles. However, in many cases they are scattered throughout the cytoplasm of the cell (Slautterback, 1963; Ledbetter and Porter, 1963; Sandborn *et al.,* 1964). They are occasionally seen even in the nucleus (Behnke and Forer, 1966). Their function has not been determined but their presence in abundance at the periphery of the cells having peculiar shapes, e.g., the red blood cell, makes it likely that they serve as skeletal elements. However, their presence as "fibers" in the division spindle of cells suggests that they may play a contractile role as well. The fact that the "fibers" of a cilium (in their 9 + 2 arrangement) are such microtubules lends some strength to the argument. Another suggestion common in the literature but still without experimental support is that microtubules serve for intracellular transport of ions and molecules.

Microtubules can be liberated from cells. Thus, when hemolyzed red blood cells (cells treated with distilled water to remove hemoglobin) are carefully added to the surface of a water-filled trough (Langmuir trough) the materials from the peripheral cytoplasm of the red cell are spread in a thin film at the surface. The

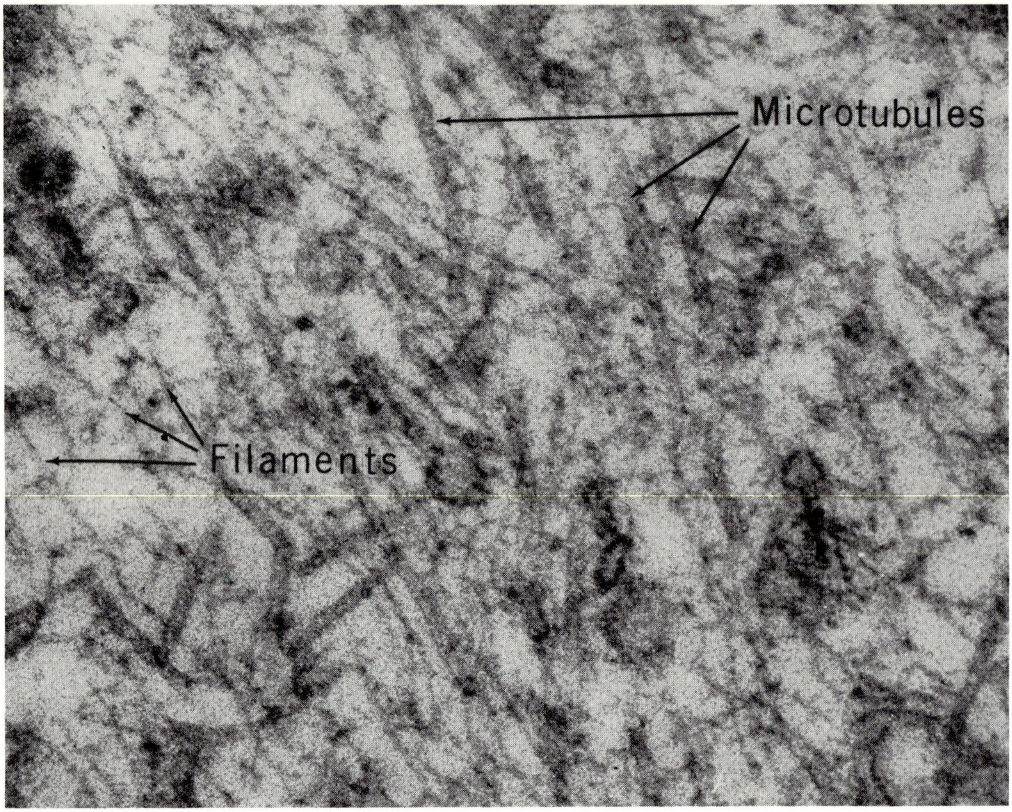


***Figure 5.12.*** Microtubules and microfilaments. Cells of rat kidney glomeruli, glutaraldehyde and osmium tetroxide fixation; lead citrate staining (×120,000). (From Fawcett, 1966: The cell: Its Organelles and Inclusions. W. B. Saunders Co., p. 229; by courtesy of Dr. Arnold Schechter).

film is swept and mounted on a grid, shadowed with heavy metal and studied by electron microscopy. The individual microtubules are readily seen (Wolfe, 1965; see Fig. 5.12).

*Microfilaments* are about 60 Å in diameter and their function, like that of microtubules, is uncertain. Some workers ascribe a contractile function to microfilaments because of their abundance in the cytoplasm of contractile plasmodia of slime molds (McManus and Roth, 1965).

## PLASTIDS; CHLOROPLASTS

Plastids are microscopic organelles found in plant cells such as leukoplasts in which starch granules develop, chloroplasts which contain the chlorophyll (green pigment) of plants, and chromoplasts which contain other pigments (e.g., the red carotenoid in the tomato). Best known are the chloroplasts, since they have been of interest in the extensive studies on photosynthesis.

Chloroplasts are of various shapes—spherical, discoidal or ovoidal and sometimes even disc shaped. The discoidal ones frequently have a colorless center containing a starch granule. Chloroplasts in cells of shade plants are larger than those of sun plants, but the commonest ones are 4 to 6 μ in diameter. The number of chloroplasts in a cell ranges from one—as in some algal cells—to a very large number in cells of higher plants. The single chloroplast in an algal cell may form a network, a spiral band, a stellar plate or a large massive structure occupying much of the cell.

Chloroplasts, isolated by centrifugation of mechanically disrupted spinach leaf cells, contain about 56 per cent protein, 32 per cent lipids and 8 per cent chlorophyll. Among the lipids are carotenoids and xanthophylls, as well as triglycerides, steroids, phospholipids and waxes. The chlorophyll probably exists in combination with some of the protein (Mercer *et al.,* 1955).

When chloroplasts are microsectioned they appear to be lamellar. The lamellae, being on the border of resolution under

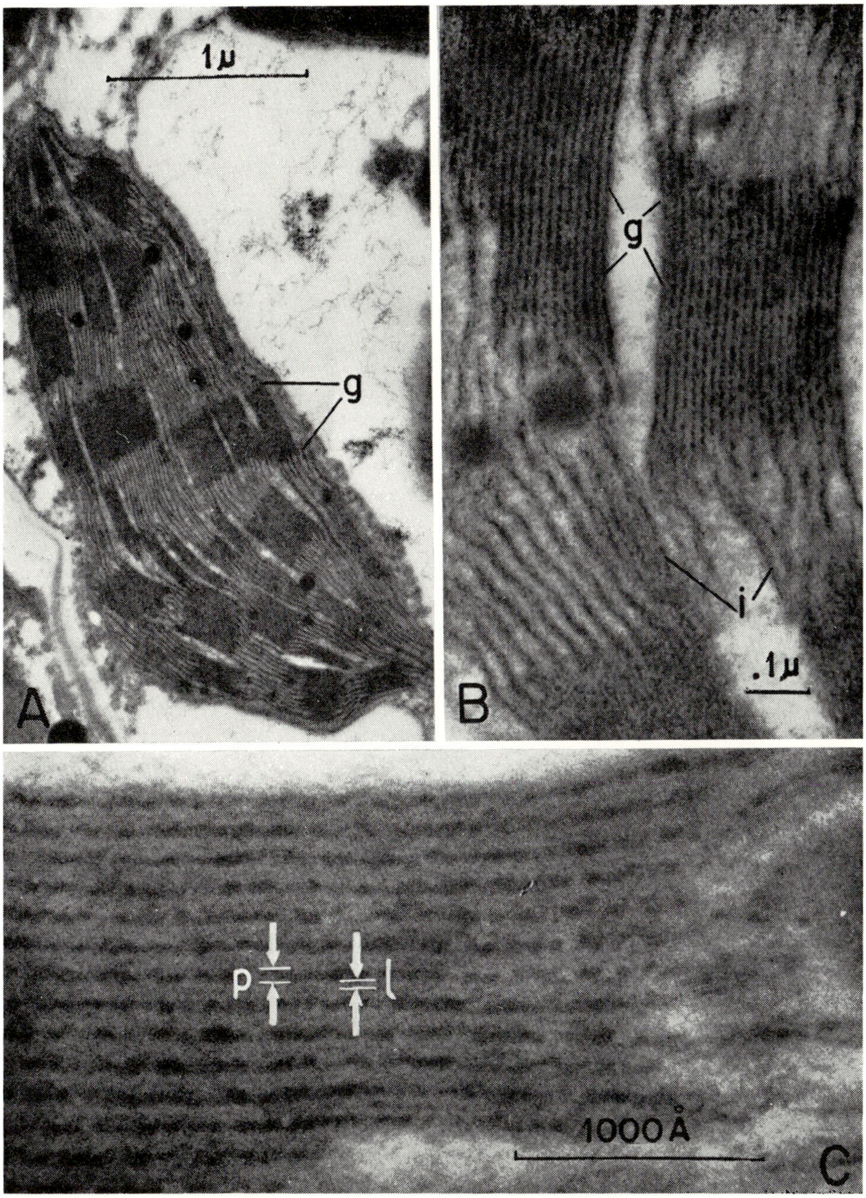


***Figure 5.13.*** *A,* Chloroplast of a mesophyll cell of maize showing numerous grana (g) (×30,000). *B,* higher magnification of a portion of a chloroplast showing two grana (g), and lamellae (i) between grana (×105,000). *C,* Higher magnification of a portion of one granum; P, protein lamellae; l, lipid (×370,000). (By courtesy of A. J. Hodge.)

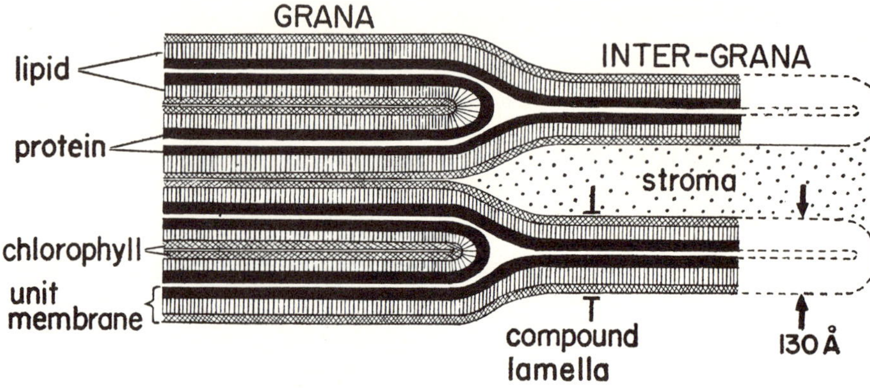


***Figure 5.14.*** Diagram of probable macromolecular structure of grana of chlorophyll. (By courtesy of A. J. Hodge.)

the ordinary microscope, can be seen quite plainly with the electron microscope (Fig. 5.13). In addition to the general lamellation, some chloroplasts show denser green bodies called *grana* (Fig. 5.13). In grana the lamellae are numerous and closely packed so that the grana appear to be distinct bodies suspended in the matrix of the chloroplast which has less dense lamellations. The grana are formed by the localized bifurcation of the lamellae containing the chlorophyll, thus doubling the chlorophyll content and increasing the optical opacity (Fig. 5.14). In a given species grana may be present in chloroplasts in some locations and absent from those in others; for example, in maize the mesophyll chloroplasts have grana but sheath parenchyma chloroplasts do not (Hodge *et al.,* 1955).

When chloroplasts, containing grana, are fragmented (by mechanical means or by detergents), the grana may be separated from the rest of the lamellae as individual bodies. Grana vary in size from 0.3 to 1.7 $\mu$ in various species; in spinach chloroplasts they are about 0.6 $\mu$ in diameter.

The structure of lamellae and grana has been determined from studies with

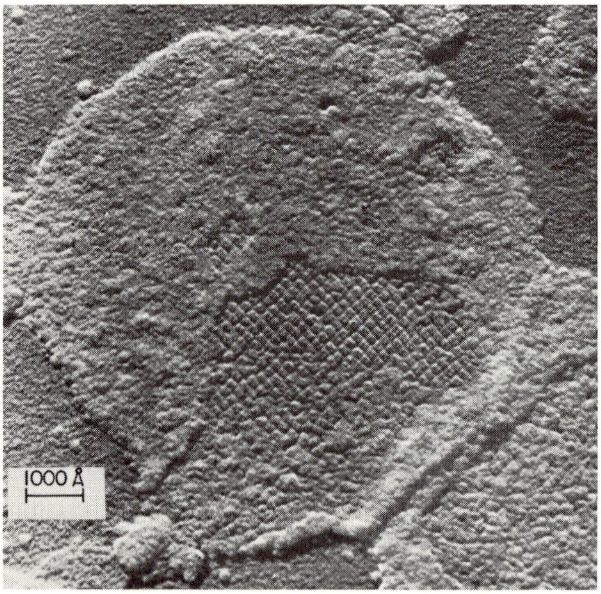


***Figure 5.15.*** A single lamella of a granum of a spinach leaf cell showing the regularly spaced particles called quantasomes. Osmium tetroxide fixed, heavy metal shadowed presparation. Each quantasome is about 155 Å long, 180 Å wide and 100 Å thick; a quantasome has a molecular weight of about $2 \times 10^6$ and contains 230 chlorophyll molecules. (From Park and Biggins, 1964: Science *144*:1010, by courtesy of Dr. R. Park.)

polarizing and electron microscopes. The latter show that the various layers making up a single membrane of a granum differ in electron opacity. Chemical analysis of isolated grana indicates that, per unit weight, they contain more lipids of various types and more chlorophyll than the other portions of the chloroplast. Using inferences from all available data, the structure of lamellae and grana has been reconstructed as shown in Figure 5.14.

Still smaller bodies are present inside the grana (see Fig. 5.15). These are called the *quantasomes* (Park, 1966) and are believed to be the units of photosynthesis, each consisting of 250 molecules of chlorophyll, the minimum amount found necessary for photosynthesis according to flashing light experiments (see Chapter 19).

The importance of this highly organized chlorophyll structure to photosynthesis is discussed in Chapter 19. It need

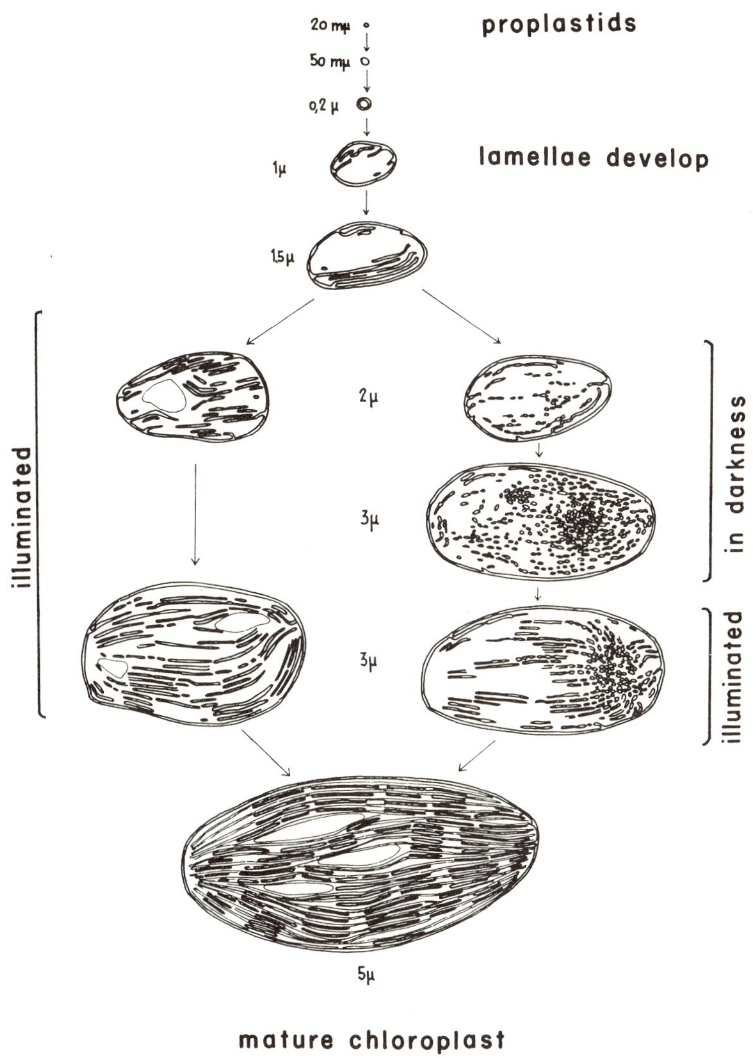


***Figure 5.16.*** The development of a chloroplast from a submicroscopic proplastid. When the illuminated proplastid is 1 μ in diameter, its inner surface membrane begins to invaginate to form lamellae. The proplastids grow and, in light, progressively develop lamellae to form the mature chloroplast (left). In darkness, however, the lamellae break up into vesicles (right), but no reillumination the vesicles unite to reform the lamellae. (From Mühlethaler and Frey-Wyssling, 1959: J. Biophys. Biochem. Cytol. *6*:509.)

only be mentioned here that when a culture of *Euglena* is kept in the dark the lamellae of the chloroplast and the chlorophyll disappear. Re-placing the cells in the light is followed by the simultaneous recovery of chlorophyll content and lamellation. The chloroplasts remain sensitive to light and their structure continues to be controlled by exposure to light (Packer and Siegenthaler, 1966).

Chloroplasts swell in hypotonic solutions and shrink in hypertonic solutions, indicating that the properties of the limiting membrane of the chloroplast appear to be similar to those of the external cell membrane (Mercer *et al.,* 1955).

Chloroplasts originate from minute submicroscopic ameboid proplastids. When the proplastid reaches a diameter of 1 $\mu$, its inner membrane invaginates to form lamellae which, in light, continue developing and differentiating to the mature condition found in the chloroplast (Fig. 5.16). Because the lamellae develop in the same way as they do in the bodies which give rise to mitochondria, some workers believe the two structures (which are also the major loci of cellular energy conversions) to be closely related (Mühlethaler and Frey-Wyssling, 1959).

Chloroplasts contain DNA and appear to be capable of self-duplication. The chloroplast DNA is distinct in its properties from that of the nuclear DNA, e.g., in bouyant density (Ray and Hanawalt, 1965). The *Euglena* chloroplast DNA appears to be more sensitive to ultraviolet radiation than nuclear DNA in this organism (Lyman *et al.,* 1961).

## VACUOLES

A vacuolar system of very small dimensions is found in the endoplasmic reticulum of all cells and in the Golgi complex of the cells that possess this organelle. However, both plant and animal cells may have much larger vacuoles. For example, many plant cells have large central sap vacuoles, which contain various salts and organic molecules and which maintain the turgor or rigidity of the plant cell upon uptake of water. In some animal cells (e.g., in protozoans) various types of vacuoles may occur. Sometimes such cells form minute vacuoles following degenerative changes or by "cell drinking"—the engulfment of some of the surrounding fluid forming *pinosomes* (see pinocytosis, Chapter 14). Food vacuoles (*phagosomes*) are formed by evagination of the cell membrane surrounding food particles and enclosing some of the fluid or by the invaginated membrane of the mouth. Contractile vacuoles may be formed in either of two ways—by coalescence of many spherical vacuoles or by filling with fluid from channels in the cytoplasm (Kitching, 1956). Contractile vacuoles are of special interest because of their dynamic nature and cyclic appearance and disappearance. Minute vacuoles appear as a result of cell drinking in some cells of vertebrates as well.

Little is known of the structure of the membranes surrounding the kinds of vacuoles just described. In general they appear to be similar to the plasma membrane (see Chapter 11), but their permeability properties may be different.

## CILIA AND FLAGELLA

*Cilia* (and flagella) are not universal cell organelles but are found in many animal and plant cells. Cilia are short motile organelles, generally present in large numbers, covering the surface of some cells (e.g., the ciliates). *Flagella* (characteristic of the flagellates, for example) are similar but larger and less numerous, as well as more complex in their movements. The physiology of ciliary and flagellar motion is dealt with in Chapter 24.

Little structure is visible in cilia and flagella under a light microscope or under a polarizing microscope, unless the cilium is frayed by chemical treatment. However, the detailed structure of cilia and flagella is clearly resolved by the electron microscope. Nine sets of paired peripheral fibers appear to surround two larger centrally placed fibers, as seen in cross section in Figure 5.17. This pattern, with slight

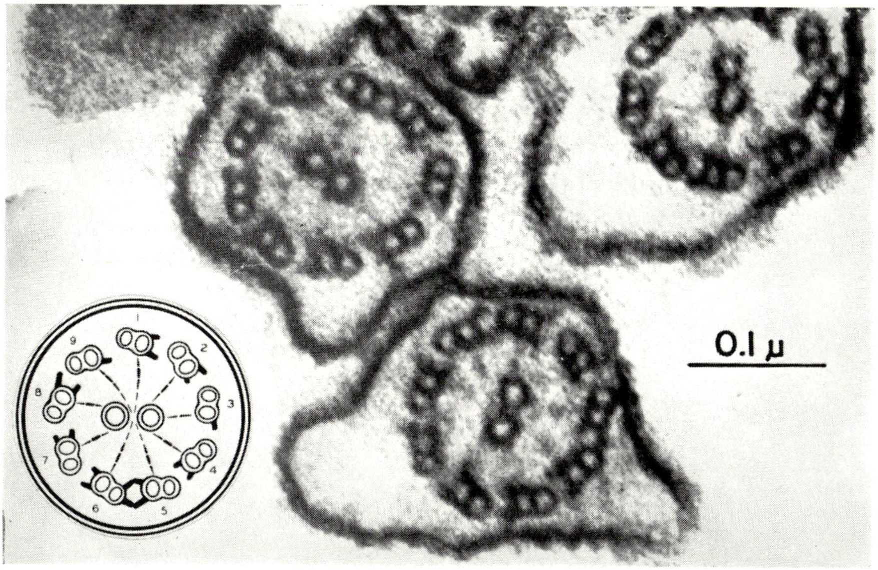


***Figure 5.17.*** Electron micrograph of a cross section through the tails of three spermatozoans with an arrangement of the filaments similar to that indicated in the inset. The inset shows the radial projections and the arms that each pair of spermatozoan filaments possesses. (By courtesy of B. Afzelius.)

variations, recurs in the tails of various spermatozoa, in cilia from ciliated epithelia, in ciliate and flagellate protozoans and in the cilia of plant cells (e.g., zoospores of algae). Occasionally a deviation from the 9 + 2 pattern is found for individual cells (Afzelius, 1963).

Spermatozoan tails have an additional element, a helical fibril around the periphery of the cilium. Finlike structures have also been found projecting in the direction of the central fibers in spermatozoan tails of some species (Fawcett and Porter, 1954).

The surface of a cilium (or flagellum) is an extension of the cell membrane. Studies on the development of a cilium indicate that it takes origin below the surface of the cell in a *basal granule* connected to the cilium by the ciliary root. Electron micrographs reveal a basic structural resemblance between cilium, ciliary root and basal granule, the cilium being essentially an extension of the basal granule (Fawcett, 1966). The importance of the basal granules to the ciliature has been the subject of extensive studies in the protozoa (Lwoff, 1950).

The flagella of bacteria are somewhat different in plan of organization, consisting of single fibers. These fibers are made up of long chains of molecules of a fibrous protein, flagellin, belonging to the keratin-myosin-elastin-fibrinogen group. In some cases three such fibers may be intertwined together as they are in the body of a spirochete (Stanier and Gunsalus, 1960).

## THE CENTROSOME AND THE CENTRIOLE

The *centrosome* consists of two bodies, the *centrioles* and the surrounding material. Each centriole is a structure about 0.2 μ in diameter. It is present in all cells but not always visible, except possibly during cell division. At that time the centriole divides, forming the two poles for the division spindle. When asters are present as part of the division figure they radiate from the centriole.

Electron micrographs show that the centriole is a short cylinder which has a structural plan essentially like that of the basal granule and the cilium (Fig. 5.18). Centrioles are generally paired, the two cylinders always being oriented at right

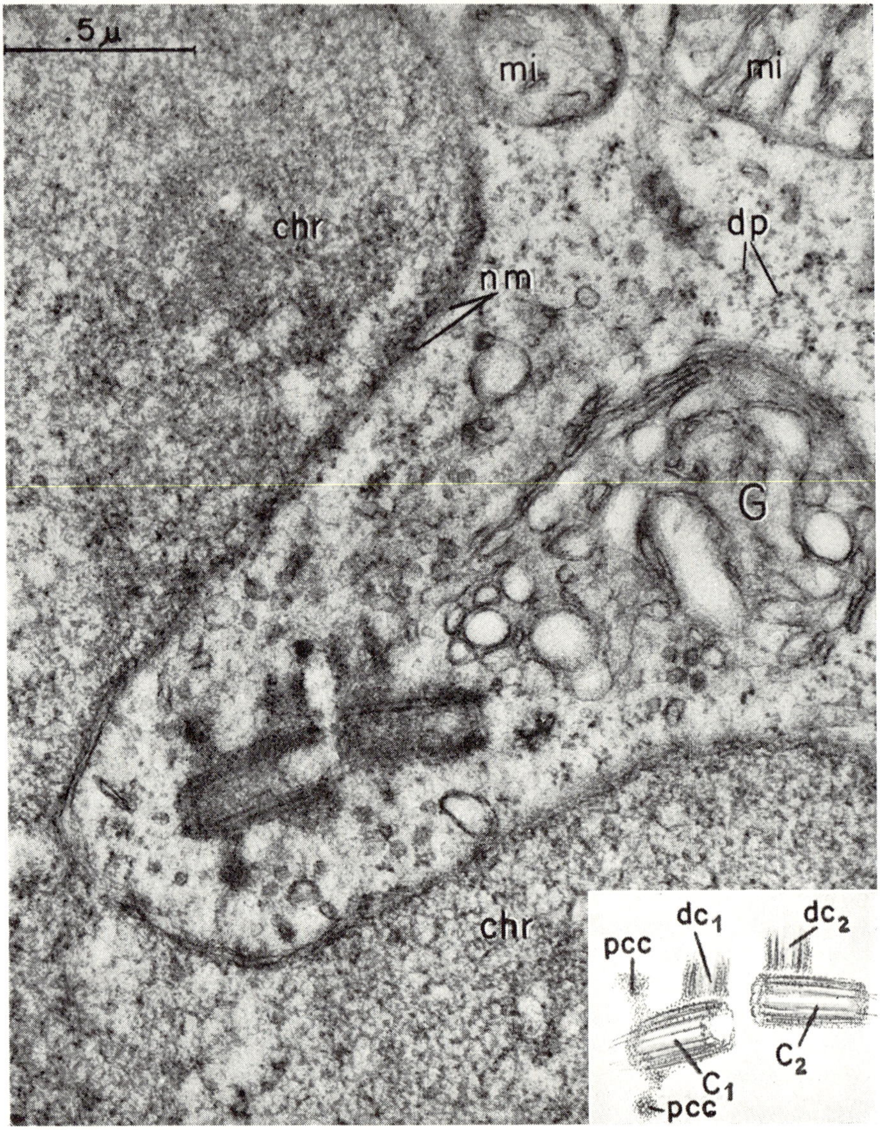


*Figure 5.18.* Electronmicrograph of two centrioles of an embryonic cell in chicken spleen. The two centrioles ($C_1$) and ($C_2$) are in the invaginated portion of the nuclear membrane near the Golgi complex (G); chr, chromatin with filamentous structure; pcc, pericentriolar bodies; $dc_1$ and $dc_2$ are daughter centrioles of $C_1$ and $C_2$, respectively; mi, mitochondria; dp, dense particles. (×60,000.) (By courtesy of E. DeHarven and W. Bernhard.)

angles to one another. Each centriole shows nine peripheral paired filaments in a circle, as in cilia except that the nine units consist of three fibers instead of two.

In spermatozoa the tail filaments are directly in contact with the tubule of one centriole, the other centriole remaining separate. Because of this it is suggested that the basal granules of cilia take origin in the centriole (Fawcett, 1966).

## RETINAL ROD CELLS

The receptor cells of dim vision in the eye are called *rods.* These cells consist of three parts, an outer segment containing the visual pigment, a thin middle connecting segment and an inner segment containing the nucleus of the cell, mitochondria and so forth (see Chapter 9). Studies on the embryological develop-

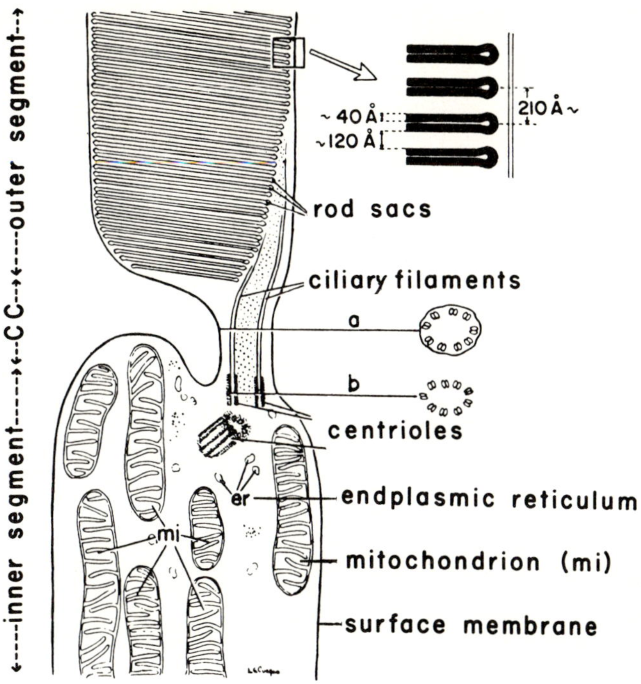


***Figure 5.19.*** Diagram of a retinal rod cell in the rabbit. The upper portion with the rod sacs is the outer segment, connected to the inner segment (lower portion) by a connecting cilium (CC), in which can be seen the ciliary filaments, lengthwise (vertical) as well as in cross section at (a). The two centrioles are shown in typical position, approximately at right angles to one another and in cross section at (b), in which the origin of the ciliary filaments is observed. The inner segment has the organelles of a typical cell—endoplasmic reticulum, mitochondria, and so forth (nucleus not shown). (From DeRobertis, Nowinski and Saez, 1965: Cell Biology. 4th Ed. W. B. Saunders Co., Philadelphia.)

ment and differentiation of a rod from an embryonic cell indicate similarities between the outer segment of the rod and a cilium. Electron micrographs of the middle segment of the rod reveal the unmistakable pattern of a cilium—nine paired filaments forming a circle, as shown in Figure 5.19. During histogenesis a cilium-like bulge appears on the embryonic cell. This projection enlarges and becomes filled with vesicles. These vesicles, initially displaced to one side, gradually flatten, enlarge and reorient, coming to occupy the major part of the terminal segment. These form the discs found in the external segment rod of the mature retinal cell. Only the narrow connecting segment then retains the structural pattern of a cilium.

The photoreceptor appears to be built on the cilium model in the Coelenterata, Echinodermata, Protochordata and Vertebrata but a proliferation of cell membranes without ciliary affinities is found in the photoreceptor in the Annelida, Arthropoda and Mollusca (Eakin, 1963). Since these two groups of animals differ in fundamental embryology and in other ways, a correlated difference in origin of the eyes in the two groups is of considerable evolutionary interest.

The photoreceptors are not the only receptors originating from a cilium. The hair cells in the auditory organ of vertebrates and the hair cells in the gravity receptors are other examples.

## SUMMARY

Knowledge of the organization and activity of the cell organelles is ever widening. It is to be noted, of course, that structural analysis of cell organelles is limited by the resolving power of the various microscopes now available, and the study of their functions is beset with technical difficulties. Clearly, further study depends upon the development of new and ingenious methods for coordinated approaches to both structure and functions of the cell organelles.

## *LITERATURE CITED AND GENERAL REFERENCES*

Afzelius, B. A., 1963: Cilia and flagella that do not conform to the 9 + 2 pattern. 1. Aberrant members within normal populations. J. Ultrastruc. Res. *9*:381–392.

Baudhuin, P., Miller, M., Poole, B., and DeDuve, C., 1965: Non-mitochondrial oxidizing particles (microbodies) in rat liver and kidney and in *Tetrahymena pyriformis*. Biochem. Biophys. Res. Commun. *20*:53–59.

Bayley, S. T., 1964. Physical studies on ribosomes from pea seedlings. J. Molec. Biol. *8*:231–238.

Behnke, O., and Forer, A., 1966: Intranuclear microtubules. Science *153*:1536–1537.

Bouck, G. B., 1963: Stratification and subsequent behavior of plant cell organelles. J. Cell Biol. *18*: 441–457.

Brachet, J., and Mirsky, A. E., 1959–1961: The Cell. 5 vol. Academic Press, New York.

Brandes, D., 1965: Observation on the apparent mode of formation of "pure" lysosomes. J. Ultrastruc. Res. *12*:63–80.

Branton, D., 1964: Fine structure in freeze-etched *Allium cepa*. J. Ultrastruc. Res. *11*:401–411.

Brieger, E. M., 1963: Structure and Ultrastructure of Microorganisms. Academic Press, New York.

Brierley, G. P., Murer, E., Bachmann, E., Green, D. E., and Hadley, D. G., 1963: Ion transport. II. The accumulation of inorganic phosphate and magnesium ions by heart mitochondria. J. Biol. Chem. *238*:3482–3489.

Campbell, P. N., 1966: The Structure and Function of Animal Cell Components. Pergamon Press, Oxford.

Cohn, Z. A., and Benson, B., 1965: The *in vitro* differentiation of mononuclear phagocytes. II. The influence of serum on granule formation, hydrolase production, and pinocytosis. J. Exp. Med. *121*:835–848.

Coimbra, A., and Leblond, C. P., 1966: Sites of glycogen synthesis in rat liver cells as shown in electron microscope radioautography after administration of glucose-$H^3$. J. Cell Biol. *30*:151–175.

Copeland, D. E., and Dalton, A. J., 1959: An association between mitochondria and the endoplasmic reticulum in cells of the pseudobranch gland of a teleost. J. Biophys. Biochem. Cytol. *5*:393–396.

Criddle, R. S., Bock, R. M., and Green, D. E., and Trisdale, H., 1962: Physical characteristics of proteins of the electron transfer system and interpretation of the structure of the mitochondrion. Biochemistry *1*:827–842.

Dallmer, G., Siekevitz, P., and Palade, G. E., 1966: Biogenesis of endoplasmic reticulum. J. Cell Biol. *30*:73–118.

Dalton, A. J., 1961: Golgi apparatus and secretion granules. *In* The Cell. Brachet and Mirsky, eds. Academic Press, New York, Vol. 2, pp. 603–619.

DeDuve, C., 1965: The separation and characterization of subcellular particle. Harvey Lect. *59*:49–87.

DeDuve, C., Berthet, J., and Beaufay, H., 1959: Gradient centrifugation of cell particles, theory and applications. Progr. Biophys. *9*:325–373.

DeDuve, C., and Wattiaux, R., 1966: Functions of lysosomes. Ann. Rev. Physiol. *28*:435–492.

*DeRobertis, E. D. P., Nowinski, W. N., and Saez, F. A., 1965: Cell Biology. 4th Ed. W. B. Saunders Co., Philadelphia.

DeRobertis, R., 1956: Morphogenesis of the retinal rods. J. Biophys. Biochem. Cytol. Suppl. *2*:209–218.

Eakin, R. M., 1963: Lines of evolution of photoreceptors. *In* The General Physiology of Cell Specialization. Mazia and Tyler, eds. McGraw-Hill Book Co., New York, pp. 393–425.

Engstrom, A., and Finean, J. B., 1958: Biological Ultrastructure. Academic Press, New York.

Fawcett, D. W., 1965: Surface specializations of absorbing cells. J. Histochem. Cytochem. *13*:75–91.

*Fawcett, D. W., 1966: The Cell: Its Organelles and Inclusions. W. B. Saunders Co., Philadelphia.

Fawcett, D. W., and Porter, K. R., 1954: A study of the fine structure of ciliated epithelia. J. Morph. *94*:221–281.

Fletcher, M. J., and Sanadi, D. R., 1961: Turnover of rat-liver mitochondria. Biochim. Biophys. Acta *51*:356–360.

Frederic, S., 1958: Recherches cytologiques sur le chondriome normal ou soumis à l'expérimentation dans des cellules vivantes cultivées *in vitro*. Arch. Biol. *69*:167–349.

Freeman, J. A., 1966: Goblet cell fine structure. Anat. Rec. *154*:121–147.

Frey-Wyssling, A., and Mühlethaler, K., 1965: Ultrastructural Plant Cytology. Elsevier Publishing Co., Amsterdam.

Goodwin, T. W., and Lindberg, O. (eds.), 1961: Biological Structure and Function. Academic Press, New York.

Granick, S., 1961: The chloroplasts; inheritance, structure and function. *In* The Cell. Brachet and Mirsky, eds. Academic Press, New York, Vol. 2, pp. 489–602.

Green, D. E., Jan. 1964: The mitochondrion. Sci. Am. *210*:63–74.

Green, D. E., and Hatef, Y., 1961: The mitochondrion and biochemical machines. Science *133*: 13–19.

Greenawalt, J. W., Rossi, C. S., and Lehninger, A. L., 1964: Effect of active accumulation of calcium and phosphate ions on the structure of rat liver mitochondria. J. Cell Biol. *23*:21–38.

Haguenau, F., 1958: The ergastoplasm: its history, ultrastructure and biochemistry. Intern. Rev. Cytol. *7*:425–483.

Halder, D., Freeman, K., and Work, T. S., 1966: Biogenesis of mitochondria. Nature *211*:9–12.

Hayashi, T. (ed.), 1959: Subcellular Particles. Ronald Press Co., New York.

Hodge, A. J., McLean, J. D., and Mercer, F. V., 1955: Ultrastructure of the lamellae and grana in the chloroplasts of *Zea mays* L. J. Biophys. Biochem. Cytol. *1*:605–614.

International Review of Cytology.

Iwamura, T. 1966: Nucleic acid in chloroplasts and metabolic DNA. Progr. Nucleic Acid Res. *5*:133–155.

Journal of Cell Biology (formerly Journal of Biophysical and Biochemical Cytology).

Journal of Cell Science. Vol. 1, No. 2, 1966.

Journal of Ultrastructure Research.

Kitching, J. A., 1956: Contractile Vacuoles. Protoplasmatologia D.3.a. Springer-Verlag, Vienna.

Kuff, E. L., and Dalton, A. L., 1959: Biochemical

studies of isolated Golgi membranes. *In* Subcellular Particles. Hayashi, ed. Ronald Press Co., New York, pp. 114–127.

Ledbetter, M. C., and Porter, K. R., 1963: A "microtubule" in plant cell fine structure. J. Cell Biol. *19*:239–250.

Lehninger, A. L., 1965: The Mitochondrion: Molecular Basis of Structure and Function. W. A. Benjamin, New York.

Levry, G. A., and Conchie, J., 1964: The subcellular localization of the "lysosomal" enzymes and its biological significance. Progr. Biophys. Molec. Biol. *14*:105–129.

Luck, D. J. L., 1963: Formation of mitochondria in *Neurospora crassa.* A quantitative autoradiographic study. J. Cell Biol. *16*:483–499.

Lwoff, A., 1950: Problems of Morphogenesis in the Ciliates. John Wiley & Sons, New York.

Lyman, H., Epstein, H. T., and Schiff, J. A., 1961: Studies of chloroplast development in *Euglena.* I. Inactivation of green colony formation by UV light. Biochim. Biophys. Acta *50*:301–309.

Mazia, D., and Tyler, A. (eds.), 1963: The General Physiology of Cell Specialization. McGraw-Hill Book Co., New York.

McManus, M. A., and Roth, L. E., 1965: Fibrillar differentiation in myxomycete plasmodia. J. Cell Biol. *25*:305–318.

Mercer, F. V., Hodge, N. J., Hope, A. N., and McLean, J. D., 1955: The structure and swelling properties of *Nitella* chloroplasts. Austral. J. Biol. Sci. *8*:1–18.

Moor, H., and Mühlethaler, K., 1963: Fine structure in frozen-etched yeast cells. J. Cell Biol. *17*:609–620.

Mühlethaler, K., and Frey-Wyssling, A., 1959: Entwicklung von Struktur der Protoplasten. J. Biophys. Biochem. Cytol. *6*:507–512.

Neutra, M., and Leblond, C. P., 1966: Radioautographic comparison of the uptake of galactose-$H^3$ and glucose-$H^3$ in the Golgi region of various cells secreting glycoproteins or mucopolysaccharides. J. Cell Biol. *30*:137–150.

Onishi, T., and Onishi, T., 1962: A contractile protein of mitochondria. J. Biochem. (Japan) *51*:380–381.

Ohnishi, T., and Ohnishi, J., 1962: Extraction of actin- and myosin-like proteins from liver mitochondria. J. Biochem. (Japan) *52*:230–231.

Orrenius, S., and Ericsson, J. L. E., 1966: Enzyme-membrane relationships in phenobarbital induction of drug-inhibiting enzyme system and proliferation of endoplasmic membranes. J. Cell Biol. *28*:181–198.

Packer, L., and Siegenthaler, P., 1966: Control of chloroplast structure by light. Intern. Rev. Cytol. *20*:97–124.

Palade, G. E., 1956: Intracisternal granules in the exocrine cells of the pancreas. J. Biophys. Biochem. Cytol. *2*:417–422.

Pappas, G. D., and Brandt, P. W., 1959: Mitochondria. I. Fine structure of the complex pattern in the mitochondria of *Pelomyxa carolinensis* Wilson (*Chaos chaos*). J. Biophys. Biochem. Cytol. *6*:85–90.

Park, R. B., 1966: Subunits of chloroplast structure and quantum conversion in photosynthesis. Intern. Rev. Cytol. *20*:67–95.

Parsons, D. F., 1963: Mitochondrial structure: two types of subunits on negative stained mitochondrial membranes. Science *140*:985–987.

Perry, R., 1966: The nucleolus and the synthesis of ribosomes. Nat. Cancer Inst. Monogr. *18*:325–340.

Petermann, M. L., 1964: The Physical and Chemical Properties of Ribosomes. American Elsevier Publishing Co., New York.

Pitelka, D. R., 1963: Electron-microscope Structure of Protozoa. Pergamon Press, London.

Pollister, A. W., and Pollister, P. F., 1957: The structure of the Golgi apparatus. Intern. Rev. Cytol. *6*:85–106.

Porter, K. R., 1961: The ground substance; observations from electron microscopy. *In* The Cell. Brachet and Mirsky eds. Academic Press, New York, Vol. 2, pp. 621–675.

Porter, K. R., and Bonneville, M. A., 1963: An Introduction to the Fine Structure of Cells and Tissues. Lea & Febiger, Philadelphia.

Porter, K. R., Claude, A., and Fullman, E. F., 1945: A study of tissue culture cells by electron microscopy. J. Exp. Med. *81*:233–244.

Ray, D. S. and Hanawalt, P. C., 1965: Satellite DNA components in *Euglena gracilis* cells lacking chloroplasts. J. Molec. Biol. *11*:760–768.

Rebhun, L. T., 1965: Freeze substitution: fine structure as a function of water concentration in cells. Fed. Proc. *24*:S217–S231.

*Rhodin, J. A. C., 1963: An Atlas of Ultrastructure. W. B. Saunders Co., Philadelphia.

Rich, A., Dec. 1963: Polyribosomes. Sci. Am. *209*: 44–53.

Rich, A., Penman, S., Becker, Y., Darnell, J., and Hall, C., 1963: Drug-induced changes in liver endoplasmic reticulum: association with drug-metabolizing enzymes. Science *142*:1657–1663.

Robertson, J. D., 1959: The ultrastructure of cell membranes and their derivatives. Symp. Biochem. Soc. London. *16*:3–43.

Roodyn, D. B. (ed.), 1967. Enzyme Cytology. Academic Press, New York.

Rottenberg, H., and Solomon, A. K., 1965: Energy linked K uptake in mitochondria. Biochem. Biophys. Res. Commun. *20*:85–92.

Rouiller, C., 1960: Physiological and pathological changes in mitochondrial morphology. Intern. Rev. Cytol. *9*:227–292.

Sandborn, E., Koen, P. F., McNabb, J. D., and Moore, G., 1964: Cytoplasmic microtubules in mammalian cells. J. Ultrastruc. Res. *11*:123–138.

Scanga, F., 1964: Atlas of Electron Microscopy and Biological Applications. American Elsevier Publishing Co., New York.

Schneider, W. C., and Kuff, E. L., 1964: Centrifugal isolation of subcellular components. *In* Cytology and Cell Physiology. Bourne, ed. Academic Press, New York, pp. 19–89.

Siekevitz, P., July 1957: Powerhouse of the cell. Sci. Am. *197*:131–140.

Siekevitz, P., 1963: Protoplasm: endoplasmic reticulum and microsomes and their properties. Ann. Rev. Physiol. *25*:15–40.

Siekevitz, P., 1965: Origin and tractorial nature of microsomes. Fed. Proc. *24*:1153–1155.

Siekevitz, P., and Palade, G. E., 1959: A cytochemical study of the pancreas of the guinea pig. IV. Chemical and metabolic investigation of RNP particles. J. Biophys. Biochem. Cytol. *5*:1–10.

Sjöstrand, F. S., 1964: The endoplasmic reticulum. *In* Cytology and Cell Physiology. Bourne, ed. Academic Press, New York, pp. 311–375.

Slater, E. C., Kaniuga, Z., and Wojtczak, L., 1967: Biochemistry of Mitochondria. Academic Press, New York.

Slautterback, D. B., 1963: Cytoplasmic microtubules. I. Hydra. J. Cell. Biol. *18*:367–388.

Stanier, R. Y., and Gunsalus, I. C. (eds.), 1960–1964: The Bacteria; a Treatise on Structure and Function. 5 vol. Academic Press, New York.

Tedeschi, H., 1959: The structure of the mitochondrial membrane: inference from permeability properties. J. Biophys. Biochem. Cytol. *6*:241–252.

Vincent, W. S., and Miller, O. L. Jr. (eds.), 1966: International Symposium on the Nucleolus, Its Structure and Function. National Cancer Institute Monograph No. 23, U. S. Government Printing Office, Washington, D. C., pp. 1–610.

Weibull, C., 1960: Movement. *In* The Bacteria. Gunsalus and Stanier, eds. Vol. 1, Structure, pp. 153–205.

Weissman, G., 1964: Labilization and stabilization of lysosomes. Fed. Proc. *23*:1038–1044.

Wettstein, F., and Noll, H., 1965. Binding of transfer RNA to ribosomes engaged in protein synthesis: number and properties of ribosomal binding sites. J. Molec. Biol. *11*:35–53.

Wilson, J. E., and Dove, J. L., 1965: Turnover of mitochondria in rat liver, kidney and heart. J. Elisha Mitchell Sci. Soc. *81*:21–24.

Wolfe, S. L., 1965: Isolated microtubules. J. Cell Biol. *25*:408–413.

Woodward, D., 1966: Personal communication.

Zamecnik, P. C., March 1958: The microsome. Sci. Am. *198*:118–124.

# CHAPTER 6 / THE CHEMICAL ORGANIZATION AND FUNCTION OF THE CELL NUCLEUS

Ever since its discovery by Robert Brown in 1835, the nucleus has been recognized as the most important organelle of the cell. Today it is understood to be the major seat of heredity and of the control over all synthetic reactions in the cell. Cells having a well defined nucleus are called *eukaryotic* cells. Bacteria and blue-green algae possess no true nucleus although their nuclear material may be aggregated and distinguishable in electron micrographs during cell division. No nuclear envelope separates the nuclear material of such *prokaryotic* cells from the surrounding "cytoplasm" and the nuclear material may, in fact, consist of only a chromosome, as will be discussed later. Thus, when the "nucleus" is considered in this chapter, that of a eukaryotic cell is implied. Although it is impossible to cover here all studies on the nucleus, some of the most basic features of this research are considered in this chapter (Mitchison, 1966; Moses, 1964).

## RELATION BETWEEN NUCLEUS AND CYTOPLASM

There exists a well knit and delicately balanced relationship between the nucleus and the cytoplasm of a cell. This can be seen best if we study the cell after the nucleus is removed. A cell may be enucleated in any one of a number of ways—for example, by removing the nucleus from a cell with a hooked needle or a micropipette, or by centrifuging a cell until it breaks into two pieces, one with and one without a nucleus. An even simpler operation for this purpose, which may be done freehand with a glass needle, is to cut an *Amoeba proteus* into two portions, one with and one without the nucleus (see Fig. 6.1). The enucleate half continues moving for a while but soon rounds up and becomes inactive and unresponsive to its environment, although some chemical activity continues. Thus it will digest food previously ingested but will not take in any more, and it will continue to absorb and bind radioactive phosphorus when tested (Mazia and Hirshfield, 1950), but at a lesser and continually decreasing rate, until the cell dies. The nuclear portion, in contrast, reacts normally, eats, grows and in time divides (Danielli, 1959, 1963).

When the nucleus of an ameba is removed, the ameba behaves as the enucleate half in the experiment just described. However, its activities are restored to normal

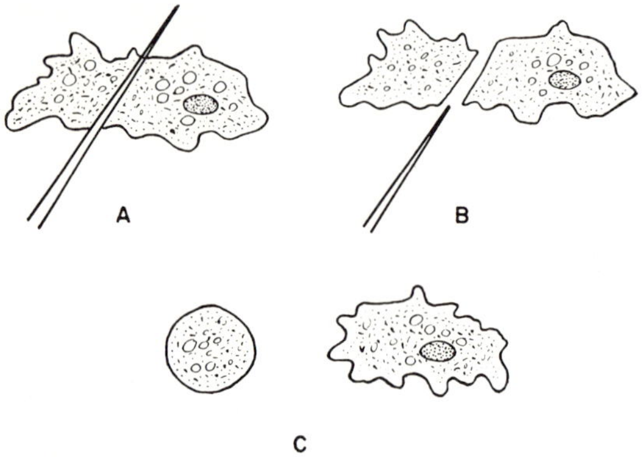


***Figure 6.1.*** Fate of nucleate and enucleate pieces of an ameba.

when it is renucleated. Since the nucleus deteriorates even if kept briefly outside the cytoplasm, renucleation is usually done by placing donor and recipient amebas of the same species close together and with the hooked needle quickly pulling the nucleus out of one and inserting it into the other. In this manner an ameba may be renucleated successfully even several days after its enucleation.

These experiments have shown that the cytoplasm of a cell cannot survive alone and that its activities are in some manner initiated and regulated by the nucleus.

Enucleation and renucleation of amphibian eggs have also been successfully accomplished in a similar manner. In fact, a nucleus taken from one of the cells of the blastula or gastrula stage gives rise to a normal embryo if it is inserted into an enucleated and unfertilized egg of the same species. It is interesting to note here that a nucleus from an older gastrula in some species is no longer effective, indicating that the nucleus has lost its totipotency between early and late gastrula stages (King and Briggs, 1956). However, in some species, as in the larva of the toad *Xenopus,* differentiation does not become irreversible until the cell from which the nucleus is taken is about to become an organ (Moore, 1960; Fischberg and Blackler, 1963).

If the nucleus of one species of ameba is introduced into an enucleated specimen of a different, but closely related species, the nucleus and cytoplasm are generally not quite compatible, and although the newly restored ameba may continue to move about, eat and divide, its progeny usually die. However, if the new strain continues to multiply for a while, its descendants .take on a character intermediate between the two strains from which they sprang, which indicates that the nucleus seems to influence cytoplasmic characteristics (Danielli, 1959).

Similar nuclear transfers have been made between species of *Acetabularia,* a sizable single celled alga which has a foot, a stalk and a cap (Brachet, 1965; Gibor, 1966) (see Fig. 6.2). The cap has a characteristic shape for each species, and if removed it is easily regenerated. When a nucleus from an individual of one species of *Acetabularia* is transferred to an enucleate individual of a different species,

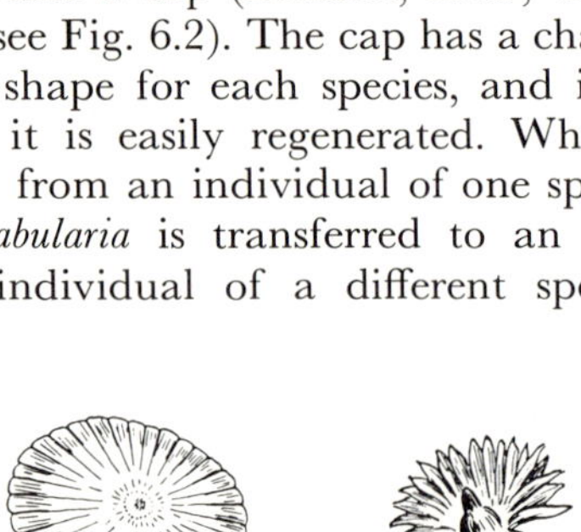


***Figure 6.2.*** Two species of *Acetabularia* used in experiments. The caps are about a centimeter in diameter, and the stalks are usually 2 to 6 cm. in length. (After Hammerling, 1953: Internat. Rev. Cyt. *2*:475.)

from which the cap has also been removed, the cap when regenerated is intermediate in form between the two species. If more than one nucleus is introduced from the donor species, the cap which regenerates has more resemblance to that of the donor than to that of the recipient species. The nucleus therefore exerts quantitative morphogenetic influence (Hammerling, 1963).

The nucleus of *Acetabularia* does not divide until just before gamete formation, at the end of its life cycle (Fig. 6.3). Division of a nucleus from a "young" cell can be induced, however, much before its time, if the nucleus of such a cell is introduced into an enucleated "old" cell, of the same species, which is about to form gametes. Apparently, in the chain of reactions the cytoplasm gradually develops substances which induce nuclear division. A two-way influence seems to exist—nucleus upon cytoplasm and cytoplasm upon nucleus (Hammerling, 1963).

An unfertilized sea urchin egg, enucleated by being cut or centrifuged, may be stimulated to divide, without fertilization, by brief immersion in hypertonic solutions. In the total absence of a nucleus it undergoes divisions to form a multicellular embryonic form which later degenerates (see Fig. 6.4). Apparently, division of cytoplasm, for a limited time, can continue without the presence of a nucleus, but a nucleus is necessary for continued and normal functioning and differentiation of the cytoplasm of a cell (Harvey, 1936).

On the other hand, a nucleus without its cytoplasm is helpless. If it is removed from the cytoplasm of an ameba and kept in physiological balanced salt solution it dies within a few hours. Even the thin layer of cytoplasm covering the nucleus of a sperm enables the sperm to live and to take up and use nutrients from its medium. Sperm of choice stock animals are thus kept alive at low temperatures for long periods and used for artificial insemination. Nucleus and cytoplasm of

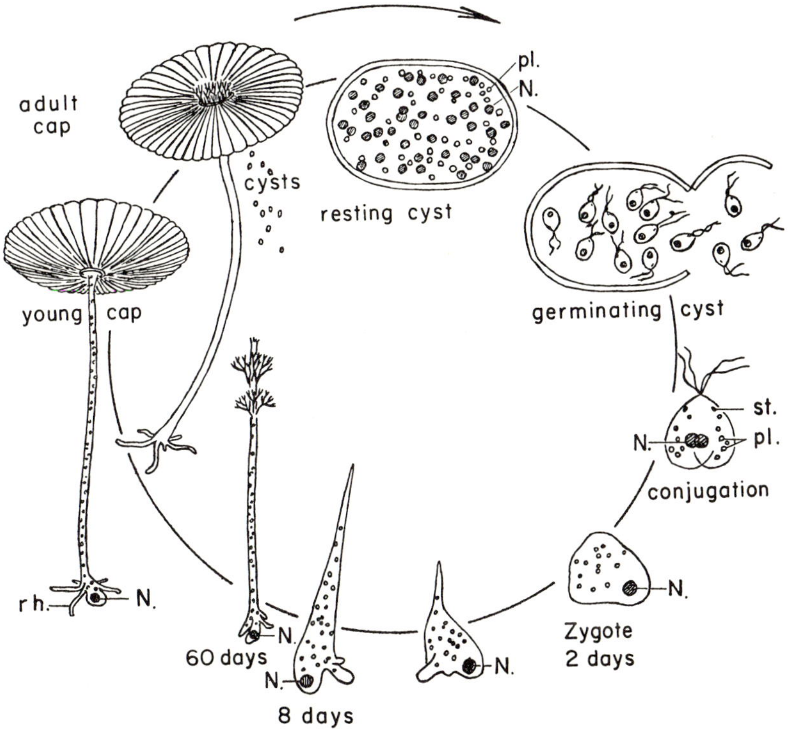


**Figure 6.3.** Life cycle of *Acetabularia mediterranea.* N refers to nucleus, rh to rhizoid, st to stigma on the isogametes and pl to the plastids. (From Brachet, 1957: Biochemical Cytology. Academic Press, New York.)

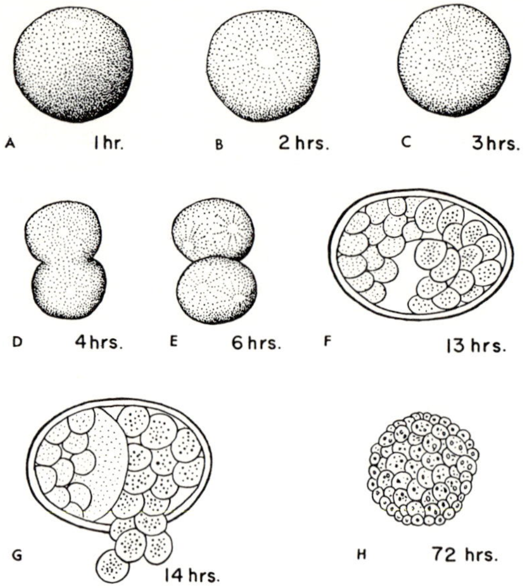


***Figure 6.4.*** Development of an enucleated *Arbacia* egg, artificially stimulated to divide. Note aster formation in figures *B* and *C,* division *D* and *E,* and the many cells that developed later. No further division occurred after the stage shown in figure *H,* although the cells vacuolated and lived for a month. (After Harvey, 1936: Biol. Bull. *71*:101.)

the living cell, therefore, constitute a reciprocal system in which each requires the other (see Goldstein, 1967).

The cytology of the nucleus of eukaryotic cells and the sequence of events which almost miraculously take place, step by step, in the nucleus during cell division, have been the subject of thorough study. The nuclear cycle includes the events by which haploid gametes form the diploid zygote, which divides by mitosis to form the cells of the adult and, during reproduction in the adult, the haploid gametes. This and the correlations between chromosomal behavior and the distribution of genes are outlined in most elementary textbooks of biology and in more detail in textbooks of cytology (e.g., DeRobertis *et al.,* 1965). Further consideration of these topics is therefore omitted here.

During division in bacteria and blue-green algae the chromosomes are presumably partitioned equally between the daughter cells because the genes of bacteria behave like those of eukaryotic cells, even to the extent of mutability and genetic recombination during sexual reproduction (Cairns, 1966).

## CHEMICAL COMPOSITION OF THE NUCLEUS—CYTOCHEMISTRY

Under the microscope, the resting nucleus of a eukaryotic cell usually shows a nucleolus and the chromosomal chromatin, sometimes in threads, suspended in the nuclear sap, the entire mass being surrounded by a nuclear envelope. The nuclear envelope, made up of two membranes, appears to be a portion of the endoplasmic reticulum applied to the nucleus, possibly with subsequent modifications. Each membrane of the nuclear envelope is therefore essentially a unit membrane (see Chapters 5 and 11). The two unit membranes of the nuclear envelope are separated by a space about 140 Å wide. The nuclear envelope contains pores large enough to let large molecules through. A nuclear envelope is seen in the electronmicrograph of Figure 6.5 and its permeability properties are discussed in Chapter 11.

The electron micrograph of an intermitotic nucleus shown in Figure 6.6 shows a rather irregular accumulation of granular material, probably representing DNA and protein. Some workers have reported minute fibrils, presumably in the chromosomes, but upon this there is no general agreement (DuPraw, 1965).

The chemistry of nuclei can be studied readily because, like other cell organelles, the nucleus can now be separated from other cell constituents by fractional centrifugation. The cells are first disrupted, by means of various methods such as treatment with distilled water (for red cells) or freezing and thawing, or applying lytic agents such as saponin. The fragmented cells are placed in solutions of sucrose and calcium chloride or a weak acid, such as citric acid, and differentially centrifuged and washed with the solution. Unfortunately, the procedures cause a loss of some of the material from the nucleus, but addition of appropriate salts and adjustment of concentration, salts and pH of the medium minimize such losses. Isolated

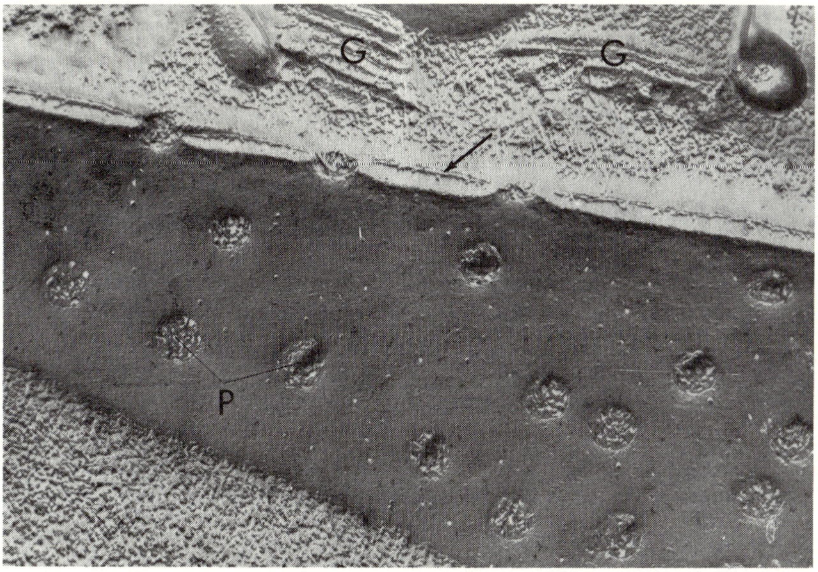


**Figure 6.5.** Surface view of the inner nuclear membrane of an onion root tip cell, freeze-etched. Note the numerous pores (P) through which the cytoplasm may be seen. In the lower left corner may be seen the nucleoplasm. The outer unit membrane of the nuclear envelope is indicated by the arrow above. Also seen in the upper part of the illustration are two cross-fractured Golgi complexes (G) (×84,000). (By courtesy of Dr. D. Branton.)

nuclei in buffered sucrose are found to retain some metabolic ability and can incorporate radioactive precursors into their nucleic acids and proteins (Siebert and Smellie, 1957; Brachet, 1960; Cohen, 1966).

Nuclei have also been isolated from frozen, dried tissues in nonaqueous solvents of graded density. Such nuclei retain all their water and acid-soluble constituents and nuclear proteins (Siebert and Smellie, 1957). With carefully designed procedures this is also possible with aqueous solvents, e.g., buffered sucrose (Allfrey and Mirsky, 1959). The two types of analysis may be used to complement each other.

Nuclear mass varies from one cell type to another and from one species to another. For example, the nucleus of a mammalian liver cell constitutes 10 to 18 per cent of the cell mass, whereas that of a thymus cell of the same individual may represent 60 per cent of the cell mass.

Certain chemical constituents are almost always present in nuclei of all cells, such as: DNA (deoxyribonucleic acid), RNA (ribonucleic acid), lipids, a basic protein (either histone or protamine), more complex proteins including enzymes, other phosphorus-containing organic compounds and various inorganic compounds, mostly salts (Davidson, 1960).

The chemistry of the nucleus is also studied by cytochemical techniques. The study of *cytochemistry* aims to determine not only the kinds of various materials present in the cell, but also their distribution in the differentiated structures of the cell and the amount present in each structure. To some extent differential centrifugation of the cell particulates and their chemical analysis permits localization of chemicals. However, cytochemical methods give additional information, especially detailed localization of the materials in a structure, providing the procedures do not destroy the materials or cause them to diffuse to new locations. The two methods —cytochemical and biochemical—are especially effective when used in conjunction with one another.

The location of some chemical constituents of the nucleus, e.g., the nucleic

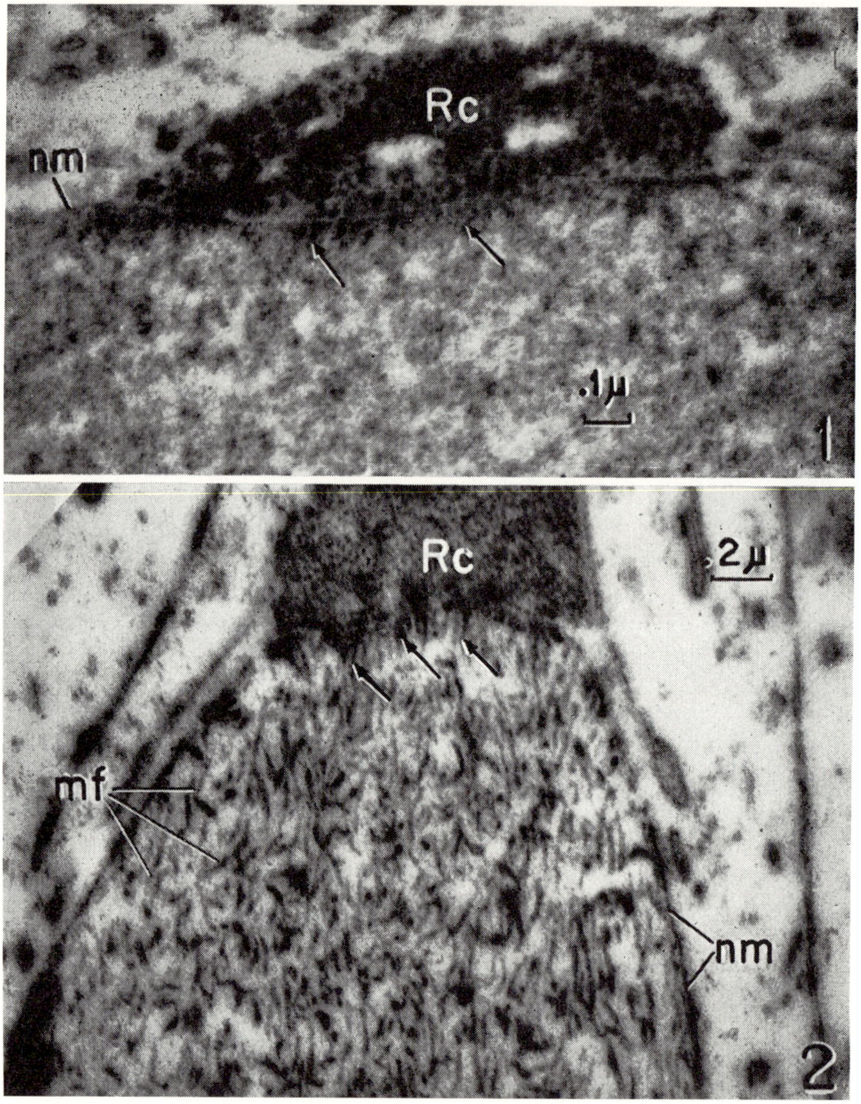


**Figure 6.6.** *1,* Electron micrograph of one region in the nucleus of a young locust spermatid. Region near the ring centriole, Rc; nuclear membrane, nm. The nucleus has a fine, fibrillar, dispersed appearance. The microfibrils, indicated by the arrows, are oriented in the vicinity of the centriole. (×60,000.) *2,* Later stage in the development of the spermatid. The microfibrils (mf) are now thicker and generally oriented in the axis of the spermatid head. (×40,000.) (From DeRobertis, 1955.)

acids, can also be determined by ultraviolet photomicrography (Caspersson, 1950). This method is based on the fact that no other protoplasmic constituents absorb ultraviolet light significantly at the absorption peak of nucleic acid at 260 nm. (see Chapter 3, Fig. 3.20); therefore, the density of photomicrographs at this wavelength measures the relative content of nucleic acid in the cell. The measurements of nucleic acid in cells may be rendered quantitative by comparison with the optical density of a known sample of nucleic acid in solution. The measuring equipment is illustrated in Figure 6.7. A graph showing the absorption spectra for nucleus and cytoplasm appears in Figure 6.8. Because the method involves inherent approximations and errors, the measurements cannot be expected to give precision

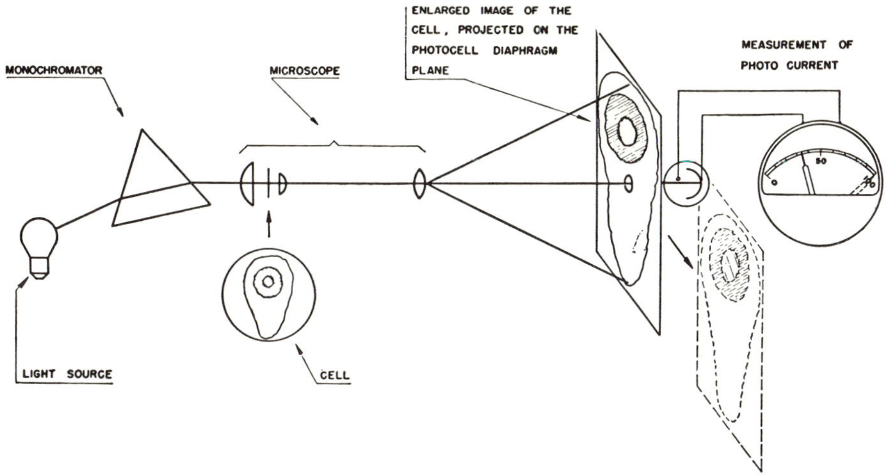


*Figure 6.7.* Principle of photoelectric microspectrophotometry. Microspectrophotometric equipment of Caspersson (diagrammatic). (From Caspersson, 1950: Cell Growth and Cell Function. W. W. Norton & Co., New York.)

as to absolute amounts of nucleic acids present.

Some staining reactions may also be used as chemical analyses. For example, by the Feulgen reaction, sections of material, fixed in the usual manner, are hydrolyzed by hydrochloric acid to liberate the aldehyde group of the pentose sugar of DNA and are then treated with leucofuchsin (basic fuchsin in sulfurous acid). The reaction between leucofuchsin and the aldehyde in deoxyribose results in the appearance of a purplish blue color. The depth of color can be measured under a microscope spectrophotometrically and used for quantitative determinations. DNA is Feulgen positive and RNA is Feulgen negative. Polymerized DNA may also be stained specifically with methyl green, which does not stain the RNA in cells (Kurnick and Mirsky, 1950; Allfrey *et al.,* 1955; Davidson, 1960). However, methyl green will not stain depolymerized DNA even though it is still present in its original location in the cell. Acridine orange differentially stains single stranded DNA (present in some viruses), double stranded DNA and RNA (Mayor and Diwan, 1961; Mayor, 1961).

Studies of nuclear sap, removed by microdissection from amphibian oocytes, showed it to contain proteins and salts in solution and no detectable nucleic acid (Davidson, 1960).

The nuclear enzymes have been studied but controversial findings have been reported, some workers claiming a large number of different types of enzymes to be present in their nuclear preparations, others claiming far fewer. These differences in findings probably result from differences in technique. If cytoplasmic granules are not carefully removed from the surface of the nucleus, enzymes of the granules will be added to those of the nucleus. On the other hand, very careful washing without proper precautions may remove nuclear enzymes normally present. There now seems to be general agreement that the nucleus lacks some of the enzymes for aerobic metabolism, such as are found in the mitochondria (see Chapter 5), but that it contains enzymes for anaerobic metabolism and those enzymes which are involved in the formation of high energy phosphates, transferring the energy liberated, for example, during anaerobic metabolism. It also contains enzymes for protein synthesis, as well as coenzyme synthesis (nicotinamide adenine dinucleotide, or NAD, see Chapter 16). The latter two syntheses require the presence of oxygen.

The nucleus contains some lipids such as might be expected in any cellular ma-

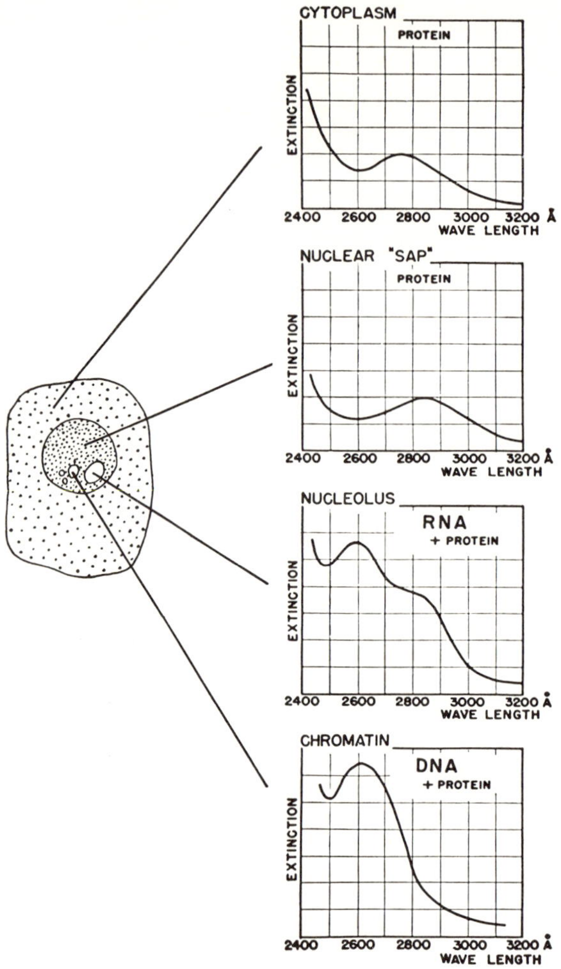


***Figure 6.8.*** Absorption spectra of constituents of nucleus and cytoplasm measured on a section of a cell. (Adapted from Caspersson, 1950: Cell Growth and Cell Function. W. W. Norton & Co., New York.)

terials and in a membrane. The lipids of the nucleus do not seem to have been the subject of any extensive and pointed study, but there does not seem to be any lipid unique to the nucleus.

The most characteristic constituents of the nucleus are the nucleic acids, DNA and RNA (Brachet, 1960). The amount of DNA in salmon sperm heads (being mostly nucleus) may constitute 48.5 per cent of the dry, fat-free material. Generally, DNA constitutes considerably less than that percentage of the dry weight. The absolute amount of this nucleic acid varies from species to species, but for a given species it is constant for a haploid set of chromosomes. When diploid, tetraploid and octaploid cells are present, as in the liver, the absolute amount of DNA per cell varies accordingly. DNA, as we have seen, is Feulgen positive, and its concentration can also be measured by the depth of color thus obtained. Although the total amount of DNA in a cell varies by as much as 15 per cent as determined by quantitative analysis, the larger variations reported by cytologists on the basis of staining reactions are probably erroneous. DNA is attached to residual proteins which vary extensively, especially with nutritional conditions, falling markedly on starvation. Since staining often depends upon the DNA-protein complex, an error in estimation by this means could easily be introduced (Allfrey *et al.,* 1955; Davidson, 1960).

In still another test for DNA, enzymes specific for the hydrolysis of DNA or RNA have been used. For example, treatment of fixed cells with deoxyribonuclease removes the DNA, and a Feulgen test on the residue is negative. Ribonuclease, on the other hand, removes the RNA and does not affect Feulgen-positive material. The two treatments thus complement each other (Davidson, 1960; Kurnick and Mirsky, 1950).

Ribonucleic acid may be selectively stained with pyronin. The amount of RNA may then be estimated from the depth of color by spectrophotometric methods in the same way as DNA. Treatment of a section of tissue with ribonuclease before staining removes the pyronin-reacting material (RNA) from a cell and the test is then negative (Kurnick and Mirsky, 1950).

Since DNA and RNA can be selectively stained by this method in a fixed cell or tissue, their distribution in the cell can be ascertained. DNA is apparently present chiefly in bands on the metaphase chromosome, most clearly seen in the multiple (polytene) chromosomes of the salivary gland cells of flies (Fig. 6.9). In interphase nuclei the DNA is visible primarily in the granules of the nucleolus, which gives a positive reaction with the Feulgen reagents.

RNA is present in large amounts in the nucleolus. It is also found around the chromosomes. However, it will be recalled (see Chapter 5) that the major part of the RNA of the cell occurs in the ribosomes, both those free and those attached to the endoplasmic reticulum. RNA is also present in a diffuse form (in solution) in the ground substance and a small amount is found in the mitochondria.

By cytochemical methods all the RNA molecules appear to be alike. Chemical methods, however, detect differences. The

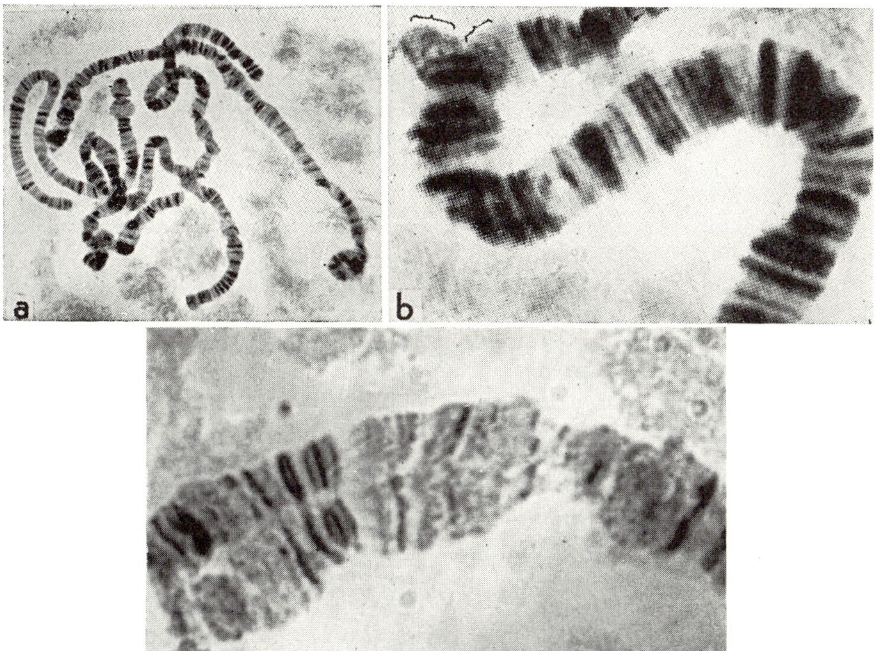


***Figure 6.9.*** Polytene chromosomes of *Drosophila melanogaster*. *a*, View of the chromosomes of one nucleus, showing the dark bands and the clear interbands. Preparation by crushing in acetocarmine. *b*, Part of the sex chromosome X, as it is observed with ultraviolet light; photograph taken at 2750 Å (after Schultz, 1941). *c*, Absence of pairing in chromosome III of a female hybrid of *Drosophila pananaensis* X. (From DeRobertis *et al.*, 1965: Cell Biology. 4th Ed. W. B. Saunders Co., Philadelphia; by courtesy of Dr. Rosina de Barros.)

primary difference that is stressed in most reports is a difference in base ratios and molecular weight. It will be recalled (Chapter 3) that RNA contains the nucleotides of the purines, adenine and guanine, and of the pyrimidines, cytosine and uracil (Table 3.8). In a given RNA the ratios of these bases is always constant. However, the RNAs from the ribosomes, the nucleolus, the chromosomes and the soluble RNAs separated from one another by fractional centrifugation all show different base ratios from one another, indicating a different chemical composition. Differences in molecular weight have already been mentioned in Chapter 3. Briefly, transfer RNA has a low molecular weight of about 25,000, ribosomal RNA is quite large, its molecular weight running into many millions and messenger RNA is variable but large, probably from 500,000 to several million. There are, presumably, other differences in properties less readily detectable at the present time, such as affinities for other molecules (Edstrom, 1960, 1964).

The chemistry of the chromosomes has also been extensively studied. Chromosome-like threads can be separated from nuclei fragmented in a Waring Blendor and their chemical composition can be determined. By this means "isolated chromosomes" from lymphocytes, from fish and fowl erythrocytes and from thymus cells and liver cells have been studied. Lymphocyte chromosomes in neutral, molar sodium chloride disperse to form a viscous suspension which, after centrifugation at 18,000 to 19,000 rpm for 1 to 2 hours, separates into a highly viscous, somewhat opalescent supernatant and a tightly packed sediment. The fluid consists of practically all of the DNA and of nucleohistone which constitutes 90 to 92 per cent of the mass of the chromosome. Pouring the fluid into water yields a fibrous precipitate containing 45 per cent DNA and 55 per cent histone (Davidson, 1960).

The sediment, which does not dissolve in sodium chloride, consists of a mass of coiled threads resembling chromosomes but somewhat smaller. They have been called residual chromosomes and account for about 8 to 10 per cent of the mass of the original chromosomes and show longitudinal differentiations and variations in diameter corresponding to similar regions on the original chromosomes. The nucleic acid content of the residual chromosomes is about 4 per cent of the total originally in the chromosome and is chiefly RNA, there being only a small amount of DNA left. Removal of the nucleic acid with hot trichloracetic acid leaves a residual protein, of complex nature, containing aromatic amino acids and different from histone and protamine (Allfrey and Mirsky, 1959).

The chromosome thus appears to be made up of residual protein and DNA, to which is attached the histone (in all somatic cells) or the protamine (in sperm). The histone or protamine can be removed from isolated metaphase chromosomes without altering the morphology of the latter. Only when the DNA is removed from the residual protein, or the protein removed from the DNA, is the chromosome structure destroyed. Neither the combination nor the configuration can be restored by mixing the two once they have been separated. Residual protein, therefore, shares in some way with DNA the architectural framework of the chromosomes (Allfrey *et al.,* 1955).

The histone present in association with DNA in all eukaryotic cells, but absent from prokaryotic cells, may possibly play a part in regulating either nucleic acid activity or synthesis (Irvin *et al.,* 1963) or in maintaining chromatin structure (Littau *et al.,* 1965).

The relative proportions of these constituents vary in chromosomes from cells of different tissues and for cells from the same tissue in different species. Only 45 per cent of the mammalian liver cell chromosome consists of nucleohistone, 12 per cent of the nucleic acid is RNA and 39 per cent consists of residual protein (Busch *et al.,* 1964). A small amount of lipid is also present (Gahan, 1965).

## THE NUCLEOLUS

A nucleolus is generally present in the nucleus of most cells, although it is inconspicuous or absent in some cells, e.g., sperm and muscle fibers (Sirlin, 1961, 1962). It is conspicuous in highly active cells, such as neurons, or cells carrying on active syntheses, such as secretory cells. Some cells possess multiple nucleoli. When the nucleoli are removed by experimental manipulation the cell dies (Beermann, 1960).

Studies of the nucleolus under the electron microscope show no outer limiting membrane and some fine coiled material in an amorphous background (Sirlin *et al.*, 1963; Vincent, 1955; Stevens, 1965). The nucleolus contains most of the RNA of a resting nucleus. Thus, the nucleolus stains deeply with basic dyes, but after action of ribonuclease this reaction is lost in the central part of the nucleolus. The outer envelope of the nucleolus, not affected by ribonuclease, has been shown to contain a small amount of DNA. Isolated nucleoli, which may have lost some of their constituents during the procedure, have been found to have 3 to 5 per cent RNA. Some of the proteins of the nucleolus are of the complex type, for example, enzymes. Relatively few enzymes have been identified in the nucleolus, e.g., enzymes having to do with RNA synthesis and the manufacture of a coenzyme (nicotinamide adenine dinucleotide, see Chapter 16). As shown by autoradiography, tagged molecules are incorporated readily into protein and RNA in the nucleolus, suggesting that the nucleolus plays a role in the synthesis or accumulation of these compounds. The nucleolus becomes enlarged during periods of synthetic activity in cells and is less well developed in quiescent stages (Davidson, 1960; DeRobertis, 1964).

The nucleolus is essentially Feulgen negative, indicating lack of DNA, except for some peripheral chromatin from the chromosome which serves as the nucleolus organizer. The nucleolus has been isolated from cells and its chemical composition has been determined. It contains a high percentage of RNA and protein.

The function of the nucleolus has been debated for a long time and is still not clearly defined (Vincent and Miller, 1966, e.g., p. 235 vs. p. 449). Since the nucleolus contains, in addition to a varying mass of fibrils, a larger mass of granular material resembling ribosomes in size and shape, it has been thought to be the site of ribosomal manufacture or assembly (Allfrey, 1963; Elson, 1964; Flamm and Birnstiel, 1964a, b; Perry, 1962, 1964; Prescott, 1964). The base ratio of nucleolar RNA resembles closely that of the cytoplasm (Edstrom, 1964). One conception is that the message for the ribosomes comes from the nucleolus organizer region of the chromosome at which the ribosomes are formed. These then come out of the nucleolus by way of the pores in the nuclear envelope (DeRobertis *et al.,* 1965). The nucleolus may also be the site of histone synthesis (Birnstiel *et al.,* 1963; Birnstiel and Flamm, 1964).

## THE MOLECULAR STRUCTURE OF THE CHROMOSOME

In the giant chromosomes of the salivary gland cells of fly larvae, e.g., *Drosophila* (Fig. 6.9), the individual deoxyribonucleohistone units believed to be the fundamental units of the chromosomes have multiplied without cell division to the point that the polyteny (multiplicity) may reach 1000 (as determined by DNA content) and the length, 2000 $\mu$, compared to a length of 7.5 $\mu$ in normal tissue cells of *Drosophila.* Because the chromatids are lined up in phase with one another, the Feulgen-positive DNA-containing regions form bands in register which are readily seen with the light microscope. The bands can be correlated with positions of various genes as determined from crossover data. Consequently, they have been favorite objects for chromosome researches. These bands can be readily removed by action of the enzyme deoxyribonuclease, yet a matrix remains, a matrix shown by color tests to be composed of protein. The protein matrix may also be removed selectively from these chromosomes by the action of pepsin and trypsin. If the nucleic acid is first rendered insoluble by the action of lanthanum salts and then the protein of the chromosomes is digested by such proteolytic enzymes as pepsin or trypsin, the DNA remains as a series of discs. Similar experiments have also been performed with chromosomes from other animals, and similar results have been obtained (Caspersson, 1950).

The organization of the chromosome has been conceived to be either (1) a sort of protein backbone to which discs or masses of DNA or deoxyribonucleoprotein are applied; or (2) a string of protein and nucleoprotein masses such as were described earlier. Although the integrity of chromosomes as indicated by their resistance to various physical and chemical procedures appears to support the first of the possibilities, treatment of chromosomes with chelating agents (which bind calcium and magnesium and so remove these ions from the chromosome), making possible easy fragmentation of the chromosomes, supports the second. It would seem that calcium and magnesium may normally bind these units together in the chromosomes. The union between these large units is therefore not easily broken by various chemical procedures applied during separation of chromosomes from other materials, during all of which the chromosomes maintain their indentity (Mazia, 1954).

The experiments reviewed here are not definitive, however. In the absence of proteolytic enzymes but in the presence of sufficient salt, chromosomes are disrupted into small pieces, suggesting breakage of hydrogen bonds essential to chromosomal integrity. In solutions of low ionic strength (approximately equal to the solutions used for their isolation), it was found that solutions of trypsin and ribonuclease were without action but that deoxyribonuclease fragmented the chromosome into many small fragments. The larger Feulgen-staining bodies (chromomeres) that are seen under the light microscope in giant chromosomes may be connected to one another by a fine Feulgen-positive fibril too thin to be resolved by the light microscope (Callan, 1963). It would therefore seem more likely that a chromosome consists of either a continuous molecule of DNA, the enlargements of which perhaps represent multiple loopings of the molecule, or a series of molecules of DNA connected to one another linearly. Protein is attached to the DNA, forming a nucleohistone, and RNA is also present. The histone is removed by trypsin and the RNA by ribonuclease action but the integrity of the chromosome is not destroyed except by deoxyribonuclease.

Decisive evidence for the presence of DNA in the regions between the chromomeres has been obtained from staining with acridine orange. The molecules of acridine orange align themselves parallel to the purine and pyrimidine bases in nucleic acids. Exposure to ultraviolet radiation evokes fluorescence, orange in RNA and yellow-green in DNA (Mayor, 1961; Sheriff, 1964). The fluorescence measurements indicate that 5 per cent as much DNA is present between chromomeres as in the chromomeres. It therefore seems likely that the thin segments between chromomeres are not seen in Feulgen-stained preparations because the color is just too faint to be seen under the light microscope.

At times giant polytene chromosomes of fly larval organs show enlargements called "puffing" (see Fig. 6.10). The enlargements contain much RNA as shown by RNA-specific staining reactions. They are regions of active RNA synthesis as shown by rapid incorporation of $^3$H-uridine into the RNA. An inhibitor such as actinomycin D, which stops DNA-dependent RNA synthesis, prevents the development of the puffs. At different times during development, and at the same time in different tissues, different regions of chromosomes develop puffs (Beermann, 1964). Analysis of base ratios in the RNA from the puffs in different regions of the chromosome shows that each has a unique chemical composition (Beermann and Clever, 1964). Since the chromosomes are disrupted by deoxyribonuclease but not by ribonuclease and trypsin, it is probable that the continuum of each chromatid consists of DNA. It is thought that the puffing results from intensive transcription of RNA in response to the pressing developmental needs of the larval insect at the particular time.

Specific regions of the chromosome may be excited to activity by injecting into the larva the insect-molting hormone, ecdysone. It is thus likely that the hormone activates the specific genes involved and that considerable messenger RNA is formed in response (Beermann and Clever, 1964; Davidson, 1965).

Another type of enlarged chromosome which has been helpful in a study of chromosome structure is the *lampbrush chromo-*

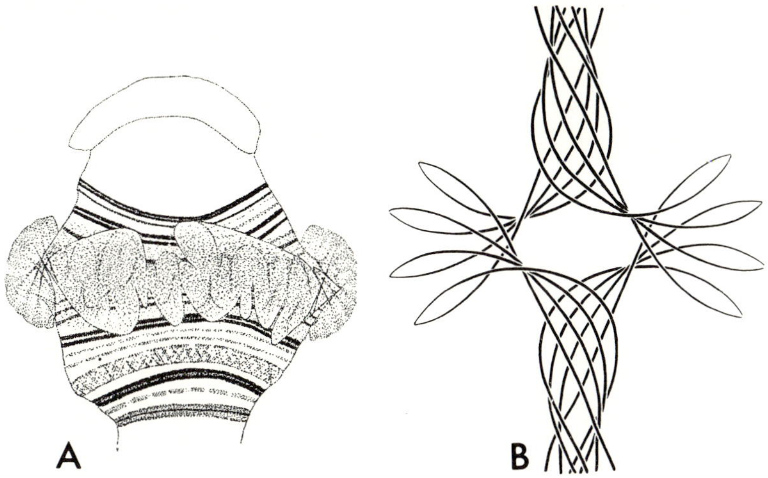


***Figure 6.10.*** *A,* Puffing in a segment of a giant chromosome of a midge (*Chironomus*) salivary gland cell. *B,* Interpretive diagram of the chromosome chromatids in the region of a puff. There would actually be a thousand or more chromatids in a giant chromosome rather than the few shown here. (From Beermann, 1952: Chromosoma *5*:139.)

*some.* Lampbrush chromosomes are very much elongated (5900 $\mu$ in some salamander oocytes) and have a tinsel-like appearance; they are found in oocytes of many animals and have been studied particularly in salamander oocytes. They are generally present before the first meiotic division of the oocyte at the time of maximal rate of yolk synthesis. The central axis of the lampbrush chromosome consists of at least four chromatids from which the lateral loops protrude (Fig. 6.11).

To ascertain the fine structure of the chromosomes electron micrographs have been made (Fig. 6.12). In such studies the chromosomes usually are visible as accumulations of dense granules, each chromosomal mass being without a visible limiting membrane. Only in a few specialized cells are more details of structure evident. For example, in some spermatocytes thread-like elements with delicate side chains the size of nucleic acid helices, extending laterally, are reported (Bennett, 1959).

A survey of the literature on the fine structure of the chromosome suggests that

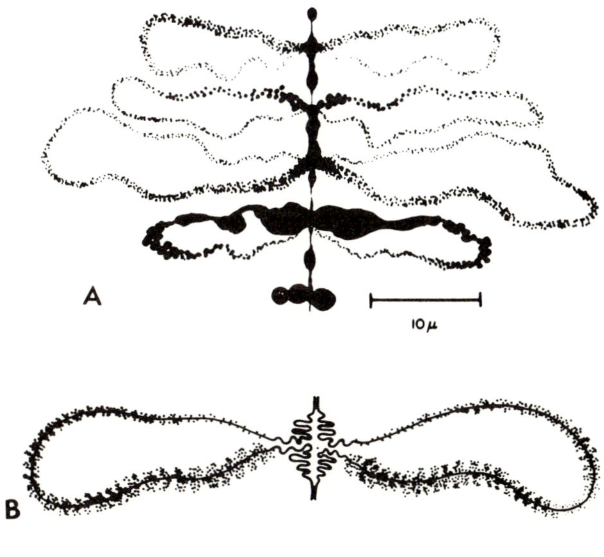


***Figure 6.11.*** *A,* Segment of a lampbrush chromosome of a newt (*Triturus*), showing characteristic paired loops projecting from the main axis of Feulgen-positive granules. *B,* Interpretation of the chromosome structure of a pair of loops in terms of two continuous chromatids. It is thought that the loop consists of a very thin DNA strand surrounded by ribonucleoprotein material. From Gall, 1963: Nature *198*:36.)

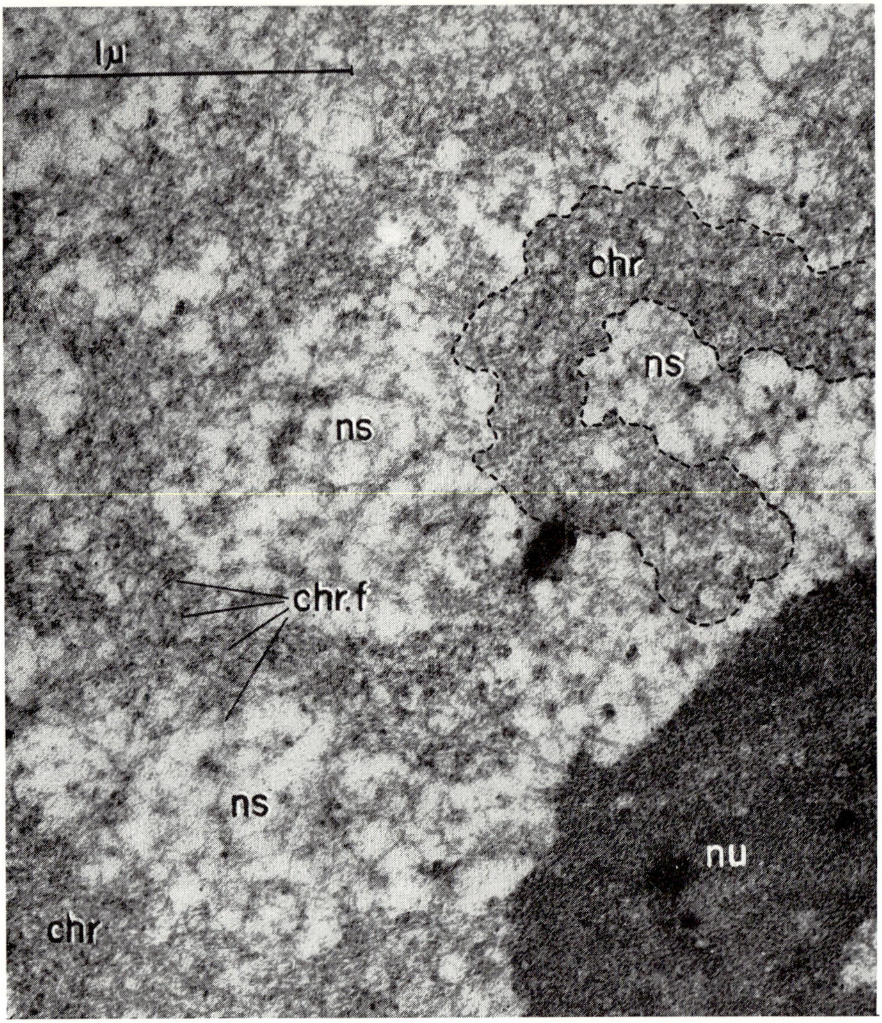


***Figure 6.12.*** Electron micrograph of a group of chromosomes (chr) in the nucleoplasm (ns) of an onion root tip nucleus. Microfibrils (chr.f) constitute the ultrastructure of the chromosome. The chromosome on the right has been outlined to make its identity clearer. The compact nucleolus (nu) has a characteristic macromolecular structure. (By courtesy of Peraz del Cerro and Solari.)

the linear unit of 50 to 120 Å in diameter is common to most cells (e.g., Fig. 6.13). Variations in diameter are probably interpretable as variations in the degree of coiling (Wettstein and Sotelo, 1965; Wolfe, 1965). It would appear that the linear strand is actually composed of two substrands of nucleohistone diameter (20 to 30 Å) which are helically arranged. Although the details are not clear, "nevertheless, experimental studies of the fine structure in lampbrush chromosomes and thin sections of meiotic chromosomes of both plants and animals increasingly suggest that the chromatid* consists of one strand" (Hyde, 1965, p. 143). Data from polarization microscope studies on DNA in

*At certain stages a chromosome may be composed of two chromatids; the paired chromosomes of a meiotic cell are thus tetrads, being composed of four chromatids. The chromatid as used here, therefore, implies the fundamental linear structural unit of the chromosome (DuPraw, 1965). The giant salivary gland chromosomes of fly larvae may have a thousand such units. This use of the terms, although convenient for descriptive purposes here, does not meet with the approval of all cytologists.

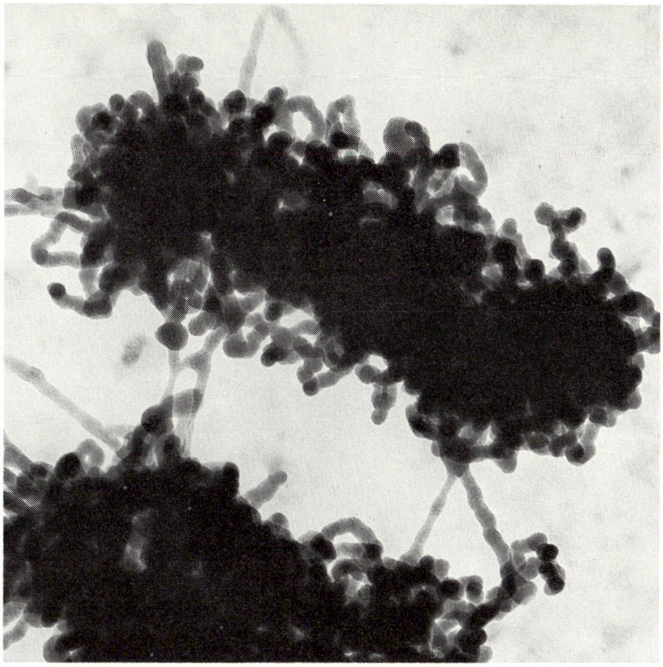

**Figure 6.13.** Fibrous fine structure of a pair of honey bee metaphase chromatids, after surface spreading and critical point drying (× 18,600). (By courtesy of Dr. E. DuPraw; see DuPraw and Rae, 1966: Nature *212*:598–600).

sperm are in agreement with this view (Inoué and Sato, 1962). The possible arrangement of the deoxyribonucleohistone in the chromosome of a eukaryotic cell is given in Figure 6.14.

In bacteria and viruses the electron microscope also reveals the more primitive hereditary material, analogous to the chromosome, to be threadlike in organization (see Fig. 6.15). It consists of a single two-stranded DNA molecule forming a circle (Fig. 6.16). The DNA thread of phages is about 20 Å thick, and in the nucleoids of the bacteria, filaments of like macromolecular dimensions are seen. Larger helices are reported in *Amoeba.* (It should be mentioned that the helices seen by the cytologist viewing stained chromosomes under the light microscope are of a larger order of magnitude and are not to be confused with the much smaller macromolecular helices seen under the electron microscope.)

A method for the study of the structure of the chromosome of bacteria was devised by Kleinschmidt *et al.* (1961), and has been applied to the study of eukaryotic cells by Gall (1963b) and others. The bacteria, whose cell walls have been removed by lysozyme, are applied to the surface of a trough containing distilled water. The protoplasts of the bacteria rupture as a result of uptake of water and the contents are spread on the surface by surface forces.

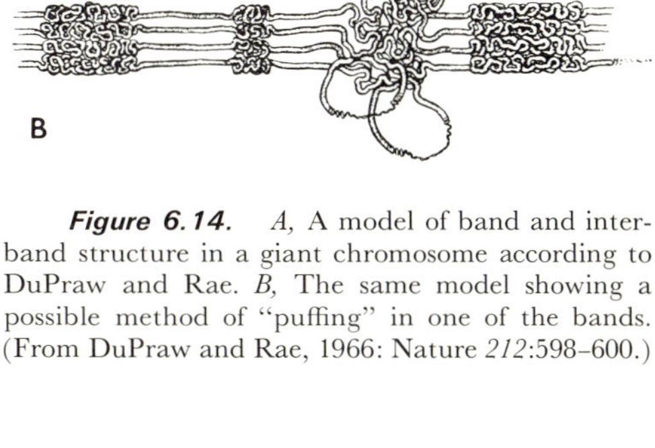


**Figure 6.14.** *A,* A model of band and interband structure in a giant chromosome according to DuPraw and Rae. *B,* The same model showing a possible method of "puffing" in one of the bands. (From DuPraw and Rae, 1966: Nature *212*:598–600.)

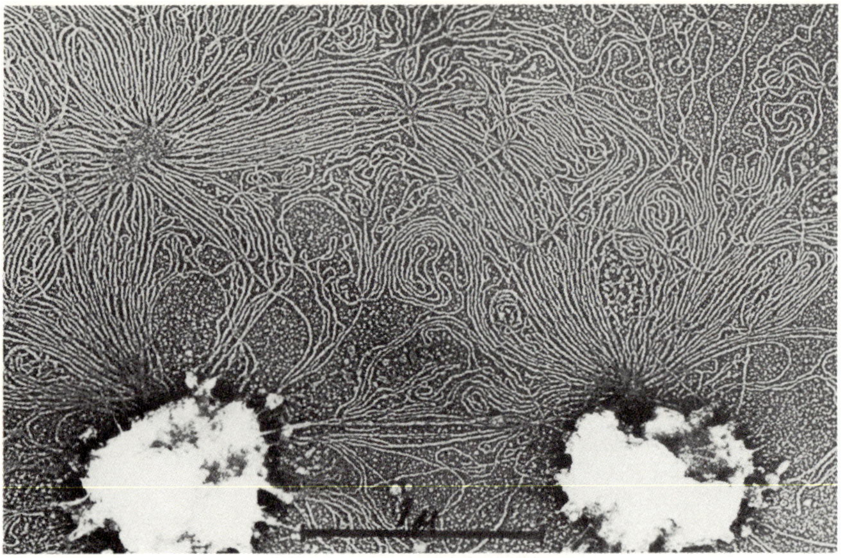

***Figure 6.15.*** Portion of two spread protoplasts with liberated DNA fibers that run together in the middle of the figure. Platinum shadow cast; × 30,000. (From Kleinschmidt and Lang, 1962: Proc. 5th Intern. Congress Electron Microscopy *2*:1–8.)

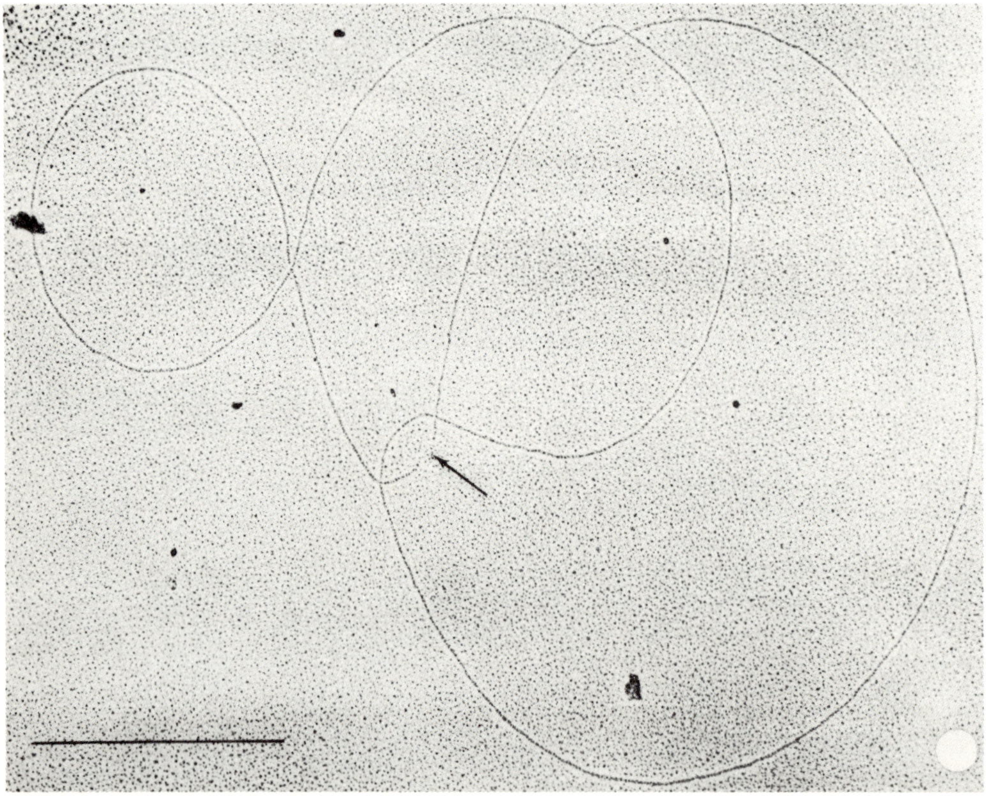

***Figure 6.16.*** Electron micrograph of lambda phage DNA. The material was phenol extracted and spread as a monolayer on a surface of water from a solution of DNA and the basic protein cytochrome C. The arrow indicates the discontinuity, perhaps the place where replication begins. (From Ris and Chandler, 1963: Cold Spring Harbor Symp. Quant. Biol. *28*:1.)

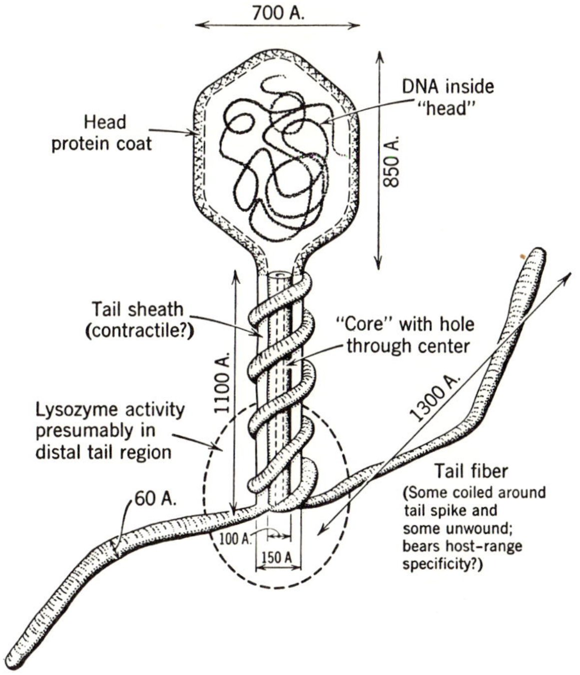


***Figure 6.17.*** Schematic reconstruction of the morphology of a bacteriophage particle. The dimensions shown are approximate and are averaged from several sources. In keeping with present concepts of the DNA molecule the two ends should be joined to form a circle. (From Anfinsen, 1959: The Molecular Basis of Evolution. John Wiley & Sons, New York.)

The cell contents picked up on a grid and studied by electron microscopy disclose a naked filament of DNA (dissolvable by deoxyribonuclease) many times the length of the bacterium (Fig. 6.15). Erythrocytes, similarly treated, disclose much thicker lumpy fibers (200 Å in average diameter, 300 Å at the lumps) with a core about 150 Å in diameter. Similar experiments have been done with cells of ascites tumors and other cells. (Cairns, 1966).

Whether the fibers isolated from eukaryotic cells in this manner are made up of a single, highly packed DNA molecule of the Watson-Crick type, and covered with a protein coat which adds to its diameter, is not detectable by electron microscopy. However, labeling with tracers reveals a single string of incorporated radioactive atoms. Because a eukaryotic chromosome shows incorporation of labeled thymidine in many places simultaneously, it is probably constructed of a string of DNA units, each perhaps the size of an *E. coli* chromosome. When a labeled eukaryotic chromosome is replicated, the labeled DNA is passed intact to only one of the progeny (semiconservative replication). In contrast, tracer labels in both RNA and protein are lost in the course of four generations. Therefore, neither protein nor RNA appears to be the backbone of the chromosome.

It is not clear what role the protein may play in the structure of the chromosome. However, electron microscopic studies of condensing chromosomes show impacting of the fibers into compact chromosomes. Digestion with trypsin causes the impacted strands of DNA to "spring out" as if they had been bound by the protein. The point of attachment of the chromosome to a spindle fiber is thinned out by digestion with trypsin but still shows presence of DNA (DuPraw, 1965).

The linear model of chromatid molecular organization has more appeal because it is easier to see how it might explain the partition of the hereditary materials between daughter chromatids during cell division. In a cell about to undergo division the core splits lengthwise and separates into two units, each of which carries one of the members of each of the pairs of DNA molecules; each helix may then re-form its complementary helix as required by the Watson-Crick model of DNA. The structure of chromosomes in spermatocytes cited earlier suggests that such a structure might very well be possible.

This model for the molecular structure of the chromosome is also in keeping with the division of chromosomes as revealed in experiments with labeled preparations (see Fig. 3.21). For example, seedlings of the broad bean (*Vicia faba*) were incubated in a medium containing tritiated ($^3H$) thymidine as a marker (tritium, emitting a low energy beta particle, gives a sharp photographic image). Most of the 12 chromosomes present in cells of *Vicia faba* became labeled, as evidenced by the correspondence between the location of silver grains in the photographic emulsion and the microscopically visible chromosomes in the control root cells. The labeled roots were then transferred to solutions with nonradioactive thymidine containing colchicine and incubated for various times. Colchicine causes the chromosomes to contract to the metaphase condition in which sister chromatids (the two strands in a chromosome duplex) are spread and are more readily observed than in controls, but it prevents anaphase. Since colchicine does not prevent the cells from growing or the chromosomes from dividing—although it prevents the cell from dividing—a count of the number of chromosomes in each cell is a measure of the number of rounds of chromosome replication. When 24 chromosomes are present each chromosome has divided once; when 48 are present each has divided twice (Taylor, 1959; Anfinsen, 1959).

The results indicate clearly the presence in each chromosome of two morphological halves held together at the centrosome and containing two equivalent strands. During the initial labeling period each chromatid becomes radioactive and metaphase figures in cells with 12 chromosomes show them all to be equally effective in blackening the test photographic emulsion. On division, however, only one of the two strands of each chromosome pair is labeled, indicating that the new one, presumably complementary to it, is assembled from nonradioactive material. When there were 48 chromosomes, only one of each four strands of each type of chromosome was labeled. Occasionally, a "second generation" chromosome was labeled over part of its length; in that case its partner was labeled over the remainder of the length, indicating a crossover. These experiments suggest that DNA is synthesized as a unit extending the length of the chromosome, the units remaining intact throughout subsequent replications (Taylor, 1959; Taylor *et al.*, 1957).

A series of somewhat similar experiments was carried out on *Escherichia coli.* The bacteria were grown on a medium containing labeled nitrogen ($^{15}N$) which appeared in the DNA. They were then grown on a medium with ordinary nitrogen and samples of the bacteria were withdrawn each generation thereafter. The DNA was isolated and it was shown that after one generation on unlabeled nitrogen, only one-half the DNA was labeled; after two generations, only one-fourth. This shows that the original DNA remains intact, the new DNA being synthesized from the unlabeled material (Meselson and Stahl, 1958).

Even more convincing are experiments on *E. coli* in which the chromosome is labeled with $^3H$-thymidine and then extended for autoradiography. This is done as follows: The bacteria are lysed by treatment with a mild detergent. The material is then mounted on a membrane through which the fluid can be withdrawn, leaving the material from the lysed bacteria spread on the membrane. Over this material is spread a layer of photographic emulsion and the preparation is placed in the dark for an appropriate time to allow the radiation from the incorporated $^3H$-thymidine to affect the silver in the photographic emulsion. Development with photographic developer discloses a circular chromosome, often with a replicating portion (see Fig. 6.18). It has been shown that the growing point of the bacterial chromosome is attached to a point of the plasma membrane called the mesosome (Ganesan and Lederberg, 1965). The mesosome is perhaps a region in which oxidative phosphorylation is occurring

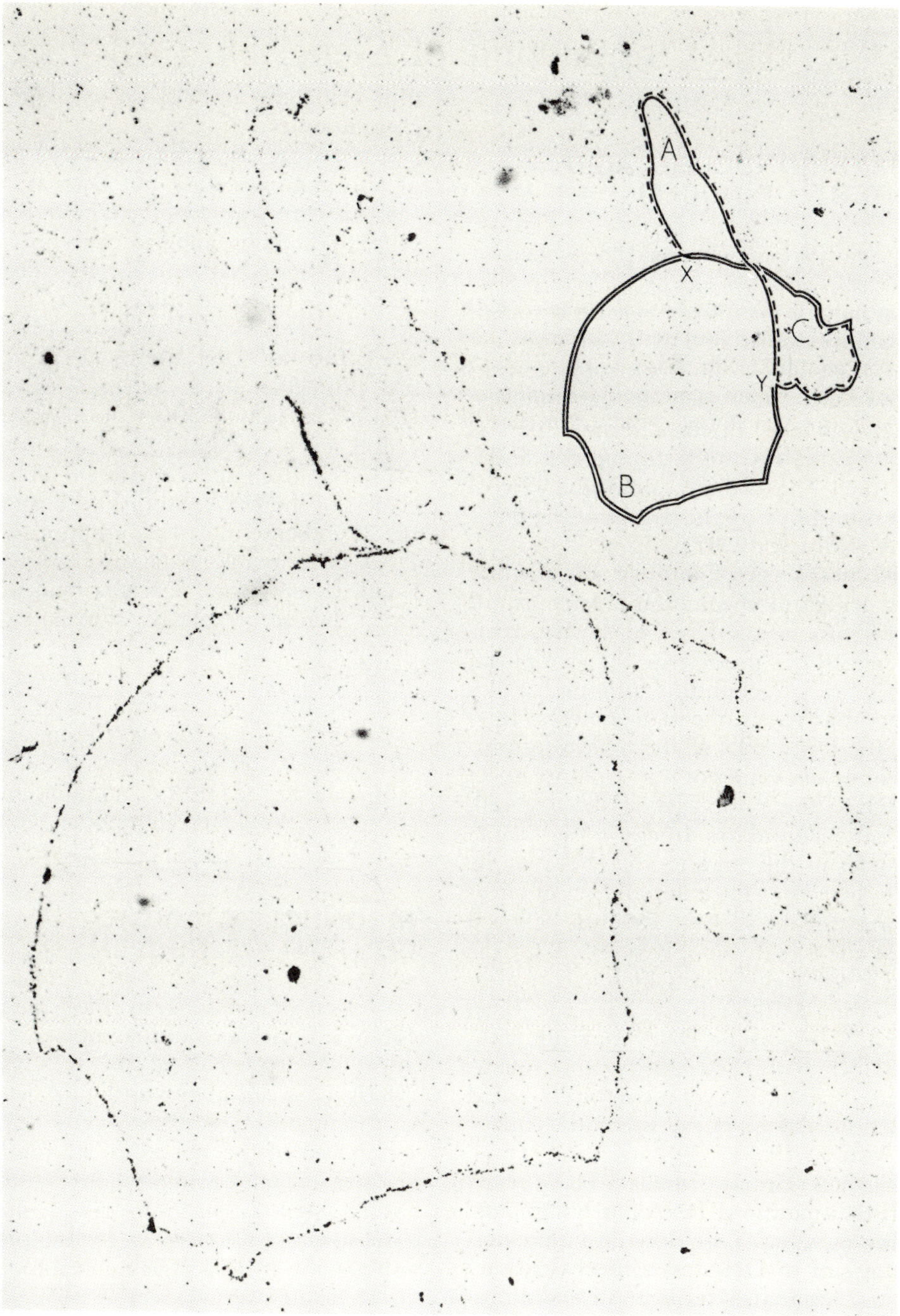


***Figure 6.18.*** Autoradiograph of the chromosome of *E. coli* (strain K12 Hfr) labeled with tritiated thymidine for two generations and extracted by treatment of the cells with lysozyme. The inset shows the same chromosome in diagram. A predominantly half "hot" chromosome that has completed two-thirds of its second round of duplication is shown. Part of the still-unduplicated section is half marked with tracer (Y to C) and part is doubly marked with tracer (C to X). (From Cairns, 1963: Cold Spring Harbor Symp. Quant. Biol. *28*:44.)

and where energy for DNA replication is made available in the form of phosphorylated compounds (Cairns, 1963).

## DEOXYRIBONUCLEIC ACID CONTENT OF NUCLEI

Regardless of whether the nucleus of a liver cell, an erythrocyte, a nerve cell or any other cell of a given species is tested, the same amount of DNA is found per haploid set of chromosomes of a resting cell nucleus. Cells such as the gametes, with only one set of chromosomes, possess only half the DNA of a diploid cell. Polyploid nuclei have the haploid quantity of DNA multiplied by the degree of the polyploidy (Mirsky, 1950).

Although the quantity of nuclear DNA per set of chromosomes is constant, that of cytoplasmic DNA may vary. Usually so little is present in the cytoplasm (mostly in mitochondria and in nucleoplasm) that it is difficult to detect. However, the nurse cells of the ovaries of animals or of the ovules of many plants inject DNA into the cytoplasm of the egg cell, where it accumulates. In egg cells of some echinoderms the quantity of DNA may be many times that for a haploid set of chromosomes for the species. This was believed to provide a ready supply of DNA for the rapid series of divisions which occurs after fertilization although it has not been proved. Cytoplasmic DNA may not stain noticeably with Feulgen because of its dispersed condition, yet it can be extracted by standard biochemical methods (Allfrey and Mirsky, 1959).

Each species of organism has a characteristic amount of DNA per haploid set of chromosomes. It should be noted that the amount of DNA per cell of a species does not necessarily indicate its evolutionary position; even though in invertebrates DNA content per cell progressively increases with increasing degree of complexity, in vertebrates little relation is found between DNA content and phylogenetic position (Mirsky, 1950).

The amount of DNA in a dividing cell is duplicated in the later interphase (or in very early prophase in some cases) long before the cell is ready to divide. The DNA therefore varies in discrete amounts, the unit being the amount per haploid set of chromosomes.

It is interesting that the DNA of a set of chromosomes does not decrease when the cells are starved or exposed to other unfavorable physiological conditions, although the RNA and protein decrease markedly. It would thus appear that the genes, necessary for instituting growth when favorable conditions again appear, are protected in some manner (Swift, 1962). This may be an evolutionary adaptation even though it seems to result primarily from the general metabolic inertness of the DNA. Studies with phosphorus-32 indicate that the phosphorus of the DNA is not readily exchanged or incorporated, except when DNA is being duplicated before a cell division and when phosphorus-32, and carbon-14, used to label compounds serving as purine precursors, are incorporated into the new DNA (Mirsky, 1950). However, some of the other labels are exchanged to some extent, even in the DNA of nondividing cells (Brachet, 1960), although not as rapidly as in RNA.

## RIBONUCLEIC ACID CONTENT OF NUCLEI

Cyclic variations in the content of RNA in nucleus (and cytoplasm) occur with activity cycles of the cell (Crippa, 1966). In a cell about to divide the RNA of the nucleolus gradually passes into the cytoplasm of the cell as the chromosomes re-form during prophase, and the nucleolus as a separate structure disappears. The nucleolus is re-formed at the telophase of mitosis and meiosis. The nucleolus appears to be formed from a section of a chromosome (nucleolus organizer), and some of the chromatin of this chromosome remains directly associated with the nucleolus during the interphase. If the nucleolar organizer is damaged, e.g., by radiation, multiple small nucleoli may form. The nucleolar cycle is shown in Figure 6.19. The RNA of the cytoplasm increases in amount dur-

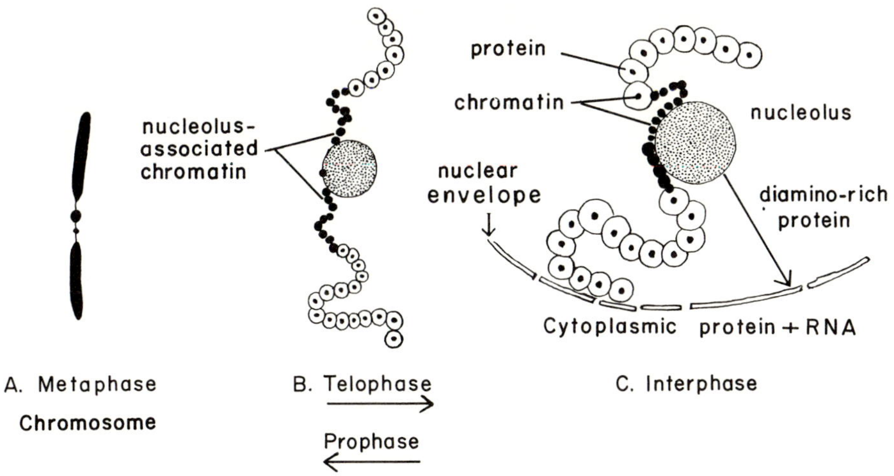


***Figure 6.19.*** Nucleolus-associated chromatin in the nucleus, and formation of the nucleolus. *A*, Chromosome with a centromere in the center, as it would appear in metaphase. *B*, Chromosome as it would appear in prophase or telophase. *C*, Chromosome as it would appear in the interphase. As protein forms about the chromatin in telophase and interphase, the chromatin loses its characteristic staining properties except for the small amount called the nucleolus-associated chromatin, which remains about the nucleolus. The nucleolus, in which RNA accumulates, serves as a center of production of diamino-rich protein, while at the border of the nuclear envelope, RNA and cytoplasmic proteins, rich in aromatic amino acids, are formed. Usually only one nucleolus is present per nucleus, although in some species several may be present. (After Caspersson, 1950: Cell Growth and Cell Function. W. W. Norton & Co., New York, p. 95; and after Darlington, 1942: Nature *149*:66; see also Brown, 1966.)

ing the growth of the cell preceding mitosis and is partitioned equally during cell division (DeRobertis, 1964).

Wherever high metabolic activity or active growth occurs, RNA also accumulates in both nucleus (especially in the nucleolus) and cytoplasm. For example, embryonic eggs contain accumulations of RNA, and similar stores occur in regenerating nerve cells, active neurons, gland cells, cells infected with virus and cells of tumors. Actively metabolizing yeast cells contain a large amount of RNA, but those in a starved culture possess little (Caspersson, 1950). Starved cells in general show a depletion of RNA. RNA also varies with other physiological conditions such as lack of oxygen and presence of metabolic poisons.

Since RNA varies so markedly with the conditions and metabolic state of cells one might expect it to be quite labile, that is, to be degraded and rebuilt readily. This is indeed indicated by the quick incorporation of radioactive tracers such as phosphorus-32 in phosphate and carbon-14 in precursors of the purine and pyrimidine bases into RNA. It is interesting that the RNA is labile not just in dividing cells but also in cells which are not dividing. Consequently, the RNA turnover must accompany various cell activities. Nuclear RNA (chiefly nucleolar) appears to be more readily turned over than cytoplasmic, but this may be an artifact of the methods used (Davidson, 1960).

## CHEMICAL NATURE OF THE GENE

At one time it was thought that DNA was too simple to code the information transmitted in heredity. The present consensus, however, is that the helical structure, proposed for DNA by Watson and Crick, allows for considerable specificity (Watson and Crick, 1953; Watson, 1965). It is now evident that a code made up only of the four main nucleotides could give sufficiently complex information to the cytoplasm to account for inheritance (Anfinson, 1959; Davidson, 1960). Some of the evidence which has accumulated to support this postulate is summarized here.

No cell can long exist without nuclear DNA after enucleation. As we have already seen, a cell can continue limited activities, but synthesis of enzymes and other needed substances is not continued and the cell deteriorates, some types of cells dying much sooner than others. It is considered that DNA directs the most vital syntheses of the cell, such as the synthesis of enzymes and other proteins. Presumably, these are synthesized on the ribosomes, but the information as to what kind of protein is to be synthesized comes from the nucleus. The constant quantity of DNA among all the other varying materials in the cell bespeaks of its importance as well.

More direct information was gained from experiments using radioactive tracers for study of the syntheses induced by phages. About 45 per cent of the dry weight of a given phage (T-2 of *Escherichia coli*) consists of DNA. Shortly after adsorption of an active phage particle to the surface of a bacterium, the DNA portion of the phage enters into the cell and replicates (Fig. 6.20), ultimately forming many phage particles, even if the phage protein portion originally adhering to the surface of the bacteria is washed off, leaving not more than 1 per cent of the $^{35}$S-labeled phage protein with the bacterium. On the other hand, phage "ghosts," prepared by subjecting active phage particles to osmotic shock, during which the DNA portion is lost, still retain their ability to adsorb to the host and to modify the surface of the host in certain ways, but they have lost the power of reproduction (Taylor, 1946; Hershey and Chase, 1952).

Since phage genetic material shows a variety of characters apparently arranged in linear order, as in a chromosome, it is clear that the genetic information is here carried by DNA alone. Somewhat similar experiments have been done with plant viruses in which, apparently, RNA replaces DNA as the hereditary material. In these experiments the virus protein and the virus RNA were separated from one another in each of two strains of virus. When the protein from a mutant form was mixed with the wild-type virus RNA, a new "hybrid" virus was formed. Such a virus would infect plants and reproduce, but when the protein formed under the influence of the hybrid virus was isolated, it proved to be like that formed by the wild-type RNA. The protein which had come in with the virus evidently had had no genetic effect (Fraenkel-Conrat, 1959).

Equally direct evidence for the importance of DNA in heredity is the *transformation* (Fig. 6.21*B*) of *Pneumococcus* from one type to another (Avery *et al.,* 1944; Zinder, 1955). Such studies were possible since the smooth and rough colony types of this bacterium are quite distinct. Addition of extract of the smooth colony transforms a rough strain into a smooth (virulent) strain. The provocative constituent in the extract is the DNA of the smooth strain, since incubation with deoxyribonuclease prevents the transformation because the enzyme destroys the DNA. It would appear that the added DNA of the extract of the smooth strain replaces the DNA of the rough strain and induces manufacture of more material like itself, since the change is hereditary. (In this instance the DNA of the rough strain does not transform the smooth strain.) Similar changes have been induced as well in other microorganisms (Braun and Whallom, 1954; Hotchkiss and Weiss, 1956; Zinder, 1955), including a protozoan (Honigberg and Read, 1960). In some cases biochemical traits such as specific nutritional requirements have been transmitted from one strain of bacteria to another. Evidence has accumulated that such transformations consist of incorporation of the new DNA into the chromosomes of the recipient bacterium, which, being haploid, breeds true to the new DNA (Fig. 6.21*B*). The transforming principle in one case has been "mutated" by nitrous acid outside the cell and the bacterium transformed by the mutated DNA was altered in its hereditary characteristics by the process (Gierer and Mundry, 1958).

Lysogeny, already considered briefly in Chapter 2, is another instance in which genetic material—in this case the DNA of the phage—has been attached to the chromosome of a bacterium (Fig. 6.22) which, accepting the phage DNA, is thereby changed in one of several ways, the most

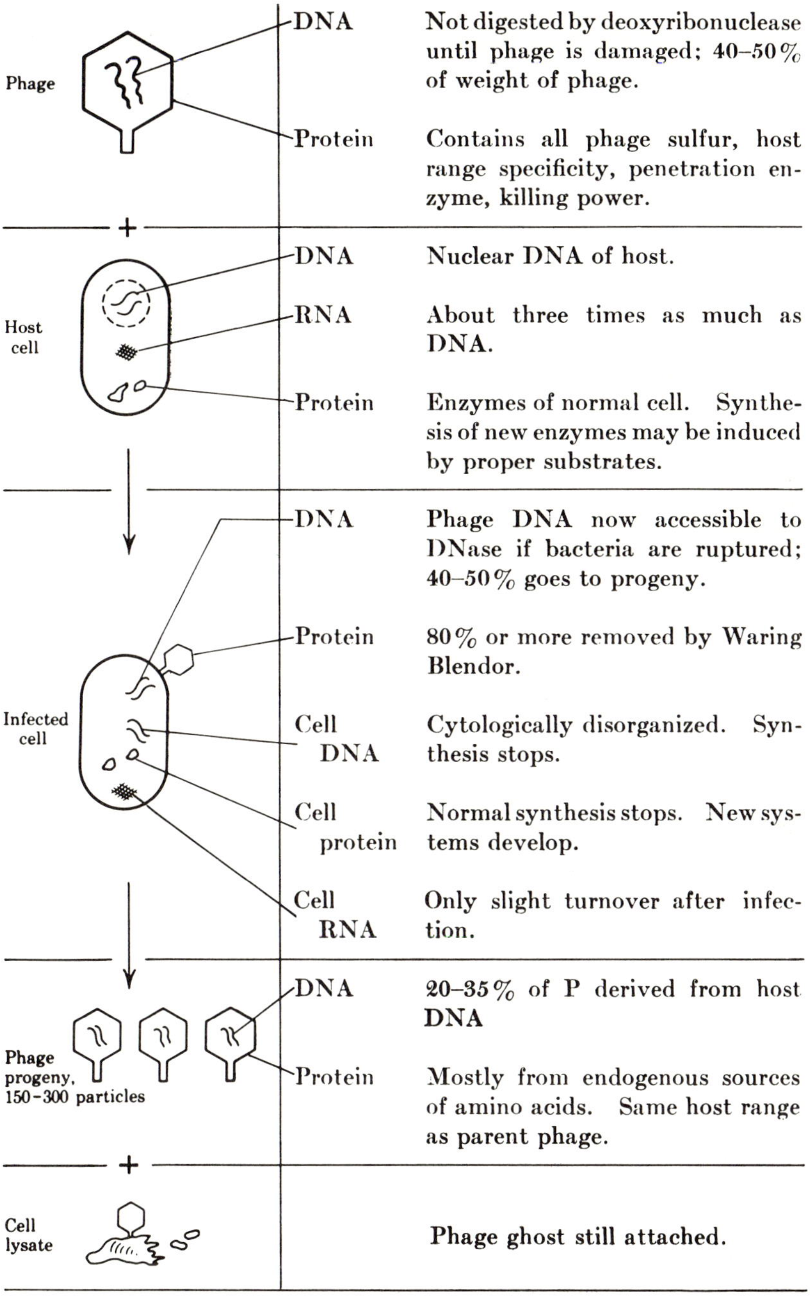


***Figure 6.20.*** The events in multiplication of phage in a bacteria cell. Specifically the diagram shows what happens in the course of infection of the colon bacillus with bacteriophages of the T series. (Redrawn after Hotchkiss, R. D., 1954: The Nucleic Acids. Vol. 2. Academic Press, New York; from Anfinson, 1959: The Molecular Basis of Evolution. John Wiley & Sons, New York.)

noticeable being its increased resistance to the infection by the lytic variety of the phage (Anfinsen, 1959).

Akin to lysogeny is transduction in bacteria, in which genetic material from one strain of bacteria is transferred to another strain of the same species by phages (Wollman and Jacobs, 1956; Zinder, 1958). For example, each of two strains of *Salmonella typhimurium* lacked the ability to synthesize one of the amino acids, but each one had a deficiency for a different amino acid. Had the bacteria mated, a certain percentage would have transferred genes, enabling the offspring of the recipient to synthesize all the amino acids. How-

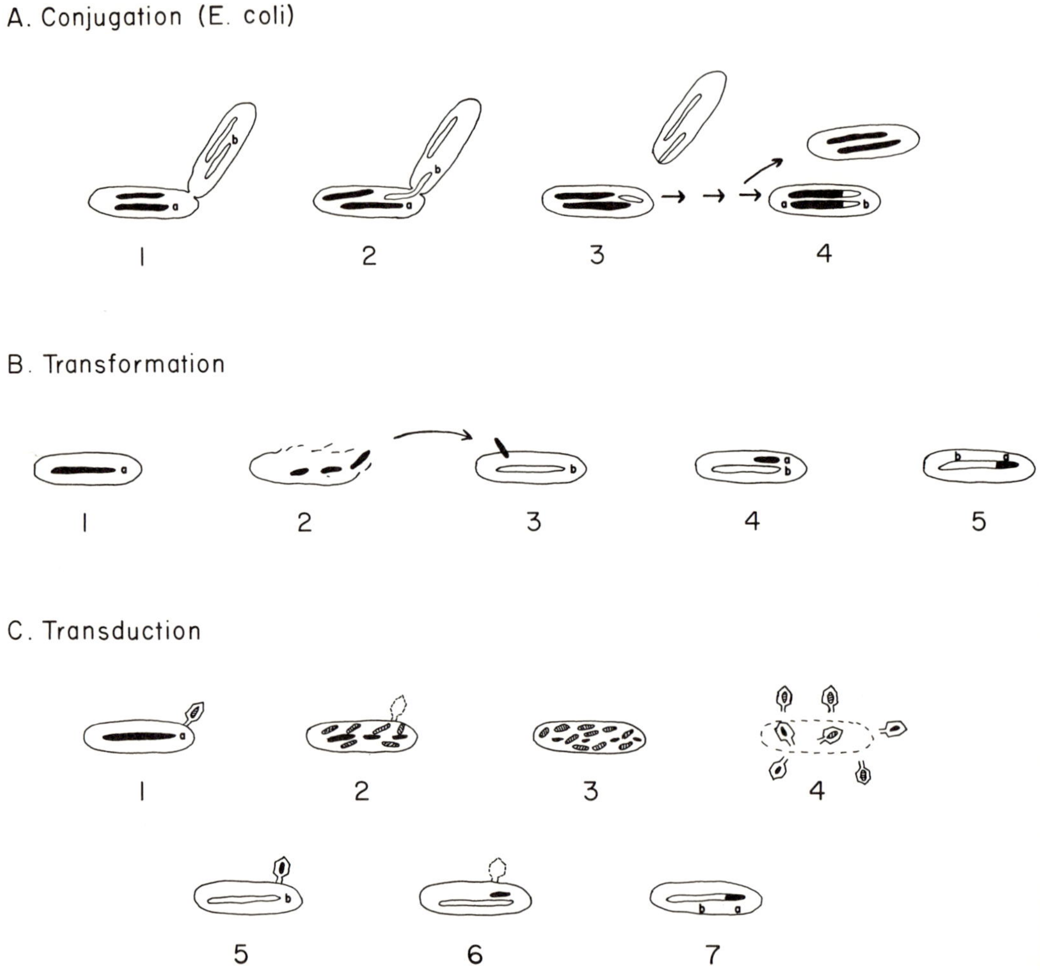


***Figure 6.21.*** *A*, Conjugation in bacteria, with special reference to the colon bacillus. The conjugating strains shown (a and b) are multinucleate. The conjugants remain linked together for an hour or two by a bridge of cytoplasm during which the male cell contributes some genetic material to the female (passive) cell. By labeling, it is shown that passage of material is in one direction only. Fertilization is progressive and can be stopped by shearing the bacteria apart, for example, with a Waring blendor. *B*, Transformation in bacteria. Resistance to drugs and biochemical variations can be transmitted by nucleic acid from one strain (a) to another (b). While pieces of the chromosome are shown to be transmitted in the diagram, this has been accomplished experimentally for specific characters by solutions of the nucleic acid of a given strain applied to the other (see text). *C*, Transduction in bacteria. Infection by a phage may bring in genetic material picked up by the phage from another strain of bacteria; such introduced genetic material may become incorporated into the genome of the bacterium. Note that in each of the three cases, *A*, *B* and *C*, a piece of the new chromosomal material from another bacterium is at first loosely added but later may become incorporated into the genome of the bacterium, probably as part of its chromosome. The resultant products of each of the processes have some of the characters of each of the two strains of bacteria.

Lysogeny

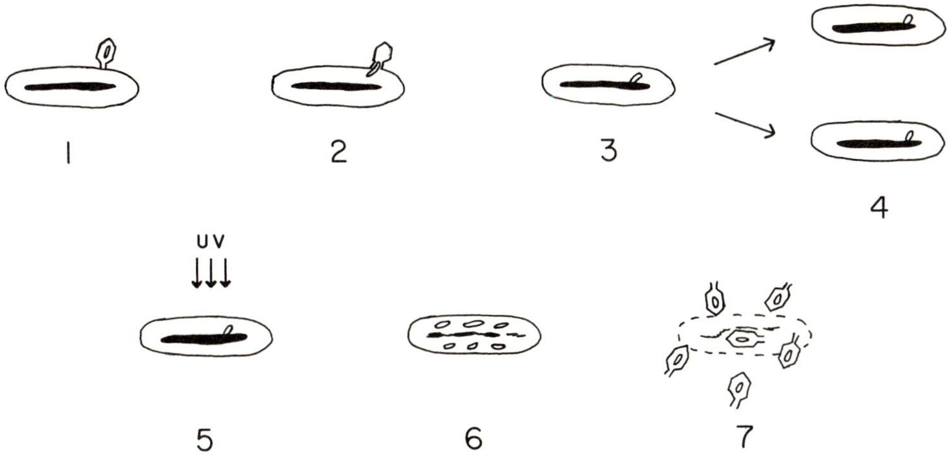


***Figure 6.22.*** Lysogeny in bacteria. The phage genome does not replicate independently but, added to the bacterial genome, replicates when the latter divides. Injury of the bacterium by ultraviolet radiation or other harmful agents, however, activates the phage, which then multiplies rapidly like a lytic phage and lyses the bacterium.

ever, even when mating was prevented by interposing between suspensions of the two strains a membrane through which they could not move, new strains of bacteria appeared which were able to produce the respective amino acids for which the two original strains were deficient. The only possible carrier of the hereditary material which could pass the membrane is a phage. It has now been shown that the phage, which transmits the information from one strain of bacterium to another, is incapable of multiplication and cannot lyse the infected bacteria (Fig. 6.21*C*). Presumably the genetic material of the phage has become defective and can no longer replicate, perhaps because by a mistake during its replication in the preceding bacterium, a part of its chromosome has been replaced by a gene block of the bacterium. The phage's very effectiveness in transduction might thus prevent its further effectiveness as a phage.

Since the transduced bacteria breed true, it must be assumed that the apparent place on the chromosome occupied by genic material, which previously produced the deficiency in amino acid manufacture, has now been replaced by material brought in by the phage, making it competent in this respect. Experiments in which radioactive tracers were used show that only DNA has been transferred by the phage. Transduction is thus another evidence that DNA is the genetic material.

It can now be stated unequivocally that the gene is a subdivision of the DNA molecule which codes for a polypeptide, except in RNA viruses, in which RNA has this role (Shapiro and August, 1966). Since a polypeptide consists of one to several hundred amino acids, each amino acid coded by a set of three nucleotide bases, an average gene consists of about 1500 nucleotide pairs (molecular weight $10^6$) (Watson, 1965).

It is of interest that some proteins consist of several polypeptide chains. For example, the hemoglobin molecule consists of two $\alpha$ and two $\beta$ polypeptide chains. Sometimes the respective polypeptides which form a given molecule are encoded on successive portions of the genome, but not in the case of hemoglobin (Nance, 1963; Ingram, 1963).

Since a mutation is now thought to represent a substitution, deletion or addition of one of the nucleotides in one of the sets of three bases coding for an amino acid, and since about 1500 nucleotides containing such bases are present in a gene, one might expect that crossover and recombination might occur within a gene. Data proving that this does occur have been obtained for the so-called rII gene of the bacterial virus T-4 of *E. coli.* From crossover data obtained on about 1000 mutants it is possible to construct a gene map and it is found that the arrangement of mutations in a gene is linear, resembling in this respect the arrangement of genes in a chromosome (Benzer, 1961; Watson, 1965).

## FUNCTIONS OF THE NUCLEIC ACIDS

The functions of the nucleic acids have always been of great interest, because the prominence of DNA in the chromosomes and in transmission of hereditary characteristics, and the presence of RNA in the nucleolus and ribosomes suggests that the nucleic acids are closely connected with cell syntheses (Davidson, 1960; Hoagland, 1959; Gay, 1960). It is also likely that DNA is involved in cell differentiation during development, but less is known about its function in this regard (Swift, 1962).

The synthesis of DNA *in vitro* by Kornberg, who received the Nobel prize for these studies in 1959, indicates the probable mechanism by which this synthesis occurs in the cell (Kornberg, 1960). The requirements for this purpose are certain enzymes, notably the DNA polymerase (in Kornberg's studies extracted from *Escherichia coli*), the nucleoside triphosphates (to provide the material necessary for the four nucleotides usually found in DNA and the energy required for the synthesis) and a DNA primer, or template, which serves as a model for the synthesis of the new DNA. It will be recalled from Chapter 3 (Fig. 3.16) that the Watson and Crick DNA model consists of two strands in a helix, bonded from purine to pyrimidine residues, so that thymine (T) attracts adenine (A) and cytosine (C) attracts guanine (G); hence the linkages across the two strands are T-A, C-G, A-T and G-C.

In the presence of the DNA polymerase the triphosphates of the nucleosides become linked, in accordance with the aforementioned attractions of bases, and in the processes of the formation of bonds between the nucleotides, two phosphates (as pyrophosphate) are split out of each triphosphate. In this way a strand of DNA which is complementary to the template in its placement of bases is formed. Since each strand forms its own complement, the net effect of the synthesis is the formation of a new complete unit of DNA similar to the template (Fig. 20.5). In the synthesis of DNA the specificity is determined solely by the template DNA presented. The same enzyme when presented with the same four nucleoside triphosphates was found capable of duplicating a variety of DNA models. The identity of the DNA produced was determined by comparing its base ratios with those of the DNA used as a primer.

In the cell DNA *replication* occurs during that part of the interphase known as the S (for synthetic) phase; this lies between $G_1$ (Gap 1) after telophase and $G_2$ (Gap 2) which precedes prophase (see Fig. 27.1) (DeRobertis *et al.,* 1965). Presumably, DNA synthesis occurs only in the chromosomes but the little DNA present in the mitochondria, and, in green plants in the chloroplasts, is also presumably self-replicating. It is apparent that DNA replication by the method described here ensures that replication of DNA present in a cell will continue from cell generation to cell generation, unless a "mistake" in copying (mutation) is made.

It is well known that DNA is in control of protein synthesis, since removal of the nucleus or the removal of DNA by deoxyribonuclease stops protein synthesis in the cell (as indicated by failure in incorporation of labeled amino acid precursors of proteins). Cytoplasmic changes accompanying DNA removal include the gradual loss of basophilia, indicating the loss of RNA in ribosomes which are the centers of protein synthesis in the cytoplasm.

RNA always is present, in one form or another, when protein synthesis occurs. This is true even in phage which has DNA as the hereditary material but no RNA.

The DNA of phage, however, induces in its bacterial host the production of a new RNA, the messenger RNA or mRNA, under the influence of which is synthesized the protein which the phage needs for coating its new DNA molecules. In general, protein synthesis in cells is also accompanied by a turnover of RNA as is shown in experiments with incorporation of labeled precursors. Apparently, the mRNA is first formed in the nucleus since in autoradiographs the label appears over the nucleus (chromosomes) in 5 minutes, the nucleolar count being low until after 10 minutes have passed. The label appears in the cytoplasm much later (Goldstein and Micou, 1959; Goldstein, 1964, 1965; Prescott and Goldstein, 1967). In the cytoplasm the mRNA becomes attached to the ribosome where the *translation* of the message is carried out. It has been said that the original information is kept in the nucleus and only the copy is allowed to move into the cytoplasm. Although protein production depends upon RNA directly, the RNA involved in protein synthesis is formed by *transcription* from DNA (Atwood, 1965). DNA production is, therefore, the master reaction and DNA is the ultimate agent indirectly controlling protein synthesis.

Protein is synthesized primarily in the cytoplasm and more specifically on the ribosomes, which contain 80 per cent of the RNA of the cell. In gland cells with a well developed endoplasmic reticulum, protein globules may be readily identified within the canals (Hoagland, 1959).

The RNA, like DNA, consists of four types of nucleotides, except that uracil replaces thymine, and the sugar molecules constituting part of the nucleotides have hydroxy groups (which are absent in the deoxy sugars of DNA). Ideas concerning transcription of RNA on DNA templates are largely speculative, although some proof is accumulating. Presumably, as in the case of DNA replication, the ribonucleoside triphosphates are attracted in the same manner to the template DNA, guanine to cytosine, adenine to thymine, cytosine to guanine and uracil to adenine, in accordance with the pattern on the template. In this manner would be formed the *messenger RNA* molecule, which leaves the nucleus and goes to the cytoplasm, where it translates the information as to what protein is to be made. There it attaches itself to a ribosome, presumably by complementation of bases. The ribosomal RNA is presumably made directly or under the influence of DNA since it disappears if the DNA is removed by deoxyribonuclease or the cell is enucleated (Crick, 1954; Hoagland, 1959; Hurwitz and Furth, 1962).

The clearest evidence for a messenger RNA is perhaps that from work on T-2 and T-4 phages. Soon after infection of *E. coli* a new RNA is produced in which the adenine-uracil/guanine-cytosine ratio duplicates the phage DNA adenine-thymine/guanine-cytosine ratio. The RNA is rapidly formed after infection and just as rapidly destroyed (Hurwitz and Furth, 1962). The purpose of the mRNA is to inform the bacterial cell ribosomes as to what kind of protein to produce as a coating for the phage DNA.

In the ground substance of the cytoplasm occurs another form of RNA, also made under the influence of DNA, but having free terminal groups by which it can attach amino acids. This RNA, originally thought to be a part of a ribosome which had been solubilized by treatment, is different in chemical properties from the ribosomal RNA. The RNA present in the ground substance of the cytoplasm, called *soluble* or *transfer* RNA (sRNA or tRNA), consists of relatively small molecules (low molecular weight) which combine with amino acids before the latter can be united to form a protein.* For this purpose the amino acids must be in a phosphorylated (high energy) form. Such compounds of RNA with amino acids have been isolated from the cell; there are some 20 of them and they are being purified. Presumably, by attaching to a transfer RNA the amino acids obtain an identity which the RNA templates attached to the ribosome can recognize. The amino acid–RNA complexes are, there-

*See Chapter 20 for further discussion and diagrams of protein synthesis.

fore, attached to the appropriate places on the template, fitting by the method of complementation already described for DNA. Once the RNA–amino acid complexes are affixed to the surface of the ribosome the respective amino acids, which are in an activated or energetic form, combine to form peptide bonds with one another, at the same time detaching from the RNA nucleotide, which detaches from the ribosome and presumably goes on for another round of transfers (Crick, 1954; Hoagland, 1959).

For protein synthesis it is essential to have a supply of guanidine triphosphate (GTP). Except as a source of energy the exact function of GTP is not understood, but it cannot be replaced by adenosine triphosphate (ATP), cytidine triphosphate (CTP) or uridine triphosphate (UTP). During protein synthesis GTP is decomposed to guanidine monophosphate (GMP) and pyrophosphate (POP). Pyrophosphatase present in the cell decomposes POP to inorganic phosphate (Ingram, 1966). Some data suggest that GTP is necessary for peptide bond formation, possibly in translocation of the nascent peptide chain on the ribosome from one site to the next.

That existing ribosomes of the bacterial cell are used in the synthesis of phage proteins is shown by growing bacteria on nutrients labeled with heavy carbon and nitrogen until these bacteria are infected with phage. At this time the bacteria are returned to unlabeled nitrogen and carbon sources, but are briefly supplied with phosphorus-32 to label newly formed messenger RNA. Density gradient separation of the light and heavy ribosomes shows all the new RNA (marked with phosphorus-32) associated with the old (heavy) ribosomes (Hurwitz and Furth, 1962).

It was also demonstrated in parallel experiments, in which sulfur-35 was supplied in protein precursors, that the new proteins ($^{35}$S-labeled) were associated with the old ribosomes before they appeared elsewhere in the cell (Brenner *et al.,* 1961).

The enzymes which catalyze union of RNA carriers and amino acids are probably produced in the same way as other enzymes, i.e., on an RNA template, but are probably always present in "starter" proportions at the beginning of any protein synthesis in the cell (see Chapter 22).

Amino acid activation, attachment of amino acids to transfer RNA, and the final building of protein in ribosomes are facts supported by good data. Evidence is also good that RNA originates in the nucleus and is under the control of the DNA of the chromosomes. However, some of the hypothetical steps outlined here require further documentation (Harris, 1960; Hogness, 1962).

## THE GENETIC CODE

Since usually only four purine and pyrimidine bases (sometimes a fifth, and even a sixth) are present in RNA and DNA, it would, at first glance, seem difficult to make a code from just four units sufficient to locate 20 amino acids on a ribosome as is done in the synthesis of proteins. However, it was found possible to develop codes which are specific to each one of the 20 amino acids by using only four nucleotides. Thus, designating the nucleotides 1, 2, 3 and 4 for purposes of argument, one can make the following sequences of three each, each to designate one and only one amino acid: 112, 212, 131, 132, 133, 231, 232, 233, 141, 142, 143, 144, 241, 242, 243, 244, 341, 342, 343 and 344.* If transfer RNA nucleotides with one of these sequences were to pick up one amino acid each, and unite with the complementary spaces on the messenger RNA, it would be possible to line up the amino acids to build any protein, even with this relatively simple code (Watson and Crick, 1953, 1962; Hoagland, 1959).

Experimental support for DNA coding of *in vitro* protein synthesis was first provided by the brilliant discovery that protein-like molecules are produced in response to artificial RNA-like polyribo-

---

* Perhaps 111, 222, 333, and 444 might figure in this series as well.

nucleotides of known composition [for example, a polyuridine made up only of uridine (U)]. Thus, when an RNA-like polyribonucleotide composed of three repeating units of uridine [the U-U-U sequence of the polyuridine being complementary to the adenine (A) sequence, A-A-A, of the DNA] is supplied, the synthesized protein-like polypeptide molecule consists of repeating phenylalanine residues (polyphenylalanine) (Nirenberg and Matthei, 1962). When other synthetic RNA-like compounds are used [e.g., polyribonucleotides made up of uridine (U), cytosine (C), guanine (G) and adenine (A), in each case consisting of units of three—UUA, UGG, UUC, UCA, and so forth], the protein-like polypeptides synthesized are characteristic for each type of RNA-like polyribonucleotide supplied. Experimental data suggesting the RNA code (and so also the complementary DNA code) for all of the amino acids have now been gathered (Ochoa *et al.*, 1961; Lengyel, 1962; Yanofsky, 1963).

When it proved possible to synthesize regular polymers, e.g., CUCUCU, additional code words were discovered. Others were determined by the specific sRNA binding to ribosome-mRNA complexes. The RNA code words for amino acids determined in these various ways are summarized in Table 6.1.

It is evident from Table 6.1 that there are several code words for many individual amino acids; to this extent the code is said to be degenerate (lacking specificity). However, it is possible that lack of specificity may in itself give a certain amount of flexibility to synthesis of proteins in the cell.

Although a given gene may code for one polypeptide, one mRNA molecule may code for several polypeptide chains. Therefore, in addition to coding for amino acids, it is possible that code signals must be given to start and stop polypeptide chain growth. When such information is lacking, for example, in a UUU polynucleotide (mixed with ribosomes and phenylalanine), the polypeptide produced (polyphenylalanine) does not become detached from the polynucleotide. Information for beginning and stopping polypeptide synthesis is probably present in mRNA. It is not clear, however, whether the code words for beginning and stopping polypeptide synthesis are similar to the code words for amino acids, e.g., triplets of bases (Watson, 1965). It is possible that the odd bases of infrequent occurrence in nucleic acids serve this purpose (a sort of punctuation).

When a codon for one amino acid is mutated to that for another amino acid, the mutation is called a *missense mutation.* For example, abnormal hemoglobins are produced by substitutions of a single wrong amino acid in one location, as already discussed (Ingram, 1963; Nance, 1963). When a mutation occurs in which one codon for a particular amino acid is changed in such a way that it no longer corresponds to any amino acid, the mutation is called a *nonsense mutation,* and it too results in abnormal protein synthesis. It is important to point out that in the case of *in vitro* protein synthesis small changes in conditions, e.g., concentration of magnesium ions, may result in misreading of the message and in mistakes in the protein synthesized. In the cell, the latter type of mutations would presumably be minimized by maintenance of conditions favorable for normal protein synthesis (Watson, 1965).

The genetic code appears to be universal. Thus poly-U promotes synthesis of polyphenylalanine in cell-free extracts of many organisms, and poly-C for production of polyproline, and so forth, provided all the components of the system are included in each case. Ribosomes from one organism serve for mRNA originating in another organism to produce proteins in conformity with the mRNA message. This is another instance that at the level of "biochemical universals" there is "little difference between cabbages and kings" (Crick, 1963, 1966; Jukes, 1963, 1965; McCarthy, 1965; Nirenberg *et al.,* 1963; Watson, 1965; Marshall *et al.,* 1967).

The analysis of the genetic code will be incomplete until the colinearity between amino acids and nucleotides in messenger mRNA, and again between the

**Table 6.1.** The Genetic Code*

SECOND LETTER

| FIRST LETTER | U | | C | | A | | G | | THIRD LETTER |
|---|---|---|---|---|---|---|---|---|---|
| U | UUU | Phe | UCU | Ser | UAU | Tyr | UGU | Cys | U |
| | UUC | | UCC | | UAC | | UGC | | C |
| | UUA | Leu | UCA | | UAA | OCHRE | UGA | ? | A |
| | UUG | | UCG | | UAG | AMBER | UGG | Tryp | G |
| C | CUU | Leu | CCU | Pro | CAU | His | CGU | Arg | U |
| | CUC | | CCC | | CAC | | CGC | | C |
| | CUA | | CCA | | CAA | GluN | CGA | | A |
| | CUG | | CCG | | CAG | | CGG | | G |
| A | AUU | Ileu | ACU | Thr | AAU | AspN | AGU | Ser | U |
| | AUC | | ACC | | AAC | | AGC | | C |
| | AUA | | ACA | | AAA | Lys | AGA | Arg | A |
| | AUG | Met | ACG | | AAG | | AGG | | G |
| G | GUU | Val | GCU | Ala | GAU | Asp | GGU | Gly | U |
| | GUC | | GCC | | GAC | | GGC | | C |
| | GUA | | GCA | | GAA | Glu | GGA | | A |
| | GUG | | GCG | | GAG | | GGG | | G |

GENETIC CODE, consisting of 64 triplet combinations and their corresponding amino acids, is shown in its most likely version. The importance of the first two letters in each triplet is readily apparent. Some of the allocations are still not completely certain, particularly for organisms other than the colon bacillus (*Escherichia coli*). "Amber" and "ochre" are terms that referred originally to certain mutant strains of bacteria. They designate two triplets, UAA and UAG, that may act as signals for terminating polypeptide chains.



messenger RNA and DNA, has been shown (Watson, 1965; Thomas, 1966). Now that nucleotide sequences have been determined in some small RNA molecules (Holley, 1966; Holley *et al.,* 1965), such correlations are becoming possible and progress has been made in a number of instances (e.g., Yanofsky, 1963; Yanofsky *et al.,* 1964; Yanofsky and Ito, 1967; Yanofsky, 1967).

The kind of protein produced from the same pool of amino acids would then depend upon the mRNA presented. The mRNA would depend upon what gene had produced it, since presumably many types of genes are present on the chromosomes of a cell, each capable of producing a corresponding messenger RNA. On this basis, a cell can have only as many types of polypeptides as it has types of genes (Crick, 1954; Hoagland, 1959; Loftfield, 1957).

## EXTRACHROMOSOMAL INHERITANCE

Extrachromosomal inheritance has been known for a long time under the name of maternal inheritance (Jinks,

1964). It is now clear that DNA is present in both mitochondria and chloroplasts and might serve as the carrier of extrachromosomal inheritance. Such DNA is self-replicating and apparently codes for protein synthesis which occurs in these organelles.

Extrachromosomal mutants are not produced by the mutagens which cause chromosomal mutations; therefore, study of such inheritance has been slow. Recently some chemicals have been found suitable for the purpose (e.g., sublethal concentrations of streptomycin) and a fairly large number of mutations have been accumulated in the single celled alga *Chlamydomonas reinhardii* (Sager, 1966). The mutations have been observed to occur in linkage groups, with crossing over between them; therefore, the behavior of the extrachromosomal DNA appears to be much like that of the chromosome. Since two such linkage groups have been found it is suggested, but not yet proved, that one linkage group corresponds to the DNA in the chloroplast and the other to the DNA in the mitochondria (Gillham, 1965).

## CODING FOR mRNA, rRNA AND sRNA ON DNA

Since all the information for synthesis in the cell resides in its DNA, some portions of the chromosome must code for the various types of cellular RNA. Anticodes to the DNA must therefore be present on each type of RNA present in the cell. It has been possible to show this by hybridization between DNA and various types of cellular RNA. Hybridization is the pairing of strands of RNA and DNA (see Chapter 3). For hybridization to occur, the hydrogen bonds between the two strands of DNA and within the RNA strand must be broken by heat (Marmur *et al.,* 1963). Then the mixture is slowly cooled. In locations where RNA and DNA are complementary they will hybridize upon cooling (Spiegelman, 1964).

Hybridization reveals that approximately 0.32 per cent of the chromosome of *E. coli* codes for the ribosomal RNA. An even smaller portion, approximately 0.025 per cent, codes for sRNA. However, there are at least 20 sRNA molecules encoded. The amount of DNA coding for one mRNA is also a relatively small part of the entire chromosome. However, there are at least a thousand kinds of mRNA coding for the thousand (or more) types of proteins in the cell.

It is of interest to determine whether both strands of DNA, or only one, serves for coding the various types of RNA. Hybridization techniques with single stranded DNA of φX174 bacteriophage have provided some information (Sinsheimer, 1962). It will be recalled that a virus inhibits syntheses normally occurring in the host cell, but initiates and directs syntheses in the cell to produce more of its own DNA and the proteins required for its replication and packaging (Chapter 2). The virus therefore induces mRNA formation in the bacterium. The mRNA isolated from the φX174 virus-infected *E. coli* does not hybridize with the DNA isolated from the mature φX174 phages. However, when the mRNA induced in the host cell was mixed with the active virus in the bacteria before packaging had occurred, hybridization was obtained. It has been shown that the replicating form of the virus is double stranded, the second strand being formed by complementation from the single strand. Hybridization therefore appears to involve only the replicated strand. From this and other experiments it can be inferred that only one of the two strands of DNA carries the codes for the various types of RNA (Cairns, 1962; Hayashi *et al.,* 1965; Spiegelman, 1962).

## COMPARISON OF THE GENOME IN BACTERIA, VIRUSES AND EUKARYOTIC CELLS

The genome of *Escherichia coli* is better known and more completely studied than any other genome (Watson, 1965). There are thought to be about 2000 to 3000 genes coding for the various types of proteins necessary to the life of *E. coli* (or any other cell), about a third of which have been described. The DNA molecule of *E. coli* makes up only 1 per cent of the mass of the cell and is present as a single double-stranded molecule about 1000 times the length of the cell, presumably coiled up like a tangled skein of yarn. Each gene has a molecular weight of about $10^6$ and contains about 1500 base pairs (and nucleotides). During replication, it is thought that the growing point is accompanied by a separation of the two strands in the DNA molecule (see Fig. 6.17). Since replication in *E. coli* under favorable conditions takes only 20 minutes, the growing point must move at a rate of about 40 to 50 $\mu$ per minute.

*Escherichia coli* (wild type) can live in the absence of other organisms on a diet of glucose and a variety of salts, including a source of nitrogen and sulfur. It is, therefore, capable of synthesizing all of its other organic cellular requirements. The pleuropneumonia-like organisms (PPLO; see Chapter 2) are minute organisms which usually live as parasites on other organisms. The genome of a PPLO, such as the free-living PPLO *Mycoplasma laidlawii,* is estimated to contain about 700 genes coding for as many proteins. As pointed out before, the PPLO organisms, when free-living, must be supplied with many vitamins, amino acids, purine and pyrimidine bases, ribose and deoxyribose and other complex molecules which they cannot synthesize themselves. The genome of this species, which is much smaller than in *E. coli,* thus suffices to code for the remaining proteins, presumably enzymes involved in growth and reproduction of their kind (Watson, 1965; Vinograd and Lebowitz, 1966).

Even more reduced in size is the genome of viruses. Viruses are parasites which depend upon the host cell for most of their synthetic machinery. The virus is shown to consist of the DNA molecule (chromosome), a coat of one or more proteins and sometimes one or more enzymes. However, it has the code for synthesis of mRNA but depends upon the host's sRNA and rRNA. The T-2 phage, for example, probably possesses several hundred genes. It can therefore code for the enzymes necessary for its replication and packaging. The genome of the phage may overlap with that of the bacterium. For example, the T-2 phage infecting a thymineless (thymine-minus) mutant of *E. coli* will induce in the thymineless host the synthesis of thymine, the code for which the phage contains in its genome (Watson, 1965).

Some phages have much smaller genomes than the T-2 phage, for example, lambda phage which has only about 15 to 20 genes; none of its genic information overlaps that of the bacterium. Although in many cases phage infection represses host DNA synthesis, lambda phage infection does not. It often even joins the genome of the host and becomes dormant—it is therefore a lysogenic phage.

The smallest viruses contain only three to five genes. Thus the F2 RNA virus of *E. coli* has a nucleic acid chain which contains only 3000 nucleotides and so can code for approximately 1000 amino acids; of these, 100 are used to construct the proteins of the shell of the virus. Probably about 100 of the remaining 900 amino acids are present in the F2-specific RNA synthetase. The remainder of the amino acids probably form several proteins (enzymes?) involved in the reproduction of the virus.

An even smaller virus than the F2 is the one which is found in tobacco cells infected with tobacco necrosis virus. Its nucleic acid contains only about 1000 nucleotides, which would be sufficient to code for one protein of average size. Necrosis virus is never successful in invading a host cell alone and multiplies only when the cells are also infected with a different, larger (6000 nucleotide) RNA virus. Presumably the small virus uses the enzymes of the larger virus (e.g., the RNA synthetase) for its own replication (Watson, 1965).

The smallest DNA virus is polyoma, a virus that causes tumors in the mouse. The polyoma virus inhabits the nucleus of the cell. It has about 5000 nucleotide pairs in its DNA molecule, sufficient to code for several proteins. These are probably enzymes and the proteins for the virus coat (Watson, 1965).

In summary, the bacterial and virus chromosome consists of a DNA molecule of varied size, depending on the size of the genome. The DNA is generally double stranded; exceptions are the single stranded virus $\varphi$X174 (DNA) and tobacco mosaic virus (RNA).

The chromosomes of eukaryotic cells are vastly more complex than those of bacteria or viruses. This is apparent even from the number of genes present in eukaryotic cells. Thus *Drosophila* is estimated to possess 30,000 genes, and a single chromosome of man has about 10 times as many. The haploid genome of man, as found in a sperm, is estimated, on the basis of DNA content, to contain three million genes. However, it is thought that the chromosome consists of a single double-stranded DNA molecule.

It will be recalled that the eukaryotic cell chromosome contains about as much protein as DNA, weight for weight. The protein is mostly histone, largely made up of basic amino acids such as arginine and lysine. The histone is strongly bound to the acidic nucleic acid. The histone is combined, largely to the chromosome; it is not found in the cytoplasm.

## LITERATURE CITED AND GENERAL REFERENCES


Allfrey, V. G., 1963: Nuclear ribosomes, messenger RNA and protein synthesis. Exp. Cell Res. Suppl. *9*:183–212.

Allfrey, V. G., Luttau, V. C., and Mirsky, A. E., 1963: On the role of histones in regulating RNA synthesis in the cell nucleus. Proc. Natl. Acad. Sci. U.S.A. *49*:414–421.

Allfrey, V. G., and Mirsky, A. E., 1959: Biochemical properties of the isolated nucleus. *In* Subcellular Particles. Hayashi, ed. Ronald Press Co., New York, pp. 186–207.

Allfrey, V. G., Mirsky, A. E., and Stern, H., 1955: Chemistry of the cell nucleus. Advances Enzymol. *16*:411–500.

Allfrey, V. G., Osawa, S., and Mirsky, A. E., 1957: Protein synthesis in isolated cell nuclei. J. Gen. Physiol. *40*:451–490.

Anfinsen, C. B., 1959: The Chemical Basis of Evolution. John Wiley & Sons, New York.

Atwood, K. C., 1965: Transcription and translation of genes. *In* Reproduction: Molecular, Subcellular and Cellular. Locke, ed. Academic Press, New York, pp. 17–38.

Avery, O. T., MacLeod, C. M., and McCarty, M., 1944: Studies on the chemical nature of the substance inducing transformation of pneumococcal types. Induction of transformation of a DNA fraction isolated from *Pneumococcus* Type III. J. Exp. Med. *79*:137–158.

Bassel, A., Hayashi, M., and Spiegelman, S., 1964: The enzymatic synthesis of a circular DNA-RNA hybrid. Proc. Natl. Acad. Sci. U.S.A. *52*:796–804.

Beadle, G., and Beadle, M., 1966: The Language of Life. Doubleday & Co., Garden City, N. Y.

Beermann, W., 1960: Der Nukleolus als lebenswichtiger Bestandteil des Zellkerns. Chromosoma *11*:263–296.

Beermann, W., 1964: Control of differentiation at the chromosomal level. J. Exp. Zool. *157*:49–62.

Beermann, W., and Clever, U., Apr. 1964: Chromosome puffs. Sci. Am. *210*:50–58.

Bennett, H. S., 1959: Fine structure of cell nucleus, chromosomes, nucleoli, and membrane. *In* Biophysical Science, a Study Program. Oncley, ed. John Wiley & Sons, New York, pp. 297–300.

Benzer, S., 1961: Genetic fine structure. Harvey Lect. *56*:1–21.

Benzer, S., Jan. 1962: The fine structure of the gene. Sci. Am. *206*:70–84.

Bernfield, M. R., and Nirenberg, M. W., 1965: RNA codewords and protein synthesis. Science *147*: 479–483.

Birnstiel, M. L., and Flamm, W. G., 1964: Intranuclear site of histone synthesis. Science *145*: 1435–1437.

Birnstiel, M. L., Fleissner, E., and Borek, E., 1963: Nucleolus: a center of RNA methylation. Science *142*:1577–1580.

Block, D. P., and Brack, S. D., 1964: Evidence for the cytoplasmic synthesis of nuclear histone during spermiogenesis in the grasshopper *Chortaphaga viridifasciata* (DeGreer). J. Cell Biol. *22*:327–340.

Brachet, J., 1960: The Biological Role of Ribonucleic Acids. Elsevier Publishing Co., Amsterdam.

Brachet, J. A., 1965: *Acetabularia.* Endeavour *24*: 155–161.

Brachet, J., 1965: The role of nucleic acids in morphogenesis. Progr. Biophys. Molec. Biol. *15*:99–127.

Braun, W., and Whallom, J., 1954: The effects of DNA and enzyme-treated DNA on bacterial population changes. Proc. Natl. Acad. Sci. U.S.A. *40*:162–165.

Brenner, S., Jacob, F. A., and Meselson, M., 1961: An unstable intermediate carrying information from genes to ribosomes for protein synthesis. Nature *190*:576–581.

Brown, S. W., 1966: Heterochromatin. Science *151*: 417–426.

Busch, H., Starbach, W. G., Singh, E. J., and Ro, T. S., 1964: Chromosomal proteins. *In* The Role of Chromosomes in Development. Locke, ed. Academic Press, New York, pp. 51–71.

Cairns, J., 1962: Proof that the replication of DNA involves separation of the strands. Nature *194*: 1274.

Cairns, J., 1963: The bacterial chromosome and its manner of replication as seen by autoradiography. J. Molec. Biol. *6*:208–213.

Cairns, J., Jan. 1966: The bacterial chromosome. Sci. Am. *214*:37–44.

Callan, H. G., 1963: The nature of lampbrush chromosomes. Intern. Rev. Cytol. *15*:1–34.

*Carlson, E. A., 1966: The Gene: A Critical History. W. B. Saunders Co., Philadelphia.

Caspersson, T. O., 1950: Cell Growth and Cell Function. W. W. Norton & Co., New York.

Cohen, D., 1966: The Biological Role of the Nucleic Acids. American Elsevier Publishing Co., New York.

Crick, F. H. C., Oct. 1954: The structure of the hereditary material. Sci. Am. *194*:54–66.
Crick, F. H. C., 1963a: On the genetic code. Science *139*:461–464.
Crick, F. H. C., 1963b: The recent excitement in the coding problem. Progr. Nucleic Acid Res. Molec. Biol. *1*:164–217.
*Crick, F. H. C., Oct. 1966: The genetic code: III. Sci. Am. *215*:55–60. (Also Oct. 1962, *207*:66–74.)
Crippa, M., 1966: The rate of RNA synthesis during the cell cycle. Exp. Cell Res. *42*:371–375.
Danielli, J. F., 1959: The cell-to-cell transfer of nuclei in amoebae and a comprehensive cell theory. Ann. N. Y. Acad. Sci. *78*:675–687.
Danielli, J. F., 1963: The theory of cells in relation to the study of cytoplasmic inheritance in amoebae by nuclear transfer. Harvey Lect. *58*:217–231.
Davidson, E. H., June 1965: Hormones and genes. Sci. Am. *212*:35–45.
Davidson, J. N., 1960: The Biochemistry of the Nucleic Acids. 4th Ed. John Wiley & Sons, New York.
DeRobertis, E., 1964: Advances in the ultrastructure of the nucleus and chromosomes. Natl. Cancer Inst. Monogr. *14*:35–55.
*DeRobertis, E. D. P., Nowinski, V. W., and Saez, F. A., 1965: Cell Biology. 4th Ed. W. B. Saunders Co., Philadelphia.
Doudney, C. O., 1966: Requirement of RNA synthesis for DNA replication in bacteria. Nature *211*:39–41.
DuPraw, E. J., 1965: Macromolecular organization of nuclei and chromosomes: a folded fiber model based on whole-mount electron microscopy. Nature *206*:338–343.
Edstrom, J., 1960: Extraction, hydrolysis and electrophoretic analysis of RNA from microscopic tissue units (microphoresis). J. Biochem. Biophys. Cytol. *8*:39–46.
Edstrom, J., 1964: Chromosomal RNA and other nuclear RNA fractions. *In* The Role of Chromosomes in Development. Locke, ed. Academic Press, New York, pp. 137–152.
Elson, D., 1964: The ribosome factors of the cell. Discovery *25*:24–29.
Fischberg, M., and Blackler, A. W., 1963: Nuclear changes during the differentiation of animal cells. Symp. Soc. Exp. Biol. *17*:138–156.
Flamm, W. G., and Birnstiel, M. L., 1964a: Inhibiting DNA replication and its effect on histone synthesis. Exp. Cell Res. *33*:616–619.
Flamm, W. G., and Birnstiel, M. L., 1964b: The nuclear synthesis of ribosomes in cell cultures. Biochim. Biophys. Acta *87*:101–110.
Fraenkel-Conrat, H., 1959: Structure and infectivity of tobacco mosaic virus. Harvey Lect. *53*:56–58.
Fraenkel-Conrat, H., Oct. 1964: The genetic code of a virus. Sci. Am. *211*:47–54.
Gahan, P. B., 1965: Histochemical evidence for the presence of lipid on the chromosomes in animal cells. Exp. Cell Res. *39*:136–144.
Gahan, P. B., and Chayen, J., 1965: Cytoplasmic deoxyribonucleic acid. Intern. Rev. Cytol. *18*:223–247.
Gall, J. G., 1963a: Chromosomes from an interphase nucleus. Science *139*:120–121.
Gall, J. G., 1963b: Chromosomes and cytodifferentiations. *In* Cytodifferentiation and Macromolecular Synthesis. Locke, ed. Academic Press, New York, pp. 119–143.
Ganesan, A. T., and Lederberg, J., 1965: A cell-membrane bound fraction of bacterial DNA. Biochem. Biophys. Res. Commun. *18*:824–835.
Gay, H., Jan. 1960: Nuclear control of the cell. Sci. Am. *202*:126–136.
Genetic Code. 1967: Cold Spring Harbor Symp. Quant. Biol. Vol. 31, 750 pp.
Gibor, A., Nov. 1966: *Acetabularia:* a useful giant cell. Sci. Am. *25*:118–124.
Gierer, A., and Mundry, K. W., 1958: Production of mutants of tobacco mosaic virus by chemical alteration of RNA *in vitro.* Nature *182*:1457–1458.
Gillham, N. W., 1965: Linkage and recombination between non-chromosomal mutations in *Chlamydomonas reinhardii.* Proc. Natl. Acad. Sci. U.S.A. *54*:1560–1567.
Goldstein, L., 1964: Nucleocytoplasmic relationship. *In* Cytology and Cell Physiology. 3rd Ed. Bourne, ed. Academic Press, New York, pp. 559–635.
Goldstein, L., 1965: Proteins in nucleo-cytoplasmic interactions in *Ameba proteus.* J. Cell Biol. *27*:35A–36A.
Goldstein, L. (ed.), 1967: The Control of Nuclear Activity. Prentice-Hall, Englewood Cliffs, N. J.
Goldstein, L., and Micou, J., 1959: On the primary site of RNA synthesis in the cell nucleus. Anat. Rec. *134*:570 (Abstr. 145).
Green, D. E., and Goldberger, R. F., 1966: Molecular Insights into the Living Processes. Academic Press, New York.
Gudbjarnzson, S., and De Schryver, C., 1964: Nuclear ribosomes, an early factor in tissue reparation. Biochem. Biophys. Res. Commun. *14*:12–16.
Guest, J. R., and Yanofsky, C., 1966: Relative orientation of gene, messenger, and polypeptide chain. Nature *210*:799–802.
Gurdon, J. B., 1966: The cytoplasmic control of gene activity. Endeavour *25*:95–99.
Hammerling, J., 1963: Nucleo-cytoplasmic interactions in *Acetabularia* and other cells. Ann. Rev. Plant Physiol. *14*:65–92.
Harris, R. J. C. (ed.), 1960: Protein Biosynthesis. Academic Press, New York.
Harvey, E. B., 1936: Parthenogenetic merogony or cleavage without nuclei in *Arbacia punctulata.* Biol. Bull. *71*:101–121.
Hayashi, M., Hayashi, M. N., and Spiegelman, S., 1963: Restriction of *in vivo* genetic transcription to one of the complementary strands of DNA. Proc. Natl. Acad. Sci. U.S.A. *50*:664–672.
Hershey, A. D., and Chase, M., 1952: Independent function of viral protein and nucleic acid in growth of bacteriophage. J. Gen. Physiol. *36*: 39–56.
Hoagland, M. B., Dec. 1959: Nucleic acids and proteins. Sci. Am. *201*:55–61.
Hogness, D. S., 1962: The activity of viral DNA in the genesis of proteins in the bacterial cell. *In* The Molecular Control of Biological Activity. Allen, ed. McGraw-Hill Book Co., New York, pp. 189–220.

Holley, R. W., Feb. 1966: The nucleotide sequence of a nucleic acid. Sci. Am. *214*:30–39.

Holley, R. W., Apgar, J., Everett, G. A., Madison, J. T., Marquisee, N., Merrill, S. H., Penswick, J. R., and Zamir, A., 1965: Structure of a ribonucleic acid. Science *147*:1462–1465.

Honigberg, B. M., and Read, C. P., 1960: Virulence transformation of a trichomonad protozoan. Science *131*:352–353.

Hotchkiss, R. D., 1955a: The genetic material of the pneumococcal transformations. Harvey Lect. *49*: 124–144.

Hotchkiss, R. D., 1955b: Bacterial transformation. J. Cell. Comp. Physiol. *45*(Suppl. 2):1–22.

Hotchkiss, R. D., and Weiss, E., Nov. 1956: Transformed bacteria. Sci. Am. *195*:48–53.

Hurwitz, J., and August, J. T., 1963: The role of DNA in RNA synthesis. Progr. Nucleic Acid Res. Molec. Biol. *1*:59–92.

Hurwitz, J., and Furth, I. J., Feb. 1962. Messenger RNA. Sci. Am. *206*:41–49.

Hyde, B. B., 1965: Ultrastructure in chromatin. Prog. Biophys. Molec. Biol. *15*:131–148.

*Ingram, V. M., 1963: The Hemoglobins in Genetics and Evolution. Columbia University Press, New York.

*Ingram, V. M., 1966: The Biosynthesis of Macromolecules. W. A. Benjamin, New York.

Inoué, S., and Sato, H., 1962: Arrangement of DNA in living sperm: a biophysical analysis. Science *136*:1122–1124.

Irvin, J. L., Holbrook, H. J., Jr., Evans, J. H., McAllister, H. C., and Stiles, E. P., 1963: Possible role of histones in regulation of nucleic acid synthesis. Exp. Cell Res. Suppl. *9*:359–366.

Jacob, F., 1966: Genetics of the bacterial cell. Science *152*:1470–1478.

Jinks, J. L., 1964: Extrachromosomal Inheritance. Prentice-Hall, Englewood Cliffs, N. J.

Jukes, T. H., 1963, 1965: The genetic code. Am. Sci. *51*:227–244; *53*:477–494.

Kaji, H., and Kaji, A., 1964: Specific binding of sRNA with the template ribosome complex. Proc. Natl. Acad. Sci. U.S.A. *52*:1541–1547.

Kendrew, J. C., 1966: The Thread of Life. Harvard University Press, Cambridge, Mass.

King, T. J., and Briggs, R., 1956: Serial transplantation of embryonic nuclei. Cold Spring Harbor Symp. Quant. Biol. *21*:271–290.

Kleinschmidt, A. K., Lang, D., Plesher, C., Hellmann, W., Haas, V., Zahn, P. K., and Hagedorn, A., 1961: Über die intrazellulare Formation von Bakterien—DNS. Z. Naturforsch. *16b*:730–739.

Kornberg, A., 1960: Biologic synthesis of DNA. Science *131*:1503–1508.

Kurnick, N. B., and Mirsky, A. E., 1950: Methyl green staining. II. Stoichiometry of reaction with nucleic acids. J. Gen. Physiol. *33*:265–274.

Lengyel, P., Speyer, J. F., Basilio, C., and Ochoa, S., 1962: Synthetic polynucleotides and the amino acid code. Proc. Natl. Acad. Sci. U.S.A. *48*:282–284.

Lipser, D., and Perrin, 1963: Complementation on ribosomes. Cold Spring Harbor Symp. Quant. Biol. *28*:533–537.

Littau, V. C., Burdick, C. J., Alfrey, V. G., and Mirsky, A. E., 1965: The role of histones in the maintenance of chromatin structure. Proc. Natl. Acad. Sci. U.S.A. *54*:1204–1212.

Loftfield, R. B., 1957: The biosynthesis of protein. Progr. Biophys. *8*:348–386.

Maggio, R., Siekevitz, P., and Palade, G. E., 1963a: Studies on isolated nuclei: II. Isolation and chemical characterization of nucleolar and nucleoplasmic subfractions. J. Cell Biol. *18*:293–312.

Maggio, R., Siekevitz, P., and Palade, G. E., 1963b: Studies on cell nuclei. I. Isolation and chemical characterization of a nuclear fraction from guinea pig liver. J. Cell Biol. *18*:267–291.

Marmur, J., Rownd, R. and Schildkraut, C. L., 1963: Denaturation and renaturation of DNA. Progr. Nucleic Acid Res. Molec. Biol. *1*:232–300.

Marshall, R. E., Caskey, C. T., and Nirenberg, M., 1967: Fine structure of RNA codewords recognized by bacterial, amphibian and mammalian transfer RNA. Science *155*:820–826.

Mayor, H. D., 1961: Differential staining of single stranded DNA, double stranded DNA, and RNA. Proc. Soc. Exp. Biol. Med. *108*:103–105.

Mayor, H. D., and Diwan, A. R., 1961: Studies on the acridine orange staining of two purified RNA viruses: polio virus and TMV. Virology *14*:74–82.

Mazia, D., 1954: The particulate organization of the chromosome. Proc. Natl. Acad. Sci. U.S.A. *40*:521–527.

Mazia, D., and Hirshfield, H., 1950: Prenuclear dependence of $P^{32}$ uptake by the cell. Science *112*: 297–299.

McCarthy, J., 1965: The evolution of base sequences in polynucleotides. Progr. Nucleic Acid Res. and Molec. Biol. *4*:129–160.

*Meselson, M., and Stahl, F. W., 1958: The replication of DNA in *E. coli.* Proc. Natl. Acad. Sci. U. S. A. *44*:671–682.

Mirsky, A. E., 1950–51: The chemical composition of chromosomes. Harvey Lect. *46*:98–115.

Mitchison, J. M., 1966: Some functions of the nucleus. Intern. Rev. Cytol. *19*:97–110.

Moore, J. A., 1960. Nuclear transfer of embryonic cells of the amphibia. *In* New Approaches in Cell Biology. Walker, ed. Academic Press, New York.

Moses, M. J., 1964: The nucleus and chromosomes: a cytological perspective. *In* Cytology and Cell Physiology. 3rd Ed. Bourne, ed. Academic Press, New York, pp. 423–558.

Nance, W. E., 1963: Genetic control of hemoglobin synthesis. Science *141*:123–141.

*Nirenberg, M. W., March 1963: The genetic code: II. Sci. Am. *208*:80–94.

Nirenberg, M. W., Jones, O. W., Leder, P., Clark, B. F. C., Sly, W. S., and Pestka, S., 1963: On the coding of genetic information. Cold Spring Harbor Symp. Quant. Biol. *23*:549–557.

Nirenberg, M., and Leder, P., 1964: RNA codewords and protein synthesis. Science *145*:1399–1407.

Nirenberg, M., and Matthei, J. H., 1962: An intermediate in the biosynthesis of polyphenylalanine directed by synthetic template RNA. Proc. Natl. Acad. Sci. U.S.A. *48*:104–109.

Ochoa, S., Burma, D. P., Kröger, H., and Weill, J. D., 1961: DNA-dependent incorporation of nu-

cleotides from mucleoside triphosphates into RNA. Proc. Natl. Acad. Sci. U. S. A. *47*:670–680.
Osawa, S., 1965: Biosynthesis of ribosomes in bacterial cells. Progr. Nucleic Acid Res. Molec. Biol. *4*:161–188.
Peacock, W. J., 1963: Chromosome duplication and structure as determined by autoradiography. Proc. Natl. Acad. Sci. U.S.A. *49*:793–801.
Perry, R. P., 1962: The cellular sites of synthesis of ribosomal and 4S RNA. Proc. Natl. Acad. Sci. U. S. A. *48*:2179–2186.
Perry, R. P., 1964: The role of the nucleolus in RNA metabolism and other cellular processes. Nat. Cancer Inst. Monogr. *14*:73–89.
Prescott, D. M., 1964: Cellular site of RNA synthesis. Progr. Nucleic Acid Res. Molec. Biol. *3*:33–57.
Prescott, D. M., and Bender, M. A., 1963: Autoradiographic study of chromatid distribution of labeled DNA in two types of mammalian cells *in vitro*. Exp. Cell Res. *29*:430–442.
Prescott, D. M., and Goldstein, L., 1967. Nuclear-cytoplasmic interaction in DNA synthesis. Science *155*:469–470.
Reich, E., and Goldbert, I. H., 1964: Actinomycin and nucleic acid function. Progr. Nucleic Acid Res. and Molec. Biol. *3*:183–234.
Roels, H., 1966: Metabolic DNA: a cytochemical study. Intern. Rev. Cytol. *19*:1–34.
Sager, R., 1966: Mendelian and non-Mendelian heredity: a reappraisal. Proc. Roy. Soc. London *B164*:290–297.
Shapiro, L., and August, T. T., 1966: RNA virus replication. Bact. Rev. *30*:279–287.
Sherif, M., 1964: Intravital fluorescence microscopy in the qualitative evaluation of the interaction of acridine orange with nucleic acid in intact living human organs. Nature *204*:390–391.
Siebert, G., and Smellie, R. M. S., 1957: Enzymatic and metabolic studies on isolated nuclei. Intern. Rev. Cytol. *6*:383–424.
*Sinsheimer, R. L., July 1962: Single stranded DNA. Sci. Am. *207*:109–116.
Sirlin, J. L., 1960: Cell sites of RNA and protein synthesis in the salivary gland of *Smittia* (Chironomidae). Exp. Cell Res. *19*:177–180.
Sirlin, J. L., 1961: The nucleolus of the cell nucleus. Endeavour *20*:146–153.
Sirlin, J., 1962: The nucleolus. Progr. Biophys. *12*:25–66.
Sirlin, J. L., Jacob, J., and Tandler, C. J., 1963: Transfer of the methyl group of methionine to nucleolar RNA. Biochem. J. *89*:447–452.
*Spiegelman, S., May 1964: Hybrid nucleic acids. Sci. Am. *210*:48–56.
Spirin, A. S., 1963: Some problems concerning the macromolecular structure of ribonucleic acid. Progr. Nucleic Acid Res. Molec. Biol. *1*:301–345.
Stevens, B. J., 1965: The fine structure of the nucleolus during mitosis in the grasshopper neuroblast cell. J. Cell Biol. *29*:349–368.
Swift, H., 1962: Nucleic acids and cell morphology in dipteran salivary glands. *In* Molecular Control of Cellular Activity. Allen, ed. McGraw-Hill Book Co., New York, pp. 73–125.
Taylor, A. R., 1946: Chemical analysis of the $T_2$ phage and its host, *E. coli* (strain B). J. Biol. Chem. *165*:271–284.
Taylor, J. H., 1959: Autoradiographic studies of nucleic acids and proteins during meiosis in *Lillium longiflorum*. Am. J. Bot. *46*:477–484.
Taylor, J. H., 1964: The arrangement of chromosomes in the mature sperm of the grasshopper. J. Cell Biol. *21*:286–289.
Taylor, J. H., Woods, P. S., and Hughes, W. L., 1957: The organization and duplication of chromosomes as revealed by autoradiographic studies using tritium labelled thymidine. Proc. Natl. Acad. Sci. U.S.A. *42*:122–128.
Thomas, C. A., Jr., 1966: The arrangement of information in DNA molecules. J. Gen. Physiol. *49*:143–169.
Vincent, W. S., 1955: Structure and chemistry of nucleoli. Intern. Rev. Cytol. *4*:269–298.
Vincent, W. S., and Miller, O. R., Jr. (eds.), 1966: International Symposium on the Nucleolus, Its Structure and Function. National Cancer Institute Monograph No. 23. U. S. Government Printing Office, Washington, D. C.
Vinograd, J., and Lebowitz, J., 1966: Physical and topological properties of circular DNA. J. Gen. Physiol. *49*:103–125.
Wagner, R. P., and Mitchell, H. K., 1963: Genetics and Metabolism. 2nd Ed. John Wiley & Sons, New York.
Watson, J. D., 1963: Involvement of RNA in the synthesis of proteins. Science *140*:17–26.
*Watson, J. D., 1965: The Molecular Biology of the Gene. W. A. Benjamin, New York.
Watson, J. D., and Crick, F. H. C., 1953: Molecular structure of nucleic acids. A structure for deoxyribose nucleic acid. Nature *171*:737–738.
Watson, M. L., 1959: Further observations on the nuclear envelope of animal cells. J. Biophys. Biochem. Cytol. *6*:147–156.
Wettstein, R., and Sotelo, J. R., 1965: Fine structure of meiotic chromosomes. J. Ultrastruc. Res. *13*:367–381.
Wolfe, S. L., 1965: The fine structure of isolated chromosomes. J. Ultrastruc. Res. *12*:104–112.
Wollman, E. L., and Jacob, F., July 1956: Sexuality in bacteria. Sci. Am. *195*:109–118.
Yanofsky, C., 1963: Genetic control of protein structure. *In* Cytodifferentiation and Macromolecular Syntheses. Locke, ed. Academic Press, New York, pp. 15–29.
Yanofsky, C., May 1967a: Gene structure and protein structure. Sci. Am. *216*:90–94.
Yanofsky, C., 1967b: Gene structure and protein structure. Harvey Lect. *61*:145–168.
Yanofsky, C., Carlton, B. O., Guest, J. R., Helinski, D. R., and Henning, U., 1964: On the colinearity of gene structure and protein structure. Proc. Natl. Acad. Sci. U.S.A. *51*:266–272.
Yanofsky, C., and Ito, J., 1967: Orientation of antipolarity in the tryptophan operon of *Escherichia coli*. J. Molec. Biol. *24*:143–145.
Zinder, N. D., 1955: Bacterial transduction. J. Cell. Comp. Physiol. *45*(Suppl. 2):23–49.
Zinder, N. D., Nov. 1958: Transduction in bacteria. Sci. Am. *199*:38–43.

SECTION
II

# The Cell Environment

Microorganisms and single celled animals and plants are subjected directly to all the changes in the external environment: changes in concentration of oxygen, carbon dioxide, salts and hydrogen ions and changes in temperature and in the quality and intensity of light. Cells of multicellular animals and plants are also subjected to all such changes, but their cells are protected to various degrees from the external environment by protective coatings (e.g., skin), by the proximity of neighboring cells and by the development of an internal environment in the form of sap, body fluid or blood, in which considerable constancy of conditions is maintained. The greatest control of the internal environment of cells is reached in the warm blooded animals, in which the oxygen, hydrogen ion concentration, salts, foods, proteins and temperature of the blood which bathes all the cells of the animal are maintained within narrow limits. Such control has been called *homeostasis* (Cannon, 1932).

It will be interesting to consider the degree of variation in the factors of the natural environments and the degree to which these variations are tolerated by various forms of life, because from this inquiry certain fundamental characteristics of life will become apparent. Furthermore, it will be of interest to see to what extent an organism which controls its internal environment has been able to invade and occupy successfully an environment inaccessible to an organism without such control (Prosser and Brown, 1961). Only the major environmental factors will be considered here; e.g., water, carbon dioxide, oxygen, salts, pressure, hydrogen ion concentration, radiations (light, ultraviolet rays and ionizing radiations) and temperature.

# CHAPTER 7 / WATER, GASES AND PRESSURE IN THE CELL ENVIRONMENT

## WATER AS A FACTOR IN THE CELL ENVIRONMENT

A cell must always possess direct or indirect water frontage, inasmuch as all cell reactions occur in aqueous solutions. In simpler forms of life this is achieved by direct contact with environmental water. The more complex plant or animal may become covered with an outer coating impervious to water, but each of its cells has contact with the aqueous environment of the internal medium (the *milieu interieur* of Claude Bernard, 1879).

Water makes up 80 per cent of the cell. When the cell loses its water, life is suspended or extinguished. Activity reappears in desiccated, dormant cells when water again becomes available. The cells of some dried mosses, lichens, rotifers, tardigrades and nematodes resist unfavorable conditions for years, only to resume activity quickly when moistened, provided other conditions are favorable. Many protozoa encyst and some bacteria sporulate when deprived of water. Such dried resistant stages are especially characteristic of fresh water and terrestrial organisms which are subjected to seasonal drying.

Many properties of water make it suitable not only as the main constituent of the living cell, but also as the main factor in the cell environment (Henderson, 1927).

***Water as a Temperature Stabilizer.*** Among the most important properties of water are its *temperature-stabilizing* properties, especially its high specific heat, its high heat of vaporization and its high heat of fusion (Edsall and Wyman, 1958).

The *specific heat* of a substance is defined as the number of calories required to raise the temperature of 1 gram of that substance 1° C. The specific heat of water is 1.0, a value exceeded only by that of liquid ammonia.

The *latent heat of vaporization* is the number of calories required to change 1 gram of liquid to vapor. For water at 100° C. this is 539 calories, one of the highest values known. Close to the heat of vaporization of water is that of hydrofluoric acid (360 calories); that of ammonia is less.

The *heat of fusion* is the number of calories required to convert 1 gram of solid at the freezing point to 1 gram of liquid at the same temperature. For water it is 80 calories and for ammonia, 108 calories per gram. Although the heat of fusion of water is less than that of ammonia, it is greater than that of most other organic or inorganic materials tested.

How do these properties of water provide a better control of temperature on the surface of the earth as well as in the external and internal environment of the cell? This can be illustrated best by a number of examples. Radiant energy of the order of $10^{15}$ cal./km.$^2$ a year falls on the surface of the earth at the equator. This is equivalent to the energy liberated by the fission of 100 kg. of uranium-235 or by the burning of 700,000 tons of coal. If a dry land mass made up of rock and soil with a low specific heat were to absorb this much radiant energy, it would rise to a lethal temperature by day and fall below the freezing point of water by night, as it apparently does on the moon. However, the water on earth and water vapor in the atmosphere of the earth absorb some of the incoming radiation in sunlight, preventing overheating in daylight. Some of the sunlight which gets through the atmospheric water vapor "blanket" is absorbed and heats the land masses (Fig. 7.1). As the heat is reradiated from the land by night, the atmospheric water vapor absorbs the outgoing infrared rays, preventing the earth from cooling at night as rapidly as it otherwise would. This so-called "greenhouse effect" of the atmospheric water vapor, the moderation of the temperature of the land masses, is of special importance to life (Henderson, 1927; Blum, 1955; Plass, 1958, 1959).

Because of the high specific heat of water, the temperature of the great bodies of water covering the earth's crust is increased much less than that of the land. A body of water can absorb or lose a considerable amount of heat without much change in its temperature; that is, its heat capacity is high. The *heat capacity* of a body may be defined as the number of calories

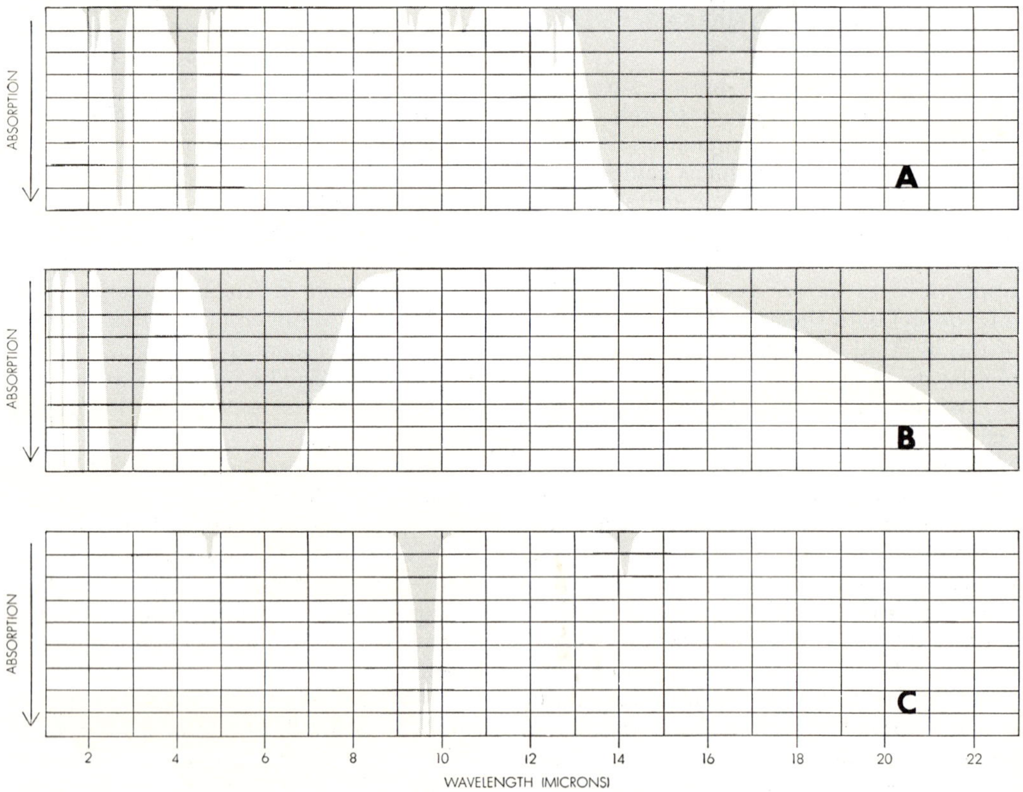


***Figure 7.1.*** Absorption of infrared radiations in the earth's atmosphere by carbon dioxide (*A*), water vapor (*B*) and ozone (*C*). (From Carbon Dioxide and Climate, by Gilbert N. Plass. Copyright © 1959 by Scientific American, Inc. All rights reserved.)

required to raise the temperature of the entire body 1° C. The larger a body is and the higher its specific heat, the greater is its heat capacity. The vastness of the oceans makes them of especial importance as thermostats resisting temperature change. Consequently, the climate of land masses is tempered by nearby water.

The oceans moderate the temperature on earth in yet another way. Because of the high heat of vaporization of water, a tremendous amount of the sun's heat is absorbed during evaporation from the vast surface of the oceans. The vapor holds the heat until the warm, water-laden air is chilled in a cooler part of the earth. Then the vapor condenses and liberates the heat, thereby counteracting the coolness of the region in which precipitation occurs. The water-laden air may move far from its original source before it condenses and gives off its heat; therefore, the heat absorbed at the surface of the ocean may moderate climatic conditions far away. For example, heat absorbed in the tropics may be carried to the northern polar regions (Davis and Day, 1961).

The high heat of fusion of water is another factor serving to reduce radical changes in temperature, since ice is formed from water only after a considerable loss of heat, and it melts only when this amount of heat is replaced. The system:

$$[\text{water}]_{0^\circ\text{C.}} \underset{+\text{heat}}{\overset{-\text{heat}}{\rightleftharpoons}} [\text{ice}]_{0^\circ\text{C.}} \qquad (7.1)$$

remains at 0° C. during the withdrawal of heat, accompanied by ice formation, or the addition of heat, melting the ice.

An additional thermostabilizing property of water is its high specific gravity (weight per unit volume) at 4° C. Because of this, pure water cooled below 4° C. rises and freezes only at the surface. If it were not for this, bodies of water such as lakes might freeze solid and become uninhabitable (Henderson, 1927).

Finally, water is a very much better heat conductor than any other common liquid. Consequently the heat it absorbs is readily distributed. This tends to equalize the temperature, not only in the external and internal environment but also inside cells.

***Water as a Solvent.*** Besides serving as a thermostabilizer, *water is the best solvent known*—more substances dissolve in it than in any other solvent. Electrolytes dissolved in water readily ionize because water molecules, by virtue of their own charged nature (see Fig. 7.2), act as small dipoles attracting the ions of the solute, thereby weakening the attraction between them. In more technical terms it may be said that water is a polar solvent of high *dielectric constant.* Such a constant can be measured by placing the vapor or solution of a substance between the plates of a condenser and comparing the attraction between the plates with that which occurs in a vacuum. For example, water has a dielectric constant of 78.5, ethyl alcohol 24.3 and benzene 2.28. Ionization and ionic reactions are of great importance to life, as will be shown in later chapters. For all its importance as a solvent, water is remarkably inert and innocuous and most substances are unchanged by being dissolved in it. Therefore, they may remain in solution in the cell, unaltered until such time as they may be utilized.

Water is also necessary for *hydrolyses* and for biological *oxidations,* the two chemical reactions most fundamental to the living cell.

***Special Properties of Water.*** Water has several other properties which make it particularly fit as part of the external and internal environment. It is *transparent to light,* enabling the specialized photosynthetic organelles, the chloroplasts, inside the plant cells to absorb the sunlight—a process upon which depends all life on earth. The transparency of the ocean also enables floating plants and plants attached to rocks to grow below the surface. This property is of considerable importance since marine plants are the ultimate source of food for most marine animals. The *high surface tension* of water is also an important property since it enables the water to move closer to the surface of the soil than would otherwise be possible, making it available to life.

The properties of water and ice continue to be an area of active study (Frank, 1965; Runnels, 1966; Claussen, 1967). Interest has been focused on the properties of water in the interpretation of resistance of cells to freezing (Meryman, 1966).

Attention has recently been directed to some anomalies in the curve relating to temperature the surface tension and viscosity of water, which shows changes in its slope at 15, 30, 45 and 60° C. Since several properties of biological systems also show changes at these temperatures, such changes in properties of water are thought attributable to the changes in structure in water at these transition points (Drost-Hansen, 1965a, b). However, some of the measurements indicating discontinuities in the properties of water are considered by other investigators to be within the limits of error (Falk and Kell, 1966).

***Structure of Water.*** Some of the remarkable properties of water are now interpretable on the basis of its structure. The shape of a water molecule is that of an isosceles triangle; the intermolecular O—H distance is nearly 0.99 Å, and the H—O—H angle is not far from 105°. The powerful attraction of the oxygen nucleus tends to draw electrons away from the protons (H nuclei), leaving the region around them with a net positive charge. Since only two pairs of electrons around the oxygen atom are shared with the protons, the two other pairs in the eight-electron shell point outward from the O—H bond, attracted to the net positive charge of other water molecules with a force of about 4500 cal./mole. A group of water molecules therefore describes a tetrahedron around the oxygen atom, the positively charged region of one water molecule tending to orient itself toward a negatively charged region of one of its neighbors. This is shown in Figure 7.2. Each water molecule therefore tends to have four near neighbors and each oxygen atom is the center of a tetrahedron of other oxygens, the O—O distance, as determined by x-ray diffraction studies, being 2.76 Å (Edsall and Wyman, 1958).

When ice melts about 15 per cent of the highly regular arrangements found in ice break. Even at higher temperatures, however, some of the structure remains and the high heat of vaporization of water indicates some quasicrystalline structure even at 100° C.

The type of linkage indicated in Figure 7.2 by a dotted line between the proton and the neighboring oxygen atoms is called a *hydrogen bond.* As already pointed out in Chapter 3, hydrogen bonds are of great importance in aqueous systems and occur in solutions of proteins and other

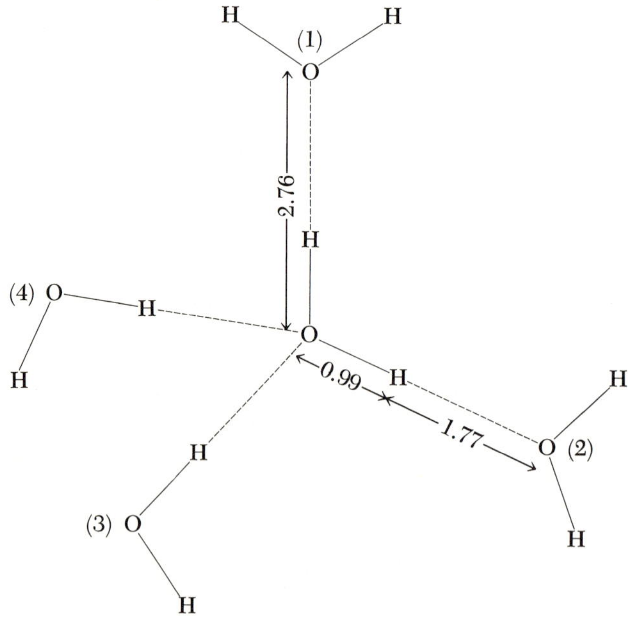


***Figure 7.2.*** The dipolar nature of the water molecule. The diagram shows the orientation of water molecules through tetrahedral coordination in ice. Molecules (1) and (2), as well as the central water molecule, lie entirely in the plane of the paper; molecule (3) lies above this plane, and molecule (4) below it, so that oxygens (1), (2), (3) and (4) lie at the corners of a regular tetrahedron. In water the orientation is not as regular, but multiples occur bound by hydrogen bonds between the molecules (shown in dashes). The distances are given in Angstrom units ($10^{-8}$ cm.). (From Edsall and Wyman, 1958: Biophysical Chemistry. Academic Press, New York.)

relatively polar substances. Formation and breakage of hydrogen bonds (about 4500 cal./mole) give water most of its peculiar properties. For example, the heat of vaporization of water is high because all the hydrogen bonds between a water molecule and its neighbors must be broken before it can escape from the surface of the liquid. The high heat of fusion results from similar attraction (Buswell and Rodebush, 1956).

It is difficult to conceive of life without water. For this reason one of the first questions exobiologists ask in considering whether life exists on other planets is: does water occur on the surface or in the atmosphere of the planet? (Lederberg, 1960; Horowitz, 1966; Pittendrigh *et al.,* 1966). On the other hand, there is always the question whether life in the dry state or in some other solvent (e.g., ammonia) is truly impossible and inconceivable (Wunder, 1966).

## CARBON DIOXIDE IN THE CELL ENVIRONMENT

Carbon dioxide is present in the crust of the earth, in the air and in all natural waters. It is difficult to displace carbon dioxide from water; even after water is boiled, it is rapidly acidified by the reentry of carbon dioxide which dissolves in it. The great solubility of carbon dioxide in water is evident when it is compared with that of other gases present in the atmosphere of the earth. For example, 1.715 liters of carbon dioxide are absorbed at 0° C. by 1 liter of distilled water when the pressure of the carbon dioxide is 1 atmosphere, as compared with 0.049 liters for oxygen and 0.024 liters for nitrogen (see Table 7.1). Increases in temperature decrease the solubility of carbon dioxide in water, as shown in Table 7.1. Increases in salt content of the water, as measured by the chlorinity, have a like effect (see Table 7.2). Since carbon dioxide enters water so readily, an equilibrium always exists between the carbon dioxide in water and in air; water can never completely remove it from air, nor can air remove it from the water. At room temperature the amount of carbon dioxide dissolved in water is about equal to that in the air with which the water is in contact.

Forming an acid as it dissolves in water, carbon dioxide increases the solvent powers of water for many substances. As a weak acid, it forms one of the primary buffers of the internal fluids of organisms and of terrestrial water (Chapter 8).

**Table 7.1.** Solubility or Absorption Coefficients of Some Gases

| TEMPERATURE °C. | SOLUBILITY COEFFICIENTS | | |
|---|---|---|---|
| | $CO_2$ | $O_2$ | $N_2$ |
| 0 | 1.713 | 0.0489 | 0.0235 |
| 5 | 1.424 | 0.0428 | 0.0209 |
| 10 | 1.194 | 0.0380 | 0.0186 |
| 15 | 1.019 | 0.0341 | 0.0168 |
| 20 | 0.878 | 0.0310 | 0.0154 |
| 25 | 0.759 | 0.0283 | 0.0143 |

The solubility or absorption coefficient of a gas is the volume of gas reduced to 0°C. and 760 mm. Hg dissolved per volume of distilled water when the partial pressure of the gas is 1 atmosphere. To obtain milliliters of gas per milliliters water in equilibrium with air, multiply by the partial pressure of the gas in air: 0.03 per cent for $CO_2$, 20.99 per cent for $O_2$ and 78.03 per cent for $N_2$. (Data from Handbook of Chemistry and Physics. 1964. 45th Ed. Chemical Rubber Co., Cleveland.)

**Table 7.2.** Solubility Coefficients of Carbon Dioxide, Showing Variation with Temperature and Chlorinity

| TEMPERATURE °C. | CHLORINITY IN GM./KG. | | |
|---|---|---|---|
| | 0 | 16 | 20 |
| | SOLUBILITY COEFFICIENTS OF $CO_2$ | | |
| 0 | 1.715 | 1.489 | 1.438 |
| 12 | 1.118 | 0.980 | 0.947 |
| 24 | 0.782 | 0.695 | 0.677 |

Chlorinity is the halide content of sea water in grams per kilogram. Salinity = 0.03 + 1.805 × chlorinity. For other terms see Table 7.1. For conversion of data to milliliters of carbon dioxide in equilibrium with air multiply by the partial pressure of carbon dioxide in air, 0.03 per cent. (Data from Sverdrup, Johnson and Fleming, 1942.)

The importance of carbon dioxide as a raw material for photosynthesis in plant cells and as one of the wastes from the respiratory processes of plant and animal cells will also be discussed later (Chapters 18 and 19). It has been calculated that the carbon dioxide in the earth's atmosphere is renewed every 300 to 400 years, that is, each carbon dioxide molecule is incorporated into a plant cell every 300 to 400 years (Rabinowitsch, 1945).

Carbon dioxide in the air absorbs some of the sunlight entering the atmosphere at wavelengths between 2 and 5 $\mu$, and in this manner tempers the heating of the surface of the earth's crust. It also absorbs strongly the radiations between 13 and 17 $\mu$, near the wavelengths at which the infrared radiations reradiated from the earth's crust are most intense (Fig. 7.1), thereby delaying the cooling of the surface of the earth with nightfall much as in the "greenhouse effect" described for water vapor.

A current theory ascribes an important role to carbon dioxide in regulating the overall climate of the earth (Plass, 1958, 1959). Thus, the increase in carbon dioxide content of the atmosphere (11 per cent since 1870), from burning of fossil fuels is thought to account for the rising temperature at the surface of the earth for the last 90 years. If the carbon dioxide concentration in the earth's atmosphere should double, the average temperature increase of the earth's surface would be 3.6° C. If the concentration of carbon dioxide should decrease to one-half of its present value, the average temperature would fall by 3.6° C., enough to cause great ice sheets to form again, as in the great glacial epochs of the earth's history. However, some workers postulate that variations in the output of the sun's radiation may also account for the great changes in climate (Opik, 1958).

## OXYGEN IN THE CELL ENVIRONMENT

Because most cells of organisms are aerobic, requiring oxygen to carry on life activities, oxygen is an important factor of the cellular environment. Oxygen is present in the air and constitutes 20.99 per cent of the atmosphere of the earth. The oxygen in the atmosphere becomes available to living cells as it is dissolved in the liquid medium outside and inside the cell. The oxygen of the atmosphere is continually being replenished as it is liberated during photosynthesis of green cells.

Oxygen is much less soluble in water than is carbon dioxide and diffuses slowly in water. If oxygen were to be distributed by diffusion alone, the oxygen content of a body of water would fall off rapidly below the surface because of the slowness of the diffusion process and aerobic life would be impossible in the depths. However, turbulence caused by currents or welling up mixes a body of water with water containing oxygen and thereby carries oxygen down from the surface waters, greatly facilitating its distribution.

The photosynthetic process of plant cells present in water is also a major source of oxygen and water may even become saturated or supersaturated with oxygen in the neighborhood of photosynthetically active plant cells. On the other hand, if photosynthetic plants are absent and organic material is available, water may become devoid of oxygen as a consequence of the respiration of microorganisms of decay. In muds containing much organic matter, as for example those under brine pools, oxygen is absent within less than a millimeter of the surface. Wet, mucky soil is usually more or less lacking in oxygen, but loose soil contains air pockets with oxygen equal in concentration to that of air.

The solubility of oxygen in water falls with increase in temperature (see Table 7.1). The deleterious effect of high temperature on animals confined in an aquarium is often the result of lack of oxygen in the water, and care must be exercised to provide an adequate supply.

The solubility of oxygen in water also decreases with increase in salinity (see Table 7.3). The oxygen content of sea water is therefore slightly less than that of fresh water at the same temperature. The oxygen content of brine, which may

**Table 7.3.** Solubility Coefficients of Oxygen, Showing Variation with Temperature and Chlorinity

| TEMPERATURE °C. | CHLORINITY IN GM./KG. 0 | 15 | 17 | 20 |
|---|---|---|---|---|
| | SOLUBILITY COEFFICIENTS OF $O_2$ | | | |
| 0 | .0489 | .0406 | .0395 | .0378 |
| 5 | .0429 | .0359 | .0350 | .0336 |
| 10 | .0380 | .0321 | .0313 | .0301 |
| 15 | .0342 | .0292 | .0284 | .0274 |
| 20 | .0310 | .0267 | .0262 | .0253 |
| 25 | .0283 | .0246 | .0240 | .0231 |

For definition of terms see Tables 7.1 and 7.2. For conversion of data to milliliters of oxygen in equilibrium with air multiply by the partial pressure of oxygen in air, 20.99 per cent. (Data from Sverdrup, Johnson and Fleming, 1942; and Handbook of Chemistry and Physics. 1964. 45th Ed.)

contain 10 times as much salt as sea water, is very much less (Sverdrup *et al.,* 1942).

The gaseous content of liquids is presented in a variety of ways. In Tables 7.1 and 7.3 it is given in terms of the absorption or *solubility coefficient,* which is the volume of gas dissolved per volume of liquid when the partial pressure of the gas in question is 1 atmosphere. It is of greater biological interest to know how much of a particular gas is dissolved in a given volume of liquid in equilibrium with air at 1 atmosphere pressure rather than with the pure gas. According to Dalton's law, the total pressure of a mixture of gases is equal to the sum of the partial pressures of the gases constituting the mixture.

If a liquid is exposed to a mixture of gases, each gas dissolves in proportion to its own *partial pressure* and independently of the other gases present. The volume of a gas dissolved per unit volume of water is determined by multiplying the solubility coefficient of the gas by the ratio of its partial pressure to the total gas pressure (e.g., at 1 atmosphere). To obtain the volume of a gas dissolved in any volume of water, the volume of gas dissolved per unit volume of water need only be multiplied by the volume of the water. For example, to determine the oxygen content of 100 ml. of water in equilibrium with air at 20° C. and 1 atmosphere pressure, multiply the absorption coefficient, 0.031 (see Table 7.1), by 0.2099, since air contains 20.99 per cent oxygen, and multiply the product by 100. This gives a value of 0.65 ml. oxygen per 100 ml. distilled water.

In physiological literature the volume of the dissolved gas per 100 volumes of liquid is often spoken of as "volumes per cent." However, the concentration of dissolved oxygen, from a physical-chemical standpoint, is better expressed in terms of the partial pressure of oxygen with which the liquid would be in equilibrium, a value which some physiologists call the *oxygen tension* of the liquid (Krogh, 1941).

The partial pressure of the oxygen is readily determined, if the solubility coefficient of the oxygen and the volume dissolved in the liquid under the particular experimental conditions are known, by the use of the equation:

$$\text{volume} = \text{partial pressure} \times \text{solubility coefficient} \quad (7.2)$$

where the volume of dissolved oxygen is given as volumes of gas per unit volume of liquid, the partial pressure in atmospheres (converted to mm. Hg by multiplying the fraction found by 760 mm.) and the solubility coefficient in volumes of oxygen per volume of liquid under the conditions of the experiment (Guyton, 1966).

The problem of an adequate oxygen supply for all the cells of a multicellular animal is one which in the course of evolution has been solved by the development of oxygen-transporting pigments such as the hemoglobin in our blood. When these are lacking the body fluid contains only as much oxygen as an equal volume of water under the same conditions of temperature and pressure (e.g., in many echinoderms). However, when an oxygen-transporting pigment is present the amount of oxygen rises in some mollusks to 1 or 2 volumes per cent; in cephalopod mollusks, chironomid larvae (larvae of midges in pond mud) and annelid worms, 2 to 10 volumes per cent; in the blood of fishes, amphibians and reptiles, 5 to 15 volumes per cent; and in the blood of birds and mammals,

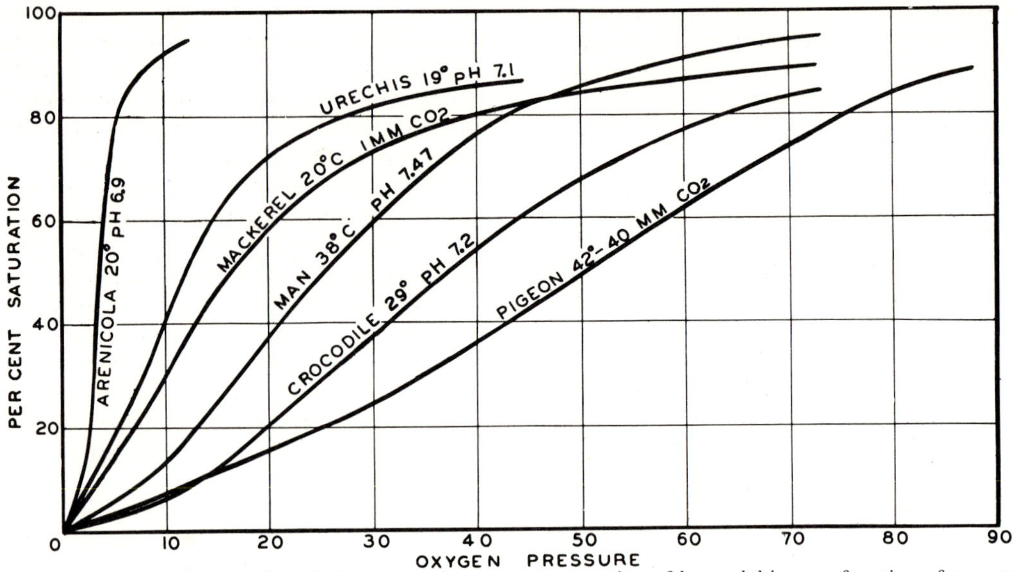


***Figure 7.3.*** Oxygen dissociation curves in per cent saturation of hemoglobin as a function of oxygen pressure (in mm. Hg) in a variety of animals. (From Prosser (ed.), 1950, Comparative Animal Physiology, Saunders, p. 310.)

15 to 20 volumes per cent. The blood of some diving mammals has an oxygen capacity of 40 volumes per cent; that is, almost twice as much oxygen is combined with the hemoglobin of the blood as is present in an equal volume of air (Krogh, 1941).

The oxygen loosely combined with the oxygen-transporting (respiratory) pigment is carried to all the cells. There a quantity of oxygen is released for use in metabolism in proportion to the difference between the higher oxygen tension in the blood and the lower one in the cells, in accordance with the relationship shown in Figure 7.3.

Some respiratory pigments take up oxygen readily and almost saturate at relatively low oxygen tensions (e.g., the blood of the worm *Arenicola* and the eel), whereas others saturate only at much higher oxygen tensions (e.g., man and the pigeon) (Manwell, 1960; Guyton, 1966). The advantage of the blood pigment such as that of *Arenicola* is the ready uptake of oxygen even when the oxygen tension of the environment is relatively low, but such a pigment will not release the bound oxygen until the oxygen tension of the tissue cells is even less than that of the blood (which in this case would have to be below 10 mm. Hg). In the pigeon, on the other hand, an oxygen tension a little below 80 mm. Hg is sufficient to withdraw oxygen from the blood.

In the cells of green plants, the oxygen is supplied during photosynthesis all day. At night the cells must draw upon the oxygen in the atmosphere reaching them by diffusion. According to Rabinowitsch (1945) the oxygen in the atmosphere is renewed every 2000 years through photosynthesis. If utilization of oxygen is faster than can be supplied by diffusion, the plant cells are capable of anaerobic metabolism which can tide them over until oxygen becomes available in daylight (see Chapter 16). Most organisms, including plants, have a capacity for varying degrees of anaerobic existence and anaerobic metabolism is considered by some as the primeval type of energy liberation (von Brand, 1946).

## PRESSURE AS AN ENVIRONMENTAL FACTOR

Because dredgings from deep waters of the Mediterranean Sea disclosed no life, it was once thought that no living things existed in the depths of the ocean. Early in the last century, however, several peculiar animals were brought up on cables laid in the deep waters of the Atlantic, and later some peculiar fishes were caught where water wells up from the depths of the ocean near Madeira off the northwest coast of Africa. Since that time, towing, dredging and trawling have disclosed life at various levels of the ocean, including the greatest depths such as the Marianas Trench near the Philippines.

Life is not only present in deep waters but it is diverse, bacteria and representatives of all the animal phyla being found. Living plants, however, are largely confined to the upper 100 meters of the ocean but some microscopical forms occur in the unlighted deep waters where they presumably live on nutrients released by the decomposition of dead organisms. The failure to find living organisms in the Mediterranean Sea has been attributed to lack of oxygen in the particular regions initially tested. Later they were found (Marshall, 1954) in other parts of the sea floor sampled.

The cells of organisms in deep waters are under great hydrostatic pressures. For every 10 meters down (slightly over 5 fathoms or 33.4 feet) the pressure increases by about 1 atmosphere (0.987 atmosphere). At the bottom of the Marianas Trench, which is over 10,000 meters deep, the pressure is over 1000 atmospheres. The cells of these organisms of the depths therefore live in a unique environment quite different from that of surface forms.

Since hydrostatic pressure is distributed equally over the surface of organisms, it has been argued that deep water organisms need no special adaptations except, perhaps, receptors which enable them to perceive each other in the dark. Sensitive, large, stalked or tubular eyes, luminous organs and other receptors are common in deep water fishes and squids. However, when coastal organisms (mussels, clams, snails, worms and crustaceans) are subjected to pressures gradually increasing to 600 atmospheres, they are first stimulated to greater activity, then depressed and inactivated. Some (starfish and coelenterates) tolerated for one hour pressures up to 1000 atmospheres, which is equivalent to the pressures of the deepest oceans. At this pressure they swelled, but several hours after their return to atmospheric pressure, they recovered. Frogs and fishes were more sensitive and were killed by 300 to 400 atmospheres of pressure (Cattell, 1936). It would seem that organisms of the deep seas have developed some special adaptations enabling them to withstand the great pressures at these depths and still carry on their activities.

The nature of the pressure tolerance adaptations at the cell level have been investigated to only a limited extent and primarily with marine eggs and microorganisms. Sea urchin eggs (*Arbacia punctulata*), beginning to cleave, revert back to the one-celled stage under pressures of more than 400 atmospheres. The contraction and gelation required for cleavage are suppressed. If the pressure is released within a short time, cleavage is resumed. About the same amount of pressure suppresses formation of pseudopodia in *Amoeba proteus;* the plasmagel appears to solate and the viscosity of the cytoplasm falls to a value so low that the food vacuoles and the nucleus sink to the bottom of the cell. It is thus apparent that pressure has a striking effect on sol-gel relationships in the cell (Marsland, 1958).

It might be expected that chemical reactions would also be affected by pressure, and investigations have demonstrated suppression of bacterial luminescence under certain conditions. The luminescence is a result of oxidation of a substrate in the presence of a specific enzyme, luciferase. When pressure is applied, luminescence is gradually quenched until the suspension of bacteria (or a mixture of the enzyme and substrate) becomes dark. If at this point the temperature is raised, luminescence reappears. This has been interpreted as a result of renaturation by heat of the enzyme luciferase denatured by pressure (Brown *et al.,* 1942).

Presumably, growth also is suppressed by pressure because cultures of such bacteria as *Bacillus subtilis* and *Pseudomonas fluorescens* fail to grow at pressures of 300 atmospheres or more. Replication of phage in *Escherichia coli* is also prevented by 300 atmospheres of pressure, but as soon as the pressure is released, growth is resumed. Yet there exist marine bacteria, such as *Bacillus submarinus,* isolated from the ocean depths where pressure approximates 500 atmospheres, which grow readily at a pressure of even 600 atmospheres (Zobell and Johnson, 1949; Johnson *et al.,* 1959).

Since most of the reactions of the cell are catalyzed by enzymes, it is thought that the suppression of cell activities by pressure is a result of action of pressure upon cell enzymes which are proteinaceous in nature. Pressure is found to reduce or prevent the incorporation of precursors into proteins and nucleic acids (Landau, 1966). Some evidence has already been presented (Chapter 3) that pressure reverses the denaturing effect of temperature on proteins. It will be recalled that an increase in temperature influences the equilibrium between native and denatured protein in the direction of the denatured form. Denaturation in most proteins is accompanied by an increase in volume as shown by dilatometric measurements. Since pressure and temperature act in opposition to one another, one would therefore expect pressure to reverse thermal denaturation by suppressing volume increase in accordance with the principle of Le Chatelier, which states that if a system is subjected to a stress it will react in such a way as to reduce the stress. This is indeed found to be the case. Enzymes which are necessary for life activities are catalytic in only a particular configuration, neither fully unfolded nor folded (Johnson, 1959).

Thus, surface bacteria which cannot tolerate the pressures of deep waters at low temperatures are shown to tolerate them and grow if the temperature is increased, as seen in the data of Table 7.4. Perhaps the truly barophilic (literally, pressure-loving) bacteria and other organisms are those in which the enzymes have the proper configuration even under great pressures and at relatively low temperatures.

It is not known by what mechanism pressure causes sex neversal in a copepod

**Table 7.4.** Effect of Temperature and Hydrostatic Pressure on Bacterial Growth

| CULTURE | 300 ATM (4410 PSI) | | | 400 ATM (5880 PSI) | | | 500 ATM (7350 PSI) | | | 600 ATM (8820 PSI) | | |
|---|---|---|---|---|---|---|---|---|---|---|---|---|
| | 20° | 30° | 40° | 20° | 30° | 40° | 20° | 30° | 40° | 20° | 30° | 40° |
| Terrestrial: | | | | | | | | | | | | |
| *Clostridium sporogenes* | — | 4+ | 4+ | — | 2+ | 3+ | — | — | — | — | — | — |
| *Bacillus subtilis* | — | 3+ | 4+ | — | 2+ | 4+ | — | — | 3+ | — | — | 2+ |
| *Pseudomonas fluorescens* | 2+ | 3+ | 4+ | — | 2+ | 4+ | — | — | 3+ | — | — | — |
| Marine: | | | | | | | | | | | | |
| *Achromobacter thalassius* | — | 4+ | 2+ | — | — | 3+ | — | — | — | — | — | — |
| *Bacillus submarinus* | 2+ | 4+ | 4+ | 1+ | 4+ | 4+ | — | 4+ | 4+ | — | 4+ | 4+ |
| *Pseudomonas xanthocrus* | 4+ | 4+ | — | 4+ | 3+ | 1+ | 2+ | 2+ | 2+ | 1+ | 1+ | — |

All cultures, except *P. xanthocrus,* showed 4+ growth at all three temperatures at atmospheric pressure; this culture grew at 20°C. and 30°C., but not at 40°C. ATM signifies atmospheres; PSI, pounds per square inch. (After Zobell and Johnson, 1949; from Oginsky and Umbreit, 1954: Introduction to Bacterial Physiology. W. H. Freeman & Co., San Francisco.)

(Vacquier and Belser, 1965) or mutation in *Euglena* (Gross, 1965).

## LITERATURE CITED AND GENERAL REFERENCES


Benson, O. O., Jr., and Stughold, H., 1960: Physics and Medicine of the Atmosphere and Space. John Wiley & Sons, New York.

Bernal, J. D., Aug. 1960: The structure of liquids. Sci. Am. *203*:125–134.

Bernard, C., 1879: Leçons sur les Phénomènes de la Vie aux Animaus et aux Végéteau. 2 Vol. Paris.

Blum, H. F., 1955: Time's Arrow and Evolution. Princeton University Press, Princeton, N. J.

Brown, D. E., Johnson, F. H., and Marsland, D. A., 1942: The pressure, temperature relations of bacterial luminescence. J. Cell. Comp. Physiol. *20*:151–168.

Buswell, A. M., and Rodebush, W. H., Apr. 1956: Water. Sci. Am. *194*:77–89.

Cannon, W. B., 1932: The Wisdom of the Body. W. W. Norton & Co., New York.

Cattell, J. M., 1936: The physiological effects of pressure. Biol. Rev. *11*:441–476.

Claussen, W. F., 1967: Surface tension and surface structure of water. Science *156*:1226–1227.

Davis, K. S., and Day, J. A., 1961: Water, the Mirror of Science. Doubleday & Co., Garden City, N. Y.

Dick, D. A. T., 1966: Cell Water. Butterworth, Washington, D. C.

Drost-Hansen, W., 1965a: The effects of biologic systems of higher-order phase transitions in water. Ann. N. Y. Acad. Sci. *125*:471–501.

Drost-Hansen, W., 1965b: Aqueous interfaces. Industr. Eng. Chem. *57*:38–44; 18–37.

*Edsall, J. T., and Wyman, J., 1958; Biophysical Chemistry. Vol. 1. Academic Press, New York, pp. 27–46.

Falk, M., and Kell, G. S., 1966: Thermal properties of water: discontinuities questioned. Science *154*: 1013–1015.

Frank, K. S., 1965: The structure of water. Fed. Proc. *24*:S1–S11.

Goldberger, E., 1965: A Primer of Water, Electrolyte and Acid-base Syndromes. 3rd Ed. Lea & Febiger, Philadelphia.

Gross, J. A., 1965: Pressure-induced color mutation of *Euglena gracilis*. Science *147*:741–742.

Guyton, A. C., 1966: Textbook of Medical Physiology. 3rd Ed. W. B. Saunders Co., Philadelphia.

*Henderson, L. J., 1927: Fitness of the Environment. Macmillan Co., New York.

Horowitz, N. H., 1966: The search for extraterrestrial life. Science *151*:789–792.

James, J. N., July 1963: The voyage of Mariner II. Sci. Am. *209*:70–84.

Johnson, F. H., Eyring, H., and Polissar, M., 1954: The Kinetic Basis of Molecular Biology. John Wiley & Sons, New York.

Johnson, F. S., 1961: Satellite Environment Handbook. Stanford University Press, Stanford, Calif.

Kaplan, J., 1953: The earth's atmosphere. Am. Sci. *41*:49–64.

Krogh, A., 1941: The Comparative Physiology of Respiratory Mechanisms. University of Pennsylvania Press, Philadelphia.

Landau, J. V., 1966: Protein and nucleic acid synthesis in *Escherichia coli:* pressure and temperature effects. Science *153*:1273–1274.

Lederberg, J., 1960: Exobiology: approaches to life beyond the earth. Science *132*:393–400.

Manwell, C., 1960: Comparative physiology: blood pigments. Ann. Rev. Physiol. *22*:191–244.

Marshall, N. B., 1954: Aspects of Deep Sea Biology. Philosophical Library, New York.

*Marsland, D., Oct. 1958: Cells at high pressure. Sci. Am. *199*:36–43.

Meryman, H. T. (ed.), 1966: Cryobiology. Academic Press, New York.

Morita, Y., 1957: Effect of hydrostatic pressure on succinic, formic and malic dehydrogenases in *Escherichia coli.* J. Bact. *74*:251–255.

Olmstead, E. G., 1966: Mammalian Cell Water. Lea & Febiger, Philadelphia.

Opik, E. J., June 1958: Climate and the changing sun. Sci. Am. *198*:85–92.

Pittendrigh, C. S., Vishniac, W., and Pearson, J. P. T., 1966: Biology and Exploration of Mars. National Academy of Science, Washington, D. C.

Plass, G. N., 1958: Carbon dioxide and the climate. Am. Sci. *44*:302–316.

Plass, G. N., July 1959: Carbon dioxide and climate. Sci. Am. *201*:41–47.

Prosser, C. L., and Brown, F. A., 1961: Comparative Animal Physiology. 2nd Ed. W. B. Saunders & Co., Philadelphia.

Rabinowitsch, E. I., 1945: Photosynthesis and Related Processes. Vol. 1. Interscience Publishers, New York.

Richard, D. W., 1954: Homeostasis versus hyperexis: or St. George and the Dragon. Sci. Monthly *77*: 289–294.

Runnels, L. K., Dec. 1966: Ice. Sci. Am. *215*:118–126.

Sverdrup, H. U., Johnson, M. W., and Fleming, R. H., 1942: The Oceans. Prentice-Hall, New York.

Vacquier, V. D., and Belser, W. L., 1965: Sex conversion induced by hydrostatic pressure in the marine copepod *Tigriopus californicus.* Science *150*:1619–1621.

von Brand, T., 1946. Anaerobiosis in Invertebrates. Biodynamics, Normandy, Mo.

Wunder, C. C., 1966: Life into Space; an Introduction to Space Biology. F. A. Davis Co., Philadelphia.

Zimmerman, A. M., 1963: Incorporation of $^3H$-thymidine in the eggs of *Arbacia punctulata.* A pressure study. Exp. Cell Res. *31*:39–51.

Zobell, C. E., 1952: Bacterial life at the bottom of the Philippine trench. Science *115*:507–508.

Zobell, C. E., and Johnson, F. H., 1949: The influence of hydrostatic pressure on the growth and viability of terrestrial and marine bacteria. J. Bact. *57*:179–189.

# CHAPTER 8 / ELECTROLYTES IN THE CELL ENVIRONMENT

Electrolytes in the cell environment as well as in the inside of the cell play an important role in all cell activities. Many ions affect the consistency of protoplasm, the selectivity of the cell membrane and the activity of enzymes in cells. Contractility of muscle cells requires magnesium, potassium and other ions. Excitability of cells depends upon the presence of various monovalent and divalent cations. The hydrogen ion concentration determines not only the degree of dissociation of a protein but also its charge. Since, next to water, protein is the most prevalent substance in cytoplasm, the hydrogen ion concentration exerts a profound effect, not only on the structural relations of protein molecules in cytoplasm but also on the activity of proteins as enzymes in cells.

## ACIDS, BASES AND HYDROGEN IONS

An *acid* may be defined as a substance which forms hydrogen ions in solution and from which hydrogen may be displaced by a metal to form a salt. A *hydrogen ion* is a positively charged hydrogen atom or *proton.* A *base* may be defined as a substance which yields *hydroxyl ions* in solution and reacts with an acid to form a salt and water. A *salt* may be defined as a compound formed when the hydrogen of an acid has been replaced by a metal.

Acids and bases are present in natural waters and in the internal media of plant and animal cells. The total acid or base content of a given quantity of medium can be determined by titration (see Table 8.1), but we are much more interested in the concentration of the hydrogen ions present than in the titratable acid or base, since the hydrogen ion is such an important catalyst in biological reactions (Bull, 1964; Pearson, 1966).

## DISSOCIATION OF WEAK ACIDS, WEAK BASES AND WATER

Strong acids and strong bases are completely ionized and dissociated in water, but weak acids and weak bases are only partially dissociated in water, the degree of dissociation depending upon the strength of the acid or base. The degree of dissociation of weak electrolytes may be determined by the conductivity of solutions, as well as by other methods.

When a weak acid, HA, is dissolved in water, it dissociates, and if the velocity of the reaction in the direction of dissociation is given by $V_1$, and the velocity of association of the ions to re-form molecules of acid is given by $V_2$, at equilibrium we have the following relationship:

$$HA \underset{V_2}{\overset{V_1}{\rightleftharpoons}} H^+ + A^- \qquad (8.1)$$

The value of $V_1$ increases as the acid dissociates more completely, and the equilibrium is then said to lie to the far right. Under a given set of conditions the velocity of the reaction to the right depends primarily on the concentration of HA:

$$V_1 = k_1 [HA] \qquad (8.2)$$

where $k_1$ represents a constant.

The velocity of the reaction to the left depends upon the concentration of $H^+$ and $A^-$.

$$V_2 = k_2 [H^+][A^-] \qquad (8.3)$$

where $k_2$ represents another constant.

When equilibrium is reached the quantity of acid dissociating is equal to that reassociating; therefore, $V_1$ equals $V_2$, and

$$k_2[H^+][A^-] = k_1[HA]. \qquad (8.4)$$

By dividing each side by $k_2$ and by [HA], expression of the law of mass action is obtained:

$$\frac{[H^+][A^-]}{[HA]} = \frac{k_1}{k_2} = K_a \qquad (8.5)$$

where $K_a$ is the dissociation constant of the weak acid or the ratio of dissociation to association. At room temperature for acetic acid, $K_a$ is $1.85 \times 10^{-5}$ (or 43 molecules in 10,000); for formic acid, it is $2.4 \times 10^{-4}$; and for boric acid, $6.4 \times 10^{-10}$. It should be apparent that the smaller the dissociation constant the weaker the acid. A parallel formulation might be developed for dissociation of weak bases (Bull, 1964).

Water also dissociates, though very slightly, and at all times is present in the two states, dissociated and undissociated. Water is unique in that, on dissociation, it gives rise to both hydrogen ions (like an acid) and hydroxyl ions (like a base). For the dissociation of water the following equations may be written:

$$H_2O \rightleftharpoons H^+ + OH^- \qquad (8.6)$$

$$\frac{[H^+][OH^-]}{[H_2O]} = K \qquad (8.7)$$

where K is the dissociation constant of water.

Then,

$$[H^+][OH^-] = K[H_2O] \qquad (8.8)$$

Since the degree of dissociation is so slight, $[H_2O]$ may be considered as approximately constant. Therefore it may be combined with K, giving $K_w$:

$$[H^+][OH^-] = K_w: \qquad (8.9)$$

The experimentally determined electrical conductivity of pure water at room temperature indicates that the concentration of hydrogen (or hydroxyl) ions is approximately $10^{-7}$N.

Therefore,

$$K_w = 10^{-7} \times 10^{-7} = 10^{-14} \qquad (8.10)$$

The dissociation of water, as is true of dissociation in general, depends upon the temperature. At lower temperatures it is somewhat less than the value given for room temperature, and at higher temperatures somewhat greater. For example, $K_w$ is $0.115 \times 10^{-14}$ at 0° C. and $9.614 \times 10^{-14}$ at 60° C. Since the hydrogen ion is one of the important biological catalysts, the increase in hydrogen ion concentration with rise in temperature may contribute to the increased activity of reactions in organisms when the temperature is raised (Harned and Hamer, 1933), although effects of temperature on the dissociation constants of buffer systems are probably of greater importance.

## THE pH SCALE

Although the hydrogen ion concentration can be expressed in terms of nor-

**Table 8.1.** Difference Between Normality and pH of Acetic and Hydrochloric Acids

| N OF SOLUTION | pH OF HAc | pH OF HCl |
|---|---|---|
| .20 | 2.69 | 0.81 |
| .10 | 2.85 | 1.09 |
| .04 | 3.08 | 1.46 |
| .02 | 3.24 | 1.77 |
| .01 | 3.39 | 2.06 |

**Table 8.2.** The Relation Between Normality and pH

| $[H^+]$ | | $[OH^-]$ | LOG $[H^+]$ | $-$LOG $[H^+]$ = pH |
|---|---|---|---|---|
| 1N | $10^0$N | $10^{-14}$N | 0 | 0 |
| 0.1N | $10^{-1}$N | $10^{-13}$N | $-1$ | 1 |
| 0.001N | $10^{-3}$N | $10^{-11}$N | $-3$ | 3 |
| 0.000,01N | $10^{-5}$N | $10^{-9}$N | $-5$ | 5 |
| 0.000,000,1N | $10^{-7}$N | $10^{-7}$N | $-7$ | 7 |
| 0.000,000,001N | $10^{-9}$N | $10^{-5}$N | $-9$ | 9 |
| 0.000,000,000,01N | $10^{-11}$N | $10^{-3}$N | $-11$ | 11 |
| 0.000,000,000,000,01N | $10^{-14}$N | $10^{-0}$N | $-14$ | 14 |

mality (see Tables 8.1 and 8.2), this is inconvenient because either exponents or small decimals must be employed. To avoid these difficulties Søensen chose to express the concentration of hydrogen ions by negative logarithms of the hydrogen ion concentration. This function he called the pH:

$$pH = -\log_{10}[H^+] \quad (8.11)$$

The relation between pH and hydrogen ion concentration is given in Table 8.2. The pH scale ranges from 0 to 14 or from 1N $[H^+]$ to 1N $[OH^-]$. A greater range of acidity and alkalinity is possible but is not likely to be encountered in biological systems or in the normal environment of cells. Even the present scale is beyond the limits normally encountered in nature.

Because, in water, the concentration of hydrogen ions equals the concentration of hydroxyl ions, water is considered neutral. Since the hydrogen ion concentration of water is $10^{-7}$, at room temperature the pH of water is 7.0. Therefore a pH lower than 7.0 indicates a concentration of hydrogen ions greater than that found in water and the solution is considered acid. The acidity increases as the pH falls (see Table 8.2). When the pH is greater than 7.0, the hydrogen ion concentration is less than that in water and the solution is considered alkaline or basic. The higher the pH, the more alkaline the solution. The product of the concentrations of the hydrogen and the hydroxyl ions is always $10^{-14}$N at 20° C.

The effects of pH upon cells and cell reactions have been extensively explored and references to these effects will be encountered in succeeding chapters as various biological activities are considered.

Although biologists were the first to emphasize the importance of pH, chemists also soon became aware of the need to control the hydrogen ion concentration, especially in industrial operations.

## MEASUREMENT OF THE STRENGTH OF A WEAK ACID BY THE pK$_a$

By applying the law of mass action to dissociation of a weak acid, HA, the following expressions may be written:

$$HA \rightleftharpoons H^+ + A^- \quad (8.1)$$

$$\frac{[H^+][A^-]}{[HA]} = K_a, \text{ the acid dissociation constant} \quad (8.5)$$

Solving for $[H^+]$:

$$[H^+] = \frac{K_a[HA]}{[A^-]} \quad (8.12)$$

Taking the negative logarithm of each side,

$$-\log_{10}[H^+] = -\log_{10}K_a - \log_{10}\frac{[HA]}{[A^-]} \quad (8.$$

By definition, $-\log_{10}[H^+] = pH$, and $-\log_{10}K_a = pK_a$; and since

$$-\log_{10}\frac{[HA]}{[A^-]} = +\log_{10}\frac{[A^-]}{[HA]}$$

substitution of these values in equation 8.13 gives:

$$pH = pK_a + \log_{10}\frac{[A^-]}{[HA]} \quad (8.14)$$

When the salt of a weak acid is added to a solution of that acid, the dissociation of the acid is suppressed by the common ion effect (law of mass action). Consequently, most of the acid is present in the undissociated form and the concentration of anion in solution becomes approximately equal to the concentration of dissolved salt. By applying equation 8.14 to this case it will be seen that the term for anion concentration may be replaced by a term for salt concentration, and the acid concentration may be considered approximately equal to the total acid. The equation may now be written:

$$pH = pK_a + \log_{10}\frac{[\text{salt}]}{[\text{acid}]} \quad (8.15)$$

This equation is known as the Henderson-Hasselbalch equation for buffer action.

When the concentrations of salt and acid are equal, the second term drops out of the Henderson-Hasselbalch equation, since the $\log_{10}$ of 1 is zero. The pH then equals the $pK_a$. By this method it is relatively easy to determine the $pK_a$ values for acids.

It should be noted that a large $pK_a$ indicates a weak acid, a small $pK_a$ a strong acid. Multivalent acids possess several dissociation constants, for each of which a $pK_a$ may be determined (Bull, 1964).

Once the $pK_a$ of an acid is known, the ratio of acid to salt required to attain a desired pH may be readily calculated. Conversely, if the concentrations of salt and acid are known, the pH of the mixture may be readily calculated. However, it must be remembered that the Henderson-Hasselbalch equation is an approximation only and that it holds best when salt and acid are in nearly equal concentration. It is most in error when applied to widely divergent ratios of acid and salt. As the $pK_a$ is also a function of the total salt concentration, an experimentally determined $pK_a$ may differ significantly from a handbook value which is cited for the weak acid and its salt at a given concentration in the absence of extraneous salts.

## BUFFERS

The commonest *buffer* is a mixture of a weak acid and its salt. When acid is added to a buffer the pH changes slowly since, according to the Henderson-Hasselbalch equation, the pH depends upon the ratio of the salt to the undissociated acid and this ratio changes slowly when acid is added, because the salt suppresses the dissociation of the acid by the common ion effect.

The $pK_a$ of the acid used in a buffer represents the midpoint of the effective buffer range because at this point acid and salt are present in equal concentrations. If acid is added the pH slowly falls; if base is added the pH slowly rises. The range at which a buffer is most useful lies between one pH unit on either side of the $pK_a$. The change in pH with addition of acid or base is shown in Figures 8.1,

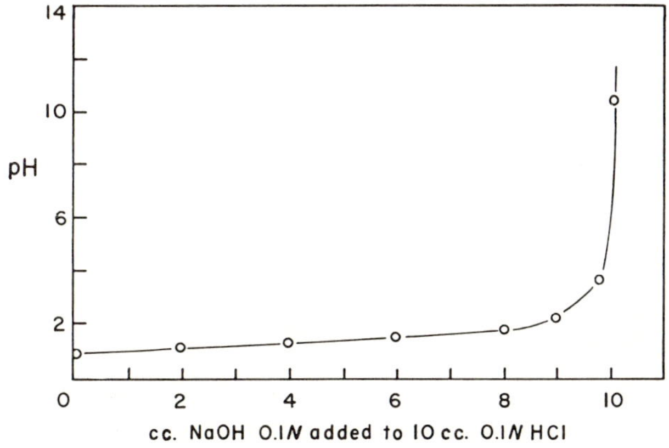


*Figure 8.1.* Change in pH when a strong acid is titrated with a base.

8.2 and 8.3. When the acid is completely neutralized the pH will sweep to the alkaline range.

Often weak acids with several dissociation constants are used as buffers. Each $pK_a$ of a multivalent acid represents the midpoint of one of the buffer ranges. The effective buffer ranges of such a system are about one pH unit to either side of each of these $pK_a$ values. When the dissociation constants are close together, as in citric acid, a wide pH range may be covered by buffers made from the acid. Citric acid, in fact, is one of the favorite acids used in buffers (see McIlvaine buffer series, Table 8.3), although like phosphate it removes calcium ions, a factor which must be taken into account in making up physiological media.

When a strong acid and one of its salts are used to make a buffer system the buffering action in restricted to the low pH range, because no matter how much salt is present the acid remains completely dissociated and the pH low. The $pK_a$ of a strong acid cannot be determined because in absence of the common ion effect suppressing ionization of the acid, the Henderson-Hasselbalch equation does not apply. A buffer made from a strong acid and its salt also resists pH change on addition of base, changing no more than one pH unit until about 80 per cent of the acid is neutralized (see Fig. 8.1).

All that has been said of acids and their salts as buffering systems may also be said of bases and their salts (West, 1942; Bull, 1964) (see Appendix 8.1).

***Buffer Capacity.*** A buffer's capacity is measured by the degree of change of pH following addition of a given amount of acid or base. A buffer of a given concentration has greatest buffer capacity at a pH near the $pK_a$ of the buffer acid. The concentration of the buffer determines its capacity to absorb base or acid at any pH. Dilution of a buffer usually changes the pH slightly and also proportionally decreases the buffer capacity. Some characteristics of buffers are given in Table 8.3.

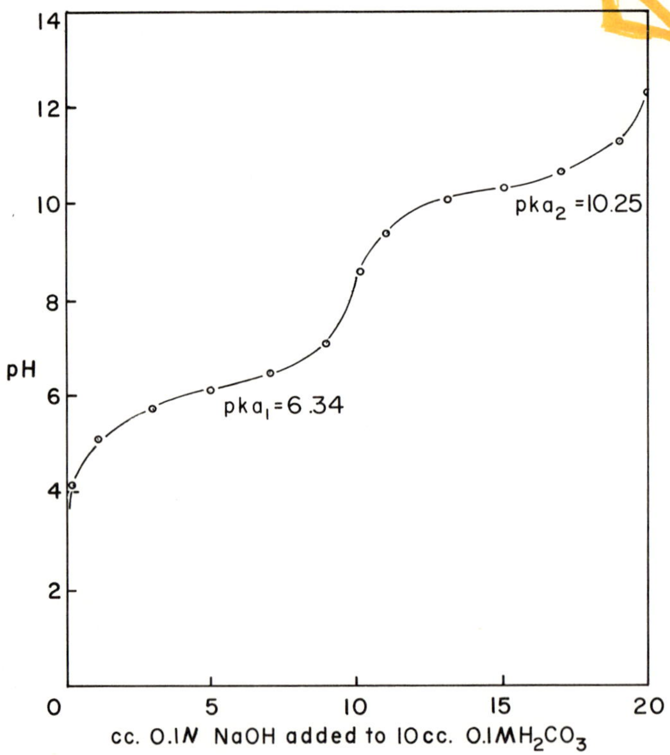


***Figure 8.2.*** Change in pH when a weak acid, carbonic acid, is titrated with a base. Note the two dissociation constants of carbonic acid (25° C.).

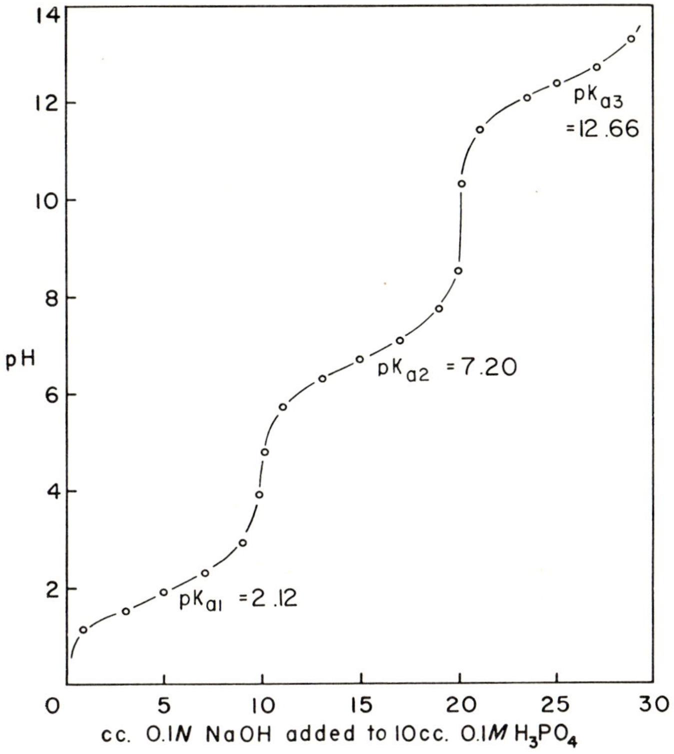


***Figure 8.3.*** Change in pH when a weak acid, phosphoric acid, is titrated with a base. Note the three dissociation constants of phosphoric acid (25° C.).

**Table 8.3.** Standard Buffer Mixtures

| NAME OF BUFFER | CONSTITUENTS | pH RANGE |
|---|---|---|
| Clark and Lubs | Phthalate-HCl | 2.2 – 3.8 (20° C.) |
| Clark and Lubs | Phthalate-NaOH | 4.0 – 6.2 (20° C.) |
| Clark and Lubs | $KH_2PO_4$-NaOH | 5.8 – 8.0 (20° C.) |
| Clark and Lubs | Boric acid, KCl-NaOH | 7.8 –10.0 (20° C.) |
| Clark and Lubs | HCl-KCl | 1.0 – 2.2 (20° C.) |
| Sφrensen | Glycine-NaCl-NaOH | 8.58–12.97 (18° C.) |
| Sφrensen | Borate-NaOH | 9.24–12.38 (18° C.) |
| Sφrensen | Borate-HCl | 7.61– 9.23 (20° C.) |
| Sφrensen | Citrate-HCl | 1.04– 4.96 (18° C.) |
| Sφrensen | Glycine-HCl | 1.04– 3.68 (18° C.) |
| Sφrensen | $Na_2HPO_4$-$KH_2PO_4$ | 5.29– 8.04 (18° C.) |
| Sφrensen | Citrate-NaOH | 4.96– 6.69 (20° C.) |
| Palitzsch | Borax-borate | 8.69– 6.77 (18° C.) |
| McIlvaine | Citric-$NaH_2PO_4$ | 2.2 – 8.0 |
| Tris(hydroxymethyl) aminomethane | Tris + HCl | 7.0 – 9.0 (20° C.) |
| Kolthoff and Vleeschhouwer | $Na_2CO_3$-borax | 9.2 –11.0 (18° C.) |
| Kolthoff and Vleeschhouwer | $Na_2HPO_4$-NaOH | 11.0 –12.0 (18° C.) |

Data from Clark, 1928. Data for Tris(hydroxymethyl) aminomethane from Whitehead, 1959: J. Chem. Educ. *36*:297; and Bulletin 106, Sigma Chemical Co., St. Louis.

**Table 8.4.** pH Values of Environmental Media and Some Organisms Found Therein

| TYPE OF HABITAT | pH RANGE | MEDIUM | CHARACTERISTIC ORGANISMS |
|---|---|---|---|
| Very acid | 0.0 | Sulfur and salts | *Thiobacillus thiooxidans.* |
| | 1.0– 3.0 | Organic | The fungus, *Merulius lacrymans.* |
| | 2.0 | Organic | Acetate flagellates after adaptation. |
| | 3.2– 4.6 | Peat bogs | Mosses: *Sphagnum* and *Drepanocladus.* |
| | 3.4 | Vinegar | Vinegar eels (nematodes), vinegar bacteria. |
| | 3.2 | Soil water | High moor vegetation. |
| | 3.3– 3.7 | Organic | The flagellate, *Astasia,* in tryptone (lower limit). |
| Somewhat acid | 5.2 | Sour milk | Lactic acid bacteria. |
| | 5.7– 5.8 | Distilled water<br>Fresh rain | |
| Somewhat acid to alkaline | 5.2– 9.0 | Soil water | Peat pit mosses. |
| | 6.2– 9.2 | Neutro-alkaline lakes | Mosses at edge. |
| | 4.5– 8.5 | Soil water in most soils | Various forest and field plants. |
| Nearly neutral | 6.8– 7.0 | Freshly boiled (cooled) distilled water | |
| | 6.5– 8.0 | Drinking water | |
| | 6.8– 8.6 | River water | River plants and animals. |
| | 6.2– 8.2 | Springs | Algae around springs. |
| | 7.8– 8.6 | Sea water | Marine plants and animals. |
| Alkaline | 7.5– 9.1 | Soil water | Greasewood, rabbit bush, salt grass and brome grass growing on alkaline flats. |
| | 9.0– 9.6 | Organic medium | *Astasia,* extreme alkaline range. |
| | 10.0 | Alkaline desert pool | Cyprinid fishes (Denio, Nevada). |
| | 10.5 | Organic medium | Dermatophytes. |
| | 10.5 | Inorganic | Larvae of *Aedes matronius.* |
| | 9.3–11.1 | Organic medium | Several species of *Penicillium,* one of which tolerates pH 11.1; also some species of *Aspergillus.* |

Data for *Thiobacillus* from Stanier, Doudoroff and Adelberg, 1963: The Microbial World; for fungi from Wolf, 1947: The Fungi, II, and Leise and James, 1946: Arch. Dermat. & Syph. 53:481; for mosses from Sørensen, 1948: Dansk Botanisk Arkiv., *12*:5, and Richards and Verdoorn, 1932: Manual of Bryology; for Salt Lake from Flowers, 1933: The Bryologist, 34; for acetate flagellates, Hutner and Provasoli, 1951, *In* Lwoff: Biochem. & Physiol. of the Protozoa, I; for *Astasia* from Schoenborn, 1949: J. Exp. Zool., *111;* for desert pool near Denio, personal communication from Eminger Stewart; for *Aedes* from Beadle, 1932: J. Linn. Soc. London, Zool. *38;* others from the Handbook of Chemistry and Physics. 1964. 45th Ed., and various other sources.

*Note:* More examples are given for the extremes because endurance of these conditions is so unusual. Most organisms occur in an environment slightly to the acid or alkaline side of neutrality.

## BUFFERING SYSTEMS IN THE CELL ENVIRONMENT

Environmental waters free of decomposing organic matter are generally alkaline or neutral, as may be seen from the data in Table 8.4. For example, the pH of ocean water is generally 7.8 to 8.2. Fresh water is more variable, and if any decomposing material containing organisms is present, the pH falls because of the organic acids produced.

In the ocean, rivers and lakes the primary buffering is accomplished by the carbonate-bicarbonate system (see Fig. 8.2). Because of the ubiquity of carbon dioxide in natural waters this buffer system is unavoidable in nature; in the presence of such ions as sodium, potassium, calcium and magnesium, carbon dioxide forms bicarbonates, thereby keeping natural waters on the slightly alkaline side of neutrality. Whenever living organisms are present in natural waters they constantly contribute some carbon dioxide, but unless many are crowded in a small space this is not likely to have much effect upon the pH, unless at the same time organic acids are being produced, as occurs in bogs covered by mosses.

Since carbon dioxide is one of the main wastes of cell metabolism and is constantly being formed in the organism, it is always available in the internal environment and inside the cells. Therefore, the carbonate-bicarbonate buffer system is also the main buffer of cells and of plasma and body fluids of organisms. The $pK_{a1}$ of carbonic acid is 6.34; the $pK_{a2}$ is 10.25 at 25° C. MacInnes and Belcher, 1933, 1935).

Another buffer system of considerable importance to the cell is the phosphate buffer system (Bull, 1964). At 25° C. phosphoric acid of very low concentration has a $pK_{a1}$ value of 2.12, a $pK_{a2}$ value of 7.12 and a $pK_{a3}$ value of 12.66 (see Fig. 8.3). The values differ with change of temperature and with change in concentration, as has already been pointed out, and these factors should be taken into account when buffers are used in experiments. The middle $pK_a$ value lies in the pH range which is of importance in biological systems and helps maintain the pH of blood and body fluids. In this range the buffer salts present are usually $NaH_2PO_4$ and $Na_2HPO_4$. Phosphate is especially important as a buffer inside cells.

Proteins, amino acids, fatty acids and various organic acids probably also serve as buffers in the cell. Some of the organic acids like citric acid are especially suitable. However, the role of such buffers is decidedly secondary to the carbonate and phosphate systems in the cell environment.

Inside the living cell phosphate (see Fig. 3.1) is present in relatively large amounts, together with the carbonate and protein buffers. The pH readings inside cells depend upon the method used, the part of the cell tested and the species. For example, animal tissue cells and *Amoeba* varied between pH 6.5 and pH 7.5, whereas yeast and plant cells were below pH 6.0. However, none of the methods of measurement appears satisfactory because most require piercing the cell in order to insert microelectrodes or to inject pH indicator dyes, procedures which are likely to be injurious (Caldwell, 1956). Furthermore, as discussed in Chapter 5, cytoplasm is a heterogeneous assembly of organelles and it is therefore difficult to know just what portion of the cytoplasm is being measured for its pH in such experiments. Inside some cells there are present natural dyes, such as the yellow pigment of *Physarum,* which may be used as a pH indicator in the ground substance, where it is presumably found (Seifriz and Zetzmann, 1935).

## AMPHOLYTES OR ZWITTERIONS

*Ampholytes* are inorganic and organic substances which can act as acids or as bases; that is, they can either give off or take up hydrogen ions, depending upon the conditions under which they are placed. If an acid is added to an ampholyte, the latter will dissociate as a base. If a base is added, the ampholyte will dissociate as an acid. Perhaps the concept may be illustrated best with amino acids (and proteins), which are of special importance to the cell.

An amino acid has no net charge at its *isoelectric pH.* The isoelectric pH is defined as the pH at which there is no net accumulation of amino acid at either anode or cathode of an electric field, and the ions are therefore considered to be equally charged negatively and positively.

Several lines of evidence indicate that amino acids exist as zwitterions rather than as uncharged molecules. For instance, they have a high solubility in water; they have a highly polar nature, shown by their behavior in an electromagnetic field; and they have relatively high boiling points—properties which cannot be explained if uncharged molecules are assumed. The isoelectric pH is different for different amino acids (6.1 for glycine, 5.7 for serine, 3.0 for aspartic acid), and it depends upon the number and relative strength of acid and basic groups in the amino acid, i.e., their tendency to ionize. For a monocarboxylic, monoamino amino acid the isoelectric pH (pI) is the sum of its acid pK and its basic pK divided by 2. On the acid side of the isoelectric pH an amino acid is positively charged, and on the alkaline side, negatively charged:

$$\underset{\text{zwitterion}}{H_3\overset{+}{N}-R-COO^-} + H^+ \rightleftharpoons H_3\overset{+}{N}-R-COOH \quad (8.16)$$

$$HO^- + \underset{\text{zwitterion}}{H_3\overset{+}{N}-R-COO^-} \rightleftharpoons H_2N-R-COO^- + H_2O \quad (8.17)$$

It must not be construed that proteins dissociate as if they were a multitude of amino acids. The peptide backbone of the protein binds most of the amino and carboxyl groups of the constituent amino acids. Only the end groups of the peptide chains and the extra amino and carboxyl groups of the polyamino and polycarboxy amino acids and special groups such as the hydroxy group of tyrosine are free to dissociate as the pH of the solution containing the protein is changed.

## RESISTANCE OF CELLS TO pH CHANGES

An ameba tolerates an environmental pH range of 4.2 to 8.2, and yeast tolerates a range of about 4.5 to 8.5. Since each pH unit is equal to a tenfold change in hydrogen ion concentration, this range represents a ten thousand fold change. As seen from the data in Table 8.4, cells of organisms such as vinegar bacteria, vinegar eels (nematodes), certain mosses and some fungi tolerate much more acid conditions; others, such as the cells of dermatophyte fungi and plants of the alkali regions, tolerate more alkaline conditions; cells of most organisms exist in a range of pH 6.0 to 8.0.

The hydrogen ion does not penetrate into cells rapidly, but under certain conditions hydrogen ions may drive a weak acid into cells. For example, carbonic acid at a low pH is largely in the molecular form, which enters cells readily. Inside the cell it dissociates, acidifying the cell. Conversely, under alkaline conditions carbonic acid may be withdrawn from the cell, combining with the alkali to form bicarbonate. The pH may therefore influence the entry and exit of various weak acids to and from the cell. (For the converse effects obtained with a weak base such as ammonia, see Chapter 13.)

Although the pH of environmental waters containing decomposing material is highly variable, the pH of body fluids and bloods is found to be held within narrow limits as seen in Table 8.5. For example, the blood of man is very closely regulated at a pH of 7.4 and if the pH of blood falls to 7.0, acidotic coma and death ensue; at a pH of 7.8 life terminates in tetany.

The body fluids of a number of invertebrates lie in the same pH range; however, a few are neutral—for example, the body fluid of ascidians—and others are as alkaline as pH 7.8. In fact, the fluid inside a coelenterate may have the same pH as sea water, approximately 8.2. On the other hand, the body fluid of some insects undergoing metamorphosis may be as acid as pH 6.8. Acidity in such cases probably

**Table 8.5.** pH Values of Body Fluids, Blood and Plant Fluids

| pH RANGE | FLUIDS |
|---|---|
| 7.3–7.5 | Blood of man |
| 7.3–7.5 | Spinal fluid of man |
| 6.9–7.2 | Whole blood, dog |
| 6.7–7.9 | Body fluids or blood of various invertebrates |
| 6.6–7.6 | Milk, human |
| 6.9–7.5 | Saliva, human |
| 1.0–3.0 | Gastric contents, man |
| 4.8–8.2 | Duodenal contents, man |
| 2.9–3.3 | Apple juice freshly extracted |
| 2.2–2.4 | Lemon juice |
| 1.8–2.0 | Lime juice |
| 5.0–5.2 | Expressed juice of *Sedum* leaf |

Data from Handbook of Chemistry and Physics. 1964. 45th Ed., and from Prosser and Brown, 1961.

results from the accumulation of metabolic carbon dioxide and organic acids which are incompletely voided when the excretory system is not functioning.

Some plant saps and fruit juices are highly acid (see Table 8.5), but it must be remembered that most of their organic acids are stored in the vacuoles of the cells, not in the cytoplasm. The pH of sap obtained by squeezing fruits is probably the pH of the vacuolar contents, not necessarily that of the cytoplasm of the cells.

The pH inside cells is variable but is usually in a narrow zone on either side of neutrality (Small, 1955; Wiercinski, 1955).

## METHODS OF MEASURING pH

Because of its simplicity and inexpensiveness, the most readily available method for measuring pH is the *colorimetric method.* Dyes, usually weak acids and bases which are either differently or more intensely colored in their ionic form than in their undissociated form, are used for this purpose. Standard solutions are made up in carefully prepared buffers against which unknown solutions, containing the same amount of dye as the standard, may be matched.

The $pK_a$ value of the dye acid (or base) gives the point of half dissociation. The approximate range over which the dye may be used effectively is roughly one pH unit to either side of the $pK_a$ value. Various dyes used for this purpose are described by Clark (1928). Some data for dyes frequently used in the biological range are given in Table 8.6.

One disadvantage of the dye method is that it depends upon the hue sensitivity of the human eye, a variable trait. Another disadvantage is that some opaque or light-scattering material may obscure color matching. Furthermore, when a solution to be tested has a color (e.g., blood), colorimetric methods are inapplicable. In the presence of proteins, or in the presence of salt, additional errors arise. With the exercise of proper precautions the colorimetric method can be accurate to about 0.1 pH unit for solutions not subject to the errors mentioned, but under crude conditions it is probably accurate to only one pH unit. Nonetheless,. it is effective for field studies when a pH meter is not available.

The electrometric method is simpler and more accurate than the colorimetric method and is in common use at present. In the commercial pH meter two electrodes, a glass electrode and a calomel electrode, are dipped into the unknown solution, as shown in Figure 8.4. The glass electrode consists of a bubble of glass containing a solution of known concentration of hydrochloric acid, usually 0.1 normal with respect to hydrogen ions, into which extends a silver–silver chloride wire connected to the potentiometer. The calomel electrode contains calomel ($Hg_2Cl_2$) in contact on one side with saturated potassium chloride and on the other with mercury from which a platinum wire connects to the potentiometer.

A potential arises because of a difference in concentration between the hydrogen ions inside the glass electrode and in the unknown solution, the two being separated by the glass membrane through which only the hydrogen ions can pass. The potential developed is proportional to

**Table 8.6.** Characteristics of Indicators for Measuring pH

| CONCENTRATION OF INDICATOR IN SOLUTION, % | pH RANGE COVERED | NAME AND ABBREVIATION | $pK_a$ | COLOR CHANGE |
|---|---|---|---|---|
| 0.04 | 1.2–2.8 | Thymol blue (TB) | 1.5 | Red–yellow |
| 0.04 | 3 –4.6 | Brom phenol blue (BPB) | 3.98 | Yellow–blue |
| 0.01 | 3.2–5 | Congo red (CoR) | — | Blue–violet–scarlet |
| 0.016 | 3.8–5.4 | Brom cresol green (BCG) | 4.67 | Yellow–blue |
| 0.02 in 60% alc. | 4 –6 | Methyl red (MR) | 5.1 | Red–yellow |
| 0.010 | 4.8–6.4 | Chlor phenol red (CPR) | 5.98 | Yellow–red |
| 0.04 | 5.2–6.8 | Brom cresol purple (BCP) | 6.3 | Yellow–purple |
| 0.04 | 6.0–7.6 | Brom thymol blue (BTB) | 7.0 | Yellow–blue |
| 0.01 in 50% alc. | 7 –8 | Neutral red (NR) | — | Yellow–red |
| 0.02 | 6.8–8.4 | Phenol red (PR) | 7.9 | Yellow–red |
| 0.02 | 7.2–8.8 | Cresol red (CR) | 8.3 | Yellow–red |
| 0.04 | 8.0–9.6 | Thymol blue (TB) | 8.9 | Yellow–blue |
| 0.05 in 50% alc. | 8 –10 | Phenolphthalein (P) | 9.7 | Colorless–red |

Data mostly from Clark, 1928. Aqueous solutions are used except as indicated. Congo red is considered undesirable because of large salt errors. Details for making up solutions may be found on pp. 91 to 94 of Clark's treatise. Most of the aqueous indicators are made up in 0.01N NaOH, appropriately diluted. Salts of the dyes, however, are soluble in water. Many of them are now available in this form.

the difference in concentration between the two solutions. While hydrogen ions, in sufficient numbers to set up a potential, pass from the solution in which they are more concentrated through the glass electrode* toward the other solution, actual leakage of acid does not occur, because anions cannot accompany the hydrogen ions.

In a potentiometer the potential developed between the two solutions inside and outside the glass electrode is balanced by that from a battery, the point of balance indicated by a null point on the dial of the galvanometer. In the pH meter the calomel electrode is used as a reference electrode against which the potential between the two solutions inside and outside the glass electrode is measured (see

*A glass electrode useful in determining sodium and potassium ions has been described (Friedman, 1962), as well as one for divalent ions (Garrels *et al.*, 1962) and for calcium in sea water (Thompson and Ross, 1966).

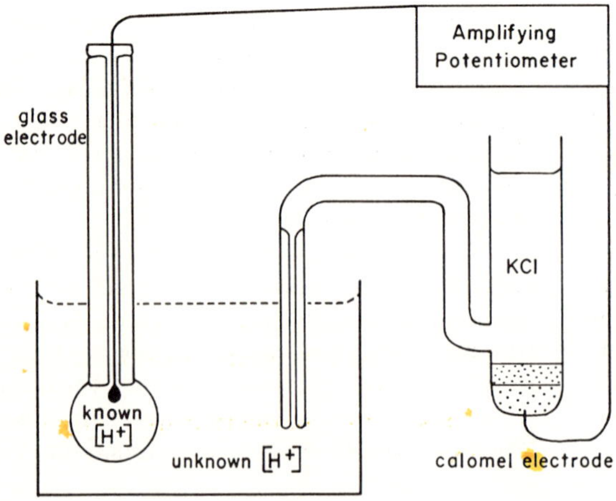


***Figure 8.4.*** A simplified diagram of the electrometric determination of pH with a glass electrode. The sample whose pH is to be determined is placed in the vessel marked "unknown [$H^+$]" into which dip the glass electrode (containing 0.1N $H^+$) and the calomel electrode. A potential is developed between the solution in the vessel and the glass electrode because of the difference in hydrogen ion concentrations between the two. This potential is amplified and measured by a potentiometer. The dial on the instrument is generally read directly in pH units. For details of the calomel electrode see Figure 17.2.

Chapter 17). In practice, the potential is amplified, and the dial of the potentiometer slide wire, which then measures the potential, is calibrated to read directly in pH units (Eisenman *et al.,* 1966).

For measurement of the pH inside the living cell, special micromethods (injection of indicator dyes and microelectrodes) have been devised (Small, 1955; Wiercinski, 1955; Caldwell, 1956).

## RELATION BETWEEN CONCENTRATION AND ACTIVITY OF ELECTROLYTES

The pH of a weak acid is less than would be expected on the basis of the normality of the acid, because dissociation is incomplete. However, the concentration of hydrogen ions of even a strong acid like hydrochloric, which is considered to be completely dissociated, is also somewhat less than would be expected on the basis of its normality. This is because the electrostatic attraction between oppositely charged ions in solutions prevents free activity of the ions. Some hydrogen ions are therefore not free to act at any given moment. If strong acid is sufficiently diluted, however, the concentration of hydrogen ions as measured by pH approaches the normality of the acid, because as interionic distance increases, ionic interference decreases to the vanishing point. This is strictly true only at infinite dilution.

Since in solutions of electrolytes, ionic interference reduces the effectiveness of ions so that their concentration seems lower than it is, the concept of activity has been introduced. *Activity* is the effective concentration of an ion in solution. The *activity coefficient* is defined as the ratio of the activity, as measured by some property such as the depression of the freezing point of a solution, to the true concentration (molality):

$$\text{activity coefficient } (\gamma) = \frac{\text{activity}}{\text{concentration}} \quad (8.18)$$

The activity coefficient, which is usually less than one, increases as the solution becomes more dilute, and approaches unity at infinite dilution when the attractive forces between oppositely charged ions become negligible.

Since pH measures the hydrogen ions free to act, it is a measure of the hydrogen ion activity, and pH should therefore be more strictly defined as:

$$\text{pH} = -\log_{10}[\text{hydrogen ion activity}] \quad (8.19)$$

or pH is the negative logarithm of the effective hydrogen ion concentration. When this is done the $pK_a$ obtained for a weak acid is designated $pK_a'$. This added precision may not affect the general discussion of the pH relationships or the Henderson-Hasselbalch equation, but it must be borne in mind should it become necessary to calculate data with this degree of accuracy.

In a like manner the activity of bases is restricted by electrostatic attraction between oppositely charged ions. The pOH is therefore the negative logarithm of the hydroxyl ion activity rather than the negative logarithm of the hydroxyl ion concentration.

## HYDRONIUM AND HYDROGEN IONS

According to Brönsted (1928), an *acid* is a substance capable of giving off a proton (hydrogen ion), and a *base* is a substance capable of combining with a proton. By this definition the chloride ion is a base because it is capable of combining with a hydrogen ion. Since water combines with hydrogen ions to form *hydronium* ions, $H_3O^+$, it is a base, but since $H_3O^+$ may yield a proton, the hydronium ion is an acid. Since $NH_3$ combines with a proton to form $NH_4^+$, it is a base, whereas $NH_4^+$ is an acid since it is capable of yielding a proton. This method of defining acids and bases has special use for nonaqueous systems.

The *hydronium ion* is always formed by combination of a proton with water whenever free protons are dissociated from acids; it is not probable that free hydrogen ions exist. Proof of the presence of $H_3O^+$ is obtained from x-ray diffraction study of the crystalline hydrate of perchloric acid. The distances between the ions identified indicate $H_3O^+$ and $ClO_4^-$. The hydronium rather than the hydrogen ion is also demonstrated by electrolysis of hydrogen bromide and water dissolved in liquid sulfur dioxide. Neither hydrogen bromide nor water alone conducts electricity in sulfur dioxide. Yet in the mixture of hydrogen bromide and water in liquid sulfur dioxide, water and hydrogen are obtained at the negative pole, not hydrogen alone, as would be the case if hydrogen ions were migrating as such. The presence of hydronium ions rather than hydrogen ions has been assumed throughout the discussion here and in subsequent chapters, but following custom, all formulations (since they apply equally well to hydronium or hydrogen ions) are given in terms of hydrogen ions.

## SALTS

Most cells do not tolerate distilled water and can live only in a medium which contains some salts. Melted snow and freshly fallen rain water are sometimes devoid of salt and seldom contain more than 0.003 per cent salt, yet some organisms are resistant to this dilution. Hard fresh water contains about 0.1 per cent salt, calcium being the predominant cation and carbonate the predominant anion. Sea water contains about 3 per cent salt (30 gm. per liter) and salt lakes in which salt is beginning to crystallize, about 30 per cent, which some halophilic organisms can tolerate (Larsen, 1962). The principle salts in sea water are the chlorides, sulfates and carbonates of sodium, potassium, calcium and magnesium. Sodium is the predominant cation, and chloride the predominant anion (see Table 8.7).

The salt content of natural waters is determined by a variety of methods, e.g., the total salt content by evaporation under vacuum, and individual salt ions by quantitative analytical methods (Kolthoff and Belcher, 1957).

Because chloride is the commonest ion in salt waters, the chlorinity, as determined by silver nitrate titration, is often used to indicate the relative salinity of a salt water. The chlorinity is related to the salinity of sea water as follows: Salinity $= 0.03 + 1.805 \times$ chlorinity, when both are expressed in grams per kilogram of sea water. (Sverdrup *et al.*, 1942).

Every cell contains salts (see Chapter 2), and salt is always present in the internal environment of the cells. The salt concentration of the blood of most marine invertebrates is equal to that of sea water (3 per cent). The salt concentration in blood of all other animals is less than that of sea water, being 0.68 to 0.9 per cent in vertebrates, whether fresh water, marine or terrestrial, and even less, 0.3 to 0.7 per cent, in fresh water and terrestrial invertebrates (Prosser and Brown, 1961).

Life can exist only in the presence of certain salts, and it is interesting to note that the relative proportions of the various ions most favorable for biological functions are approximately those characteristic of sea water (see Table 8.8). Perhaps this can be understood best from an example. The heart of a frog stops beating when placed in pure sodium chloride solution equal in concentration to the total salt content in blood. If a soluble calcium salt is added, the heart may resume beating, but the beat becomes normal only when potassium is also added and only

**Table 8.7.** Composition of Some Typical Natural Waters in Grams per Liter

| WATER | Na | K | Ca | Mg | Cl | $SO_4$ | $CO_3$ |
|---|---|---|---|---|---|---|---|
| Sea water | 10.7 | 0.39 | 0.42 | 1.34 | 19.3 | 2.69 | 0.073 |
| Hard fresh water | 0.021 | 0.016 | 0.065 | 0.014 | 0.041 | 0.025 | 0.119 |
| Soft fresh water | 0.016 | — | 0.010 | 0.00053 | 0.019 | 0.007 | 0.012 |

Data from Baldwin, 1948: An Introduction to Comparative Biochemistry. Cambridge University Press, London.

**Table 8.8.** Relative Ionic Composition of Sea Water as Compared with the Relative Ionic Composition of the Blood or Body Fluid of Various Animals

(The Na ion in all cases is taken as 100.)

A

| LOCATION OF SEA WATER | ION | | | | | |
|---|---|---|---|---|---|---|
| | Na | K | Ca | Mg | Cl | $SO_4$ |
| Woods Hole | 100 | 2.74 | 2.79 | 13.94 | 136.8 | 7.10 |
| Japan | 100 | 2.14 | 2.28 | 11.95 | 119.0 | 5.95 |

B

| ANIMAL | CLASSIFICATION | Na | K | Ca | Mg | Cl | $SO_4$ |
|---|---|---|---|---|---|---|---|
| Aurelia | Coelenterate | 100 | 2.90 | 2.15 | 10.18 | 113.05 | 5.15 |
| Strongylocentrotus | Echinoderm | 100 | 2.30 | 2.28 | 11.21 | 116.1 | 5.71 |
| Phascolosoma | Sipunculid | 100 | 10.07 | 2.78 | — | 114.06 | — |
| Venus | Mollusk | 100 | 1.66 | 2.17 | 5.70 | 117.3 | 5.84 |
| Carcinus | Crustacean | 100 | 2.32 | 2.51 | 3.70 | 105.2 | 3.90 |
| Cambarus | Crustacean | 100 | 3.09 | 2.60 | 6.70 | 30.9 | — |
| Hydrophilus | Insect | 100 | 11.1 | 0.92 | 16.8 | 33.6 | 0.12 |
| Lophius | Fish | 100 | 2.85 | 1.01 | 1.61 | 71.9 | — |
| Frog | Amphibian | 100 | 2.40 | 1.92 | 1.15 | 71.4 | — |
| Man | Mammal | 100 | 3.99 | 1.78 | 0.66 | 83.97 | 1.73 |

Calculated from data in Prosser and Brown, 1961.

when the three salts are present in the relative proportions characteristic of blood. The cations are said to act antagonistically to one another (see Chapter 13). It must be emphasized, therefore, that not only the total concentration of salt (osmotic factor) but also the concentration of each ion relative to that of the others (antagonism) is important in making a *balanced medium* such as blood or sea water.

Although neither the relative nor the absolute concentrations of salts are exactly the same in blood as in sea water, a fundamental relation between sea water and blood is suggested by the striking similarity in the relative concentrations of the ions found in each. Paleontologic evidence indicates that life probably originated in the sea, and in evolution an internal environment, the blood, may have simulated the medium in which life arose. With this general concept almost everyone seems to agree (Prosser and Brown, 1961).

The question, however, arises: If blood resembles sea water because of its ultimate derivation from a marine environment, why is the blood of the vertebrate so dilute compared with present-day sea water? (Fig. 8.5) Also, why is it that potassium, for example, is often present in slightly larger proportions, and magnesium in considerably smaller proportions in blood than in sea water? This has been explained by MacCallum (1926) as a relic of the ancient and perhaps more dilute seas in which potassium was more concentrated and magnesium less so than in present-day sea water. Although this theory is appealing, newer evidence indicates that the primeval sea was probably at least twice as concentrated as present-day vertebrate blood (Edsall and Wyman, 1958).

On the other hand, Pantin (1931) suggests that in the course of evolution the presence of some ions in blood in proportions different from those in sea water has been found to be more effective for biological functions. For example, it is advantageous for an active species to elim-

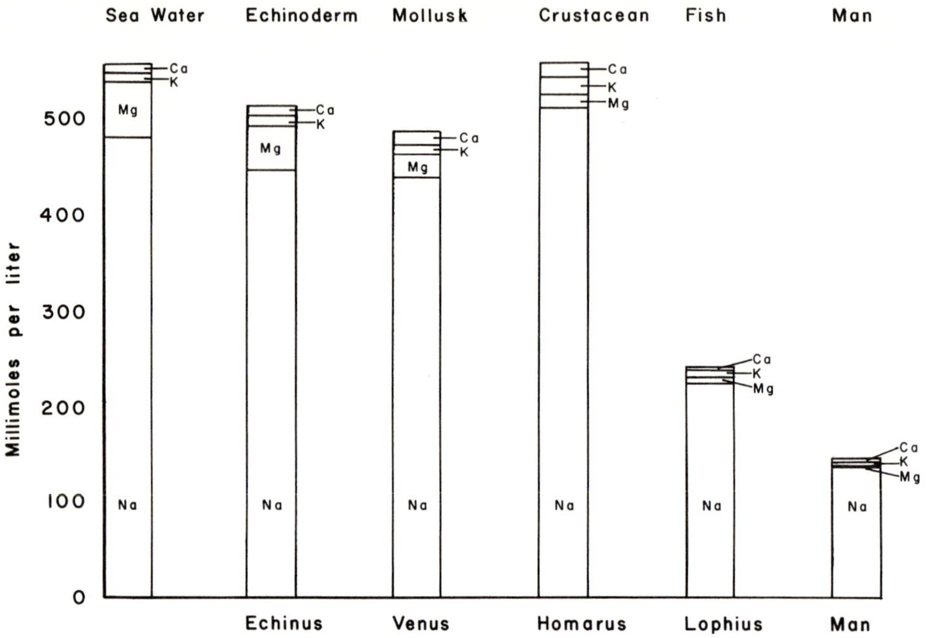


***Figure 8.5.*** Cations in sea water, body fluid and blood. Note the decreased amount of magnesium in the three forms on the right and the relative increase of potassium. The much lower concentration of total salt in the vertebrate is also readily noticeable. *Echinus* is a sea urchin, *Venus* a clam, *Homarus* a lobster and *Lophius* a fish.

inate a substance with narcotic action like the magnesium ion. Such differences in the proportions of ions in blood and sea water could be maintained by marine animals only as kidneys developed, because the differences are the result of a balance between the input and outgo of salts as determined by kidney function. A steady state is achieved, characteristic of the blood of each species.

It is to be noticed in Figure 8.5 that organisms (e.g., echinoderms) whose body fluids are almost identical to sea water in absolute and in relative concentration of the various ions, have no kidneys or nephric organs, whereas those with good nephric organs (e.g., mollusks, crustaceans and vertebrates) have blood and body fluid most unlike sea water, either in relative concentration of the various ions or absolute concentration or both.

In hard fresh water the concentration of calcium relative to sodium and other cations is higher than in sea water. This is important to life in fresh waters since calcium retards the transfer of materials across the plasma membrane of the cell, protecting the cells of the organism from the loss of useful salts to the dilute external medium (see Chapter 13).

X-ray diffraction studies show that salts are completely ionized even in the dry, crystalline state. On solution in water the ions of the salt become freely movable because of the high dielectric constant of water (see Chapter 7).

Salts, then, are completely dissociated in solution, but the effective concentration of the salt ions, as measured by some property which depends upon the number of active particles (e.g., the depression of the freezing point of a solution), is less than the value given by the concentration (in molality) of the salt ions in solution (see Chapter 12). The apparent decrease in concentration results mainly from attractions between salt ions of opposite sign, but attractions between solvent and solute molecules and even attractions between solvent molecules themselves may effectively reduce the activity of the ions present. It is therefore desirable to measure the *activity* rather than the concentration of the salts, as in the case of acids and bases discussed earlier.

## APPENDIX

### 8.1 BUFFER MIXTURES USING A WEAK BASE AND ITS SALT

For making buffers it is possible to calculate the ratio of the salt of a weak base to the base (BOH) in much the same way as is done for a weak acid. From the relationship,

$$BOH \rightleftharpoons B^+ + OH^- \quad (8.20)$$

the Henderson-Hasselbalch equation is derived

$$pOH = pK_b + \log_{10}\frac{[\text{salt}]}{[\text{base}]} \quad (8.21)$$

Since $[H^+] \cdot [OH^-] = 10^{-14}$ (8.22)

and $pH + pOH = 14$ (8.23)

or $pOH = 14 - pH$ (8.24)

Then, substitution of equation 8.24 for pOH in equation 8.21 gives:

$$14 - pH = pK_b + \log_{10}\frac{[\text{salt}]}{[\text{base}]} \quad (8.25)$$

or

$$-pH = -14 + pK_b + \log_{10}\frac{[\text{salt}]}{[\text{base}]} \quad (8.26)$$

or $$pH = 14 - pK_b - \log_{10}\frac{[\text{salt}]}{[\text{base}]} \quad (8.27)$$

## LITERATURE CITED AND GENERAL REFERENCES


Bates, R. G., 1954: Electrometric pH Determination. John Wiley & Sons, New York.

Bayliss, L. E., 1959: Principles of General Physiology. John Wiley & Sons, New York, Vol. 1, pp. 76–168.

Bittar, E. E., 1964: Cell pH. Butterworth, Washington, D. C.

Brönsted, J. N., 1928: Acid and basic catalysis. Chem. Rev. *5*:231–338.

*Bull, H. B., 1964: An Introduction to Physical Biochemistry. F. A. Davis Co., Philadelphia, Chapters 3 and 5.

Caldwell, P. C., 1956: Intracellular pH. Intern. Rev. Cytol. *5*:229–277.

Christensen, H. N., 1963: pH and Dissociation. A Learning Program for Students of the Biological and Medical Sciences. W. B. Saunders Co., Philadelphia.

Clark, W. M., 1928: The Determination of Hydrogen Ions. 3rd Ed. Williams & Wilkins, Baltimore.

Clark, W. M., 1952: Topics in Physical Chemistry. 2nd Ed. Williams & Wilkins, Chapters 14–16.

Davenport, H. W., 1958: The ABC of Acid-Base Chemistry. 4th Ed. University of Chicago Press, Chicago.

Edsall, J. T., and Wyman, J., 1958: Biophysical Chemistry. Academic Press, New York.

Eisenman, G., Baker, R., Mattock, G., and Friedman, S. M., 1966: The Glass Electrode. Interscience Publishers, New York.

Friedman, S. M., 1962: Measurement of sodium and potassium by glass electrodes. Meth. Biochem. Anal. *10*:71–106.

Garrels, R. M., Sato, M., Thompson, M. E., and Truesdell, A. H., 1962: Glass electrodes sensitive to divalent ions. Science *135*:1045.

Gold, V., 1956: pH Measurements. John Wiley & Sons, New York.

Harned, H. S., and Hamer, W. J., 1933: The ionization constant of water and the dissociation of water in potassium chloride solutions from electromotive forces of cells without liquid junction. J. Am. Chem. Soc. *55*:2194–2206.

Hastetler, P. B., Truesdell, A. H., and Christ, C. L., 1967: Activity coefficients of aqueous potassium chloride measured with a potassium sensitive glass electrode. Science *155*:1537–1539.

Joseph, N. R., 1958: A pH calculator based on linear transformations of the Henderson-Hasselbalch equation. Science *128*:1207.

Kolthoff, I. M., and Belcher, R., 1957: Volumetric Analysis. Interscience Publishers, New York.

Larsen, H., 1962: Halophilism. *In* The Bacteria. Gunsalus and Stanier, eds. Academic Press, New York, Vol. 4, pp. 297–342.

Lockwood, A. P. M., 1962: Ringer solution and some notes on the physical basis of their ionic composition. Comp. Biochem. Physiol. *2*:241–289.

MacCallum, A. B., 1926: The paleochemistry of the body fluids and tissues. Biol. Rev. *6*:316–357.

MacInnes, D. A., Jan. 1951: pH. Sci. Am. *184*:40–43.

MacInnes, D. A., and Belcher, D., 1933: The thermodynamic ionization constants of carbonic acid. J. Am. Chem. Soc. *55*:2630–2646.

MacInnes, D. A., and Belcher, D., 1935: The thermodynamic ionization constants of carbonic acid at 38° from electromotive force measurements. J. Am. Chem. Soc. *57*:1683–1685.

Pantin, C. F. A., 1931: The origin and composition of the body fluids in animals. Biol. Rev. *6*:459–482.

Pearson, R. G., 1966: Acids and bases. Science *151*: 172–177.

Prosser, C. L., and Brown, F. A., 1961: Comparative Animal Physiology. 2nd Ed. W. B. Saunders Co., Philadelphia.

Seifriz, W., and Zetzmann, M., 1935: A slime mold pigment as indicator of acidity. Protoplasma *23*: 175–179.

Small, J., 1955: The pH of plant cells. Protoplasmatologia II.B.2. Springer-Verlag, Vienna.

Sverdrup, H. U., Johnson, M. W., and Fleming, R. H., 1942: The Oceans. Prentice-Hall, New York.

Thompson, M. E., and Ross, J. W. Jr., 1966: Calcium in sea water by electrode measurement. Science *154*:1643–1644.

West, E. S. 1942: Physical Chemistry for Students of Biochemistry and Medicine. Macmillan Co., New York.

Wiercinski, F., 1955: The pH of Animal Cells. Protoplasmatologia II.B.2.c. Springer-Verlag, Vienna.

# CHAPTER 9 / RADIATIONS IN THE CELL ENVIRONMENT

Every one knows that all life is ultimately dependent upon the sun's radiant energy that is stored by green plant cells during photosynthesis (see Chapter 19). In addition to this primary role of light in the living economy, a persistent environment of various radiations produces other effects, reactions and adaptations, which may influence the life activities of the organisms living within this environment.

## NATURAL RADIATIONS

In addition to the radiation of the sun's spectrum, natural radiations include radio waves from the sun and the stars, ionizing radiations from the breakdown of the radioactive elements in the earth's crust and the cosmic rays coming from the sun and from the depths of space. Although natural ionizing radiations are usually of a minor consequence on the earth's surface, protons of solar flares and the van Allen belts of ionizing radiation around the earth (as shown in Fig. 9.1) and probably other planets become hazards in space travel (O'Brien, 1963; Dye and Wilkinson, 1965; Rice, 1966).

***Radiations of Sunlight.*** Sunlight is the major source of radiations on earth. Its spectrum (see Fig. 9.2) includes ultraviolet radiation, visible light, infrared rays and radio waves, the latter only recently discovered as a part of the sun's spectrum. Solar flares are a source of x-rays which on absorption produce ionization 60 miles high in the earth's atmosphere (Frier and Webber, 1963). Although the visible wavelengths are the most noticeable, about 60 per cent of the energy of sunlight is in the invisible infrared span, the limit of which is indefinite in sunlight. The sun is the major source of the earth's heat without which life as we know it would be impossible on earth. The emission from the sun in the visible and infrared parts of the spectrum suggests that the sun is a body with a surface temperature of 6000° absolute (Kiepenheuer, 1959; Menzel, 1959; Robinson, 1966).

At the surface of the earth little ultraviolet radiation of wavelengths shorter than 290 nm. is measured (The abbreviation nm. = nanometer, or $10^{-9}$ meter). The shortest wavelength recorded on a clear day with the sun at the zenith is 287 nm. Such a source of radiation as the sun, theoretically, should emit ultraviolet radiations shorter than 287 nm. To determine whether it does, spectrographs were first placed on V-2 rockets soon after World War II and sent over 50 miles up into the atmosphere of the earth. Records from these rockets showed the presence of ultraviolet light of wavelengths as short as

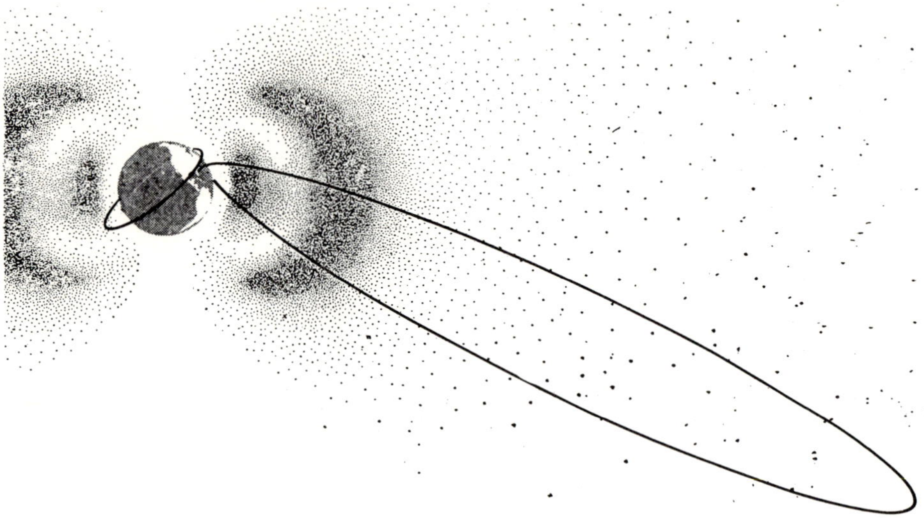

***Figure 9.1.*** Van Allen belts of ionizing radiation around the earth. The satellites Explorer IV and Pioneer III gave the first detailed information of the radiation belts. Explorer IV (short ellipse) monitored radiation levels for nearly 2 months at altitudes up to 1300 miles. Pioneer III, a lunar probe (long ellipse) provided data out to 65,000 miles. (From Radiation Belts around the Earth, by James A. Van Allen. Copyright © 1959 by Scientific American, Inc. All rights reserved.)

220 nm. Recent records from double spectrographs in Aerobee-High rockets have demonstrated the presence of radiation of ultraviolet wavelengths so short as to overlap with the x-ray region (Tousey, 1961) (see Table 9.1). Apparently sunlight radiations of wavelengths shorter than 287 nm. are absorbed in the earth's atmosphere before these radiations reach the surface of the earth. It has been demonstrated that high in the atmosphere oxygen absorbs ultraviolet radiation, forming a layer of

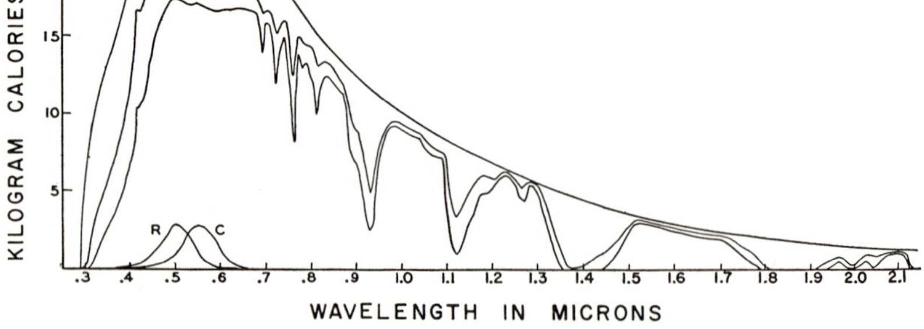


***Figure 9.2.*** Spectral distribution of sunlight: 0, outside the atmosphere; 1, with the sun at zenith; 2, with the sun at 60° from zenith. Curves 1 and 2 are for 20 mm. $H_2O$, 2.8 mm. ozone, and 300 dust particles per cc. Curves R and C indicate, respectively, the spectral sensitivity of the human rods and cones; the ordinate units are arbitrary. (From Blum, 1945: J. Clin. Invest. *24*:713.)

**Table 9.1.** Wavelengths of Radiations

| TYPE OF RADIATION | SUBDIVISION | WAVELENGTH IN nm.* |
|---|---|---|
| Gamma | | 0.0001–0.14 |
| X-rays | | 0.0005–20 |
| | hard | 0.0005–0.1 |
| | soft | 0.1–20 |
| Ultraviolet | | 15–390 |
| | very short | 15–200 |
| | short (far) | 200–300 |
| | long (near) | 300–390 |
| Visible | | 390–780 |
| | violet | 390–430 |
| | blue | 430–470 |
| | blue green | 470–500 |
| | green | 500–530 |
| | yellow green | 530–560 |
| | yellow | 560–590 |
| | orange | 590–620 |
| | red | 620–780 |
| Infrared | | $780–4 \times 10^5$ |
| | near | $780–2 \times 10^3$ |
| | far | $2 \times 10^3–4 \times 10^5$ |
| Hertzian waves | | $10^5–3 \times 10^{13}$ |
| | space heating | $10^5–10^6$ |
| | radio | $10^6–10^{12}$ |
| | radar | $10^6–10^9$ |
| | television | $10^9–10^{11}$ |
| Power A.C. | | $10^{15}$ |

* 1 nm. (nanometer) = $10^{-9}$ meter; 1 nm. = 1 m$\mu$ = 10 Å (angstrom). Although various units of length are used to measure wavelength of light, the nanometer is being recommended.

ozone. The ozone in turn absorbs somewhat longer ultraviolet rays, re-forming oxygen. The ozone umbrella, as this layer has sometimes been called, exists about 25 to 30 miles high in the atmosphere and is apparently the agent which removes the ultraviolet radiations of very short wavelengths (Kiepenheuer, 1959).

*Radioactive emanations* consist of three components: (1) gamma rays, which are penetrating radiations of very short wavelength but otherwise like x-rays; (2) alpha particles, positively charged helium nuclei; and (3) beta particles, rapidly moving electrons. Besides the naturally occurring radioactive elements such as radium and polonium, many elements have been made radioactive by bombardment with high energy particles such as helium nuclei. These are known as artificial radioactive elements (Glasstone, 1958; Shilling, 1964).

***Cosmic Rays.*** Cosmic rays were once thought to be produced by the creation and annihilation of matter in the depths of space. It is now thought that they may obtain their energy by passing through enormous magnetic fields (galactic accelerators) that drift among the stars in the Milky Way galaxy (Glasstone, 1958). Also, explosions occurring in the central parts of some galaxies are believed to contribute to the cosmic rays (Sandage, 1964). On entering the earth's atmosphere cosmic rays consist of charged particles, mainly protons (positively charged hydrogen nuclei), and in addition some heavier particles. After collisions of these particles with atmospheric gases, cosmic rays are found to include also uncharged particles (neutrons), gamma rays and charged particles other than protons, such as mesons (positively or negatively charged particles intermediate in mass between protons and electrons) and positive and negative electrons (Korff, 1957; Ginsburg and Razoryonov, 1959; Anderson, 1960; Hafner, 1964; Bowyer *et al.,* 1965; Burbridge, 1966).

Several cosmic rays strike the human body every second. Even organisms which escape the effect of visible and ultraviolet radiation are reached by cosmic radiations, some components of which are able to penetrate many centimeters of lead. Because of the difficulty of screening them out, the effects of cosmic rays on life have not yet been fully ascertained.

***Measurement of Radiation.*** The intensity of all radiation, other than ionizing radiation, is measured in terms of amounts of energy per unit time per unit area. It may be given in terms of any energy unit, for example, the erg (the work done by a force of 1 dyne acting through a distance of 1 centimeter) or the calorie. One calorie equals $4.186 \times 10^7$ ergs.

When radiation is measured in terms of its heating power, it is only necessary to absorb all the incident radiation on a black surface and convert the radiation to heat which may be taken up in water and measured by a thermometer (as in a heliometer used to measure the energy of sunlight). If a small amount of radiation is to be measured, thermocouples are placed in the water or on the black receiving surface. A thermocouple consists of a junction of two metals between which a thermoelectric current is induced by a change in temperature. This current is measured by a galvanometer (see Fig. 12.5). If a minute amount of radiation is to be measured, a thermopile, which consists of a series of thermocouples, is the instrument of choice because the current passing from the thermopile to the high sensitivity galvanometer is the sum of the currents induced in each of the thermocouples.

Because measurement of radiation by a thermopile usually requires a high sensitivity galvanometer, itself a delicate instrument, photoelectric cells are more frequently used for this purpose. However, unlike the thermopile, which absorbs all wavelengths of radiation equally, photoelectric cells absorb some wavelengths more than others. Therefore, to be useful for measuring the intensity of radiation at different wavelengths, photoelectric cells must be calibrated wavelength for wavelength, generally with a thermopile or a bolometer (Forsythe, 1937).

Ionizing radiation is usually measured by its ionizing power. Although it is also possible to measure the intensity of ionizing radiation by converting it to heat, the amount of heat produced by a strong dose of x-rays is rather small and difficult to measure with accuracy. One unit of ionizing radiation is the *roentgen,* abbreviated "r." The ICRU (International Commission on Radiation Units) defines the roentgen (r) as the "unit of exposure to x- or $\gamma$-radiation such that the associated corpuscular emission per 0.001293 g of air (1 cc at 0° C and standard atmospheric pressure) produces in air, ions carrying 1 electrostatic unit of quantity of electricity of either sign" (Morgan, 1963, p. 567). "It corresponds to the liberation of $2.082 \times 10^9$ ion pairs per cc. of air at 0° C. and 760 mm. pressure, and involves an energy dissipation of 0.1083 ergs/cc. of air . . . " (Lea, 1955). Because of differences in relative biological effectiveness of various ionizing radiations, units other than the roentgen are also used to measure dose rate.*

***Nature of Radiations.*** All radiations, from radio waves to the shortest

---

*"The ICRU defines the *rad* as the unit of absorbed dose and 1 rad corresponds to 100 ergs per gram medium. In radiation protection work (unless otherwise specified) 1 rad corresponds to the absorption per gram soft tissue of 100 ergs of energy delivered by ionizing radiation." To take account of differences in biological effectiveness of different kinds of ionizing radiation, "the ICRU introduced the term *rem* as a unit of RBE (relative biological effectiveness) dose. By definition the RBE dose in rems equals the dose in rads multiplied by the relative biological effectiveness." (Morgan, 1963, p. 567.)

gamma rays, are electromagnetic in nature and travel at the same speed, the speed of light ($3 \times 10^{10}$ cm./sec.); they differ only in wavelength. The wavelength distribution in the radiation spectrum is shown in Table 9.1. Alpha particles, electrons, protons and other *particulate radiations* travel at various speeds, depending upon the voltage gradient. However, they are accompanied by electromagnetic radiations which travel with the speed of light.

Although radiation travels as a wave and many of its properties are best explained by its behavior as a wave, radiation acts as if it were produced and delivered in discrete amounts of energy called quanta. The energy in the quantum, q, is related to the wavelength of radiation, being greater the shorter the wavelength, as shown by the following law (Planck's law):

$$\text{q(quantum of energy)} = h\nu = \frac{hc}{\lambda} \quad (9.1)$$

where h is Planck's constant, $6.624 \times 10^{-27}$ erg-seconds;
$\nu$ is the frequency of light in waves per second, equal to $c/\lambda$;
c is the velocity of light, $3 \times 10^{10}$ cm./sec.;
$\lambda$ is the wavelength expressed in centimeters.

As may be seen by dimensional analysis of the right-hand term, the energy in a quantum is expressed in ergs.

According to the quantum theory, quanta of light are absorbed at random during irradiation. The absorption of 1 quantum of light by a molecule (or atom) produces only one activated molecule (or atom). This is known as the Einstein-Starck law of photochemical equivalence. Absorption of light constitutes the *primary reaction*, which is independent of temperature since the addition of kinetic energy does not increase the probability that a molecule will absorb light. The primary reaction is followed by any one of a number of *secondary reactions*. For example, the energy absorbed may be reradiated at the same wavelength (resonance) or at longer wavelengths (fluorescence), or degraded to heat. On the other hand, the energy absorbed may render the activated molecule capable of reacting with other molecules as, for example, in photodynamic sensitization (Livingston, 1955). These secondary reactions are affected by temperature (see Chapter 10).

## PHOTODYNAMIC SENSITIZATION OF CELLS

In 1898 Raab found that protozoans were killed when exposed to strong visible light in the presence of dilute solutions of some fluorescent dyes, but that they were unaffected in the dark in the same solutions. On the other hand, light, even intense visible light, had no effect in the absence of the dyes. Dyes of similar concentration illuminated in the absence of paramecia are not injurious to paramecia subsequently immersed in them (Blum, 1964; Spikes and Glad, 1964; Spikes, 1968). Protozoans, such as paramecia, which possess no pigment, are translucent and appear white because the particles in their cytoplasm pass or scatter all wavelengths of visible light. They are unaffected even by fairly intense visible light, because they do not absorb it. However, when a dye which absorbs light, such as erythrosin, is present in the medium, it becomes absorbed to the surface of the paramecia, and the energy of the light absorbed by the dye is transferred to the cytoplasm, resulting in injury or death. Such a dye is spoken of as a *photosensitizer* and its action is *photodynamic* if molecular oxygen is required for expression of the effect (Spikes, 1968). This set of experiments is an illustration of the *Grotthus-Draper law*, which states that only absorbed light produces photochemical (or photobiological) reaction.

If the light intensity and the time of exposure are varied in such a way that the product of the two is always the same, the photochemical effect is the same. This principle is called the *Bunsen-Roscoe reciprocity law* and is well known in photography. The photodynamic effect obeys the reciprocity law but deviates from it if the intensity of the light is reduced to a very low level, in which case the effect diminishes, perhaps because the organism has a chance to carry on some recovery simultaneously with injury.

Photodynamic action occasionally occurs in nature. The protozoan *Blepharisma* if grown in the dark produces a red pigment which accumulates in the outer part of the protoplasm. Deeply pigmented cells are killed if exposed to intense visible light. The pigment extracted from the colored *Blepharisma* cells also sensitizes colorless ones to visible light (Giese, 1946; Giese and Zeuthen, 1949).

Another case of natural photosensitization is that observed in the carotenoidless mutant strain of the bacterium *Rhodopseudomonas spheroides,* which can carry on photosynthesis in absence of oxygen but is killed by light when oxygen is present whereas the wild-type parental form, which contains carotenoid, is not affected. The action spectrum for this photosensitization resembles the absorption spectrum of chlorophyll and it is thought that chlorophyll, in the absence of carotenoid, photosensitizes the destructive photo-oxidation. When cultures of the carotenoidless strain are grown in the dark in the presence of oxygen, they become essentially chlorophyll free (nonphotosynthetic) and at the same time lose their photosensitivity. These experiments are of interest because they suggest that carotenoid has a protective role in photosynthesis, preventing oxidation of important cell materials by oxidants produced in the process (Stanier, 1960; Krinsky, 1968).

A photodynamic effect is occasionally found in the cells of higher organisms. For instance, cells in the unpigmented skin of cattle eating certain weeds (St.-John's-wort, *Hypericum crispum*) are sensitized by a dyestuff which is present in the leaves. The dye is absorbed during digestion and is carried by the blood to the cells of the skin. Exposure to the sun then results in a rash, an edematous swelling or, in extreme cases, convulsions and death. This dye has been extracted from the leaves, purified and injected into animals with the same result. Similar effects have been demonstrated with dyes from various parts of other plants on various animals. In man a photodynamic effect is observed when abnormal metabolic reactions occasionally form products which sensitize the body to sunlight (Blum, 1964).

Photodynamic action by definition is that type of photosensitization which occurs only in the presence of oxygen. When oxygen is withdrawn no injury is observed, even when cells in dye solutions are exposed to bright visible light. These experiments, and the increased uptake of oxygen during illumination of the cells in the presence of dye, indicate that during photodynamic action photo-induced oxidation occurs, and evidence is available that the proteins of the cell are selectively oxidized. The aromatic amino acid residues in the proteins seem to be most susceptible (Blum, 1964). Some dyes, however, appear to combine selectively with nucleic acids (Mayor and Melnick, 1961). Irradiation in the presence of such dyes has been found to produce mutations (Zetterberg, 1964).

## RESPONSE IN PHOTORECEPTOR CELLS—VISION

The entire cell of a unicellular animal, plant or microbe, for example, ameba, paramecium or zoospores of some algae, may be sensitive to light and the cell may move toward the light or away from it (Bendix, 1960). The streaming inside plant cells is affected by light so that, even though the entire cell cannot move, the cytoplasm is sensitive to light. It is probable that the cytoplasm of cells is generally sensitive to light, but it is not possible to observe the reaction in all cells (Virgin, 1964).

Some unicellular organisms have developed a receptor, a region of the cell especially sensitive to light. Flagellates, both photosynthetic, plantlike ones and the nonphotosynthetic animal-like forms, usually display a stigma containing carotenoid pigments and a photoreceptor. Such a receptor enables an organism to make a response to smaller changes in intensity of light and sometimes to even detect the direction of the light (Bendix, 1960).

Although some multicellular organisms have scattered photoreceptive cells in many parts of the body, such as the interesting isolated photoreceptive cells found in the nervous system of some invertebrates, most multicellular animals have photoreceptors, e.g., eyespots and eyes of different degrees of complexity. (These are considered in textbooks of comparative animal physiology.) Multicellular plants have also developed photoreceptors which usually consist of cells near the growing tip that contain a pigment enabling them to absorb light which results in differential auxin transport and differential growth at a distance from the receptors (Briggs, 1964). (Plant photoreceptors are covered in textbooks of plant physiology.)

The focus of interest here is analysis of the nature of the cellular responses to light. This can be illustrated by the studies on vision. The visual process in its essential features is a cellular process and is much the same in eyes throughout the animal kingdom. The chemistry of visual reactions has been studied and analyzed most carefully in the photosensitive cells of eyes of vertebrates.

The inside of the vertebrate eye is lined by the cells of the *retina*, which receives the light after it has passed through the cornea, aqueous humor, lens and vitreous humor. The cornea and the lens are the refracting parts of the eye; the humors are the bathing media; the retina is the seat of photochemical and excitatory processes. In the retina of many vertebrates there are two types of photoreceptive cells, the *rods*, which are narrow and spindle shaped, and the *cones*, which are cone shaped. (see Fig. 9.3). Cones, exclusively, are present in the central fovea of man; both rods and cones are present in the rest of the retina.

The cones are concerned with color vision and the fovea is the center of acuity or detailed form vision. Although all cones appear alike histologically, physiological evidence (electric recording after stimulation by light) indicates the presence of several types of cones, each spe-

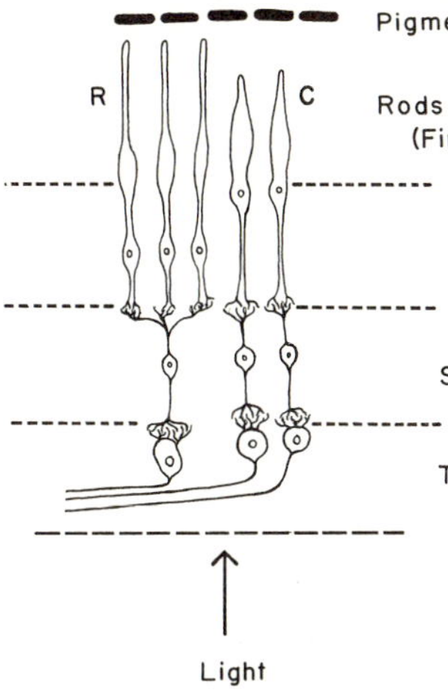


***Figure 9.3.*** Diagram of rods (upper left) and cones (upper right) in a vertebrate retina.

cialized for reception of light from a certain span of the spectrum. Although rods are incapable of distinguishing colors, they are sensitive to much lower intensities of light than the cones; therefore, they serve as photoreceptors in a much lower intensity range. At night a very dim light can be seen from the corner of the eye because of its action on the rods, which are more concentrated at the periphery of the retina. If the head is turned so that the light falls directly on the fovea, the light seems to disappear because the cones are insensitive to such dim light. This difference in sensitivity is best illustrated by the dark adaptation curves (see Fig. 9.4).

The ultrastructure of the rod in the vertebrate eye which suggests its origin from a cilium has been considered in Chapter 5. However, ultrastructural studies of eyes in representatives of many phyla of animals indicate perhaps a different origin in some of them (Eakin, 1963).

Though rods do not register a sense of color, both rods and cones are differentially sensitive to the various wavelengths in the visible spectrum. To determine the relative effectiveness of light at different wavelengths the dark-adapted individual in a dark chamber determines the lowest intensity of light of a given wavelength which he can see (Fig. 9.5). The visual

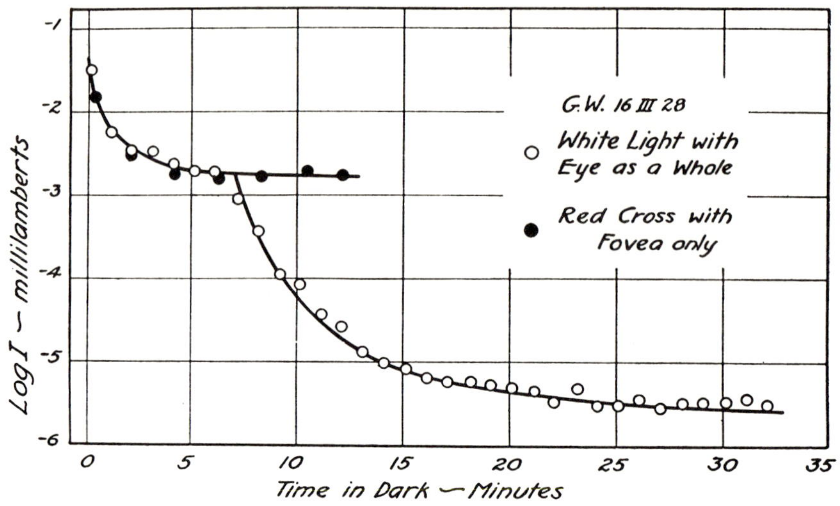


***Figure 9.4.*** The dark adaptation of the human eye. The points are single measurements of the just perceptible intensity at different times in the dark after high light adaptation. The white circles are for white light. The black circles are for the extreme red end of the spectrum which stimulates the cones alone. Dark adaptation is thus in two stages; first, cone adaptation; and second, rod adaptation. (From Hecht, 1929: J. Op. Soc. Am. *18*:269.)

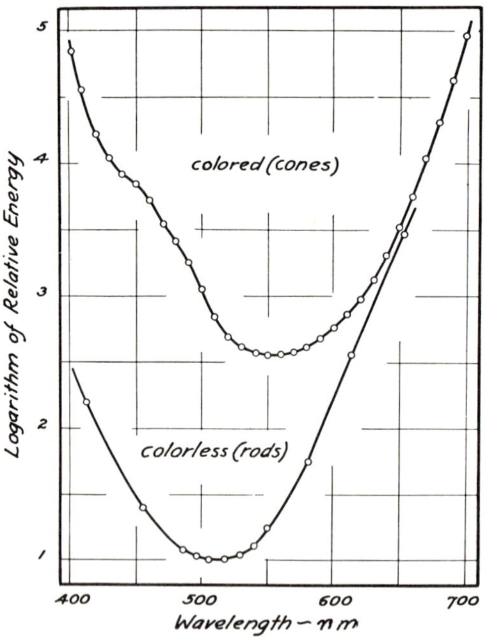


***Figure 9.5.*** Spectrum sensibility curves for rods and cone vision on a real energy basis. The position of the two curves on the ordinates corresponds to the fact that after complete dark adaptation, any region of the retina outside the fovea sees red light of 650 nm. as colorless at the threshold, and as colored only above the threshold. (From Hecht and Hsia, 1945: J. Op. Soc. Am. *35*:262.)

efficiency of a wavelength of light is accordingly given as the reciprocal of the intensity required to stimulate the eye. In Figure 9.6 the relative efficiencies of different wavelengths for rod and cone vision are plotted. A diagram plotting the relative efficiencies of different wavelengths in producing a biological effect is spoken of as an *action spectrum.* It is apparent that the action spectrum has a maximum for rod vision at 505 nm. and for foveal cone vision at 562 nm. This suggests that the rods and the cones each possess distinctive photosensitive pigments.

That the retina is bleached by light is shown by experiments obtaining an optogram formed when the dark-adapted eye of an animal is briefly exposed to a brilliantly illuminated window. If the eye is quickly removed and the retina separated and fixed in alum, it shows an image of the bright window.

The visual purple pigment, or *rhodopsin,* may be extracted in the dark by rupturing with a detergent the rods of cells of a dark-adapted animal. Rhodopsin is a typical protein of large molecular weight with an absorption spectrum almost identical with the action spectrum for rod

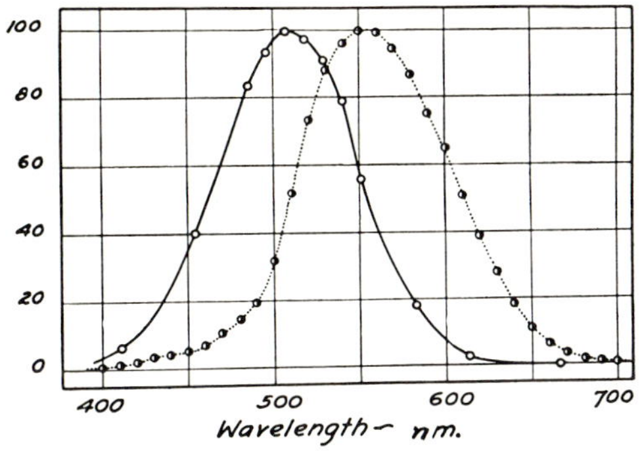


**Figure 9.6.** Luminosity curves for scotopic (rod) vision and for photopic (cone) vision. Since the maxima are arbitrarily set at 100, these curves give no information about the relative sensitivity of rods and cones for red, or for any other part of the spectrum. (From Hecht and Hsia, 1945: J. Op. Soc. Am. *35*:262.)

vision, as can be seen in Figure 9.7. It is readily bleached by light and regenerates to some extent in the dark.

The nature of the chemical reactions following absorption of light by rhodopsin in the rods of vertebrates has been investigated extensively in the last 20 years. The spectacular recovery from night blindness observed in individuals treated with vitamin A connected this vitamin with vision (Dowling, 1966). Research showed dark-adapted retinas to possess little vitamin A, but light-adapted retinas had an abundant supply. When retinas from dark-adapted animals were treated with strong fat solvents, a carotenoid called *retinene* (retinaldehyde) appeared in place of vitamin A. As this carotenoid disappeared during light adaptation, an equivalent amount of *trans*-vitamin A appeared in the excised retina. Experimentation has demonstrated that vitamin A is indeed a reduced form of retinene. During light adaptation the rhodopsin decomposes to form *trans*-retinene and the protein *opsin* (see Fig. 9.8). The released *trans*-retinene then becomes reduced to *trans*-vitamin A (Wald, 1954).

Retinene and opsin obtained experimentally by action of light on retinas or retinal extracts do not recombine to form rhodopsin, nor do vitamin A and opsin. In seeking the reason for this, optical isomers of retinene were tested and it was discovered that although the *cis*-isomer combines with opsin to form rhodopsin, the *trans*-isomer does not. Evidence has been gathered which indicates that the isomerization from *trans*- to *cis*-vitamin A normally occurs under the influence of isomerizing enzymes acting only on vitamin A, not retinene; consequently, the reduction of retinene to vitamin A and its reoxidation to retinene subsequent to isomerization are necessary for completion of the cycle (Hubbard *et al.,* 1965).

The cycle of transformations involved in the breakdown and resynthesis of rhodopsin is called the *visual cycle* (see Fig. 9.9). For his elucidation of the events in the visual cycle Wald was awarded the Nobel

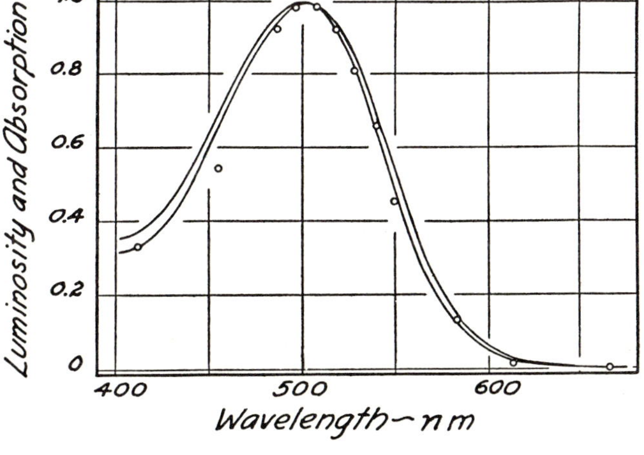


**Figure 9.7.** Comparison of scotopic luminosity at the retina with visual purple absorption. The points are corrected for quantum effectiveness and ocular media transmission. The curves are the percentage absorption spectra of visual purple; the upper curve represents 20 per cent maximal absorption and the lower one 5 per cent maximal absorption. All curves have been made equal to 1 at the maximum, 500 nm., for ease in comparison. (From Hecht, 1942: J. Gen. Physiol. *25*:831.)

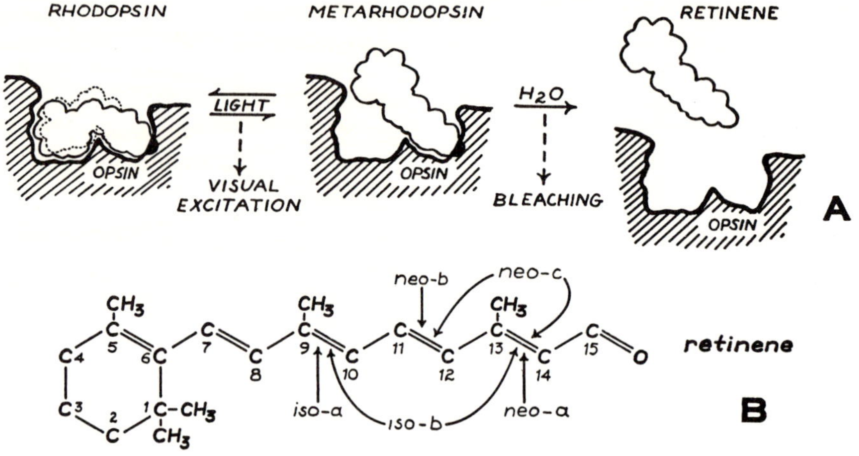


*Figure 9.8.* *A*, Hypothetical scheme showing the effect of light on vertebrate rhodopsins. In rhodopsin, the neo-b (11-*cis*) chromophore fits into the chromophoric site on opsin. Light isomerizes the *cis* chromophore to the all-*trans* configuration, thus decreasing the interaction between the chromophore and opsin. The result is all-*trans* metarhodopsin, which hydrolyzes readily to retinene and opsin. The isomerization of rhodopsin to metarhodopsin is probably responsible for visual excitation, but bleaching is due to the hydrolysis of metarhodopsin. *B*, Structural formula of retinene, showing the system of numbering carbon atoms. The structure as shown has the all-*trans* configuration. Arrows indicate the double bonds which are in *cis* configuration in the various *cis* isomers. (From Kropf and Hubbard, 1958: Ann. N.Y. Acad. Sci. *74*:266.)

prize in 1967. Recently methods have been developed to measure the bleaching of rhodopsin in the intact eye (Rushton, 1959). These studies indicate that only a very small proportion (0.1 per cent) of the rhodopsin is bleached even when the retina is fully light adapted.

Just which of these chemical changes, if any, enables the rods to excite the neurons connected with them to carry messages to the central nervous system is not yet known. The amount of energy required to stimulate the dark-adapted eye of man is exceedingly small. Hecht and his co-workers (1942) have shown that only 5 to 14 quanta need be delivered to the retina. Since each of these is probably absorbed by a different rod, a single rod probably absorbs only a single quantum. Therefore only a single light-activated molecule is probably produced in a rod and it seems most likely that the rod is already set to discharge. The light-activated molecule probably acts only as a trigger which upsets a delicately balanced system (Wald, 1964). It is also possible

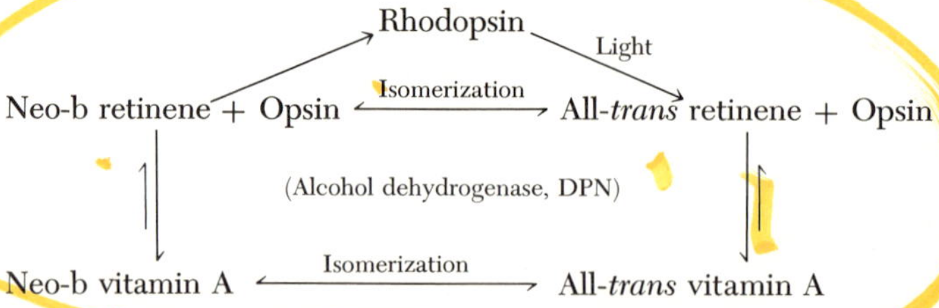


*Figure 9.9.* The visual cycle. The *cis* form of retinene present, conjugated with opsin (protein) in rhodopsin, is transformed by light to the all-*trans* retinene. *Trans*-retinene is most readily isomerized back to the *cis* form by an indirect route (lower path), but direct isomerization may occur under the influence of blue light (upper path). The *trans*-retinene is first reduced to *trans*-vitamin A which is isomerized to the *cis*-form, neo-b vitamin A, in the presence of an isomerase. When the bond between opsin and retinene is broken the retina becomes yellow. Reduction of retinene to vitamin A results in loss of color (After Wald, 1961: Light and Life. McElroy and Glass, eds. Johns Hopkins Press.)

that on illumination rhodopsin induces excitation by acting as an adenosine triphosphatase (McConnell and Scarpelli, 1963). Consideration of such excitability problems is reserved for Chapters 22 to 25.

The pigment *iodopsin,* in the cones of the eye, has never been extracted free of rhodopsin and has been studied only in mixtures with the latter. The absorption spectrum of iodopsin, determined by indirect means, corresponds to the action spectrum for cone vision. There is some evidence that vitamin A is also involved in visual responses of the cones (Wald, 1964). In general, however, relatively little is known of the biochemistry of cone vision at present although the response of cones in intact eyes is being extensively studied (Rushton, 1959, 1962, 1964; Land, 1959, 1964; Wald and Brown, 1965; MacNichol, 1966).

Carotenoids and a pigment resembling rhodopsin are also present in the cells of the eyes of various types of invertebrates, e.g., the squid, the blue crab and the king crab. However, visual processes in these organisms may be somewhat different since when the squid eye is exposed to light, vitamin A is not formed from retinene (Wald, 1954). Even in flagellates with eyespots, e.g., *Euglena,* carotenoids are associated with responses to visible light, and the light absorbed most effectively by these substances is most effective in phototaxic responses. The phototropism, or the reaction of plants to light, is probably effected by different mechanisms, and is not as well analyzed, since sometimes even the role of carotenoids in this process has been questioned (see Briggs, 1964; Page, 1968).

## CIRCADIAN RHYTHMS

Much attention has been focused on the effect of light on daily rhythms which occur even in the life activities of unicellular plants and animals and in cells of multicellular organisms. It had long been supposed that the natural alternating periods of day and night impose this rhythm, but if the rhythm is related to environmental periodicity in this simple causative way it should disappear when the plant or animal is placed in continuous light or darkness. Experiments show that in some nonphotosynthetic cells the rhythm usually persists in continuous darkness (or continuous dim light in some photosynthetic cells).

Significantly, the rhythm in continuous darkness is never exactly 24 hours, being slightly more or slightly less than this figure. For this reason these rhythms have been called *circadian* (about a day) by Halberg and the name has been generally accepted (Halberg and Howard, 1958; Aschoff, 1963). Circadian rhythms are found in the cell activities of many microorganisms, plants and animals. Such an endogenous rhythm within cells suggests some mechanism which times the rhythm; hence the name *biological clock.* It would be impossible here to discuss or appraise the vast amount of literature on the subject which includes considerable information on biochemical and cellular rhythms in man (Aschoff, 1965). Instead, only one example of an endogenous rhythm and its characteristics in a unicellular dinoflagellate, *Gonyaulax polyhedra,* will be considered in detail here (Sweeney and Hastings, 1958; Hastings, 1964; Hastings and Keyman, 1965).

*Gonyaulax* is a photosynthetic cell which emits a brief flash of light (90 msec.) when stimulated by agitation. Cultures grown in a daily cycle of day and night display rhythmically a greater luminescence upon agitation during the dark period than during the day, as shown in Figure 9.10*A*. When such cultures are transferred to a dark chamber, the rhythm continues, but its amplitude decreases progressively because of lack of nutrients, since photosynthesis ceases in the dark. Bright light permits photosynthesis but inhibits the rhythm; dim light permits just enough photosynthesis to supply necessary nutrients to the cell but does not inhibit the rhythmic bioluminescent response which persists in dim light indefinitely without decrease in amplitude. Even cultures grown in bright light for a year, during which time the rhythm disappears, resume the rhythm

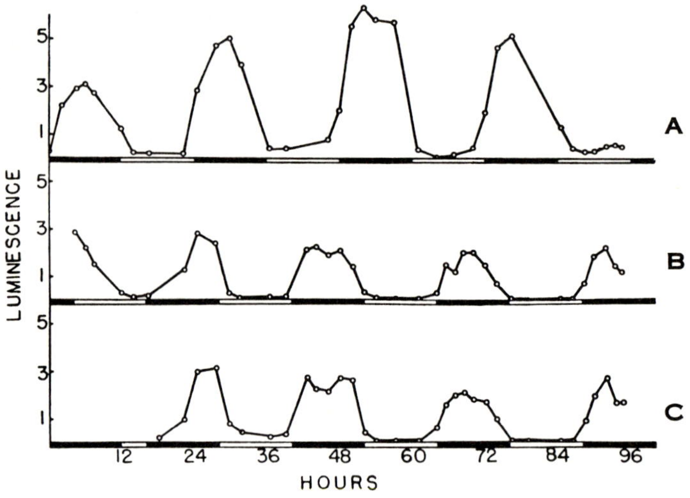


***Figure 9.10.*** Phase shift. This experiment illustrates the effect of changing the solar time at which the light and dark periods occur. Curve *A* shows the pattern of luminescence changes in the dinoflagellate, *Gonyaulax,* in an LD culture—one exposed to a diurnal rhythm of 12 hours light (L) and 12 hours of darkness (D)—which had been on the schedule indicated for some time. The black bars on the time axis indicate dark periods. The lower two graphs, *B* and *C,* illustrate the effect of imposing upon cultures (which were previously on the schedule shown in the top graph) an LD schedule in which the light and dark periods were at a different time of day. The new schedules were started at zero hours on the graph. Temperature, about 26° C. Light intensities used, about 250 foot-candles. (From Hastings and Sweeney, 1958: Biol. Bull. *115*:443.)

when again grown in dim light. The *innate period* of the rhythm, that is, the time from one peak to another, under constant dim illumination is somewhat less than 24 hours under some conditions (lower temperatures), somewhat more under others (higher temperatures), but always almost 24 hours.

The *phase* or beginning of the rhythm of luminescence in *Gonyaulax* can be shifted, as shown in Figure 9.10*B* and *C.* For example, if a culture of *Gonyaulax,* which has been grown under natural periods of night and day, is exposed to bright light during the night, and then placed in continuous dim light, it will once again show an endogenous rhythm but with a *phase shift* of several hours, the shift span depending especially upon the intensity of illumination and its duration. The rhythm will now operate from this new base line and without reference to the solar day, since the organisms are in continuous dim light. Clearly, the biological clock of *Gonyaulax* is set by bright light, but its innate period is set by endogenous factors, and so long as conditions are constant it will continue with its innate rhythm.

*Gonyaulax* taken from a rhythmic culture with an innate period of 24 hours and exposed to alternate periods of 7 hours of darkness and 7 hours of light for a while become *entrained* to a new rhythm (Fig. 9.11). However, as soon as they are again placed under continuous dim illumination they revert to the innate period of almost 24 hours. Similar entraining periods of 6, 8 and 16 hours give like results. It appears that the light, as a timer, has a powerful effect but not a persistent one* (Sweeney and Hastings, 1960; Hastings, 1964).

Besides the luminous rhythm, a circadian rhythmic outburst of divisions was also observed in cultures of *Gonyaulax.* These show some of the same features as the luminous rhythm, including a phase shift (Sweeney and Hastings, 1958).

It is important to know what factors other than light play a part in endogenous circadian rhythms. Experiments with sus-

*In other species of organisms it has usually been found possible to entrain periods only when they are very close to the natural periods of the organisms.

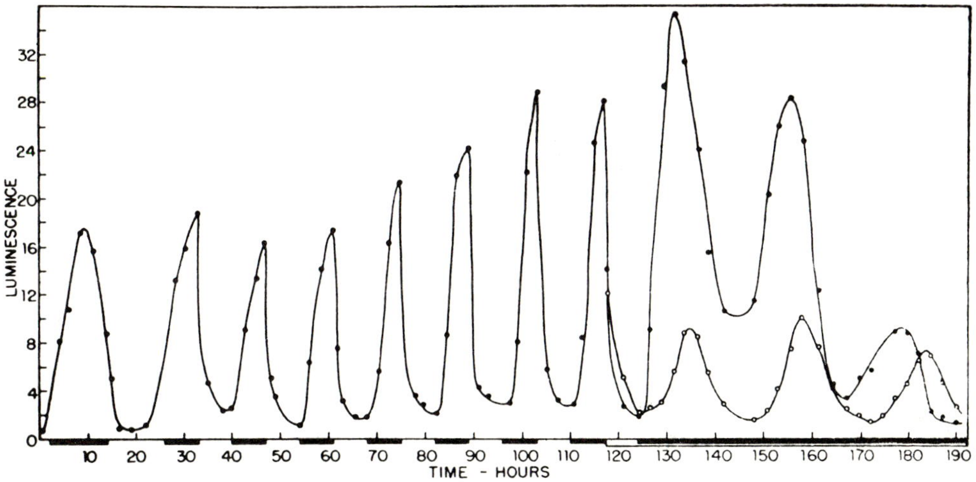


***Figure 9.11.*** This illustrates the entrainment of the luminescence rhythm in the dinoflagellate, *Gonyaulax* to a 14 hour cycle and the manifestation of an endogenous diurnal rhythm when the cells are placed in constant conditions subsequent to the treatment. Dark periods are indicated by black bars on the time axis. The cells were on an LD—12 hours light (L), 12 hours darkness (D)—schedule previous to the time when the 14 hour cycle was started (at 26 hours). Light intensity throughout the 14 hour cycling was 800 foot-candles. At 117 hours some aliquots were removed from the dark and placed in constant dim light at 230 foot-candles. The luminescence changes in these cultures are shown by the open circles. From 124 hours on, the other aliquots were left in the dark. Luminescence changes are plotted with solid dots. Note rapid decline of peak luminescence in darkness. Temperature, 21° C. (From Hastings and Sweeney, 1958: Biol. Bull. *115*:444.)

pensions of *Gonyaulax* have shown that there is apparently no relation between mechanical disturbance of any kind and the innate period (Hastings, 1964). (In birds, however, entrainment of circadian rhythms has been reported with sound—Menaker and Eskin, 1966). It has been shown also that, with changes in temperature, some changes in period are almost always observed, an increase in temperature increasing the period (Sweeney and Hastings, 1960) (Fig. 9.12). Thus at 16° C. the period was 22.8 hours; at 26.7° C. it was 26.5 hours. At 11.5° C. and 32° C. rhythms practically ceased. Relative independence of temperature has generally been observed for endogenous rhythms in many organisms.

It is not possible at present to do more than make a consistent verbal framework for the description and interpretation of existing data on rhythms since specific cellular components have not yet been identified as parts of the mechanism of the clock. The circadian rhythm is considered to be unaffected by the input of information coming to the cell from the environment in the form of pressure (including osmotic pressure), mechanical disturbance, ionizing radiation (e.g., cosmic rays) and to a considerable extent temperature. The endogenous period depends upon a metabolic supply of energy, as already shown, but it is unlikely that the period is determined by metabolic reactions in a simple manner, since the rate of most metabolic reactions is doubled by a 10° rise in temperature within the viable range (see Chapter 10), whereas such a change in temperature has little effect on the endogenous period, as we have seen. It is postulated by some that two coupled systems are involved, one of them light sensitive and temperature independent which serves as the pacesetter and a second one temperature sensitive but light insensitive, phased by the first.

Shocks of ultraviolet light which affect the nucleic acids of the cell set the biological clock back (in *Paramecium*). However, visible light which affects the clock in all organisms does not affect nucleic

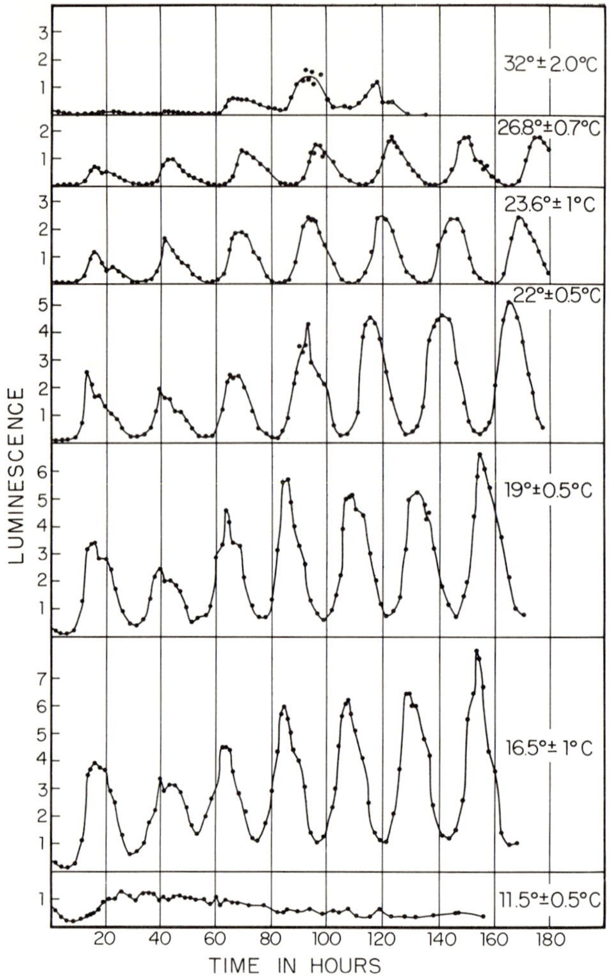


***Figure 9.12.*** Characteristics of the persistant rhythm of luminescence in the dinoflagellate *Gonyaulax* at several different temperatures. The cells were kept at 22° C. in LD (i.e., alternate 12 hours light and 12 hours dark) conditions prior to the beginning of the experiment shown on the graph. At that time the cells were transferred from the dark at 22° to the various temperatures, all in dim light, and left in these conditions throughout the experiment. The luminescence of the cells was determined approximately every 2 hours. At the dim light intensity used, there is little cell division, even at the optimal temperature for growth. The number of cells therefore remained essentially the same in all tubes at all temperatures. (From Hastings and Sweeney, 1957: Proc. Nat. Acad. Sci. U.S.A. *43*:804.)

acids. The action spectrum for visible light-induced shifts in period in *Gonyaulax* is similar to that of chlorophyll, but obviously chlorophyll is absent from colorless cells whose rhythms are also set by visible light. Therefore some substance other than chlorophyll must absorb the light which sets the endogenous rhythms, but clear-cut evidence of the nature of the pacemaker reaction is not available (Hastings and Keynan, 1965).

Many cellular activities in man and other mammals are also rhythmic. Although normally set to a diurnal rhythm by the day-night cycle, they persist under constant conditions as circadian rhythms, being nearly, but not exactly, of 24 hour periodicity. Man in a cave or in an underground isolated bunker, for example, continues to show a circadian temperature cycle and maxima in excretion of sodium, calcium and ketosteroids in the urine (Aschoff, 1965). There is evidence that numbers of white blood cells, the concentration of various hormones and certain nutrients in the blood, the synthesis of nucleic acids in some cells of the body and the susceptibility of the cells to drugs and toxins also show circadian rhythms (Halberg and Howard, 1958). Even susceptibility of cells in mice to ionizing radiation is said to display circadian rhythmicity (Pizzarello *et al.,* 1964).

The period of mammalian circadian

rhythms is sensitive to light intensity; it is shortened by high intensity and lengthened by low in diurnal species, and inversely affected by light in nocturnal species. The phase of circadian rhythms has in some instances been shifted by appropriate lighting regimes (e.g., light by night, darkness by day) (Aschoff, 1965). The malaise experienced for several days by jet travelers on arrival at a distant point in east-west or west-east travel is partly the result of displacement of the phase of their circadian rhythms since north and south flights of equal distance without shift in latitude do not produce comparable effects (Hauty and Adams, 1965). Although circadian rhythms persist under constant conditions, they are disrupted by random lighting (Holmquist *et al.,* 1966).

Circadian rhythms in animals may take origin in certain cells of the body which then serve as pacesetters for others, as has been claimed for certain cells in the nervous system of the cockroach (Harker, 1964) and in the mollusk *Aplysia* (Strumwasser, 1965). Such pacesetter cells may possibly exist in the mammal and may evoke circadian secretion of hormones which in turn may affect distantly located cells. On the other hand, endogenous circadian rhythms of secretion of corticosteroid hormones from the adrenal cortex cells of mammals have been observed and such cells continue to show a circadian rhythm of secretion in tissue culture as well (Halberg, personal communication). The adrenocorticosteroids might in turn pace other biochemical activities in the body but the details have not been elucidated. However, some endogenous rhythms (blood serum level of certain hormones) in mammals are not abolished by adrenalectomy (Halberg *et al.,* 1959, 1965). The pineal gland in the brain is thought by some to be the major pacesetter in circadian rhythms of vertebrates but there is some uncertainty as to whether it is independent of control by other centers (e.g., hypothalamus). Furthermore, some of the activities of the pineal gland are under exogenous control by light (Wurtman and Axelrod, 1965).

## PHOTOPERIODISM

Photoperiodism is the growth response of cells in plants and animals to alternate periods of light and darkness and is correlated with the length of the light and dark periods. Thus, some plants (long-night plants) bloom only when exposed to a number of short days; others (short-night plants) bloom only if exposed to a number of long days. Experiments have demonstrated that it is the cells in the young leaves of a plant that receive the stimulus from exposure to light and that these cells produce a diffusible substance (hormone) that arouses the floral organs and results in their growth. The hormone can pass by direct contact from illuminated cells to others not treated, but will not pass through agar or cellophane. Variations in the intensity of the light (above threshold), such as are caused by clouding of the skies, do not affect the response, which is regulated only by the number of hours of exposure.

The period of darkness is also of primary importance and its interruption, even with a brief flash of light, alters the response. In other words definite, alternate periods of light and darkness are required. Many other growth responses, including germination of seeds, are regulated by the same rhythms of day and night (Borthwick *et al.,* 1956). In some cases, by manipulation of the light periods, several flowering cycles have been induced in one calendar year, although under natural conditions only one cycle occurs.

The red end of the spectrum, particularly at wavelength 660 nm., is the region of maximal stimulation for photoperiodic response in plant cells. Blue light is relatively ineffective.

The photoperiodic response induced by red light (660 nm.) can be reversed by illumination with far red (730 nm.); it can be reactivated by another exposure to red (660 nm.) and again reversed by far red (730 nm.). As a result of these experiments the hormone is considered to be a pigment called *phytochrome.* The pigment was conceived to exist in cells in two forms, one absorbing strongly in the red, and another in the far red (Fig. 9.13):

$$P_r \underset{730 \text{ nm.}}{\overset{660 \text{ nm.}}{\rightleftharpoons}} P_{fr} \qquad (9.2)$$

$P_{fr}$ is considered the active form, which is perhaps an enzyme inducing the photoperiodic responses, probably by affecting some key enzymatic reaction. $P_{fr}$ gradually and spontaneously reverses to form $P_r$, quickly in cells of long-day plants and slowly in cells of short-day plants. The dark conversion is somewhat affected by temperature.

Although phytochrome is present in exceedingly small amounts (1 part in 10 million), it has been extracted from the cells of the tips of young albino plants and was shown to be a bluish compound consisting of a protein and another molecule attached to it. It has all the properties predicted for it from photoperiodic studies. It shows maximal absorption in one form at 660 nm., being converted to the other form with maximal absorption at 730 nm. (Fig. 9.13). The latter wavelength reconverts it to the form with maximal absorption at 660 nm. The conversions are entirely photochemical and occur at very low temperatures as readily as at high. Phytochrome is

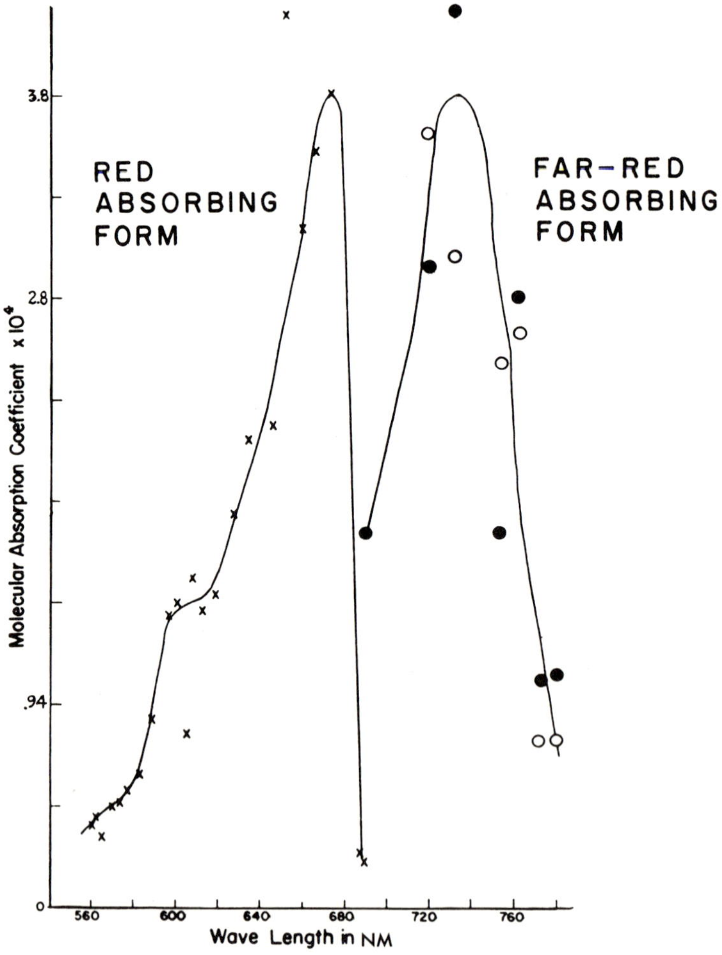


***Figure 9.13.*** Molecular absorption coefficients of the red and far-red absorbing forms of phytochrome (After Hendricks, 1959: Photoperiodism and Related Phenomena in Plants and Animals. Withrow, Ed. American Association for the Advancement of Science, Washington.)

as widely distributed as chlorophyll and is to photoperiodism as chlorophyll is to photosynthesis (Hendricks, 1964).

Photoperiodic phenomena similar to those just discussed for plant life have been observed for animal life also. For example, many birds breed in the spring when days are getting longer, and precocious breeding can be induced in winter by artificially lengthening the days, even when the temperature is below freezing. The converse pattern is found in some mammals that breed in the fall when the days are short and produce their young in the spring. Here the stimulus of the light apparently acts upon cells in the hypophysis. When the hypophysis is activated it stimulates growth of gonadal cells. Some evidence exists that photoperiodic control regulates breeding in many higher and lower forms of life as well as fat deposition, migration and other behavior patterns (Farner, 1961). In some cases (e.g., some insects) reproduction can be controlled practically at will by manipulation of the photoperiod, although nutrition and temperature play a role as modifiers (Beck, 1963; Lees, 1968). The analysis of the cellular mechanisms involved in these responses is as yet rudimentary (Solberger, 1965).

## EFFECTS OF ULTRAVIOLET LIGHT ON CELLS

One effect of ultraviolet rays that everyone has experienced is *sunburn.* The ultraviolet radiations in sunlight injure cells in the epidermis (see Fig. 9.14). The injured cells (prickle cell layer) liberate chemicals which diffuse out and cause a relaxation of the walls of blood vessels in the dermis, resulting in a reddening of the skin (erythema). If the injury is slight, the red or pink color soon disappears, and in most individuals after a few hours' delay a small amount of pigment (melanin) develops (*tanning*). If the injury is more severe, the prickle cells may die, whereupon the layer is invaded by white cells and serum accumulates, causing a blister which later dries, and the skin "peels" off as new epidermis takes the place of the old (Blum, 1955).

The shortest ultraviolet wavelengths in sunlight which reach the surface of the earth are most effective in producing sun-

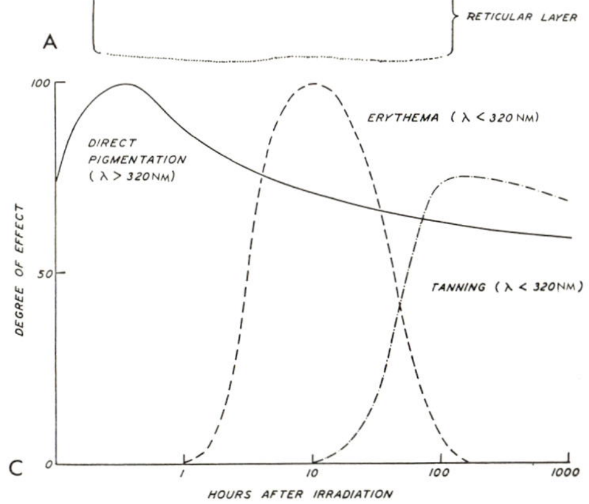


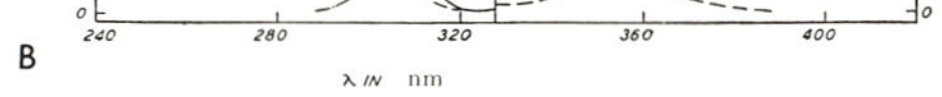


**Figure 9.14.** *A*, Diagram of a section of human skin. Sunburn, or killing of the cell, occurs in the prickle cell layer; erythema (reddening) occurs in the dermis, especially the papillary layer, as a result of dilation of blood vessels. Pigment develops chiefly in the cylindrical cell layer. *B*, Erythema and tanning curves. The question mark on the tanning curve at short wavelengths indicates lack of accurate information. The vertical line indicates a change in ordinate to the percentage scale on the right. *C*, Time of appearance of erythema and pigmentation. (From Giese and Wells, 1946: Scient. Monthly *62*:464.

burn, but shorter ultraviolet wavelengths produced experimentally are also effective. The relative efficiency of different wavelengths (the action spectrum, see Fig. 9.14*B*) in producing sunburn is similar to that in producing tanning. However, long ultraviolet wavelengths, which do not produce erythema, may evoke tanning by affecting precursors of melanin already present in the skin (see Fig. 9.14*B* and *C*).

Another injurious effect of the sun is eyeburn or snow blindness. Ultraviolet radiation in sunlight directly striking the eye, reinforced by similar radiation reflected from snow (or the surface of water), may kill the superficial layers of cells covering the cornea. These cells become opaque, blinding the individual until the layer of dead cells is shed (after a few days). Injury resulting from ultraviolet radiation is superficial and leaves no permanent damage to the eye.

Although ultraviolet radiation does not penetrate deeply and its effects are confined to the surface layers of organisms, prolonged exposure induces superficial skin cancer in experimental animals such as rats and mice (Blum, 1955).

The effect of ultraviolet radiation of sunlight on cells and unicellular organisms is much more drastic. Ever since Downes and Blunt, in 1877, discovered that ultraviolet radiation kills bacteria, all types of cells—microbes, protozoans, eggs of marine animals, algae, fungi and cells in tissue culture—have been shown to be killed by ultraviolet radiations of wavelengths shorter than 310 nm. Doses of ultraviolet radiation sufficient to prevent division of cells (that is, to produce reproductive death) induce mutations in surviving cells. Consequently, ultraviolet radiation has been extensively used by geneticists to produce mutations in microorganisms. Doses of the radiation milder than necessary for reproductive death retard cell division, decrease respiration and affect most cellular functions, especially synthetic processes (Giese, 1964).

The relative efficiency of different wavelengths of ultraviolet radiation in producing a certain cellular effect (e.g., retarding cell division) is illustrated in Figure 9.15. Scrutiny of the figure shows that it resembles the absorption spectrum of nucleic acid (Fig. 3.20). This suggests that retardation of division, induction of mutations, killing of bacteria and other

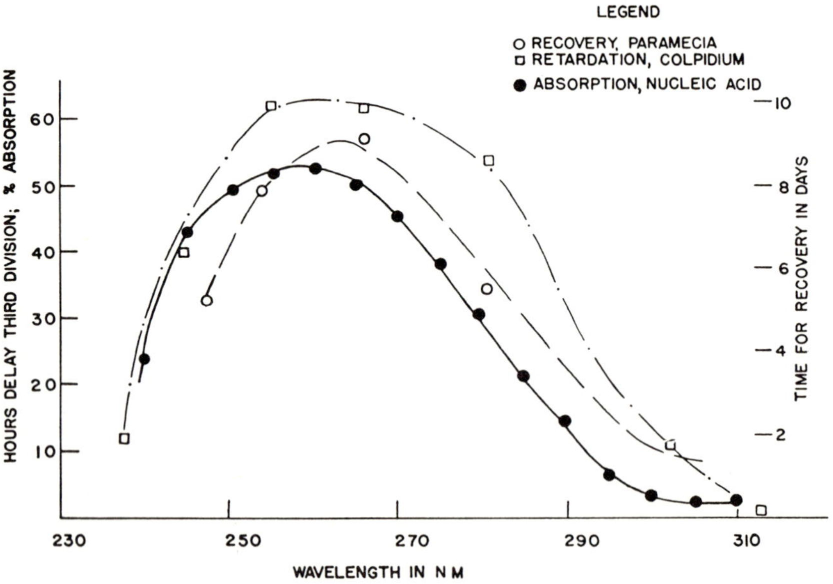


***Figure 9.15.*** Action spectra for retardation of division of paramecia and colpidia by ultraviolet radiation, and absorption spectrum of nucleic acid. (From Giese, 1953: Physiol. Zool. *26*:12.)

effects of ultraviolet radiation with similar action spectra are caused by the effects of this radiation on nucleic acids in cells.

Experiments have shown that inhibition of DNA synthesis by ultraviolet radiation stops cell division. Larger doses progressively inhibit RNA synthesis and protein synthesis in that order. Inhibition of synthetic processes probably underlies the various deleterious effects of ultraviolet radiation (Giese, 1964, Volume 2, Chapter 17).

Ultraviolet radiation is strongly and selectively absorbed by many organic chemicals and the absorption spectra are so characteristic that many compounds may be identified by their ultraviolet absorption spectra (Sinsheimer, 1955). The distinctive absorption spectra for nucleic acid and serum albumin were already given in Figure 3.20 and serve as examples of selective uptake of ultraviolet radiation. Thousands of compounds have been studied in a similar manner (for laws of absorption see Appendix 9.1).

Much excitement was aroused when Kelner (1949) found that visible light, either accompanying or immediately following ultraviolet radiation, will, to a considerable extent, reverse the injurious effects of the latter (*photoreversal* or *photoreactivation;* see Fig. 9.16). Fungus spores, bacteria, animal and plant cells and viruses inactivated by ultraviolet radiation are photoreactivated by treatment with visible light. Mutations and cytological changes induced by ultraviolet radiation are also reversed by treatment with visible light, as in ultraviolet tumor induction (Jagger, 1958, 1960). Recently it has been shown that even erythema in the skin of man is subject to photoreactivation (van der Leun, 1966).

Photoreactivation is never complete, and cells act as if they had been given a smaller dose of ultraviolet than that actually given, (dose-reduction principle).

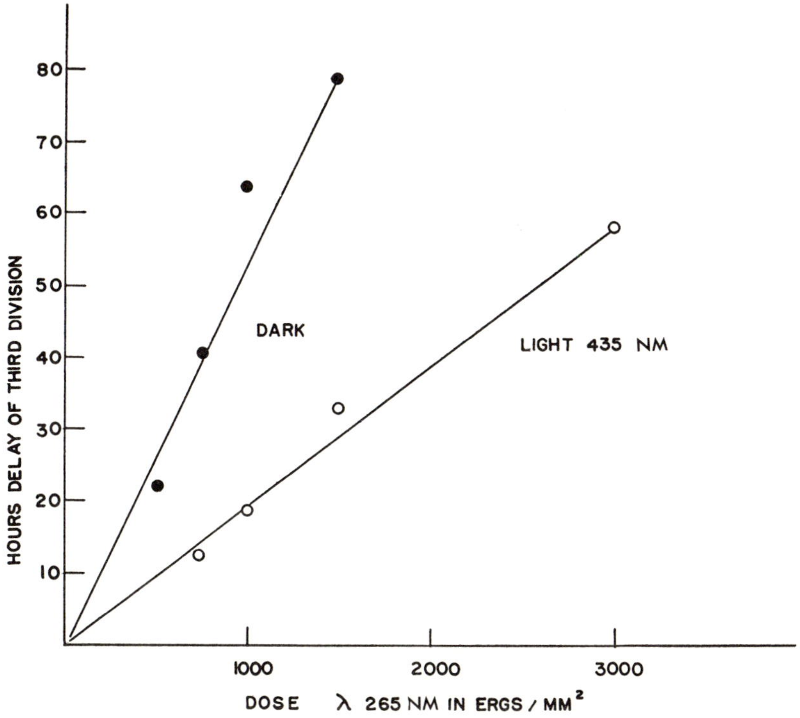


***Figure 9.16.*** Photoreversal recovery of *Colpidium colpoda* treated with various doses of λ265 nm. followed by treatment with blue light of 435 nm. (From Giese, 1953: Physiol. Zool. *26*:14).

A residual damage from ultraviolet always remains after treatment with visible light, damage which increases with the dose and with the repetition of these two treatments in sequence (e.g., Fig. 9.17). Whether the residual damage occurs to the same systems which are being photoreactivated or to others which are not subject to photoreactivation is not known with certainty.

Only the blue end of the visible spectrum and the bordering long ultraviolet region are effective in photoreversal. Theoretically it should be possible to determine the substance absorbing the visible light from the action spectrum of photoreactivation. Unfortunately, available determinations of the action spectrum are not in agreement, and also the action spectra are not detailed enough to identify the absorbing material, although riboflavin and a porphyrin have been suggested (Jagger, 1958).

Photoreactivation appears to involve a photochemical reaction (absorption of visible and long ultraviolet light in the chromophore), followed by a thermochemical reaction, as indicated by a marked temperature effect (photoreversal even ceases entirely when the cells are placed on ice before and during illumination) (Mayor and Melnick, 1961).

DNA synthesis, RNA synthesis and protein synthesis, in that order, are sensitive to ultraviolet radiation and photoreactivation. Since DNA seems to be most readily affected by action of ultraviolet radiation, attempts were made to determine whether DNA itself was altered by ultraviolet radiation and whether this alteration could be reversed by visible light. For this purpose, DNA-transforming reaction (Fig. 6.21*B*) was used as a sensitive biological test of damage to DNA. It is known that in the bacterium *Hemophilus influenzae,* for example, some strains are streptomycin resistant and others are not. By addition of DNA extracted from the resistant strain, a sensitive strain can be transformed into a streptomycin-resistant strain (Chapter 6). It was found, moreover, that the DNA of the streptomycin-resistant strain is altered when subjected to ultraviolet radiation and apparently loses its capacity to transform the streptomycin-sensitive strain into a streptomycin-resistant form. Also, experiments show that if the altered DNA is illuminated with visible light before it is added to the streptomycin-sensitive strain, there is no photoreversing effect in the DNA-altering process (Fig. 9.18).

However, if a crude extract from yeast (or some other organism) that shows photoreactivation is added to the irradiation-altered transforming DNA before illumination, then photoreactivation is obtained. It was found by fractionation that yeast, for example, has an enzyme (photoenzyme) which facilitates photoreactivation of the ultraviolet-treated DNA from the streptomycin-resistant *Hemophilus.* It is inferred that the enzyme of yeast which facilitates photoreversal combines only with the

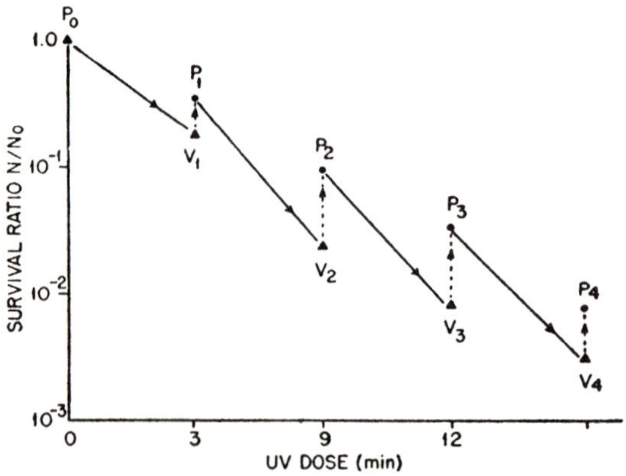


***Figure 9.17.*** Four cycles of ultraviolet inactivation and photoreactivation in a bacterium (*Escherichia coli*). The slopes of the ultraviolet curves do not differ significantly. $V_1$, After first ultraviolet treatment; $P_1$, after first photoreactivation, and so forth. (From Nishiwaki, 1954: Yokohama Med. Bull. *5*:21.)

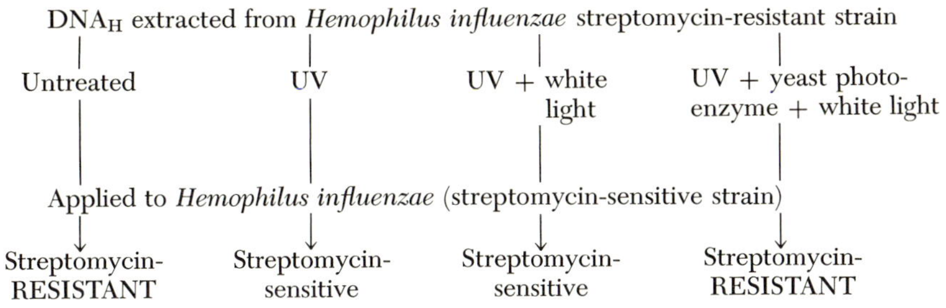


***Figure 9.18.*** Photoreactivation of the transforming principle of *Hemophilus.*

ultraviolet-treated DNA (in some way recognizing the ultraviolet lesion), and the DNA-enzyme complex absorbs visible light which photoreverses the ultraviolet injury of the DNA. Ultraviolet radiation, then, alters the DNA molecule, making possible combination with the yeast photoenzyme; unirradiated DNA (from *Hemophilus* or other sources) does not combine with the enzyme (see Fig. 9.18 and equations 9.3 to 9.6). It was also found that an extraneous DNA, irradiated with ultraviolet, acts as a competitive poison in the system, uniting with the yeast enzyme and interfering with photoreactivation of the *Hemophilus* transforming DNA. This is taken to indicate that irradiation alters some bond, presumably in every DNA molecule, making it capable of uniting with the photoreversing enzyme in yeast (Rupert, 1964):

$$DNA_H + E \rightarrow DNA_H + E \text{ (no combination)} \quad (9.3)$$

$$DNA_H + UV \rightarrow (DNA_H)_{uv} \text{(alteration)} \quad (9.4)$$

$$(DNA_H)_{uv} + E \rightarrow (DNA_H)_{uv} \cdot E \text{ (enzyme complex)} \quad (9.5)$$

$$(DNA_H)_{uv} \cdot E + \text{visible light} \rightarrow DNA_H + E \quad (9.6)$$

The molecular basis of ultraviolet radiation injury to DNA and its photoreversal is, of course, of great interest. It was first found that thymine molecules in frozen solution on irradiation with ultraviolet become linked to one another forming double molecules or *dimers.* Soon afterwards it was shown that adjacent pyrimidines in a strand of DNA also form dimers on exposure of DNA to ultraviolet radiation shorter than 300 nm. The first dimers in DNA to be discovered were thymine dimers which are also the most common, but it is now known that cytosine-thymine mixed dimers and cytosine dimers are also formed in irradiated DNA in decreasing frequency in the order given. Such dimers can be removed from DNA by hydrolyzing the molecules and can be identified by chromatography. Dimers of thymine and other pyrimidines have also been identified in cellular DNA of irradiated bacteria, protozoa and mammalian cells (Wacker, 1963; Trosko *et al.,* 1965). It has also been shown in bacteria that dimerization of thymine or other pyrimidines interferes with replication, thus resulting in death of the cells (Setlow, 1966).

Thymine dimers are also induced in transforming DNA by ultraviolet radiation. When the irradiated transforming DNA was illuminated with visible light, the thymine dimers (and other dimers) were found to be split by visible light in the presence of yeast photoreactivating enzyme, the energy presumably being supplied by the visible light. Chromatography of the hydrolyzed DNA in this case showed no dimers but only thymine and other pyrimidines. The ultraviolet radiation-induced damage to transforming DNA is photoreversed only about 90 per cent, under the most favorable conditions, indicating that damage to DNA other than dimerization may also have occurred.

When irradiated with ultraviolet, cells capable of photoreactivation by visible light, such as *E. coli,* produce thymine dimers, whereas those irradiated with ultraviolet and subsequently exposed to visible light produce only thymine and no thymine dimers. Here photoreactivation occurs to the extent of about 50 to 80 per cent, indicating that photoproducts other than pyrimidine dimers are also formed in ultraviolet irradiated cells. The nature of these photoproducts has not been fully determined. Cross linkages between two DNA strands are known to be formed by ultraviolet irradiation of cells but the doses needed are larger than those required to inactivate cells. Linkages are also known to occur between irradiated DNA and proteins in mixtures, and presumably in cells, but their contribution to cell injury has not yet been assessed (Smith, 1964; Setlow, 1966). Some photoproducts, other than thymine dimers, have been found in extracts of ultraviolet-treated bacterial spores but they have not yet been identified (Donnelan and Setlow, 1965).

Because photoreactivation occurs in tobacco mosaic virus possessing an RNA genome, and dimerization follows exposure of RNA to ultraviolet radiation, it is presumed that these dimers are split by visible light in the presence of a photoenzyme (Setlow, 1966).

The ecological significance of photoreactivation should be self-evident. Ultraviolet radiations in sunlight injure small organisms or cells and induce mutations, but the accompanying visible light greatly reduces its injurious action.

## DARK RECOVERY

When cultures of *E. coli* which had been exposed to ultraviolet radiation were in one case immediately plated on nutrient agar and in another case allowed to remain for a number of hours in buffered salt solution in the dark before plating, the number of colonies in the latter were very much more numerous than in the former. This phenomenon was labeled *liquid holding recovery* (Hollaender and Claus, 1936). Since it occurs in the dark it has recently been renamed *dark recovery*.

Different mutants of *E. coli* show different degrees of recovery from ultraviolet radiation. Witkin in 1946 isolated a mutant B/r which proved much more resistant to these radiations as well as to x-rays. In 1958 Ruth Hill isolated another mutant, $B_{s-1}$, which is much more sensitive to ultraviolet radiation than the parent strain B. The suggestion was made that the kind of repair seen in liquid holding recovery occurred to the greatest degree in B/r, less in B and perhaps not at all in $B_{s-1}$. Howard-Flanders and his collaborators (1962) showed that in cross mating the character for resistance could be transferred to the sensitive strain, at least three genes being involved.

Setlow and Carrier (1964) demonstrated that during dark repair, dimers produced by ultraviolet radiation were excised. They had grown the bacteria on radioactive thymine to label the DNA. Ultraviolet irradiation of the sensitive strain showed DNA containing the original tracer in the intact DNA; the other strains showed labeled thymine-containing fragments which had been split out of the DNA. Since the pieces split out of the DNA contained three types of bases it seemed likely that the dimer plus other material was split out of the DNA molecule. Repair would then consist of reinstatement of the missing portion of the DNA molecule using the information in the complementary strand of DNA. The enzyme recognizing the damage and repairing it may be the polymerase catalyzing the formation of DNA or it may be another enzyme; in fact, several enzymes may be involved since ultraviolet resistance is dependent upon several genes, possibly one for each protein. The repair mechanism has been named the *cut and patch mechanism*.

Direct evidence has been obtained for the repair step by labeling the DNA during the repair process with tritiated 5-bromouracil which, although heavier than thymine, is used by the cell in place of thymine. It was found that the DNA molecules replicated during dark repair were labeled with tritium but not sufficiently labeled with the 5-bromouracil to make them heavy enough to separate from normal DNA by centrifugation (Pettijohn and Hanawalt, 1964). This indicates that only small pieces of the DNA were labeled during repair, supporting the idea of the cut and patch mechanism.

Dark recovery follows injuries by agents other than ultraviolet radiation, e.g., ionizing radiation and chemical agents such as the mutagen nitrosoguanidine and the antibiotic mitomycin C. Strains of cells with strong, intermediate and weak recovery systems are, however, equally sensitive to heat, which has no selective effect on DNA at temperatures known to denature proteins (Setlow, 1966; Hanawalt and Haynes, 1967).

It has been suggested that recovery of this type makes use incidentally of a repair system in the cell whose primary role is the guidance of DNA replication. During replication of DNA, breaks occurring in a strand may be followed by excision of adjacent nucleotides and repair by the enzymes described. The extraordinarily infrequent failure (estimated to be 1 in $10^8$) in replication of bacterial genes attests to the effectiveness of repair. Whatever the primary function of the repair system, it plays an important role in maintaining intact the genome of the cell (Setlow, 1966; Hanawalt and Haynes, 1967).

## CHARACTERISTICS OF DIFFERENT KINDS OF IONIZING RADIATIONS

The effects of ionizing radiations upon cells are dependent upon the nature of the radiations used—alpha or beta particles, fast or slow neutrons, gamma rays, hard or soft x-rays. As background, each of these types of ionizing radiation is first characterized.

Alpha particles are positively charged helium particles, produced by radioactive disintegration of natural or artificially radioactive elements. Polonium gives off only alpha particles and is therefore an ideal source of these radiations. Alpha particles produce intense ionization, especially at the end (tail) of the ionization trail, and are absorbed superficially. Alpha particles may have a velocity of about $3 \times 10^9$ cm. per second and energy corresponding to about 5 MeV (where M is $10^6$ and e is the charge on an electron and V stands for volts). They are especially effective in causing chromosome breaks. Because they produce such a high density of ionization, reducing the oxygen tension does not reduce their effectiveness.

Beta particles are negatively charged particles. They vary in speed and energy, depending on the voltage by which they are accelerated. The ionization produced by beta particles is much sparser than that produced by alpha particles. As a consequence of their charge and resulting interaction with matter, beta particles are superficially absorbed in tissue. Usually they are polyenergetic, but monoenergetic particles can be produced. In tissue, less than 1 per cent of the beta particles are absorbed with the emission of an x-ray.

Fast neutrons are uncharged, highly energetic particles produced by bombarding low molecular weight substances with protons or deuterons. They may give rise to gamma rays on inelastic collision with some matter, but in tissue of organisms which consists of nuclei of atoms of low atomic weight, collisions give rise to protons. Protons are positively charged particles of about the same mass. Protons pass through tissue, forming a high density of ions. Because of the density of ionization, neutron effects are independent of oxygen. Since neutrons are absorbed in proportion to the number of atomic nuclei, light materials like paraffin or water are more effective absorbers of neutrons than sheets of lead because they have larger numbers of nuclei per unit mass. Most sources of neutrons are therefore kept in huge vats of water. Since neutrons are themselves uncharged they may penetrate deeply into tissue of organisms before absorption brings about ionization.

Slow neutrons, also called thermal neutrons, are much less energetic (about 0.03 MeV); consequently they do not cause ionization or dislodge atomic nuclei but react with atomic nuclei to form a new isotope which may disintegrate to give rise to ionizing particles.

X-rays and gamma rays are photons. Gamma rays produced by radioactive disintegration of elements have energies of between 0.3 and 5 MeV, and x-rays have now been produced as energetic as gamma radiations. It is therefore best to speak of the properties of both, but with the understanding that the energy of both varies considerably with the voltage. Ordinary x-rays have much lower voltages than gamma rays. If the energy is less than 0.3 MeV the photoelectric effect is observed, and the absorption of the quantum causes ionization of the atomic nucleus with the emission of an electron containing the energy of the quantum minus the binding energy of the electron. When more energy is available the photon and free or loose electrons collide (Compton effect), resulting in deflection of the electron and the photon, the latter now possessing a longer wavelength, having given up some of its energy. For quanta of energy of 5 MeV or more, absorption is usually followed by positron and electron formation (Glasstone, 1958).

Gamma rays and x-rays penetrate deeply into material of low density, the depth depending upon the voltage. Consequently the density of ion pairs per unit path length varies, being quite sparse for hard or short wavelength radiations. These radiations penetrate deeper than the charged particles because they do not react at once with matter into which they penetrate. Once a photoelectron is ejected from a particle of absorbing material, its energy is quickly dissipated in formation of ion pairs, the density of which is greatest at the end (tail) of the ionization trail.

## EFFECTS OF IONIZING RADIATIONS ON CELLS

Brief exposure of the cells of the skin to ionizing radiation leads to a reddening, or erythema, which is followed by slight tanning; such brief exposures are used in the treatment of various skin diseases. More prolonged exposure, however, leads to injuries to the skin which are quite different from those caused by ultraviolet light.

Many ionizing radiations (e.g., hard x-rays) readily penetrate through a considerable thickness of living tissue and some of them will even go through a mass of concrete or lead. The effect of ionizing radiation is not confined to the surface (except in the action of alpha rays or very soft x-rays); therefore, cells deep in the body may be injured. Certain kinds of cells are especially sensitive to ionizing radiations, e.g., lymphocytes, synthetically active cells and proliferating cells. The cells of the germinative layer in the skin (Fig. 9.14) are more readily destroyed than the surrounding connective tissue cells, so that injuries may be healed only by the deposit of scar tissue (keloids). Similarly, blood-forming cells are selectively killed; consequently there is a change in the number and the proportion of the various types of cells. Actively growing, undifferentiated cells such as those found in tumors are more readily affected than the normal, differentiated cells about them, so that ionizing radiations have been especially useful in retarding or inhibiting the growth of tumors. Gametogenic cells, even those deep in a gonad, are selectively affected by ionizing radiations. As a result, mutations are induced in gametes subsequently formed; therefore these radiations are the choice for genetic experiments. Doses greater than mutagenic ones inhibit gametogenesis completely, thereby sterilizing the organism (Alexander, 1965). Extensive tests on plant cells make it appear likely that these differential sensitivities of cells are correlated with the nuclear DNA or nucleotide content—the greater their DNA content the greater their sensitivity to ionizing radiations (Fig. 9.19) (Sparrow and Miksche, 1961; Kaplan and Moses, 1964).

Differences in potency, penetration and manner of action between ultraviolet and x-ray radiations are largely attributable to a difference in energy content of the quanta of these radiations (see Table 9.2). Quanta of short ultraviolet radiations are only about twice the magnitude of visible light quanta, but x-ray quanta are 2000 to 200,000 times greater (Alexander, 1965).

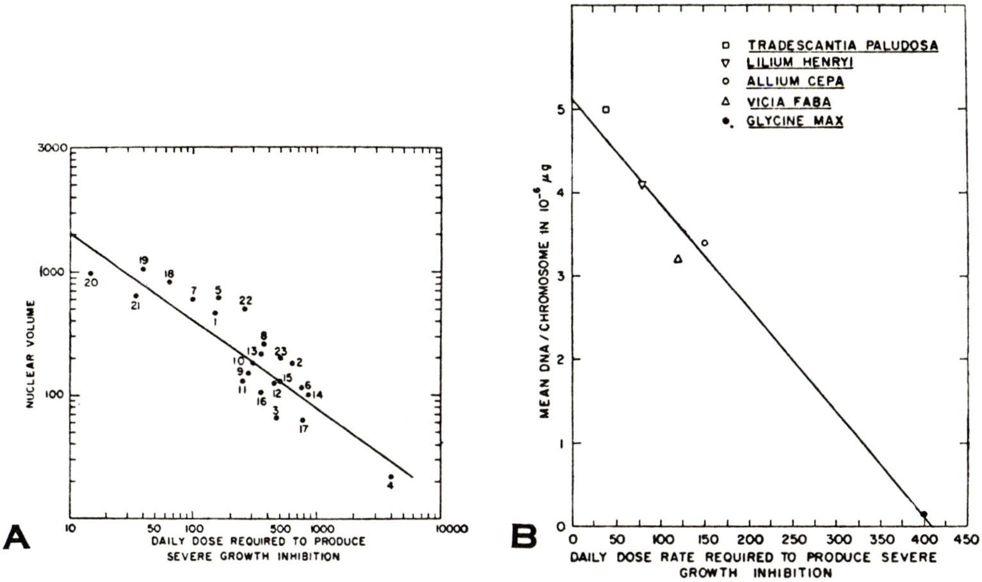


***Figure 9.19.*** Relationship between DNA content and sensitivity to ionizing radiations. In *A* is shown the relationship between nuclear volume in cubic microns and radiosensitivity in 23 species of plants. 1, *Allium cepa.* 2, *Anethum graveolens.* 3, *Antirrhinum majus.* 4, *Arabidopsis thaliana.* 5, *Brodiaea bridgesii.* 6, *Graptopetalum bartramii.* 7, *Haworthia attenuata.* 8, *Helianthus annuus.* 9, *Impatiens sultanii.* 10, *Luzula purpurea.* 11, *Nicotiana glauca.* 12, *Oxalis stricta.* 13, *Pisum sativum.* 14, *Raphanus sativus.* 15, *Ricinus communis.* 16, *Saintpaulia ionantha.* 17, *Sedum oryzifolium.* 18, *Tradescantia ohiensis.* 19, *Tradescantia paludosa.* 20, *Trillium grandiflorum.* 21, *Tulbaghia violacea.* 22, *Vicia faba.* 23, *Zea mays.* *B*, Relationship between mean DNA values per chromosome and radiosensitivity in five species of plants. (From Sparrow and Miksche, 1961: Science *134*:282.)

Ionizing radiations are absorbed in proportion to the density of the absorbing materials. Thus, metal is most opaque to x-rays, bone is next, cells in tissue are still less opaque and air least. Of these, metal has the greatest density (the density of gold is 19.3); that of bone is slightly over 3 (apatite, the mineral present in bone, has an average density of 3.2); that of tissue or water is close to 1.0; and of air, 0.00129.

Two solutions of the same density but chemically differing in solute will absorb ionizing radiation to the same extent. Therefore, x-rays cannot be used as a tool in detecting the presence of various chemical substances in solution by obtaining specific absorption spectra in the same manner as was possible with infrared, visible or ultraviolet radiation because absorption of x-rays is determined more by the physical properties of a substance than

**Table 9.2.** Comparison of Energy Content of Quanta of Various Wavelengths

| WAVELENGTH IN Å ($10^{-8}$ CM.) | QUANTUM* $q = \frac{hc}{\lambda}$ | RATIO OF ENERGY IN ONE QUANTUM TO THAT IN ONE OF GREEN LIGHT |
|---|---|---|
| Green (5000 Å) | $hc/5000 \times 10^{-8}$ | 1 |
| Far ultraviolet (2500 Å) | $hc/2500 \times 10^{-8}$ | 2 |
| Soft x-ray (2.5 Å) | $hc/2.5 \times 10^{-8}$ | 2000 |
| Hard x-ray (0.025 Å) | $hc/0.025 \times 10^{-8}$ | 200,000 |

* Where $h = 6.62 \times 10^{-27}$ erg-sec.; $c = 3 \times 10^{10}$ cm./sec.; and $\lambda$, the wavelength, is given in centimeters.

by its chemical nature (Fano, 1955). However, x-rays can be used for identification of some elements and compounds (Barrieau, 1957).

When a molecule absorbs a quantum of moderately hard x-rays it ionizes with the ejection of an electron which carries most of the energy of the x-ray quantum. This fast moving photoelectron collides with other molecules, giving to each of them some of its energy, as a consequence of which each of them ionizes. For every 34 electron volts (the charge on an electron times the voltage; an energy unit) of energy in the original x-ray beam, a pair of ions is formed (Lea, 1955). A 100,000 volt x-ray, therefore, produces about 3000 ion pairs.

Such a shower of ions produced by absorption of x-rays can be made visible in the following manner. An x-ray is passed through a Wilson cloud chamber saturated with water vapor. The volume of the chamber is suddenly increased whcn thc piston forming the floor of the chamber is pulled. Around each ion formed by the x-ray beam, water vapor then condenses, forming a cloud which can be photographed.

Very hard x-rays, on the other hand, transmit their energy to molecules in other ways; their action is considered in the references cited (Lea, 1955).

The length of the ionizing pathway produced by an x-ray is directly proportional to a power function of the velocity of the electron ejected from some molecule ionized by the x-ray, and inversely proportional to the density of the absorbing material. Consequently, its length will be short in a very dense material like gold, considerably longer in bone and tissue, and much longer in air. The shorter the wavelength of the x-ray (and the higher the energy content) the greater will be the velocity of the primary electron. Therefore, other things being equal, a hard x-ray (short wavelength) will have a proportionally longer pathway in tissue than a softer ray (Fano, 1955) (see Appendix 9.3).

Some of the effects of x-rays, gamma rays, radioactive emanations and other ionizing radiations are understandable on the basis of the facts already presented. The effects of x-rays on cells are largely independent of wavelength and produce effects primarily related to the amount of ionization they induce in cells. Sometimes the density of ionization is of great importance. For example, a pathway of very dense ionization is most effective in splitting chromosomes and for this purpose, radiations such as alpha particles, which produce dense clusters of ions, are more effective than hard x-rays or other radiations which have a greater range but produce a lesser density of ions.

Because the number of mutations produced by ionizing radiations is proportional to the dose, and so is related to the probability of a definite structure being "hit," a target or "direct action" theory or ionizing radiation was proposed (Alexander, 1965). However, when it was discovered that addition of a reducing compound, such as cysteine, protects the cells against radiation action whilc substances promoting oxidation—like oxygen itself—potentiate the effect of the radiations upon cells,* indirect action of the radiations had to be taken into account (Hollaender, 1955). Radiochemical studies with enzymes, for example, also indicate that the environment affects the radiochemical efficiency of ionizing radiations upon molecules, whereas if the action were direct, the effect of the environment should be negligible. However, if the organisms or enzyme solutions are first dried, the effects of the environment are essentially eliminated (Lea, 1955).

Under such conditions it is found that the effects of radiations are in accordance with the direct action theory. A single hit usually produces mutation or inactivation (reproductive death) of a cell. Because the slope of the line relating the logarithm of the survival ratio to the dose indicates the size of the target, it has been possible to use this biophysical method to define the approximate size of biological entities such

*However, increasing the oxygen tension of the culture medium containing the cells above that of the surrounding air does not potentiate effects of ionizing radiations.

as viruses, genes and molecules. However, when water is present, both direct and indirect effects occur and the relative proportion of the effect resulting from direct and indirect action may be assessed by varying the environment.

Since the cell is essentially an aqueous system containing a matrix of proteins, most of the ionizing radiations falling upon the cell are absorbed by water. Water is ionized and decomposed with the formation of oxidizing free radicals like the hydroxyl and perhydroxyl, and oxidizing compounds like hydrogen peroxide or organic peroxides, as shown in Table 9.3 (Blois, 1961).

A free radical is unstable inasmuch as it has an unpaired electron. Thus the hydrogen radical H· must lose an electron to become a hydrogen ion and the hydroxyl radical OH· must add an electron to form the hydroxyl ion, or each must become associated with a molecule forming another free radical in the process, or combine with another free radical forming a normal paired electron configuration of a stable compound (Table 9.3). Free radicals react very readily with organic compounds. Because the unpaired electron of a free radical can make a transition to a higher energy state as a result of resonant absorption in an electromagnetic field of appropriate frequency, free radicals produced in biological systems by ionizing radiations or in other ways can be detected and measured (see Chapter 17) (Allen, 1961; Blois, 1961).

There is considerable evidence to indicate that the aforementioned oxidizing substances are responsible for many of the effects of ionizing radiations upon cells. The strongest evidence is perhaps the following: (1) the formation of such oxidizing materials during irradiation is decreased when the oxygen tension is lowered, and it occurs at only a very low rate when oxygen is lacking; (2) in apparent correlation with this, biological effects of ionizing radiation on cells, such as induction of mutations, fragmentation of chromosomes and lethal effects, are decreased correspondingly when the oxygen tension is reduced; (3) pretreatment of organisms, particularly microorganisms, with reducing substances also protects them to a considerable extent from ionizing radiation, even in the presence of air (Alexander, 1965).

Radiation protection has become an important study and it is now known that protective agents mitigate many of the effects of ionizing radiation, including shortening of the life span, mutations, cataract formation and radiation sickness, but not cancer induction. Cell death in critical organs such as the bone marrow is reduced by radioprotective substances and the few surviving viable stem cells (after a minimal lethal dose) are able to repopulate the marrow. Radioprotective substances, the best of which appear to be the sulfhydryl compounds cysteine and cysteamine (betamercaptoethanol) and their derivatives, are most effective if administered just shortly before exposure to ionizing radiation. They have little protective effect in the absence of oxygen (absence of oxygen alone acts as a dose reduction factor), nor against dense ionizing radiation such as alpha particles or proteins (which show no oxygen effect). This suggests that the protective effect is probably a reduction of the potent oxidizing free radicals formed immediately on

**Table 9.3.** Formation of Free Radicals in water as a Result of Absorption of Ionizing Radiations*

$H_2O$ + ionizing radiation → $H_2O^+$ + e (electron)
$H_2O + e \rightarrow H_2O^-$
$H_2O^+ \rightarrow H^+ + OH\cdot$ (hydroxyl radical)
$H_2O^- \rightarrow OH^- + H\cdot$ (hydrogen radical)

. . . . . . . . . . . . . . . . . . . .

$H\cdot + OH\cdot \rightarrow H_2O$
$OH\cdot + OH\cdot \rightarrow H_2O_2 \rightarrow H_2O + \frac{1}{2}O_2$
$H\cdot + H\cdot \rightarrow H_2$
$OH\cdot + H_2 \rightarrow H_2O + H\cdot$
$H\cdot + H_2O_2 \rightarrow H_2O + OH\cdot$

. . . . . . . . . . . . . . . . . . . .

In presence of oxygen:
$H\cdot + O_2 \rightarrow HO_2\cdot$ (perhydroxyl radical)
$HO_2\cdot + H\cdot \rightarrow H_2O_2$

* The unpaired electron of the radical is shown as a dot. (From Alexander, 1965: Atomic Radiation and Life. Revised. Penguin Books, Baltimore.)

irradiation of cells. Consequently, introduction of the protective agent even as soon as 0.001 sec. after irradiation is without protective action. However, the mechanism by which protection occurs is not conclusively demonstrated (Hollaender *et al.,* 1959; Bacq and Alexander, 1961; Alexander, 1965).

Many pioneer investigators and medical men suffered radiation burns, sometimes a long time after exposure, before the hazards of ionizing radiation were recognized. As a consequence, an International Commission on Radiation Protection (ICRP) and national commissions, in the United States the National Commission on Radiation Protection (NCRP), were formed to regulate conditions under which workers might handle the dangerous radiations, to decide on units of dosage and to determine the maximal permissible exposure. The maximal permissible exposure has been successively reduced as it was realized that there may be undesirable subthreshold effects of radiation (Morgan, 1963).

The converse of radiation protection is sensitization to radiation. For example, iodoacetamide, an antagonist to sulfhydryl compounds, sensitizes bacteria to ionizing radiation. Also, incorporation of certain analogues to thymine or thymidine, e.g., 5-bromodeoxyuracil or 5-bromodeoxyuridine, into the DNA of cells renders the cells radiosensitive (Bacq and Alexander, 1961; Alexander, 1965), probably by interfering with dark repair of the DNA in cells irradiated with ionizing radiation.

Recently, cells in colonies originating from single cells of human tissue have been grown on a synthetic medium. Such a culture consists of a relatively homogeneous population of cells and is useful for testing radiation effects. The experiments show that cells from several different types of tissues have about the same degree of resistance to ionizing radiations, as measured by reproductive death, when they

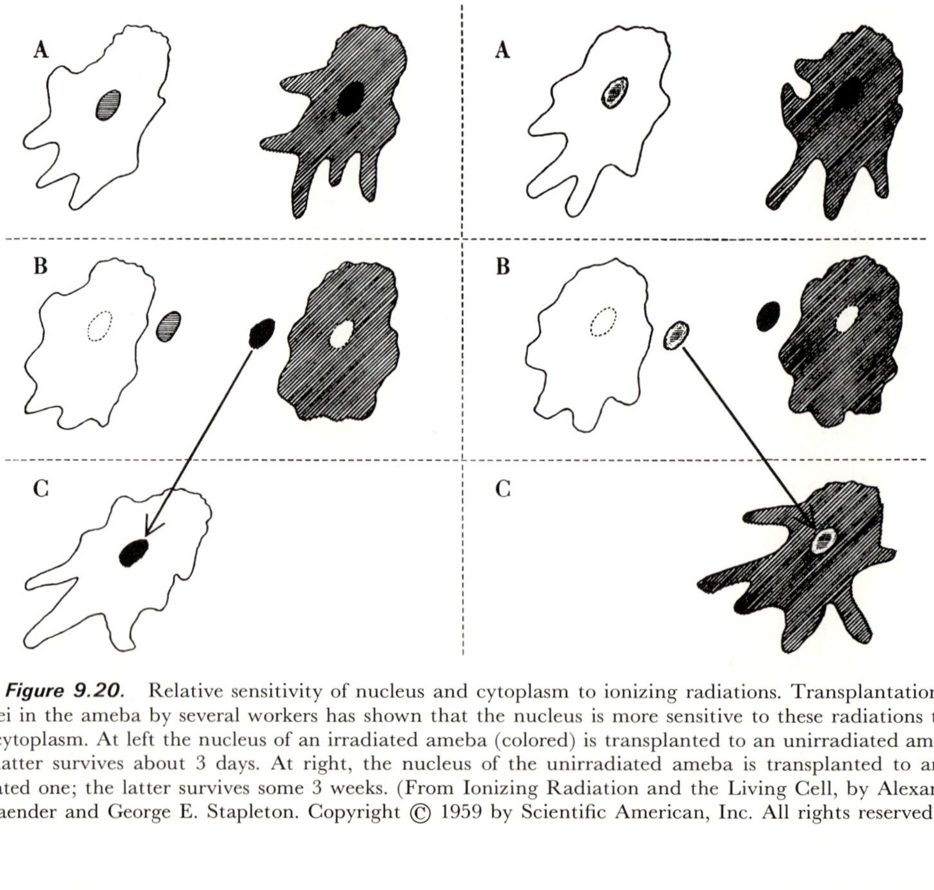


***Figure 9.20.*** Relative sensitivity of nucleus and cytoplasm to ionizing radiations. Transplantations of nuclei in the ameba by several workers has shown that the nucleus is more sensitive to these radiations than the cytoplasm. At left the nucleus of an irradiated ameba (colored) is transplanted to an unirradiated ameba; the latter survives about 3 days. At right, the nucleus of the unirradiated ameba is transplanted to an irradiated one; the latter survives some 3 weeks. (From Ionizing Radiation and the Living Cell, by Alexander Hollaender and George E. Stapleton. Copyright © 1959 by Scientific American, Inc. All rights reserved.)

are all growing and dividing (about every 18 hours). The dose required to produce reproductive death is only about 50 r. This is closely correlated with the doses needed for the production of a chromosome break, resulting in abnormal division (Alexander, 1965).

Radiosensitivity varies markedly for cells of different species—the same effect being produced in the cells of mammalian tissue by 50 r as in *E. coli* by 10,000 r, in yeast by 30,000 r, in ameba by 100,000 r and in paramecia by 300,000 r—all of the cells being in approximately the same physiological state at the time of treatment (Alexander, 1965).

The nucleus of the cell is more sensitive to ionizing radiations than the cytoplasm. When wasp eggs, in which the nucleus is asymmetrically placed, are irradiated, a single alpha particle delivered to the nucleus prevents division, whereas a million particles are required in the cytoplasm to do the same. Also, when a nucleus is transferred from an irradiated *Amoeba proteus* to an unirradiated one, the effect is much greater than when a nucleus is transferred from an unirradiated ameba to an irradiated cytoplasm (Fig. 9.19).

The locus of action of ionizing radiations in the cell has not been identified with certainty. The overall metabolism of the cell, as measured by oxygen consumption and carbon dioxide output, is not altered in a consistent way for all types of cells studied. Also, no evidence has been found for damage by ionizing radiation to specific metabolic enzymes in the cell. Some *in vitro* studies with enzymes which require a free SH group suggested their special vulnerability, but this has not been borne out in studies with SH-containing enzymes extracted from irradiated cells, these enzymes being no more sensitive when irradiated *in situ* in the cell than any of the others tested.

Doses which produce reproductive death do not necessarily stop DNA synthesis. The minimal doses that are required for reproductive death of the cells, if applied before synthesis of DNA has begun, inhibit incorporation of phosphorus-32 into DNA precursors. However, once the DNA synthesis has begun, much larger doses are necessary to stop it. It has been suggested that the process most sensitive to ionizing radiation is the coupling of enzymes for energy transfer from the oxidative energy-liberating reactions of the nucleus to enzymes during phosphate bond formation. Nucleotides and nucleotide precursors of nucleic acid then accumulate, possibly because the DNA template has been damaged. A possible alternative explanation is that the Kornberg enzyme that bears the template has been damaged (Alexander, 1965).

Synthesis of RNA and protein is not much affected by a dose that inhibits cell division (reproductive death). Cells in tissue culture given 10 to 100 times the reproductive death dose may grow up to a millimeter in diameter. The increase of cytoplasm in such a giant cell indicates that many of the metabolic activities continue in a reproductively dead cell, including uptake of nutrients, liberation of energy and RNA and protein synthesis, although the growth is unbalanced inasmuch as the cell DNA is not replicated. Such an irradiated cell can be subsequently infected with a virus which then grows and replicates in much the same manner as in the normal cell. From all such experiments it is apparent that the reproductive capacity of a cell is infinitely more sensitive to radiations than are its other functions (Puck, 1960).

No photoreactivation by illumination with visible light has been demonstrated with tissues injured by x-ray. However, as in the case of injury by a variety of agents, x-ray injury is followed by dark repair.

## BIOLOGICAL EFFECTS OF THE ATOMIC BOMB

An atomic bomb explosion lasts about $10^{-6}$ seconds and develops in its center a temperature of about 1,000,000° C. Its destructive effects are the result of its extreme heat, the ultraviolet flash, shock waves and ionizing radiations. The temperature of the surface of an object such as human skin at 400 yards from the bomb may be raised to about 50° C. The ultraviolet flash burn may be severe. Therefore, surface cells are readily damaged whether they are microorganisms or cells in the surface of the organism. The shock waves rip the capillaries of the sense organs and lungs and cause hemorrhages throughout the entire digestive system. The ionizing radiations consist of about 15 per cent neutrons and about 15 per cent alpha and beta particles; the remainder consists of short gamma rays. Cells, such as protozoans and bacteria as well as those in the body of an animal or plant, are injured in proportion to their sensitivity to these radiations.

A characteristic set of symptoms in a person injured by ionizing radiations is nausea accompanied by vomiting, prostration and, after a latent period, fever, bloody diarrhea, loss of hair and appearance of purple hemorrhagic spots in the skin. Should the individual survive, various skin lesions appear in areas where the germinative epithelium has been killed by the radiations. If infection is prevented and the wounds heal, great sausage-like growths of connective tissue known as keloids develop over severely burned areas of skin. The connective tissue of the keloids may contract enough to make it impossible for the victim to use hands, legs or other parts affected. Delayed effects include mutations in the progeny of the exposed organisms (Los Alamos Laboratory, 1950).

Some idea of the destructiveness of the bomb's radiations may be gathered from the fact that it takes only between 350 and 530 r of ionizing radiations to kill a man. At 1250 meters from the center of the old type bomb, the calculated intensity is about 473 r. Anyone caught in this range would receive a lethal dose unless hidden underground or behind thick concrete walls (Glasstone, 1958; Los Alamos Laboratory, 1950; Shilling, 1964).

## APPENDIX

### 9.1 ABSORPTION OF LIGHT IN A SOLUTION

Equal fractions of incident light are absorbed by successive layers of equal thickness of light-absorbing substance. The fraction of light transmitted (the total incident light minus that absorbed) by a solution is dependent upon the concentration of the absorbing substance, c (usually expressed in moles per liter), the thickness of the solution, l (expressed in centimeters), and the nature of the absorbing substance summarized in $\epsilon$ (the molecular extinction coefficient). The relation between these factors is given in Beer's law:

$$T = 10^{-\epsilon cl} \quad (9.7)$$

$$\text{or } \log_{10} T = -\epsilon cl \quad (9.8)$$

$$\text{or } \log_{10}\frac{1}{T} = \epsilon cl \quad (9.9)$$

Since by definition, D, the optical density, equals $\log_{10} 1/T$,

$$D = \epsilon cl \quad (9.10)$$

$$\epsilon = \frac{D}{cl} \quad (9.11)$$

Since the optical density of a solution is measured by a spectrophotometer, the extinction coefficient is readily determined.

Not all substances obey Beer's law. Association or dissociation of molecules may affect absorption. It is therefore desirable to study the absorption of a solution at different concentrations to test for obedience to the law.

The extinction coefficient, $\epsilon$, is inversely proportional to the transmission. For purposes of illustration, various degrees of transmission are assumed when light is passed through 1 cm. of an 0.1M solution of a substance, S, and the values of $\epsilon$ are calculated as follows:

| HYPOTHETICAL TRANSMISSION (0.1 MOLAR SOLUTION) | $\epsilon$ |
|---|---|
| 90% | .46 |
| 75% | 1.25 |
| 50% | 3.02 |
| 25% | 6.03 |
| 10% | 10 |
| 1% | 20 |
| 0.1% | 30 |

## 9.2 RELATION OF X-RAY WAVELENGTH TO VOLTAGE

A quantum of x-rays is given by Planck's law: $q = h\nu$

$h\nu = Ve$ where V is the voltage through which the electron (e) falls in producing the x-ray, and e is the charge on the electron, $1.6 \times 10^{-19}$ coulomb.

Since $h\nu = hc/\lambda = Ve$

$$\lambda = \frac{hc}{Ve} = \frac{hc/e}{V} = \frac{12.34}{V}$$

where V is given in kilovolts.

By this relation it is possible to calculate the minimal wavelength to be expected for a given voltage.

## 9.3 THE RANGE OF AN ELECTRON

The range, R, of an electron is related in the following way to its velocity and to the density of the material into which it enters:

$$R = \frac{V_0^4}{k \times P_\rho \times 10^{43}}$$

where R is the range in centimeters;
$V_0$ is the initial velocity of the photoelectron in centimeters per second;
k is the constant equal to 0.6 cm.$^6$ sec.$^{-4}$ gm.$^{-1}$;
$\rho$ is the density of the material.

The range is inversely proportional to the density and directly proportional to the initial velocity of the photoelectron. The initial velocity, $V_0$, may be determined from the relation,

$$Ve = \tfrac{1}{2}\, mv^2$$

where V is the voltage;
e is the charge on the electron ($1.6 \times 10^{-19}$ coulomb);
Ve is in volt coulombs or joules (1 joule = $10^7$ ergs or dyne-cm.);
m is the mass of an electron ($9.1066 \times 10^{-28}$ gm.);
v is the velocity of the electron in centimeters per second;
and the dyne is a gram cm. sec.$^{-2}$.

The initial velocity of an electron produced by absorption of a maximum quantum of 100,000 volt x-ray is $1.87 \times 10^{10}$ cm. per second; the range in air (density 0.00125 gm. per cubic centimeter) is 16 cm., in tissue or water (density 1) about 0.02 cm., in bone about 0.007 cm. and in gold (density 19.3) about 0.00108 cm.

The density of ionization will be greater the shorter the range. Since soft x-rays produce primary photoelectrons of low velocity, the density of ionization is greater than for a hard x-ray in the same material, e.g., in tissue.

## *LITERATURE CITED AND GENERAL REFERENCES*


Alexander, P., June 1960: Radiation-imitating chemicals. Sci. Am. *202*:99–108.

*Alexander, P., 1965: Atomic Radiation and Life. Revised. Penguin Books, Baltimore.

Allen, A. O., 1961: Radiation Chemistry of Water and Aqueous Solutions. D. Van Nostrand Co., New York.

Anderson, K. A., June 1960: Solar particles and cosmic rays. Sci. Am. *202*:64–71.

Aschoff, J., 1963: Diurnal rhythms. Ann. Rev. Physiol. *25*:581–600.

Aschoff, J., 1965a: Circadian rhythms in man (a summary). Science *148*:1427–1432.

Aschoff, J. (ed.), 1965b: Circadian Clocks. North Holland Publishing Co., Amsterdam.

Augenstein, L. G., Mason, R., and Quastler, H. (eds.), 1964: Advances in Radiation Biology. Vol. 1. Academic Press, New York.

Bacq, Z. M., and Alexander, P., 1961: Fundamentals of Radiology. Pergamon Press, New York.

Barrieau, R. E., 1957: X-ray absorption edge spectrometry as an analytical tool. Determination of molybdenum and zinc. Chem. *29*:348–352.

*Beck, S., 1963: Animal Photoperiodism: Relationship of Daylength to Animal Growth, Development and Behavior. Holt, Rinehart & Winston, New York.

Bendix, S., 1960: Pigments in phototaxis. *In* Comparative Biochemistry of Photoreactive Systems. Allen, ed. Academic Press, New York, pp. 107–128.

Biologic Effects of Laser Radiation. 1965. Fed. Am. Soc. Exp. Biol., Washington, D. C.

Blois, M. S. (ed.), 1961: Free Radicals in Biological Systems. Academic Press, New York.

Blum, H. F. 1955: Sunburn. *In* Radiation Biology. Hollaender, ed. McGraw-Hill Book Co., New York, Vol. 2, pp. 487–528.

Blum, H. F., 1959: Carcinogenesis by Ultraviolet Light. Princeton University Press, Princeton.

Blum, H. F., 1964: Photodynamic Action and Diseases Caused by Light. Rev. ed. Hafner Publishing Co., New York.

Borthwick, H. A., Hendricks, S. B., and Parker, M. W., 1956: Photoperiodism. *In* Radiation Biology. Hollaender, ed. McGraw-Hill Book Co., New York, Vol. 3, pp. 479–517.

Bowyer, S., Byram, E. T., Chubb, T. A., and Friedman, H., 1965: Cosmic x-ray sources. Science *147*:394–398.
Bridges, C. D. B., 1962: The flash photolysis of visual pigments. Vision Res. *2*:201–214.
Briggs, W. R., 1964: Phototropism in higher plants. *In* Photophysiology. Giese, ed. Academic Press, New York, Vol. 1, pp. 223–271.
Bunning, E., 1964: The Physiological Clock. Endogenous Diurnal Rhythms and Biological Chronometry. Springer-Verlag, New York.
Burbridge, G., Aug. 1966: The origin of cosmic rays. Sci. Am. *215*:32–38.
Butler, S. T., Dec. 1962: Atmosphere tides. Sci. Am. *207*:49–55.
Cellular Radiation Biology. 1965. Williams & Wilkins, Baltimore.
Cole, L. J., and Nowell, P. C., 1965: Radiation carcinogenesis: the sequence of events. Science *150*: 1782–1786.
Comar, C. L., 1962: Biological aspects of nuclear weapons. Am. Sci. *50*:339–353.
Davson, H. (ed.), 1962: The Eye. 4 vol. Academic Press, New York.
Donnellan, J. E., Jr., and Setlow, R. B., 1965: Thymine photoproducts but not thymine dimers found in UV irradiated bacterial spores. Science *149*:308–310.
Dowling, J. E., Oct. 1966: Night blindness: Sci. Am. *215*:78–84.
Dye, D. L., and Wilkinson, M., 1965: Radiation hazards in space. Science *147*:19–24.
Eakin, R. M., 1963: Lines of evolution of photoreceptors. *In* The General Physiology of Cell Specialization. Mazia and Tyler, eds. McGraw-Hill Book Co., New York, pp. 383–425.
Fano, U., 1954: Principles of radiation physics. *In* Radiation Biology. Hollaender, ed. McGraw-Hill Book Co., Vol. 1, pp. 1–144.
Farner, D. S., 1961: Comparative physiology: photoperiodicity. Ann. Rev. Physiol. *23*:71–96.
Forsythe, W. E., 1937: The Measurement of Radiant Energy. McGraw-Hill Book Co., New York.
Frier, P., and Webber, W. R., 1963: Radiation hazards in space from solar flares, Science *142*:1587–1592.
Gates, D. M., 1966: Spectral distribution of solar radiation at the earth's surface. Science *151*:523–529.
Giese, A. C., 1945: Ultraviolet radiations and Life. Physiol. Zool. *18*:223–250.
Giese, A. C., 1946: An intracellular photodynamic sensitizer in *Blepharisma.* J. Cell. Comp. Physiol. *28*:119–127.
Giese, A. C., 1947: Radiations and cell division. Quart. Rev. Biol. *22*:253–282.
*Giese, A. C. (ed.), 1964: Photophysiology. 2 vol., Academic Press, New York.
Giese, A. C., 1967: Effects of radiation on protozoa. *In* Research in Protozoology. Chen, ed. Academic Press, New York, Vol. 2, pp. 281–356.
Giese, A. C., and Wells, J. M., 1946: Sunburn protection, natural and artificial. Sci. Monthly *62*: 458–464.
Giese, A. C., and Zeuthen, E., 1949: Photooxidations in pigmented *Blepharisma.* J. Gen. Physiol. *32*: 525–535.
Ginsburg, V., and Razoryonov, L., 1959: New data on the chemical composition of cosmic rays. Am. Sci. *47*:562–566.
Glasstone, S., 1958: Sourcebook on Atomic Energy. 2nd Ed. D. Van Nostrand Co., New York.
Hafner, E. M., 1964: Cosmic electromagnetic radiation. Science *145*:1263–1271.
Halberg, F., Engeli, M., Hamburger, C., and Hillman, D., 1965: Spectral resolution of low frequency, small amplitude rhythms in excreted 17-ketosteroids; probably androgen-induced circaseptan desynchronization. Acta Endocrinol. *50*(Suppl. 103):5–54.
Halberg, F., Halberg, E., Barnum, C. P., and Bittner, J. J., 1959: Physiologic 24-hour periodicity in human beings and mice, the lighting regime and daily routine. *In* Photoperiodism and Related Phenomena in Plants and Animals. Withrow, ed. American Association for the Advancement of Science, Washington, D. C., pp. 803–878.
Halberg, F., and Howard, R. B., 1958: 24 hour periodicity and experimental medicine, examples and interpretations. Postgrad. Med. *24*:349–358.
Hanawalt, P. C., 1965: Repair replication of DNA in bacteria: irrelevance of chemical nature of base defect. Biochem. Biophys. Res. Commun. *19*: 462–467.
Hanawalt, P. C., 1966: The UV sensitivity of bacteria: its relation to the DNA replication cycle. Photochem. Photobiol. *5*:1–12.
Hanawalt, P. C., 1968: Cellular recovery from photochemical damage. *In* Photophysiology: Current Topics. Giese, ed. Academic Press, New York, Vol. 4, pp. 203–251.
*Hanawalt, P. C. and Haynes, R. H., Feb. 1967: The repair of DNA. Sci. Am. *216*:36–43.
Harker, J., 1964: The Physiology of Diurnal Rhythms. Cambridge University Press, New York.
Hastings, J. W., 1964: The role of light in persistent daily rhythms. *In* Photophysiology. Giese, ed. Academic Press, New York, Vol. 1, pp. 333–361.
Hastings, J. W., and Keynan, A., 1965: Molecular aspects of circadian rhythms. *In* Circadian Clocks. Aschoff, ed. North Holland Publishing Co., Amsterdam, pp. 167–182.
Hauty, G. T., and Adams, T., 1965: Phase shifting of the human circadian system. *In* Circadian Clocks. Aschoff, ed. North Holland Publishing Co., Amsterdam, pp. 413–425.
Haynes, R. H., Wolff, S., and Till, J. E. (eds.), 1967: Structural defects in DNA and their repair in microorganisms. Radiation Res. Suppl. 6, pp. 1–243.
Hecht, S., Shlaer, S., and Pirenne, M. H., 1942: Energy, quanta and vision. J. Gen. Physiol. *25*: 819–840.
Hendricks, S., 1964: Photochemical aspects of plant photoperiodicity. *In* Photophysiology. Giese, ed. Academic Press, New York, Vol. 1, pp. 305–331.
Hollaender, A. (ed.), 1954–1956: Radiation Biology. 3 Vol. McGraw-Hill Book Co., New York.
Hollaender, A., and Claus, W. D., 1936: The bactericidal effect of ultraviolet radiation on *Escher-*

*ichia coli* in liquid suspensions. J. Gen. Physiol. *19*:753–765.

Hollaender, A., Congdon, C. C., Doherty, D. G., Makinodan, T., and Upton, A. C., 1959: New developments in radiation protection and recovery. Progress in Nuclear Energy, Series VII, Vol. 2, Medical Sciences, Pergamon Press, London.

Holmquist, D. L., Retiene, K., and Lipscomb, H. S., 1966: Circadian rhythms in rats: effects of random lighting. Science *152*:662–664.

Howard-Flanders, P., Boyce, R. P., Simson, E., and Theriot, L., 1962: A genetic locus in *E. coli* K12 that controls the reactivation of UV-photoproducts associated with thymine in DNA. Proc. Natl. Acad. Sci. U. S. A. *48*:2109–2115.

Hubbard, R., and Kropf, A., June 1967: Molecular isomers in vision. Sci. Am. *216*:62–74.

Hubbard, R., Bownds, D., and Yoshizawa, T., 1965: The chemistry of visual photoreception. Cold Spring Harbor Symposia Quant. Biol. *30*:301–315.

Jagger, J., 1958: Photoactivation. Bact. Rev. *22*:99–142.

Jagger, J., 1960: Photoreactivation. Rad. Res. Suppl. *2*:75–90.

Jagger, J., 1967: Introduction to Research in Ultraviolet Photobiology. Prentice-Hall, Englewood Cliffs, N. J.

Jagger, J., and Stafford, R. S., 1965: Evidence for two mechanisms of photoreactivation in *Escherichia coli B.* Biophysical J. *5*:75–88.

Kaplan, H. S., and Moses, L. E., 1964: Biological complexity and radiosensitivity. Science *145*:21–25.

Kelner, A., 1949: Effect of visible light on recovery of *Streptomyces griseus* conidia from UV irradiation injury. Proc. Natl. Acad. Sci. U. S. A. *35*:73–79.

Kiepenheuer, K., 1959: The Sun. University of Michigan Press, Ann Arbor.

Korff, S. A., 1957: The origin and implications of the cosmic radiation. Am. Sci. *45*:281–300.

Krinsky, N. I., 1968: The protective function of carotenoid pigments. *In* Photophysiology. Giese, ed. Academic Press, New York, Vol. 3, pp. 123–195.

Land, E. H., May 1959: Experiments in color vision. Sci. Am. *200*:84–100.

Land, E. H., 1964: The retinex. Am. Sci. *52*:247–264.

Lea, D. E., 1955: Actions of Radiations on Living Cells. Macmillan Co., New York.

Lees, A. D., 1968: Photoperiodism in insects. *In* Photophysiology. Giese, ed. Academic Press, New York, Vol. 4, pp. 47–137.

Livingston, R., 1955: Photochemistry. *In* Radiation Biology. Hollaender, ed. McGraw-Hill Book Co., Vol. 2, pp. 1–40.

Los Alamos Scientific Laboratory, 1950: The Effects of Atomic Weapons. U. S. Govt. Printing Office, Washington, D. C.

MacNichol, E. F., Jr., Dec. 1964: Three pigment color vision. Sci. Am. *211*:48–56.

Mayor, H. D., and Melnick, J. L., 1961: Studies on the acridine orange staining of the purified RNA viruses: polio virus and TMV. Virology *14*:74–82.

McConnell, D. G., and Scarpelli, D. G., 1963: Rhodopsin: an enzyme. Science *139*:848.

McElroy, W. D. and Glass, B. (eds.), 1961: Light and Life. Johns Hopkins Press, Baltimore.

McIlwain, C. E., 1963: The radiation belts, natural and artificial. Science *142*:355–361.

McLaren, A. D., and Shugar, D., 1964: Photochemistry of Proteins and Nucleic Acids. Macmillan Co., New York.

Menaker, M., and Eskin, A., 1966: Entrainment of circadian rhythms by sound in *Passer domesticus.* Science *154*:1579–1581.

Menzel, D. H., 1959: The Sun. 2nd Ed. Harvard University Press, Cambridge, Mass.

Morgan, K. Z., 1963: Permissible exposure to ionizing radiation. Science *139*:565–571.

Muhammed, A., 1966: Studies on the yeast photoreactivating enzyme. I. A method for the large scale purification and some properties of the enzyme. J. Biol. Chem. *241*:516–523.

Muller, H. J., 1954: Manner of production of mutations by radiation. *In* Radiation Biology. Hollaender, ed. McGraw-Hill Book Co., New York, pp. 475–626.

O'Brien, B. J., May 1963: Radiation belts. Sci. Am. *208*:84–96.

Page, R. M., 1968: Phototropism in fungi. *In* Photophysiology. Giese, ed. Academic Press, New York, Vol. 3, pp. 65–90.

Patt, H. M., and Brues, C. T., 1954: The pathological physiology of radiation injury in the mammal. 1. Physical and biological factors in radiation injury. 2. Specific aspects of the physiology of radiation injury. *In* Radiation Biology. Hollaender, ed. McGraw-Hill Book Co., New York, Vol. 1, pp. 919–1028.

Pettijohn, D., and Hanawalt, P. C., 1964: Evidence for repair-replication of UV damaged DNA in bacteria. J. Molec. Biol. *9*:395–410.

Photochemistry and Photobiology. 1962, Vol. 1.

Pittendrigh, C. S., 1965: On the mechanism of the entrainment of a circadian rhythm by light cycles. *In* Circadian Clocks. Aschoff, ed. North Holland Publishing Co., Amsterdam, pp. 277–297.

Pizzarello, D. J., Isaak, D., and Chua, K. E., 1964: Circadian rhythmicity in the sensitivity of mice to whole body radiation. Science *145*:286–291.

Pizzarello, D. J., and Witcofski, R. L., 1967: Basic Radiation Biology. Lea & Febiger, Philadelphia.

Puck, T. T., April 1960: Radiation and the human cell. Sci. Am. *202*:142–153.

Radiation Research. Current issues and supplements.

Reid, C., 1957: Excited States in Chemistry and Biology. Academic Press, New York.

Reinberg, A., Ghata, J., and Cameron, D. T., 1964: Biological Rhythms. Walker & Co., New York.

Rice, F. O., 1966: Production and reactions of free radicals in outer space. Am. Sci. *54*:158–169.

Robinson, N. (ed.), 1966: Solar Radiation. American Elsevier Publishing Co., New York.

Rupert, C. S., 1964: Photoreactivization of ultraviolet damage. *In* Photophysiology. Giese, ed. Academic Press, New York, Vol. 2, pp. 283–327.

Rushton, W. A. H., 1959: Visual pigments in man and animals and their relation to seeing. Progr. Biophys. *9*:240–283.

Rushton, W. A. H., Nov. 1962: Visual pigments of man. Sci. Am. *207*:120–132.

Rushton, W. A. H., 1964: Vision as a photic process. *In* Photophysiology. Giese, ed. Academic Press, New York, Vol. 2, pp. 123–162.

Sandage, A. R., Nov. 1964: Exploding galaxies. Sci. Am. *211*:38–47.

Sanderson, J. A., and Hurlburt, E. O., 1955: Sunlight as a source of radiation. *In* Radiation Biology. Hollaender, ed. McGraw-Hill Book Co., New York, Vol. 2, pp. 95–118.

Scientific American. Sept. 1959. Issue on Ionizing Radiation, Vol. 201, No. 3.

Setlow, J. K., 1966: The molecular basis of biological effects of ultraviolet radiation and photoreactivation. *In* Current Topics in Radiation Research. Ebert and Howard, eds. North Holland Publishing Co., Amsterdam, Vol. 2, pp. 195–248.

Setlow, R. B., 1964: Physical change and mutagenesis. J. Cell. Comp. Physiol. *64*(suppl. 1):51–68.

Setlow, R. B., and Carrier, W. L., 1964: The disappearance of thymine dimers from DNA: an error correcting mechanism. Proc. Natl. Acad. Sci. U. S. A. *51*:226–231.

Setlow, R. B., Carrier, W. L., and Williams, R. W., 1966: Three levels of DNA repair in *E. coli.* Science *152*:676.

Shilling, C. W., 1964: Atomic Energy Encyclopedia in the Life Sciences. W. B. Saunders Co., Philadelphia.

Sinsheimer, R. I., 1955: Ultraviolet absorption spectra. *In* Radiation Biology. Hollaender, ed. McGraw-Hill Book Co., Vol. 2, pp. 165–201.

Smith, K. C., 1964: Photochemistry of the nucleic acids. *In* Photophysiology. Giese, ed. Academic Press, New York, Vol. 2, pp. 329–388.

Solberger, A., 1965: Biological Rhythm Research. American Elsevier & Publishing Co., New York.

Sparrow, A. H., and Miksche, J. P., 1961: Correlation of nuclear volume-DNA content with a higher plant tolerance to chronic radiation. Science *134*: 282–283.

Spikes, J. D., 1968: Photodynamic action. *In* Photophysiology. Giese, ed. Academic Press, New York, Vol. 3, pp. 33–64.

Spikes, J. D., and Glad, B. W., 1964: Photodynamic action. Photochem. & Photobiol. *3*:471–487.

Stanier, R., 1960: Carotenoid pigments: problems of synthesis and function. Harvey Lect. *54*:219–255.

Strumwasser, F., 1965: The demonstration and manipulation of a circadian rhythm in a single neuron. *In* Circadian Clocks. Aschoff, ed. North Holland Publishing Co., Amsterdam, pp. 442–462.

Sweeney, B. M., and Hastings, J. W., 1958: Rhythmic cell division in populations of *Gonyaulax polyhedra.* J. Protozool. *5*:217–224.

Sweeney, B. M., and Hastings, J. W., 1960: Effects of temperature upon diurnal rhythms. Cold Spring Harbor Symp. Quant. Biol. *25*:87–104.

Tousey, R., 1961: Solar Research from rockets. Science *134*:441–448.

Trosko, J. E., Chu, E. H. Y., and Carrier, W. L., 1965: The induction of thymine dimers in ultraviolet-irradiated mammalian cells. Rad. Res. *24*: 667–672.

Van Allen, J., March 1959: Radiation belts around the earth. Sci. Am. *200*:39–47.

Van der Leun, J. C., 1966: Ultraviolet Erythema, a Study on Diffusion Processes in Human Skin. Doctoral Dissertation, Utrecht.

Virgin, H. I., 1964: Some effects of light on chloroplasts and plant protoplasm. *In* Photophysiology. Giese, ed. Academic Press, New York, Vol. 1, pp. 273–303.

Wacker, A., 1963: Molecular mechanisms of radiation effects. Progr. Nucleic Acid Res. Molec. Biol. *1*:369–399.

Wald, G., 1954: The molecular basis of visual excitation. Am. Sci. *42*:73–95.

*Wald, G., Oct. 1959: Life and light. Sci. Am. *201*: 92–108.

Wald, G., 1964: The receptors of human color vision. Science *145*:1007–1017.

Wald, G., 1965: Radiation and life. *In* Recent Progress in Photobiology. Bowen, ed. Blackwell Scientific Publications, Oxford, pp. 333–350.

Wald, G., and Brown, P. K., 1965: Human color vision and color blindness. Cold Spring Harbor Symp. Quant. Biol. *30*:345–361.

Weale, R. A., 1965: High light intensities and photochemical reactions of human visual pigments in situ. Cold Spring Harbor Symp. Quant. Biol. *30*:335–343.

Wilkins, M., 1965: The influence of temperature and temperature changes in biological clocks. *In* Circadian Clocks. Aschoff, ed. North Holland Publishing Co., Amsterdam, pp. 146–163.

Wurtman, R. J., and Axelrod, J., July 1965: The pineal gland. Sci. Am. *213*:50–60.

*Zetterberg, G., 1964: Mutagenic effects of ultraviolet and visible light. *In* Photophysiology. Giese, ed. Academic Press, New York, Vol. 2, pp. 247–281.

# CHAPTER 10 / TEMPERATURE AS A FACTOR IN THE CELL ENVIRONMENT

## THE BIOKINETIC ZONE

All living cells are exposed to some thermal variation in their environment. Some are exposed to much less variation than others, since the temperature of the ocean varies much less than that of land and fresh water (see Chapter 7). Even when the temperature of the blood or body fluid bathing the cells is closely controlled, as in warm blooded animals, slight temperature changes occur. Because of its effect on life processes and because most cells have only a small range of adaptation to thermal changes, temperature is an important factor in the distribution of living things.

A temperature which is most favorable for one cell function, such as respiration, may not be most favorable for another, such as cell growth. Furthermore, a temperature which may appear to be favorable for a given function during brief exposure of a cell may prove harmful on longer exposure. Temperature effects must therefore be examined rather fully, and it is a good experimental procedure when studying a given biological process to determine the effect of a temperature range on the process in order to select the temperature most favorable for the experiments.

The narrow range of temperatures within which a living cell carries on its life activities, called the *biokinetic zone,* lies approximately between 10° and 45° C. There are some organisms, however, that tolerate temperatures much higher than 45° C., and there are other organisms that tolerate temperatures much lower than 10° C. Most organisms when in a dormant state can withstand greater temperature variations than when they are in an active state (Bělehrádek, 1935, 1957). Some organisms have been successfully adapted to heat by culturing at progressively higher temperatures. For example, Dallinger in 1887 showed that over a period of seven years, flagellates could become acclimated to a temperature of 70° C.

## HEAT RESISTANCE AND HEAT DEATH OF ACTIVE CELLS

*Thermophilic* (heat-loving) algae and bacteria are those which normally live at higher temperatures than their *mesophilic* relatives which live in the temperature range of the biokinetic zone. For example, some algae and bacteria live in hot springs at temperatures of 60° and 70° C., and the alga *Phormidium* has been reported to exist even at 89° C. These are temperatures which would kill most active cells (Brues, 1927; Bělehrádek, 1935). Recent studies on hot springs of Yellowstone National Park disclosed some thermophilic bacteria that can grow in superheated pools at 93.5° to 95.5° C., which is above the boiling point of water (92° C.) at this altitude (Brock, 1967). This suggests that some form of life is possible regardless of temperature, provided water remains in the liquid state.

It is common knowledge that the activities of organisms are primarily catalyzed by enzymes which are made up of proteins and that many proteins are altered by heat; for instance, egg albumin is quickly coagulated at temperatures between 60° and 100° C. Many enzymes are inactivated by these temperatures.

It is thought that some thermophilic bacteria withstand high temperatures because they appear to synthesize enzymes faster than the high temperature can destroy them, since these bacteria are resistant to heat only when nutrients are available. In some cases at high temperature they require the amino acids and vitamins which they are able to make at lower temperatures, suggesting a partial breakdown of the synthetic mechanisms. In the absence of appropriate nutrients, such bacteria are no more resistant to heat than mesophilic strains of the same species.

On the other hand, thermophilic bacteria are generally more resistant to heat because they possess enzymes which continue to function normally at high temperatures (Militzer *et al.*, 1951). Stronger hydrogen bonding of the proteins of such cells is one of the reasons for their greater resistance to heat. As a consequence, such thermophiles are unable to live at lower temperatures, presumably because the bonding then becomes too strong for normal catalytic action of the enzymes (Allen, 1950, 1953, 1960). Acclimation presumably occurs by development of suitable enzyme systems.

The high temperature required to kill a living cell depends upon the temperature range to which it has been previously adapted, and upon the length of exposure to that high temperature. For example, some polar fish may die when the temperature is raised to 6° C.; others, when it is raised to 10° or 15° C. Marine animals of temperate zones may be killed when the temperature is raised to about 30° or 35° C., whereas fresh water and terrestrial animals may resist temperatures as high as 40° to 45° C. Detailed data on lethal temperatures and exposures for a wide variety of organisms are given by Bělehrádek (1935).

Lethal temperatures also depend upon the physiological state of the cell. Those in an active state are much more readily affected than those in a dormant state.

Heat death of a cell may conceivably result from thermal inactivation of its enzymes. It is well known that many enzymes which are reversibly inactivated by mild heat are irreversibly inactivated by high temperatures. The action of heat on an enzyme is perhaps most clearly illustrated with the bioluminescent reaction in which the heat-stable substrate, luciferin, is oxidized in the presence of the enzyme, luciferase with the emission of light. The intensity of luminescence is a measure of the rate of this reaction. The intensity of luminescence increases when the temperature of the mixture of enzyme and substrate is raised up to a certain point. When the temperature is raised still higher, luminescence declines and is ultimately quenched. If the temperature is lowered, luminescence reappears, provided neither the temperature nor the time of exposure has been excessive. If left at 60° C. for a sufficient period, the mixture will not luminesce on cooling, indicating that the enzyme has been inactivated (luciferin is heat stable).

Similar experiments have been performed with luminous bacteria. It is interesting that suppression of luminescence of bacteria by heat can be reversed by application of hydrostatic pressure; conversely, suppression of luminescence by pressure can be reversed by raising the temperature. The explanation offered is that the pressure reduces the volume of the enzyme molecule while heat increases it by breaking hydrogen bonds, causing the secondary and tertiary structures to unfold, exposing hydrophobic groups to the aqueous medium (Johnson *et al.,* 1954).

Inactivation of the luminous system in the bacteria by heat is concomitant with inactivation of some other enzymes as well. High temperatures shift the equilibrium of each thermolabile enzyme in favor of the denatured, inactive form.

$$\text{Native enzyme} \underset{}{\overset{\text{heat}}{\rightleftharpoons}} \text{denatured enzyme} \quad (10.1)$$

The time of exposure to heat is also a factor and when the heat treatment is prolonged, even if the temperature is one at which the enzyme is not irreversibly altered (e.g., at a temperature of 45° C. or lower), the processes required for life are probably suspended, and the cell dies.

Various enzymes of a cell have different temperature sensitivities, some even withstanding boiling, e.g., ribonuclease. Of course it is possible that heat affects enzymes indirectly by dissociating the aggregates of which they are a part. The subunits may reassociate in a manner different from that normally present, and the newly formed aggregate may be less effective catalytically than the original enzyme. It is often difficult experimentally to differentiate such an effect from a direct effect (Prosser, 1958, 1967).

The possibility that derangements of the lipids of the cell are the cause of heat injury has also been suggested. As has already been shown (Chapter 3), lipids are important structural components of the cell and their disruption might well lead to death. The following evidence supports this interpretation: (1) high resistance of an organism to heat is correlated with a high melting point of the main lipids laid down in the organism, and (2) the melting point of lipids stored in animals adapted to higher temperatures is higher than that of lipids stored in control animals (for example, in fishes) (Hoar and Dorchester, 1949). Lipids are important constituents of cell membranes (see Chapter 11). Disruption of the cell membranes, either internal or bounding the cell, results in cell chaos and death. There is increasing evidence of disruption of cell membranes as the possible cause of heat and cold death (Brock, 1967).

The liberation of a coagulating enzyme by heat is suggested as a third possible explanation for heat injury. Heat is found to release calcium from the outer portion of the cytoplasm of some cells, and this calcium is said to liberate a clotting enzyme which gels the cell (Heilbrunn, 1954). It is interesting in this respect that bacterial spores which are highly resistant to heat may have ten times as much calcium as the vegetative stages (Allen, 1960).

Curves for killing bacteria by heat indicate that heat death is the result of a single event, probably inactivation of some factor necessary for replication or cell division. Since diploid yeast is much more resistant to heat than haploid yeast, the critical factor(s) present in both chromosome sets must probably be inactivated before cell reproduction is blocked. It is conceivable that such heat death is the result of a lethal mutation, that is, alteration of the bases in DNA making replication impossible. The greater complexity of the killing curves for diploid yeast than those for haploid yeast or bacteria suggests that other factors, perhaps in the cytoplasm, also play a role in determining heat resistance (Wood, 1956).

## COLD RESISTANCE AND COLD DEATH OF ACTIVE CELLS

Most cells become relatively inactive as the temperature falls to the lower limits they tolerate. Few remain active at the freezing point of water, although they may survive. Hypothermia, both naturally occurring (e.g., in hibernators) and induced, results in decreased metabolism (Hock and Covino, 1958). However, *cryophilic,* or cold-resistant organisms, polar fishes and invertebrates carry on their life activities at a temperature of 0° C. or below (Elkus and Finean, 1953; Sailer, 1950). For example, some molds grow slowly even at −4° to −6.7° C. and bacteria grow slowly in ice cream at −10° C. (Wood, 1956).

The Antarctic polar region is the coldest part of the earth, the temperature in

August falling to −83° C., because it is a large land mass and land has a low heat capacity compared to water. Sea water cannot fall below −1.9° C. without freezing and it therefore acts as a great reservoir of heat. The heat of fusion of water warms the region as freezing occurs. Because of this the Arctic, which is a comparatively small land mass around which water is seldom more than 5 meters below the surface ice, is considerably warmer than the Antarctic and richer in land organisms. Both Arctic and Antarctic, however, have a large and diversified marine fauna although many of the marine forms must live at −1.9° C. essentially all of their lives (Rubin, 1962). At McMurdo in the Antarctic, for example, the temperature of the sea water varies only about 0.1° C. annually (Somero, 1967).

There is apparently an increase in rate of respiration of cells connected with cold adaptation of organisms, but the physiological changes which underlie such adaptation to cold have not been determined (Prosser, 1958). Some of the enzymes concerned with energy liberation are increased in activity (increase in quantity or increase in proportion of a more effective isozyme) during acclimation to low temperature, e.g., succinic dehydrogenase (Somero, 1967) and lactic dehydrogenase (Hochachka, 1965, 1967). Others show no change and some become more active only with increase in temperature (e.g., aldolase) (Kruger, 1962).

Some organisms are injured or killed even when the temperature falls to points well above the freezing point of water. The cause of death in such cases is not known, but some cells so injured by cold above freezing show altered respiration. This suggests that electrolyte balance is affected, since respiration furnishes energy for the exclusion of certain ions (see Chapter 14) (Fuhrman and Fuhrman, 1959). It is a well known phenomenon that water in contact with hydrocarbons forms "ice" at temperatures well above freezing, but as yet no evidence exists for such relations in the living cell (Buswell and Rodebush, 1956).

As an organism is slowly exposed to freezing temperatures, water freezes outside the cells, thereby concentrating the solution surrounding the cells. Consequently, water diffuses out of the cells into the bathing fluid, and the result is an increase in the concentration of solutes inside the cell, which lowers the freezing point inside the cell. However, the abnormally increased concentration of solutes is injurious to the cell as shown by direct application of equally concentrated solutions to the cells (Meryman, 1959, 1966). In some cases as much as 60 to 90 per cent of the free cell water may be withdrawn. Many cells, however, can endure being nearly frozen at −5° C., reimbibing water on thawing without apparent injury (Scholander *et al.,* 1950, 1953).

If the cells are slowly cooled, ice crystals may form inside them when the temperature falls below −40° C. because ice crystals outside the cell serve as nucleators through channels in the cell membrane and induce ice formation inside. However, if the water inside the cell has been reduced to less than 10 per cent of its original content, crystals will not form since this concentrated residue is uncrystallizable (Mazur, 1966). Withdrawal of water from the cell during freezing has been followed with nuclear magnetic resonance techniques (Sussman and Chin, 1966).

The time factor here is also of great consideration. Rapid freezing results in ice formation (as small crystals) inside as well as outside cells. This may concentrate the dissolved materials to the point of injury and the formation of ice crystals inside the cells may by itself disrupt the cytoplasmic structures and those of the nucleus beyond the point of recovery. This is especially true if the cells are stored for a time, allowing for growth of large ice crystals from the small ones.

Studies on the effects of freezing and thawing on the ultrastructure of the cell have shown that there is considerable damage to membranes of the endoplasmic reticulum, mitochondria and the plasma membrane of the cell, although remarkable recovery often occurs (Stowell *et al.,* 1965;

Trump *et al.,* 1965). It is possible that freezing removes lipid from the membranes, making them so fragile that they disintegrate on warming (Mazur, 1965). It is also possible that enzymes inside the cells are the foci of cold injury. When some enzymes in solution are frozen, they may be split into their composite polypeptide chains by the change in concentration of the salt and the accompanying pH change. Warming results in reassociation during which the subunits may interact in such a way as to give abnormal polymers with lesser or no catalytic activity (Chilson *et al.,* 1965).

After both slow and rapid freezing the nature of the thawing process proves especially important. For example, if the ice melts slowly, the cell will be exposed to a high electrolyte concentration at a higher temperature than that applied during freezing, allowing for more ready denaturing of proteins. Also during slow thawing, growth of ice crystals, especially inside cells, may kill them. Thus even if the cell initially survived freezing it might still be damaged severely on thawing. Rapid thawing is therefore especially important (Meryman, 1956, 1960, 1966).

If glycerol is added to the suspending

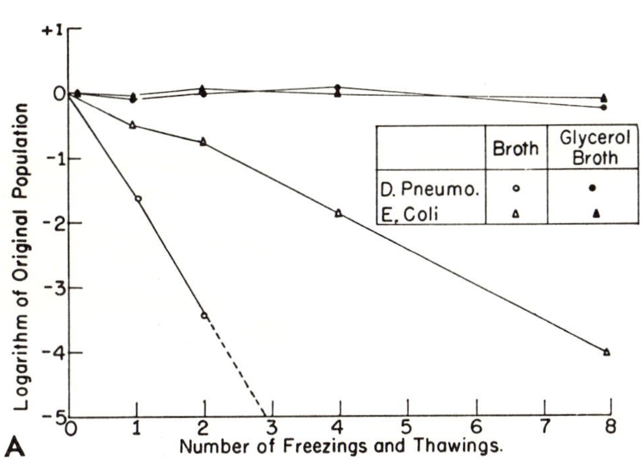


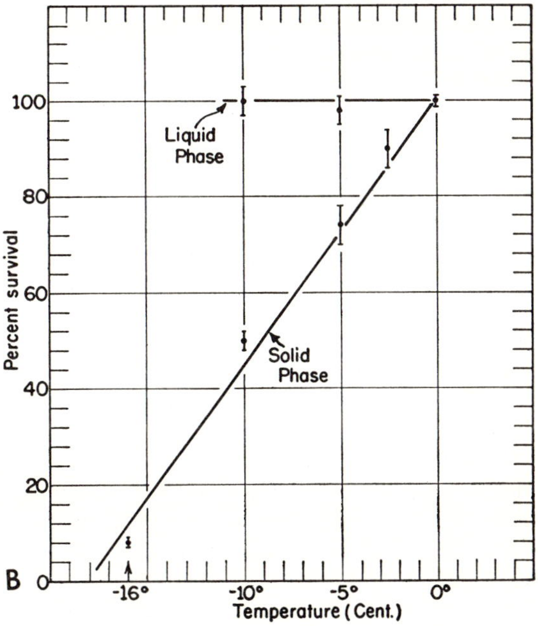


***Figure 10.1.*** *A,* Survival of the colon bacillus (*Escherichia coli*) and the pneumonia coccus (*Diplococcus pneumoniae*) after repeated freezing (5 minutes at −80° C.) and thawing, with and without glycerol (15 per cent). (After Hollander and Nell, 1954.) *B,* Survival of T-2 coliphage exposed to various temperatures for 45 minutes in both liquid (supercooled) and solid (frozen) phases. Note that when freezing is prevented and the phage remains in the liquid phase, little if any damage occurs, but when ice is formed (solid phase) damage is progressive with lowering of temperature. Since phage does not respire, the damage would appear to be direct, rather than from the accumulation of unnatural products. (From Wood, 1956: Advances Biol. & Med. Phys. *4*:119.)

medium, some of it penetrates into the suspended cells and it lowers the freezing point of both the bathing medium and the fluid in the cells and there acts as an "antifreeze." But even if freezing occurs, less ice is produced than in the absence of glycerol, and the crystals formed are smaller because of the strong hydrogen binding of glycerol for water. Concentration of electrolytes by freezing is thereby reduced; for this reason glycerol is sometimes spoken of as an "electrolyte buffer." Bacteria and phages which are readily killed by freezing in broth have been shown to survive eight cycles of freezing and thawing (Fig. 10.1) in 15 per cent glycerol. However, since glycerol itself is toxic, it must be removed by dialysis after thawing. Glycerol is especially useful in slow freezing procedures. Treated in this manner, erythrocytes and other cells remain viable and can be stored at low temperatures for long periods. Even eggs in ovaries have been kept for months at temperatures well below freezing and then successfully transplanted (Parkes, 1956). When cells are frozen with glycerol the rate of thawing matters little (Meryman, 1956; Nash, 1966).

Arctic organisms are able to reduce susceptibility to cold death by dehydration before the onset of winter, some retaining between 80 and 90 per cent of the summer moisture, others only as little as 25 per cent. However, if subjected to very low temperatures, simpler forms of Arctic organisms may be frozen, yet survive. Respiration continues at a much reduced rate. Ice crystals can be seen (e.g., in tissue spaces of midge larvae) or demonstrated in the organisms by a decrease in the specific gravity of the frozen material (Scholander *et al.*, 1953). Debris which allows contact for the organism, preventing physical disruption on freezing, as well as a line of entry for oxygen, may help preserve life in ice.

A fish completely frozen in ice has never revived, despite the stories to the contrary. It would thus appear that even the cells of Arctic organisms are not free from injury by freezing, though they may be more resistant than temperate forms; however, hamsters have been frozen for an hour and revived (Parkes, 1956).

## THERMAL RESISTANCE OF DORMANT CELLS

When unfavorable thermal conditions occur in the environment, many cells suspend life activities and become dormant. Bacteria and lower plants form *spores,* protozoans and some invertebrates form *cysts,* and the cells in seeds of higher plants become dormant. The cold season of the polar, subpolar and north and south temperate zones is a period of dormancy for many organisms. The hot, dry season in desert and other arid habitats is a period of dormancy for many organisms which may dry up in the absence of water; these organisms renew their activity only when water becomes available again (e.g., rotifers, tardigrades, nematodes and mosses).

Cells of organisms in dormant states become dehydrated—in some cases to a considerable extent—and their greater resistance to both heat and cold during their dormant stages is, in part, probably the result of the dehydrated state of their cells. Cysts and spores, for example, resist exposure to dry heat of over 100° C. Dry sterilization (as a means of killing dormant organisms), therefore, requires a temperature of 150° to 170° C. for half an hour or so. Moist sterilization, on the other hand, requires only 15 minutes at 115° C. (15 pounds pressure in an autoclave or pressure cooker). Boiling (100° C.) kills some cysts and spores, but it only induces germination of others, and following a time lapse a second boiling is necessary to kill the bacteria which have emerged from the spores. Repeated boiling for sterilization is called tyndallization. Pasteurization, or heating at 60° C. for half an hour, kills vegetative stages only, not dormant ones (spores).

The greater thermal resistance of organisms in their dormant stages is not yet fully explained. It is known that the temperature required to denature proteins is higher the lower their water content. For example, at a relative humidity of 0 per cent, half of a given quantity of egg albumin is denatured when a sample is kept at a temperature of 140° C. for an hour; at a relative humidity of 20 per cent, an hour at 124° C. suffices; at 47

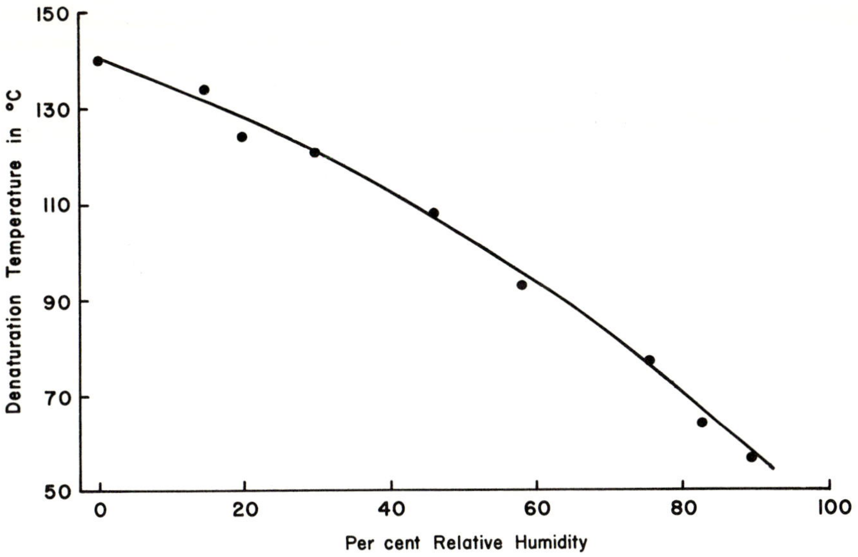


*Figure 10.2.* Relation between relative humidity and temperature for denaturation of egg albumin. The data are for denaturation of half of the egg albumin in 60 minutes. (From Barker, 1933: J. Gen. Physiol. *17*:27.)

per cent, an hour at 99° C.; at 75 per cent, an hour at 77° C.; and at 93 per cent, an hour at 57° C. (see Fig. 10.2). Similarly, a cyst, or seed, each highly resistant to heat, contains very little water; for example, only 5 per cent of water is present in seeds of some desert plants, whereas the active organism, highly sensitive to heat, contains 80 to 90 per cent water (Christophersen and Precht, 1953).

Some spores and cysts have been found capable of germinating after exposure to temperatures of between −250° C. and −272° C., which is close to absolute zero (−273.18° C.), yet cells in active states do not survive even −183° C (Scholander, 1953). In such dehydrated cells, little free water is available to form ice. This may account for their resistance to such low temperatures (Fuhrman and Fuhrman, 1959). The fact that hibernating insects and cocoons of insects resist low temperatures even though they are not dehydrated suggests that factors other than dehydration may be involved (e.g., antifreeze).

Because life on earth exists within a rather limited range of temperature, exobiologists consider determination of the temperatures of other planets (by means of probes such as Mariner II to Venus) of utmost importance. Life seems unlikely on Venus because of its high temperature. Conversely, life is questionable on Mars and all of the distant planets because of their low temperatures (Sagan, 1961; Salisbury, 1962).

## THE RATE OF THERMOCHEMICAL REACTIONS

The rate of reactions in the cell increases with the temperature up to a maximum; with further increase in temperature, the rate of activity declines as lethal thermal limits are approached. Therefore, for meaningful comparisons of the effects of temperature on the rates of various biological processes, the rates of reaction must be compared for the same temperature interval. Usually one determines the ratio of the rate of activity at a given temperature to the rate at a temperature 10° C. lower. This ratio, called the *temperature coefficient*, is designated $Q_{10}$. Four relatively different kinds of temperature coefficients have been found for cell reactions, each corresponding to a distinctive mechanism of action: those which are about 2, those which are considerably higher than 2, those which are considerably less than 2 and those which are about 1. Illustrative data for a number of reactions are given in Table 10.1

Examination of this table will disclose that, in comparable thermobiological reactions, as is true in thermochemical reactions, the rate increases at least twofold (i.e., 100 per cent) for each 10° C. rise in temperature, a relation which has sometimes been called van't Hoff's rule (see Appendix, 10.1 for equation).

The fundamental cause underlying this principle is perhaps the same for both types of reactions. One possibility suggests itself, namely, a comparable rise in kinetic energy of the system. However, calculation shows that a rise of 10° C. in temperature increases the average kinetic energy of the molecules only to a very limited extent—equal to 10/T, where T is the absolute temperature. For instance, at room temperature the absolute temperature is about 300°; therefore, 10/T would be 10/300 or 3.3 per cent. This trifling increase in the average kinetic

**Table 10.1.** Increase in Rate of Reaction with a 10° Rise in Temperature

| TYPES OF REACTION | REACTIONS IN INANIMATE SYSTEM | $Q_{10}$ | TEMP. RANGE °C. | REACTION IN ORGANISMS | $Q_{10}$ | TEMP. RANGE °C. |
|---|---|---|---|---|---|---|
| Thermo-chemical | Reaction between peroxide and hydriotic acid | 2.08 | 6–10 | Photosynthesis at high light intensity | 1.6–4.3 | 4–30 |
| | Activity of malt amylase | 2.2 | 10–20 | Rate of cleavage of *Arbacia eggs* | 3.9<br>3.3<br>2.6<br>1.7 | 7–17<br>8–19<br>15–25<br>20–30 |
| | Most thermochem. reactions (van't Hoff rule) | 2.3 | | Contractility of duodenum | 2.42 | 28–38 |
| | | | | Respiration of pea seedlings | 2.4 | 10–20 |
| | Protein coagulation: | | | Heat killing: | | |
| | egg albumin | 635 | 69–76.3 | spores | 2–10 | 40–140 |
| | | | | bacteria | 12–136 | 48–59 |
| | | | | protozoans | 891–1000 | 36–43 |
| | hemoglobin | 13.8 | 60–70.4 | rabbit leucocytes | 28.8 | 38–60 |
| Diffusion and con-ductivity | Diffusion: | | | Penetration of molecules and ions into cells: | | |
| | strong electrolytes | 1.27 | 0–50 | Dyes into *Arbacia* | 4.7 | 20–25 |
| | sugar | 1.37 | | Accumulation of bromine in *Nitella* | 1.57 | 10–20 |
| | dextrin | 1.41 | | same in light | 2.4–2.7 | 10–20 |
| | | | | Water into *Arbacia* eggs | 2.4 | 11–24.8 |
| | Conductivity: KCl in water | 1.22–1.24 | 10–25 | Conductivity of living *Laminaria* | 1.32–1.34 | 10–20 |
| Photo-chemical | Photographic plate | 1.05 | —85–15 | Bleaching of retinal pigment | 1.0 | 5–36.1 |
| | Decomposition of ozone | 1.15 | 18–35 | Killing of *Staphylococcus* by ultraviolet light | 1.06 | 5–36 |
| | $Cl_2 + H_2 \rightarrow 2HCl$ | 1.17 | 10–40 | Killing infusoria by ultraviolet light | 1.5–1.6 | 11–21 |
| | Bromination of ethyl maleate ester | 1.07<br>1.02 | 21.5–24.5<br>4–14 | Photosynthesis, low light intensity | 1.06 | 15–25 |

The data for coagulation of proteins are from Chick and Martin, 1910: J. Physiol. *40;* for photochemical reactions from Dhar, 1931: The Chemical Action of Light; for photosynthesis from Warburg, 1919: Biochem. Zeitschr. *100;* for killing of Staphylococcus from Gates, 1929: J. Gen. Physol. *13;* all other data from Bělehrádek (1935) or directly from references cited by him.

energy of the molecules by itself is quite inadequate to explain the twofold rise in the rate of reaction which follows a 10° C. rise in temperature.

Arrhenius in 1889, however, proposed a plausible explanation for a twofold (or greater) rise in the rate of a thermochemical reaction with an increase in temperature of 10° C. He pointed out that not all the molecules in a population of a given kind of molecules, at a given temperature, have the same amount of kinetic energy. Following random collisions, some of the molecules accumulate more energy than others. Some become energy rich, some energy poor, and most are in the middle class between. Arrhenius suggested that it is the energy-rich molecules which are most likely to react on collision, and that the rate of reaction might therefore depend upon the concentration of such molecules. A molecule must have a certain amount of kinetic energy to undergo a thermochemical reaction. This critical amount of energy has been called the *energy of activation.* Arrhenius found means for determining experimentally this energy of activation.

He measured the rate of a thermochemical reaction at different temperatures and plotted the natural logarithm (ln) of the velocity, k, at each temperature against the reciprocal of the absolute temperature (T). He obtained a straight line. The equation for this relation is:

$$\frac{d\,ln\,k}{dT} = \frac{a}{T^2} \qquad (10.2)$$

or

$$ln\,k = c - \frac{a}{T} \qquad (10.3)$$

where a is the slope and c is the intercept on the temperature axis, 1/T.

On the basis of theoretical considerations he considered that the rate of reaction is dependent on the number of molecules possessing a given amount of energy of activation, E. The slope, a, measures the rate of reaction. On the basis of theoretical deductions, the relation of a to E was given by Arrhenius as: a = E/R, where R is the gas constant. Substitution into equation 10.2 gives:

$$\frac{d\,ln\,k}{dT} = \frac{E}{RT^2} \qquad (10.4)$$

Integrated, this becomes:

$$\frac{k_2}{k_1} = e^{-\frac{E}{R}\left[\frac{1}{T_1} - \frac{1}{T_2}\right]} \qquad (10.5)$$

or,

$$ln\,\frac{k_2}{k_1} = \frac{E}{R}\left[\frac{1}{T_1} - \frac{1}{T_2}\right]$$

where E is the energy of activation; $k_1$ and $k_2$ are the velocity constants for the rates of the reaction at two different absolute temperatures, $T_1$ and $T_2$; ln is the natural logarithm; and R is the gas constant. E may be determined from the plot of the logarithm of the rate of the reaction against 1/T (Arrhenius, 1915; see Fig. 10.4).

Arrhenius then used the Maxwell-Boltzmann distribution law to determine the number of molecules possessing the energy of activation before and after a 10° C. rise in temperature. The calculation shows that several times as many activated molecules are present at the higher temperature as at a temperature 10° C. lower (see Appendix 10.2 and Fig. 10.3). It seems most likely, therefore, that the increase in the rate of thermochemical reactions with a rise in temperature is attributable to the disproportionate increase in the fraction of molecules with the required energy of activation for a particular reaction (Lehninger, 1965).

Strictly speaking, the E value determined in this manner is actually a much more complicated entity according to the transition state theory of Eyring and is preferably spoken of as the *temperature characteristic,* $\mu$, rather than E, the energy of activation. The relationship is briefly considered in Appendix 10.2, but the original literature must be consulted for a full treatment. Fig. 10.4 shows the type of curve obtained when a given function affected by temperature is plotted against the absolute temperature.

Biologists have often used the Arrhenius equation instead of the van't Hoff equation for presenting their data on relations between temperature and the rate of a

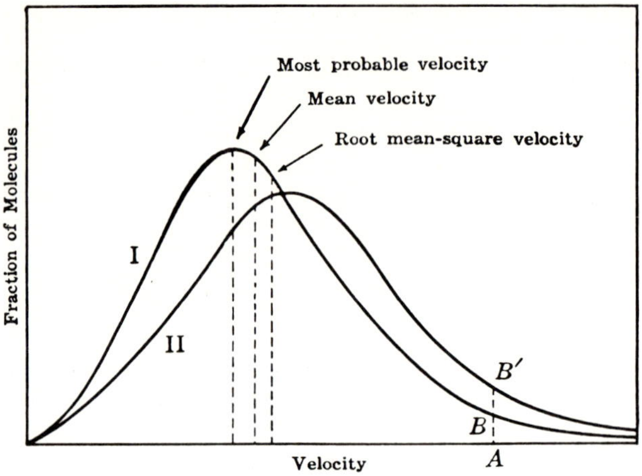


*Figure 10.3.* Maxwell-Bolzmann distribution diagram (schematic) for increase in the number of molecules having energy equivalent to (or more than) the energy of activation (*ABB′*) with increase in temperature. The ordinate measures the fraction of molecules (or the probability for a given molecule) having the velocity (a function of the kinetic energy) corresponding to the abscissa. The abscissa measures the velocity of the molecules. Curve I represents the lower temperature; curve II, the higher temperature. The line *ABB′* marks the minimal activation energy. (Modified from Glasstone, 1946: The Elements of Physical Chemistry. D. Van Nostrand Co., New York.)

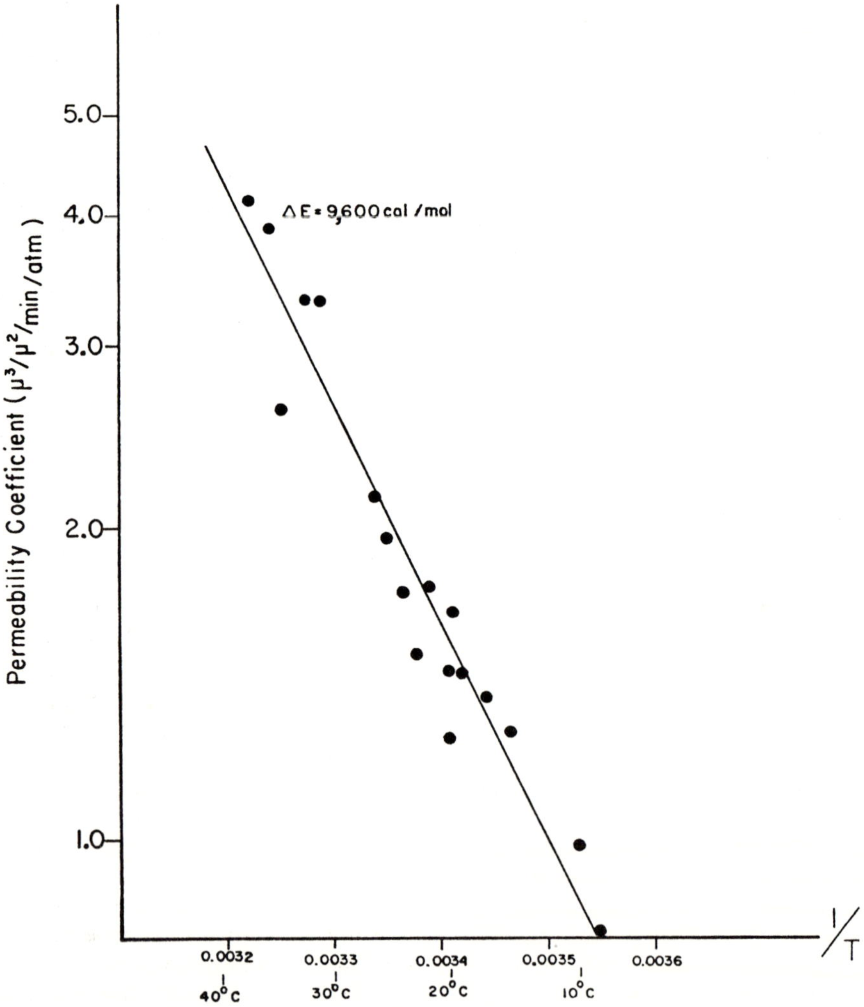


*Figure 10.4.* Relationship between the logarithm of the water permeability coefficient of Ehrlich ascites tumor cells and the reciprocal of the absolute temperature; the energy of activation determined by the plot is 9600 cal. per mole. (From Hempling, 1960: J. Gen. Physiol. *44*:365.)

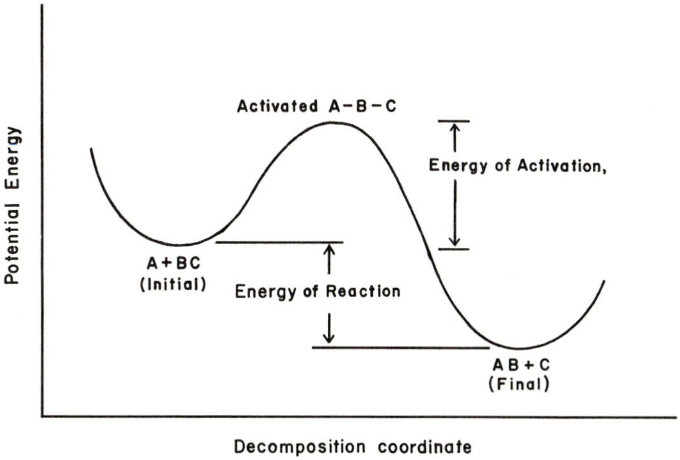


***Figure 10.5.*** Potential energy diagram for a thermochemical reaction. Before reacting, the molecules must accumulate sufficient kinetic energy (energy of activation) to pass over the energy barrier represented by the rising curve, even though the reaction occurs with a loss in free energy (approximately indicated by the energy of reaction).

reaction. The $\mu$ or E is then considered to be the energy of activation for the limiting reaction, and is therefore a valuable index of the order of magnitude of the energy of activation required for a particular reaction. E values are sometimes found fairly constant over relatively wide temperature ranges in particular studies, which probably indicates that particular reactions are limiting the rates over those ranges. Although such an index may not identify a limiting reaction, it may serve as the basis of inferences leading to fruitful experimentation by other procedures (Johnson *et al.,* 1954). Molecules, on absorbing energy, show several types of activation: increased movement, more rapid rotation and more rapid vibration and higher electronic states. Absorption of heat usually results in the first two types of activation and may result in the third as well. The velocity of a reaction is proportional to the frequency with which the state of activation is achieved by the molecules in a given system. Since an increase in temperature increases the chances for molecules to acquire a higher energy state, it increases the speed of thermochemical reactions.

As an example, let us consider a spontaneous reaction, that is, one which goes with a decrease in free energy. It will proceed rapidly if the molecules need to accumulate little energy before they are activated. If, on the other hand, the energy of activation is large, the reaction will proceed slowly. Anything (e.g., heat or catalysis) that increases the probability that the molecules will acquire the necessary energy of activation will increase the rate of the reaction.

These concepts might be summarized in the form of a diagram (Fig. 10.5). When a molecule moves from one energy level to another in a reaction, it must pass over an energy "barrier," even though the overall effect is to decrease the free energy, and the reaction is, therefore, thermodynamically possible. The "barrier" is the energy of activation which a molecule must first accumulate. Anything which increases the chances of the molecule passing over this barrier increases its chances of reaction (Johnson *et al.,* 1954). The lowering of the energy of activation in the presence of catalysts (Fig. 15.1) is discussed in Chapter 15.

## TEMPERATURE COEFFICIENTS FOR DENATURATION OF PROTEIN AND THERMAL DEATH OF CELLS

The temperature coefficient ($Q_{10}$) for denaturation of proteins is of a different order of magnitude from that for the thermochemical reactions just discussed. As seen in Table 10.1, the coefficient is very high, being higher at the upper range of temperatures used to denature the proteins. It must be realized that proteins are highly organized molecules and that denaturation involves the change from this highly organized structure to a much more random one. For this reason, even a small rise in temperature in the range which causes denaturation has a marked effect out of all proportion to the increase in the kinetic energy of the system (see Chapter 1).

Exceedingly large temperature coefficients are also observed for thermal death. Values of the $Q_{10}$ for thermal death range from about 10 to 100 or even 1000 for cells as diverse as bacteria, protozoa, eggs, sperm, vertebrate leukocytes and erythrocytes and cells of invertebrates and plants. (See Table 10.1.) These findings have been considered by many workers as corroboration of the hypothesis that heat death of organisms is attributable to the denaturation of certain critical proteins (Johnson *et al.,* 1954).

## TEMPERATURE EFFECTS ON THE RATE OF DIFFUSION AND CONDUCTIVITY

The $Q_{10}$ for the rate of diffusion of nonelectrolytes and of electrolytes is less than 2.0; for example, it is 1.27 for electrolytes, 1.37 for sugar and 1.41 for dextrin (see Table 10.1). Since the kinetic energy of a system increases 3 to 4 per cent for each 10° C. rise in temperature, the increase of 30 to 40 per cent in rate of diffusion and of conductivity (movement of electrolytes in an electric field) must be attributed to some factor other than increase in kinetic energy of the system only. It has been suggested that perhaps changes in the viscosity of water may account for it, since the viscosity of water falls with a rise in temperature and the diffusion of molecules or ions dissolved in it is therefore enhanced. The values, as shown in Table 10.1, are of the same order of magnitude for both diffusion and conductivity.

Passage of molecules or ions through the cell membrane at times appears to be a simple diffusion process. Thus the $Q_{10}$ for conductivity through the tissue of the alga *Laminaria* is similar to the $Q_{10}$ for conductivity of a solution *in vitro* containing a concentration of electrolytes equal to that in the algal tissue. However, in other cases given in Table 10.1, the $Q_{10}$ for passage of some materials into cells is often 2 or higher. A possible explanation of the high $Q_{10}$ values for these examples is a change in the viscosity of the cell membrane through which the ions or molecules must pass. However, there may be other explanations, since passage of molecules and ions through cell membranes is not always governed by the same mechanism (see Chapter 13).

## EFFECT OF TEMPERATURE ON PHOTOCHEMICAL AND PHOTOBIOLOGICAL REACTIONS

The $Q_{10}$ for photochemical reactions is only slightly greater than 1.0, indicating that the rate of a photochemical reaction of small molecules is little influenced by heat. As seen in Table 10.1, the $Q_{10}$ for the bromination of ethyl maleate ester by light is 1.02. Similar low values are found for most of the photobiological reactions listed in the table; for instance, bleaching of visual purple in the cells of the eye by yellow light is 1.0. However, photochemical reactions of macromolecules (e.g., enzymes) may show higher values (McClaren and Shugar, 1964).

Absorption of light by a molecule leads to more rapid vibration or electronic excitation. The energy required for this purpose is larger than that which is usually supplied by heat. Absorption of light occurs, as already seen, in discrete units called quanta. The quantum is related to the wavelengths of radiations, being larger the shorter the wavelength (see Chapter 9). A molecule absorbs a quantum at a time; therefore, the rate of photochemical reactions is determined by the rate of uptake of light quanta. The probability that a molecule will capture a light quantum is not noticeably enhanced at all by increasing its kinetic energy (heat). Therefore, an increase in temperature has little effect on photochemical reactions. Consequently, the temperature coefficient for a photobiological reaction is low when the overall process is limited by the rate of strictly photochemical reactions.

## TEMPERATURE COEFFICIENTS AND LIMITING REACTIONS

Most biological reactions occur in a series of steps. Depending upon the conditions, one or another of these stepwise reactions may become the "bottleneck" or *rate-limiting reaction* (master reaction). The temperature coefficient for a biological process will therefore depend upon the conditions of the experiment.

The temperature coefficient for photosynthesis is an interesting example. At low light intensities it is about unity. At high light intensities, however, it is found to be 2 or more (see Table 10.1). At low light intensity the rate at which the quanta of light are being absorbed (photochemical reaction) is the bottleneck that determines the rate of the overall reaction. At high light intensity, however, when the chlorophyll molecules of the plant are saturated with light, the rate at which the light energy can be utilized in subsequent thermochemical reactions (such as the synthesis of carbohydrate) then becomes the bottleneck, limiting the rate of the overall reaction. The nature of the photochemical (light) and the thermochemical (dark) reactions is considered in Chapter 19.

In like manner it is possible to distinguish whether a thermochemical reaction, a diffusion process or a protein denaturation is the bottleneck in a series of reactions in cells, because the differences in the temperature coefficients for these types of reactions are large enough to be measurable. Examples of the usefulness of temperature coefficients in analysis of reactions occurring in cellular activities will appear in subsequent chapters.

Crozier (1925) attempted to make even wider use of temperature coefficients for identification of specific thermochemical reactions in cells. It is well known that biological thermochemical reactions (e.g., those which liberate energy from food in the cell) occur in steps. If one step in a series becomes rate-limiting over some temperature range and has a constant temperature coefficient, it might be possible to match the coefficient for each temperature range with that of a biochemical reaction *in vitro* (in the test tube) and in this manner to identify each of the bottleneck or master thermochemical reactions, which are assumed to be occurring in the cell.

Although temperature coefficients for chemical reactions in solution are fairly constant (see Table 10.2), temperature coefficients for enzymatic reactions *in vitro* (and presumably also in the cell) are quite variable with change in temperature

**Table 10.2.** Variation in Temperature Coefficients over a Temperature Range

| TEMPERATURE INTERVAL | REACTION BETWEEN $H_2O_2$ AND HI | ACTIVITY OF A MALT AMYLASE | $CO_2$ PRODUCED BY PEA SEEDLINGS | FREQUENCY OF HEART BEAT (CAT) | PROTOPLASMIC STREAMING IN *Vallisneria* | DIVISION OF *Arbacia* EGG |
|---|---|---|---|---|---|---|
| 0–10° | 2.080 | 3.2 | 3.0 | — | 4.0 | — |
| 10–20 | 2.078 | 2.2 | 2.4 | 5.4 | 2.3 | 3.9 |
| 20–30 | 1.940 | 2.0 | 1.8 | 2.15 | 1.6 | 1.7 |
| 30–40 | 1.932 | 1.5 | 1.4 | 1.84 | 1.3 | — |
| 40–45 | 1.920 | — | — | 1.35 | — | — |

Data from Bělehrádek (1935) except for *Arbacia* which are from Harvey, 1932: Biol. Bull. *42*.

(Table 10.2). Since biological reactions are usually enzymatic, it seems unlikely that their identification by this method would be possible even if the more refined methods suggested by Crozier were used. The possibility of identification of specific thermochemical reactions *in vivo* by temperature coefficients has great inherent appeal, but it has not been considered valid by most workers, even though the $Q_{10}$ allows one to distinguish a thermochemical reaction from a protein denaturation, a photochemical reaction or a diffusion process (Burton, 1939; Morales, 1947; Johnson *et al.*, 1954; Bull, 1964).

## APPENDIX

### 10.1 VAN'T HOFF EQUATION FOR $Q_{10}$

When it is inconvenient to perform experiments at temperature intervals 10° apart, and data are gathered for smaller or larger temperature intervals, the temperature coefficient may be determined by use of the van't Hoff equation (Bělehrádek, 1935):

$$Q_{10} = \left(\frac{k_2}{k_1}\right)^{10/(t_2-t_1)} \qquad (10.7)$$

where $k_2$ is the rate of the reaction at temperature $t_2$, and $k_1$ is the rate at $t_1$. This equation is more convenient to handle in the form:

$$\log Q_{10} = \frac{10}{t_2 - t_1} \log \frac{k_2}{k_1} \qquad (10.8)$$

### 10.2 CALCULATION OF NUMBER OF ACTIVATED MOLECULES

From the Maxwell-Boltzmann distribution law one can calculate the relative number of molecules in a given population which possess a given amount of energy. $N_1$, the number of molecules having energy equal to or in excess of E (the energy of activation or the energy required for a given reaction), can be determined from the equation:

$$\frac{N_1}{N_0} = e^{-(E/RT)} \qquad (10.9)$$

where $N_0$ is the total number of molecules, e the base of the natural system of logarithms, R the gas constant in calories and T the absolute temperature. E for a given reaction is determined experimentally as explained earlier in this chapter (Arrhenius, 1915).

### 10.3 REACTION RATE THEORY

Reaction rates may be formulated according to the collision theory as already shown or by another approach, that of *transition states.* The collision theory is more familiar, but the newer theory has the advantage that by it all types of reactions may be formulated in a single way. The basic concept of the transition state theory (or absolute reaction rate theory) of Eyring and his collaborators is that before molecules can react they must pass through the activated state which has an energy content greater than that of the normal reactants and a very short life of activation (about $10^{-13}$ sec.). The average of this energy increment is the activation energy (see Figs. 10.3 and 10.5). An equilibrium is established between the normal reactants and the activated complex, which is disturbed by the continuous passage of some activated molecules into the final state (lower trough, Fig. 10.5) (Glasstone, 1960).

The specific reaction velocity, Vs, for the crossing of the energy barrier (the hump in Fig. 10.5) is given by the equation:

$$Vs = \frac{\alpha kT}{h} K^* \qquad (10.10)$$

where K* is the equilibrium constant between the normal reactants and the activated complex; kT/h is a frequency, k being the Boltzmann constant (the gas constant, R, per molecule, or R/N where N is Avogadro's number), h is Planck's constant (see Chapter 9), T is the absolute temperature and $\alpha$ is the transition coefficient and represents the probability of the system in the activated state (at the top of the hump) returning to the normal state or to the final state. The transition approaches unity in many reactions although it is a complicated function which cannot be considered here.

Since according to principles of thermodynamics:

$$RT \ln K^* = -\Delta F^* \quad (10.11)$$

where F* is the free energy or maximum work content of the system

or,
$$\ln K^* = -\frac{\Delta F^*}{RT} \quad (10.12)$$

Taking the antilogarithms to the base, e, gives:

$$K^* = e^{-\Delta F^*/RT} \quad (10.13)$$

Substituting this value of K* into equation 10.9, gives:

$$Vs = \frac{\alpha KT}{h} \cdot e^{-\Delta F^*/RT} \quad (10.14)$$

It is readily observable that equation 10.5 is much like equation 10.14, both being formulations for rates of reaction but on different grounds. However, the ΔF*, the free energy of activation, is not exactly equal to E, the temperature characteristic, the latter incorporating many other factors which are separately considered in equation 10.14.

The reaction rate theory has shown usefulness in application to data on bacterial luminescence, growth, dehydrogenase activity and so forth and, in fact, is also applicable to rates of any molecular-rate process. However, it does not fit in some cases and the error may be large (Bělehrádek, 1957).

## LITERATURE CITED AND GENERAL REFERENCES


Allen, M. B., 1950: The dynamic nature of thermophily. J. Gen. Physiol. *33*:205–214.

Allen, M. B., 1953: Thermophilic aerobic spore-forming bacteria. Bacteriol. Rev. *17*:125–173.

*Allen, M. B., 1960: Utilization of thermal energy by living organisms. Comp. Biochem. *1*:487–514.

Arrhenius, S., 1915: Quantitative Laws in Biological Chemistry. Bell & Sons, London.

Asakina, E., 1965: Freezing process and injury in isolated animal cells. Fed. Proc. *24*:S183–S187.

Bělehrádek, J., 1935: Temperature and Living Matter. Protoplasma-Monographien. Vol. 8. Verlag von Gebrüder Borntraeger, Berlin.

Bělehrádek, J., 1957a: Physiological aspects of heat and cold. Ann. Rev. Physiol. *19*:59–82.

Bělehrádek, J., 1957b: A unified theory of cellular rate processes based upon an analysis of temperature action. Protoplasma *48*:53–71.

Bender, M. A., Tran, P. T., and Smith, L. H., 1960: Preservation of viable bone marrow cells by freezing. J. Appl. Physiol. *15*:520–524.

Berry, J. A., and Magoon, C. A., 1934: Growth of microorganisms at and below 0° C. Phytopath. *24*:780–796.

Biology of the Antarctic Seas. 1964. American Geophysical Union, Washington, D. C.

Brock, T. D., 1967: Life at high temperatures. Science *158*:1012–1019.

Brues, C. T., 1927: Animal life in hot springs. Quart. Rev. Biol. *2*:181–203.

Bull, H. B., 1964: An Introduction to Physical Biochemistry. F. A. Davis Co., Philadelphia.

Burton, A. C., 1939: The properties of the steady state compared to those of equilibrium as shown in characteristic biological behavior. J. Cell. Comp. Physiol. *14*:327–349.

Buswell, A. M., and Rodebush, W. H., April 1956: Water. Sci. Am. *194*:77–89.

Chilson, O. P., Costello, L. A., and Kaplan, N. O., 1965: Effects of freezing in enzymes. Fed. Proc. *24*:S55–S65.

Christophersen, J., and Precht, H., 1953: Die Bedeutung des wassergehaltes der Zelle fur Temperaturanpassungen. Biol. Zent. *72*:104–119; Chem. Abstracts *47*:8117 (1953).

Cooney, D. G., and Emerson, R., 1964: Thermophilic Fungi, An Account of Their Biology, Activities, and Classification. W. H. Freeman & Co., San Francisco.

Crozier, W. J., 1925–1926: Note on the distribution of critical temperatures for biological processes. J. Gen. Physiol. *9*:525–529.

Cryobiology Symposium. Fed. Proc. *24*:S1–S324.

Dallinger, W. H., 1887: On a series of experiments made to determine the thermal death point of known Monad germs when the heat is endured in a fluid. J. Royal Micr. Soc., London, pp. 185–199.

Dyer, D. L., and Gafford, R. D., 1961: Some characteristics of a thermophilic blue-green alga. Science *134*:616–617.

Elkus, J., and Finnian, J. B., 1953: X-ray diffraction studies on the effect of temperature on the structure of myelin in the sciatic nerve of the frog. Exp. Cell Res. *4*:69–82.

Farr, D. A., and Fuhrman, F. A., 1965: Role of diffusion of oxygen in the respiration of tissues at different temperatures. J. Applied Physiol. *20*:637–646.

Fuhrman, F. A., and Farr, D. A., 1964: The role of diffusion of oxygen in the respiration of rat diaphragm at different temperatures. Helgol. Wiss. Meeresuntersuch. *9*:324–329.

Fuhrman, G. J., and Fuhrman, F. A., 1959: Oxygen

consumption of animal and tissues as a function of temperature. J. Gen. Physiol. *42*:715–722.

Glasstone, S., 1960: The Elements of Physical Chemistry. D. Van Nostrand Co., New York.

Glasstone, S., Laidler, K. J., and Eyring, H., 1941: The Theory of Rate Processes. McGraw-Hill Book Co., New York.

Gordienko, P. A., May 1961: The Arctic Ocean. Sci. Am. *204*:88–102.

Harris, R. J. C. (ed.), 1954: Biological Applications of Freezing and Drying. Academic Press, New York.

Heilbrunn, L. V., April 1954: Heat death. Sci. Am. *190*:70–75.

Heilbrunn, L. V., Harris, D. L., LeFevre, P. G., Wilson, W. L. and Woodward, A. A., 1946: Heat death, heat injury and toxic factor. Physiol. Zool. *19*:404–429.

Hoar, W. S., and Dorchester, J. E. C., 1949: The effect of dietary fat on the heat tolerance of Goldfish (*Carassius auratus*). Can. J. Res. Sect. D. *27*: 85-91.

Hochachka, P. W., 1965: Isozymes in metabolic adaptation of a poikilotherm; subunit relationships in lactic dehydrogenase of goldfish. Arch. Biochem. Biophys. *111*:96–103.

Hochachka, P. W., 1967: Organization of metabolism during temperature compensation. *In* Molecular Mechanisms of Temperature Adaptation. Prosser, ed. American Association for the Advancement of Science, Washington, D. C. pp. 177–203.

Hock, R. J., and Covino, B. C., March 1958: Hypothermia. Sci. Am. *198*:104–114.

Johnson, F. H. (ed.), 1957: Influence of Temperature on Biological Systems. American Physiological Society, Washington, D. C.

*Johnson, F. H., Eyring, H., and Polissar, M., 1954: The Kinetic Basis of Molecular Biology. John Wiley & Sons, New York.

Kruger, G., 1962: Uber die Temperaturadaptation des Bitterlings (Rhodeus amarus Bloch). A. wiss Zool. *167*:87–104.

*Lehninger, A. L., 1965: Bioenergetics. W. A. Benjamin, New York.

Levitt, J., 1958: Frost, drought and heat resistance. Protoplasmatologia VIII.6. Springer-Verlag, Vienna.

Llano, G. A., Sept. 1962: The terrestrial life of the Antarctic. Sci. Am. *207*:213–230.

Lovelock, J. E., 1954: Physical instability and thermal shock in cells. Nature *173*:659–661.

Lozina-Lozinsky, L. K., 1965: Survival of some insects and cells following intracellular ice formation. Fed. Proc. *24*:S206–S211.

Luyet, B. J., and Gehenio, P. M., 1940: Life and Death at Low Temperatures. Biodynamica, Normandy, Mo.

Mazur, P., 1963: Kinetics of water loss from cells at subzero temperatures and the likelihood of intracellular freezing. J. Gen. Physiol. *47*:347–369.

Mazur, P., 1965a: The role of cell membranes in the freezing of yeast and other single cells. Ann. N. Y. Acad. Sci. *125*:658–676.

*Mazur, P., 1965b: Causes of injury in frozen and thawed cells. Fed. Proc. *24*:S175–S182.

Mazur, P., 1966: Physical and chemical basis of injury in single celled microorganisms subjected to freezing and thawing. *In* Cryobiology. Meryman, ed. Academic Press, New York, pp. 213–315.

McClaren, A. D., and Shugar, D., 1964: Photochemistry of Proteins and Nucleic Acids. Macmillan Co., New York.

Melnick, P. J., 1965: Effect of freezing rate on the histochemical identification of enzyme activity. Fed. Proc. *24*:S259–S267.

Meryman, H. T., 1956: Mechanics of freezing in living cells and tissues. Science *124*:515–521.

Meryman, H. T., 1960: General principles of freezing and freezing injury in cellular materials. Ann. N. Y. Acad. Sci. *85*:503–509.

Meryman, H. T. (ed.), 1966: Cryobiology. Academic Press, New York.

Militzer, W., Tuttle, L. C., and Georgi, C. E., 1951: Thermal enzymes. III. Apyrase from a thermophilic bacterium. Arch. Biochem. Biophys. *31*: 416–423.

Morales, M. E., 1947: A note on limiting reactions and temperature coefficients. J. Cell. Comp. Physiol. *30*:303–313.

Murphy, R. C., Sept. 1962: The oceanic life of the Antarctic. Sci. Am. *207*:187–210.

Nash, T., 1966: Chemical constitution and physical properties of compounds able to protect cells against damage due to freezing and thawing. *In* Cryobiology. Meryman, ed. Academic Press, New York, pp. 179–211.

Parkes, A. S., June 1956: The freezing of living cells. Sci. Am. *194*:105–114.

Parkes, A. S., and Smith, A. U., 1951: Regeneration of rat ovarian tissue grafted after exposure to low temperature. Proc. Roy. Soc. London *B140*:455–478.

Patton, A. R., 1965: Biochemical Energetics and Kinetics. W. B. Saunders Co., Philadelphia.

Prosser, C. L. (ed.), 1958: Physiological Adaptation. American Association for the Advancement of Science, Washington, D. C.

Prosser, C. L. (ed.), 1967: Molecular Mechanisms of Temperature Adaptation. American Association for the Advancement of Science, Washington, D.C.

Rose, P. H. (ed.), 1967: Thermobiology. Academic Press, New York.

Rubin, M. J., Sept. 1962: The Antarctic and the weather. Sci. Am. *207*:84–94.

Sagan, C., 1961: The planet Venus. Science *133*: 849–858.

Sailer, R. I., 1950: A thermophobic insect. Science *112*:743.

Salisbury, F. B., 1962: Martian biology. Science *136*:17–26.

Scholander, P. F., Flagg, W., Hock, R. J., and Irving, L., 1953: Studies on physiology of frozen plants and animals in Arctic. J. Cell. Comp. Physiol. *42*(Suppl. 1):1–56.

Scholander, P. F., Hock, R., Walter, V., and Irving, L., 1950: Adaptation to cold in Arctic and tropical mammals and birds in relation to body temperature, insulation, and basal metabolic rate. Biol. Bull. *99*:259–271.

Sherman, J. K., 1965: Practical applications and chemical problems of preserving spermatozoa by freezing. Fed. Proc. *24*:S288–S296.

Smith, A. U., 1958: The resistance of animals to cooling and freezing. Biol. Rev. *33*:197–253.

Smith, A. U., 1961: Biological Effects of Freezing and Supercooling. Williams & Wilkins, Baltimore.

Smith, R. E., and Hoijer, D. J., 1962: Metabolism and cellular function in cold acclimation. Physiol. Rev. *42*:60–142.

Solomon, A. K., 1953: The kinetics of biological processes. Advances Biol. Med. Phys. *3*:65–97.

Somero, G., 1967: Mechanisms of cold adaptation in some Antarctic fishes. Doctoral Dissertation, Stanford University, Stanford, Calif.

Stowell, R. E., Young, D. E., Arnold, E. A., and Trump, B. F., 1965: Structural, chemical, physical and functional alterations in mammalian nucleus following different conditions of freezing, storage and thawing. Fed. Proc. *24*:S115–S141.

Sussman, M. V., and Chin, L., 1966: Liquid water in frozen tissue studied by nuclear magnetic resonance. Science *151*:324–325.

Trump, B. F., Young, D. E., Arnold, E. A., and Stowell, R. E., 1965: Effects of freezing and thawing on the structure, chemical composition and function of cytoplasmic structure. Fed. Proc. *24*: S144–S168.

Wang, G. T., and Marquardt, W. C., 1966: Survival of *Tetrahymena pyriformis* and *Paramecium aurelia* following freezing. J. Protozool. *13*:123–128.

Wohlschlag, D. E., 1964: Respiratory metabolism and ecological characteristics of some fishes in McMurdo Sound, Antarctica. Antarctic Res. Series *1*:33–62.

*Wood, T. H., 1956: Lethal effects of high and low temperature on unicellular organisms. Advances Biol. and Med. Phys. *4*:119–165.

# SECTION III

# Exchange of Materials across Cell Membranes

The plasma membrane must permit the entry of cell foods and oxygen and the exit of cell wastes. The properties which enable the plasma membrane to regulate entry and exit of materials are of prime importance because maintenance of life depends upon such exchanges. The movement of water and of dissolved materials through the plasma membrane is a dynamic process; the living cell is never in equilibrium with the environment in terms of materials across its membrane. In fact, a cell in complete equilibrium with the solutes in its environment is dead.

Many of the cell organelles, such as the nucleus, the endoplasmic reticulum, the Golgi complex, the lysosomes and the mitochondria, are separated from one another and from the cytoplasmic ground substance by membranes. These membranes resemble the plasma membrane in showing selective permeability and permit separation of the cell into compartments; within each compartment composition can be regulated by what passes into and out of the compartment through the limiting membrane. All the membranes (including the plasma membrane) together constitute the *cell membranes*. The physical and chemical nature of cell membranes is discussed in Chapter 11.

At one time it was thought that movement of substances through the plasma membrane was determined solely by the concentration gradient; i.e., a substance moved inward only when it was present outside in a concentration higher than inside, and vice versa. This is probably true in some instances, and several cases will be considered in Chapters 12 and 13, but movement against a concentration gradient has been observed in some other cases; that is, a substance which is present outside the cell in smaller amounts than inside is nevertheless absorbed. For example, potassium is usually accumulated in plant and animal cells to a concentration many times that in the medium surrounding the cells. Such transport requires expenditure of energy by the cell and has been called active transport. It is considered in Chapter 14 along with cell drinking or pinocytosis.

Inasmuch as the topics discussed in each of the following chapters in this section are artificially separated from one another to simplify discussion, a certain amount of overlap is bound to occur. Such overlap is permitted only because it helps to elucidate the topic under discussion.

# CHAPTER 11 / THE CHEMICAL NATURE AND STRUCTURE OF CELL MEMBRANES

## EXTRANEOUS MEMBRANES ON CELLS

Although it is often stated that the cytoplasm of a cell is bounded by a *cell membrane* or *plasma membrane,* extraneous coats in addition may surround the delicate plasma membrane and protect it. In a sea urchin egg a *jelly layer* covers the *vitelline membrane* within which is the plasma membrane, as shown in Figure 11.1. Both of these layers are secreted by the ovary and both may be removed without damaging the plasma membrane. The jelly coating may be dissolved by treatment with acidified sea water. The vitelline membrane, which is a tough elastic coating, may be lifted and stretched with a needle and can be removed by microdissection or by washing in isotonic potassium chloride solution. An egg, denuded of the jelly and the vitelline membrane, becomes sticky and is believed to be separated from the medium by the plasma membrane. The "naked" egg is much more susceptible to injury than when it is protected by the vitelline membrane and the jelly, but it nevertheless develops normally when fertilized (Chambers, 1940; Kopac, 1940; Brandt, 1962).

When an egg is fertilized the vitelline membrane is raised as a *fertilization membrane,* and a space then separates it from the surface of the egg (Runnstrom, 1952). This space is filled with fluid except for some strands connecting the membrane to the surface of the egg. About the surface of the egg is then secreted a *hyaline plasma layer.* This becomes prominent as the cement which binds the blastomeres together when the egg undergoes cleavage.

The presence of a plasma membrane on a naked sea urchin egg can be demonstrated by microdissection. When the surface is depressed by the application of a microdissection needle, it visibly crinkles (Chambers, 1949). If the egg, freed of its extraneous membranes, is sawed in two with a needle, two spheres are obtained, indicating that the cut surface has developed a membrane which holds the cell material. Even when the egg in sea water is torn with a needle, a new membrane forms (see Fig. 11.2*A*). However, if the membrane is torn in calcium-free sea water, the contents of the egg disperse in the solution (see Fig. 11.2*B*). Calcium appears to be essential for the formation of the membrane, which is presumably formed by the precipitation of the cell

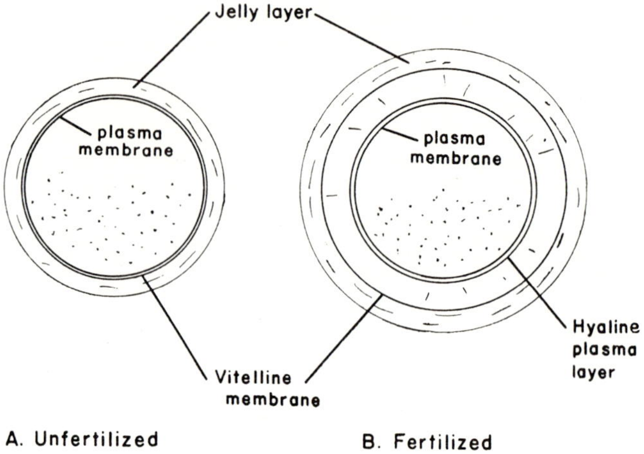


***Figure 11.1.*** Diagrams of unfertilized and fertilized sea urchin eggs to show extraneous layers. The plasma membrane is indicated, although it is not actually possible to see it. The vitelline membrane is raised during fertilization, becoming the fertilization membrane.

material in the presence of calcium. Heilbrunn (1930) has called this reaction the *surface precipitation reaction.*

As the cells differentiate into tissues they are cemented together by a material much like the hyaline plasma layer which binds together the blastomeres of an embryo. In capillaries the *cellular cement* acts like a filter through which blood plasma passes into the surrounding tissue spaces. The intercellular cement is probably primarily acid mucopolysaccharide, the prevalent one being hyaluronic acid. Hyaluronic acid contains D-glucuronic acid and N-acetyl-D-glucosamine in equivalent quantities, each alternating with the

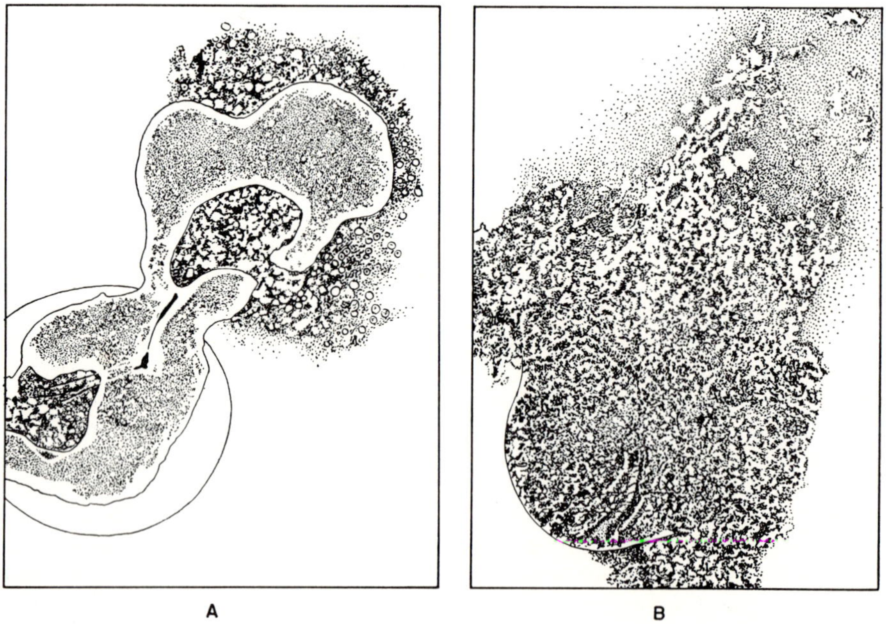


***Figure 11.2.*** *A*, Surface precipitation reaction in an egg of a starfish (*Partiria miniata*) crushed in sea water. *B*, The membrane fails to develop when the egg is crushed in calcium-free sea water. (After a photograph by Richard Boolootian.)

other in a linear chain. Even in low concentration hyaluronic acid forms a firm gel (Cantarow and Schepartz, 1967).

In ciliate protozoans the plasma membrane is protected by a tough proteinaceous pellicle which, superficially, resembles the cell wall of a plant cell. However, when a protozoan is placed in hypertonic solution the pellicle decreases in size to fit the shrunken protozoan, which indicates that the pellicle is elastic in nature. Details of the organization of the pellicle have been studied with the electron microscope (Pitelka, 1963).

The pellicular material can be partially removed from some protozoan cells. Thus, when the ciliate *Blepharisma* is placed in a solution of 0.05 per cent strychnine sulfate, a layer of protein and pigment the shape of the cell is shed (Kennedy, 1966). The *Blepharisma* withdraws from this mantle and swims out of it through the mouth or the contractile vacuole opening. The *Blepharisma* then becomes pear shaped, as if unable to maintain its long slender form without the rigid pellicle. A rigid pellicle is regenerated within a day or two and the cell gradually resumes its normal shape (Nadler, 1929).

In lieu of the pellicle an ameba is coated with a *slime,* made up chiefly of polysaccharides (O'Neill, 1964b), as are even some cells with a pellicle or a cell wall (e.g., desmids and diatoms). An ameba pulled with a microneedle and released may snap back to its original position as a result of contraction of the elastic slime (Chambers and Chambers, 1961).

**Plant Cell Walls.** In a plant cell a *cell wall* is always present, protecting the cell membrane. The cell wall is usually composed chiefly of cellulose secreted by the cell. The adjacent cell walls of abutting cells are cemented together by *pectin.* Old cell walls may be thickened by *lignin,* a ring compound (see Chapter 3), and surfaces of epidermal cells may be covered with a *waxy* covering which reduces water loss (Eglinton and Hamilton, 1967).

Extensive studies by x-ray diffraction, polarized microscopy and electron microscopy have been made of the architecture of the plant cell wall, and it appears that the cytoplasm lays down an interwoven network of fibers (primary wall) to which lamellar deposits (secondary wall) are later added. Spaces remain between the fibers, and the network design gives maximal strength for a minimum of material.

The first rudiment of a future cell wall in a cell about to divide is seen as a series of droplets on the midline between the two daughter cells and at right angles to the spindle. Half an hour later birefringence is observed, probably at the moment of formation of the cellulose fibrils on either side of the pectic materials forming the middle lamella (Picken, 1960).

In *Valonia,* the unit fibers are composed of $\alpha$ cellulose and may be quite long, about 10 $\mu$, and astonishingly constant in diameter—150 to 300 Å. The fibers are ribbon-like and lie with the broader face in the plane of the cell wall. X-ray evidence indicates "periods" of 500 glucose residues, each one 2500 Å long (Preston, 1959). X-ray mapping before and after stripping of these fibers revealed two sets of fibers: one meridional and the other following a shallow, spiral path; the two kinds of fibers were laid down in layers following a definite pattern of succession, a total of 200 to 300 layers being present.

The primary cell wall of many other plant cells, however, has fibers which are distinctly interwoven, not in plane sheets as in *Valonia,* indicating that the secreting cytoplasm penetrates the fibril meshwork. The secondary cell wall has parallel fibers of constant thickness laid down in close texture and forming lamellae (Fig. 11.3). The intimacy of contact between cytoplasm and cell wall is shown also by the adhesion of the cytoplasm upon plasmolysis to the growing tips of cell wall fibers (Mühlethaler, 1961; Northcote, 1963; Albersheim, 1965).

A concentrated preparation of the enzyme cellulase removes the cell walls from cells in the root, leaf and other parts of higher plants and liberates the naked protoplasts bounded by the plasma membrane. The protoplasts of plant cells act

***Figure 11.3.*** Electron micrograph of the arrangements of cellulose fibers in the cell walls of *Cladophora rupestris* (green alga). (×24,000) Note the three directions in which the fibers run. Smaller units, of which the fibers are composed, may be seen in the isolated fibers at the top and the bottom. (From Preston, 1961: Proc. Roy. Soc. *B 154*:70–94.)

like any other naked cells; they round up and swell or shrink depending upon the concentration of solutes in the surrounding medium (Ruesink and Thimann, 1966).

The cell wall may have a degree of plasticity. For example, the stamen hair cells of *Tradescantia,* placed in hypotonic solutions, may extend 100 per cent and again return to their original size on replacement into isotonic solutions. In the growing tip the cell walls must be plasticized before they can elongate, there being as much as a sixfold increase in dry weight of the wall substance during maximal extension (Picken, 1960).

The cell wall composition is not always the same. For example, although the cotton hair is 90 per cent (by dry weight) cellulose, only 2 per cent of a red algal cell wall is cellulose, other polysaccharides of complex nature making up the remainder. The cell walls of fungi, including yeast, are made up largely of chitin, a glucosamine polymer (see Table 3.1). In chitinous cell walls of some invertebrate animal cells self-tanning proteins are also present (Picken, 1960).

In bacteria the separateness of the cell wall and the cell membrane is readily demonstrated by plasmolysis (Fig. 2.6). The structure of the cell wall, as shown by electron microscopy, in most cases appears to be made up of minute rectangular or hexagonal units about 120 to 140 Å in dimension (Fig. 11.4). In some cases, however, it consists of many stranded curving cables. The chemical composition of the cell wall is somewhat different in the two major divisions of bacteria.

In gram-positive bacteria, e.g., *Staphylococcus,* the wall amounts to 20 per cent of the dry weight of the bacterium and is made up of a small number of amino acids, including diaminopimelic acid (unique to bacteria and blue-green algae), hexosamine and 1 to 3 per cent of lipid. The cell walls of gram-positive bacteria also contain teichoic acids which are polymers containing chains of glycerol or ribityl residues joined to each other by

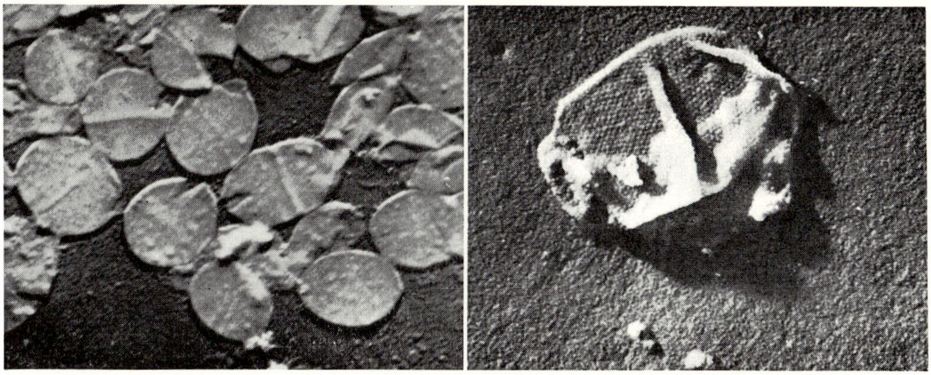

**Figure 11.4.** Electron micrographs of cell walls of bacteria. *Left,* Cell walls of *Streptococcus fecalis,* after the cells have been split by grinding to permit the contents to escape. (×12,000) *Right,* A portion of a *Spirillum rubrum* cell, showing the regular pattern of bodies of which this wall is constructed. (×42,000) (Salton and Williams; Salton and Horne. From Burrows, Textbook of Microbiology, 1959.)

phosphate groups. Sugars are attached to the glycerol or ribityl, and D-glucose residues are in ester linkage with some of the hydroxyl groups of these compounds (Braddiley, 1964). The cell walls of gram-negative bacteria, e.g., *Escherichia coli,* are thinner and contain much more lipid, proportionally, and are chemically more elaborate than in the gram-positive group. Two layers are found in the cell wall of *E. coli.* One is extractable with phenol and is made up of lipoprotein, accounting for 80 per cent of the dry weight of the wall; the other is chemically much like the wall of the gram-positive types, containing lipopolysaccharide, lysine and hexosamine (Robinow, 1960; Rogers, 1963; Nickerson, 1964).

Penicillin is bacteriostatic to gram-positive bacteria because it inhibits development of the cell wall in the growing bacteria. Streptomycin is bacteriostatic to many gram-negative forms (Carpenter, 1961).

The cell walls of bacteria can be removed with the aid of the enzyme lysozyme. The protoplast so liberated, like any other naked cell, shrinks or swells depending upon the concentration of solutes in the medium. Bacteria without cell walls will not divide even in a favorable nutrient medium.

Outside the cell wall of most, if not all, bacteria is found an encapsulating layer of slime or firmer matter made up of polysaccharides, lipid and sometimes other materials. These encapsulating materials are important to bacteriologists as antigens, but to bacteria they are primarily additional mechanical protection (Carpenter, 1961).

Cell walls, vitelline membranes, fertilization membranes arising from vitelline membranes, and the hyaline plasma layer have a porosity such that most materials (except very large molecules) can pass through them. It is the plasma membrane that determines what shall enter or leave the cell. An analysis of the plasma membrane is therefore most pertinent to an understanding of the transport of materials into and out of cells.

## FINE STRUCTURE OF CELL MEMBRANES

Much of the basic electron microscopic work leading to the present concept of cell membrane structure has come from a study of the myelin sheath of the neuron (nerve cell). As shown diagramatically in Figure 11.5, the axon of the neuron depresses the plasma membrane of the Schwann cell (sheath cell) and, rotating in the cell (or being spirally wrapped in the cell), it gradually accumulates around it a number of double membranes in which it lies. The myelin sheath, therefore,

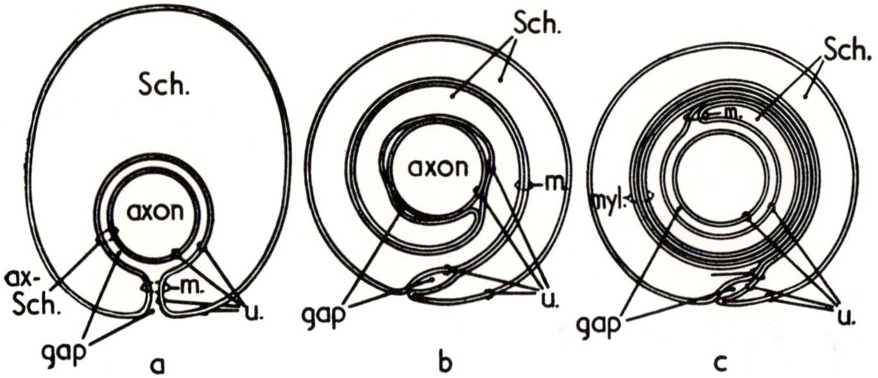


***Figure 11.5.*** Three stages in the development of nerve myelin are indicated. The earliest recognized stage is shown in (*a*). Here a single axon is embedded in a Schwann cell (Sch.). The enveloping lips of the Schwann cell come together around the axon to form a double membrane known as the mesaxon (m.). There is usually a small gap between the Schwann cell membranes (u.) forming the mesaxon, which is continuous with the gap between the axon and the Schwann cell membranes making up the axon-Schwann membrane (ax.-Sch.). At a later stage of myelination represented by (*b*) the mesaxon is elongated in a spiral around the axon and its central gap is largely obliterated. The gap of the axon-Schwann membrane is also partially obliterated. At a later stage (*c*) the mesaxon is elongated further and the cytoplasmic surfaces of the spiral mesaxon loops come together to form the major dense lines of compact myelin (myl.). (From Robertson, 1959: Biochem. Soc. Symp. *16*:3.)

consists essentially of numerous plasma membranes derived from the Schwann cell and wound around the axon of the neuron (Robertson, 1960, 1964).

Analysis of the sheath by methods of x-ray diffraction and polarizing microscopy had already led to an understanding of the existence of repeating units of great regularity. Electron micrographs confirmed and extended these studies. High resolution on ultrathin cross sections of the neuron and its myelin sheath indicates a unit membrane of 75 Å consisting of two osmophilic layers (i.e., staining with osmium tetroxide) each about 25 to 40 Å wide and a less dense band between them about 20 to 25 Å wide (Fig. 11.6). Because of the way the myelin sheath is formed during development (Fig. 11.5), such unit membranes occur in pairs, in apposition to each other within the sheath. Many such pairs with some material between them (shown between membranes in Fig. 11.5) make up the myelin sheath.

Such a basic unit membrane as seen in the sheath appears to be general for a wide variety of cells, e.g., cells of intestinal epithelium, nerves, skin, liver, pancreas, kidney, endothelium, blood and muscle of a variety of species of vertebrates and invertebrates including ciliates and amebas. Even more remarkable is the finding of a similar unit in plant cell membranes as well (Robertson, 1960, 1964). Since the membranes of the endoplasmic reticulum and mitochondria and even the Golgi complex show the same unit structure, it would appear to be a fundamental unit common to all cell membranes (Robertson, 1960; see Fig. 5.12).

Electron microscopic studies also reveal extensions of the plasma membranes by folds. The intestinal epithelium cell may have as many as 3000 microvilli, and similar structures are seen in brush border cells of the kidney tubules. These microvilli (Fig. 11.7) are believed to increase the surface of the cell for exchange with the materials in the bathing medium. In other cells, invaginations and irregularities also occur, but not to the spectacular degree as they do in the cells in which a major activity is the exchange of materials with the environment. All these extensions have approximately the thickness (75 Å) of a unit membrane (Robertson, 1959).

It has been claimed that in some fungal cells connections between the plasma mem-

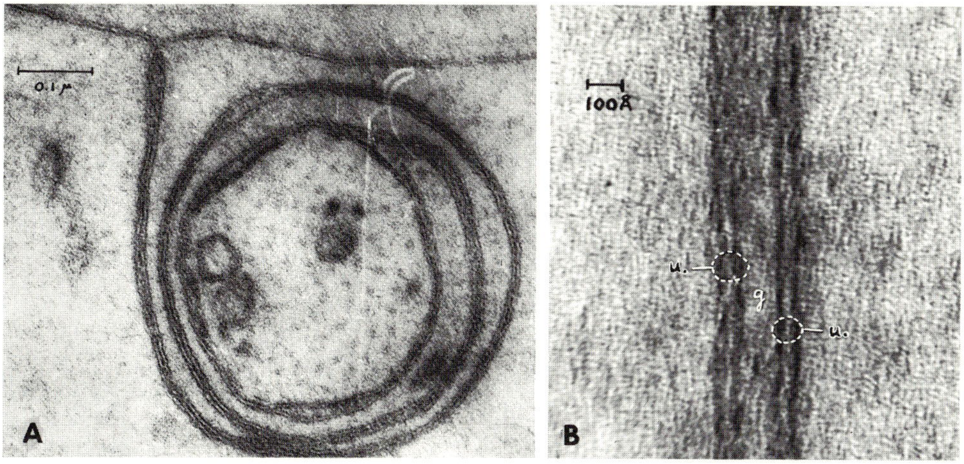


***Figure 11.6.*** Electron micrographs of unit membranes in the myelin sheath. *A*, Unit membranes are clearly seen at the top of the figure and pairs of such membranes are seen beyond the point at which the unit membranes meet. Each unit membrane is made up of three layers. *B*, Higher magnification of a pair of unit membranes, each member of the pair being indicated by u; g shows the material between the pair of unit membranes. The three layers of the unit membrane probably correspond to the three layers shown in Fig. 11.10, the outer layers being, perhaps, protein; the inner, lipid material. (Robertson, 1959: Biochem. Soc. Symp. *16*:3.)

brane and the nuclear membrane can be seen (McAlear and Edwards, 1958). It is also thought that possibly all the membrane systems in the cell have their origins as invaginations of the plasma membrane, from which their similarity is derived. The infoldings are then considered as a means for communication from surface to endoplasmic reticulum and nucleus. Because the membranous systems in the cytoplasm vary with conditions it is assumed that they are in a state of dynamic equilibrium, forming and breaking down continually. Because all of these studies were made with fixed materials, and not on the living cells, it is difficult to pass judgment on these provocative ideas (Robertson, 1964).

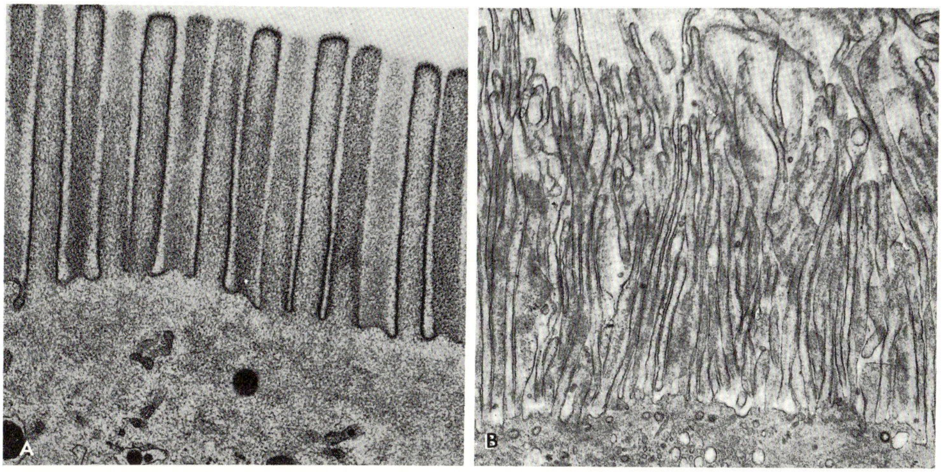


***Figure 11.7.*** Microvilli on cells of *A*, intestine of the rat (phosphate buffered osmium fixation, uranyl acetate and Karnovsky A lead staining, ×22,000); and *B*, epithelium of rabbit epididymis (collidine buffered glutaraldehyde and osmium fixation, lead citrate staining, ×16,000). (From Fawcett, 1966: The Cell: Its Organelles and Inclusions. W. B. Saunders Co., Philadelphia.)

## CHEMICAL CONTENT OF CELL MEMBRANES

A single cell membrane is too small for microchemical analysis, but such an analysis is possible with a mass of cell membranes. The ghosts of erythrocytes (Rogers, 1963; Nickerson, 1964) and bacterial protoplasts (Robinow, 1960) have been subjected to biochemical analysis in an attempt to determine some of their constituents. More recently such studies have been made upon ameba (O'Neill, 1964a, b), microsomal membranes (Kamat and Wallach, 1965) and mitochondrial membranes (Lehninger, 1965).

Upon hemolysis, an erythrocyte liberates hemoglobin, and the ghost which is left behind appears to consist of nothing but the plasma membrane, although the membrane when returned to Ringer's solution may continue to swell and shrink in response to osmotic stresses (Ponder, 1948). Analysis of a mass of ghosts indicates that they consist of lipid and protein in the ratio of 1.0 to 1.7 by weight. The protein, called stromatin, has a high molecular weight and is fibrous in nature. The lipids are mainly cholesterol and phospholipids. Cholesterol makes up between 15 and 32 per cent of the total lipid content. The phospholipids, which make up between 55 and 75 per cent of the total lipid content, consist primarily of *lecithin* and *cephalin.* On the average the membrane has 75 to 90 molecules of lipid for every molecule of protein, enough lipid to cover the surface of the red blood corpuscle with a layer two to three molecules thick (Parpart and Ballantine, 1952).

Some investigators question these analyses of the chemical contents of the cell membrane on the basis that perhaps more than the plasma membrane may be present in the ghost, especially since the nature of the ghost varies with its mode of preparation (Ponder, 1952). For instance, it is possible that the erythrocyte ghost includes some of the internal structure upon which the hemoglobin has been bound.

The data from ghosts of bacteria corroborate those on erythrocytes in general, although the ratio of protein to lipid varies considerably in different kinds of plasma membranes. Comparisons given here have been made on the basis of areas occupied by proteins and lipids in a surface film. Thus myelin, consisting of multiple membranes of Schwann cells wrapped about axons (see Figs. 11.5 and 11.6), has a protein to lipid area of 0.43; the erythrocyte, 2.5; *Bacillus megatherium,* 5.4; and *Mycoplasma laidlawii,* 0.4. The proportion of cholesterol to phospholipid varies in membranes of various vertebrate cells and cholesterol is completely lacking in bacteria. Nor are the phospholipids in the membranes of different kinds of cells the same (Korn, 1966). Furthermore, some plasma membranes contain considerable polysaccharide, e.g., *Amoeba proteus,* in which it constitutes 16 per cent of the total dissolved material in the membrane (O'Neill, 1966a, b). Carbohydrates are probably generally present in the plasma membrane conjugated to proteins but usually in lesser amounts than in *Amoeba.*

Some of the proteins present in unit membranes of the mitochondria and chloroplasts have enzymatic activity, among which is the splitting of adenosine triphosphate. Furthermore, claims have been made that this protein in chloroplasts and mitochondria of some cells is contractile. The plasma membranes of some cells also have adenosine triphosphatase activity but contractility has not been reported in any of them (Onishi and Onishi, 1962).

Some workers, using staining methods, have identified nucleic acid in the membrane of the *Arbacia* egg (Lansing and Rosenthal, 1952), and the same has been claimed in other instances. It is difficult to be sure that some of the endoplasmic reticulum with ribosomes is not being included in these instances.

Salts are also present in cell membranes, and microincineration tests suggest that some of these are present in higher concentration in the membranes than elsewhere in cells (Scott, 1933). Water present in cell membranes forms part of the structure of the membranes, as it does in all cell materials (Hechter, 1965).

## LIPIDS IN THE CELL MEMBRANES

The general correlation between lipid solubility and entry of various compounds into the cell (see Fig. 13.4) is strong evidence for the presence of lipoidal material in the cell membrane. Other experiments demonstrate the presence of lipoidal material in the plasma membrane, but none of them excludes the possibility that other materials are also present.

The high electrical resistance of cells (e.g., 100,000 ohms per square centimeter of surface in *Nitella,* and 1000 ohms in the squid giant axon) is attributable to the presence of a lipoidal layer at the surface of the cell. Ions penetrate through the lipoidal layer with difficulty; consequently, conduction of electricity through the membrane is slight and the resistance large. It has been calculated that a layer of lipids about 50 Å thick, consisting of two layers of molecules, is sufficient to account for such resistance (Danielli, 1954).

The coalescence of oil with the cell membrane is also explained by the presence of lipid in the cell membrane. Thus, oil, forced out of a micropipette under the microscope, when brought next to the cell to be tested, may completely engulf a cell such as an erythrocyte. A small drop of oil brought next to *Amoeba dubia* forms an oil cap on it, but because of the rigid nature of the ectoplasm, the oil cap is not taken into the *Amoeba.*

An oil cap also forms on an *Arbacia* egg freed of extraneous coats (see Fig. 11.8), but if no protein is present in the sea water, the oil droplet pops inside the cell because the surface tension within the cell is less than at the surface of the oil. If the vitelline membrane is present the oil does not coalesce with an *Arbacia* egg, since the lipoidal layer of the plasma membrane of the egg cannot make contact with the oil through the vitelline membrane. When an egg, from which the extraneous membranes have been removed is fertilized, the hyaline plasma layer is secreted over the plasma membrane, and oil again will not cap the egg.

The lysis (breakdown) of the plasma membrane by lipolytic agents such as benzene and carbon tetrachloride, both of which are lipoid solvents, is thought to be a result of their action on the lipids in the plasma membrane. Digitonin, sodium and potassium salts of fatty acids, and paraffin sulfonates, all of which lyse the plasma membrane, act as lysins primarily because of their effect on cell lipids. Digitonin, for example, is thought to form a complex with membrane sterols. Much more specific is the lysis of the plasma membrane by lecithinase, an enzyme which catalyzes the hydrolysis of lecithin in the plasma membrane (Ponder, 1953).

The permeability of various substances through a similar but artificial layer of lipid 50 Å thick is of the same order of

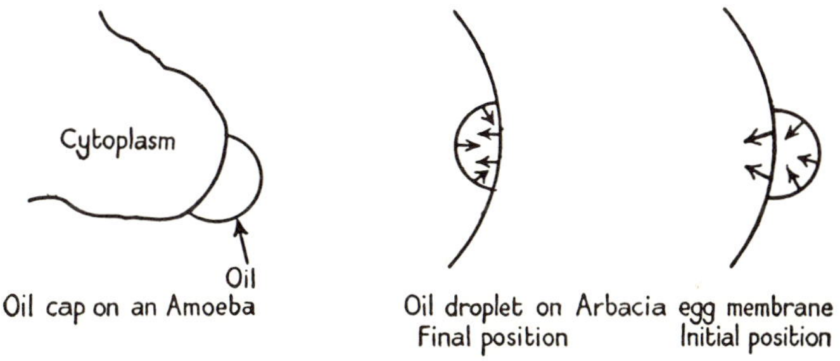


***Figure 11.8.*** Diagrams illustrating the behavior of oil drops brought into contact with the cell surface. The arrows indicate the internally directed force in the droplets due to interfacial tension. (From Danielli, 1951: Cytology and Cell Physiology. Bourne, ed. Oxford University Press, London.)

**Table 11.1.** Comparison of the Permeability of Water and Oil Films and a Variety of Cells to Three Substances

| SUBSTANCE | 5 mμ WATER | 5 mμ OIL | ARBACIA EGG | *Chara ceratophylla* | *Pylaiella litoralis* | OX ERYTHROCYTE |
|---|---|---|---|---|---|---|
| Ethylene glycol | $1.7 \times 10^9$ | 0.1 | 0.73 | 1.2 | 0.1 | 0.2 |
| Glycerol | $1.4 \times 10^9$ | 0.002 | 0.005 | 0.02 | 0.002 | 0.002 |
| Erythritol | $1.2 \times 10^9$ | 0.00005 | — | 0.001 | 0.0001 | — |

Data from Danielli, 1952. The permeability is given in moles of a substance entering a square micron of surface area per second for a unit mole per liter difference in concentration between the inside and the outside of the cell (multiplied by $10^{16}$ to give a number which can be handled more conveniently than can small fractions). The permeability was calculated for water and oil films and was taken from experimental data for the cells. Note that the permeability of cells to these compounds is of the same magnitude as the permeability of an oil film to them.

magnitude as the permeability of the same substances penetrating living cells. Calculation of the penetration of the same substances through an aqueous layer of this thickness gives permeability constants of an entirely different order of magnitude, as seen in Table 11.1.

## PROTEIN IN THE CELL MEMBRANES

The elasticity and strength of the plasma membrane of the cell strongly suggest that the lipids in the plasma membrane are bound together by some other type of compound, since a membrane made up exclusively of lipids would lack mechanical strength. In erythrocyte ghosts are found fibrous proteins of large molecular weight containing hydrophobic and hydrophilic groups which could well perform such a role, the hydrophobic groups binding lipids and the hydrophilic groups binding water-soluble substances. Proteins at the surface could thus bind the lipids to the aqueous cytoplasm and at the same time provide the elasticity of the plasma membrane observed and measured (Mitchison, 1953).

The low tension between the plasma membrane and the surrounding medium also suggests the presence of protein in the plasma membrane. By measuring the force necessary to distort an egg with a gold fiber it is possible to determine the tension at the surface of the egg. Determination of the centrifugal force required to break an egg in two may also be used for this purpose. Regardless of the type of cell tested and of the method used, the tension of the plasma membrane appears to be low. The values found for *Arbacia* are between 0.2 and 0.08 dynes per centimeter; the value for the *Triturus* erythrocyte is 0.1; for *Amoeba dubia,* 1.3; for the slime mold *Physarum,* 0.45; for the egg of *Triturus,* 0.1.

The surface tension of oil against water is about 15 dynes per centimeter. However, when some protein is spread on an oil, the surface tension against water falls to values of 1.0 or less. The low tension at the surface of cells suggests the presence of protein (Davson and Danielli, 1952). Some uncertainty exists, however, as to whether the term surface tension really applies in making measurements on an elastic membrane such as the cell membrane.

Lysis of the plasma membrane by agents which act specifically upon proteins offers corroborative evidence of the presence of proteins in the surface of the cell. Furthermore, cells injected into an animal act as antigens against which antibodies are formed in the serum and lysis of the foreign cells by such antibodies implies action on their membranes. Antibodies are not known to be produced against lipids, but they are demonstrable against proteins and carbohydrates. Since both are present in plasma membranes, the evidence is indecisive (Brown and Danielli, 1964a). On the other hand, the polymerized polyhydroxyphenols are known to denature proteins but not to affect carbohydrates or lipids. When such a substance is applied to a cell, the structure of the plasma membrane is lysed (Ponder, 1954), suggesting the presence of protein. Evidence from electron micrographs of erythrocyte ghosts also indicated the presence of proteins in the plasma membrane (Robertson, 1964).

Chemical determinations on suspensions of plasma membranes indicate that sufficient protein is present to coat the bimolecular lipid leaflet with protein on both sides.

Among the proteins of the plasma membrane are glycoproteins and mucoproteins, conjugates of carbohydrates and proteins. They are widespread and may form a considerable component of the membrane protein. These are important as antigens. An example is the agglutination (precipitation) of red blood cells by influenza virus, as a result of reaction between enzymatically active groups on the surface of the virus particles with the surface mucopolysaccharides on the red blood cells. The widespread nature of immunological reactions to all types of cells indicates the general presence of such mucoids in their surfaces. The mucoids of the foreign cells serve as antigens against which the host organism develops antibodies (Robertson, 1960).

The plasma membrane must also be the site of receptor proteins in bacteria containing permease systems. In any cells in which active transport occurs, and this is probably universal, it is likely that enzymes occur in the surface of the cell. Perhaps part of the structural protein of a plasma membrane is enzymatic protein (Chambers and Chambers, 1961). Some evidence indicates that lipids in cell membranes may also influence the catalytic activity of enzymes (Rothfield *et al.*, 1966).

## THICKNESS OF CELL MEMBRANES

Since the dimensions of the cell membrane are within the limits of resolution by the electron microscope, considerable effort has been made to study the cell membrane by this means. Such studies made of flattened entire cell membranes and of thin sections of cell membranes indicate a thickness of about 75 Å for the cell membrane (see Fig. 11.6). This is the dimension for the unit membrane already considered in Chapter 5. However, the unit membranes from different parts of the cell are not necessarily of exactly the same dimensions. By measuring the peak to peak density on unit membranes of electron micrographs of bullfrog cells it was found that the membranes of the Golgi vesicles and of synaptic vesicles of neurons are of about the same dimensions as the plasma membrane, whereas the unit membranes of the nuclear envelope, the endoplasmic reticulum and the Golgi lamellae are only 85 to 90 per cent as thick (Yamamoto, 1963; see Fig. 11.9). Sometimes the plasma membrane is thicker than 75 Å and may be as large as 100 Å (Korn, 1966). The cross-sectional thickness of the cell membranes depends in part on the method of fixation, being thicker when fixed in permanganate than when fixed in osmium tetroxide (Sjöstrand, 1963). Some evidence has also been obtained for smaller subunits in these membranes, especially by freeze-etching (Branton, 1964).

All the other methods for estimating the thickness of plasma membranes are indirect. Before the advent of the electron microscope the interference with light reflectance from a mirror by erythrocyte ghosts and layers of barium stearate (of known thickness) was used to determine the plasma membrane dimensions. Estimates determined in this manner for the thickness of plasma membranes of different types of erythrocytes vary from 140 to 250 Å. About half this thickness is left after the lipids are extracted, indicating a residue of protein.

Estimates obtained by dividing the total amount of protein and lipid in the membrane by the area of the membrane indicate a thickness of the same order of magnitude as was obtained by the preceding method; namely, about 120 to 140 Å for the cell membrane of the erythrocyte (Parpart and Ballantine, 1952).

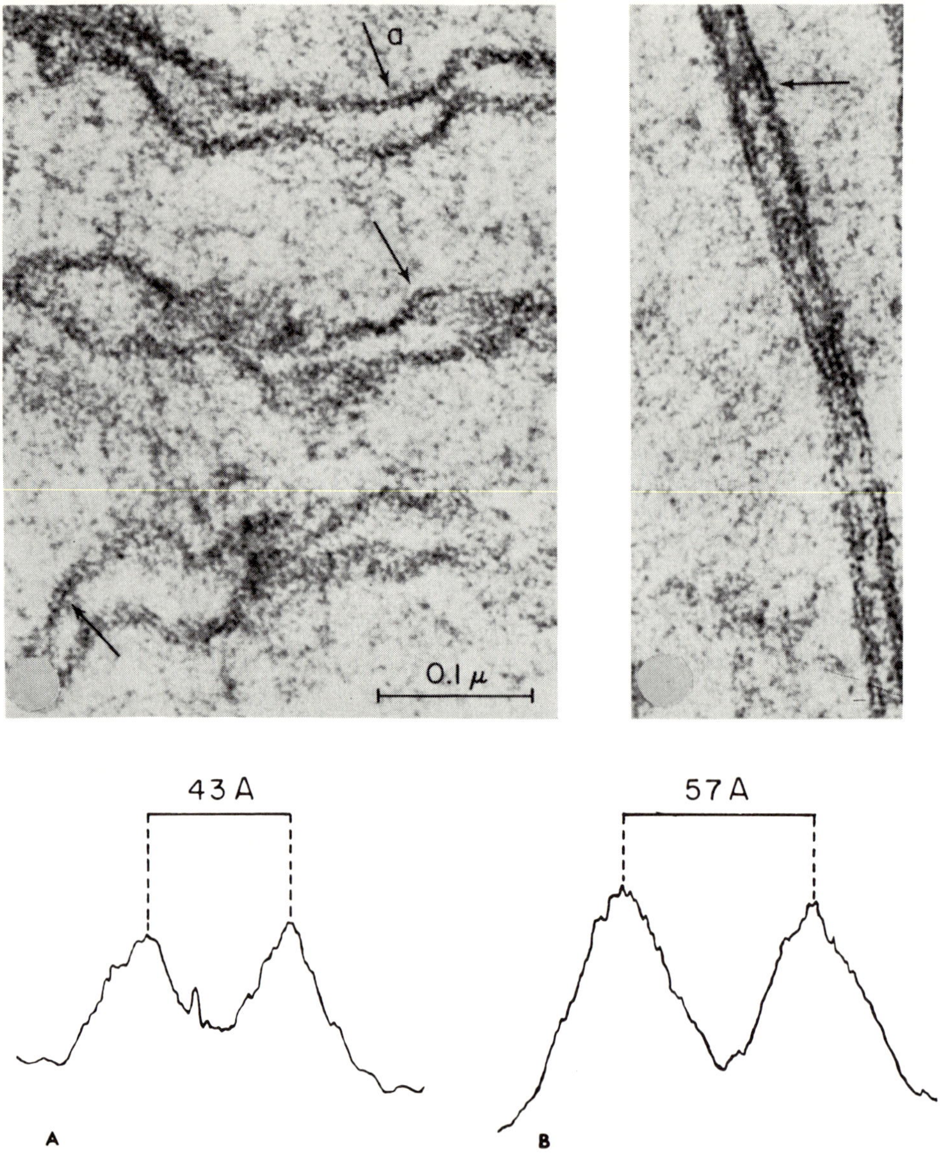


*Figure 11.9.* Thickness of the unit membrane as determined by densimetric recording on electron micrographs of various membranes. *A*, Endoplasmic reticulum membrane of bullfrog ganglion cell, and *B*, plasma membrane of the same cell; magnification of the electronmicrographs is 225,000×. The value in Å is the peak to peak value; the thickness of the membrane should be measured from the initial increase in density over the background on either side. Peak to peak measurement is more accurate. (From Yamamoto, 1963: J. Cell Biol. *17*:413.)

From experiments with polarized light somewhat thicker cell membranes are estimated, but still of the same order of magnitude as determined by other methods (Waugh, 1950). Still another approach used for estimating the thickness of the plasma membranes of a variety of cells is that of measuring the thickness of an oil film required to duplicate the electric properties characteristic of the plasma membrane. The electrical resistance of a 50 Å oil film is found to be of the same order of magnitude as that of the cell membrane (Danielli, 1952).

Some confidence, therefore, may be placed in the values found, since all the experimental techniques have yielded estimates of the same order of magnitude.

## STRUCTURE OF CELL MEMBRANES

No technique exists as yet that permits direct observation of the molecular arrangements of proteins and lipids in the cell membranes. All information on membrane structure has been inferred from indirect evidence (Smith, 1962; Jacobs, 1962; Kavanau, 1964; Finean, 1966). Little consideration has been given to the polysaccharides and their possible structural role.

The striking correlation between lipoid solubility and permeability led Overton, more than 50 years ago, to postulate a cell covered essentially with lipid. Langmuir had shown that fatty acid molecules at an interface stand at right angles to the surface of the water, and the same was thought to be likely in the plasma membrane. When, in 1925, Gorter and Grendel showed that all the extracted lipids from a red blood cell occupy an area twice that for a single molecular film, it was thought that there was a membrane with a double layer of lipid molecules, each with its hydrophobic end to the other, standing at right angles to the surface. However, because the tension at the surface of an egg plasma membrane is much lower than that expected for a lipid, but approximately equal to the value for a lipid with a coating of proteins, it was postulated by many workers that proteins may cover, though not necessarily completely, the outer surface of the plasma membrane (Davson and Danielli, 1952; Davson, 1962).

Most of the evidence for the arrangement of lipids and proteins in the plasma membrane comes from studies using x-ray diffraction (Schmitt and Palmer, 1940) and polarized light (Mitchison, 1953). Studies with the polarizing microscope suggest that the protein molecules in the plasma membrane are parallel to the surface of the cell and that the lipids are radially arranged. However, according to the investigators presenting the data, this cannot be considered an established fact because the evidence may be interpreted in other, though less likely, ways (Waugh and Schmitt, 1940; Schmitt and Palmer, 1940).

The plasma membrane as visualized by Danielli and Davson consists of two layers of lipid molecules arranged radially with their hydrocarbon chains toward each other, and with their respective polar groups arranged outwardly and inwardly, the entire double layer of lipid molecules being sandwiched between two layers of protein (see Figs. 11.6 and 11.10).

Electron micrographs of all cell membranes fixed in permanganate, which binds protein, consistently show the tripartite structure and thus would appear to support the Davson-Danielli "sandwich" model. The outer layers of protein of the membrane become electron opaque on fixation with permanganate, whereas the inner lipid "filling" does not. Some workers do not accept this evidence (Korn, 1966).

The layers and pores in the plasma membrane are considered to be dynamic, not static, and at times a lipid chain appears to make contact with the outside of the cell, at least in some spots, because lipases (e.g., lecithinase) lyse cell membranes, indicating that they are able to reach the lipids. Conversely, some investigators propose that in this dynamic membrane protein-lined "pores" periodically appear through the lipid layers, giving rise to a protein continuum from outer to inner protein layers in the cell membrane (Danielli, 1954). Such pores, however, if they occur, are below the limit of resolution of the electron microscope. Pores of small size, about 7 Å in diameter, have been postulated from experiments on penetration of water and salts into cells (Solomon, 1960). Larger molecules probably must dissolve into the structure of the membrane to gain entry.

Another hypothesis suggests that the protein of the plasma membrane may form a lattice of fibrous protein between the meshes of which the lipid component

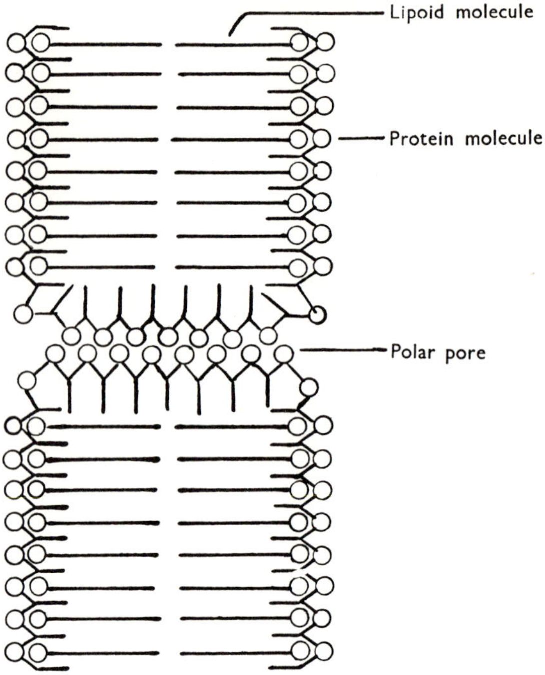


*Figure 11.10.* Hypothetical molecular structure of the plasma membrane of the cell. (From Danielli, 1954: Colston Papers *7*:1.)

is distributed (Hillier and Hoffman, 1953). For instance, electron micrographs show that erythrocyte ghosts from which the lipids had been extracted with solvents have a fibrous structure interrupted by holes which are assumed to be the places where the lipids had previously been located. Untreated ghosts show no such cavities. Therefore, since both polar and nonpolar groups are found in proteins, the plasma membrane may be visualized as a mosaic of patches in which polar groups of lipids are attached to polar groups of proteins, and other patches in which the nonpolar groups of proteins are attached to nonpolar groups of lipids. The first type of patch would permit water-soluble materials to enter the plasma membrane, and the second type would permit oil-soluble materials to enter the membrane.

Several other models of plasma membranes have been described (Ponder, 1954). For example, an alternate view is that of an all-lipid membrane with protein pores. However, studies of thin cross sections of plasma membranes by electron microscopy do not show a homogeneous membrane with radial heterogeneity such as might be expected of this structure, although the radii of protein might conceivably be below the limit of resolution. Still another view of the plasma membrane is a series of cylinders of lipid between layers of protein (Korn, 1966). However, the tripartite nature of the unit membrane seen in electron micrographs favors, although it does not prove, the validity of the Danielli-Davson model shown in Figure 11.10 (Picken, 1960; Robertson, 1960, 1964).

## THE NUCLEAR ENVELOPE

The *nuclear envelope* is of profound importance because during interphase the nucleus can pass information to the cytoplasm only through the nuclear envelope in controlling or in directing the synthetic activities occurring in the cytoplasm of the cell.

Early studies suggested that the nuclear envelope was a selectively permeable membrane, much like the plasma membrane of the cell (Churney, 1942).

Isolated envelopes from nuclei of amphibian cells show an outer layer of protein and lipid and an inner one containing a relatively insoluble protein. The protein in both appears to be elastin-like and capable of being stretched reversibly (Picken, 1960).

The nucleus accumulates labeled sodium ($^{22}Na$) (Allfrey *et al.*, 1961). Supporting this view were observations on changes in nuclear volumes of cells placed in hypertonic and hypotonic solutions. For instance, when marine eggs are placed in hypertonic solutions, the nucleus shrinks along with the general shrinkage of the cell. In hypotonic solutions, the nucleus swells as does the cytoplasm. It was later recognized that these experiments do not constitute adequate support for this view because a highly porous nuclear membrane would also permit the nucleus to swell or shrink when the cytoplasm does (Picken, 1960).

Meantime some experiments indicated that the nuclear envelope was more porous than the plasma membrane. For example, heparin and sulfated polymanuronic acid, added to a suspension of isolated nuclei from rat liver cells, displace highly polymerized DNA, a very large molecule, which then passes out of the nucleus through the nuclear membrane. Furthermore, nucleases are able to penetrate isolated nuclei of rat liver cells and to digest the nucleic acids in very short periods, indicating that these large molecules have entered the nuclei rapidly. Partially hydrolyzed gelatin and albumin, both of which are large molecules, appear to penetrate such nuclei as well (Anderson, 1953). Also, isolated nuclei of amphibian eggs immersed in a solution of hemoglobin accumulate hemoglobin (Picken, 1960).

However, it can be argued that such nuclei removed from the cytoplasm of cells are abnormal (Allfrey *et al.,* 1961). Thus in nuclear transfer experiments, great care must be taken to prevent contact of the nuclei with the medium, lest the nuclei be injured. Even a short stay of the nuclei in the medium prevents their normal functioning. The ready passage of macromolecules through the membranes of isolated nuclei cannot, therefore, be considered conclusive evidence that the normal nuclear envelope is equally porous.

However, experiments performed with nuclei still inside the cytoplasm avoid the criticism leveled against those performed on isolated nuclei. Such experiments have demonstrated that nuclear envelopes of some cells are more porous than plasma membranes and that they permit macromolecules to enter and leave the nuclei. Thus, in cases where ribonuclease and deoxyribonuclease are able to penetrate an intact cell, they also enter the nucleus and hydrolyze the nucleic acids in the nucleus. Changes in function and in the staining properties of the nuclei are then observed (Brachet, 1957). The appearance of fluorescent antibodies (proteins) in nuclei of embryos also indicates that the nuclear membrane has a greater porosity than the cell membrane (Picken, 1960). It is difficult to accumulate such evidence for all types of cells since the large molecules to be tested do not penetrate the plasma membrane except presumably by pinocytosis, a process which may not be general (see Chapter 14).

The most direct evidence on the nature of the nuclear envelope is from electron micrographs (Fig. 6.5). These clearly show two unit membranes similar to the unit plasma membranes. According to some workers the *nuclear membranes* might be considered modified extensions of the endoplasmic reticulum or the cisterna surrounding the nucleus (Watson, 1959). The outer nuclear membrane has pores 400 to 700 Å in diameter and these have been seen in cells of a variety of tissues to extend through the inner membrane as well. On the other hand, in some electron micrographs such pores are not seen as clearly in the inner membrane, and, considering the differences in techniques employed by different workers, the status of these pores in some cells must be considered uncertain until resolved by further study (Picken, 1960). The pores in the outer nuclear membrane often appear to be plugged with material. Electron micrographs do not permit one to tell how effective the plug may be against entry of small molecules, nor whether the plug remains tight at all times. Pinocytosis or evidence for it from cytoplasm to nucleus has not been described (Watson, 1959).

The possibility of another kind of communication between nucleus and cytoplasm has been suggested, namely, outpouching from chromosomes forming blebs on the nuclear membrane which separate into the cytoplasm as vesicles or laminae. These structures are covered by membranes derived from the nuclear envelope (Gay, 1960).

A potential of 13 mv. between nucleoplasm (negative) and cytoplasm (probably the ground substance) has been observed in gland cells of *Drosophila,* as well as a high resistance to passage of an electric current across the nuclear envelope. This could be a formidable barrier to ions even as small as $K^+$, $Na^+$ and $Cl^-$. The pores seen in electron micrographs of these cells are perhaps artifacts or they have become plugged with material equally resistant to passage of electricity as the plasma membrane. On the other hand, no potential and low resistance to passage of electric current were observed across the nuclear membrane of amphibian oocytes (Loewenstein and Kanno, 1963). The potential across the nuclear envelope of the midge *Chironomus* salivary gland cell showed change in response to injection of insect molting hormone ecdysone, suggesting changes in permeability (Loewenstein, 1965).

## CELL TO CELL CONTACTS

Cells in a tissue, though in contact with one another, are most often separated from one another by a distance equal to one or more unit membranes, the space between them being filled with some amorphous substance. At most contacts between nerve cells (synapses) a similar spacing is also found. However, at a few synapses (e.g., in the giant fibers of the crayfish mediating a rapid tail flip and in the Mauthner cells of fishes mediating a rapid escape response) there are found *tight junctions* or closed contacts where the two plasma membranes touch one another. Tight junctions may occur between smooth muscle fibers and in amphibian epithelium (Farquhar and Palade, 1960). Also, when motile connective tissue cells (fibroblasts) touch each other and show contact inhibition, such closed contacts are developed. Some neuroglia cells (supporting cells of the central nervous system) may also form tight junctions, especially those in the blood-brain barrier (Gray, 1964). Cell to cell contacts may cause low electrical resistance coupling of the cells, permitting salts and perhaps other substances to move freely from cell to cell, as in the case of *Drosophila* salivary gland cells (Kanno and Loewenstein, 1964). Yet such cells have a typical high electrical resistance across the surface exposed to the bathing medium.

Cells also may have specialized contacts, *desmosomes,* where the unit membranes are separated by the usual distance but the innermost layer of each of the apposed plasma membranes is thickened by amorphous material, subjacent to which is a layer of fibrils (Fig. 11.11). The fibrils seem to anchor the desmosome to the cell. Desmosomes occurring on adjacent cells are always present in pairs. However, when a cell borders on a basement membrane (secreted by the cells) of a tissue, a single unit of the desmosome may be present (Kelly, 1966). Desmosomes are common in epithelial cells but they may be present in cells of the nervous system as well, and perhaps elsewhere.

Cells are sometimes joined by bridges of cytoplasm passing between pores in the plasma membranes of the adjacent cells. Such junctions are called *plasmodesmata* and they are found only in some plant cells (Picken, 1960).

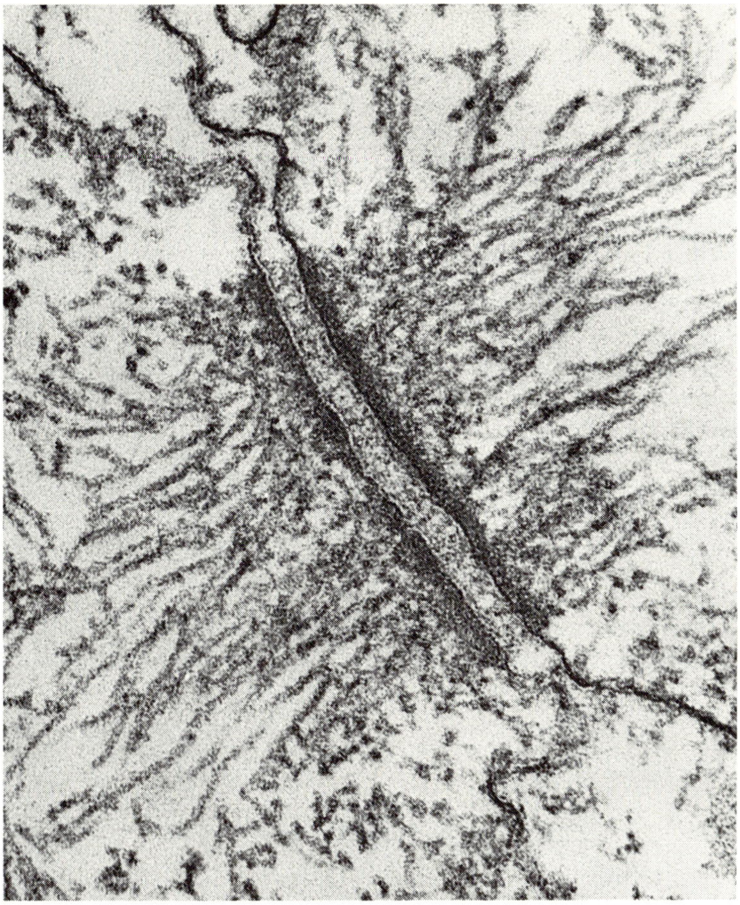

***Figure 11.11.*** A desmosome from a larval *Triturus torosa* epidermis fixed in s-collidine buffered osmium tetroxide to which the dye ruthenium red was added. The section was subsequently stained with lead and uranyl acetate. (×125,000) Note the thickening of the unit cell membranes of the two apposed cells, the amorphous material between and the fibrils leading from the thickening into the cytoplasm of each cell. (From Kelly, 1966: J. Cell Biol. *28*:51–72.)

## COMPARTMENTALIZATION OF THE CELL BY MEMBRANES

As already discussed, membranes are present on many of the cell organelles—double-unit membranes on the mitochondria, single-unit membranes on the endoplasmic reticulum, Golgi complex and lysosomes, and a nuclear envelope consisting of two unit membranes. The membranes serve to compartmentalize the cell, each compartment presumably being able to regulate its contents by virtue of the selective permeability of its membrane. As seen in Chapter 5, mitochondria are able to concentrate magnesium, potassium and phosphate against a concentration gradient at the expense of metabolic energy (Rottenberg and Solomon, 1965; Brierley, 1963; Greenawalt *et al.,* 1964). The endoplasmic (sarcoplasmic) reticulum of muscle cells concentrates calcium ions against a concentration gradient at the expense of metabolic energy (Hasselbach, 1965). The Golgi complex also displays a capacity for such selective activity. The lysosomes contain a ready supply of hydrolytic enzymes but the membrane confines them and prevents their action on the cytoplasm until released. The membranes about these organelles thus serve as "buffers" protecting the reactions going on within them from the changes which may be occurring in the ground substance of the cytoplasm. The cytoplasm is in turn itself buffered from the changes occurring in the bathing fluid by the plasma membrane.

The nucleus within a nuclear envelope may be similarly protected from changes occurring in the cytoplasm. However, it has been argued that even large molecules have access to the nucleus because of the large pores seen in the nuclear envelope (Fig. 6.5). The maintenance of a potential difference between nucleus and cytoplasm has already been mentioned. When found, such a potential indicates that there is ionic selectivity; therefore, there should also be selectivity for large molecules. Since plugs have been seen in pores of the nuclear envelope (Branton, 1964), it is possible that they serve to stopper the pores, at least at times. This subject is of great interest and needs further study.

## LITERATURE CITED AND GENERAL REFERENCES

Albersheim, P., 1965: The substructure and function of the cell wall. *In* Plant Biochemistry. Bonner and Varner, eds. Academic Press, New York, pp. 151–186.

Allfrey, V. G., Meudt, R., Hopkins, J. W., and Mirsky, A. E., 1961: Sodium-dependent transport reactions and their role in protein and nucleic acid synthesis. Proc. Natl. Acad. Sci. U.S.A. *47*:907–932.

Anderson, N. G., 1953: On the nuclear envelope. Science *117*:517–521.

Brachet, J., 1957: Biochemical Cytology. Academic Press, New York.

Braddiley, J., 1964: Teichoic acids and the bacterial cell wall. Endeavour *23*:33–37.

Brandt, P. W., 1962: A consideration of the extraneous coats of the plasma membrane. Circulation *26*:1075–1091.

Branton, D., 1964: Fine structure in freeze-etched *Allium cepa*. J. Ultrastruc. Res. *11*:401–411.

Branton, D., and Park, R. B. (eds.), 1968: Papers on Biological Membrane Structure. Little, Brown & Co., Boston.

Brierley, G. P., Murer, E., Bachman, E., Green, D. E., and Hadley, D. G., 1963: Ion transport. II. The accumulation of inorganic phosphate and magnesium by heart mitochondria. J. Biol. Chem. *238*: 3482–3489.

Brown, F., and Danielli, J. F., 1964a: The cell surface and cell physiology. *In* Cytology and Cell Physiology. Bourne, ed. Academic Press, New York.

Brown, F., and Danielli, J. F., 1964b: Plasma membrane as essential to life as nucleus. *In* Cytology and Cell Physiology. Bourne, ed. Academic Press, New York.

Cantarow, A., and Schepartz, B., 1967: Biochemistry. 4th Ed. W. B. Saunders Co., Philadelphia.

Carpenter, P. L., 1967: Microbiology. 2nd Ed. W. B. Saunders Co., Philadelphia.

Chambers, R., 1940: The relation of extraneous coats to the organization and permeability of cellular membranes. Cold Spring Harbor Symp. Quant. Biol. *8*:144–153.

Chambers, R., 1949: Micrurgical studies on protoplasm. Biol. Rev. *24*:246–265.

Chambers, R., and Chambers, E. L., 1961: Explorations into the Nature of the Living Cell. Harvard University Press, Cambridge, Mass.

Churney, L., 1942: The osmotic properties of the nucleus. Biol. Bull. *82*:52–67.

Danielli, J. F., 1952: Structural factors in cell permeability and secretion. Symp. Soc. Exp. Biol. *6*:1–15.

Danielli, J. F., 1954: Morphological and molecular aspects of active transport. Symp. Soc. Exp. Biol. *8*:502–516.

Davson, H., 1962: Growth of the concept of the paucimolecular membrane. Circulation *26*:1022–1037.

Davson, H., and Danielli, J. F., 1952: Permeability of Natural Membranes. 2nd Ed. Cambridge University Press, London.

Eglinton, G., and Hamilton, R. J., 1967: Leaf epicuticular waxes. Science *156*:1322–1335.

Farquhar, M. G., and Palade, G. E., 1965: Cell junctions in amphibian skin. J. Cell Biol. *26*:263–291.

Fawcett, D. W., 1962: Physiologically significant specializations of the cell surface. Circulation *26*:1105–1132.

Finean, J. B., 1966: The molecular organization of cell membranes. Progr. Biophys. *16*:143–170.

Gates, G. G., 1964: Physiological aspects of membrane function in plant cells during development. *In* Cellular Membranes in Development. Locke, ed. Academic Press, New York, pp. 299–320.

Gay, H., Jan. 1960: Nuclear control of the cell. Sci. Am. *202*:126–136.

Gray, E. G., 1964: Electron microscopy of the cell surface. Endeavour *23*:61–65.

Greenawalt, J. W., Rossi, C. S., and Lehninger, A. L., 1964: Effect of active accumulation of calcium phosphate on the structure of rat liver mitochondria. J. Cell Biol. *23*:21–38.

Hasselbach, W., 1965: Relaxing factor and the relaxation of muscle. Progr. Biophys. *14*:167–222.

Hechter, P., 1965: Role of water structure in the molecular organization of cell membrane. Fed. Proc. *24*:S91–S102.

Heilbrunn, L. V., 1930: The surface precipitation reaction of living cells. Am. Phil. Soc. Proc. *69*: 295–301.

Hillier, J., and Hoffman, J. F., 1953: On the ultrastructure of the plasma membrane as determined by the electron microscope. J. Cell. Comp. Physiol. *42*:203–247.

Jacobs, M. H., 1962: Early osmotic history of the plasma membrane. Circulation *26*:1013–1021.

Kamat, V. B., and Hoelzl Wallach, D. F., 1965: Separation and partial purification of plasma-membrane fragments from Ehrlich ascites carcinoma microsomes. Science *148*:343-2345.

Kanno, Y., and Loewenstein, W. R., 1964: Low resistance coupling between gland cells. Some observations on intercellular contact membranes and intercellular space. Nature *201*:194-195.

Kavanau, J. L., 1964: Structure and Function in Biological Membranes. Holden-Day, San Francisco.

Kelly, D. E., 1966: Fine structure of desmosomes, hemidesmosomes and an adepidermal globular layer in developing newt epidermis. J. Cell Biol. *28*:51-72.

Kennedy, J. R., Jr., 1967: The effect of strychnine and light on pigmentation in *Blepharisma undulans* Stein. J. Cell Biol. *28*:145-153.

Kopac, M. J., 1940: The physical properties of the extraneous coats of living cells. Cold Spring Harbor Symp. Quant. Biol. *8*:154-170.

Korn, E. D., 1966: Structure of biological membranes. Science *153*:1491-1498.

Lansing, A. L., and Rosenthal, T. B., 1952: The relationship between RNA and ionic transport across the cell surface. J. Cell. Comp. Physiol. *40*:337-345.

Lehninger, A. L., 1965: The Mitochondrion. W. A. Benjamin, New York.

Loewenstein, W. R., 1965: Permeability of a nuclear membrane: changes during normal development and changes induced by growth hormone. Science *150*:909-910.

Loewenstein, W. R., and Kanno, V., 1963: The electrical conductance and potential across the membrane of some cell nuclei. J. Cell Biol. *16*:421-425.

Maddy, A. H., 1966: The chemical organization of the plasma membrane of animal cells. Int. Rev. Cytol. *20*:1-65.

McAlear, J. H., and Edwards, G. A., 1958: Continuity of plasma membrane and nuclear membrane. Exp. Cell Res. *16*:689-692.

Mitchison, J. M., 1953: A polarized light analysis of the human red cell ghost. J. Exp. Biol. *30*:397-432.

Mühlethaler, K., 1961: Plant cell walls. *In* The Cell. Brachet and Mirsky, eds. Academic Press, New York, Vol. 2, pp. 85-134.

Nadler, J. E., 1929: Notes on the loss and regeneration of the pellicle in *Blepharisma undulans.* Biol. Bull. *56*:327-330.

Nickerson, W. J., 1964: Composition and function of microbial cell walls. *In* Cellular Membranes in Development. Locke, ed. Academic Press, New York, pp. 281-297.

Northcote, D. H., 1963: The nature of plant cell surfaces. Biochem. Soc. Symp. *22*:105-125.

O'Neill, C. H., 1964a: Isolation and properties of the cell surface membrane of *Amoeba.* Exp. Cell Res. *35*:477-496.

O'Neill, C. H., 1964b: Preparation and properties of cell surface membranes. *In* Recent Progress in Surface Science. Danielli *et al.,* eds. Academic Press, New York, pp. 427-442.

Onishi, T., and Onishi, T., 1962: Extraction of actin- and myosin-like proteins from liver mitochondria. J. Biochem. (Japan) *52*:230-231.

Overton, E., 1902: Beitrage zur allgemeinen Muskel- und Nervenphysiologie. Arch. ges. Physiol. *92*: 115-280.

Pappenheimer, J. R., 1953: Passage of molecules through capillary walls. Physiol. Rev. *33*:387-423.

Parpart, A. K., and Ballantine, R., 1952: Molecular anatomy of the red cell plasma membrane. Modern Trends in Physiology and Biochemistry. Barron, ed. Academic Press, New York, pp. 135-148.

Picken, L., 1960: The Organization of Cells. Clarendon Press, Oxford.

Pitelka, D. R., 1963: Electron-microscopic Structure of Protozoa. Pergamon Press, Long Island City, N. Y.

Ponder, E., 1948: Hemolysis and Related Problems. Grune & Stratton, New York.

Ponder, E., 1952: Various types of ghosts derived from human red cells: heat fragmentation and phase optic studies. J. Exp. Biol. *29*:605-619.

Ponder, E., 1953: The kinetics of progressive reactions in systems containing saponin, digitonin and sodium taurocholate. J. Gen. Physiol. *36*:723-732.

Ponder, E., 1954: Present concepts of the structure of the mammalian red cell. Blood *9*:227-235.

Preston, R. D., 1959: Wall organization in plant cells. Int. Rev. Cyt. *8*:33-60.

Robertson, J. D., 1960: The molecular structure and contact relationships of cell membranes. Progr. Biophys. *10*:343-418.

Robertson, J. D., Apr. 1962: The membrane of the living cell. Sci. Am. *206*:65-82.

Robertson, J. D., 1964: Unit membranes: a review with recent new studies of experimental alterations and a new subunit structure in synaptic membranes. *In* Cellular Membranes in Development. Locke, ed. Academic Press, New York, pp. 1-81.

Robinow, C. F., 1960: Outline of the visible organization of bacteria. *In* The Cell. Brachet and Mirsky, eds. Academic Press, New York, Vol. 4, pp. 45-108.

Roche, P. (ed.), 1960: Physiology of the cell membrane. J. Gen. Physiol. *43*(Suppl.):1-189.

Rogers, H. J., 1963: The surface structure of bacteria. Biochem. Soc. Symp. *22*:55-100.

Rothfield, L., Takeshita, M., Pearlman, M., and Hovac, R. W., 1966: Role of phospholipids in the enzymatic synthesis of the bacterial cell envelope. Fed. Proc. *25*:1495-1502.

Rottenberg, H., and Solomon, A. K., 1965: Energy linked $K^+$-uptake in mitochondria. Biochem. Biophys. Res. Commun. *20*:85-92.

Ruesink, A. W., and Thimann, K. V., 1966: Protoplasts: preparation from higher plants. Science *154*:280-281.

Runnstrom, J., 1952: The cell surface in relation to fertilization. Symp. Soc. Exp. Biol. *6*:39-88.

Schmitt, F. O., and Palmer, K. J., 1940: X-ray diffraction studies of lipide and lipide protein systems. Cold Spring Harbor Symp. Quant. Biol. *8*:94-101.

Scott, G. H., 1933: The localization of mineral salts in cells of some mammalian tissues by microincineration. Am. J. Anat. *53*:243-287.

Sjöstrand, F. S., 1963: A comparison of plasma membrane, cytomembrane and mitochondiral membrane elements with respect to ultrastructural features. J. Ultrastruc. Res. *9*:561–580.

Smith, H. W., 1962: The plasma membrane, with notes on the history of botany. Circulation *26*: 987–1012.

Solomon, A. K., Dec. 1960: Pores in the cell membrane. Sci. Am. *203*:146–156.

Symposium on the plasma membrane. 1962. Circulation *26*:983–1227.

Teorell, T., 1952: Permeability properties of erythrocyte ghosts. J. Gen. Physiol. *35*:669–701.

Thompson, T. E., 1964: The properties of biomolecular phospholipid membranes. *In* Cellular Membranes in Development. Locke, ed. Academic Press, New York, pp. 83–96.

Watson, M. L., 1959: Further observations on the nuclear envelope of the animal cell. J. Biophys. Biochem. Cytol. *6*:147–156.

Waugh, D. F., 1950: The ultrastructure of the envelope of mammalian erythrocytes. Ann. N. Y. Acad. Sci. *50*:835–853.

Waugh, D. F., and Schmitt, F. O., 1940: Investigations of the thickness and ultrastructure of cellular membrane by the analytical leptoscope. Cold Spring Harbor Symp. Quant. Biol. *8*:233–241.

Weiss, L., 1960: The adhesions of cells. Internat. Rev. Cyt. *9*:187–225.

Whittam, R., 1964: Transport and diffusion in red blood cells. Williams & Wilkins, Baltimore.

Yamamoto, T., 1963: On the thickness of the unit membrane. J. Cell Biol. *17*:413–421.

# CHAPTER 12 / MOVEMENT OF WATER ACROSS CELL MEMBRANES: OSMOSIS

Water moves in and out through the plasma membrane at all times, but no effect is demonstrable if the solute concentration is the same on both sides of the membrane. It is when the concentrations of solute are different on the two sides of the plasma membrane that movement becomes apparent by shrinkage or swelling of the cell. Water moves in or out of the cell until the concentration of solute is equalized between the inside and the outside of the cell. *Osmosis* is the diffusion of water (or more generally, a solvent) through a membrane in response to a concentration gradient.

The principles governing osmotic transfers have been studied with artificial, natural and living cell membranes.

## PASSAGE OF WATER ACROSS ARTIFICIAL MEMBRANES

Many artificial membranes have been devised as analogues to the plasma membrane to facilitate study of osmotic transfers. A membrane of copper ferrocyanide, formed by the interaction of potassium ferrocyanide and copper sulfate, has been a favorite artificial membrane since it was first introduced by Traube in 1867. When a crystal of copper sulfate in a piece of gauze is suspended in a test tube filled with a solution of potassium ferrocyanide, the copper sulfate dissolves and interacts with the ferrocyanide, forming a membrane. Water enters this membrane causing the bag or "cell" to swell and burst. A new membrane precipitates in a new position as more of the copper sulfate is dissolved and additional water is taken up, and the process repeats itself until the copper sulfate is completely dissolved.

A similar membrane for experimental study was devised by Pfeffer in 1877. Potassium ferrocyanide solution is placed inside an unglazed pottery jar surrounded by a solution of copper sulfate. An electric current is then passed between the two solutions. The membrane of copper ferrocyanide which is precipitated in the small pores of the unglazed jar allows only water to pass through it (see Fig. 12.1). If the jar is then filled with 1 molal NaCl and is suspended in a beaker of distilled water, water molecules pass through the membrane but salts cannot, and since the water molecules cannot get out as readily as they enter, because they collide with sodium and chloride ions inside the jar, water accumulates inside the jar. If a tube is fitted to the mouth of the jar, water rises in the tube until the pressure developed by the rise of water equals the *osmotic pressure*

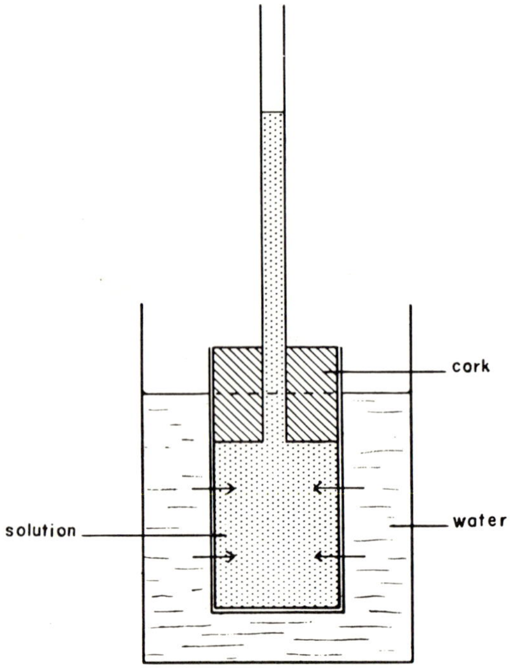


***Figure 12.1.*** A simple osmometer. The pores in the jar have been coated with a precipitate of copper ferrocyanide, which is permeable to water but not to solutes. Solute is dissolved in the jar, and after the cork with the tube is inserted, water enters faster than it leaves the solution; consequently, the level rises in the tube and indicates the level of osmotic pressure.

(the tendency of the water molecules to enter the jar in response to the concentration gradient between the inside and the outside of the jar). Such a membrane, which is permeable to solvent only—in this case, water—is said to be *semipermeable.* In aqueous systems a semipermeable membrane implies a membrane permeable to water only.

A better way of measuring the osmotic pressure than by the height of water in a tube is by means of the mercury osmometer introduced by Pfeffer and illustrated in Figure 12.2. An *osmometer* is a device used to measure osmotic pressure. It must have a semipermeable membrane which allows water (solvent) but not the solute to pass.

The physical chemist employs still other devices to determine osmotic pressure. One such method is measuring the pressure that just prevents the entry of water (solvent) through the semipermeable membrane; this is taken as a measure of the osmotic pressure. Another method is the Hepp osmometer in which very small quantities of material can be used and a direct reading of the osmotic pressure may be made. Or, alternatively, the osmotic pressure registered on a diaphragm may be transduced to an electrical change readily measured and recorded (Peters and Saslow, 1939; Rehm, 1944; Mauro, 1965).

An artificial membrane often used in biological studies is cellophane dialysis tubing. If such a membrane is fastened over a thistle tube and the tube is filled with concentrated sodium chloride solution and immersed in distilled water, water will enter and the volume of solution in the thistle tube will increase as a consequence, indicating that cellophane is permeable to water. Tests for chloride in the distilled water outside, however, show that sodium chloride is also coming through the cellophane. But the rise in the level of the solution in the tube proves that entry of water is greater than outward movement of salt. If glucose is used instead of sodium chloride similar results are obtained, except that the outward movement of glucose is slower than that of salt. To show that passage of solute is not the result of a mechanical leak, Congo red, a colloidal dye, is added to the solution in the thistle tube during such an experiment. It will be recalled that colloids do not pass through cellophane (Chapter 4); therefore, the dye will be retained by an intact membrane, but the appearance of a pink color in the water surrounding the tube will indicate a tear in the membrane. A membrane which lets one material (water) pass through more readily than another (salt or sugar) is said to be *selectively permeable.*

Natural membranes, such as goldbeater's skin (outside membrane of the ox intestine) and "skin" (epidermis) from a scale of an onion bulb, can be used to study movement of water in the same manner as artificial membranes. Since the results for natural membranes are simi-

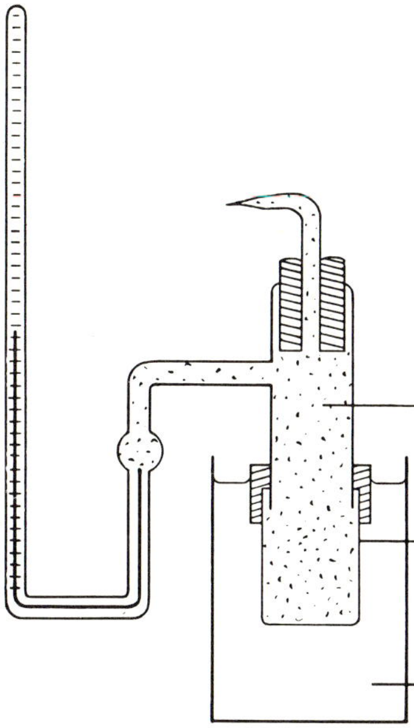


***Figure 12.2.*** Pfeffer's osmometer with mercury manometer to measure the osmotic pressure. The test solution is kept inside the jar, in the walls of which has been precipitated a copper ferrocyanide membrane. Entry of water occurs until the osmotic pressure is counterbalanced by the pressure of the column of mercury.

lar to those for cellophane, further discussion is omitted here.

## PERMEABILITY OF THE PLASMA MEMBRANE TO WATER

Cells act as osmometers to the extent that they act as systems attaining equilibrium for water. Experiments indicate that cells are generally highly permeable to water. For example, when a sea urchin egg is placed in sea water diluted with distilled water it swells. The swelling increases with greater dilution of the sea water. Diluted sea water is said to be *hypotonic* to the sea urchin egg, which is accustomed to undiluted sea water. Swelling continues with dilution until the membrane bursts and cytolysis occurs. When the egg is placed in sea water concentrated by evaporation, it shrinks, and the shrinkage increases as the medium becomes more concentrated. Such a medium is said to be *hypertonic* to the sea urchin egg. The egg in unaltered sea water does not swell or shrink, and the sea water is said to be *isotonic* (Lucké, 1940).

Experiments with red blood cells (erythrocytes) indicate that they also are permeable to water. If the erythrocytes are placed in a hypertonic solution they shrink in a characteristic way and are said to be crenated. When placed in a hypotonic salt solution they swell and may become spherical instead of discoidal. Erythrocytes become hemolyzed and lose their hemoglobin when placed in excessively dilute salt solutions (Ponder, 1948, 1955). A concentration of sodium chloride of approximately 0.7 per cent is isotonic to a red blood cell of a frog, and a concentration of 0.9 per cent is isotonic to mammalian erythrocytes. Many of the problems of water permeability in animal cells are discussed by Dick (1959, 1966).

Plant cells, too, are quite permeable to water. The cells of a fresh water plant like *Elodea* are normally turgid; that is, the protoplast is held firmly against the cell wall, and the entire leaf has a certain rigidity as a result of the collective rigidity of the individual cells. If the plant cells are placed in a concentrated salt solution, they become plasmolyzed (see Fig. 12.3), and the protoplast of each cell contracts about the shrunken vacuole. If the hypertonic salt solution is replaced by fresh water, each cell regains water and the protoplast

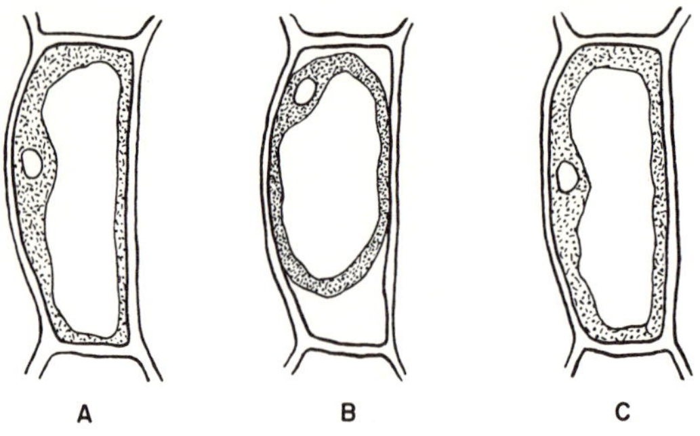

***Figure 12.3.*** A plant cell (*A*) undergoes plasmolysis (*B*) when it is placed in a hypertonic solution. Deplasmolysis (*C*) occurs after a period of time, but only if the solute can penetrate. Time for deplasmolysis depends upon the rate of entry of the solute.

again fills the space within the cell wall. When *Elodea* is placed in a series of solutions containing increasing concentrations of salt, plasmolysis occurs in its cells when the concentration of a salt, such as potassium nitrate, reaches about 0.6 per cent. Less salt than this is present in the fresh water of ponds and streams in which *Elodea* is found. The rigid cell wall, limiting the volume, prevents excessive uptake of water. However, if the cell wall is first digested by cellulase then the naked protoplast will swell with uptake of water (Ruesink and Thimann, 1966).

Cells plasmolyzed by hypertonic solutions return in time to their initial volume (deplasmolysis). Deplasmolysis is rapid in solutions of solutes which penetrate rapidly, and it is slow in solutions of solutes which penetrate slowly. For example, the initial shrinkage of an egg or of *Elodea* cells, when in hypertonic solutions, indicates that water leaves the cell more readily than the solute molecules enter, but subsequent swelling shows that the solute molecules have penetrated the cells. Solute molecules accumulate in the cell and water reenters, causing the cell to swell. The membrane of a cell is therefore not semipermeable but selectively permeable; that is, it is more permeable to water than to other substances. This is found to be true of all living cells studied. (For permeability to solutes, see Chapter 13.) It would appear that the plasma membrane is freely permeable to water and permits water to enter when the cell is placed in hypotonic solutions and permits water to leave when the cell is placed in hypertonic solutions (Dick, 1966).

Bacteria, like plant cells, are also held within cell walls and do not appear to swell in hypotonic solutions. However, if the cell wall of the bacteria is removed by lysozyme (Acker and Hartsell, 1960), the protoplast swells and shrinks like any other cell.

The cell organelles, as we have seen, are covered by unit membranes, which appear to be selectively permeable like the plasma membrane. Thus mitochondria shrink or swell in response to the concentration of solute in the bathing media. Less is known of the membranes of the endoplasmic reticulum, Golgi complex and other organelles because they are less susceptible to experimental study in this regard. However, it seems likely that these membranes are also freely permeable to water. Their capabilities in concentration of certain ions at the expense of metabolic energy is dealt with elsewhere (Chapters 14, 16 and 17).

## OSMOTIC PRESSURE OF NONELECTROLYTES

As already discussed, when a solution of a nonelectrolyte in a thistle tube is separated from distilled water by a semipermeable membrane, water will pass through the membrane into the solution causing an increase in volume and a rise of solution in the tube. The pressure exerted by the solution in the tube is proportional to the height (h) of the column and is called the osmotic pressure ($\pi$). Water will continue to increase the volume of solution until the pressure of the column on the membrane just prevents the entry of water into the membrane. At that time equilibrium will have been established, and

$$\pi = h \times k \qquad (12.1)$$

where k is a constant of proportionality for a given temperature and concentration. The higher the concentration of solute, the higher the rise of the water will be.

The plant physiologist, Pfeffer (1877), made measurements of osmotic pressure of various solutes by using a copper ferrocyanide osmometer and found that the osmotic pressure divided by the concentration of solute is constant (see Table 12.1), but he did not realize the full significance of his data. The physical chemist, van't Hoff, showed that Pfeffer's data could be taken as evidence of the applicability of the gas laws to osmotic pressure. A gram molecular weight of a gas at atmospheric pressure occupies 22.4 liters of space. If it is compressed into the confines of a liter it exerts 22.4 atmospheres pressure. If a solute is compared to a gas, then a gram molecular weight of the solute in a liter of solution, if it acts like a gas, should exert a pressure of 22.4 atmospheres. When analyzed, Pfeffer's data showed that a gram molecular weight of a nonelectrolyte dissolved in a liter of water exerts an osmotic pressure of approximately 22.4 atmospheres, confirming van't Hoff's hypothesis. The agreement is only approximate, since exact agreement would occur only if the molecules of solute occupied no volume and the attractions between them were nil.

**Table 12.1.** Pfeffer's Data for Osmotic Pressure at Different Concentrations of Sucrose

| CONCENTRATION (C) IN PER CENT | OSMOTIC PRESSURE ($\pi$) IN ATMOSPHERES | $\frac{\pi}{C}$ |
|---|---|---|
| 1 | 0.70 | 0.7 |
| 2 | 1.34 | 0.67 |
| 4 | 2.74 | 0.68 |
| 6 | 4.10 | 0.68 |

To determine whether the rise in osmotic pressure is proportional to the concentration of the solution, Pfeffer divided the osmotic pressure ($\pi$) by the concentration (C) of the solution being tested. As seen in the third column, $\pi$/C is fairly constant. (After Glasstone, 1960.)

The gas law may be written PV = nRT, where P is the pressure in atmospheres, V the volume in liters, n the number of moles, R the gas constant (equal to 0.082 liter atmospheres per degree per mole) and T the absolute temperature. By rearrangement,

$$P = \frac{n}{V} \times RT \qquad (12.2)$$

Since the number of moles of solute divided by the number of liters of solvent in which it is dissolved is equal to the molal concentration (moles in a kilogram of solvent), C may be substituted for n/V. Then,

$$P = CRT \qquad (12.3)$$

For osmotic pressure this is usually written:

$$\pi = CRT \qquad (12.4)$$

in which $\pi$ denotes the osmotic pressure. This relationship can be derived from basic thermodynamic principles (see Dick, 1966).

For a gram molecular weight of osmotically active ideal nonelectrolyte, C is 1, R is 0.082 liter atmospheres per degree per mole, and the absolute temperature at 0° C. is 273°. Then

$$\pi = (1 \text{ mole/liter}) \cdot (0.082 \text{ liter atm./degree/mole}) \cdot (273°) \qquad (12.5)$$

$$\text{or } \pi = 22.4 \text{ atmospheres}$$

Since the gas laws were developed for perfect gases whose molecules have no volume and do not attract one another, their application to solutions is even more of an approximation than their application to gases. The gas law can be applied only to weak solutions at best. Deviations become marked at higher concentrations (Glasstone, 1960; Dick, 1966).

## OSMOTIC PRESSURE OF ELECTROLYTES

Since electrolytes dissociate, each molecule gives rise to a number of ions. Because the osmotic pressure is determined by the number of particles present in the solution, for a given *molal** concentration an electrolyte will exert a greater osmotic pressure than a nonelectrolyte. The magnitude of the difference depends on the degree of dissociation and the number of ions produced.

Therefore, the osmotic pressure of an electrolyte, calculated by the gas law, must be multiplied by a factor correcting for dissociation. The correction factor is less than the total number of ions formed from a molecule of an electrolyte, because the activity of the ions, rather than their concentration, determines the osmotic pressure. Ionic interference reduces the effective concentration of the ions; e.g., unlike ions attract one another, reducing the number of ions free to act. The activity of a given electrolyte varies at different concentrations, being generally greater at low concentrations in which interference between ions is least.

*The physical chemist uses molal concentrations in dealing with solutions. A *molal* (m) solution contains 1 gram molecular weight of a solute in 1000 grams of solvent, whereas a *molar* (M) solution is a liter of solution containing 1 gram molecular weight of a solute. Use of molal solutions has the advantage that molal solutions of different substances contain the same ratio of solute to solvent molecules. At very low concentrations, molar becomes equivalent to molal, since the effect of molecular volume of the solute becomes negligible.

The determination of the activity for each concentration of an electrolyte is most easily done by cryoscopy (freezing point determination), which measures the number of particles active, not just their concentration. Then, by comparing the freezing point depression of an electrolyte with that of an ideal nonelectrolyte of the same concentration (usually 1 molal of each is used) a factor called the *cryoscopic coefficient,* G, may be obtained:

$$G = \frac{\Delta T_{fp}}{\Delta T_{\overline{fp}}} \qquad (12.6)$$

where $\Delta T_{fp}$ is the freezing point depression produced by a given molal concentration of the electrolyte in question and $\Delta T_{\overline{fp}}$ is the lowering of the freezing point by the same molal concentration of an ideal nonelectrolyte. A few cryoscopic coefficients are given in Table 12.2. The osmotic pressure of an electrolyte can be calculated from the gas law equation (12.4), but the concentration term in the gas law equation must be multiplied by the cryoscopic coefficient, G, for each concentration of electrolyte

$$\pi = (C \times G)RT \qquad (12.7)$$

since G measures the relative increase of number of particles by ionization and includes an activity correction (Glasstone, 1960).

**Table 12.2.** Cryoscopic Coefficients (G)* at Various Molal Concentrations

| ELECTROLYTE | 0.02 | 0.05 | 0.1 | 0.2 | 0.5 |
|---|---|---|---|---|---|
| $MgCl_2$ | 2.708† | 2.677 | 2.658 | 2.679 | 2.896 |
| $MgSO_4$ | 1.393† | 1.302 | 1.212 | 1.125 | — |
| $CaCl_2$ | 2.673† | 2.630 | 2.601 | 2.573 | 2.680 |
| LiCl | 1.928 | 1.912 | 1.895 | 1.884 | 1.927 |
| NaCl | 1.921 | — | 1.872 | 1.843 | — |
| KCl | 1.919 | 1.885 | 1.857 | 1.827 | 1.784 |
| $KNO_3$ | 1.904 | 1.847 | 1.784 | 1.698 | 1.551 |

Data from Heilbrunn, 1952: An Outline of General Physiology. 3rd Ed. W. B. Saunders Co., Philadelphia.

* $G = \frac{T_{fp}}{T_{\overline{fp}}}$, as defined in equation 12.6.

† 0.025 molal.

## COLLIGATIVE PROPERTIES AND THE DETERMINATION OF OSMOTIC PRESSURE

The methods already discussed for the direct measurement of osmotic pressure are not very suitable for the measurement of the osmotic pressure of cell contents because the materials are available only in relatively small quantities. Therefore, use has to be made of some properties of solutions that are functions of the same variable, namely, the *number* of particles in solution and not the *kind* of particles.

Those properties of a solution that depend upon the number of particles present per unit volume of solution are called its *colligative properties,* and they are a manifestation of a phenomenon common to all of them; i.e., solute molecules decrease the tendency of water molecules to move from one solution to another or from one phase to another. The *colligative properties* of solutions are (1) the osmotic pressure, (2) the freezing point depression, (3) the boiling point elevation and (4) the vapor pressure.

The dependence of osmotic pressure upon the number of active particles has already been discussed. The greater the number of particles, assuming that they are free to move, the greater will be their interference with movement of water molecules from a solution through a membrane. This interference with movement of water molecules also affects the freezing point, boiling point and vapor pressure in a like manner. The freezing point is lowered because solute molecules interfere with the formation of ice crystals. The boiling point is raised because the solute particles in solution interfere with the establishment of an equilibrium between the water molecules and the atmospheric pressure. Similarly, the vapor pressure, which depends upon the evaporation of water from the solution, is decreased by the interference of the solute particles with the escape of water molecules, and the decrease is proportional to the concentration of solute. It is self-evident, therefore, that if one of the colligative properties is known, any of the others can be calculated, since they are all related through the same variable.

The vapor pressure determination is perhaps the simplest, as seen from the following example. When two beakers containing aqueous solutions of different concentrations of solute are placed in an airtight enclosure (see Fig. 12.4), the water molecules leave and reenter both solutions continually, but because exit of water molecules is less impeded in the dilute solution, more water will leave it than leaves the more concentrated solution. Thus, the dilute solution loses more water than it receives and the concentrated solution receives more water than it loses until the two become equal in concentration. Since the temperature falls in the solution losing water, and since the change in temperature is proportional to the difference in concentrations of the two solutions, the vapor pressure can be calculated from the change in temperature at equilibrium.

Using this principle, an elegant method for determining osmotic pressures of very small samples of fluid has been developed. This method employs thermocouples to measure the temperature, because thermometers are not sensitive enough for such small samples of fluid. A thermocouple consists of a junction of two metals in which a change in temperature in-

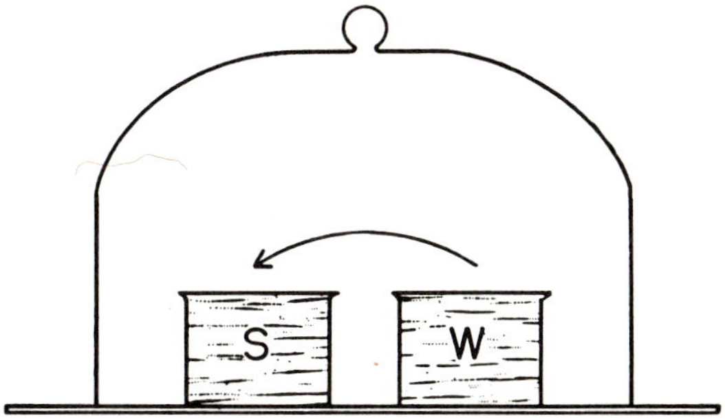


***Figure 12.4.*** Water distils over into a solution because of a difference in vapor tension between pure water (W) and the solution (S). The solute molecules reduce the escaping tendency of the water molecules in the solution. The temperature of distilled water is lowered by evaporation from its surface. This lowering of temperature can be measured by a thermocouple (see Fig. 12.5).

duces an electromotive force creating a current measurable with a galvanometer (see Fig. 12.5). First the junctions are calibrated when one is placed in a solution of known concentration (and osmotic pressure) and the other in distilled water. A series of solutions of known osmotic pressures are tested in this manner, and a graph is constructed that relates galvanometric deflection to osmotic pressure. The galvanometric deflection produced by a sample of unknown osmotic pressure

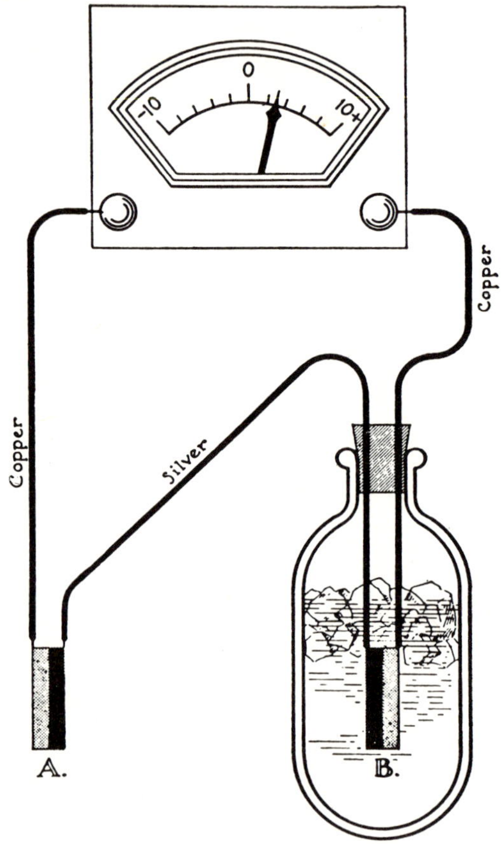


*Figure 12.5.* An elementary form of thermocouple for measuring change in temperature. Junction B is always kept at some convenient known temperature, e.g., on ice. When junction A is placed on a substance of different temperature, the thermoelectric current produced deflects the galvanometer. Galvanometric deflection is calibrated with A at several known temperatures, and a calibration curve is constructed to relate the deflection to the temperaature at A. Then junction A can be used to measure unknown temperatures within the range of the scale. (From Amberson and Smith, 1948: Outline of Physiology. Appleton-Century-Crofts, New York.)

is then determined in the same way and under the same conditions. This galvanometric deflection, placed on the standard curve, at once enables one to determine the osmotic pressure of the sample (Lifson and Visscher, 1944). Commercially produced instruments for vapor pressure measurements are now available.

Another method for determining osmotic pressure from vapor pressure is Barger's method. The fluid to be tested is taken up into one end of each of a series of capillary tubes. On the other end of each tube, a sample from a series of known concentrations is introduced, an air space remaining between the two solutions in the tube. The initial length of each solution in each tube is marked. The unknown solution is tested in this way against the series of solutions of known concentrations. Each tube is examined periodically under a microscope to see if the test sample has gained or lost water. This is continued until it is definitely established that a pair is found in which no water exchange occurs. If no water is exchanged between the two samples, the vapor pressures in the two solutions are equal. Then the osmotic pressures also are equal in the two, and that of the unknown can be calculated in terms of the known. This method has the advantage of requiring a minimum of equipment, but it is somewhat tedious in operation (Krogh, 1939).

In biological work, determination of the freezing point of a solution has been used most extensively for determining the osmotic pressure, because the equipment is generally available and the measurement can be made without altering the nature of the solution. It is known that the freezing point of an aqueous solution containing a gram molecular weight (1 molal) of ideal nonelectrolyte is depressed by 1.86° C. The solution is placed in a freezing mixture with a thermometer (or thermistor, in the Fiske osmometer), and the temperature is taken at the time ice crystals appear. The temperature is taken again when the crystals disappear on warming. From the depression of the freezing point the osmotic pressure is readily calculated with the following equation:

$$\frac{\Delta_x}{\Delta_m} = \frac{x}{22.4 \text{ atm.}} \quad (12.8)$$

$$x = \frac{\Delta_x}{\Delta_m} \times 22.4 \text{ atm.} \quad (12.9)$$

$$\text{or} \quad x = \frac{\Delta_x \cdot 22.4}{1.86} = 12.06\ \Delta_x \quad (12.10)$$

Here $\Delta_x$ = the depression of the freezing point produced by the "unknown" solution whose osmotic pressure is sought. $\Delta_m$ = the depression of the freezing point produced by a 1 molal solution of ideal nonelectrolyte and x = the osmotic pressure of the "unknown" solution. At present methods are available for freezing point determinations by this method on microquantities of fluid (Prager and Bowman, 1963).

A variant of the freezing point method may be used for very small quantities of fluid. Test solutions are drawn up into capillary tubes and frozen. Their melting is then observed through crossed polaroids, under a microscope. Standards of known concentrations of sodium chloride (or other solute) are prepared in the same manner so that standard and test solutions can be examined simultaneously for comparison of the times required for the melting of the ice crystals (see Fig. 12.6). The different times required for melting the ice in the known sodium chloride solutions are plotted against the concentrations of the solutions. From this standard graph the osmotic pressure of an unknown can be

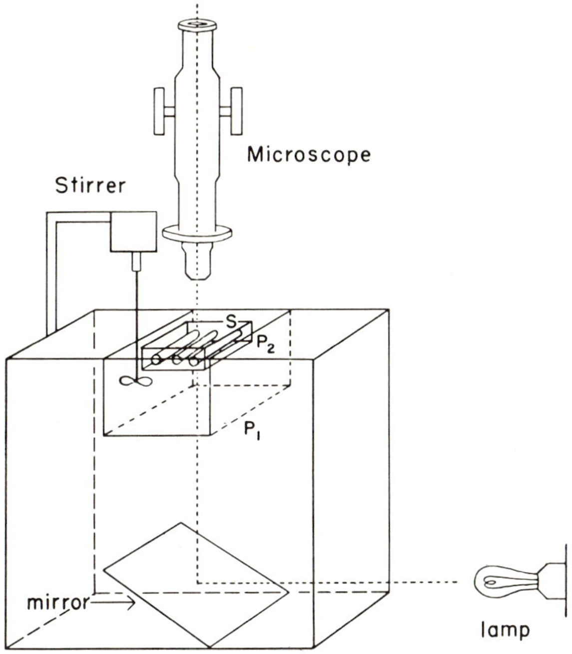


***Figure 12.6.*** Determining the melting point of minute quantities of biological materials. $P_1$ and $P_2$ are polaroid sheets suitably oriented so that the ice crystals appear bright against a dark background. Samples of several known concentrations of sodium chloride (S) are taken up into small capillary tubes and sealed with Vaseline. Samples of unknown solutions (S) are taken up into similar tubes and sealed. Both are then placed in a rack and frozen. They are then placed in a water bath chilled with dry ice, and the time for melting the ice crystals is determined, the water in the bath being stirred the while. The crossed polaroids enable one to see the ice crystals clearly against a black field, since the crystals polarize the light. Disappearance of the ice crystals is indicated by the sudden blackening of the field. The melting points for all samples are plotted as in Figure 12.7, and the values of the unknown samples are determined from the curve. (After Gross, 1954: J. Exp. Biol. *31*:402.)

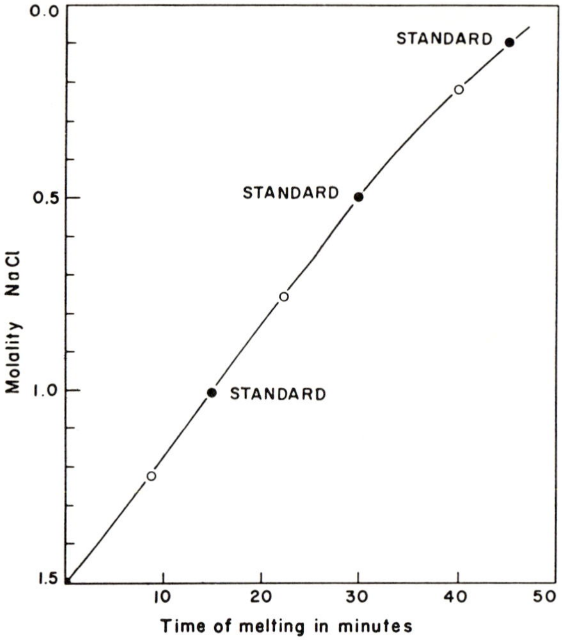


***Figure 12.7.*** Use of the melting method (shown in Fig. 12.6) for determination of osmotic pressure. A standard curve is first established by plotting data for the time required to melt frozen solutions of known concentrations of NaCl against the concentrations of the solutions. Three such points are shown as black dots, and the curve is drawn through these points. The concentration of an unknown solution, in terms of the standards, can be determined if the melting time (open circle) for a sample is placed on the curve.

determined in terms of the osmotic pressures of known concentrations of sodium chloride or other solute (see Fig. 12.7) (Gross, 1954).

Osmotic pressure can also be determined by the rise in boiling point of a solution, since it is well known that when a nonelectrolyte is dissolved in water, the rise in boiling point is 0.52° C. per molal dissolved nonelectrolyte. However, this method is unsatisfactory for determining the osmotic pressures of biological solutions, because proteins coagulate when they are boiled.

## OSMOLAL SCALE FOR BIOLOGICAL STUDIES

It is the concentration of osmotically active particles inside cells or in the fluids bathing the cells that is of consequence to movement of water between the medium and the cell, not whether the particles are ionic or nonionic, organic or inorganic. It is, therefore, useful to have terminology in terms of osmotic activity. For example, the *osmole* is a gram molecular weight of osmotically active particles in 1000 gm. of solvent (water, in biological systems). An osmole exerts 22.4 atm. of osmotic pressure and an osmolal solution lowers the freezing point 1.86° C. Frequently osmolal concentrations are given in milliosmolal units (a milliosmole is 1/1000 osmole) because of the low concentration of solutes in biological materials and media. The osmolal concentration of a solution is determined by any one of the colligative properties (Prosser and Brown, 1961; Dick, 1966).

## SOURCE OF ENERGY FOR WATER MOVEMENT

Traversing the plasma membrane by any molecule requires energy. Molecules at a temperature consistent with life processes possess kinetic energy. Some molecules in a given population possess more than the average amount of energy. Such molecules with at least the required energy of activation (Fig. 10.5) can surmount the hurdles (Fig. 13.10) required for their entry by diffusion. When water moves against a gradient metabolic energy is required (see Chapter 14).

## CELLS AS OSMOMETERS

The living cell, having a selectively permeable membrane rather than a semipermeable membrane, could hardly be

**Table 12.3** *Arbacia* Egg as an Osmometer

| RELATIVE PRESSURE | VOLUME OBSERVED* | $\pi V$ | $\pi(V - b)$ |
|---|---|---|---|
| 1.0 | 2,121 | 2,121 | 1,881 |
| 0.9 | 2,316 | 2,084 | 1,868 |
| 0.8 | 2,570 | 2,056 | 1,864 |
| 0.7 | 2,922 | 2,045 | 1,878 |
| 0.6 | 3,420 | 2,053 | 1,909 |
| 0.5 | 4,002 | 2,002 | 1,881 |

After Lucké and McCutcheon, 1932.
* Volume in arbitrary units.

expected to behave like a perfect osmometer. However, a cell might perhaps show some approximation to an osmometer if in response to a concentration gradient water were to pass through the membrane much more readily than would the solute. However, one correction must be made before such a test can be applied with fairness. The cell contains in its structures much protein and lipid, which add to the total volume of the cell, thereby excluding from that volume a corresponding volume of solution. The volume occupied by cell structures and called the *nonsolvent volume* (usually designated b), must be subtracted from the term for total volume in the equation for osmotic pressure: $\pi V = nRT$. The equation therefore becomes:

$$\pi(V - b) = nRT \qquad (12.11)$$

where b represents the nonsolvent volume and the other terms have the meanings previously assigned. In Table 12.3 the nonsolvent volume is given as 12.5 per cent of the total volume of the cell, or 0.125. When correction for the nonsolvent volume is made, the sea urchin egg in diluted sea water is observed to behave reasonably well as an osmometer, the values in the fourth column being more nearly alike than those in the third column, where no correction was made. In other cells, where solutes pass through the membrane more rapidly than they do in the sea urchin egg, the cells do not act even approximately like osmometers.

To compare the relative permeabilities of different kinds of cells to water, it is necessary to obtain the volume of water entering per unit area of cell surface per unit time per unit difference in osmotic pressure between the inside and the outside of the cell. A permeability constant, k, for water is given in these terms by the following equation (Lucké and McCutcheon, 1932):

$$\frac{dV}{dt} = k\,A(\pi_{in} - \pi_{out}) \qquad (12.12)$$

where V is the volume of the cell, dV, a small change in volume; t is the time, dt, a small change in time, A is the area of the cell surface; $\pi_{in}$ is the osmotic pressure of the cell; and $\pi_{out}$ is the osmotic pressure of the external medium.

The low permeability rates to water of different cells, as determined primarily by swelling of cells in hypotonic media, are given in Table 12.4. Almost a two hundred fold variation in permeability to water is observed in these cells. Cells with a higher surface to volume ratio have a higher permeability to water than those with a lower ratio (Dick, 1966). It is perhaps significant that the fresh water forms, in which the greatest gradient is found between the inside and the outside of the cell, have the lowest permeability constants for entry of water. Protozoan cells in fresh water are forced to void the constant excess of incoming water by means of contractile vacuoles. A low permeability to water reduces this work to a considerable degree.

Water movement into and out of a cell

**Table 12.4.** Comparison of Permeabilities of Some Cells in Water

| SPECIES | PERMEABILITY CONSTANT* |
|---|---|
| *Amoeba proteus* | 0.026–0.031 |
| *Pelomyxa carolinensis* | 0.023 |
| Fresh water peritrichs | 0.125–0.25 |
| *Arbacia* egg | 0.4 |
| Human erythrocyte | 3.0 |

From Prosser and Brown, 1961: Comparative Animal Physiology. 2nd Ed. W. B. Saunders Co., Philadelphia.
* In cubic microns of water, per square micron of surface area of cell, per atmosphere difference in pressure between the inside and outside of the cell, per minute.

is not always as simple as this discussion would lead one to believe (Robinson, 1953). It is important to point out that the permeability constant of a cell to water is only an approximation. For example, difference in osmotic pressure cannot be the only factor governing entry or exit of water, since in some cells, even the inward and outward permeability to water is different, indicating operation of some biological factor as well (Jacobs, 1952). Furthermore, depending upon its physiological state, the same cell may show changes in its permeability to water. For example, the permeability of some echinoderm eggs to water doubles, triples or quadruples on fertilization of the eggs. After cell division occurs, permeability to water may fall back to the original value (Lillie, 1918; Höber, 1945). It is curious that some cells are much more permeable to water than others and some cells even secrete water; for example, the gland cells in the carnivorous bladder plant *Utricularia,* set the trap by this means (Bennett-Clark, 1959). All of these are interesting and fundamental problems involving the movement of water across the cell membrane.

## LITERATURE CITED AND GENERAL REFERENCES


Acker, R. F., and Hartsell, S. E., June 1960: Flemming's lysozyme. Sci. Am. *202*:132–142.

Brown, F., and Danielli, J. F., 1964: The cell surface and cell physiology. *In* Cytology and Cell Physiology. Bourne, ed. Academic Press, New York, pp. 239–310.

Brown, F., and Stein, W. D., 1960: Balance of water, electrolytes and nonelectrolytes. Comp. Biochem. *2*:403–470.

Bennett-Clark, T. A., 1959: Water relations of cells. *In* Plant Physiology, A Treatise. Steward, ed. Academic Press, New York, pp. 105–191.

Blum, G., 1957: Osmotischer Wert, Saugkraft, Turgor. Protoplasmatologia II. Springer-Verlag, Vienna.

Bull, H. B., 1964: An Introduction to Physical Biochemistry. F. A. Davis Co., Philadelphia.

*Davson, H., 1964: Textbook of General Physiology. 3rd Ed. Little, Brown & Co., Boston.

Davson, H., and Danielli, J. F., 1952: The Permeability of Natural Membranes. 2nd Ed. Macmillan Co., New York.

Dick, D. A. T., 1959: Osmotic properties of living cells. Int. Rev. Cytol. *8*:388–448.

*Dick, D. A. T., 1966: Cell Water. Butterworth, Washington, D. C.

Fiske Osmometer, Fiske Associates, Inc., Bethel, Conn.

Glasstone, S., 1960: Elements of Physical Chemistry. 2nd Ed. D. Van Nostrand Co., New York.

Grim, E., and Sollner, K., 1957: The contribution of normal and anomalous osmosis to the osmotic effects arising across charged membranes with solutions of electrolytes. J. Gen. Physiol. *40*:887–899.

Gross, W. J., 1954: Osmotic responses in the sipunculid *Dendrostomum zostericolum.* J. Exp. Biol. *31*: 402–423.

Höber, R. (ed.), 1945: Physical Chemistry of Cells and Tissues. Blakiston Co., Philadelphia, Chapter 14.

Jacobs, M. H., 1952: The measurement of cell permeability with particular reference to the erythrocyte. *In* Modern Trends in Physiology and Biochemistry. Barron, ed. Academic Press, New York, pp. 149–171.

Kitching, J. A., 1956: Contractile Vacuoles of Protozoa. Protoplasmatologia III.D.3.a. Springer-Verlag, Vienna.

Krogh, A., 1939: Osmotic Regulation in Aquatic Animals. Cambridge University Press, London.

Lifson, N., and Visscher, M. B., 1944: Osmosis in living systems. *In* Medical Physics. Glasser, ed. Year Book Medical Publishers, Chicago, Vol. 1, pp. 869–892.

Lillie, R. S., 1918: The increase of permeability of water of sea urchin eggs and the influence of cyanide and anaesthetics on this change. Am. J. Physiol. *45*:406–430.

Lucké, B., 1940: The living cell in osmotic systems and its permeability to water. Cold Spring Harbor Symp. Quant. Biol. *8*:123–132.

Lucké, B., and McCutcheon, M., 1932: The living cell as an osmotic system and its permeability to water. Physiol. Rev. *12*:68–139.

Mauro, A., 1965: Osmotic flow in a rigid porous membrane. Science *149*:867–869.

Milicevic, B., 1964: Application of thermodynamics of irreversible processes to the theory of vapor-membrane osmometers. Helv. Chim. Acta *47*: 2059–2067.

Mitchell, P., and Moyle, J., 1956: Osmotic function and structure in bacteria. *In* Bacterial Anatomy. Spooner and Stocker, eds. Cambridge University Press, London, pp. 150–180.

Olmstead, E. G., 1966: Mammalian Cell Water. Lea & Febiger, Philadelphia.

Osterhout, W. J. V., 1956: The role of water in protoplasmic permeability and in antagonism. J. Gen. Physiol. *39*:963–976.

Peters, E., and Saslow, G., 1939–40: Performance of the Hepp micro-osmometer. J. Gen. Physiol. *23*: 177–184.

Pfeffer, W., 1877: Osmotische Undersuchungen. Engelmann, Leipzig.

Ponder, E., 1948: Hemolysis and Related Problems. Grune & Stratton, New York.

Ponder, E., 1955: Red Blood Cells, Structure and Function. Protoplasmatologia X.2. Springer-Verlag, Vienna.

Prager, D. J., and Bowman, R. L., 1963: Freezing point depression: new method for measuring ultra-micro quantities of fluids. Science *142*:237–239.

Prosser, C. L., and Brown, F., 1961: Comparative Animal Physiology. 2nd Ed. W. B. Saunders Co., Philadelphia, Chapters 2 and 3.

Rehm, W. S., 1944: An easily constructed Hepp osmometer. Science *100*:364.

Robinson, J. R., 1953: The active transport of water in living systems. Biol. Rev. *28*:158.

*Robinson, J. R., 1960: Metabolism of intracellular water. Physiol. Rev. *40*:112–149.

Ruesink, A. W., and Thimann, K. V., 1966: Protoplasts: preparation from higher plants. Science *154*:280–281.

Solomon, A. K., Dec. 1960: Pores in the cell membrane. Sci. Am. *203*:146–156.

Teorell, T., 1955: A contribution to the knowledge of rhythmical transport processes of water and salts. Exp. Cell Res. *3* (suppl.): 339–345.

# CHAPTER 13 / MOVEMENT OF SOLUTES THROUGH CELL MEMBRANES IN RESPONSE TO A CONCENTRATION GRADIENT

Many substances other than water move in and out of cells in response to a concentration gradient, although only a few of them (carbon dioxide and oxygen) are able to move as freely as water. If a cell membrane allows a substance to move in or out through it, it is said to be permeable to that substance.

## METHODS FOR STUDYING PERMEABILITY OF CELL MEMBRANES TO SOLUTES

Only a few of the methods employed in permeability studies will be discussed here. Some methods are only qualitative. For example, entry of dyes, such as neutral red, into cells like paramecia can be observed by the appearance of a pink color inside the cells (Brooks and Brooks, 1941). Sometimes use is made of such dye penetration as an indicator. Thus the entry of a base like ammonia into paramecia stained with neutral red may be observed because the pink paramecia become yellowish as soon as the base has entered. However, it is virtually impossible to make quantitative measurements because if the penetration is slow, or if only a minute quantity enters, no color change is noticed. Furthermore, the method depends upon a personal element—judgment of color—which is likely to vary from one person to another. The same criticism applies also to other visual methods, such as the formation of an insoluble precipitate inside cells (e.g., the formation of calcium oxalate crystals in cells of *Elodea* as an indication of the penetration of oxalic acid).

***Volume Change.*** A quantitative method much more satisfactory for studying the permeability of cell membranes is the observation of the rate of deplasmolysis of plasmolyzed cells (see Chapter 12). The relative rates of entry into cells of different substances used in hypertonic solutions may be tested by a study of rates of deplasmolysis as the cells reabsorb water. For example, when equimolal hypertonic solutions of several substances like ethylene glycol, glycerol or glucose are applied to cells, they will plasmolyze the cells (see Fig. 12.3). Deplasmolysis will occur first among the cells in ethylene glycol and last among those in glucose (Fig. 13.1). This means that ethylene glycol penetrates the fastest of the three and glucose the slowest (Baptiste, 1935). If the cell remains shrunken in a solution, the solute has failed to enter the cell or changes in the cell structure have occurred.

A similar method, particularly useful for the study of the permeability of the plasma membranes of minute cells, is the determination of the volume of a mass of cells before and after immersion in hypertonic solutions. The volume of the cell mass is determined after the cells have been packed together by centrifugation. This is a favorite method for the study of permeability of red blood corpuscles. A special centrifuge tube called the *hematocrit,* with a calibrated fine cylindrical portion 1 or 2 mm. in diameter, is used for this purpose. If the series of experiments with different substances is carried out under similar conditions and with the same centrifugal force applied for the same time, quantitative data on the rate of entry of the various substances can be readily obtained by the volume changes of the packed cells. However, it is virtually impossible to pack the cells without any intercellular fluid; therefore such experiments should be designed to compare relative packing of a control and experimental sample (Ponder and Saslow, 1930, 1931).

***Hemolysis.*** Hemolysis, indicated by the sudden appearance of a clear red solution instead of the previously murky suspension of red blood cells, has also been used as a measure of the rate of penetration of substances into erythrocytes. For this purpose red blood cells are placed in hypertonic solutions such as ethylene glycol. At first water leaves the cells, causing crenation, but in time ethylene glycol molecules enter. This entry increases the store of osmotically active molecules inside the cells, which results in a corresponding uptake of water. The cell membrane can stretch only to a limited extent without altering its properties, however, and the cells undergo *hemolysis* when this limit has been passed; that is, the membrane has become porous enough to permit the hemoglobin molecules to leave the cell.

In solutions of glucose, erythrocytes are hemolyzed only after a much longer

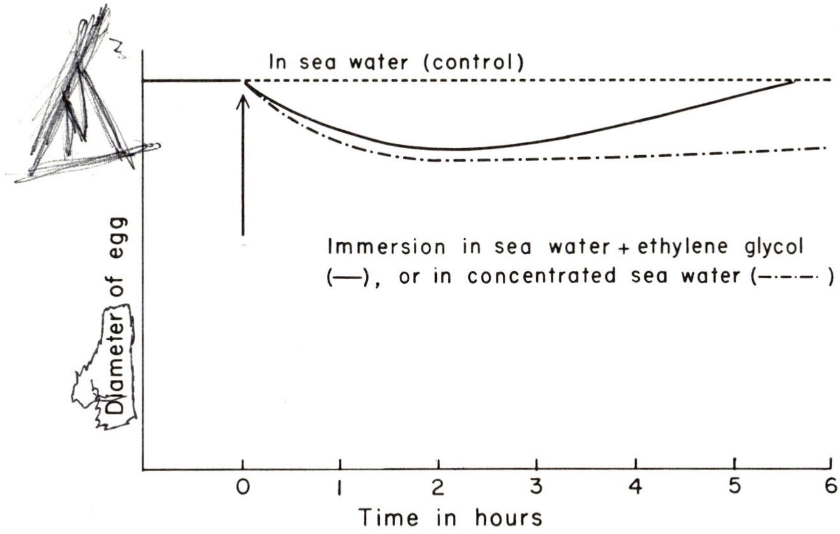


***Figure 13.1.*** Return to normal volume as a measure of permeability. Sea urchin eggs placed in sea water rendered hypertonic by addition of ethylene glycol in one case and by evaporation of water in the other case. Ethylene glycol enters more rapidly than salts; therefore, the eggs return sooner to their initial volume in ethylene glycol than in sea water.

time than when they are in ethylene glycol. The relative rates of entry of various kinds of molecules may therefore be measured by the rates of hemolysis. However, some question exists whether this method can be used for comparisons of hemolysis of red blood cells in different species, because the strength of their membranes may vary and thus hemolysis rates may not have the same meaning.

***Quantitative Analyses.*** Perhaps the most direct method for measuring permeability of cells is the quantitative analysis of the internal contents of a cell exposed to a solute for a known period. For this purpose large coenocytic (multinucleate) plant cells, such as *Valonia, Chara* and *Halicystis,* are ideal, because the uptake of the materials can be measured by analysis of the contents of the large central sap vacuoles. The vacuole of *Valonia* or *Halicystis* often contains more than 1 ml. of sap, making possible microchemical tests on a single cell. Few animal cells are large enough to permit chemical analyses of a single cell, but the squid giant axon has been so employed. The chemical method is especially valuable because the data lend themselves to calculation of permeability constants in terms of the quantity of material transported per unit time across a unit area of cell membrane per unit osmotic gradient. Data obtained on one cell for a variety of substances may therefore be quantitatively compared with data on another cell (Collander and Bärlund, 1933).

Still another method used to measure permeability of a cell to a given substance is the rate of uptake of that substance from solution after a suspension of cells is added. This is an especially valuable quantitative method for study of the rate of entry of solutes into yeast or other small, active cells, because the internal contents of such cells are too minute for quantitative analysis. The cells are readily centrifuged off from the test solution, leaving a supernatant fluid which may be subjected to quantitative analysis. To illustrate, a suspension of $10^8$ yeast cells per milliliter removes about 6 mg. of sugar from solution in the first hour. The rate of removal may be measured by sampling the supernatant fluid 15, 30 and 60 minutes after yeast is first exposed to a known concentration of the glucose solution (in phosphate buffer). A small correction has to be made for the glucose respired during the course of the experiment. Another possible source of error is adsorption of some solutes to the surfaces of the cells (classroom experiment).

Another important quantitative method is that employing radioactive tracers. Cells may be exposed for a given period to solutions containing radioactive isotopes, e.g., $^{24}Na$. After the test suspension is centrifuged, the supernatant is removed and the cells are rinsed off. The amount of radioactive ion which has been taken up may be measured with a Geiger counter (see Fig. 13.2). By making readings after varying the exposure time to radioactive sodium, one may determine the rate of ion uptake. Moreover, periodic readings taken after the cells have been removed from the radioactive solutions and suspended in nonradioactive ("cold") medium, indicate the rate of loss of ions. Valuable quantitative data have been obtained by this method. For instance,

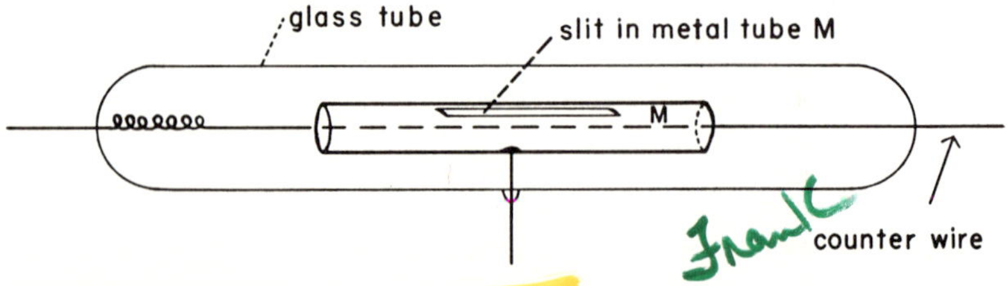


***Figure 13.2.*** A Geiger-Müller counter (diagrammatic). The counter wire and the metal tube are at a difference in potential of about 1000 to 1500 volts, so that absorption of even a single electron or quantum by the metal leads to an electric discharge from the metal to the wire. This is registered as a "plunk" by the relay-driven counter, which in simplest form resembles a cyclometer or tally counter.

unique data not obtainable by other methods have thus been gathered for sodium. Previous experiments suggested that sodium does not enter cells, or enters so slowly that it cannot be measured. Exposure of cells to radioactive sodium, however, demonstrated that sodium ions enter and leave cells at the same readily measurable rate; therefore, no net change occurs in the cell content of sodium ions. Consequently, movement of sodium ions could not have been demonstrated by even the most precise quantitative chemical determinations (Brooks, 1938).

The radioactive isotope method has been confined largely to study of the entry of a number of electrolytes which have radioactive isotopes with sufficiently long half-life periods to permit measurement. $^{3}H$, $^{14}C$, $^{24}Na$, $^{32}P$, $^{35}S$, $^{36}Cl$, $^{42}K$, $^{45}Ca$, $^{59}Fe$, $^{64}Cu$, $^{65}Zn$, $^{82}Br$, $^{89}Sr$ and $^{131}I$ are some of the isotopes available for tracer studies. However, the tracer method need not be confined to electrolytes, since organic molecules may be labeled with radioactive carbon, etc. (Ussing, 1953).

## THE CORRELATION OF PERMEABILITY WITH PARTITION COEFFICIENTS

A definite rationale lies behind the groups of compounds tested in permeability studies. By testing various groups of compounds, each with certain properties in common, it has been possible to find some correlation between these properties and movements of each solute through the cell membrane.

*Nonpolar compounds* are compounds in which electrons are equally shared by the two atoms forming a bond (see Fig. 13.3), as in the paraffins, olefins and cyclic compounds. They have a high solubility in fats or fat solvents, but a low solubility in water.

In *nonionic polar compounds* the electrons of the bond between two atoms are attracted more to one of the atoms (e.g., to the strongly electronegative atoms, such as oxygen, nitrogen and chlorine) than to the other. As a consequence, the molecule acts as a dipole and orients in an electric field. Nonionic polar compounds are derivatives of nonpolar compounds into which electronegative atoms or radicals have been introduced. Thus, methyl alcohol is a derivative of methane, ethyl alcohol a derivative of ethane, and so on (see Fig. 13.3).

In *ionic compounds* the electrons of a valency bond are held by only one of the atoms or radicals, so that ions are present even in the crystalline state of such compounds. In the crystal the ions are bound together by electrostatic forces. A liquid of high dielectric constant, like water,

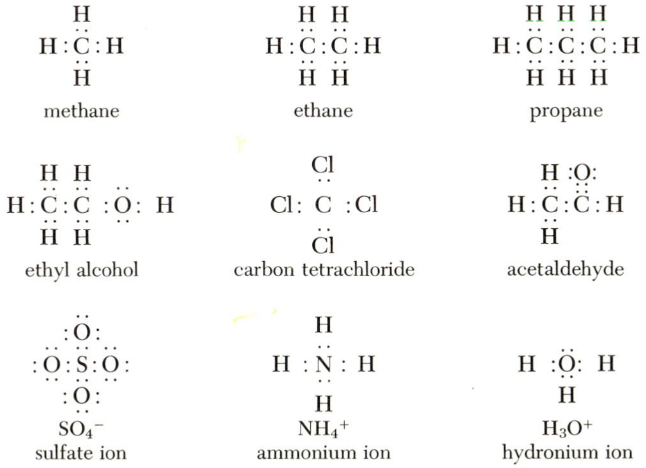


***Figure 13.3.*** Electron configuration in nonpolar and polar compounds. Note that in polar compounds, both nonionic and ionic, the electrons are closer to the electronegative atoms. (From West, 1942: Physical Chemistry for Students of Biochemistry and Medicine. Macmillan Co., New York.)

stabilizes the attractive forces between pairs of ions, allowing them to separate from one another and act as individual particles. Ionic compounds are of two kinds—*weak electrolytes,* such as weak acids and bases, and *strong electrolytes,* such as salts and strong acids and bases. The stronger the electrolyte, the greater will be its solubility in water and the less its solubility in oil and oil solvents. In a homologous series such as the organic acids, the strength (ionization) of the acids decreases as the chain length increases; for instance, formic acid is stronger than acetic, the latter is stronger than propionic, and so on. As the degree of ionization declines, the oil solubility increases, and the higher fatty acids are freely soluble in oils or oil solvents but only slightly soluble in water (see Table 13.1).

The ratio of solubility of a compound in oil or fat solvent to its solubility in water is called the *partition coefficient.* The more nonpolar a compound is, the more soluble it is in fats or fat solvents, i.e., the more *lipophilic* it is. Such compounds usually have slight, if any, solubility in water. Their partition coefficients are therefore high.

On the other hand, polar derivatives of nonpolar compounds are usually less soluble in fats or fat solvents, in correspondence to their degree of polarity. An increase in polarity results in greater solubility in water. To illustrate, lower alcohols like methyl alcohol and ethyl alcohol are highly polar, and they are only slightly soluble in fats or fat solvents but completely miscible with water. Conversely, higher alcohols like butyl alcohol and amyl alcohol are sparingly soluble in water but miscible with fat solvents in all proportions. Lower alcohols have low partition coefficients; higher alcohols have high ones.

Ionized compounds are polar and correspondingly less soluble in fats but usually highly soluble in water. Weak electrolytes with lipophilic radicals, like the long chains of the higher fatty acids mentioned before, are soluble in fats and have high partition coefficients, whereas the lower fatty acids (e.g., formic acid, acetic acid) are highly soluble in water and have low partition coefficients. Strong electrolytes are sparingly soluble in fats but highly soluble in water. Their partition coefficients are therefore low.

The entry of many compounds into the large cells of *Chara* was studied by Collander and Bärlund (1933), and some of their results are cited in detail here because they constitute the most extensive study of its kind. The amount of a substance entering a *Chara* cell per unit time was determined by chemical analysis, and the permeability constant in terms of the quantity of compound entering per unit time per unit area of cell surface per unit osmotic gradient was then calculated. These data are given graphically in Figure 13.4. The approximate molecular diameters of the compounds are indicated by the size of the circles in the figure, the largest circles designating the largest molecules.

**Table 13.1.** Partition Coefficients of a Homologous Series of Acids

| ACID | NUMBER OF CARBONS | MOLECULAR WEIGHT | DISSOCIATION CONSTANT | WATER SOLUBILITY IN GRAMS/100 ML. | PARTITION COEFFICIENT, BENZENE : WATER |
|---|---|---|---|---|---|
| Formic | 1 | 46.03 | $1.77 \times 10^{-4}$ | very high | — |
| Acetic | 2 | 60.05 | $1.76 \times 10^{-5}$ | very high | 0.055 |
| Propionic | 3 | 74.08 | $1.34 \times 10^{-5}$ | very high | 0.28 |
| Butyric | 4 | 88.10 | $1.48 \times 10^{-5}$ | 5.62 at −1.1° C. | 1.08 |
| Caproic | 6 | 116.16 | $1.43 \times 10^{-5}$ | 0.4 | 3.73 |
| Caprylic | 8 | 144.22 | — | 0.25 at 100° C. | 16.10 |

Data from Handbook of Chemistry and Physics, 1968. 49th Ed. Chemical Rubber Co., Cleveland; and Höber, 1945.

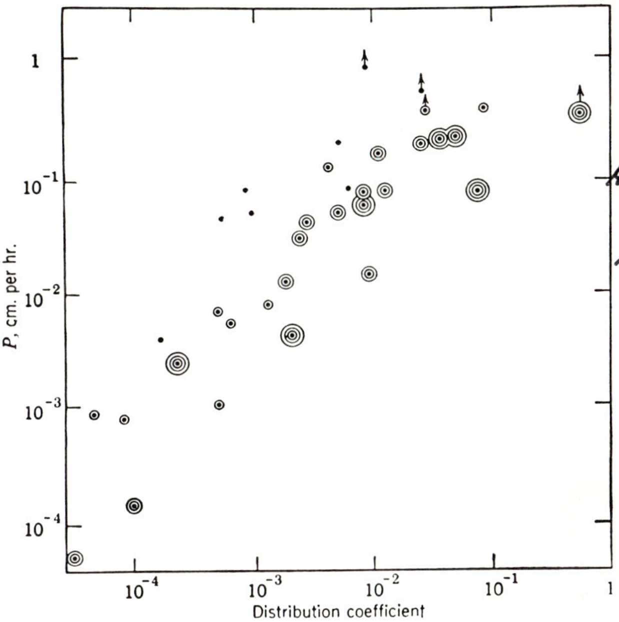


***Figure 13.4.*** Relation between partition coefficient (distribution coefficient) of a substance between oil and water and its penetration (*P*) into the *Chara* cell. The small dots indicate small molecules, and successive circles about these indicate corresponding increases in molecular size. Note that even large molecules of high solubility in oil penetrate the cell readily. (From Collander, 1937: Trans. Faraday Soc. *33*:986.)

Scrutiny of Fig. 13.4 discloses that usually molecules with the largest partition coefficients enter the cell most readily, regardless of molecular size. Thus if in a homologous series such as the alcohols the narcotic action of the alcohols on respiration is used to measure their penetration, higher alcohols with greater oil solubility are observed to enter cells much more readily than lower alcohols of less oil solubility, even though the molecular sizes of the higher alcohols are considerably larger (Höber, 1945) (see Table 13.2).

Another example showing that entry of molecules into cells increases with increase in partition coefficient, even though there is an increase in molecular weight, is that of the urethans. Comparing the effects of urethans (alkyl carbamates) on the resting potential of frog nerves as a measure of their penetration into cells, one can demonstrate that urethan (ethyl carbamate) penetrates less readily than the substituted urethans in which methyl, ethyl or other radicals are substituted for some of the hydrogen atoms. The substituted urethans have higher oil to water partition coefficients than ethyl carbamate (Crescitelli, 1948).

Since the lipoid solubility of a mole-

**Table 13.2.** Partition Coefficients (Ratio of Solubility in Oil to Solubility in Water) of a Series of Homologous Alcohols

| ALCOHOL | MOLECULAR WEIGHT | SOLUBILITY IN GM. PER 100 ML. WATER | PARTITION COEFFICIENT, OIL : WATER | "NARCOTIC CONCENTRATION" |
|---|---|---|---|---|
| Methyl | 31.06 | ∞ | 0.0097 | 5.0 |
| Ethyl | 46.07 | ∞ | 0.0357 | 1.6 |
| Propyl | 60.09 | ∞ | 0.156 | 0.8 |
| n-Butyl | 74.12 | 7.9 | 0.588 | 0.15 |
| i-Amyl | 88.15 | slight | 2.13 | 0.045 |

Data from Handbook of Chemistry and Physics, 1968. 49th Ed.; and Höber, 1945. The narcotic concentration is in moles needed to give 50 per cent inhibition of oxygen consumption by goose erythrocytes.

cule is a factor of greater moment than its size in determining rate of entry into a cell, the question arises whether partition coefficient alone determines if a molecule shall pass through the plasma membrane. However, further experiments demonstrate that molecular size does influence the passage of substances through the plasma membrane and that many other factors, such as hydrogen ion concentration, salts, temperature and physiological state, also influence the permeability of the plasma membrane.

## THE EFFECTS OF MOLECULAR SIZE UPON PERMEABILITY

In studying the effects of molecular size on permeability, one should note that in general the greater the molecular weight, the larger will be the molecule. Molccular diameters also depend upon steric configuration, especially in organic molecules. Symmetry, or its lack, also influences the effective diameter of a molecule, an asymmetrical molecule having a larger effective diameter than a symmetrical one.

It is understood that very large molecules probably have only a slight chance of entering a cell through the plasma membrane. Sucrose, for example, penetrates cells in very small amounts, if at all. Larger carbohydrate molecules, such as starch, glycogen or inulin, have not been shown to diffuse through plasma membranes; for this reason they serve effectively as food stores. The probability that a protein molecule will pass through a cell membrane is also very slight, although the absorption of antigens indicates that penetration occurs in some instances, perhaps by pinocytosis (see Chapter 14).

On the other hand, it has been found that, provided the partition coefficients of the compounds are the same, the molecular size has some effect upon permeability. For example, cyanamide, propionamide, succinamide and diethylmalonamide have about the same partition coefficients, but their molecular diameters increase in the order given (see Fig. 13.5). When these compounds were tested on *Chara* it was found that their ability to penetrate the cell membrane decreased with increase in molecular diameter. Size, then, is also a factor, perhaps secondary, in determining whether or not a molecule will pass into a cell and in determining the rate of penetration.

However, there are cases in which molecular size of small molecules may be of even greater importance than the partition coefficient in determining the penetration of molecules into some cells. This is illustrated by the data in Figure 13.6 for *Beggiatoa*, a large sulfur bacterium, tested with compounds similar to those used with *Chara*. Here the correlation between permeation and molecular size is much more decisive than the correlation between permeation and partition coefficient (see Fig. 13.7).

Another example of the influence of molecular size on permeability is shown in the penetration of aldehydes into a dark mutant strain of luminous bacteria, which lacks the ability to form its own aldehydes needed for luminescence, but will luminesce if aldehyde is supplied. By using the first phase of luminescence (when luminous intensity increases line-

Cyanamide $CN \cdot NH_2$

Propionamide $CH_3 \cdot CH_2 \cdot CO \cdot NH_2$

Succinamide $NH_2 \cdot CO \cdot CH_2 \cdot CH_2 \cdot CO \cdot NH_2$

Diethylmalonamide $C_2H_5—C(CO \cdot NH_2)_2—C_2H_5$

$$\begin{array}{c} CO \cdot NH_2 \\ | \\ C_2H_5—C—C_2H_5 \\ | \\ CO \cdot NH_2 \end{array}$$

***Figure 13.5.*** Compounds of like partition coefficient but of different molecular size. Penetration into *Chara* is inversely correlated with molecular size, the permeability decreasing in the order given. (After Collander and Bärlund, 1933: Acta Bot. Fennica *11*:1.)

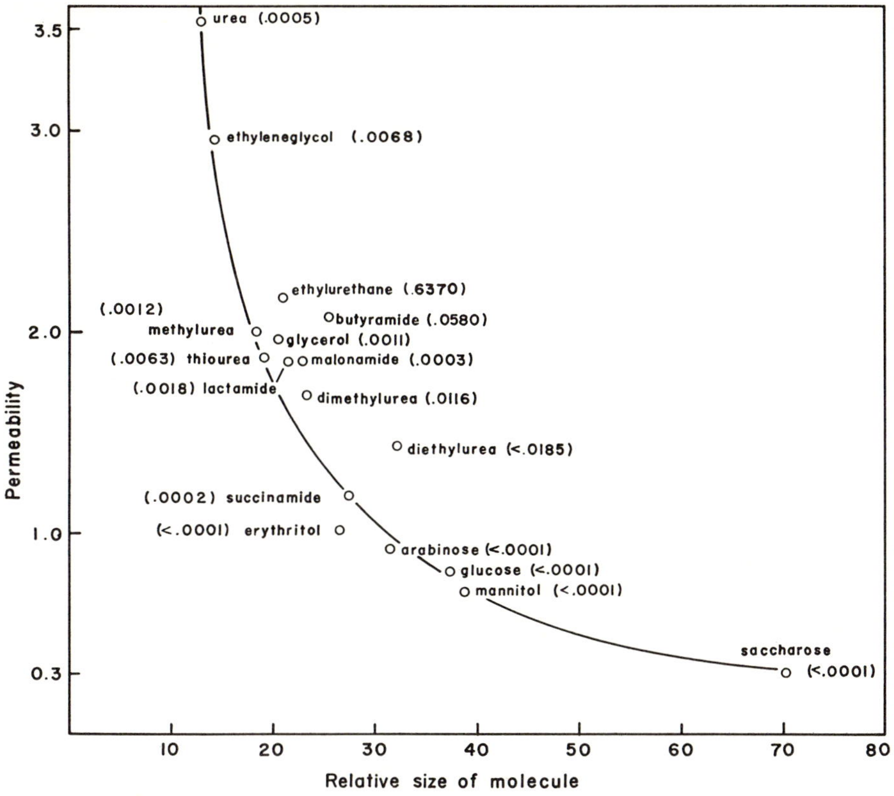


***Figure 13.6.*** Relation between molecular size and permeability in *Beggiatoa*. Molecular size is measured by molecular refraction in yellow light. Permeability is measured by the $\log_{10}$ ($10^4 \times$ the threshold plasmolytic concentration moles per liter). Partition coefficients in ether:water are given in parentheses. Correlation is clearly with molecular size. (Data from Ruhland and Hoffmann, 1925: Planta *1*:1.)

arly with the concentration of aldehyde) as a measure of entry, the data for rate of penetration of a homologous series of aldehydes was obtained, as shown in Table 13.3. The permeability constants calculated for the rate of entry of aldehydes into bacteria are of the same order of magnitude as for entry of other nonelectrolytes into plant cells (Rogers and McElroy, 1958 a, b).

It is apparent that it is impossible for macromolecules such as polypeptides, proteins and polysaccharides to pass through the pores in (or dissolve into) a plasma membrane through which smaller molecules enter and leave the cell because of their large size (Solomon, 1960). The cell has developed another method for uptake of such macromolecules, i.e., *pinocytosis*, which is discussed in Chapter 14.

**Table 13.3.** Rate of Penetration of Aldehydes into Luminous Bacteria

| NUMBER OF CARBONS | NAME | PENETRATION IN CM. PER SEC. $\times 10^{-6}$ | $Q_{10}$ |
|---|---|---|---|
| 7 | heptanal | 10.6 | 1.13 |
| 8 | octanal | 6.4 | 1.19 |
| 9 | nonanal | 1.68 | 1.26 |
| 10 | decanal | 0.95 | 1.26 |
| 11 | undecanal | 0.85 | 1.42 |
| 12 | dodecanal | 0.7 | 1.89 |
| 14 | tetradecanal | 0.31 | 1.78 |

Data from Rogers and McElroy, 1958a, b.

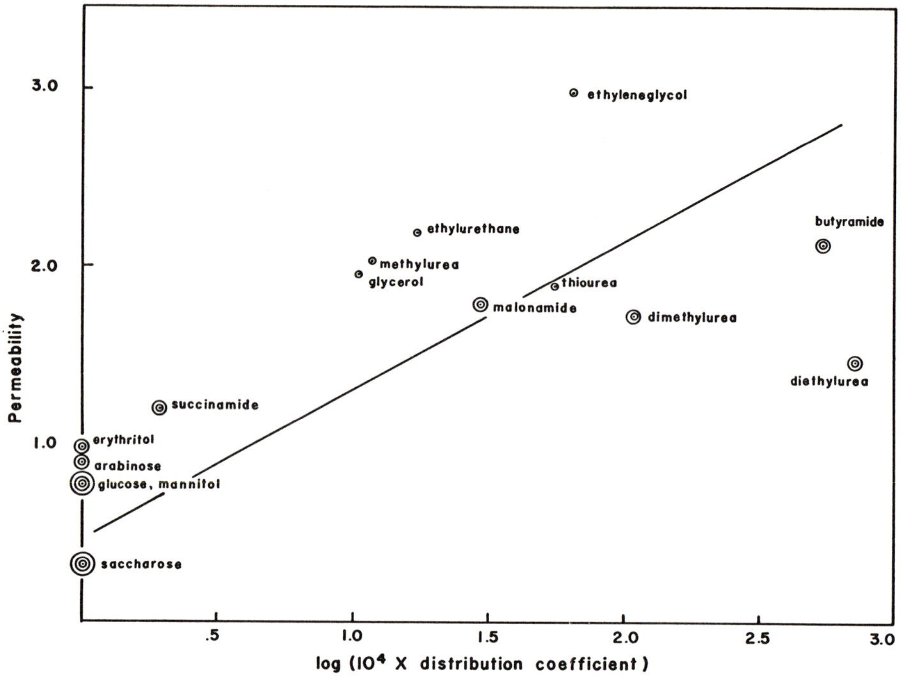


***Figure 13.7.*** Relation between partition coefficient and permeability of *Beggiatoa* to various organic substances. Although the permeability increases with the partition coefficient, the correlation is not nearly so good as that found in *Chara* (see Fig. 13.4). Permeability is measured by $\log_{10}$ ($10^4 \times$ threshold plasmolytic concentration in moles per liter). The abscissa is $\log_{10}$ ($10^4 \times$ partition coefficient between ether and water). (Data from Ruhland and Hoffmann, 1925: Planta *1*:1.)

## THE EFFECT OF IONIZATION UPON PERMEABILITY

Electrolytes enter cells more slowly, in general, than nonelectrolytes of similar molecular dimensions, and weaker electrolytes enter more rapidly than strong ones. The relation between charge and penetration is best seen by studying the rate of entry of weak electrolytes as affected by their degree of ionization. The degree of ionization of a weak electrolyte is determined by the pH. By varying the pH, within limits safe for the living cell, one may determine what effect the charge on an electrolyte has upon its ability to penetrate a plasma membrane.

In Figure 13.8 is shown the effect of pH on the penetration of carbon dioxide or carbonic acid. At pH 6.34 one-half is dissociated. As the pH rises (see Fig. 8.2), the dissociation increases but the entry of carbon dioxide decreases. Conversely, with fall in pH the amount of dissociation decreases and the entry of carbon dioxide into the cell increases (Osterhout and Dorcas, 1926). Similar relations hold for hydrogen sulfide, hydrocyanic acid, auxins (plant growth hormones), dinitrophenol (metabolic stimulant) and other biologically active weak acids. The entry of weak bases such as the amines and alkaloids is governed by the same principles, except that in their case the rise in pH suppresses dissociation and enhances penetration.

The effect of degree of ionization on the entry of electrolytes into cells is further illustrated by comparing the rates of entry of members of a homologous series such as the weak organic acids illustrated in Table 13.1. As judged by criteria such as color changes in cells with natural pigments, penetration of the anions was shown to increase in the order given in the table. It appears that in spite

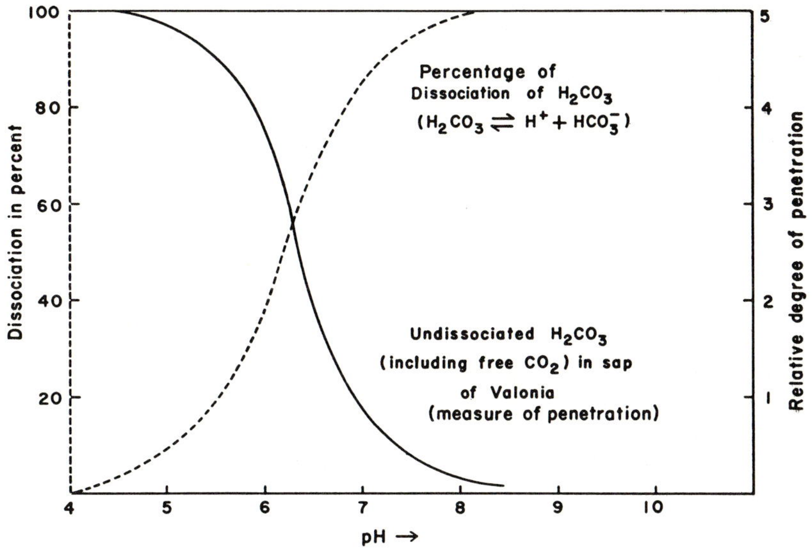


***Figure 13.8.*** Relation between pH and penetration of a weak electrolyte in *Valonia.* (After Osterhout and Dorcas, 1926: J. Gen. Physiol. *9*:259.)

of the increase in molecular size, the entry of the acids is more rapid, the smaller their degree of dissociation. However, lipoid solubility plays a role here also, since the rate of entry increases with a rise in the partition coefficient, a factor previously emphasized as of paramount importance (Höber, 1945). Presumably, permeation is proportional to the number of undissociated molecules, and a change in pH in the direction which increases the proportion of undissociated molecules will enhance penetration. The data indicate that the presence of a charge on an ion decreases its chance of entry.

If the presence of an electric charge is indeed a disadvantage for entry of ions into the cell, then the stronger the charge on an ion the less probable should be the ion's entry into a cell. This is found to be true. In studies on a series of strong electrolytes (completely dissociated) some ions are found to enter more readily than others. Monovalent cations such as $Na^+$ and $K^+$ enter more readily than divalent cations such as $Ca^{++}$ or $Mg^{++}$, and divalent cations enter more readily than trivalent cations such as $Fe^{+++}$. Similarly, monovalent anions such as $I^-$ or $Cl^-$ enter more readily than divalent anions such as $SO_4^{=}$ or tartrate (Ussing, 1954).

Not all ions of the same valency enter cells at the same rate. For example, the ammonium ion enters more rapidly than potassium, potassium more rapidly than sodium, and sodium more rapidly than lithium. This might be explained on the basis of the relative size of the hydrated ions. As was explained in Chapter 4, ions bear charges which attract and orient water molecules about themselves in hydration spheres. The resultant particle is therefore much larger than the ion itself. Hydration is not equal in all ions. For instance, the nucleus of potassium is surrounded by several shells of electrons; therefore its net charge density is lower than that of lithium, which has one shell of electrons. Consequently, less water is attracted to potassium than to lithium.

Hydration must always be taken into account when one considers the effective size of an ion, since the size of a hydrated ion determines the diffusion rate as much as it determines the ease of the ion's passage through the plasma membrane (see Table 13.4). For the entry of anions into cells, a similar explanation has been ap-

**Table 13.4.** Penetration and Properties of Some Cations

| PROPERTY | ION | | |
|---|---|---|---|
| | Li | Na | K |
| Atomic weight | 6.939 | 22.997 | 39.096 |
| Degree of hydration | most | intermediate | least |
| Effective diameter | largest | intermediate | smallest |
| Entry into cells | slowest | intermediate | fastest |

plied to explain the differential entry of ions of the same valency, e.g., $SCN > I > NO_3 > Cl$ (Brooks and Brooks, 1941). Thus permeability to ions is dependent to some extent upon a "sieve" action of the plasma membrane. However, it can be demonstrated that the hydration sphere does not electrically neutralize the ion. The residual charge still has its effect upon penetration, as does, in addition, the particle size resulting from hydration.

The fact that uncharged particles enter a cell more readily than ions do suggests that the plasma membrane is charged. Anions are attracted to positively charged areas and cations to negatively charged ones. Since both anions and cations are affected, the plasma membrane is considered to be a *mosaic* of negative and positive charges. However, the overall charge on the plasma membrane is positive, so that anions would be expected to enter cells more readily than cations. This is found to be the case (Höber, 1936).

Some cells appear to be freely permeable to anions while at the same time they largely exclude cations. Other cells may be freely permeable to cations but may largely exclude anions. The passage of cations and anions from an external medium into a cell is thought to be in exchange for an ion of like charge inside the cell. Hydrogen ions resulting from metabolic activity are always available for exchange with cations (cation exchange diffusion), and bicarbonate ions of similar origin are available for exchange with anions (anion exchange diffusion). For maintenance of the electric potential across the plasma membrane, equal numbers of anions and cations need not be taken up by the cell, but the exchange of ion for ion must be equal (Ussing 1953, 1954; Finean, 1966).

All of these considerations on the permeability of cells to ions apply only to situations in which no marked accumulation of some ions occurs and when no active transport mechanism seems to be involved. When marked accumulation occurs, as for example, that of potassium in most cells, an active transport system is suggested (Chapter 14).

## OSMOTIC RELATIONS IN THE PRESENCE OF PROTEIN

Proteins form ions that penetrate neither natural nor artificial membranes with small pores. When a potassium proteinate solution, at the appropriate pH, is placed in a cellophane bag and immersed in a solution of potassium chloride, the initial distribution of ions is similar to that shown in Figure 13.9*A*. At this time the initial concentration of potassium ions inside is designated as $C_i$, the concentration of chloride is zero, and the concentration of proteinate is designated as $C_i$. The initial concentration of potassium ions outside at this time is designated as $C_o$, and the concentration of chloride ions is designated as $C_o$.

Since chloride ions penetrate the membrane, they diffuse inward in response to the concentration gradient. To maintain electric neutrality, potassium ions accompany them. At equilibrium, therefore, a certain concentration, X, of potassium and chloride ions will have entered the

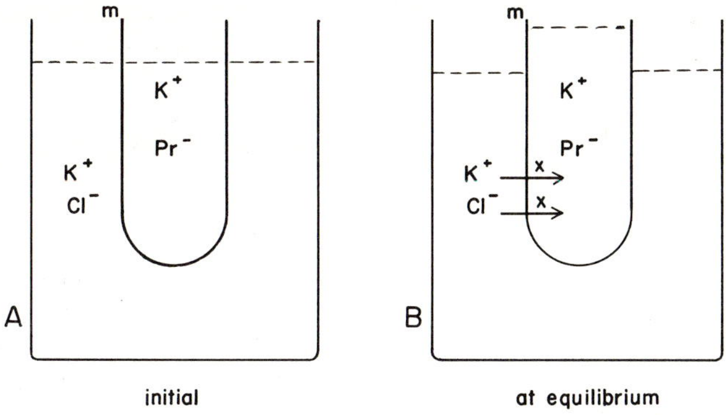


| Initial concentration | Final concentration |
|---|---|
| inside | inside |
| $K^+ = C_i$ | $K^+ = C_i + X$ |
| $Pr^- = C_i$ | $Pr^- = C_i$ |
| $Cl^- = 0$ | $Cl^- = X$ |
| outside | outside |
| $K^+ = C_o$ | $K^+ = C_o - X$ |
| $Cl^- = C_o$ | $Cl^- = C_o - X$ |

***Figure 13.9.*** Donnan equilibrium. The membrane (m) within which the protein solution is contained is impermeable to protein ions but permits all the other ions to pass. Note: Because salts are taken up into the tube containing the protein, the osmotic pressure will rise and water will enter.

membrane (Fig. 13.9*B*). At that time the concentration of potassium ions inside the membrane will be ($C_i + X$); that of the chloride ions inside will be X. The concentration of potassium ions outside the cellophane bag will be ($C_o - X$) and that of the chloride ions outside the bag will be ($C_o - X$).

Donnan (1927), by applying the law of mass action, found that at equilibrium the product of the concentrations of the diffusible ions inside the bag equals the product of the concentrations of the diffusible ions outside the bag:

$$[K^+]_{in}[Cl^-]_{in} = [K^+]_{out}[Cl^-]_{out} \quad (13.1)$$

The physical chemistry and thermodynamics of the Donnan equilibrium are dealt with elsewhere (e.g., Bull, 1964).) Substituting for these symbols the concentration terms used before, for the simplest case, in which the compartments are equal, one can write the following equation:

$$(C_i + X)X = (C_o - X)(C_o - X) \quad (13.2)$$

Solving for X:

$$X^2 + C_iX = C_o^2 - 2C_oX + X^2 \quad (13.3)$$

$$X = \frac{C_o^2}{C_i + 2C_o} \quad (13.4)$$

Knowing the initial concentrations of ions inside and outside a membrane, one can use this equation to solve for X, the concentration (at equilibrium) of the diffusible ions which have entered the membrane. The equilibrium is called the *Donnan equilibrium* in honor of its discoverer.

If a mixture of uni-univalent, diffusible and membrane-penetrating electrolytes such as NaCl, $KNO_3$, LiBr and HI is used instead of KCl inside and outside the tube containing the protein (Fig. 13.9), the following relations hold:

$$\frac{[Na^+]_i}{[Na^+]_o} = \frac{[K^+]_i}{[K^+]_o} = \frac{[Li^+]_i}{[Li^+]_o} = \frac{[H^+]_i}{[H^+]_o} =$$

$$\frac{[Cl^-]_o}{[Cl^-]_i} = \frac{[NO_3^-]_o}{[NO_3^-]_i} = \frac{[Br^-]_o}{[Br^-]_i} = \frac{[I^-]_o}{[I^-]_i}$$

## THE EFFECTS OF TEMPERATURE ON PERMEABILITY—ENERGY OF ACTIVATION

If certain substances enter a cell in response to a concentration gradient between the outside and inside of the cell, and if the molecules are freely able to enter and leave the plasma membrane, the temperature coefficient for entry of these substances into cells should be like that found for free diffusion, namely, 1.3 to 1.4 (Table 10.1). Experiments show this to be true in a few cases. For example, a temperature coefficient of this order is found for the entry of propyl alcohol into erythrocytes and of some aldehydes into luminous bacteria (Tables 10.1 and 13.3), suggesting that these molecules move freely through these cell membranes.

However, in many other cases, e.g., the entry of dyes into *Arbacia* or into frog skin (Table 10.1), the rate of entry of compounds into a cell is increased several fold by a 10° C rise in temperature, indicating that most molecules have to be activated to cross some barrier in the plasma membrane (Dawson and Danielli, 1952) (see Chapter 11).

Assuming that the plasma membrane is a sandwich of two layers of lipid between an outer and an inner layer of protein, as was previously discussed, the following illustration will perhaps clarify some of the problems involved.

The plasma membrane of the cell (see Fig. 13.10) presents the main hurdles to any entering molecule which must pass from the aqueous external medium into the cell membrane, through the plasma membrane and, finally, from the plasma membrane into the aqueous medium of the cytoplasm within the cell. The plasma membrane, as was shown, contains a considerable amount of lipoidal material. If a molecule entering a cell has a high lipoid solubility (that is, a high partition coefficient), neither its entry into nor its transport through the membrane presents an obstacle. However, entry of the molecule from the membrane into the aqueous cytoplasm does present an obstacle because the cohesive bonds it has formed with the lipoid molecules of the membrane must be broken before it can leave the membrane.

If, on the other hand, a molecule entering a cell membrane is highly soluble in water (that is, it has a low partition coefficient), its strong hydrogen bonds with the water in which it is dissolved outside the cell must be broken before the molecule can enter the lipoid membrane. Its movement through the lipoid membrane also presents an obstacle. However, such a molecule passes readily from the membrane into the aqueous

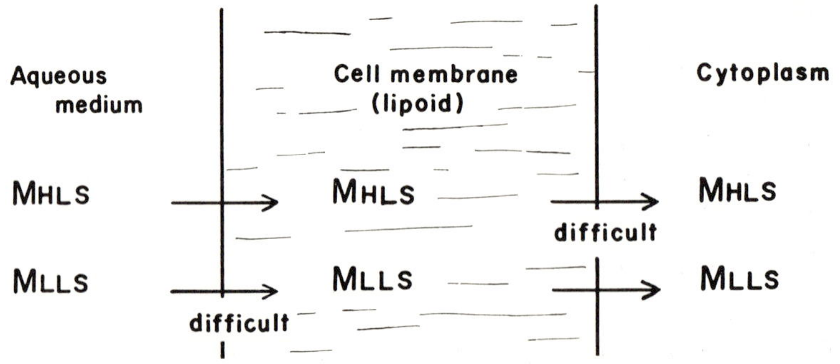


*Figure 13.10.* Hurdles encountered by a molecule with high lipoid solubility ($M_{HLS}$) and a molecule with low lipoid solubility ($M_{LLS}$) when passing through the cell membrane to the cytoplasm. (After Davson and Danielli, 1952: The Permeability of Natural Membranes. 2nd Ed. Cambridge University Press, London.)

cytoplasm inside the cell where it forms new hydrogen bonds. In this case, therefore, passage into and through the cell membrane is the most difficult hurdle.

If the energy required to get across any of these hurdles is large, only the molecules with the required energy of activation are successful in moving through the membrane (Davson and Danielli, 1952; Rogers and McElroy, 1958b). For this reason, temperature coefficients for permeability may be quite large even when active transport is not involved.

## EFFECT OF RADIATION ON PERMEABILITY

The effects of visible light, ultraviolet radiations and ionizing radiations upon the penetration of substances into cells has been studied, and some literature on this topic exists. Some of the findings are controversial, and therefore they will not be reviewed here, but it is known that when cells are sufficiently injured by radiations their permeability to various solutes increases as it does following any other injury (Green, 1954; Shepard and Stewart, 1952; Giese, 1968).

## THE EFFECTS OF SALTS ON PERMEABILITY—SALT ANTAGONISM

The salt environment to which a cell is subjected affects the permeability of the plasma membrane to other substances. To study these effects, the electrical conductivity of cells has been measured, as an index of entry of ions. With a cork borer, discs of tissue cells of the kelp *Laminaria* are cut from the blade of the plant. These are stacked like a roll of coins between two electrodes, and the electrical resistance (the reciprocal of the conductivity) is determined (see Fig. 13.11). A decrease in resistance means that the ions are passing through the cells more readily and that the permeability has therefore been increased. A few samples of the many interesting studies performed are cited in the following paragraphs.

When *Laminaria* discs are immersed in sodium chloride solutions isotonic with sea water (0.52 molar), their electrical resistance gradually decreases, indicating an increase in permeability. After extended exposure the cells, having become injured, are somewhat freely permeable to sodium chloride, suggesting that the plasma membrane of the cells now presents little resistance to entry and exit of the sodium and chloride ions. If not subjected to too long an exposure to the sodium chloride, the *Laminaria* discs recover when immersed again in sea water, and their electrical resistance increases to some extent (see Fig. 13.12).

When this experiment is repeated with calcium chloride isotonic to sea water (0.278 molar solution), the electrical resistance of the discs first increases, indicating decreased permeability to the salt (see Fig. 13.13), and then falls, show-

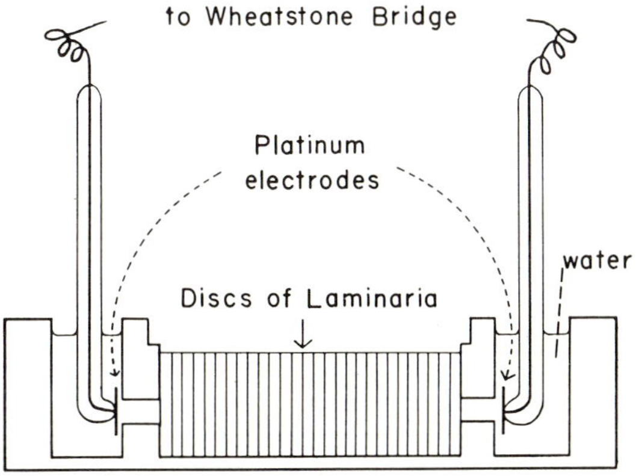


***Figure 13.11.*** Apparatus for conductivity studies. (After Osterhout, 1922: Injury, Recovery and Death in Relation to Conductivity and Permeability. J. B. Lippincott Co., Philadelphia.)

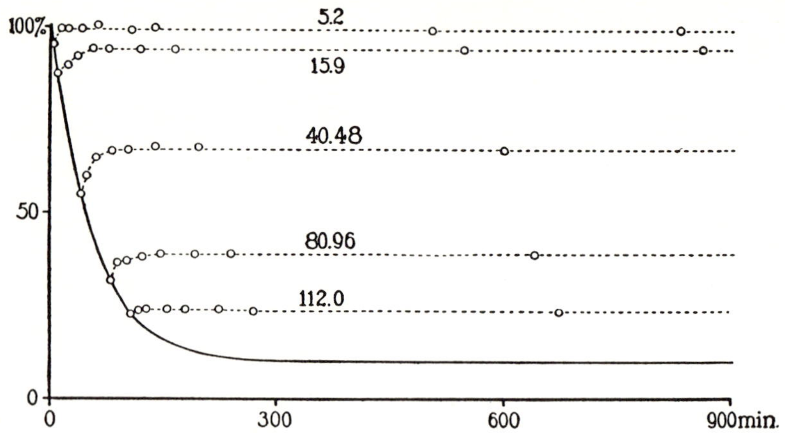


***Figure 13.12.*** The effects of isotonic NaCl on the permeability of disks cut out of the kelp *Laminaria.* The curves show the net electric resistance (ordinate) in per cent of the initial resistance of *Laminaria agardhii* in 0.52 molar NaCl (unbroken line) and recovery in sea water (dotted lines). The figure attached to each recovery curve denotes the time of exposure (in minutes) to the solution of NaCl. (From Osterhout, 1922: Injury, Recovery and Death in Relation to Conductivity and Permeability. J. B. Lippincott Co., Philadelphia.)

ing that the cells, having become injured are now more permeable to the ions.

Isotonic potassium chloride produces effects similar to those of sodium chloride, and isotonic magnesium chloride produces effects similar to those of calcium chloride.

When *Laminaria* discs are placed in an isotonic mixture of sodium chloride and calcium chloride, in the same proportions as present in sea water, the electrical resistance remains about the same as that of the control in sea water (17

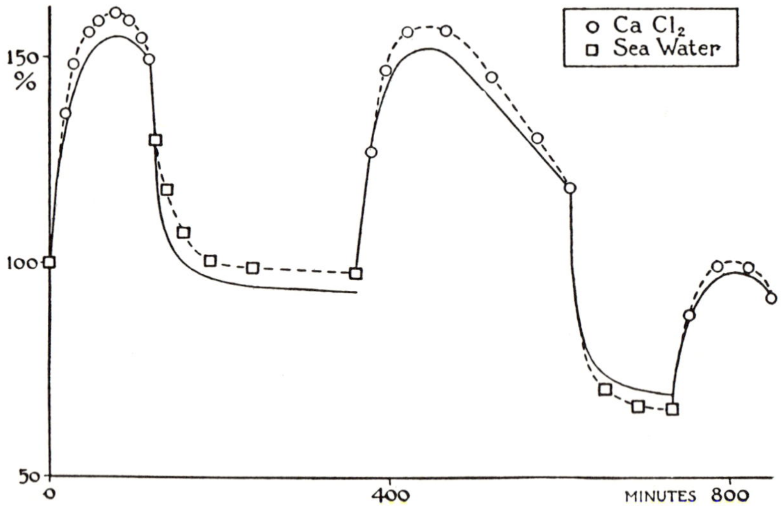


***Figure 13.13.*** The effect of isotonic $CaCl_2$ on the permeability of disks cut out of the kelp *Laminaria.* The curves show the net electric resistance (ordinate) in per cent of the initial resistance of *Laminaria agardhii* in 0.278 molar $CaCl_2$ and in sea water. The unbroken line represents calculated values, and the broken line represents observed values. The curves are based on an average of 10 or more experiments. (From Osterhout, 1922: Injury, Recovery and Death in Relation to Conductivity and Permeability. J. B. Lippincott Co., Philadelphia.)

hours, in one test), indicating that the permeability is practically normal and that the cells have not been injured. Since calcium and sodium chlorides have opposite initial effects on the permeability of the plasma membrane, calcium chloride decreasing it and sodium chloride increasing it, these two ions are said to antagonize each other. The permeability of the membrane is completely normal if potassium and magnesium chlorides are also added to the solution of sodium and calcium chlorides in the proportions in which these cations are found in sea water. Salt antagonism appears to be the antagonistic action of different salts in maintaining normal permeability of the plasma membrane.

Similar salt antagonism is also observed on animal cells. For instance, a frog heart placed in pure sodium chloride solution ceases beating, but if calcium chloride is added in the proportion found in blood (or sea water), the heart revives. However, the beat is not normal unless potassium chloride and a buffer are added also (see Fig. 13.14). Ordinarily glucose is also added to the solution to supply nutrient to the heart muscle cells. Such a balanced salt solution was developed for frog heart by Sidney Ringer, a British physician, and is called *Ringer's solution* in his honor (Ringer, 1880, 1887). However, it is not known whether the salts exert their action by affecting the permeability of heart muscle cells or for other reasons; e.g., calcium may be required in the bathing medium because of its inciting action on muscle contraction. This topic is discussed in Chapter 25. Many similar solutions have been developed since for experimental use with protozoans, marine eggs, blood cells, tissues, organs and tissue cultures (Prosser and Brown, 1961; Lockwood, 1961).

The fact that calcium decreases the permeability of the plasma membrane to water and to other substances is of great importance, especially in the water relations of cells of fresh water organisms. Such organisms must maintain within them a certain amount of salt for normal functioning, even though little salt is available in the bathing medium and even though such cells continually take up water and lose salts. Experiments show that calcium ions in water reduce the permeability of gill cells to both water and salts; some gill cells have developed the capacity to take up salt even against a gradient (see Chapter 14).

The problem of regulating the permeability of gill cells is acute for brackish water animals like crabs, which migrate into highly diluted sea water or even fresh water. When such animals are placed in distilled water they may lose salts and gain water so rapidly that they die. In hard fresh water, on the other hand, they are able to get along because calcium reduces the permeability of the plasma membranes of gill cells to both solutes and water.

The antagonistic relations between various cations are more complex than would be suggested by these discussions. They have been extensively studied, but a satisfactory interpretation in terms of molecular interactions in the cell mem-

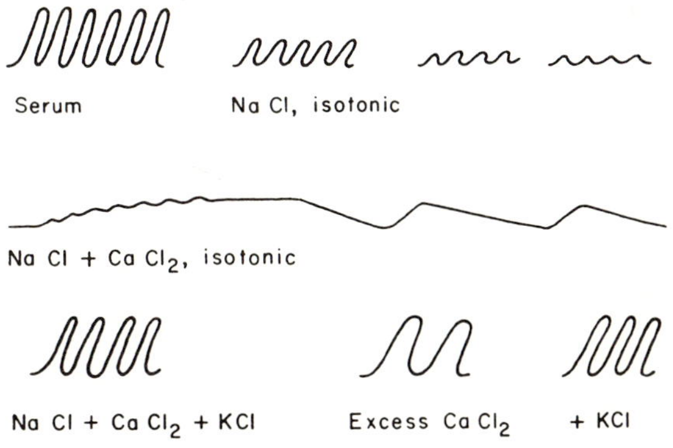


***Figure 13.14.*** The effects of salts on a frog's heart. The heart stops in isotonic NaCl, but beats slowly when $CaCl_2$ is added in the proportions found in blood (the heart must be stimulated once). The beat is almost normal when KCl is added in the proportion found in blood and when the mixture contains a buffer. Excess $CaCl_2$ prolongs the beat of the heart and is antagonized by KCl. (After Ringer, 1882–83: J. Physiol., *4*:29, 422.)

brane has not been found. Although anions of salts, used in balanced media, probably also affect the life of the cells suspended in them, salt antagonism is thought to be essentially a cationic phenomenon (Heilbrunn, 1952; Lockwood, 1961).

## THE EFFECT OF NARCOTICS AND ANESTHETICS UPON PERMEABILITY

Narcotics and anesthetics generally seem to exert their characteristic effects on cell sensitivity and respiration in proportion to their lipoid solubility, indicating that they may need to enter the cell to affect it. Some values for the narcotic action of a number of compounds, correlated with their solubility in olive oil, are given in Table 13.5.

Anesthetics also affect the permeability of cells in an interesting manner. For example, butyl alcohol decreases the penetration of glycerol into the red blood cells of human beings, rats, rabbits, guinea pigs and birds, but it increases the penetration of glycerol into the red blood cells of sheep, horses, dogs, bats, pigs, reptiles and fishes. Thiourea, on the other hand, has effects opposite to that of butyl alcohol on the penetration of glycerol into the red blood cells of the two series of animals tested (Jacobs, 1952). Furthermore, the same anesthetic, at different concentrations, may have the reverse effects on the permeability in a cell. For example, urethan in low concentration decreases the permeability of discs of algal tissue, whereas at high concentrations it has the opposite effect when first applied. In both instances prolonged exposure results in injury and increased permeability (Osterhout, 1922).

**Table 13.5.** Relation Between Narcotic Action and Lipoid Solubility

| COMPOUND | VOL. % REQUIRED TO JUST IMMOBILIZE MICE | SOLUBILITY IN OLIVE OIL AT 37° C. |
|---|---|---|
| Ethane | 80 | 1.3 |
| Acetylene | 65 | 1.8 |
| Dimethyl ether | 12 | 11.6 |
| Methylchloride | 6.5 | 14.0 |
| Dimethyl acetal | 1.9 | 100 |
| Chloroform | 0.5 | 265 |

Data from Höber, 1945.

Such diverse effects of anesthetics on the permeability of cells are difficult to explain. Perhaps the anesthetic molecules, being highly lipid soluble, accumulate at the surface of the cell, thus altering the properties of the membrane by either increasing or decreasing the ease of entry of other molecules. For example, if the solution of anesthetic in the membrane makes the membrane more lipoidal in nature, molecules of high partition coefficient will enter more readily than before, whereas those of low partition coefficients will enter less readily. The converse would also be possible (Danielli, 1950).

## THE EFFECT OF THE PHYSIOLOGICAL STATE

Movement of solutes through the plasma membrane is dependent primarily on the physiological state of the cells. Active cells utilize cell foods and accessory materials, thus maintaining a continual diffusion gradient for these substances, enabling the substances to enter the cell readily.

When a muscle cell or nerve cell is shocked and stimulated, its permeability to ions is increased. An active muscle cell is permeable to amino acids, glucose and other materials, whereas a resting one is not (Steinbach, 1954). Electric stimulation of various plant and animal cells results in an increase in their permeability to solutes at the negative electrode, as shown by increased conductivity (decreased resistance). The leakage of potassium ions at this electrode in stimulated *Nitella* and the squid giant axon may be sufficient even for measurement by microchemical methods (Hodgkin and Keynes, 1953). As shown by experiments with radioactive sodium, sodium ions also pass through the membrane of a stimulated cell more freely than they do through the membrane of a resting cell.

Injury of the cell by heat, radiations,

overstimulation, anesthetics, pH change, salt unbalance or any other means results in an increased permeability of the cell. By use of the conductivity method Osterhout (1922) has made an extensive study of the effects various degrees of injury have on permeability of *Laminaria* and other algae. If the injury is slight there is only a slight increase in the conductivity, indicating that the permeability of the cells has increased slightly. If the injury is large, the increase in conductivity is greater. If exposure to any of the injurious factors just listed has not been excessive before the cells are returned to normal conditions, complete recovery of normal permeability often follows. However, as shown in Figures 13.12 and 13.13, only partial recovery may occur, and the permeability may be permanently increased by injury. All gradations of change are possible, depending upon the damage to the cells. Studies of permeability changes following injury of a variety of cells indicate a similar action. Death, as measured by permeability, is a quantitative phenomenon.

## FACILITATED DIFFUSION

When the rate of entry of a solute into a cell, e.g., galactose into some red blood cells, increases proportionally with increase in solute concentration in the medium only to a limit, beyond which there is little further change in rate of entry with further increase in concentration, permeation of the solute is considered to be limited by saturation of a *carrier system.* The carrier is perhaps some substance in the plasma membrane which combines with the solute in question, transporting it into the cell. Presumably the carrier liberates the solute inside the cell and itself returns to "pick up" another molecule (or ion). This type of transport is often called *facilitated diffusion.* Facilitated diffusion occurs at the expense of kinetic energy alone; only those solute molecules with the critical increment of kinetic energy are able to penetrate through the plasma membrane. Facilitated diffusion is to be distinguished from active transport in which a carrier may also be involved, but passage of the solute through the plasma membrane occurs at the expense of metabolic energy (see Chapter 14). Facilitated diffusion is therefore not affected by metabolic poisons, whereas active transport is.

Facilitated diffusion may show a high temperature coefficient. This is to be expected because passage of the solute through the plasma membrane is dependent upon the number of solute carrier-complexes having sufficient energy for the purpose. The proportion of complexes having the critical energy may increase several fold with an increase in temperature of 10° C. (see Chapter 10).

Although facilitated diffusion does not alter the final equilibrium which could be attained even in the absence of a carrier system, it does greatly speed the attainment of the equilibrium. It is, therefore, advantageous to the cell (see Crone, 1965; Levene, 1965).

## EXCHANGE DIFFUSION

When the plasma membrane of a cell is permeable to given molecules or ions it is conceivable that some will come out of the cell at the same time as some go in, especially if a carrier mechanism transports the molecules (or ions). Thus, when some cells are kept in a solution of $^{24}Na$ for a period, then placed in "cold" sodium ion solution, the radioactive sodium leaves the cell in proportion to its relative concentration in the cells. The loss of $^{24}Na$ presumably occurs by exchange diffusion, a radioactive sodium ion being exchanged at random for a cold sodium ion, because if the bathing solution containing sodium is replaced by one of isotonic mannitol, the exchange ceases since no sodium ion is available for exchange. The exchange is at the expense of kinetic energy, not metabolic energy, as shown by its independence of metabolic poisons (Davson, 1964; Keynes, 1966).

Exchanges of one molecular (or ionic) species for another are also known to occur when the two species are transported by a common carrier. For example, the amino acids histidine and tryptophan are readily exchanged for each other by Ehrlich ascites tumor cells. These exchanges do not require metabolic energy nor are they subject to the inhibitors which affect metabolically linked transport (Quastel, 1965).

Some cation-for-cation and anion-for-anion exchanges require metabolic energy. Such exchanges are discussed in Chapter 14.

## COMPARISON OF THE PERMEABILITY OF DIFFERENT KINDS OF CELLS

The relative permeability of different kinds of cells to a particular substance may be compared by any method which is equally applicable to all of the cells. For example, the rate of hemolysis has been used for comparison of the permeability of different erythrocytes. Erythrocytes of various species of mammals, when tested in this manner for their permeability to glycerol, show differences covering a range of 240 fold (see Table 13.6). Similar differences are observed for entry of sodium ions, potassium ions, urea and glucose.

When comparing the permeability of various kinds of cells to a given substance, and the same method cannot be applied to all of the cells, then it is necessary to measure the amount of the substance entering a unit area per unit of time and per unit concentration difference between the inside and outside of the cells. Calculation of a permeability constant incorporating these quantitative data has been made possible by means of certain equations (Davson, 1964). These are based on Fick's law of diffusion, which states that a substance, S, will diffuse through an area, A, at a rate, dS/dt (amount, dS, per unit time, dt), which is dependent upon the difference in concentration between the substances at a certain distance apart.

$$\frac{dS}{dt} = DA\frac{dC}{dx} \qquad (13.6)$$

In this equation D is the diffusion coefficient and is given in moles per unit area per unit concentration gradient, and dC is the difference in concentrations $(C_1 - C_2)$ over the distance dx.

For passage through a membrane such as that of a cell, equation 13.6 becomes

$$\frac{dS}{dt} = \frac{DA}{M}(C_{out} - C_{in}) \qquad (13.7)$$

Here $C_{out}$ represents the concentration of the substance, S, outside, and $C_{in}$ is its concentration inside the cell. A is the area of the membrane, and M—which is substituted for dx in equation 13.6—represents the thickness of the membrane.

Since the amount of substance (S) per volume (V) of the cell is $C_{in}$, $C_{in}V$ may be substituted for S. When k, the permeability constant, is substituted for D/M, the equation is written as follows:

$$\frac{dC_{in}}{dt} = k\frac{A}{V}(C_{out} - C_{in}) \qquad (13.8)$$

If the volume of fluid outside the cell is large, integration of equation 13.8 and solution for k gives

$$k = \frac{V}{At}\, ln\frac{C_{out} - C_{in}}{C_{out} - C^1_{in}} \qquad (13.9)$$

Here $C_{in}$ and $C^1_{in}$ are the concentrations inside the cell before and after time t. The permeability constant, k, is therefore given in moles of the substance entering per square micron of cell surface per second per mole difference in concentration of substance S outside and inside the cell (Davson, 1964).

The permeability constants for entry of a variety of substances into a given cell are different because some substances enter readily, others slowly. Although for any one substance tested the permeability con-

**Table 13.6.** Variation in Permeability of Erythrocytes of Various Mammals, Measured by Time for 75 Per Cent Hemolysis in 0.3 Molar Glycerol in 0.12 Per Cent NaCl

| ANIMAL | TIME | ANIMAL | TIME |
|---|---|---|---|
| Rat | 3.5 | dog | 253 |
| Man | 5.1 | pig | 340 |
| Mouse | 12.9 | cat | 459 |
| Rabbit | 21.8 | ox | 612 |
| Guinea pig | 38.2 | sheep | 850 |

Data from Höber, 1945.

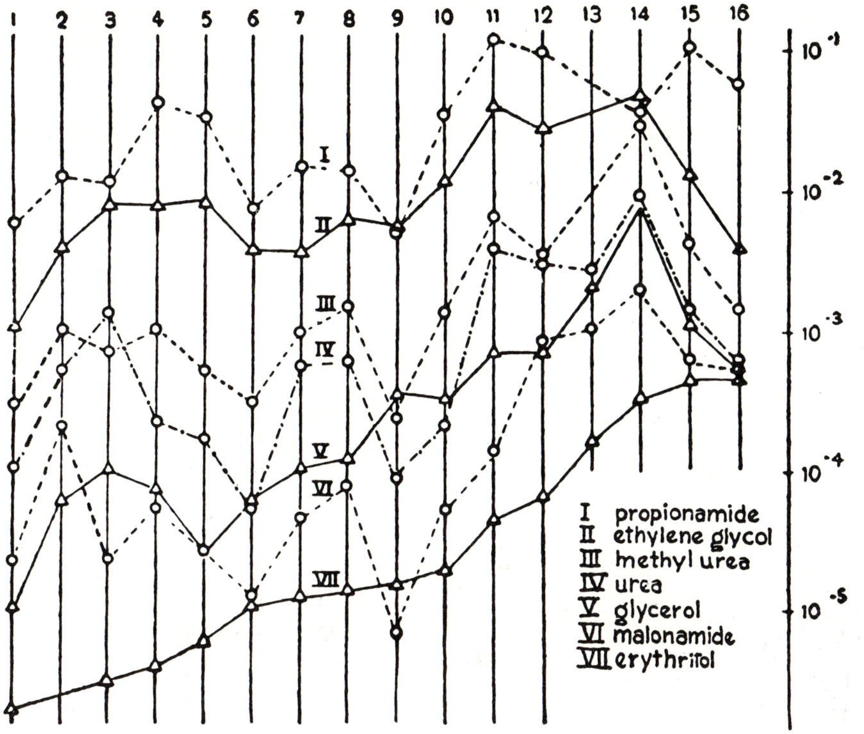


***Figure 13.15.*** Permeability constants of plant cells to nonelectrolytes. Cells from 16 species are represented: 1, Leaf cells of *Plagiothecum denticulatum;* 2, *Oedogonium sp.;* 3, root cells of *Lemna minor;* 4, *Pylaiella litoralis;* 5, *Zygnema cyanosporum;* 6, subepidermal cells of *Curcuma rubricaulis;* 7, *Spirogyra sp.;* 8, leaf cells of *Elodea densa;* 9, epidermal cells of *Rhoeo discolor;* 10, epidermal cells of *Taraxacum pectinatiforme;* 11, "leaf cells" of *Chara ceratophylla;* 12, internodal cells of *Ceramium diaphanum;* 13, *Bacterium paracoli;* 14, *Oscillatoria princeps;* 15, *Melosira sp.;* 16, *Licmophora sp.* (From Collander, 1937: Trans. Faraday Soc. *33*:989.)

stant is different for cells of each of a number of species of organisms, the cells of each of these various species often show the same relative differences in permeability constants for different substances tested on each of them. This is illustrated in a study on penetration of a number of substances into cells of 16 species of plants illustrated in Figure 13.15. This finding gives greater confidence that the rules of permeability, developed from experiments on the cells of a fair sample of organisms, are likely to apply to other cells in a qualitative way. No method yet exists for predicting permeability quantitatively.

## INTRACELLULAR TRANSPORT

Transport also occurs between organelles in a cell. Such transport may occur by the same mechanisms as between the cell and its bathing medium, although information is at present rather limited (Warren, 1966).

## LITERATURE CITED AND GENERAL REFERENCES

Baptiste, E. C. J., 1935: The effect of cations on the permeability of cells to water. Ann. Bot. *49*:345–366.

Bělehrádek, J., 1935: Temperature and Living Matter. Protoplasma-Monographien. Verlag von Gebrüder Borntraeger, Berlin, Vol. 8, p. 31.

Brooks, S. C., 1938: The penetration of radioactive KCl into living cells. J. Cell. Comp. Physiol. *11*: 247–252.

Brooks, S. C., 1951: Penetration of radioactive isotopes $P^{32}$, $Na^{24}$ and $K^{42}$ into *Nitella.* J. Cell. Comp. Physiol. *38*:83–93.

Brooks, S. C., and Brooks, M. M., 1941: The Permeability of Living Cells. Protoplasma-Monographien. Verlag von Gebrüder Borntraeger, Berlin, Vol. 19, Chapter 12.

Bull, H. B., 1964: An Introduction to Physical Biochemistry. F. A. Davis Co., Philadelphia.

Collander, R., 1959: Cell membranes: their resistance to penetration and their capacity for transport. *In* Plant Physiology, a Treatise. Steward, ed. Academic Press, New York, Vol. 2, pp. 3–102.

Collander, R., and Bärlund, H., 1933: Permeabilitatsstudien in *Chara ceratophylla.* Acta Bot. Fennica *11*:1–114.

Coons, A. H., 1952: The cellular fate of injected antigens. Symp. Soc. Exp. Biol. *6*:166–172.

Crescitelli, F., 1948: The dual action of carbamates on the resting potential of frog nerve. J. Cell. Comp. Physiol. *32*:187–210.

Crone, C., 1965: Facilitated transfer of glucose from blood into brain tissue. J. Physiol. (London) *181*: 103–113.

Danielli, J. F., 1950: Cell Physiology and Pharmacology. Elsevier, Houston.

Davson, H., 1964: A Textbook of General Physiology. 3rd Ed. Little, Brown & Co., Boston.

Davson, H., and Danielli, J. F., 1952: The Permeability of Natural Membranes. 2nd Ed. Macmillan Co., New York, especially Chapters 16 and 17.

Donnan, F. G., 1927: Concerning the applicability of thermodynamics to the phenomena of life. J. Gen. Physiol. *8*:685–688.

Etherton, B., and Higginbotham, N., 1960: Transmembrane potential measurements of cells of higher plants as related to salt uptake. Science *131*:409–410.

Giese, A. C., 1968: Effect of ultraviolet radiation on some activities of animal cells. *In* Ultraviolet Radiation (Conference on Biological Effects of UV-Radiation) Urbach, ed. Pergamon Press, Oxford, pp. 61–76.

Gray, W. D., and Sova, C., 1956: Relation of molecular size and structure to alcohol inhibition of glucose utilization by yeast. J. Bacteriol. *72*:349–356.

Green, J. W., 1954: The action of ultra-violet invisible light on human red cells. Anat. Rec. *120*: 707.

Heilbrunn, L. V., June 1951: Calcium and life. Sci. Am. *184*:60–63.

Heilbrunn, L. V., 1952: An Outline of General Physiology. 3rd Ed. W. B. Saunders Co., Philadelphia, Chapter 33.

Heilbrunn, L. V., and Mazia, D., 1936: The action of radiation on living protoplasm. *In* Biological Effects of Radiation. Duggar, ed. McGraw-Hill Book Co., New York, pp. 625–658.

Höber, R., 1936: Membrane permeability to solutes in its relation to cellular physiology. Physiol. Rev. *16*:52–102.

Höber, R. (ed.), 1945: Physical Chemistry of Cells and Tissues. Blakiston Co., Philadelphia.

Hodgkin, A. L., and Keynes, R. D., 1953: The mobility and diffusion coefficient of potassium in giant axons from *Sepia.* J. Physiol. *119*:513–528.

Jacobs, M. H., 1952: The measurement of cell permeability with particular reference to the erythrocyte. *In* Modern Trends in Physiology and Biochemistry. Barron, ed. Academic Press, New York, pp. 149–171.

Keynes, R. D., 1966: Exchange diffusion of sodium in frog muscle. J. Physiol. (London) *184*:318–328.

Krogh, A., 1939: Osmotic Regulation in Aquatic Animals. Cambridge University Press, London.

Levene, M., 1965: The substrate-facilitated transport of glucose carrier across the human erythrocyte membrane. Biochim. Biophys. Acta *109*:151–163.

Mazia, D., and Clark, J. M., 1936: Free calcium in the action of stimulating agents on *Elodea* cells. Biol. Bull. *71*:306–323.

Mullins, L. J., 1949: Ion transfer in cells under high gas tensions. Exp. Cell. Res. *1*(Suppl.):328–330.

Osterhout, W. J. V., 1922: Injury, Recovery, and Death in Relation to Conductivity and Permeability. J. B. Lippincott Co., Philadelphia.

Osterhout, W. J. V., and Dorcas, M. J., 1926: The penetration of carbon dioxide into living protoplasma. J. Gen. Physiol. *9*:255–267.

Overbeek, J. T., 1956: The Donnan equilibrium. Progr. Biophys. *6*:57–84.

Ponder, E., 1948: Hemolysis and Related Problems. Grune & Stratton, New York.

Ponder, E., and Saslow, G., 1930: Alterations of cell volume in solutions of various tonicities. J. Physiol. *70*:169–181.

Ponder, E., and Saslow, G., 1931: Alterations of cell volume in extremely hypertonic solutions. J. Physiol. *73*:267–296.

Prescott, D. M., and Mazia, D., 1954: Permeability of nucleated and enucleated fragments of *Amoeba proteus* to $D_2O$. Exp. Cell Res. *6*:117–126.

Prosser, C. L., and Brown, F. H., 1961: Comparative Animal Physiology. W. B. Saunders Co., Philadelphia, pp. 60–62.

Quastel, J. H., 1965: Molecular transport at cell membranes. Proc. Roy. Soc. (London), Ser. B. *163*:169–196.

Ringer, S., 1880: Concerning the influence exerted by each of the constituents of the blood on the contraction of the ventricle. J. Physiol. *3*:380–392.

Ringer, S., 1887: Further experiments regarding the influence of lime, potassium and other salts on muscular tissue. J. Physiol. *7*:291–308.

Rogers, P., and McElroy, W. D., 1958a: Enzymatic determination of aldehyde permeability in luminous bacteria. I. Effect of chain length on light emission and penetration. Arch. Biochem. Biophys. *75*:87–105.

Rogers, P., and McElroy, W. D., 1958b: Enzymatic determination of aldehyde permeability in luminous bacteria. II. Effect of temperature on enzymatic activity and penetration. Arch. Biochem. Biophys. *75*:106–116.

Ruhland, W., and Hoffman, C., 1925: Die Permeabilitat von *Beggiatoa mirabilis.* Ein Beitrag ultrafiltertheorie des Plasmas. Planta *1*:1–83.

Scholander, P., 1965: Sap pressure in relation to osmotic potential. Science *148*:344–346.

Shepard, C. W., and Stewart, M., 1952: The direct effect of radiation on erythrocytes. J. Cell. Comp. Physiol. *39*(Suppl. 2):189–215.

Solomon, A. K., Dec. 1960: Pores in the cell membrane. Sci. Am. *203*:146–156.

Steinbach, H. B., 1954: The regulation of sodium and potassium in muscle fibers. Symp. Soc. Exp. Biol. *8*:438–452.

Takada, H., 1954: Protoplasmatological studies on the yeast cells adapted to high osmotic environment. J. Inst. Polytechnics. Osaka City Univ. *4*:17.

Ussing, H. H., 1952: Some aspects of the application of tracers in permeability studies. Advances Enzymol. *13*:21–65.

Ussing, H. H., 1953: Transport through biological membrane. Ann. Rev. Physiol. *15*:1–20.

Ussing, H. H., 1954: Active transport of inorganic ions. Symp. Soc. Exp. Biol. *8*:407–422.

Warren, K. B., 1966: Intracellular Transport. Symp. Soc. Cell Biol. *5*:1–325.

# CHAPTER 14 / METABOLICALLY LINKED TRANSPORT THROUGH CELL MEMBRANES AND PINOCYTOSIS

## ACTIVE TRANSPORT

When a substance moves through a cell membrane against a concentration gradient, either into or out of a cell, it is at once apparent that such movement could not be by diffusion powered by kinetic energy alone. To move the molecules of a substance against a concentration gradient and (or) against an electrochemical gradient requires a mechanism other than—or in addition to—diffusion, and a source of energy for such movement. The cell must do work in moving the substance through the cell membrane against a concentration gradient. Any movement of substances through the cell membrane which requires work on the part of the cell is called *active transport** or more accurately, *metabolically-linked transport.*

The phenomenon of active transport is perhaps best illustrated by the activity of the cells of the kidney tubule. Pieces of kidney tubule, separated with a micromanipulator from a chick embryo, often round up into spheres one cell layer thick (see Fig. 14.1). When such spheres are immersed in dilute solutions of phenol red, dissolved in embryonic tissue fluid, each sphere accumulates the dye in its internal cavity so that the cavity becomes deep pink in color. The dye may be seen passing through the cells of the tubule in which it lingers in transit and where it becomes orange in color, a change to be expected of phenol red at the pH (about 6.7 to 6.8) inside the cells. If phenol red were diffusing through the cells in response to the difference between the concentrations of the dye in the bathing solution and in the vesicle, it would move inward until equilibrium was established. The appearance of the dye in much higher concentration inside the cavity of the vesicle than elsewhere means that the dye is passing through the cells and is being extruded or secreted into the cavity against a concentration gradient (Chambers *et al.*, 1935).

If the kidney cell preparation is placed on ice at the time phenol red is added, it

*The term active transport is used in different ways by various authors, no definition being at present satisfactory to everyone. For a discussion consult Rosenberg (1954) and Anderson and Ussing (1960).

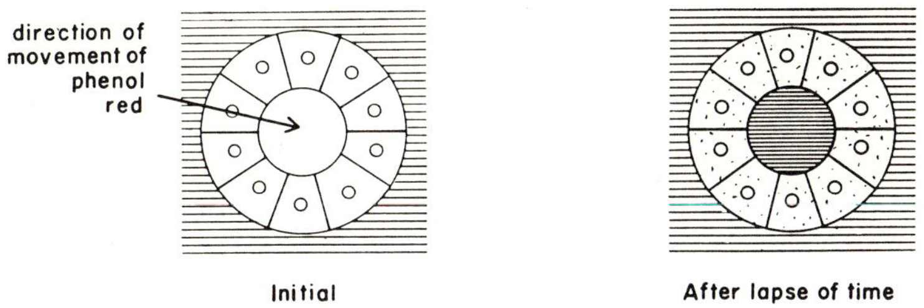


***Figure 14.1.*** Accumulation of phenol red in sections of proximal kidney tubules of embryo chick (diagrammatic). Phenol red moves inward until it becomes more concentrated in the inside of the vesicle than on the outside.

fails to accumulate the dye. If it is placed on ice after accumulation of the dye inside the vesicle, the dye then diffuses out and distributes itself equally between the medium and the lumen of the vesicle. Warmed up to chick body temperature (39° C.), the vesicle again accumulates the dye. The experiment demonstrates that although the dye continues to diffuse through tubule cells even at low temperature, it is no longer accumulated, probably because some process essential to accumulation is inhibited. Measurements show that the respiration of the kidney tubule cells comes virtually to a halt when the tubule is placed on ice. Since respiration is the main source of the energy by which cellular work is accomplished, the inference is drawn that the movement of the dye against a concentration gradient requires cellular work. This inference is corroborated by experiments, again using dyes, in which respiration of the kidney tubule cells, even though at 39° C., is inhibited by a metabolic poison. When respiration virtually ceases, dye is no longer accumulated. Diffusion, however, continues in response to the concentration gradient, and a preparation which is poisoned after it has already accumulated dye is still able to lose it to the medium.

These experiments demonstrate that kidney tubule cells can do work against a concentration gradient only if they are able to liberate energy. Energy, then, is needed to carry molecules of dye against a concentration gradient, just as energy is needed to climb a hill. The work done in transporting a dye against a concentration gradient can be calculated. It is proportional to a constant multiplied by the natural logarithm of the ratio of the concentration of the dye inside and outside the cell, the difference in concentration representing the height of the concentration "hill" (see Fig. 14.2). (When ions are being transported, a term for the work done against the electrochemical gradient must be added to this.) From the respiratory rate of the kidney cells it is also possible to determine how much energy has been released by the cell during the time that a given amount of solute has been transported. The calculations show that the energy released in respiration is more than enough to do this work.

Two criteria have often been used to determine whether a substance moves across the plasma membrane by active transport: first, whether it moves against a concentration gradient; second, whether it requires metabolic energy for the movement.

If a substance moves against a concentration gradient, transport is probably active, although not all active transport need be against a concentration gradient. However, it is usually necessary to demonstrate that the substance in question is osmotically active inside the cell and that its movement really occurs against a concentration gradient. That some substances are present inside cells in a concentration greater than outside (Fig. 3.1) does not necessarily imply that they enter the cell by an active process against a concentration gradient. For instance, a substance may be rendered osmotically inactive inside the cell in any of a number of ways. If ionic, it might unite with another ion to form an insoluble compound (e.g., calcium oxalate). Or it might possibly be dis-

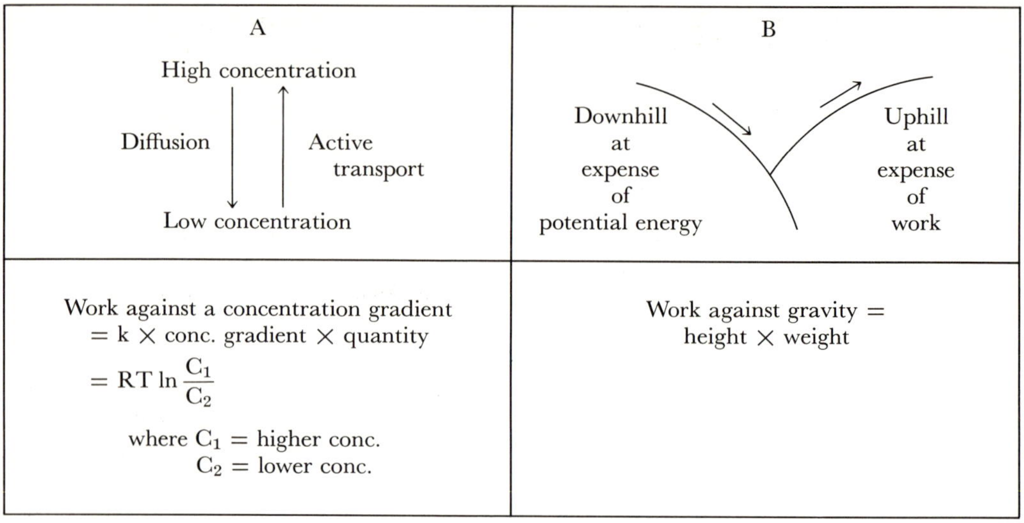


***Figure 14.2.*** Analogy between concentration gradient (left half of A) and potential gradient (left half of B) and between movement against a concentration gradient (right half of A) and work done in moving uphill (right half of B). Equations for work against a concentration gradient (osmotic work) and for work in lifting a weight up a height are given at the bottom of the figure. When work is done against a potential (electrochemical) gradient another term must be added to the equation given. This term (for electrochemical work) consists of the moles of charge transported, multiplied by the potential gradient: $W = n\mathcal{F}E$, where n is the number of equivalents in one stoichiometric occurrence, $\mathcal{F}$ the Faraday or gram molecular equivalent of unit charge and E the potential gradient in volts (Cirillo, 1966).

solved in lipid and therefore be unavailable for osmotic action in the aqueous continuum of the cell. Organic substances like glucose may be polymerized to form insoluble glycogen or starch, and glucose molecules could thus be trapped, that is, taken out of solution, and a concentration gradient favorable for uptake might be maintained in spite of extensive entry.

Since active transport depends on release of energy from metabolic reactions, the effect of respiratory activity on uptake of the materials by a cell should be tested. If the accumulation of a substance ceases after respiration is stopped by a respiratory inhibitor (e.g., cyanide), active transport is highly probable. Also, if accumulation is slowed by lowering the temperature (which decreases the rate of respiration), active transport is probable. Likewise, when the oxygen tension is greatly reduced, the aerobic respiration of the cell decreases and usually active transport declines as well. In cases in which blocking aerobic respiration does not inhibit active transport, the energy for the process might be supplied by anaerobic metabolism. This could be tested by blocking the anaerobic metabolism, for example, with iodoacetic acid, to determine whether this will stop active transport. It is likely that active transport always occurs at the expense of either aerobic or anaerobic metabolism, but ceases when both are stopped (Harris, 1956; Prankerd, 1961).

As much as 30 per cent of the metabolism of a mammalian red blood cell and 15 per cent of the metabolism of an ascites (tumor) cell is thought to be expended in active transport processes (Stein, 1960).

It is of interest at this point to compare active transport and facilitated diffusion. In both processes a carrier mechanism is involved. Therefore, both show the phenomenon of saturation; that is, the permeability increases with increasing concentration of the compound until the carrier is translocating all it can (much like an enzyme saturated with substrate; see Fig. 15.11). In both processes related substrates compete for the carrier, and stereospecificity (e.g., of D- and L- compounds) is of consequence in complexing of the substrate with the carrier. In both cases the $Q_{10}$ for transport may be 2 or more. In both cases, also, permeation is greatly accelerated over what could occur by ordinary diffusion. However, a temperature low enough to stop metabolism stops active transport but only decreases facilitated diffusion. Finally, active transport can occur against an electrochemical gradient whereas facilitated diffusion cannot.

## EXAMPLES OF ACTIVE TRANSPORT

Not much is known of the mechanism of active transport, but the process seems to be quite general in plant cells, animal cells and microorganisms. It is thought that probably most solutes enter cells actively (Le Fevre, 1955). A number of interesting cases are cited here.

***Accumulation of Potassium in Cells.*** A great accumulation of potassium is found in many plant and animal cells and in microorganisms. For example, *Nitella,* a fresh water plant with large cells, has 1065 times as much potassium as the water in which it grows; *Valonia,* another large plant cell, has 41.6 times as much; and *Chara* has 63 times as much (Table 23.2). Although the concentration of potassium is not always this much greater inside than outside the cell, such concentrations are characteristic of plant cells generally. A somewhat similar difference in concentration of potassium between the inside of the cell and the bathing medium is found in muscle and nerve cells (Fig. 3.1) and in microorganisms such as yeast.

That potassium is present in cells as an ion and is osmotically active is demonstrated by measurement of the electrical conductivity. The electrical conductivity of cells is of such a magnitude as to indicate that potassium exists in the cells in dissolved, ionic form; otherwise, the conductivity would be much lower since potassium is the main cation of the cell. The potassium gradient between the inside of the cells and the medium therefore could be maintained only by active transport. If this is the case, potassium should leak out of the cells, in response to the concentration gradient, when metabolism is inhibited. This occurs; for example, if either a squid giant axon or a *Halicystis* cell is placed under anaerobic conditions, potassium leaks out because the energy liberated by aerobic metabolism is no longer available to keep it inside. If oxygen is bubbled through the water, it enables the cells to take up potassium again. In these cells anaerobic metabolism apparently does not provide the energy required for active uptake of potassium (Blinks, 1949).

Lowering the temperature to near the freezing point permits potassium to escape from and sodium to enter human erythrocytes, presumably because the metabolic reactions required for potassium uptake are suspended. When the temperature is raised, potassium is again taken up. On the other hand, the human erythrocyte continues to hold its potassium even when the oxygen tension is greatly decreased at room or body temperature. This implies that in this case an anaerobic process (called glycolysis in animal cells) liberates the energy for active transport of potassium. Such a postulate is corroborated by the finding that poisoning the anaerobic glycolysis with fluoride causes potassium to leak out of the cells, even at body temperature (Prankerd, 1961).

Yeast holds its potassium under anaerobic conditions, provided glucose is present. It is unable to liberate energy anaerobically (by fermentation) unless nutrient is present in the bathing medium; i.e., it does not use its stored food for this purpose. Therefore, when yeast is washed free of glucose, and kept in the absence of oxygen, it loses its potassium. When glucose is restored, potassium also is recovered. A similar phenomenon is observed for *E. coli.*

A normal, healthy cell membrane is necessary for maintenance of the difference in concentrations of potassium between the inside and the outside of the cell. Alteration or injury of the membrane or of the cell leads to potassium leakage. For example, when the giant axon of the squid is stimulated, potassium leaks out (Huxley and Stämpfli, 1951). The same is true of *Nitella.* In other words, when the membrane is temporarily altered, ions can get through. On recovery, however, potassium is reabsorbed.

Potassium leakage following stimulation of cells has been the subject of a number of studies. However, it has been found that potassium is not the only ion which moves through the membrane of a stimulated cell. By carefully timing the movements of radioactive sodium and potassium in separate experiments one can observe that in a stimulated cell sodium moves in, then potassium leaks out. The sodium subsequently leaves and the potassium once again enters. Analysis of this occurrence suggested, as a possible explanation, that potassium is not accumulated by active transport, but rather that sodium is extruded by active transport; this process is referred to as the *sodium pump.* Elimination of sodium in this manner presumably goes on in all cells which accumulate potassium, and supposedly the elimination of the sodium creates a cation gradient favorable for the passive uptake of potassium. In such a case, when potassium leaks out of cells which have been inhibited by metabolic poisons, low temperature or anaerobiosis, it could be considered that the sodium pump is temporarily out of action and sodium enters the cells by diffusion. Sodium is thought to be extruded only when metabolic energy for this purpose is liberated by the cell (in the absence of poisons and at appropriate temperatures) (Conway, 1953).

However, evidence indicates that the sodium pump explanation alone is not in keeping with the facts (Rummel *et al.*, 1961). Thus, not only is sodium actively extruded in many cases, but potassium is actively transported into the cell as well, since sodium extrusion is linked to potassium uptake. For example, in the red blood cell the electrochemical gradient is inadequate to bring in the potassium, and active transport is physically necessary. It would also appear likely, in some cases, that one mechanism handles both sodium and potassium because changing the concentration of one of the ions outside the cell also affects the transport of the other ion (Stein, 1960). Evidence from use of tracers indicates that potassium is actively transported in muscle and nerve cells as well, even though the electrochemical gradient is favorable for passive entry (Keynes, 1961; Solomon, 1962; Baker, 1966). Somewhat similar data are found for the cells of the red alga *Porphyra perforata* (Eppley, 1959). In yeast, potassium is taken up by active transport in exchange for hydrogen ions (Andersen and Ussing, 1960).

It is probable that the ion transport is not the same in cells of different species; for example, the erythrocytes of a few species of mammals, such as the dog and cat, present an exception, accumulating sodium instead of potassium (Ponder, 1948). Nor is ion transport the same even in cells of different tissues in the same species. For example, the epithelial cells of the small intestines transport actively both sodium and chloride ions whereas cells of the kidney tubules transport sodium actively and chloride moves passively in response to concentration gradients (Andersen and Ussing, 1960).

***Root Pressure and Guttation.*** Considerable information has accumulated on the development of root pressure in plants. Experimental analyses of the sap indicate that the root hair cells are capable of taking up, against a concentration gradient, the various necessary salt ions such as potassium, magnesium, calcium and nitrate (Biddulph, 1959). From the root hair cells the salts are transferred to the cortical cells and from the cortical cells they are actively secreted against a concentration gradient into the xylem ducts. From there the salts may readily be distributed by diffusion to any cells where they are required in the plant. Accumulation of salts in the root cells, thus leading to a high osmotic pressure, accounts for bleeding of a cut stem and for guttation, the phenomenon in which, under humid conditions, droplets of water appear from water-secreting glands along the edge of a leaf. Although a detailed discussion of bleeding and guttation is out of place here, such phenomena indicate that the plant cells actively transport some materials against concentration gradients.

***Vacuolar Function.*** It is possible that fresh water protozoans make use of active transport in voiding water. The water surrounding them contains less salt than the protozoan protoplasm does; therefore, water enters in response to the concentration gradient but must leave the protozoans against the concentration gradient. The magnitude of the elimination of water becomes evident when it is realized that every 15 or 20 minutes a *Paramecium* empties through its contractile vacuole a quantity of water equal to its entire volume (Kitching, 1954, 1956).

On the other hand, a brackish water protozoan in slightly diluted sea water may take in little water, in which case the activity of its contractile vacuole may be very low. However, if the medium is greatly diluted, as after a heavy rain, the rate of vacuolar contraction may be greatly increased.

When a fresh water protozoan is poisoned with cyanide, and its respiration thus suspended, the contractile vacuoles fail to function and the cell swells (Fig. 14.3). If it is not washed free of cyanide, it bursts.

Active transport is likely in these examples, but to determine the mechanism it was necessary to determine whether, in the formation of the vacuole, salts are first secreted against a concentration gradient so that then water enters passively into the vacuole, or whether water is actively passed into the vacuole directly against a concentration gradient. Evidence from micropuncture and analysis of the fluid withdrawn from the contractile vacuole of *Amoeba* at various times in the vacuolar cycle suggests that the fluid secreted into the contractile vacuole is isosmotic with the cytoplasm. The salt is subsequently actively absorbed against a concentration gradient, leaving the fluid hypoosmotic to the cytoplasm (Schmidt-Nielsen and Schrauger, 1963).

***Active Transport in Cells of the Gut.*** When there is a high concentration of sugars in the gut of a dog, a concentration gradient exists from the cells of the lumen of the gut to the blood stream favorable to movement of sugar by simple diffusion through the plasma membranes. To test whether such movement does occur,

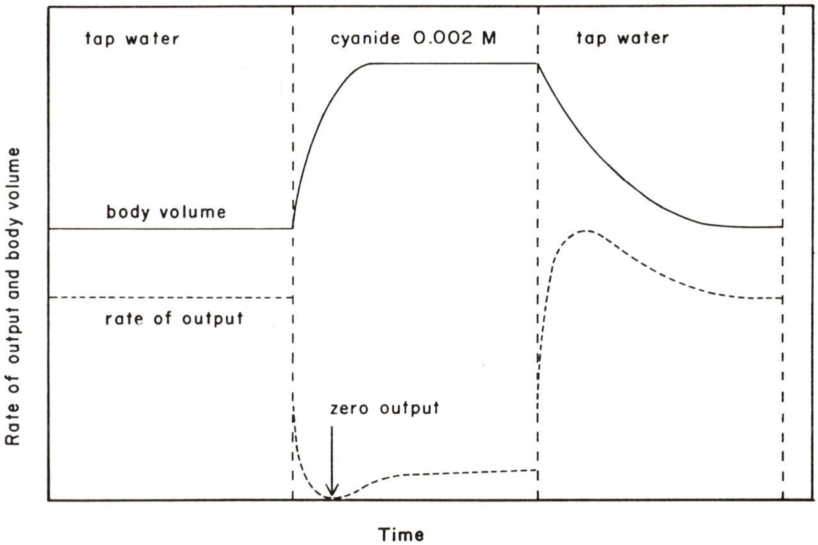


***Figure 14.3.*** The effects of cyanide on the swelling and vacuolar output of the fresh water protozoan *Zoothamnion.* The time of a complete experiment is about 2 hours. (After Kitching, 1938: Biol. Rev. *13*:423.)

the metabolism of the gut cells must be inhibited by appropriate metabolic poisons, so that active transport can be stopped. When this is done in isolated loops of the intestine, it was found that pentoses move as freely as they did before, but though some glucose and galactose still move into the blood, they do so at a much lower rate than in the absence of poison. This experiment suggests that the movement of glucose and of galactose depends in part upon active transport. To test this hypothesis, the blood stream is loaded with glucose (or galactose) so that any movement of glucose from the unpoisoned gut cells to the blood must then occur against a concentration gradient. The fact that it continues at a relatively rapid rate indicates that active transport is involved. A similar experiment is one in which a concentration of glucose (or galactose) less than that in the blood is placed in the gut so that any movement from gut to blood must occur against a concentration gradient. It still does. The transporting mechanism however, has a limited capacity. For example, if the glucose concentration in the gut is increased, the rate of its transport into the blood rises until a definite concentration limit is reached, beyond which further increase in its concentration has no effect on the rate of its transport. The rate of transport is seemingly limited by that amount which the carrier mechanism can transport per unit time (saturation). The transport of sugars requires sodium and appears to be linked to sodium transport (Schultz and Zalusky, 1964; Alvarado, 1966).

The conclusion from all of these experiments is that some sugars (e.g., some pentoses), probably move from gut to blood by simple diffusion through the plasma membranes of the gut cells and therefore only when the concentration gradient in that direction is favorable. So also do glucose and galactose to a small extent, but in addition they are actively transported even against a concentration gradient, much of the transport of glucose occurring apparently in this manner (Davson, 1964).

It is interesting that when two sugars, e.g., D-glucose and D-galactose, which are known to be actively transported, are tested at the same time, they compete with each other. This indicates a limited amount of carrier to which either sugar but not both can be bound at the same time. The binding is stereospecific and neither L-glucose nor L-galactose can be actively transported. Of some 49 sugars tested, only 14 were found to be actively transported in the intestine (Crane, 1962).

***Active Transport in the Cells of the Nephric Organs.*** Active transport against a concentration gradient is most readily demonstrable in the tubules (nephrons) of the aglomerular kidney found in only a few marine fishes (such as the pipefish and the goosefish), because fluid can enter such a tubule only through the membranes of the cells. The only way in which the tubule of a nephron can form urine in this case is by transport of matcrials, across thc cclls of the tubule, from the blood into the lumen (Smith 1956; see Fig. 14.4). If the concentration of a given compound in the tubule is less than that in the blood, entry of that compound into the tubule can occur by simple diffusion through the cell. On the other hand, the urine of aglomerular fishes is found to contain creatine, creatinine, uric acid, magnesium, potassium, chloride and sulfate, each of them often in a concentration considerably greater than in the blood. The aglomerular kidney also concentrates in the urine a variety of foreign materials, such as nitrate, thiosulfate, sulfocyanides, indigo carmine, neutral red and phenol red, that have been experimentally injected into the blood. Moreover, this activity ceases when oxygen is withheld or when the metabolism of the kidney cells is inhibited by appropriate poisons. All evidence, therefore, indicates that active transport against a concentration gradient, requires metabolic energy.

The glomerular kidney, which is characteristic of all vertebrates except the fishes cited, contains a glomerulus or tuft of blood capillaries in an epithelial cup for filtration of water and of dissolved

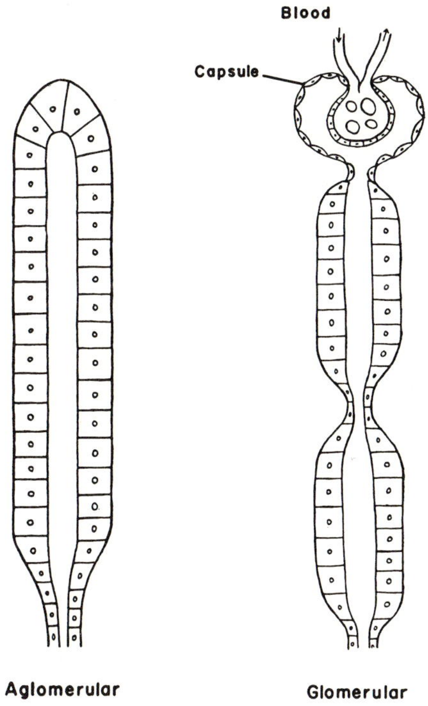


*Figure 14.4.* Diagram of aglomerular and glomerular kidney tubules (nephrons). Both are copiously supplied with blood capillaries along the outside. Only the glomerular tubule has a capillary tuft in a capsule.

constituents of the blood other than the proteins (see Fig. 14.4). While the fluid passes along the length of the tubule, part of the water and some of the other filtered constituents are reabsorbed (Robinson, 1953, 1954). Although active transport is more difficult to demonstrate in the tubules of glomerular kidneys, it occurs in the reabsorption process (which often takes place against a concentration gradient) as well as in the secretion of certain wastes into the tubules. Inhibition of the metabolic processes in the cells of the kidney inhibits both of these reactions. Also, some wastes are found concentrated in the urine to a greater extent than would have been possible only by reabsorption of water from the filtrate. These must have been voided into the urine by active transport against the concentration gradient. Decisive and direct proof of such secretion against a gradient has already been shown in sections of the kidney tubules of a chick embryo (Smith, 1956; Ruch and Patton, 1965).

Accumulating evidence indicates that active transport occurs also in the cells of tear glands of marine birds (Schmidt-Nielsen, 1959), and in cells of the nephric organs of some invertebrates. For example, in some fresh water invertebrates the urine is found to be hypotonic to the blood. This indicates either the elimination of water against a concentration gradient or the selective uptake of salts against a concentration gradient. When certain species of crabs (or other invertebrates), capable of tolerating brackish water, are transferred from sea water to diluted sea water, they void more urine, and their respiration increases correspondingly. The increase in each case is larger if the dilution is greater. The increased respiration has sometimes been interpreted as a result of the increased amount of energy required for eliminating the excess water taken up osmotically by the gills, but proof is lacking (Potts and Parry, 1964).

In general, studies on the movement of various materials against concentration gradients in the kidney cells indicate that the materials move at different rates. The rules governing such movements resemble the rules of permeability previously discussed. Thus, very large molecules are not eliminated (e.g., an aglomerular kidney never passes sucrose, inulin or albumin); movement is related to lipoid solubility, size playing a lesser role; and so on.

## ACTIVE TRANSPORT OF ORGANIC NUTRIENTS INTO CELLS

Glucose appears to enter yeast and erythrocytes only when energy is expended, even in the presence of a favorable concentration gradient. However, the evidence for this is indirect and consists in part of the finding that heavy metals, such as copper and uranium, inhibit glucose uptake in these cells. It is postulated that enzymes at the cell surface are necessary for uptake of glucose, and that these enzymes are selectively poisoned by copper, uranium and other heavy metals (Le Fevre, 1954; Rothstein, 1954).

Evidence of a chemically and metabolically coupled transport system for uptake of some substances, such as amino acids, into bacteria has also been presented (Gale, 1954; Quastel, 1965). The entry of these substances shows pH dependency, it is poisoned by typical metabolic poisons (see Chapter 17) and the transport is competitive with analogues. Furthermore, the transport system can be saturated. Such a system is feasible for bacteria in which all the enzymes required for aerobic metabolism are located in the surface of the protoplast (see Chapters 2 and 5). In such cases an enzyme would probably transport a molecule by coupling with it at the surface and handing it on to another enzyme in the metabolic main line (Mitchell, 1959; Quastel, 1965; see Chapter 17).

## ACTIVE TRANSPORT IN MITOCHONDRIA AND ENDOPLASMIC RETICULUM

Active transport at the expense of metabolic energy occurs not only through the plasma membrane but also through the membranes of the mitochondria, and in some cells (if not generally) in the endoplasmic reticulum and possibly in other organelles. Thus active accumulation of calcium phosphate against a concentration gradient has been observed many times (Greenawalt *et al.*, 1964; Brierley *et al.*, 1963; Lehninger, 1964). As much as a 50-fold increase in calcium of rat kidney mitochondria has been recorded, leading to practically complete exhaustion of calcium in the medium. Such uptake occurs only if an oxidizable substrate is present and stops if the respiration is blocked by a metabolic inhibitor (e.g., cyanide). Also required is the high energy phosphate compound adenosine triphosphate (see Chapter 17) and inorganic phosphate as well as magnesium ion. Uptake does not occur at 0° C. (Lehninger, 1964). Inorganic phosphate accompanies calcium ion in being absorbed and is also accumulated.

The endoplasmic reticulum of striated muscle cells has also been found an active site of calcium accumulation (Hasselbach, 1964, 1965) which obeys all the criteria of active transport. The uptake has been traced autoradiographically to those elements of the muscle cells which are portions of the endoplasmic (sarcoplasmic) reticulum (Shanes and Bianchi, 1960; Costantin *et al.*, 1965). The rapid and active uptake of calcium removes one of the agents needed to continue muscular contraction and so brings about relaxation (Winegrad, 1965).

## THE PERMEASE SYSTEM

In some bacteria, e.g., *E. coli,* sugar transport occurs through a special transport system known as the *permease system.* Such a system can be demonstrated most readily in cases in which it is not a *constitutive* enzyme system (one always present and demonstrable) but rather develops or is *induced* after the addition of a substrate. The following examples serve to illustrate the phenomenon. The wild-type strain of *E. coli,* placed in a solution of lactose, a disaccharide, does not use this sugar at once but only after a lapse of time. This delay in utilization could be interpreted as the time required either for the entry of the sugar into the cell or for the induction of the enzyme ($\beta$-galactosidase) necessary to digest the disaccharide into monosaccharides before its use in metabolism. It can readily be demonstrated that in wild-type *E. coli* the digestive enzyme is always present in the cell. The cells need only be lysed and the digestive enzyme can be isolated. The time lapse in utilization of lactose must be due then to failure of the lactose to enter the cell until some change occurs. The nature of this change has been analyzed by the use of a number of mutant strains.

A mutant of *E. coli* was isolated which apparently lacks the constitutive digestive enzyme since the enzyme could not be detected in extracts of the bacteria. As a consequence, the mutant presumably cannot use lactose in its metabolism. Yet after an induction period it is able to remove lactose from the medium, as shown by quantitative determinations before and after the induction period. It was found that if the cell walls are removed (by lysozyme), the mutants accumulate lactose to the point that the protoplasts swell extensively and even burst, taking up water in response to the increased osmotic pressure. As much as 22 per cent of the dry weight of the bacteria consists of lactose at this time. Entry of lactose in this case is not passive but rather enzymatically controlled.

Still another mutant of *E. coli* has been found which possesses the digestive enzyme (as demonstrated in extracts of the bacteria) but is unable to use the lactose because, even after an inductive period, it is unable to develop the permease system. There is thus no doubt that separate systems are involved in uptake and in digestion of lactose (Cohen and Monod, 1957).

The separateness of uptake mechanism and utilization can be shown in yet another way, i.e., by the use of a metabolically inactive analogue such as thiomethyl-D-$\beta$-galactoside which is structurally similar to lactose. This lactose analogue is not normally utilized by wild type *E. coli,* but when supplied to the wild type divested of cell walls, the protoplasts take up the analogue to the point of swelling and even bursting. Thus they can neither digest nor utilize the lactose analogue, but they can accumulate it—the permease system fails to distinguish between it and lactose.

Even more convincing are experiments demonstrating that induction of the permease system, as that of any other induced system, is poisoned by inhibitors of protein synthesis such as chloramphenicol and by ultraviolet radiation, which inhibits various cell syntheses including enzyme inductions (see Chapter 9).

It is apparent in the example just given that uptake of lactose occurs against a concentration gradient and must require work. One might therefore expect it to be blocked by metabolic poisons which generally block active transport. Azide and dinitrophenol are indeed found to stop accumulation of nutrients in permease systems. Since these poisons prevent the formation of high energy phosphate bonds during respiration, it is possible that such bonds are involved in uptake of materials by permease systems (Cohen and Monod, 1957).

Monod has concluded that the permease system is a specialized enzyme transport system which, once formed, is independent of and does not participate in metabolism—it only brings the sugar (or other nutrient) through the cell membrane. Any coupling with metabolism is, therefore, physical, not chemical. Permeases have been postulated for a variety of substrates, each one specific to a substrate. *E. coli* is thought to have 30 to 60 such systems. It is conceivable that some, if not all, are coupled to energy donors. The permease presumably combines with the substrate molecule, aiding it through the osmotic barrier.

Permeases have been described for a number of species of bacteria and for yeast. They may occur in plant and animal cells but are much more difficult to demonstrate in such materials (Horecker *et al.*, 1961).

Although active transport of water has been considered a possibility by some workers, others are unwilling to accept the evidence presented. True, water is voided in hypotonic urine, but filtration of the blood and active reabsorption of the solutes in the filtrate during the countercurrent flow between uriniferous tubules and the parallel small blood vessels, into which the solutes are absorbed, would account for this satisfactorily in the glomerular kidney. However, the metabolic cost to the cell of active extrusion of the quantity of water involved alone is considered prohibitive. (Andersen and Ussing, 1960; Robinson, 1953, 1954, 1960).

The various examples cited give some idea of the great variety of biological processes which are dependent upon active uptake of compounds or ions against concentration gradients. Such active uptake occurs in plant cells, animal cells and microorganisms. It may appear surprising that cells bring in many substances by active transport when some of them might enter on the basis of their own kinetic energy, i.e., when present outside the cell in higher concentration than inside. However, active transport makes it possible for the cell to take in a particular substance at a rate determined solely by the needs of the cell at any time, regardless of how favorable or unfavorable the concentration gradient may be for its uptake at that time. Thus, to a cell, active transport of a substance has considerable advantage over its diffusion through the cell membrane in answer to a concentration gradient, in spite of its cost in metabolic energy to the cell.

## THE EFFECT OF ENVIRONMENT UPON ACTIVE TRANSPORT

Because temperature affects metabolic processes it is expected to influence the rate of active transport. If a substance is carried through a cell membrane by active transport, the rate of liberation of energy by respiration is likely to be the limiting factor, and a temperature coefficient characteristic of thermochemical reactions is to be expected; that is, the rate of entry of such a compound should be doubled or tripled by a rise of 10° C. (see Chapter 10). However, a high temperature coefficient obtained for uptake of any substance would not by itself prove that the substance is taken in by an active process. It must be corroborated by other means as well.

Other environmental factors which affect the metabolic system are likely to affect active transport also; for example, narcotics and anesthetics, heavy metals, radiations, salts and pH. Analysis of the effects of the environment on active transport, although necessary, has not yet been extensive enough for a fruitful discussion.

## POSSIBLE MECHANISM OF ACTIVE TRANSPORT

Many suggestions for the actual mechanisms of active transport, including a working model for ion transport (Teorell, 1953), have appeared in the literature during recent years, but without general agreement. Among these are suggested the following:

1. A contractile protein molecule is conceived to attach to the membrane, extending and contracting (coiling). Uncoiling involves the breaking of intramolecular bonds, thus forming affinities which can be satisfied by adsorption of foreign molecules or ions. When the protein is recoiled the original bonds are re-formed and the foreign matter desorbed inside the cell. Such protein molecules perhaps may also act as enzymes capable of liberating energy from high energy phosphate compounds by which the work necessary for active transport is accomplished (Goldacre, 1952).

2. Nucleic acid molecules attach themselves to incoming molecules or ions (Lansing and Rosenthal, 1952).

3. Electroendosmosis or mass movement of fluid (fluid circuit) occurs through a fixed membrane as a result of a difference in potential on both sides of it.

4. Enzyme carriers in the reduced state act as carriers and when oxidized release the ions (Andersen and Ussing, 1960; Stein, 1960). This has also been called electron-linked transport and the prototype is Lundegard's theory (1954). Electrons are transferred along a chain of enzymes in the main line of oxidations (see Chapter 16) as the iron in the cytochromes and cytochrome oxidase changes in valency from two to three. Cation transport is supposed to be accomplished by the exchange of hydrogen ions for the cations in question. The source of energy is the drop in potential of the electrons. Although it is appealing, the theory is not in quantitative agreement with other findings.

A variation of the enzyme transport hypothesis postulates a $Mg^{++}$-requiring enzyme system (an adenosinetriphosphatase) present in the plasma membrane with two closely associated sites, an outer end and an inner one (Skou, 1965). When the enzyme transport system is activated by adenosine triphosphate the outer site has a greater affinity for potassium ion than for sodium ion, whereas the inner site has a greater affinity for sodium ion than for potassium ion. Splitting of a high energy phosphate bond from the adenosine triphosphate (with the formation of adenosine diphosphate and inorganic phosphate) reverses these affinities, as a consequence of which sodium ion is expelled from the inside of the cell while potassium ion is taken in, completing the cycle. The enzyme system is sensitive to low concentrations of the cardiac glycoside G-strophanthin (or ouabain, as it is more generally called). It requires $Mg^{++}$ and both $Na^+$ and $K^+$ for its operation; in absence of $K^+$, $Na^+$ is not "pumped out." Sodium was localized in the cell by electron microscopic studies as sodium pyroantimonate in ouabain-treated corneal cells (Kaye and Cole, 1965). Evidence for such a system has been found in the cells of 29 of 36 tissues in the cat tested and in 10 other species. Although the quantitative details vary, the essential qualitative features remain the same (Skou, 1965).

5. Carriers other than respiratory enzymes are responsible for active transport.

Consider Figure 14.5 as an example of a generalized scheme for active transport across cells of a tubule or across a cell membrane. A cell of a nephric tubule is bathed with a solution in which a certain compound (A) is present in lesser concentration on the outside than in the lumen of the tubule. The problem is to get A through the cell and into the lumen of the tubule against the concentration gradient. Suppose that in the outer surface of the cell (*O*) there is present a compound with a high energy bond ($\sim$B), continually produced by metabolic reactions. When the compounds A and $\sim$B meet in the presence of a certain enzyme ($E_1$), they form a complex (AB). As the complex forms at the outer surface of the cell it diffuses through the cell down a concentration gradient. When AB reaches the inner surface (*i*) of the cell, bordering the lumen, it meets another enzyme ($E_2$) which decomposes it into A and B. Compound A, present in high concentration locally at the surface of the cell along the lumen of the tubule, diffuses into the lumen of the tubule. The complex AB is thus continually produced at *O* and decomposed at *i* by enzymatic reactions. Compound B, the molecular carrier, is conceived of as a large molecule which cannot readily pass through the cell membrane; consequently it and its complex with A remain inside the cell.

This formulation has the appeal of simplicity since the complex formed at the cellular surface farthest from the lumen diffuses through the cell in a concentration gradient, and A diffuses from the cell in a concentration gradient when it

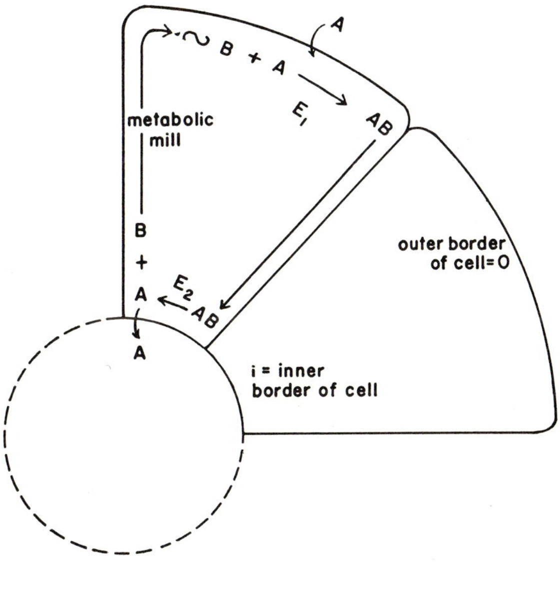


***Figure 14.5.*** Diagram of a possible mechanism of active transport. See text for explanation.

is liberated from the complex at the inner surface of the cell where it is locally present in high concentration. This scheme requires polar localization of enzymes, such as $E_1$ and $E_2$, of which there is some evidence in both kidney tissues and gut. Formation of some substances like $\sim$B is well known in the metabolic cycles of cells. However, identification of enzymes or compounds involved in active transport has so far not met with success.

The same hypothesis might explain unidirectional active transport through the cell membrane against a concentration gradient, as in the accumulation of ions in cells. In that case *O* (see Fig. 14.5) will be thought of as the outside surface of the cell membrane and *i* as the inner one, bordering on the cytoplasm.

Some attempts have been made to determine the identity of the carrier molecule (or molecules) in active transport of this type. From studies on the salt-secreting cells of glands in the eyes of marine birds (e.g., the albatross) good evidence has accumulated that phosphatidic acid (glycerol esterfied with two molecules of fatty acid and one of phosphoric acid, Fig. 3.12) might be the carrier of sodium and possibly of hydrophilic molecules and ions generally. Phosphatidic acid has the property of ready solubility in the lipid phase, while at the same time it has a minus charge which can bind the sodium. It does not leave the lipid phase but makes contact with the aqueous phase, there releasing the aqueous soluble material. The energy required for the movement is believed to be supplied by the high energy phosphate bond (Hokin and Hokin, 1960, 1963, 1965). From studies with other cells a protein carrier is suggested, presumably enzymatic (Alvarado, 1966; Skou, 1965).

## PINOCYTOSIS

Edwards, in 1925, working with *Amoeba,* noticed the formation of small channels in the cell membrane followed by the separation of small vacuoles, or *pinosomes,* in the cytoplasm (see Fig. 14.6). Lewis in 1931 described the same phenomenon which he saw in cells of tissue culture and gave it the name *pinocytosis.* Some more detailed observations of the process were made by Mast and Doyle on *Amoeba proteus* in 1934 (see Fig. 14.6). Little work then appeared upon the topic until recently when workers in a number of laboratories developed an intense interest in the process.

***Occurrence.*** The study of pinocytosis has been much facilitated by modern technological methods. For example, the development of the electron microscope has made possible resolution of much smaller channels and pinosomes. Also, some molecules, such as proteins labeled with dyes which fluoresce in the ultraviolet, when used to induce pinocytosis experimentally, can be identified readily inside the cell by fluorescence microscopy (Holter and Holtzer, 1959). They can be traced, following pinocytosis, in the large number of minute vacuoles scattered in the cytoplasm. Organic molecules labeled with radioactive tracers such as $^{14}C$ may also be employed. In this case, autoradiographs are made of the cells and if pinocytosis has occurred, small vacuoles containing radioactive material are evident. Ferritin, a protein containing much iron has also been used. The iron, having a high density, is opaque to electrons and, following pinocytosis, electron micrographs reveal it in small vacuoles in the cytoplasm. Thorium dioxide is also useful in the same way.

It has been shown that pinocytosis occurs in ameba and ameboid cells, such as leukocytes (white blood cells), brush border cells of the kidney (active in fluid exchanges), epithelial cells of the intestine, reticuloendothelial cells, Kupffer (phagocytic) cells of the liver, cells lining the walls of the capillaries, macrophages, fibroblasts and some malignant cells in tissue culture and some other cells. Pinocytosis was not observed (using labeled protein) in cardiac, smooth or skeletal muscle cells or in cells of a number of other tissue culture cells, including human amnion cells (Holter and Holtzer, 1959).

In cells of plant roots, little convincing

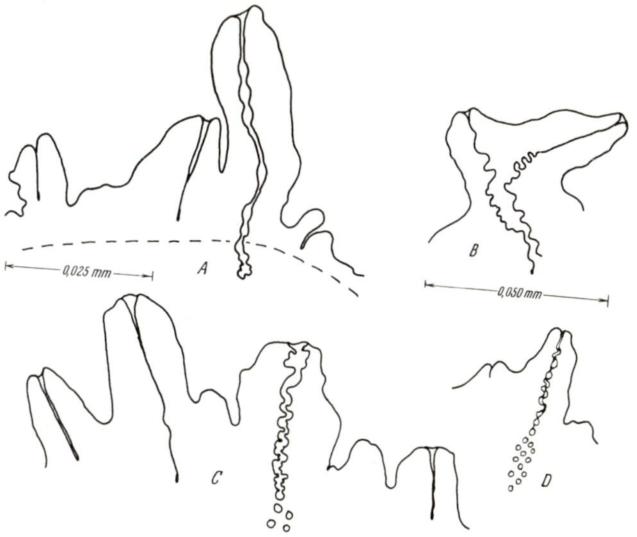


***Figure 14.6.*** Pinocytosis channels in *Amoeba proteus: A*, channels in small pseudopods in various stages of formation; *B*, convoluted channels; *C*, channel beginning to disintegrate at the inner end; *D*, further disintegration showing that many drops of fluid are ingested in the formation of each channel. (After Mast and Doyle, 1934: Protoplasma *20*:555.)

evidence has been found for pinocytosis, although labeled proteins and enzymes were found to penetrate into these cells from the outside and to appear in the cytoplasm around the nucleus. Electron micrographs of plant cells in some instances suggest canals leading from the surface into the cytoplasm which might be interpreted as pinocytosis channels, but reevaluation of these results has led to the conclusion that pinocytosis is questionable in plant cells (Bradfute *et al.*, 1964).

Indirect evidence also indicates that pinocytosis probably occurs on a wide scale in cells of embryos, since many proteins are known to pass through the placenta and immunologically specific proteins have been found to enter a wide variety of tissue cells, even into the cell nuclei (Schechtman, 1956). The entry of the enzyme ribonuclease into many types of cells has also been demonstrated indirectly by its hydrolysis of the cellular RNA, as revealed by the subsequent decreased staining by basic dyes and the decreased absorption of ultraviolet radiation by the cells. Pinocytosis has also been observed in the developing eggs of *Lymnaea stagnalis* (Brachet, 1957).

It is thought that pinocytosis is perhaps a general phenomenon and the negative evidence is discounted on the basis that the appropriate conditions may not yet have been supplied to the cells. The phenomenon is perhaps limited to cells capable of forming pseudopodia or at least having highly mobile membranes.

***Induction of Pinocytosis.*** Mast and Doyle (1934) showed that in appropriately starved cells *Amoeba proteus* pinocytosis is induced by salts and proteins. It is now known that amino acids (Fig. 14.7) and viruses are also effective, but that neither carbohydrates nor nucleic acids can induce pinocytosis. In some instances basic dyes have also induced pinocytosis (Rustad, 1961). In general it appears that the ion must be positively charged to induce pinocytosis.

Chapman-Andresen (1962), in an extensive study on *Amoeba proteus,* made quantitative comparisons between various conditions for inducing pinocytosis by counting the number of channels induced in 10 amebas (starved two days) in 20 minutes. Some data gathered in this manner are cited from her monograph in Figures 14.7 to 14.10.

The proteins found most effective for induction of pinocytosis in *Amoeba* are shown in Figure 14.7. For leukocytes, solutions of 1 to 4 per cent albumin and globulin were effective, although even

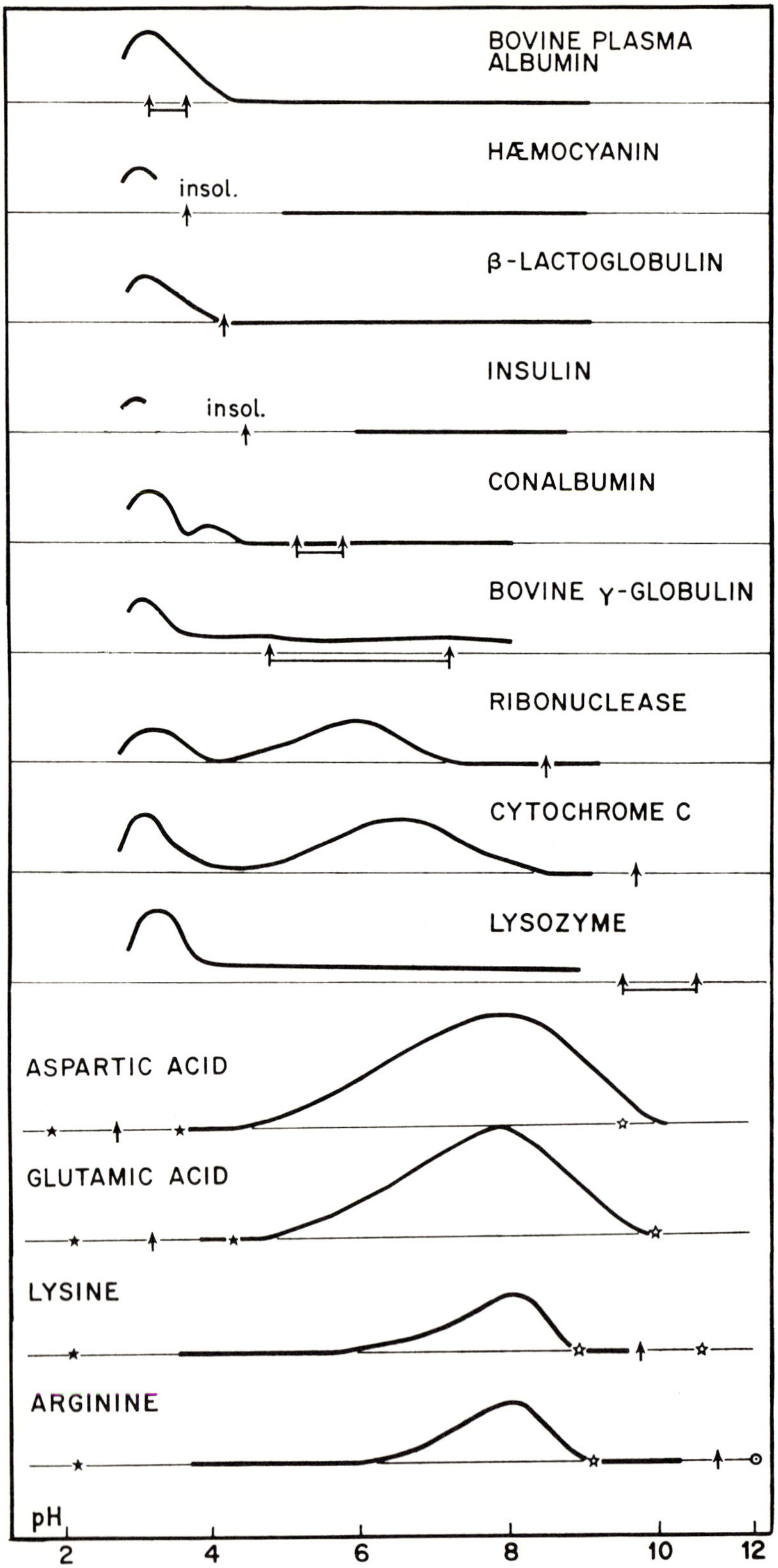


*Figure 14.7.* See opposite page for legend.

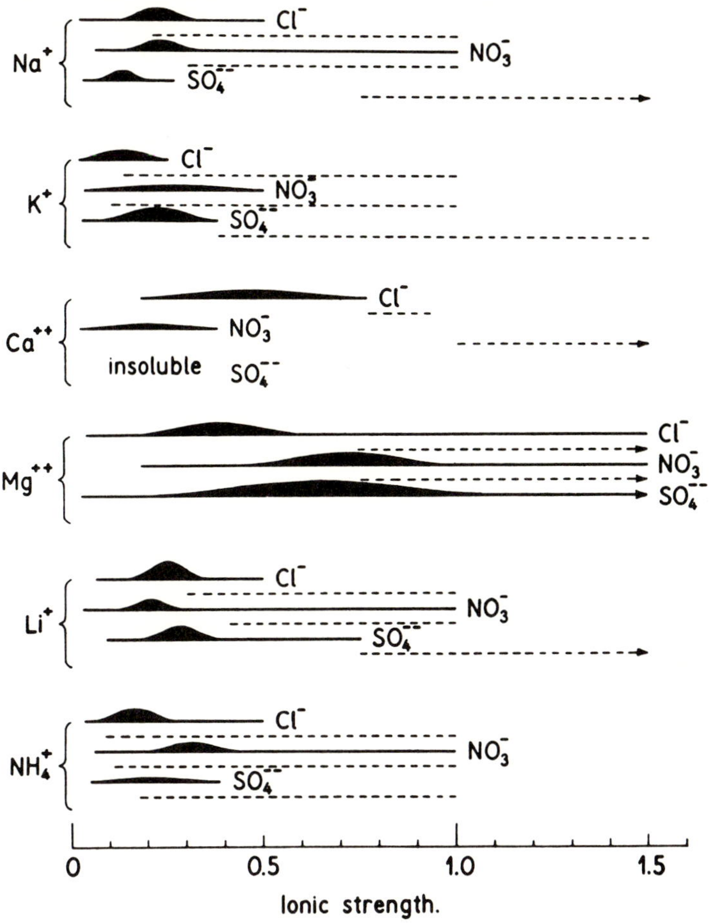


***Figure 14.8.*** Pinocytosis and toxicity in relation to concentrations of salts expressed in terms of ionic strength (the ionic strength is half of the sum of the terms obtained by multiplying the molality of each ion in the solution by the square of its valence). Dotted line indicates toxicity, solid line indicates pinocytosis, the width of the lines giving an indication of the intensity of the reaction. (From Chapman-Andresen, 1958: Comp. Rend. Lab. Carlsberg *31*:77.)

less than 1 per cent would suffice (Chapman-Andresen, 1958, 1962). It is necessary to set the pH of the solutions of proteins and amino acids on the acid side of the isoelectric pH, where they are positively charged, as it is only positively charged ions that are effective in inducing pinocytosis.

Salts induce pinocytosis also, cations being more effective than anions, as shown in Figure 14.8. Potassium, sodium and magnesium salts cause induction more readily than calcium. Sodium and magnesium induce narrower channels than potassium. A pH of 6.5 to 8.0 (Fig. 14.9) and a temperature between 12 and

***Figure 14.7.*** Effectiveness of induction of pinocytosis, as judged by the mean number of channels induced in 10 amebas in 20 minutes, of solutions of various amino acids (above) and proteins (below). Cysteine, asparagine, triglycine, glutamine, methionine, glycine and histidine were ineffective except for a very slight action at very high pH values. Pepsin, hemocyanin, $\beta$-lactoglobulin and insulin were ineffective except for slight activity at very low pH.

The isoelectric pH's are indicated by arrows. For amino acids the dark star indicates the pK(COOH), the light star the pK($NH_3^+$) and the circle with a dot, the pK of the guanidine. (From Chapman-Andresen; 1962: Comp. Rend. Lab. Carlsberg *33*:172, 186.)

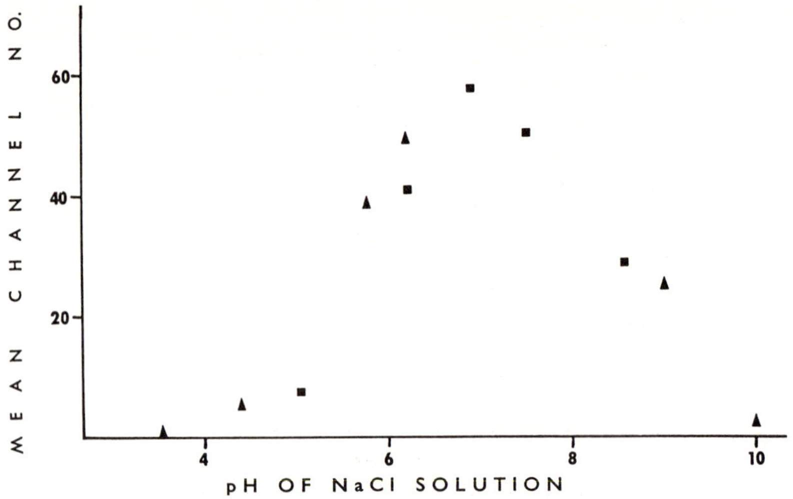


***Figure 14.9.*** Effect of pH of inducing solution on response of *Amoeba proteus* to 0.125M NaCl in distilled water. (From Chapman-Andresen, 1962: Comp. Rend. Lab. Carlsberg *33*:147.)

22° C. are most favorable (DeTerra and Rustad, 1959; Fig. 14.10). The optimal concentration is different for each salt, although 0.1 M solutions are adequate in most cases. The upper limit of concentration, however, is set by the toxic action of the salt upon the cytoplasm and lies between 0.25 and 1.0 M. Since pinocytosis has been seen in the eleocytes of the sea urchin (Holter, 1959), it would be interesting if the salt relations in such a form were to be tested since the cells are already accommodated to a concentration of about 0.5 M.

It was found that more pinocytotic channels were induced in enucleated than in nucleated pieces of *Amoeba proteus* (in 0.1 M NaCl) 90 minutes after cutting. However, if the tests are made 5 days after the cutting (during which time the enucleated pieces have presumably deteriorated) many more channels are induced in the nucleate than in the enucleate pieces.

Visible light has no effect on pinocytosis, but ultraviolet induces it (Rinaldi, 1959). There also appears to be no seasonal variation (Chapman-Andresen, 1962).

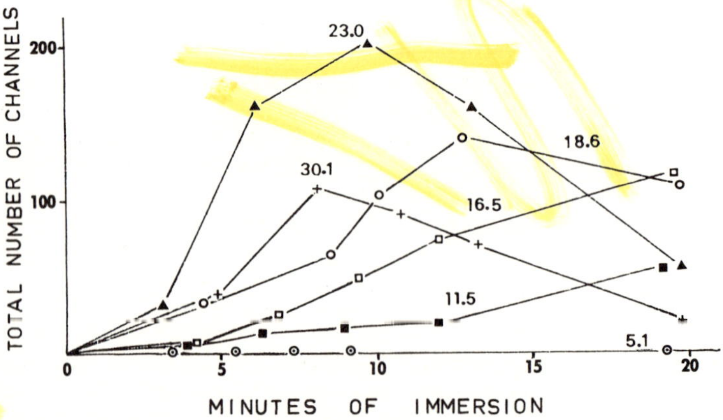


***Figure 14.10.*** Effect of temperature on pinocytosis in *Amoeba proteus* in 0.125M NaCl at pH 6.5. The temperature is indicated to the side of each of the curves. No pinocytosis occurred at 5.1° C. (From Chapman-Andresen, 1962: Comp. Rend. Lab. Carlsberg *33*:137.)

Once pinocytosis has been induced, in amebas, it goes on actively for 15 to 45 minutes, and then decreases. It was found necessary to wash the amebas free of ionic medium and then to reimmerse them in the salt or protein solutions in order to induce another period of activity (Holter, 1959).

Leukocytes appear to be able to continue pinocytosis for longer periods than *Amoeba*. It is presumed that only a portion of the surface of the leukocyte is active at any one time; therefore, pinocytotic action continues perhaps at all times.

Inducers of pinocytosis may cause a 50-fold decrease in electrical resistance of the plasma membrane prior to formation of typical channels and vacuoles. At this time structural changes in the unit membrane can be observed—the middle electron-transparent layer increases about twofold. Such changes in the *Amoeba* membrane are presumably dependent upon the calcium concentration in the external medium (Brandt and Freeman, 1967).

Pinocytosis in Sarcoma 180 II tumor cells resembles the process observed in *Amoeba* (Ryser, 1968). Also, basic polyamino acid inducers are more effective the higher their molecular weight, and are ineffective if smaller than 1000. Steric factors play a role since poly-D-lysine is more effective than poly-L-lysine. Acidic polyamino acids may be inhibitory (e.g., polyglutamic acid). Pinosomes combine with lysosomes resulting in digestion of foreign proteins.

## THE PINOSOMES

When an ameba is placed into a solution favorable for induction of pinocytosis it becomes detached from the substrate, the contractile vacuole temporarily ceases operation and it forms a rosette, most of its surface being covered by small active pseudopodia, each made up almost entirely of clear ectoplasm (Fig. 14.11). In the pseudopodia (Fig. 14.12) may be seen the clear channels which are the forerunners of the pinosomes. By the use of proteins marked with fluorescent dyes, the pinosomes, formed by the pinching off of the vacuoles, may be readily seen inside the ameba.

The pinocytotic vacuoles contain within them the same solutes as the external medium. Holter (1959) has shown that the vacuoles in an ameba do not centrifuge down soon after formation but

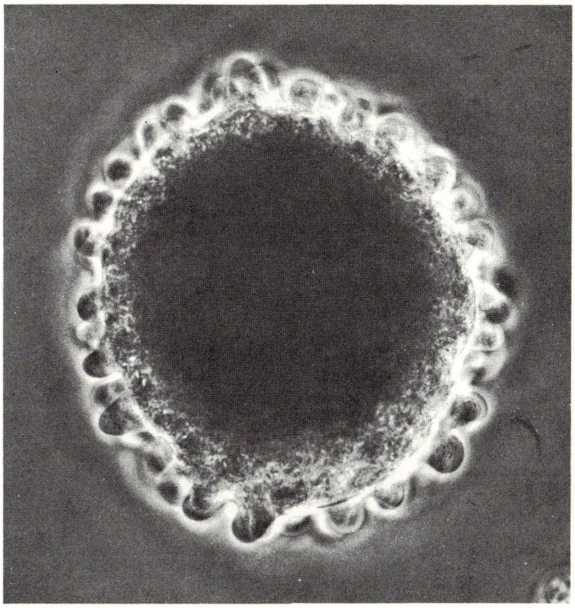

***Figure 14.11.*** Induction of many round pseudopods in *Amoeba* placed in strong protein solution. (From Pinocytosis, by Ronald C. Rustad. Copyright © 1961 by Scientific American, Inc. All rights reserved.)

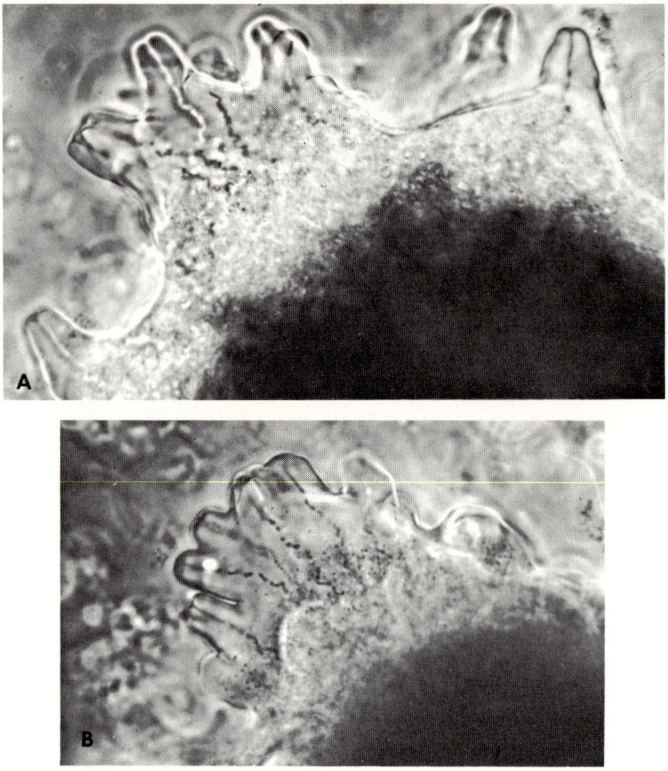

***Figure 14.12.*** *A*, Pinocytosis in *Amoeba proteus* as seen under phase contrast microscopy. The *Amoeba* was starved for 24 hours, then immersed into 1 per cent egg albumin in balanced salt solution. The pinocytotic channels are clearly visible in the pseudopodia. In some of the channels deep inside the cell small pinosomes may be seen forming from the engulfed fluid. *B*, Taken 35 seconds later during which time the ameba has moved; therefore, the pseudopodia do not match. The most prominent channels in *A* have become the most prominent chains of vacuoles (pinosomes) in *B*. (From Holter, 1960: Int. Rev. Cytol. *8*:484–485. Photographs by courtesy of Dr. David Prescott, Oak Ridge National Laboratory.)

rather stay at the centripetal end with the lighter materials of the cell. If the ameba is centrifuged after a lapse of time, the vacuoles are carried down to the centrifugal end with the heavier materials in the cell. Presumably in the meantime the material in the vacuoles has been concentrated by the withdrawal of water. This can be shown by the use of fluorescent proteins which gradually become concentrated into a dense, brilliantly fluorescent globule. Finally, only a fluorescent granule remains when most of the water is withdrawn.

It is of interest to know whether the membrane of the pinocytotic vacuole is like the plasma membrane, and whether other materials besides water can also enter the cytoplasm from the vacuole. As a test case, glucose marked with $^{14}C$ was included in the inducing solution containing protein. Autoradiographs, following pinocytosis, indicated that the glucose had entered into the pinocytotic vacuoles and later autoradiographs showed that the glucose became dispersed in the cell. Ultimately, about 15 per cent of the radioactive carbon became incorporated into the cytoskeleton of the cell and about 75 per cent of the radioactive carbon was given off by the ameba as carbon dioxide. Similar experiments carried out with methionine labeled with $^{35}S$ showed it to enter into the cytoplasm also.

Less has been learned about the fate of proteins taken into the cell. The fluorescein labeling on the protein is removed from the cell much more rapidly than the $^{14}C$ label on glucose, but whether the protein is digested or the labeling material is just voided is not known (Holter, 1959).

Pinocytotic vacuoles have been seen to shrink, coalesce and even divide, but never to disappear. Whether the membrane of a vacuole is digested is not known. It has even been inferred that pinocytosis might be a way of building the laminated structures in the cytoplasm (Holter, 1959).

## MECHANISM OF PINOCYTOSIS

Pinocytosis in the ameba, *Chaos chaos,* has been described as a result of movement of the plasma membrane, which is considered to be an elastic structure fitting closely to the underlying cytoplasm. The plasma membrane wrinkles when a flow of a pseudopod begins and the membrane is not yet freed from the gel adjoining it. This leads to invagination of parts of the plasma membrane which are consequently separated from the larger advancing portion of the membrane. It is thought that the charged solute particles, used to induce pinocytosis, may weaken the structural rigidity of the membrane, allowing parts of it during movement to be drawn into the cytoplasm by its remaining attachments to the gel sheet. These vacuoles or channels are then separated from the membrane and released into the interior of the cell as pinosomes (Brandt, 1958).

Bennett (1956) on the other hand, postulated that pinocytosis does not involve just the engulfment of fluid but that there is need for attachment of materials on the surface of the cell before the vacuole can form, as shown in Figure (14.13). Decisive evidence for this mechanism has been obtained, using protein as the inducer of pinocytosis. The amebas were

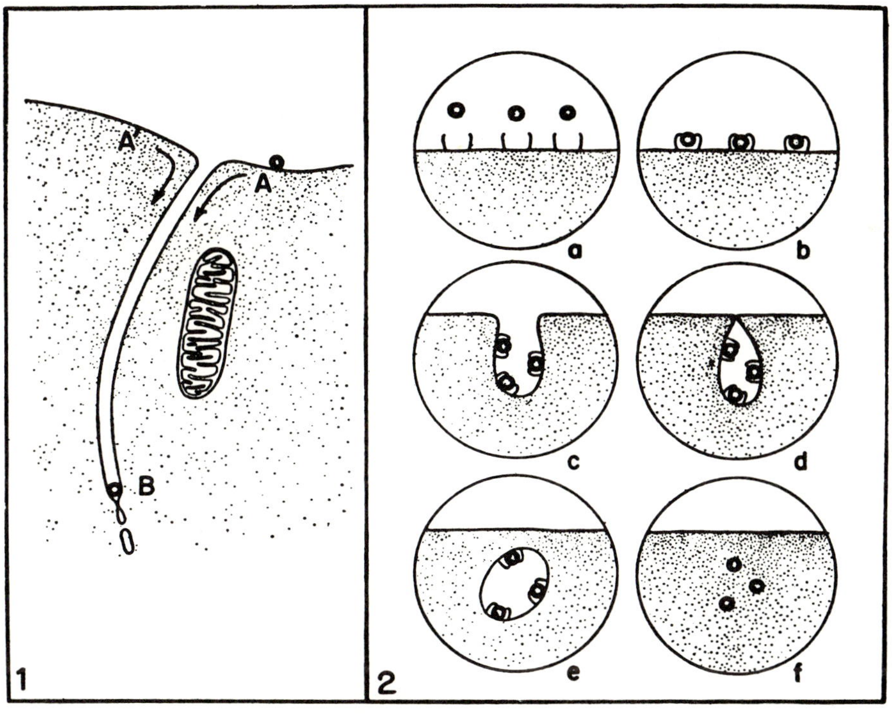


***Figure 14.13.*** Induction of pinocytosis by attachment of particles as envisaged by Bennett. A particle (*1A*) induces membrane flow, forming a channel containing the particle (*1B*) which is taken inward into a pinosome such as those illustrated forming. The series of figures illustrates the approach (*2a*) and attachment (*2b*) of a particle, invagination of the cell membrane (*2c*), formation of a pinosome or pinocytotic vaculole (*2d, e*), and, finally, incorporation of the particulate material into the cytoplasm (*2f*). (From Bennett, 1956: J. Biophys. Biochem. Cytol. *2*:Part 2 (suppl.), 99.)

immersed in an antigen protein for varying periods. They were then freeze-dried and sectioned. The sections were exposed to an antibody protein labeled with fluorescein. The sections so treated showed a brilliant fluorescence under ultraviolet radiation, indicating that the surface of the ameba was heavily coated with the antigen to which the antibody containing fluorescein had become attached. Further proof of attachment of protein came from studies using protein labeled with radioactive material. In 5 minutes an ameba binds 50 times the amount of protein contained in its own volume (of solution). It would therefore appear that such binding may play an important role in pinocytosis and may be a necessary first step, so that pinocytosis is not just "cell drinking" but a process consisting of several steps (Brandt, 1958; Marshall *et al.*, 1959; Rustad, 1961).

Basic dyes (e.g., alcian blue, toluidine blue) are found to induce pinocytosis also and can be shown to be bound to the surface of an ameba, most likely by electrostatic forces. Proteins compete for the surface of an ameba in this reaction. Some photomicrographs indicate that the surface of *Amoeba proteus* has hairlike projections of extracellular material 1000 to 2000 Å long and 60 Å wide which greatly increase the area of the surface available for binding. The extraneous membranes, composed of mucopolysaccharides, carry a negative charge reacting with the cations, which act as inducers. Experiments show that to these structures, as a first step in pinocytosis, particles become attached (Rustad, 1961).

Binding, or surface adsorption, of protein on the surface of the cell continues even in the presence of carbon monoxide or cyanide, either of which is inhibitory to the oxidative metabolism of the cell. It also continues at low temperatures which are inhibitory to completion of pinocytosis. This indicates that the first step in the process of pinocytosis is a simple physical reaction.

The reactions which follow, however, including the formation of channels and vesicles, are stopped by metabolic inhibitors in *Amoeba* (but not in sarcoma cells) and also by low temperatures. This indicates that they are dependent upon the oxidative metabolism of the cell. Thus it seems that there are at least two kinds of processes involved in pinocytosis.

## SIGNIFICANCE OF PINOCYTOSIS

Evidence from many different kinds of cells indicates that macromolecules such as proteins, polypeptides and hormones enter cells. Embryologists have shown by delicate and specific tests that proteins also enter into cells of embryos. Pinocytosis is thought to be the mechanism by which entry of such macromolecules including hormones and antibodies, is made possible, since it is difficult to consider that the small pores in the cell membrane (measured in a number of ways) or solution in the plasma membrane could possibly permit such macromolecules to enter freely. Pinocytosis might also be a means for entry of other large molecules, e.g., DNA, which the cell might incidentally engulf along with the medium containing the pinocytosis-inciting particles or ions, possibly inducing transformation.

Pinocytosis may be of special importance in some physiological processes. For example, transfer of ferritin in reticulocytes may occur in this manner. Transport of lipids from the gut may also occur by pinocytosis. The blood "dust" consisting of fine lipid droplets may well have been liberated into the blood by the reverse of pinocytosis.

***Is Pinocytosis Active Transport?*** If active transport is defined as the selective uptake of a particular solute against a concentration gradient performed at the expense of metabolic energy, then pinocytosis does not truly qualify as active transport. In pinocytosis the active step requiring energy is the formation of a vacuole or channel and its separation from the cell membrane. This phase of the process is not selective because it is a random sampling of the environmental fluid. The only selective phase appears to be the binding of proteins and perhaps other molecules to the surface of the cell and this may or may not require metabolic energy. Perhaps surface binding is prerequisite to any active transport. When the vacuole is formed these materials are included and they may move passively against a concentration gradient. It might even be that the entering proteins differ from those inside the cell, in which case they therefore move along with a concentration gradient, not against it.

Solutes other than the particles inducing pinocytosis, included in the engulfed vacuole as a random sample of the environmental fluid, are concentrated only as a result of subsequent processes, possibly active ones, not yet studied. Whether these solutes pass by diffusion or by active transport through the portion of the plasma membrane which forms the vacuolar wall, and whether the solutes are concentrated by removal of the water from the vacuole by osmosis or active transport is not known, but some solutes are readily incorporated into the cytoskeleton or used in metabolism, e.g., $^{14}C$ glucose, mentioned before. Some cells also void secretory vacuoles or granules —presumably by the reverse of pinocytosis. It must be recognized that pinocytosis appears to be more complex than active uptake through the plasma membrane.

## LITERATURE CITED AND GENERAL REFERENCES

### ACTIVE TRANSPORT

Alvarado, F., 1966: Transport of sugars and amino acids in the intestine: evidence for a common carrier. Science *151*:1010–1013.

Andersen, B., and Ussing, H. H., 1960: Active transport. Comp. Biochem. *2*:371–402.

Baker, P. F., 1966: The sodium pump. Endeavour *25*:166–172.

Biddulph, O., 1959: Translocation of inorganic solutes. *In* Plant Physiology, A Treatise. Steward, ed. Academic Press, New York, Vol. 2, pp. 553–603.

Blinks, L. R., 1949: The source of the bioelectric potentials in large plant cells. Proc. Natl. Acad. Sci. U.S.A. *35*:566–575.

Brierley, G. P., Murer, E., Bachmann, E., Green, D. E., and Hadley, D. G., 1963: Ion transport. II. The accumulation of inorganic phosphate and magnesium ions by heart mitochondria. J. Biol. Chem. *238*:3482–3489.

Brooks, S. C., 1929: The accumulation of ions in cells—a non-equilibrium condition. Protoplasma *8*:389–412.

Chambers, R., Beck, L. V., and Belkin, M., 1935: Secretion in tissue cultures. I. Inhibition of phenol red accumulation in the chick kidney. J. Cell. Comp. Physiol. *6*:425–439; 441–455.

Christensen, H. N., 1962: Biological Transport. W. A. Benjamin, New York.

Cirillo, V. R., 1966: Membrane potentials and permeability. Bact. Rev. *30*:68–79.

Cohen, G., and Monod, J., 1957: Bacterial permeases. Bact. Rev. *21*:169–194.

Conway, E. J., 1953: A redox pump for the biological performance of osmotic work, and its relation to the kinetics of free ion diffusion across membranes. Int. Rev. Cytol. *2*:419–445.

Conway, E. J., 1955: Evidence for a redox pump in the active transport of cations. Int. Rev. Cytol. *4*:377–396.

Costantin, L., Franzini-Armstrong, C., and Podolsky, R., 1965: Localization of calcium-accumulating structures on striated muscle fibers. Science *147*: 158–160.

Crane, R. K., 1962: Hypothesis for mechanism of intestinal transport of sugars. Fed. Proc. *21*: 891–895.

Csaky, T. Z., 1963: A possible link between active transport of electrolytes and non-electrolytes. Fed. Proc. *22*:3–7.

*Davson, H., 1954: Vertebrate physiology from the point of view of active transport. Symp. Soc. Exp. Biol. *8*:16–26.

Davson, H., 1964: Textbook of General Physiology. 3rd Ed. Little, Brown & Co., Boston, p. 287.

Eppley, R. W., 1959: Potassium accumulation and sodium efflux by *Porphyra perforata* tissues in lithium and magnesium sea water. J. Gen. Physiol. *43*:29–38.

Gale, E. F., 1954: The accumulation of amino acids within staphyloccocal cells. Symp. Soc. Exp. Biol. *8*:242–253.

Goldacre, R. J., 1952: The folding and unfolding of protein molecules as a basis of osmotic work. Int. Rev. Cytol. *1*:135–164.

Greenawalt, J. W., Rossi, C. S., and Lehninger, A. L., 1964: Effect of active accumulation of calcium and phosphate ions on the structure of rat liver mitochondria. J. Cell Biol. *23*:21–38.

Harris, E. J. (ed.), 1956: Transport and Accumulation in Biological Systems. Academic Press, New York.

Hasselbach, W., 1964: ATP-driven active transport of calcium in the membranes of the sarcoplasmic reticulum. Proc. Roy. Soc. (London) *B160*:501–504.

Hasselbach, W., 1965: Relaxing factor and the relaxation of muscle. Progr. Biophys. *14*:167–222.

Hoagland, D. R., 1944: Lectures on Inorganic

Nutrition of Plants. Chronica Botanica, Waltham, Mass.

Hoffman, J. F. (ed.), 1964: The Cellular Functions of Membrane Transport. Prentice-Hall, Englewood Cliffs, N.J.

Hokin, L. E., and Hokin, M. R., 1960: Studies on the carrier function of phosphatidic acid in sodium transport. J. Gen. Physiol. *44*:61–85.

Hokin, L. E., and Hokin, M. R., 1963: Phosphatidic acid metabolism and active transport of sodium. Fed. Proc. *22*:8–18.

Hokin, L. E., and Hokin, M. R., Oct. 1965: The chemistry of cell membranes. Sci. Am. *213*:78–86.

Horecker, B. L., Osborn, M. J., McLellan, W. L., Avigad, G., and Asensio, C., 1961: The role of bacterial permeases in metabolism. *In* Membrane Transport and Metabolism. Kleinzeller and Kotyk, eds. Academic Press, New York, pp. 378–387.

Huxley, A. F., and Stämpfli, R., 1951: Effect of potassium and sodium on resting and action potentials of single myelinated nerve fibers. J. Physiol. *112*:496–508.

Kaye, G. I., and Cole, J. D., 1965: Electron microscopy: sodium localization in normal and ouabain-treated transporting cells. Science *150*:1167–1168.

Keynes, R. D., 1961: The energy source for active transport in nerve and muscle. *In* Membrane Transport and Metabolism. Kleinzeller and Kotyk, eds. Academic Press, New York, pp. 131–139.

Kitching, J. A., 1954: Osmoregulation and ionic regulation in animals without kidneys. Symp. Soc. Exp. Biol. *8*:63–75.

Kitching, J. A., 1956: Contractile vacuoles of protozoa. Protoplasmatologia III.D.3.a. Springer-Verlag, Vienna.

Kleinzeller, A., and Kotyk, A. (eds.), 1961: Membrane Transport and Metabolism. Academic Press, New York.

Klemperer, H. G., 1957: The accumulation of iodide by *Fucus ceranoides.* Biochem. J. *67*:381–390.

Koch, H. J., 1954: Cholinesterase and active transport of sodium chloride through the isolated gill of the crab *Ereiocheir sinensis. In* Cell Physiology. Kitching, ed. Academic Press, New York, pp. 15–26.

Lansing, A. L., and Rosenthal, T. B., 1952: The relation between RNA and ionic transport across the cell surface. J. Cell. Comp. Physiol. *40*: 337–345.

LeFevre, P. G., 1954: The evidence for active transport of monosaccharides across the red cell membrane. Symp. Soc. Exp. Biol. *8*:118–135.

LeFevre, P. G., 1955: Active Transport through Cell Membranes. Protoplasmatologia III.D.3.a. Springer-Verlag, Vienna.

LeFevre, P. G., and McGinniss, G. F., 1960: Tracer exchange vs. net uptake of glucose through the human red cell surface. J. Gen. Physiol. *44*:87–103.

Lehninger, A. L., 1964: The Mitochondrion. W. A. Benjamin, New York.

Lundegard, H., 1954: Anion respiration: the experimental basis of a theory of absorption, transpiration and exudation of electrolytes by living cells and tissues. Symp. Soc. Exp. Biol. *8*:262–296.

Mitchell, P., 1959: Biochemical cytology of microorganisms. Ann. Rev. Microbiol. *13*:407–440.

Ponder, E., 1948: Hemolysis and Related Problems. Grune & Stratton, New York.

Potts, W. T. W., and Parry, G., 1964: Osmotic and Ionic Regulation in Animals. Macmillan Co., New York.

Prankerd, T. A. J., 1961: The Red Cell: An Account of Its Chemical Physiology and Pathology. Charles C Thomas, Springfield, Ill.

Quastel, J. H., 1965: Molecular transport at cell membranes. Proc. Roy. Soc. (London) *B163*: 169–196.

Robinson, J. R., 1953: The active transport of water in living systems. Biol. Rev. *28*:158–194.

Robinson, J. R., 1954: Secretion and transport of water. Symp. Soc. Exp. Biol. *8*:42–62.

Robinson, J. R., 1960: Metabolism of intracellular water. Physiol. Rev. *40*:112–149.

Rosenberg, T., 1954: The concept and definition of active transport. Symp. Soc. Exp. Biol. *8*:27–41.

Rothstein, A., 1954: Enzyme systems of the cell surface involved in the uptake of sugars by yeast. Symp. Soc. Exp. Biol. *8*:165–201.

Ruch, T. C., and Patton, H. D., 1965: Physiology and Biophysics. W. B. Saunders Co., Philadelphia.

Rummel, W., Pfleger, K., and Seifen, E., 1961: Experimental examination of the hypothesis of coupling between the sodium and potassium pump in erythrocytes. J. Theoret. Biol. *1*:318–323.

Schmidt-Nielsen, B., and Schrauger, C. R., 1963: *Amoeba proteus:* studying the contractile vacuole by micropuncture. Science *139*:606–607.

Schmidt-Nielsen, K., Jan. 1959: Salt glands (of birds). Sci. Amer. *200*:109–116.

Schultz, S. G., and Zalusky, R., 1964: Ion transport in isolated rabbit ileum. II. The interaction between sodium and active sugar transport. J. Gen. Physiol. *47*:1043–1059.

Shanes, A. M., and Bianchi, C. P., 1960: Radiocalcium release by stimulated and potassium-treated sartorius muscle of the frog. J. Gen. Physiol. *43*:481–493.

Skou, J. C., 1965: Enzymatic basis for active transport of $Na^+$ and $K^+$ across cell membrane. Physiol. Rev. *45*:596–617.

Smith, H. W., 1956: Principles of Renal Physiology. Oxford University Press, New York.

*Solomon, A. K., Aug. 1962: Pumps in the living cell. Sci. Am. *207*:100–108.

Stein, W. D., 1960: The transfer mechanism in active transport. *In* New Approaches in Cell Biology. Walker, ed. Symposium, 1958, Imperial College, London, pp. 125–138.

Steward, C. F., and Sutcliffe, J. F., 1959: Plants in relation to organic salts. *In* Plant Physiology, A Treatise. Steward, ed. Academic Press, New York, Vol. 2, pp. 253–478.

Stewart, F. C., and Miller, F. K., 1954: Salt accumulation in plants: a reconsideration of the role of growth and metabolism. Symp. Soc. Exp. Biol. *8*:367–406.

Teorell, T., 1953: Transport processes and electrical properties in ionic membranes. Prog. Biophys. *3*:305-369.

Ussing, H. H., 1952: Some aspects of the application of tracers in permeability studies. Advances Enzymol. *13*:21–65.
Ussing, H. H., 1961: Transport of ions across cellular and biological membranes. *In* Membrane Transport and Metabolism. Kleinzeller and Kotyk, eds. Academic Press, New York, pp. 115–130.
Warren, K. B., 1966: Intracellular Transport. Symp. Soc. Cell Biol. *5*:1–325.
Winegrad, S., 1965: Role of intracellular calcium movements in excitation-contraction coupling in striated muscles. Fed. Proc. *24*:1146–1152.

## PINOCYTOSIS

Bennett, H. S., 1956: The concepts of membrane flow and membrane vesiculation as mechanisms for active transport and ion pumping. J. Biophys. Biochem. Cytol. Suppl. *2*:99–103.
Brachet, J., 1957: Biochemical Cytology. Academic Press, New York.
Bradfute, O. E., Chapman-Andresen, C., and Jensen, W. A., 1964: Concerning morphological evidence for pinocytosis in higher plants. Exp. Cell Res. *36*:207–210.
Brandt, P. W., 1958: A study of the mechanism of pinocytosis. Exp. Cell Res. *15*:300–313.
Brandt, P. W., and Freeman, A. R., 1966: Plasma membrane substructure changes correlated with electrical resistance and pinocytosis. Science *155*:582–585.
Chapman-Andresen, C., 1958: Pinocytosis of inorganic salts by *Amoeba proteus* (*Chaos diffluens*). Compt. rend. Lab. Carlsberg *31*:77–92.
Chapman-Andresen, C., 1962: Studies on pinocytosis in amoebae. Comp. rend. Lab. Carlsberg *33*:73–264.
Chapman-Andresen, C., and Prescott, D. M., 1956: Studies on pinocytosis in the amoebae *Chaos chaos* and *Amoeba proteus.* Compt. rend. Lab. Carlsberg *30*:57–58.
DeTerra, N., and Rustad, R. C., 1959: The dependence of pinocytosis on temperature and aerobic respiration. Exp. Cell Res. *17*:191–195.
Edwards, J. G., 1925: Formation of food cups in Amoeba induced by chemicals. Biol. Bull. *48*: 236–239.
Elbers, P. F., and Bluemink, J. B., 1960: Pinocytosis in the developing egg of *Limnaea stagnalis* L. Exp. Cell Res. *21*:619–622.
Gropp, A., 1963: Phagocytosis and pinocytosis. *In* Cinematography in Cell Biology. Rose, ed. Academic Press, New York.
Hayward, A. F., 1963: Electron microscopy of induced pinocytosis in *Amoeba proteus.* Comp. rend. Lab. Carlsberg *33*:537–558.
Holter, H., 1959: Pinocytosis. Int. Rev. Cytol. *8*: 481–504.
Holter, H., 1961: The induction of pinocytosis. *In* Biological Approaches to Cancer Chemotherapy. Harris, ed. Academic Press, New York, pp. 77–88.
Holter, H., 1963: Pinocytosis. Proc. Fifth Intern. Congr. Biochem. *2*:248–256.
Holter, H., and Holtzer, H., 1959: Pinocytotic uptake of fluorescein-labelled proteins by various tissue cells. Exp. Cell Res. *18*:421–423.
Holter, H., and Marshall, J. M., Jr., 1954: Studies on pinocytosis in the amoeba *Chaos chaos.* Compt. rend. Lab. Carlsberg *29*:7–26.
Holtzer, H., and Holtzer, S., 1960: The *in vitro* uptake of fluorescein labelled plasma proteins. I. Mature cells. Compt. rend. Lab. Carlsberg *31*:373–408.
Ledoux, L., 1965: Uptake of DNA by living cells. Prog. Nucleic Acid Res. Molec. Biol. *4*:231–267.
Lewis, W. H., 1931: Pinocytosis. Johns Hopkins Hospital Bull. *49*:17–28.
Marshall, J. M., and Nachmias, V. T., 1965: Cell surface and pinocytosis. J. Histochem. Cytochem. *13*:92–104.
Marshall, J. M., Jr., Schumaker, V. N., and Brandt, P. W., 1959: Pinocytosis in amoebae. Ann. N. Y. Acad. Sci. *78*:515–523.
Mast, S. O., and Doyle, L., 1934: Ingestion of fluid by Amoeba. Protoplasma *20*:555–560.
Rinaldi, A. R., 1959: The induction of pinocytosis in *Amoeba proteus* by ultraviolet radiation. Exp. Cell Res. *18*:70–75.
Rudzinska, M. A., Trager, W., and Bray, R. S., 1963: Pinocytotic uptake and digestion of hemoglobin in malarial parasites. J. Protozool. *12*:563–576.
Rustad, R. C., April 1961: Pinocytosis. Sci. Am. *204*:120–130.
Rustad, R. C., 1964: The physiology of pinocytosis. *In* Recent Progress in Surface Science. Danielli, Pankhurst and Riddiford, eds. Academic Press, New York pp. 353–376.
Ryser, H.J-P., 1968: Uptake of protein by mammalian cells: an underdeveloped area. Science *159*:390–396.
Schechtman, A. M., 1956: Uptake and transfer of macromolecules by cells with special reference to growth and development. Int. Rev. Cytol. *5*: 303–322.
Straus, W., 1964: Occurrence of phagosomes and phago-lysosomes in different segments of the nephron in relation to reabsorption, transport, digestion and extrusion of intravenously injected horseradish peroxidase. J. Cell Biol. *21*:295–308.

# SECTION IV

# Conversions of Energy and Matter in the Cell

The cell can perform its work only at the expense of chemical energy obtained from nutrient molecules. All cells require certain cell foods: soluble nutrients such as glucose, amino acids and certain vitamins, and salts which can penetrate the cell membrane. All cells except anaerobes require free oxygen. It is in the cells that conversions of matter and transfers of energy occur, during which wastes are produced which must be removed if life is to be maintained.

For the unicellular organisms the problem is relatively simple. They absorb or engulf nutrients, and take in oxygen directly from their environment. They eliminate their metabolic waste products directly into it.

In a multicellular animal, in which the individual cells are often removed from immediate contact with the exterior environment, a cellular environment is provided by specialized supply systems, e.g., digestive, respiratory, circulatory and excretory systems. Through the services of these supply systems, each and every cell is supplied with the nutrients and oxygen it needs and is washed free of its metabolic waste products. The following scheme shows the conversions which occur in the course of these processes.

The nutritional pattern of green plants differs from that of animals in certain fundamental respects. Green plants contain chlorophyll, manufacturing their own carbohydrates by photosynthesis; with these and salts they synthesize the various other compounds they need. As a consequence of the overall synthetic character of plant metabolism, excretion of nitrogen compounds is limited, and much of the waste nitrogen resulting from cellular metabolism is recycled via the amides. In seed plants the xylem transports water and salts, and the phloem system distributes such cell foods as amino acids, glucose and vitamins so that each cell is adequately supplied by the sap.

The nutritional problems of an individual cell in a multicellular plant or animal organism are, in the final analysis, the same as the nutritional problems of a unicellular organism. The nature of enzymes and their

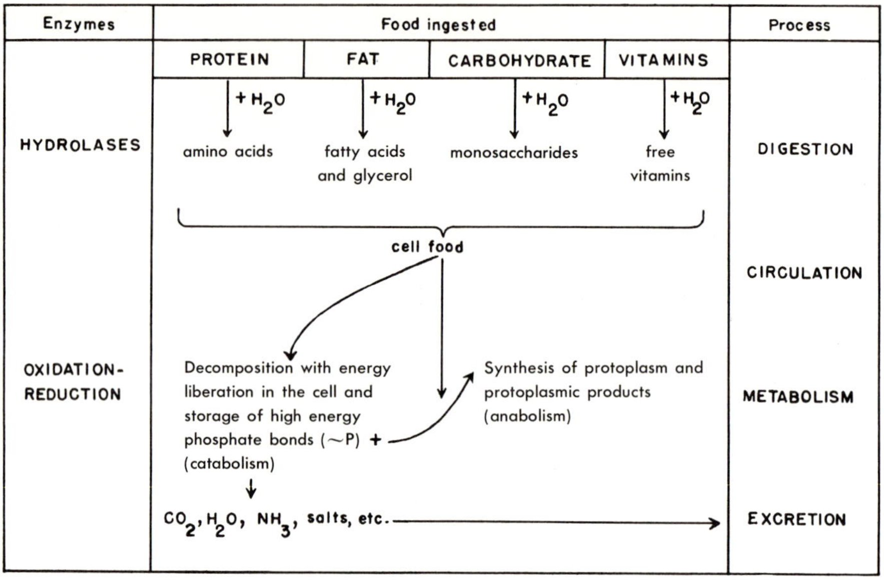


General scheme of animal nutrition. For colorless plants a similar scheme could be drawn. Such plants do not ingest food, however; instead, they secrete digestive enzymes, and they absorb the digested material. Green plants synthesize their own carbohydrates in the process of photosynthesis, and they synthesize other foods from carbohydrates and salts.

functions are considered in Chapter 15. The energy-liberating reactions of metabolism which provide the energy for maintenance of life in the cell are outlined in Chapter 16. The nature of oxidation-reduction in cells is dealt with in Chapter 17. Cell respiration, glycolysis and bioluminescence are discussed in Chapter 18. Photosynthesis in green plant cells is considered in Chapter 19, biosynthesis in Chapter 20 and regulation and control of cell metabolism in Chapter 21.

# CHAPTER 15 / CELLULAR ENZYMES

## LOWERING THE ENERGY OF ACTIVATION BY ENZYMES

An *enzyme* is a *biocatalyst.* It is made up of protein, and like other catalysts* it accelerates the rate of a reaction without itself being used up (except incidentally in side reactions). It is safe to assume that most reactions occurring in the cell are catalyzed by enzymes. Only a few reactions, some examples of which will be cited later, may be rapid enough without them. The main function of a catalyst is to lower the *energy of activation* (kinetic energy) required of a molecule before the molecule can proceed in a particular reaction. In other words, the catalyst lowers the energy requirement or energy barrier over which a molecule must pass (Fig. 15.1). It thus increases the rate of the reaction because at any given temperature it is probable that the number of molecules having a smaller amount of energy will be greater than the number having a larger amount of energy (see Fig. 10.4). Enzymes function in this manner in the cell. For example, the energy of activation for the decomposition of hydrogen peroxide to water and oxygen is 18,000 calories per mole. In the presence of the enzyme catalase, which occurs in all aerobic cells, it is only 5500 calories per mole. The advantage of enzymes to the cell, which must operate at what a chemist would consider a relatively low temperature, is self-evident (Dixon, 1964; Gutfreund, 1965).

A very important aspect of enzyme action is the channeling of reactions along certain pathways. Thus many reactions may be thermodynamically possible (spontaneous) but do not occur at a finite rate. Selected reactions may be promoted by supplying the appropriate enzymes. Thus, in a sense, by supplying the required enzymes the cell achieves selective control over a series of possible reactions (Wilkie, 1960).

It is not known whether it is possible for the cell to make enzymes to catalyze any chosen reaction, or whether they can be made only for certain types of reactions. If the latter is true, then the metabolic chain of reactions not only must satisfy the energetic requirements, but it also must be channeled by means of such enzymes as are possible (Wilkie, 1960).

Although an enzyme cannot increase the amount of energy obtainable from a given chemical reaction, it can enormously increase the efficiency with which the energy is obtained. The effect of the enzyme is thus to raise what might be called the chemical conductance, thereby permitting a given rate of energy production with a lower dissipation of free energy.

*The term "catalyst" is used by the physical chemist for a substance which alters the rate of a reaction. Consequently, negative catalysis is possible. The word is used here in the popular sense, i.e., a substance which accelerates a reaction.

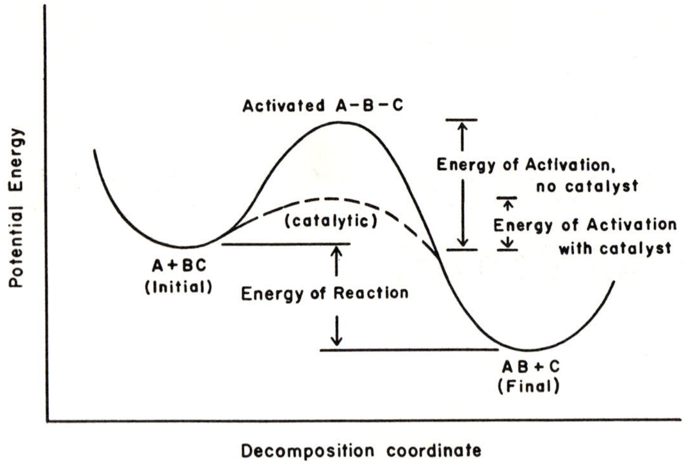


***Figure 15.1.*** Potential energy diagram for a reaction occurring in one case in the absence and in another case in the presence of a catalyst. Before reacting, the molecules must accumulate sufficient kinetic energy (energy of activation) to pass over the energy barrier represented by the rising curve, even though the reaction occurs with a loss in free energy (approximately indicated by the heat of reaction). In the presence of a catalyst the energy of activation is considerably less than in its absence.

## HYDROLYTIC ENZYMES—HYDROLYSIS

*Hydrolysis* is the process by which a compound is split into fragments by the addition of water. The manner in which proteins, fats and polysaccharides are hydrolyzed is of special importance because all these compounds are necessary for maintenance of life, yet they are too large to penetrate through the cell membrane. Such hydrolyses are illustrated in Figure 15.2.

Hydrolyses are spontaneous reactions in which a small amount of energy is liberated; that is, they occur with a net decrease in free energy. However, in the absence of a catalyst, they often proceed too slowly to be measured. For instance, it is an everyday experience that starch does not seemingly hydrolyze to sugar when suspended in water (provided that growth of microorganisms is prevented). This means that the number of molecules having the requisite energy of activation is too small to give a perceptible change. However, addition of a small amount of acid catalyzes the hydrolysis, which occurs especially rapidly if heat is also applied to supply the energy of activation to more of the molecules. In cells hydrolyses are catalyzed by enzymes.

Enzymes which help to hydrolyze or split proteins, carbohydrates and fats into smaller residues are known as hydrolytic enzymes (i.e., *hydrolases*) or digestive enzymes, and the process of hydrolysis is also spoken of as digestion. Hydrolases are present inside all cells (in lysosomes) and make possible intracellular movements of proteins, lipids and carbohydrates.

In unicellular organisms, hydrolysis or digestion is often intracellular. For example, *Amoeba* and *Paramecium* form food vacuoles that enclose the food which is hydrolyzed there; the cell foods resulting from hydrolysis pass through the vacuolar membrane into the cell, and the undigestible residues are voided by the cell into the environmental medium. Bacteria, however, digest their foods by secreting enzymes onto the substrate, which is then liquefied during hydrolysis, and the cell foods are absorbed by the bacteria.

In multicellular animals, digestion of a food may occur in steps, each of which may be localized in a separate part of a very specialized and compartmented digestive system. For example, starch digestion in man (and a few other vertebrates)

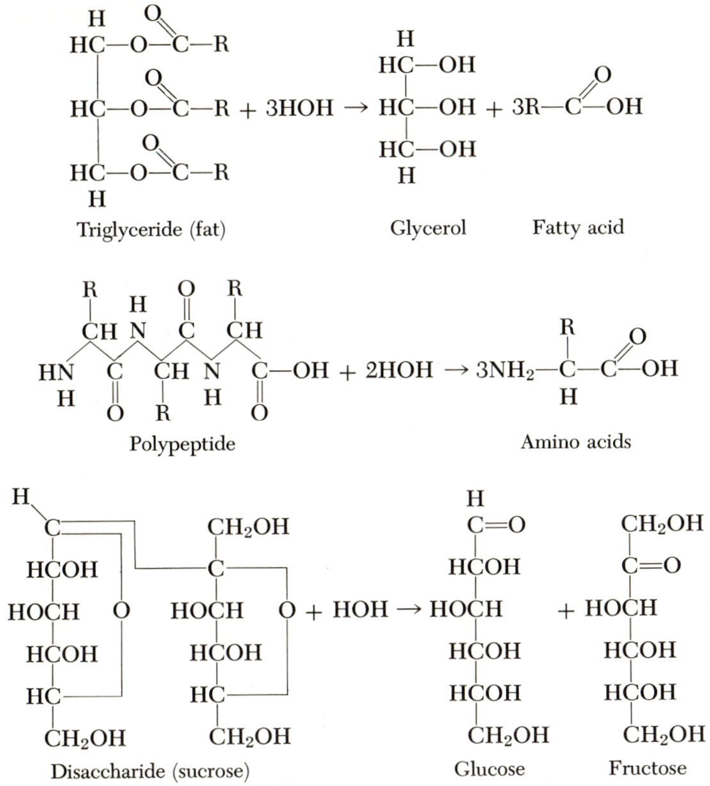


***Figure 15.2.*** Diagrams of the digestion of food by hydrolysis.

begins in the mouth, where it is catalyzed by an enzyme secreted by the cells of the salivary glands, continues, aided by enzymes from the cells of the pancreas, and finally is completed by enzymes from cells in the walls of the small intestine. Protein digestion in vertebrates begins in the stomach (acid phase) and then continues in the small intestine (alkaline phase), assisted by the action of enzymes in the pancreatic juice and intestinal wall juices. Fat digestion, however, occurs in one step, catalyzed in the small intestine by enzymes from cells of the pancreas. In the lower invertebrates digestion may occur in a single stomach-intestine or it may occur intracellularly or by amebocytes.

Digested cell foods pass into the blood and are distributed to all the cells of the body. Cell foods in excess over immediate needs are deposited in places of storage in insoluble form, e.g., glycogen (in liver and muscle), fat (in adipose tissue) and protein (in muscle). Furthermore, by the activities of hydrolytic enzymes, foods in storage depots can be mobilized when need arises. They then move into the blood for intercellular relocation to places where foods are required.

In multicellular green plants, digestion is usually intracellular, and digested cell foods move into the sap and are thus carried to each cell. Here, too, excess of manufactured food goes into storage, e.g., sugar (in fruit), starch (in potatoes), fat (in nuts), protein (in beans) and so forth. Digestion is concerned mainly with hydrolysis of foods into a form suitable for transport from regions of storage to regions of need and growth. An exception is found in carnivorous plants, in which digestion, resembling the process in animals, may occur at the tips of glandular hairs, as in the sundew *Drosera,* or in a cavity, as in a modified leaf of the pitcher plant *Darlingtonia.*

## SPECIFICITY OF HYDROLYTIC ENZYMES

Proteins are essentially polymers of amino acids; carbohydrates, such as glycogen and starch, are polymers of monosaccharides; and fats are conjugates of glycerol and fatty acids. Large polymeric molecules are usually hydrolyzed in several steps, each step being catalyzed by a specific enzyme.

For example, a protein molecule is split first by the *endopeptidases* into smaller fragments from the ends of which amino acids are hydrolyzed by the *exopeptidases.* The exopeptidases perform their work more effectively after the action of endopeptidases, since more ends of protein fragments then become available upon which they may act.

Three familiar endopeptidases are: pepsin, secreted by cells of the vertebrate stomach, and chymotrypsin and trypsin, present in vertebrate pancreatic juice. Their high degree of specificity is shown by the fact that each one attacks a particular linkage between amino acids. Pepsin attacks a protein, or protein fragment, on the amino side of the peptide bond between tyrosine (or phenylalanine) and other amino acids. Chymotrypsin attacks the carboxyl end of the bond between tyrosine (or phenylalanine, tryptophan and methionine) and other amino acids (see Fig. 15.3). Trypsin acts at the carboxyl end of lysine or arginine residue (Cantarow and Schepartz, 1967).

Exopeptidases are also specific in action. Aminopolypeptidases attack a polypeptide on the end where a free amino group exists; carboxypolypeptidases attack a polypeptide on the end where a free carboxyl group exists. The result of each reaction is the production of a free amino acid and a polypeptide molecule reduced in molecular weight. Dipeptidases attack only dipeptides, producing amino acids as products. Many dipeptidases probably exist, each one specific to a particular dipeptide (Smith, 1960).

The *carbohydrases,* which catalyze hydrolysis of carbohydrates, are also specific. Thus the $\alpha$-amylases attack the bonds between glucose residues in starch randomly; $\beta$-amylases attack only the second link from the nonreducing end of the chain and liberate maltose. The net effect of digestion of the various hexosans (polymerized hexoses, see Table 3.1) is the formation of monosaccharides, primarily glucose. However, plant cells also produce and store pentosans (polymerized pentoses) and have enzymes by which these are hydrolyzed into pentoses (Baldwin, 1963).

Certain animal and plant cells have a variety of special carbohydrases which en-

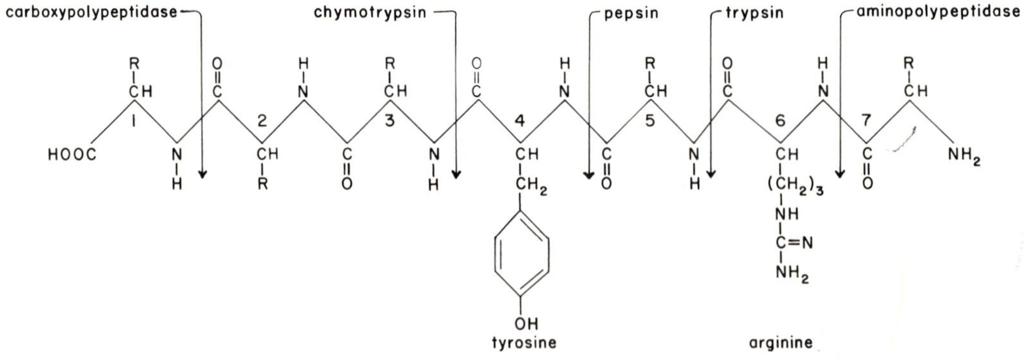


***Figure 15.3.*** A diagram to indicate the specificity of various proteolytic enzymes. The numbers refer to the amino acid residues only two of which, tyrosine and arginine, are labeled below. The polypeptidases are specific, one to the free carboxyl end (left) of a protein molecule or peptide, the other to the free amino end (right) of such molecules. Pepsin is specific to the amino side of tyrosine (or phenylalanine) residues inside a protein molecule; chymotrypsin is specific to the carboxyl side of such residues; and trypsin is specific to the carboxyl side of arginine or lysine residues.

able them to cope with hexosans or pentosans. Some protozoans in termites possess a cellulase that enables them to digest wood cellulose to glucose. Some of the bacteria in cattle and other herbivores also secrete cellulases (Hungate, 1966). An enzyme, inulase, secreted by cells in the tubers of the Jerusalem artichoke, enables them to digest the stores of inulin, a fructose polymer, to fructose whenever needs arise.

*Lipases,* which hydrolyze fats by attacking the ester linkages between alcohols (such as glycerol) and fatty acids, are relatively unspecific in action. Provided the acid is organic, its chemical identity and that of the alcohol radical with which it is combined do not appear to matter so that one lipase usually acts catalytically upon all of them, presumably because the steric configuration of the bonds is similar in all of them. However, lipases from different organisms are not quite identical since their pH optima vary. For example, the pancreatic lipase of man acts best at pH 7.0, whereas that of the castor bean has an optimum at pH 5.0. But in either case the products of the complete digestion of fats are fatty acids and glycerol. The products of digestion of phospholipids, on the other hand, include, in addition to fatty acids and glycerol, phosphoric acid and an organic base. For example, lecithin (see Fig. 3.12) yields, on hydrolysis, fatty acids, glycerol, phosphoric acid and choline (Baldwin, 1963).

An enzyme only accelerates a reaction; it cannot make a reaction proceed which is otherwise impossible. If a reaction is reversible the enzyme accelerates the reaction in both directions, and at a given time the direction of the reaction depends upon the law of mass action. In other words, the equilibrium constant for the reaction is not changed by an enzyme. For example, it has been demonstrated that the hydrolysis of fat by a lipase proceeds to an equilibrium, at which time there is a mixture of fat, fatty acid and glycerol. If more fatty acids and glycerol are then added to this mixture, the same lipase will synthesize fat until a new equilibrium is reached (see Fig. 15.4). The hydrolyses of carbohydrates and proteins, however, are not reversible. For instance, disaccharides and polysaccharides can be synthesized only by phosphorylases from monosaccharides possessing a high energy phosphate bond, indicating that energy is necessary for the synthesis as well as an enzyme other than a hydrolase (Hassid and Doudoroff, 1950). Enzymes other than the hydrolases are necessary to catalyze such syntheses since polymerization occurs by a pathway entirely different from that of hydrolysis. Similarly, the synthesis of proteins requires phosphorylated amino

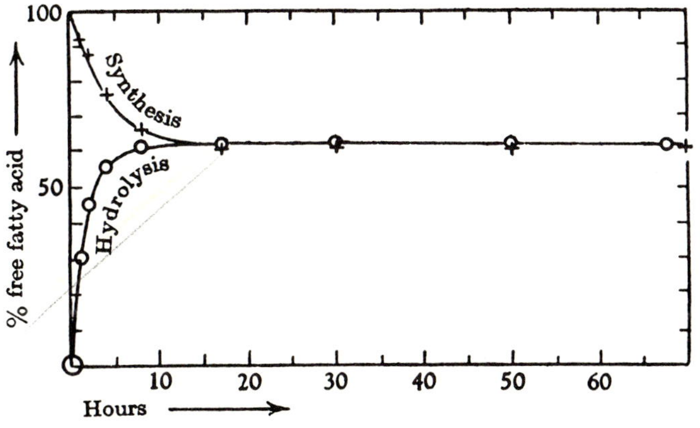


***Figure 15.4.*** Hydrolysis and synthesis of triolein by *Ricinus* lipase. (After Parsons; and data of Armstrong and Gosney.)

acids and enzymes other than the hydrolases (see Chapter 20).

## THE CHEMICAL NATURE OF HYDROLYTIC ENZYMES

Sumner, in 1926, first succeeded in crystallizing urease from jack bean seeds. Urease is a hydrolytic enzyme which decomposes urea to carbon dioxide and ammonia. In 1930, Northrop crystallized pepsin from the gastric juice of the cow, and he and Kunitz soon thereafter crystallized trypsin from the pancreas of the same animal. Since then many hydrolytic and other types of enzymes have been crystallized and purified (Northrop, 1948; Florkin and Stolz, 1964). Over 900 enzymes are now identified and about 75 have been crystallized (Dixon, 1964; Lehninger, 1965).

Several methods have been used in the purification of enzymes. For example, hydrolytic enzymes may be purified by salting out the proteins from digestive juices or from a mash of digestive tissue, and subsequently testing the activities of both the supernatant and the precipitated material. Since some proteins in a mixture are precipitated by a low concentration of a salt (like ammonium sulfate), and others by a higher concentration, separation of the proteins in a mixture is possible only by successive salting out operations. Thus each fraction obtained in this manner is tested for enzymatic activity. The more active fraction is dissolved and the precipitation procedure is repeated until the enzyme activity is confined largely to one of the fractions. The salt may then be removed from the protein by dialysis (e.g., through cellophane) and the active concentrated material is crystallized. At present a variety of chromatographic procedures are being employed for enzyme purification (see Appendix 19.3). Ionic resins which selectively remove molecules of particular sizes permit even quicker separation of enzymes from other materials than classic chromatography (see Colowick and Kaplan, 1955–1959).

Even crystalline enzymes often contain impurities, but they may be further freed of minute quantities of contaminating protein impurities by the Tiselius electrophoresis apparatus. This method takes advantage of the fact that a protein molecule is positively charged on the acid side of its isoelectric pH and negatively charged on the alkaline side; therefore, it migrates in an electric field except at its isoelectric pH. Three factors affect the mobility of a protein: its size, symmetry and charge density. Movement is therefore hindered by large size, by asymmetry and by decrease in charge density. As a result, proteins in a mixture moving in response to an electric field are "sorted out," by these three variables and the concentration of each one in a gradient may be measured optically, e.g., with ultraviolet light or, more commonly, by the change in the index of refraction (Schlieren optics). Each protein may be removed separately from the electrophoresis tube or from the appropriate region of the filter paper when paper electrophoresis is used. Each fraction can then be tested for enzyme activity. Even electrophoretic separation is not final proof of purity since two proteins may form an aggregate and move together.

Proof of enzyme purity can be obtained, however, by the method called *end-group analysis*. This is based on the knowledge that any single peptide chain has only one terminal amino and one terminal carboxyl group. The terminal amino group can be labeled, for example, with the dye 2,4-dinitrofluorobenzene (and comparable methods are available for identifying the carboxyl group). If the protein consists of a single polypeptide strand, only one labeled amino acid should appear per molecule when the protein is hydrolyzed and chromatographed (or separated on an ion exchange column). If the protein consists of two (or more) chains, two (or more) terminal amino acids should appear per molecule, but always in the same stoichiometric ratios in each sample tested. If a mixture of proteins occurs in an aggregate, an exact stoichiometric ratio is unlikely.

All hydrolytic enzymes purified by

these methods show properties characteristic of proteins: color reactions, ultraviolet absorption spectra, inactivation by heat and radiations, and sensitivity to pH and salts and so forth. Therefore the specific properties of an enzyme probably depend upon the kind of protein of which it is composed. However, in some enzymes the proteins are complexed with other ions or molecules.

## ACTIVATION OF HYDROLYTIC ENZYMES

Many hydrolytic enzymes appear to associate with salts which are important to the catalytic activity of these enzymes. For example, when ptyalin (salivary amylase) is freed of sodium chloride by dialysis, neither the protein portion nor the sodium chloride alone can effect the hydrolysis of starch. When the two are mixed together again, the preparation once more becomes catalytically active. It has been shown that in this case it is the chloride ion in sodium chloride which is essential for the enzyme activation. Many of the elements needed for enzymatic reactions are present in minute quantities in cells of plants and animals, and may be associated in a similar manner with such enzymes (see Tables 3.4 and 15.1). However, not all hydrolytic enzymes are activated by salts. Some, like pepsin, trypsin and chymotrypsin, are not known to require salt ions for their catalytic action (Gutfreund, 1965).

Hydrolases are often in an inactive form when they leave the cells in which they are manufactured. In such a case they must be activated before they can catalyze hydrolyses. For example, *pepsinogen,* the inactive forerunner of pepsin, is secreted from cells (chief cells) in the stomach lining into the lumen, where it is activated by hydrochloric acid ($H^+$ ions) coming from other secretory cells (parietal cells) nearby and also by the autocatalytic action of pepsin itself. Activation here involves removal of a polypeptide, since the molecular weight of pepsinogen differs from that of pepsin by an amount equivalent to the molecular weight (4000) of a small polypeptide, as shown in Fig. 15.5 (Boyer *et al.*, 1960). The suggestion has been made that the polypeptide masks the active spots of the enzyme.

Enzymes which require free sulfhydryl groups (—SH) for catalytic activity must be reduced after purification to restore their catalytic activity since during the purification process these sulfhydryl groups are oxidized. For example, the protease *papain,* as extracted from the papaya, is inactive until it is treated with reducing agents such as glutathione, hydrogen cyanide or hydrogen sulfide, which reduce the disulfide bonds (S—S) in the enzyme to thiol or sulfhydryl bonds (—SH). The enzyme is active only when one of these bonds is present in the reduced state.

**Table 15.1.** Specific Metal Activation of Various Hydrolytic Enzymes

| ENZYME | HYDROLYSES CATALYZED BY ENZYMES | METAL REQUIRED |
|---|---|---|
| Prolidase | Glycylproline + HOH → proline + glycine | Mn |
| Dehydropeptidase | Glycyldehydrophenylalanine + HOH → glycine + $NH_3$ + phenylpyruvic acid | Zn |
| Carboxypeptidase | Carbobenzoxyglycyl-l-leucine + HOH → carbobenzoxyglycine + leucine | Mg |
| Glycylglycine dipeptidase | Glycylglycine + HOH → 2 glycine | Zn |

These hydrolyses will proceed only if the metals are present in small quantities, indicating that the metals are catalytic in action. The peculiar substrates of the second and third reactions are used because analytic procedures are easier with such synthetic compounds. (Data from McElroy and Nason, 1954, and from Smith and Hanson, 1949: J. Biol. Chem. *179.*)

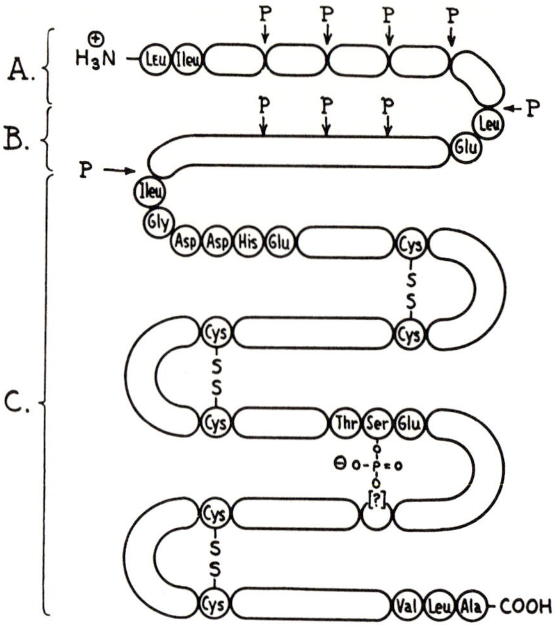


***Figure 15.5.*** The structure of pepsinogen. The principle attack by hydrogen ions (of HCl) and pepsin is at points marked *P*, releasing miscellaneous peptides (*A*), pepsin inhibitor (*B*), and pepsin (*C*). The undetermined sequences are actually much larger in relation to the known sequences than is indicated. The location of the phosphoserine residue with respect to the disulfides is also uncertain. (From Boyer, Lardy and Myrbäck, 1960: The Enzymes. Vol. 4. Academic Press, New York.)

Cathepsin, a protein-digesting enzyme in the liver, is also active only in its reduced state. Such enzymes are inactivated when oxidized; oxidation may occur even by exposure to air during extraction or purification.

## ENZYMES INVOLVED IN CELLULAR OXIDATION-REDUCTIONS

In addition to the hydrolases which catalyze hydrolysis of organic cell foods, a number of other kinds of enzymes—isomerases, transferring enzymes, adding and splitting enzymes, oxidases and dehydrogenases—take part in the synthesis and degradation of nutrients in the cell (Gutfreund, 1965). Although some of these enzymes are known only in impure state in cell fragments, others have been crystallized and purified (Boyer *et al.*, 1960).

The *isomerases* catalyze internal changes in the molecules of compounds, such as the shift from glucose-1-phosphate to fructose-6-phosphate. Quite a number of isomerases catalyze such changes in both the anaerobic and aerobic phases of metabolism.

*Lyases* are enzymes that remove a group of atoms from a substrate with the formation of a double bond. *Ligases* are enzymes that combine adenosine triphosphate (ATP) hydrolysis with a bonding of two substrates (1961 Union of Biochemistry).

*Adding* and *splitting enzymes* catalyze addition of a fragment to a molecule or splitting of a fragment from a molecule, e.g., the addition of carbon dioxide to a molecule, with the formation of a new carboxyl group, or the splitting out of carbon dioxide from a carboxyl group (decarboxylation). As an illustration, pyruvic acid may be carboxylated to form oxaloacetic acid or decarboxylated to form an acetyl fragment and carbon dioxide. Some transferring enzymes catalyze the addition or removal of water but their action differs from that of hydrolases because they do not split molecules.

*Transferring enzymes* catalyze transference of various radicals, such as the high energy phosphates, methyl groups, acetyl groups and amino groups. For instance, the enzyme hexokinase, a phosphorylase, catalyzes the phosphorylation of glucose, in the course of which a high energy phosphate bond is transferred from an ATP (see Fig. 16.3) to a glucose molecule. Dehydrogenases and related enzymes (e.g., flavoproteins) are a special group of transferring enzymes by which hydrogen is transferred.

*Oxidases* and *dehydrogenases* are concerned with electron and hydrogen ion transfers. Examples are cytochrome oxidase, which transfers electrons, and coenzyme I (nicotinamide adenine dinucleotide or NAD; see Fig. 15.6), which transfers hydrogen ions and electrons.

Although a large number of different and specific types of enzymes are concerned with oxidation-reduction reactions in the cell, all of them have some properties in common. Each oxidative enzyme studied has been found to consist of a protein carrier and a prosthetic group. The *prosthetic group* is an organic radical, sometimes of considerable complexity, which is attached to the protein. Neither the protein nor the prosthetic group alone is enzymatically active, as can be demonstrated by separating the two components. For example, when the flavoprotein enzyme (the yellow enzyme) is dialyzed

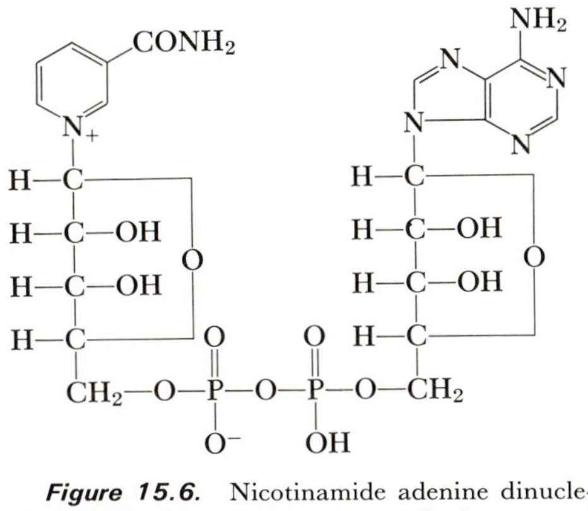


***Figure 15.6.*** Nicotinamide adenine dinucleotide (NAD), the coenzyme or prosthetic group of some of the dehydrogenases. Nicotinamide adenine dinucleotide phosphate (NADP) is similar in structure except that it has an extra phosphoric acid residue attached to the pentose of adenosine (right side of the molecule). The reduced form of NAD may be designated $NADH_2$ or NADH + $H^+$ when it is desirable to show the release of an $H^+$ ion in the reduction. If the latter designation is employed then the oxidized form of NAD should be given as $NAD^+$. The same conventions are to be used with NADP (Thompson, 1962).

against distilled water, a yellow compound diffuses out, leaving the protein in the cellophane bag. The yellow compound, containing riboflavin or vitamin $B_2$, is the prosthetic group of the enzyme (see Fig. 15.7). Neither the protein part inside the cellophane bag nor the prosthetic group outside shows enzymatic activity. The two when mixed become active, together forming the enzyme. In the cell the two are probably present mostly in the associated form, since the dissociation constant for this enzyme is small.

In some enzymes, like the *dehydrogenases,* the prosthetic group is usually dissociated from the protein and associates with it only when acting on the substrate. When the dissociation constant is large, as in this case, the prosthetic group is generally called a *coenzyme.* Two prosthetic groups (see Fig. 15.6) are known for dehydrogenases—coenzyme I, or nicotinamide adenine dinucleotide ($NAD^+$),* and coenzyme II, or nicotinamide adenine dinucleotide phosphate ($NADP^+$).† $NADP^+$ is most frequently the coenzyme in synthetic reactions and represents a "reduction reservoir." It also functions in the pentose oxidative shunt (Fig. 16.12), whereas $NAD^+$ is the coenzyme for most degradative dehydrogenations. All the dehydrogenases formed by union of these prosthetic groups with various proteins are highly specific as to substrate. Dissociation appears advantageous since the same prosthetic group can thus associate with various proteins, each combination being specific to some substrate. The specificity of a dehydrogenase, therefore, must be due largely to the protein moiety of the enzyme with which the coenzyme associates (Cantarow and Schepartz, 1967).

Cytochromes are enzymes which transfer electrons from flavoproteins or other carrier enzymes to cytochrome oxidase. The cytochrome oxidase in turn passes the electrons to oxygen, activating it and enabling it to combine with the hydrogen ions and form water. The cytochromes

*Coenzyme I was previously designated diphosphopyridine nucleotide (DPN).

†Coenzyme II was previously designated triphosphopyridine nucleotide (TPN).

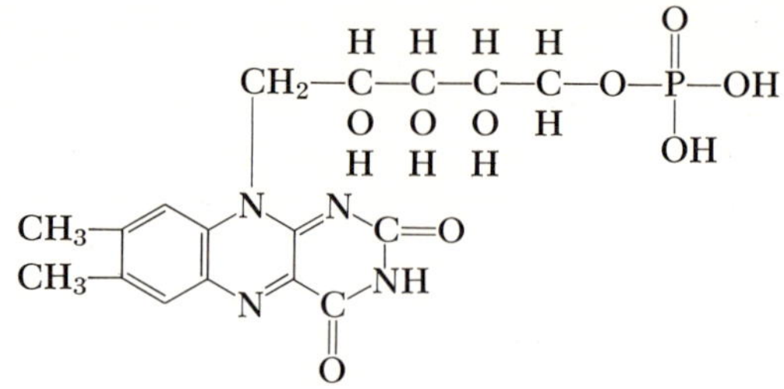

Riboflavin phosphate
(flavin mononucleotide)

↓ + ATP

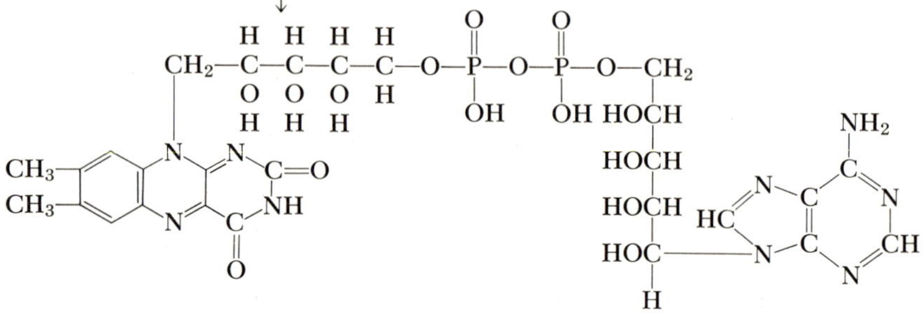

Flavin adenine dinucleotide (FAD)

$$\text{DPNH} + \underset{\text{FAD}}{\overset{\text{FMN}}{\text{or}}} \longrightarrow \text{DPN} + \underset{\text{FADH}_2}{\overset{\text{FMNH}_2}{\text{or}}}$$

***Figure 15.7.*** Riboflavin prosthetic groups of "yellow" enzymes. Both of the yellow enzymes participate primarily in transport of hydrogen ions and electrons from prosthetic groups of one of the dehydrogenases (e.g., nicotinamide adenine dinucleotide) to the cytochromes. Each prosthetic group has specific functions, e.g., FMN is the prosthetic group of lactic dehydrogenase and FAD is the prosthetic group of aldehyde oxidase, D-amino oxidase and fatty acyl-CoA dehydrogenases. (From McElroy, 1964: Cellular Physiology and Biochemistry. 2nd Ed. Prentice-Hall, Englewood Cliffs, N. J.)

(Fig. 15.8), of which there are a number in all aerobic cells, are cell pigments with a prosthetic group resembling the heme of hemoglobin. Cytochrome oxidase also has a similar prosthetic group. In cytochromes and cytochrome oxidase the iron changes from the oxidized ($Fe^{+++}$) to the reduced ($Fe^{++}$) state as an electron is received, and from the reduced to the oxidized state as an electron is passed on to oxygen. In hemoglobin the iron is always in the reduced state and adds oxygen only loosely (becomes oxygenated) without changing in valence.

The cytochromes and cytochrome oxidase are tetrapyrrole compounds, four pyrroles surrounding the iron. Tetrapyrroles without the metal are called *porphyrins.* When a metal is present they are called *metalloporphyrins.* Cytochrome and heme are examples of metalloporphyrins in which iron is present. It is interesting to note that the plant pigment chlorophyll (see Fig. 19.10) is a metalloporphyrin in which magnesium is present.

As already pointed out, many of the enzymes catalyzing the oxidation-reduction reactions are quite specific in action. Thus, succinic acid dehydrogenase catalyzes dehydrogenation of succinic acid only, and malic acid dehydrogenase catalyzes dehydrogenation of malic acid only. However, enzymes exist which catalyze a number of reactions, being specific not to an individual molecule but to a particular steric configuration of atoms, which may

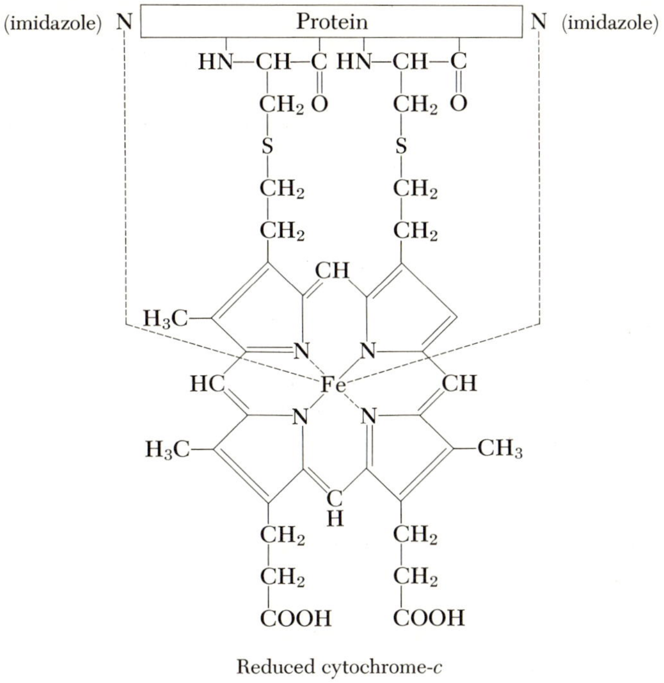

*Figure 15.8.* Reduced cytochrome C. (From Florkin, 1960: Unity and Diversity in Biochemistry. Pergamon Press, London.)

occur in a number of different molecules. For instance, isocitric dehydrogenase catalyzes dehydrogenation of both isocitric and oxalosuccinic acids in the Krebs cycle (Baldwin, 1963).

Some of the enzymes participating in oxidation-reduction reactions are activated by various elements. Some of these elements are listed in Table 15.2, along with the enzymes for which they are necessary (McElroy and Nason, 1954; Gutfreund, 1965).

It is interesting to note the presence of members of the vitamin B complex as components of the prosthetic groups of the enzymes involved in cellular oxidations and syntheses. This accounts for the presence of certain B vitamins in cells of all organisms examined, e.g., vitamin $B_1$, or thiamine, which forms part of the prosthetic group of cocarboxylase (see Fig. 15.9), an enzyme important in the metabolism of pyruvic acid. Some of the pertinent information on the metabolic role

**Table 15.2.** Elements Necessary for Some Oxidative Enzymes

| ENZYME | REACTION | METAL |
|---|---|---|
| Carbonic anhydrase | $CO_2 + H_2O \rightleftharpoons H_2CO_3$ | Zn |
| Inorganic pyrophosphatase | Pyrophosphate + $H_2O \rightarrow PO_4$ | Mg* |
| Catalase | $2H_2O_2 \rightarrow 2H_2O + O_2$ | Fe |
| Cytochromes | Electron transport | Fe |
| Tyrosinase | Tyrosine + $\frac{1}{2}O_2 \rightarrow$ hallachrome | Cu |
| Laccase | Phenols → ortho and paraquinones | Cu |
| Ascorbic acid oxidase | Ascorbic acid → dehydroascorbic acid | Cu |

From McElroy and Swanson, Jan. 1953: Sci. Am. *188.* A much more complete account is given in McElroy and Nason, 1954.

* Magnesium is indicated to be an activator for a large variety of enzymes—kinases, mutases and so forth.

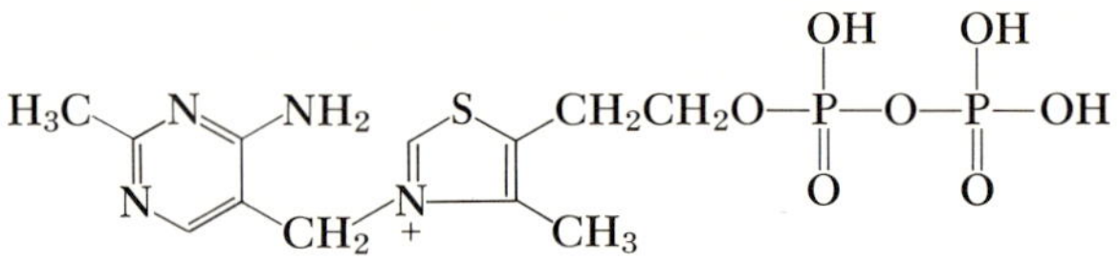

**Figure 15.9.** Thiamine pyrophosphate, or cocarboxylase.

of these vitamins is summarized in Table 3.9, along with a summary of the gross superficial symptoms which result from lack of the vitamins in the diet. These pathological symptoms are the cumulative effects of prolonged metabolic disturbances (Szent-Györgyi, 1939; Sebrell and Harris, 1954).

## EFFECT OF ENVIRONMENT ON ENZYME ACTIVITY

The rate of an enzyme catalyzed reaction, in most cases, increases proportionally with increasing concentration of the enzyme, provided the substrate is present in excess and is therefore not rate limiting.

The relation between the rate of the reaction and substrate concentration has been formulated in most general terms as follows:

$$\text{Rate} = k[S]^n \qquad (15.1)$$

In this equation k and n are constants and S is the concentration of the substrate. In some cases increasing the substrate concentration (in the lower part of the concentration range) proportionally increases the rate of the reaction. In that case the exponent n in equation 15.1 has the value of unity (first order). When a high concentration of substrate proves inhibitory, as it does in some cases, the exponent n is said to have a negative value. In some other cases, much more complex relations than these exist between the concentration of substrate and the rate of the enzyme-catalyzed reaction (Bull, 1964).

When the concentration of enzyme is low enough so that it is rate limiting, and the concentration of substrate is varied, the rate of the reaction may prove independent of the substrate concentration. In that case n in equation 15.1 is zero, and the term for the substrate drops out of the equation. Qualitatively speaking, the enzyme becomes saturated with substrate or nearly so. This becomes the "bottleneck" determining the rate of the reaction, and further increase in substrate concentration has a negligible effect.

If the substrate concentration is varied, but the enzyme concentration kept constant, a limiting rate is reached which is the maximal turnover rate or work rate of the available enzyme at the particular temperature and pH of the experiment. Examples of the effect of concentration of enzyme and substrate on rate of reaction are given in Figures 15.10 and 15.11. These data suggest that an enzyme-substrate complex is formed as the first step in any enzyme-catalyzed reactions (Bull, 1964).

Temperature also is a factor in enzyme activity. Experiments with hydrolytic enzymes indicate that the greatest rate of hydrolysis occurs at the temperature optimal for the hydrolase. The optimal temperature is different for enzymes from different organisms. It is usually low for enzymes present in cells of plants and cold blooded animals of the polar regions, very high for enzymes from some thermophilic bacteria and intermediate for enzymes from cells of other forms. Seemingly, the optimal temperature for the catalytic activities of hydrolytic enzymes is close to the normal temperature of the particular cell.

A rise in temperature up to the optimum is accompanied by a corresponding increase in the rate of hydrolysis (Fig. 15.12) because of the increase in the number of molecules with the required energy of activation, as shown in Figure 10.4. An increase in temperature above the optimum is accompanied by a decrease in

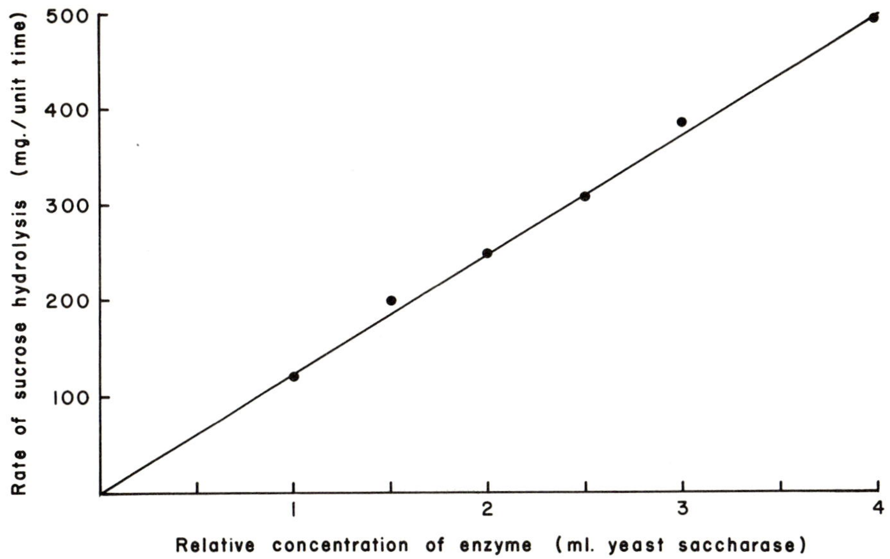


***Figure 15.10.*** Effect of enzyme concentration on the rate of a reaction when an excess of substrate is present. (After Baldwin, 1963: Dynamic Aspects of Biochemistry. 4th Ed. Cambridge University Press, London.)

the rate of hydrolysis (Fig. 15.12). This has been explained in the following manner. Only the native or unaltered form of an enzyme is catalytically active; therefore, other things being equal, the rate of a hydrolytic reaction will depend in part upon the concentration of the native enzyme. At all temperatures, within a temperature range characteristic of the organism, an equilibrium exists between native and denatured enzyme. This equilibrium can be shifted in either direction. As the temperature rises, more of the enzyme goes into the denatured state, as shown in Table 15.3. Therefore, as the quantity of native enzyme decreases, the rate of the hydrolytic reaction catalyzed by the enzyme declines. The temperature

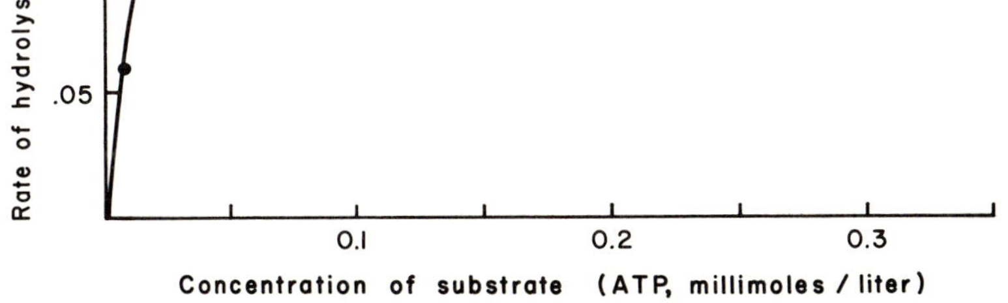


***Figure 15.11.*** Effect of substrate concentration on the rate of reaction when a limited amount of enzyme is present. Note that the enzyme becomes saturated with substrate so that further increase in concentration of substrate has no effect. (After Ouellet *et al.,* 1952: Arch. Biochem. Biophys. *39*:41.)

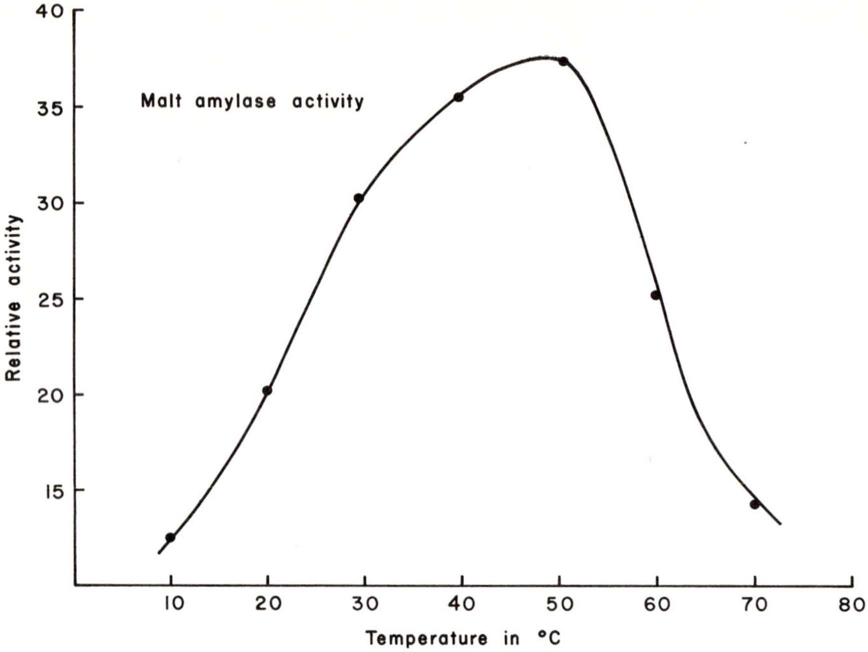


*Figure 15.12.* Effect of temperature on the rate of enzyme activity. (Data from Gortner, 1949: Outlines of Biochemistry. 3rd Ed. John Wiley & Sons, New York.)

coefficient, $Q_{10}$, for denaturation of enzymes between 70° and 80° C. is several hundred, indicating a tremendous rate of denaturation at these temperatures. Although the conformity of the peptide chain(s) in the enzyme represents the condition of lowest free energy, the entropy is low since not only is the sequence of amino acids specified but the secondary and tertiary structure is also quite specific. Such structure is maintained primarily by hydrogen bonds and some sulfur-to-sulfur linkages. Many of these bonds are readily broken, leading to a more random arrangement of higher entropy (Lehninger, 1965).

**Table 15.3.** Denaturation of Trypsin by Heat

| T° C. | PER CENT DENATURED | T° C. | PER CENT DENATURED |
|---|---|---|---|
| 42 | 32.8 | 45 | 57.4 |
| 43 | 39.2 | 48 | 80.4 |
| 44 | 50.0 | 50 | 87.8 |

Data from Northrop, Kunitz and Herriott: Crystalline Enzymes.

Most enzymes are denatured irreversibly at a temperature between 50° and 80° C. Others, however, are very resistant to high temperature. For example, crystalline trypsin is reversibly denatured even at 100° C. and so is Taka-Diastase. The latter is irreversibly denatured only above 140° C. Each enzyme is inactive catalytically while at the limit of its high temperature tolerance but resumes its activity when cooled (Johnson *et al.*, 1954).

Enzymes are also inactivated by salts of heavy metals such as copper or mercury. These metals are often used in fixing agents for preserving cell proteins, because they act as denaturants and protein precipitants. Their action on enzymes is, therefore, not unexpected. The same might be said of an alkaloidal reagent such as phosphotungstic acid, which is often used to precipitate proteins in biochemical procedures.

One of the most interesting environmental influences on enzymatic activity

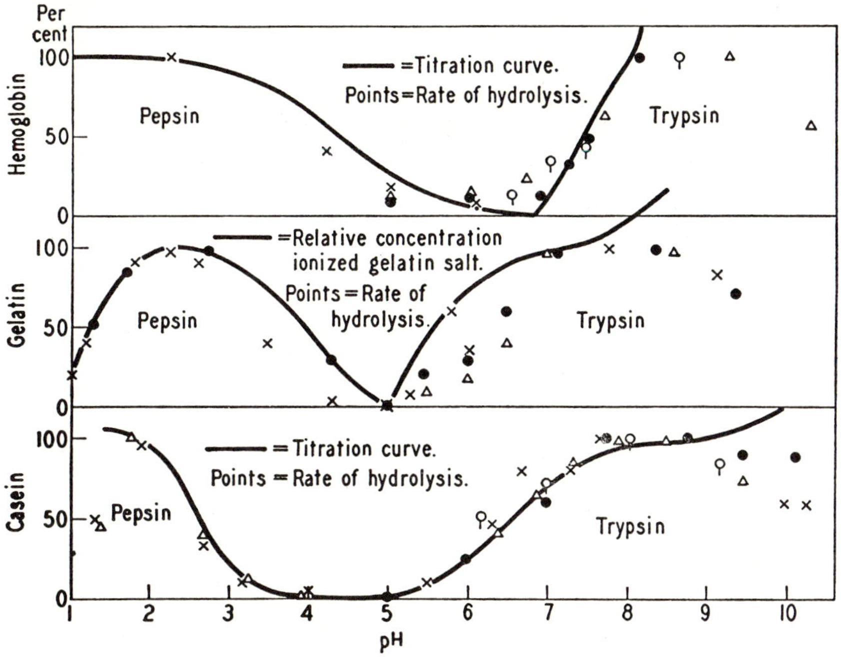


***Figure 15.13.*** The effect of pH upon the activity of pepsin and trypsin. The ordinate measures the relative rates of hydrolysis. Note that the pH effect varies with the protein. (From Northrop, 1939: Crystalline Enzymes. Columbia University Press, New York.)

is the effect of pH. Each enzyme is catalytically active within a limited range of pH (see Fig. 15.13). For example, the optimal range for the activity of pepsin is between pH 1.5 and 2.5; for trypsin, between 8 and 11; for salivary amylase, between 6.7 and 6.8; for malt amylase, at 5.2; for pancreatic lipase, and 7; for castor bean lipase, at 5. In all these cases activity declines gradually on either side of the optimum. The pH sensitivity of enzymes is thought to result from the necessity of a particular charge distribution about the active center for optimal activity. Since different enzymes are composed of amino acids in different proportions, the optimal pH of an enzyme is apt to be a characteristic property dependent upon its composition.

## MECHANISM OF ENZYME ACTION

An enzyme appears to act in much the same way as an inorganic catalyst, except that it is usually much more specific. It is known that an enzyme adsorbs the substrate upon which it acts, forming an intimate union with it, as shown diagrammatically for pepsin in Figure 15.14 and for an amylase in Figure 15.15 (Frieden, 1959). In this case apparently the positive and negative charges present on the protein of the enzyme attract the oppositely charged radicals of the substrate. The precise nature of the union here is not known with certainty and may vary with different enzymes.

As we have seen in Chapter 10, the rate of a thermochemical reaction is proportional to the number of molecules possessing the required energy of activation (Fig. 10.4). If at a given temperature the number of activated molecules is small, the rate of the reaction will be low. A catalyst is thought to increase the rate of a reaction because the energy of activation level is lowered (Fig. 15.1) when the reactant molecules are adsorbed to its surface and subjected to the stress of adsorption, which may amount to a com-

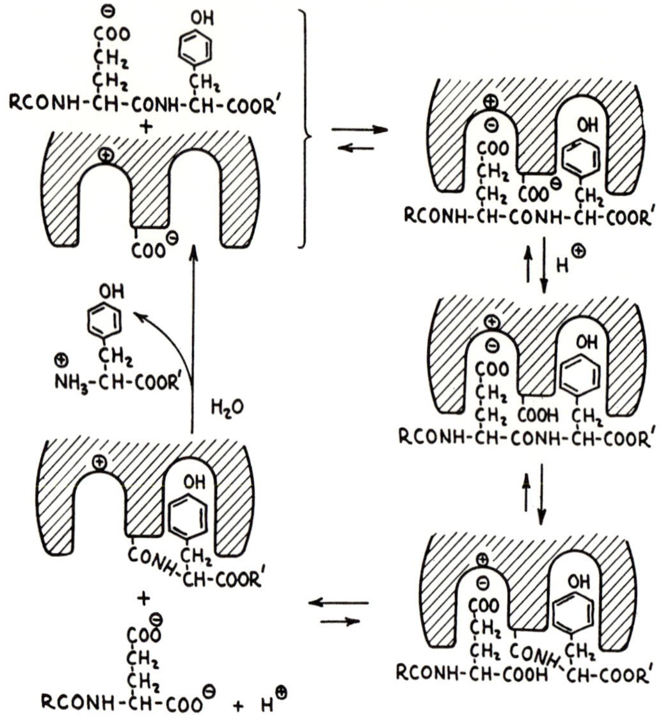


*Figure 15.14.* A proposed hypothetical mechanism of action of pepsin; N-acyl-L-glutamyl-L-tyrosine ester represented as substrate. (From Boyer, Lardy and Myrbäck, 1960: The Enzymes. Vol. 4. Academic Press, New York.)

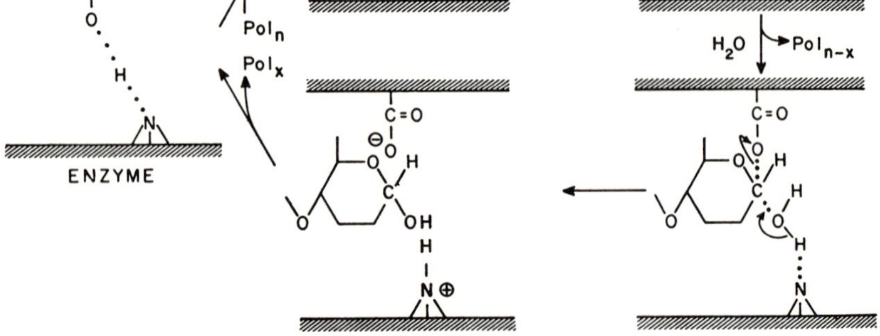


Scheme I

*Figure 15.15.* Proposed hypothetical mechanism of action of amylases. (From Boyer, Lardy and Myrbäck, 1960: The Enzymes. Vol. 4. Academic Press, New York.)

pression of several hundred atmospheres pressure. Consequently, even at low temperatures, in the presence of a catalyst the number of molecules having energy sufficient to react is vastly increased. After the adsorbed molecules react they are less attracted to the catalyst, and as they leave the adsorbing surface, room becomes available on it for more reactant molecules.

In general, enzymes lower the energy of activation for cellular reactions, making possible, at a temperature and pressure present in cells, reactions which would otherwise require very high temperature or pressure, or both and would thus be impossible in the cell. For example, the energy of activation for hydrolysis of cane sugar by acid is 25,600 calories per mole, whereas for its hydrolysis in the presence of the enzyme invertase, it is only 9080 calories per mole (Johnson *et al.*, 1954).

The hydrolysis of a polypeptide by a peptidase, in which a metal ion is often associated with a protein (see Tables 15.1 and 15.2), has been envisaged in the following manner. The metal ion attaches itself by residual valency to a polypeptide molecule on one side of the peptide bond, while the protein portion of the enzyme attaches itself on the other side of the same peptide bond. A stress is thereby placed on the peptide bond, weakening the bond between the amino and carboxyl groups of two adjacent amino acid residues of the polypeptide. Thereby the energy required for breaking the peptide bond is reduced, and the hydrogen ions, always present, then cause its hydrolysis, even at room temperature (Smith, 1960). In the absence of an enzyme a high temperature is required for this hydrolysis (Phillips, 1966).

It has been possible to resolve the three-dimensional detail of an enzyme crystal at the 2 Å level with x-ray diffraction methods, which permit analysis of the positions of the amino acid residues by joining lines (contours) of equal electron density in the x-ray diffraction pictures. Such a study has been made for the hydrolytic enzyme lysozyme, which digests the cell walls of bacteria, by the computer analysis of some 10,000 contours.

Lysozyme appears to be a single polypeptide chain of 129 residues of 20 different kinds of amino acids. The sequence of amino acids in lysozyme has been identified by the methods already discussed for insulin (Chapter 3). From the x-ray diffraction studies it is evident that amino acid residues numbered 5 through 15, 24 through 34 and 88 through 96 represent three lengths of alpha helix structure, although the helical sections are somewhat distorted. The amino acid residues with polar hydrophilic acid and basic groups are on the surface of the molecule; those with hydrophobic nonpolar groups (e.g., the methyl groups in leucine and isoleucine) are on the inside of the molecule. The crystal of lysozyme contains water of crystallization to the extent of about 35 per cent of weight. Parts of the polypeptide chain, especially near the amino terminal end, are folded into stable conformations and appear to act as centers around which the remainder of the molecule is in turn folded. The end result is a lysozyme molecule which consists of a helix in a gap between two "wings" formed by other residues with their hydrophobic groups turned deep into the molecule. The gap is not filled and a deep cleft running up on the side of the molecule appears to be the active site on the enzyme molecule (Phillips, 1966). The remaining amino acid residues are wound around the globular unit formed by the terminal amino acid end of the polypeptide.

By using a polymer of an amino sugar (e.g., N-acetyl muramic acid) which lysozyme does not appear to attack, but which it binds (by competitive inhibition), and studying the x-ray diffraction pattern of the enzyme-sugar complex, it was found that the substrate binds to the enzyme in a number of places within the cleft of the lysozyme molecule. Activation presumably occurs by distortion resulting from the binding. Although x-ray determinations do not permit study of the enzyme in action, they supply three-dimensional data which give stronger evidence for the mechanism of enzyme action than had previously been available (Koshland, 1963). Such a cleft also appears in some other enzymes which have been under intensive study, e.g., chymotrypsin.

Not all portions of an enzyme molecule are necessary for its function as an enzyme. Thus papain may lose 120 of its 180 amino acid residues without losing its enzyme activity. On the other hand, a small change in a critical part of an enzyme molecule may alter or destroy its activity. Thus, blocking the sulfhydryl groups of the protein of phosphoglyceraldehyde dehydrogenase (after removal of the coenzyme NAD, which is necessary for dehydrogenase activity) induces in the same molecule phosphatase activity. Glutamic dehydrogenase can be fragmented into subunits (molecular weight 250,000) which catalyze the reversible deamination of alanine. Removal of one histidine residue destroys the activity of chymotrypsin (Niemann, 1964).

Experimental evidence indicates that oxidation-reduction enzymes, like hydrolytic enzymes, combine with the substrate. For example, peroxidase, which absorbs light maximally in the visible spectrum at wavelengths 498, 538, 548 and 645 nm., on addition of peroxide shows maximal absorption at 561 and 630.5 nm. Such a change in absorption spectrum indicates that the union between substrate and enzyme must be rather intimate, since the absorption spectrum of a compound is not so altered except by a chemical change (Baldwin, 1963).

Evidence of union between enzyme and substrate is also indicated by the following experiments on bioluminescence. An enzyme, luciferase, catalyzes the oxidation of a substrate, luciferin, with the emission of light. If the luciferin and luciferase are mixed before oxygen is admitted, the oxidation will take less time than if the oxygen is first added to each separately and the two are then mixed (see Fig. 15.16), indicating that first there is the formation of a complex of luciferin and luciferase followed by the oxidation of the complex by atmospheric oxygen:

$$\text{luciferin} + \text{luciferase} \rightarrow \text{luciferin-luciferase complex} \quad (15.2)$$

$$\text{luciferin-luciferase complex} + O_2 \rightarrow \text{oxyluciferin} + \text{luciferase} + \text{light} \quad (15.3)$$

Because the formation of the complex,

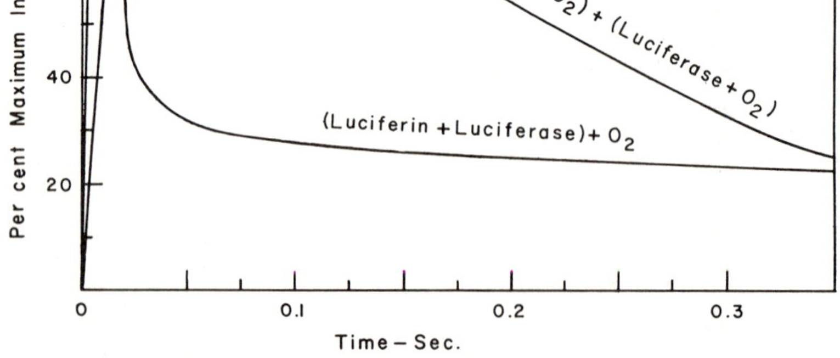


***Figure 15.16.*** Rate of emission of light when *Cypridina* luciferin (oxidizable substrate) is mixed with luciferase (enzyme) before admission of oxygen, and when each solution is mixed with oxygen before the two are mixed together. (After Chance *et al.*, 1940: J. Cell. & Comp. Physiol. *15*:195.)

luciferin-luciferase, is a slower reaction than its oxidation, a mixture of luciferin and luciferase in which the complex has already formed will give off light more rapidly on admission of oxygen then will the reaction upon addition of luciferin and oxygen to luciferase and oxygen. The difference in time in the two cases measures the time required for the association of luciferin and luciferase (Chance *et al.*, 1940).

From such evidence as this it is thought that the substrates involved in oxidations and reductions are bound selectively to the surfaces of the enzymes concerned. As a result of such adsorption, the energy of activation required for dehydrogenation or other oxidative enzymatic reactions is lowered (see Fig. 15.1) so that the reactions can proceed at temperatures in the biokinetic zone (see Chapter 10).

## THE MICHAELIS-MENTEN LAW

Assuming that the enzyme and the substrate combine reversibly to form a complex in enzyme-catalyzed reactions: enzyme + substrate → complex, or

$$E + S \underset{k_2}{\overset{k_1}{\rightleftharpoons}} ES \qquad (15.4)$$

where $k_1$ is the velocity constant of the formation of complex (ES), and $k_2$ is the velocity constant for dissociation of (ES). If complex → enzyme + products, or

$$ES \xrightarrow{k_3} E + P \qquad (15.5)$$

then $k_3$ is the velocity constant for decomposition of the complex (ES) with the formation of reaction products (P). Reaction 15.5 is considered to be the slowest; therefore it sets the pace for the overall reaction since E cannot be recovered until complex ES is decomposed.

Under certain conditions, the rate of the enzymatic reaction will also depend upon the concentration of the substrate. This relation was formulated quantitatively by Michaelis and Menten in 1913 on the basis of the law of mass action. The following derivation follows that given by Bull (1964).

If $E_0$ represents the total concentration of enzyme, E the total concentration of free enzyme, and ES the concentration of bound enzyme, then

$$E = E_0 - ES \qquad (15.6)$$

The dissociation constant of the enzyme-substrate complex is:

$$K_m = \frac{E \times S}{ES} = \frac{(E_0 - ES)S}{ES} \qquad (15.7)$$

or, by rearrangement and by making certain approximations;

$$ES = \frac{E_0 \times S}{K_m + S} \qquad (15.8)$$

The velocity of the reaction at any time for the formation of the product is:

$$v = k_3 \times ES \qquad (15.9)$$

where $k_3$ has the same meaning as in equation 15.5.

Then substitution of equation 15.9 into equation 15.8 gives:

$$v = \frac{k_3 \times E_0 \times S}{K_m + S} \qquad (15.10)$$

If the substrate concentration is increased to the point at which the surface of the enzyme is saturated with substrate, the maximum velocity of the reaction is then:

$$V = k_3 \times E_0 \qquad (15.11)$$

The Michaelis-Menten substrate-enzyme complex is not the activated complex, the latter being present in much lower concentration. The substrate-enzyme complex must pass into an activated form before it can decompose into the final products of the reaction.

Tests of this expression of the velocity on many biological reactions, as affected by substrate in the lower concentration range, indicate good quantitative agreement with prediction. This is usually done by a double reciprocal plot 1/v, where v stands for the velocity at any time, against 1/S, where S stands for the substrate concentration (Fig. 15.17). This gives assurance that the kinetics of enzymatic

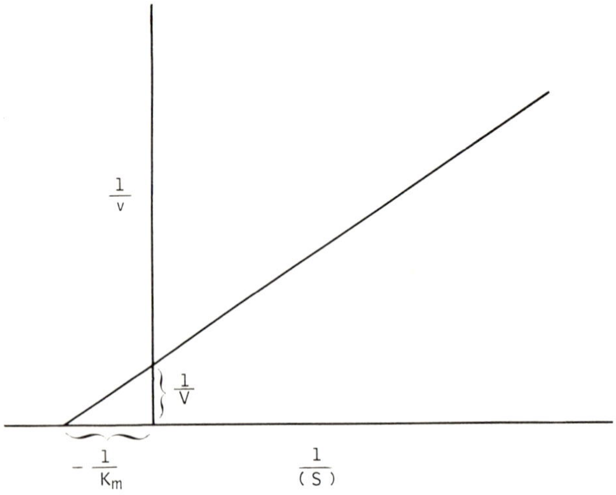


**Figure 15.17.** Lineweaver-Burk double reciprocal plot as a test of the applicability of the Michaelis-Menten formulation of the reaction kinetics. The v stands for the reaction velocity at any time, V is the maximum reaction velocity when the surface of the enzyme is saturated with substrate, S is the concentration of substrate and $K_m$ is the Michaelis-Menten constant. $K_m$ is equal numerically to the substrate concentration that permits the reaction to proceed at one-half the maximum velocity. (From Cantarow and Schepartz, 1967: Biochemistry. 4th Ed. W. B. Saunders Co., Philadelphia.)

reactions are subject to the law of mass action and indicates the incisiveness of the formulation of the relationships between enzyme and substrate by Michaelis and Menten (Patton, 1965).

## INHIBITION OF ENZYMES BY POISONS

Since enzymes have sites (one or more) at which they combine with a substrate, inactivation of those sites prevents the activity of the enzyme upon its substrate. An agent inactivating an enzyme is an *enzyme poison.* Poisons are effective to different degrees and act upon enzymes in different ways (Hochster and Quastel, 1963; Webb, 1966).

When a nonmetabolizable molecule, A, resembles a metabolizable molecule, B, sufficiently to be bound to the enzyme, then A remains attached to the enzyme and prevents the attachment of B, thus competing with B for space on the enzyme surface. Its inhibition of the reactions of B is called *competitive inhibition.* For example, malonic acid resembles succinic acid (see Fig. 15.18) because both have two carboxyl groups. Malonic acid inhibits the action of succinic dehydrogenase on succinic acid by "clogging the active spots" on the enzyme, and since the malonic acid is not metabolized it remains attached to the enzyme. However, the complex of malonic acid and succinic dehydrogenase does dissociate at a finite rate given by the dissociation constant. Therefore, an excess of succinic acid will reverse the action of malonic acid (law of mass action). In general, when competitive inhibition occurs, an excess of substrate reverses the action of the poison, as shown in Figure 15.19. An experiment in which the substrate concentration is varied generally enables one to distinguish between competitive and noncompetitive inhibition. (Sizer, 1957). The effects of competitive and noncompetitive inhibitors may also be compared on the basis of the Michaelis-Menten formulation (Fig. 15.20).

Some molecules which are metabolic antagonists to particular substrates exhibit noncompetitive inhibition of enzyme reactions. In such cases also, the inhibitor

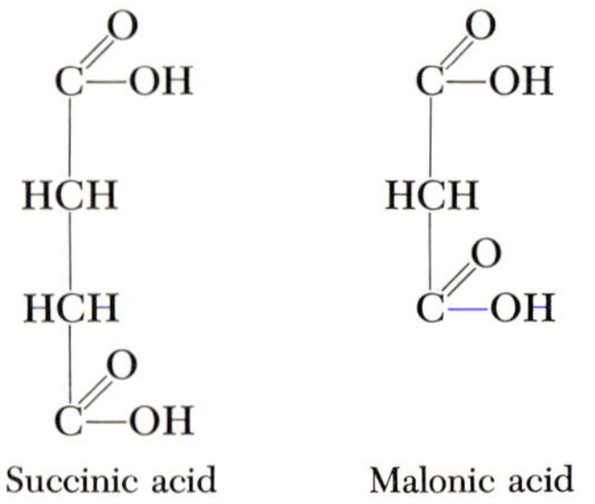


**Figure 15.18.** The structural formulas of malonic and succinic acids.

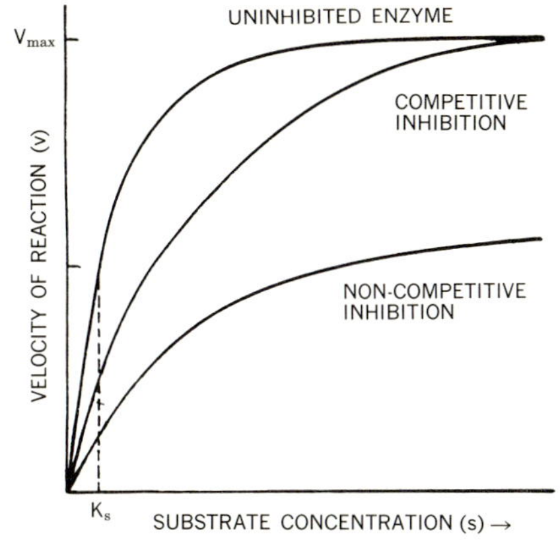


***Figure 15.19.*** The rate of a catalyzed reaction as a function of substrate concentration is plotted for an active enzyme and for an enzyme inhibited by either a competitive or a noncompetitive inhibitor. (From Sizer, 1957: Science *125*:34.)

resembles the metabolizable substance, but it is apparently attached to the enzyme at some point other than the one which binds the original substrate, and by its antagonistic action it prevents the expected enzyme reaction (see Fig. 15.21). Sometimes such inhibition is not relieved under any conditions, and high substrate concentration even at low inhibitor concentration has no effect. In other cases, however, when the inhibitor concentration is low, a high concentration of substrate reverses the inhibition. For instance, inhibition of bacterial metabolism by a small concentration of sulfonamide is relieved by a high concentration of the vitamin B factor, folic acid, the prosthetic group of the enzyme (Sizer, 1957).

Enzymes which depend upon the presence of sulfhydryl (—SH) groups for their activity are inactivated by salts of heavy metals or alkylating agents. Mercury and copper salts inactivate some hydrolases by combining with the sulfhydryls. Iodoacetic acid inhibits some oxidative enzymes by reacting with their sulfhydryl groups (alkylation). Some of the enzymes in the anaerobic phase of sugar decomposition require active sulfhydryl groups and are therefore readily inhibited by iodoacetic acid. This type of inhibition depends upon the concentration of the poison, not on the ratio of poison to substrate, and is a

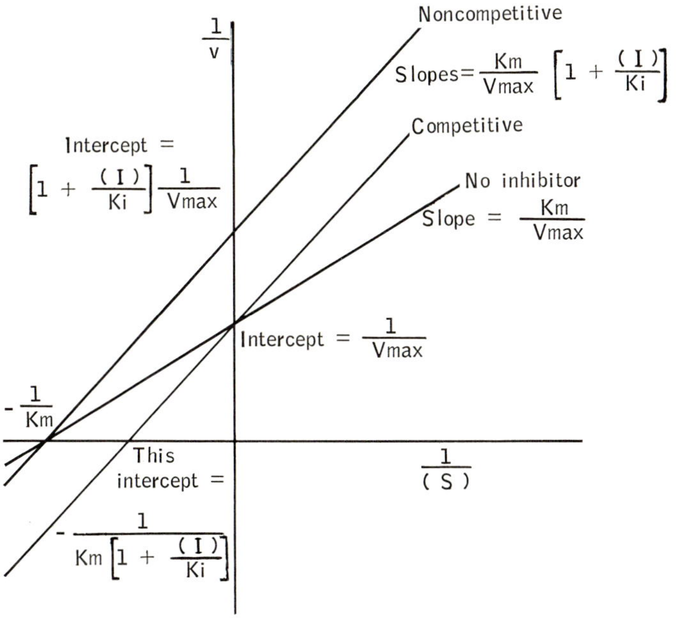


***Figure 15.20.*** Burk-Lineweaver plot of the Michaelis-Menten formulation of enzyme kinetics in the presence and absence of competitive and noncompetitive inhibitors. $K_m$ is the Michaelis constant in the absence of inhibitors, $K_i$ in the presence of inhibitors; I stands for inhibitor. (For details of the formulation see the source of this figure, Patton, 1965: Biochemical Energetics and Kinetics. W. B. Saunders Co., Philadelphia.)

*Enzyme Inhibition and Activation*

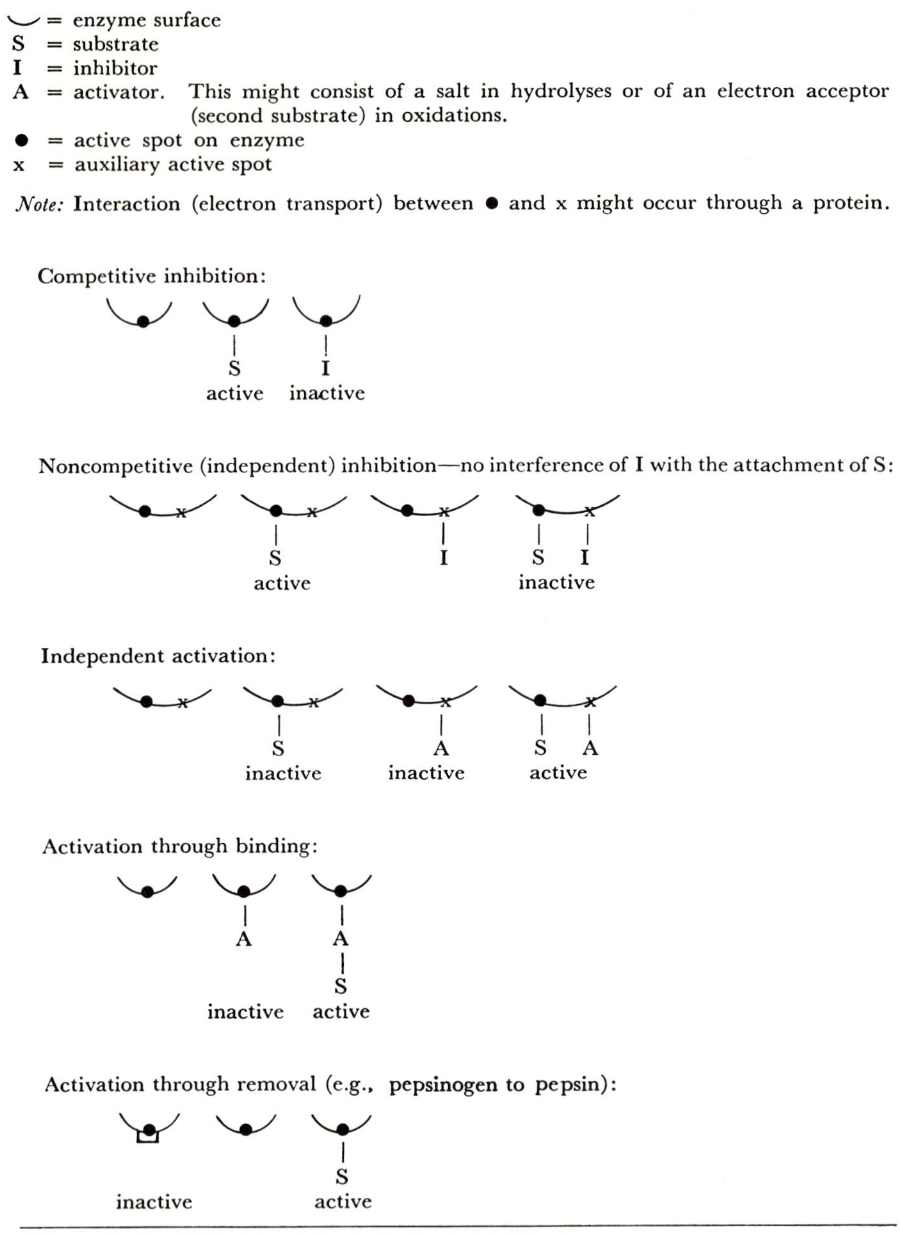


***Figure 15.21.*** Enzyme inhibition and activation. (Based on a number of sources, partly by courtesy of Dr. B. Hofstee, Palo Alto Medical Research Foundation.)

noncompetitive inhibition (Sizer, 1957; Webb, 1966).

Members of a class of enzymes which require for their activity metal ions such as $Mg^{++}$, $Mn^{++}$, $Co^{++}$, $Zn^{++}$ or $Fe^{++}$ are inactivated by substances which combine with these. For example, cyanide forms an almost nondissociable complex with the copper in the prosthetic group of polyphenol oxidase, an enzyme catalyzing the oxidation of polyphenols in plant cells. In a similar manner, cyanide also binds the iron in cytochrome oxidase, an enzyme in the main pathway of oxidation-reduction in the cell (see Chapter 16). In this case inhibition depends upon the concentration of the poison, not on the ratio of poison to substrate (Sizer, 1957).

Many other types of enzyme poisons exist besides the ones considered here. For example, some enzymes not affected by sulfhydryl poisons or metal poisons are affected by diisopropylfluorophosphate (DFP), e.g., acetylcholine esterase. Again, the coupling between oxidative reactions and formation of high energy phosphates is affected by dinitrophenol (DNP). Some of these enzyme poisons will be cited later when discussion of them is pertinent.

## LITERATURE CITED AND GENERAL REFERENCES

Acker, R. F., and Hartsell, S. E., June 1960: Flemming's lysozyme. Sci. Amer. *202*:132–142.

Baldwin, E., 1963: Dynamic Aspects of Biochemistry. 4th Ed. Cambridge University Press, London.

Bayliss, L. E., 1959: Principles of General Physiology. John Wiley & Sons, New York, Vol. 1, pp. 285–325.

Boyer, P. D., Lardy, H., and Myrbäck, K., 1960: The Enzymes. Academic Press, New York, Vol. 4, Chapters 4 and 5.

Bull, H., 1964: An Introduction to Physical Biochemistry. 2nd Ed. F. A. Davis Co., Philadelphia.

Cantarow, A., and Schepartz, B., 1967: Biochemistry. 4th Ed. W. B. Saunders Co., Philadelphia.

Chance, B., Harvey, E. N., Johnson, F. H., and Millikan, G., 1940: The kinetics of bioluminescent flashes, a study in consecutive reactions. J. Cell. Comp. Physiol. *15*:195–215.

Colowick, S. P., and Kaplan, N. D., 1955–1959: Methods in Enzymology. 6 vol. Academic Press, New York.

Dixon, M., 1964: Enzymes. 2nd Ed. Academic Press, New York.

Florkin, M., and Stotz, E. H., 1964: Enzymes—General Considerations. Vol. 12. Comprehensive Biochemistry. American Elsevier Publishing Co., New York.

Frieden, E., Aug. 1959: The enzyme substrate complex. Sci. Am. *201*:119–125.

Graebler, O. H. (ed.), 1956: Enzymes. Units of Biological Structure and Function. Academic Press, New York.

Gutfreund, H., 1965: An Introduction of the Study of Enzymes. John Wiley & Sons, New York.

Hassid, W., and Doudoroff, M., 1950: Synthesis of disaccharides with bacterial enzymes. Advances Enzymol. *10*:123–143.

Hochster, R. M., and Quastel, J. H. (eds.), 1963: Metabolic Inhibitors. Academic Press, New York.

Hungate, R. E., 1966: The Rumen and Its Microbes. Academic Press, New York.

Johnson, F. H., Eyring, H., and Polissar, M., 1954: The Kinetic Basis of Molecular Biology. John Wiley & Sons, New York.

Kaplan, N. D., Ciotti, M. M., Hamolsky, M., and Bieber, R. E., 1960: Molecular heterogeneity and the evolution of enzymes. Science *131*:392–397.

Klotz, I. M., 1967: Energy Exchanges in Biochemical Reactions, Academic Press, New York.

Koshland, D. E., 1963. Correlation of structure and function in enzyme action. Science *142*:1533–1540.

*Lehninger, A., 1965: Bioenergetics. W. A. Benjamin, New York.

McElroy, W. D., and Nason, A., 1954: Mechanism of action of micronutrient elements in enzyme systems. Ann. Rev. Plant Physiol. *5*:1–30.

Meister, A., 1955: Transamination. Advances Enzymol. *16*:185–246.

Niemann, C., 1964: Alpha chymotrypsin and the nature of enzyme catalysis. Science *143*:1287–1296.

Northrop, J. H., 1948: Crystalline Enzymes. Columbia University Press, New York.

Ochoa, S., 1951: Biological mechanisms of carboxylation and decarboxylation. Physiol. Rev. *31*:56–106.

Patton, A. R., 1965: Biochemical Energetics and Kinetics. W. B. Saunders Co., Philadelphia, Chapter 10.

*Phillips, D. C., Nov. 1966: The three dimensional structure of an enzyme molecule. Sci. Am. *215*: 78–90.

Sebrell, W. H., and Harris, R. S., 1954: The Vitamins: Chemistry, Physiology, Pathology. Academic Press, New York.

Sizer, I. W., 1957: Chemical aspects of enzyme inhibition. Science *125*:34–59.

Smith, E. L., 1960: Peptide bond cleavage (survey). *In* The Enzymes. Boyer, Lardy, and Nyoback, eds. Academic Press, New York, Vol. 4, pp. 1–10.

Szent-Györgyi, A., 1939: On Oxidation, Vitamins, Health and Disease. Williams & Wilkins, Baltimore.

Theorell, H., 1956: Nature and mode of action of oxidation enzymes. Science *124*:467–472.

Tiselius, A., 1937: A new apparatus for electrophoretic analysis of colloidal mixtures. Trans. Farad. Soc. *33*:524–531.

*Thompson, R. H. S., 1962: Classification and nomenclature of enzymes. Science *137*:405–408.

Webb, J. L. (ed.), 1966: Enzyme and Metabolic Inhibitors. Academic Press, New York.

Wilkie, D. R., 1960: Thermodynamics and the interpretation of biological heat measurements. Prog. Biophys. *10*:260–298.

# CHAPTER 16 / THE RELEASE OF ENERGY IN CELLS

Cell foods are the source of all the building materials and energy available to the cell. Upon the breakdown of these cell foods the energy, stored therein as chemical energy, is released and used for carrying out all the functions of life. The release of free energy is achieved through oxidation-reduction reactions. The elucidation of the nature of these reactions and of the pathway of oxidation-reduction reactions in the cell is one of the most exciting developments of the biochemistry and physiology of the cell in the last two decades.

The sum total of all the chemical changes occurring in the cell constitutes its *metabolism.* Although the decomposition (catabolism) of some substances in the cell is usually linked with the synthesis (anabolism) of other substances necessary to the life of the cell, in this chapter attention is centered on the decomposition of cell foods with the release of energy.

## THE NATURE OF OXIDATION-REDUCTION REACTIONS

The very word *oxidation* implies a process in which a substance combines with oxygen—for example, the union of hydrogen and oxygen in the formation of water—and the early workers so defined it. Hydrogen gives up electrons to oxygen and is thereby oxidized; oxygen, taking up the electrons, is thereby reduced:

$$2H_2 + O_2 \rightarrow 2H_2O \qquad (16.1)$$

In many cases, however, electrons are given up even when oxygen is not present; for example, in the formation of ferric from ferrous ion.

$$Fe^{++} \underset{\text{reduction}}{\overset{\text{oxidation}}{\rightleftharpoons}} Fe^{+++} + e \qquad (16.2)$$

$Fe^{++}$ is the electron donor which is oxidized to $Fe^{+++}$. In the reverse reaction $Fe^{+++}$ is the electron acceptor which is reduced to $Fe^{++}$. *Oxidation,* in its most generalized sense, is at present defined as the donation of electrons from one molecule (ion, atom, radical) to another, and *reduction* is defined as the acceptance of electrons by one molecule (ion, atom, radical) from another. The two processes always occur simultaneously.

In many of the oxidation-reduction reactions which occur in the cell, not only electron transfer but also hydrogen transfer occurs. For example, when succinic acid is oxidized to fumaric acid, two hydrogen ions and two electrons are removed. In this case, oxidation consists of the removal of the hydrogen ions and electrons from the succinic acid to the enzyme catalyzing the reaction:

$$\underset{\text{Succinic acid}}{\begin{array}{c} COOH \\ | \\ CH_2 \\ | \\ CH_2 \\ | \\ COOH \end{array}} \xrightarrow[\text{Succinic dehydrogenase}]{-(2H^+ + 2e)} \underset{\text{Fumaric acid}}{\begin{array}{c} COOH \\ | \\ CH \\ \| \\ CH \\ | \\ COOH \end{array}} \qquad (16.3)$$

The donor of the hydrogen ions and electrons is a reductant called the *hydrogen donor.* On giving up hydrogen ions and electrons it is oxidized. The enzyme catalyzing the reaction and accepting the hydrogen ions and electrons is an oxidant called the *hydrogen acceptor.* It is reduced in the process. A generalized equation for such an oxidation-reduction reaction may be written as follows:

$$\underset{\text{H-donor}}{H_2B} + \underset{\text{H-acceptor}}{A} \xrightarrow{\text{oxidation}} \underset{\substack{\text{Oxidized} \\ \text{H-donor}}}{B} + \underset{\substack{\text{Reduced} \\ \text{H-acceptor}}}{H_2A} \qquad (16.4)$$

When oxygen is present in a cell, it generally acts as the terminal hydrogen and electron acceptor, becoming reduced in the process. Oxygen readily accepts electrons from molecules (or from ions, radicals and so forth) and it is because of this property that the oxidations in which oxygen participates were studied first. In the main line of oxidation in the cell, oxygen serves as the final hydrogen acceptor, and water ($H_2O$) is the end product. When flavoproteins pass hydrogen ions and electrons to oxygen in other than the main line of oxidation, $H_2O_2$ is usually formed (Baldwin, 1963).

## STEPWISE RELEASE OF ENERGY DURING OXIDATION-REDUCTION REACTIONS

Whereas many hydrolyses (see Chapter 15) generally proceed with small change in free energy, oxidation-reduction reactions in the cell usually are accompanied by a considerable change in free energy. The work of the cell is directly or indirectly accomplished by this free energy. Chemical energy is usually stored in compounds containing so-called high energy phosphate bonds (e.g., adenosine triphosphate), from which the energy can be liberated and used for synthesis of other necessary compounds in the cell.

The energy available in a nutrient molecule is never liberated by the cell in one step. The complete oxidation of glucose into carbon dioxide and water occurs with a $\Delta F$ of $-686{,}000$ cal./mole. The cell has no way of using this much energy liberated at once; it could efficiently convert only a very small fraction of it. Rather, the cell degrades the glucose stepwise, with a small free energy change in each step, each step being generally catalyzed by a specific enzyme. The merit of a series of stepwise reactions is that they can be coupled energetically by a reactant in common, and in certain steps the energy liberated is transferred through such coupled reactions to high energy phosphate bonds, in which it is stored. A sequence of reactions involving such small free energy changes is therefore advantageous to the organism. The overall efficiency, as indicated by the formation of 38 high energy phosphate bonds during the decomposition of 1 mole of glucose (see Tables 16.1 and 16.3), is about 40 per cent (Lehninger, 1965a).

## PATHWAY OF OXIDATION-REDUCTION REACTIONS IN THE CELL

Table 16.1 shows a general scheme for the oxidation-reduction of glucose in the cell, during which glucose is gradually decomposed, with the evolution of energy, into carbon dioxide and water as the end products. Glucose is phosphorylated, and after atomic rearrangements in the molecule, it is split into two triose phosphates. Each triose phosphate, in the presence of dehydrogenases and other enzymes, is oxidized to pyruvic acid. This phase of

**Table 16.1.** General Scheme of the Decomposition of Glucose

| PROCESS | $O_2$ NEED | REACTIONS | CLASSES OF ENZYMES INVOLVED | HIGH ENERGY PHOSPHATES* |
|---|---|---|---|---|
| Glycolysis | − | Glucose (hexose) +<br>↓ ATP<br>Hexose-phosphate<br>↓<br>2 triose phosphates<br>↓<br>2 pyruvic acid +<br>energy as ~P | Phosphate-transferring enzymes<br>Phosphate-splitting enzymes<br>Isomerases<br>Dehydrogenases<br>Phosphate-transferring enzymes | ~P + ADP → ATP<br>ATP + C → ADP + CP |
| Krebs Tricarboxylic Acid Cycle | + | ↓<br>$CO_2$ + $H_2O$ +<br>energy as ~P | Dehydrogenases<br>Isomerases<br>Flavoproteins<br>Cytochromes<br>Cytochrome oxidase<br>Decarboxylase<br>Phosphate-transferring enzymes | ~P + ADP → ATP<br>ATP + C → ADP + CP |

* The phosphates are identified as follows:
ATP = adenosine triphosphate (adenine–pentose–phosphate ~ phosphate ~ phosphate)
ADP = adenosine diphosphate (adenine–pentose–phosphate ~ phosphate)
CP = creatine phosphate (creatine ~ phosphate)
C = creatine
In some cells arginine takes the place of creatine.

glucose decomposition can also occur in the absence of oxygen, and is discussed in more detail later as the anaerobic decomposition of glucose (glycolysis).

In the second phase of glucose decomposition, which occurs only in the presence of oxygen, the pyruvic acid formed as the end product of the first series of reactions described earlier is first decarboxylated (carbon dioxide being given off) and the 2-carbon fragment, called acetyl, is combined with a coenzyme called *coenzyme A,* or CoA (see Fig. 16.1). The acetyl-CoA combines with a 4-carbon acid, to form a 6-carbon acid. The 6-carbon acid is subsequently decarboxylated and oxidized again to form a 5-carbon acid, carbon dioxide and water being given off in the process. A series of decarboxylations follows, each one resulting in the liberation of carbon dioxide and water. As a result of this entire series of reactions glucose is completely decomposed to carbon dioxide and water. These reactions of the aerobic decomposition of glucose are discussed in more detail later.

Part of the free energy change in the oxidation of glucose in the cell is stored in *phosphorylated organic compounds.* The high energy phosphate bond (usually written ~P) in such compounds is easily transferred from one organic compound to another (e.g., from creatine phosphate to adenosine diphosphate) in the presence of the proper enzyme. The phosphate transfer potential of such compounds is thus high. Therefore, it serves as a convenient means of storing and distributing the energy in cells. The energy of such a phosphate bond is about 7000 cal./mole at standard conditions—at 25 C. and pH 7.0, when each reactant and

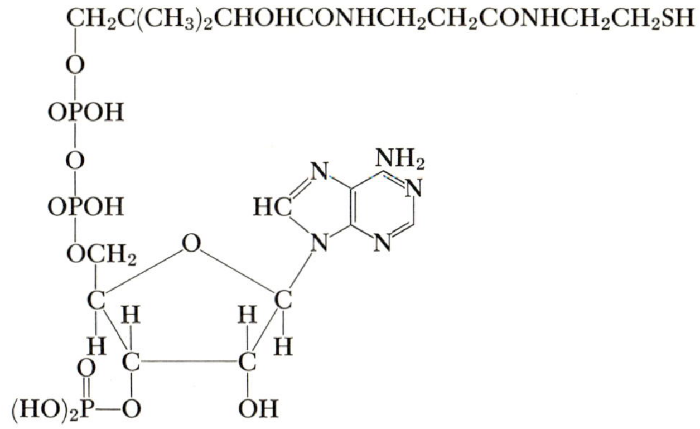


***Figure 16.1.*** Coenzyme A.

product is at thermodynamic activity, i.e., at 1.0 molal—the exact amount depending upon the conditions of the reaction, such as pH and salt environment. When all of the energy of a phosphate bond is utilized in a reaction (e.g., synthesis of a compound) inorganic phosphate (low energy content) appears as the end product (Lehninger, 1965a).

Some high energy phosphate compounds of importance are listed in Table 16.1. *Adenosine triphosphate* (ATP), a nucleotide, is composed of a pentose sugar, three phosphoric acids, and the nitrogenous base, adenine (see Fig. 16.2). The bond between the pentose and the first phosphoric acid residue is an ordinary low energy bond, but the bonds between the other two phosphoric acid residues are high energy phosphate bonds, that is, they have a high phosphate transfer potential.

In *adenosine diphosphate* (ADP), which has two phosphoric acid residues, only one high energy phosphate bond is present. Free energy release from ATP usually* results in the loss of one high energy bond and formation of one ADP. The two phosphates are therefore always present together. In muscle cells a relatively small amount of adenosine triphosphate is present, but a larger supply of another phosphate, known as creatine phosphate (CP), exists as the high energy phosphate reservoir (Fig. 16.2). Muscle cells may contain up to 0.5 per cent, wet weight, of creatine phosphate. Creatine phosphate in the presence of the appropriate enzyme (a kinase) may transfer its high energy bond to ADP, forming creatine and ATP, and conversely, creatine may accept one ~P from ATP, forming CP and ADP; ATP and ADP appear to be distributors of high energy phosphate bonds. In cells of many invertebrates *arginine phosphate* takes the place of creatine phosphate. Creatine phosphate and arginine phosphate are called *phosphagens* (Baldwin, 1963).

Lehninger (1965a) calls attention to the fact that "high energy phosphate bond" is a misnomer. The fact is that there is a large difference in free energy between reactants and the products, but the energy does not really reside in the phosphate bond. However, the terminology has become so prevalent that it is used here to indicate compounds of high phosphate transfer potential.

Organic phosphates that can transfer phosphate to other compounds with a large decrease in free energy are said to have a high phosphate transfer potential; those that do so with a lesser decrease in free energy are said to have a lower phosphate transfer potential. Compounds of varying phosphate transfer potential are shown in Figure 16.3.

Although high energy phosphate bonds are usually outlined as a category distinct from low energy bonds, this is somewhat deceptive since a whole series of bonds of different energy content occurs, as shown in Figure 16.3. The transfer po-

*Splitting of pyrophosphate from ATP with formation of AMP occurs in a number of synthetic reactions discussed in Chapter 20.

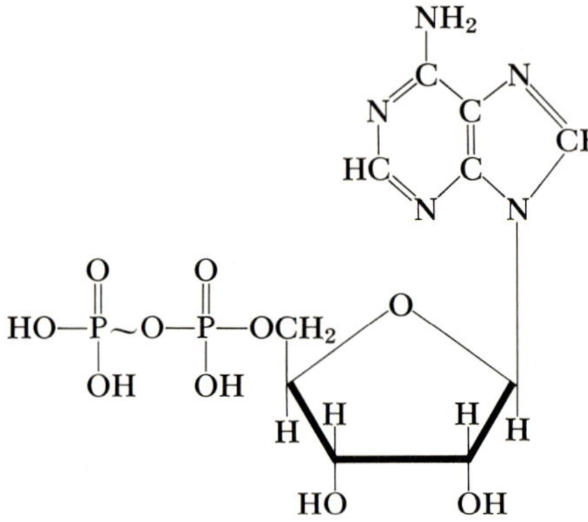

adenosine diphosphate (ADP)

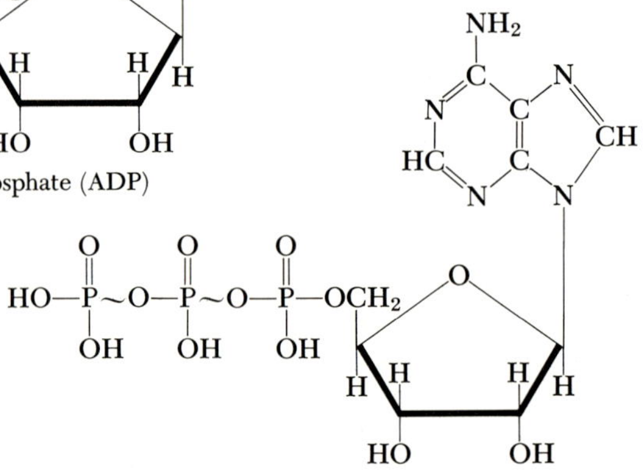

adenosine triphosphate (ATP)



Creatine phosphate (CP)

***Figure 16.2.*** Compounds of high phosphate transfer potential. In ADP and ATP the nitrogen-containing moiety is adenine; it is attached to the pentose which in turn is attached to the phosphate.

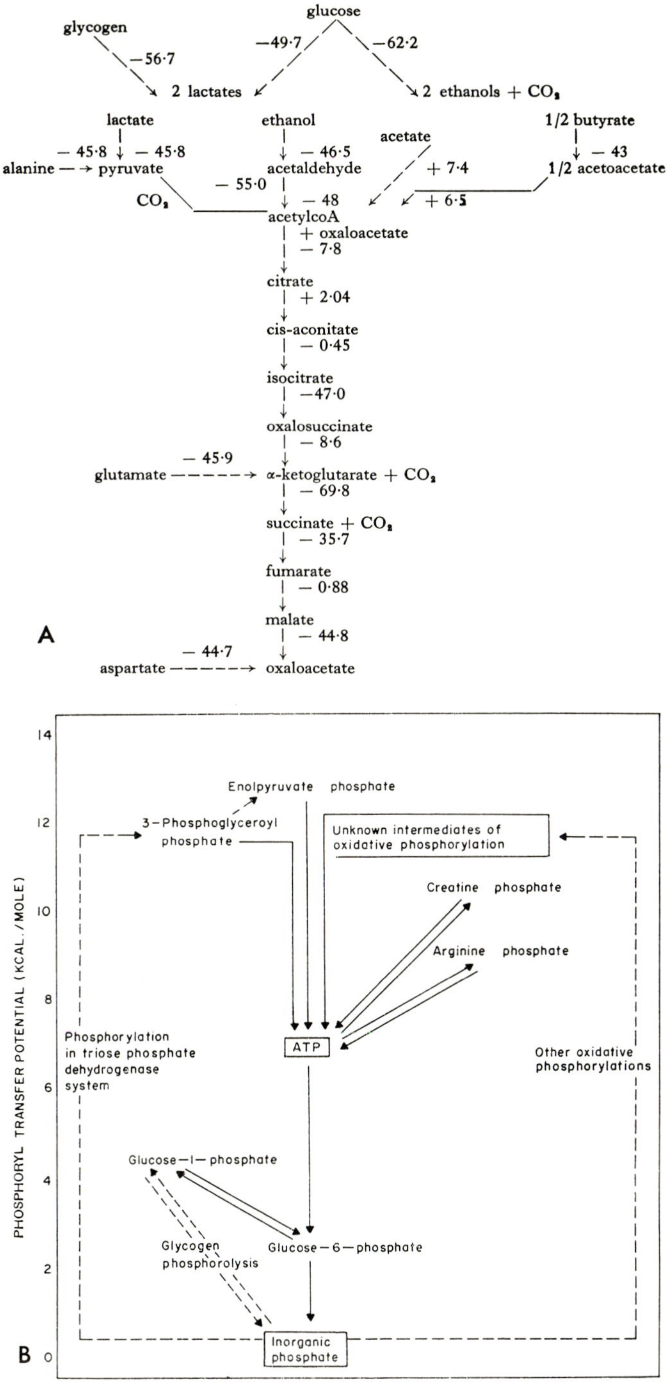


***Figure 16.3.*** *A*, changes in free energy ($\Delta F_0$) in kcal. per mole at 25°, pH 7.2, 0.2 atm. $O_2$, 0.05 atm. $CO_2$ and the other reactants having a concentration of 0.01M. The free energy changes include changes due to associated reactions notably reactions with molecular oxygen acting as an acceptor for hydrogen removed at various stages. (From Florkin, 1960: Unity and Diversity in Biochemistry. Pergamon Press, London.) *B*, Some free energy changes in the biosynthesis of phosphates. (From Atkinson and Morton, 1960: Comparative Biochemistry *2*:74.)

tential of ATP at pH 7.0 and unit molality is about 7000 cal./mole. The efficiency of the cell in producing high energy phosphates from nutrients therefore should be calculated on this basis (Lehninger, 1965b). Actually, in the cell the concentration of ATP and ADP is much less than 1 molal; furthermore, $Mg^{++}$ complexes with ATP, ADP and phosphate, shifting the equilibrium toward ATP hydrolysis. The hydrolysis of ATP has such a large $\Delta F$ because the equilibrium constant for the reaction is so large (it will be remembered that $\Delta F = -RT \ln K_{eq}$; see Chapter 1, Table 1.3).

Dinitrophenol is an inhibitor which prevents the coupling between the carrier molecule ADP and the enzymes transporting the high energy phosphate bonds. Consequently, energy liberated in oxidation is lost to the cell. For this reason, dinitrophenol was tested in a reducing diet for the obese and it was found to stimulate caloric output without causing an increase in synthetic activity. However, because it causes cataract of the eye, its medical use has been banned.

## ANAEROBIC DECOMPOSITION OF GLUCOSE

As was stated previously, in the cell, the initial steps in glucose breakdown up to the formation of pyruvic acid do not require oxygen and are the same regardless of whether or not oxygen is present. Anaerobic oxidation of glucose (called *glycolysis* in animals and *fermentation* in microorganisms and plants) stops at this point.

The various steps in the anaerobic decomposition of glucose are given in Figure 16.4 and are called the Embden-Meyerhof pathway after two pioneer investigators. Glucose must first be phosphorylated at the expense of a high energy phosphate from ATP (Phosphorylation I). Glycogenolysis (which is phosphorolysis catalyzed by phosphorylase), on the other hand, liberates a phosphorylated glucose which already has one high energy phosphate bond (glucose-1-phosphate). In either case a second ATP is needed (Phosphorylation II) before the hexose phosphate shown in the outline is split into two triose phosphates (3-phosphoglyceraldehyde and dihydroxyacetone phosphate). It should be noted that in the breakdown of the triose phosphates, hydrogen is removed by dehydrogenases, and after a series of reactions pyruvic acid appears. Some of the liberated energy in the anaerobic process is stored in compounds of high phosphate transfer potential (Phosphorylations III and IV, right side of Fig. 16.4) converting ADP to ATP. Since some ATP is used in phosphorylating glucose, the net gain of ATP is the total stored minus that used. In making a calculation of such an ATP gain, it should be remembered that each glucose gives rise to two trioses and that for each triose decomposed to pyruvic acid, two ADP molecules are converted to two ATP molecules. In terms of energy this means that the anaerobic oxidation of glucose nets about 14,000 cal./mole starting with glucose, or about 21,000 cal./mole starting with glycogen. A large number of the enzymes given in Table 16.2 are involved in the reactions shown, one enzyme usually catalyzing each step.

Under anaerobic conditions, pyruvic acid acts as a hydrogen acceptor, forming lactic acid, as shown in Figure 16.5. For example, muscle cells in oxygen debt may accumulate a considerable amount of lactic acid as a consequence. The lactic acid is in part oxidized to carbon dioxide and water, and part of it is reduced to triose, which forms hexose phosphate in cells of the liver; the latter is stored in the cells as glycogen. Liver cells store as much as 20 per cent glycogen per unit dry weight, whereas muscle cells store only about 2 per cent. When sugar is needed, the glycogen undergoes phosphorolysis and the sugar is released into the blood (Cantarow and Schepartz, 1967).

In yeast still another path is followed. The pyruvic acid is decarboxylated and the acetaldehyde thus formed is reduced to alcohol, as shown in Figure 15.5 in which the enzymes required for the reactions are also indicated. Alcohol dehydrogenase is sensitive to iodoacetate, addition of which interferes with fermentation.

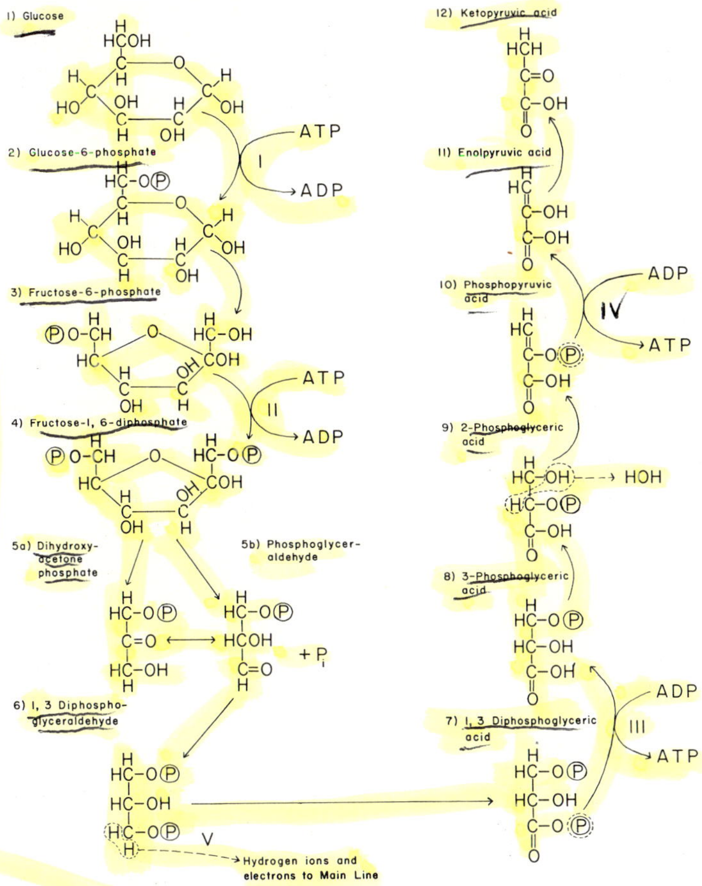


***Figure 16.4.*** Glycolysis or anaerobic decomposition of glucose. The terminal reactions are given in Figure 16.5. The enzymes concerned in the reactions given here are shown in Table 16.2. Ⓟ refers to organically found phosphate, $P_i$ to inorganic phosphate.

**Table 16.2.** Enzymes of Glycolysis

| REACTION* | ENZYME | COENZYME | INHIBITED BY |
|---|---|---|---|
| 1 - 2 | Glucokinase | ATP, $Mg^{++}$ | dialysis |
| 2 - 3 | Phosphoglucoisomerase | — | — |
| 3 - 4 | 6-Phosphofructo-1-kinase | ATP, $Mg^{++}$ | dialysis |
| 4 - 5 | Fructoaldolase | — | — |
| 5a - 5b | Trioseisomerase | — | — |
| 5b - 6 | Spontaneous? | — | — |
| 6 - 7 | Glyceraldehyde-3-phosphate dehydrogenase | $NAD^+$, $HOPO_3^=$ | iodoacetate, dialysis |
| 7 - 8 | 3-Phosphoglycerate-1-kinase | ADP, $Mg^{++}$ | dialysis |
| 8 - 9 | 2-3 Phosphoglyceric mutase | $Mg^{++}$ | — |
| 9 - 10 | Enolase | $Mg^{++}$ | NaF |
| 10 - 11 | Pyruvic kinase | ADP, $Mg^{++}$ | dialysis |
| 11 - 12 | Spontaneous? | — | — |

After Sallach and McGilvery, 1960: Gibson Medical Electronic Chart on Intermediary Metabolism, and Baldwin, 1963.

* Refer to Figure 16.4.

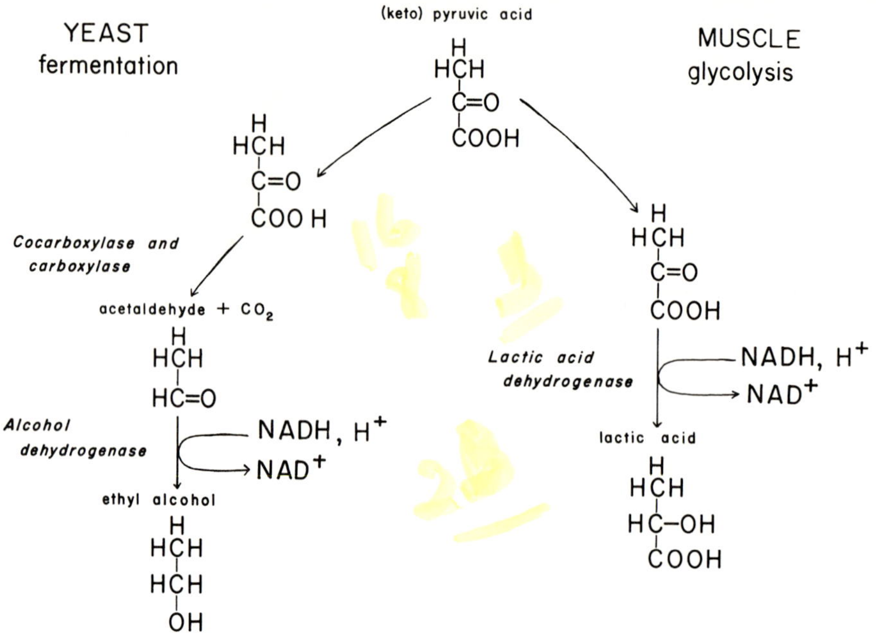


*Figure 16.5.* Terminal reactions in yeast fermentation and muscle glycolysis. These reactions follow directly after those in Figure 16.4.

Bisulfite ($NaHSO_3$), when added to trap the acetaldehyde, permits the development of glycerol by reduction of the dihydroxyacetone (Kluyver and van Niel, 1956).

Neither the reduction of pyruvic acid to lactic acid nor the decarboxylation of pyruvic acid to acetaldehyde liberates energy; this can be seen from the heat of combustion of each of these compounds given in Table 1.1. The heat of combustion of lactic acid is greater than that of pyruvic acid, as one might expect of a more highly reduced compound. The heat of combustion of acetaldehyde is the same as that of pyruvic acid.

## AEROBIC DECOMPOSITION OF GLUCOSE

Only a small fraction of the energy in a glucose molecule has been liberated by the time pyruvic acid is formed (Table 16.3). For that reason glycolysis is sometimes considered quite inefficient. However, it must be remembered that only a small part of the chemical potential in glucose is released during glycolysis ending with the formation of lactate and is still available for release in aerobic oxidation of lactate to carbon dioxide and water. If the efficiency of glycolysis is calculated on the basis of the chemical potential of ATP produced at the expense of free energy released in production of lactate from glucose, then the efficiency is 14,000 divided by 52,000 cal./mole (for two moles of ATP produced with glycolysis of one mole of glucose to lactate), or 27 per cent. The rest of the energy is liberated during the oxidation of pyruvic acid, the latter being perhaps the most important source of energy in all aerobic cells (see Fig. 16.3*A*).

The energy of pyruvic acid is liberated after a series of stepwise reactions. Pyruvate combines with coenzyme A (CoA) and is decarboxylated; that is, carbon dioxide is split out of the carboxyl group of the acid. The enzyme catalyzing decarboxylation contains thiamine phosphate (TPP) in its prosthetic group. The reactions resulting in the decarboxylation

**Table 16.3.** Yield ATP from ADP and Phosphate in Aerobic Degradation of a Mole of Glucose

| REACTION | TEXT FIGURE NO. | REACTION ON TEXT FIGURE | NET GAIN |
|---|---|---|---|
| Initial phosphorylations | 16.4 | I,II | −2 |
| Substrate phosphorylation (x2) | 16.4 | III,IV | +4 |
| Hydrogen ions and electrons through main line: | | | |
| from oxidation of 1,3-diphosphoglyceraldehyde (x2) | 16.4 | V | +6 |
| from pyruvic to citric acid (x2) | 16.7 | VI | +6 |
| from isocitric to oxalosuccinic acid (x2) | 16.7 | VII | +6 |
| from ketoglutaric to succinic acid (x2) | 16.7 | VIII | +8 |
| from succinic to fumaric acid (x2) | 16.7 | IX | +4 |
| from malic to oxaloacetic acid (x2) | 16.7 | X | +6 |
| | | Total | +38 |

of pyruvic acid and the formation of acetyl CoA are shown in Figure 16.6. The hydrogen is passed to coenzyme I—nicotinamide adenine dinucleotide (NAD). The 2-carbon fragment (acetyl) is now added to a 4-carbon acid (oxaloacetic) to form a 6-carbon acid (citric) as shown in Figure 16.7 (top). Citric acid first undergoes

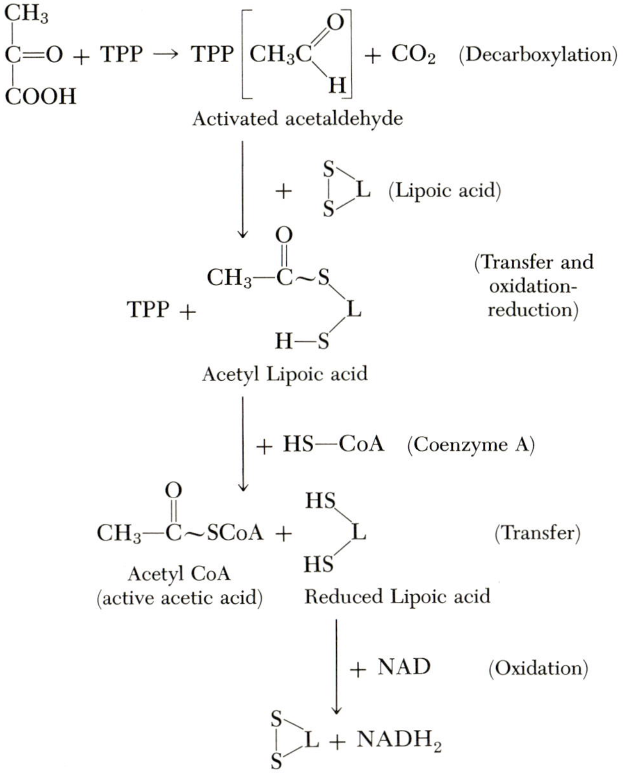


Net reaction:

$$CH_3COCOOH + CoA + NAD \rightarrow CH_3\overset{O}{\overset{\|}{C}}{\sim}S{-}CoA + NADH_2 + CO_2$$

***Figure 16.6.*** The mechanism of oxidative decarboxylation of pyruvic acid. TPP refers to thiamine pyrophosphate. (From McElroy, 1964: Cellular Physiology and Biochemistry. 2nd Ed. Prentice-Hall, Englewood Cliffs, N. J.)

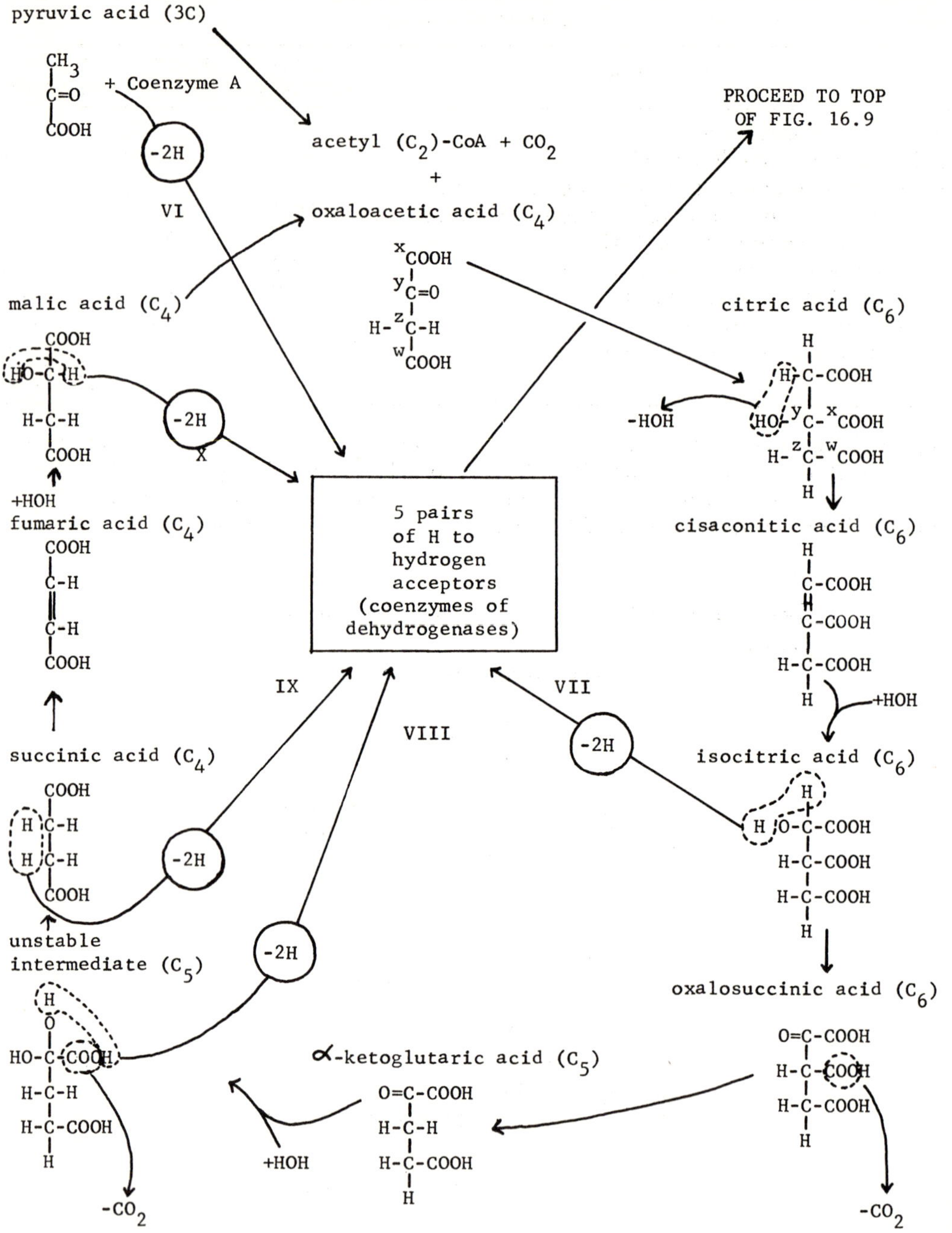


*Figure 16.7.* Aerobic breakdown of pyruvic acid. (For anaerobic reactions see Fig. 16.6.) The enzymes involved in the various steps are given in Table 16.4. Decarboxylation of an acid, with release of $CO_2$ (emphasized in this diagram), can be *reversed* under appropriate conditions in all types of cells, animal as well as green plant or microorganism, thus permitting incorporation of ("fixing") free $CO_2$, although in a different way than in photosynthesis (see Chapter 19).

This diagram shows the origin of the hydrogen ($H^+$ and electrons) from various substrates. The diagram is continued in Figure 16.9 which shows the hydrogen passing through the "main line" along with the electrons, the point at which most of the free energy from the oxidation is obtained.

internal rearrangements and then is dehydrogenated. After this it is decarboxylated, giving rise to a 5-carbon acid (α-ketoglutaric). The 5-carbon acid in turn is decarboxylated in the presence of CoA (forming succinyl CoA), giving rise to a 4-carbon acid (succinic). The 4-carbon acid undergoes a series of dehydrogenations ending with the formation of oxaloacetic acid which again becomes a receptor for an acetyl fragment. A considerable amount of energy, amounting to 40,000 or 50,000 cal./mole, is liberated in each oxidative decarboxylation (see Fig. 16.3). Part of this energy is stored in high energy phosphate bonds (see Table 16.3).

These events, first associated by Krebs (1948–1949), form a cycle which has been called the *Krebs tricarboxylic acid*

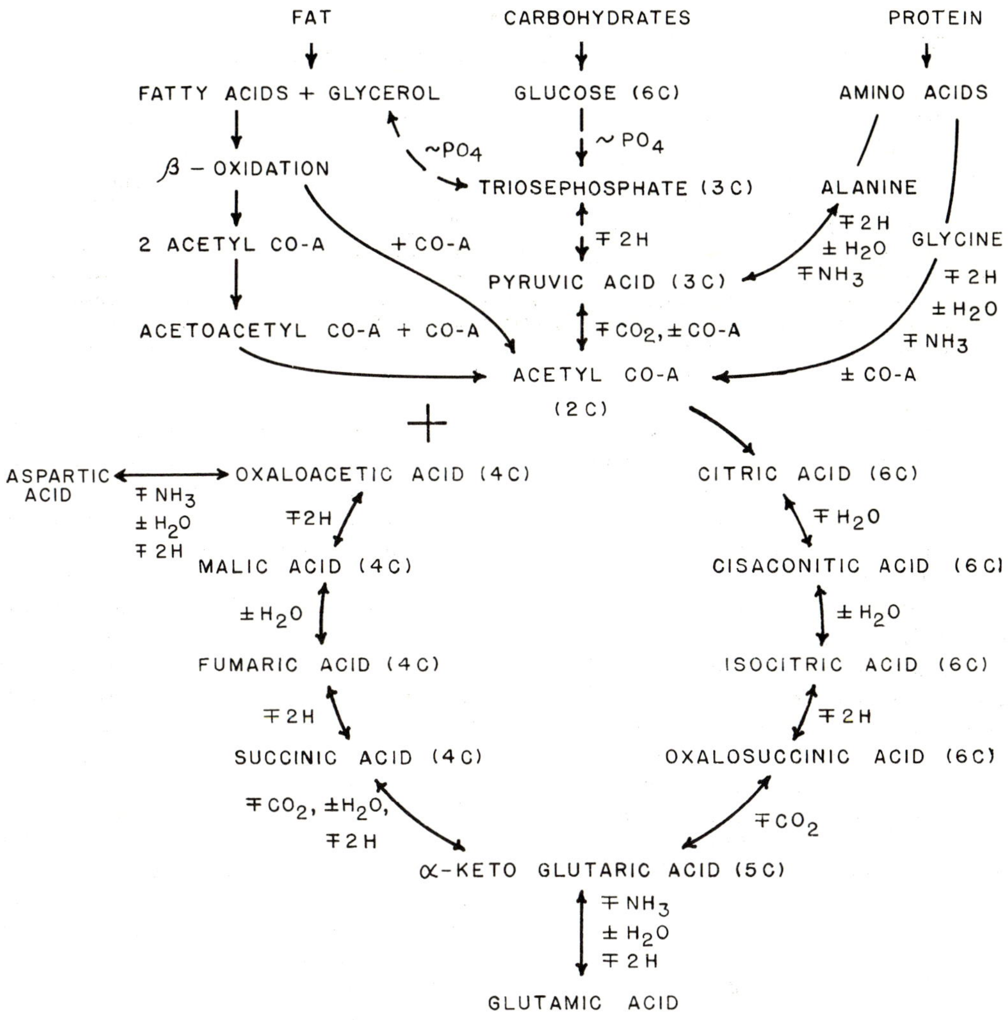


***Figure 16.8.*** The metabolic mill. Note: 1. Citric, cisaconitic, isocitric and oxalosuccinic acids are tricarboxylic.

2. Serine, threonine, cysteine and valine go by way of the pathway shown for alanine. Arginine, proline, hydroxyproline, histidine and ornithine go by way of the pathway shown for glutamic acid. Aspartic acid on deamination gives rise to oxaloacetic acid.

3. For high energy phosphate yield see Table 16.3.

4. The upper part of each plus or minus sign indicates the reaction going clockwise; the lower part indicates the converse.

For detailed structures of the acids see Figures 16.4, 16.5, 16.6 and 16.7, and for a list of the enzymes involved see Tables 16.2 and 16.4. (Modified after Krebs, 1948–1949: The Tricarboxylic Acid Cycle. Harvey Lectures, Charles C Thomas, Springfield, Ill.)

**Table 16.4.** Enzymes Involved in the Krebs Tricarboxylic Acid Cycle

| REACTION | ENZYME* | INHIBITOR |
|---|---|---|
| 1. Pyruvic acid and acetyl CoA | Oxidative decarboxylation system | — |
| 2. Acetyl CoA and oxaloacetic acid | Citrate synthetase | — |
| 3. Citric to cisaconitic acid | Aconitase | Fluorocitrate |
| 4. Cisaconitic to isocitric acid | Aconitase | — |
| 5. Isocitric to oxalosuccinic acid | Isocitric dehydrogenase; $NADP^+$ | Anaerobiosis |
| 6. Oxalosuccinic to $\alpha$-keto-glutaric acid | Oxalosuccinic decarboxylase; $Mn^{++}$ | — |
| 7. $\alpha$-Ketoglutaric to succinic acid | Oxidative decarboxylation system | Arsenite |
| 8. Succinic to fumaric acid | Succinic dehydrogenase | Malonate |
| 9. Fumaric to malic acid | Fumarase | — |
| 10. Malic to oxaloacetic acid | Malic dehydrogenase; $NAD^+$ | — |

* The coenzyme for dehydrogenases of most cells is $NAD^+$; the coenzyme for isocitric dehydrogenase is $NADP^+$ although yeast can use both $NAD^+$ and $NADP^+$. The only dehydrogenation in the Krebs cycle which does not require a coenzyme is that from succinic to fumaric acid in which the hydrogen acceptor is a cytochrome. Largely after Baldwin, 1967.

*cycle* because several of the acids appearing in it are tricarboxylic. The importance of coenzyme A in the transformation was shown by Lipmann, who elucidated the chemical nature of coenzyme A. The prosthetic group of coenzyme A contains pantothenic acid, one of the vitamins B. For their elucidation of these reactions Krebs and Lipmann shared the Nobel prize in 1953. These reactions of the Krebs cycle are presented in a simplified way in Figures 16.7 and 16.8. The enzymes concerned in the reactions are given in Table 16.4. The yield of high energy phosphates from the reactions of the Krebs cycle is given in Table 16.3 (Cheldelin *et al.*, 1962).

## THE METABOLIC MILL

The Krebs tricarboxylic acid cycle has much wider significance in the animal or plant cell than is shown by the degradation of pyruvic acid. It is the means whereby fragments of organic compounds available from various metabolic reactions involving the degradation of proteins, fats and carbohydrates can be utilized effectively. For example, when plant proteins in the food of an animal are digested into their constituent amino acids, the various amino acids appear in proportions different from the proportions needed for the synthesis of animal protein. Most of the amino acids that are present in excess or are not usable are then deaminated, and the residues enter the Krebs cycle where, upon being oxidized, they will form compounds of high phosphate transfer potential. To illustrate, glutamic acid on deamination forms $\alpha$-ketoglutaric acid; aspartic acid forms oxaloacetic acid; and alanine forms pyruvic acid. Fatty acid fragments also enter the Krebs cycle. Fatty acids are oxidized by what is called $\beta$-oxidation, in the course of which a 2-carbon acid fragment (acetyl) is split from the long chain of the fatty acid. The acetyl fragment, as described earlier, enters the Krebs cycle. These various relations are shown in Figure 16.8 (Lipmann, 1955; Krebs, 1957).

Such cyclic conversion into the compounds of high phosphate transfer potential such as creatine phosphate or into the formation of glycogen is an effective and efficient method of storing chemical potential energy. It would be infinitely more difficult for the organism to store and use the odd assortment of compounds that enter the Krebs cycle than it is to store glycogen, ATP and CP.

This diagram of the metabolic mill also serves to show synthesis in the cell of more complex compounds from fragments, for example, of amino acids from ammonia and the appropriate keto acids (e.g., glutamic acid from ammonia and $\alpha$-ketoglutaric acid). Protein synthesis from phosphorylated amino acids is dealt with elsewhere (Chapter 20). Carbohydrate in excess over that which can be stored in the glycogen depots is converted into fat by way of synthesis of fragments from the Krebs cycle. The synthesis of fatty acids is more indirect than reversal of fatty acid degradation (Fig. 16.8). It involves the incorporation of carbon dioxide into acetyl CoA at the expense of energy from ATP to form a 3-carbon compound. The latter compound is combined with another acetyl CoA to form a 5-carbon compound from which, after reduction and decarboxylation, a 4-carbon fatty acid is produced. Each addition of two carbons to the fatty acid involves a similar round of reactions, details of which may be obtained from a textbook of biochemistry. In animals fat is stored in fat bodies or adipose tissue. In plants fat is often found in fruit (avocado) or in seeds (nuts). It is this overall set of interrelations in the degradation and synthesis of proteins, fat and carbohydrates that is called the *metabolic mill.*

## HYDROGEN AND ELECTRON TRANSFER IN THE MAIN LINE OF OXIDATION-REDUCTION REACTIONS

In the major pathway, often called the *main line of oxidation-reduction* reactions

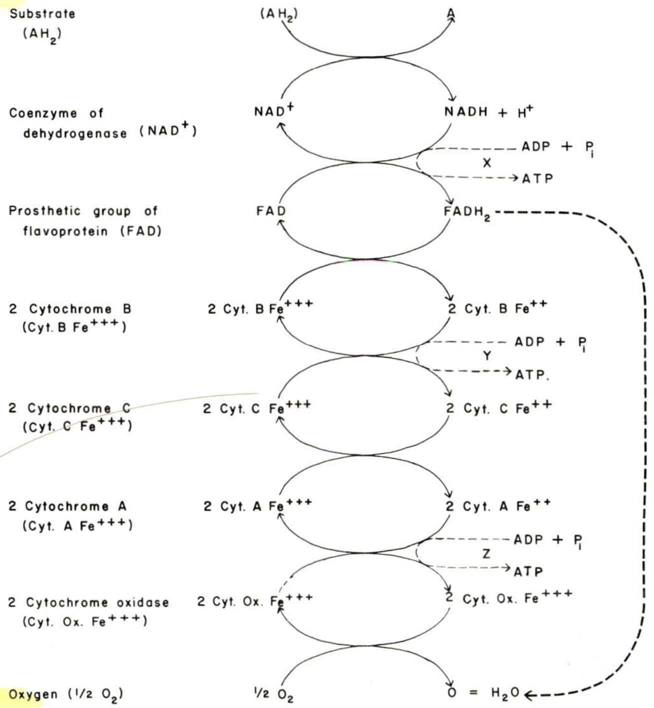


***Figure 16.9.*** The main line of electron transfers and the probable reactions in ATP synthesis from ADP and inorganic phosphorus (Pi). X, Y and Z denote coupling enzymes, the exact nature of which is still unknown. (After Lehninger, 1960: Fed. Proc. *19*:952.)

in the cell, hydrogen is removed from a substrate and transferred through a series of enzymes until it is oxidized to water, with an accompanying release of energy. As stated previously, most dehydrogenases (enzymes which catalyze hydrogen transfer) are formed by association of one of the coenzymes—coenzyme I (NAD) or coenzyme II (NADP)—with a protein molecule specific to a particular substrate. Hydrogen is usually picked up from the substrate by the coenzyme part of the dehydrogenase and carried to the next enzymes in the chain of events, the flavoproteins, which act as hydrogen carriers (e.g., flavin adenine dinucleotide, FAD). From the flavoprotein each hydrogen is discharged into the cell fluid as a hydrogen ion, and the electrons are passed on to cell pigments known as cytochromes. From the cytochromes electrons are passed to the enzyme *cytochrome oxidase,* which ultimately discharges the electrons to oxygen. The oxygen, thus excited, unites with the hydrogen ions, forming water. The whole series of reactions is shown in Figure 16.9 with the transfer of hydrogens and of electrons indicated in each of the steps (Lehninger, 1965a).

Dehydrogenases in cells can be most easily detected by observing the color change of methylene blue when the dye is mixed with the tissue extract. In the presence of active hydrogen liberated from a substrate by a dehydrogenase, the methylene blue, acting as a carrier like flavoprotein, is reduced to colorless leucomethylene blue. If the solution containing leucomethylene blue is shaken up with air, the colorless dye is again oxidized, and the blue color reappears—the hydrogen having combined with the oxygen of the air to form water.

The relative dehydrogenase content or dehydrogenase activity of cells may be estimated by the rate of decolorization of methylene blue mixed with tissue mash or *brei* in a Thunberg tube (see Fig. 16.10) from which oxygen has been withdrawn. The presence and activity of the dehydrogenases specific to each of a series of substrates may be studied if various substrates are placed in the tube one at a time (Thunberg, 1930).

The *cytochromes* (Fig. 16.11*A*) are cell

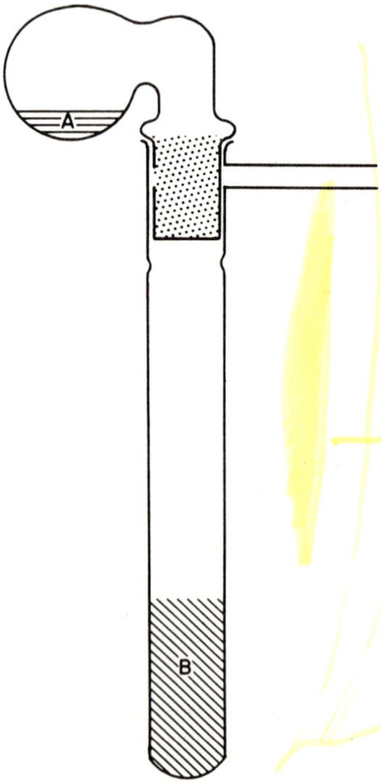


***Figure 16.10.*** The Thunberg tube. The top may be turned through 180° and the tube evacuated. An inert gas may then be added, or the top may be turned to maintain the vacuum. At any desired time nutrient or poison may be poured from A to the tissue or suspension of cells in B. The Thunberg tube has been used extensively in studies on dehydrogenases. The tissue mash in which dehydrogenase activity is to be determined is placed in B, along with methylene blue. Oxygen is withdrawn, and the rate of decolorization of the dye, which serves as an index of dehydrogenase activity, is determined. The effect of a nutrient (or poison), poured from A, upon the rate of action of the dehydrogenase, as measured by the rate of decolorization of the dye, may also be tested by this method. The kinds of dehydrogenase present in a tissue may be determined by this means since the dye is decolorized only if hydrogen is transferred to it from the nutrient by the dehydrogenase.

A

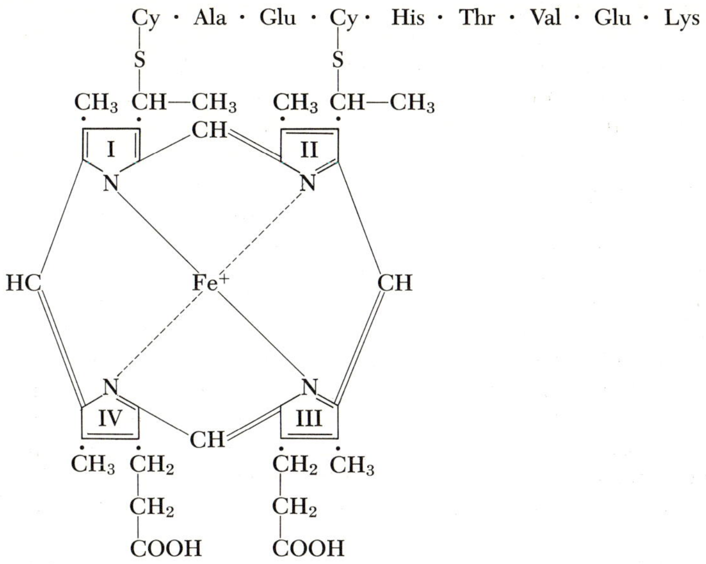


B

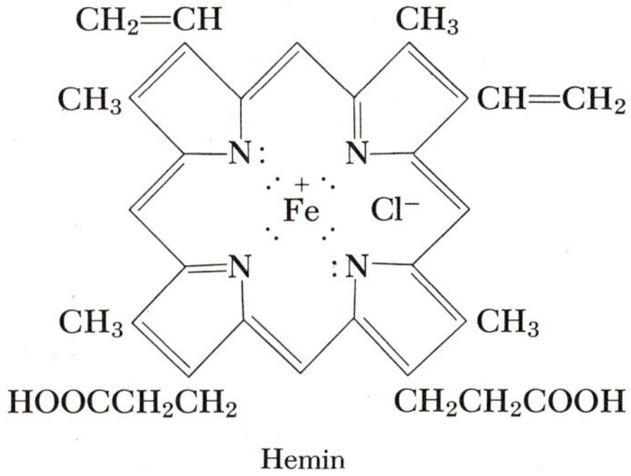


***Figure 16.11.*** *A*, Possible structure of the cytochrome C-peptide. *B*, Structure of a metalloporphyrin, hemin (the chloride of heme). A porphin is a cyclic tetrapyrrole with no substituent side chains. When substituent side chains are present, the compound is called a porphyrin. Various porphyrins have different side chains. When a metal is present in the center of the ring, the compound is called a metalloporphyrin. (From Anfinsen, 1959: The Molecular Basis of Evolution. John Wiley & Sons, New York. After Tuppy and Bodo.)

pigments universally present in aerobic cells (Keilin, 1925). They receive electrons from the flavoproteins or other carrier enzymes and pass them to cytochrome oxidase which activates oxygen. At least four cytochromes, A, $A_3$, B and C, are usually found in cells. In bacteria several new ones have also been described (Wood, 1957). In plant cells nine cytochromes have been identified. In all, 23 different cytochromes have been studied (Goddard and Bonner, 1960). A cytochrome has a prosthetic group like the heme portion of hemoglobin (see Fig. 16.11*A* and *B*). The cytochromes are distinguishable by their absorption spectra.

The cytochromes were first seen by MacMunn in 1886 in spectroscopic studies of various cells, but their role in the cell was not understood until 1926 when Keilin showed that the state of oxidation or reduction of the cytochromes may be determined by their absorption spectra. Keilin thus surmised their importance in cellular oxidation.

*Cytochrome oxidase* is an enzyme that

has as its porphyrin prosthetic group cytochrome $A_3$ (see Fig. 16.9). On receiving an electron from cytochrome A, the iron in the cytochrome oxidase molecule becomes reduced, and when the electron is passed to oxygen the iron is again oxidized. It is possible to demonstrate the presence of cytochrome oxidase by a simple color reaction called the *nadi reaction* devised by Warburg (1930). For this reaction, *p*-phenylenediamine and $\alpha$-naphthol are mixed with a suspension of cells or a tissue extract, and if cytochrome oxidase is present a bluish color soon appears. The color is the result of the formation of the dye, indophenol. By the intensity of the color one can determine roughly the amount of enzyme present and demonstrate a concentration gradient of cytochrome oxidase in cells. For example, the animal pole of a marine egg has more cytochrome oxidase than the vegetal pole, and the anterior end of the embryo has more than the posterior end (Child, 1953).

## ALTERNATE PATHWAYS OF GLUCOSE OXIDATION

***The Pentose Shunt.*** For most aerobic cells glucose degradation follows the series of reactions of the Embden-Meyerhof and Krebs cycles just described. However, alternate pathways exist, one of the most important of these being the *pentose* (*or oxidative*) *shunt* which may account for as much as 10 per cent of the glucose metabolism, and even more under some circumstances (Lehninger, 1960; Cheldelin *et al.*, 1962).

The first reaction in the pentose shunt (Fig. 16.12) is the oxidation of glucose-6-phosphate, in the presence of the enzyme glucose-6-phosphate dehydrogenase, to form 6-phosphogluconic acid by way of 6-phosphogluconolactone. The 6-phosphogluconic acid is then oxidized, in the presence of coenzyme NADP, to form NADPH (the reduced form of the coenzyme), $H^+$ and 3-keto-6-phosphogluconic acid which, in the presence of the enzyme 3-keto-6-phosphogluconic acid decarboxylase, is decarboxylated to form ribulose-5-phosphate and carbon dioxide. The NADPH and $H^+$ can then reduce some compound or be oxidized through the flavin-cytochrome system in the main line, in which case water is formed.

Ribulose-5-phosphate can be converted into other pentoses, either ribose-5-phosphate or xylulose-5-phosphate, by phosphoriboisomerase and phosphoketopentose epimerase, respectively. All three pentoses, ribose, ribulose and xylulose, can enter into a series of reactions catalyzed by the enzymes transketolase, transaldolase and aldolase, but the details of the reactions are not fully determined. However, triose phosphate is formed, possibly by a pathway which is the reverse of that in photosynthesis (see Chapter 19). In this way, for example, three molecules of ribulose-5-phosphate give rise to five molecules of triosephosphate. Some of the trioses may again form hexoses. However, a 7-carbon sugar, sedulose-7-phosphate, and a 4-carbon sugar, erythrose-4-phosphate, are also intermediates, in the presence of the enzymes transaldolase and transketolase, respectively. The reactions resemble those in photosynthetic plants concerned with the regeneration of pentose (see Table 19.3). Thus, if the course of the cycle as shown is assumed to occur, the net effect is as if one of the six molecules of glucose had been completely decomposed, the other five being reconstituted.

The enzymes necessary for these reactions occur in all cells tested (Wood, 1955). High energy phosphates are formed as in the reactions previously considered (38 ATP molecules per hexose decomposed). Since some of these compounds are involved in photosynthesis, it is not surprising that this pathway also exists in plant cells. It is also found in some microorganisms. Cells of some tumors carry on oxidation by way of the oxidative pentose shunt to a greater extent than do normal cells.

***Alternate Pathways in Microorganisms.*** Microorganisms, perhaps the most versatile "chemists" among cells, have explored many pathways for decomposition of organic compounds.

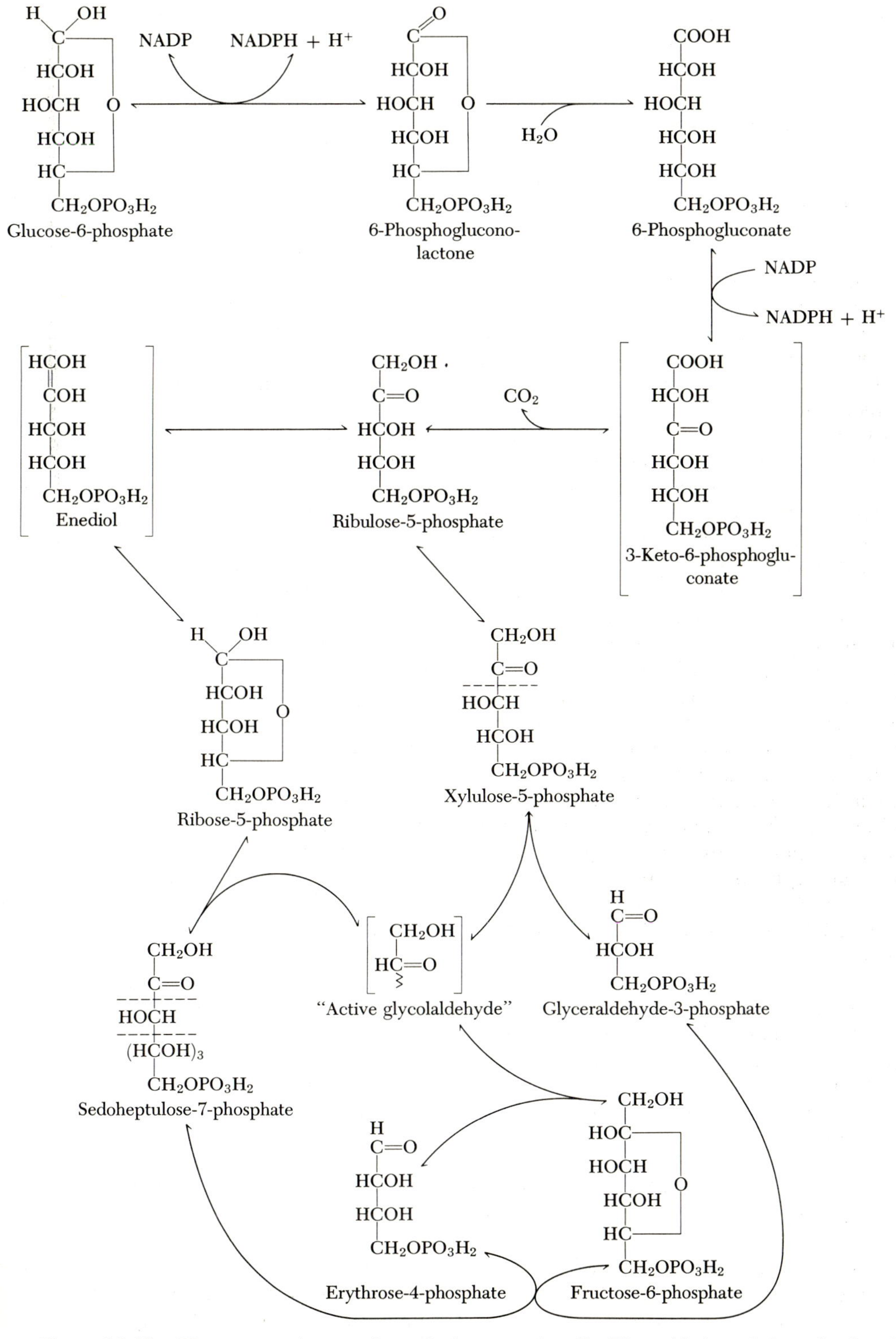


***Figure 16.12.*** The pentose shunt pathway in heterotropic cells. The oxidation of six molecules of glucose with the formation of six molecules of pentose is equivalent to complete oxidation of a glucose molecule to carbon dioxide and water. (From Cantarow and Schepartz, 1967: Biochemistry. 4th Ed. W. B. Saunders Co., Philadelphia, p. 343.)

Some bacteria—the *chemoautotrophs*—employ oxidation of inorganic compounds such as sulfur or iron as a source of the energy needed to carry on the synthesis of their own cell material from carbon dioxide, water and salts. Some autotrophic bacteria are able to synthesize their own cell material because they are photosynthetic (see Chapters 19 and 20). The majority of microorganisms are *heterotrophic,* that is, they depend upon organic materials as their sources of energy. Also found among the microbes are some peculiar fermentations with production of methane, hydrogen sulfide and various products of anaerobic carbohydrate and nitrogen metabolism foreign to animal and plant cells.

It has been stated that a microbe can probably be found to decompose any organic compound, even one resulting from the skill of the organic chemist and not known to occur in nature. When the microbe lacks the enzyme for a substance, a new enzyme may be induced in it after it has had contact with the substrate for a time. If the microbe fails to develop a new enzyme but is grown for several generations in the presence of the strange substrate, mutants may appear which are capable of using the substrate as a source of energy. This is true even of substances toxic to many cells. This subject, though fascinating in its implications, is too large for consideration here (see Kluyver and van Niel, 1956; Gunsalus and Stanier, 1961).

Even though many kinds of fermentative and oxidative processes are carried out by microorganisms, a degree of unity prevails in their biochemistry (Kluyver and Donker, 1926). All the series of reactions depend upon certain key reactions common to most life, and some similar enzymes and coenzymes, as well as compounds with a high phosphate transfer potential are found in all of them (Florkin, 1960).

***Pathways in Poisoned Cells.*** Since cyanide binds the iron in the prosthetic group of cytochrome oxidase, making it incapable of transferring electrons, the iron remains in the oxidized state and the cytochromes are unable to discharge their electrons to the enzyme (Sizer, 1957). Because the absorption spectra of reduced and oxidized cytochromes are different, the effect of cyanide on the respiratory condition of the cell can be detected spectroscopically, cyanide causing the absorption spectrum of the reduced cytochromes to appear.

Although 0.001 molar cyanide blocks the main line of oxidation, a residual respiration, constituting about 5 to 15 per cent of the total and called the cyanide-stable respiration, continues—presumably by pathways other than through cytochrome oxidase. For example, flavoproteins have been found to pass electrons slowly to oxygen, and a few oxidases specific to amino acids remove hydrogen from corresponding amino acids and pass it directly to oxygen, even in the presence of cyanide. The experiments with cyanide, therefore, demonstrate the presence of oxidative pathways other than the main line pathway of oxidation through cytochrome oxidase (Chance, 1954).

A poison such as urethan which affects many enzymes in a rather unspecific manner is not as useful in demonstrating alternate pathways. Urethan has a certain degree of selectivity for dehydrogenases, and in its presence the oxidized absorption spectrum of the cytochromes appears.

Poisons have been useful in many cases because when they are used to block a specific pathway, alternate pathways show up more clearly. In this way it has been possible to identify and characterize some of the subsidiary pathways which play a role in normal metabolism as well.

***Pathways in Tumor Cells.*** Warburg has emphasized the glycolytic nature of tumor metabolism. Regardless of whether this is a cause or an effect of the derangement of growth that is cancer, it is interesting that there is greater emphasis in many cases, though not all, upon the anaerobic metabolism. Since tumors are often rapidly dividing cells without too much organization, glycolysis could be favored because the supply of oxygen is probably inadequate. No special kind of metabolism unique to tumor cells has been found, the difference in reactions and enzymes being quantitative (Warburg, 1930; Greenstein, 1954; Aisenberg, 1961).

## HYDROGEN PEROXIDE FORMATION AS AN END PRODUCT IN CELLULAR OXIDATION

When enzymes transport the hydrogen, removed from a substrate, directly to atmospheric oxygen, hydrogen peroxide rather than water is formed. An example of such a reaction is shown by D-amino oxidase, which catalyzes the oxidative deamination of its substrates to the corresponding keto acids. The hydrogen peroxide, which is known to be an inhibitor of many enzymes and thus acts as a cell poison, is decomposed as soon as it is formed through the action of some specific enzymes. The most generally distributed enzyme of this type, *catalase,* decomposes hydrogen peroxide to water and oxygen. Catalase, which contains iron in a porphyrin nucleus, has some peroxidase-like action and is inhibited by cyanide. In many plant cells, *peroxidases,* the enzymes that serve to oxidize substrates by using peroxide as a hydrogen acceptor, dispose of the hydrogen peroxide (Baldwin, 1963).

## METHODS FOR INVESTIGATING INTERMEDIARY METABOLISM

The concepts summarized in the preceding pages were developed only after prolonged research by diverse techniques. The history of their development is involved and cannot be considered here. However, discussion of an interesting controversy illustrates the devious routes taken in making some of the discoveries.

In analyzing cell respiration, Warburg considered activation of oxygen to be the important step. The activated oxygen, he said, would combine with the substrate. At the time, many investigators sought to analyze problems of the cell by making simple analogues, or models, and Warburg found that activated blood charcoal catalyzes oxidation of many of the metabolites present in the cell by activating oxygen. The active agent in blood charcoal is iron, which is subject to poisons like cyanide and carbon monoxide. Warburg therefore searched for an iron-containing agent in the cell that activates oxygen. He obtained evidence for such an agent by the effective use of respiratory inhibitors. He found that this cellular catalyst is inhibited by cyanide and carbon monoxide. The carbon monoxide poisoning proved reversible with visible light, and from the action spectrum of this reversal Warburg determined that the respiratory enzyme was a porphyrin. He called this enzyme the "atmungsferment" or respiratory enzyme. Later, isolation of the enzyme verified his deductions (Warburg, 1930).

On the other hand, Wieland, Warburg's contemporary, emphasized the importance of activation of hydrogen, and as a model of the cell enzymes he used palladium black, which is known to adsorb hydrogen. He found that the palladium black catalyzes the oxidation of some of the organic compounds of the cell, and he postulated the presence of dehydrogenase action in the cell, with activation and removal of hydrogen. Thunberg (1930) later gathered evidence for this type of enzyme, using the method shown in Figure 16.10. By using different substrates with tissue preparations he was able to identify a large number of dehydrogenases and to measure their relative concentrations.

It was finally realized that Wieland and Warburg were both correct, that many enzymes are involved in cell respiration and that both hydrogen and oxygen have to be activated. Many of the respiratory enzymes have since been obtained in relatively pure form and their properties have been investigated. Some enzymes seem to be inseparable from membranes in the cell. At the present time the few purified oxidation-reduction enzymes of this type require co-factors (metals and coenzymes) in concentrations higher than those present in the living cell. This might imply that the structural and functional relations between the enzymes and co-factors are optimal in the cell, or perhaps that they have been altered in some way during the purification procedure (Lehninger, 1965b).

For a study of the steps in the breakdown or synthesis of an organic compound in the cell of an organism (a field of great interest at the present time), the organism is fed various diets or given certain nutrients by injection. Its urine is then examined for metabolic intermediates (i.e., intermediate stages in the breakdown of a nutrient). Many nutrients can now be labeled with "tagged" nitrogen, carbon, phosphorus, sulfur and other elements, and therefore can be much more readily traced in metabolism than was previously possible. In addition, it has been found possible to determine pathways of breakdown and synthesis by studying cells of organisms possessing genetic metabolic abnormalities, which would make them incapable of completely decomposing some compounds. As a result, intermediates accumulate in their urine in quantities sufficient for identification. Mutants lacking some enzymatic constituent have also been produced in the bread mold *Neurospora* and in cells of other organisms (see Chapter 20). Such mutants accumulate some of the metabolic intermediates in large quantities, making identification easier (Wagner and Mitchell, 1963).

*In vitro* studies of intermediary metabolism have been made with cells in tissue

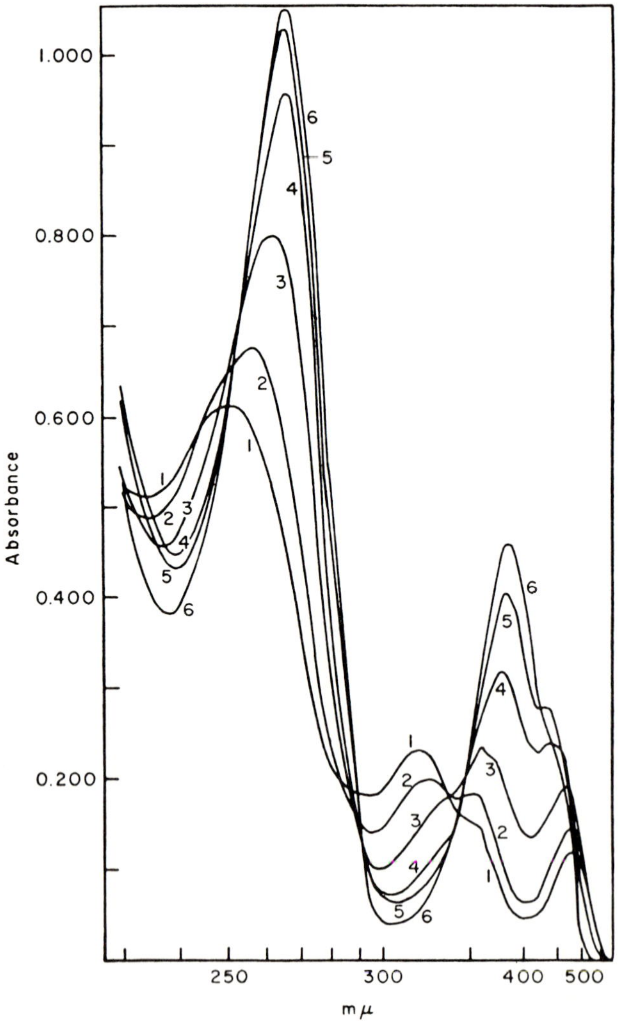


***Figure 16.13.*** Absorption spectra of flavin mononucleotide ($6.4 \times 10^{-5}$M in 1 N HCl) at successive oxidation states from about 75 per cent reduction (curve 1) to full oxidation (curve 6); light path 0.5 cm., temperature 28° C.; reduced with metallic Zn. (From Beinert, 1956: J. Am. Chem. Soc. *78*:5323.)

cultures, cells of perfused organs (e.g., liver and kidney) and cells in slices of tissues. Sometimes the cells are ground to a brei for biochemical studies. Manometric studies of gaseous exchanges in cells in tissue slices have added to the knowledge of intermediary metabolism (see Chapter 18), because comparisons can be made between respiration of normal and of pathological cells or of tissue cells from animals grown on diets omitting certain essential requirements (Duncan, 1964).

It is also important that, as each enzyme is separated and purified, the reaction it catalyzes be studied *in vitro.* Only when the various thermodynamic and kinetic properties of the isolated system have been studied under controlled conditions can the part it plays in the whole sequence of reactions in the cell be established with any certainty.

As mentioned previously, enzymes can be separated from extraneous proteins by salting out or by adsorption on resins, and they may be purified by electrophoretic procedures. Even when purification is not possible, considerable information may be obtained by use of a partially purified enzyme, provided the extraneous material does not interfere with the reaction under study. Much has been learned even by the use of heterogeneous suspensions containing many enzymes, such as entire mitochondria, fragmented mitochondria or partially purified enzyme systems from mitochondria.

The rate of an enzymatic reaction is often measured spectrophotometrically because accurate measurements can be made quickly and continually during the course of the reaction. Fortunately many of the enzymes important in the reactions discussed in this chapter have markedly different absorption spectra in their oxidized and reduced states. Consequently the relative amount of the enzyme system in each state can be readily determined and the course of the reaction can be followed by the change in proportion of each. An example is given in Figure 16.13. Methods for studying enzymatic reactions are considered in treatises on enzymes (Colowick and Kaplan, 1955–1959).

## ENZYMES IN THE ORGANELLES OF THE CELL

The enzymes involved in the anaerobic phase of sugar decomposition are apparently membrane bound, but whether on the endoplasmic reticulum or the plasma membrane is not known. On the other hand, all of the enzymes of aerobic oxidation of glucose, e.g., of the Krebs tricarboxylic acid cycle, are localized in the mitochondria. Most of the energy for the metabolism of the cell, therefore, appears to be released and transferred in the mitochondria (Siekevitz, 1957). Since the various enzymes which participate in the formation and transfer of high energy phosphate bonds are also localized there, the role of the mitochondria appears to be the formation and liberation of compounds of high phosphate transfer potential (e.g., ATP).

As already described, the mitochondrion has a complex structure, usually with lamellate cristae protruding into the central lumen, providing a large surface area. The problem is to localize and isolate each enzyme, to identify the individual reactions catalyzed by the enzyme and, finally, to reconstruct the enzyme system *in vitro,* especially the enzymes of the Krebs cycle, electron transport and phosphorylation. Attempts to separate many of the enzymes have failed, although some success has been achieved in submitochondrial systems obtained by fragmenting the mitochondria, the whole series of reactions now being obtained in pieces equal to $\frac{1}{40}$ of a mitochondrion. The difficulties suggest that there may be a biological necessity for structural organization of some of these catalysts in a moderately rigid, geometrically organized constellation in the membrane to minimize the path distance between slowly diffusing large molecules and to increase the probability of interaction.

Now it is apparent that enzymes that catalyze the various reactions of the Krebs tricarboxylic cycle are probably present on membranes, perhaps the outer mitochondrial membrane. On the other hand, the electron-transport particles appear to be on the cristae (Lehninger, 1965b). By the use of negative staining fine structures like small tennis rackets have been detected on the cristae (Fig. 16.14). The dimensions of these structures are of the right order of magnitude to serve as electron-transport particles, but the reactions occur equally well when the knobs are removed (Parsons, 1963; Chance *et al.,* 1964).

The enzymes concerned with phosphorylation may be present on the outer mitochondrial membrane. Mitochondria undergo changes in volume indicating changes in permeability (see Chapter 14). These changes in permeability accompany oxidative reactions on the membranes of the mitochondria. The mitochondrial membrane is not very permeable to ATP or ADP; therefore, it is thought that the phosphate transfer and transfer of energy occurs at the membrane surface itself, the ADP just outside the membrane being phosphorylated to ATP. The ratio of ATP to ADP at the surface determines the swelling of the mitochondria (Lehninger, 1965b).

Some evidence indicates that each mitochondrion has about 5000 to 10,000 complete sets of such enzymes. The assemblies are probably distributed equally over the cristae. It will be recalled that a cell has many mitochondria—about 500 being present in many kinds of cells (a liver cell has about 2500); therefore there is ample room to accommodate numerous sets of enzymes (Lehninger, 1965b).

Corresponding to their role in energy transfer, mitochondria are numerous in active cells and, in general, concentrate in regions of greatest activity. During cell division they may be seen actively moving about the spindle. Mitochondria appear at synapses between neurons. They are very numerous in secretory cells and in muscle cells, especially in the very active flight muscle cells of insects and birds.

Enzymes participating in the synthesis of proteins are present in the ribosomes that adhere to the outer surfaces of the walls of the endoplasmic reticulum of the cell. When cells are mechanically crushed the endoplasmic reticulum breaks up into vacuoles, on the outer surface of which are the ribosomes. Treatment with a detergent separates the ribosomes from the lipoprotein membranes of the vacuoles. The ribosomes thus separated are concentrated by centrifugation and are found

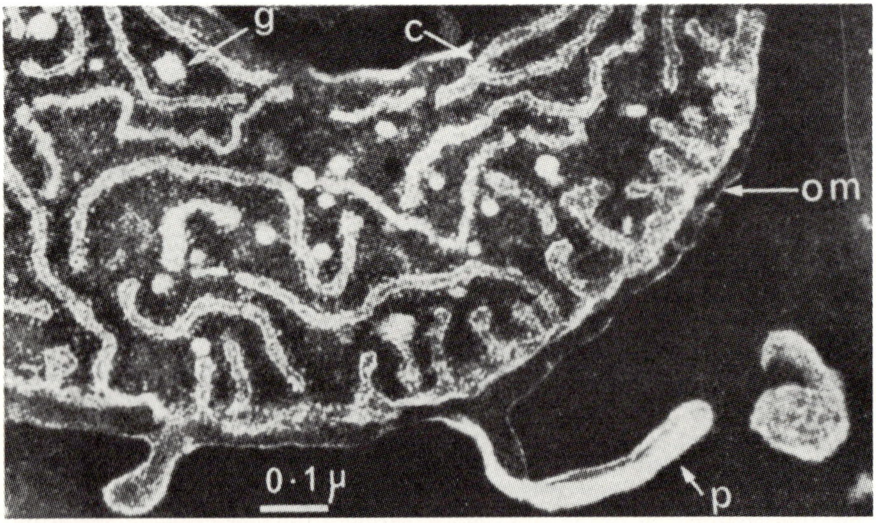


***Figure 16.14.*** Postulated electron transport particles (oxysomes) on cristae of the mitochondrion; these are the small spherical bodies on pedicels seen most clearly in the middle portion of the figure. The preparation is a negatively stained liver mitochondrion: om is the outer membrane, p a projection of the outer membrane and g a dense granule; c indicates structures that are probably cristae, according to the author of the study. (From Parsons, 1963: Science *140*:cover, 985.)

to incorporate labeled amino acids into proteins very readily when provided with mRNA, sRNA, the appropriate enzymes, high energy phosphates and the proper conditions. It is on the ribosomes then that the major part of protein synthesis of the cell is performed (see Chapter 20).

The nucleus contains enzymes for anaerobic metabolism and for some oxidative phosphorylation as well. The nucleus also contains enzymes for synthesis of DNA, RNA and some proteins. The coenzymes NAD and NADP are synthesized in the nucleus, and nuclear control of cytoplasmic metabolism is in part the result of the control the nucleus has over the production of these substances which afterwards pass through the nuclear membrane into the cytoplasm (see Chapter 6).

## DYNAMIC NATURE OF CELL CONSTITUENTS

Through the use of cell foods containing isotopes of nitrogen ($^{15}N$) and carbon ($^{14}C$) it is possible to follow their incorporation into the cell and thus to "label" cell proteins. The rate of decrease of the labeled nitrogen and carbon may then be measured. By this means it has been demonstrated that cells of all tissues lose some of their labeled material and replace it anew periodically. Even the protein in connective tissue is replaced, although less rapidly than in more active tissues. Replacement is very rapid in some cells like those in the liver (Schoenheimer, 1942; Lehninger, 1965a).

The results have been generally taken to indicate that the cellular protein is continually breaking down and being replaced. On the other hand, another interpretation of the results is also possible—it may be that individual cells in the tissues studied are dying and are being replaced (for example, in the intestine, blood and epidermis), rather than that part of the protein in all of the cells is being replaced. Studies with bacteria show that nitrogen is being incorporated into the cell at a steady rate so long as the cell is growing. There appears to be no breakdown of protein except when the cells are starved for nitrogen (Spiegelman, 1957). It thus appears necessary to test the *dynamic state* hypothesis with a variety of animal and plant cells, perhaps in tissue culture, and by a variety of methods. It may be that the increases occur only with growth (or storage) and that losses of protein attend nitrogen starvation. Some tests on tissue cultures suggest that cells which have stopped dividing do replace their cellular materials slowly (Mazia, 1961).

## LITERATURE CITED AND GENERAL REFERENCES

Aisenberg, A. C., 1961: The Glycolysis and Respiration of Tumors. Academic Press, New York.

Albritton, E. C., 1954: Standard Values in Nutrition and Metabolism. W. B. Saunders Co., Philadelphia.

Atkinson, M. R., and Morton, R. K., 1960: Free energy and the biosynthesis of phosphates. Comp. Biochem. *2*:1–95.

Baldwin, E., 1967: Dynamic Aspects of Biochemistry. 5th Ed. Cambridge University Press, London.

Barker, H. A., 1956: Bacterial Fermentations. John Wiley & Sons, New York.

Beinert, H., 1956: Spectral characteristics of flavins at the semiquinoid oxidation level. J. Am. Chem. Soc. *78*:5323–5328.

Boyer, P. D., 1963: Phosphohistidine. Science *141*: 1147–1153.

Bueding, E., and Farber, E., 1960: Comparative biochemistry of glycolysis. Comp. Biochem. *1*: 411–439.

*Cantarow, A. and Schepartz, B., 1967: Biochemistry. 4th Ed. W. B. Saunders Co., Philadelphia.

Chance, B., 1954: Spectrophotometry of intracellular respiratory pigments. Science *120*:767–774.

Chance, B., and Parsons, D. F., 1963: Cytochrome function in relation to inner membrane structure of mitochondria. Science *142*:1176–1179.

Chance, B., Parsons, D. F., and Williams, G. R., 1964: Cytochrome content of mitochondria stripped of inner membrane structure. Science *143*:136–139.

Cheldelin, V. H., Wang, C. H., and Wang, T. E., 1962: Saccharides: alternate routes of metabolism. Comp. Biochem. *3*:427–502.

Child, C. M., 1953: Indicator gradient patterns in oocytes and early developmental stages of echinoderms, a reexamination. Biol. Bull. *104*:12–27.

Colowick, S. P., and Kaplan, M. D., 1955–59: Methods in Enzymology. 6 vol. Academic Press, New York.

Duncan, G. G., 1964: Diseases of Metabolism. 5th Ed. W. B. Saunders Co., Philadelphia.

Florkin, M., 1960: Unity and Diversity in Biochemistry. Pergamon Press, Long Island City, N. Y.

Frieden, E., Aug. 1959: The enzyme-substrate complex. Sci. Am. *201*:119–125.

Goddard, D. R., and Bonner, W. D., 1960: Cellular respiration. *In* Plant Physiology, A Treatise. Steward, ed. Academic Press, New York, pp. 209–312.

Green, D. E., July 1958: Biological oxidation. Sci. Am. *199*:56–62.

Green, D. E., Feb. 1960: The synthesis of fat. Sci. Am. *202*:46–51.

Green, D. E., Jan. 1964: The mitochondrion. Sci. Am. *210*:63–74.

Green, D. E., and Hatefi, Y., 1961: The mitochondrion and biochemical machines. Science *133*: 13–23.

Greenberg, D. M. (ed.), 1961: Metabolic Pathways. Vol. I. Amino acids, nucleic acids, porphyrins, vitamins and coenzymes. Vol. II. Carbohydrates, lipids and related compounds. Academic Press, New York.

Greenstein, J. P., 1954: Biochemistry of Cancer. Academic Press, New York.

Gunsalus, I. C., and Stanier, R. Y. (eds.), 1961: The Bacteria. 5 vol. Academic Press, New York.

Hsia, D. Y., 1959: Inborn Errors of Metabolism. Yearbook Medical Publishers, Chicago.

Ingraham, L., 1961: Biochemical Mechanisms. John Wiley & Sons, New York.

James, W. O., and Leech, R. M., 1960: The plant cytochromes. Endeavour *19*:108–114.

Kamen, M. D., 1957: Isotopic Tracers in Biology. Academic Press, New York.

Kamen, M. D., Aug. 1958: A universal molecule of living matter. Sci. Am. *197*:77–82.

Keilin, D., 1925: On cytochrome, a respiratory pigment, common to animals, yeast and higher plants. Proc. Roy. Soc. London *B98*:312–339.

Kluyver, A. J., and Donker, H. J. L., 1926: Die Einheit in der Biochemie. Chemie der Zelle und Gewebe *13*:134–190.

Kluyver, A. J., and van Niel, C. B., 1956: The Microbe's Contribution to Biology. Harvard University Press, Cambridge, Mass.

Krahl, M. E., 1961: The Action of Insulin on Cells. Academic Press, New York.

Krebs, H. A., 1948–1949: The tricarboxylic acid cycle. Harvey Lect. *44*:165–199.

Krebs, H. A., 1957: Control of metabolic processes. Endeavour *16*:125–139.

Krebs, H. A., and Kornberg, H. L., 1957: A survey of the energy transformations in living matter. Ergebn. Physiol. *49*:212–298.

Lehninger, A. L., May 1960: Energy transformation in the cell. Sci. Am. *202*:102–114.

*Lehninger, A. L., 1965a: Bioenergetics. W. A. Benjamin, New York.

Lehninger, A. L., 1965b: The Mitochondrion. W. A. Benjamin, New York, Chapters 3–7.

Lipmann, F., 1955: Coenzyme A and biosynthesis. Am. Sci. *43*:37–47.

Lipmann, F., Jones, M. E., Black, S., and Flynn, R. M., 1953: The mechanism of the ATP Co-A-acetate reaction. J. Cell. Comp. Physiol. *41*(Suppl. 1):109–112.

Mazia, D., 1961: Mitosis and the physiology of cell division. *In* The Cell. Brachet and Mirsky, eds. Academic Press, New York, Vol. 3, pp. 77–412.

*McElroy, W. D., 1964: Cellular Physiology and Biochemistry. 2nd Ed. Prentice-Hall, Englewood Cliffs, N. J.

Meister, A., 1955: Transanimation. Advances Enzymol. *16*:185–246.

Ochoa, S., 1951: Biological mechanisms of carboxylation and decarboxylation. Physiol. Rev. *31*:56–106.

Parsons, D. F., 1963: Mitochondrial structure: two types of subunits on negatively stained mitochondrial membranes. Science *140*:985–987.

Racker, E., 1954: Alternate pathways of glucose and fructose metabolism. Advances Enzymol. *15*:141–182.

*Racker, E., 1957: Micro- and macrocycles in carbohydrate metabolism. Harvey Lect. *51*:143–174.

Racker, E., 1961: Mechanisms of synthesis of ATP. Advances Enzymol. *23*:323–399.

Schoenheimer, R., 1942: The Dynamic State of Body Constituents. Cambridge University Press, London.

Siekevitz, P., July 1957: Powerhouse of the cell. Sci. Am. *197*:131–140.

Sizer, I. W., 1957: Chemical aspects of enzyme inhibition. Science *125*:54–59.

Spiegelman, S., 1957: Nucleic acids and the synthesis of proteins. *In* A Symposium on the Chemical Basis of Heredity. McElroy and Glass, eds. Johns Hopkins Press, Baltimore, pp. 232–267.

Theorell, H., 1956: Nature and mode of action of oxidation enzymes. Science *124*:467–472.

Thunberg, T., 1930: The hydrogen-activating enzymes of the cells. Quart. Rev. Biol. *5*:318–347.

Wagner, R. P., and Mitchell, H. K., 1963: Genetics and Metabolism. 2nd Ed. John Wiley & Sons, New York.

Warburg, O., 1930: The Metabolism of Tumours. Constable & Co., London.

Webb, J. L., 1961: Enzymes and Metabolic Inhibitors. Academic Press, New York.

Wieland, H., 1932: On the Mechanism of Oxidation. Yale University Press, New Haven.

Wood, H. G., 1955: Significance of alternate pathways in the metabolism of glucose. Physiol. Revs. *35*:841–859.

Wood, W. A., 1957: Metabolism of carbohydrates and related compounds. Ann. Rev. Microbiol. *11*:253–282.

Wooley, D. W., 1953: A Study of Antimetabolites. John Wiley & Sons, New York.

# CHAPTER 17 / OXIDATION-REDUCTION POTENTIALS IN CELLULAR OXIDATION

In the main line of oxidation in the cell the hydrogen and electrons removed from a substance are passed to oxygen by a series of enzymes (Figure 16.9). The particular sequence of transfers from one enzyme to another is determined by the *oxidation-reduction* (redox) *potential,* e.g., its tendency to give up or take on electrons. Transfer is along a potential gradient. The mechanical analogue of this is a staircase down which a load is dropped.

The study of redox potentials of biological systems is of great significance because these potentials determine the particular sequence of reactions occurring in a cell.

## REDOX POTENTIALS AT ELECTRODES IMMERSED IN ELECTROLYTES

The redox potential of an element—its tendency to give up or take on electrons—is a characteristic property of the element, and depends upon the nuclear and electronic constitution of the element. Members of one class of elements (like hydrogen) give up electrons readily, forming ions in solution, whereas those of another class (like oxygen) take up electrons instead.

An electrode is a device for conducting electrons. It may conduct them to or from an electrolyte solution, an electric arc or a vacuum tube. It is usually made of metal, or of a metal and one of its salts; but a gas, like hydrogen, may serve as an electrode if the gas is adsorbed on the surface of an inert metal like spongy platinum. The calomel electrode is a metal electrode which is widely used because of its independence of pH and because of its stability. Another metal electrode used in oxidation-reduction studies is the platinum electrode, which is useful because it is inert and does not react with a solution in which it is immersed.

In order to measure the electrode potential of an element, it is necessary to immerse the element as the electrode into a solution of one of its salts, to connect the electrode to a potentiometer and to measure the potential difference between it and some standard electrode, also connected to the potentiometer, as a baseline. Thus zinc, which like hydrogen gives up its electrons readily, dissolves when immersed as an electrode in a solution of one of its salts. As it dissolves it forms ions in the solution, leaving the electrons on the electrode:

$$Zn \underset{\text{reduction}}{\overset{\text{oxidation}}{\rightleftharpoons}} Zn^{++} + 2e \quad (17.1)$$

or, in more general terms:

$$\text{Metal} \underset{\text{reduction}}{\overset{\text{oxidation}}{\rightleftharpoons}} \text{ion} + \text{electron} \quad (17.2)$$

An electrode made of zinc in contact with one of its salts is therefore negatively charged. The greater its tendency to go into solution, other things being equal, the greater will be the potential developed between the solution and the electrode. On the other hand, copper, like oxygen, does not dissolve when immersed as an electrode in one of its salts. Instead, it delivers electrons to the cupric ions in the salt solution, and the cupric ions plate out as copper on the metal. An electrode made up of such a metal in contact with one of its salts is therefore always positive. All metals, on the basis of similar experimental data, can be arranged in an *electromotive series* which measures their relative potentials (see Table 17.1 and Fig. 17.1).

The potential of the "normal" hydrogen electrode (see Fig. 17.2) taken at standard state is arbitrarily taken as zero and serves as a reference point for potentials of all other electrodes (Fig. 17.1). The standard state for the normal hydrogen electrode means that gaseous hydrogen at 1 atmosphere pressure is continuously bubbled past a spongy platinum electrode which is kept in a solution 1 normal with respect to hydrogen ions.

To measure the potential, at a given temperature, of a metal electrode immersed in a solution of one of its own salts 1 normal in activity, one must connect the electrode through a potentiometer with a hydrogen electrode. For example, it is found by this means that at 25° C. the normal magnesium electrode, measured

**Figure 17.1.** Electromotive potentials and a mechanical analogue.

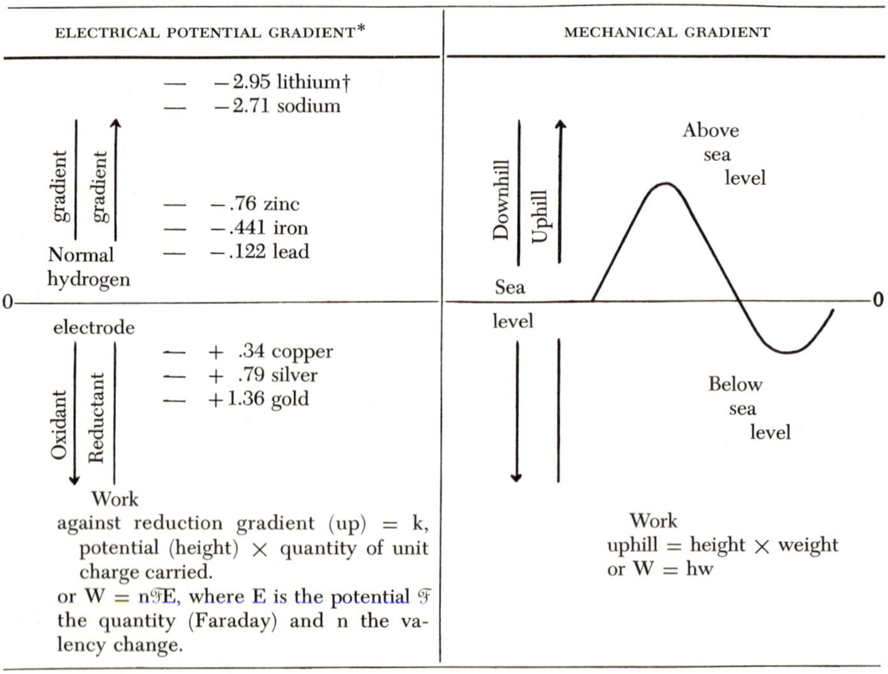


* Potentials at pH 0.0. Note that the sign of the charge is not that of the ion but of the electrode in contact with a solution of ions.

† The sign of the potential represents the reduction potential (see Table 17.1). Oxidant means oxidizing agent and reductant, reducing agent.

**Table 17.1.** Normal Electrode Potentials at 25° C.

| ELECTRODE | ELECTRODE REACTION | NORMAL* ELECTRODE POTENTIAL (VOLTS) AGAINST HYDROGEN ELECTRODE |
|---|---|---|
| $Li^+$, Li | Li $= Li^+ + e$ | −2.9595 |
| $Na^+$, Na | Na $= Na^+ + e$ | −2.7146 |
| $Zn^{++}$, Zn | Zn $= Zn^{++} + 2e$ | −0.7618 |
| $Fe^{++}$, Fe | Fe $= Fe^{++} + 2e$ | −0.441 |
| $Pb^{++}$, Pb | Pb $= Pb^{++} + 2e$ | −0.122 |
| $H^+$, $H_2$ (1 atm.) | $\frac{1}{2}H_2 = H^+ + e$ | 0.0000† |
| $Cu^{++}$, Cu | Cu $= Cu^{++} + 2e$ | +0.3441 |
| $Ag^+$, Ag | Ag $= Ag^+ + e$ | +0.7978 |
| $Au^{+++}$, Au | Au $= Au^{+++} + 3e$ | +1.36 |

Data from West, 1942. The electrode potential is the reduction potential; its sign is arbitrary and follows Clark, 1952. When the opposite sign is used for the electrode potentials the values represent the oxidation potentials or their readiness to be oxidized (e.g., Lewis and Randall, 1961).

* All salts 1 normal in activity (standard state).

† At pH 0.0 (1 normal $H^+$).

against the normal hydrogen electrode, registers a potential of −1.55 volts. A saturated calomel electrode, containing mercury in contact with mercurous chloride, which in turn is in contact with potassium chloride (Fig. 17.2), registers a potential of +0.2458 volts against the normal hydrogen electrode.

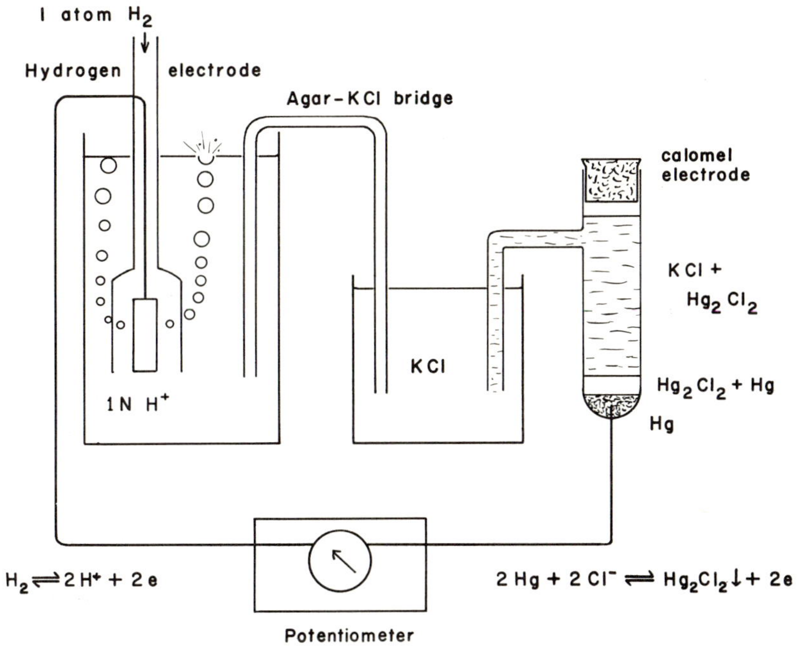


*Figure 17.2.* The hydrogen electrode as used to determine the potential of another electrode, in this case the calomel electrode. The normal hydrogen electrode is taken as 0.0, and against it the calomel electrode has a potential of +0.2458 volts. The potentials of other electrodes can be determined by substituting the electrodes, one at a time, for the calomel electrode. The potential difference between the two electrodes is measured by a potentiometer in which an unknown emf is matched against a known emf (divided as desired by passing it through a slide wire resistance) from a battery. The matching is registered by a null point on a galvanometer. In this way the data in Table 17.1 and in Figure 17.1 were determined.

It should be emphasized that biological convention places the strongest reducing agents at the tops of tables and graphs (see Table 17.1 and Fig. 17.1) instead of at the bottom. During oxidation-reductions when electrons are passed in response to an oxidation gradient, that is, from reducing potential to oxidizing potential (downward in Fig. 17.1), the free energy change is the product of the change in potential through which the electrons fall, multiplied by the number of electrons:

$$\Delta F_0 = -n\mathcal{F}\,\Delta E \qquad (17.3)$$

where $\Delta F_0$ is the standard free energy change; n is the number of equivalents in one stoichiometric occurrence; $\mathcal{F}$ is the Faraday, 96,494 coulombs. This constant is a product of the charge on an electron ($1.592 \times 10^{-19}$ coulombs) and the gram mole number of electrons (Avogadro's number, $6.024 \times 10^{23}$, the number of molecules in a gram molecular weight). $\Delta E$ is the change in potential in volts.

The product is in volt-coulombs or joules. To convert to calories, a more familiar unit, it is necessary to divide by 4.185 because a calorie is equal to 4.185 joules. It will be recalled that the free energy for a reaction represents the maximal work that can be done by that reaction, that is, its chemical potential.

Conversely, when a gram mole of electrons moves along a reduction gradient from oxidizing potential to reducing potential, the same amount of free energy (work) must be expended on the system (upward in Fig. 17.1).

It is instructive to calculate the free energy available from moving a gram mole of electrons from $NADPH_2$ (− 0.324 volts) to the level of oxygen (the oxygen electrode, +0.815 volts, at pH 7.0):

$$\Delta F = -n\mathcal{F}\,\Delta E = (1)(96{,}500)(1.14)/4.185 = -26{,}300 \text{ cal.}$$

The energy in calories for any voltage change in the series of enzymes and carriers in the cell may be calculated in the same way.

In order to emphasize the reducing capacity of a substance, that is, its capacity to donate electrons, the biological convention followed here assigns a negative sign to electrode potentials on the reducing side of the normal hydrogen electrode (Clark *et al.,* 1928; Clark, 1952, 1960). Chemical convention, emphasizing the tendency of an element to be oxidized, assigns the opposite sign to the electrode (Glasstone and Lewis, 1960; Lewis and Randall, 1961).

An atom which gives up electrons and becomes a positively charged ion is a reducing agent (reductant), and the greater its tendency to give up electrons, the greater is its reducing potential. The greater the negative electrode potential, the greater is the reducing tendency of the element. Elements (or ions) with highly negative electrode potentials reduce those less negative, and, on the other side of the hydrogen electrode potential, the elements with less positive electrode potentials reduce the more positive ones.

Conversely, anything which takes on electrons, like an oxygen molecule or a copper ion, is an oxidizing agent (oxidant). The greater the tendency of a substance to take on electrons, the greater is its oxidizing potential. Therefore, the more positive the electrode potential, the greater is the oxidizing potential of the element. Elements or ions with highly positive potentials oxidize those less positive, and on the other side of the hydrogen electrode potential the less negative elements oxidize the more negative ones.

To summarize, then, in the electromotive series as listed in Table 17.1, any element is capable of oxidizing the one above it and of reducing the one below it. Each element is therefore an oxidizing agent to those above it in the electromotive series and a reducing agent to those below it. Therefore, the electromotive series represents an *oxidation-reduction series,* and the electrode potential of each element is a measure of the oxidation and reduction capacity of the element.

## MEASUREMENT OF REDOX POTENTIALS OF ORGANIC COMPOUNDS

Organic materials cannot be made to serve as electrodes for measurement of oxidation-reduction potentials in the laboratory, yet in the cell there are oxidation-reduction systems similar to those discussed. For example, in biological oxidation-reduction reactions one often encounters changes in the valencies of iron in iron-prophyrin enzymes, such as cytochrome and cytochrome oxidase:

$$\underset{\text{Reductant}}{Fe^{++}} \underset{\text{reduction}}{\overset{\text{oxidation}}{\rightleftharpoons}} \underset{\text{Oxidant}}{Fe^{+++}} + e \qquad (17.4)$$

One also encounters organic systems in which hydrogen and electrons are transferred (in pairs) in oxidation-reduction reactions:

$$\underset{\text{hydroquinone}}{H_2Q} \underset{\text{reduction}}{\overset{\text{oxidation}}{\rightleftharpoons}} \underset{\text{quinone}}{Q} + 2H^+ + 2e \qquad (17.5)$$

Fortunately, however, such redox potentials can be measured indirectly by the use of the platinum electrode. Platinum, an inert metal, does not dissolve when placed in contact with a solution and does not form an oxidation-reduction system with it. Although chemically inert, platinum is a good conductor of electrons, which it transfers from the solution to the potentiometer. The redox potential of an organic system, like that of the inorganic system, is always measured against the normal hydrogen electrode as a reference standard (Clark, 1960).

The apparatus for such a determination is illustrated in Figure 17.3. The reductant must be kept reduced by gassing with nitrogen free of oxygen. Aliquots of oxidant are added to the reductant (titrated), the mixture is stirred after each addition and the potential of the mixture is measured with the potentiometer. Redox potentials for different ratios of oxidant to reductant are plotted in Figure 17.4, giving a curve for each redox system. The curves resemble titration curves for weak acids. It is interesting to note that near the middle of the curves, addi-

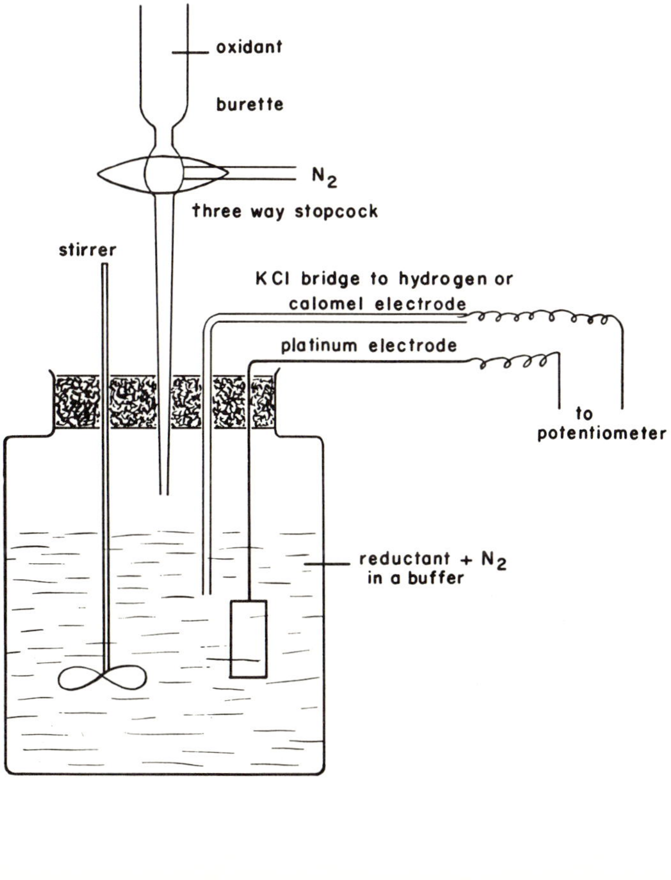


***Figure 17.3.*** Apparatus for measuring oxidation-reduction potentials. Although a redox potential is always given as between a platinum and a hydrogen electrode (the latter being the standard of reference), a calomel electrode is found in practice to be more convenient than the hydrogen electrode for a laboratory determination. The potential of the calomel electrode against the hydrogen electrode must be taken into account.

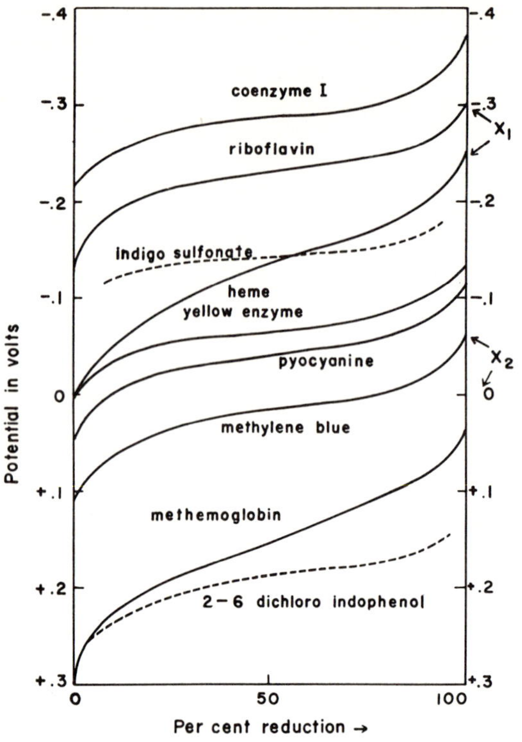


*Figure 17.4.* Oxidation-reduction titration curves for dyes and cellular constituents. $X_1$ indicates the state of an ameba in nitrogen; $X_2$ represents its state in air.

*Note:* The convention of placing the potential of greater reduction capacity at the top is followed here and in succeeding graphs. (Data at pH 7.0 from Clark, 1938: J. Appl. Physics *9*:102. Data on indigo sulfonate and indophenol at pH 7.4 from Clark, 1928: U. S. Hygienic Lab. Bul. *151*:308.)

tion of either oxidant or reductant has little effect on the redox potential, a phenomenon resembling the buffer action of a weak acid (Clark, 1960).

Although such measurements are made without too much difficulty, it is more convenient to determine the redox potential (E) of a particular mixture indirectly, through the use of Peter's equation (for derivation see Appendix 17.1). For a temperature of 30° C. and a valency change of 2, and converting from natural logarithms (*ln*) to $\log_{10}$, the equation is:

$$E = E_0 + 0.030 \log_{10} \frac{[\text{oxidant}]}{[\text{reductant}]} \quad (17.6)$$

Here [oxidant] and [reductant] are the given concentrations of the particular oxidant and reductant studied, and $E_0$, which is experimentally determined, is the redox potential of a mixture containing an equimolal concentration of the given oxidant and reductant. The factor 0.030 includes a term for the absolute temperature; therefore, it is always necessary to specify the temperature when any data on redox potentials are presented.

It is clear that once the $E_0$ for any given redox system (e.g., ferrous to ferric, or hydroquinone to quinone) is found, the redox potential at any other ratio of concentrations for the same redox system can be quickly calculated (Clark, 1952).

## THE EFFECT OF pH ON REDOX POTENTIALS

When an oxidation-reduction reaction involves a hydrogen atom of an organic molecule, the hydrogen ion concentration of the solution becomes of great importance. For example, in the oxidation of hydroquinone mentioned before (see equation 17.5) a change in hydrogen ion concentration is found to influence the equilibrium between hydroquinone and quinone (law of mass action). This influences the redox potential by a definite factor (see Appendix 17.2). This fact is of great importance to the cell since many

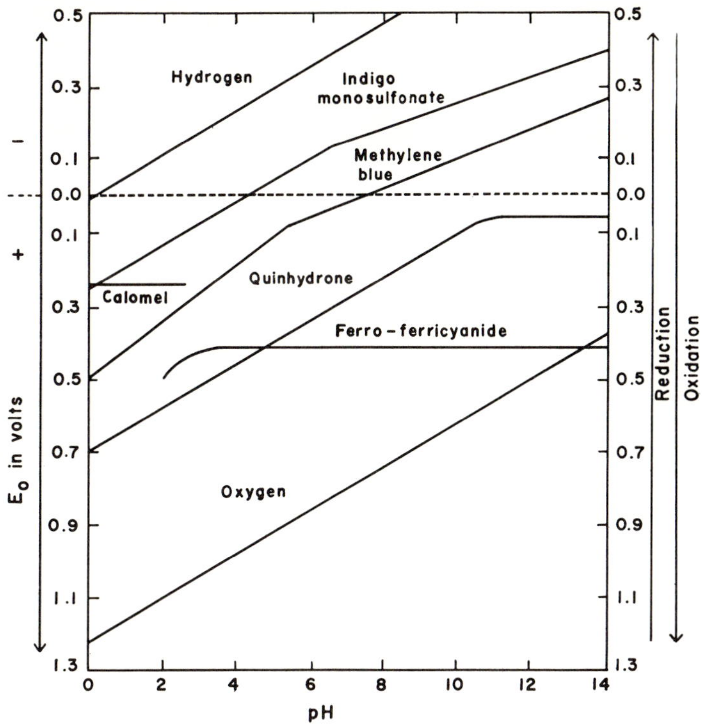


***Figure 17.5.*** The effect of pH on oxidation-reduction potential when concentration of oxidant equals concentration of reductant. (Data from Gortner, 1949: Outlines of Biochemistry. 3rd Ed. John Wiley & Sons. New York.)

oxidation-reduction reactions in the cell involve not merely electron transfer but also hydrogen transfer. It should be noted that decreasing the hydrogen ion concentration increases the reducing powers of most of the redox systems, as shown in Figure 17.5. This is to be expected since the hydrogen ion itself is an oxidant:

$$\underset{\text{Reductant}}{\tfrac{1}{2}H_2} \underset{\text{reduction}}{\overset{\text{oxidation}}{\rightleftharpoons}} \underset{\text{Oxidant}}{H^+} + e \qquad (17.7)$$

Furthermore, it is important to know the pH of a redox system because the redox potential of enzyme systems cannot be measured at the pH of the normal hydrogen electrode (pH 0.0 or 1 normal in respect to hydrogen ions). At such a pH, enzymes would be denatured and the results would not have biological significance. Cellular reactions are carried on at pH 7 (or near it), as was shown in Chapter 16. The redox potentials of biological systems are, however, always referred to the normal hydrogen electrode as a standard. A hydrogen electrode at pH 7.0 and at 30° C. has a potential of −0.42 volt against the normal standard hydrogen electrode.

Therefore, when comparisons of redox potentials are made at a pH other than 0.0, the pH and the temperature should be indicated. The redox potential of a substance under such conditions is designated E′ rather than E, and the potential of an equimolal mixture of reductant and oxidant under such conditions is designated $E_0'$ rather than $E_0$.

## POTENTIALS DEVELOPED BY BIOLOGICAL REDOX SYSTEMS

Why does a dehydrogenase pass the electrons it has removed from a substrate to flavoproteins? Why, in turn, do the flavoproteins pass the electrons to each of a number of cytochromes which donate them to cytochrome oxidase, from which they are transferred to oxygen? On the basis of the principles developed in this chapter, and on the basis of the redox potential measurements, it is now possible to give a reasonable explanation for such sequences. Values (at the indicated temperatures) of redox potentials for sample extracted systems are given in Table 17.2, along with reference values for the hydrogen and the oxygen electrode potentials. These values are $E'_0$ and were determined at or near pH 7.0. It will be seen from this table that each enzyme system develops a characteristic redox potential, the entire range of values lying between the values of the hydrogen electrode at one end and the oxygen electrode at the other. The enzyme systems and substrates in a cell therefore form a "biological electromotive series" in which each system is theoretically capable of reducing the one below it and oxidizing the one above it.

It has been found that the most rapid pathway for the oxidation-reduction reactions in the cell is that which makes use of the entire sequence of steps from dehydrogenases to cytochrome oxidase. When this pathway is abridged—for example, when flavoprotein passes electrons to oxygen directly, instead of by way of cytochrome and cytochrome oxidase—the pathway is slow. This is demonstrated when a respiring cell is poisoned with a low concentration of cyanide, which inactivates the cytochrome oxidase. Respiration continues but only at a small frac-

**Table 17.2.** Normal Oxidation-Reduction Potentials of Some Biologically Important Systems at pH 7.0

| SYSTEM | $E'_0$ | T IN °C. |
|---|---|---|
| Ketoglutarate $\rightleftharpoons$ succinate + $CO_2$ + $2H^+$ + 2e | −0.68 | —‡ |
| Ferredoxin | −0.432 | —§ |
| Formate $\rightleftharpoons$ $CO_2$ + $H_2$ | −0.420 | 38 |
| $H_2 \rightleftharpoons 2H^+ + 2e$ | −0.414 | 25 |
| NADH + $H^+$ $\rightleftharpoons$ $NAD^+$ + $2H^+$ + 2e | −0.317 | 30† |
| NADPH + $H^+$ $\rightleftharpoons$ $NADP^+$ + $2H^+$ + 2e | −0.316 | 30† |
| Horseradish oxidase | −0.27 | —† |
| $FADH_2$ $\rightleftharpoons$ FAD + $2H^+$ + 2e | −0.219 | 30† |
| $FMNH_2$ $\rightleftharpoons$ FMM + $2H^+$ + 2e | −0.219 | 30† |
| Lactate $\rightleftharpoons$ pyruvate + $2H^+$ + 2e | −0.180 | 35 |
| Malate $\rightleftharpoons$ oxaloacetate + $2H^+$ + 2e | −0.102 | 37 |
| Reduced flavin enzyme $\rightleftharpoons$ flavin enzyme + $2H^+$ + 2e | −0.063 | 38 |
| Luciferin* $\rightleftharpoons$ oxyluciferin + $2H^+$ + 2e | −0.050 | ?* |
| Ferrocytochrome B $\rightleftharpoons$ ferricytochrome B + e | −0.04 | 25 |
| Succinate $\rightleftharpoons$ fumarate + $2H^+$ + 2e | −0.015 | 30 |
| Decarboxylase | +0.19 | —† |
| Ferrocytochrome C $\rightleftharpoons$ ferricytochrome C + e | +0.26 | 25 |
| Ferrocytochrome A $\rightleftharpoons$ ferricytochrome A + e | +0.29 | 25 |
| Ferrocytochrome $A_3$ $\rightleftharpoons$ ferricytochrome $A_3$ + e | ? | —‡ |
| $H_2O \rightleftharpoons \frac{1}{2}O_2 + 2H^+ + 2e$ | +0.815 | 25 |

Data from Goddard, 1945. Potentials in all cases are at or near neutrality.

* From McElroy and Strehler, 1954: Bact. Rev. *18*.

† From Clark, 1960.

‡ From Goddard and Bonner, 1960: *In* Plant Physiology, a Treatise. Steward, ed. Academic Press, New York. Goddard and Bonner give the NADPH/$NADP^+$ system as −0.324, and NADH/$NAD^+$ as −0.320.

§ From Tagawa and Arnon, 1962: Nature *195*:537–543. The value cited is for spinach ferredoxin.

tion of its former rate (Michaelis, 1930; Goddard, 1945).

Such a series of reactions, in which each reaction involves a small amount of energy exchange and a small difference in redox potential, has great thermodynamic advantage. The larger the number of steps in each of which the energy exchange is small, the closer the system comes to being reversible, and the greater will be the thermodynamic efficiency. It is to be noted that the steps in decomposition more or less correspond to the free energy required to make 1 ATP; hence the steps are correlated with the number of ATP molecules formed from ADP and inorganic phosphate. Consider, for example, the decomposition of one molecule of glucose by way of glycolysis and the Krebs cycle. This process may yield 38 high energy phosphate bonds (in ATP and CP, see Chapter 16). If for purposes of calculation we take the standard free energy change for one gram mole of ATP being converted to ADP and inorganic phosphate as 7000 cal./mole, the yield per gram mole of glucose decomposed will be 266,000 cal. The $\Delta F$ for degradation of glucose to $CO_2$ and $H_2O$ is $-686{,}000$ cal./mole. The efficiency of the process on this basis is therefore about 39 per cent, a testimonial to the success of a cell in solving its thermodynamic problem, as compared with a heat engine, in which the maximal possible efficiency is 34 per cent. It is probable that the efficiency in the cell is even greater than this (see Chapter 16).

The presence of at least four cytochromes, B, C, A and $A_3$, the redox potential of each being successively closer to the redox potential of oxygen, in the order given, is probably interpretable on these grounds. Since the cytochromes have the same prosthetic group, an iron-porphyrin compound, the specificity of each cytochrome probably depends upon the protein to which the prosthetic group is attached. It is the nature of the protein and its union to the prosthetic group which gives each of these protein-porphyrin complexes a characteristic $E_0'$ value.

Altogether about 15 biological redox systems with metabolic function have been studied. However, few if any of them are amenable to exact theoretical treatment, because the derivation of the oxidation-reduction equation (Appendix 17.1) depends upon a system in equilibrium, and living systems develop steady states approaching, but never reaching, equilibrium (Bull, 1964).

## COUPLING OF UNIVALENT AND DIVALENT ENZYMATIC REDOX SYSTEMS

During an oxidation-reduction reaction the flavoprotein molecule transfers two electrons. Originally it was thought necessary for the flavoprotein molecule to transfer both electrons at once, one to each of two ferricytochrome B molecules, since the iron in the ferricytochrome B is capable of receiving only one electron. But the reaction of one flavoprotein molecule with two cytochrome B molecules at one time requires a triple collision. Such an occurrence is much less probable than the reaction of one flavoprotein molecule with two cytochrome B molecules, one at a time. As a two-step reaction, each step would require a collision between only two molecules.

This problem, whether oxidation-reduction reactions involving the exchange of two electrons occur in one step or in two steps, formed the basis of a series of brilliant studies by Michaelis (1935, 1940). His solution of the problem can best be illustrated by his study of the oxidation of hydroquinone, as a model system. Michaelis demonstrated that curves (see Fig. 17.6) obtained for the redox potentials when increments of oxidizing material are added to hydroquinone indicate first the loss of one electron, then another, in successive steps. The loss of the first electron is thought to correspond to the formation of a semiquinone, which is a type of *free radical.* (The free radical formed by loss of the first electron from hydroquinone is highly stable, existing as an insoluble substance called quinhy-

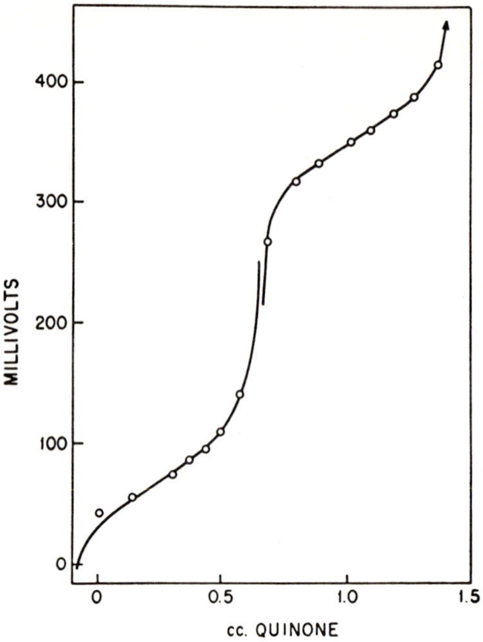


***Figure 17.6.*** Potentiometric demonstration of single electron transfer in an oxidation-reduction. Potential values in millivolts (ordinate) are plotted for increasing amounts of oxidant (quinone) added to the reductant ($\alpha$-oxyphenazine in 1N HCl, pH 0.08). The temperature is 30° C. Note that for each of the two electrons transferred from reductant to oxidant a separate curve (inflection) is obtained, indicating the transfer of a single electron at a time. (From Michaelis, 1931: J. Biol. Chem. *92*:221.)

drone, which separates from the solution of hydroquinone in the form of dark green crystals.) The second electron is lost from the semiquinone, and the semiquinone becomes quinone, the fully oxidized form (Fig. 17.7). Perhaps semiquinones, or free radicals, are always formed whenever two hydrogens and two electrons are transferred in oxidation-reduction reactions (Michaelis, 1930), but are too short lived to be demonstrable by these methods (see paramagnetic methods, in the next section).

An example of biological interest is the formation of a semiquinone during the oxidation-reduction of riboflavin (vitamin $B_2$). In this reaction, two hydrogen ions and two electrons are transferred stepwise. Riboflavin is colorless when reduced in acid solution by a reducing agent, but when it is exposed to air by shaking, it turns a bright orange color until it is fully oxidized, at which stage it becomes yellow. In this example, loss of a single electron results in the formation of an orange colored semiquinone; loss of the second electron results in the yellow color of the quinone form of riboflavin (see Fig. 17.8). It is evident in this case that the electrons are transferred one at a time from a riboflavin molecule to an electron acceptor. The color changes can be followed spectrophotometrically and the concentration of the free radical ascertained by this means (Blois, 1961).

A similar oxidation-reduction reaction occurs in the prosthetic group of the flavoprotein enzyme. This enzyme is colorless in the reduced form and yellow when oxidized. Here too it is suggested that electrons need be transferred one at a time from flavoprotein enzymes to an electron acceptor (cytochrome B). Since proteins stabilize free radicals of compounds attached to them, flavoprotein is particularly favorable as an electron transporting and coupling mechanism.

No problem is posed by transfers among cytochromes B, C, A and $A_3$ (cytochrome oxidase), because in each case a single electron is passed and only two molecules need collide. The activation of oxygen, however, involves two electrons, but at a given time only one oxygen is coupled with cytochrome oxidase (see Chapter 16 and Fig. 16.9). If the organized systems of enzymes on the cristae of the mitochondria were to serve as semi-

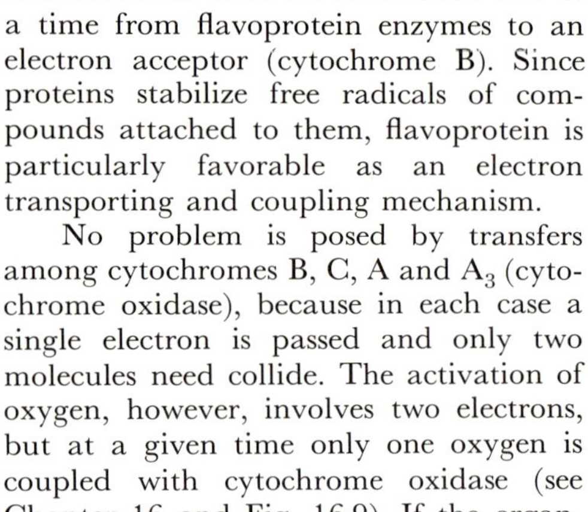


***Figure 17.7.*** Free radical formation (semiquinone) in the oxidation of hydroquinone to quinone. The oxidation occurs in two steps, one hydrogen ion and electron coming off at a time, the semiquinone having an unpaired electron which is paired after the second hydrogen ion and electron are removed.

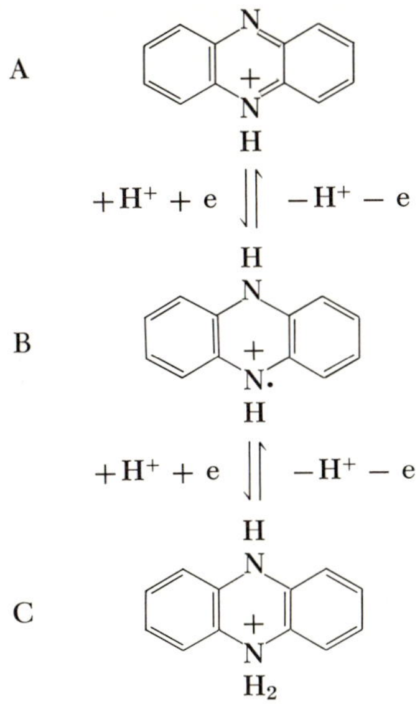

*Figure 17.8.* Free radical formation with single electron and hydrogen ion transfer in riboflavin. Note the change in double bonds in the middle ring as electrons and hydrogen ions are transferred. *A*, The quinone; *B*, the sequinone; and *C*, the hydroquinone. (After Leach, 1954: Advances Enzymol., *15*:1.)

conductors, it would be unnecessary to postulate sequential collisions between enzymes and substrate molecules (Leach, 1956).

## DETECTION OF FREE RADICALS BY THEIR MAGNETIC PROPERTIES

A spinning electron generates a magnetic field. In atoms or molecules the electrons are paired and spinning in opposite directions; therefore their magnetic fields cancel out. A free radical, having an unpaired electron, is paramagnetic; that is, it has a magnetic moment because of the unpaired electron spin. Application of a magnetic field of appropriate strength to such a substance aligns its magnetic axes in the direction of the field at the expense of energy from the magnetic field. By measuring the absorption of energy from the magnetic field, information can be obtained about the kinds of free radicals present and their relative concentrations.

Thus, in 1938, Michaelis brought a reaction mixture suspected of having free radicals between the poles of an electromagnet (Fig. 17.9), and after bringing it to a balance point on a sensitive balance, he turned on the current to the electromagnet. The balance point was displaced. The degree of displacement of the balance point, as measured by the change in

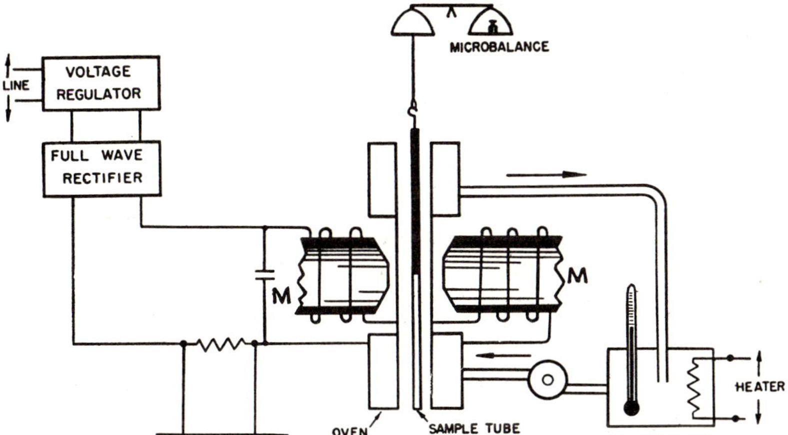


*Figure 17.9.* Balance method for measurement of paramagnetism of free radicals. The Gouy apparatus has a thermostat for the sample and means for control and measurement of the magnetizing current. M refers to the electromagnet. (From Blois, 1956: *In* Physical Techniques in Biological Research. Vol. 2. Oster and Pollister, eds. Academic Press, New York.)

weight of the reaction mixture before and after application of the magnetic field, is proportional to the concentration of free radicals present. This method is suitable only when free radicals are produced in slow reactions and in relatively high concentrations (Chance, 1961). However, it is possible to refine the method for use in rapid reactions and the method, so modified, is still in use (Brill, 1961; Chance *et al.*, 1962).

A more sensitive and more rapid method for detecting the magnetic moments of unpaired electrons of free radicals is the electron paramagnetic resonance (EPR) or electron spin resonance (ESR) method. Here detection of free radicals depends upon the fact that in a magnetic field the orientation of a free electron with the field differs by a significant and measurable amount of energy from its orientation against the field. The electron, spinning as it does continually, generates a magnetic field. Application of an external magnet causes the electron to precess, just as tipping from the vertical causes a gyroscope (or a top) to precess rather than tip over. The precession of the electron is in the microwave frequency range (about 3 cm. or 10,000 megacycles per second). When microwaves travel down a rectangular waveguide (the tube used to guide such waves) they produce a rotating magnetic field at any fixed point. The material to be tested is placed in a side wall of the waveguide (Fig. 17.10), the radio waves are turned on and the external magnetic field is applied to make the electrons precess. When the precession rate reaches the resonance value and the electrons flip to align their fields with the applied magnetic field, they extract energy from the radio waves and the reading on a receiver at the end of the tube dips accordingly (Fig. 17.11).

The energy involved in causing the flip is given by the following equation:

$$h\nu = g\beta H \qquad (17.8)$$

where $h\nu$ is the quantum of electromagnetic energy; h is Planck's constant ($6.624 \times 10^{-27}$ erg seconds); $\nu$ is the frequency per second; g is the spectroscopic splitting factor which is 2.0023 for an electron exhibiting pure "spin only" paramagnetism; $\beta$ is the Bohr magneton $= 0.927 \times 10^{-20}$ erg/gauss; and H is the magnetic field strength in gauss.

In most organic free radicals a small interaction between the electron spin and the electron orbit causes the g-value to deviate from the spin-only value of 2.0023. The g-value varies in the narrow range of 2.000 to 2.0070 for the materials of greatest importance to the cell. Each type of free radical shows a characteristic g-value within this range.

The main pattern traced on the record of ESR on the receiver is from the unpaired electron in the free radical, but the fine structure is the result of the interaction of its field with the fields of

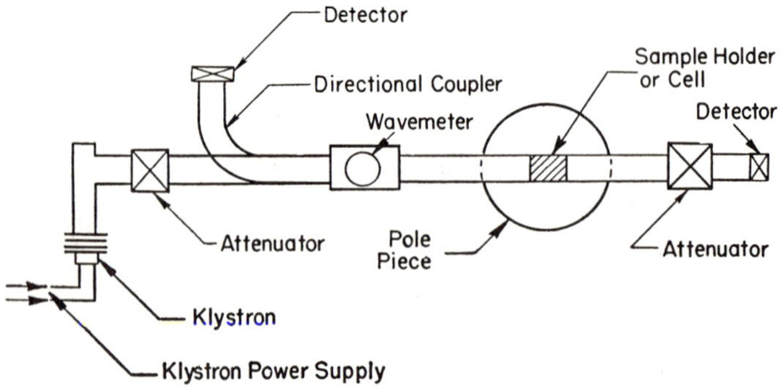


***Figure 17.10.*** Typical experimental arrangement used to observe electron paramagnetic resonance at microwave frequencies. The klystron is the source of the waves. The sample holder is in the side wall of the waveguide. (From Blois, 1956: *In* Physical Techniques in Biological Research. Vol. 2. Oster and Pollister, eds Academic Press, New York.)

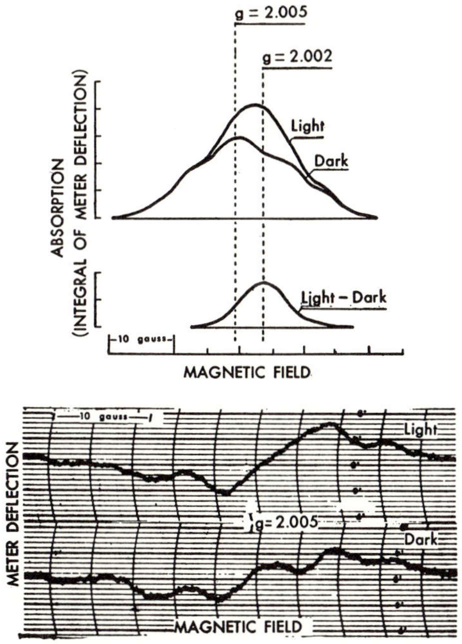


***Figure 17.11.*** Electron spin resonance from spinach chloroplasts washed twice with 0.5M sucrose solution. The lower curves represent the actual spectrometer records, and the upper curves represent the integral curves derived from them. (From Commoner, 1957: Science *126*:4.)

certain surrounding atomic nuclei. Local fields produced by the nuclei may be strong enough to "split" the electron resonance. It was hoped that this "hyperfine" structure would "fingerprint" free radicals in biological systems and permit their identification, but really well resolved patterns have not, thus far, characterized biological systems although there is consistent improvement (Weaver, personal communication).

Free radicals have been found much more widespread in chemical and biological systems than had previously been realized. Free radicals have been detected not only in oxidation-reductions but also in a large number of other reactions of biological interest such as photochemical reactions, thermal cleavage of molecules in chemical systems, photosynthesis, bioluminescence and enzymatic reactions in cells, and as a result of action of radiations upon cells (Ehrenberg and Ludwig, 1958; Commoner and Hollocher, 1960; Gutowsky and Nachod, 1958; Beinert and Sands, 1961).

Biological studies of free radicals initially had to be made with lyophilized (low temperature, vacuum-dried) cells because water absorbs so strongly the microwaves used for detection of free radicals. However, it has proved possible to make the measurements with frozen biological materials, thus enabling determination of the free radicals present within the materials at the moment of freezing. Free radicals are usually short lived, but in a frozen system they last for a longer time and concentrations of free radicals of the order of $10^{-6}$ to $10^{-7}$ M have been found in intact frozen cells (Herzfeld and Bass, 1957; Gutowsky and Nachod, 1958; Blois *et al.,* 1961). Recent improvements in physical methods have made possible the detection of free radicals in unfrozen, undried biological systems, opening the possibility of gathering data during the course of biological reactions, rather than only at the moment of freezing (Blois and Weaver, 1964).

An example of results obtained with ESR analysis is shown in Figure 17.11 for chloroplasts (from spinach leaves) carrying on photosynthesis. It seems that excitation of the photosynthetic system in chloroplasts by absorption of light results in the appearance of free radicals which interact with the magnetic field to give the resultant curve (light deflections minus dark deflections).

## DETERMINATION OF REDOX POTENTIALS IN THE LIVING CELL

Cell activities generally influence the redox potentials of the environment by reducing the oxygen tension and shifting the potential toward a more negative value. Changes in redox potentials, whether naturally or artificially produced, profoundly influence growth and differentiation of some cells (Cater, 1960).

It is of some interest to measure the

redox potential at which a whole cell, or some part of it, is poised when subjected to various environmental conditions. A number of approaches have been used to measure intracellular redox potentials. Some investigators, for instance, have inserted microelectrodes into a cell, but satisfactory electrodes are not easy to make, and it is not possible to be certain that the cell remains in healthy condition (Cater, 1960).

Another method used for measuring redox potentials of cells is to inject into a living cell some dye which might act as an oxidation-reduction indicator, and to determine from the color change whether the dye is reduced or oxidized. By injecting, in succession, dyes covering a range of $E_0'$ values, the likely $E_0'$ value of the cell may be determined. Of course, the $E_0$, $E_0'$ and titration curves must first be determined separately for the various oxidizable dyes used. For example, at pH 7.0 methylene blue has an $E_0'$ of +0.011 volt, pyocyanin of −0.034 volt and indophenol just below the $E_0'$ of cytochrome C (−0.26 volt). As might be expected, in the absence of oxygen the overall redox potential of a cell was found to be more negative than in the presence of oxygen (Hewitt, 1950).

Since the cytoplasm of the cell is a polyphasic colloid and has many particles, organelles and surfaces upon or within which many enzymatic reactions may occur independently at the same time, it would be of interest to study the potentials of such isolated fragments. A few such studies have been made, particularly with mitochondria, which are active centers of aerobic processes (Cater, 1960).

In summary, oxidation-reduction studies on biological systems indicate that the sequence of reactions in the cell occurs along a most probable biological electromotive series. Not only enzyme kinetics but also enzyme thermodynamics appear to be in accord with physical-chemical principles. The fundamental basis of enzymatic action is well established, although much remains to be learned about the manner in which enzymes carry out their functions (Anfinsen and Kielley, 1954; Green and Beinert, 1955).

## APPENDIX

### 17.1 DERIVATION OF PETER'S EQUATION FOR OXIDATION-REDUCTION POTENTIALS

To derive Peter's equation (17.6), it is necessary to equate the mechanical work to the electrical work done in reducing 1 gram mole of metal:

$$\text{Metal} \rightleftharpoons e_{\text{solution}} + \text{ion} \quad (17.2)$$

If for purposes of argument the electron e is considered to act as a gas particle, the formulation of laws applying to gases will apply to electrons which are at pressure $e_{\text{metal}}$ in the metal and $e_{\text{solution}}$ in the solution.

If a gas is allowed to expand slowly in a cylinder by pushing a piston over the distance l (an isothermal, reversible, ideal expansion), an equilibrium existing all the time, the work done (W) is the force (F) times the distance (l):

$$W = Fl \quad (17.9)$$

If the piston moves only the distance (dl) the work done (dW) equals:

$$dW = Fdl \quad (17.10)$$

Since $$P = F/\text{area} \quad (17.11)$$

then $$F = P \cdot \text{area} \quad (17.12)$$

Substituting this value of F into equation 17.10:

$$dW = P \cdot \text{area} \cdot dl = PdV \quad (17.13)$$

since area × distance = volume.

According to the gas law,

$$PV = RT \text{ or } P = RT/V \quad (17.14)$$

Substituting this value of P into equation 17.13 gives:

$$dW = \frac{RT}{V}\, dV, \text{ or } RT\,\frac{dV}{V} \quad (17.15)$$

Integrating the equation between $V_1$ and $V_2$ gives:

$$W = RT\, ln\, \frac{V_2}{V_1} \quad (17.16)$$

Here R is the gas constant (8.315 joules per degree), T the absolute temperature

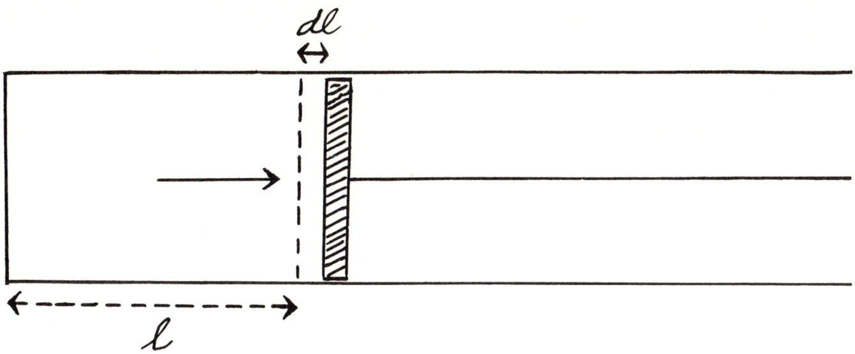


*Figure 17.12.*

and *ln* the natural logarithm (2.302 × $\log_{10}$).

But $$V_2 = \frac{RT}{P_2}$$

and $$V_1 = \frac{RT}{P_1} \qquad (17.17)$$

Substituting these values of $V_2$ and $V_1$ in equation 17.15,

$$W = RT\ ln\ \frac{RT/P_2}{RT/P_1} = RT\ ln\ \frac{P_1}{P_2} \qquad (17.18)$$

Since for purposes of the derivation the electrons are considered to act as gas particles, they should exert pressure. If the pressure of the electrons in the metal, $e_{metal}$ is taken as equal to $P_1$ and $e_{solution}$ as equal to $P_2$, substitution of these values in equation 17.18 gives:

$$W = RT\ ln\ \frac{e_{metal}}{e_{solution}} \qquad (17.19)$$

The electrical work (W) done in the same process is a product of the valency of the metal, the Faraday or quantity of electricity transferred, and the potential against which the work is done:

$$W = n\mathcal{F}E \qquad (17.20)$$

Here n is the valency of the metal, $\mathcal{F}$ the Faraday (96,000 coulombs) and E the potential in volts. Since the work for reduction of a gram mole of metal is the same whichever way it is calculated, then:

$$n\mathcal{F}E = RT\ ln\ \frac{e_{metal}}{e_{solution}} \qquad (17.21)$$

$$E = \frac{RT}{n\mathcal{F}}\ ln\ \frac{e_{metal}}{e_{solution}} \qquad (17.22)$$

$$E = \frac{RT}{n\mathcal{F}}\ ln\ e_{metal} - \frac{RT}{n\mathcal{F}}\ ln\ e_{solution} \qquad (17.23)$$

If $RT/n\mathcal{F}$ *ln* $e_{metal}$ is set at C′, since it is a property characteristic of each metal, then:

$$E = C' - \frac{RT}{n\mathcal{F}}\ ln\ e_{solution} \qquad (17.24)$$

However, since metal = $e_{solution}$ + ion, or, in more general terms:

$$\text{Reductant} = e_{solution} + \text{oxidant} \qquad (17.25)$$

then by the law of mass action:

$$\frac{[e_{sol.}][ox.]}{[red.]} = K \qquad (17.26)$$

$$[e_{sol.}] = K\frac{[red.]}{[ox.]} \qquad (17.27)$$

Substituting 17.27 into 17.24,

$$E = C' - \frac{RT}{n\mathcal{F}}\ ln\ \frac{[red.]}{[ox.]} - \frac{RT}{n\mathcal{F}}\ ln\ K \qquad (17.28)$$

Since $C' - (RT/n\mathcal{F})$ *ln* K is a constant for any metal (or oxidation-reduction system) it may be set equal to $E_0$:

$$E_0 = C' - \frac{RT}{n\mathcal{F}}\ ln\ K \qquad (17.29)$$

Then substituting this value of $E_0$ in equation 17.28 gives:

$$E = E_0 - \frac{RT}{n\mathcal{F}}\ ln\ \frac{[red.]}{[ox.]} \qquad (17.30)$$

or

$$E = E_0 + \frac{RT}{n\mathcal{F}} ln \frac{[ox.]}{[red.]} \quad (17.31)$$

Peter's equation may be simplified if it is applied under a special set of conditions, for example, at 30° C., or 303° absolute, and with n equal to 2, as was done in equation 17.6,

$$E = E_0 + 0.030 \log_{10} \frac{[oxidant]}{[reductant]} \quad (17.6)$$

Also, by setting,

$$[oxidant] = [reductant],$$

in which case,

$$\log_{10} \frac{[oxidant]}{[reductant]} = O,$$

then: $E = E_0$

$E_0$ values for various oxidation systems are given in Table 17.2. Comparisons of the redox potentials of compounds are most frequently made on this basis.

## 17.2 EQUATION FOR THE EFFECT OF pH ON OXIDATION-REDUCTION POTENTIALS

If hydrogen ions are produced in the course of an oxidation-reduction reaction, a second term must be added to Peter's equation. Thus:

$$\underset{\text{Reductant in solution}}{H_2Q} \rightleftharpoons \underset{\text{Oxidant in solution}}{Q} + 2H^+ + \underset{\text{On electrode}}{2e} \quad (17.32)$$

The equation corresponding to this reaction is:

$$E = E_0 + \frac{RT}{n\mathcal{F}} ln \frac{[oxidant]}{[reductant]} + \frac{RT}{\mathcal{F}} ln [H^+] \quad (17.33)$$

If we substitute numerical values for R, n (the valency change, taken as 2 for organic redox systems since two electrons are generally involved, and as 1 for hydrogen since only one electron is involved), $\mathcal{F}$ and T (take T as 30° C., or 303° absolute) and give the equation in terms of $\log_{10}$, the equation becomes:

$$E = E_0 + 0.03 \log_{10} \frac{[oxidant]}{[reductant]} - 0.06 (pH) \quad (17.34)$$

## 17.3 MEASUREMENT OF pH BY THE QUINHYDRONE ELECTRODE

Because of the relation between E and pH in equation 17.34, it is possible to use some oxidation-reduction systems to measure the pH. In practice this is seldom done since the glass electrode is so much more convenient. However, it is instructive to discuss the use of the quinhydrone electrode for this purpose because it illustrates the linear response of the oxidation-reduction potential to pH. Quinhydrone, an equimolar mixture of hydroquinone and quinone, is added to the solution of unknown pH to be tested, and the potential, E, between the platinum electrode and the calomel electrode is determined. $E_0$ for quinhydrone is known, and the second term in equation 17.34 is zero because the concentration of oxidant equals the concentration of reductant; therefore the pH is readily calculated:

$$E = E_0 - 0.06 (pH) \quad (17.35)$$

It is even more convenient to read the pH from a chart relating pH to E values for quinhydrone, as shown in Figure 17.5.

## 17.4 USE OF A CALOMEL ELECTRODE INSTEAD OF A HYDROGEN ELECTRODE

The hydrogen electrode is always the baseline against which all potentials are measured whether at pH 0.0 for metals or at pH 7 for systems of biological origin. In practice in the laboratory, however, a calomel electrode, with a known potential against the hydrogen electrode, is used instead of a hydrogen electrode, because the latter is so inconvenient. The saturated calomel electrode is nonpolarizable* and has a potential of +0.2458 volt against

*No gases are formed during its operation.

the normal hydrogen electrode at pH 7.0. When the potential of a biological oxidation-reduction system (X) is determined against a calomel electrode, the reading obtained for X must be corrected by adding algebraically to it the potential difference of the calomel electrode against the hydrogen electrode. For example, if the potential of X against the calomel electrode is $-0.1742$ volt, its true potential against the hydrogen electrode is $+0.2458$ volt plus $-0.1742$, or $+0.0716$ volt.

## LITERATURE CITED AND GENERAL REFERENCES


Anfinsen, C. B., and Kielly, W. W., 1954: Biological oxidations. Ann. Rev. Biochem. *23*:17–54.

Barnothy, M. F., 1964: Biological Effects of Magnetic Fields. Plenum Publishing Corp., New York.

Barron, E. S. G., 1952: The mechanism of enzymatic oxidation reduction. *In* Trends in Physiology and Biochemistry. Barron, ed. Academic Press, New York, pp. 1–24.

Beinert, H., and Sands, R. H., 1961: Semiquinone formation of flavins and flavoproteins. *In* Free Radicals in Biological Systems. Blois, ed. Academic Press, New York, pp. 17–52.

Blois, M. S., 1956: Magnetic methods. *In* Physical Techniques in Biological Research. Oster and Pollister, eds. Academic Press, New York, Vol. 2, pp. 393–440.

Blois, M. S. (ed.), 1961: Free Radicals in Biological Systems. Academic Press, New York.

Blois, M. S., and Weaver, E. C., 1964: Electron spin resonance and its application to photophysiology. *In* Photophysiology. Giese, ed. Academic Press, New York, Vol. 1, pp. 35–63.

Brill, A., 1961: The detection of free-radical intermediates in biochemical reactions by their magnetic susceptibility. *In* Free Radicals in Biological Systems. Blois, ed. Academic Press, New York, pp. 53–74.

*Bull, H. B., 1964: An Introduction to Physical Biochemistry. 3rd Ed. F. A. Davis, Philadelphia.

*Cater, D. B., 1960: Oxygen tension and oxidation-reduction potential in living tissues. Prog. Biophys. *10*:153–194.

Chance, B., 1961: Free radicals and enzyme-substrate compounds. A tribute to Lenor Michaelis. *In* Free Radicals in Biological Systems. Blois, ed., Academic Press, New York, pp. 1–16.

Chance, B., Cohen, P., Jobsis, F., and Schoener, B., 1962. Intracellular oxidation reduction states in vivo. Science *137*:499–506.

Clark, W. M., 1952: Topics in Physical Chemistry. 2nd Ed. Williams & Wilkins, Baltimore.

Clark, W. M., 1960: Oxidation Reduction Potentials of Organic Systems. Williams & Wilkins, Baltimore.

Clark, W. M., *et al.,* 1928: Studies on Oxidation-Reduction. Hygienic Lab. Bull. No. 151.

Commoner, B., and Hollocher, T. C., Jr., 1960: Free radicals in heart muscle mitochondrial particles: general characteristics and localization in electron transport system. Proc. Natl. Acad. Sci. U.S.A. *46*:405–416.

Commoner, B., Lippincott, P., Norberg, R. E., Heise, J. J., Passaneua, J. V., and Townsend, J., 1957: Biological action of free radicals. Science *126*:57–63.

Dechev, G., and Moscona, A., 1964: Possible identity of redox and bioelectric potentials in live systems (*Vicia faba*). Comp. Rend. Acad. Bulgare Sci. *17*: 159–162.

Ehrenberg, A., and Ludwig, G. D., 1958: Free radical formation in reaction between old yellow enzyme and reduced TPN. Science *127*:1177–1178.

Glasstone, S., and Lewis, D., 1960: Elements of Physical Chemistry. 2nd Ed. D. Van Nostrand Co., New York.

Goddard, D., 1945: Respiratory enzymes. *In* Physical Chemistry of Cells and Tissues. Höber, ed. Blakiston Co., Philadelphia, pp. 397–420.

Green, D. E., and Beinert, H., 1955: Biological oxidation. Ann. Rev. Biochem. *24*:1–44.

Gutowsky, H. S., and Nachod, F. C. (eds.), 1958: Nuclear magnetic resonance. Ann. N. Y. Acad. Sci. *70*:764–930.

Herzfeld, C. M., and Bass, A. M., March 1957: Frozen free radicals. Sci. Am. *196*:90–104.

Hewitt, L. F., 1950: Oxidation-Reduction Potentials in Bacteriology and Biochemistry. 2nd. Ed. Williams & Wilkins, Baltimore.

Hollocher, T. C., Jr., and Commoner, B., 1960: Mechanism of free radical formation. Proc. Natl. Acad. Sci. U.S.A. *46*:416–427.

Kochi, J. K., 1967: Mechanisms of organic oxidation and reduction by metal complexes. Science *155*:415–424.

Leach, S. J., 1956: The mechanism of enzymic oxidoreduction. Advances Enzymol. *15*:1–47.

Lewis, G. N., and Randall, M., 1961: Thermodynamics. 2nd Ed. McGraw-Hill Book Co., New York.

Michaelis, L., 1930: Oxidation-Reduction Potentials. J. B. Lippincott Co., Philadelphia.

Michaelis, L., 1935: Semiquinones, the intermediate steps of reversible organic oxidation-reduction. Chem. Rev. *16*:243–286.

Michaelis, L., 1940: Occurrence and significance of semiquinone radicals. Ann. N. Y. Acad. Sci. *40*:39–76.

Pake, G. E., 1961: Observation and interpretation of paramagnetic resonances: A review. *In* Free Radicals in Biological Systems. Blois, ed. Academic Press, New York, pp. 91–100.

Podolsky, R. J., and Morales, M. F., 1956: The enthalpy change of ATP hydrolysis. J. Biol. Chem. *218*:945–959.

Szent-Györgyi, A., 1956: Bioenergetics. Science *124*:873–875.

Weaver, E. C., and Chon, H., 1966: Spin label studies in Chlamydomonas. Science *153*:301–303.

West, E. S., 1942: Physical Chemistry for Students of Biochemistry and Medicine. Macmillan Co., New York.

# CHAPTER 18 / CELLULAR RESPIRATION, FERMENTATION AND LUMINESCENCE

*Cellular respiration* is the aerobic oxidation of nutrients in the cell with the help of atmospheric oxygen, in the course of which energy is liberated and carbon dioxide is given off. In *glycolysis* and *fermentation,* nutrients are oxidized in the absence of atmospheric oxygen, that is, anaerobically. Organisms that live in the presence of oxygen are *aerobes;* those that live in its absence are *anaerobes.*

## AEROBIC METABOLISM: OXYGEN CONSUMPTION OF CELLS

The oxygen consumption of cells is measured by the volume of oxygen consumed (at standard temperature and pressure, STP) per unit weight of organism per unit of time. This is called the $Q_{O_2}$ or rate of respiration. The units of measurement used are microliters ($\mu$l.) per milligram per hour for small samples or milliliters (ml.) per gram per hour for larger samples. Data could be cited either for wet or dry weight; the latter is approximately one-fourth of the former (see Table 18.1).

Although many methods have been used for measurement of respiration, only one will be described here. For a description of others, see Appendix 18.1. As seen in Fig. 18.1, a suspension of cells is placed in vessel A, the center well of which contains potassium hydroxide. Vessel B contains no cells and serves as a compensator. The two vessels are connected to each other through a manometer, which is set at the zero position (level) at the start of an experiment. As oxygen is consumed, the pressure in the closed system falls and the level on the right side of the manometer (as shown in the figure) rises because carbon dioxide, given off during the respiration, is absorbed by the potassium hydroxide. The volume of oxygen consumed can be determined by adding enough air from the calibrated syringe to just compensate for the change in level of the manometer. To insure equilibrium, the vessels should be shaken, preferably in a constant temperature bath.

Unicellular organisms, such as bacteria, yeast, some molds, unicellular algae, small colonial algae, protozoans, marine eggs, sperm and red blood cells, lend themselves especially well to such studies since suspensions of them are readily pre-

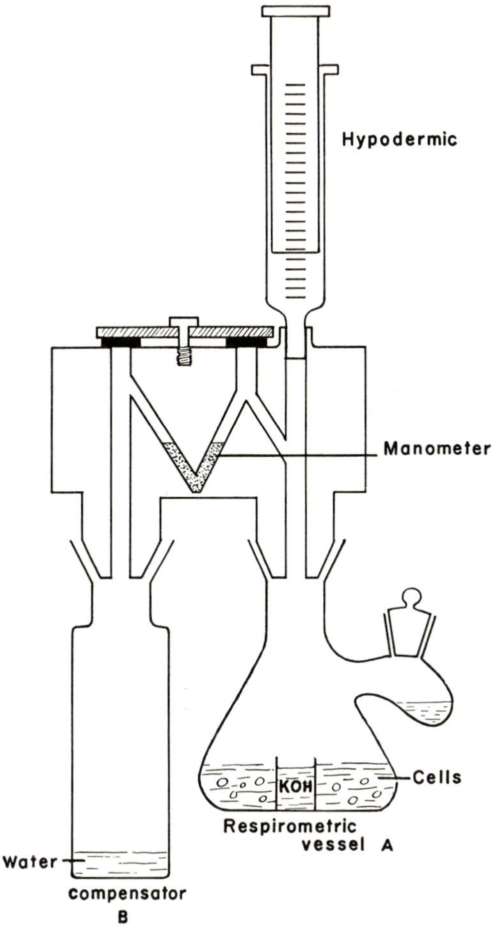


***Figure 18.1.*** The Scholander-Wennesland modification of the Fenn-Winterstein respirometer. Usually a calibrated screw plunger is used instead of a hypodermic syringe. (After Wennesland, 1951: Science *114*:100.)

pared and handled. Compact tissues, however, must be cut in thin slices before the respiration of the cells can be studied in respirometric flasks. In this manner liver and brain slices have been extensively studied, and to a lesser extent sections of leaf, stem and root.

When data gathered in this manner are compared, it becomes immediately apparent that the $Q_{O_2}$ of diverse organisms and tissues (Table 18.1) varies widely. Some microbes, like *Azotobacter* and *Bacillus fluorescens,* respire more actively than any of the others listed, and yeast and mold respire more actively than protozoans and cells from animal tissues.

It is of interest to find out what connection exists between rates of respiration and sizes of various cells. If we compare the respiratory rates of the unicellular organisms in Table 18.1, we will see that the rate of respiration varies with the ratio of surface to volume of the cell. Microbes, being of small linear dimensions, have a larger surface in relation to volume than do the larger cells. When cell size increases, the $Q_{O_2}$ declines (Zeuthen, 1953). The range in size among protozoans is ample to illustrate this point clearly (see Fig. 18.2). Thus the increase in the respiratory rate of the large multinucleate ameba *Pelomyxa carolinensis* (*Chaos chaos*) over that of the small ameba *Amoeba proteus* is only about half the increase in bulk. Probably *Pelomyxa* represents the limit of effective increase in size of a single cell. In the course of evolution the increase in mass brought about by the aggregation of small cells proved more effective in maintaining a sufficiently high respiratory rate than could a corresponding increase in the mass of a single cell (Zeuthen, 1953).

When the $Q_{O_2}$ of larger animals and plants is compared with that of microbes, the contrast is even more striking (Table 18.1). Yet the respiration of the entire organism or tissue is the sum of the respiration of its component cells. One probable reason for this difference is the method used for comparison. To illustrate, respiration of microorganisms is generally measured in the presence of nutrient favorable for maximal activity. On the other hand, respiration of cells of a larger organism is usually measured during a relatively inactive state. When measured during maximal activity, their metabolic rates are 20-fold to 100-fold greater (Table 18.1), at which time their oxygen consumption comes within the range of the oxygen consumption shown by active microorganisms (Holmes, 1937; Stafford, 1963).

If respiratory rate is plotted against the size of organism, it will be seen that respiratory rate falls with increase in size. Even when an individual of a species increases in size during its development,

**Table 18.1.** Respiration Rates of Various Cells and of Organisms and Tissues ($Q_{O_2}$ is Given in Milliliters per Gram of Wet Weight per Hour)

| GROUP | ORGANISM | °C. | $Q_{O_2}$ | REMARKS |
|---|---|---|---|---|
| Microorganisms | *Bacillus mesentericus vulgatus** | 16 | 12.1 | |
| | *Azotobacter chroococcum*† | 28 | 500–1000 | |
| | *Bacillus fluorescens non liquefaciens*‡ | | 4100 | |
| | *Neurospora crassa*§ | 26 | 6.4 | |
| | *Saccharomyces cereviseae*‖ | 26 | 8–14.5 | |
| | Paramecium¶ | 20 | 0.5 | |
| Plants | *Verbascum thapsus** | 23 | 0.093 | leaf |
| | | | 0.204 | pistil |
| | | | 0.190 | stamen |
| | *Papaver rhoeas** | 22 | 0.803 | leaf |
| | | | 0.172 | pistil |
| | | | 0.280 | stamen |
| Invertebrates | *Anemonia sulcata*‡ | 18 | 0.0134 | |
| | *Asterias rubens*‡ | 15 | 0.03 | |
| | *Nereis virens*‡ | 15 | 0.026 | |
| | Mytilus¶ (mussel) | 20 | 0.02 | |
| | Astacus¶ (crayfish) | 20 | 0.047 | |
| | Vanessa¶ (butterfly) | 20 | 0.6 | at rest |
| | | | 100.0 | in flight |
| Vertebrate | carp¶ | 20 | 0.1 | |
| | mouse¶ | 37 | 2.5 | at rest |
| | | | 20.0 | running |
| | man¶ | 37 | 0.2 | at rest |
| | | | 4.0 | maximal work |
| Animal tissues | rat liver† | 37.5 | 2.2–3.3 | |
| | rat kidney cortex† | 37.5 | 5.2–9.0 | |
| | rat brain cortex† | 37.5 | 2.7 | |
| | rat voluntary muscle# | 37 | 1.5 | at rest |
| | | 37 | 10.0 | active |
| | frog nerve# | 15 | 0.02 | at rest |
| | | 15 | 0.75 | active |
| | rabbit nerve# | 37 | 0.29 | at rest |

All $Q_{O_2}$ values given as volume of oxygen per gram of dry weight per hour have been divided by 4 to give the value for wet weight.

* Stiles and Leach, 1952.
† Tabulae Biologicae, 1934:9.
‡ Heilbrunn, 1952.
§ Giese and Tatum, 1946: Arch. Biochem. 9.
‖ Giese and Swanson, 1947: J. Cell. Comp. Physiol. *30.*
¶ Krogh, 1941.
# Holmes, 1937.

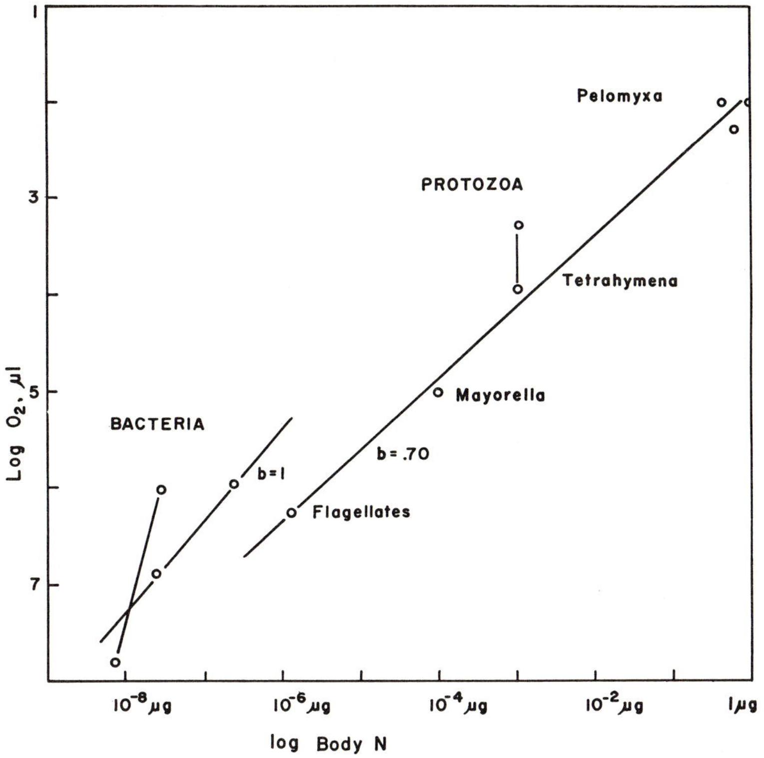


***Figure 18.2.*** The relation between cell size, as measured by the logarithm of the cell nitrogen content (N), and respiration, as measured by the logarithm of the volume of oxygen consumed by single celled organisms. Note that the points of protozoans fall along a straight line, the slope of which (b, the exponent in equation 18.1) is 0.70. This slope tells us that the rate of respiration increases much more slowly than the mass (size) of the protozoan. For bacteria, on the other hand, the slope (b) of the line is 1.0, indicating that over the small size range covered by the experiments the respiration increases proportionally with cell size. The higher value for b indicated in far left line has questionable significance. (After Zeuthen, 1953: Quart. Rev. Biol. *28*:3.)

its rate of respiration ($Q_{O_2}$) falls (see Fig. 18.3). In warm blooded animals it is clear that the greater the ratio of surface area to volume, the greater is the heat loss and the greater is the rate of respiration. For example, the shrew loses so much heat because of its large surface-to-volume ratio that it must eat continually to get enough food to keep warm. The hummingbird has solved this problem by becoming torpid during sleep, its temperature falling almost to that of the environment. Consequently, a higher metabolic rate is expected in small mammals or birds than in larger ones. A limit is thus set to the minimal size of a warm blooded form. Whether loss of heat is also the cause of the greater respiratory rates found in small invertebrates is not certain at the present time (Pearson, 1953, 1954).

The oxygen consumed by an animal per unit time can be related to its surface and volume by means of an empirical equation:

$$O_v = KW^b \qquad (18.1)$$

In this equation $O_v$ is the volume of oxygen consumed per unit time, K is a constant, W is the weight of the organism and b is an exponent. For most animals tested, $b = 0.73$. Since $O_v$ is the $Q_{O_2}$ times the weight of the animal, equation 18.1 indicates that the $Q_{O_2}$ is related to the 0.73 power of W (Krogh, 1941).

One reason for the decrease in $Q_{O_2}$

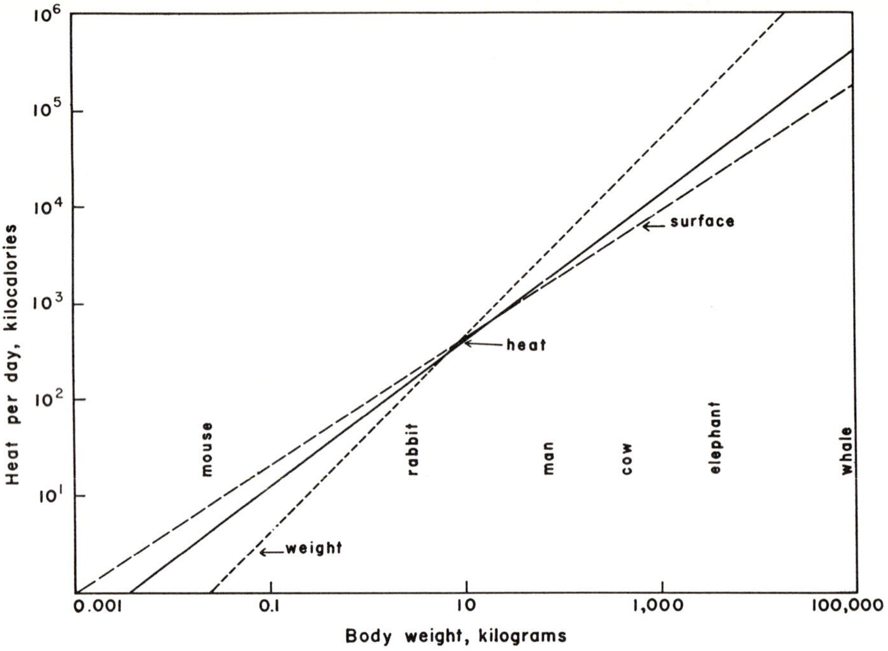


***Figure 18.3.*** Relation between respiratory rate (measured as heat per day) and size. Heat is shown by the solid line; surface is shown by the broken line with longer segments; weight is shown by the broken line with shorter segments. Note that respiratory rate is more closely correlated with surface than with weight. (After Kleiber, 1947: Physiol. Rev. *27*:530.)

with increase in size is perhaps the corresponding increase in the larger organism's content of inert materials. According to the principle of similitude (Bonner, 1952), the volume of an animal increases as the cube of its linear dimensions, whereas the strength of bones and other connective tissue increases only as the square of the animal's linear dimensions. Consequently, the mass of bone, cartilage and fibrous connective tissue increases disproportionally as the animal increases in size. Such tissues are, metabolically speaking, relatively inert and contribute little to the $Q_{O_2}$, which is primarily a function of the active tissues. Similarly, in plants, cell walls and supporting tissue and vascular tubes ("dead wood") add bulk without contributing metabolizing tissue.

If the cells could be removed from the tissues of animals (or plants) regardless of size, individual cells would probably respire at much more nearly the same rate. This is apparent from studies on respiration of cell suspensions growing actively in tissue culture (Umbreit *et al.,* 1964).

## THE EFFECT OF TEMPERATURE ON RESPIRATORY RATE

The respiratory rate of a suspension of cells increases with a rise of environmental temperature within the limits of the biokinetic zone (see Chapter 10). The *temperature coefficient* ($Q_{10}$) of respiration is generally found to be in the range of 2 to 4, indicating that a 10° C. rise in temperature increases the rate of the reaction twofold to fourfold (see Fig. 18.4). However, exposure to temperatures higher than the optimum in the biokinetic zone may result in injury, after which the respiration declines.

The respiratory rates of multicellular

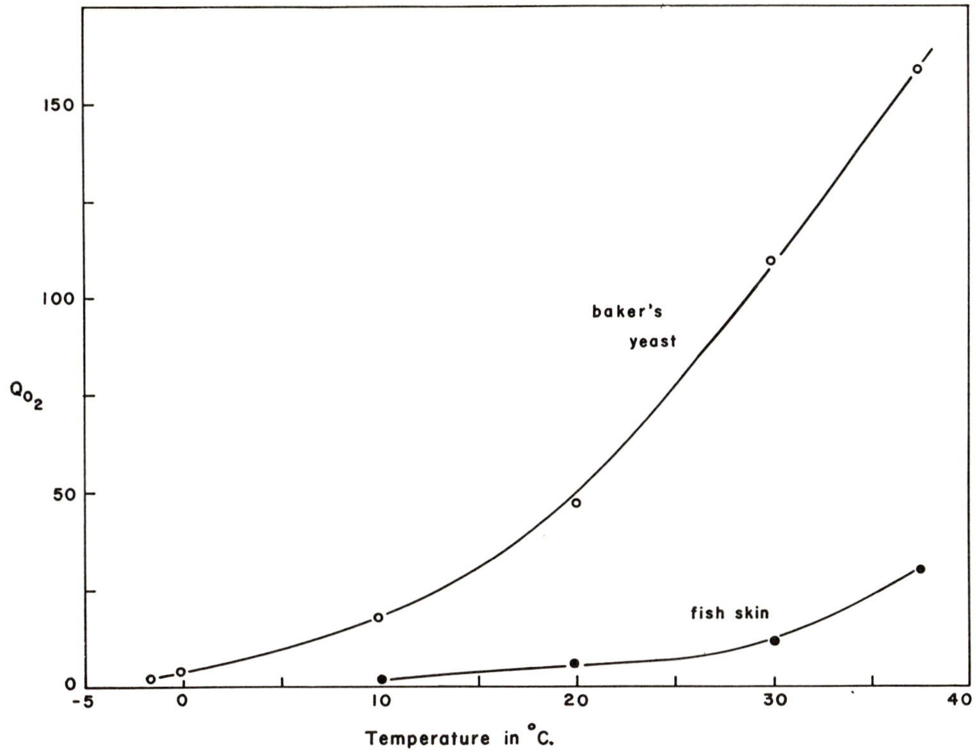


*Figure 18.4.* Relation between temperature and respiratory rate in fish skin and yeast. (Data from Tabulae Biologicae, 1934, *9*:223.)

plants and cold blooded multicellular animals increase in a characteristic way when the temperature of the environment rises within the limits of the biokinetic zone. The respiratory rates of young mammals, in which the "thermostatic" mechanism is not yet in operation, vary with temperature in the same way as the respiratory rates of cold blooded organisms. The cells in the tissues therefore appear to respond to temperature in much the same way as suspensions of cells. However, the rates of respiration of mature warm blooded animals do not vary with temperature in the same manner because their body temperature tends to remain constant, and a fall in environmental temperature may even lead to increased respiration, to release the extra heat needed to maintain a constant temperature.

## THE EFFECT OF OXYGEN TENSION ON RESPIRATION

The relation between oxygen tension (see Chapter 7) and cellular respiration is of considerable interest, since oxygen must enter all living cells in dissolved state. For example, when the oxygen tension of a liquid containing luminous bacteria in suspension falls to 3 per cent of that in the atmosphere, the respiratory rate of the bacteria begins to fall. It declines rapidly to one-half the normal rate when the oxygen tension has fallen to 0.25 per cent of atmospheric oxygen, as shown by the data in Figure 18.5. The curve plotting the relation between respiratory rate and oxygen tension is similar to the curve relating the amount of oxygen adsorbed by charcoal at different gas pressures. This suggests that adsorption of

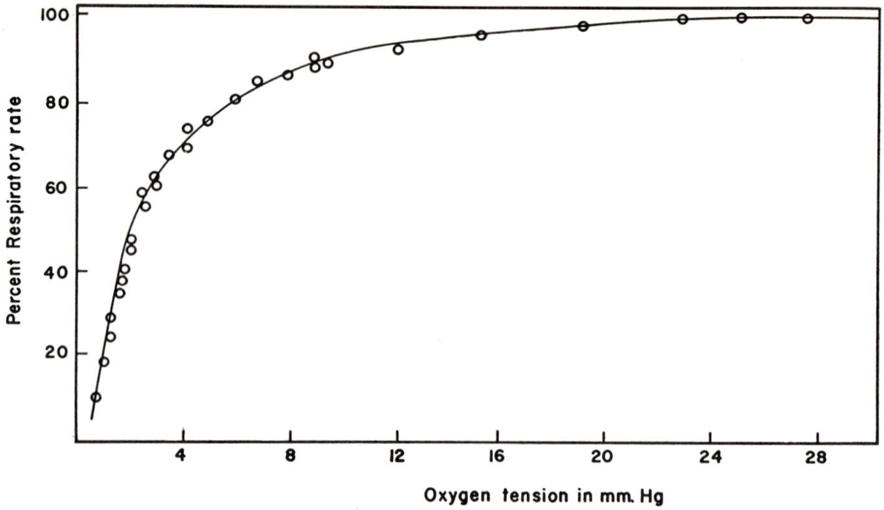


*Figure 18.5.* Relation between oxygen tension and the respiratory rate of luminous bacteria. Dimming occurs when the respiratory rate falls to 50 per cent. (From Shoup, 1929–1930: J. Gen. Physiol. *13*:41.)

oxygen perhaps underlies the relation between cell respiration and oxygen tension. It is suggested that, for maximal respiration, oxygen sufficient to saturate the active surfaces of the cytochrome oxidase must be present (Harvey, 1940).

## RESPIRATORY INHIBITORS

Many substances inhibit respiration (see Fig. 18.6). Some data on such narcotics have already been discussed in connection with permeability studies (see Chapter 13), and the data in Table 13.5 correlate the lipoid solubility of a number of narcotics with their effects on the respiration of cells. The excellent correlation suggests that a narcotic must penetrate the cell to exert its action on respiration (Danielli, 1950).

Other substances that have been studied for their effects on respiration are cyanide, azide and ethyl urethan. Cyanide binds the iron group in cytochrome oxidase, preventing transfer of electrons to oxygen, thereby preventing activation of oxygen. Consequently, it depresses respiration even when present in relatively low concentrations (about 0.001 molar). Azide resembles cyanide in its action. Ethyl urethan is much less specific and interferes with a large number of enzymes in the cell; consequently, a relatively high concentration (from 0.1 to 0.5 molar) is necessary to reduce respiration. Some enzymes, e.g., dehydrogenases and luciferases, are more sensitive to urethan than are other enzymes.

## THE EFFECT OF WATER ON RESPIRATION

Dehydration depresses respiration, and it is difficult to demonstrate oxygen consumption in dormant dehydrated stages of organisms such as cysts, spores or seeds. The decline in the respiratory rate of a single protozoan undergoing encystment, and presumably dehydration, has been studied in a microrespirometer (see Fig. (18.7). Conversely, it has been found that the respiratory rate increases as water is absorbed during excystment. Similar experiments with wheat seeds show that as the percentage of water content in the wheat kernels increases above 14.75 per cent, the respiratory rate rapidly increases. This suggests that perhaps below this value all the water is bound, whereas above it, free water is present in appreciable amounts for the resumption of cell activities (Bailey and Gurjar, 1910). Ex-

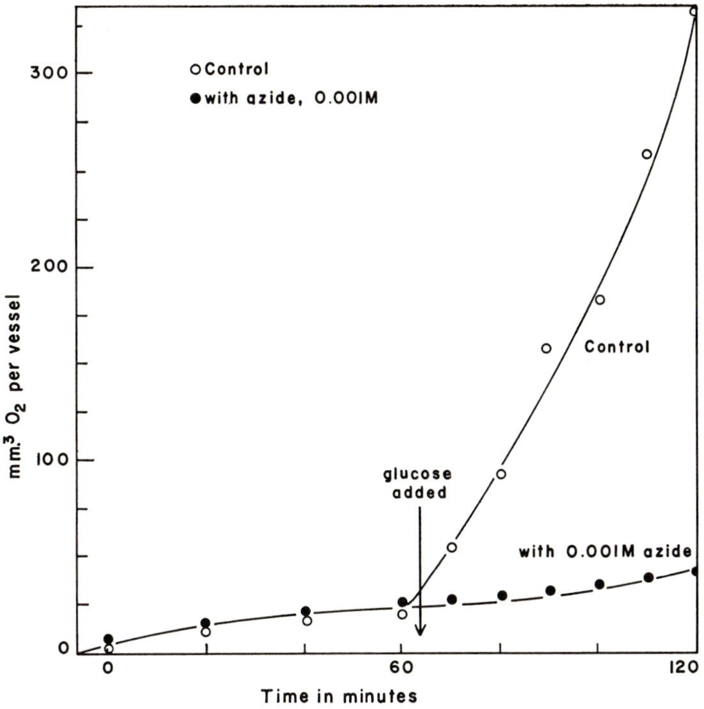


***Figure 18.6.*** The effect of a respiratory poison on the respiration of yeast. The azide, which resembles cyanide in its action on respiration, was added to one group of cultures (black circles) but not to the control (open circles). Note that the endogenous respiration (before the addition of glucose) is little affected by azide, since experimental and control points fall side by side, but note also that exogenous respiration (after the addition of glucose) is virtually abolished. In the same experiment 0.3M urethan had little effect on endogenous or exogenous respiration, although it strongly inhibited fermentation. (Data from classroom experiment.)

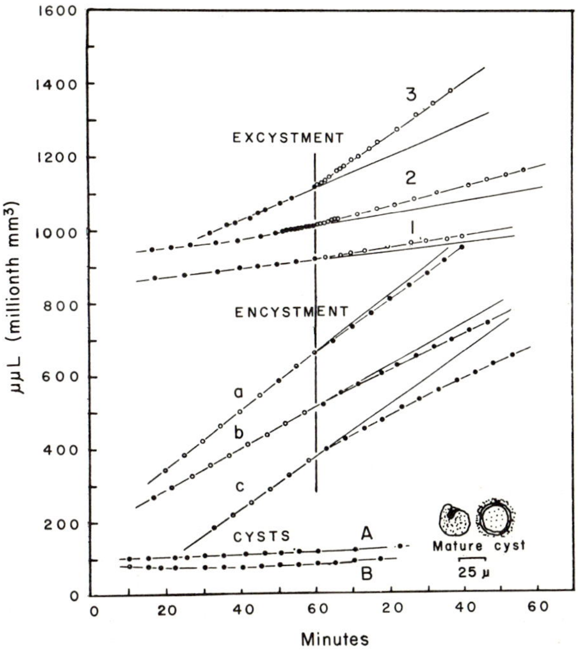


***Figure 18.7.*** Respiration of a single ciliate protozoan (*Bresslaua*) in a microrespirometer during encystment and excystment. Note that during encystment respiration progressively falls (a, b, c). Cysts (A, B) show little respiration. During excystment respiration gradually rises (1, 2, 3). (After Scholander *et. al.,* 1952: Biol. Bull. *102*:182.)

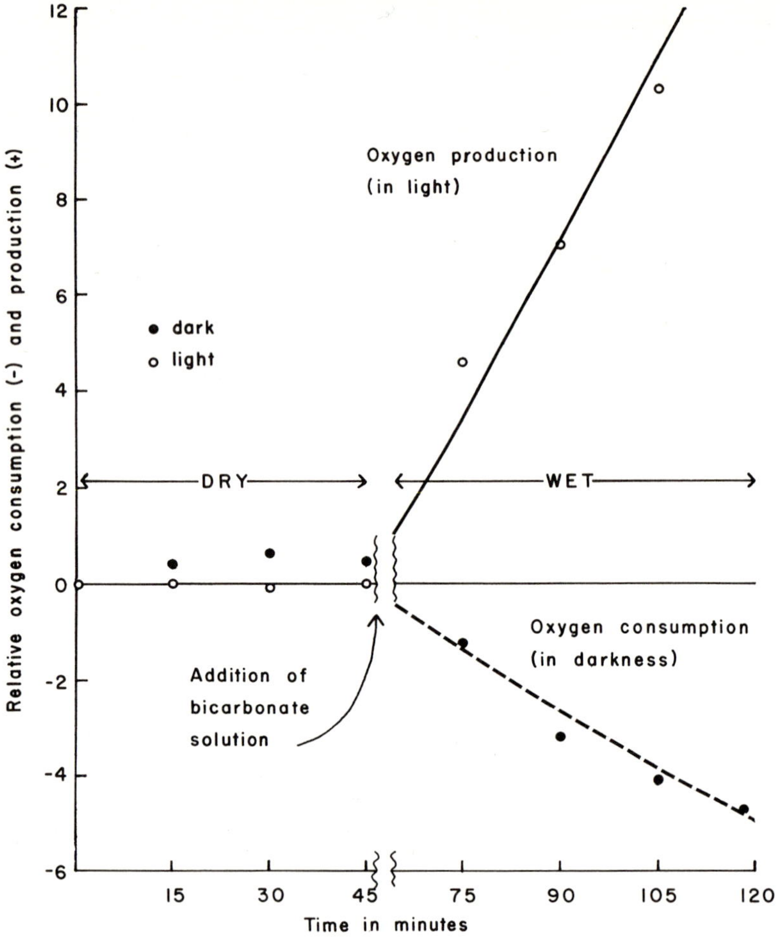


***Figure 18.8.*** Effect of hydration on respiration and photosynthesis in a lichen. About 0.5 gm. of foliose lichen freshly collected was added to each of two flasks, one of which was tightly sealed with tinfoil to exclude light, the other illuminated. The dark flask had KOH in the side well, the other none. After addition of bicarbonate buffer as a source of $CO_2$ for photosynthesis, the vessels were kept open to the atmosphere for about 20 minutes. No readings were taken during this time, because desorption of gases attached to the dry lichen occurs in both cases. Readings were resumed when desorption was about complete. (Data from classroom experiment.)

periments with cells of lichens demonstrate that the metabolically inert dried cells can be activated to a high rate of respiration (and photosynthesis in the light) about 30 minutes after wetting (Cuthberts, 1934; Butin, 1954; Fig. 18.8). These and other experiments indicate the importance of adequate free water for maintenance of cellular reactions. The water present in dormant stages of organisms is largely *bound water,* not free to participate in chemical reactions. Neither hydrolyses nor oxidations, both of which are required for maintenance of active existence, occur in the absence of water.

## RESPIRATORY QUOTIENT (R.Q.) AND REDUCTION LEVEL (R.L.)

The *respiratory quotient* (R.Q.) is the ratio of the carbon dioxide produced to the oxygen consumed:

$$\text{R.Q.} = \frac{CO_2}{O_2} \quad (18.2)$$

From this ratio considerable information can be gathered about the nutrients being metabolized.

The respiratory quotient for cells metabolizing carbohydrates with a general formula $(CH_2O)_n$ is 1, because a molecule of carbon dioxide is produced for every molecule of oxygen that is consumed during the complete combustion of a carbohydrate. For example, experiments with various microorganisms fed on carbohydrates show a respiratory quotient of 1.

$$CH_2O + O_2 \rightarrow CO_2 + H_2O;\quad R.Q. = \frac{CO_2}{O_2} = \frac{1}{1} = 1$$

The respiratory quotient of cells metabolizing fats is less than 1, because oxygen consumption is greater than carbon dioxide production. A calculation of the oxygen required for complete combustion of a fat molecule such as tripalmitin, $(C_{15}H_{31}COO)_3C_3H_5$, makes this clear:

$$(C_{15}H_{31}COO)_3C_3H_5 + 72.5\ O_2 \rightarrow 51\ CO_2 + 49\ H_2O$$

$$R.Q. = \frac{CO_2}{O_2} = \frac{51}{72.5} = 0.70$$

The respiratory quotient varies with the nature of the fat metabolized. For instance, for tributyrin, a fat with short-chain fatty acid, it is 0.8.

The respiratory quotient for cells metabolizing protein is also less than 1, because proteins are made up of amino acids, which require more oxygen for complete combustion than carbohydrates do. Since the amino acids have a short chain, an R.Q. of 0.8 is generally assigned to proteins.

If an organism is fed a mixed diet, the R.Q. depends upon the relative proportion of the types of nutrients used. For example, the R.Q. for man on an average diet of carbohydrates, fats and proteins is 0.825 (Ruch and Patton, 1965).

Occasionally the R.Q. of respiring cells is greater than 1. For example, when some of the oxygen released by the carbohydrate during its synthesis into fat is used as a hydrogen acceptor instead of atmospheric oxygen, less atmospheric oxygen is consumed. For quite a different reason, the R.Q. of cells going into oxygen debt may be temporarily greater than 1 (until the debt is paid), because more carbon dioxide is being produced by fermentative processes than can be accounted for by the cellular uptake of oxygen. The R.Q. of plant cells in the dark depends upon the nutrient being consumed (Stiles and Leach, 1952).

Another useful concept is the *reduction level* (R.L.). The reduction level of a compound that is being oxidized is the reciprocal of the respiratory quotient and is usually expressed in the following way:

$$R.L. = \frac{1}{R.Q.} = \frac{2\ n_C + \frac{1}{2}\ n_H - n_O}{2\ n_C} \qquad (18.3)$$

In this equation $n_C$, $n_H$ and $n_O$ are, respec-

**Table 18.2.** Reduction Levels of Some Carbon Compounds

| COMPOUND | FORMULA | REDUCTION LEVEL | HEAT OF COMBUSTION (MEASURED) IN KCAL./MOLE | HEAT OF COMBUSTION (CALCULATED) IN KCAL./MOLE | NUMBER OF CARBONS | HEAT OF COMBUSTION PER CARBON IN KCAL./MOLE |
|---|---|---|---|---|---|---|
| Methane | $CH_4$ | 2.0 | 210.8 | 220.0 | 1 | 220 |
| Ethyl alcohol | $C_2H_5OH$ | 1.5 | 327.6 | 330.0 | 2 | 165 |
| Carbohydrate | $(CH_2O)_n$ | | | | | |
| e.g., glucose | $(CH_2O)_6$ | 1.0 | 673.0 | 660.0 | 6 | 110 |
| Lactic acid | $C_3H_6O_3$ | 1.0 | 326.0 | 330.0 | 3 | 110 |
| Succinic acid | $C_4H_6O_4$ | 0.88 | 357.1 | 387.2 | 4 | 94.5 |
| Pyruvic acid | $C_3H_4O_3$ | 0.83 | 279.1 | 273.9 | 3 | 91.3 |
| Malic acid | $C_4H_6O_5$ | 0.75 | 320.1 | 330.0 | 4 | 82.2 |
| Formic acid | $CH_2O_2$ | 0.5 | 62.8 | 55.0 | 1 | 55 |
| Oxalic acid | $C_2H_2O_4$ | 0.25 | 60.2 | 55.0 | 2 | 27.5 |
| Carbon dioxide | $CO_2$ | 0.0 | 0.0 | 0.0 | 1 | 0.0 |

Data on the heat of combustion of pyruvic acid are from International Critical Tables, V. Data on heats of combustion of all other compounds are from Handbook of Chemistry and Physics, 1968. 49th Ed. Chemical Rubber Co., Cleveland.

tively, the numbers of carbon, hydrogen and oxygen atoms in the molecule being oxidized (Rabinowitch, 1945).

For a carbohydrate ($CH_2O$) the value of R.L. is 1:

$$\text{R.L.} = \frac{2(1) + \frac{1}{2}(2) - 1}{2(1)} = 1$$

Since the R.Q. is the reciprocal of the R.L., an R.Q. value gives information on the level of reduction of the compounds considered. When the R.Q. is low (e.g., when fats or proteins are used) the reduction level is high. A high R.Q. indicates low reduction level of the combusted compounds, e.g., carbohydrates or partially oxidized substances of lower reduction level.

The higher the reduction level of a compound, the greater are its chemical energy content and its metabolic usefulness. An approximation of the heat of combustion in kilocalories per mole can be calculated from the reduction level by multiplying the R.L. by 110 and the number of carbons in the compound.* Values calculated in this manner for a number of compounds are given in Table 18.2 alongside data determined experimentally with a calorimeter.

Because fats have a high reduction level they serve as valuable storage materials in some animal and plant cells. In a gram of fat there is almost double the energy available in a gram of protein or carbohydrate.

## HEAT EVOLVED DURING RESPIRATION

Considerable heat is developed as a by-product of cellular respiration, aerobic and anaerobic. This is readily measured with a thermometer during the activities of microorganisms in a manure pile. Heat produced by plant cells is more difficult to detect because it is so readily lost, but a temperature rise is demonstrable under controlled conditions. For example, considerable heat is produced by germinating seeds, and the cells of a spadix of an arum lily are said to become warm enough to attract insects.

Movement and exertion of all animals involving increased contraction of muscle cells is accompanied by a marked increase in the rate of respiration which increases with their temperature. For example, the temperature of the muscle cells of a fish may be several degrees higher than the temperature at the surface of the body (Krogh, 1941). Maximal exertion in man may be accompanied by the development of a fever temperature as high as 103° F. which, after exertion has ended, falls rapidly through evaporation of water from the skin. An increase in temperature, however achieved (e.g., fever), increases the cellular respiration; cooling reduces the respiratory rates of the cells.

*This figure stems from the fact that approximately 110 kcal. of energy is required to reduce 1 gram mole of carbon dioxide to the carbohydrate level. This level lies between the more highly reduced fats and the less reduced organic acids.

## INDIRECT CALORIMETRY

Energy exchanges in organisms may be determined directly with a calorimeter (see Chapter 1) or indirectly by a measurement of the gaseous exchanges, since the energy spent by the cells of an organism is ultimately derived from the oxidation of foodstuffs. Because of its convenience, the latter method is now used almost exclusively in medicine for tests of basal and of activity metabolism.

As already seen, when 1 gram molecular weight of glucose is oxidized in metabolism, 6 × 22.4 liters, or a total of 134.4 liters of oxygen is consumed, and 673,000 calories of heat are released:

$$C_6H_{12}O_6 + 6\ O_2 \rightarrow 6\ CO_2 + 6\ H_2O + 673{,}000 \text{ calories} \quad (18.4$$

per liter of oxygen used, then, 673,000/134.4 or 5007 calories of heat are liberated. Per liter of oxygen used for the combustion of fat, 4686 calories are liberated; and per liter of oxygen used for the combustion of proteins, 4500 calories are liberated. When these foodstuffs are oxidized in the proportions present in the average mixed diet of man, 1 liter of oxygen releases 4825 calories. Therefore, if in an experiment a man consumes 5 liters of oxygen in 20 minutes, he will liberate approximately 25,125 calories of heat. This rate of respiration continued for a day liberates approximately $2 \times 10^6$ calories or 2000 kilocalories (kcal.) per day. This is equivalent to the heat required to raise 20 liters of water from the freezing point to the boiling point. On the basis of the principles outlined and on the basis of the $Q_{O_2}$, it is possible to make calculations of the food requirements of cells or organisms.

If, in addition to oxygen consumption, it were possible to measure the carbon dioxide production as well, the R.Q. could be calculated. From the R.Q. and the nitrogenous waste of a suspension of cells or an animal, the energy contribution of carbohydrate, fat and protein can be determined. This has been done in a few cases. For example, the R.Q. of man on a mixed diet is 0.825. One gram of urea represents the utilization of 3 grams of protein. For each gram of protein 0.95 liters of oxygen has been consumed and 0.76 liters of carbon dioxide produced. The fraction of oxygen used in combustion of the protein portion may be determined and subtracted from the total oxygen used.

The fraction of carbon dioxide produced by protein may similarly be subtracted from the total carbon dioxide produced. The remainder represents that produced by a combination of fat and carbohydrate. From these values for oxygen and carbon dioxide it is possible to obtain an R.Q. for the nonprotein component of the food utilized. By use of Figure 18.9, relating the nonprotein R.Q. to oxygen used in carbohydrate consumption, it is possible to determine the fraction of carbohydrate used in the diet, the remainder constituting the fat component. Such studies show that generally the main source of energy in cells is carbohydrate (Ruch and Patton, 1965).

However, the cells of some animals (butterflies, hibernators and migrators) can utilize fat directly, as indicated by an R.Q. of about 0.7. For example, when a butterfly is fed sugar, the R.Q. rises to 1.5 or more, indicating that fat is being formed and deposited in the cells (Niemierko, 1959). Cells of starving animals use fats (and some protein) when reserves of carbohydrates are gone and, as expected, the R.Q. is then about 0.7.

The R.Q. of respiring plant cells varies. A plant cell using carbohydrate as its main source of nutrient has an R.Q. near unity. When dependent upon fat, e.g., a germinating seedling with large fat stores, the R.Q. may be about 0.7. In some cases

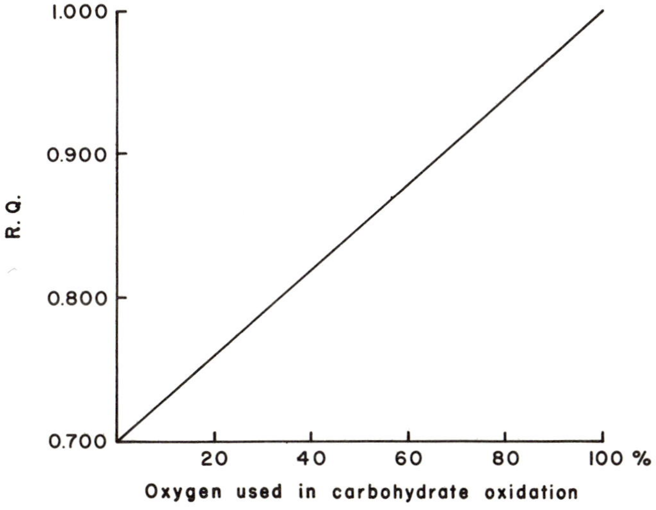


***Figure 18.9.*** Relation between nonprotein R.Q. and percentage of the nonprotein oxygen consumption which is used in oxidation of carbohydrate. (After Ruch and Patton, 1965: Physiology and Biophysics. 19th Ed. W. B. Saunders Co., Philadelphia.)

the R.Q. may be as low as 0.4 (*Linum*, fourth day after germination). This is believed to be the result of uptake of oxygen during the formation of carbohydrates from fats, the converse of what happens when plant and animal cells make fats from sugar. The carbohydrates are then presumably oxidized, since most plant cells show typical anaerobic and aerobic pathways like those described in Chapter 16. The R.Q. of succulent plant cells, which accumulate organic acids, may be quite low also (Stiles and Leach, 1952; Stafford, 1963).

## ANAEROBIC METABOLISM: GLYCOLYSIS AND FERMENTATION

The term *glycolysis* refers to fermentation or anaerobic oxidation as it occurs in animal cells; the term *fermentation* is often applied only to anaerobic processes of oxidation in other types of cells, although it was originally defined by Pasteur as "respiration in the absence of air." During anaerobic metabolism hydrogen acceptors other than atmospheric oxygen receive the hydrogen, and carbon dioxide is usually given off. Equations for the reactions in glycolysis and alcoholic fermentation are given in Chapter 16.

Anaerobic metabolism releases only about one-twentieth of the energy available in glucose and is consequently much less efficient than aerobic metabolism in utilization of a given quantity of nutrient. However, both animal and plant cells make use of it in emergencies.

All cells are capable of temporary anaerobic existence. Many bacteria and protozoans tolerate anaerobiosis for prolonged periods, some indefinitely (von Brand, 1945). Even cells in most tissues can tolerate anoxia briefly. However, the cells of the central nervous system of vertebrates suffer very quickly from oxygen lack (Ruch and Patton, 1965).

Violent exertion of muscle cells results in an oxygen debt; that is, energy is liberated largely by glycolytic processes which proceed in the muscle cells in the absence of adequate oxygen. The oxygen debt is paid off when exertion ceases, and the incompletely oxidized products of glycolysis are then completely oxidized or rebuilt into glycogen for future use (Guyton, 1966).

It is doubtful that many animals are obligate anaerobes, although a few can live under relatively anaerobic conditions. Under unfavorable conditions, cells of animal parasites lacking in free oxygen depend entirely upon anaerobic processes for release of energy. Some sea animals in the intertidal zone are subjected periodically to low oxygen in the stagnant water of pools or burrows when the tide is out (Krogh, 1941). Some symbiotic protozoans (like those constituting the cellulose-digesting fauna of the termite) live under relatively anaerobic conditions and are killed when the oxygen tension is raised (Cleveland, 1926). Protozoans in the rumen of a cow are probably under completely anaerobic conditions (Hungate, 1966).

Plant cells seem to survive an environment lacking in oxygen better than animal cells do, although a young maize seedling dies after 14 hours in nitrogen at 40° C. or after 24 hours in nitrogen at 18° C.; apparently the actively growing cells of seedlings are more sensitive than those of older plants. The nature of the anaerobic processes in plant cells, termed fermentation, has not been fully elucidated, although in many cases alcohol is formed and carbon dioxide is evolved (Stiles and Leach, 1952; Stafford, 1963).

Many bacteria are capable of anaerobic existence, and some can live only in the complete absence of oxygen. Examples of such obligate anaerobes are certain butyric and lactic acid bacteria and the denitrifying bacteria like *Bacillus denitrificans. Clostridium botulinum* is another obligate anaerobe, which produces a highly toxic poison in foods on which it grows. Many microbes are facultative anaerobes; that is, they can survive a lack of free oxygen if they are given adequate nutrients, but in the presence of free oxygen they need be provided with only the simplest substrates. Other butyric and lactic acid bacteria, as well as various thermophilic bacteria, are facultative anaerobes. Yeast is capable of anaerobic existence, utilizing the energy in sugar from which it makes alcohol and carbon dioxide. When oxygen is available, yeast is capable of oxidizing the alcohol completely to carbon dioxide and water. When the yeast strain is genetically defective and lacks the cytochrome oxidase system, however, it cannot oxidize the sugar completely to carbon dioxide and water (Harris, 1956).

Microbes have exploited various pathways of fermentation in their struggle for survival. Among these might be mentioned lactic acid, propionic acid, butyric acid, alcoholic fermentations and mixed fermentations. Many of these interesting fermentations have been of industrial importance (Gale, 1947; Barker, 1956).

When carbon dioxide is the only gas produced during glycolysis or fermentation, the rate of the process may be determined manometrically. Sometimes, however, gases other than carbon dioxide are formed in addition (e.g., hydrogen in *Escherichia coli* fermentations). When this occurs the carbon dioxide must be absorbed by potassium hydroxide, and the other gases must then be measured in the manometer and identified by chemical tests.

Speculatively speaking, the fact that all cells have some capacity for anaerobic existence, at least for a time, may indicate a possible vestige of anaerobic metabolism of early life. If life originated on an earth that lacked free oxygen, as some would have us think, it would have been anaerobic, utilizing for metabolism the stores of organic nutrient which had been formed by photochemical and thermochemical reactions.

## CELL OXIDATIONS WITH EMISSION OF LIGHT

***Occurrence.*** The flash of the firefly and the luminous wake of a boat have always attracted attention, even in ancient times (Harvey, 1957). Since then bioluminescence has been described in bacteria, fungi, protozoa and species belonging to 40 different orders in half the 25 phyla of animals (see Figs. 18.10 and 18.11). *Bioluminescence* is the emission of light as a consequence of the cellular oxidation of some substrate in the presence of an enzyme (Cormier and Totter, 1968).

Robert Boyle, in 1667, showed that luminescence of fungi on rotten wood, like cellular respiration, ceases when air (later shown to be the oxygen in air) is withdrawn, and recurs when air is readmitted. Reamur, in 1723, demonstrated that shellfish cease luminescing when dried, but resume luminescing when water is added. Dubois in 1885 demonstrated that luminescence results from the oxidation of a heat-stable substrate, which he called *luciferin,* in the presence of a heat-

***Figure 18.10.*** Colonies of the luminous bacteria, *Achromobacter fischeri,* photographed using only their own light. The brilliant strain is a variant that tolerates the acidity developed in the medium and remains bright while the other strain turns dim.

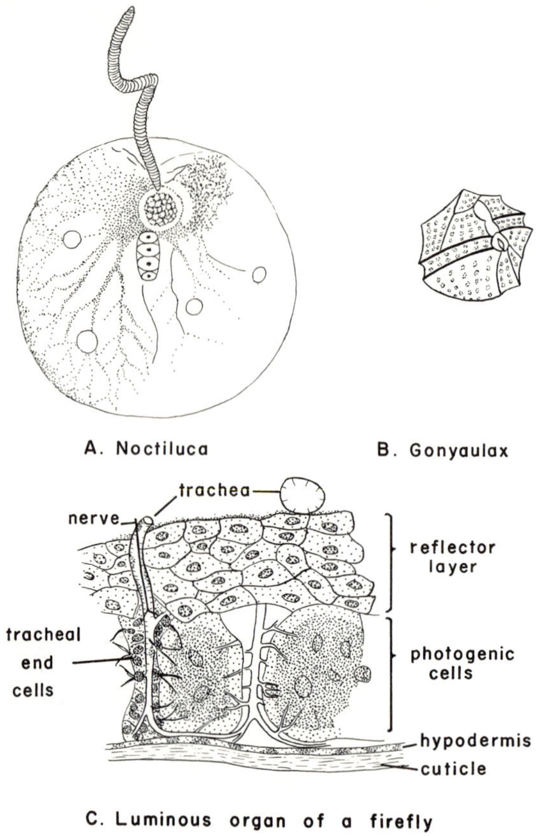


***Figure 18.11.*** *A*, *Noctiluca* (about 700 $\mu$ in diameter) and *B*, *Gonyaulax* (about 200 $\mu$ in diameter), two luminous dinoflagellates found along the Atlantic and Pacific Coasts. The wake of a boat passing through a dense population of such dinoflagellates is luminous because the movement of the water excites the protozoans to give off light. *C*, The luminous organ of a firefly, *Photinus*. (After Williams, 1916: J. Morph. *28*:145.)

sensitive enzyme, which he called *luciferase.* He found that luciferin is present in the luminous extract from a firefly in limited quantities and is soon exhausted, but that the enzyme luciferase remains essentially unchanged (Harvey, 1940, 1952, 1960).

***Chemical nature of bioluminescence.*** Investigations with extracts from luminous bacteria have demonstrated the intimate relationships between the luminescent systems and the energy-liberating enzymatic reactions of the cell. To obtain the extract, the bacteria (grown on a synthetic medium) are harvested by centrifugation and dried in acetone cooled to −15° C. The dried powder is homogenized in water at room temperature and centrifuged to remove debris. The supernatant thus obtained is luminous, but the luminescence can be inhibited at −20° C., a temperature at which the molecules are unable to attain the necessary energy of activation (Harvey, 1960). Most of the characteristics of the luminescence in this extract are similar to those of the luminous system in the bacteria. For example, for both the temperature optimum is about 25° C., and the emission spectrum is between 475 nm. and 500 nm. There is little doubt that the extract contains the luminous system of the bacteria (McElroy, 1960).

Evidence points to the probable identity of luciferin from bacteria with reduced flavin mononucleotide ($FMNH_2$) which has been shown to react with an aldehyde present in the cells to form a complex:

$$FMNH_2 + CH_3 \ldots CHO \rightarrow FMNH_2{-}CH_3 \ldots CHO \qquad (18.5)$$

The complex is oxidized in the presence of luciferase:

$$FMNH_2{-}CH_3 \ldots CHO + O_2 \rightarrow FMN{-}CH_3 \ldots COOH^* + H_2O \quad (18.6)$$

$$FMN{-}CH_3 \ldots COOH^* \rightarrow FMN + CH_3 \ldots COOH + h\nu \text{ (light)} \quad (18.7)$$

$$FMN + NADH + H^+ \rightarrow NAD^+ + FMNH_2 \quad (18.8)$$

The $NAD^+$ is then once again reduced by hydrogen from various dehydrogenations occurring in the bacterial cells (see Chapter 16), e.g., the oxidation of malate to oxaloacetate. Various aldehydes from $C_6$ to $C_{16}$ may complex *in vitro* with $FMNH_2$ to form bacterial luciferin (McElroy, 1960). However, it is not known which or how many of these aldehydes serve this purpose in the luminous bacteria. It is interesting to note that mutant strains of luminous bacteria (*Achromobacter fischeri*) which lack the ability to form aldehydes, and therefore do not luminesce, can be made luminous by adding a solution of any of these aldehydes.

The luciferase extracted from bacteria has been purified by fractional precipitation with ammonium sulfate and by separation on starch-column chromatographs. The enzyme appears to require a free sulfhydryl group for its activity and is therefore extremely sensitive to heavy metals. Its only function appears to be the luminous reaction, although it was first believed to be a respiratory enzyme since it appears in the crude extract with $NADH_2$ oxidase (McElroy, 1960).

The luciferin and luciferase systems of the East Coast fireflies have also been extracted and studied by McElroy and his associates. Firefly luciferin probably occurs in the lantern in aqueous solution and, being quite stable, can be purified in the presence of air by filter-paper chromatography or by use of appropriate resins. At a pH greater than 7.0, pure luciferin, when exposed to ultraviolet light, gives off a yellow-green fluorescence. The luciferin luminesces only in the presence of ATP and some divalent ion, such as $Mn^{++}$, $Mg^{++}$ or $Co^{++}$, in addition to firefly luciferase and oxygen. The need for ATP links the luciferin-luciferase system to carbohydrate metabolism as the source of energy. Since inorganic phosphate appears as a byproduct of firefly luminescence, a degradation of compounds with high energy phosphate bonds is indicated.

The sequence of reactions in firefly luminescence appears to be as follows:

$$LH_2 \text{ (luciferin)} + ATP \rightarrow LH_2\text{-AMP(adenyl-luciferin)} + \text{POP(pyrophosphate)} \quad (18.9)$$

Adenyl-luciferin is regarded as the active luciferin molecule that is peroxidated by reaction with atmospheric oxygen and subsequently breaks down to adenyloxyluciferin, water and adenylic acid:

$$LH_2\text{-AMP} + \tfrac{1}{2}O_2 \rightarrow L\text{-AMP}^* + H_2O \quad (18.10)$$

$$L\text{-AMP}^* \rightarrow L + AMP + h\nu \quad (18.11)$$

The enzymes (luciferases) of the two species of firefly were purified and then tested with the luciferin of one of them. It was found that the color of the emitted light depended upon the enzyme used. A similar experiment performed with the luciferases from two species of luminous bacteria and the luciferin from one of them showed that in this case it is the complex of the luciferase with the flavin mononucleotide (FMN) that is light emitting. However, since changes in ionic environment change the color of the light when purified luciferase from one species is used, it is probable that the way in which the luciferin is bound to the luciferase determines the color of light emitted (allosteric effect?). The binding of the luciferin to luciferase is in turn determined by the tertiary structure of the luciferase, which is probably different in various luciferases (Seliger and McElroy, 1965). For each luciferin molecule oxidized, 1 quantum of light is emitted (0.73 to 1.03). This extraordinary efficiency of the process suggests an energy-stabilizing role for the luciferase and great specificity of the enzyme. However, adenyloxyluciferin readily combines with luciferase, thereby acting as a potent inhibitor of the enzyme. Since coenzyme A combines with adenyloxyluciferin to form oxyluciferyl CoA

*Indicates an excited molecule.

and adenylic acid, it has a stimulating effect on luminescence (Seliger and McElroy, 1960). The pyrophosphate formed in reaction 18.9 is decomposed by the pyrophosphatase available in the cells; therefore it does not affect the reaction.

Some of the compounds postulated by McElroy and his co-workers for these reactions have been isolated. For example, adenyl-luciferin can be formed *in vitro* from ATP and crystalline luciferin, and this compound can be hydrolyzed anaerobically to form luciferin and adenylic acid (AMP). The oxyluciferin and various phosphate compounds have also been isolated from firefly extracts. Intermediate compounds (peroxides?) are being investigated. They are postulated because, as was previously pointed out, the energy yield (40 to 50 kcal.) from oxidation of one luciferin molecule is insufficient to furnish the energy for the light quanta emitted (52 to 64 kcal.). The energy yield from peroxide formation is more than enough (75 kcal.), thus making this, by analogy with peroxidation in chemiluminescence, a tempting suggestion (McElroy, 1960).

The molecular weight of crystalline firefly luciferin is 248 and its empirical formula is $C_{11}H_8N_2S_2O_3$. Oxyluciferin, which usually occurs with luciferin as a contaminant, has a molecular weight just two units less, 306. Luciferin has two pK values, one at pH 3.0 (COOH), another at pH 8.4. The structural formula of firefly luciferin has been determined and the luciferin has been synthesized.

The luciferase of fireflies, which has also been purified and crystallized, shows all the general properties of its class of enzymes. Its molecular weight is about 100,000, its isoelectric pH is 6.1 to 6.2, its pH optimum is 7.8 and its temperature optimum is 23° C. (Green and McElroy, 1956; Seliger and McElroy, 1965).

Bioluminescence has also been studied in the extracts of a number of other organisms, e.g., from *Cypridina* ("water firefly"), a Japanese ostracod crustacean, from the marine dinoflagellate *Gonyaulax* and from others (Chase, 1964). By use of pure preparations of luciferin and luciferase of *Cypridina* it has been shown that the dark-adapted eye will perceive a flash from a solution containing $10^{-11}$ M luciferin and $10^{-15}$ M luciferase (Johnson *et al.,* 1961).

It must be emphasized that the word *luciferin* refers not to a chemical entity, but rather to a class of compounds which may differ chemically among themselves, but all of which are oxidized in the presence of an enzyme and atmospheric oxygen with the emission of light. *Luciferase* is the name given to the class of enzymes which catalyze the bioluminescent reactions. They have all the common properties of enzymes (see Chapter 15). Like all other enzymes, luciferase absorbs ultraviolet light maximally at 280 nm. and at wavelengths shorter than 250 nm., resembling in this respect various unconjugated proteins. Like other enzymes it is inactivated by ultraviolet light (within the region in which it absorbs), by heat and by various other agents which denature proteins. The luciferases do not appear to contain metal ions and are not sensitive to cyanide, but some of them are markedly sensitive to urethan (Harvey, 1960; Seliger and Morton, 1968).

***Physical Nature of Biolumineseence.*** A molecule gives off light when the outer shell electrons in virtual orbits of the excited state return once more to their normal unexcited positions. When heat furnishes the energy for excitation, as in incandescent bodies, the low efficiency of light production (e.g., 0.42 per cent in the carbon filament lamp and 2.5 per cent in the tungsten filament lamp) indicates the low probability of this type of excitation. In fluorescent lamps, electrons, falling through a voltage gradient, collide with gas molecules, transferring kinetic energy to the gas molecules and thereby exciting their electrons to virtual orbits. The efficiency of conversion in this case is about five times that in the best incandescent lamps. Light may also be obtained by exciting electrons mechanically, as is done by grinding crystals.

Light is also produced by some exergonic chemical reactions called chemiluminescences, during which sufficient energy is released to raise electrons of the substrate or its products to an excited state. The chemiluminescent oxidation of

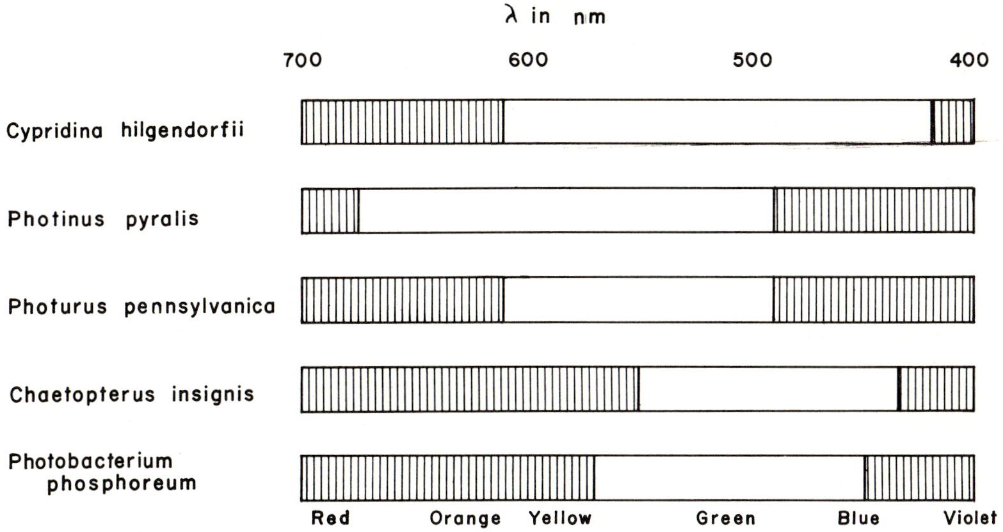


***Figure 18.12.*** Luminescent spectra from luminous secretions or cells in various species of animals. The uppermost is from the "water firefly" of Japan, a crustacean; the next two are from American fireflies; the fourth is from a luminous tube-dwelling worm; and the last is from a luminous bacterium. To the eye the light of the first appears bluish, the light of the second yellowish, the light of the last three greenish. (Data from Harvey, 1920: The Nature of Animal Light. J. B. Lippincott Co., Philadelphia, p. 47).

luminol (3-aminophthalcyclohydrazide) has been studied more extensively than most of the others (McElroy, 1960). Bioluminescence, then, is chemiluminescence occurring in living cells (Harvey, 1960).

The intensity of light emitted in bioluminescence is very low. Lode has estimated that if the entire dome of St. Peter's in Rome were covered with a layer of luminous bacteria, the intensity of light would be about equal to that from one standard candle (Harvey, 1940).

Although the intensity of the luminescence is low, the visual efficiency of the light is high compared with artificial sources, being 95 per cent for the light from the firefly *Photurus pennsylvanica,* 45 per cent for the luminous bacteria, 20 per cent for *Cypridina* and 12 to 14 per cent for fluorescent lamps. In these comparisons, account is taken of the visual efficiency of the human eye (Fig. 9.6) at the different wavelengths in the emission spectra of the luminous organisms (Fig. 18.12). Neither infrared nor ultraviolet light has been detected in the emission spectra of luminous organisms (Harvey, 1940).

## *APPENDIX*

### 18.1 VARIOUS METHODS FOR MEASURING RESPIRATION

***Chemical Methods.*** The oxygen content of a sample of water may be measured by the Winkler method. This procedure involves addition of a solution of manganous chloride and of a mixture of potassium iodide and potassium hydroxide to the sample of water, which is then gently shaken to mix the reagents. The oxygen present in the sample oxidizes some of the manganous ion to manganic ion. Strong acid (hydrochloric or sulfuric) is added when the precipitate is settled. The iodide is then oxidized to iodine in proportion to the manganic ion present, and the iodine is titrated with thiosulfate, with starch used as an indicator. This method may be used to calculate oxygen consumption (or oxygen production) of cells (Allee and Oesting, 1934).

Carbon dioxide produced during cell respiration may be measured if it is trapped in calcium hydroxide or barium hydroxide which is then titrated with oxalic acid. A considerable amount of error may result, however, unless the carbon dioxide produced during respiration has been completely absorbed by the alkali.

The expired carbon dioxide may also be absorbed by a weighed sample of soda lime which is weighed again after exposure. This method is used in calorimetry (see Fig. 1.5).

Carbon dioxide production may also be measured by the change in pH, if an indicator is used. This is feasible only if buffers are lacking in the solution and if the respiration is not too rapid. The method does not distinguish between carbon dioxide and any other metabolic acids produced during the experiment (Blinks and Skow, 1938; Baumberger and Winzler, 1939).

***Volumetric Methods.*** In the Fenn-Winterstein apparatus (see Fig. 18.13), two flasks are attached to a capillary containing a droplet of kerosene. Into one flask, containing a well filled with hydroxide, are placed respiring cells. The corresponding flask on the other side serves as a compensator or control. As the oxygen is used, carbon dioxide is formed and is absorbed by the hydroxide. The net effect is a decrease in pressure in the vessel containing the respiring cells, thus causing the kerosene drop to move toward the vessel. If the volume of the capillary is known, the oxygen consumption may be measured directly. Since a temperature change affects both flasks, its effect is canceled out, and since the system is closed to the atmosphere, it is also unaffected by barometric changes. The temperature change, however, does affect the rate of respiration of the cells (Winterstein, 1912; Umbreit *et al.,* 1964).

Another direct and convenient volumetric method for determination of oxygen consumption (see Fig. 18.1) is the Scholander-Wennesland modification of the Fenn-Winterstein method. A multiple unit differential respirometer has recently been made available by Gilson (1963), simplifying the system by using a manifold and only one reference flask (Fig. 18.14).

***Manometric Methods.*** A method developed by Barcroft for blood-gas analysis was later modified by Warburg for measurement of gaseous exchanges such as those which occur in respiration, photosynthesis or various chemical reactions. The essential pieces of equipment, a manometer and an attached vessel, are shown in Figure 18.15. The cells or tissues are placed in the flask and hydroxide is added to the center well. After thermal equilibrium has been achieved, the vessel is closed off from the atmosphere by the stopcock at the top of the manometer. As oxygen is consumed and carbon dioxide is given off, the carbon dioxide is absorbed by the hydroxide in the well, and the pressure in the system falls, causing the manometric fluid to rise in the arm toward the respiratory flask.

In making readings, the manometric fluid is first set at a fixed point on the right side of the manometer to keep the volume of the system constant. The difference between the heights of the fluid in the two arms of the manometer is then determined after a period of respiration. By use of the gas laws, the volume of oxy-

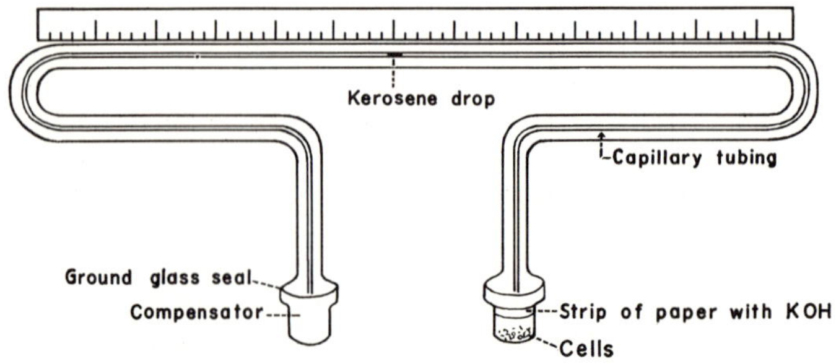


***Figure 18.13.*** Microadaptation of the Fenn-Winterstein manometer.

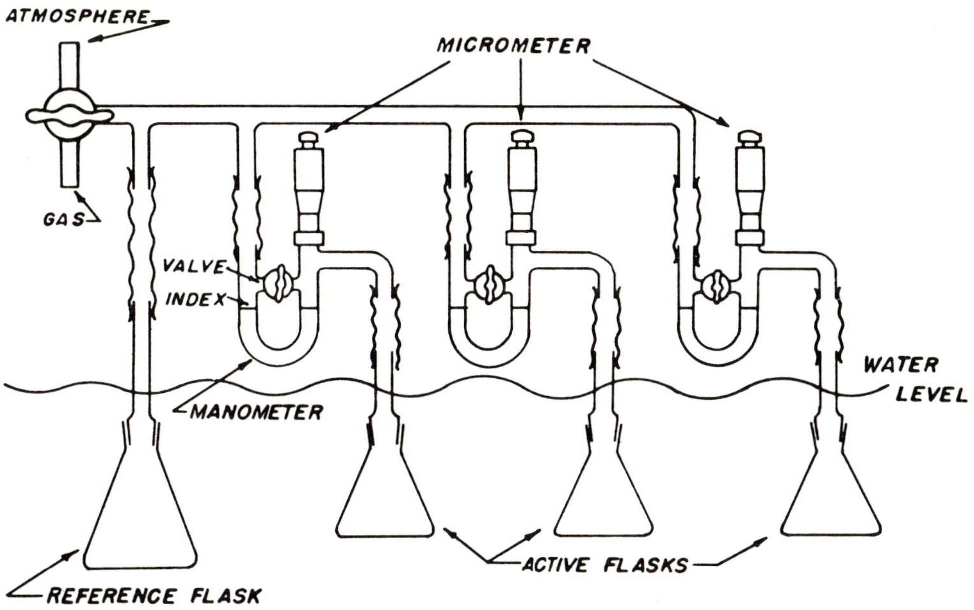


***Figure 18.14.*** Gilson multiple respirometer for direct reading of oxygen consumption. Because a single reference flask serves for all the respirometric flasks, instead of a reference flask for each respirometric flask as in Figure 18.1, the unit is compact. The vessels are attached to the manifold by flexible plastic tubing. (By courtesy of Gilson Electronics, Inc. 3000 West Beltline, Middleton, Wisconsin.)

gen consumed can be determined from the change in pressure (Myers and Matsen, 1955; Umbreit *et al.,* 1964).

Because changes in temperature and barometric pressure affect the manometers, which are open to the atmosphere on the outer side, it is necessary to have a control manometer (called a *thermobarometer*), similar to the other manometers with respiring cells, but without cells.

The main advantage of manometric methods is their versatility. By proper management, exchanges of oxygen and carbon dioxide, glycolysis or other acid production and any system yielding gas exchanges may be studied. The need to calibrate the vessels and manometers for quantitative measurements is a minor disadvantage (MacLeod and Summerson, 1940; Goldstein, 1949; David, 1965).

In either volumetric or manometric techniques, it is possible to use vessels with side arms containing nutrients, poisons or other substances of interest. These substances may be added independently of one another and at any time desired, and their effects on respiration may thus be determined. Vessels of various designs are available for special purposes (Umbreit *et al.,* 1964).

***Cartesian Diver Methods.*** For measuring the respiratory rate of a small number of cells or of a single egg or a single protozoan, the Cartesian diver method (Figs. 18.16 and 18.17) has been used (Holter and Zeuthen, 1966). (Other microrespirometers have also been studied and improved—see Halprin and Gilbert, 1965; Gregg, 1966.)

The Cartesian diver is a very small tube in which are placed in succession a drop of fluid containing the cells, a drop of hydroxide to absorb the carbon dioxide released during respiration and an oil seal or a little plug. Each of these drops is separated by a bubble of air (see Fig. 18.17). As oxygen is consumed by the cells, and as the carbon dioxide released is absorbed by the hydroxide, the diver sinks because its buoyancy decreases. The diver may be refloated if the pressure on the "diving chamber" is lowered. In practice a definite level in the diving chamber is taken as the zero position, and the diver

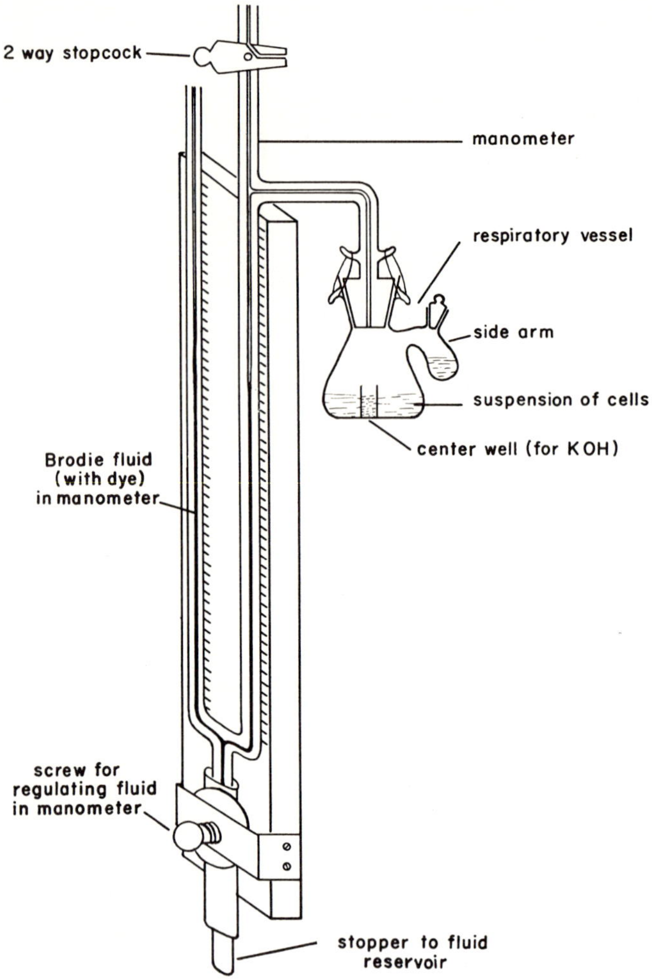


***Figure 18.15.*** A vessel and manometer for the Barcroft-Warburg respirometer. The stopcock is set in position for a measurement.

is kept at this level. In this way errors resulting from solution of gas at different pressures are avoided.

The Cartesian diver is used most frequently for measurement of the *relative* rates of oxygen consumption during the course of the experiments rather than for absolute measurements, but the latter can be obtained by calculation from the pressure changes measured. The disadvantage of the Cartesian diver method is that the techniques are sufficiently delicate to require extensive training for successful operation.

***The Oxygen Electrode.*** The oxygen (platinum) electrode was originally developed for the determination of the oxygen content of water, provided that electrolytes are present in the water, a condition always met in biological experiments since organisms require salts.

In practice, a platinum electrode (see Fig. 18.18) is attached to the negative terminal of a battery. The positive terminal of the battery is connected through a potentiometer or other voltage regulator to the experimental solution by a reference electrode. The potentiometer is generally set to deliver $-0.5$ ($-0.3$ to $-0.9$) volt and the amplified current is measured by an appropriate meter. Hydrogen ions combine with oxygen to form hydrogen peroxide at the platinum electrode surface. At a fixed voltage applied between the two electrodes the rate of discharge of hydrogen ions and the rate of formation of

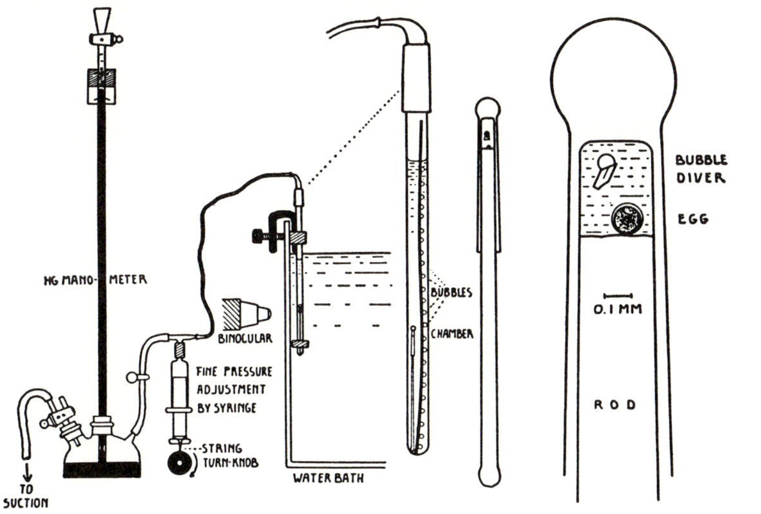


***Figure 18.16.*** The ultramicro or reference Cartesian diver. In this case the oxygen tension of the sample containing the egg is measured by the buoyancy of the minute bubble diver, itself smaller than the egg. (From Scholander *et al.,* 1952: Biol. Bull. *102*:159.)

hydrogen peroxide is proportional to the oxygen content of the water (Fig. 18.19). The study is conducted at a low voltage because electrolysis of water will occur above 2 volts (Davies, 1962; Coe and Coe, 1965; Foster, 1966).

The apparatus is calibrated as follows. First the baseline of current for zero oxygen content is determined after the water is gassed with nitrogen for a long time. Next the water is saturated with air, or any desired known mixture containing oxygen, and the oxygen content of the solution is determined by the Winkler method, or the

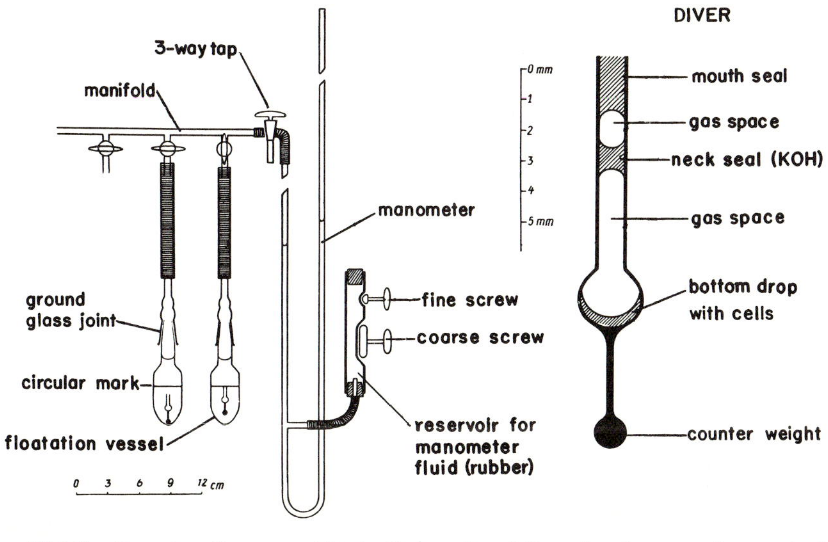


***Figure 18.17.*** Cartesian diver assembly. A whole series of diver chambers may be attached through a manifold to a single manometer. The floatation vessels containing the divers are attached by way of rubber tubing to the manifold as is the manometer. The reservoir is attached to the manometer by rubber tubing. (After Holter, 1943: Compt. Rend. Lab. Carlsberg Series Chimique *24*:403.)

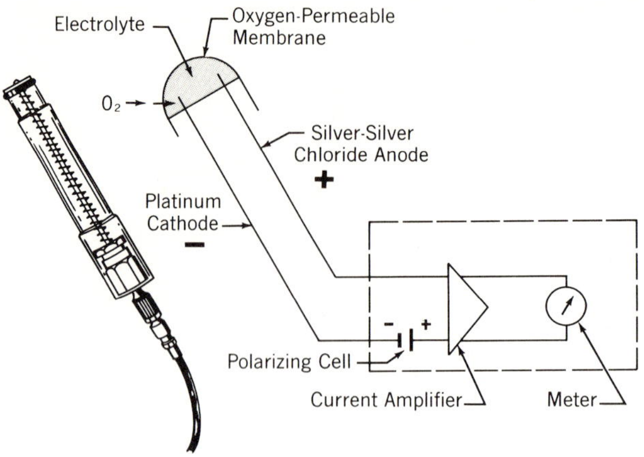


***Figure 18.18.*** The oxygen electrode (platinum electrode). The actual appearance of one type of oxygen electrode is shown to the left; a diagram of an electrode and its connections is shown to the right. (By courtesy of Beckman Instruments Inc., Fullerton, Calif.)

content of oxygen in equilibrium with air is calculated from known data. The current of the solution is then measured to give a starting point for the experiment. Next the organism is placed in the solution, and the subsequent decrease in oxygen content is measured by the decrease in current. Each decrement in current represents a proportional decrease in oxygen content. Knowing the oxygen content at the start, one can calculate the respiratory rate from the rate of change in current.

The main advantage of the platinum electrode is the rapidity and accuracy with which measurements can be made. One disadvantage is the necessity of adequate stirring within the reaction vessel to make certain of equalization of conditions throughout its volume. However, shaking is necessary in most respirometric methods to insure equilibrium between the water and air. Shaking is unnecessary when the volume is small, as in a diver, or when the entire water volume is sampled by the Winkler method.

The platinum electrode is highly

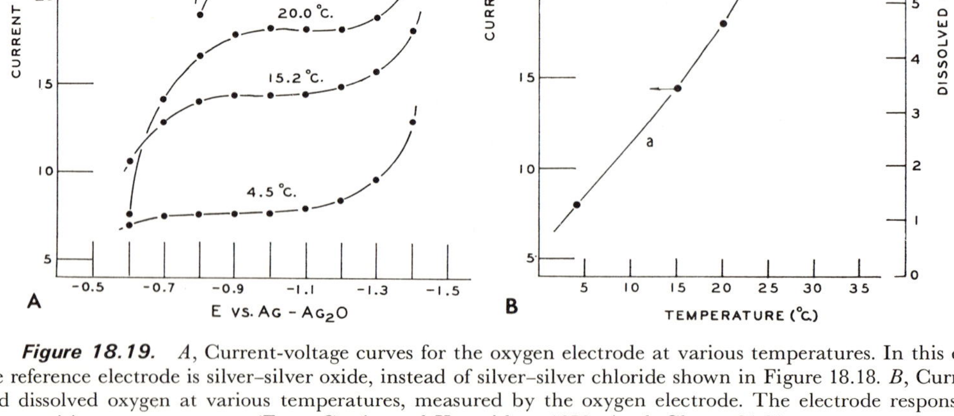


***Figure 18.19.*** *A*, Current-voltage curves for the oxygen electrode at various temperatures. In this case the reference electrode is silver–silver oxide, instead of silver–silver chloride shown in Figure 18.18. *B*, Current and dissolved oxygen at various temperatures, measured by the oxygen electrode. The electrode response is very sensitive to temperature. (From Carritt and Kanwisher, 1959: Anal. Chem. *31*:6.)

adaptable to a wide range in size of respiring organisms (Barker and Milner, 1961). It has the advantage over all the other methods in that it records the oxygen tension, which is in some cases more important to know than the oxygen content of the water or the oxygen consumed by the organism. Because it records continuously and instantaneously, it is especially valuable in following quick changes in respiration (or oxygen production) and for determining the critical oxygen tension, the tension below which the respiration of cells (or organisms) is proportional to the oxygen tension. The output of the platinum electrode may be fed into a recorder for a continuous record of oxygen tension over prolonged periods. An oscillating platinum electrode decreases the possibility of local exhaustion of oxygen (Cater, 1960). The platinum electrode is especially valuable when it is desirable or necessary in respirometric studies to avoid contact with the gas phase, a necessary factor in manometric studies (Davies, 1962).

The platinum electrode has been used extensively in photosynthetic research where layers of algal fronds one cell thick can be placed directly on the surface of the platinum electrode. In closed vessels of solution the dissolved oxygen is continuously consumed by the electrode reaction itself, so that the current drifts slowly downward. If respiring cells are present, the downward drift is faster. Conversely, if oxygen is produced photosynthetically, the current will rise. In the "steady state" modification of the method, the fluid in the vessel caused to flow, is stirred or is aerated.

***Other Methods.*** Another electrometric method for measuring respiratory rates is determination of the change in electrical resistance of a strip of moist filter paper taking up carbon dioxide from respiring organisms. The speed with which these measurements can be made allows one to study the almost immediate effects of an experimental variable. This method is so sensitive that it enables one to determine the respiratory rate of a single *Nitella* cell (Skow and Blinks, 1940).

The carbon dioxide respired by cells in air can also be measured by infrared absorption *spectroscopy* since carbon dioxide absorbs specifically in the infrared (i.e., at 4.2 $\mu$). However, special and expensive equipment that is generally not available in most laboratories is required; therefore the method will not be discussed here (McAlister, 1937).

In luminous bacteria, oxygen is required for both respiration and luminescence, but much less oxygen is used for luminescence than for respiration. When the oxygen tension falls below the limiting

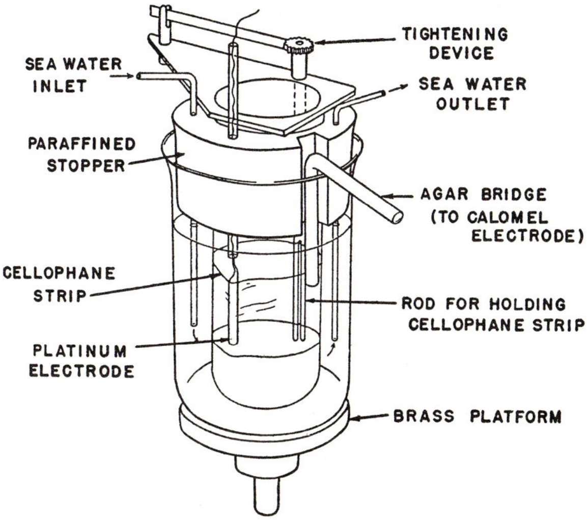


***Figure 18.20.*** The platinum electrode as used for studies with thin sheets of marine algae. The platinum electrode is cemented to the test tube, and over it is mounted the algal tissue, which is held in place by a permeable cellophane strip (tightened by a slotted rod at the rear). Sea water in the beaker may be made to flow through continuously, or aerated. (From Haxo and Blinks, 1949–50: J. Gen. Physiol. *33*:393.)

value, the bacteria suddenly cease luminescing. They "dim out" (Harvey, 1940). The lapse of time between the start of the experiment and the time when luminescence ceases could be used to measure the relative rate of respiration. The effects of various drugs, narcotics and poisons upon respiration can be quickly determined by this method, provided they have no selective effect on luminescence.

## 18.2 RELATIVE SENSITIVITIES OF VARIOUS METHODS

The Warburg method, in which accurate manometer readings can be made to 1 mm., has a sensitivity of about 1 μl. Its sensitivity can be increased by the hydraulic lever principle (Burk and Hobby, 1954). The Fenn-Winterstein manometer is more sensitive than the Warburg manometer but less sensitive than Cartesian divers. A variety of Cartesian divers have been developed. That of Zeuthen (1950) has an air volume of 1 μl. and a sensitivity of $2.5$ to $5.0 \times 10^{-5}$ μl. Larger divers (Linderstrøm-Lang, 1943) have a correspondingly lower sensitivity, whereas the minute divers used for single eggs (Scholander *et al.,* 1952a, b; Scholander and Iversen, 1958) are a hundred times more sensitive.

## *LITERATURE CITED AND GENERAL REFERENCES*


Albritton, E. C., 1954: Standard Values in Nutrition and Metabolism. W. B. Saunders Co., Philadelphia.

Allee, W. C., and Oesting, R., 1934: A critical examination of Winkler's method for determining dissolved oxygen in respiration studies with aquatic animals. Physiol. Zool. *7*:509–541.

Bailey, C. H., and Gurjar, A. M., 1910: Respiration of stored wheat. J. Agr. Res. *12*:685–713.

Barcroft, J., 1946: Features in the Architecture of Physiological Function. Cambridge University Press, London.

Barker, G. C., and Milner, G. W. C., 1961: Polarography. Endeavour *20*:26–31.

Barker, H. A., 1956: Bacterial Fermentations. John Wiley & Sons, New York.

Baumberger, J. P., and Winzler, R. J., 1939: A photometric method for the determination of carbon dioxide. Ind. and Eng. Chem. Anal. *11*:371–374.

Blinks, L. R., and Skow, R. K. 1938a: The time course of photosynthesis as shown by glass electrodes with anomalies in the acidity changes. Proc. Natl. Acad. Sci. U. S. A. *24*:413–419.

Blinks, L. R., and Skow, R. K., 1938b: The time course of photosynthesis as shown by a rapid electrode method for oxygen. Proc. Natl. Acad. Sci. U. S. A. *24*:420–427.

Bonner, J., 1952: Morphogenesis. Princeton University Press, Princeton, N. J.

Brand, T. von, 1945: The anaerobic metabolism of invertebrates. Biodynamica *5*:165–295.

Brown, F. A., Jr., Bennett, M. F., and Webb, H. M., 1955: Persistent daily and tidal rhythm of $O_2$-consumption in fiddler crabs. J. Cell. Comp. Physiol. *44*:477–505.

Burk, D., and Hobby, G., 1954: Hydraulic-leverage principle for magnification of sensitivity of gas change in free and fixed-volume manometry. Science *120*:640–648.

Butin, H., 1954: Physiologisch-okologische Untersuchengen über den Wasseraushalt und die Photosynthese bei Flechten. Biologisches Zentralbl. *73*:459–502.

Cater, D. B., 1960: Oxygen tension and oxidation-reduction potentials in living tissues. Progr. Biophys. *10*:153–194.

*Chase, A. M., 1964: Bioluminescence—production of light by organisms. *In* Photophysiology. Vol. 2. Giese, ed. Academic Press, New York, pp. 389–421.

Cleveland, L. R., 1926: Symbiosis among animals with special reference to termites and their intestinal flagellates. Quart. Rev. Biol. *1*:51–60.

Coe, E. L., and Coe, M. H., 1965: A comparison of photographic methods for detecting changes in respiratory rate in cell-suspensions. Anal. Biochem. *11*:65–72.

Cormier, M. J., and Totter, J. R., 1964: Bioluminescence. Ann. Rev. Biochem. *33*:431–458.

Cormier, M. J., and Totter, J. R., 1968: Bioluminescence: enzymatic aspects. *In* Photophysiology. Vol. 4. Current Topics. Giese, ed. Academic Press, New York, pp. 315–353.

Cuthberts, J. B., 1934: Further notes on the physiology of *Teloschistis flavicans*. Trans. Roy. Soc. So. Africa *22*:35–54.

Danielli, J. F., 1950: Cell Physiology and Pharmacology. Hafner Publishing Co., New York.

David, H., 1965: An improved apparatus for calibration of Warburg apparatus. Anal. Biochem. *11*: 351–359.

Davies, P. W., 1962: The oxygen electrode. *In* Physical Techniques in Biological Research. Nastuk, ed. Academic Press, New York, Vol. 4, pp. 137–179.

Dole, M., 1949: The history of oxygen. Science *109*: 77–81.

Fenn, W. O., and Raha, H., 1964: Respiration. *In* Handbook of Physiology. Vol. 1. Sect. B. Williams & Wilkins, Baltimore.

Foster, J. M., 1966: An oxygen electrode respirometer for measuring very low respiration rates. Anal. Biochem. *14*:22–27.

Gale, E. J., 1947: The Chemical Activities of Bacteria. University Tutorial Press, London.

Gilson, W. E., 1963: Differential respirometer of simplified and improved design. Science *141*:531–532.

Goldstein, A., 1949: A "free manometer" technique for use with the Warburg apparatus. Proc. Am. Acad. Arts Sci. *77*:235–253.

Green, A. A., and McElroy, W. D., 1956: Crystalline firefly luciferase. Biochim. Biophys. Acta *20*:170–176.

Gregg, J. A., 1966: A microrespirometer capable of quantitative substrate mixing. Exp. Cell Res. *42*:260–264.

Guyton, A., 1966: Textbook of Medical Physiology. 3rd Ed. W. B. Saunders Co., Philadelphia.

Halprin, K. M., and Gilbert, D., 1965: An improved microrespirometer for tissue slices. Anal. Biochem. *12*:542–546.

Harris, M., 1956: Occurrence of respiration-deficient mutants in baker's yeast cultivated anaerobically. J. Cell. Comp. Physiol. *48*:95–112.

Harvey, E. N., 1940: Living Light. Princeton University Press, Princeton, N. J.

Harvey, E. N., 1952: Bioluminescence. Academic Press, New York.

Harvey, E. N., 1957: A History of Bioluminescence. American Philosophical Society, Philadelphia.

*Harvey, E. N., 1960: Bioluminescence. Comp. Biochem. *2*:545–591.

Hastings, J. W., Spudich, J., and Malnic, G., 1963: The influence of aldehyde chain length upon the relative quantum yield of the bioluminescent reaction of *Achromobacter fischeri*. J. Biol. Chem. *238*:3100–3105.

Heilbrunn, L. V., 1952: A Textbook of General Physiology. 3rd Ed. W. B. Saunders Co., Philadelphia.

Holmes, E., 1937: Metabolism of Living Tissues. Cambridge University Press, London.

Holter, H., 1943: Technique of the cartesian diver. Comp. rend. Carlsberg Lab. *24*:399–478.

*Holter, H., and Zeuthen, E., 1966: Manometric techniques for single chain cells. *In* Physical Techniques in Biological Research. Pollister, ed. Academic Press, New York, pp. 251–317.

Horowitz, N. H., 1945: On the evolution of biochemical syntheses. Proc. Natl. Acad. Sci. U. S. A. *31*:153–157.

Hungate, R. E., 1966: The Rumen and Its Microbes. Academic Press, New York.

Johnson, F. H. (ed.), 1955: Luminescence in Biological Systems. American Association for the Advancement of Science, Washington, D. C.

Johnson, F. H., Shimomura, S., and Saiga, Y., 1961: Luminescence potency of the Cypridina system. Science *134*:1755–1756.

Kleiber, M., 1947: Body size and metabolic rate. Physiol. Rev. *27*:511–541.

Kluyver, A. J., and van Niel, C. B., 1956: The Microbe's Contribution to Biology. Harvard University Press, Cambridge, Mass.

*Krogh, A., 1941: The Comparative Physiology of Respiratory Mechanisms. University of Pennsylvania Press, Philadelphia.

Linderstrøm-Lang, K., 1943: On the theory of the cartesian diver micro respirometer. Comp. rend. Carlsberg Lab. *24*:333–398.

Lyman, C. P., and Chatfield, P. O., 1955: Physiology of hibernation in mammals. Physiol. Rev. *35*:403–463.

MacLeod, J., and Summerson, W. H., 1940: Graphical method for determining Warburg vessel constants at various fluid volumes. Science *91*:201–202.

McAlister, E. D., 1937: Time course of photosynthesis for a higher plant. Smith. Inst. Miscel. Publ. *95*:1–17.

McElroy, W. D., 1960: Bioluminescence. Fed. Proc. *19*:941–948.

McElroy, W. D., and Glass, B. (eds.), 1961: Light and Life. Johns Hopkins University Press, Baltimore.

Myers, J., and Matsen, F. A., 1955: Kinetic characteristics of Warburg manometry. Arch. Biochem. Biophys. *55*:373–388.

Niemierko, W., 1959: Some aspects of metabolism in insects. *In* Symposium Proceedings of the 4th International Congress in Biochemistry. Levelbook, ed. Pergamon Press, London, Vol. 12, pp. 185–200.

Nursall, J. R., 1959: Oxygen as a prerequisite to the origin of the metazoa. Nature *183*:1170–1172.

Oparin, A. I., 1957: The Origin of Life on Earth. 3rd Ed. Oliver & Boyd, Edinburgh.

Pearson, O. P., Jan. 1953: The metabolism of humming birds. Sci. Am. *188*:69–72.

Pearson, O. P., Aug. 1954: Shrews. Sci. Am. *191*:66–70.

Rabinowitch, E. I., 1945: Photosynthesis and Related Problems. Interscience Publishers, New York, Vol. 1, p. 109.

Ruch, T. C. and Patton, H. D., 1965: Physiology and Biophysics. 19th Ed. W. B. Saunders Co., Philadelphia.

Schmidt, H., and von Stackelberg, M., 1963: Modern Polarographic Methods. Academic Press, New York.

Scholander, P., Dec. 1963: The master switch of life. Sci. Am. *209*:92–106.

Scholander, P. F., Claff, C. L., and Sveinsson, S. L., 1952a: Respiratory studies on single cells. I. Methods. Biol. Bull. *102*:157–177.

Scholander, P. F., Claff, C. L., and Sveinsson, S. L., 1952b: Respiratory studies on single cells. II. Observations on the oxygen consumption in single protozoans. Biol. Bull. *102*:178–184.

Scholander, P. F., and Iversen, O., 1958: New design of volumetric respirometer. Scand. J. Clin. Lab. Invest. *10*:429–431.

Seliger, H. H., and McElroy, W. D., 1960: Spectral emission and quantum yield of firefly bioluminescence. Arch. Biochem. Biophys. *88*:136–141.

Seliger, H. H., and McElroy, W. D., 1964: The colors of firefly bioluminescence. Enzyme configuration and species specificity. Proc. Natl. Acad. Sci. U. S. A. *52*:75–81.

Seliger, H. H., and McElroy, W. D., 1965: Light: Physical and Biological Action. Academic Press, New York, Chapters 3 and 4, pp. 152–205.

Seliger, H. H., and Morton, R. A., 1968: A physical approach to bioluminescence. *In* Photophysiology. Vol. 4. Current Topics. Giese, ed. Academic Press, New York, pp. 253–314.

Skow, R. K., and Blinks, L. R., 1940: Collecting Net, Vol. 15, No. 10.

Stafford, G. A., 1963: Plant Metabolism. Harvard

University Press, Cambridge, Mass.
Stiles, W., and Leach, W., 1952: Respiration in Plants. 3rd Ed. John Wiley & Sons, New York.
Umbreit, W. W., Burris, R. H., and Stauffer, J. F., 1964: Manometric Techniques: A Manual Describing Methods Applicable to the Study of Tissue Metabolism. 4th Ed. Burgess Publishing Co., Minneapolis.
Webb, J. L., 1963: Enzyme and Metabolic Inhibitors. Vol. I. General Principles of Inhibition. Academic Press, New York.
Wennesland, R., 1951: A volumetric respirometer for studies of tissue metabolism. Science *114*: 100–103.
Winterstein, H., 1912: Ein apparat zur mikroblutgasanalyse und Mikrorespirometrie. Biochem. Zeitschr. *46*:440–449.
Winzler, R. J., and Baumberger, J. P., 1939: A photometric method for the determination of carbon dioxide. Ind. Eng. Chem. *11*:371–374.
Zeuthen, E., 1950: Cartesian diver respirometer. Biol. Bull. *98*:139–143.
*Zeuthen, E., 1953: Oxygen uptake as related to body size in organisms. Quart. Rev. Biol. *28*:1–12.

# CHAPTER 19 / PHOTOSYNTHESIS

Of all the processes in nature, photosynthesis is perhaps the most fundamental. Transferring the free energy of sunlight, the green plant combines carbon dioxide and water—the end products of metabolism in all organisms—and from them builds carbohydrates, which directly or indirectly serve as the source of free energy of all living things. It has been calculated that each carbon dioxide molecule in the atmosphere is incorporated into a plant structure once every 300 years and that all the oxygen in the air is renewed by plants every 2000 years. Yet it is only recently that a thorough, systematic study of photosynthesis has been attempted, and botanists, chemists, physicists and physiologists, working together on the photosynthetic process, have made some of the most dramatic discoveries in cell physiology (Rabinowitch, 1945; Clayton, 1965).

More than 2000 years ago Aristotle observed that sunlight was necessary for greening of plants. The requirement of light for photosynthesis was next indicated by the experiments of Stephen Hales (1677–1761) and Senebier (1742–1809). Then, in 1786, Ingenhousz, inspired by Priestley's work on oxygen, provided proof that in the light, carbon dioxide is absorbed and oxygen is given off, whereas in the dark, carbon dioxide is given off and oxygen is absorbed by green plants. In 1804 de Saussure showed that during photosynthesis, equal volumes of carbon dioxide and oxygen are exchanged and that often equal volumes are exchanged during respiration. His was the first truly quantitative study of photosynthesis (Arnon, 1961a).

Warburg (1923) was one of the first to use *Chlorella vulgaris,* a unicellular green alga (see Fig. 19.1), for studies of photosynthesis. Cultures of pedigreed *Chlorella* grown under light of known quality and

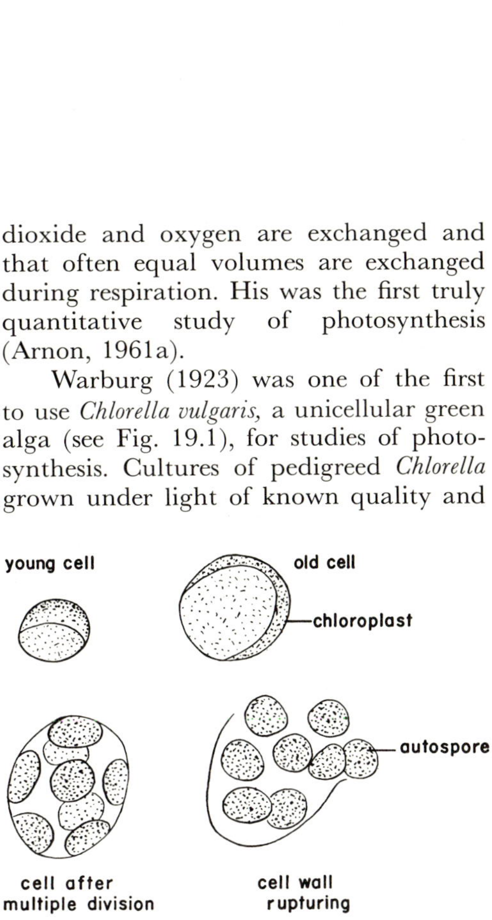


***Figure 19.1.*** *Chlorella,* an alga used extensively in studies on photosynthesis. Note the cup-shaped chloroplast and the multiple fission. Eight autospores are shown in the figure, but 4, 8 or 16 may be formed. Individual *Chlorella* cells are small (2 to 8 $\mu$ in diameter) and lacking in definitive structures, making identification of a species difficult. This may create confusion in experimental work unless subcultures of the same stock are used. (After Smith, 1933: Freshwater Algae of the United States. McGraw-Hill Book Co., New York; and after Fritsch, 1927: British Freshwater Algae. Cambridge University Press. London.)

intensity, and kept at a constant temperature and in a nutrient of standard composition, are found to be more alike than the cells of a leaf. The population of *Chlorella* can be washed any number of times and suspended in a replenishable medium of known constitution for experimentation. The temperature, the concentration of carbon dioxide in equilibrium with a bicarbonate buffer, the intensity and the quality of the light and the rate of diffusion of the materials from the cells can all be controlled during the course of experimentation (Myers, 1951).

Such cultures are used in most studies employing manometric techniques. Thin thalli of marine algae are more convenient when the platinum oxygen electrode is used to measure oxygen production because the tissue can be held directly over the electrode (Haxo and Blinks, 1950). When photosynthesis was measured by conductivity change or pH change (Blinks and Skow, 1938), single large algal cells were used. For still other studies photosynthetic sulfur bacteria are of interest, and for investigation of photophosphorylation and photoreductions occurring during photosynthesis, chloroplasts isolated from spinach leaves and other plants are most effective (Arnon, 1960; Blinks and van Niel, 1963).

## PHOTOSYNTHESIS AS AN OXIDATION-REDUCTION REACTION

The overall reaction which takes place in photosynthesis might be written as follows:

$$nCO_2 + nH_2O \xrightarrow{\text{light}} (CH_2O)_n + nO_2 \quad (19.1)$$

Here $(CH_2O)_n$ represents a carbohydrate. That the ratio of oxygen evolved to carbon dioxide consumed is 1 has been verified many times. In some plant cells, such as diatoms, fats appear as a result of photosynthesis, and some investigators have suggested that perhaps fats might be the first products of photosynthesis. However, the photosynthetic quotient in diatoms has a value of 1 (that of fat synthesis is greater than 1), and fat never accumulates except in cells which have ceased growing. In plant cells as in animal cells, fat is a food reserve.

Van Niel (1931) suggested that water is the hydrogen donor in the oxidation-reduction process which occurs in photosynthesis in plant cells. He based his argument on the reactions which occur in photosynthetic sulfur bacteria:

$$2H_2S + CO_2 \rightarrow H_2O + 2S + CH_2O \quad (19.2)$$

In the green plant cell $H_2O$ could be the hydrogen donor, playing a role similar to that of $H_2S$ in the sulfur bacteria, and, by analogy, equation 19.2 may be written as follows:

$$2H_2O + CO_2 \rightarrow H_2O + O_2 + CH_2O \quad (19.3)$$

Convincing proof of the validity of van Niel's argument came from work with isotopic tracers. Thus using water marked with the oxygen isotope $^{18}O$, Ruben *et al.* (1941) showed that when a culture of *Chlorella* is exposed to light, the gaseous oxygen that is liberated—and only the gaseous oxygen—contains the isotopic oxygen. This is taken as clear evidence that all the oxygen liberated comes from water and that water functions as a hydrogen donor in photosynthesis.

On the other hand, bacteria capable of photosynthesis use a variety of hydrogen donors, including $H_2S$ and various organic compounds. Because of this it has been proposed that organisms using organic hydrogen donors are primitive ones. Presumably, as organic donors were exhausted in some early geological era, organisms evolved that could use water as the hydrogen donor. Oxygen, according to this view, appeared in the atmosphere only after the photosynthetic pattern of the green plant cells had been established. The basic similarity of photosynthesis suggests that the process had a common origin but that photosynthesis, as it occurs in eukaryotic plant cells now, developed only later (van Niel, 1936, 1949).

## LIGHT (PHOTOCHEMICAL) REACTIONS IN PHOTOSYNTHESIS

That the primary reaction in photosynthesis is a photochemical reaction is apparent from the fact that light is necessary. However, several lines of evidence indicate that thermochemical reactions follow the photochemical reaction.

At low light intensity the photochemical reaction is the limiting reaction in photosynthesis. In a *Chlorella* suspension illuminated with low intensities of light, the temperature may be varied over fairly wide limits (15° to 32° C.) without greatly altering the rate of the reaction (see Table 19.1). The temperature coefficient ($Q_{10}$) of photosynthesis over this range is about 1.06. The rate at which the chlorophyll molecules absorb and transmit the limited number of light quanta is not increased by raising the temperature. The initial portion of the curves in Figure 19.2 brings this out clearly.

If light intensity is increased to the point at which it saturates the system, that is, the point at which each chlorophyll molecule is supplied with all the light it can absorb, and if the temperature is then increased, the rate of photosynthesis increases. At high intensity, then, it is the thermochemical reactions, which follow the photochemical reaction, that are rate limiting. In this case a temperature coefficient as high as 4.3 is found over the range of 10° to 32° C. (see Table 19.1 and Fig. 19.2).

A preparation of *Chlorella* exposed to flashing light performs more photosynthesis per unit dose of light than does the same preparation when exposed to continuous light. In other words, a given amount of light is more efficiently used if a dark period follows a light period. At low temperature, when the periods of darkness following the flashes are varied, the efficiency of a given dose of light increases with the length of the dark period up to a maximum which may be considered as the time required for completion of the dark reactions. Some of the data are shown in Figure 19.3. The lower the temperature, the longer must be the dark period, because the rate of carbon dioxide fixation is determined by the kinetic energy of the molecules (Emerson and Arnold,

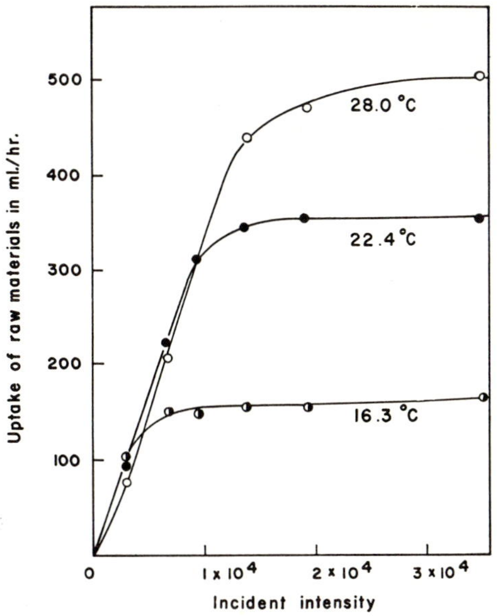


***Figure 19.2.*** Influence of temperature on rate of photosynthesis of purple sulfur bacteria (Thiorhodaceae) at different intensities of light. Note that at low light intensity, temperature has little effect on the rate of photosynthesis. (From Wassink *et al.*, 1942: Enzymologia *10*:304.)

**Table 19.1.** Effect of Temperature and Light Intensity on the Rate of Photosynthesis

| Temp. range | 4–10 | 10–20 | 20–30 | 5–10 | 16–25 | 15–25 | 25–32 |
|---|---|---|---|---|---|---|---|
| $Q_{10}$ | 4.3 | 2.1 | 1.6 | 4.7 | 2.0 | 1.06 | 1.0 |
| Rel. intensity | 45 | 45 | 45 | 16 | 16 | 1.8 | 1.0 |
| of light | | High | | Medium | | Low | |

From Warburg, 1919: Biochem. Zeit. *100*.

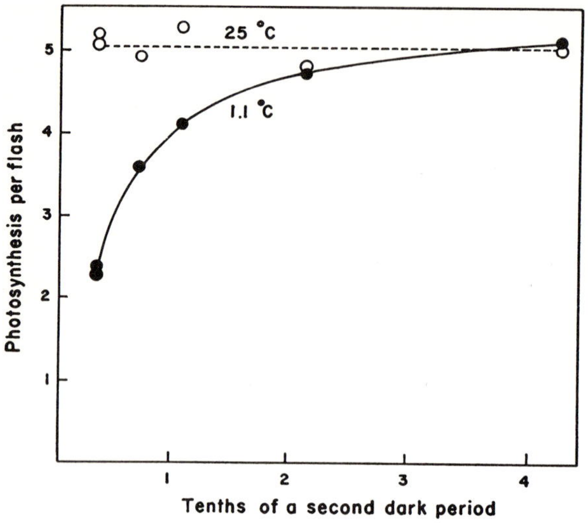


***Figure 19.3.*** Increase in efficiency of use of light at low temperature with long dark periods between flashes. (After Emerson and Arnold, 1932: J. Gen. Physiol. *15*:403.)

1932). Some supporting evidence for light and dark reactions in photosynthesis also comes from studies of the effects of ultraviolet radiations on the process of photosynthesis (Arnold, 1933).

All these experiments demonstrate that at least two types of reactions are involved in photosynthesis. The primary reaction in which light is absorbed and used by chlorophyll is called the Hill reaction. The fundamental thermochemical dark reaction in which carbon dioxide fixation takes place has been named the Blackman reaction. The names are in honor of pioneering investigators.

Robert Hill (1939) discovered that fresh leaves ground in water containing suitable hydrogen acceptors give off oxygen when exposed to light, even though the cells are crushed and photosynthesis of carbohydrate has ceased. A variety of dyes and other compounds, such as ferricyanide, when added to the crushed cells are photochemically reduced. For example, quinone is reduced to hydroquinone (see Fig. 17.7). These experiments indicate that a reduction pool of hydrogen is produced even when light is absorbed by the fragmented chloroplasts.

It is interesting to note that the number of quanta of light which must be absorbed by the chlorophyll in disrupted chloroplasts to liberate one molecule of oxygen is of the same order as the number required in intact green plant cells. This indicates that the process in disrupted chloroplasts is probably the same as in the intact green cells. The only difference is that under these conditions the hydrogen produced cannot be utilized in doing work useful to the cell since the systems are damaged.

According to van Niel (1931, 1936, 1949), the decomposition of water during photosynthesis may be expressed as follows:

$$\text{HOH} + h\nu \rightarrow (\text{H}) + (\text{OH}) \quad (19.4)$$

In this equation (H) represents a reductant, and (OH), an oxidant of unspecified nature, possibly each attached to unknown substances X and Y, respectively, and $h\nu$ represents a quantum of light. By accumulation of (H) a reduction pool is produced which reduces the carbon dioxide in the plant cells. Recombination of (OH) presumably forms water and oxygen:

$$(\text{OH}) + (\text{OH}) \rightarrow H_2O + \tfrac{1}{2}O_2 \quad (19.5)$$

As already pointed out, the origin of oxygen from the water was demonstrated by providing the plant cells with $H_2{}^{18}O$. All the label appeared in the oxygen given off (Ruben *et al.,* 1941).

Van Niel's hypothesis of water decomposition by light is one of the landmarks in photosynthetic research. It postulates that light by way of chlorophyll causes a separation of oxidizing and reducing entities: "What are these entities, how (in physical detail) are they formed and what (in chemical detail) does the plant do with them?" (Clayton, 1965, p. 10). These are questions still not answered fully.

## THE PHOTOSYNTHETIC UNIT

Emerson and Arnold (1932) found that, per unit of energy, flashing light is more effective for photosynthesis than continuous illumination, as measured by oxygen evolved from *Chlorella,* and that one oxygen molecule is evolved per 2500 molecules of chlorophyll present under conditions giving maximal efficiency of light-energy utilization (at a dark interval of 0.04 second). Since 8 quanta of light are required to reduce one carbon dioxide molecule to carbohydrate and liberate one oxygen molecule, this means that one quantum event is mediated by approximately 300 chlorophyll molecules. This is the so-called *"photosynthetic unit"* of chlorophyll.

Since a plant cell achieves maximal rate of carbon dioxide assimilation and oxygen evolution almost immediately upon illumination, it is thought likely that the energy absorbed by the chlorophyll is collected at a reaction center in the photosynthetic unit. This implies that although all of the 300 chlorophyll molecules may collect and harvest light quanta, they are not equivalent and only certain ones presumably participate in the photochemical reaction.

It is interesting that evidence other than flashing light data points to the likelihood of such a photosynthetic unit. Thus the cytochromes unique to chloroplasts (cytochromes F and $B_6$) are present in the proportion of approximately 1 to 300 chlorophyll molecules. Also, the herbicide dichlorophenol dimethylurea inhibits photosynthesis almost completely when present in a concentration of one molecule for every 300 chlorophyll molecules (Clayton, 1964, 1965). Evidence has also been obtained by electron microscopic studies made with the freeze-etching technique on preparations split through chloroplast lamellae. Attached to such lamellae are structures about 100 Å in diameter (Fig. 5.15). It is estimated that a piece of the chloroplast lamella accommodating about 300 chlorophyll molecules would be approximately this size. These units have been called *quantasomes,* and have tentatively been identified as the possible photosynthetic units (Park and Biggins, 1964).

## PHOTOPHOSPHORYLATION

Evidence has accumulated that electrons in long-lived excited states (triplet states), as well as free radicals, are formed during photosynthesis and that there is a loss of electrons from chlorophyll during photic excitation (Sogo *et al.,* 1961).

Arnon and his collaborators have obtained direct evidence of conversion in chloroplasts of the energy of photic excitation into ATP (from ADP and inorganic phosphate) and of reduction of NADP to $NADPH_2$ (Arnon, 1961, 1965). In oxidative phosphorylation, electrons pass from a large negative (reducing) potential through a series of cytochromes (the "electron cascade") to a positive (oxidizing) potential, the change in free energy during this process being stored in production of ATP from ADP and inorganic phosphate (Fig. 16.9).

In chlorophyll of chloroplasts electrons are excited to high energy states by light absorption. If the "electron cascade" were coupled with the cytochromes present in the cell, then light energy could be converted to the chemical energy of ATP and other triphosphates in a manner similar to that in oxidative phosphorylation. Investigations showed that photosynthetic phosphorylation does indeed occur when isolated chloroplasts are illuminated, provided appropriate factors and co-factors, removed during isolation procedures, are once again added. Such photophosphorylation, requiring no oxygen, is independent of oxidative phosphorylation which, as will be recalled, occurs in mitochondria but not in chloroplasts (Chapter 16). The green plant cell can form 30 times as much ATP by photophosphorylation as the cell kept in the dark can by oxidative phosphorylation.

It is possible to separate the photochemical reaction completely from the thermochemical (dark) reactions. When chloroplasts are ground and centrifuged, three components are obtained: a green layer, a supernatant and a lipid layer. The green portion, consisting of *grana,* can carry out the entire photochemical reaction, e.g., the anaerobic phosphorylation of ADP to form ATP, and the reduction of coenzyme NADP to $NADPH_2$ (Arnon, 1960, 1961a, b, 1963) provided catalytic amounts of several factors are supplied to the mixture: among these are vitamin K, FMN (flavin mononucleotide, one of the flavin coenzymes) and chloride ion. The other two components alone can not carry on photochemical reactions but when the products of photochemical phosphorylation and coenzyme reduction (reduction pool) are added to a mixture of these two chloroplast fractions the result is a reduction of carbon dioxide to carbohydrate. This indicates the presence of enzymes for the dark reactions of photosynthesis in the colorless fractions.

The absorption spectra of the cytochromes in the mixture change as these reactions proceed, suggesting that the cytochromes are indeed serving as intermediates in electron transport. Furthermore, it was found that photosynthetic green plant cells have unique cytochromes ($B_6$ and F) not found in animal cells or microorganisms.

In photosynthesis two types of photophosphorylation, cyclic and noncyclic, have been described (Arnon, 1960). In *cyclic photophosphorylation* ATP formation is coupled with liberation of the energy arising from electronic excitation following the absorption of light by chlorophyll; no net reducing power is stored, nor is oxygen liberated. Presumably the electron in chlorophyll which has been raised to an excited state by light absorption returns to the ground state with the free energy stored in ATP; no external source of electrons is required to return chlorophyll to its original state. Thus the electron in chlorophyll completes a cycle.

In *noncyclic photophosphorylation* both the electrons of chlorophyll excited by light absorption and the $H^+$ ions from water are used to reduce NADP to $NADPH_2$, which serves as the reductant for the dark reactions of photosynthesis (see later). In this case, the electron contributed from the chlorophyll molecule to the reduction of NADP is *not returned* and must be obtained from some other source, the hydroxyl ion of water: the two hydroxyl ions then combine to form water and oxygen, $OH^- + OH^- \rightarrow H_2O$ and $\frac{1}{2}O_2$ (see Fig. 19.7). Since in this type of photophosphorylation the electron does not complete a cycle it is called noncyclic photophosphorylation. The reaction of the hydroxyl ions is hypothetical but plausible (Arnon, 1960, 1961).

## THE FUNCTION OF PIGMENTS IN PHOTOSYNTHESIS

Chlorophyll is present in all photosynthetic cells, although it is sometimes masked by the presence of other pigments. Although several varieties of chlorophyll have been identified, all have the same essential porphyrin structure, or tetrapyrrole nucleus, shown in Figure 19.4. Magnesium, in nonionic form, is present in this nucleus and is attached to the ends of the pyrroles. In chemical terms, chlorophylls are methyl phytol esters of the parent dicarboxylic acids, the chlorophyllins. Phytol is a long chain alcohol, containing one double bond. It can be thought of as derivable from vitamin A by hydrogenation.

The various chlorophylls differ from one another only in the side chains attached to the outer ends of the tetrapyrrole nucleus. Green plant cells have chlorophylls A and B, brown algal cells and diatoms have chlorophylls A and C and red algal cells have chlorophylls A and D. The purple bacteria have a chlorophyll called bacteriochlorophyll, which resembles

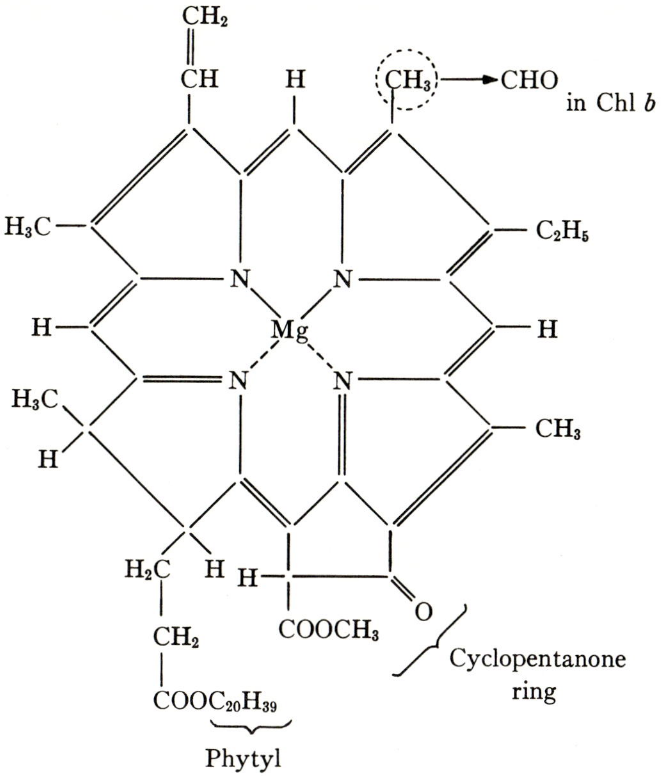


Structure of chlorophyll *a*

***Figure 19.4.*** The tetrapyrrole structure of chlorophyll. The difference in structure between chlorophyll *a* and chlorophyll *b* is shown at the top right of the figure. Bacteriochlorophyll is different in a number of ways; the cyclopentanone ring is the same but the radical replacing the phytyl has not yet been determined with certainty. (From Clayton, 1965: Modern Physics in Photosynthesis. Blaisdell Publishing Co., Waltham, Mass.)

chlorophyll A. The green sulfur bacteria have a so-called *Chlorobium* chlorophyll, the exact structure of which has not been determined. Cells of etiolated green plants (sprouted in the dark) contain protochlorophyll which is quickly reduced to chlorophyll when the leaf is illuminated. However, "blue" mutant* forms of purple sulfur bacteria, lacking carotenoid, are capable of photosynthesis in absence of oxygen (Clayton, 1965). Evidence exists that the chlorophylls in chloroplasts are attached to proteins in the grana.

Most of the chlorophylls absorb in the red and blue-violet parts of the spectrum (see Fig. 19.5), except for bacteriochlorophyll, which absorbs in the infrared and the blue violet (French, 1960). When illuminated with various wavelengths of light (including ultraviolet), chlorophyll shows a brilliant fluorescent band in the red. This is one of its most characteristic properties. It also shows a weaker band in the far red. In addition, after illumination, chlorophyll emits light in the dark for as long as 2 minutes, a property which is

*By the use of x-rays, green mutant strains of *Chlorella* have been produced which absorb light and evolve oxygen but are incapable of performing the photosynthesis of carbohydrates. Such mutants can be grown in a solution of glucose and salts, indicating that they are capable of manufacturing their other requirements (Davis, 1948).

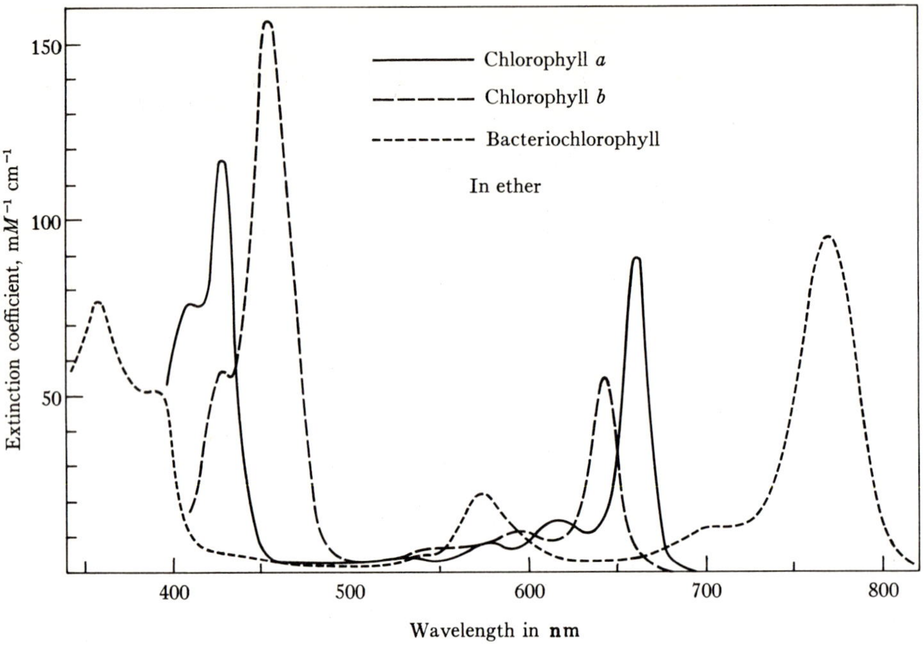


***Figure 19.5.*** Absorption spectra of various chlorophylls. The absorption spectra in the red and infrared portions of the spectrum from 600 to 800 nm., are of special interest. Note that chlorophyll *a* absorbs at longer red wavelengths than chlorophyll *b* and that bacteriochlorophyll absorbs in the infrared portion, just beyond the limit of vision in the red. (From Clayton, 1965: Modern Physics in Photosynthesis. Blaisdell Publishing Co., Waltham, Mass.)

inhibited when agents such as cyanide or azide are present. Since these agents also inhibit photosynthesis, light emission by chlorophyll is thought to be the result of a reversal of some step in photosynthesis (Arnold and Davidson, 1954; Krasnovskii, 1959).

In all plant cells and in the purple sulfur bacteria (which are capable of photosynthesis), in addition to chlorophyll, carotenoids (yellowish pigments) are present. The carotenoids are essentially hydrocarbons (see Fig. 19.6) and fall into two main groups, the carotenes and the xanthophylls, the latter being hydroxy compounds otherwise similar to carotenes. A large

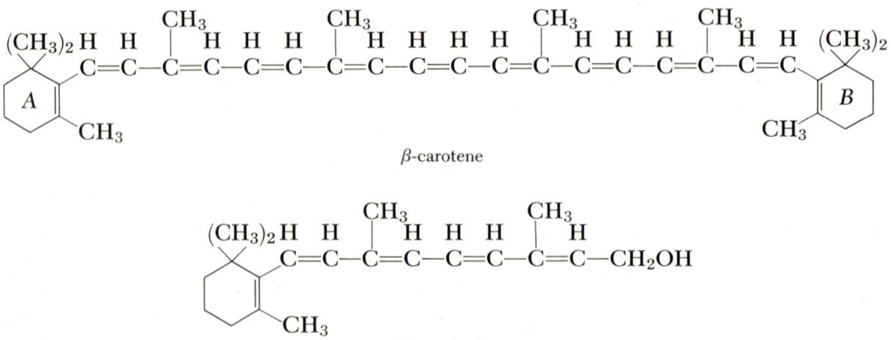


β-carotene

Vitamin $A_1$

***Figure 19.6.*** β-Carotene and vitamin $A_1$. Note that vitamin A is essentially one-half of a β-carotene molecule in which units *A* and *B* are mirror images of each other.

number of carotenes and xanthophylls exist, but they will not be enumerated here. The carotenoids absorb strongly in the blue-violet and ultraviolet ends of the spectrum. In the cells of each plant group, characteristic carotenoids are found. A few of the main ones are listed in Table 19.2. The carotenoids, like the chlorophylls, seem to be attached to proteins in the chloroplasts of cells. In some anaerobic photosynthetic bacteria, they appear to protect chlorophyll from photooxidation (Sistrom *et al.,* 1956). They may play a similar role in other cells (Krinsky, 1968).

The action spectrum for photosynthesis in many green plants shows highest efficiency in the red end of the spectrum, somewhat lower efficiency in the blue end and least efficiency in the green portion of the spectrum. Chlorophyll shows maximal absorption in the blue end with somewhat lower absorption in the red end of the spectrum and virtually none in the green (see Fig. 19.5); carotenoids absorb little, if at all, in the red end but absorb strongly in the blue end of the spectrum. This suggests that light absorbed by carotenoids in many green plant cells is, perhaps, not useful in photosynthesis (Blinks, 1964). However, the carotenoids of brown algae and some green algae do participate in the process of photosynthesis, since the action spectrum of photosynthesis here corresponds to the absorption spectrum of a mixture of extracted chlorophyll and the carotenoid (see Fig. 19.7).

The phycoerythrin (red pigment) and phycocyanin (blue pigment) found in the red and blue-green algae are protein-linked pigments (phycobilins), destroyed by heat. The prosthetic group in each of these pigments is related to the bile pigments, being in each case an open tetrapyrrole (see Fig. 19.8). This prosthetic group differs from that of chlorophyll in that the pyrroles are not joined to form a ring, and magnesium is lacking. The pigments in red and blue-green algae take an active part in photosynthesis, and the highest efficiency for photosynthesis in the red algae is in the green part of the spectrum (see Fig. 19.10), where the red pigment absorbs strongly (see Fig. 19.9) and where chlorophyll absorbs least (Blinks, 1964).

Although in the photosynthesizing cell, pigments other than chlorophyll, such as the phycobilins (in cells of red and blue-green algae) and some carotenoids (e.g., fucoxanthin in cells of brown algae), absorb light used in photosynthesis, the fluorescence spectrum obtained on illu-

**Table 19.2.** Pigments in Photosynthetic Bacteria and Plants

| GROUP OF PLANTS | CHLOROPHYLLS | CAROTENOIDS* | OTHER PIGMENTS |
|---|---|---|---|
| Green plants | A and B | several, e.g., β-carotene and lutein | anthocyans (nonphotosynthetic) |
| Green sulfur bacteria | bacterioviridin | several | — |
| Purple sulfur bacteria | bacteriochlorophyll | several, e.g., spirilloxanthin | — |
| Brown algae | A and C | several, e.g., fucoxanthin | — |
| Diatoms | A and C | several, e.g., fucoxanthin | — |
| Yellow-green algae | A and E | several | — |
| Red algae | A and D | several, e.g., β-carotene and lutein | phycobilins (phycoerythrin, phycocyanin) |
| Blue-green algae | A | several | phycobilins |

After Strain, H. H., 1951: Manual of Phycology. Smith, ed. Chronica Botanica Co., pp. 243–262.
* All plants and bacteria contain both carotenes and xanthophylls; several kinds of each are often present. β-carotene is found in all plants.

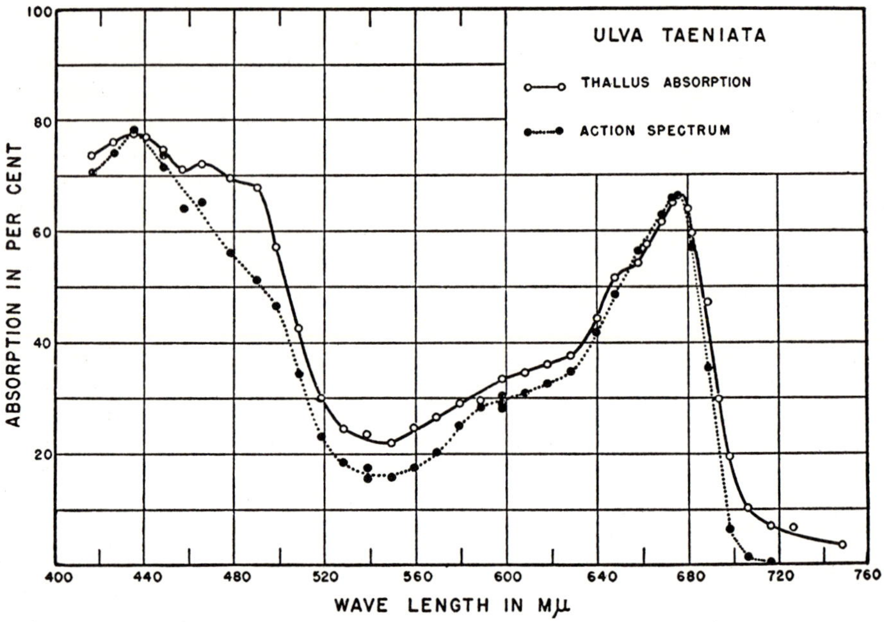


***Figure 19.7.*** Correspondence between absorption spectrum and action spectrum for a green alga, *Ulva taeniata.* Note that in this plant the efficiency in blue light is very high. (From Haxo and Blinks, 1950: J. Gen. Physiol. *33*:408.) In this case photosynthetic efficiency is high in the blue-violet, unlike that in many green plants. Presumably energy absorbed in the short wavelengths is being used in photosynthesis.

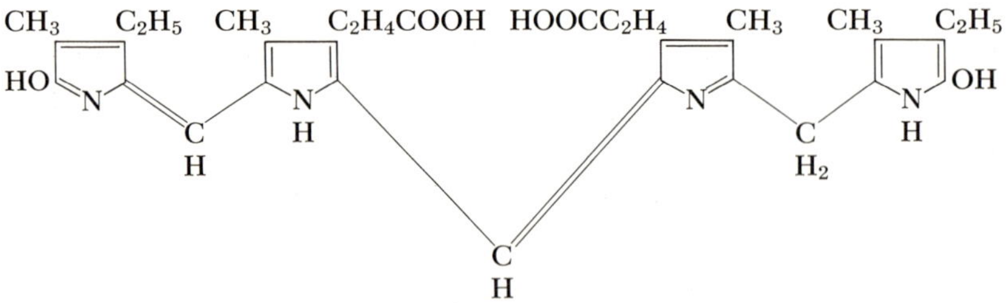


***Figure 19.8.*** The structure of the prosthetic group of phycocyanin, a pigment found in red and blue-green algae. Note that it is an open pyrrole instead of a closed one like chlorophyll, and it resembles the bile pigments resulting from a breakdown of heme in hemoglobin. Phycocyanin has the chemical name mesobiliviolin. Phycoerythrin, the red pigment of red algae, is similar in general structure to phycocyanin and is called mesobilierythrin. These prosthetic groups are attached to proteins.

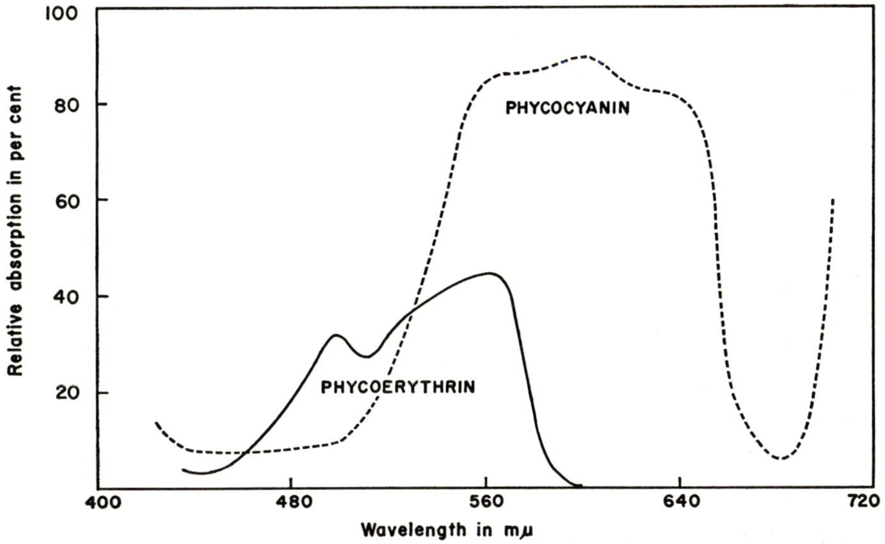


***Figure 19.9.*** Absorption spectra of phycocyanin and phycoerythrin. (Modified after Emerson and Lewis, 1942: J. Gen. Physiol. *25*:579; and from Haxo and Blinks, 1950: J. Gen. Physiol. *33*:389, respectively.)

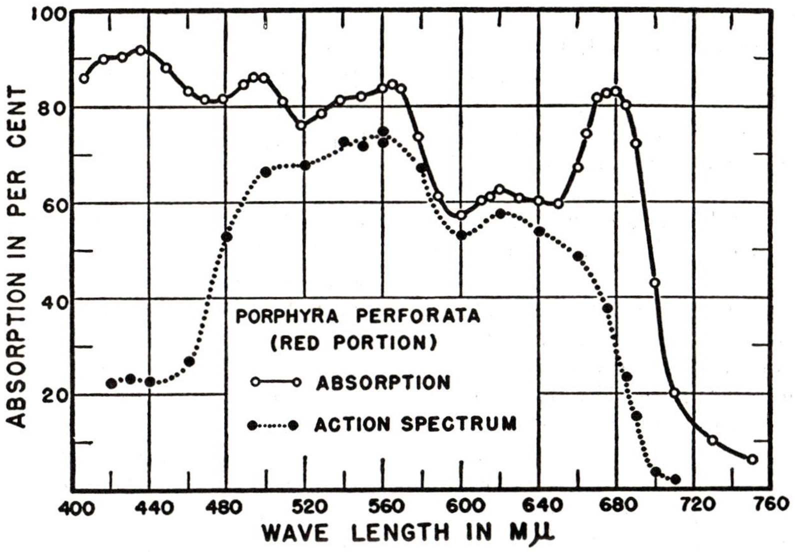


***Figure 19.10.*** Action and absorption spectra of a red alga, *Porphyra.* Note that the region of the spectrum absorbed most by the phycoerythrin (compare with Fig. 19.9) is most effective in photosynthesis. (From Haxo and Blinks, 1950: J. Gen. Physiol. *33*:414.)

mination is always that of chlorophyll A, even when plants are illuminated with wavelengths which are little absorbed by chlorophyll. Absorption of light by chlorophylls B, C and D also excites chlorophyll A to fluorescence. This suggests that light absorbed by all the other pigments is transferred to chlorophyll A, which perhaps may use the energy to excite electrons.

## EVIDENCE FOR TWO SYSTEMS OF PHOTOCHEMICAL REACTIONS

In 1950 Blinks demonstrated with the platinum electrode that there is a brief (transient) enhancement in photosynthesis which follows shifts from a long wavelength (e.g., 700 nm.) to a shorter wavelength. This effect has been called the *Blinks effect* (French, 1961). In 1957 Emerson showed that the photosynthetic efficiency of a long wavelength of light is enhanced by simultaneous exposure of photosynthetic plant cells to shorter wavelengths of light. In essence, if $x$ is the rate of photosynthesis when one wavelength (e.g., long red) alone is applied, $y$ is the rate when a shorter wavelength alone is applied and $z$ is the rate when both are superimposed, then $z$ is greater than the sum, $x + y$, when enhancement occurs. The fraction $z/(x + y)$ is a measure of the enhancement (Clayton, 1965). This effect has been called the *Emerson effect* (French, 1961).

For both of these enhancement effects, the action spectrum obtained by constant illumination of green plant cells with a long wavelength of light and simultaneous exposure also to shorter wavelengths, one wavelength at a time, resembles the absorption spectrum of chlorophyll B (Fig. 19.11). Conversely, if exposure to various long wavelengths, one at a time, is superimposed over a constant illumination with a short wavelength, the action spectrum obtained is that of chlorophyll A (Myers and Graham, 1963). Chlorophyll B absorbs more strongly in the short red and chlorophyll A absorbs more strongly

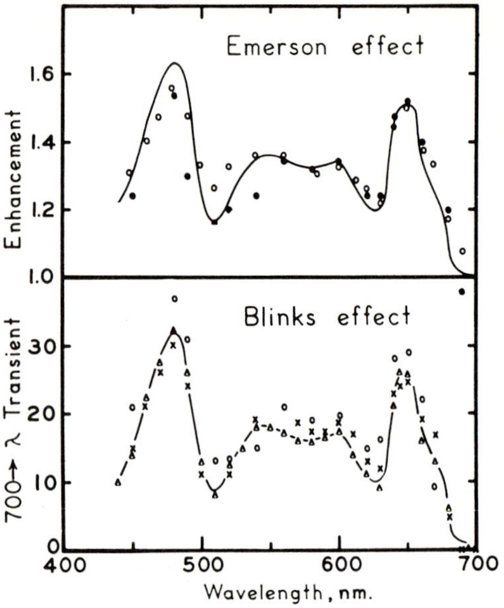


***Figure 19.11.*** The Emerson enhancement effect and the Blinks transient effect. The action spectrum for the enhancement effect (Emerson effect) was obtained by illuminating a plant exposed to 700 nm. with each of a number of shorter wavelengths. The action spectrum for the Blinks effect was obtained by recording the transient increase in photosynthesis following the shift from 700 nm. to each of a number of shorter wavelengths of light. Each wavelength (λ) was adjusted in intensity to give a rate of photosynthesis equal to that obtained with 700 nm. alone. Then, the two wavelengths were used simultaneously and the enhancement of the transient was determined. (From French, 1961: *In* Life and Light. McElroy and Glass, eds. Johns Hopkins Press, Baltimore.)

in the longer red wavelengths (Fig. 19.5); simultaneous use of illumination with long and short wavelengths brought not only chlorophyll B but also chlorophyll A into photosynthetic activity, thus increasing the effectiveness of the photosynthetic process. These two reactions are spoken of as the two *photic reaction systems* in photosynthesis. The action spectra for the Emerson and Blinks effects obtained with plants containing phycobilins (red and blue-green algae) and those with fucoxanthins (diatoms) in addition to chlorophylls suggest that these accessory pigments are also involved in photosynthetic enhancement.

If only one of the pigments in a photosynthetic cell is excited by simultaneous exposure to two wavelengths of light absorbed by the pigment, no enhancement is observed. That is, the amount of photosynthesis induced is equal to the sum of the photosynthesis induced by each alone. Thus, 650 nm. is the peak of absorption by chlorophyll B in *Chlorella* (Fig. 19.5). There is no significant enhancement with superimposition of exposure to any other wavelength in the region 600 to 680 nm. (the long wavelength limit of absorption by chlorophyll B) (Myers and Graham, 1963).

It is interesting that Emerson had earlier observed a decline in efficiency of photosynthesis in monochromatic long wavelength red light (long red wavelength "dropoff"). This was also a result of excitation of only one of the photic systems. He later showed that simultaneous illumination with short wavelengths permits the same efficiency of photosynthesis in long red wavelengths as was found over the remainder of the photosynthetic spectrum, owing to enhancement.

Many of the compounds involved in photosynthesis in the chloroplast are capable of oxidation and reduction and show characteristic redox potentials. Many of them also show absorption spectra which are characteristic of their oxidized and reduced states, making it possible to determine spectrophotometrically whenever illumination or other procedures bring about their oxidation or reduction. Measurements made with the spectrophotometer are rapid (oscillographic recording) so that an immediate change in absorbance at any given wavelength may be detected and its decline followed (Fig. 19.12). Often the absorbance changes are of a rather small order of magnitude, e.g., 1 to 2.5 per cent, and for this purpose a special technique of differential spectrophotometry has been used. The differential, or rate of change of absorption (da/dt), is plotted instead of the absorption. By this means small differences are greatly magnified and the decay constants can be measured with much greater accuracy (Fig. 19.13; Clayton, 1965).

Electron excitation resulting from

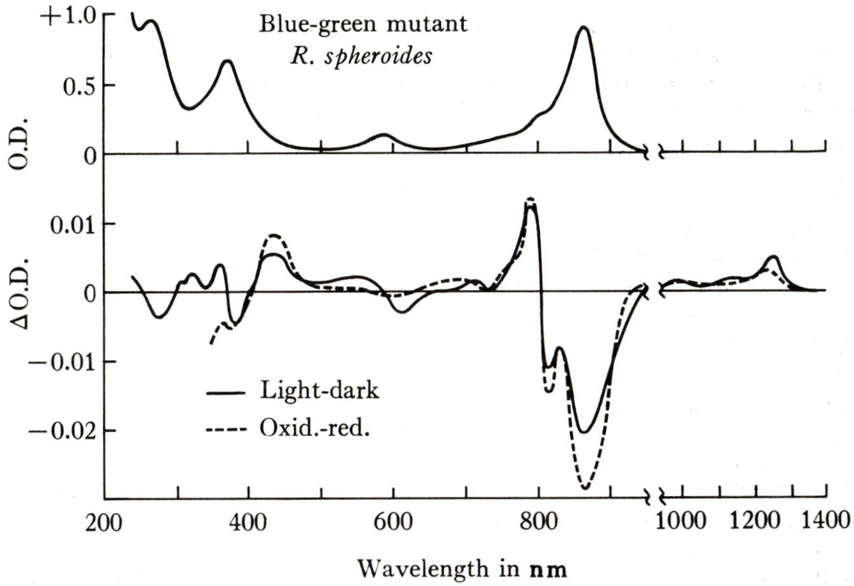


***Figure 19.12.*** Light-induced changes in absorption spectra of chromatophores of the blue-green mutant of the purple photosynthetic bacterium *Rhodopseudomonas spheroides.* Changes reflecting cytochrome oxidation were eliminated in this measurement by saturating the chromatophore suspension with the cytochrome poison *p*-chloromercuribenzoate. The solid line shows the effects of illumination, the dashed line the effect of chemical oxidation (ferri/ferrocyanide, potential 470 mv., or $KIO_4$, potential 500 mv.). (From Clayton, 1965: Modern Physics in Photosynthesis. Blaisdell Publishing Co., Waltham, Mass.)

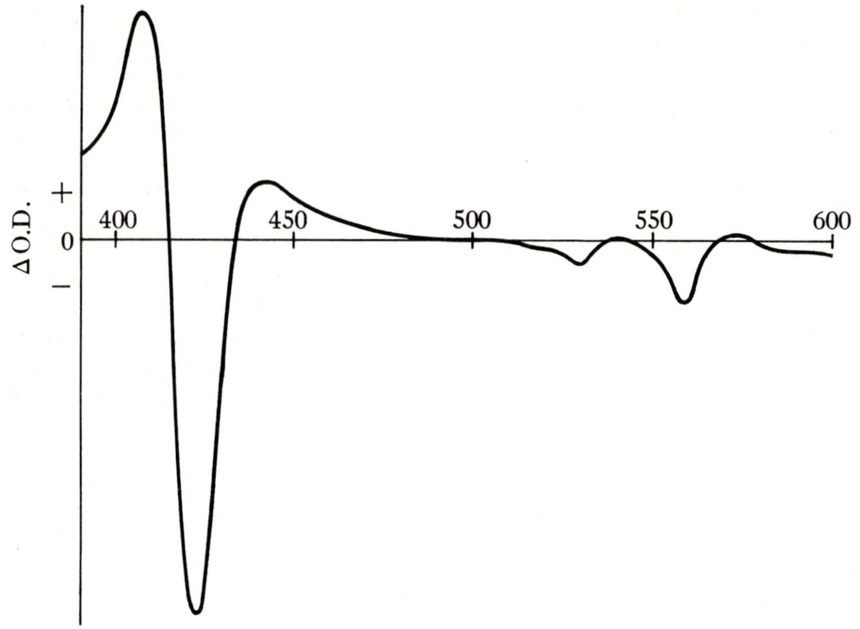


***Figure 19.13.*** Difference spectra showing the light-induced oxidation of cytochrome in *Chromatium* chromatophores. (From Clayton, 1965: Modern Physics in Photosynthesis. Blaisdell Publishing Co., Waltham, Mass.)

absorption of light by chlorophyll is clearly indicated by electron spin resonance studies. Such studies also provide evidence for two photic reactions in photosynthesis, a characteristic signal being associated with each one (Weaver and Bishop, 1963; Weaver and Chon, 1966).

On the basis of these types of experiments the scheme shown in Figure 19.14 has been developed. This is called the *series formulation* of photosynthesis because it pictures the two photic reactions in *System I* and *System II* as succeeding each other. The alternate scheme of parallel reactions is conceivable and has been proposed, but it appears to have more inconsistencies than the series formulation (see Gaffron, 1960; Clayton, 1965).

According to the series formulation, shorter wavelengths of light absorbed by chlorophyll B in green plants (by phycobilins in red and blue-green algae, and by fucoxanthin in diatoms) enter System II and excite chlorophyll from its ground state of +800 mv. to a reduced state of about 0.0 mv. The excited electron cascades down the potential gradient through a number of steps, i.e., a quinone (plastoquinone; see Bishop, 1961), cytochrome $B_6$ and cytochrome F (at + 360 mv.; see Duysens, 1964), and finally to the pigment (chlorophyll A) of the longer wavelength System I, usually designated as P700.

Absorption of light by System I (P700) results in excitation of the electron from the level it had reached in chlorophyll A to about −420 mv. (possibly even more negative in the presence of other unidentified carriers) at ferredoxin, from which it passes to NADP, reducing it to $NADPH_2$ in the presence of $H^+$.

It will be noted that just as in metabolic oxidation in the mitochondria the free energy of an electron at a reducing potential is utilized to phosphorylate ADP, so in photosynthesis the light energy of excited electrons results in photophosphorylation. The consequence of the operation of this scheme is the production of both a reduction pool (as $NADPH_2$) and a supply of high energy phosphate com-

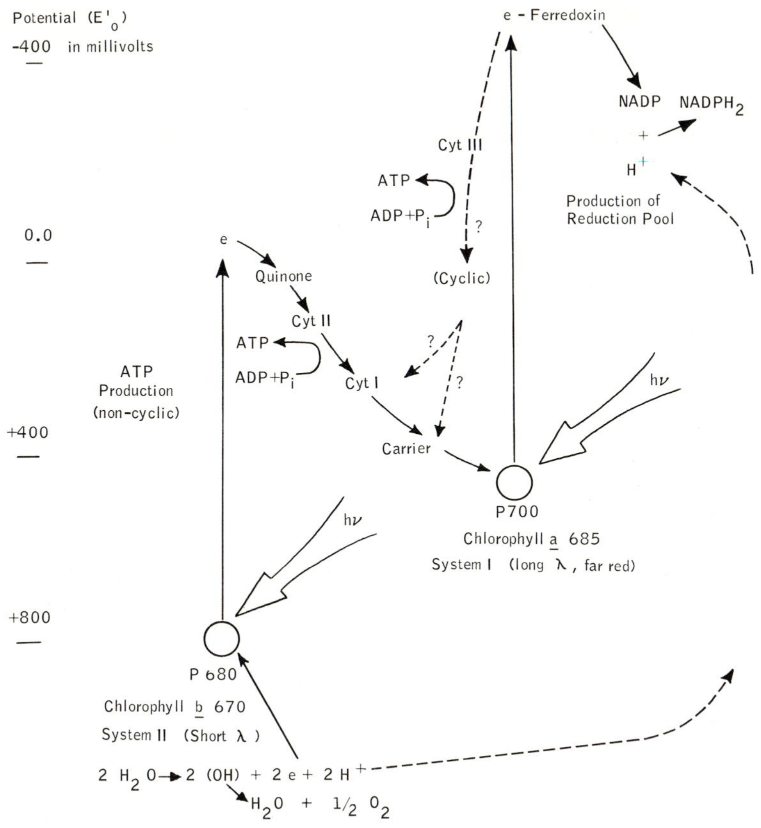


***Figure 19.14.*** Generalized series scheme (tentative) for the two systems of light reactions in photosynthesis. System II has pigment 680 nm. at the reaction center but most of the chlorophyll in this system is chlorophyll *b* with maximum absorption at 670 nm. in green plants. In red and blue-green algae phycobilins are also present, and in diatoms, fucoxanthin. Some carotenoids are present in all photosynthetic forms. System II is concerned mainly with formation of a store of ATP from ADP and inorganic phosphate, and an oxidant (OH) leading to formation of $O_2$.

System I has pigment 700 nm. at the reaction center, but most of the chlorophyll *a* present absorbs at 685 nm. System I is presumably concerned with production of a reduction pool, as indicated by the reduction of NADP to $NADPH_2$. $NADPH_2$ and ATP are used in the dark reactions as shown in Figure 19.16. Cyclic ATP production also occurs but the exact pathway is not known.

Of the cytochromes: I is probably cyt $f_3$, II is probably cyt f and III is probably cyt $b_6$. The "other carrier" is probably plastocyanin, a copper-containing protein. From many sources.

pounds, both necessary for completion of the thermochemical (dark) reactions in photosynthesis.

It is thought that most of the chlorophyll molecules in the photosynthetic unit serve only to harvest light and that the energy conversion takes place in only one part of the unit called the *reaction center.* In a photosynthetic unit of the purple bacteria all of the chlorophyll molecules except one at the reaction center can be deprived of their magnesium (which in chlorophyll lies in a strategic position like that of iron in hemoglobin) and yet the photosynthetic unit continues to collect light for use at the reaction center. On the other hand, even if the other molecules are intact, when the chlorophyll at the reaction center is altered, photosynthesis ceases.

In System I of green plants the reaction center also consists of a single chlorophyll molecule with absorption in the very long red: for this reason it is called P700 (pigment 700 absorbing maximally at wavelength 700 nm.). Only one molecule in the 300 molecules constituting the photosynthetic unit is P700. Its peculiar properties and its characteristic absorption spectrum apparently result from its specific binding within the photosynthetic unit (a type of allosterism?) since in extracts it cannot, on the basis of its properties, be separated from other chlorophyll present in the unit. It alone of all the light-absorbing chlorophyll molecules in the unit is reversibly bleached (oxidized) by long wavelength red light.

In System II of green plants chlorophyll B predominates, although some chlorophyll A is also present. The pigment P680 (chlorophyll B absorbing strongly at wavelength 680 nm.) is present at the reaction center. In blue-green algae the phycobilins (phycocyanin and phycoerythrin) are present in System II, and in diatoms the carotenoid fucoxanthin is also present in addition to some chlorophyll A (Rabinowitch and Govindjee, 1965).

The photosynthetic purple bacteria which use a variety of hydrogen donors in place of water (e.g., hydrogen sulfide or various organic compounds)* lack System II, which in green plants is involved in oxygen production (from $OH^-$), among other things. For this reason the bacteria do not produce oxygen, show no enhancement (Blinks and van Niel, 1963) and are of special interest as more primitive photosynthetic systems (Fuller and Conti, 1963; Vernon, 1964; Clayton, 1965).

The scheme of Figure 19.14 does not require direct photolysis of water by light. Water is slightly ionized. The $OH^-$ ion serves in System II as electron donor (combining therewith to form water and oxygen) and the $H^+$ ion is used in System I to combine with $NADP^+$ on acceptance of electrons from ferredoxin. Light in System II excites the electrons made available by $OH^-$, only indirectly causing separation of water into $OH^-$ and $H^+$ by continued withdrawal of electrons from $OH^-$.

Despite the many important discoveries in photosynthesis, the nature of the light reactions and their sequence during photophosphorylation and photoreduction in the chloroplast remain uncertain. However, many new techniques are being applied to study photosynthesis, some of them with high temporal resolving power. These will enable the investigator to follow even reactions that are completed in a very brief period. There is therefore reason to expect that the sequence of reactions by which the photic reactions occur will be elucidated in as clear a manner as the dark reactions.

## THE DARK REACTIONS IN PHOTOSYNTHESIS

Some workers have considered the possibility that photosynthesis is a reversal of the Krebs cycle, that the hydrogen made available from water during photosynthesis might be used to reverse each step in the dehydrogenation and decarboxylation occurring in respiration (see Chapter 16; Ochoa and Vishniac, 1952). The reactions of the Krebs cycle (Fig. 16.7) are reversible in both plant and animal cells: the reverse of decarboxylation is carbon dioxide fixation; the reverse of dehydrogenation is reduction. Thus, carbon dioxide is split out of isocitric acid to form α-ketoglutaric acid (a 5-carbon acid) in one of the steps of the Krebs cycle. But carbon dioxide can also be incorporated into α-ketoglutaric acid, provided isocitric dehydrogenase, manganous ion and reduced coenzyme II ($NADPH_2$), are present. In the dark, carbon dioxide in all cells is combined with pyruvic acid to form oxaloacetic acid (Whittingham, 1955). In fact, for reversals of the oxidations and decarboxylations of the Krebs cycle, it is necessary only to reduce the coenzymes of the dehydrogenases and to supply ATP and the appropriate enzymes. In animal cells, the coenzymes of the dehydrogenases are reduced only in the presence of a suitable hydrogen donor from which hydrogen and en-

*In some anaerobic photosynthetic bacteria using hydrogen donors other than hydrogen or water, e.g., succinate, not only is NADP reduced to $NADPH_2$, but also atmospheric nitrogen is reduced to ammonia (fixed). Such nitrogen fixation occurs at the expense of photic energy. This fact is of considerable importance in view of the great importance of nitrogen fixation in the economy of nature (see nitrogen cycle, Fig. 1.7).

ergy are liberated. In green plant cells, sunlight absorbed by the grana of chlorophyll supplies most of the energy for production of ATP and for reduction of coenzymes since few mitochondria are present. The simplicity of the idea is appealing, but experimental data are not in keeping with this explanation. It has been shown that reversal of the Krebs cycle occurs in plant cells most conspicuously in the dark. Furthermore, all the reactions of photosynthesis can occur in the chloroplast, while the reactions of the Krebs cycle occur only in the mitochondria (Arnon, 1955).

Several techniques have been responsible for the successful analysis of the thermochemical "dark" reactions in photosynthesis. One of these is the use of radioactive carbon, making it possible to tag the carbon dioxide ($^{14}CO_2$) which is incorporated into organic compounds during photosynthesis. Another technique is paper chromatography (see Appendix 19.2), by which the exceedingly small amounts of substances which appear even after a second of illumination can be identified. Each compound containing $^{14}C$ can then be determined quantitatively if sensitized photographic films are exposed to the radioactive spots on the paper chromatogram.* From the relative density of spots on the developed films the relative concentrations of the various compounds containing $^{14}C$ can be determined, and if the film is calibrated by a simultaneous exposure to a compound with a known amount of $^{14}C$, the actual concentration of each compound can be determined quantitatively. By increasing the length of exposure of cells to light, one can chart the spread of radioactive carbon into a variety of organic compounds. It is also possible to compare the pathway by which plants incorporate $^{14}C$ into compounds in the dark with the pathway taken during illumination (Calvin, 1950–1951). Still another technique which has proved very fruitful in analyzing thermochemical reactions is the isolation of the enzymes involved in each of the thermochemical steps in photosynthesis (Racker, 1955).

*An alternate method is to cut out the paper containing the radioactive compound and measure its radioactivity directly with a Geiger counter.

In most of the experiments tracing the incorporation of $^{14}C$ into photosynthetic products, the radioactive carbon is provided in a solution of bicarbonate to cells which have been brought to a steady state of photosynthesis in the presence of adequate carbon dioxide. At a determined time the cells are killed and analyzed quantitatively by chromatography and sensitized photographic film for various compounds containing $^{14}C$ (see Appendix 19.2). In cells killed immediately after a 1 second exposure to light the $^{14}C$ is found mainly in phosphoglyceric acid. In cells killed after 5 seconds of illumination, traces of phosphoglyceric acid, alanine, asparic acid and malic acid containing $^{14}C$ appear. In cells killed after 15 seconds of illumination, in addition to the other compounds listed, faint traces of sucrose containing $^{14}C$ may be detected and also some pentose, glucose phosphate and fructose phosphate, all containing $^{14}C$.

After a minute of illumination has elapsed, before the cells are killed, abundant radioactive glucose phosphate and fructose phosphate are found. At the end of 5 minutes of illumination radioactivity appears even in the lipids of the cell, and by way of the amino acids such radioactivity may be able to reach the proteins. It is quite evident that the radioactive carbon obtained from carbon dioxide is rapidly spread throughout the various classes of compounds present in the plant cell (Calvin, 1950–1951, 1962; Bassham and Calvin, 1962; Bassham *et al.,* 1955).

Since the 3-carbon acid, phosphoglyceric acid, is the first compound so far found even after very brief illumination to incorporate radioactive carbon, its genesis becomes of prime importance. About half of its radioactivity resides in the carboxyl group, suggesting that the $^{14}CO_2$ had been added to a larger molecule from which the phosphoglyceric acid has been split off. Since radioactivity accumulates rapidly in the center of a pentose sugar, the pentose sugar is the molecule which most likely incorporates the carbon dioxide. The enzymes involved in this transfer are now known (see Table 19.3).

**Table 19.3.** Photosynthetic Dark Reactions

Reactions Fixing Carbon Dioxide into Pentose Forming Hexose

| NUMBER REACTANTS | PRODUCTS | ENZYMES |
|---|---|---|
| 1. Pentose phosphate + ATP | Ribulose diphosphate + ADP | Phosphopentokinase |
| 2. Ribulose diphosphate + $CO_2$ + $H_2O$ | 2 Phosphoglyceric acid | Pentose diphosphate carboxylase |
| 3. 2 Phosphoglyceric acid + 2ATP | 2 Diphosphoglyceric acid + 2ADP | Phosphoglyceric acid kinase |
| 4. 2 Diphosphoglyceric acid + 2NADPH + $2H^+$ | 2 Triose phosphate + $2NADP^+$ + 2 phosphate (inorganic) | Triose phosphate dehydrogenase |
| 5. 2 Triose phosphate | Hexose diphosphate | Aldolase, triose phosphate isomerase |
| Reactions Regenerating Pentose | | |
| 6. Hexose diphosphate + $H_2O$ | Hexose monophosphate + phosphate (inorganic) | Hexose diphosphatase (magnesium dependent) |
| 7. Hexose monophosphate + triose phosphate | Pentose phosphate + erythrose (tetrose) phosphate | Transketolase |
| 8. Hexose monophosphate + erythrose phosphate | Sedoheptulose phosphate + triose phosphate | Transaldolase |
| 9. Sedoheptulose phosphate + triose phosphate | 2 Pentose phosphate | Transketolase |
| *Sum* of (1 to 9) × 3: $3CO_2$ + 9ATP + $5H_2O$ + 6NADPH + $6H^+$ | 1 triose phosphate + 9ADP + $6NADP^+$ + 8 inorganic phosphate | |
| Reactions Regenerating NADPH and ATP and Providing $H^+$ | | |
| 10. $H_2O$ + $NADP^+$ | NADPH + $H^+$ + ½$O_2$ + 2e | Splitting of water. |
| 11. NADPH + 3ADP + 3 inorganic phosphate + $H^+$ + ½$O_2$ + 2e | $NADP^+$ + 3ATP + $4H_2O$ | |
| *Sum* of (1 to 11) × 3: $3CO_2$ + $2H_2O$ + inorganic phosphate | 1 Triose phosphate + $3O_2$ | |

ATP = adenosine triphosphate; ADP = adenosine diphosphate; $NADP^+$ = nicotinamide adenine dinucleotide phosphate oxidized, NADPH + $H^+$ (or $NADPH_2$), reduced; e = electron. (Data from Racker, 1955.)

After the pentose combines with carbon dioxide it almost immediately splits to form two molecules of phosphoglyceric acid. Following the pathway of $^{14}CO_2$, the molecules of phosphoglyceric acid are further phosphorylated to form diphosphoglyceric acid. In the presence of reduced coenzyme II ($NADPH_2$), diphosphoglyceric acid is reduced to form triose phosphate and inorganic phosphate. Two triose phosphate molecules unite to form the sugar hexose (fructose) diphosphate, as indicated by the fact that a higher percentage of the labeling in this molecule appears in the centrally located carbons 3 or 4.

From hexose (fructose) diphosphate, hexose phosphate is formed by loss of a phosphate, and from it glucose phosphate is formed in the presence of an isomerase.

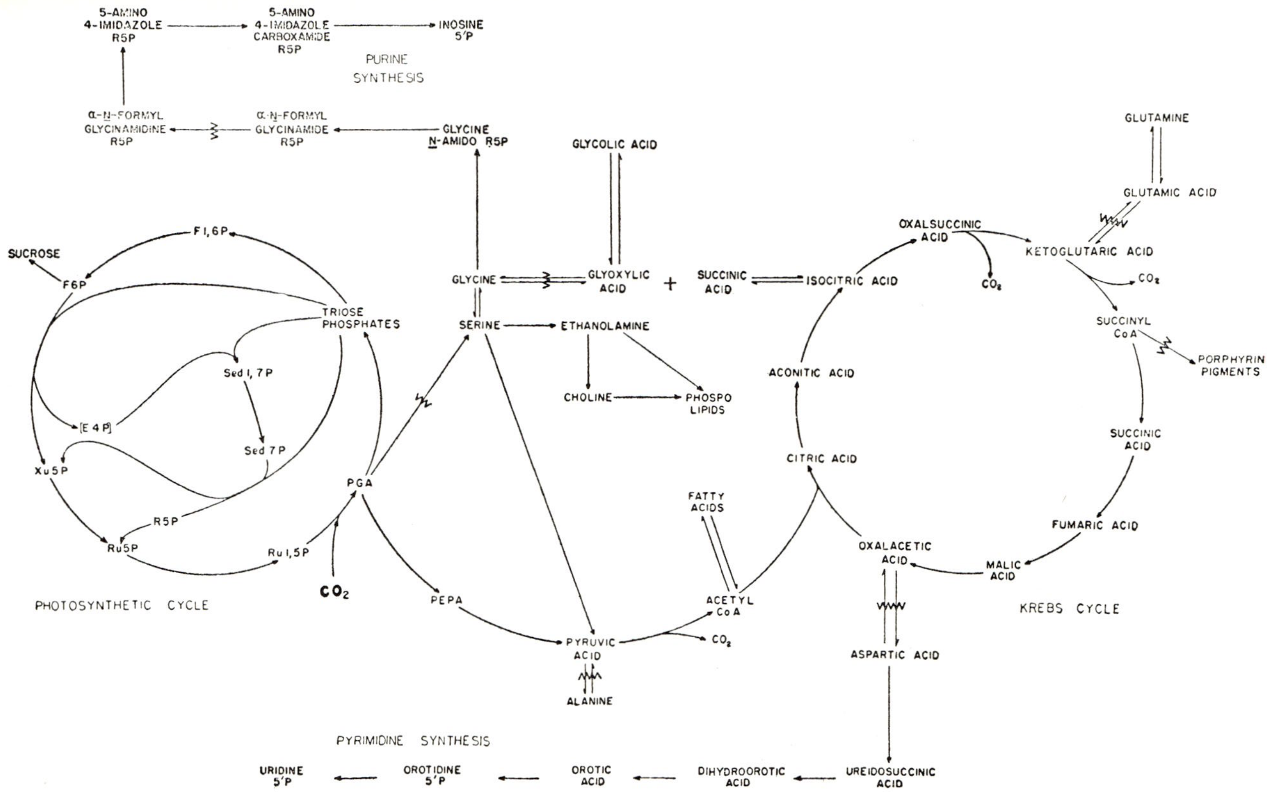


***Figure 19.15.*** Interrelationships between photosynthesis, respiration and biosyntheses in the green plant cell. The following abbreviations are used in the photosynthetic cycle:

(F 6P), fructose 6-phosphate (fructose monophosphate)
(F 1, 6P), fructose 1,6-diphosphate
(Sed 1, 7P), sedoheptulose 1,7-diphosphate
(Sed 7P), sedoheptulose 7-phosphate
(PGA), phosphoglyceric acid
(Ru 1, 5P), ribulose 1,5-diphosphate
(Ru 5P), ribulose 5-phosphate
(R 5P), ribose 5-phosphate
(Xu 5P), xylulose 5-phosphate
(E 4P), erythrose (tetrose) 4-phosphate

Leading from the photosynthetic cycle toward the Krebs cycle on the lowest arc in the diagram are phosphoenol pyruvic acid (PEPA) and coenzyme A (CoA). (From Bassham and Calvin, 1957: The Path of Carbon in Photosynthesis. Prentice-Hall, Englewood Cliffs, N. J. A more detailed photosynthetic cycle is given in Calvin, 1962.)

From these molecules various disaccharides and polysaccharides may then be synthesized. The acids of the Krebs cycle containing $^{14}C$ are formed from the glucose used in respiration (primarily in the dark, since illumination inhibits these reactions). Carbon-14 appears in amino acids when some of the keto acids are aminated, and it appears in lipids and proteins when these compounds are built from fragments in the Krebs cycle (see Fig. 16.8).

Pentose, which appears to serve as the carbon dioxide acceptor in the thermochemical reactions of photosynthesis, can be produced by the oxidation of glucose to gluconic acid, which can be oxidatively decarboxylated in small amounts. However, since this would constitute reversals of photosynthetic carbon fixation, another source of pentose seems more likely.

Tracing the pathway of $^{14}C$ in compounds suggests the following scheme for the synthesis of pentose. It is similar to the pentose shunt already discussed for heterotropic cells (Fig. 16.12). Hexose phosphate combines with a triose in the presence of transketolase to form a pentose and a tetrose (4-carbon) sugar. The tetrose combines with a hexose under the influence of transaldolase and the compound formed then splits to form a 7-carbon sugar, heptulose, and a triose. The heptulose may then combine with triose in the presence of transketolase to form two molecules of pentose.

In this manner continuous supplies of pentose are maintained. These supplies are adequate to fix carbon dioxide in the thermochemical reactions of photosynthesis (see Table 19.3 and Figs. 19.15 and 19.16). This scheme is supported by the fact that pentose is always isolated in illuminated cells along with the 7-carbon sugar, heptulose. The latter contains about half of its radioactivity in the central carbons, indicating that it is formed by the combination of smaller fragments. For elucidating the dark reactions in photosynthesis, Calvin received the Nobel Prize in 1961.

***The Chloroplast as a Complete Photosynthetic Entity.*** Intact chloroplasts are capable of complete photosynthesis outside the cell, provided all the co-factors are supplied (Arnon, 1955).

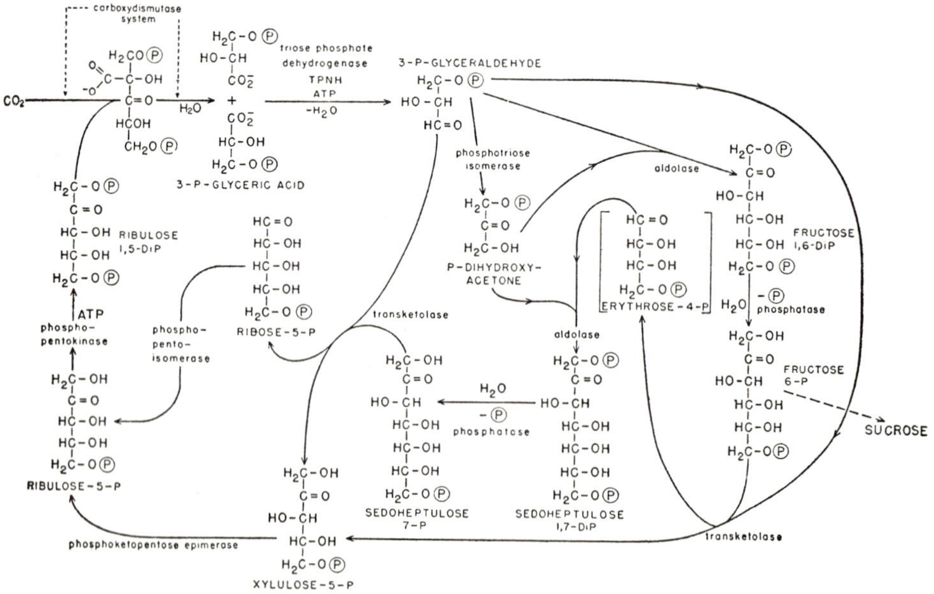


*Figure 19.16.* Details of the dark reactions of photosynthesis and the pentose cycle. Compare the pentose cycle in photosynthetic plants to the pentose shunt in cells of heterotrophic microorganisms and animals (Fig. 16.12). TPNH is now designated NADPH. (From Bassham and Calvin, 1957: The Path of Carbon in Photosynthesis. Prentice-Hall, Englewood Cliffs, N. J.)

The chloroplasts in the cells of green plants contain highly organized units of chlorophyll called *grana* (see Fig. 5.13). These grana can be seen by electron microscopy to be organized into lamellae, as shown in Figures 5.14 and 5.15. By chemical analysis grana have been shown to consist of lipids, proteins and other compounds which are thought to be arranged in a characteristic manner as diagramed in Figure 5.14. The organization of a granum is destroyed when chloroplasts are disrupted by grinding or other methods.

Grana comparable to those in cells of green plants are lacking in algal cells. Free grana or *chromatophores* are found in photosynthetic bacteria (Clayton, 1965).

Three sets of reactions are observed in intact chloroplasts: (1) the Hill reaction, following light absorption; (2) photosynthetic phosphorylation, during which adenosine diphosphate forms adenosine triphosphate—a reaction occurring in the absence of oxygen and distinct from metabolic phosphorylation; and (3) the dark reaction or carbon dioxide fixation. That they are three separate reactions can be demonstrated as follows. Carbon dioxide fixation ceases when iodoacetamide is added, but photosynthetic phosphorylation and the Hill reaction continue; methylene blue inhibits both carbon dioxide fixation and photosynthetic phosphorylation but not the Hill reaction; *o*-phenanthroline inhibits all three reactions (Arnon, 1955, 1963). Presumably these reactions all occur in the chloroplasts inside the living cell. In the cytoplasm of the cell occur many other reactions and during the course of these reactions the photosynthetic products are built into other derivatives (see Fig. 19.15).

## APPENDIX

### 19.1 QUANTUM EFFICIENCY OF PHOTOSYNTHESIS

Formation of 1 gram mole of glucose requires 709,000 calories (free energy change), that is, this amount of energy is required to reduce 6 gram moles of carbon dioxide. To reduce 1 gram mole of carbon dioxide requires one-sixth of this energy or 118,000 calories. A mole quantum at 660 nm. (red light) is equivalent to 43,000 calories (see Chapter 9). Theoretically, therefore, at least 3 quanta of red light are necessary to bring about the reduction of one molecule of carbon dioxide.

Warburg and Negelein (1923) determined the amount of light required by *Chlorella* to bring about the reduction of one carbon dioxide molecule. They used a dense suspension of *Chlorella* in a manometric vessel, the sides and top of which were silvered to reflect the light. From their experimental data they found that about 4 quanta were necessary (Table 19.4).

Subsequent investigators have found 8 to 12 quanta necessary to reduce a molecule of carbon dioxide in a variety of green, brown and red algae, and in photosynthetic sulfur bacteria (Wassink, 1948; Franck, 1951; Nishimura *et al.*, 1951). Furthermore, the same values were obtained by manometric, electrometric and calorimetric techniques (see Table 19.4). The types of errors characteristic of each of these techniques are quite different; therefore, it seems unlikely that because of systematic errors they would all deviate consistently in the same direction.

In view of these findings Warburg and his co-workers redetermined the quantum requirements for reduction of one carbon dioxide molecule, using a number of plant cells, and they reaffirmed their previous value (Warburg *et al.*, 1949; Burk, 1953). In these studies, respiration is usually compensated for by white light. Once the compensation point is found, the suspension of cells is illuminated with an experimental source of light of any desired and known wavelength or intensity.

It is difficult to assign reasons for the differences between the results of the two groups of workers. The following possibilities have been suggested: (1) photosynthesis cannot be completely disentangled from other concomitant reactions, especially those of respiration (Brown, 1953; Bassham *et al.*, 1955; Whittingham, 1955); (2) the exact end product of photo-

**Table 19.4.** Quantum Efficiency of Photosynthesis in Representative Plants

| GROUP | ORGANISM | *Number of quanta required to reduce one molecule of $CO_2$*: DATA | ORDER OF MAGNITUDE |
|---|---|---|---|
| Green sulfur bacteria | *Chlorobium thiosulfatophilum** | White light: 7.8 to 11.8 | 8–12 |
| Purple sulfur bacteria | Chromatium strain D† | Band: 540 to 600 nm. 8.5 to 11.8 (on thiosulfate); 8.5 to 16 (on hydrogen) | 8–12 |
| Green algae | *Chlorella pyrenodosa*‡ | Green (546 nm.): Wisconsin strain: 8.8 to 9.1; Emerson strain: 7.3 to 9.3; Burk strain: 8.1 to 9.7 | 8–10 |
| | Chlorella§ | Blue (436 nm.): 5.7 to 5.1; Yellow (578 nm.): 3.8 to 4.3; Red (660 nm.): 4.1 to 4.4 | 4–6 |

*Number of quanta required to reduce one molecule of $CO_2$*

| GROUP | ORGANISM | WAVELENGTH IN NM. 436 | 500 | 560 | 620 | 675 | ORDER OF MAGNITUDE |
|---|---|---|---|---|---|---|---|
| Green algae | *Ulva lobata*‖ | 14.3 | 18.1 | 20.0 | 16.6 | 12.5 | 12–20 |
| Brown algae | *Coiladesme californica*‖ | 14.5 | 13.7 | 14.3 | 13.2 | 13.3 | 13–14 |
| Red algae | *Porphyra naiadum*‖ | 35.6 | 17.2 | 13.3 | 14.5 | 23.2 | 14–36 |
| | *Delesseria decipiens*‖ | 29.4 | 15.8 | 15.8 | 19.2 | 32.2 | 16–32 |
| Flowering plants | *Phyllospadix scouleri*‖ | 17.0 | 20.8 | 17.9 | 14.7 | 13.3 | 14–21 |

* Larsen *et al.*, 1952.

† Wassink *et al.*, 1942: Enzymologia *10*.

‡ Yuan *et al.*, 1955: Biochem. et Biophys. Acta *17*. This article summarizes many of the experiments performed on quantum efficiency in Chlorella.

§ Warburg and Negelein, 1923. See also Burk *et al.*, 1949: Science *110;* and Burk, 1955: Science *121,* reaffirming these low values.

‖ Yocum and Blinks, 1954: J. Gen. Physiol. *38*. Quoted in detail because it covers such a variety of forms, 436 nm. is absorbed by both chlorophyll and carotenoids, 500 by carotenoids, 560 especially by phycoerythrin, 620 especially by phycocyanin, and 675 especially by chlorophyll A. The lowest quantum values are given in the table.

synthesis is not known—if different workers alter conditions to vary the end metabolic product, their results will differ; (3) most of the measurements require that the available instruments be pushed to the very limits of their sensitivity. The difficulty will be resolved only when workers from the opposite camps pool their resources and attack the problem jointly.

Even if the exact number of quanta required for the reduction of one molecule of carbon dioxide cannot be given with certainty at the present time, significant findings have emerged from the quantum efficiency studies. Investigators have consistently found the same quantum requirement (within the limits of error) for the reduction of one molecule of carbon dioxide, regardless of the type of cells studied—whether green plant cells; blue-green, red or brown algae; purple or green sulfur bacteria (Larsen *et al.,* 1952).

The photosynthetic sulfur bacteria, some of which use hydrogen sulfide as a hydrogen donor, are especially interesting since, on thermodynamic grounds, a lesser energy requirement might be expected, hydrogen sulfide being a highly reduced compound. However, experiments show that the same number of quanta are necessary for the reduction of one molecule of carbon dioxide. The same is true for strains of algae or bacteria which are adapted to use hydrogen gas. In certain purple bacteria various organic compounds containing potential chemical energy can be utilized as hydrogen donors, but the quantum requirement is still 8 to 12. This points again to the similarity of the photosynthetic mechanism wherever it is present.

If the efficiency of the photosynthetic conversion of light energy to chemical energy is calculated on the basis that 8 to 12 quanta (43,000 cal./mole quantum) of red light (660 nm.) are required to reduce 1 gram mole of carbon dioxide, the theoretical requirement being 118,000 cal./mole, a value of 23 to 43 per cent is obtained. The efficiency is 118,000 calories, divided by the product of the number of quanta required to reduce one carbon dioxide molecule, times the mole quantum at the particular wavelength. Considering the number of steps in the process of photosynthesis, the value of 23 to 43 per cent is a relatively high order of efficiency.

## 19.2 PAPER CHROMATOGRAPHY

In the general chromatographic method as introduced into biology by Tswett in 1906, various adsorbants—e.g., starch, aluminum oxide—are poured as a slurry into a tube (see Fig. 19.17*a*). An unknown is poured on the widened top of the tube and then "developed" by the slow trickle of an appropriate solvent through the column. The rate at which any compound migrates through the column depends upon the balance between its affinity for the solvent and its adsorption on the particles of the column. This method of separating compounds is spoken of as adsorption chromatography.

Another chromatographic method depends upon partition of a solute between two solvents in which the solute has differential solubility. If a solute has a greater solubility in one solvent than in another, the two solvents being immiscible, it is possible to remove and concentrate the solute in the solvent in which it has greater solubility.

Something like this separation can be achieved on a column of some powdered material if only one of the solvents wets the powder. The solvent wetting the powder

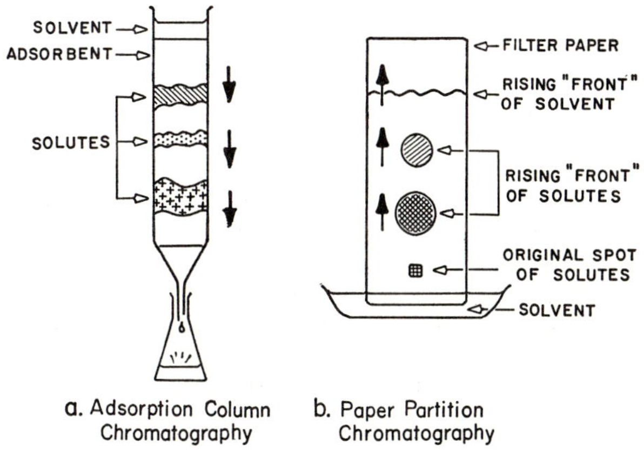


***Figure 19.17.*** Column and paper chromatography. (From Cantarow and Schepartz, 1967: Biochemistry. 4th Ed. W. B. Saunders Co., Philadelphia.)

will form a stationary solvent phase, and the powder in this case holds the solvent which wets it. In practice, the solute in the appropriate organic solvent is added to the top of the wet column and developed by washing with additional aliquots of the solvent. As the second solvent, immiscible with the first, moves through the column, the solute is partitioned between the two solvents, as if an infinite number of separatory funnels were being used in series to extract the solute. The solute moves as a band on the column. If instead of one solute, several have been dissolved in the solution poured on the top of the column, they will be separated as several bands on the column because of their differences in partition between the two solvents (Strain, 1958).

Paper chromatography, as developed in 1941 by Martin and Synge, is a variation of Tswett's partition chromatographic method. It is a method based also on the differential solubility of compounds in solvents, and it achieves separation by partition of the solute between the two solvents. The paper acts as a support for one of the solvents (in theory at least).

The solution to be tested is placed as a spot on a large sheet of filter paper. When the spot is dry the paper is placed with one edge dipping into a tray containing the mixture of solvents, the choice of solvents depending upon the solubility of the solutes to be chromatographed. After the solvent front has moved the desired distance beyond the sample spot, the solvent front is marked, and the paper is dried and sprayed with some agent to "develop" the spots of materials which have migrated from the original sample spot.

For example, a mixture of amino acids is spotted on filter paper, and the paper is dipped into a butyl alcohol–water mixture. Diverse amino acids are differentially soluble in these two solvents, some being more soluble in water, others more soluble in butyl alcohol. The alcohol-water mixture migrates along the filter paper fibers upward through the sample spot, and the various substances with differential solubility migrate to different distances from the original spot, those most soluble in butyl alcohol moving farthest. After the solvent front has advanced past the spot to a distance considered from experience to be adequate, the line of the solvent front is marked and the paper is dried. For detection of amino acids the paper is then sprayed with ninhydrin, which forms a purplish color on reacting with an amino acid. Control chromatograms, made from known mixtures of amino acids, are run in parallel with the unknown. From the positions of known amino acids the unknown acids can be determined (see Fig. 19.17*b*). From the depth of the color reaction quantitative data may be gathered (Block *et al.*, 1955).

The ratio of the distance traveled by any particular substance to the distance covered by the solvent is known as the *ratio of the fronts* or $R_f$. This ratio is usually constant for a given substance in a particular solvent and under given conditions (e.g., temperature). Thus the $R_f$ values determined from controls of known amino acids under similar conditions may be used as standards for identification of unknown amino acids.

When a "one-dimensional" chromatogram does not adequately separate the components of a mixture, the once-developed paper is placed, after it has dried, in a pan of another mixture of solvents in such a way that the movement of the second mixture of solvents occurs at right angles to the movement of the first, producing a "two-dimensional" chromatogram (Block *et al.*, 1955).

When the substances separated by paper chromatography are radioactive, they can be visualized by photography. The filter paper is pressed against a sensitive photographic film. The ionizing radiations from the radioactive material will so affect the film that on development the film will show darkening in proportion to the radiation, the amount of radiation depending upon the concentration of the radioactive substance. From the $R_f$ values of controls and from the density of the darkened areas on film exposed to known concentrations of radioactive material, identification and quantitative analysis of an unknown material are possible, even when the unknown is present only in traces, as, for example, after a few seconds of photosynthesis (Bassham and Calvin, 1962).

## *LITERATURE CITED AND GENERAL REFERENCES*


Arnold, W. A., 1933: Effect of ultraviolet light on photosynthesis. J. Gen. Physiol. *17*:135–143.

Arnold, W. A., 1965: An electron hole picture of photosynthesis. J. Phys. Chem. *69*:788–791.

Arnold, W., and Davidson, J. B., 1954: The identity of the fluorescent and delayed light emission spectra in *Chlorella.* J. Gen. Physiol. *37*:677–684.

Arnon, D. I., 1955: The chloroplast as a complete photosynthetic unit. Science *122*:9–16.

*Arnon, D. I., Nov. 1960: The role of light in photosynthesis. Sci. Am. *203*:105–118.

Arnon, D. I., 1961a: Changing concepts of photosynthesis. Bull. Torrey Bot. Club *88*:215–259.

Arnon, D. I., 1961b: Cell-free photosynthesis and the energy conversion process. *In* Life and Light. McElroy and Glass, eds. Johns Hopkins Press, Baltimore, pp. 489–566.

Arnon, D. I., 1963: Photosynthetic phosphorylation and a unified concept of photosynthesis. Proceedings of the Fifth International Congress of Biochemistry. Pergamon Press, Oxford, pp. 201–232.

Arnon, D. I., 1965: Ferredoxin and photosynthesis. Science *149*:1460–1470.

Bassham, J. A., and Calvin, M., 1962: The Photosynthesis of Carbon Compounds. W. A. Benjamin, New York.

Bassham, J. A., Shibata, K., and Calvin, M., 1955: Quantum requirement in photosynthesis related to respiration. Biochim. Biophys. Acta *17*:332–340.

Bishop, N. I., 1961: The possible role of plastoquinone (Q-254) in the electron transport system of photosynthesis. *In* Ciba Foundation Symposium on Quinones in Electron Transport. Wolstenholme and O'Connor, eds. Little, Brown & Co., Boston, pp. 385–404.

Bishop, N., 1966: Partial reactions of photosynthesis and photoreduction. Ann. Rev. Plant Physiol. *17*:185–208.

Blinks, L. R., 1950: Action spectra of chromatic transients and the Emerson effect in marine algae. Proc. Natl. Acad. Sci. U.S.A. *46*:327–333.

*Blinks, L. R., 1964: Accessory pigments and photosynthesis. *In* Photophysiology. Giese, ed. Academic Press, New York, Vol. 1, pp. 199–221.

Blinks, L. R., and van Niel, C. B., 1963: The absence of enhancement (Emerson effect) in the photosynthesis of *Rhodospirillum rubrum. In* Microalgae and Photosynthetic Bacteria. Special issue of Plant and Cell Physiology, pp. 297–307.

Blinks, L. R., and Skow, R. K., 1938: The time course of photosynthesis as shown by a rapid electrode method for oxygen. Proc. Natl. Acad. Sci. U.S.A. *24*:420–427.

Block, R. J., Durrum, E. L., and Zweig, G., 1955: Paper Chromatography and Paper Electrophoresis. 2nd Ed. Academic Press, New York.

Branton, D., 1968: Structure of the photosynthetic apparatus. *In* Photophysiology. Vol. 3. Current Topics. Giese, ed. Academic Press, New York, pp. 197–224.

Brown, A. H., 1953: The effect of light upon respiration using isotopically enriched oxygen. Am. J. Bot. *40*:719–729.

Burk, D., 1953: Photosynthesis: a thermodynamic perfection of nature. Fed. Proc. *12*:611–625.

Calvin, M., 1950–1951: The path of carbon in photosynthesis. Harvey Lect. *46*:218–251.

*Calvin, M., 1962: The path of carbon in photosynthesis. Science *135*:879–889 (Nobel lecture).

Clayton, R. K., 1962: Recent developments in photosynthesis. Bact. Rev. *26*:151–164.

Clayton, R. K., 1964: Physical aspects of the light reaction in photosynthesis. *In* Photophysiology. Giese, ed. Academic Press, New York, Vol. 1, pp. 144–197.

*Clayton, R. K., 1965: Modern Physics in Photosynthesis. Blaisdell Publishing Co., Waltham, Mass.

Davis, E. A., 1948: Photosynthetic studies with mutant strains of *Chlorella.* Science *108*:110–111.

*Duysens, L. N. M., 1964: Photosynthesis. Prog. Biophys. *14*:1–104.

Emerson, R., and Arnold, W., 1932: A separation of the reactions in photosynthesis by means of intermittent light. J. Gen. Physiol. *15*:391–420.

Emerson, R., Chalmers, R., and Cederstrand, C., 1957: Some factors influencing the long wave limit of photosynthesis. Proc. Natl. Acad. Sci. U.S.A. *43*:133–143.

Franck, J., 1951: A critical survey of the physical background of photosynthesis. Ann. Rev. Plant Physiol. *2*:53–86.

French, C. S., 1960: The chlorophylls in vivo and in vitro. *In* Encyclopedia of Plant Physiology. (Ruhland *et al.,* eds.), Springer-Verlag, Berlin, pp. 252–297.

French, C. S., 1961: Light, pigments and photosynthesis. *In* Life and Light. McElroy and Glass, eds. Johns Hopkins Press, Baltimore, pp. 447–474.

Fuller, R. C., and Conti, S. F., 1963: The microbial photosynthetic apparatus. *In* Studies on Microalgae and Photosynthetic Bacteria. Special issue of Plant and Cell Physiology, pp. 49–63.

Gaffron, H., 1960: Energy storage: photosynthesis. *In* Plant Physiology, a Treatise. Steward, ed. Academic Press, New York, Vol. 1B, pp. 3–277.

Goodwin, T. W. (ed.), 1966: Biochemistry of Chloroplasts. Academic Press, New York.

Haxo, F. T., and Blinks, L. R., 1950: Photosynthetic action spectra of marine algae. J. Gen. Physiol. *33*:389–422.

Hill, R., 1939: Oxygen produced by isolated chloroplasts. Proc. Roy. Soc. London *B127*:192–210.

Hoch, C., and Knox, R. S., 1968: Primary processes in photosynthesis. *In* Photophysiology. Vol. 3. Current Topics. Giese, ed. Academic Press, New York, pp. 225–251.

Iwamura, T., 1966: Nucleic acids in chloroplasts and metabolic DNA. Progr. Nucleic Acid Res. *5*:133–155.

Jagendorf, A. T., 1962: Biochemistry of energy transformations during photosynthesis. *In* Survey of Biological Progress. Glass, ed. Academic Press, New York, Vol. 4, pp. 181–344.

Kamen, M. D., 1963: Primary Processes in Photosynthesis. Academic Press, New York.

Kok, B., 1963: Correlation between P700, C700 and the low temperature emission bands. *In* Photosynthetic Mechanisms in Green Plants. Bishop and Gaffron, eds. Publication 1145, National Academy of Sciences of the U.S.A., National Research Council, Washington, pp. 45–55.

Kok, B., 1965: Photosynthesis: the path of energy. *In* Plant Biochemistry. Bonner and Varner, eds. Academic Press, New York, pp. 903–960.

Krasnovskii, A. A., 1959: The luminescence of chlorophyll in relation to photosynthesis. Biophysica *4*:1–16.

Krinsky, N. I., 1968: The protective function of carotenoid pigments. *In* Photophysiology. Vol. 3. Current Topics. Giese, ed. Academic Press, New York, pp. 123–195.

Larsen, H., Yocum, C. S., and van Niel, C. B., 1952: On energetics of photosyntheses in green sulfur bacteria. J. Gen. Physiol. *36*:161–171.

Martin, A. J. P., and Synge, R. L. M., 1941: A new form of chromatogram employing two liquid phases. 1. A theory of chromatography. 2. Application to the micro-determination of the higher monoamino-acids in proteins. Biochem. J. *35*:1358–1368.

Myers, J., 1951: On the mass culture of algae. Plant Physiol. *26*:539–548.

Myers, J., and Graham, R., 1963: Enhancement in *Chlorella.* Plant Physiol. *38*:105–116.

Nishimura, M. S., Whittingham, C. P., and Emerson, R., 1951: The maximum efficiency of photosynthesis. Symp. Soc. Exp. Biol. *5*:176–210.

Ochoa, S., and Vishniac, W., 1952: Carboxylation reactions and photosynthesis. Science *115*:297–301.

Packer, L., and Deamer, D. W., 1968: Studies on the effect of light on chloroplast structure. *In* Photophysiology. Vol. 3. Current Topics. Giese, ed. Academic Press, New York, pp. 91–122.

Park, R. B., and Biggins, J., 1964: Quantasome: size and composition. Science *144*:1009–1011.

Rabinowitch, E. I., 1945: Photosynthesis and Related Processes. Vol. 1. Chemistry of Photosynthesis. Interscience Publishers, New York.

Rabinowitch, E. I., 1951: Photosynthesis and Related Processes. Vol. 2. Part 1. Problems of Spectroscopy and Fluorescence of Photosynthetic Pigments. Interscience Publishers, New York.

Rabinowitch, E. I., 1956: Photosynthesis and Related Processes. Vol. 2, Part 2. Kinetics of Photosynthesis. Interscience Publishers, New York.

*Rabinowitch, E. I., and Govindjee. July 1965: The role of chlorophyll in photosynthesis. Sci. Am. *213*:74–83.

Racker, E., 1955: Synthesis of carbohydrate from $CO_2$ and $H_2$ in a cell free system. Nature *175*: 249–251.

Ruben, S., Randall, M., Kamen, M., and Hyde, J. L., 1941: Heavy oxygen ($O^{18}$) as a tracer in the study of photosynthesis. J. Am. Chem. Soc. *63*:877–879.

San Pietro, A., 1965: Non-heme iron proteins. *In* Energy Metabolism. Antioch Press, Yellow Springs, Ohio.

San Pietro, A., and Black, C. C., 1965: Enzymology of energy conversion in photosynthesis. Ann. Rev. Plant Physiol. *16*:155–174.

San Pietro, A., Greer, F. A., and Army, T. J. (eds.), 1967: Harvesting the Sun. Academic Press, New York.

Sistrom, W. R., Griffiths, M., and Stanier, R. Y., 1956: The biology of a photosynthetic bacterium which lacks colored carotenoids. J. Cell. Comp. Physiol. *48*:473–515.

Sogo, P. B., Carter, L. A., and Calvin, M., 1961: Electron spin resonance studies on photosynthetic materials. *In* Free Radicals in Biological Systems. Blois, ed. Academic Press, New York, pp. 311–335.

Strain, H. H., 1958: Chloroplast Pigments and Chromatographic Analysis. Pennsylvania State University, University Park, Pa.

Tswett, M., 1906: Physikalisch-chemische Studien uber das Chlorophyll. Die Adsorptionen. Ber. Deutsch. Botan. Ges. *24*:316–323.

van Niel, C. B., 1931: On the morphology and physiology of the purple and green sulphur bacteria. Arch. Mikrobiol. *3*:1–112.

van Niel, C. B., 1936: On the metabolism of the Thiorhodaceae. Arch. Mikrobiol. *7*:323–358.

van Niel, C. B., 1949: The comparative biochemistry of photosynthesis. *In* Photosynthesis in Plants. Frank and Loomis, eds. Iowa State College Press, Ames, Iowa, pp. 437–495.

Vernon, L. P., 1964: Bacterial photosynthesis. Ann. Rev. Plant Physiol. *15*:73–100.

Vernon, L. P., and Earhart, D. J., 1967: Current literature survey of photosynthesis—X. Photochem. Photobiol. *6*:163–175.

Vernon, L. P., and Seely, G. R., 1966: The Chlorophylls. Academic Press, New York.

Warburg, O., and Negelein, E., 1923: Uber den einfluss der Wellenlange auf den Energieumsatz bei der Kohlensaureassimilation. Ztschr. physik. Chemie *106*:191–218.

Warburg, O., Schocken, T., Burk, D., Hendricks, S., and Korzenovsky, M., 1949: The maximum efficiency of photosynthesis—a rediscovery. Science *110*:225–229.

Wassink, E. C., 1948: Photosynthesis. Ann. Rev. Biochem. *17*:559–578.

Weaver, E. C., and Bishop, N. I., 1963: Photosynthetic mutants separate electron paramagnetic resonance signals of *Scenedesmus.* Science *140*: 1095–1097.

Weaver, E. C., and Chon, H., 1966: Spin label studies in *Chlamydomonas.* Science *153*:301–303.

Whatley, F. R., and Losada, M., 1964: The photochemical reactions of photosynthesis. *In* Photophysiology. Giese, ed. Academic Press, New York, Vol. 1, pp. 111-154.

Whittingham, C. P., 1955: Energy transformations in photosynthesis and the relation of photosynthesis to respiration. Biol. Rev. *30*:40–64.

# CHAPTER 20 / BIOSYNTHESIS

## NUTRITIONAL PATTERNS

Photosynthesis represents only one aspect of the synthetic activity of a chlorophyll-containing plant cell. The cell can use the products of photosynthesis to synthesize the complex spectrum of small molecules (monosaccharides, oligosaccharides, fatty acids, amino acids, vitamins) and macromolecules (lipids, polysaccharides, proteins, nucleic acids) which it needs for life and reproduction.

Colorless plant cells, animal cells and bacteria ultimately depend upon the products of the green plant cell for their organic supplies—organic carbon, nitrogen and some vitamins. With these, and inorganic salts, they too can synthesize the small molecules and macromolecules which are required for maintaining life, and for growth and cell division. If it is assumed that the green plant cells (*photoautotrophic* cells) represent the pattern of nutrition with maximal synthetic activity, inasmuch as they synthesize all their organic requirements from the low energy compounds (carbon dioxide, water and salts) at the expense of light energy, all other types of cells (*heterotrophic* cells) may be considered cells in which the capacity for synthetic activities unique to the photoautotrophic cells have been progressively lost, the loss being greatest in heterotrophic cells requiring the most complex media for their growth.

There are different degrees of dependence of heterotrophic cells upon supply of organic nutrients. Some organisms (e.g., wild-type *Escherichia coli*), require only organic carbon (usually glucose) and various salts to synthesize all their other organic requirements for life. Others show progressive loss of capacity to synthesize various common molecules needed for functioning or growth and must obtain these directly from their nutrients. In this respect, animal cells are generally much more dependent upon their nutrients than other cells. This should not be taken to mean that animal cells are incapable of synthetic activity, as witness the genome of such cells (Chapter 6) which codes for a tremendous number of proteins, but rather that here syntheses start from the level of complex organic molecules which the cells obtain in their nutrients, and from these they synthesize their characteristic complex structures.

In some cases, however, complex nutrients are required by the cells but complex structures are not synthesized by them, as, for example, in the pleuropneumonia-producing organisms (PPLO); hence, the general synthetic activity has declined and the genome has in this case become correspondingly simplified. The noncellular phages represent replicating systems with even more simplified genomes, much of the synthetic activity required for replication of their own kind being dependent upon that present in the host cell. This reaches its extreme in some viruses which have only two genes coding for two proteins (see Chapter 6).

Among the organic requirements of many cells are vitamins needed in small amounts as components of various enzyme systems (Chapters 3 and 15). Such vitamins, like many other common organic compounds, are interchangeable in cells of different organisms which have similar enzyme systems. A discussion of the nutritional requirements of different kinds of cells and a classification of modes of nutrition is beyond the scope of this book but can be found in textbooks of microbiology and nutrition.

On the other hand, some bacteria (e.g., *chemoautotrophic* cells) synthesize all their organic cell requirements from only carbon dioxide, water and salts at the expense of oxidation of inorganic compounds. For example, colorless sulfur bacteria utilize energy liberated by oxidation of hydrogen sulfide to sulfate, and iron bacteria utilize the energy liberated by oxidation of ferrous to ferric iron. Extraordinary as such metabolism may appear to be, and interesting as it may be from an evolutionary standpoint, it represents a mode of nutrition exploited by only a small group of organisms and will not be considered in detail here (see Stanier and Gunsalus, 1960–1964).

## BIOSYNTHESIS OF SMALL MOLECULES

Perhaps the most successful attack on the problem of biosynthesis of small molecules is that pioneered by Beadle and Tatum (1941), largely with the bread mold *Neurospora,* for which they received the Nobel Prize in 1958. Wild-type *Neurospora* supplied with a base medium of sucrose, certain salts and the B vitamin biotin produces all the other growth factors and all the amino acids and other compounds it needs. However, when wild-type *Neurospora* is irradiated with ultraviolet radiations or x-rays, gene "blocks" may occur; that is, mutants may appear that are unable to synthesize some growth requirements synthesized by the wild type. These mutants grow only if the particular required compound is added to the base medium (see Figs. 20.1 and 20.2). This finding provides a tool for studying the synthesis of many cell requirements and has yielded results that would have been difficult by other means.

To illustrate the method described above, a number of mutants of *Neurospora* have been found that fail to produce either tryptophan (an aromatic amino acid) or some precursors of tryptophan. All these mutants will grow if tryptophan is added to the base medium. But by adding a variety of possible precursors of tryptophan to the medium of the mutants, and by studying the metabolic intermediates that accumulate in the medium, the course of the tryptophan synthesis can be analyzed. In this manner, as illustrated in Figure 20.3, it has been found that after mutation of gene 1, a culture of the mutant *Neurospora* synthesizes tyrosine or phenylalanine but is unable to make anthranilic acid from these. If it is supplied with anthranilic acid, it forms indole, which is then coupled with serine to form tryptophan. Mutation of gene 2 "blocks" synthesis of indole from anthranilic acid, although all the preceding syntheses occur. Provided with indole, *Neurospora* with a mutated gene 2 readily synthesizes tryptophan. In a culture of *Neurospora* with a mutation at gene 2, anthranilic acid accumulates (Beadle, 1949).

Many naturally occurring metabolic defects have been found that lead to the accumulation of peculiar intermediates. By identifying these intermediates it has been possible to map the pathways of biosynthesis and degradation of a great number of compounds. The agreement between data gathered on biosynthesis in animal cells and data on biosynthesis in *Neurospora* is impressive and leads one to a realization of the unity in the biochemistry of the cell, be it in the plant, animal or microorganism. Such unity for degradation of nutrients was realized long ago (Kluyver and Donker, 1926) but is now also apparent for the synthetic processes as well. The oldest homologies in life are

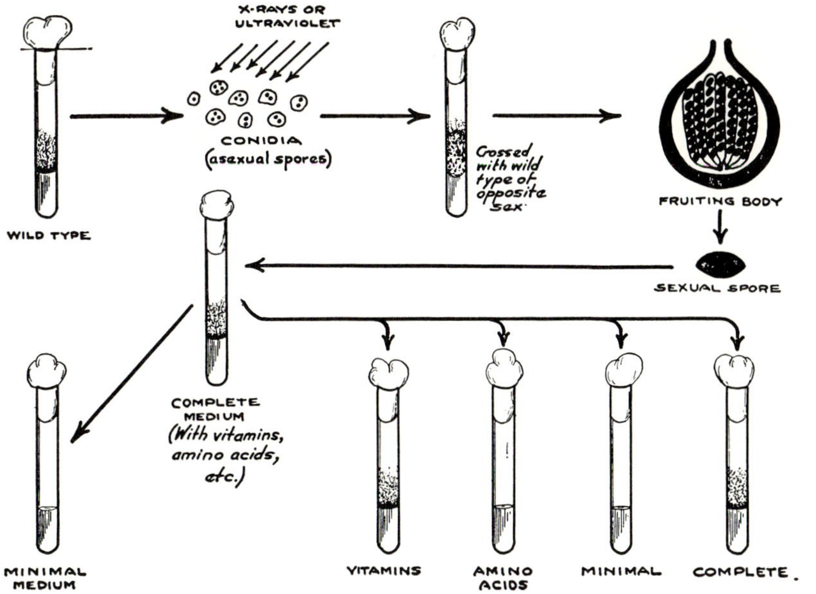


*Figure 20.1.* Experimental method for testing biochemical mutants in *Neurospora.* In practice the irradiated asexual spores must be sprouted to form a mycelium, which is fused with another mycelium developed from a nonirradiated asexual spore (see life cycle in Fig. 20.2). The fused mycelia form a fruiting body in which are developed the ascospores. It is the ascospores which are isolated for the tests shown. (From Beadle, 1947: Science in Progress *5*:176.)

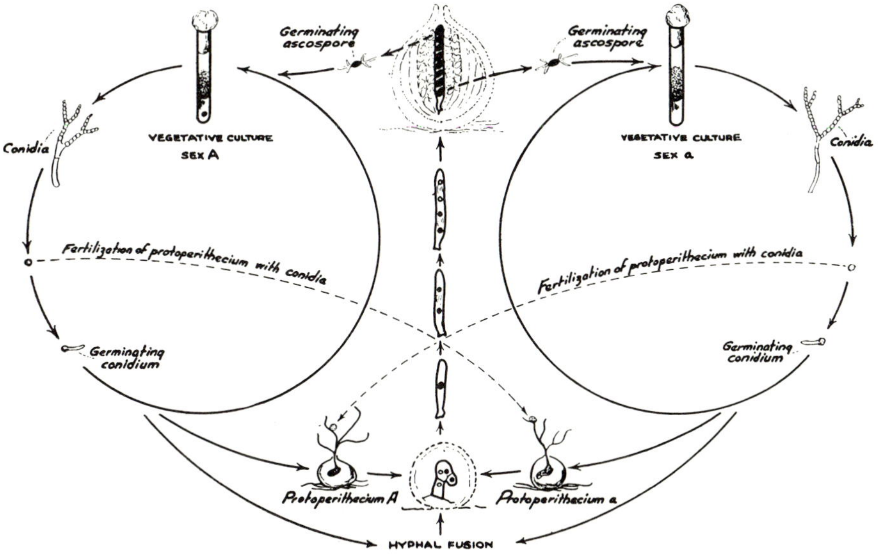


*Figure 20.2.* Diagram of the life cycle of *Neurospora crassa.* In addition to the sexual cycle shown, the mold is able to multiply asexually by microconidia and by means of fragments of the mycelium. (From Beadle, 1946: Am. Scientist *34*:36.)

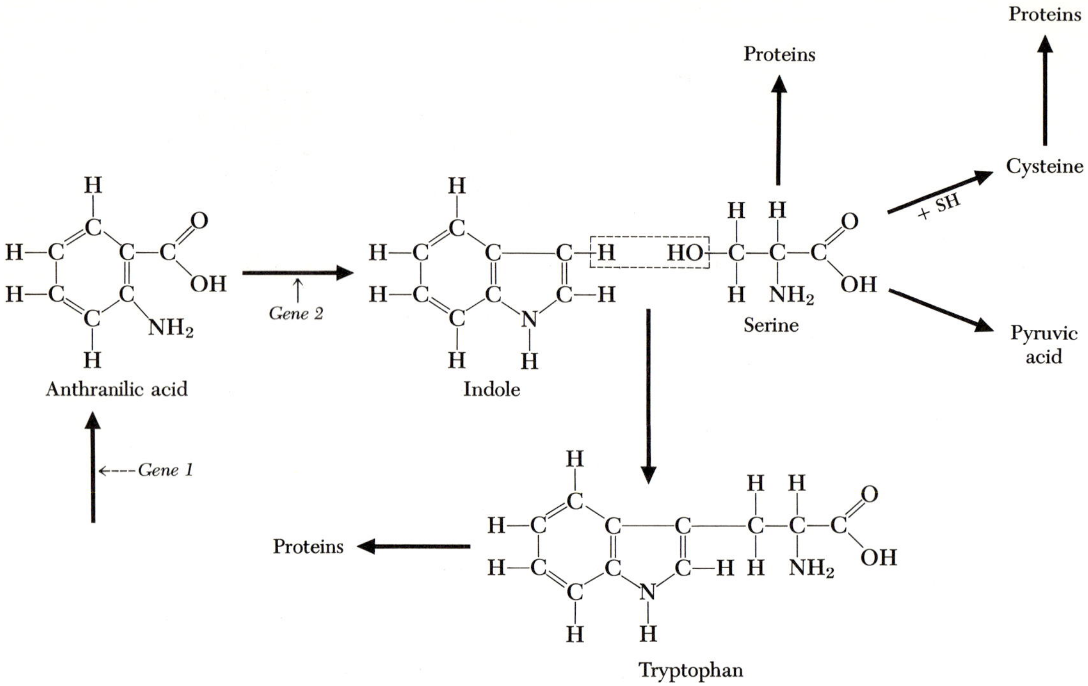


*Figure 20.3.* The action of various genes on tryptophan synthesis in *Neurospora.* (After Tatum and Bonner, from Beadle, 1946: Am. Scientist *34*:37.)

therefore at the cellular level (Szent-Györgyi, 1940).

The biosynthetic method used by Beadle and Tatum on *Neurospora* has been applied by many workers to other organisms, especially bacteria, e.g., the colon bacillus, *Escherichia coli.* By this method much has been learned about biosyntheses in cells of various vitamins, purines, pyrimidines and amino acids, and also about the isolation of the enzymes concerned.

## MACROMOLECULAR SYNTHESES

### BIOSYNTHESIS OF DNA

The biosynthesis of deoxyribonucleic acid (DNA) *in vitro,* through polymerization of nucleoside triphosphates in the presence of the appropriate enzyme, was successfully completed by Kornberg, and for this he received the Nobel Prize in 1959. The nucleic acid DNA consists of four kinds of nucleotides (units of purine or pyrimidine attached to a phosphoric acid residue through a pentose sugar) attached to one another, the phosphoric acid and pentose forming the backbone of the spiraling structure of the Watson-Crick DNA model (see Fig. 3.16). The two complementary DNA fibers are connected to one another by hydrogen bonds between their respective purine and pyrimidine bases—thymine (T) to adenine (A), and guanine (G) to cytosine (C).

The first step in the synthesis of DNA is the formation of the nucleoside triphosphates (activation) by the union of an ATP molecule with each of the four nucleoside monophosphates (phosphorylation) (Kornberg, 1960). This mechanism is not unique—it is the means whereby amino acids are activated before protein synthesis and fatty acids are activated before any of their synthetic reactions. It is also the means whereby coenzymes are phosphorylated.

For the four deoxyribonucleoside phosphates, the reactions are:

$$dAMP + ATP \rightarrow dAPPP + P_i \quad (20.1)$$
$$dGMP + ATP \rightarrow dGPPP + P_i \quad (20.2)$$
$$dCMP + ATP \rightarrow dCPPP + P_i \quad (20.3)$$
$$TMP + ATP \rightarrow TPPP + P_i \quad (20.4)$$

where d refers to the pentose deoxyribose; ATP refers to adenosine triphosphate; dAMP, to deoxyadenosine monophosphate; dGMP, to deoxyguanosine monophosphate; dCMP, to deoxycytidine monophosphate; TMP, to thymidine monophosphate; dAPPP, to deoxyadenosine triphosphate; dGPPP, to deoxyguanosine triphosphate; dCPPP, to deoxycytidine triphosphate; TPPP, to thymidine triphosphate; and $P_i$, to inorganic phosphate. An enzyme (*phosphorylase*) must be present to catalyze the reaction in each case. The nature of the reaction of the ATP with each nucleoside phosphate is shown in Figure 20.4*A*.

The next step in the synthesis of DNA is polymerization of the activated nucleoside triphosphates, as they are added one at a time to the end of a DNA molecule supplied to serve as a primer or template in the synthesis. If one of the triphosphates is omitted, synthesis is reduced by about 100-fold. Addition of a nucleoside triphosphate is accompanied by splitting off of a pyrophosphate. The pyrophosphate is in turn decomposed by the enzyme pyrophosphatase which is always present, with the liberation of two molecules of inorganic phosphate. These reactions are shown in Figure 20.4*B*. The high phosphate transfer potential (about 7000 cal./mole) of the triphosphates supplies the energy required for the synthesis. This occurs only in the presence of an enzyme, *nucleotide polymerase,* which in this instance was extracted from *E. coli.* It also proceeds only if some DNA is supplied as a primer. A similar polymerase has been isolated from mammalian cells (Bollum, 1963). The overall reaction is shown in Figure 20.4*C.*

DNA of high molecular weight serves best as primer. Denaturing the DNA by heat increases its efficiency, perhaps because breaking the hydrogen bonds holding the two antiparallel chains of nucleotides characterizing the DNA molecule makes the information more accessible. During replication each strand is presumably acted upon by the polymerase separately, as shown in Figure 20.5. The material enzymatically synthesized *in vitro* has been shown to have many physical and chemical properties indistinguishable from the natural primer; for example, it shows hyperchromicity on heating and melting (denaturation), although it renatures more readily than the primer after heating, and it requires the same temperature for 50 per cent melting (Tm). The sedimentation coefficient and intrinsic viscosity of the product also approach that of the primer (Ingram, 1965).

The function of the DNA primer in the synthesis of DNA is to supply a template. The exactness of the duplication is perhaps best illustrated by the nature of the DNA produced when different template models are supplied. Each DNA, it will be recalled, always has an equal number of adenine and thymine residues ($A = T$) and an equal number of guanine and cytosine ($G = C$) residues, but the ratio of $A + T$ to $G + C$ is a characteristic property of each type of DNA. In Table 20.1 are listed the primers used and the base analysis of the primer and product. The agreement is remarkably close. The enzyme, in some unknown and perhaps unique way, takes directions from the template and adds the nucleotides in the order and in the numbers appropriate to the primer supplied. It is always the double-stranded DNA that is produced by the enzyme, even when a single-stranded DNA, such as that from the tiny $\varphi$x174 virus (Sinsheimer, 1959), is provided as template. When DNA, which is normally double stranded but which had uncoiled and become single stranded in heating, was used as a template, the formation of only the normal double-stranded DNA was obtained. Evidently enough information is available in a single strand of DNA to enable the enzyme to synthesize the appropriate double-stranded DNA. With highly purified enzymes the single-stranded DNA (obtained by heating the double-stranded one) alone is effective (Kornberg, 1960).

Kornberg used an ingenious method consisting of labeling the nucleosides

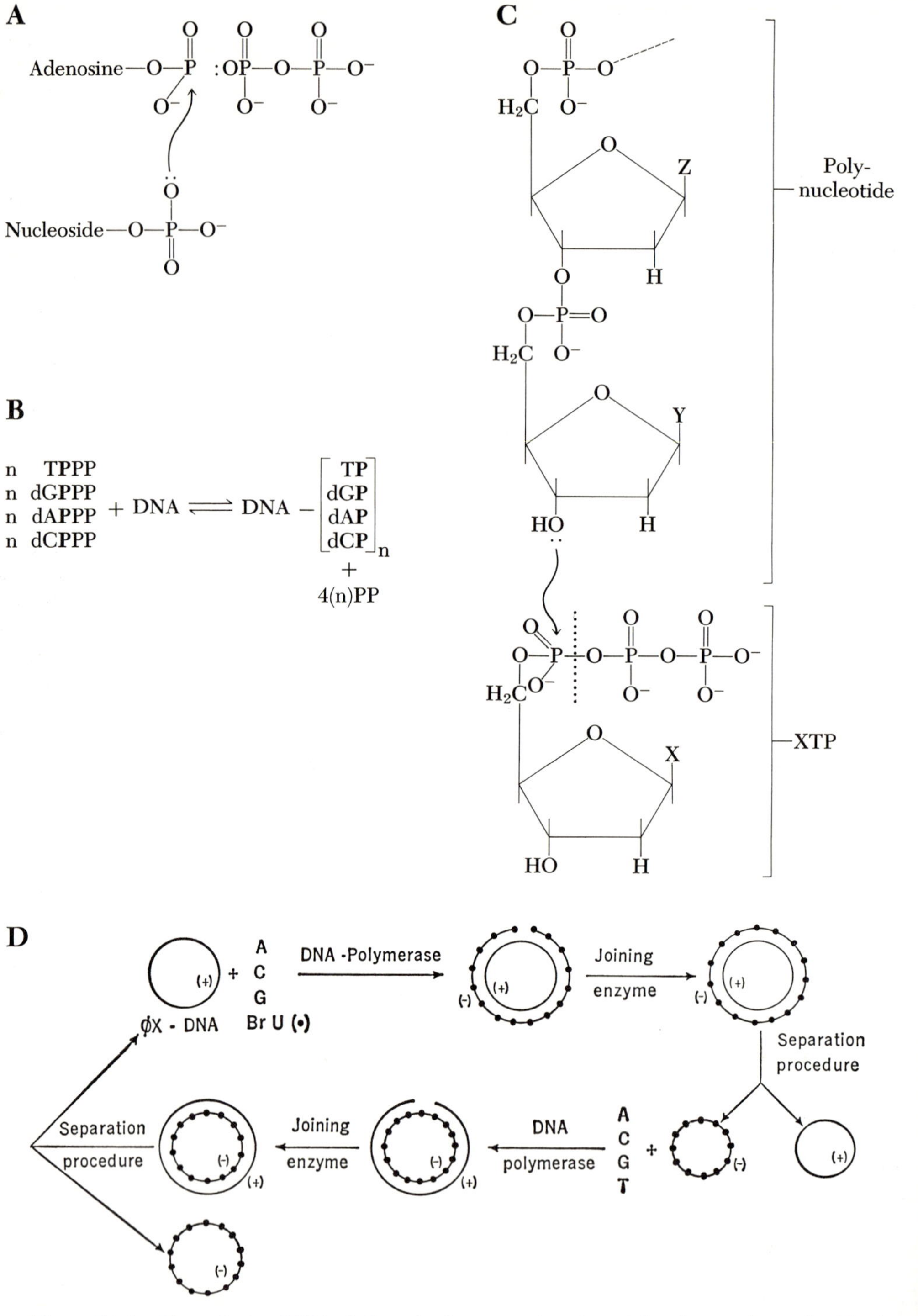


*Figure 20.4.* Biosynthesis of DNA. *A*, Reaction between adenosine triphosphate and a nucleoside phosphate. *B*, Overall equation for polymerization of the nucleoside triphosphates formed as a result of the reaction shown at *A*. *C*, Mechanism of attachment of a nucleoside to the polynucleotide chain. (*A*, *B* and *C* from Kornberg, 1960: Science *131*:1505.) *D*, Synthesis of infective viral DNA. (After Goulian, Kornberg and Sinsheimer, 1967.)

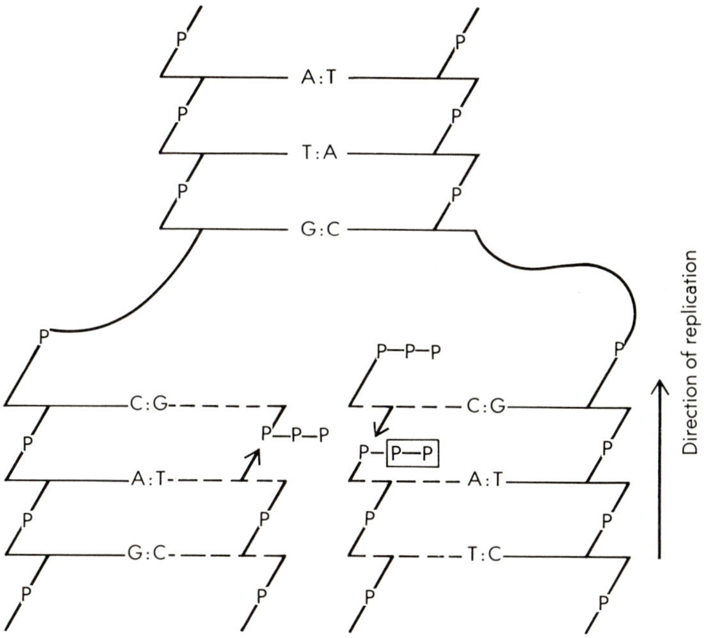


***Figure 20.5.*** Diagram showing the "growing point" in a replicating DNA molecule if both strands are duplicated simultaneously. (From Ingram, 1965: The Biosynthesis of Macromolecules, W. A. Benjamin, New York.)

**Table 20.1.** Chemical Composition of DNA Enzymatically Synthesized with Different Primers*

| | A | T | G | C | $\frac{A+G}{T+C}$ | $\frac{A+T}{G+C}$ |
|---|---|---|---|---|---|---|
| *Mycobacterium phlei* | | | | | | |
| Primer | 0.65 | 0.66 | 1.35 | 1.34 | 1.01 | 0.49 |
| Product | 0.66 | 0.65 | 1.34 | 1.37 | 0.99 | 0.48 |
| *Escherichia coli* | | | | | | |
| Primer | 1.00 | 0.97 | 0.98 | 1.05 | 0.98 | 0.97 |
| Product | 1.04 | 1.00 | 0.97 | 0.98 | 1.01 | 1.02 |
| Calf thymus | | | | | | |
| Primer | 1.14 | 1.05 | 0.90 | 0.85 | 1.05 | 1.25 |
| Product | 1.12 | 1.08 | 0.85 | 0.85 | 1.02 | 1.29 |
| Bacteriophage T2 | | | | | | |
| Primer | 1.31 | 1.32 | 0.67 | 0.70 | 0.98 | 1.92 |
| Product | 1.33 | 1.29 | 0.69 | 0.70 | 1.02 | 1.90 |
| A-T copolymer | 1.99 | 1.93 | $<0.05$ | $<0.05$ | 1.03 | 40 |

* *A*, adenine; *T*, thymine; *G*, guanine; *C*, cytosine.
From Kornberg, 1960: Science *131*.

with $^{35}P$ and incorporating them at one bond of the pentose moiety and then selectively splitting them at another bond, by which it proved possible to determine the nucleoside sequence (nearest neighbor analysis) in a DNA molecule. Then, by using as a primer a sample of synthesized DNA in which the sequence of nucleotide bases was known and again analyzing the product for nucleotide sequence, it was demonstrated that the sequence of nucleotides in the primer was duplicated in the newly synthesized DNA. Furthermore, it was possible to distinguish between two general models of DNA, one in which the strands were parallel with respect to their linkages and one in which they were antiparallel. Apparently, the two strands in a DNA molecule are antiparallel (see review in Ingram, 1965).

When a copolymer, made up only of two nucleotides—deoxyadenylate and deoxythymidilate—is supplied as a template to the polymerase and activated nucleotide triphosphates, only the copolymer is duplicated even though all four nucleoside triphosphates are present in the reaction mixture. This polymer (Table 20.1, last line) indicates the extent to which the enzyme copies what is presented as a template and also shows that the hydrogen bonding between pyrimidine and purine must be of considerable importance in the structure of the polymer produced. It has recently proved possible, with the highly purified DNA polymerase in an *in vitro* system containing the four nucleoside triphosphates, to synthesize φx174 virus DNA, which is infective to *E. coli* (Goulian, Kornberg and Sinsheimer, 1967). This is a major advance over the previous attempts in which the artificially synthesized DNA (e.g., the transforming principle) was biologically inactive even though biochemically similar to the template provided (Ingram, 1965). It will be remembered that the φx174 virus is a single-stranded circular DNA which is infective only in the circular form. When the DNA strand (arbitrarily designated the + strand) enters the bacterium it induces by complementation the formation of the − strand, which is also circular DNA. In the *in vitro* experiment it is necessary to include a polynucleotide-joining enzyme, recently discovered in *E. coli* by Martin Gellert, in addition to the template DNA, the DNA polymerase and the four nucleoside triphosphates. The polynucleotide-joining enzyme connects the two ends of the DNA molecule. By supplying the *in vitro* system with a thymidylic acid analogue (5-bromodeoxyuridylic acid) the DNA made in complementation to the +DNA template becomes heavier than the DNA template, thus making it possible to separate the two by density gradient centrifugation. Possible contamination of one DNA strand with the other was checked by labeling the +DNA with $^{32}P$ and the components of the −DNA with $^{3}H$. The synthetic DNA was obtained free of template DNA and, in circular form, was infective to *E. coli* (supplied in protoplast form, that is, free of cell walls) and induced the formation of +DNA. The +DNA was then separated from the −DNA and also was found infective (Fig. 20.4*D*). These experiments have many important implications.

## BIOSYNTHESIS OF RNA

Three types of RNA are found in cells: *ribosomal RNA* (rRNA), the major RNA component, comprises 70 to 80 per cent of the total; *messenger RNA* (mRNA) makes up 5 to 10 per cent of the total; and *soluble* or *transfer RNA* (sRNA) makes up the remainder. It will be recalled (see Chapters 3, 6) that all three RNAs conjugate with DNA of the same species, indicating complementation and sharing of information between the two molecules. Messenger RNA, when so conjugated, occupies a larger portion of the DNA than rRNA and sRNA. It is therefore no surprise that the synthesis of all three types of RNA is DNA dependent (Smellie, 1963).

A single enzyme, RNA polymerase, is responsible for incorporating the four ribonucleoside triphosphates, ATP, GTP, CTP and UTP, which contain the bases adenine, guanine, cytosine and uracil, respectively, in the highly energetic triphosphate form. The presence of DNA is necessary for RNA synthesis and the synthesized RNA is specific to the DNA (template) used for its induction. Thus when the DNA is obtained from *E. coli,* the ratio of adenine plus uracil to guanine plus cytosine $\left(\text{i.e., } \frac{A + U}{G + C}\right)$ is 1.0 in the RNA synthesized, which is characteristic of the colon bacillus RNA. When the DNA used is obtained from calf thymus, the $\left(\frac{A + U}{G + C}\right)$ ratio in the RNA synthesized is found to be 1.3, which is characteristic of calf thymus RNA.

Additional evidence of the specificity of DNA as an inducer of RNA synthesis was obtained by using a synthetic DNA-like polydeoxyribonucleotide, e.g., one containing adenine and thymine deoxyribonucleotides (A-T copolymer). In this case the RNA-like polyribonucleotide, synthesized under DNA influence, consisted of polymers of uracil and adenine polyribonucleotides in alternating sequence. When polythymine deoxyribonucleotide was used as the template, adenine polyribonucleotide consisting of only adenine ribonucleotide units was synthesized (Horwitz and Furth, 1962).

*The DNA-dependent RNA polymerase* appears to be distributed in bacteria and the nuclei of most cells. Either magnesium or manganese ion is required for functioning of this polymerase. Since of the two ions magnesium is found in higher concentration in the cells, it is probably the ion which generally performs this catalytic function. The DNA-dependent RNA polymerase requires also an active sulfhydryl group for functioning and is therefore readily poisoned by substances which react with this group. During its purification a sulfhydryl reagent is generally added to insure activity of this group in the enzyme (Horwitz and August, 1963).

As polymerization proceeds in the presence of all four ribonucleoside triphosphates, pyrophosphate is formed. The pyrophosphate is split into inorganic phosphate by the enzyme pyrophosphatase, which is always present in cells. Omission of one of the triphosphates decreases the rate of the polymerization to nearly zero (Ingram, 1965).

The DNA template used in the RNA syntheses is not used up in the reaction and remains separate from the RNA synthesized. For example, a sample of transforming DNA has the same transforming activity as it had before it was used in synthesizing six to ten times its weight of RNA (Ingram, 1965).

It appears that both strands of the DNA serve as templates during polymerization *in vitro,* but there is some evidence that only one strand is used *in vivo.* However, the DNA from the single-stranded DNA of phage φx174 serves as a primer for RNA polymerase *in vitro,* although it is less active than double-stranded DNA. The RNA strands produced by polymerization have been shown to be antiparallel by the same procedures used to demonstrate the antiparallel nature of the two strands in a DNA molecule (review in Ingram, 1965).

A "terminal adding" enzyme completes the synthesis of sRNA begun by the polymerase by adding to its amino acid–acceptor end the characteristic three nucleotides pCpCpA (p refers to phosphate). The precursors of these nucleotides are the corresponding nucleoside triphosphates.

The synthesis of all three types of RNA (DNA-dependent) is inhibited by the antibiotic actinomycin D which probably binds the guanine-containing regions of the RNA polymerase system. In partial inhibition, there is preferential inhibition of the G- and C-copying sequences in DNA (Reich and Goldberg, 1964).

RNA is also synthesized from nucleoside diphosphates in the presence of polynucleotide phosphorylase, an enzyme discovered in extracts of the microbe *Azotobacter vinelandii* (DNA-independent) in Ochoa's laboratory by Grunberg-Manago *et al.* (1956). The enzyme has since been extracted and purified from a variety of microorganisms, plant cells and pig liver cell nuclei (Grunberg-Manago, 1963). The enzyme catalyzes the synthesis of high molecular weight polyribonucleotides with the release of inorganic phosphate:

$$\text{nBRPP} \rightarrow (\text{BRP})_n + \text{nP}_1 \quad (20.5)$$

where B stands for base (adenine, hypoxanthine, uracil or cytosine); R, for ribose; and $P_i$, for inorganic phosphate.

Single polymers containing one of the nucleotides—AMP (adenosine monophosphate), UMP (uridine monophosphate), CMP (cytidine monophosphate) or IMP (inosine monophosphate)—as the only basic units have been obtained by incubating the enzyme with the corresponding nucleoside diphosphate in each case, and even nucleosides not normally present in RNA are polymerized. Mixed polymers of adenine and uridine ribonucleoside phosphates or of all four nucleoside phosphates of RNA can be obtained if two or four of the nucleoside diphosphates, respectively, are supplied.

The polymers have molecular weights of the order of 30,000 to 2,000,000; therefore they are in the range of molecular weights for RNA, as obtained from different sources. They display many RNA-like properties. For example, the polynucleotides even react with tobacco mosaic virus protein to form noninfective virus-like particles. It is interesting that no template RNA molecule is required for the polymerization of RNA in presence of polynucleotide phosphorylase. However, no specific nucleotide sequence is found in the synthesized RNA, the sequence appearing to be a matter of chance. Such synthetic polynucleotides formed from various mixtures of nucleoside diphosphates have played an important part in deciphering the genetic code for protein synthesis.

The function of polynucleotide phosphorylase* in the cell is not known but it may function to decompose mRNA, which in bacterial cells generally has a short life span. Neither rRNA nor sRNA is decomposed at an appreciable rate by the enzyme (Ingram, 1965).

Three kinds of rRNA are now recognized: one in the large (50S) and one in the small (30S) ribosomal subunits, and one (low MW) of unknown function.

*It will be recalled that enzymes catalyze the reaction in both directions to equilibrium (*in vitro*). Therefore, the polynucleotide phosphorylase in presence of RNA alone will decompose it until an equilibrium between the RNA and the products is reached.

## BIOSYNTHESIS OF PROTEIN

Proteins have been of great interest for many years. A great deal is known of protein composition and much has been learned about protein synthesis. An important early observation that dinitrophenol, which is known to uncouple phosphorylations, also inhibits protein synthesis led to the realization that not the amino acids but rather their "activated" phosphorylated forms are probably used in synthesis of proteins in the cell. Use of amino acids labeled with radioactive substances has made it possible to measure their rate of incorporation into proteins, and thus to follow the steps of protein synthesis in the living cell. Histochemical and cytological studies, especially by Brachet and Caspersson (see Chapters 3 and 6), have implicated RNA of the ribosomes of the cytoplasm in protein synthesis.

Although the pathway by which the present knowledge of protein synthesis has been achieved is somewhat devious (Zamecnik, 1960; Nisman and Pelmart, 1964), it is now clear that the synthesis of protein in the cell includes a number of steps and the participation of several forms of RNA.

The first step in protein synthesis involves activation of the amino acid by reaction with ATP in the presence of an enzyme, forming an *amino acyl–adenosine monophosphate complex* (aa-AMP) (Step 1, Fig. 20.6*A*). The aa-AMP remains bound to the activating enzyme. The next step is the preparation of an acceptor site for the amino acid on soluble RNA, usually written sRNA,* by reaction with cytidine triphosphate (CTP) and adenosine triphosphate (ATP) (Step 2, Fig. 20.6*B*). The aa-AMP–enzyme complex is next attached to the phosphorylated sRNA, and AMP and the enzyme are released, as shown in Step 3 in Figure 20.6*C*.

Presumably at least 20 such sRNA molecules occur. They are the low molecular weight fraction of RNA dissolved in the ground substance, liberated into the

*sRNA (soluble RNA) is the same as tRNA (transfer RNA) and adapter RNA.

supernatant of disintegrated cells, and precipitated at pH 5—therefore also called the pH 5 fraction. When a particular amino acid (labeled) is added until the saturation level is reached, addition of any other labeled amino acid at this time leads to further labeling of the protein, indicating a lack of competition between the two amino acids (Hoagland, 1959; Davidson, 1960). This could best be explained by postulating for the second amino acid a separate soluble RNA molecule different from that involved with the first amino acid. These sRNA molecules are now being fractionated and identified. A specific enzyme is probably involved in

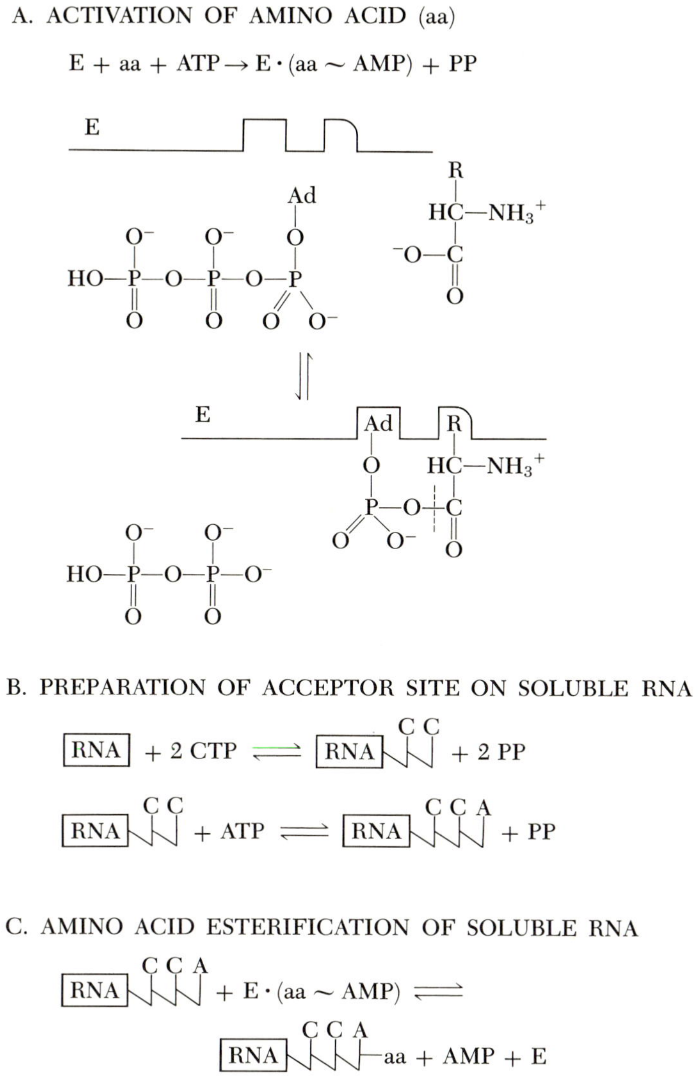


***Figure 20.6.*** Schematic representation of preparatory steps in protein synthesis concerned with amino acid activation. *A*, Phosphorylation. ATP represents adenosine triphosphate; AMP, adenosine monophosphate; aa, amino acid, reaction and schematic representation of mechanism; E, enzyme surface; Ad, adenine; R, a radical on the amino acid; PP, pyrophosphate. Only the alpha amino group and the carboxyl group are shown in detail. (From Hoagland, Keller and Zamecnik, 1956: J. Biol. Chem. *218*:345.) *B*, Preparation of the acceptor site on soluble RNA by phosphorylation. CTP represents cytidine triphosphate; C, cytidine. *C*, Addition of amino acid to the soluble RNA. Note that the enzyme is released for further activity after esterification is completed. (From Zamecnik, 1960: Harvey Lect. *54*:265.)

the attachment of each of the 20 amino acids to the respective molecule of soluble RNA (see Figs. 20.8 and 20.10).

It has been found that a given amino acid may attach to any one of several different sRNA molecules. This might be expected in view of the degeneracy of the code, indicated by multiple coding for a specific amino acid (Chapter 6). Soluble RNA recognizes its location by base pairing on mRNA and if several base triplets code for an amino acid one might expect that the corresponding sRNAs with complementation to each such triplet might also do so (Ingram, 1965).

On heating, soluble RNA shows hyperchromicity (increased absorption at 260 nm.), indicating that there is some base pairing which lowers absorption at 260 nm. in its native state. It is suggested that the single strand of which sRNA is composed loops back upon itself, attachment between adjacent portions of the loop occurring where base complementarity occurs (Ingram, 1965). Four sites of activity are expected on sRNA: the site recognizing the amino acid (CpCpA) which is alike in all sRNA molecules, the site recognizing the specific amino acid–activating enzyme, the coding recognition site and the ribosome recognition site. A hypothetical structure for sRNA is shown in Figure 20.7.

It was also observed that not only soluble RNA but also ribosomal RNA is needed for amino acid incorporation into a protein (Allfrey and Mirsky, 1961). Presumably, sRNA carries the activated amino acid to the rRNA, and in the presence of guanosine triphosphate (GTP) as a source of energy, the amino acid is transferred to the rRNA while the sRNA is liberated (see Fig. 20.9). The ribosomes,

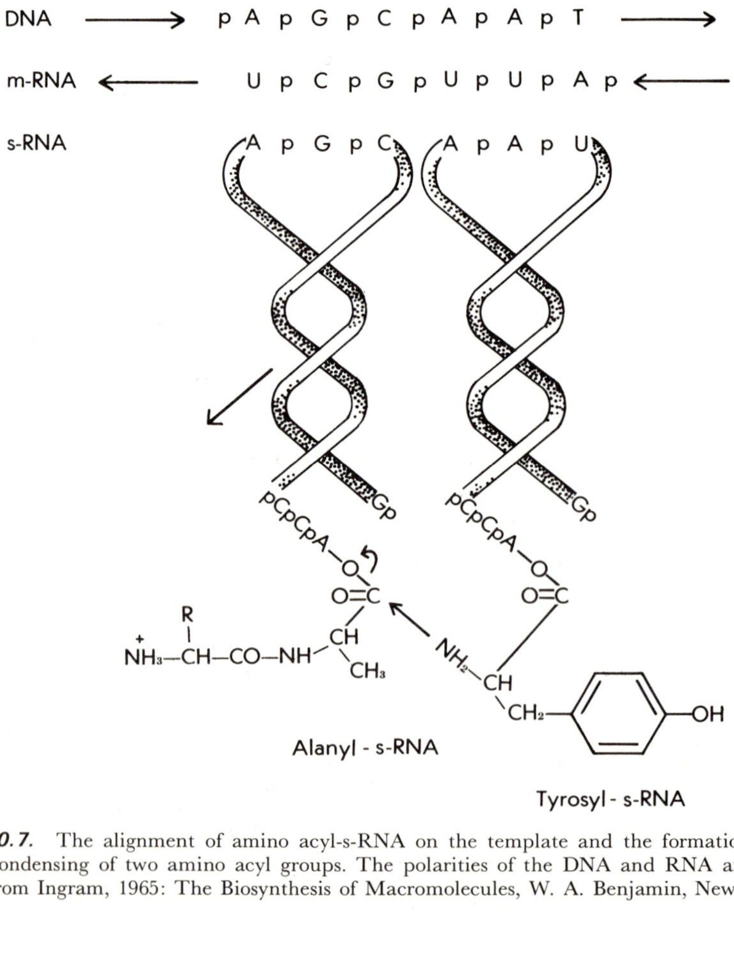


***Figure 20.7.*** The alignment of amino acyl-s-RNA on the template and the formation of a peptide bond by the condensing of two amino acyl groups. The polarities of the DNA and RNA are indicated by the arrows. (From Ingram, 1965: The Biosynthesis of Macromolecules, W. A. Benjamin, New York.)

however, are not active in this respect until they are first combined with messenger RNA (mRNA). Magnesium ions are also required in order to obtain a stable bonding of mRNA to rRNA and for the stability of the ribosomes themselves. It is the message on mRNA which is translated by the orderly arrangement of the acyl-amino molecules, one by one by complementation of the base sequences on sRNA (see Fig. 20.7). The ribosome presumably acts only as a carrier and will function for any mRNA supplied to it.

The idea of a messenger RNA and its functioning in this manner was first developed by Jacob and Monod (1961). The strongest early evidence for the concept was obtained by Volkin and Astrachan (1956), who infected *E. coli* with T-2 phage and found that synthesis of protein and nucleic acids of the bacterium ceased and that, during the lag before the beginning of synthesis of the protein characteristic of the T-2 coat, a rapid RNA synthesis took place. The synthesized RNA was complementary in its base composition to the phage DNA but not to that of the bacterial cell, and had a short life span. Since then, more conclusive evidence from tracer experiments has accumulated so that at present the concept is well documented and accepted. Messenger RNA is short-lived in bacteria but may have a long life in some cells, as in reticulocytes which produce hemoglobin and give rise to red blood cells (Lipmann, 1963).

Ribosomes act in groups in cells or extracts which carry on active protein synthesis. Such groups are called polyribosomes or *polysomes* (see Fig. 5.6) (Rich, 1963). Such aggregation was discovered when the ribosomes from actively synthesizing extracts were separated by centrifugation. The sedimentation constants of the "ribosomes" were found to be much larger than those of preparations in which no protein synthesis was occurring and in which a low concentration of magnesium ions was present. Electron micrographs of the sediment confirmed the presence of aggregates rather than single ribosomes. This is thought to indicate that several ribosomes transcribe the message at the same time; as one drops off after completing its protein molecule, another slips on (Fig. 20.8).

Protein synthesis can be readily followed by using ascites tumor cells at

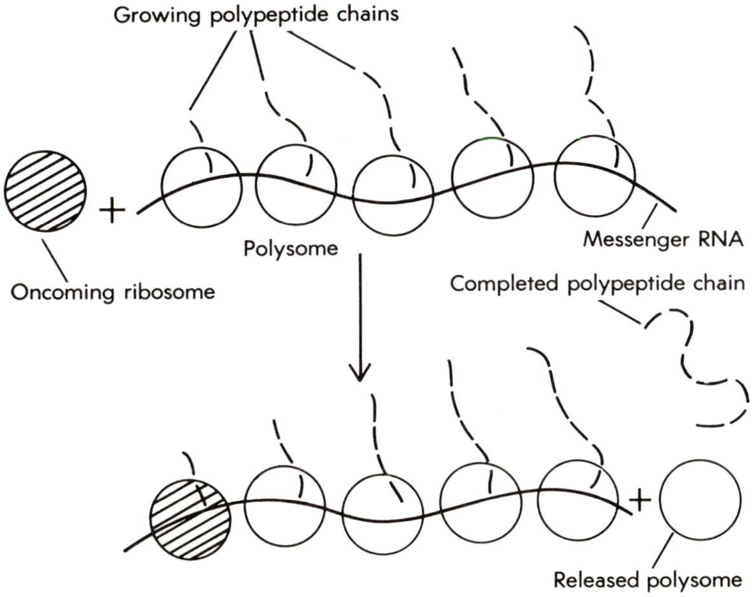


***Figure 20.8.*** A schematic model of polysome function. (After Warner, Knopf and Rich, from Ingram, 1965: The Biosynthesis of Macromolecules, W. A. Benjamin, New York.)

20° C. Incubation of tumor cells with a $^{14}C$-labeled amino acid is followed by the appearance of radioactivity, first in the transfer RNA fraction, more slowly in the ribosomal RNA, then in the ribosome protein and finally in the soluble protein of the cell. For these reactions guanosine triphosphate is needed, probably as a source of high energy phosphate bonds. GTP probably serves to attach the amino-acid sRNA complex to the ribosome.

Peptide bonding between successive amino acids attached to the ribosome then takes place by what has been described as a "zipper action," apparently starting with the amino acid valine at the N terminus of the chain and closing bond after bond until the protein molecule is finished in accordance with the template presented, the C-terminal amino acid being the last one added. Meanwhile, each sRNA is liberated as its amino acid joins in the formation of protein (Fig. 20.9). New amino acid molecules are added to the growing chain at the rate of about two a second, and a protein molecule of 150 amino acids is finished in 1.5 minutes. The amino acids are in an activated state and presumably need only to be juxtaposed in the appropriate steric relationships for this to happen.

Following protein synthesis, the next step would be the separation or "peeling off" of the protein molecule from its template on the ribosome (Fig. 20.9) (Zamecnik, 1960). The overall scheme of protein synthesis is shown in Figure 20.10. It is probable that condensing enzymes and releasing enzymes are also required for protein synthesis, although little is known about these at present. That the peptide chain is formed in linear fashion from one end is apparent from the studies of Dintzis (1961). Dintzis used a cell-free system derived from reticulocytes which carry on an active synthesis of primarily one protein (hemoglobin). He incubated the cells with $^{14}C$-labeled amino acids for a long time to mark the peptide chains evenly. He then exposed the preparations to $^{3}H$-labeled amino acids for graded pulse periods. It is argued that if synthesis is sequential there should be a progressively greater labeling with $^{3}H$ relative to $^{14}C$ with increased pulse periods. Brief labeling was found to mark only the por-

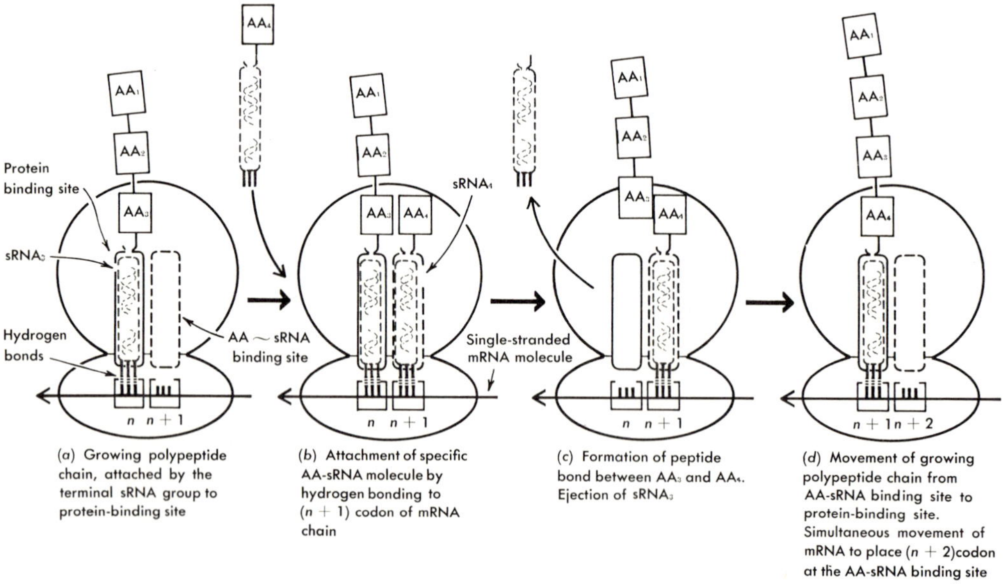


***Figure 20.9.*** Diagrammatic representation of the stepwise growth of the polypeptide chain. The two subunits of the ribosome are shown; the protein binding site (to the left) and the AA-sRNA binding site are shown on the ribosome. (From Watson, 1965: The Molecular Biology of the Gene. W. A. Benjamin, New York.)

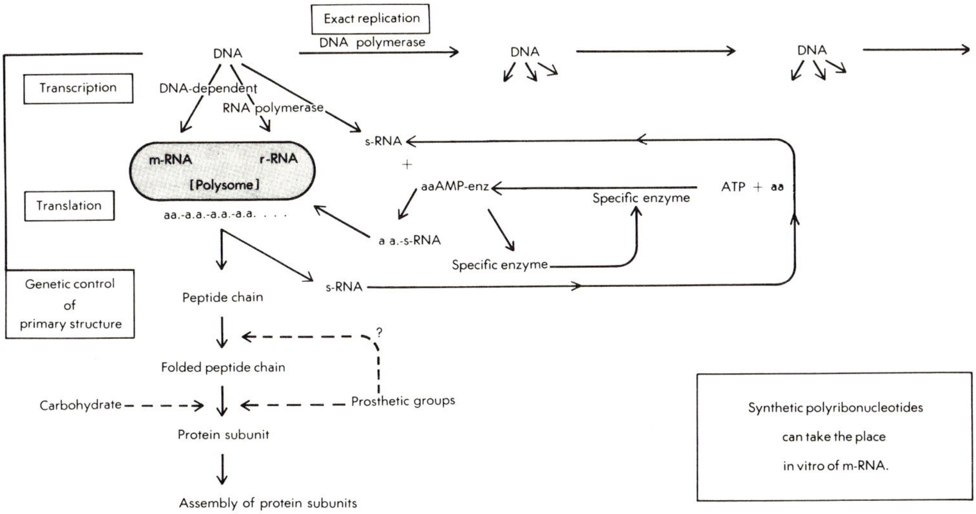


***Figure 20.10.*** The overall scheme for DNA, RNA and protein synthesis. (From Ingram, 1965: The Biosynthesis of Macromolecules, W. A. Benjamin, New York.)

tion of the polypeptide chain nearest the carboxyl (terminal) end, whereas longer periods labeled proportionally larger portions away from the terminus until the entire chain was labeled.

Protein synthesis occurs to some extent also in the nucleus, as is shown by uptake of labeled amino acids into proteins of nuclei. The nucleus, like the cytoplasm, contains the enzymes for activation of amino acids and for their incorporation into proteins. Nuclei depleted of DNA by treatment with deoxyribonuclease lose their capacity for incorporation of ATP and of amino acids, but upon addition of DNA, synthesis is restored* (Crick, 1958).

It is of great importance to note that specific protein syntheses may be induced in cells. For example, it will be recalled that enzymes can be induced in microorganisms in response to the presence of a specific nutrient for which the organism does not have a constitutive enzyme. Thus *E. coli* develops a $\beta$-galactosidase in response to the presence of lactose in the medium (see Chapter 14). Another case of induced protein synthesis is the development of immunity, that is, the production of antibodies in response to antigens. The active group (hapten) in the antigen is generally a carbohydrate (polysaccharide) or a protein, and the antibody protein is highly discriminatory in its action, indicating its "custom-built" synthesis.

Protein synthesis in cells is readily inhibited by the antibiotic chloramphenicol, which has a structure resembling phenylalanine and presumably takes its place on the sRNA and is transferred to the polypeptide in process, terminating its growth. However, RNA and DNA synthesis may continue under these conditions. Radiation can inhibit DNA synthesis without immediately affecting other syntheses, indicating partial separation of the three processes. Stopping both protein and RNA synthesis, however, also stops DNA synthesis (Doudney and Billen, 1961).

The antibiotic puromycin has a structure resembling that of amino-acyl-sRNA and therefore blocks protein synthesis at a relatively late stage. Experiments indicate that in its presence small peptide chains are released from the ribosomes, protein synthesis failing to go to completion. One puromycin molecule was found to be bound to each N-terminal end, presum-

**Acetabularia* continues to synthesize protein for several weeks after enucleation. This apparent paradox is resolved if the report that DNA is present in some chloroplasts is found to be general, in view of the role of DNA in protein synthesis.

ably forming a premature C-terminal end because puromycin lacks the sRNA bonding to maintain attachment of the peptide chain to the ribosome (Ingram, 1965).

## BIOSYNTHESIS OF POLYSACCHARIDES

In the "informational" molecules such as DNA, RNA and proteins, there is distinct ordering of the units forming the information-carrying sequences that form the polymer. In polysaccharides the molecules are generally not ordered into information-carrying sequences, nor are they of a definite chain length. This is perhaps understandable since polysaccharides serve either as storage molecules or as structural components. Polysaccharides, as the name implies, are composed of many subunits. Synthesis of polysaccharides requires that individual subunits be in an activated state. For example, synthesis of sucrose from fructose and glucose in organisms is a result of three enzymatic reactions, the source of energy being compounds of high phosphate transfer potential, as shown in Figure 20.11. However, it will be recalled that sucrose is hydrolyzed by invertase with a small change in free energy (Chapter 15). Theoretically, by increasing the concentration of the products one should be able to reverse the hydrolytic reaction. This occurs only to some extent in the laboratory in a mixture of lipase and fats, not in the case of carbohydrase action.

In the synthesis of *glycogen,* the enzyme *glycogen synthetase* serves to catalyze the transfer of glucose residues in activated state from UDP-glucose to the nonreducing ends of a glycogen primer. As a result, the chain is lengthened and UDP (uridine diphosphate) is released in stoichiometric ratio. The reaction proceeds with a decrease in free energy of approximately 3000 cal./mole. Glycogen consists of a branching system of linearly arranged glucose units. For the formation of branch points, enzymes other than the synthetase are necessary. Thus, the *liver-branching enzyme* forms branch points on the straight polyglucose chains by transferring to it segments of straight chains (Lehninger, 1965; Ingram, 1965).

Starch is a polysaccharide whose main constituent is the unbranched polysaccharide, amylose. Synthesis of starch also requires activated glucose units, ATP and UTP serving as energy sources. Cellulose is a polysaccharide in which the glucose units are linked by $\beta$-D-(1,4)glucosidic bonds. The chains of cellulose are of indefinite length, and resistance of cellulose to reagents may stem in part from the close packing in parallel alignment of such linear units. Synthesis of cellulose requires the presence of GTP, the guanine analog of UTP, but the mechanism by which this occurs is still under investigation (Ingram, 1965).

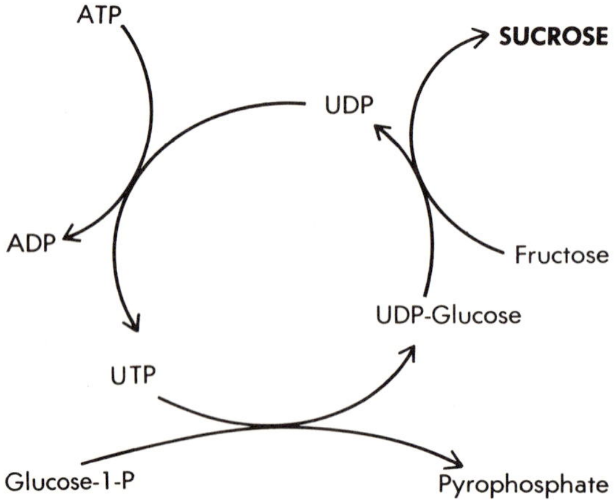


***Figure 20.11.*** Scheme of sucrose synthesis involving three enzymatic reactions. UDP is uridine diphosphate; UTP is uridine triphosphate. (From Ingram, 1965: The Biosynthesis of Macromolecules, W. A. Benjamin, New York.)

**Table 20.2.** Sequential Reactions in the Biosynthesis of Phosphatidyl Choline

| |
|---|
| *Activation of fatty acids* |
| RCOOH + ATP + CoASH → RCOSCoA + AMP + pyrophosphate |
| R′COOH + ATP + CoASH → R′COSCoA + AMP + pyrophosphate |
| 2 AMP + 2 ATP → 4 ADP |
| 2 pyrophosphate + 2 $H_2O$ → 4 phosphate |
| *Activation of glycerol* |
| glycerol + ATP → 3-phosphoglycerol + ADP |
| *Activation of choline* |
| choline + ATP → phosphorylcholine + ADP |
| phosphorylcholine + CTP → CDP-choline + pyrophosphate |
| pyrophosphate + $H_2O$ → 2 phosphate |
| *Assembly reactions* |
| 3-phosphoglycerol + RCOSCoA → 1-acyl 3-phosphoglycerol + CoASH |
| 1-acyl 3-phosphoglycerol + R′COSCoA → α-phosphatidic acid + CoASH |
| α-phosphatidic acid + $H_2O$ → 1,2-diglyceride + phosphate |
| 1,2-diglyceride + CDP-choline → phosphatidyl choline + CMP |
| CMP + ATP → CDP + ADP |
| CDP + ATP → CTP + ADP |
| Sum: 2 fatty acids + glycerol + choline + 8 ATP → phosphatidyl choline + 8 ADP + 8 phosphate |

From Lehninger, A. L., 1965: Bioenergetics. W. A. Benjamin, New York.

## BIOSYNTHESIS OF LIPIDS

*Lipids* are generally molecules of small molecular weight, seldom over 1000. Since they are often assembled from several types of molecules, the synthesis of lipids appears somewhat more complex than that of carbohydrates. However, in essence, synthesis consists of activation of each of the reacting molecules through reaction with ATP and CTP. The series of reactions involved in the formation of a phospholipid, phosphatidyl choline or lecithin, is shown in Table 20.2.

## RATE OF BIOSYNTHESIS

Lehninger (1965) has made some interesting calculations of the rate of biosyntheses in the *Escherichia coli* cell. The chemical constitution of *E. coli* is well known and the rate of duplication under optimal conditions is about once every 20 minutes. On the basis of these data it can be shown that 12,500 molecules of lipid and about 32 molecules of polysaccharide are synthesized per second by a single *E. coli* cell. The *E. coli* cell produces some 1400 molecules of protein per second, each one containing some 500 amino acids of 20 different kinds, arranged in some precise order as dictated by the genetic code. About 12 RNA molecules are synthesized per second but only one duplication of the DNA molecules occurs during the division interval (time between divisions) in *E. coli.*

The rate of production of molecules occurs more slowly in cells with a long division interval (such as 24 hours for some mammalian cells in tissue culture).

Lipids have been synthesized in the laboratory by man but only slowly. Only recently has a true protein of relatively large molecular weight (insulin) been synthesized in the laboratory (Katsoyannis, 1966; Bodanszky and Bodanszky, 1967). This emphasizes the efficiency of the methods employed by the cell.

The rate of ATP turnover is remarka-

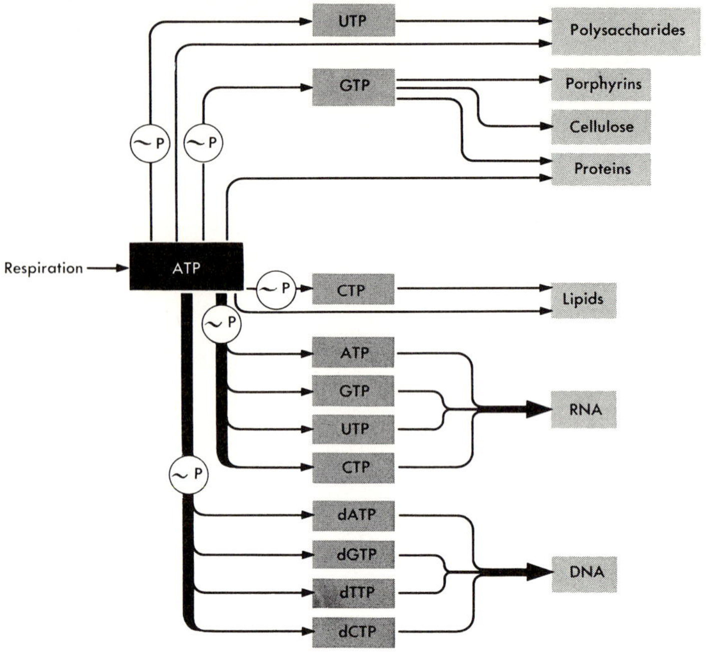


***Figure 20.12.*** Channeling of phosphate bond energy of ATP into specific biosynthetic routes. (From Lehninger, 1965: Bioenergetics. W. A. Benjamin, New York.)

ble in an active *E. coli* cell. Lehninger (1965) has calculated that about two and a half million molecules of ATP are used per second in an *E. coli* cell growing at its maximal rate. About 88 per cent of the energy released by the ATP is used in the synthesis of protein, 3.7 per cent in lipid synthesis, 3.1 per cent in RNA synthesis, 2.5 per cent in DNA synthesis and 2.7 per cent in carbohydrate synthesis.

As discussed previously, the terminal phosphate of the ATP is split off with a decrease of free energy of about 7000 cals./mole (*orthophosphate cleavage*). However, in many of the biosynthetic reactions considered here, pyrophosphate is split from ATP, forming AMP and pyrophosphate (POP) (*pyrophosphate cleavage*). The change in free energy is about the same in both pyrophosphate and orthophosphate cleavage. The pyrophosphate still retains one high energy phosphate bond but this energy cannot be used to recharge AMP or ADP. The pyrophosphate is split by the enzyme pyrophosphatase with a decrease of free energy about 7000 cals./mole; therefore, the total cost in loss of free energy for biosynthetic reactions in which POP is formed is about $2 \times 7000$ cals./mole. There is, therefore, what Lehninger calls a greater thermodynamic "pull" in such coupled reactions than when only one high energy phosphate is removed.

It is interesting to note that ATP energy is channeled through many other triphosphates during the synthesis of various compounds. These relationships are summarized in Figure 20.12. No satisfactory explanation has been suggested for such channeling.

It is evident that the synthesis of the highly ordered macromolecules proceeds with a decrease in entropy. The system as a whole, however, shows an increase in entropy, the ordered energy in various triphosphates having been expended in increasing the orderliness of the macromolecules.

## CHANGES IN BIOSYNTHESES—MUTATIONS

A mutation is a sudden but relatively permanent change in heredity. It is to be expected, on the basis of probabilities, that in the course of a long series of replications of genic DNA a mistake will occur. Such a mistake might well produce changes in synthesis of the specific protein with profound effects on the organism; the frequency of mistakes is greatly accelerated by some chemicals (Lawley, 1966) and by radiations (see Chapter 9).

The method of "fingerprinting" proteins has made it possible to document some mutations and to demonstrate that a mutation with gross effects on the cell or organism may rest upon a simple alteration in synthesis of a protein molecule, perhaps changes in only one amino acid at a given location in the protein molecule. For example, the hereditary disease of man known as sickle cell anemia results from the substitution of a molecule of the amino acid valine for glutamic acid, the fourth amino acid of the $\beta$ chain of the hemoglobin molecule. As a result, the hemoglobin of the mutant (diseased individual) is so altered in its physical properties that the entire erythrocyte takes on a new form—that of a sickle instead of a disc—which is much less effective in transporting oxygen. In another mutant type, replacement of the glutamic acid by the amino acid lysine at the same position on the hemoglobin molecule gives rise to still another pathological condition. Several other amino acid substitutions on hemoglobin have been found to have corresponding effects on the protein molecule and on the red blood cell in the mutants (Ingram, 1958; Duncan, 1964).

Many other mutations have also been analyzed. In *Neurospora crassa* several mutations affect the enzymes tyrosinase and tryptophan synthetase, with over 100 available mutants in the latter case. A single gene mutation involves a single amino acid alteration in the enzymic proteins produced, and this causes drastic effects on the properties of the enzyme. In *Escherichia coli,* tryptophan synthetase, alkaline phosphatase and $\beta$-galactosidase have all been similarly studied after mutation and in all cases the fully analyzed "fingerprint" shows substitution of an amino acid at one locus on the enzyme protein molecule (Yanofsky and St. Lawrence, 1960; Carlton and Yanofsky, 1965; Immamoto and Yanofsky, 1967).

There is considerable evidence to support the concept of "one gene–one polypeptide chain." The gene expresses itself by the formation of a polypeptide chain which, if enzymatic, carries out the syntheses designated by the code in the DNA. A slight change in this polypeptide chain—even the replacement of a single amino acid at a critical point on the molecule—results in such a change of its properties that the reaction it catalyzes is carried out more slowly or not at all. This deficiency in the genic DNA template is thus reflected in a deficiency of the polypeptide chain produced under direction of the particular gene. The resultant deficiency shows up as a mutation in the cell or the organism (Yanofsky, 1960, 1967).

What now remains is an exciting prospect, almost within reach—namely, the determination of the sequence of bases in the DNA of the gene, which will elucidate the nature of the genic coded message to the enzymes of the cell (Yanofsky and St. Lawrence, 1960; Yanofsky, 1967a,b,c).

## DEVELOPMENT OF CELL ORGANELLES

Little is known about the origin and development of the various organelles and other structures in the cell. It is likely that to a considerable extent such structures owe their origin to operation of weak cohesive forces attracting molecules into the most stable conformation. Proteins, for example, show, in addition to the primary peptide bond structure, secondary and tertiary (even quaternary) structure mostly as a result of hydrogen bonds and other types of weak forces, upon which depend some of their specific properties. Denaturation by heat (and other agents) ruptures the bonds that form such structures. For example, an enzyme loses its specific catalytic properties when it is heated but on cooling the catalytic activity reappears, essentially at the original level of activity. This indicates that on cooling the higher order of structure destroyed by heat has re-formed spontaneously, since it occurs without input of energy.

Chemical disaggregation of protein also is sometimes reversible; for example, hemoglobin is split by urea into two half-molecules of molecular weight 34,000. When urea is removed, the active hemoglobin molecule is re-formed.

This principle of "self-assembly" (Watson, 1965) may be extended to more complex structures, such as cell membranes. Thus, when a cell membrane is ruptured a new membrane is quickly re-formed (e.g., Fig. 11.2). It is thought that lipid molecules, held in reserve or newly synthesized, are quickly attracted to the membrane in existence, forming cohesive bonds with it more readily than with the polar cytoplasm, thus extending it (see also Woodward and Munkres, 1967).

Some attempts have been made to gain insight into the processes involved in the formation of various cell organelles by studying what appear to be analogous reactions in colloidal systems. The interaction of colloids of opposite charge in the formation of tactoids and coacervates (formed by various weak forces) has already been considered in Chapter 4. Some globular inclusions in the cell have been likened to coacervates, and spindle-shaped fibers, and even structures such as the division spindle, have been likened to tactoids.

Alternatively, it is thought that many of the organelles in the living cell may arise *de novo* as outgrowths of the cell membrane (Robertson, 1960); it is also postulated that they result from replication of existing organelles which act as templates, guiding the enzymes in depositing the new material (DeRobertis *et al.,* 1965).

## ABIOTIC ORGANIC SYNTHESES AND THE ORIGIN OF LIFE

It is inevitable that the question of the origin of life should arise whenever the spectacular advances in biosynthesis of cellular constituents are reviewed. As the syntheses of various organic compounds—vitamins, sugars, fatty acids, amino acids, polypeptides, proteins and nucleic acids—are studied, one asks how these substances were first formed on the lifeless planet, and how the aggregations of such materials were enclosed in a membrane and attained life and the ability to replicate. Much has been learned about the synthesis of various organic compounds from precursors such as might have been present in the primeval atmosphere of the abiotic earth by subjecting them to heat, radiations and electric discharges which simulate conditions on the surface of the primeval earth. The viruses which duplicate themselves may represent an experiment in replication, but their inability to liberate energy from organic compounds makes them dependent upon host cells and consequently they do not appeal as the first forms of life. It is possible to consider here only a few of the many experiments and speculations present in the extensive literature (Oparin, 1957; Florkin, 1960).

Reproduction, growth, nutrition and response, together, are considered primary properties of life, although each of these properties may be found separately in lifeless physical systems. It is difficult to conceive of replication without either DNA or RNA, or at least RNA, although it has been speculated that perhaps enough information is present in the protein molecule for replication by some mechanism other than the one finally adopted by cells in the course of evolution. Watson (1965) points out that the accuracy of a protein template in replication would be so low as to be useless.

It is equally difficult to conceive of growth, which involves reducing the entropy in the cell or organism (not of the entire system, of course), without coupled reactions involving transfer of energy from compounds of high phosphate transfer potential. Even the synthesis of small molecules involves energy from such compounds, and macromolecular synthesis, as we have seen, does not proceed without them. However, Calvin and Calvin (1964, p. 168) have stated, "Not only can the building blocks of today's organisms be generated by abiogenic processes, but the basic 'energy currency' (ATP) used by all organisms can be formed in a similar abiogenic conversion of the prime energy sources, ionizing radiation and light."

It appears necessary also that the reactions in the primeval cell be confined inside a compartment, separated from the environment by a selectively permeable membrane, presumably containing lipid. Finally, it is an indication of its orderliness that the living cell produces only one of two possible optical isomers, e.g., of the D and L forms of amino acids (Wald, 1957).

According to Oparin in his book *The Origin of Life on Earth* (1957), the original organic substances on the cooling surface of the earth were probably highly reduced compounds, e.g., methane and other hydrocarbons, water and ammonia. The hydrocarbons and their derivatives offer immense chemical possibilities. Using them, almost all the organic substances present in nature can be reproduced in the laboratory. For example, by oxidation of acetylene, aldehyde may be produced. Aldehydes can be polymerized to form sugars. Organic acids may be formed by oxidations, and from them, amino acids, in the presence of ammonia. By polymerization of amino acids, proteins are probably formed. In the relative absence of oxygen in the primitive earth's atmosphere it is likely that photochemical reactions were of greater importance than oxidations requiring oxygen.

It is thought that although the reduced compounds of the earth's primitive atmosphere absorb in the short ultraviolet portion of the spectrum—methane at wavelengths shorter than 150 nm.; water, 170 nm.; and ammonia, less than 220 nm. —it is likely that sufficient short ultraviolet radiation still passed through the atmosphere to effect various organic syntheses. However, spark discharges were also available. In an atmosphere lacking in oxygen, and on an earth devoid of life, the organic chemicals, formed under the energetic action of electric discharges and the short ultraviolet radiations, were stable and remained for long periods, giving ample opportunity for interaction with one another.

The primitive atmosphere of the earth was presumably well supplied with organic materials, mostly hydrocarbons and their products, which could have served as sources of energy for nutrition. While such materials were available, the primitive cell did not have to synthesize its organic constituents from inorganic matter, but enzymes (proteins) were needed for catalyzing the reactions for energy-yielding metabolism.

The question arises, could the chemicals of which the cell is composed have arisen in the primeval atmosphere? This can be answered affirmatively most clearly in the following manner: "In a sense organisms demonstrate to us what organic reactions and products are *possible.* We can be certain that, given time, all of these must occur. Every substance that has been found in an organism displays thereby the finite probability of its occurrence. Hence, given time, it should arise spontaneously" (Wald, 1954, p. 49). It is also clear from what has already been said (Chapter 1) that reactions in the cell occur in accordance with the principles of thermodynamics, that the cell channels reactions and accelerates them by possessing enzymes, but it cannot make a thermodynamically impossible reaction go, although it does couple endergonic and exergonic reactions to provide a source of free energy for synthesis. The constitution of a possible primeval atmosphere and possible pathway of chemical evolution is shown in Figure 20.13.

In the presence of one or another energy source such as ultraviolet light, spark discharge (high voltage) and ionizing radiation, methane, water and ammonia have been shown to produce amino acids (Ponnamperuma, 1967). Without ammonia, but with carbon dioxide, such compounds as sugars and fatty acids have been demonstrated (Calvin and Calvin, 1964). The very important biocatalysts, porphyrins, have been produced from succinic acid and glycine (Calvin, 1956). Recently even lipids have been identified in irradiated mixtures of methane, water and carbon

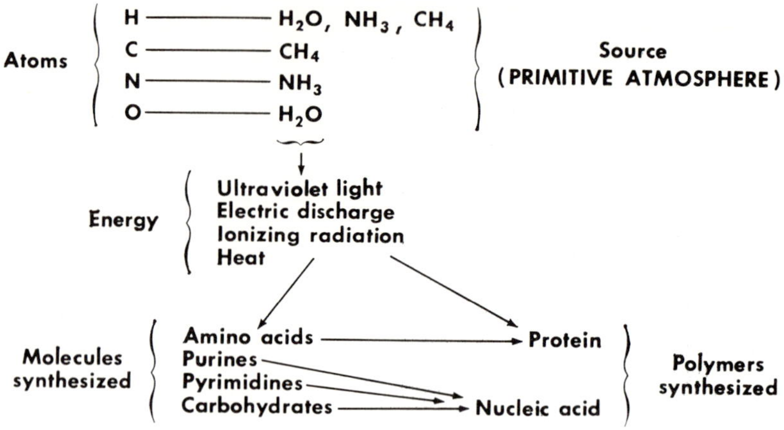


*Figure 20.13.* Chemical evolution: compounds in the primeval atmosphere and possible synthesis of organic compounds on the primitive earth. (From Young and Ponnamperuma, 1964: Early Evolution of Life. D. C. Heath & Co., Boston (BSCS Pamphlets #11) by permission of the Biological Sciences Curriculum Study.)

dioxide (Allen and Ponnamperuma, 1967). When hydrocyanic acid is used as a substrate, irradiation produces purine and pyrimidine bases, and with appropriate additions, nucleosides. Irradiation of cyanide to which phosphate compounds have been added produces nucleotides (Ponnamperuma, 1967). Neither polynucleotides nor proteins have been obtained as a result of action of radiations or electric discharges on the primeval atmosphere, but it is conceivable that in the presence of the appropriate catalysts or conditions, even these syntheses might also have occurred. Calvin and Calvin (1964) indicate that hydrocyanic acid, abundant on the early earth, may have served for the dehydration synthesis of polymers from the monomers in the case of carbohydrates, proteins and nucleic acids.

Polypeptide and protein syntheses have, however, been achieved by heat acting on mixtures of amino acids under conditions which might have prevailed on the surface of the primeval earth. Fox (1960, 1964) has shown that application of heat, especially to drying or dry mixtures of amino acids, can induce polymerization; dissolving the material in water then yields globules of proteinoids. Such proteinoids have many of the properties of proteins, e.g., similar color reactions, infrared spectra, ranges of molecular weights, solubility properties and electrophoretic properties, and function as proteolytic substrates and nutrient for bacteria. Some amino acids are much more readily incorporated into the proteinoids than others, indicating that the process as it occurs in the abiotic manner is not random but somewhat selective. The proteinoids do not serve as antigens, differing in this respect from natural proteins. Some critics have objected to the experiments on the basis that proteinoids are produced only at high temperatures, for example, heating a mixture of amino acids at 170° C. for 6 hours, but it has been shown that in the presence of polyphosphoric acid a polymer is readily formed from 15 amino acids at temperatures as low as 70° C. Such a mixture could have been present on the primitive earth and the temperature is not unreasonable. It is interesting that hydrolysis of the polymers yields both D and L forms of amino acids, a finding again at variance with natural proteins, which usually consist of the L-acids, although a few bacteria incorporate some D-acids also (Fox, 1960).

In the absence of life on the primeval earth degradation of organic compounds would have been very slow, because neither organisms nor oxidation would have affected them in an abiotic and nearly anaerobic planet. Therefore, mixtures of various types of organic chemicals might have persisted for prolonged periods. Cosmologists estimate that our galaxy took origin from the primordial hydrogen gas cloud about 20 billion years ago and our immediate solar system (including the earth) arose about 4.7 billion years ago. Fossil algae have been reported dating 2 to 2.7 billion years ago (Holmes, 1954), which would indicate that life had already come into being and developed to the stage of photosynthetic organisms. The inception of life, therefore, dates to between 4.5 and 2 billion years ago (Greenstein, 1961), sufficient time for a large number of trials and errors.

It appears likely that primitive life was heterotrophic, dependent upon the organic materials present in the atmosphere or produced from these constituents by the action of radiations, electric discharges and heat. Furthermore, it is likely that such life was facultatively anaerobic since oxygen was not a prominent constituent of the primitive atmosphere. Nurtured by the organic compounds which it used to duplicate itself, life soon depleted the environment of an available supply of such materials. But inherent in life is the capacity for change, and presumably some forms of life early developed a capacity to utilize the energy in sunlight for the reduction of carbon dioxide. According to Horowitz (1945), the synthesizing capacity of cells increased stepwise as organic nutrients dwindled one by one in the environment, because then each genetic acquisition would have survival value. Thus, cells capable of photosynthesis appeared only after the organic nutrient available for life had dwindled or disappeared, at which time the capacity for photosynthesis had supreme survival value. According to Oparin (1957), free oxygen appeared in the atmosphere only after forms of life capable of photosynthesis had developed. When oxygen became available, efficient aerobic reactions were in turn added to the anaerobic processes utilized by the cells (Nursall, 1959). On the other hand, the content of oxygen isotopes in sea water is different from that of the atmosphere, suggesting an origin of the oxygen in the atmosphere by some means other than or in addition to photosynthesis, since the latter liberates oxygen only from water (Dole, 1949; see Chapter 19).

Once polymers became available, as well as the high energy phosphates and a number of catalysts (based upon porphyrin and metal ions), a good deal of the machinery of life was in being. This is the more so because the secondary, tertiary and quaternary structures of proteins required in the cell structure are the most stable state thermodynamically, and therefore form spontaneously once the primary structure has been formed. The same reasoning applies to nucleic acids. Even such structures as membranes in many instances will re-form spontaneously, the membrane representing the state of greatest stability of its components. Such considerations do not minimize the problems yet remaining, e.g., to explain the initial spark of life, but they simplify the requirements. Although "spontaneous generation" is an improbable event, if it occurred once it should be possible again, once the necessary conditions are provided.

## LITERATURE CITED AND GENERAL REFERENCES

### BIOSYNTHESIS

Allfrey, V. C., and Mirsky, A. E., Sept. 1961: How cells make molecules, Sci. Am. *205*:74–82.

Allen, J. M., 1961: The Molecular Control of Cellular Activity. McGraw-Hill Book Co., New York.

Anfinsen, C. B., 1959: The Molecular Basis of Evolution. John Wiley & Sons, New York.

Beadle, G. W., 1949: Genes and biological enigmas. *In* Science in Progress. Baitsell, ed. Yale University Press, New Haven, Vol. 6, pp. 184–265.

*Beadle, G. W., 1959: Genes and chemical reactions in *Neurospora.* Science *129*:1715–1719.

Beadle, G. W., and Tatum, E. L., 1941: Genetic control of biochemical reactions in *Neurospora.* Proc. Natl. Acad. Sci. U.S.A. *27*:499–506.

Bodanszky, M., and Bodanszky, A. A., 1967: From peptide synthesis to protein synthesis. Am. Sci. *55*:185–196.

Bollum, F. J., 1963: "Primer" in DNA polymerase reactions. Progr. Nucleic Acid Res. Molec. Biol. *1*:1–26.

Campbell, P. N., 1965: The biosynthesis of proteins. Progr. Biophy. Molec. Biol. *15*:1–38.

Carlton, B. C., and Yanofsky, C., 1965: Amino acid substitutions in the A proteins of tryptophan synthetase mutants and revertants. J. Biol. Chem. *240*:690–693.

Crick, F. H. C., 1958: On protein synthesis. Symp. Soc. Exper. Biol. *12*:138–163.

Davidson, J. N., 1960: The Biochemistry of the Nucleic Acids. 4th Ed. Methuen & Co., London.

DeRobertis, E. D. P., Nowinski, W. W., and Saez, F. A., 1965: Cell Biology. 4th Ed. W. B. Saunders Co., Philadelphia.

Dintzis, H. M., 1961: Assembly of the peptide chains of hemoglobin. Proc. Natl. Acad. Sci. U.S.A. *47*:247–261.

Doudney, C. O., and Billen, D., 1961: Incorporation of thymidine into DNA by extracts from bacteria exposed to U.V. light. Nature *190*:545–546.

Duncan, G. G. (ed.), 1964: Diseases of Metabolism. 5th Ed. W. B. Saunders Co., Philadelphia.

Goulian, M., Kornberg, A., and Sinsheimer, R., 1967: Enzymatic synthesis of DNA. XXIV. Synthesis of infectious phage φx174 DNA. Proc. Nat. Acad. Sci. U.S.A. *58*:2321–2328.

Green, D. E., and Goldberger, R. F., 1967: Molecular Insights into the Living Process. Academic Press, New York.

Grunberg-Manago, M., 1963: Polynucleotide phosphorylase. Progr. Nucleic Acid Res. Molec. Biol. *1*:93–133.

Grunberg-Manago, M., Ortiz, M. P. J., and Ochoa, S., 1956: Enzymic synthesis of polynucleotides. I. Polynucleotide phosphorylase of *Azotobacter vinelandii.* Biochim. Biophys. Acta *20*:269–285.

Hoagland, M. B., Dec. 1959: Nucleic acids and proteins. Sci. Am. *201*:55–61.

Horwitz, J., and August, J. T., 1963: The role of DNA in RNA synthesis. Progr. Nucleic Acid Res. Molec. Biol. *1*:59–92.

Horwitz, J., and Furth, J. S., Feb. 1962: Messenger RNA. Sci. Am. *206*:41–49.

Immamoto, F., and Yanofsky, C., 1967: Transcription of the tryptophan operon in polarity mutants of *Escherichia coli.* II. Evidence for normal production of tryp-mRNA molecules and for premature termination of transcription. J. Molec. Biol. *28*: 25–35.

Ingram, V. M., Jan. 1958: How do genes act? Sci. Am. *198*:68–74.

Ingram, V. M., 1963: The Hemoglobins in Genetics and Evolution. Columbia University Press, New York.

*Ingram, V. M., 1965: The Biosynthesis of Macromolecules. W. A. Benjamin, New York.

Jacob, F., and Monod, J., 1961: Genetic regulatory mechanisms in the synthesis of proteins. J. Molec. Biol. *3*:318–356.

Jacob, F., and Wollman, E. L., June 1961: Viruses and genes. Sci. Am. *204*:93–106.

Jones, O. W., Jr., and Nirenberg, M. W., 1962: Qualitative survey of RNA codewords. Proc. Natl. Acad. Sci. U.S.A. *48*:2115–2123.

Jukes, T. H., 1962: Relations between mutations and base sequences in the amino acid code. Proc. Natl. Acad. Sci. U.S.A. *48*:1809–1815.

Katsoyannis, P. G, 1966: Synthesis of insulir Science *154*:1509–1514.

Keir, H. M., 1966: DNA polymerases from mammalia cells. Prog. Nucleic Acid Res. Molec. Biol. *4*:82 128.

Kluyver, A. J., and Donker, H. J. L., 1926: Di Einheit in der Biochemie. Chemie der Zelle un Gewebe *13*:134–190.

Kornberg, A., 1960: Biologic synthesis of deoxyribo nucleic acid. Science *131*:1503–1508 (Nobel lecture)

Lawley, P. D., 1966: Effects of some chemica mutagens and carcinogens on nucleic acids. Progr Nucleic Acid Res. Molec. Biol. *5*:89–131.

*Lehninger, A. L., 1965: Bioenergetics. W. A. Benjamin, New York.

Lengyel, P., 1966: Problems in protein biosynthesis J. Gen. Physiol. *49*:305–330.

Lipmann, F., 1963: Messenger RNA. Progr. Nuclei Acid Res. Molec. Biol. *1*:135–161.

Nisman, B., and Pelmart, J., 1964: *De novo* protei synthesis *in vitro.* Progr. Nucleic Acid Res. Molec Biol. *3*:235–297.

Ochoa, S., and Heppel, L. A., 1957: Polynucleotid synthesis. *In* Symposium on the Chemical Basi of Heredity. McElroy and Glass, eds. John Hopkins Press, Baltimore, pp. 615–638.

Reich, E., and Goldberg, I. H., 1964: Actinomycir and nucleic acid function. Progr. Nucleic Acic Res. Molec. Biol. *3*:183–234.

Rich. A., Dec. 1963: Polyribosomes. Sci. Am. *209* 44–53.

Richardson, C. C., and Weiss, B., 1966: The enzymatic phosphorylation of nucleic acids and it application to end group analysis. J. Gen. Physiol *49*:81–97.

Robertson, J. D., 1960: The molecular structure anc contact relationships of cell membranes. Progr Biophys. *10*:343–418.

Sinsheimer, R. L., 1959: A single stranded DNA from bacteriophage φx174. J. Molec. Biol. *1*:43–53

Smellie, R. M. S., 1963: The biosynthesis of RNA in animal systems. Progr. Nucleic Acid Res. Molec. Biol. *1*:27–58.

Stanier, R. Y., and Gunsalus, I. C. (eds.), 1960–1964: The Bacteria: A Treatise on Structure and Functions. 5 vol. Academic Press, New York.

Synthesis and structure of macromolecules, 1963. Cold Spring Harbor Symp. Quant. Biol. Vol. 28.

Szent-Györgyi, A., 1940: Vitamins. *In* The Cell and Protoplasm. Taylor, ed. Science Press, New York, pp. 159–165.

Tissieres, A., Schlesinger, D., and Gros, F., 1960: Amino acid incorporation into proteins by *Escherichia coli* ribosomes. Proc. Natl. Acad. Sci. U.S.A. *46*:1450–1463.

Volkin, E., and Astrachan, L., 1956: Phosphorus incorporation in *Escherichia coli* RNA after infection with bacteriophage T2. Virology *2*:149–161.

Watson, J. D., 1965: The Molecular Biology of the Gene. W. A. Benjamin, New York.

Woodward, D. O., and Munkres, K. D., 1967: Genetic control, function and assembly of a structural protein. *In* Organizational Biosynthesis. Vogel, Lampen, and Bryson, eds. Academic Press, New York, pp. 489–502.

Yanofsky, C., 1960: The tryptophane synthetase system. Bact. Rev. *24*:221–245.

Yanofsky, C., 1967a: Gene structure and protein structure. Harvey Lect. *61*:145–168.

Yanofsky, C., 1967b: Structural relationships between gene and protein. Ann. Rev. Genetics *1*:117–138.

*Yanofsky, C., May 1967c: Gene structure and protein structure. Sci. Am. *216*:80–93.

Yanofsky, C., and Ito, J., 1966: Nonsense codons and polarity in the tryptophan operan. J. Molec. Biol. *21*:213–234.

Yanofsky, C., and St. Lawrence, P., 1960: Gene action. Ann. Rev. Microbiol. *14*:311–340.

Zamecnik, P. C., 1960: Historical and current aspects of the problem of protein synthesis. Harvey Lect. *54*:256–281.

## ORIGIN OF LIFE

Abelson, P. H., 1957: Some principles of paleobiochemistry. Ann. N. Y. Acad. Sci. *69*:255–376.

Allen, W. V., and Ponnamperuma, C., 1967: A possible prebiotic synthesis of monocarboxylic acids. Currents Modern Biol. *1*:24–28.

Calvin, M., 1956: Chemical evolution and the origin of life. Am. Sci. *44*:248–263.

Calvin, M., 1962: The origin of life on earth and elsewhere. Biol. Med. Phys. *8*:315–342.

Calvin, M., 1967: Chemical evolution. Progr. Theoret. Biol. *1*:1–34.

Calvin, M., and Calvin, C. J., 1964: Atom to Adam. Am. Sci. *52*:163–186.

Dole, M., 1949: The history of oxygen. Science *109*:77–81.

Ehrensvard, G., 1962: Life: Origin and Development. University of Chicago Press, Chicago.

Florkin, M. (ed.), 1960: Aspects of the Origin of Life. Pergamon Press, New York.

Fox, S. W., 1957: Chemical problems of spontaneous generation. J. Chem. Educ. *34*:472–479.

Fox, S. W., 1960: How did life begin? Science *132*:200–208.

Fox, S. W. (ed.), 1964: The Origins of Prebiological Systems and of Their Molecular Matrices. Academic Press, New York.

Greenstein, J. L., 1961: Stellar evolution and the origin of the chemical elements. Am. Sci. *49*: 449–472.

Holmes, A., 1954: The oldest dated materials of the Rhodesian shield. Nature *173*:612–614.

Horowitz, N. H., 1945: On the evolution of biochemical syntheses. Proc. Natl. Acad. Sci. U.S.A. *31*:153–157.

Keosian, J., 1964: The Origin of Life. Reinhold Publishing Corp., New York.

Miller, S. L., 1953: A production of amino acids under possible primitive conditions. Science *117*: 528–529.

Nursall, J. R., 1959: Oxygen as a prerequisite to the origin of the metazoa. Nature *183*:1170–1172.

Oparin, A. I., 1957: The Origin of Life on Earth. 3rd Ed. Academic Press, New York.

Ponnamperuma, C., 1968: Ultraviolet radiation and the origin of life. *In* Photophysiology. Vol. 3. Current Topics. Giese, ed. Academic Press, New York, pp. 253–267.

Sanchey, R., Ferris, J. P., and Orgel, L. E., 1966: Conditions for purine synthesis: did prebiotic synthesis occur at low temperatures? Science *153*:72–73.

Wald, G., Aug. 1954: The origin of life. Sci. Am. *191*: 45–53.

Wald, G., 1957: The origin of optical activity. Ann. N. Y. Acad. Sci. *69*:255–376.

Watson, J. D., 1965: The Molecular Biology of the Gene. W. A. Benjamin, New York, p. 181.

Young, R. S., and Ponnamperuma, C., 1964: Life: origin and evolution. Science *143*:384–388.

# CHAPTER 21 / REGULATION AND CONTROL OF CELL METABOLISM

In normal cells few soluble metabolic products accumulate. Only in cells in which mutation has blocked some step in a synthetic or degradative pathway or in cells which have been exposed to appropriate metabolic inhibitors is there accumulation of such products. Studies on such cells have been of great value in elucidation of intermediary metabolism and synthetic pathways. The fleeting nature of the intermediates of metabolic reactions in normal cells suggests that cellular metabolism is regulated in such a manner that various substances are present in the cell in concentrations optimal for cell functions (Bonner, 1961; Chance *et al.,* 1965; Eakin, 1963; Touster, 1963; Weber, 1963–1965; Wright, 1963).

## TYPES OF REGULATION

Some metabolic regulation may result from operation of the law of mass action and some may be due to competition of enzymes for a common substrate. However, most metabolic regulation is of a more complex nature, and occurs by affecting either enzyme activity or enzyme synthesis.

Enzyme activity is dependent upon many factors. The manner in which the polypeptide chain folds (secondary and tertiary structures), the attachment of prosthetic groups to the protein and, in enzymes made up of several polypeptide chains, the way the subunits complex (quaternary structure) affect the enzyme activity. Also, metabolic end products may cause feedback activation or feedback inhibition of the enzyme.

The rate of enzyme action depends in part upon the concentration of the enzyme. The concentration of the enzyme is determined by the balance between the rate of synthesis and the rate of degradation of the enzyme.

Enzyme synthesis can be induced or repressed. The rate of polypeptide synthesis is determined in part by the concentration of the mRNA specific to the polypeptide. The concentration of the mRNA is determined by the balance between its synthesis and degradation. The rate of mRNA synthesis is determined by the rate of assembly of the nucleoside phosphates on the template and the rate of release of the finished product from the template.

Finally, the rate of translation of the mRNA message is affected by the rate of initiation, synthesis, ribosomal movement and release of the finished polypeptide chain from the ribosomes. Conceivably regulation could occur at any one, several or all of these levels. Some of these regulatory mechanisms are considered in more detail in the following paragraphs.

## MASS ACTION IN REVERSIBLE REACTIONS

When two substances, A and B, react to form C and D and the process is reversible, then when C and D begin to accumulate, the rate of formation of A and B from C and D increases:

$$A + B \rightleftharpoons C + D \qquad (21.1)$$

Thus, unless C and D are used up or diffuse away, the reaction reaches equilibrium in which as much C and D are formed as A and B. The Lohmann reaction, in which the high energy phosphate bond of ATP is transferred to creatine, forming creatin phosphate, is an example of such a simple regulation:

$$ATP + C \rightleftharpoons CP + ADP \qquad (21.2)$$

Here CP will accumulate as long as there is an excess of ATP. When ATP is in short supply, high energy phosphate will pass from the CP to ADP (Cantarow and Schepartz, 1967).

## COMPETITION FOR SHARED SUBSTRATE

A "bottleneck" occurs when each of two metabolic pathways compete for a common metabolite or cofactor. An example is $NADH_2$ produced during the oxidative step in muscle glycolysis. $NADH_2$ can be oxidized either by lactic dehydrogenase or by oxidative reactions. When aerobic conditions prevail the latter reactions predominate.

Such competition is commonplace in intermediary metabolism. For example, $\alpha$-ketoglutarate is a participant in the Krebs cycle (Fig. 16.8) as well as in the glutamate dehydrogenase cycle. Glutamate is formed in the presence of ammonia and $\alpha$-ketoglutarate. In other words, in the presence of $NH_3$, the Krebs cycle may be suppressed.

It is possible that the *Pasteur effect* is of this nature also. Many years ago Pasteur noticed that fermentation (glycolysis) in yeast is suspended when oxygen is available. It is well known that both oxidative and glycolytic pathways require inorganic phosphate and ADP. Since phosphorylation proceeds at a much higher rate in mitochondrial oxidative metabolism than in glycolysis, the shift from glycolysis to oxidative metabolism may rest upon this fact.

It is also possible that the *Crabtree effect,* in which addition of glucose to some tissues suppresses aerobic oxidative processes, is explainable by competition for a common substrate. Tissues in which this effect is noticed are ones with an exaggerated glycolytic pathway. Aerobic glycolysis depletes the inorganic phosphate and adenine nucleotides, thus interfering with oxidative metabolism (Cantarow and Schepartz, 1967).

## ALLOSTERIC INHIBITION AND ACTIVATION

One of the most prevalent controls for a series of reactions is by *negative feedback inhibition* (Magasanik, 1961; Gerhart and Pardee, 1962, 1964; Henderson, 1962; Umbarger, 1964). Suppose that in the series of reactions:

$$A \xrightarrow{E_1} B$$

$$B \xrightarrow{E_2} C$$

$$C \xrightarrow{E_3} D$$

$$D \xrightarrow{E_4} F$$

it is found that the end product inhibits the first of the series of reactions catalyzed by enzyme 1 ($E_1$); then the whole reaction series is slowed down. For example, it has been found that the end product sometimes reacts with the enzyme, reducing its activity (Magasanik, 1961). The presence of F on enzyme 1, in addition to A,

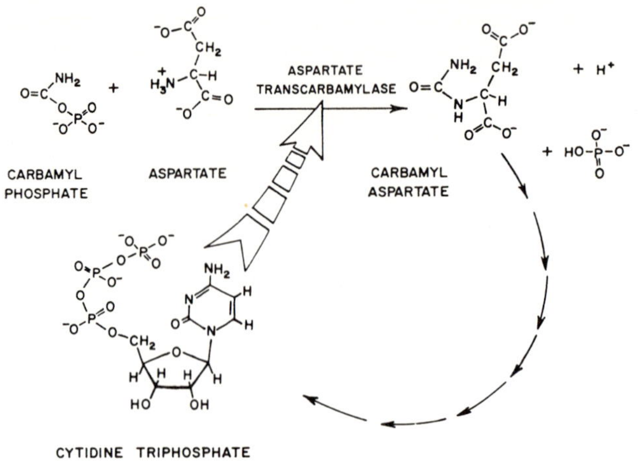


***Figure 21.1.*** Schematic diagram of feedback inhibition in the pyrimidine pathway of *Escherichia coli* (the end product, cytidine triphosphate, inhibits the activity of the enzyme aspartate transcarbamylase). (From Gebhardt and Pardee, 1964: Fed. Proc. *23*:728.)

causes a change called the *allosteric effect,* presumably a configurational alteration of the enzyme, making it less catalytically active. As soon as the excess of the end product F is used up in subsequent reactions, the reaction series accelerates again, because F will have then become detached from $E_1$.

For example, consider the inhibition of aspartate transcarbamylase by CTP (cytidine triphosphate). The normal substrate, aspartate, has little in common in its structure with CTP, the inhibitor (see Fig. 21.1). The probability that this enzyme might accept the two substrates at two different points, e.g., at two different *sites,* was indicated by heat treatment of the enzyme: it was found that heat treat-

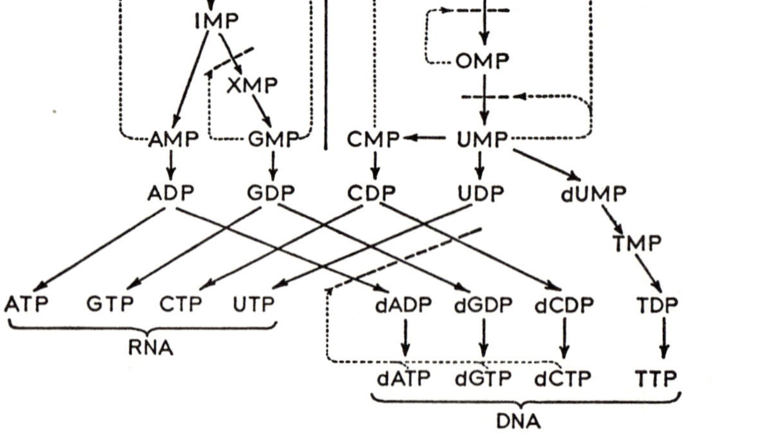


***Figure 21.2.*** Some feedback relationships involved in the regulation of RNA and DNA synthesis. (I, inosine; X, xanthosine; A, adenosine; G, guanosine; O, orotidine; U, uridine; C, cytidine; T, thymidine.) (From Paul, 1964: Cell Biology. Stanford University Press, Stanford, Calif.)

ment could destroy the inhibitor site without affecting the catalytic site. However, heat treatment caused dissociation of the enzyme into four subunits, each subunit still possessing catalytic activity. It is thought that aspartate reduces the reactions between the subunits, causing looser association, whereas CTP increases them (Gerhart and Pardee, 1964).

An example of feedback inhibition in a more complex series of reactions is shown in Figure 21.2 for some of the reactions involved in RNA and DNA synthesis. In this series of reactions end products affect enzymes at ten different places (Paul, 1964). Some similar examples are found in pyrimidine synthesis (Yates and Pardee, 1956) and purine synthesis (Henderson, 1962). Such regulation is also found in many reactions during development and differentiation of embryos (Walker, 1965).

In a fairly large number of cases an enzyme catalyzing a reaction which is recognized as an initial step in a sequence of reactions leading to a specific end product has been found to be inhibited by the end product (Umbarger, 1964).

The allosteric effect of metabolite interaction with the enzyme of the primary reaction need not always be inhibitory. A configurational change might well increase the activity of the enzyme. Thus, it has been found in many cases that a small molecule which does not take part in a reaction is necessary for the reaction; in other words, it acts as an enzyme activator (Koshland, 1964). Similarly, if a metabolite had such an activating effect upon an enzyme in the sequence, the net effect would be to accelerate the series of reactions. An example is observed in the reactions involved in glycogen control. Glycogen is made from glucose-6-phosphate in three enzymatic steps. First glucose-6-

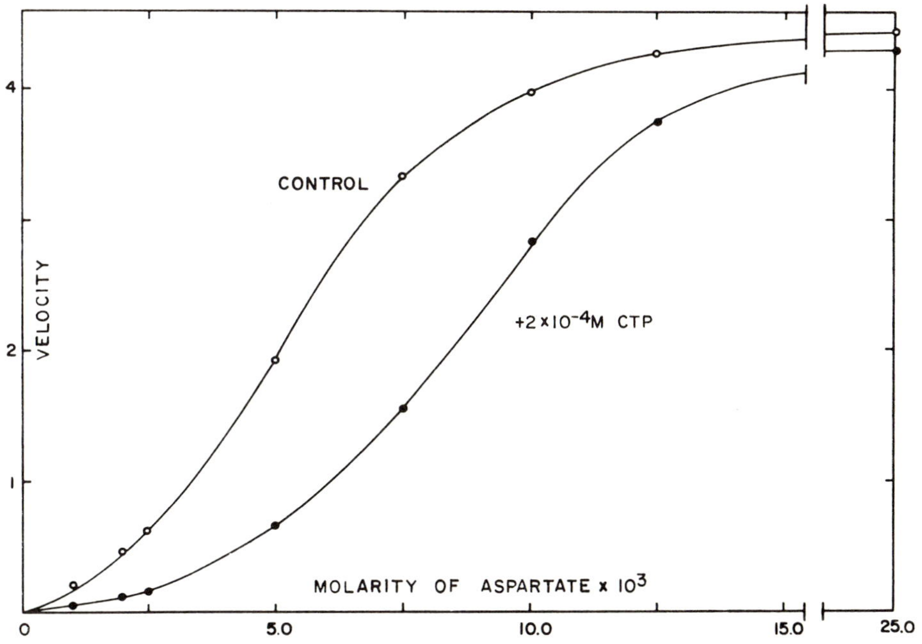


***Figure 21.3.*** Effect of increase in concentration of substrate on rate of reaction of a regulator enzyme (control). The aspartic transcarbamylase is a regulator enzyme and its rate shows a characteristic sigmoid curve. Shown also is the competitive inhibition between the end product cytidine triphosphate and the substrate with increasing substrate (aspartate) concentration. (Velocity = units of activity per milligram of protein $\times 10^{-3}$. The reaction mixture contained $3.6 \times 10^{-3}$M carbamyl phosphate, with aspartate varied as indicated; $2 \times 10^{-4}$M CTP when used; 0.04M potassium phosphate buffer, pH 7.0: and $9.0 \times 10^{-2}$ $\mu$g. of enzyme protein per ml.) (From Gerhard and Pardee, 1964: Fed. Proc. *23*:731.)

phosphate is isomerized to glucose-1-phosphate, and then glucose-1-phosphate reacts with uridine triphosphate (UTP) to form uridine diphosphate-D-glucose (UDPG), whereupon the glucose moiety from UDPG is incorporated into glycogen. A good supply of glucose-6-phosphate stimulates the synthesis of glycogen in a cell by activating the enzyme which catalyzes the conversion of UDPG to glycogen. On the other hand, when the sugar in the cell decreases to a low level, glycogen phosphorylase is activated and catalyzes splitting of sugar from glycogen. The molecule activating phosphorylase appears to be adenosine monophosphate (AMP). AMP is derived from ATP by removal of two high energy phosphate bonds during various reactions in the cell.

The advantage to the cell of end product control is the speed with which changes in rate of reaction may be instituted. In this way an equilibrium state may readily be maintained by constant adjustment and readjustment of rates of reaction.

The allosteric effect is also observed in the action of *regulator enzymes.* The curve for the effect of substrate concentration on the rate of reaction for a regulator enzyme is different from that for most enzymes; for most enzymes the curve is hyperbolic and the rate of reaction falls off as the substrate concentration increases, presumably because the active sites on the enzyme have been saturated (Fig. 15.11). For a regulator enzyme like aspartate transcarbamylase, however, the curve for the rate of reaction with rising substrate concentration is sigmoid (Fig. 21.3). This is thought to indicate that binding one substrate molecule to the enzyme makes binding the next one easier, and the catalytic effect of the enzyme is greatest when four substrate molecules are bound, the allosteric effect presumably reaching its optimal state. This resembles the uptake of oxygen by hemoglobin which also is made up of four subunits (two $\alpha$ and two $\beta$), each with a heme group (see Fig. 7.3). In both cases this represents a regulation of the function of the enzyme activity by an allosteric effect induced by the substrate (Changeux, 1965).

## REGULATION BY ISOZYME FORMATION

*Isozymes* (isoenzymes) are enzymes in multiple forms, all performing the same general function at different rates. They are different to some extent in chemical composition so that they are separable electrophoretically. Thus lactate dehydrogenase is found in five electrophoretically distinct fractions. Each of these electrophoretic species of lactic dehydrogenase is a tetramer consisting of two polypeptide chain units, H and M, present in different proportions: $H_4$, $H_3M$, $H_2M_2$, $HM_3$, and $M_4$. The five lactate dehydrogenase isozymes differ in many properties: catalytic activity, amino acid composition, sensitivity to heat and immunological responses.

During development of an organism different proportions of the various isozymes are produced in cells of differentiating tissues, presumably in relation to the specific requirements of the cells of each of the tissues. For example, species of lactate dehydrogenase with high content of H are most active at low pyruvate concentrations and are inhibited by high concentrations of pyruvate. Such species are found in the muscle cells of the heart. The properties of the isozyme high in H would tend to channel the lactate into the aerobic reactions of the Krebs cycle rather than to the formation of lactate, a regulation obviously of adaptive significance for heart cells. Conversely, lactate dehydrogenases with a high M content remain active at high pyruvate

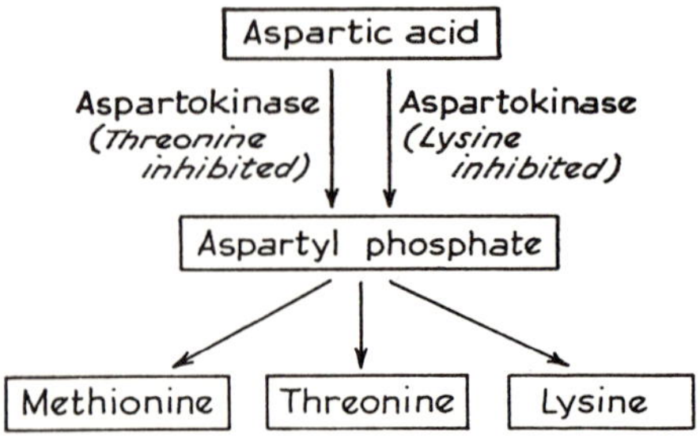


**Figure 21.4.** The way in which the synthesis of aspartyl phosphate is regulated by different end products through two isozymes of aspartokinase. (From Paul, 1964: Cell Biology. Stanford University Press, Stanford, Calif.)

concentrations and therefore channel the lactate metabolism toward formation of lactic acid. Such lactate dehydrogenase isozymes are found in skeletal muscle cells. They also represent regulations of adaptive value because skeletal muscle cells undergo periods of intense activity during which they go into oxygen debt and at such times lactic acid formation permits rapid release of energy. The lactic acid is rapidly removed from the muscle cells by the blood (Cantarow and Schepartz, 1967).

Isozymes may also form in response to stresses upon cells. Fish exposed to low temperatures acclimate in such a manner that the respiratory rate increases at the low temperature over the initial rate at that temperature. Isozymes have been found in the muscle cells of such fish and it has also been noticed that the energy of activation for some of the reactions involved in metabolic activity is decreased. The change in respiratory rate presumably occurs as a result of development of isozymes with decreased energy of activation (Hochachka, 1966, 1967).

An interesting case of isozyme formation coupled with negative feedback control is that of aspartokinase. This enzyme catalyzes the reaction between aspartic acid and ATP to form aspartyl phosphate. In a series of subsequent reactions aspartyl phosphate forms the amino acids lysine, threonine and methionine. In a number of species of bacteria the aspartokinase exists in two isozymes, one of them sensitive to feedback inhibition by lysine and the other by threonine. In this way if either of the amino acids accumulates there is a reduction in the formation of aspartyl phosphate, but the reaction is not stopped (Fig. 21.4; Paul, 1964).

## COMPARTMENTATION

Certain enzymes involved in particular reactions are segregated within membrane-bound units. Such compartmentation may serve as regulatory control in the cell. Thus the reactions of the Krebs cycle can occur only in the mitochondria within which are the required enzymes. Enzymes involved in lipid synthesis appear to be confined to the membranes of the endoplasmic reticulum. Protein synthesis is shown to occur on the surface of ribosomes. On the other hand, enzymes involved in glycolysis seem to be in the particle-free cytoplasm. This, perhaps, helps to explain both the Pasteur effect and the Crabtree effect since in both cases one of the competing enzyme systems is in the mitochondria, and the other is in the particle-free cytoplasm.

The rate of a reaction involving an enzyme free in solution may be different from that involving an enzyme bound to a membrane, as has been shown for a number of mitochondrial enzymes. Thus the various membranes in the cell may play a role in regulation of rate of enzymatic reactions. Although the way in which this occurs is not known, it is thought to be probably akin to the allosteric effect.

It is known that the plasma membrane and other membranes in the cell are sites of active transport, which depends upon metabolic energy. The transport of various metal ions by the membranes may play a part in regulation of activity of the enzymes contained within the membrane-bound structures. Thus in muscle cells the calcium ion concentration is dependent upon the activity of the endoplasmic (sarcoplasmic) reticulum. The calcium ion is required for the functioning of actomyosin as an ATPase.

Membranes act in one other form of regulation—as the targets of hormones, which may alter their permeability properties. Since such regulation is primarily in the realm of intercellular regulation it will not be considered further here; the phenomena are discussed in textbooks of animal and plant physiology.

## ENZYME INDUCTION AND REPRESSION

When *E. coli* cells are placed in a medium containing a galactoside sugar (e.g., lactose) as the sole source of carbon, they are at first incapable of utilizing the sugar because they lack the enzyme $\beta$-galactosidase. After a few minutes, however, the lactose has induced the *E. coli* to form $\beta$-galactosidase, which is capable of digesting lactose into the usable products glucose and galactose. After induction of $\beta$-galactosidase the enzyme can be extracted from the lysed cells. The induced enzyme disappears again if the cells are kept on glucose without a $\beta$-galactoside for a while. The enzyme must again be induced when such cells are replaced in a $\beta$-galactoside.

This experiment indicates that the bacteria may possess enzymes of two kinds: constitutive and induced. Constitutive enzymes are always present, at least in small quantities. Induced enzymes are absent from the cell and appear only on presentation of an inducing substrate (Pollock, 1959; Halvorson, 1960), but they fail to appear if such protein synthesis inhibitors as puromycin or chloramphenicol are added to the substrate. This indicates that protein synthesis is necessary for induction of such enzymes. However, mutants of *E. coli* are known which are incapable of developing the enzyme under conditions favorable for induction in inducible strains.

The experiments on enzyme induction led to far-reaching findings in the hands of Jacob and Monod (both of whom, with Lwoff, received the Nobel prize in 1966), because they opened the way for an analysis of the genetic basis of induction and repression (for details of these experiments, see Jacob and Monod, 1961a, b, 1963; Wilson and Pardee, 1964).

From what has been considered in Chapters 6 and 20, it is evident that an enzyme is synthesized from a message transcribed from a gene by mRNA. This means that a gene must encode the information for each inducible enzyme just as another gene encodes the information for synthesis of a constitutive enzyme. Since no inducible enzyme is present in the absence of the inducing substance, the gene for this enzyme, if present, must in some manner be repressed, because it is being neither transcribed nor translated. A possible mechanism for explaining the phenomenon is given here.

Adjacent genes concerned with some common activity may be grouped into units of the chromosome called *operons* (Stent, 1963). In *E. coli* the gene for the permease concerned with the active transport of the $\beta$-galactoside into the cells and the gene for the $\beta$-galactosidase are in the same operon next to the $\beta$-galactosidase operator. The genes of the operon are under the control of the operator, a short segment of the chromosome which turns the $\beta$-galactosidase gene on and off, depending on the presence or absence of an inducer. Present also is a regulatory gene which codes for a relatively low molecular weight protein which can combine with the operator to turn off the production of $\beta$-galactosidase mRNA. In the presence of an inducer (a $\beta$-galactoside), the repressor substance is inactivated by reaction with the inducer. The inactivated repressor cannot combine with the operator and once again allows the production of the $\beta$-galactosidase. All of these ideas are summarized in Figure 21.5.

It is seen in Figure 21.5 that all the genes in the operon are affected by the inducer and the repressor; in this case both the permease and the $\beta$-galactosidase are affected at the same time. However, as discussed in Chapter 14, the permease and the $\beta$-galactosidase systems may be uncoupled by mutations. Thus some mutants are found which can accumulate a $\beta$-galactoside but cannot digest it, whereas other mutants possess the capacity to synthesize $\beta$-galactosidase, as can be demonstrated by disrupting the cells and liberating the enzyme, but they cannot obtain the sugar because they lack the permease system.

An interplay between repressors and inducers enables the cell to synthesize enzymes in relation to needs and availability of nutrients in the medium. This is important for survival because it would be difficult to accommodate the variety of enzymes which might possibly be useful in the confines of a small cell.

Many cases of induction have now been found in animal and plant cells, as well as in many bacteria, and the phenomenon of induction and repression appears to be general although there may well be differences in details. For example, animal and plant cells are diploid in contrast to most bacteria, which are haploid, so that

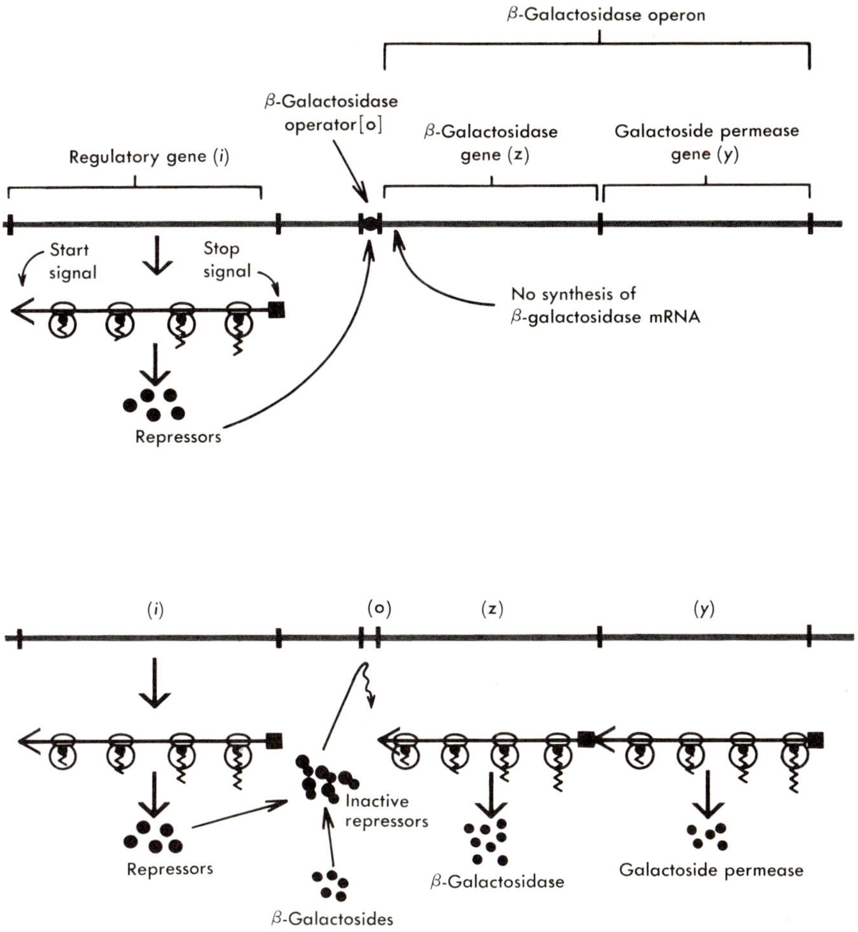


***Figure 21.5.*** The operon and its functioning. The hypothesis currently favored for how the interaction of repressor, corepressor, and operator controls the synthesis of the *E. coli* proteins β-galactosidase and β-galactoside permease. Repressors are shown combining with DNA. This point has not yet been proved, and the alternative hypothesis exists that repressors act by combining with mRNA, thereby preventing its attachment to ribosomes, and so allowing its rapid enzymatic breakdown to nucleotides. It is important to note that, under both sets of hypotheses, a specific repressor decreases the amount of a specific mRNA molecule. (From Watson, 1965: The Molecular Biology of the Gene. W. A. Benjamin, New York.)

recessive mutants will not be expressed unless there is homozygosity. This makes their detection more difficult in eukaryotic cells than in prokaryotic cells (Cantarow and Schepartz, 1967).

Although most inborn errors of metabolism in man have in the past been considered mutations in structural genes, it is now necessary to reassess them in view of inducer and repressor regulations found in bacteria. Some 55 metabolic disorders of man have been shown to be hereditary and probably variations of regulation (Hsia, 1966; Stanbury *et al.*, 1966). A hopeful approach to metabolic studies and controls in mutant cells of man is the use of tissue culture (especially skin) derived from individuals with metabolic disorders. Such studies are only beginning (Bell, 1965).

If a product from a reaction is normal in structure and biological activity but is variable quantitatively, it is likely to be under control of an operator gene. When the operator fails in its function, probably by mutation, the products may accumulate. Excessive production of porphyrin followed by its leakage into the urine and elsewhere in the body may be an example of this type of mutation in man, which expresses itself by a failure to repress the production of the porphyrin. Sometimes a reaction appears to be super-repressed, as in some cases of jaundice (Cantarow and Schepartz, 1967).

It has sometimes been pointed out that genetic controls are slow to act, so that regulation of metabolism by induction and repression presents a rather slow mechanism for metabolic control. Enzyme induction, for example, requires several minutes or more, compared with feedback inhibition of a reaction which may be immediate. Repression is also slow. Furthermore, even if repression had been induced, the enzymes in the cell, which are degraded only slowly, enable the reaction to continue as long as a supply of enzymes and substrates is available and conditions are otherwise favorable. However, although repression is rather slow, it is often accompanied by end product inhibition of activity of the first enzyme in the series of reactions. This makes possible very quick reduction of enzymatic activity, in addition to repression.

## REGULATION OF PROTEIN SYNTHESIS

In some cells a given protein is made in large quantity and for a prolonged time, e.g., the production of hemoglobin in reticulocytes (progenitors of red blood cells). In such cells a long-lived mRNA would be of value for survival. It is not surprising, therefore, to find that in reticulocytes the mRNA may last for a number of days.

In bacteria undergoing rapid growth and division it is likely that only a few protein molecules of each kind are needed per cell. For example, some data indicate that only about 15 copies of some of the enzymatic proteins are made per cell. Usually, when the required number of proteins has been synthesized the mRNA is hydrolyzed. Such a short-lived mRNA suits the bacteria, which have no need for continued synthesis of one protein. A bacterium that did continue synthesis would soon be eliminated in competition with well regulated cells. It has been shown that the *Escherichia coli* cell has only about 1000 mRNA molecules, which at any one instant must meet all the varying synthetic and catabolic needs of the cell (Yanagisawa, 1963).

## REGULATION OF NUCLEIC ACID SYNTHESIS

In bacteria, DNA synthesis goes on almost continuously (for about 80 to 95 per cent of the generation time), and the growing point of the DNA molecule must move rather fast to enable the cell to replicate the entire chromosome before the cell divides (see Chapter 6). In eukaryotic cells, DNA synthesis occurs for about half the division interval during the synthetic (or S) period. The synthetic period lasts 7.5 hours in the cells of the broad bean *Vicia faba,* 10.5 hours in the cells of *Tradescantia paludosa,* 9.9 hours in fibroblast cells of the mouse and 5.8 hours in fibroblasts of the Chinese hamster (Kihlman, 1966). During the remainder of the division interval time the DNA does not appear to be synthesized (see Fig. 27.1). The nature of the signal given for beginning and ending of DNA replication has not been determined. Presumably replication stops when the genome has been duplicated.

The nature of DNA synthesis regulation in certain cells such as the salivary gland cells of the flies *Drosophila* and *Chironomus,* in which the chromosomes multiply to form many parallel units, is not known. The duplicated chromosomes become easily visible under the light microscope and the location of various genes can easily be spotted because all the strands are in register with one another (Fig. 6.9).

Transcription of the genes to various types of RNA poses certain problems (Neihardt, 1964). Presumably rRNA and sRNA are formed continuously in the bacterial cell and they remain in the cell, being much more resistant to nucleases than mRNA. Messenger RNA for production of constitutive enzymes is formed continuously but in bacteria it is rapidly broken down. The life span of an mRNA molecule is, apparently, related to the needs for its translational activity. The short life span of mRNA in *E. coli* and the long life span of mRNA in the reticulocytes of a mammal probably represent two extremes. However, precise data do not seem to be available about the life span of mRNA in other types of eukaryotic cells. Messenger RNA for synthesis of induced enzymes is formed only in the presence of the inducer. When the inducer is absent repression occurs, as already discussed.

Although ribosomes are normally rather stable in active cells, they disappear rapidly when the cell is starved, perhaps because active ribosomes with attached mRNA appear to be resistant to nuclease, whereas inactive ones are not and consequently are degraded. A starved cell therefore has fewer ribosomes. Addition of nutrient leads to rapid synthesis of rRNA and the formation of ribosomes. Thus the quantity of rRNA is quickly adjusted to the needs of the cell (Koningsberger and Bosch, 1967).

## REGULATION BY HORMONES—INTERCELLULAR REGULATION

In the multicellular organism, metabolism and growth may be regulated in a complex manner from some coordinating center. Since these are topics concerned with the entire organism they are more appropriately considered in texts of plant physiology and animal physiology. Only two examples will be given to illustrate the nature of the cellular reactions to hormones.

The giant chromosomes of insects, studied especially in tissues of *Drosophila* and *Chironomus* (particularly the salivary glands), are shown to be the result of multiple and progressive replication of the chromatids without mitotic separation. Such chromosomes may measure ten times the length and 10,000 times the cross-sectional area of the original chromatids. The bands seen on such chromosomes (the *chromomeres*) are probably each made up of the tightly folded-up portion of the constituent double DNA-histone chromatid strand. On such giant chromosomes, puffs (Balbiani rings) or enlargements are often seen; the puffs may be different in different tissues. It has been found that injection into the larva of *ecdysone,* the molting-metamorphosis hormone of insects, results in development of puffs where none were seen before. Such puffs develop within 30 minutes of injection in one of the chromosomes (No. 1) and 30 minutes later in another (No. 4). The hormone is effective in a concentration as low as $10^{-7}$ $\mu$g. per milligram of larva for the first puff. Once puffing has been induced, the condition persists until after molting is completed, although the details depend upon the stage at which the injection is made.

It will be recalled (Chapter 6) that the puff consists of extruded sections of DNA-histone filament upon which the mRNA develops. The mRNA then translates the message into protein. The hormone apparently activates or derepresses the genes at a particular locus on the chromosome. Different genes may be activated in different tissues by the same hormone. Different hormones may activate different genes in the same chromosome. In this manner hormones may serve to coordinate the various activities concerned with a particular function (e.g., in the case of ecdysone, molting and metamorphosis) (Beermann, 1965).

The cells of the liver of a mammal are known to store a considerable amount of glycogen, which serves to regulate the blood glucose level, largely through the interplay of various hormones upon the enzymes in the liver cells. Thus after a rich meal there is an abundance of glucose in the blood. Such a high glucose level leads to secretion of insulin into the blood from the islets of Langerhans in the pancreas. Insulin not only increases the permeability of cells to glucose but also stimulates the formation of glycogen (glycogenesis) in the liver cells, possibly by action on glucokinase, the enzyme that transfers high energy phosphate from ATP to glucose. As soon as the level of blood glucose has fallen, secretion of insulin essentially stops, as does the synthesis of glycogen. Thyroxine, the secretion of the thyroid gland, on the other hand, activates a phosphorylase to form glucose phosphate

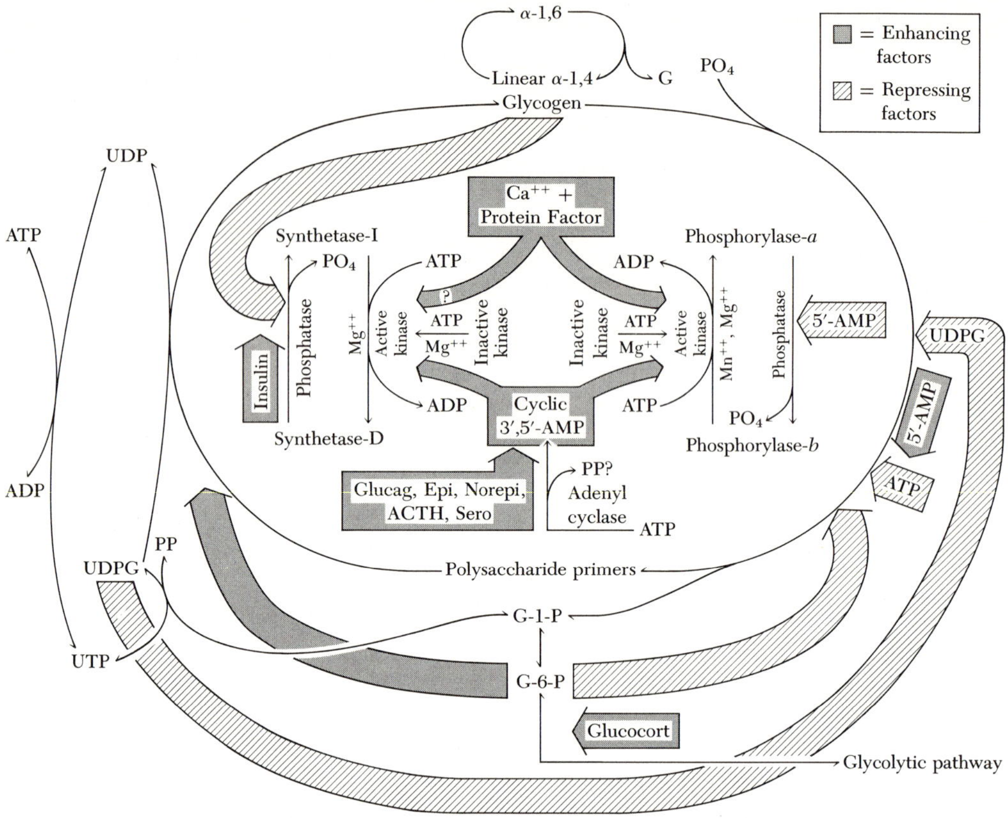


***Figure 21.6.*** Pathways and regulatory factors in glycogen metabolism. ACTH, Adrenocorticotropic hormone; Epi, epinephrine; G, glucose; G-1-P and G-6-P, glucose-1- and glucose-6-phosphate; Glucag, glucagon; Glucocort, glucocorticoid hormones; Norepi, norepinephrine; $PO_4$, phosphate; PP, pyrophosphate; Sero, serotonin; UDP and UTP, uridine di- and triphosphate. The hormones epinephrine and norepinephrine are produced by the adrenal medulla and stimulate glycogenolysis. Glucagon, produced by certain cells (alpha cells) of the islets of Langerhans in the pancreas, activates liver phosphorylase, thus stimulating glycogenolysis. Glucocorticoid hormones are produced by the adrenal cortex and regulate carbohydrate metabolism. The phosphorylases exist in two forms in all tissues studied, the a form being physiologically more active. (From Cantarow and Schepartz, 1967: Biochemistry. 4th Ed. W. B. Saunders Co.)

from glycogen. The glucose phosphate is decomposed to glucose and inorganic phosphate in the presence of a phosphatase. Secretion of thyroxine may be induced by exposure of the individual to cold. Similarly, epinephrine, appearing in the blood as a result of excitement, leads to greater glycogen breakdown by phosphorolysis (glycogenolysis), as does muscular exercise, perhaps by the same mechanism.

When the supply of carbohydrate in the diet is low, glycogen is still formed, but now from amino acids (see the metabolic mill, Fig. 16.8). The enzymes involved in *glyconeogenesis,* as the process is called, are presumably under control of some of the ketosteroid hormones of the adrenal cortex. Some of these interrelationships are shown in Figure 21.6.

## LITERATURE CITED AND GENERAL REFERENCES


Bayliss, L. E., 1966: Living Control Systems. Freeman, San Francisco.

Beermann, W., 1965: Cytological aspects of information transfer in cellular differentiation. *In* Molecular and Cellular Aspects of Development. Bell, ed. Harper & Row, New York, pp. 204–212.

Bell, E., 1965: Molecular and Cellular Aspects of Development. Harper & Row, New York, especially Sections 5, 7–10.

Bonner, D. M. (ed.), 1961: Control Mechanisms in Cellular Processes. John Wiley & Sons, New York.

Cantarow, A., and Schepartz, B., 1967: Biochemistry. 4th Ed. W. B. Saunders Co., Philadelphia, pp. 758–772.

Cellular Regulatory Processes. 1961. Cold Spring Harbor Symp. Quant. Biol., Vol. 26.

Chance, B., Estabrook, R. W., and Williamson, J. R. (eds.), 1965: Control of Energy Metabolism. Academic Press, New York.

*Changeoux, J., April 1965: The control of biochemical reactions. Sci. Am. *212*:36–45.

Eagle, H., 1965: Metabolic controls in cultured mammalian cells. Science *148*:42–51.

Eakin, R. E., 1963: An approach to the evolution of metabolism. Proc. Natl. Acad. Sci. U.S.A. *49*: 360–366.

Gerhart, J. C., and Pardee, A. B., 1962: The enzymology of control by feedback inhibition. J. Biol. Chem. *237*:891–896.

Gerhart, J. C., and Pardee, A. B., 1964: Aspartate transcarbamylase, an enzyme designed for feedback inhibition. Fed. Proc. *23*:727–735.

Halvorson, H. O., 1960: The induced synthesis of protein. Advances Enzymol. *22*:90–156.

Henderson, J. F., 1962: Feedback inhibition of purine biosynthesis in ascites tumor cells. J. Biol. Chem. *237*:2631–2635.

Hochachka, P. W., 1966: Lactic dehydrogenases in poikilotherms: definition of a complex isoenzyme system. Comp. Biochem. Physiol. *18*:261–269.

Hochachka, P. W., 1967: Organization of metabolism during temperature compensation. *In* Molecular Mechanisms of Temperature Adaptation. Prosser, ed. American Association for the Advancement of Science, Washington, D.C., pp. 177–203.

Hsia, D. Y., 1966: Inborn Errors of Metabolism. 2nd Ed. Part I. Clinical Aspects. Year Book Medical Publishers, Chicago.

Ingram, V. M., 1963: The Hemoglobins in Genetics and Evolution. Columbia University Press, New York.

Jacob, F., and Monod, J., 1961a: Genetic regulatory mechanisms in the synthesis of proteins. J. Molec. Biol. *3*:318–356.

Jacob, F., and Monod, J., 1961b: On the regulation of gene activity. Cold Spring Harbor Symp. Quant. Biol. *26*:193–211.

Jacob, F., and Monod, J., 1963: Genetic repression, allosteric inhibition and cellular differentiation. *In* Cytodifferentiation and Macromolecular Syntheses. Locke, ed. Academic Press, New York, pp. 30–64.

Jacob, F., Perrin, D., Sanchez, C., and Monod, J., 1960: L'opéron: groupe de gènes à expression coordonnée par un opérateur. Comp. Rend. Acad. Sci. Paris *250*:1727–1729.

Karlson, P., 1965: Biochemical studies of ecdysone control of chromosomal activity. J. Cell. Comp. Physiol. *66*(Suppl.1):69–75.

Kennedy, F. T., Wicks, W. D., and Greenman, D. L., 1965: Hydrocortisone stimulation of RNA synthesis in induction of hepatic enzymes. J. Cell. Comp. Physiol. *66*(Suppl. 1)125–151.

Kihlman, B. A., 1966: Action of Chemicals on Dividing Cells. Prentice-Hall, Englewood Cliffs, N. J.

Koningsberger, V. V., and Bosch, L. (ed.), 1967: Regulation of Nucleic Acid and Protein Biosynthesis. Biochimica Biophysica Acta Library, Vol. 10. American Elsevier Publishing Co., New York.

Korner, A., 1965: Growth hormone effects on RNA and protein synthesis in liver. J. Cell. Comp. Physiol. *66*:(Suppl. 1):153–162.

Koshland, D. E., Jr., 1964: Conformation changes at the active site during enzyme action. Fed. Proc. *23*:719–726.

Maaløe, O., and Kjeldgaard, N. O., 1966: Control of Macromolecular Synthesis. W. A. Benjamin, New York.

Magasanik, B., 1961: Feedback mechanisms in biological syntheses. *In* Biological Approaches to Cancer Chemotherapy. Harris, ed. Academic Press, New York, pp. 39–46.

Neihardt, F. C., 1964: The regulation of RNA synthesis in bacteria. Progr. Nucleic Acid Res. *3*:145–181.

Paul, J., 1964: Cell Biology. Stanford University Press, Stanford, Calif.

Pollock, M. R., 1959: Induced formation of enzymes. *In* The Enzymes. 2nd Ed. Boyer, Hardy and Myrback, H., eds. Academic Press, New York, Vol. 1, pp. 619–680.

Stanbury, J. B., Wyngaarden, J. B., and Frederickson, D. C. (eds.), 1966: The Metabolic Basis of Inherited Disease. Blakiston Division, McGraw-Hill Book Co., New York.

*Stent, G. D., 1963: The operon: on its third anniversary. Science *144*:816–820.

Stulberg, M. P. (ed.), 1967: International Symposium on Enzymatic Aspects of Metabolic Regulation. National Cancer Institute Monograph 27. U.S. Government Printing Office, Washington, D.C.

Touster, O., 1963: Biological control mechanisms. Science *139*:926–928.

Umbarger, H. E., 1964: Intracellular regulatory mechanisms. Science *145*:674–679.

Walker, J. B., 1965: End product repression in the creative pathway of the developing chick embryo. *In* Molecular and Cellular Aspects of Development. Bell, ed. Harper & Row, New York, pp. 285–296.

*Watson, J. D., 1965: The Molecular Biology of the Gene. W. A. Benjamin, New York.

Weber, G. (ed.), 1963–1965: Advances in Enzyme Regulation. 3 vol. Pergamon Press, Oxford.

Wilson, A. C., and Pardee, A. B., 1964: Comparative aspects of metabolic control. *In* Comparative Biochemistry. Florkin and Mason, eds. Academic Press, New York, Vol. 6, pp. 73–118.

Wooley, D. W., 1959: Antimetabolites. Science *129*: 615–621.

Wright, B. (ed.), 1963: Control Mechanisms in Respiration and Fermentations. Ronald Press Co., New York.

Yanagisawa, K., 1963: Genetic regulation of protein biosynthesis at the level of the ribosome. Biophys. Biochem. Res. Commun. *10*:226–231.

Yates, R. A., and Pardee, A. B., 1956: Control of pyrimidine biosynthesis in *Escherichia coli* by a feed-back mechanism. J. Biol. Chem. *221*:757–770.

SECTION V

# Excitability and Contractility

All life is continually challenged by the demands of the environment. If an organism is to survive it must obtain nutrients, resist changes in environment and inimical attacks and at all costs reproduce its own kind. Chance plays a large part in the struggle for existence but those organisms that survive must respond in an adaptive way.

The responses of all living things to their environment ultimately depend upon cellular *excitability,* one of the fundamental properties of life. Perhaps more than anyone else, Francis Glisson (1587–1677) of Cambridge was the first to consider that excitability, manifesting itself in the appearance of alteration or movement, is a general property of life and a result of possession of energy. Another early figure of note was John Brown (1735–1788), who said that living things are stimulated by external factors, such as heat, food, foreign matter and poisons, as well as by interactions of internal organs. Brown's is a commonplace conception today but one that was quite novel at the time. Excitability, as defined today, is the capacity to react to changes in the environment, internal and external. Energy is continuously expended by the cell in its efforts to maintain a condition of readiness to react.

The living organism appears to respond to the environment as a whole, rather than to a single environmental factor at a time. For experimental purposes, however, it is more convenient to vary each factor of the environment singly and to study its effect. If the change in quantity or quality of an environmental factor is large enough to be detected by an organism, it constitutes a *stimulus.* The stimulus evokes in the living organism a change which constitutes the *response*. The cytoplasmic change evoked by a particular stimulus is probably fundamentally the same in all cells, but the mechanisms it sets into action, culminating in the response of the organism as a whole, depends upon the type of organism under consideration.

Unicellular organisms incorporate in a single cell all they need for an adaptive response to stimuli. Some of them have no specialized structures (*receptors*) sensitive to stimuli or special motile structures (*effectors*) and can respond, so far as can be seen, as a whole. The entire *Amoeba,* for example, is sensitive to light. Some unicellular organisms, however, have a receptor, e.g., the eyespot of *Chlamydomonas* (Fig. 2.1). Special effectors may also be present, such as pseudopodia in ameboid cells and cilia (or flagella) in ciliated and flagellated protozoans, various bacteria, plant zoospores and gametes of multicellular organisms.

In multicellular plants the main adaptive response to most stimuli is by *growth movements,* although some plants respond also to certain stimuli by *turgor movements*. In each case, the cellular change is transmitted to the effector cells and the plant reacts in a particular adaptive manner.

Turgor movements are strikingly shown by the sensitive plant *Mimosa,* which may be stimulated by touch or in other ways (Datta and Dutt, 1957; Gernand and Ehrig, 1957). When the electrical impulse generated by the stimulus reaches the cells of the pulvinus, as the effector at the base of each leaflet (and petiole) is called, turgor is lost from cells at the base, causing the leaflets to close together. As turgor is gradually reestablished in the cells of the pulvinus, the leaflets recover their original position. Turgor movements are also found in floral organs of some plants and in special structures on insect-trapping (carnivorous) plants (Lloyd, 1942).

Growth movements, in response to a stimulus, also called *growth curvatures* or *tropisms,* are too slow to be detected by the human eye but are clearly visible in time-lapse photographic records. Differential growth underlies responses to gravity, temperature (thermotropism), light (phototropism), water (hydrotropism), chemicals (chemotropism) and touch (thigmotropism).

The concept of tropism was introduced in 1832 by the botanist Augustin de Candolle (1778–1841). The general features of a tropistic response are perhaps best illustrated if we consider phototropism as an example. Some time after a plant is exposed to light, more active growth (in the region of undifferentiated cells) takes place on the side away from the light, which turns the plant toward the light. Darwin, long ago, showed that when a tinfoil cap is placed over the tip of an oat coleoptile, the plant fails to respond to light. A receptor for light, present in the growing tip, can be demonstrated in other plants as well. The action spectrum of phototropism indicates activity of a pigment absorbing strongly in the blue-violet end of the visible spectrum, presumably a carotenoid (Bonner and Galston, 1952).

It can be demonstrated that a *growth hormone,* called *auxin,* is produced at the growing tip. Auxins are relatively small molecules (the main one, indole acetic acid, has a molecular weight of 175.18); therefore, they presumably diffuse readily. Auxin is known to be transported downward away from the tip and it is found to accumulate on the side away from the light. Only cells which have not yet differentiated, such as those in the region of elongation of the stem, respond to auxin by lengthening. Asymmetric distribution of auxins is thought to account for many other plant responses, e.g., tip dominance, setting of fruit, shedding of leaves and, along with other hormones (e.g., phytochrome, Chapter 9), coordination of plant activities. The effect of auxin can be counteracted by specific growth inhibitors (e.g., maleic anhydride) (Audus, 1959).

Multicellular animals on the whole respond adaptively to the various internal and external environmental factors. Some of the cells involved in these specialized activities are greatly modified in accordance with their functions and grouped in organs of complex structure. Multicellular animals respond to a stimulus chiefly by movement, usually muscular movement, which may be very rapid compared to the slow responses that characterize plants, but they may also respond to stimuli by secretion from glands; e.g., when a hungry dog is confronted with food, saliva will flow freely. Some animals, however, will respond in other ways. Those with luminous organs (Fig. 18.11) may give off light upon stimulation, and electric fish stun their prey by an electric shock.

The simplest response pattern in animals is the *reflex arc,* which is the automatic response of an effector (or a group of effectors) to a stimulus (Fig. 23.8). In a reflex arc the excitation resulting from the action of the stimulus on the receptor cells generates a disturbance called the nerve impulse, which is passed by nerve cells (neurons) called *afferent* (sensory) *neurons* to the central nervous system where it is relayed by *interneurons* to *efferent* (motor) *neurons* connecting with the effectors (generally muscles). The response of an animal as a whole may consist of a series of such reflexes in a chain, the first reflex tripping the second, the second the third, and so on, or it may also respond by complex reactions which originate in the higher centers of the nervous system. However, there are some effectors which are independent of neuronal connections. For example, leukocytes, which are widespread in the blood or body fluid of animals, and most ciliated epithelial cells respond directly to chemicals and not through nervous control (Sleigh 1966; but see also Bone, 1958).

Ductless glands of internal secretion (endocrine glands) secrete hormones into the blood which aid in coordinating responses, as well as in growth and metabolic processes of the cells of animals. Such glands are called into action sometimes by the nervous system—at other times, by chemical changes in the body cells or fluids. The central nervous system plays a major role in coordination of immediate responses. Coordinative action of endocrines and nerves is considered in textbooks of animal physiology (Ruch and Patton, 1965).

A brief discussion of cellular excitability as manifested by the electrical changes induced in cells by stimulation is given in Chapters 22 and 23. Contractility manifested in cells as a consequence of stimulation is considered in Chapter 24 and the chemistry of contractile processes of cells in Chapter 25.

# CHAPTER 22 / ACTION POTENTIALS OF CELLS

An *action potential* is usually defined as the temporary change in electric potential between stimulated and resting portions of a cell. It follows stimulation and passes from one end of the cell to the other. Because it moves as a wave of excitation, it is also spoken of as a propagated potential. Action potentials are especially prominent on nerve and muscle cells (Erlanger and Gasser, 1937), but they are also found to occur in many other excitable plant and animal cells (Blinks, 1955; Hisada, 1957; Kinosita *et al.,* 1964; Eckert, 1965, 1966) and during fertilization of an egg (Tyler *et al.,* 1955). Since an action potential is best developed on a nerve cell, where it is called a *nerve impulse,* it will be discussed in greater detail later in this chapter.

## THE NATURE OF THE NERVE IMPULSE

*Neurons,* as nerve cells are usually called, are fundamentally similar to other cells in that they have a cell body with cytoplasm, a nucleus and a cell membrane. Therefore, they are subject to the same principles that govern the behavior of all living cells. But by adaptation to their specialized function, they possess structures admirably fitting them for the conduction of nerve impulses. *Dendrites,* short outgrowths from the cell body, carry impulses to the cell body, and *axons,* long outgrowths, carry the impulses to the next neuron. An axon is often referred to as a nerve fiber, but, in a loose way, nerve fiber may also refer to the whole neuron.

As seen in Figure 22.1, the *axon* of the vertebrate myelinated neuron is wrapped in a *myelin sheath* consisting of multiple membranes of the *Schwann cells,* the nuclei of which occur occasionally along the sheath (see also Figs. 11.5 and 11.6). Recently Schwann cells in nerve tissue culture were actually observed to wrap around the neuron, thus laying the myelin sheath as had been postulated for many years (Ernyei and Young, 1966). The myelin sheath varies, being thinner on invertebrate neurons where it may consist of only a single layer; it is practically absent in the finer fibers, called *nonmyelinated* fibers, of the vertebrate autonomic nervous system. The myelin sheath does not enclose the dendrites, the cell body or the terminations of the axon. *Myelin,* containing a considerable amount of lipid (see Chapter 11), seems to act as an insulator, and myelinated neurons transmit the nerve impulse more rapidly than nonmyelinated ones of the same diameter (see Tables 22.1 and 22.2).

Neurons vary in size and form (Bodian, 1952, 1962). The largest nerve fibers in a mammalian nerve are designated the A fibers, those of intermediate size, the B fibers and the small ones, the C fibers

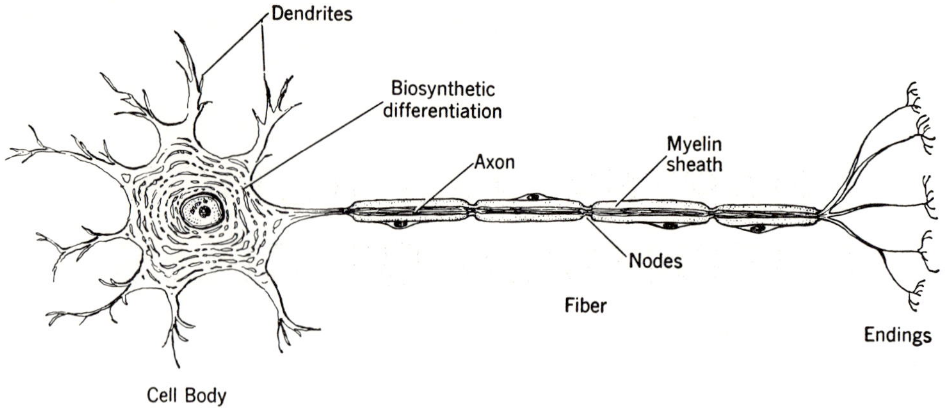


***Figure 22.1.*** Diagram of a myelinated neuron. The nuclei along the axon are in Schwann cells which form the myelin sheath. Such cells cover the axon except at the nodes of Ranvier and at the terminal branching portion. The terminal endings make synaptic connections with peripheral tissue. (From Schmitt, 1959: Biophysical Science. A Study Program. Oncley, ed. John Wiley & Sons, New York.) Radioactive tracers disclose movement of components from the cell body through the axon to the muscles (Korr *et al.,* 1967).

(see Table 22.1). These are further classified into subgroups which are given Greek letter designations (Erlanger and Gasser, 1937). Fibers of different size vary considerably in their properties.

A nerve usually consists of a bundle of many neurons. However, single fibers can be dissected out of a bundle and studied individually, and much of the information of action potentials has been obtained in this manner (Tasaki, 1953).

*Glial* cells (supporting cells of the central nervous system) in amphibians do not discharge like neurons when subjected to depolarizing (or hyperpolarizing) currents, although they show resting potentials like most cells. Glial cells connect to one another by low resistance bridges but they are separated from nerve cells by intercellular clefts (Kuffler *et al.,* 1966).

The twitching of the muscles in the leg of a frog when the sciatic nerve is cut or otherwise stimulated is a common observation. Pioneer investigators used the muscle twitch as evidence of a nerve impulse traveling along the nerve from the point of stimulation to the muscle, and the nerve-muscle preparation became a classic tool in the study of nerve transmission. Direct measurements of nerve activity, however, were initiated in 1786 by Galvani, who obtained evidence for the presence of electric currents in nerve as well as in muscle, even when stimuli other than electricity (e.g., pinching) were used. In 1848 Emil Du Bois-Reymond

**Table 22.1.** Properties of Neurons of Different Sizes (Cat and Rabbit Saphenous Nerves)

| | GROUP | | |
|---|---|---|---|
| PROPERTIES | A | B | C |
| Diameter of fiber ($\mu$) | 20–1 | 3 | — |
| Conduction velocity (m./sec.) | 100–15 | 14–3 | 2 |
| Duration of action potential (msec.) | 0.4–0.5 | 1.2 | 2.0 |
| Absolute refractory period (msec.) | 0.4–1.0 | 1.2 | 2.0 |

After Grundfest, 1940: Ann. Rev. Physiol. 2.

published a book called *Researches on Animal Electricity*. However, ready means for detecting electric currents in excited nerve cells were not available until much later.

The introduction of the cathode ray oscillograph by Gasser and Erlanger in 1922 provided an instrument which is now generally used to measure action potentials and other electrical manifestations of living things, because the electron beam, having very slight inertia, responds to an imposed potential with practically no delay. In the oscilloscope the stream of electrons coming from the hot filament of the tube impinges upon a fluorescent screen covered with a transparent plastic marked like graph paper (see Fig. 22.2). Activated by an oscillating circuit, the beam sweeps at any desired frequency across the screen from side to side between two vertical plates in the tube. An action potential from a nerve fiber can be passed through an amplifier and imposed upon two horizontal plates, causing the beam to move upward or downward according to the potential. The screen of the oscilloscope may be calibrated for time, by imposing a known sweep frequency upon the electron beam, and for voltage, by means of a standard cell. Because the image has afterglow (phosphorescence), it can be readily photographed for a permanent record. The zero position may be altered at will so that the entire screen may be used to measure either negative or positive variations only, as dictated by the experiment at hand, and the electrical connections may be made in such a way as to have upward movement of the beam indicate either positive or negative deflection.

The surface of a neuron excited by pinching or by other stimuli becomes negative to the surface of an unexcited portion, as shown by recording the electrical changes (see Fig. 22.3). If two pickup electrodes are applied at some distance from the point of excitation, the electrode nearer to the stimulated part of the neuron at first becomes negative to the other electrode; then, after a brief time, it becomes positive to the other electrode. The brief negativity can be interpreted as the arrival of a moving wave of excitation (action potential) on the nerve cell at the position of the first electrode, and the subsequent positivity can be interpreted as the neuron's recovery from the excitation at the first electrode as the wave of excitation passes to the second electrode, and the latter is rendered negative to the first. Because the electrode proximal to the stimulus is first negative and then positive, the action potential, or the nerve impulse, is said to be *diphasic* in character. A whole series of electrodes, laid along the nerve fiber, will necessarily show the same phenomenon as the impulse travels to the end of the fiber.

Once an impulse has been initiated, it passes to the end of the neuron. Regardless of the intensity of stimulation beyond the threshold, the magnitude of the action potential or nerve impulse on a

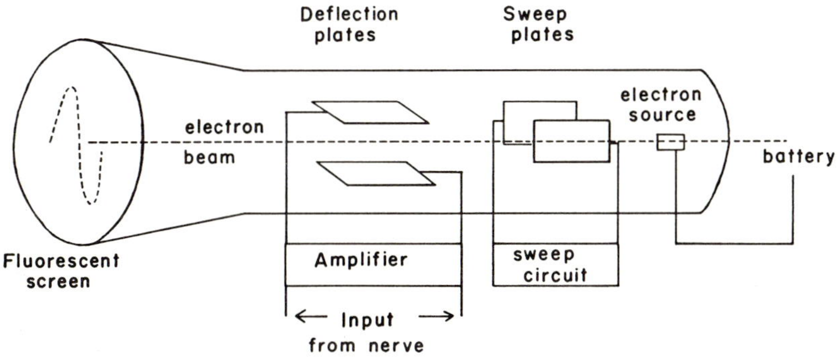


***Figure 22.2.*** Diagram of a cathode ray oscilloscope showing a trace of a diphasic spike on the fluorescent screen.

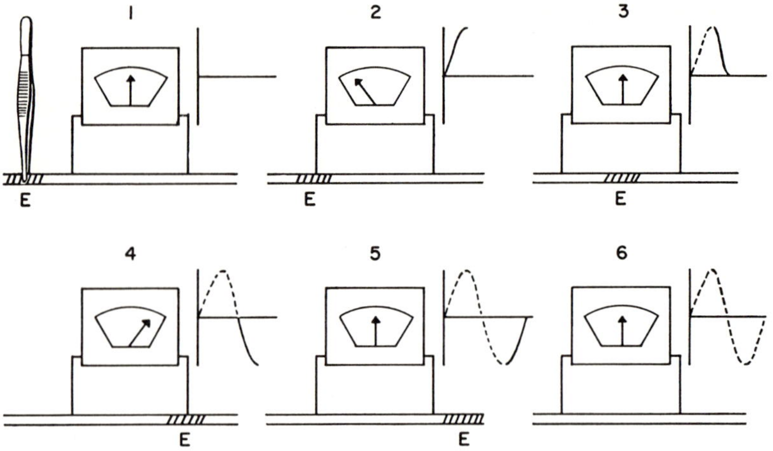


*Figure 22.3.* The diphasic nature of the action potential on a nerve. A pinch of the forceps stimulates the nerve, and a wave of excitation (E) passes the left pickup electrode (that is, the one nearest the point of stimulation), making it temporarily negative to the right one. Later, when the impulse reaches it, the right pickup electrode becomes temporarily negative to the left one. Finally, after the wave of excitation has passed both pickup electrodes, the hand on the dial returns to zero. If the movement of the hand on the dial is plotted, a graph similar to those at the upper right of these diagrams is obtained.

single neuron under the same conditions is always the same whether measured near the stimulus or far from it (see Fig. 22.4). This characteristic of the nerve impulse is referred to as the *all-or-none* principle of nerve action. The nerve impulse has been likened to a trail of gunpowder, along which each bit is set off by the fire next to it, regardless of whether a match or a blowtorch was first used to ignite it. The neuron appears to be ready to discharge, needing only to be triggered.

Cold and anesthetics decrease excitability of a neuron, but the action potential is measured in front of and behind an anesthetized (or chilled) portion of a neuron, its magnitude remaining the same provided the nerve impulse is able to pass. Transmission therefore occurs *without decrement* (Brazier, 1960, 1964).

Once a nerve impulse has passed over any point of a neuron, a *refractory period* ensues, a sort of "busy signal" period, during which the neuron is recovering its capacity to react. At the beginning of the refractory period no stimulus, regardless of its strength, will evoke a response at the point of stimulation; then a response may be elicited but only by a stimulus much stronger than is normally needed; gradually the neuron recovers its normal sensitivity. If the neuron has been excited by repeated stimulation to carry a train or *volley* of impulses, its recovery is slower than after a single impulse.

These changes in excitability of a neuron are correlated with the phases of the action potential as illustrated in Figures 22.5 and 22.6. The data for these curves are the potentials between the two recording electrodes as described earlier. The *spike* is the peak, or the maximal, action potential. For any neuron the spike is of constant magnitude and seems to pass unaltered along the length of the neuron. No additional activity can be induced by a shock of any strength until the spike at the point of stimulation has

*Figure 22.4.* Conduction without a decrement. On the oscilloscope screen the diphasic spike for a given fiber is equal in height at short and long distances from the point of stimulation.

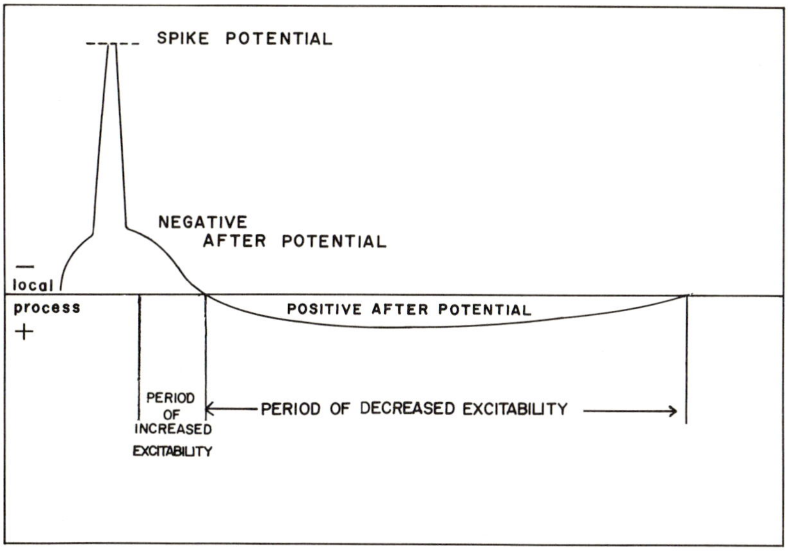


***Figure 22.5.*** The excitability cycle of a neuron. Excitation leads to a brief local change which fires into a spike if the threshold has been passed. The spike is followed by negative after-potential, and the negative after-potential is followed by a slow positive after-potential. If the spike were drawn in proportion to the local potential and the after-potentials, it would extend beyond the page edge by several feet. The prepotentials which precede the spike are not shown adequately on this diagram. See also Figure 22.9. (After Morgan, 1943: Physiological Psychology. McGraw-Hill Book Co., New York.)

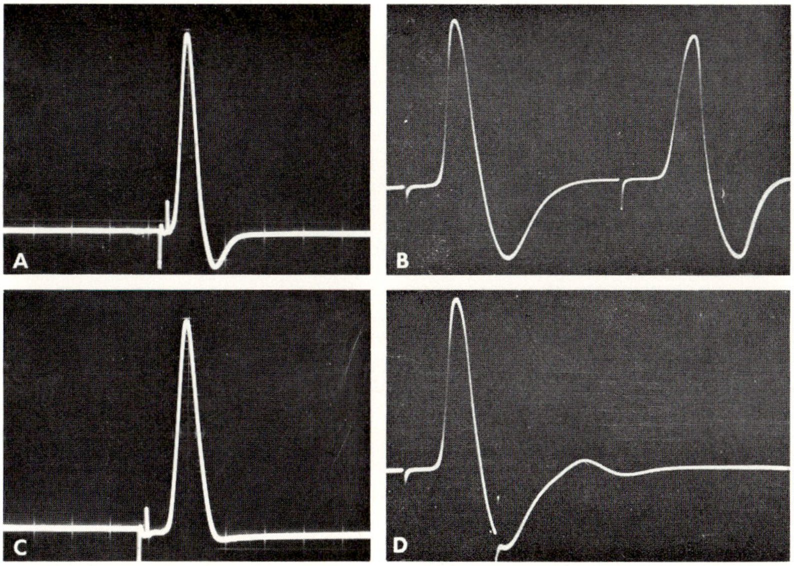


***Figure 22.6.*** Action potentials (spikes) on a frog sciatic nerve. *A,* Compound diphasic action potential from the fastest (alpha) fibers of the frog sciatic nerve, recorded by a pair of electrodes 1 cm. apart, located 1 cm. from the stimulating electrode. *B,* Two stimuli applied to the frog nerve in rapid succession 0.007 second apart. Two diphasic spikes appear. *C,* Compound monophasic action potential recorded from the nerve in *A,* at a pair of electrodes 1 cm. apart located 2 cm. from the stimulating electrodes. The nerve has been crushed between the recording electrodes. Note that the first phase of the action potential is now broader than in *A,* because of the elimination of the second electrode response, which in *A* hastens the falling phase. Markers for *A* and *C* indicate 1 millisecond. *D,* Two stimuli given 0.003 second apart. The nerve is unable to respond to the second stimulus because it is still in the refractory phase. In all the records the small break in the record is the stimulus artifact. (*A* and *C* by courtesy of Dr. D. Kennedy. *B* and *D* from, "From The Nerve Impulse", by Bernhard Katz. Copyright © 1952 by Scientific American, Inc. All rights reserved.)

been completed. It is thought that the *absolute refractory period* corresponds to the time during which the spike appears at the particular point on a neuron. For the short period during the tail of the spike, a stimulus much stronger than normal is needed to excite the neuron. This period is called the *relative refractory period.** This is followed by the *negative after-potential,* which is the broad base at the end of the tail of the spike and is a period of increased excitability (lowered threshold) resulting, presumably, from the negative after-potential. This in turn is followed by the *positive after-potential,* which is seen below the zero line of the graph (Fig. 22.5); this is a period of decreased excitability (higher threshold), presumably, a consequence of the excess positive charge (Erlanger and Gasser, 1937). Finally, the neuron resumes normal sensitivity.

The after-potentials are not always the same in all nerve fibers. Sometimes only positive after-potentials appear, as in small mammalian C fibers. The after-potentials also depend in part upon the treatment to which the neuron has been subjected (Erlanger and Gasser, 1937). The voltages and the time for the spike and the various stages of the refractory period are different for different types of neurons (Table 22.2). In large mammalian fibers (A fibers) the spike reaches a magnitude of a few hundredths of a volt, whereas the negative after-potential reaches about one-twentieth this voltage, and the positive after-potential about one-five-hundredth. The spike lasts about 1 millisecond, the negative after-potential about 15 milliseconds and the positive after-potential as long as 70 milliseconds. The spikes might be considered the message carriers of the neuron, whereas the after-potentials in contrast might be considered the indicators of the readiness with which messages will be accepted.

Figure 22.6 shows an actual oscilloscopic record of a nerve action spike. It will be noticed that the two arms of the diphasic spike are not symmetrical. The first one is larger because, when the excitation reaches the first electrode, the potential suddenly changes from zero to the maximal value. However, when the excitation reaches the second electrode, the first electrode has not yet fallen to zero but is still negative. The second spike can arise only after the second electrode has become more negative than the first, the time required for it to do this being usually taken as a measure of the time it takes the nerve impulse to travel from the first electrode to the second. Also, the second spike never reaches the magnitude of the first one because, by the time

*Sometimes the after-potentials are included in the relative refractory period, but present usage is against this.

**Table 22.2.** Velocity of Propagation of the Nerve Impulse as Affected by Axon Diameter and Sheath Thickness*

| NERVE CELL | FIBER DIAM. (FD) IN $\mu$ | AXON DIAM. (AD) IN $\mu$ | $\frac{AD}{FD}$ | BIREFRINGENCE (B) | VELOCITY (CA) (M./SEC.) | B × FD |
|---|---|---|---|---|---|---|
| Squid, giant | 650 | 637 | .98 | −.0001 | 25 | — |
| Earthworm, giant | 100 | 90 | .90 | .0010 | 25 | .10 |
| Shrimp, giant | 50 | 43 | .87 | .0024 | 25 | .12 |
| Frog, sciatic | 10 | 7.5 | .75 | .0105 | 25 | .105 |
| Cat, saphenous (20° C.) | 8.7 | 6.6 | .76 | .014 | 25 | .12 |
| Catfish | 8.8 | 5.8 | .58 | .012 | 25 | .105 |

* Sheath thickness is determined by birefringence. Birefringence is measured in terms of the degree of rotation of the analyzer prism in a polarizing microscope required to extinguish the polarized light from the nerve. The thicker the layer of myelin, the greater the birefringence, as seen in the vertebrate nerves compared with the invertebrate nerves. For use of a polarizing microscope see Chapter 2. (Modified from Taylor, 1942: J. Cell. Comp. Physiol. *20.*)

the nerve excitation has reached its maximum at the second electrode, the region of the nerve at the first electrode will not yet have fallen to zero, and the electrodes can measure only potential differences, not absolute values. The area of excitation (negativity) on the nerve at any one time may occupy a space of several millimeters to many centimeters.

It is also noteworthy that after-potentials are not recorded because they are too small to register with the amplification (gain) used. The small pip which precedes the spike is a *stimulus artifact* and is a record of the electric wave passing along the nerve fibers in purely physical fashion. A piece of string soaked in balanced salt solution will serve as well as the nerve fiber for propagation of this artifact (Stevens, 1966).

## THE RATE OF CONDUCTION OF THE NERVE IMPULSE

Although the nerve impulse travels so fast that one prominent physiologist (J. Müller, 1801–1858) despaired of its ever being measured, its speed was recorded later by one of his students, Hermann von Helmholtz (1821–1894), who used a smoked drum, a coil and a pendulum, but whose crude method nonetheless gave values close to those accepted today. The speed of the nerve impulse on large mammalian nerve fibers has since been shown to be seldom over 100 meters per second (Brazier, 1960). This shows that the impulse is not a simple electrical wave (which would travel at an extremely high velocity) but rather some reaction in the nerve cell which is accompanied by an electrical wave.

The rate of conduction of the nerve impulse increases with an increase in cross-sectional diameter of the neuron and with increasing thickness of the myelin sheath. The thickness of the myelin sheath may be measured by its birefringence. When the product of the birefringence multiplied by the diameter is nearly constant, the rate of propagation is the same even for neurons of diverse origins, as shown in the last column of Table

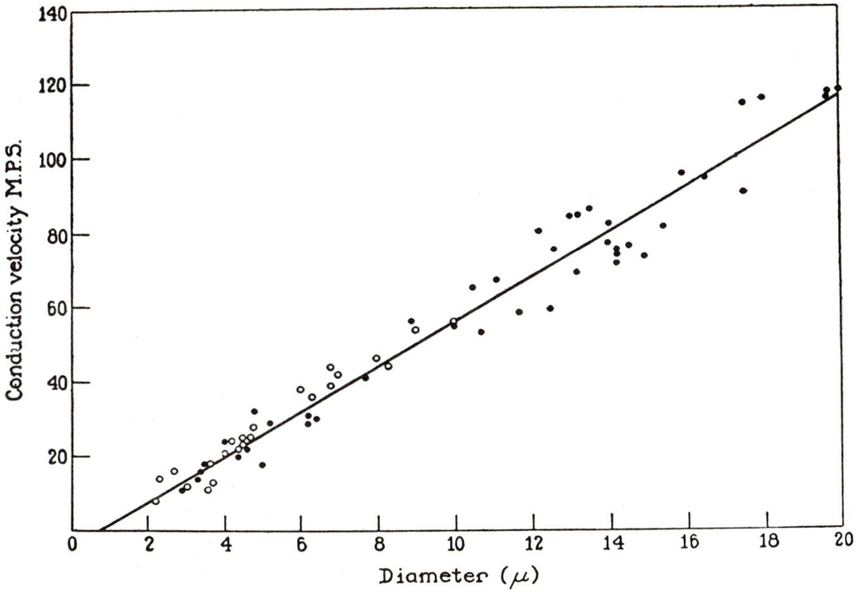


***Figure 22.7.*** The linear relation between diameter and conduction velocity of mammalian myelinated nerve fibers. Each point represents a determination of the maximum conduction velocity of a given nerve, plotted against the diameter of the largest fiber in that nerve. Dots represent adult nerves; circles represent immature nerves. (After Hursh, from Gasser, 1941: Ohio J. Sci. *41*:145.)

**Table 22.3.** Comparison of Conduction Rates of Nerve Impulses in Vertebrate and Invertebrate Nerves

| TYPE OF NERVOUS STRUCTURE | GROUP OF ANIMALS | GENUS AND TYPE OF NERVE | TEMPERATURE IN °C. | VELOCITY IN METERS/SECOND |
|---|---|---|---|---|
| Nerve net: | Coelenterate | Anemone (*Metridium*) | 21 | 0.121–0.146 |
| | | *Calliactis,* net | — | 0.04–0.15 |
| | | *Calliactis,* through tracts | — | 1.2 |
| Ganglionic cord: | Annelid | Earthworm (*Lumbricus*) | — | 0.6 |
| | Arthropod | Myriapod (*Scolopendra*) | — | 2.5 |
| | | Crayfish (*Cambarus*) | — | 1.2 |
| Nerve fibers: | Mollusk | Snail (*Helix*), $< 1\ \mu$ diam. | 15–18 | 0.4–0.05 |
| | | *Mytilus,* pedal nerve | — | 0.64 |
| | | *Sepia,* $< 50\ \mu$ diam. | — | 3.5, 2.26 |
| | Arthropod | *Limulus,* leg nerve | — | 4.6, 1.3 |
| | | Lobster, leg nerve | — | 9.2, 1.8 |
| | Vertebrate | Turtle, vagus, n.m. | — | 0.8, 0.3 |
| | | Frog, nonmyelinated | 21.5 | 0.5–0.4 |
| | | Frog, dorsal root | 21.5 | 42.0 |
| | | Rabbit, depressor, 2–4 $\mu$ | 38 | 5.0 |
| | | Dog, saphenous, 17 $\mu$ | 38 | 83.3 |
| Giant fibers: | Mollusk | *Loligo,* 718 $\mu$ diam. | — | 22.3 |
| | Annelid | Earthworm (*Lumbricus*), | | |
| | | lateral | 10–12 | 7–12 |
| | | median | 10–12 | 17–25 |
| | Arthropod | *Leander,* 35 $\mu$ diam. | 17 | 18–23 |
| | Vertebrate | *Ameirus,* 22–43 $\mu$ | 10–15 | 50–60 |

After Prosser, 1946: Physiol. Rev. *26.*

22.2. For myelinated nerve fibers of mammals, the rate of conduction of the nerve impulse varies with the cross section diameter, as illustrated in Figure 22.7. The rate of transmission for a given neuron is a constant; it may be slowed at a partial block, but beyond the block it continues at its characteristic rate. The rate of conduction of the nerve impulse at a specified temperature in nerve cells of a wide variety of animals is shown in Table 22.3.

## THE STRENGTH-DURATION RELATIONSHIP

Although changes in pressure, heat, chemicals, and other stimuli may also evoke a nerve impulse, these stimuli cannot be conveniently applied without damaging the nerve cell, nor can they be readily assessed. Ever since the pioneer experimental studies of Galvani, electrical stimulation has been used almost exclusively for experimental studies of the nerve impulse. For this purpose an inductorium is employed, and high voltage shocks at low currents can be given as stimuli or, better still, an electronic stimulator can be used which develops the desired voltage almost instantly and in which the duration of current flow can be precisely controlled (Gerebtzoff and Schoffeniels, 1960).

When a neuron is stimulated, the cathode of the stimulating electrode is placed toward the pickup electrodes of the recording instrument, since at the anode the threshold required for stimulation rises even to the point of blocking the nerve impulse arising at the cathode. In the absence of an anodal block, a nerve cell will transmit impulses in both directions from a spot stimulated electrically (Brazier, 1960).

A minimal electric current or *threshold current* is necessary to stimulate neurons. The threshold current that, after operating for a long period, is just able to stimulate the neuron is called the *rheobase*. Below the rheobase there is no response because the neuron recovers as rapidly as it is affected and no nerve impulse can be initiated, even with continuous stimulation. The time required to stimulate a neuron with a current of rheobasic strength is called the *utilization time*.

If a current stronger than rheobase is used to stimulate a nerve fiber, it need not be applied for as long a time as the utilization time. This is most clearly illustrated in a curve relating the strength of current and the duration of time it must act in order to initiate a nerve impulse. Such a curve is called the *strength-duration curve* (see Fig. 22.8). The strength-duration relationship is usually described in terms of the current rather than in terms of the voltage, and this convention is followed here.* On the basis of this curve, provided the product of the current and the square root of the time is maintained constant, the neuron is always stimulated. This product is referred to as *threshold stimulation*. Any product less than this constitutes a *subthreshold stimulation*. Though the quantitative values may be quite different for diverse excitable systems, the general shape of the strength-duration curve is usually similar (Lapique, 1926).

A current above rheobase but applied for a time less than that "required" on the basis of the strength-duration curve fails to evoke a response in a neuron. However, if the same stimulus is repeated rapidly enough (e.g., every 0.15 milliseconds) for a sufficient number of times,

*Since by Ohm's law the current in a circuit is directly proportional to the electromotive force and inversely proportional to the resistance of the circuit, the current in amperes is directly proportional to the voltage in volts.

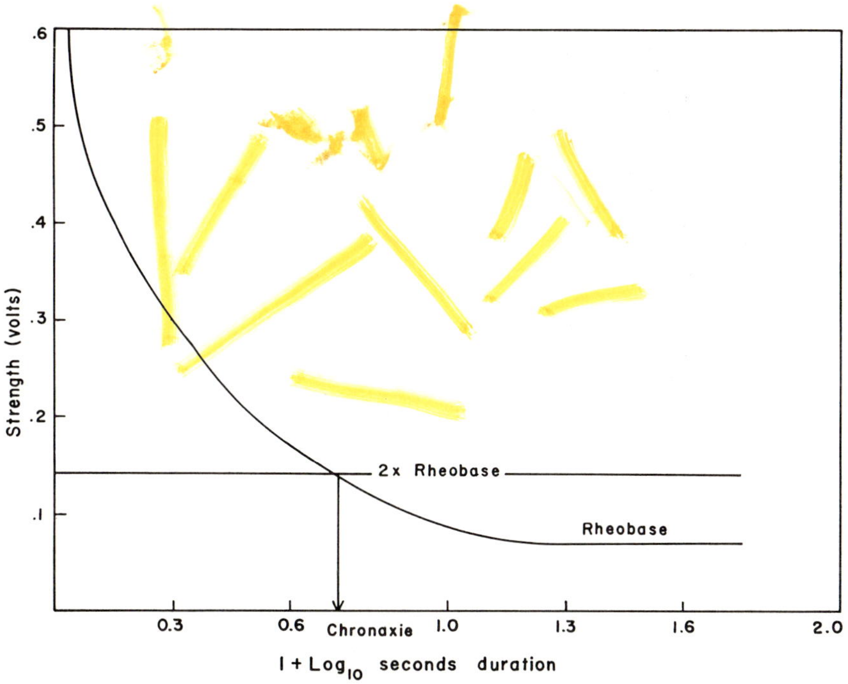


***Figure 22.8.*** Strength-duration curve for a nerve from the crab *Cancer productus*. Similar curves are obtained for any excitable system although absolute values vary. The chronaxie is the time required to stimulate when the current is twice rheobase. It is marked at the base of the arrow on the time scale.

stimulation will occur by temporal *summation,** provided the sum of all the applications of the stimulus is greater than the minimum demanded by the strength-duration curve. Stimulation under such conditions therefore does not violate the strength-duration principle. Summation, even when adequate, will fail to stimulate if the time interval between stimuli is increased beyond a certain limit, because the system recovers from each stimulus. To trigger the excitatory process, both the strength of the current and its application time must be sufficient to overcome the tendency of the neuron to recover (Johnson *et al.,* 1954; Katz, 1966).

In the special case in which the voltage is steadily applied and gradually increased up to, and kept at, a value which would normally stimulate a nerve fiber if applied suddenly, it is found that the nerve fiber will not respond. For this experiment an instrument called a rheochord is used. Although it seems to violate the strength-duration principle, this failure of the nerve fiber to respond is explainable, presumably, on the basis that the nerve fiber is capable of recovery during the time of the gradual rate of change in voltage. A new base level is established in each of these cases from which the added current is still too small a change to cause stimulation. A sudden rate of voltage change is therefore still another factor necessary to

**Spatial summation* occurs in synapses. Impulses impinging on a synapse from many neurons, each impulse alone inadequate to stimulate the postsynaptic neuron, may, if simultaneous, together stimulate it.

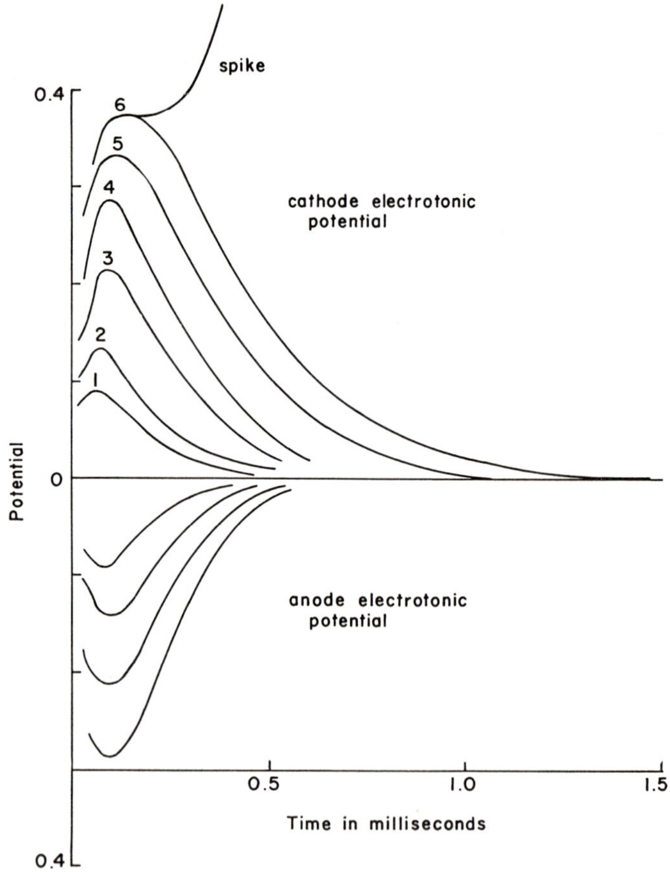


***Figure 22.9.*** Local excitatory state on a crab nerve following shocks of increasing intensity at the stimulating electrodes. The local excitation arises at both anode and cathode and decays in a millisecond. When cathodal local excitation passes threshold, the neuron suddenly fires and the excitation is then propagated the full length of the fiber. (After Hodgkin, 1938: Proc. Roy. Soc. *B126*:95.)

bring about a response, even when the strength-duration relationships are satisfied (Brazier, 1960).

Defining excitation of a nerve cell has been likened to "defining quantitatively the effort needed to bring a leaky vessel to overflowing" (Johnson *et al.,* 1954, p. 611). Although a subthreshold stimulus is incapable of evoking a propagated spike potential, it does alter the neuron, and a measurable, local, nonpropagated potential appears, indicating a mild state of excitation (see Fig. 22.9). This excitation declines exponentially since a recovery process always counters excitation. Both excitation and recovery processes have been formulated mathematically, but the subject is too complex for discussion here (Ruch and Patton, 1965).

Usually a current greater than rheobase is found more convenient experimentally for a comparison of the excitability of diverse excitable cells. Thus, the *chronaxie,* which is the time required for a current twice the rheobase to just stimulate an excitable cell has been used to compare the excitability of nerve cells, muscle cells, protozoans and plant cells (Lapique, 1926). Although the chronaxie is used less today, determination of the strength-duration curves is valuable because it illustrates most clearly the quantitative relations between excitation and stimulus.

## ACCOMMODATION

According to the strength-duration principle, a neuron responds by a propagated action spike potential only if an adequate current is applied for a sufficient time, that is, if *threshold stimulation* has been applied. This is true only at the cathode; at the anode no propagated response occurs even if stimuli of much larger magnitude are applied. That the nerve fiber is affected by anodal stimulation, but is able to accommodate to it as it does to all stimulation, is shown by the state of electrotonus induced by anodic shocks of any magnitude (a state of *anelectrotonus*). For example, after any anodal shock, the subsequent cathodal threshold required for stimulation is always greater than normal, demonstrating that a partial block has been established by the anelectrotonus. The block is greater the larger the anodal shock and the shorter the time intervening before application of cathodal stimulation, as if the anelectrotonus induced by anodal stimulation must first be overcome before stimulation can occur. On the other hand, after subthreshold cathodal shock which induces a state of *catelectrotonus,* the threshold at the cathode is lowered to the extent that a subsequent stimulus of less than normal threshold stimulation is adequate if applied soon afterward (in a fraction of a millisecond), as if the cathodal shock were being added (summation) to a partially excited neuron in a state of *catelectrotonus* (Brazier, 1960).

Reaction of a nerve fiber, therefore, depends upon its preceding experience. The threshold of a neuron is altered not only by subthreshold stimuli but also by the after-potentials following passage of a nerve impulse. Furthermore, after-potentials are much more persistent after repetitive stimulation of a neuron. They are also prolonged by some drugs such as veratrine (Katz, 1966).

## CORE CONDUCTOR THEORY OF THE NERVE IMPULSE

On applying electrodes to the surface of a nerve fiber, electrotonic potentials develop on the fiber at the electrodes and in their vicinity. These potentials indicate that current is flowing along specific lines as shown in Figure 22.10. The potential change is greatest immediately underneath the electrodes and falls off exponentially at points progressively farther away. It is proposed that the development of these potentials and their associated currents can be best accounted for if the nerve fiber is considered to act as a *core conductor,* that is, as a cylinder of conducting fluid material (axoplasm, the cytoplasm of the axon) with a sheath (cell membrane) of high electrical resistance, surrounded by a layer of conducting medium (balanced salt solution). The cell membrane which becomes polarized

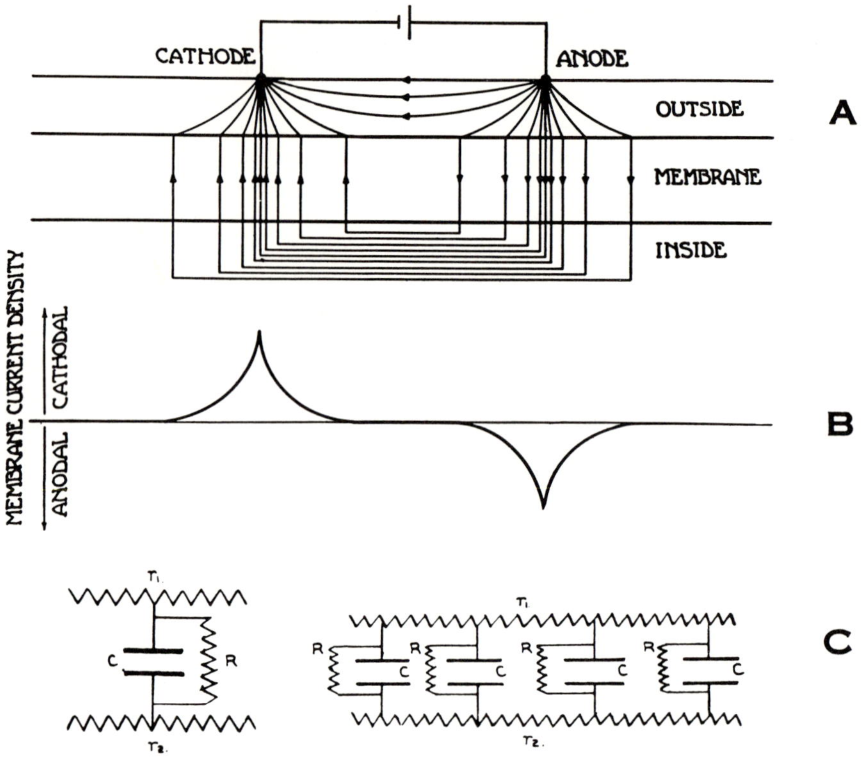


*Figure 22.10.* Core conductor theory of nerve fiber excitation. *A* shows the current distribution in a nerve fiber during passage of a constant current. Current lines represent spatial spread and direction of current flow outside, through, and inside the membrane of the nerve fiber. *B,* Lateral spread of current shown by plotting the density of penetrating current (membrane current) against the length of the nerve fiber. The current falls off in an approximately exponential manner with distance from the electrode. *C,* Electrical models of a nerve fiber. C represents the capacitance, R represents the transverse resistance and $r_1$ and $r_2$ represent the longitudinal resistances. The model to the left is the simplest electrical circuit having some of the properties of a nerve fiber. The series of such models shown to the right also account for the time characteristics in the build-up of electrotonic potential. (After Ruch and Patton, 1965: Physiology and Biophysics. 19th Ed. W. B. Saunders Co., Philadelphia.)

is thus presumed to impose a capacity on the system such that the nerve fiber may be viewed as a condenser capable of being charged on application of a potential difference across it. Such an electrical model of a nerve fiber has been constructed and studied. However, under steady state conditions in the living nerve fiber during application of a difference in potential, a slow flow (leak) of current is found to occur. To allow for this slow leak of current in the electrical model it is necessary to couple a resistance with the condenser through which such slow leakage can take place before it can be used.

A measurable time is usually required for the electrotonic potential to develop both at the electrodes applied to a nerve fiber and in their immediate neighborhood. To account for this time requirement, it is necessary to consider the nerve fiber as being made up of a series of such condenser-resistance units (Fig. 22.10*C*), each one like the one just described. When a potential is applied through a pair of electrodes on the nerve fiber, its structures, acting like a condenser, are charged, with the result that the potential at and near the electrodes builds up and tends to oppose the further flow of current.

The resistance and capacitance of the nerve fiber membrane and the resistance of its cell membrane sheath and core (axoplasm) can be calculated from experimentally determinable properties of nerve fibers. From such data it is possible to predict quantitatively the development of electrotonus, with time as a factor, in a nerve fiber, if the nerve fiber is considered to act like the electrical model described here. When the steady state condition has been attained, the distribution of electrotonic potentials around the electrodes can also be computed. The predictions made on the basis of such a model of the nerve fiber are well in agreement with the facts, and the core conductor formulation has been useful in planning experimental studies, especially in defining more clearly the properties of parts of the nerve fiber.

For passive conduction, such as occurs in an electric wire, a cable made like a nerve fiber would be of little use because its losses are great: the surface leakage and the resistivity of its core are $10^8$ greater and its sheath capacity $10^6$ greater than commercial cables. In it a weak signal (just sufficient to produce a spike) fades out within a few millimeters. The "cable" model would fail to conduct an impulse were it not for the regenerative processes on a nerve fiber that reamplify a signal above threshold along each point of the line (Katz, 1959, 1966; Clark and Plonsey, 1966).

The peculiar cable properties of a nerve cell are, however, of importance in permitting subthreshold stimulation to remain sufficiently long for summation. Especially in the dendrites of a neuron is this of importance since here the excitability is local, graded and decremental. Succeeding increments of depolarization may therefore summate to the threshold required to initiate the propagated all-or-none spike on the axon.

## SALTATORY CONDUCTION OF THE NERVE IMPULSE

Conduction in myelinated nerve fibers is exceedingly rapid compared with that in nonmyelinated nerve fibers of similar size. The myelin sheath is interrupted periodically by the *nodes of Ranvier* (Fig. 22.1), at which the membrane of a neuron makes contact with the bathing fluid. The nodes are further apart in neurons of large diameter than in neurons of small diameter. For example, the distance between nodes is about 0.2 mm. for a bullfrog neuron of 4 $\mu$ diameter, about 1.5 mm. for a neuron of 12 $\mu$ diameter and 2.5 mm. for one of 15 $\mu$ diameter (Tasaki, 1953). It has been suggested that perhaps conduction in myelinated neurons is so rapid because the action potential skips from node to node rather than having to travel along the full length of the membrane. Such skipping from node to node is called *saltatory conduction* (see Fig. 23.7*D*).

Nodal threshold for stimulation is much lower than the internodal threshold (see Fig. 22.11). Since upon stimulation of a nerve cell an electric field is formed which has been shown to operate from the point of excitation over a span of about two nodes, this electric field is much more likely to evoke a change in potential in a region of low threshold (node) than in a region of high threshold (internode). When a node is cathodally stimulated and the two nodes on either side of it are anesthetized, conduction of the impulse is still achieved across the neighboring nodes, but the same stimulus fails if applied to the internode. Also, the impulse does not pass across two anesthetized neighboring nodes even if stimulation is nodal, because in this case the third node is beyond the operation of the electric field. A squid giant axon fails to conduct the impulse across a dead section unless a metal wire is laid across this section. The electric field is not strong enough in air but is adequate in the better conductor furnished by the wire bridge (Tasaki, 1953).

One objection to the theory of *saltatory conduction* is that nodes had not generally been seen in the central nervous system, yet conduction in the brain and cord equals that in the peripheral nerve fibers. However, improved techniques have revealed nodes in preparations of the central nervous system, and at present most workers appear to be convinced of the generality of saltatory conduction.

It is interesting to note that Lillie

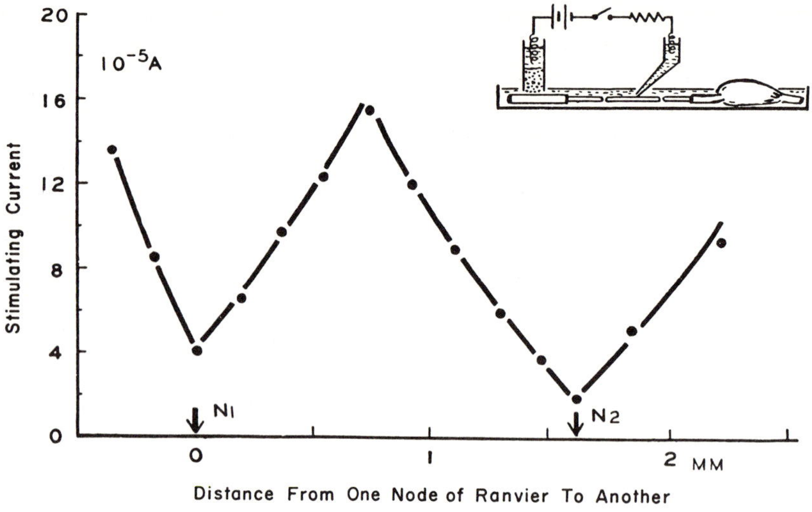


***Figure 22.11.*** A graph demonstrating that nodal stimulation is lower than internodal stimulation. (After Tasaki, 1953: Nervous Transmission. Charles C Thomas, Springfield, Ill.)

(1922) predicted saltatory conduction in nerve fibers on the basis of an iron wire analogue to the neuron. An iron wire, when placed in a solution of concentrated nitric acid, becomes coated with oxide. When the coated wire is placed in more dilute nitric acid, nothing happens unless it is stimulated by heat, electricity or mechanically by a scratch. Stimulation sets off a wave of oxidation-reduction which runs the length of the wire (all-or-none), followed by formation of a new coating. After a "refractory period," stimulation sets off another wave of oxidation-reduction. When the wire is encased in sections of glass tubing, the wave of oxidation-reduction passes from exposed "node" to "node" in saltatory fashion (Hodgkin, 1964).

## RESTING METABOLISM OF NERVE CELLS

Nerve cells at rest respire at about the same rate per unit of mass as resting muscle cells. Improvements in respirometry and especially the development of microrespirometers have made possible the measurement of gaseous exchange, and a respiratory quotient of approximately 0.8 has been found for the neuron, indicating that several nutrients are being oxidized (Fenn, 1929; Gerard, 1932, 1946; Brink *et al.,* 1952; Connelly, 1959; Arshavski, 1960). Although glucose is believed to be the main source of energy, because it is lost rapidly in the respiring nerve cells, presence of creatine, ammonia and phosphate, as metabolic wastes produced in respiration, indicates that protein and lipid are also being consumed. When nerve cells are subjected to anaerobiosis, lactic acid accumulates, indicating that during anaerobic metabolism the source of energy for maintaining excitability is mainly glycogen. The tolerance of a nerve cell to anaerobic conditions, in fact, depends partly upon the amount of glycogen reserve available. For example, crab nerve fibers, which naturally have much glycogen, conduct without oxygen for 5 hours, whereas mammalian nerve fibers, with little glycogen, stop in 20 minutes (Brazier, 1960).

The resting metabolism of nerve cells resembles the endogenous metabolism of microorganisms in that it is not susceptible to the metabolic poison azide, even in concentrations which block the activity metabolism in nerve cells. Many of the enzymes participating in aerobic and anaerobic metabolism in other cells, and also enzymes required for phosphorylations, have been identified in nerve cells, but, partly because the material is limited and so much supporting connective tissue surround a nerve, enzymatic studies on nerve cells have not been extensive (Abood and Gerard, 1954).

## ACTIVITY METABOLISM OF NERVE CELLS

On stimulation, the metabolic rate of myelinated nerve cells increases from 1.3 to 4.0 times the rate of resting metabolism. The respiratory quotient of stimulated nerve cells is 1.0, suggesting carbohydrate utilization (Gerard, 1932). Biochemical evidence indicates that glucose is oxidized in neurons of the central nervous system and in nonmyelinated nerve fibers. However, the nutrients used in myelinated nerve cells have not been identified, although decreases in ribonucleic acid content have been noticed after extensive nerve activity (Hyden, 1960) (see Chapter 6). The RNA may, however, act as a mediator in the manufacture of other materials (Katz, 1959).

It has been claimed that RNA is involved in learning and that such learning can be transferred from one individual to another by transfer of RNA (Babich *et al.,* 1965). Synthesis of protein appears to be involved in memory (Agranoff, 1967). That learning has a physical basis seems reasonable (Schmitt, 1965; Deutsch and Leibowitz, 1966), but it is probably too early to attempt a resolution of the complex problem (Young, 1965; McConnell, 1966).

It is thought that the immediate source of energy for conduction of the nerve impulse in vertebrate nerve cells is probably creatine phosphate. Creatine and inorganic phosphate, in addition to ammonia, are among the metabolic products which appear in greater amounts in a perfusate of tetanized nerve fibers. The breakdown of creatine phosphate is more rapid during nerve activity than when the neurons are at rest, and presumably the high energy of the phosphate bond is in some way involved in the transmission of the nerve impulse. In invertebrate nerve cells other phosphates have been shown to be utilized (Brink *et al.,* 1952).

Another metabolic change observed in active nerve cells is the breakdown of acetylcholine:

$$\underset{\text{acetylcholine}}{[(CH_3)_3\overset{+}{N}CH_2CH_2OCOCH_3]OH^-} \xrightarrow[\text{acetylcholine esterase}]{HOH} \underset{\text{choline}}{[(CH_3)_3\overset{+}{N}CH_2CH_2OH]OH^-} + \underset{\text{acetic acid}}{[CH_3COO^-]H^+} \quad (22.1)$$

Acetylcholine is widespread in the nervous system, and from the cut ends of stimulated nerve cells acetylcholine leaks into the perfusate. This ester is synthesized in the presence of the enzyme choline acetylase, which is also widespread in nerve cells. Acetylcholine is spontaneously hydrolyzed into the inactive compounds choline and acetic acid, a reaction greatly speeded by the enzyme acetylcholine esterase, found in all nerve cells, as well as many other kinds of cells. Acetylcholine occurs in highest concentration on the surface of neurons. The action of acetylcholine is inhibited by the drug eserine. Nachmansohn (1950b) has suggested that acetylcholine is the trigger which initiates the nerve impulse.

A clear picture of the rapid sequence of metabolic events in the nerve fiber is not yet available. Although instruments exist for measuring the evolution of heat and the physical changes in electric charge, resistance and capacitance (i.e., function as a condenser) during and after passage of the nervous impulse, no methods of comparable speed have been developed to measure the chemical changes preceding or concomitant with the nervous impulse.

## HEAT PRODUCTION DURING NERVE ACTIVITY

It is difficult to measure directly the heat evolved during the passage of a single nerve impulse. However, when the prepara-

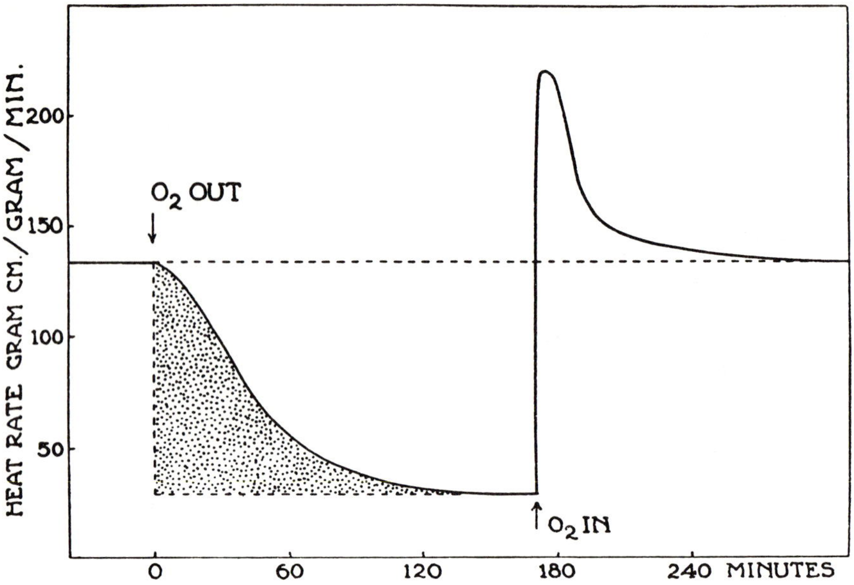


*Figure 22.12.* Resting heat production of frog nerve. At the first arrow the nerve is placed in nitrogen, with resulting slow fall in heat rate. At the second arrow the nerve is replaced in oxygen. (After Beresina, modified from Feng, 1936: Ergbn. Physiol. *38*:73.)

tion is treated with veratrine, which exaggerates after-potentials (Brink *et al.,* 1952), the volley of nerve impulses resulting from tetanic stimulation produces measurable heat. From the measurement with a thermopile of the total amount of heat produced by a volley of nerve impulses, the heat evolved in a single impulse is estimated to be $10^{-7}$ calories per gram of nerve fiber. Only a small fraction (about $\frac{1}{30}$) of the heat is apparently given off during the initial action potential; the remainder of the heat develops during the recovery process. It is thought that the initial spike of the action potential may correspond to the period of breakdown of high energy phosphate bonds. The recovery period may thus be the time during which glucose and other compounds are decomposed and the high energy phosphate bonds are rebuilt (Hill, 1933; Abbott *et al.,* 1958).

The more frequently the nerve cell is stimulated, the smaller is the increment of heat production and oxygen consumption per nerve impulse. This is an adaptive response, and the greater efficiency of the frequently stimulated neuron is thought to be a result of the better physical-chemical state of the excited cell.

In an atmosphere of pure nitrogen the rate of heat production of resting nerve cells declines over a period of 2 hours, but even after this time heat continues to be evolved, though at about one-fifth the rate in air. Metabolism continues anaerobically, accumulating lactic acid. When, on return to air, the nerve cells recover from the oxygen debt, they evolve extra heat before resuming normal heat production (see Fig. 22.12; Gerard, 1946; Schallek, 1949).

## EFFECTS OF VARIOUS FACTORS ON THE ACTION POTENTIAL OF NERVE CELLS

***Fatigue.*** Because the refractory period permits neurons to recover their normal excitability, it is practically impossible to fatigue nerve cells in the presence of an adequate supply of oxygen and nutrients. Experimentally it is possible to obtain 1000 full-sized responses per second from a nerve fiber (A fibers from a frog), and, with stronger shock, up to 2000 have been obtained for a short time (Adrian, 1950). With normal stimulation of receptor cells *in vivo,* volleys up to 160 impulses per second are found. For example, the pad of the foot of a cat sends in 9 to 100 impulses per second in response to tactile stimulation. Such a relatively low frequency of impulses allows for ready recovery. If stimulated to discharge at abnormally rapid rates, a nerve fiber conducts more slowly, its refractory period becomes longer, the initial potential declines and the threshold rises—all signs of a failure to return completely to normal. To this extent a nerve fiber may be fatigued (Adrian, 1950).

***Anoxia.*** Lack of oxygen ultimately blocks transmission of impulses in a nerve cell. For example, the action potential of a nerve fiber so stimulated that the nerve impulse has to travel through a section kept in pure nitrogen shows progressively smaller spikes in that section for about an hour (i.e., for a frog nerve fiber in nitrogen, a 3 per cent loss in spike size occurred in 15 minutes, 13 per cent in 30 and 49 per cent in 40 minutes). At the same time the rate of transmission of the nerve impulse declines (Schmitt and Cori, 1937; Fenn and Gerschman, 1950). Beyond the section in nitrogen, the spikes are of normal size, as expected on the basis of the all-or-none principle. After an hour, transmission of the nerve impulse in the section kept in nitrogen ceases, indicating that a block has been established. The higher the temperature, within the limits of viability of the nerve fiber, the more rapidly the block appears.

A nerve fiber kept anoxic for a time accumulates an oxygen debt. When it is returned to air the nerve fiber takes up oxygen more rapidly for a while until it has paid its debt. Presumably, anaerobic processes are unable to bring the nerve cell into complete recovery (Lundberg and Oscarsson, 1953). If a nerve in a state of anoxia is poisoned by iodoacetate so that glycolysis also ceases, conduction ceases much sooner. Such an experiment indicates that in the absence of oxygen, glycolysis, at least in part, supplies the energy required to synthesize the compounds that maintain the nerve cell in a condition of readiness to respond (Gerard, 1946). This is suggested also by the heat changes during anaerobic metabolism, as seen before.

Respiration in the presence of oxygen is necessary for continued maintenance of the readiness of the nerve cell to respond to stimulation, indicating a continuous expenditure of energy to maintain excitability.

***Effect of Anesthetics on the Nerve Impulse.*** Many anesthetics and narcotics block conduction of the nerve impulse. The time required to block conduction is dependent upon the concentration of anesthetic; the higher the concentration, the shorter will be the time required. The span or area of the nerve fiber to which anesthetic is applied also affects the speed with which the anesthetic acts, conduction being blocked more rapidly the larger the area treated. Some anesthetics (e.g., cocaine) produce a definite effect which is reversible; others, like chloral hydrate, produce an effect that becomes progressively more injurious and less reversible the longer the time of action. It is not possible to review here the complexities of the pharmacology of nerve cells (Brink, 1951; Salmoiraghi and Bloom, 1964).

Application of an anesthetic raises the threshold for stimulation of a nerve fiber, and larger stimuli are then required to evoke a spike. There are other changes in properties of the nerve fibers as anesthesia progresses; e.g., the sensitivity to a chemical stimulus such as citrate is reduced, the velocity of conduction of an action potential decreases and the size of the spike evoked by a stimulating shock successively declines, while the refractory period increases in length. Anesthesia, therefore, induces a progressive decrease in the excitability of neurons (Tasaki, 1953; Stevens, 1966).

## ACTION POTENTIALS IN MUSCLE FIBERS

Action potentials, as indicated previously, are not confined to nerve cells alone. Similar changes in potential occur

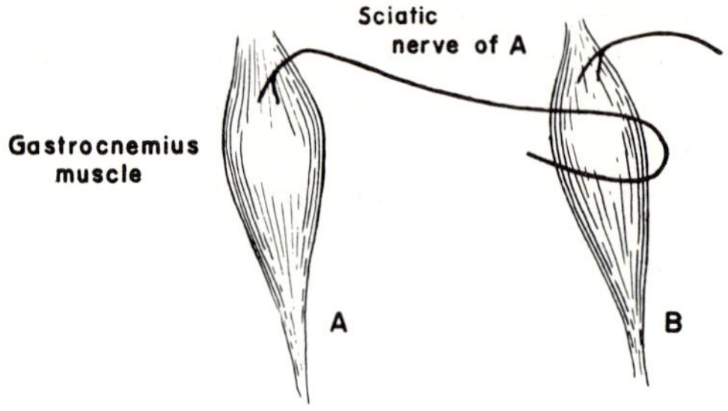


***Figure 22.13.*** Excitation of a nerve by a muscle. When the sciatic nerve that makes its ending on muscle B is pinched, muscle B is excited. Action currents on the surface of muscle B then excite the sciatic nerve leading to muscle A.

also in various muscle fibers. Thus the excited portion of the muscle fiber also becomes negative to the resting portion, and an action potential as high as 100 millivolts may be recorded on a muscle fiber. Many years ago Galvani demonstrated the electrical nature and similarity of muscle and nerve excitation by connecting two sciatic nerve–gastrocnemius muscle preparations in series (see Fig. 22.13). When the first muscle was excited to contraction, for example, by pinching the nerve connected to it, it excited the sciatic nerve of the second muscle, which contracted soon after the contraction of the first.

Action potentials are demonstrable in muscle cells of the gut, the heart, the ovi-

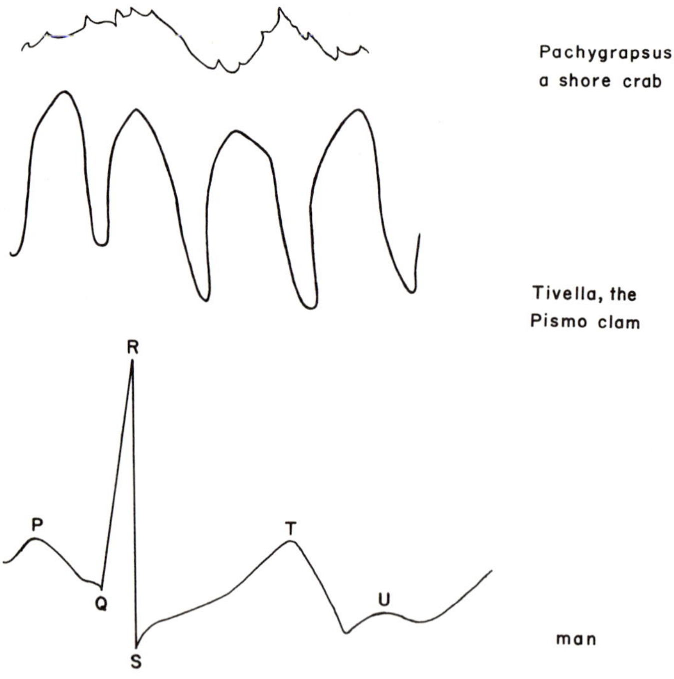


***Figure 22.14.*** Electrocardiograms of hearts of several animals. Note that the electrocardiogram for the crab, *Pachygrapsus,* shows a number of sharp peaks indicating discharge of ganglion cells in this *neurogenic* heart (that is, a heart in which the beat originates in discharges of nerve cells), whereas the electrocardiograms of the clam, *Tivella,* and man are smooth, as expected in a *myogenic* heart (that is, one in which the beat is initiated in the heart muscle itself). Both kinds of hearts may be regulated by nerves and hormones, however.

In the human heart the P wave precedes the contraction of the auricle and is probably the wave of excitation of this structure. The R wave precedes the contraction of the ventricle, T, representing the end of the ventricular systole. The events correlating with Q, S and U are not determined.

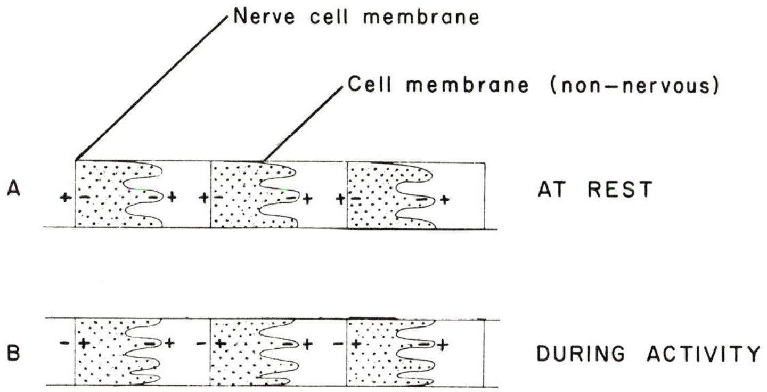


***Figure 22.15.*** Diagrammatic representation of the changes in the electroplaxes (single units) of the electric organ of a fish. At rest (*A*) both the nerve cell membrane and the ordinary cell membrane have a positive charge, the inside of the cell being negative. When excited the nerve cell membrane is not only depolarized, but becomes negatively charged (*B*). Under such conditions, because of the structure of the electric organ, the voltages of the series of electroplaxes add up, like the voltages in a series of little batteries. (After Gerebtzoff and Schoffeniels, 1960: Comp. Biochem. *2*:519.)

duct, the uterus and other organs. The tracing of the action potential of the heart is called an *electrocardiogram* (Fig. 22.14). The remarkable feature of the electrocardiogram of a large mammal is the fact that the contracting heart has a large enough action potential to be picked up on the surface of the body in spite of the activities of an enormous amount of intervening tissue (Brazier, 1960).

The most unique action potentials are those which appear in the electric fish. In electric fishes a considerable number of muscle fibers have become modified to form the electric organs, which are composed of a series of electric units. Each such electric unit, known as an *electroplax,* begins during early embryonic development as a normal skeletal muscle fiber, but later develops into what looks like a flat plate attached to a myoneural junction (see Fig. 22.15). Like a myoneural junction, the electroplax fatigues rapidly and is susceptible to the drug curare, which readily blocks the passage of the nerve impulse to the electric organ. A single electroplax gives off a peak voltage of 50 to 150 millivolts. The large voltage measured in electric fishes is made possible by the arrangement of the electroplaxes in an organ as a series of batteries, summing the voltages of the individual electroplaxes. By parallel connections between series of electroplaxes, large currents can be obtained. The electric eel *Electrophorus electricus,* for example, has been found to produce electric shocks of several hundred volts (Cox, 1943).

It is interesting to point out that Alessandro Volta (1745–1827) designed the first voltaic cell, using the electric organ of the electric fish as a model. Just as many electroplaxes in series give off a high voltage, so did his "pile" of copper and zinc discs in which the voltage was summated as in a series of batteries. The example illustrates a case in which physics received a great impetus from investigation of a biological phenomenon (Keynes *et al.,* 1953, 1956).

## ACTION POTENTIALS IN EXCITABLE PLANT CELLS

Excitable plant cells also develop action potentials when stimulated. For example, in the sensitive plant *Mimosa* or in the carnivorous plant Venus's flytrap, the action potentials that develop immediately after stimulation of the sensitive cells (see Fig. 22.16) and pass to the pulvini may be recorded by a galvanometer (Greulach, 1955; Scott, 1962). The action potentials are followed by the turgor movement.

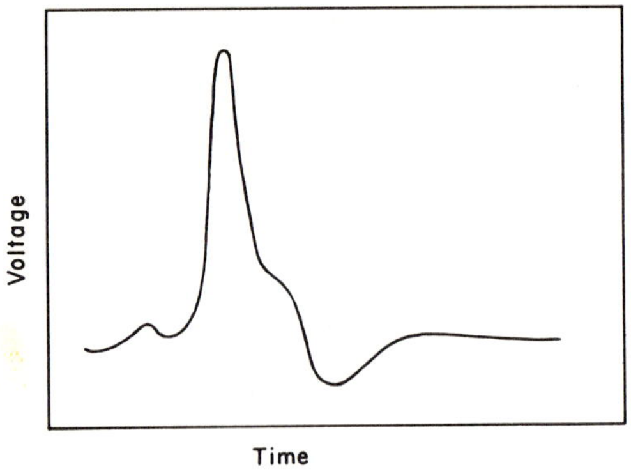


***Figure 22.16.*** An oscilloscopic trace of an electrical action potential released by the bending of a trigger hair on Venus flytrap. A maximum potential of 130 millivolts was recorded. (After Plant Movements, by Victor A. Gruelach. Copyright © 1955 by Scientific American, Inc. All rights reserved.)

Even more remarkable are the action potentials in *Nitella,* a fresh water alga with very large multinucleate cells (Osterhout, 1936; Blinks, 1949, 1955; Umrath, 1954). When a *Nitella* cell is stimulated by pressure, heat, light, chemicals or electricity, an action potential is evoked which reaches a value of about a tenth of a volt and is propagated at the rate of several centimeters per second in air, and about ten times this in water with electrolyte (Sibaoka, 1966). Like the nerve action potential, it is also diphasic. The action potential usually passes to the ends of the cell; however, it may die out after having traversed only a part of it, a behavior contrary to the all-or-none principle characteristic of the nerve impulse. Like nerve cells, however, *Nitella* cells show temporal summation of stimuli. After excitation a *Nitella* cell becomes refractory or incapable of being stimulated for a brief time. The refractory period may last 30 seconds to several minutes, or after rough handling, as during isolation of an internode cell for an experiment, *Nitella* may be refractory for hours (Blinks, 1955).

The action potential of a *Nitella* cell is not just an electrical wave, since it travels more slowly than electricity, but it is apparently a change in the cell membrane accompanied by an electrical disturbance. That the electrical wave is important for propagation of the membrane change is demonstrated experimentally. Thus when a portion of a *Nitella* cell is blocked by cold or anesthetic, the action potential does not pass the block, yet if a **U**-shaped piece of filter paper, or string, moistened with balanced salt solution bridges the block, the action potential is conducted through the bridge (see Fig. 22.17).

Passage of an action potential in *Nitella* kept in the dark is accompanied by increased oxygen consumption and carbon dioxide production. Heat is also given off. These metabolic accompaniments of the action potential are thought to be the result of reactions involved in regenerating the cellular membrane altered by excitation.

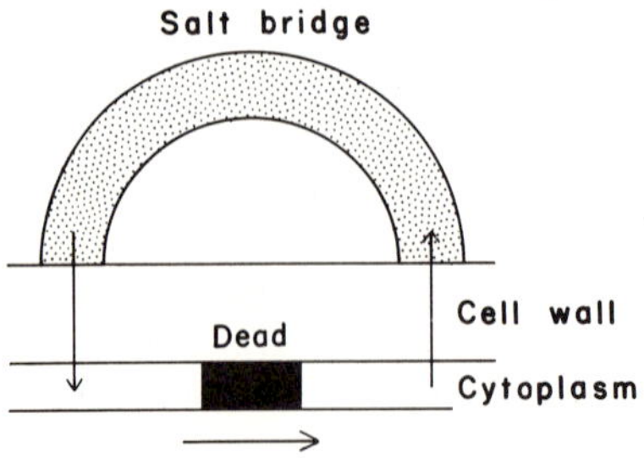


***Figure 22.17.*** Diagram of a by-pass for a block in a *Nitella* cell. Part of the cytoplasm has been killed with chloroform. The salt bridge, consisting of filter paper or string soaked in salt solution, completes the circuit and makes transmission possible (After Osterhout and Hill, 1929–30: J. Gen. Physiol. *13*:548.)

## MONOPHASIC ACTION POTENTIALS

When a nerve fiber, a *Nitella* cell or other excitable cell is crushed between one of the pickup electrodes, the action potential becomes *monophasic* (see Figs. 22.18 and 22.19), since only when the action potential passes one of the electrodes will a change in potential be recorded, the crushed part being incapable of propagating the potential. This arrangement is convenient for some types of study and is most generally used.

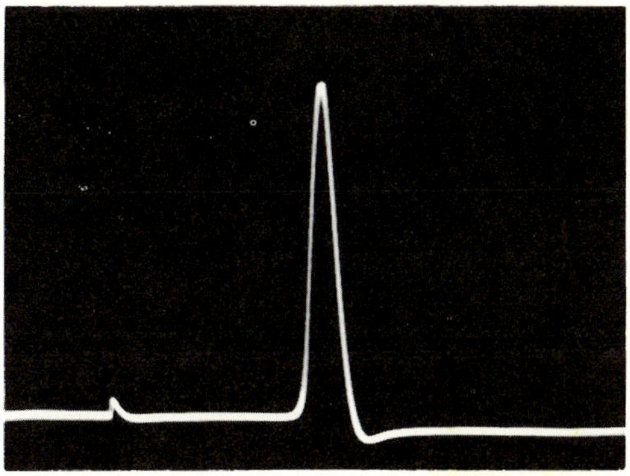

***Figure 22.19.*** Single fiber (central interneuron) isolated from the crayfish ventral nerve cord and recorded in oil. Monophasic recording leads. Time mark (below spike), 1 msec. Preceding the spike is seen the stimulus artifact. (By courtesy of Dr. D. Kennedy.)

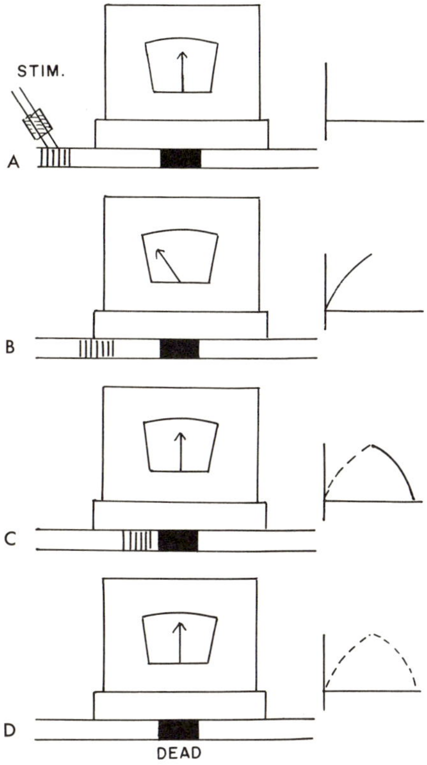


***Figure 22.18.*** Monophasic action potentials in an excitable cell with a dead section (shown in black) between the electrodes. *A,* Stimulation (stim.) excites the cell. *B,* When the excitation (shown by cross hatching) reaches the electrode, the needle of the galvanometer is deflected because the excited area is negative to the unexcited one. *C,* When the excitation passes the electrode the needle falls to zero, since both electrodes are then on unexcited areas at the same potential. *D,* Because the excitation cannot pass the dead region, the needle is not deflected again and the spike is said to be monophasic.

By placing the pickup electrodes far apart on an uninjured nerve fiber it is possible also to record what are essentially monophasic action potentials. However, the electrodes must be so far apart as to make this procedure impractical when recording from the surface of the nerve fiber.

## *LITERATURE CITED AND GENERAL REFERENCES*

Abbott, B. C., Hill. A. V., and Howarth, J. F., 1958: The positive and negative heat production associated with a nerve impulse. Proc. Roy. Soc. London *B148*:149–183.

Abood, L. G., and Gerard, R. W., 1954: Enzyme distribution in isolated particulates of rat peripheral nerve. J. Cell. Comp. Physiol. *43*:379–392.

Adrian, E. D., 1950: The control of nerve-cell activity. Symp. Soc. Exp. Biol. *4*:85–91.

Agranoff, B. W., June 1967: Memory and protein synthesis. Sci. Am. *216*:115–122.

Arshavski, Y. I., 1960: The role of metabolism in the production of bioelectric potentials. Russian Rev. Biol. *50*:55–68.

Audus, L., 1959: Plant Growth Substances. Hill & Son, Newcastle-upon-Tyne.

Babich, F. R., Jacobson, A. J., Bubash, S., and Jacobson, A., 1965: Transfer of a response to

naive rats by injection of RNA extracted from trained rats. Science *149*:656–657.
Blinks, L. R., 1949: The source of the bioelectric potentials in large plant cells. Proc. Natl. Acad. Sci. U.S.A. *35*:566–575.
Blinks, L. R., 1955: Some electrical properties of large plant cells. *In* Electrochemistry in Biology and Medicine. Shedlovsky, ed., John Wiley & Sons, New York, pp. 187–212.
Bodian, D., 1952: Introductory survey of neurons. Cold Spring Harbor Symp. Quant. Biol. *17*:1–13.
*Bodian, D., 1962: The generalized vertebrate neuron. Science *137*:323–326.
Bone, Q., 1958: Nervous control of cilia in amphioxus (*Branchiostoma*). Nature *181*:193–194.
Bonner, J., and Galston, A., 1952: Plant Physiology. W. H. Freeman & Co., San Francisco.
Brazier, M. A. B., 1960: The Electrical Activity of the Nervous System. 2nd Ed. Macmillan Co., New York.
Brazier, M. A. B., 1964: The electrical activity of the nervous system. Science *146*:1423–1428.
Brink, F., Jr., 1951: Anesthetizing action. *In* Second Conference on Nerve Impulse Transactions. Nachmansohn, ed. Josiah Macy, Jr., Foundation, New York, pp. 124–175.
Brink, F., Jr., Bronk, D. W., Carlson, F. D., and Connelly, C., 1952: The oxygen uptake of active axons. Cold Spring Harbor Symp. Quant. Biol. *17*:53–67.
Clark, J., and Plonsey, R., 1966: A mathematical evaluation of the core conductor model. Biophys. J. *6*:95–112.
Connelly, C. M., 1959: Recovery processes and metabolism of nerve. In Biophysical Science—A Study Program. Oncley, ed. John Wiley & Sons, New York, pp. 504–514.
Cox, R. T., 1943: Electric fish. Am. J. Phys. *11*:13–22.
Datta, M., and Dutt, A. D., 1957: Vacuoles and movement in the pulvinus of *Mimosa pudica.* Nature *179*:253–254.
Davson, H., 1960: A Textbook of General Physiology. 4th. Ed. Little, Brown & Co., Boston, Section 4.
Deutsch, J. A., and Leibowitz, S. F., 1966: Amnesia or reversal of forgetting by anticholinesterase depending simply on time of injection. Science *153*:1017–1018.
DiPalma, J., McMichael, R., and DiPalma, M., 1967: Touch receptor of Venus fly trap, *Dionaea muscipula.* Science *152*:539–540.
Eckert, R., 1965: Bioelectric control of bioluminescence in the dinoflagellate *Noctiluca.* Science *147*:1140–1145.
Eckert, R., 1966: Excitation and luminescence in *Noctiluca miliaris. In* Bioluminescence in Progress. Johnson and Haneda, eds. Princeton University Press, Princeton, N. J., pp. 269–300.
Erlanger, J., and Gasser, H. S., 1937: Electrical Signs of Nervous Activity. University of Pennsylvania Press, Philadelphia.
Ernyei, S., and Young, M. R., 1966: Pulsatile and myelin-forming activities of Schwann cells in vitro. J. Physiol. *183*:469–480.
Fenn, W. O., 1929: The metabolism of nerves. Harvey Lect. *23*:115–150.
Fenn, W. O., and Gerschman, R., 1950: The loss of potassium from frog nerve in anoxia and other conditions. J. Gen. Physiol. *33*:105–204.
Fernandez-Moran, H., and Brown, R. (eds.), 1958: Submicroscopic organization and function of nerve cells. Exp. Cell Res. Suppl. 5.
Frank, K., and Fuortes, M. G. F., 1961: Excitation and conduction. Ann. Rev. Physiol. *23*:357–387.
Gerard, R. W., 1932: Nerve metabolism. Physiol. Rev. *12*:469–592.
Gerard, R. W., 1946: Nerve metabolism and function. A critique of the role of acetylcholine. Ann. N. Y. Acad. Sci. *47*:575–600.
Gerebtzoff, M. A., and Schoffeniels, E., 1960: Nerve conduction and electrical discharge. Comp. Biochem. *2*:519–544.
Gernand, von K., and Ehrig, H., 1957: Untersuchungen über die elektrische Reizbarkeit von *Mimosa pudica.* Biol. Zentr. *76*:181–185.
Greulach, V. A., Feb. 1955: Plant movements. Sci. Am. *192*:100–106.
Hakansson, C. H., 1957: Action potential and mechanical response of isolated cross striated frog muscle fibres at different degrees of stretch. Acta Physiol. Scand. *41*:199–216.
Hamburger, V., 1957: The life cycle of a nerve cell. Am. Sci. *45*:263–277.
Hill, A. V., 1933: The three phases of nerve heat production. Proc. Roy. Soc. London *B113*:345–365.
Hisada, M., 1957: Membrane resting and action potentials from protozoan, *Noctiluca scintillans.* J. Cell. Comp. Physiol. *50*:57-71.
Hodgkin, A. L., 1964: The conduction of the nervous impulse. Liverpool University Press, Liverpool.
Hyden, H., 1960: The neuron. *In:* The Cell. Brachet and Mirsky, eds. Academic Press, New York, Vol. 4, pp. 215–323.
Johnson, F. H., Eyring, H., and Polissar, M., 1954: The Kinetic Basis of Molecular Biology. John Wiley & Sons, New York.
Katz, B., Nov. 1952: The nerve impulse. Sci. Am. *187*: 55–64.
Katz, B., 1959: Nature of the nerve impulse. *In* Biophysical Sciences—A Study Program. Oncley, ed. John Wiley & Sons, New York, pp. 466–474.
*Katz, B., 1966: Nerves, Muscle and Synapse. McGraw-Hill Book Co., New York.
Keynes, R. D., 1956: The generation of electricity by fishes. Endeavour *15*:215–222.
Keynes, R. D., 1957: Electric organs. *In*: Physiology of Fishes. Brown, ed. Academic Press, New York, Vol. 2, pp. 323–340.
Keynes, R. D., and Martins-Ferreira, H., 1953: Membrane potentials in the electroplexes of the electric eel. J. Physiol. *119*:315–351.
Kinosita, H., Dryl, S., and Naitoh, Y., 1964: Changes in membrane potential and the responses to stimuli in *Paramecium.* J. Fac. Sci. Univ. Tokyo *10*:291–301.
Kishimoto, U., 1957: Studies on the electrical properties of a single plant cell. Internode cell of *Nitella.* J. Gen. Physiol. *40*:663–682.
Korr, I. M., Wilkinson, P. N., and Chornock, F. W., 1967: Axonal delivery of neuroplasmic components to muscle cells. Science *155*:342–345.
Kuffler, S. W., Nicholls, J. G., and Oakland, R. K.,

1966: Physiological properties of glial cells in the central nervous system of amphibia. J. Neurophysiol. *19*:768–787.
Lapique, L., 1926: L'excitabilite en function du temps. Université de France, Paris.
Lillie, R. S., 1922: Transmission of physiological influence in protoplasmic systems, especially nerve. Physiol. Rev. *2*:1–37.
Lloyd, F. E., 1942: The Carnivorous Plants. Chronica Botanica, Waltham, Mass.
Lundberg, A., and Oscarsson, O., 1953: Anoxic depolarization of mammalian nerve fibers. Acta Physiol. Scand. *30*(Suppl. *111*):99–110.
Luttges, M., Johnson, T., Buck, C., Holland, J., and McGaugh, J., 1966: An examination of "transfer of learning" by nucleic acid. Science *151*:834–837.
McConnell, J. V., 1966: Comparative physiology: learning in invertebrates. Ann. Rev. Physiol. *28*: 107–136.
Nachmansohn, D., 1950a: The Nerve Impulse. Transactions of the First Macy Foundation Conference, New York.
Nachmansohn, D., 1950b: Studies on permeability in relation to nerve function. I. Axonal conduction and synaptic transmission. Biochim. Biophys. Acta *4*:78–95.
Osterhout, W. J. V., 1936: Electrical phenomena in large plant cells. Physiol. Rev. *16*:216–237.
Roth, L. E., 1958: Ciliary coordination in the protozoa. Exp. Cell Res. *5*(Suppl.):573–585.
Ruch, T. C., and Patton, H. D. (eds.), 1965: Physiology and Biophysics. 19th Ed. W. B. Saunders Co., Philadelphia.
Salmoiraghi, G. S., and Bloom, F. E., 1964: Pharmacology of individual neurons. Science *144*: 493–499.
Schallek, W., 1949: The glycogen content of some invertebrate nerves. Biol. Bull. *97*:252–253.
Schmitt, F. O., 1965: The physical basis of life and learning. Science *149*:931–943.
Schmitt, F. O., and Cori, C. F., 1937: Lactic acid formation in medullated nerve. Am. J. Physiol. *106*:339–349.
Scott, B. I. H., Oct. 1962: Electricity in plants. Sci. Am. *207*:107–117.
Sibaoka, T., 1966: Action potentials in plant organs. Symp. Soc. Exp. Biol. *20*:49–73.
Sleigh, M. A., 1966: The coordination and control of cilia. Symp. Soc. Exp. Biol. *20*:11–31.
Stevens, C. F., 1966: Neurophysiology: A Primer. John Wiley & Sons, New York.
Suckling, E. E., 1961: Bioelectricity. McGraw-Hill Book Co., New York.
Tasaki, I., 1953: Nervous Transmission. Charles C Thomas, Springfield, Ill.
Tasaki, I., and Spyropoulos, C. S., 1957: Influence of changes in temperature and pressure on the nerve fiber. *In* Influence of Temperature on Biological Systems. Johnson, ed. American Physiological Society, Washington, pp. 201–220.
Taylor, C. V., 1920: Demonstration of the function of the neuromotor apparatus in *Euplotes* by the method of microdissection. Univ. Calif. Publ. Zool. *19*:403–471.
Tyler, A., Monroy, A., Kao, C. Y., and Grundfest, H., 1955: Electrical potential changes upon fertilization of the starfish egg. Biol. Bull. *109*: 352–353. (Abstract)
Umrath, K., 1954: Ueber die elektrische Polarisierbarkeit von *Nitella mucronata* und *Nitella opaca.* Protoplasma *43*:237–252.
Watson, E. C., 1945: The first electric battery. Am. J. Phys. *13*:397–406.
Young, J. Z., 1965: Two memory stores in one brain. Endeavour *24*:13–26.

# CHAPTER 23 / DEVELOPMENT, PROPAGATION AND TRANSMISSION OF THE ACTION POTENTIAL

Since the action potential is apparently the "language" by which excitable cells communicate information and in this manner evoke response, its nature, its origin and the mechanism of its propagation and transmission are of profound interest. In this connection several additional questions arise for consideration: Does a potential difference already exist between the outside and the inside of a cell not undergoing stimulation? If so, what causes this resting potential, and how is it altered during the development and propagation of an action potential following stimulation of the cell?

## RESTING POTENTIALS ACROSS THE CELL MEMBRANE

It has been clearly demonstrated experimentally that a potential difference, called the *resting potential,* exists at all times between the outside and inside of all normal cells at rest (e.g., not stimulated). The experiment is performed by inserting one electrode inside a cell and placing another on its surface (see Fig. 23.1). Cells such as the squid giant axon, large muscle fibers and large plant cells like *Valonia, Halicystis* or *Nitella* are best for this purpose because of their size. However, for small cells minute transverse electrodes are being employed with success. A resting potential of approximately 10 to 100 millivolts (mv.) is registered in most cells. In the squid giant axon and frog muscle fiber the potential is about 50 to 75 mv.; in *Halicystis,* 70 to 80 mv.; in a *Nitella* cell, 100 to 200 mv. In all cases the outside is positive to the inside (Blinks, 1955; Grundfest, 1963), except in one species of *Valonia,* on which the outside is 10 to 20 mv. negative to the inside. However, this is not a contradiction and has been satisfactorily explained (Blinks, 1940).

An approximation of the potential difference between the outside and the inside of a cell may be obtained by injuring or crushing one part of the cell to "gain entry" to the inside. One electrode is

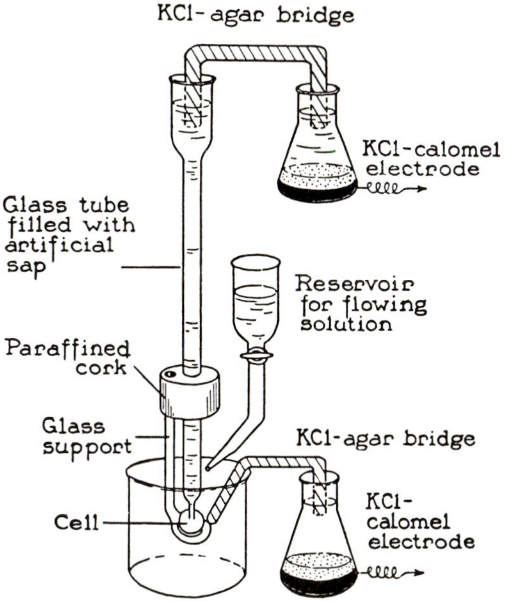


***Figure 23.1.*** Method of determining the resting potential on a plant cell. One electrode is placed inside a *Halicystis* cell and another electrode is placed on its surface. (From Blinks, 1933: J. Gen. Physiol. *17*:111.)

placed on the injured portion, and the other on an intact portion and an *injury potential* is measured. Its magnitude is 50 mv. or less for most cells tested, the outside always being positive to the injured region (inside).

It is probable that the true potential difference between the outside and the inside of the cells is larger than measured by all these methods, because the cells are injured during penetration of the electrode into the cell and because chances for electrical shorting are considerable.

In attempting to explain the resting potential of a cell it is of importance to consider in some detail an inanimate system as an analogy. It is shown that a potential difference arises when a membrane is selectively permeable to either anions or cations, and a difference in concentration of some nondiffusible ion exists between the inside and the outside of the membrane. Such a difference in potential is called a *concentration chain potential.* A potential difference also arises when cations (or anions) of different diffusibilities are present on opposite sides of a membrane, even if they are at the same concentration; it is called a *chemical chain potential* (see Appendix 23.1).

***Concentration Chain Potentials across an Artificial Membrane.*** When an artificial membrane separates solutions containing different concentrations of the same electrolyte, the nature of the potential depends upon the permeability of the membrane to the cations and anions of this electrolyte. Four different cases are described in the following paragraphs and in Figure 23.2.

1. If the membrane is equally permeable to both the cation and anion, e.g., $K^+$ and $Cl^-$, the potential is transitory, since the ions ultimately distribute themselves equally between the two solutions. The mobility of the various ions determines how long the potential will last. $K^+$ and $Cl^-$ have almost equal mobilities; therefore the potential would be slight and would fall much sooner than if NaCl were used on each side of the membrane, since $Na^+$ is less mobile than $Cl^-$. Therefore the side of the membrane with a higher concentration of NaCl would remain positive to the other for a brief but finite time. The greater the difference in mobilities, the longer would the temporary potential last.

2. If the membrane is permeable to cations but not to anions, e.g., to $K^+$ but not to $Cl^-$, then a potential develops be-

tween the two sides, and the solution of higher concentration becomes negative to the one of lower concentration. The magnitude of the potential can be calculated from the Nernst equation:

$$E = \frac{RT}{n\mathcal{F}} \ln \frac{C_1}{C_2} \qquad (23.1)$$

In this equation E is the potential in volts, R the gas constant (taken as 8.312 joules per degree per mole), T the absolute temperature in degrees, n the valency change, $\mathcal{F}$ the Faraday (96,500 coulombs per gram equivalent), *ln* the natural logarithm ($2.3 \times \log_{10}$), $C_1$ the higher concentration and $C_2$ the lower concentration (molal) of the electrolyte.

The equation can be simplified by combining the constants R, n and $\mathcal{F}$ and converting *ln* to $\log_{10}$. Then setting up the equation for room temperature, 27° C. (300° absolute), and converting to millivolts:

$$E = 59.5 \log_{10} \frac{C_1}{C_2} \qquad (23.2)$$

For a tenfold difference in concentration, $\log_{10} C_1/C_2$ equals 1; therefore, E is 59.5 mv.

An example of such a membrane is one made of celloidin (cellulose nitrate), which bears a negative charge. A potential is established across such a membrane separating two solutions of KCl because $K^+$ penetrates to some extent, whereas $Cl^-$ does not (Michaelis, 1929; Teorell, 1962). The tendency of $K^+$ to penetrate is expressed by the magnitude of the potential developed.

In general, whenever an indiffusible ion is present on one side of the membrane only, and diffusible ions are present on both sides, the situation is much like the Donnan equilibrium previously formulated for proteins (see Chapter 13).

3. If the membrane is permeable to anions but not to cations, e.g., to $Cl^-$ but not to $K^+$, a potential develops, but the solution of higher concentration in this case becomes positive to the one of lower concentration because of the movement of the anions. The tendency for the anions to diffuse is expressed in the potential developed. An example is celloidin membrane made up with the dye rhodamine B, or coated with protamine, which is positively charged and permeable to anions only.

4. If a mosaic membrane which is permeable to anions in certain portions and to cations in other portions separates solutions of different concentrations, both anions and cations will enter. For example, a membrane composed partly of rhodamine B-celloidin and partly of celloidin will permit anions (e.g., $Cl^-$) to penetrate the rhodamine B-containing portion and cations (e.g., $K^+$) to penetrate the celloidin portion. The magnitude and sign of the potential developed depend upon the ratio of the cation-permeable portion of the membrane to the anion-permeable portion of the membrane and upon the rate of penetration of the two ions and also upon their relative mobility. The greater the mobility of the ions and the greater their penetration of the membrane, and the more equal the areas permeable to them, the more transitory will be the charge. The solution of high concentration will be positive to the one of low concentration if the total membrane area permeable to anions is large, and negative if the area permeable to cations is large. The potential will be less than the calculated concentration chain potential between the two solutions, because both ions pass from the higher to the lower concentration, reducing the concentration difference. Furthermore, the potential will fall to zero when the two solutions become equalized in concentration.

***Origin of Resting Potential across the Cell Membrane.*** The resting potential is presumably maintained as long as the cell is alive and active. This was found when the potential was tested periodically for over a week on an impaled living *Halicystis* (Blinks, 1940, 1949, 1955). This indicates that the source of the potential is also persistent. It is postulated that such a potential could result from a sustained difference in either cation or anion concentration between inside and outside of cells, as discussed earlier (cases 2 and 3), but not when both ions penetrate the membrane, because then the potential would decline as equilibrium is reached. If the potential across the plasma membrane is due to such a concentration chain, some persistent concentration difference in either anions or cations should be demonstrable. Furthermore, if the concentration of the ion responsible for the resting potential is equalized on the two sides of the plasma membrane, either by leaching the cell free of it or by matching the concentration of the ion inside the cell with a solution applied to the outside of the cell, the resting potential should fall to zero.

The concentration of various ions inside cells and in their bathing medium is given in Table 23.1. It will be noted that $K^+$ is present in a much higher concentration inside the cells than outside (e.g., a *Nitella* cell contains $K^+$ in the central sap vacuole in a concentration 1065 times as great as in the pond water in which it grows). Although it has been claimed that the potassium in cells may be bound and not exchangeable with that outside, the high conductivity of the cells (in which potassium is the main cation) and the ready diffusion of labeled potassium from one end of the cell to the other indicate that, in cells tested, most of the potassium is in ionic form (Hodgkin, 1951; Katz, 1959a; Strumwasser, 1965).

In *Nitella* $Na^+$ and $Cl^-$ are also present in much higher concentration in the central sap vacuole than in pond water, but in the frog nerve fiber, the squid giant axon and in most cells both of these ions are present in much lower concentration than in the bathing medium.

Whenever indiffusible anions (or indiffusible cations) are present inside the cell and diffusible anions (or cations) are outside, a steady state develops which resembles the Donnan equilibrium for proteins. The Donnan formulation (see Chapter 13) has been applied to predict the movements of $Na^+$, $K^+$ and $Cl^-$ into and out of living cells. Since several ions are present in different concentrations on the two sides of the cell membrane, in theory, all of the diffusible ions present in the cell and in the bathing fluid to which the cell is exposed should be considered in any equation purporting to relate the concentration differences of ions outside and inside the cell to its membrane potential. However, equation 23.3 (derived by Hodgkin and Katz (1949) from the Goldman (1943) constant field equation) is a version of the Nernst equation (23.1) modified to take into consideration the development of the resting potential of the excitable cell in terms of only the three ions present in highest concentration in the cell and the bathing medium (Hodgkin, 1951; Chandler and Hodgkin, 1965).

This formulation assumes that the

**Table 23.1.** Concentration of Some Ions in Cells and Bathing Media*

| CELL | $K^+$ | $Na^+$ | $Cl^-$ | ORGANIC ANIONS |
|---|---|---|---|---|
| Nitella[1] | 54.3 | 10.0 | 90.7 | — |
| Pond water[1] | 0.05 | 0.2 | 0.9 | — |
| Squid giant axon[2] | 400 | 50 | 40 | 381.4 |
| Squid blood[2] | 20 | 440 | 560 | — |
| Sea water[2] | 10 | 460 | 540 | traces |
| Crab nerve[3] fiber | 112 | 54 | 52 | — |
| Crab blood[3] | 12.1 | 469 | 524 | — |
| Frog muscle fiber[4] | 124 | 10.4 | 1.5 | 70 |
| Frog plasma[4] | 2.25 | 109 | 77.5 | 12.9† |

* Data are in mM. per liter (1) and mM. per kg. water (2, 3, 4).
† Accounting incomplete.
[1] Hoagland and Davis, 1922–23: J. Gen. Physiol. 5:629.
[2] Hodgkin, 1958: Proc. Roy. Soc. *B148*:1. Slicks of sea water may contain considerable organic matter.
[3] Shaw, J. Exp. Biol. 32.
[4] Conway, 1957: Metabolic Aspects of Transport Across Membranes. Murphy, ed.

structure of the membrane is stable and that changes in the ionic environment in no way affect the permeability characteristics of the membrane. It is therefore an approximation, since it is known that certain ions do affect the plasma membrane. Increase in outside $K^+$ concentration results in a small decrease in the resistance of the cell membrane; increase in $Ca^+$ has the reverse effect (Gerebtzoff and Schoffeniels, 1960).

Equation 23.3 may be simplified by dropping out the term for $Na^+$ since it is shown that in the resting state most cells have permeability for $Na^+$ equal only to about 1 per cent of the permeability for $K^+$. Furthermore, when the $Na^+$ concentration of the bathing medium is varied over a wide range, the resting potential of the cell is not affected, indicating that the $Na^+$ gradient across the cell membrane plays little if any role in establishment and maintenance of the resting potential (Katz, 1959). If the permeability of the cell membrane to $K^+$ and $Cl^-$ is assumed to be about equal in both directions,* the permeability constants in the equation cancel out, leaving only the terms for concentrations of $K^+$ and $Cl^-$ inside and outside the cell, the ones for $K^+$ being of greater importance in most cells (Baker *et al.*, 1961; Tasaki and Singer, 1965; Baker, 1966).

Increasing the concentrations of $K^+$ in the medium of *Nitella* causes a progressive decline in its resting potential which is abolished when the $K^+$ concentration outside the cell equals that inside. It is even possible to reverse the sign of the resting potential if the concentration of $K^+$ becomes greater on the outside than inside the cell (Blinks, 1955). The effects are completely reversible, indicating that the cells remain healthy during the tests (see Fig. 23.3).

Since $K^+$ is present in most cells in a concentration higher than in the medium bathing them, it is tempting to postulate that it acts in other cells as in *Nitella.* In fact, experiments have shown that increasing concentrations of $K^+$ applied to the outside of the squid giant axon or frog sartorius muscle fiber or frog myelinated nerve fiber (Hodgkin *et al.*, 1949) reduce the resting potentials in these cells as in *Nitella,* but the experimental results, especially for muscle fibers, are in best agreement with theory if the $Cl^-$ concentrations are also taken into account, indicating that the $Cl^-$ contributes to the net potential, even though its share in most cells is small.

In an experiment a solution containing $K^+$ and anions was used to replace the axoplasm of a squid giant axon and the same solution was used as the outsidc medium. When the gradient of $K^+$ between the inside and the outside of the axon approximated that between axoplasm and blood, the resting potential was measured. It was found to be only slightly less than that of an isolated but unaltered axon. Also a reduction of the $K^+$ gradient, however achieved, caused a proportional decline in this "resting potential" as predicted by the Nernst equation, and the potential could even be reversed by making the $K^+$ concentration outside greater than that inside the axon. Changing the kind of anion (but not the concentration), on the other hand, had little effect (Baker *et al.*, 1961; Baker and Connelly, 1966).

In most cells $K^+$ and $Cl^-$ appear to be very nearly in electrochemical equilibrium, the large concentration difference between the inside and the outside of the cell for each of these ions being very nearly

*Under certain conditions it is not.

$$E = \frac{RT}{\mathcal{F}} \ln \left( \frac{P_{K^+}(K^+)\text{ in} + P_{Na^+}(Na^+)\text{ in} + P_{Cl^-}(Cl^-)\text{ out}}{P_{K^+}(K^+)\text{ out} + P_{Na^+}(Na^+)\text{ out} + P_{Cl^-}(Cl^-)\text{ in}} \right) \qquad (23.3)$$

Where E is the potential across the cell membrane, $P_{K^+}$, $P_{Na^+}$ and $P_{Cl^-}$ refer to the permeability coefficients of $K^+$, $Na^+$ and $Cl^-$ respectively, and "in" and "out" refer to the inside and outside of the cell, respectively.

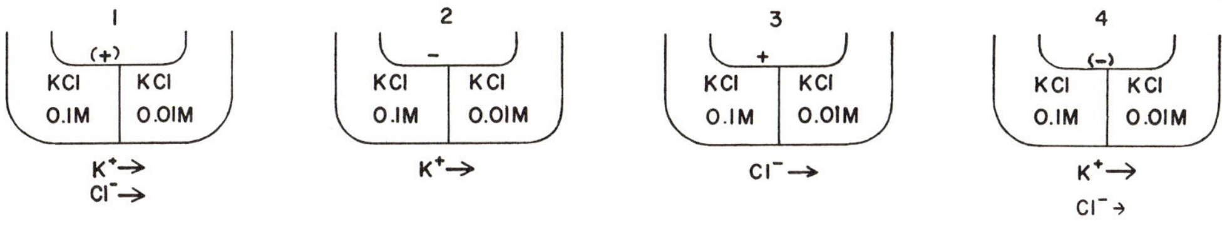


***Figure 23.2.*** Concentration chain potentials. Potentials developed when solutions of the same electrolyte are present in different concentrations on opposite sides of a membrane. The arrows show only the overall movement, since the movement of a given ion is not restricted entirely to passage through the membrane in one direction. The charge on the membrane may be temporary (as indicated by parentheses) or continued, depending upon the conditions. When the membrane is permeable to both ions, as in case 1, the potential is fleeting. When it is permeable only to cations, the solution of higher concentration becomes negative (case 2). When it is permeable only to anions, the solution of higher concentration becomes positive to the other solution (case 3). When the membrane is a mosaic of areas, some permeable to cations and some permeable to anions, the charge depends upon the ratio of the two kinds of areas. If the areas permeable to cations predominate (case 4), the solution of higher concentration becomes negative to the other solution, but only briefly since the concentrations of the two solutions eventually become equalized by the movement of cations and anions from the solution of higher concentration to the solution of lower concentration.

balanced by the charge on the cell membrane. However, for $Na^+$, both the concentration gradient and electrochemical gradient tend to bring the ion into the cell. Unless the cell membrane is infinitely impermeable to $Na^+$, a mechanism other than the one based on these factors is necessary to keep $Na^+$ at a steady low level inside the cell. It will be recalled that both $K^+$ and $Na^+$ have been shown to enter the cell membrane, and that entry of the $Na^+$, although slow, is nonetheless measurable. Yet only $K^+$ can accumulate inside the cell. It is postulated, therefore, that $Na^+$ is actively extruded by cells (see Chapter 14).

In many cases $Na^+$ extrusion from cells (by the sodium pump) appears to be linked with active uptake of $K^+$ from the medium by what appears to be a type of cation-exchange process. It has been shown, for example, that extrusion of $Na^+$ becomes reduced greatly when $K^+$ content is reduced, and in general, the rates of both processes change simultaneously and in a parallel manner (Katz, 1959).

In barnacle muscle fibers it is the calcium ion that diffuses out of the cell and sets up the potential, even though potassium is present much as in other cells (Hagiwara *et al.,* 1964; Hagiwara and Naka, 1964). Some cells may therefore have utilized cationic gradients other than potassium.

It is of interest to note that in one species of *Halicystis* the concentration of potassium in the sap is about the same as that in sea water, yet the outside of the cell is 70 mv. positive to the inside, and the potential persists even if the vacuolar content is removed and sea water is substituted. It is evident that resting potentials in such cells must be explained on some basis other than by analogy with the potassium concentration chain potential.

To determine whether anion gradients explain the resting potential of such cells, tests have been made with diverse anions. The anions inside a living cell are largely the organic ions resulting from metabolism—such as, acetate, pyruvate, lactate and amino acids—plus some inorganic ions, like sulfate and phosphate, all of which pass through the cell membranes very slowly. These ions, collectively called $X^-$, are absent from the natural bathing medium or present there in negligible quantities. Instead, the bathing medium is rich in chloride ion, which penetrates through cell membranes relatively readily but is always present inside cells in low concentration. A greater concentration of chloride ion outside rather than inside the cell favors inward penetration of chloride ion from the medium, leaving the outside of the cell positively charged. If the reasoning is correct, substitution of chloride ion in the bathing medium with a very slowly diffusing or nondiffusing anion should reduce or abolish the potential. This is verified in experiments on *Halicystis* (Blinks, 1955; see Fig. 23.4).

***Relation of the Resting Potential to Metabolism.*** Maintenance of the resting potential of the cell has been found to depend upon the presence of oxygen. Thus, when free oxygen is lacking in *Halicystis,* the resting potential falls (Fig. 23.4), but it rises again when oxygen is readmitted or the plant is illuminated (Blinks, 1955; Gaffey and Mullins, 1959). Lack of oxygen reduces the resting potential of the squid giant axon and other neurons, and the reduction is accompanied by a loss of potassium (Fenn, 1950; Hodgkin and Keynes, 1953). Since glycolytic metabolism occurs in the absence of oxygen (see Chapter 16), the nerve cell continues to liberate energy, but does so at a reduced rate, as indicated by the decrease in liberation of heat. Inhibition of glycolysis by poisons specific to glycolytic processes (e.g., iodoacetic acid, fluoride or phlorhizin*), should reduce the resting potential even more. Experiments verify this prediction, although in no case is the resting potential abolished completely so long as the nerve cell remains alive. The residual potential probably depends upon physical phenomena such as the Donnan equilibrium of the proteins. On the other hand, in an anoxic muscle fiber poisoned with iodoacetic acid, the potential falls to zero in 2 to 3 hours (Hodgkin, 1951).

The very gradual decline in the membrane potential of cells in which metabolic poisons have eliminated energy sources for the sodium pump indicates how slow transmembrane leakage really is. The decrease in potential is attrib-

*These poisons act in very different ways. Iodoacetic acid inactivates sulfhydryl groups—specifically, in this case, triosephosphate dehydrogenase (see Table 16.2). Fluoride inhibits enolase. Phlorizin inhibits glucose uptake.

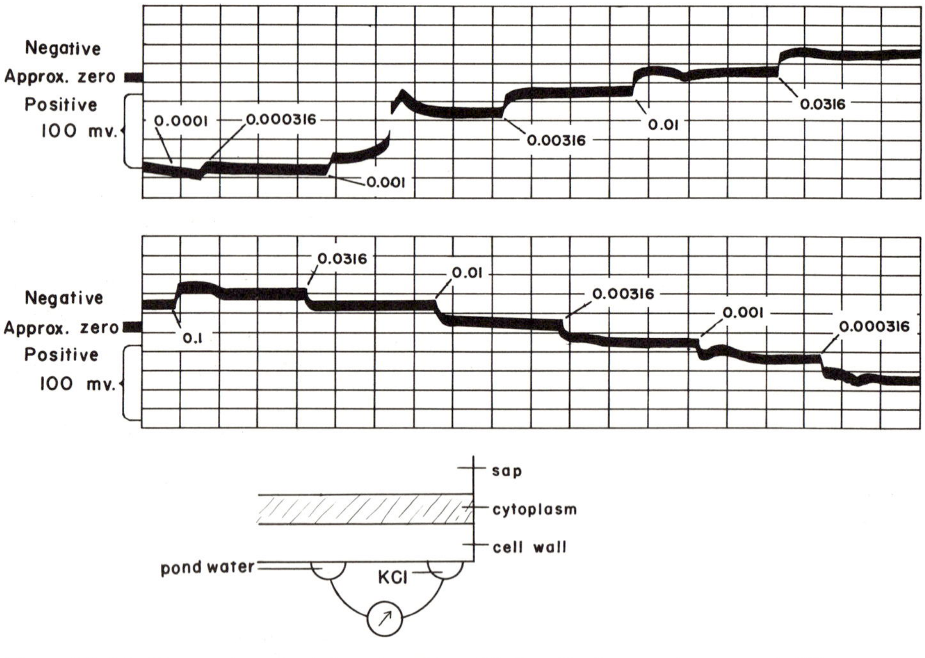


***Figure 23.3.*** Changes in potential produced in a *Nitella* cell when pond water and different concentrations of KCl are applied to the cell wall in the manner shown in the diagram beneath the graphs. The figures on the graphs indicate the molal concentrations of KCl applied. In the upper graph the potential is progressively decreased as the concentration of KCl is progressively increased, until at 0.0316M the potential is reversed. The lower graph shows the reverse effect as the concentration of KCl on the outside is lowered. (After Hill and Osterhout, 1937–38: J. Gen. Physiol. *21*:541.)

utable to the gradual diminution of the ionic gradients (Connelly, 1959; Shanes, 1963).

To sum up, some cells apparently maintain across the cell membrane a resting potential between the inside and the outside of the cell on the basis of differences in concentration of some cations, whereas others do so on the basis of differences in concentration of some anions. Since metabolism appears necessary to maintain the resting potential of a cell, it is inferred that metabolic energy is needed to maintain the difference in concentration of ions between the outside and the inside of the cell. Inasmuch as the polarized state is necessary for excitability, the cell must pay for maintaining its state of excitability by continuously expending energy.

The cost of maintaining an irritable state is suggested by the following calculation, based on the Nernst equation (23.1). The energy (W) required to move 1 mole of $Na^+$ from inside (in) to outside (out) of a two compartment system, and 1 mole of $K^+$ in the opposite direction is:

$$W = \frac{RT}{\mathcal{F}}\left(ln\,\frac{Na^{+}{}_{out}}{Na^{+}{}_{in}} + ln\,\frac{K^{+}{}_{in}}{K^{+}{}_{out}}\right) \quad (23.4)$$

where $Na^{+}{}_{out}$, $Na^{+}{}_{in}$, $K^{+}{}_{out}$ and $K^{+}{}_{in}$ are ion concentrations (or activities, strictly speaking). It is found that for the concentration gradient of $Na^+$ and $K^+$ found in the nerve fiber, about 3000 cal./mole are required to move a mole of $Na^+$ and $K^+$. If about 7000 cal./mole are available from the hydrolysis of 1 mole of ATP to ADP, it should be possible to move several moles of $Na^+$ for each mole of ADP formed. For frog nerve fibers at rest, the oxygen uptake is 1.5 $\mu$mole/gram (wet)/hour, and the $Na^+$ flux is about 6.6 $\mu$mole/gram (wet)/hour. Thus, the ratio of Na: oxygen atom (O) is 2.2. The ratio of ADP converted to ATP per oxygen atom reduced (P:O) in mitochondria is about 3. It would thus appear that only about 30 per cent of the resting metabolism is available for work other than ion transport (Connelly, 1959).

A similar calculation for nerve fibers repeatedly stimulated to produce action potentials for an hour at 50 volleys per second shows that even with the severalfold increase in respiratory metabolism, more than two-thirds of the energy made available (as high energy phosphate) from oxygen consumption is probably devoted to ion transport (Connelly, 1959).

## ORIGIN OF THE PROPAGATED ACTION POTENTIAL

An excitable cell, e.g., a neuron, a muscle fiber or a *Nitella* cell, like all other cells, has a resting potential at all times between the inside and outside of the cell, across the cell membrane. But, in addition, it can develop and propagate an action potential upon stimulation (Hoyle, 1962).

How, then, is the resting potential of an excitable cell altered during the development of an action potential? The positive charge on the outer surface of the cell membrane, because of the presence of positively charged ions, is delicately balanced by the negatively charged anions of salts, organic acids and so forth inside the cell. Application of an electrode as a stimulus disturbs the resting arrangement and distribution of these ions. If the membrane is slightly disturbed, only a local potential develops, the disturbed area becoming negative to the surrounding more positive region on the cell membrane. It is thought that at this time sodium ions enter the cell in small numbers. This local potential remains near the point of stimulation only briefly and the ions soon redistribute themselves to the steady state pattern of a resting cell. The stronger the stimulus, the larger is the local potential developed and the greater the area of disturbance on the cell membrane. If a local potential on a neuron reaches one-fifth to one-third the height of a spike, it develops into a spike which passes the full length of the axon in the all-or-none manner of a propagated action potential (see Fig. 23.5). It is estimated that about $6 \times 10^9$ electrons are needed to stimulate a nerve cell. The neuron (or other excitable cell) is then said to have been stimulated. This in essence is a statement of the Bernstein membrane theory, which

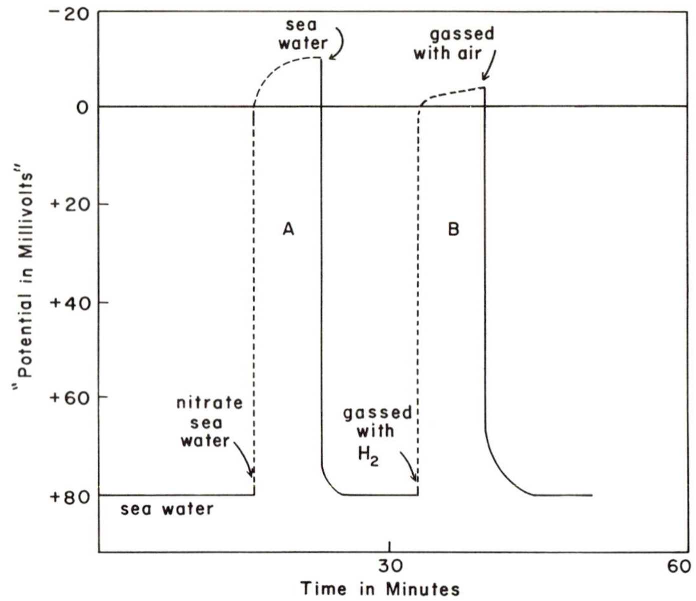


***Figure 23.4.*** *A,* The potential of *Halicystis* as affected by the replacement of the chloride of sea water with nitrate. *B,* The potential of *Halicystis* as affected by the removal of oxygen. The dotted line in *A* shows the fall in potential when *Halicystis* is placed in nitrate sea water. The dotted line in *B* shows the fall in potential when *Halicystis* is freed of oxygen by being gassed with hydrogen. The solid vertical line in *A* shows the return of the potential to the original resting value when the cell is returned to sea water. The solid vertical line in *B* shows the return of the potential when the cell is supplied with oxygen by being gassed with air. In each case the cell was set up as illustrated in Figure 23.1. (After Blinks, 1940: Cold Spring Harbor Symposia *8*:204.)

is a development of ideas first introduced by Du Bois-Reymond (Katz, 1959a, b; Katz and Melidi, 1965).

That stimulation of an excitable cell produces a profound change is seen by the altered electrical properties of the cell's membrane. Thus, upon stimulation the electrical resistance in a *Nitella* cell falls from 100,000 to 500 ohms per square centimeter and in the squid giant axon it falls from 1000 to 25 ohms per square centimeter (Hodgkin, 1951). Furthermore, potassium is lost in measurable quantities from squid giant axons and *Nitella.* Both lines of evidence indicate that on stimulation the cell membranes of these excitable cells have become much more freely permeable to certain ions. A free interchange of ions between the outside and the inside of a cell at the point of stimulation would depolarize and abolish the resting potential in that portion of the cell. Recording with electrodes inside some excitable cells has demonstrated that the point of stimulation on the membrane of the excited cell not only loses its charge (depolarizes) but changes charge and becomes negative to the inside (Fig. 23.6). Measurements on the squid giant axon indicate that the stimulated portion of the nerve fiber becomes almost as negative as it was previously positive. This has been verified on many animal cells but does not occur in cells of *Nitella* in which only depolarization occurs upon stimulation (Hodgkin, 1951).

By use of radioactive tracers the movements of sodium and potassium ions across the membrane of the nerve fiber have been followed during stimulation and recovery of the axon (Hodgkin *et al.,* 1949; Keynes and Lewis, 1951; Hodgkin and Huxley, 1952). Thus, at the point of stimulation the permeability to $Na^+$ is suddenly increased several hundredfold and reaches its peak in 100 microseconds. The potential gradient across the cell membrane is thus temporarily abolished at the point of stimulation, which be-

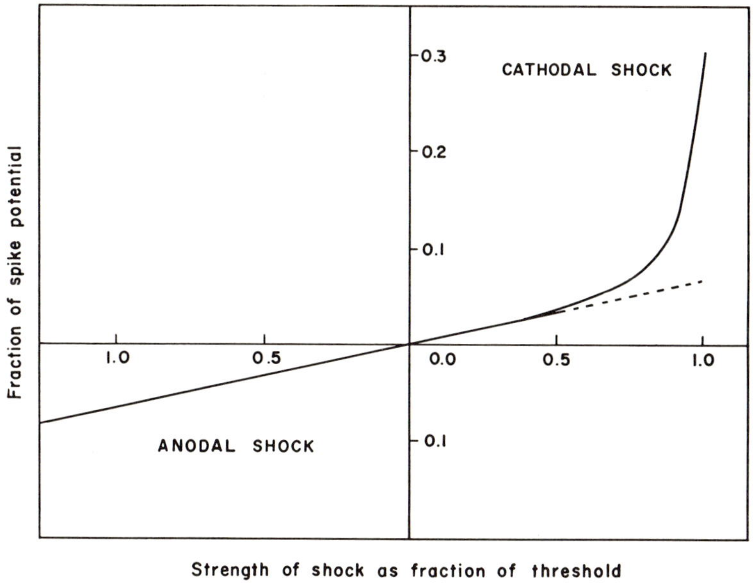


*Figure 23.5.* Relations between local potentials and spike potentials on a nerve cell. The spike does not arise from anodal shock even when the shock is very large, but it does arise from cathodal shock when a given fraction (about 0.3) of the spike potential has been reached. Each potential was measured 0.29 millisecond after the shock. The dotted line in the cathodal shock quadrant is an extrapolation showing the potential that would be developed if a spike did not arise. (After Hodgkin, 1938: Proc. Roy. Soc. *B126*:96.)

comes first depolarized and then develops a negative charge by entry of $Na^+$ into the cell. At the close of this period (the rise of the spike) the membrane again becomes essentially impermeable to sodium, but its $K^+$ permeability increases and, at the fall of the spike potential, $K^+$ leaks out from the point of stimulation and repolarizes the cell. Since $K^+$ leaks out at a slower rate than $Na^+$ moves inward, the repolarization of the membrane to resting potential is slower than its depolarization. Presumably, at the end of the spike, the sodium pump once more comes into operation, as does the coupled active transport of $K^+$ into the cell, and the membrane begins to recover its normal permeability to ions at this point (Cross *et al.*, 1965; Adrian and Slayman, 1966).

The spike cannot be explained by temporary suspension of the sodium pump because, when the latter is suspended by metabolic poisons, the action potential of the cell does change but only very slowly and gradually. The events which transpire during spike development involve large numbers of ions and are extremely rapid, but neither the nature of the changes in the cell membrane resulting in increased permeability nor the nature of the restorative processes is fully understood (Katz, 1959a).

However, it should not be inferred that sodium ions which have entered the cell will at once be expelled or that potassium ions which have left the cell will have to be regained by the nerve fiber before it can recover. The quantity of ions lost (about $10^{-12}$ M) during a single impulse is but a small fraction (about 0.0001 per cent in the squid giant axon) of that present in the nerve fiber. Even if the sodium pump and the associated uptake of potassium ions did not become active again, the squid giant axon could be stimulated thousands of times before the ion store would become exhausted (Katz, 1959a). But to recover its sensitivity, the membrane of the nerve fiber must first shut off the sodium influx as the potassium efflux repolarizes it. If the nerve fiber is not stimulated again, the sodium pump presumably resumes operation and the accumulated $Na^+$ is voided while the lost

$K^+$ is simultaneously regained. For these and other fundamental studies on nerve action potentials Hodgkin (1964a) and Huxley (1964), jointly with Eccles (1964), received the Nobel prize in 1963.

Nerve cell dendrites generally show local, nonpropagating, decremental potentials. In the dendrites both $Na^+$ entry and $K^+$ loss occur simultaneously, in contrast to the ionic movements which occur at the nerve axon. The degree of response is determined by the number of channels open for $Na^+$ and $K^+$ and the duration and extent of each channel opening. Such graded, nonpropagating potentials are of importance in receptor-sensory neuron systems as well as in synapses, making possible the effective passage of information as well as integration of the information.

That sodium is the conductor of current during the rise of the action potential is demonstrated by substituting for it a cation that does not penetrate the cell as readily as $Na^+$, such as choline chloride. Choline, an organic base with a positive charge, does not penetrate cells readily and the decrease in height of the spike is proportional to the decrease of $Na^+$ in the bathing fluid. The sodium ion is thus as vital to the height of the spike in most cells as is the potassium ion to the resting potential of cells. However, $Na^+$ is apparently not always the ion that carries the current. In some algae (e.g., *Chara*) it appears that leakage of $Cl^-$ from the cell serves the same purpose. In some crustacean muscle fibers there is evidence for influx of divalent ions instead of sodium ions.

Tetrodotoxin, a neurotoxin from the Japanese puffer fish (Mosher *et al.*, 1964), in low concentration blocks $Na^+$ permeability in nerve and muscle cells but does not affect $K^+$ movement. Consequently, it blocks spike formation in muscle and nerve cells which depend upon $Na^+$ for carrying the inward current during excitation. It is without effect in those cells in which calcium plays this role (Nakamura *et al.*, 1964a, b; Hajiwara and Nakajima, 1965; Nishi and Sato, 1966; Ozeki *et al.*, 1966). The receptors for tetrodotoxin appear to be on the outer surface of the excitatory cells since injection of the poison inside such cells does not produce a block (Narahashi *et al.*, 1966). The separateness of the $Na^+$ and $K^+$ gates, observed in studies already discussed, is thus corroborated.

## THEORY OF THE PROPAGATION OF THE ACTION POTENTIAL

It is to be expected that local circuits will flow from the unstimulated, positively charged areas of the cell membrane to the stimulated, depolarized or negative portion, and as each new area becomes depolarized or negative, it in turn acts as the sink toward which the current flows from the adjacent area (see Fig. 23.7). Progressive depolarization, or reversal of charge, of the entire length of the neuron therefore follows from the point of stimulation outwardly as if an electrode were traveling along the nerve fiber at the rate of conduction characteristic of the fiber (Brazier, 1960). It has been demonstrated that the source of the current is the flow of $Na^+$ into the cell (Hodgkin and Huxley, 1952). This, briefly, is the so-called *local circuit theory* of propagation of action potential.

Repolarization of the surface of an excited fiber with a positive charge sets in as soon as the absolute refractory period is over, and progresses so rapidly that never more than a fraction of a nerve fiber is depolarized at a time. However, this may be a region of a few millimeters to a few centimeters, depending upon the duration of the spike and its speed of propagation. Recovery occurs most likely at the expense of high energy phosphate bonds, since the concentration of creatine phosphate declines after stimulation. Because a nerve can continue to discharge for some time in the absence of free oxygen, even when glycolysis is inhibited by a poison, a store of high energy phosphate bonds must be available for the purpose. These are presumably rebuilt at the expense of the metabolic activities of the neuron—slowly in the complete absence of oxygen and much more rapidly in the presence of oxygen. Experiments with tracers indicate that, during recovery, $Na^+$ leaves the cell and $K^+$ reenters

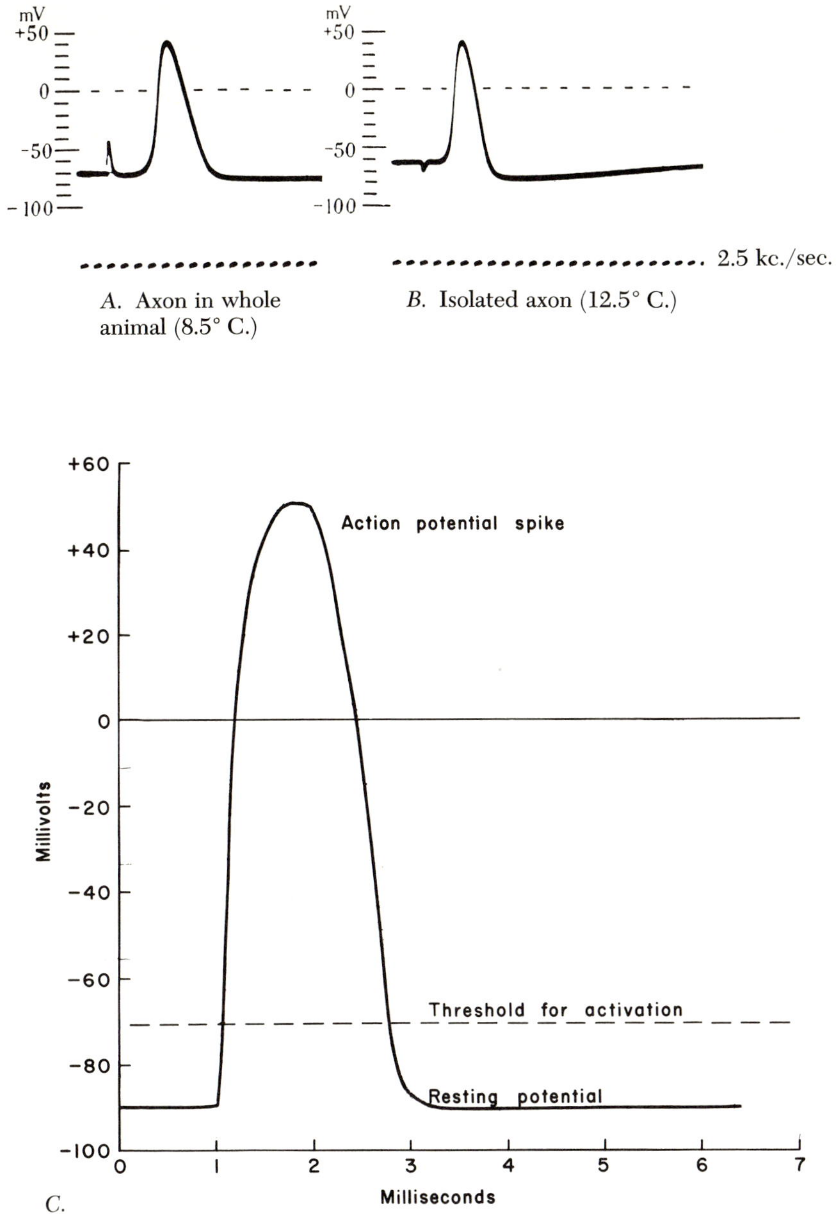


**Figure 23.6.** Potential between the inside and the outside of an excitable cell during spike formation. *A* is from an intact giant axon of the squid (*Loligo forbesi*), *B* is from an isolated squid giant axon and *C* is from a single isolated electroplax of the electric fish (*Electrophorus electricus*). (*A* and *B* from Hodgkin, 1958: Proc. Roy. Soc. *B148*:1; *C* from Gerebtzoff and Schoffeniels, 1960: Comp. Biochem. *2*:525.)

(Keynes, 1949). That work, requiring metabolic energy, must be done during recovery of a stimulated cell is indicated by the evolution of heat.

It is quite evident that various properties of the propagated action potential (nerve impulse) may be interpreted on the basis of the local circuit theory of propagation. Thus, when either a stimulus below rheobase is used or a stimulus above rheobase is applied for inadequate length of time, the membrane is not altered enough to produce a sink, and the ions soon redistribute themselves into the normal pattern for the resting potential of the unstimulated cell. All-or-none response and conduction without a decrement are to be expected of a nerve fiber if the membrane of the entire fiber is initially equally polarized (charged). The reaction is self-generating if the same amount of electrical change occurs along the entire length of the nerve fiber. The absolute refractory period represents the time during which the membrane is no longer selectively permeable, or, as is often stated, the "*sodium gate* is wide open." Sodium enters the nerve fiber during this period, and the neuron is therefore inexcitable. The initial state of polarization is presumably regained during the relative refractory period when the cell membrane once again becomes impermeable to $Na^+$, and $K^+$ leaks out in response to the concentration gradient. The states of increased and, later, of decreased excitability (as seen in vertebrate A fibers) occur probably because the membrane has not fully regained normal permeability. Consequently, a response may be elicited, but a different threshold of stimulation is required. A quantitative formulation of the theory in terms of the distribution of specific ions is not yet possible (Shanes, 1963; Pickard, 1966).

The neuron cell membrane registers many events, and its exact state and response are largely dependent upon its history. For example, in A fibers, during tetanic stimulation, the after-potentials become more prominent if the stimuli arrive before the neuron has recovered entirely. If they should arrive when the neuron displays a negative after-potential, the spike appears to be higher by the amount of the negative after-potential. The positive after-potential is also gradually augmented by frequent stimulation and the neuron becomes less excitable because a greater stimulus is required for depolarization (Erlanger and Gasser, 1937). Furthermore, subthreshold stimulation by a steady cathodal current lowers the excitability because it leads to accommodation when the rate of displacement of ions is equaled by that of recovery. When a neuron is depolarized by asphyxia, it is rendered incapable of response, yet it can be repolarized by a battery current and induced to respond at least once (Bishop, 1951). A state of polarization seems to be necessary for maintaining excitability.

Saltatory conduction, as discussed previously for myelinated nerve fibers (see Fig. 22.11 and 23.7), is thought to be the result of local circuits occurring only from node to node. The amount of naked axon is very slight even at the nodes, perhaps a few hundred angstrom units at most, because, according to electron micrograph studies, the Schwann cells of the myelin sheath send finger-like growths along the axon. The amount of $Na^+$ and $K^+$ exchange is thus greatly reduced and the net work required of the cell for response is thereby much lowered (Schmitt, 1959).

## SENSORY FIBER ACTION POTENTIALS

Receptor cells are nerve cells or epithelial cells which are much more sensitive to certain stimuli than are other cells; e.g., a pressure receptor is much more sensitive to pressure than neurons in a nerve trunk. Sometimes the distal branches of the afferent neuron are modified to serve as receptors, e.g., muscle sense organs and pain receptors.

Receptor cells when stimulated excite adjacent afferent (sensory) neurons which inform the central nervous system about the strength of the stimulus (Fig. 23.8). Since Adrian's classic experiments 36 years ago it has been recognized that varying the strength of a stimulus (e.g., pressure on a pressure receptor) affects only the frequency of discharge, not the size of the impulses sent into the central nervous system. The code, then, appears to be frequency modulation of a single type of message for a given neuron (Hartline, 1959).

In 1950 Katz demonstrated that stretching a muscle mechanoreceptor, which is a modified part of the fibers of an afferent neuron, generates in the sensitive

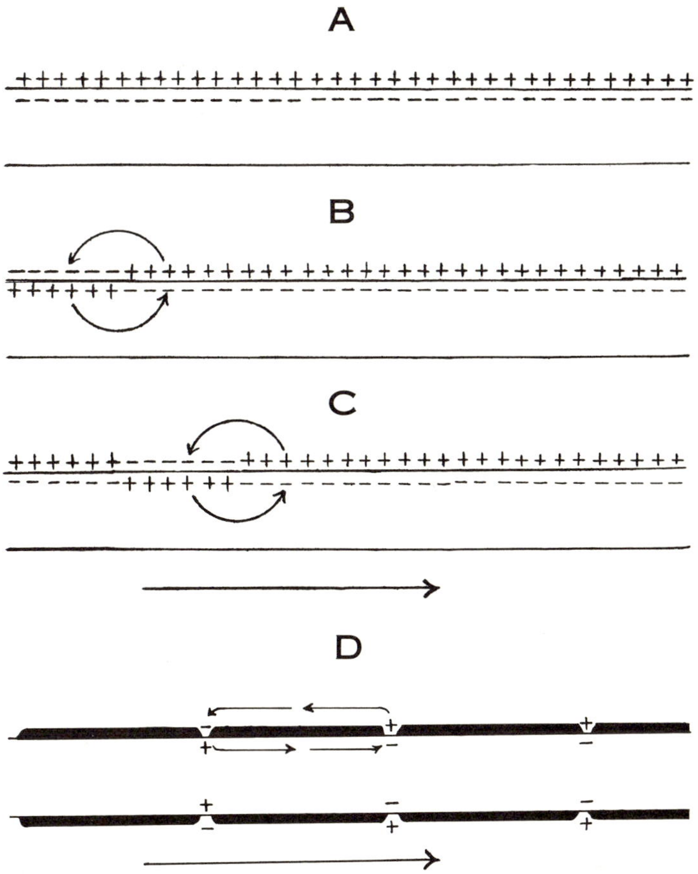


***Figure 23.7.*** Diagram to illustrate the local circuit theory of propagation of the action potential (*A, B* and *C*) in unmyelinated neurons and muscle fibers as compared to the saltatory conduction in myelinated neurons (*D*). *A* is the membrane of an unexcited nerve (or muscle) fiber; *B* shows the cell membrane excited at one end; *C* shows the movement of the action potential, followed by recovery; *D* shows node-to-node saltatory conduction. In large nerve fibers less than one-hundredth as much ionic exchange occurs during an impulse in saltatory conduction as compared to conduction in an unmyelinated nerve fiber. The arrows in *C* and *D* show the direction of impulse propagation. (After Hodgkin, 1957: Proc. Roy. Soc. *B148*:1.)

fibers a local, graded, decremental potential, the magnitude and duration of which depend upon the intensity and duration of stimulation. This local potential, also called the *generator potential,* when above threshold, sets off on the axon of the afferent neuron a volley of propagated action potentials. Discharges continue on the afferent axon as long as the depolarization of the receptor fibers remains above threshold. It was soon demonstrated that upon stimulation similar local generator potentials occur in other types of receptors as well (Davis, 1961).

A pressure receptor consists of a structure enclosing the distal fibers of an afferent neuron. By removing parts of the receptor, an attempt has been made to determine its sensitive region. It was found that when only a core of the receptor remains about the nerve fiber, stretching or deforming the core induces volleys of impulses along the attached fiber but that when the enclosed nerve fiber ending itself is damaged, discharge stops (Loewenstein, 1960, 1965; Katz, 1961). It would appear that the *biological transducer* which transforms the mechanical energy into electrical energy and the nerve impulse is possibly the terminal portion of the neuron.

There is evidence that following stimulation of such a receptor neuron, the electrical membrane resistance of its distal portion is reduced, permitting ions to enter, partially depolarizing its membrane. When depolarization has reached threshold, propagated action potentials are initiated on the afferent nerve fiber (axon). The local depolarization characteristic of a receptor is spoken of as a *generator potential,* because its generates the propagated action potential on the axon (Katz, 1961; Kennedy, 1962; Lissmann, 1963; Grundfest, 1963, 1965).

The excitation of an afferent axon does not seem to differ significantly from excitation of any nerve fiber or excitable cell. However, a fundamental difference apparently exists between electrical stimulation of a nerve fiber and stimulation through receptor cells. A single electrical stimulation above threshold of an afferent nerve fiber results in a single spike, but even a brief stimulation of a receptor almost invariably results in a volley of impulses along the afferent nerve fiber (Fig. 23.9). For a given kind of receptor, the frequency and duration of the volley depend upon the duration and intensity of stimulation of the receptor because these factors largely determine how long the generator potential is maintained at a sufficient level to continue to excite the afferent fiber. If the stimulation has been weak and brief, then the subsequent generator potential on the fine terminations of the afferent fiber will be small and will decay rapidly. Consequently, only a few spikes will be induced in the volley. If stimulation is intense, the corresponding generator potential will be greater both in electrical charge and in the area of the neuron altered, and decay will be slower. As a consequence, the resultant volley of impulses will be more prolonged (Bullock, 1959; Hartline, 1959; Katz, 1961; Ochs, 1965).

It is everyday experience that our sense receptors "get used" to a given stimulus continuously applied, unless it is increased in intensity. The probable basis for such change in sensitivity of receptor cells is the failure of the receptor to induce discharges along the afferent neuron. It can readily be demonstrated, by electrical stimulation, that the afferent neuron is not fatigued, but rather that the stimulus fails to maintain a generator potential sufficient to produce discharge; *accommodation* has thus been achieved. Accommodation occurs at different rates in diverse receptors (Fig. 23.10). For example, touch receptors accommodate fairly readily but pressure receptors in the carotid artery show almost no accommodation and continue to discharge as long as pressure is applied (Davson, 1964). In a receptor accommodated to a given intensity of stimulation, impulses can again be induced by sufficiently increasing the strength of the stimulus. In this case a generator potential has again been induced, but the accommodation will

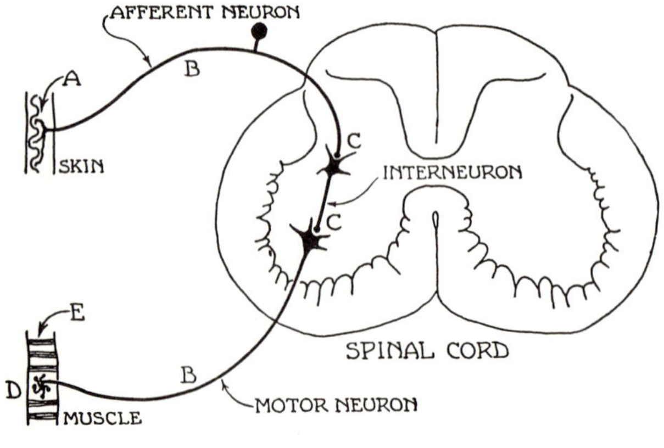


***Figure 23.8.*** Diagram of a simple reflex arc. A shows the receptor at which excitation of afferent neurons occurs. The impulses generated at A are conducted by the axons (B) to the central nervous system by the afferent neuron, in the cord by interneurons and out to the effector (muscle) (E) by the motor neuron (efferent). Synaptic transmission occurs at C and neuromuscular transmission at D. (From Ruch and Patton, 1965: Physiology and Biophysics. 19th Ed. W. B. Saunders Co., Philadelphia.)

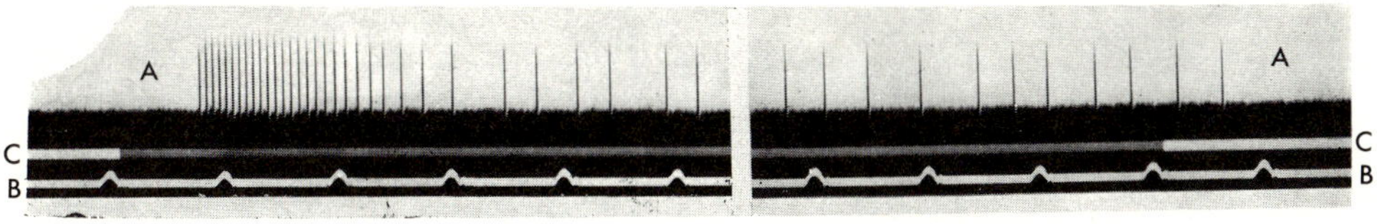


***Figure 23.9.*** Volley of impulses (A) on an afferent neuron. Single nerve fiber from an ommatidium of *Limulus.* The lower notched markings (B) are fifths of a second. The line above the time signal (C) shows when the light went on and off. The vertical break represents a time lapse of 2.4 seconds. (From Hartline and Graham, 1932: J. Cell. Comp. Physiol. *1*:285.)

occur again at a rate characteristic for the receptor (Bullock, 1959; Loewenstein, 1960).

Ultimately a receptor cell may react feebly or fail to respond when it becomes fatigued. This suggests that some material necessary for action of receptors is temporarily exhausted and the receptor is less responsive (see Prosser and Brown, 1961).

## SYNAPTIC TRANSMISSION OF ACTION POTENTIALS

The synapse is defined as the functional connection between two neurons. It was so named in 1898 by Sherrington who considered the synapse to be a region of contact, not confluence, of two excitable cells. Electron microscopy has since demonstrated that the cells at a synapse are indeed separate—not only are presynaptic and postsynaptic cell membranes present, but between them there generally exists a synaptic cleft* several hundred angstrom units wide (see Figs. 23.8 and 23.11; DeRobertis, 1958, 1967).

Synaptic contacts vary in structure. For example, a crossing of cells with similar fiber membranes lying alongside one another (*en passant* synapses) is found in coelenterates. A more prolonged contact of fiber membranes in the *loop synapses* (neuropile connections) is found in many invertebrates. The most highly differentiated contacts between neurons occur in vertebrates in which the modified ends of the nerve fibers are applied to dendrites or to the nerve cell body. Many of the sensory nerve connections are of this type, and the nerve fiber endings may be cuplike, clublike or basket-like. *End feet synapses* (boutons termineaux) are endings found on the motor neurons (see Fig. 23.11). It is interesting that in the vertebrate central nervous system, however, the synapses of special cells such as the Purkinje cells are of a rather simple type, nerve fiber abutting against nerve fiber (*en passant*). The *myoneural junction* (muscle end plate) is the synapse between a motor neuron and a muscle fiber, and is an area of considerable contact between nerve fiber and muscle cell (Fig. 23.12). These

*Except in tight junctions, already considered in Chapter 11.

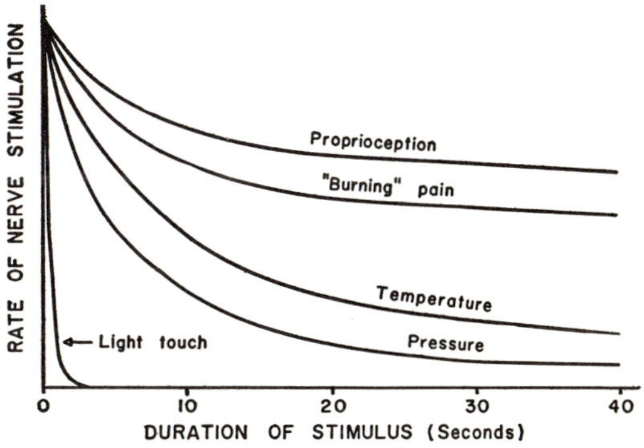


***Figure 23.10.*** Accommodation in various receptors as measured by frequency of impulse discharge after continuous application of a stimulus of the same intensity. (From Guyton, 1956: Textbook of Medical Physiology. W. B. Saunders Co., Philadelphia.)

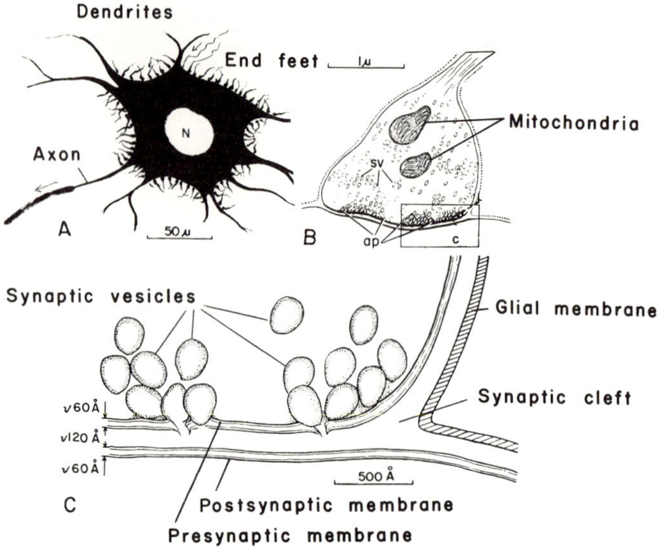


***Figure 23.11.*** Diagram showing a synaptic junction at various magnifications. *A,* A motor neuron under medium power of a light microscope. Note the numerous end feet—the terminations of axons of interneurons. *N* represents the nucleus of the motor neuron. *B,* One end foot under an electron microscope, magnified 60 times over *A*. In addition to the mitochondria, note the synaptic vesicles (sv) and the active points (ap) where the synaptic vesicles appear to aggregate and open onto the synapse. The dotted line represents the glial membrane. *C,* One sector of *B* as seen using high resolution electron microscopy. Note that some of the synaptic vesicles are rupturing into the synaptic cleft through the presynaptic membrane. Observe that the membranes bordering the cleft are much like any other cell membranes. The synaptic cleft is a real separation between the cells enclosed by the membrane of a supporting or glial cell. (Adapted from DeRobertis *et al.,* 1965: Cell Biology. 4th Ed. W. B. Saunders. Co., Philadelphia.)

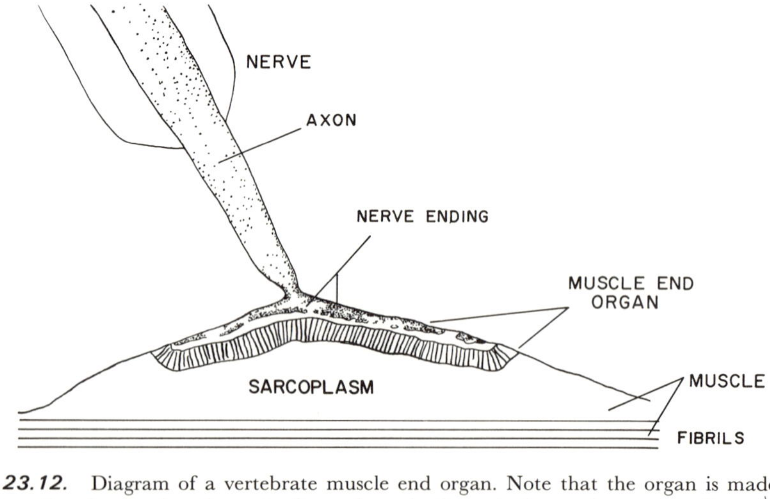


***Figure 23.12.*** Diagram of a vertebrate muscle end organ. Note that the organ is made up partly of nervous and partly of muscular elements. Sarcoplasm is the cytoplasm of a muscle fiber. (Adapted from Acheson, 1948: Fed. Proc. 7:447.)

varied types of synapses have functional significance and are discussed in textbooks of neurophysiology (Magoun, 1959).

When the surface of contact between two neurons is similar (e.g., in the *en passant* synapses of coelenterates), the synapse may transmit in both directions. When neurons interconnect in a complex manner at a synapse, the synapse transmits in one direction only, that is, it is *polarized* (Rosenblith, 1961). Polarization probably results from differentiation in neurosecretory activity at presynaptic and postsynaptic membranes and it may occur even where the contacts between these membranes are apparently structurally undifferentiated, e.g., in the central nervous system of the vertebrate.

Two theories have been proposed to explain synaptic transmission and are discussed here. Both have had their proponents, and the problem is not fully resolved (Katz, 1959b, 1961).

***Electrical Theory of Synaptic Transmission.*** Synaptic transmission by electrotonic spread from the presynaptic to the postsynaptic membrane has been demonstrated in the crayfish giant motor synapse. Here the synaptic delay is very short—0.1 millisecond (ms.) from beginning of prespike (local potential) to beginning of the excitatory postsynaptic potential. Presynaptic depolarization actually spreads across the synapse and is measurable as an appreciable potential difference in the postsynaptic membrane. The postsynaptic depolarization does not in turn cause appreciable hyperpolarization of the presynaptic fiber, indicating a unidirectional action (rectifier action) of the spread of polarization. It is proposed that the synaptic membrane acts as a rectifier of positive current from the presynaptic to the postsynaptic fiber, with high resistance to flow in the opposite direction (Furshpan and Potter, 1959a, b). The synaptic rectifier is oriented in the right direction to allow the presynaptic action current to stimulate the postfiber. The electrotonic current resulting from the spike on the presynaptic fiber is sufficient to account for normal transmission. It seems unlikely that this type of transmission is limited to neurons in the crayfish abdomen (Katz, 1959b; Furshpan and Potter, 1959a, b). It is found that where transmission occurs in this manner the junction is a tight junction; that is, the cleft separating one unit cell membrane from the next is missing (see Chapter 11 for discussion and references).

***Chemical Theory of Synaptic Transmission.*** In all other cases studied the synaptic delay is much greater than in the crayfish giant motor synapse, and varies between about 0.3 and 10 ms. Consequently it is thought unlikely that electrotonic spread occurs from presynaptic to postsynaptic fibers because the field on the presynaptic fiber is disappearing or gone before the postsynaptic local potential develops. Internal electrodes fail to indicate induction of electrotonus at the time of maximal electric field on the presynaptic fiber (Eccles, 1963, 1965). It is suggested that the presynaptic potential on reaching the synapse activates a chain of events which results in the secretion of a *transmitter substance,* which, when released, attaches itself to special receptor molecules on the surface of the contacting postsynaptic excitable cell (Florey, 1961; 1966). This chemical combination leads to a membrane change which depolarizes the postsynaptic cell locally. When depolarization exceeds the threshold level, a propagated action potential is initiated. A rise in temperature decreases synaptic delay, a finding in keeping with the theory of chemical transmission across the synapse. A rise in temperature should have little effect on electrical transmission (Katz and Melidi, 1965).

The neuromuscular junction is an example of a synapse at which chemical transmission is well supported by experimental data (Dale *et al.,* 1936). It has been possible to demonstrate that a transmitter substance (acetylcholine) is produced at the presynaptic nerve fibers which, when applied to the end plate of the muscle fiber, causes a local depolarization of the cell membrane and sets up action potentials followed by contraction of the muscle fiber. The chemical effect is localized to the synaptic area of the muscle fiber. In this junction there is an irreducible synaptic delay of 0.5 to 1 ms. between the arrival of the nerve impulse and the start of the local (end plate) potential, part of this time being necessary for secretion of the acetylcholine. Histochemical studies have demonstrated a localization at this junction of the enzyme *acetylcholine esterase,* the function of which is the decomposition of the acetylcholine to choline and acetic acid, thus preventing a prolonged depolarized state. When this enzyme is poisoned (e.g., by the drug *eserine*), stimulation of the motor nerve is followed by multiple contractions (Katz, 1959b).

Stimulation of the motor neuron produces no measurable potentials at the end plate at the time the spike reaches its terminal portion. When potential changes of either polarity are imposed on the terminal portion of the motor nerve fiber, they do not spread beyond the nerve terminal, but they increase the rate of acetylcholine discharge. There is thus no evidence of cable transfer of electric current from motor fiber to muscle fiber, but there is adequate evidence of chemical changes leading to action potentials and contractions. It has been possible to apply acetylcholine solutions directly to the motor end plate (to reach the surface of the receptor areas) and under these conditions as little as $10^{-16}$ gram equivalents of acetylcholine gives rise to an effective depolarization of the muscle fiber followed by a spike and contraction. Application of a similar concentration of acetylcholine inside the muscle fiber at the postsynaptic membrane has no effect. The muscle end organ thus acts much like a chemoreceptor, a small quantity of the substance sufficing to depolarize the cell (Katz, 1959; Zacks, 1964; del Castillo and Katz, 1967).

The way in which the transmitter substance causes depolarization is not entirely understood, but the most likely mechanism is the opening of some ionic channel in the cell membrane. There is evidence that acetylcholine simultaneously increases the membrane permeability of the motor end plate to several monovalent ions (e.g., sodium, potassium, ammonium) and possibly opens up an ionic channel to all small ions on both sides of the membrane. The result is a partial depolarization to a level of about 10 to 20 mv. (negative inside) which corresponds to a free diffusion (liquid junction) potential between the cytoplasm and the external fluid. This local nonpropagated depolarization sets off the propagated spike potential (Katz, 1959b).

When potentials are recorded on the end plate (postsynaptic) during rest, minute potential changes of about 0.5 mv. are observed on oscillographic records, with a rapid rise (1 ms.) and a slower decay (about 20 ms.), indicating momentary depolarizations of the end plate (Dudel and Kuffler, 1961). This "miniature" end plate potential appears to indicate spontaneous release of "packets" of acetylcholine by the motor neuron, since gradual application of a very low concentration of acetylcholine gives a smooth change, not the sudden changes seen in the records. The miniature end plate potentials are reduced in size by curare, and their amplitude and duration increase when the action of the acetylcholine esterase at the end plate is blocked with a poison, preventing local hydrolysis of the acetylcholine. All these experiments suggest that acetylcholine is continually manufactured and expelled from the nerve cell endings, but the spontaneous release inducing the small potentials occurs not continually but in packets. Presumably an impulse reaching the synapse causes a much larger number of such packets to be released at once and is accompanied by a much larger potential change which serves to depolarize the cell. It is interesting that in the crustacean myoneural junction, where another (unidentified) chemical transmitter is active, miniature potentials are also recorded (Dudel and Kuffler, 1961).

Electron microscopy has shown that the synaptic vesicles are almost always seen at synapses, groups of them moving to active spots (Fig. 23.11*B*) where they appear to rupture into the synaptic cleft. Although confirming evidence connecting secretion from these vesicles to transmission has not yet been presented, the suggestion is an appealing one (DeRobertis, 1964; Gray and Guillery, 1966).

Acetylcholine probably plays a role other than myoneural in synapses. For example, in autonomic ganglia where nerve fibers (preganglionic) from the central nervous system synapse with other (postganglionic) nerve fibers passing directly to effectors in the viscera, experiments show that when the preganglionic nerve fibers are stimulated, and their synapses are perfused with balanced salt solution during the excitation, the ester acetylcholine appears in the perfusate. This perfusate has been found to produce electrical changes leading to excitation and an action potential when applied at other autonomic synapses. However, experimental application of acetylcholine is usually ineffective in exciting synapses of the central nervous system (Prosser and Brown, 1961; Florey, 1966).

Other transmitter substances also play a part in synaptic transmission. For example, sympathin (norepinephrine) is important in the postganglionic fibers of the sympathetic nervous system. There is some evidence that packets of norepinephrine are secreted in such synapses much as in the case of acetylcholine in most synapses (Wolfe *et al.,* 1962; DeRobertis, 1964). A "sensory" substance of unknown chemical nature has been postulated for the synapses of afferent nerve fibers (Florey, 1961; 1966). Transmitter substances are found in the central nervous system and have been isolated and tested but not identified chemically with certainty. The potential changes that they produce, however, have been measured and correlated with the existing ionic gradients. At the synapses between interneurons and motor neurons, for example, the application of transmitter substance increases the permeability of the postsynaptic region to all ions, resulting in depolarization (Katz, 1959b).

Transmitter substances have been demonstrated in synapses of invertebrates, but the biochemical identity of these substances is uncertain. There is suggestive evidence that 5-hydroxytryptamine is one such transmitter. Acetylcholine probably functions in a few species, but evidence for norepinephrine as a transmitter substance in invertebrates is slight (Florey, 1961, 1966). Glutamate is excitatory on crayfish neuromuscular junctions but there is little evidence for it as a normal transmitter (Takeuchi and Takeuchi, 1966a).

Although biochemical identity of the transmitter substances remains doubtful in many instances, the fact that a transmitter substance is involved in transmission at synapses seems to be rather generally accepted. Thus it is considered that in most neurons a chemical change produced by a transmitter substance results in an electrical change (local generator potential) which sets off a conductile process (propagated action potential) ending in an effector action—the secretion of another transmitter substance. The action of the stimulus upon receptor cells probably also produces secretion of a transmitter substance which induces the depolarization of the afferent neuron. From this point of view the molecular mechanism for activation of various synapses may be quite similar in principle, despite diversity of transmitter substances (Florey, 1966).

***Inhibition at Synapses.*** Inhibition of synaptic transmission is part of the mechanism of integration of responses to stimuli. Attempts to explain synaptic inhibition on the basis of electrotonic spread from presynaptic to postsynaptic membranes have been as unsuccessful as those purporting to explain synaptic transmission of excitation by electrotonic spread, and for the same reasons. However, inhibitory effects of some secretions and drugs have been known for a long time.

An example of inhibition is the action of the vagus nerve (tenth cranial nerve) upon the heart beat. It is known that stimulation of the vagus releases acetylcholine which causes hyperpolarization or stabilization of the resting potential of the heart muscle fibers and results in slowing or stopping the heart beat. The mechanism, seemingly, is a change in ionic conductance, but the channel opened in this case is restricted to $K^+$ and does not include $Na^+$. As a consequence the mem-

**Table 23.2.** Changes Induced in Excitable Cells by Stimulation[1]

| EXCITABLE CELLS | EXCITATION INDUCED | EXCITOR | IONIC CHANGES AT MEMBRANE |
|---|---|---|---|
| Dendrites | Local (non-propagated), graded, decremental depolarization | Acetylcholine in some, X[2] in others | Simultaneous $Na^+$ influx and $K^+$ efflux |
| Axon of neuron | Propagated all-or-none action potential (spike), with overshoot | see text | $Na^+$ influx followed by $K^+$ efflux |
| Vertebrate skeletal muscle fiber | Spike with overshoot | Acetylcholine | $Na^+$ influx followed by $K^+$ efflux |
| Invertebrate muscle fiber | Local, graded, decremental depolarization (slow fiber) or | X | Ionic flux[3] increased |
| | spike with overshoot (fast fiber) | X | Ionic flux increased sequentially for different ions |
| Receptor cell and afferent neuron junction[4] | Local, graded, decremental, depolarizing generator potential (at transducer) | Sensory substance? | Ionic flux increased |
| Synapse: excitatory | Local, graded, decremental depolarization | In some, acetylcholine; in others, norepinephrine; in still others, X | Simultaneous $Na^+$ influx and $K^+$ efflux; possibly other ions |
| Synapse: inhibitory | Local, graded, decremental hyperpolarization; or stabilization of resting potential | Substance "I" | Change in ionic flux of some ions[5] |
| Synapse of vagus on vertebrate heart | Local, graded, decremental hyperpolarization | Acetylcholine | $K^+$ efflux |
| *Nitella* cell | Depolarization and spike which may be all-or-none or decremental, no overshoot | ? | Ionic flux changed |

[1] Data mainly from Katz, 1959, 1961; Florey, 1961; and Frank and Fuortes, 1961.

[2] X—an unknown substance present in extracts; electrical transmission in some crayfish synapses (Furshpan and Potter, 1959a, b).

[3] May involve entry of divalent cations in some, sodium and other ions in others, see Katz (1959b).

[4] The axon of the afferent neuron acts like any other axon.

[5] Hyperpolarization could occur by entry of $Cl^-$.

brane of the muscle cells affected tends to be held at the $K^+$ equilibrium potential which is somewhat greater than the existing resting potential (Katz, 1959b; Florey, 1966).

In the synapses of the vertebrate central nervous system some evidence of an inhibitory substance ("I" substance) has been gathered on the basis of the action of perfusates from inhibited synapses. Such substances generally act as hypopolarizers of the postsynaptic junctions (Table 23.2). An inhibitory substance is also found at synapses to crayfish muscle fibers and is thought to be γ-aminobutyric acid (Takeuchi and Takeuchi, 1966b.) It is likely that different inhibitory substances occur in different groups of animals. It is also possible that different inhibitory substances act in different parts of the nervous system of the same species. These problems are the basis of much current research (Florey, 1961; 1966).

Drugs may also inhibit the passage of the nerve impulse across a synapse. Perhaps the best known action of this type is *curare,* the active ingredient in South American arrow poison, which paralyzes animals by preventing transmission of the nerve impulse across the myoneural junction. Curare competes with acetylcholine for the receptor sites on the postsynaptic membrane and lowers the end plate potential, presumably by increasing the permeability to $Na^+$ and $K^+$. Acetylcholine liberated on stimulation of the presynaptic neuron cannot then lower the potential on the curarized postsynaptic membrane to the level required for excitation of a propagated action potential on the muscle fiber.

***Facilitation at Synapses.*** At most synapses more than one incoming impulse is required to excite a postsynaptic neuron to a propagated action potential. When each successive impulse induces an increment of local postsynaptic potential greater than the preceding one, rather than merely adding to the preceding one (as in summation), *facilitation* is said to have occurred. On the basis of the chemical transmitter theory of synaptic transmission, facilitation could be explained as the release of successively larger amounts of the transmitter substance with each successive impulse. When the transmitter substance has induced sufficient depolarization of the postsynaptic junction, a propagated axon potential is initiated (Florey, 1961). Facilitation thus prevents impulses generated by minor stimuli from spreading immediately through the nervous system and serves as a device to keep responses proportional to stimulation (Bullock and Horridge, 1965). In this way, facilitation, like inhibition, serves as a mechanism for coordination of responses of an organism. In some giant fiber systems, there are cases in which one presynaptic impulse induces one postsynaptic response, but they are unusual (Bullock, 1959; Bullock and Horridge, 1965).

## *APPENDIX*

### 23.1 CHEMICAL CHAIN POTENTIALS

When different kinds of electrolytes are used on the two sides of a membrane, the magnitude and sign of the potential developed depend upon a number of factors: whether the membrane is permeable to anion or cation; the difference in concentration, if any, of anions and cations on the two sides of the membrane; and the relative mobilities of the ions. Four cases are considered in the following paragraphs and in Figure 23.13.

1A. If a membrane permeable to cations but not to anions separates two electrolytes, e.g., NaCl and KCl, of *equal* molal concentration, a potential develops because of the difference in mobilities of the two cations (e.g., the mobility of $K^+$ is $7.61 \times 10^{-4}$ cm./sec., whereas that of $Na^+$ is $5.19 \times 10^{-4}$ cm./sec.). The solution with the cation of greater mobility becomes negative to the other solution and then the potential gradually declines as the ions distribute themselves on the two sides of the membrane. The magnitude of the potential reached depends upon the ratio of the mobility of the cations.

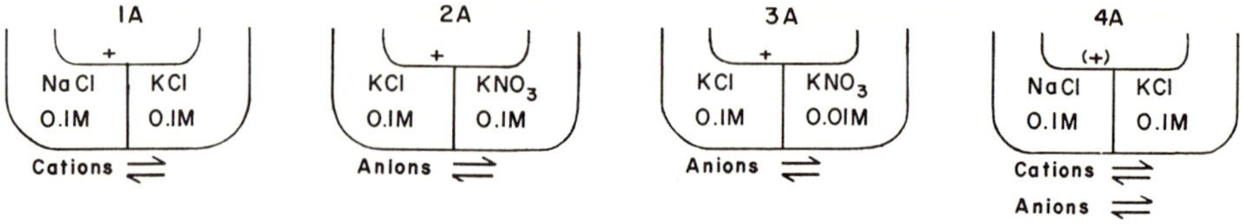


***Figure 23.13.*** Chemical chain potentials developed across a membrane separating electrolytic solutions of the same, or different, molal concentrations but with ions of different mobilities. In this case chemical "chain" potentials are developed, the sign and duration of a potential depending upon the permeability of the membrane to the ions. Case 1A illustrates what happens when the membrane is permeable to cations only. Cases 2A and 3A show what happens when the membrane is permeable to anions only. (Note that in case 3A the two solutions separated by the membrane differ in concentration as well as in the mobilities of the penetrating ions.) Case 4A shows what happens when a membrane is equally permeable to anions and cations. In case 4A the sign, which is only temporary, depends upon the relative mobilities of the ions. A fifth case might be added, namely, one in which anions *and* cations differ, with both solutions at the same concentration (e.g., 0.1M NaCl and 0.1M $KNO_3$). What would happen in this case if the membrane were permeable to anions? Cations? Both?

2A. If a membrane permeable to anions but not to cations separates two electrolytes of equal molal concentration, the relations are the converse of those holding in case 1A, and the solution with the anion of greater mobility becomes positive to the other solution. The magnitude and duration of the potential reached depend upon the ratio of the mobilities of the anions.

3A. If the salts on the two sides of the membrane are present in different concentrations and contain different kinds of cations or anions, the potential developed is the sum of the potential due to differences in concentrations plus the potential due to differences in mobilities of the ions. In this case it is assumed that the membrane is selectively permeable to anions *or* to cations, but not to both.

4A. When a membrane permeable to both anions and cations, but more permeable to cations than to anions, separates two solutions of different kinds of ions present in equal concentrations, the side with the cation of greater mobility becomes negative to the side with the cation of lesser mobility. Conversely, if the membrane is more permeable to anions than to cations, the side with the anion of greater mobility becomes positive to the one of lesser mobility. In both cases the potential then declines as fast as the various ions reach an equilibrium in their distribution.

Since chemical chain potentials are not found to contribute significantly to cell membrane potentials, quantitative formulations on the basis of relative mobilities of the various ions and of their concentration is omitted (see MacInnes, 1947).

## LITERATURE CITED AND GENERAL REFERENCES


Adrian, R. H., and Slayman, C. L., 1966: Membrane potential and conductance during transport of sodium, potassium and rubidium in frog muscle. J. Physiol. *184*:970–1014.

*Baker, P. F., March 1966: The nerve axon. Sci. Am. *214*:74–82.

Baker, P. F., and Connelly, C. M., 1966: Some properties of the external activation site of the sodium pump. J. Physiol. *185*:270–297.

Baker, P. F., Hodgkin, A. L., and Shaw, T. I., 1961: Replacement of the protoplasm of a giant nerve fiber with artificial solutions. Nature *190*:885–886.

Bennett, M. V. L., 1965: Electroreceptors in Mormyrids. Cold Spring Harbor Symp. Quant. Biol. *30*:245–262.

Bern, H. A., 1966: On the production of hormones by neurones and the role of neurosecretion in neuroendocrine mechanisms. Symp. Soc. Exp. Biol. *20*:325–344.

Bishop, G. H., 1951: The nerve impulse. *In* Transactions of the Second Macy Foundation Conference, New York.

Blinks, L. R., 1940: The relation of the bioelectric phenomena to ionic permeability and to metabolism in large plant cells. Cold Spring Harbor Symp. Quant. Biol. *8*:204–215.

Blinks, L. R., 1949: The source of bioelectric potentials in large plant cells. Proc. Natl. Acad. Sci. U.S.A. *35*:566–575.

Blinks, L. R., 1955: Some electrical properties of

large plant cells. *In* Electrochemistry in Biology and Medicine. Shedlovsky, ed. John Wiley & Sons, New York, pp. 187–212.

Brazier, M. A. B., 1960: The Electrical Activity of the Nervous System. 2nd Ed. Macmillan Co., New York.

Brazier, M. A. B., 1964: The electrical activity of the nervous system. Science *146*:1423–1428.

Brzin, M., Dettbarn, W. D., Rosenberg, P., and Nachmansohn, D., 1965: Cholinesterase activity per unit surface area of conducting membranes. J. Cell Biol. *26*:353–364.

Bullock, T. H., 1959: Initiation of nerve impulses in receptor and central neurons. *In* Biophysical Science—A Study Program. Oncley, ed. John Wiley & Sons, New York, pp. 504–514.

Bullock, T. H., 1965: In search of principles in integrative biology. Am. Zool. *5*:745–755.

Bullock, T. H., and Horridge, G. A., 1965: Structure and Function in the Nervous System of Invertebrates. W. H. Freeman & Co., San Francisco.

Chandler, W. K., and Hodgkin, A. L., 1965: The effect of internal sodium on the action potential in the presence of different internal anions. J. Physiol. *181*:595–611.

Cold Spring Harbor Symp. Quant. Biol. 1966: Sensory Receptors. Vol. 30.

Connelly, C. M., 1959: Recovery processes and metabolism of nerve. *In* Biophysical Science—A Study Program. Oncley, ed. John Wiley & Sons, New York, pp. 475–484.

Conway, E. J., 1954: Some aspects of ion transport through membranes. Symp. Soc. Exp. Biol. *8*:297–324.

Cross, S. B., Keynes, R. D., and Rybora, R., 1965: The coupling of sodium efflux and potassium influx in frog muscle. J. Physiol. *181*:865–880.

Dale, H. H., Feldberg, W., and Vogt, M., 1936: The release of acetylcholine at voluntary nerve endings. J. Physiol. *86*:353–380.

Davis, H., 1961: Some principles of sensory receptor action. Physiol. Rev. *41*:391–416.

Davson, H., 1964: A Textbook of General Physiology. 3rd Ed. Little, Brown & Co., Boston.

del Castillo, J., and Katz, B., 1967: Biophysical aspects of neuromuscular transmission. Progr. Biophys. *6*:121–170.

DeRobertis, E. D. P., 1958: Submicroscopic morphology and function of synapse. Exp. Cell Res. *5*(Suppl.)347–369.

DeRobertis, E. D. P., 1964: Histophysiology of Synapses and Neurosecretion. Macmillan Co., New York.

DeRobertis, E. D. P., 1967: Ultrastructure and cytochemistry of the synaptic region. Science *156*:907–914.

Dudel, J., and Kuffler, S. W., 1961: The quantal nature of transmission and spontaneous miniature potentials at the crayfish neuromuscular junction. J. Physiol. *155*:514–529.

Eccles, J. C., 1963: The Physiology of Synapses. Academic Press, New York.

Eccles, J. C., 1964: Ionic mechanism of postsynaptic inhibition. Science *145*:1140–1147.

*Eccles. J. C., Jan. 1965: The synapse. Sci. Am. *212*:56–66.

Eckert, R., 1965: Bioelectric control of bioluminescence in the dinoflagellate *Noctiluca*. Science *147*:1140–1145.

Erlanger, J., and Gasser, H. S., 1937: Electrical Signs of Nervous Activity. University of Pennsylvania Press, Philadelphia.

Fenn, W. O., 1950: The loss of potassium from nerves in anoxia and other conditions. J. Gen. Physiol. *33*:195–203.

Fernandez-Moran, H., and Brown, R. (eds.), 1958: Submicroscoptic organization and function of nerve cells. Exp. Cell Res. *5*(Suppl.)1–644.

Florey, E., 1961: Transmitter substances. Ann. Rev. Physiol. *23*:501–528.

Florey, E., 1966: An Introduction to General and Comparative Animal Physiology. W. B. Saunders Co., Philadelphia, Chapters 16, 18 and 21.

Frank, K., and Fuortes, M. G. F., 1957: Presynaptic and postsynaptic inhibition of monosynaptic reflexes. Fed. Proc. *16*:39–40.

Frank, K., and Fuortes, M. G. F., 1961: Excitation and conduction. Ann. Rev. Physiol. *23*:357–387.

Fuhrman, F. A., Aug. 1967: Tetrodotoxin. Sci. Am. *217*:61–71.

Furshpan, E. J., and Potter, D. D., 1959a: Transmission at the giant motor synapses of the crayfish. J. Physiol. *145*:289–325.

Furshpan, E. J., and Potter, D. D., 1959b: Slow post-synaptic potentials recorded from the giant motor fiber of the crayfish. J. Physiol. *145*:326–335.

Gabel, N. W. 1965: Excitability and the origin of life: a hypothesis. Life Sci. *5*:2085–2096.

Gaffey, C. T., and Mullins, L. J., 1959: Ion fluxes during the action potential in *Chara*. J. Physiol. *144*:505–520.

Gerebtzoff, M. A., and Schoffeniels, E., 1960: Nerve conduction and electrical discharge, Comp. Biochem. *2*:519–544.

Goldman, D. L., 1943: Potential, impedance and rectification in membranes. J. Gen. Physiol. *27*: 37–60.

Gray, E. G., and Guillery, R. W., 1966: Synaptic morphology in the normal and degenerating nervous system. Intern. Rev. Cytol. *19*:111–182.

Grundfest, H., 1957: The mechanism of discharge of the electric organ in relation to general and comparative electrophysiology. Progr. Biophys. *7*:1–85.

Grundfest, H., 1963: Impulse conducting properties of cells. *In* The General Physiology of Cell Specialization. Mazia and Tyler, eds. McGraw-Hill Book Co., New York, pp. 277–322.

Grundfest, H., 1965: Electrophysiology and pharmacology of different components of bioelectric transducers. Cold Spring Harbor Symp. Quant. Biol. *30*:1–14.

Hagiwara, S., Chichibu, S., and Naka, L. I., 1964: The effects of various ions on resting and spike potentials of barnacle muscle fibers. J. Gen. Physiol. *48*:163–179.

Hagiwara, S., and Nakajima, S., 1965: Tetrodotoxin and manganese ions: effects on action potential

of the frog heart. Science *149*:1254–1255.
Hagiwara, S., and Naka, K., 1964: The initiation of spike potential in barnacle muscle fibers under low intracellular $Ca^{++}$. J. Gen. Physiol. *48*:141–162.
Hartline, H. K., 1959: Receptor mechanisms and the integration of sensory information in the eye. *In* Biophysical Science—A Study Outline. Oncley, ed. John Wiley & Sons, New York, pp. 515–523.
*Hodgkin, A. L., 1951: The ionic basis of electrical action in nerve and muscle. Biol. Rev. *26*:339–409.
Hodgkin, A. L., 1964a: The ionic basis of nervous conduction. Science *145*:1148–1153.
Hodgkin, A. L., 1964b: The Conduction of the Nervous Impulse. Charles C Thomas, Springfield, Ill.
Hodgkin, A. L., and Huxley, A. F., 1945a: Resting and action potentials in single nerve fibres. J. Physiol. *104*:176–195.
Hodgkin, A. L., and Huxley, A. F., 1945b: Action potentials recorded from inside a nerve fibre. Nature *144*:710–711.
Hodgkin, A. L., and Huxley, A. F., 1952: Movement of sodium and potassium ions during nervous activity. Cold Spring Harbor Symp. Quant. Biol. *17*:43–52.
Hodgkin, A. L., Huxley, A. F., and Katz, B., 1949: Ionic currents underlying activity in the giant axon of the squid. Arch. Sci. Physiol. *3*:129–150.
Hodgkin, A. L., and Katz, B., 1949: The effect of sodium ions on the electrical activity of the giant axon on the squid. J. Physiol. *108*:37–77.
Hodgkin, A. L., and Keynes, R. D., 1953: The mobility and diffusion coefficient of potassium in giant axons from *Sepia.* J. Physiol. *119*:513–528.
Hoyle, G., 1962: Comparative physiology of conduction in nerve and muscle. Am. Zool. *2*:5–25.
Huxley, A. F., 1964: Excitation and conduction in nerve: quantitative analysis. Science *145*:1154–1159.
Katz, B., 1959a: Nature of the nerve impulse. *In* Biophysical Science—A Study Outline. Oncley, ed. John Wiley & Sons, New York, pp. 466–474.
Katz, B., 1959b: Mechanisms of synaptic transmission. *In* Biophysical Science—A Study Outline. Oncley, ed. John Wiley & Sons, New York, pp. 524–531.
Katz, B., Sept. 1961: How cells communicate. Sci. Am. *205*:209–238.
*Katz, B., 1966: Nerve, Muscle and Synapse. Macmillan Co., New York.
Katz, B., and Melidi, R., 1965: The effect of temperature on the synaptic delay at the neuromuscular junction. J. Physiol. *181*:656–670.
*Kennedy, D., 1962: The initiation of impulses in receptors. Am. Zool. *2*:27–43.
Keynes, R. D., 1949: The movements of radioactive ions in resting and stimulated nerve. Arch. Sci. Physiol. *3*:165–175.
Keynes, R. D., 1951: The ionic movements during nervous activity. J. Physiol. *114*:119–150.
Keynes, R. D., Dec. 1958: The nerve impulse and the squid. Sci. Am. *199*:83–90.
Keynes, R. D., and Lewis, P. R., 1951: The resting exchange of radioactive potassium in crab nerve. J. Physiol. *113*:99–114.
Krnjevic, K., 1966: Chemical transmission in the central nervous system. Endeavour *25*:8–12.
Lissmann, H. W., March 1963: Electric location by fishes. Sci. Am. *208*:50–59.
*Loewenstein, W. R., Aug. 1960: Biological transducers. Sci. Am. *203*:99–108.
Loewenstein, W. R., 1965: Facets of a transducer process. Cold Spring Harbor Symp. Quant. Biol. *30*:29–43.
Loewenstein, W. R., Terzuolo, C. A., and Washizu, Y., 1963: Separation of transducer and impulse-generating processes in sensory receptors. Science *142*:1180–1181.
MacInnes, D. A., 1947: Principles of Electrochemistry. 2nd Ed. Dover Publications, New York.
Mackie, G. O., 1965: Conduction in the nerve-free epithelia of siphonophores. Am. Zool. *5*:439–453.
Magoun, H. W. (ed.), 1959: Neurophysiology. *In* Handbook of Physiology. Sect. 1. Williams & Wilkins Co., Baltimore.
McLennan, H., 1963: Synaptic Transmission. W. B. Saunders Co., Philadelphia.
Michaelis, L., 1929: Molecular sieve membranes. Bull. Nat. Res. Council No. 69, p. 119.
Mosher, H. S., Furhman, F. A., Buchwald, H. D., and Fisher, H. G., 1964: Tarichatoxin-tetrodotoxin: a potent neurotoxin. Science *144*:1100–1110.
Nachmansohn, D. (ed.), 1950: The Nerve Impulse. Transactions of the First Macy Foundation Conference, New York.
Nakamura, Y., Nakajima, S., and Grundfest, H., 1964a: Eel electroplaques: Spike electrogenesis without potassium activation. Science *146*:266–268.
Nakamura, Y., Nakajima, S., and Grundfest, H., 1964b: The action of tetrodotoxin on electrogenic components of squid giant axon. J. Gen. Physiol. *48*:985–996.
Narahashi, T., Anderson, N. C., and Moore, J. W., 1966: Tetrodotoxin does not block excitation from inside the nerve membrane. Science *153*:765–767.
Narahashi, T., Moore, J. W., and Scott, W. R., 1964: Tetrodotoxin blockage of sodium conductance increase in lobster giant axons. J. Gen. Physiol. *47*:965–974.
Nishi, K., and Sato, M., 1966: Blocking of the impulse and depression of the receptor potential by tetrodotoxin in non-myelenated nerve terminals in Pacinian corpuscles. J. Physiol. *184*:370–386.
Ochs, S., 1965: Elements of Neurophysiology. John Wiley & Sons, New York.
Osterhout, W. J. V., 1936a: The absorption of electrolytes in large plant cells. Biol. Rev. *2*: 283–315.
Osterhout, W. J. V., 1936b: Electrical phenomena in large plant cells. Physiol. Rev. *16*:216–237.
Ozeki, M., Freeman, A. R., and Grundfest, H., 1966: The membrane components of crustacean neuromuscular systems. I. Immunity of different electrogenic components to tetrodotoxin and sanitoxin. J. Gen. Physiol. *49*:1319–1334.
Pickard, W. F., 1966: On the propagation of the nervous impulse down medulated and unmedulated fibers. J. Theoret. Biol. *11*:30–45.
Prosser, C. L., and Brown, F. A., 1961: Comparative Animal Physiology. 2nd Ed. W. B. Saunders Co., Philadelphia.
Robertson, J. D., 1955: Recent electronmiscroscope observations on the ultrastructure of the crayfish

median to motor giant synapse. Exp. Cell Res. *8*:226–229.

Rosenblith, W. A., 1961: Sensory Communication. John Wiley & Sons, New York.

Salmoiraghi, G. C., and Bloom, F. E., 1964: Pharmacology of individual neurons. Science *144*:493–499.

Schmitt, F. O., 1959: Molecular organization of the nerve fiber. *In* Biophysical Science—A Study Program. Oncley, ed. John Wiley & Sons, New York, pp. 455–465.

Shanes, A. M., 1963: Quantitative molecular approach to the permeability changes of excitation. Science *140*:51–53.

Strumwasser, F., 1965: Nervous function at the cellular level. Ann. Rev. Physiol. *27*:451–476.

Suckling, E. E., 1961: Bioelectricity. McGraw-Hill Book Co., New York.

Suckling, E. E., 1962: Bioelectricity. Am. Inst. Biol. Sci. Pamphlets, D. C. Heath & Co., Boston.

Takeuchi, A., and Takeuchi, N., 1964: The effect on crayfish muscle of iontophoretically applied glutamate. J. Physiol. *170*:296–317.

Takeuchi, A., and Takeuchi, N., 1966: A study of the inhibitory action of aminobutyric acid on neuromuscular transmission in the crayfish. J. Physiol. *183*:418–432.

Tasaki, I., and Singer, I., 1965: A macromolecular approach to the excitable membrane. J. Cell. Comp. Physiol. *66*(Suppl. 2):137–145.

Teorell, T., 1962: Excitability phenomena in artificial membranes. Biophys. J. *2*:27–52.

von Euler, U. S., 1961: Neurotransmission in the adrenergic nervous system. Harvey Lect. *55*:43–65.

Wolfe, D. E., Potter, L. T., and Richardson, K. C., 1962: Localizing tritiated norepinephrine in sympathetic axons by electron microscope autoradiography. Science *138*:440–442.

Zacks, S. I., 1964: The Motor Endplate. W. B. Saunders Co., Philadelphia.

# CHAPTER 24 / CONTRACTILITY

*Contractility,* that is, the capacity for movement, is one of the fundamental properties of the cell. Although some cells lack the power of independent motion, their cellular structures move actively in at least some stage of their development (e.g., mitosis) and probably slowly at all times. Two kinds of cell contractility are manifested: one by which the cytoplasm moves about inside the cell, e.g., cytoplasmic streaming, and the other by which the cell makes a movement relative to its surroundings. Four types of motion relative to the surroundings are recognized in living cells: ameboid, ciliary, flagellar and muscular movement or muscular contraction. Ameboid, ciliary and flagellar movements are the primary means of locomotion in protozoans, larval invertebrates and zoospores and gametes of some plants. In mature plants, movement relative to the surroundings is primarily by growth or by turgor changes. In mature animals only gametes, ciliated epithelia, wandering ameboid cells and muscle cells perform easily visible movements. Embryonic cells in animals move from places of formation to places often distantly located where the cells become parts of organs. Cells in tissue culture and healing wounds and some cancer cells also show free movement.

## CYTOPLASMIC STREAMING

*Cytoplasmic streaming* (*cyclosis*) is readily seen in a great number of cells, e.g., *Nitella,* mycelia of molds (*Neurospora*), slime molds and protozoans (*Amoeba, Paramecium*). It is a well known fact that during mitosis the contents of both plant and animal cells move actively and in a well defined pattern. Streaming is also readily seen in epidermal cells, hair cells and phloem cells of the seed plant.

Cytoplasmic streaming has long been the subject of study, and the effects of various environmental conditions upon streaming have been extensively investigated. It has been found that streaming can be initiated in some cells by visible light and also by certain chemicals. Within the biokinetic zone, streaming slows when the cell is chilled and increases when the temperature is raised. It is stopped by various stimuli, such as electric shock and ultraviolet radiation. The rate of streaming is affected by various physiological and chemical factors in much the same manner as the viscosity of cytoplasm (Ewart, 1903; Smirnova, 1955; Allen and Kamiya, 1964; Wolpert, 1965).

Although it is inherently a fascinating problem, relatively little is known of the mechanism by which cytoplasmic movement or streaming occurs. It is likened to a type of ameboid movement. Pressure is apparently "not the motile force; instead a shearing force is generated in the sol-gel interface" (Allen, 1964, p. 409). Fibrous elements are found in the cytoplasm of plant cells, and extruded endoplasm alone shows no streaming (Kuroda, 1964). It has been postulated that these fibrils are oriented in an organized fashion in the inner surface of the cortical gel and that streaming is brought about by interaction of endoplasm and these fibrils.

## AMEBOID MOVEMENT

*Ameboid movement* is the characteristic motion of amebas, hence the name. It is the motion also of various wandering ameboid cells present in the blood and body fluid of diverse animals. Ameboid wandering cells serve as scavengers and regenerate injured regions. Phagocytosis of foreign particles by ameboid white blood cells is important as a defense mechanism in multicellular animals (Karnovsky, 1962). In organogenesis during embryonic development, some cells disperse to their definitive locations in organs by ameboid movement. In rare instances, as in *Ascaris,* spermatozoa are ameboid. Ameboid motion is found rarely in plant cells; an example is the plasmodium of the slime mold, a huge multicellular ameboid mass formed by the aggregation and coalescence of vast numbers of small ameboid cells. The plasmodium, as a unit, then moves about like a giant ameba (Bonner, 1952).

Despite a considerable body of research, a satisfactory analysis of ameboid movement is still not available (Seravin, 1964). It has been shown that in an ameba, contraction is preceded by a spike potential. The potentials are variable but in some preparations they are equivalent to those shown in muscle cells (Kamiya, 1964). The most generally accepted explanation of the mechanism of ameboid movement is the conception of a reversible gel-sol transformation (Yagi, 1961; Marsland, 1964). Solation of the anterior end of the ameba accompanied by the contraction of the cortical gel at the posterior end ("tail") results in propulsion of a pseudopod (see Fig. 24.1*A, B, C*). This theory has also been used to explain the accumulation of some basic dyes (e.g., neutral red) in the interior of an ameba. According to this view, protein molecules present in the cell membrane (when in the unfolded state so that active combining sites are exposed) take up the dye molecules, but they liberate the dye molecules when they contract, at which time the combining sites are covered (Goldacre, 1964). This mechanism supposedly could serve for active transport of materials into the cell as well as for propulsion. However, it has been shown that the ectoplasm at the posterior end of an ameba has a lower viscosity than that at the anterior end, as judged by the movement of particles in the ameba during centrifugation (Allen, 1961b). Furthermore, it was found that when amebas are sucked into small glass capillaries, which are then broken under oil (or water) in such a way as to tear off the ectoplasm, vigorous streaming in the isolated endoplasm continues for an hour afterward in what is essentially the subcellular contractile apparatus of the ameba (Allen, 1961a). It would appear that the theory of ameboid movement by contraction of the posterior ectoplasm as just outlined is not in keeping with the facts (Allen and Roslansky, 1958; Allen, 1960; Griffin and Allen, 1960).

Allen and his co-workers suggest instead that an ameba literally pulls itself forward by contraction in the endoplasm at the anterior end. They found that when amebas are subjected to precisely controlled centrifugal exposures, the particulates in the axial endoplasm move only after a certain yield value has been reached, indicating structure such as is found in some colloids (see Fig. 4.5). The only regions which show a low viscosity and where the particles move with high velocity are the "shear zone," surrounding axial endoplasm, and the posterior endoplasm or "recruitment zone" (Fig.

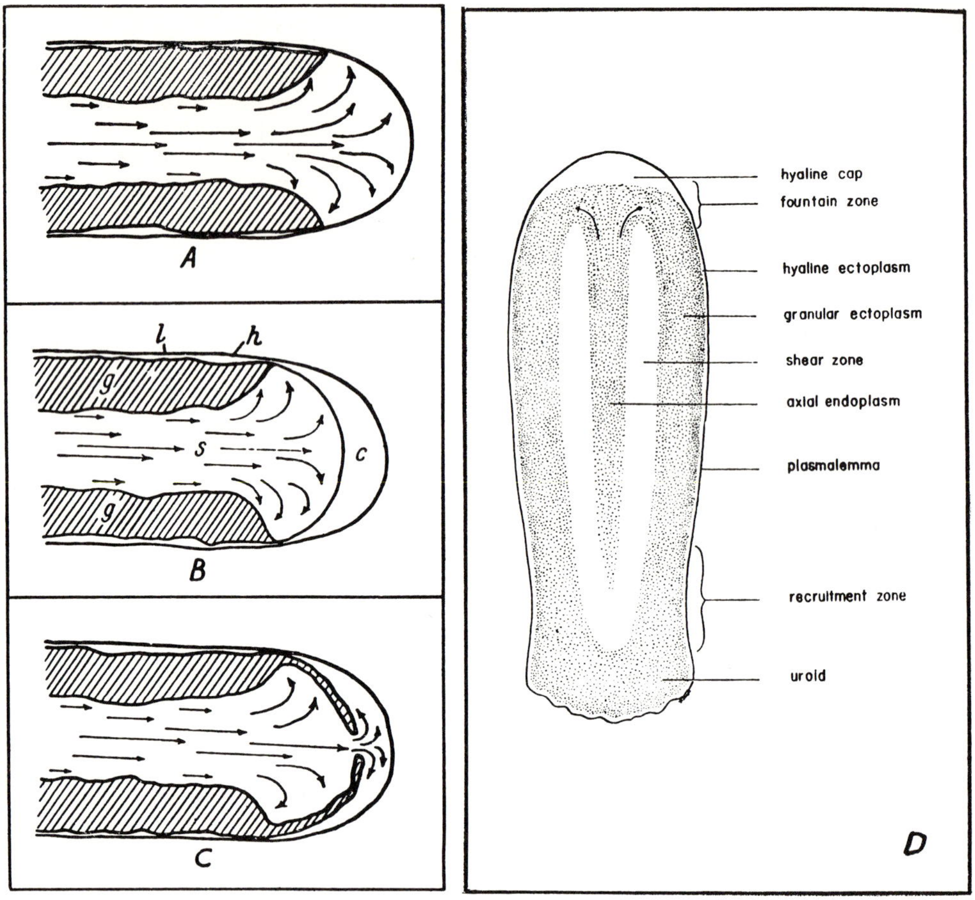


***Figure 24.1.*** *A, B* and *C,* Ameboid movement illustrating the growth of a pseudopod. The shaded region represents gelled cytoplasm. In *A* the flowing endoplasm reaches the plasmalemma (membrane) at the tip of the advancing pseudopod. In *B* there is a well defined hyaline cap (c). In *C* the hyaline cap has broken. s, Plasmasol; g, plasmagel; h, hyaline layer; l, plasmalemma. (After Mast, 1931: Protoplasma *14*:323.) *D,* A new scheme for ameboid structure and movement based on studies of cytoplasmic flow. A new terminology is proposed for regions of the cell which are differentiated with respect to consistency. *Uroid* refers to the tail process. (From Allen, 1960: J. Biophys. Biochem. Cytol. *8*:395.)

24.1*D*). The latter is so named because endoplasm is recruited from the walls of the ectoplasm in the posterior third of the cell. The endoplasm appears to contract actively as it enters the "fountain zone" (see Fig. 24.1*D*) just posterior to the hyaline cap, where the axial endoplasm becomes everted to form part of the ectoplasmic tube. The tension developed between the anchored advancing rim of the ectoplasmic tube and the movable endoplasm displaces the latter anteriorly. Continuous streaming is maintained by propagation of the contraction posteriorly along the ectoplasm at the same velocity as that at which the endoplasm advances relative to the pseudopodial tip. At the anterior end of the cell the contraction phase of the contractility cycle expresses water and solutes to form the hyaline cap (Allen, 1961a, b; Allen and Kamiya, 1964). This view has been criticized on the basis of cinematographic studies of particle movements (Rinaldi, 1963, 1964; Rinaldi and Jahn, 1963).

A third theory postulates gel-endoplasm sliding in the shear zone (Fig. 24.1). A chemical reaction is envisaged to occur between molecules in the gel and in the solated endoplasm such that the fluid endoplasm is moved forward by a mechanism perhaps akin to that proposed for muscular contraction (Allen, 1961a, b). Movement in *Amoeba proteus* also appears to involve the breakdown of ATP and glycerinated ameba models have been reactivated with ATP (Simard-Duquesne and Couillard, 1962).

A type of "ameboid" movement has been described for foraminiferan protozoans, e.g., *Allogromia,* which possess *reticulopodia,* forming a network instead of simple lobose or threadlike pseudopodia. In each of these relatively stiff filamentous reticulopodia, which may extend for many millimeters about the body of the shelled protozoan, the streaming is always in two directions simultaneously—toward and away from the body—as can be seen by watching the ever moving particles adhering to the moving surfaces on the two sides of the reticulopod (Fig. 24.2). The reticulopod is unlike a pseudopodium inasmuch as no gel-sol differences between its outer and inner cytoplasm can be detected, such as are described in a pseudopod (see earlier). When the cell is injured the reticulopod breaks into two filaments. It is postulated that this type of movement consists of shearing forces in opposite directions on the part of the two opposing faces of the filament, except at the tip where the filaments appear to be continuous as if one filament turned back upon itself (Fig. 24.2*B*). This concept has been extended to explain cytoplasmic movement in *Nitella* and other cells as well. This movement is thought to be similar, on a gross scale, to muscular contraction at the molecular level. The filaments in a reticulopodium suggest the two types of protein filaments present in muscle cells which have been demonstrated by electron microscopy to move along one another in the course of contraction (see Chapter 25) but on a different scale (Jahn and Rinaldi, 1959). Unfortunately, electron microscopic studies do not disclose any structure which supports such a mechanism (Wohlfarth-Bottermann, 1961, 1964a, b). However, membrane-bound protoplasmic fibrils containing filaments are present, several of them making up a reticulopodium. These may

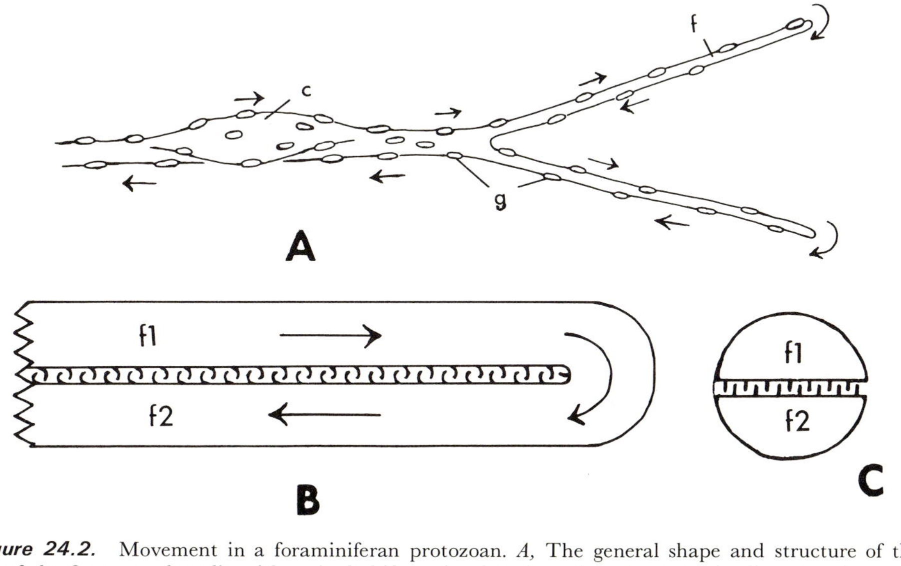


***Figure 24.2.*** Movement in a foraminiferan protozoan. *A,* The general shape and structure of the tip of one of the finer pseudopodia with a single bifurcation into branches about 1 $\mu$ in diameter. Arrows show movement of the granules (g), and of a small cytoplasmic mass (c), all of which are attached to the actively moving filament (f). *B,* Arrangement of actively moving filaments (f1, f2) of pseudopod as seen in side view. The direction of movement is shown by arrows. The moving material is assumed to be in the form of a semicylindrical filament turned back upon itself at the tip with the flat surfaces opposed. The shearing force is assumed to be between the adjacent surfaces and is designated by the short curved lines. *C,* Cross section of a filament. (From Jahn and Rinaldi, 1959: Biol. Bull. *117*:103, 111.)

play a role in the movement observed. In contraction, tension may be applied to one side of the fiber loop while compression is applied to the other, as a result of which the two sides may be forced in opposite directions. Such a model permits directive forces without requiring structures not present in the electron micrographs. Any model must be considered tentative pending accumulation of decisive data (Allen, 1964; Jarosch, 1964).

Perhaps akin to ameboid movement is the movement of slime molds, e.g., *Physarum polycephalum,* the "many-headed" one, which produces many pseudopodia, sometimes moving in several directions

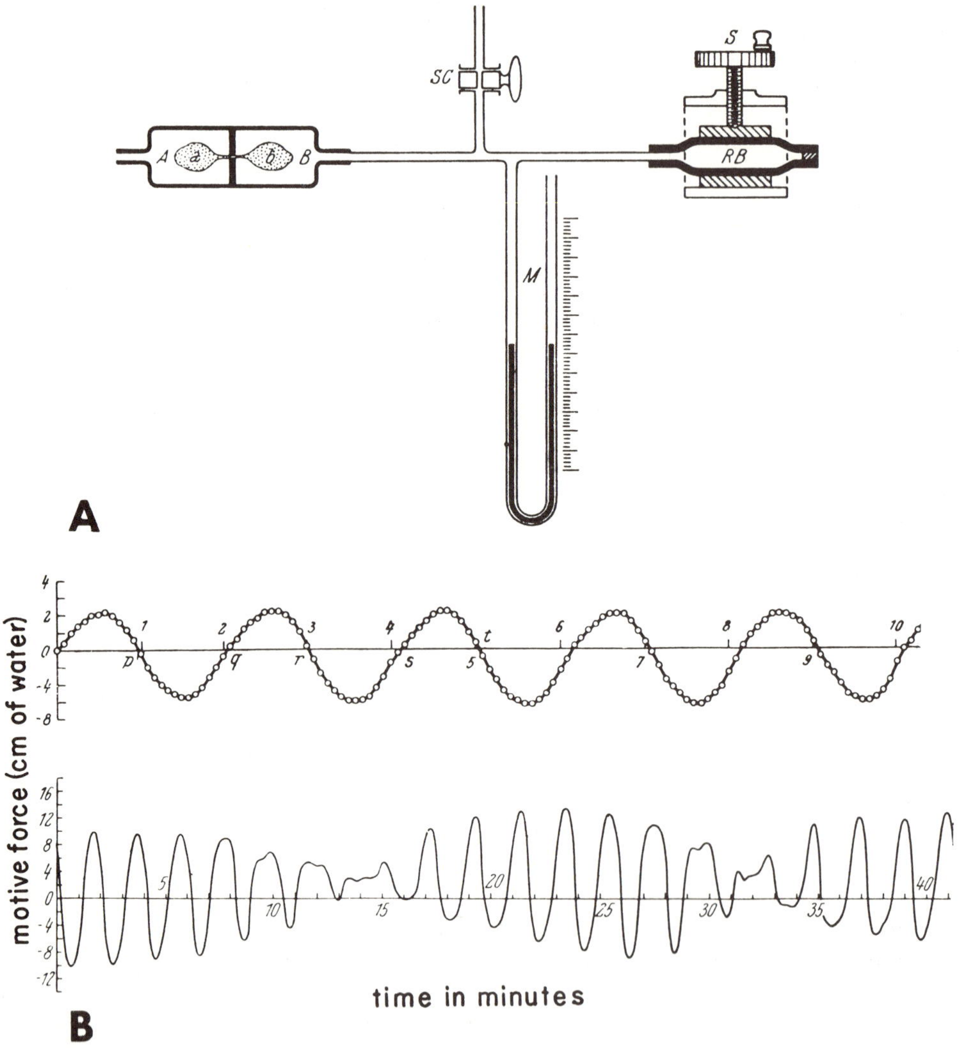


***Figure 24.3.*** *A,* Diagram showing the general arrangement for measuring the motive force of protoplasmic streaming in a myxomycete plasmodium. Note double chamber (A, B) with myxomycete (a, b) connected through a small hole between the chambers. M is a water-filled manometer, SC is a stopcock and S is a screw to control the pressure applied to the bulb RB. *B,* A recording of the motive force in centimeters of water. Note the beatlike wave pattern under normal conditions. The upper line is a recording showing all the points recorded several times a minute. The lower line shows a record in which the points are left out and the abscissa is compressed to make possible the showing of data for a longer period of time. It is possible to study the effect of temperature, anesthetics and various injurious agents on the wave pattern using this equipment. (From Kamiya, 1959: Protoplasmatologia *8*: sect. 3a, pp. 41, 43.)

(Kamiya, 1959, 1960; Rinaldi and Jahn, 1964; Steward, 1964). An interesting aspect of the movement is its pulsating nature, the pseudopodium advancing for a while, then retreating. Movement is usually greater in one direction than in the other; therefore the slime mold as a whole advances or retreats. The nature of the pulsations and the motive force has been recorded by measuring the pressure just necessary to prevent movement (see Fig. 24.3). A contractile protein has been isolated from *Physarum,* and some evidence indicates that adenosine triphosphate may play a role in movement here as in muscle cells (Loewy, 1942; Nakajima, 1964). Bursts of heat evolution synchronized with the rhythmic contractions of the plasmodium have also been recorded (Allen *et al.,* 1963).

## MOVEMENT OF PIGMENT GRANULES IN CHROMATOPHORES

Pigment cells (*chromatophores*) are of great interest in the present discussion because of the remarkable movements which may be seen in them. These cells are found in the skin of many types of animals, and enable the animal to simulate the background in many cases as in the chameleon. Although chromatophores are classified in a variety of ways with regard to the pigment they contain, there are essentially three main structural types: (1) individual reticulate cells such as those found in vertebrates (Fig. 24.4*A*, *B*); (2) larger reticulate units made up of several fused cells (syncytia), such as those found in Crustacea; and (3) multicellular

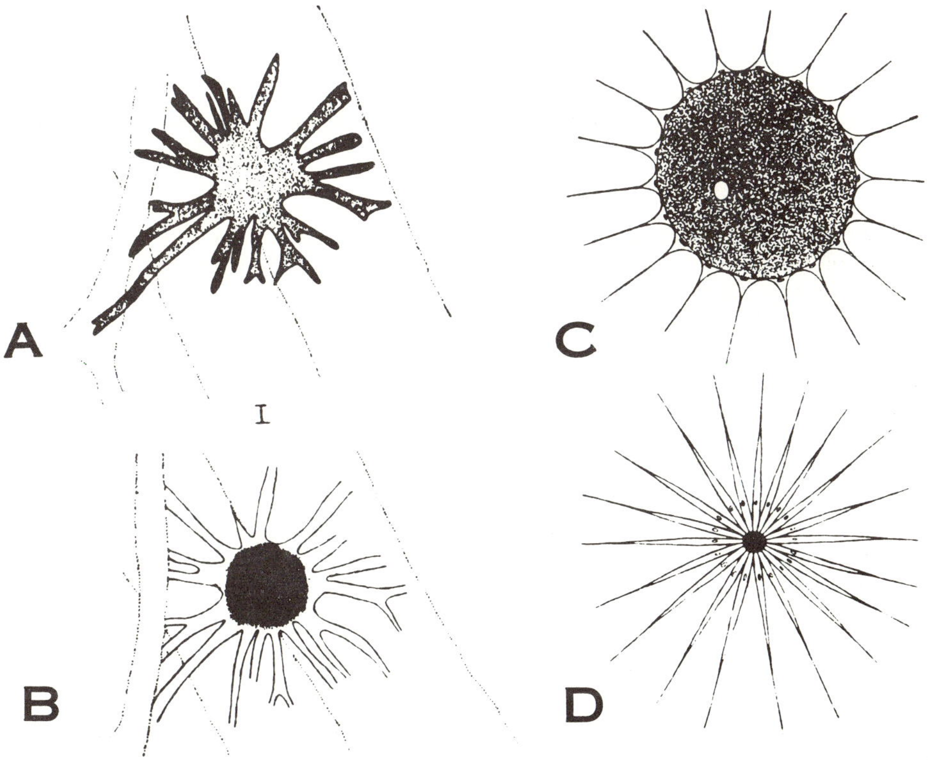


***Figure 24.4.*** Chromatophores. *A* and *B* represent camera lucida drawings of a melanophore from a scale of the killifish *Fundulus* in the expanded and contracted state. *C* and *D* represent the expanded and contracted states of the pigment in the chromatophore of the squid *Loligo.* Expansion of the pigment in this case results from contraction of the circle of muscle cells surrounding the pigment cell. (*A* and *B* from Matthews, 1931: J. Exp. Zool. *58*:471; *C* and *D* from Bozler, 1928: Z. Verg. Physiol. *7*:379.)

units consisting of a central pigment-containing cell surrounded by radially arranged muscle cells, found in some cephalopods (Fig. 24.4*C, D*; Kinosita, 1953; Nicol, 1960; Fingerman, 1959, 1966).

The pigments in chromatophores are of diverse colors and chemical nature and in most cases the pigments move inside the cell. The black and brown pigments are melanins, the red and yellow are carotenoids, the blue are protein-carotenoid complexes and the reflecting white ones are guanine crystals. The latter type of pigment granules are usually immobile and the reflecting chromatophores are deeper seated than the other kinds. A single chromatophore may contain several types of pigment granules that move independently of one another, each pigment granule individually moving from the central mass into the branching ramifications of the cell or flowing back into the central zone. Three stages of pigment spread are recognized: punctate or contracted, stellate or intermediate and reticulate or fully expanded. Some chromatophores respond directly to light, electricity and hormones. For example, the squid chromatophore showed a response much like that obtained for the twitches and summation of a muscle fiber (Hill and Solandt, 1934).

The mechanism by which the pigment moves in the reticulate chromatophore cell is not fully understood. The pigment granules appear to flow into the arborizations or channels of the cells. The channels seem to be permanent structures clearly visible, at least in tissue cultures of such cells, even when the cell is in punctate form and the pigment is in the fully contracted stage. However, it has been stated that in some cases the channels collapse when the punctate state is assumed (Matthews, 1931; Fingerman, 1966).

Experiments show that application of pressure to the chromatophores of the killifish *Fundulus* brings about pigment dispersion in the chromatophores. Since the pressure required to do this (7000 to 8000 pounds per square inch) usually induces a sol-like condition in cytoplasm, it was thought that perhaps dispersion of pigment is caused by solation, whereas, conversely, the contraction of pigment is caused by gelation. To test this hypothesis Marsland (1944) centrifuged the chromatophores in the punctate state and was unable to stratify the pigment granules, even with a force 70,000 times gravity, but he was easily able to stratify the pigment granules in the reticulate state with less centrifugal force. However, if solation and gelation do accompany pigmentation, the exact method by which this achieves the movement of the pigment granules is not clear.

Kinosita (1963) found that the pigment granules in chromatophores of the medaka fish *Oryzias* are negatively charged. Application of electric fields therefore causes displacement of the granules (electrophoresis). He postulated that when the agent which produces the contracted state (*punctate* state) of the chromatophore pigment is applied, the central part of the chromatophore becomes more positively charged than the periphery of the cell, attracting the negatively charged particles to the center. On removal of the stimulus and loss of charge difference (or possibly, the development of negativity in the central portion of the chromatophore) the pigment particles once again disperse and reach a *reticulate* state. Other theories are considered by Fingerman (1966).

In cephalopods the pigment cell is in an expanded state when the radial muscle cells are contracted, and in a punctate state when these muscle cells are relaxed. The muscle cells are under the control of the nervous system. The chromatophores of crustaceans are coordinated by hormones whereas those of fishes are controlled by both hormones and nerve impulses (see Fingerman, 1966).

## CILIARY AND FLAGELLAR MOVEMENTS

*Cilia,* the numerous hairlike motile processes on a cell, occur in all animal phyla except in nematode worms and arthropods, but they are especially prominent in members of the protozoan class Ciliata and in free-swimming larvae of many marine animals (see Fig. 24.5). A *flagellum* is a whiplike motile organelle, present singly or in small numbers at one end of a cell. It is present in members of the protozoan class Flagellata, in the spermatozoa of most animals and plants and in the flagellated cells of sponges and coelenterates, zoospores of algae and so forth.

Cilia and flagella have many characteristics in common, but flagellar motion is often much more complicated than ciliary motion. The flagellum may be used like an oar, or it may move only at the tip—like a propeller, whereas the cilium is always stroked like an oar although not always in the same manner (Fawcett, 1961; Sleigh, 1962). Cilia and flagella serve many purposes and their movements propel the organism. They usually beat continuously and act as independent effectors. However, in a few invertebrate embryos they are probably under nervous control since their movement may be stopped upon stimulation of the embryo. In ciliates they are thought to be coordinated by a neuromotor center near the mouth since destruction of the fibers connecting the center to the cilia results in uncoordinated movement (Taylor, 1920). However, Okajima (1966) reports coordinated movement in *Euplotes* even after complete dissection of the neuromotor fibers. He attributes coordination to a quick spread of electrical "events" on the surface of the cell.

Numerous studies of ciliary and flagellar movements have been made (Fawcett, 1961; Sleigh, 1962; Pitelka, 1963), and a great deal of information is available on the effect of various environmental conditions on the motion of cilia and flagella. It is interesting to note that on electron micrographs of all cells tested each cilium and flagellum has a characteristic pattern of nine circumferential and two central paired filaments, the latter filaments being somewhat more deli-

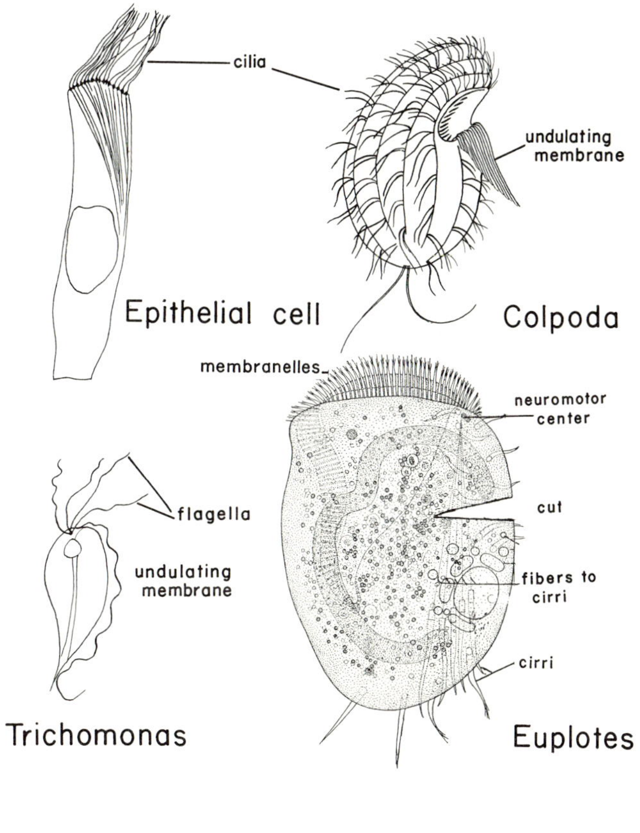


***Figure 24.5.*** Cilia and flagella of various types of cells. The epithelial cell is from the hepatic duct of a snail. (After Gray, 1928: Ciliary Movement. Macmillan Co., New York). In *Colpoda duodenaria,* a ciliate, note the fusion of cilia to form an undulating membrane and the orderly arrangement of cilia in rows. In *Trichomonas,* from the intestine of a termite, note especially the undulating membrane made by fusion of one flagellum to the cell membrane and the supporting skeletal rod (the endostyle). *Euplotes* is a ciliate with cilia fused to form membranelles at the top and along the groove leading to the mouth and into cirri, or walking appendages, on the ventral surface of the animal. Each *cirrus* is innervated by a fiber from the *neuromotor center* near the anterior end of the animal. A cut across this set of fibers causes uncoordinated movement of the cirri to which the fibers connect. A similar cut elsewhere in the body has no effect on coordination. The large sickle-shaped body is the macronucleus. Note also the numerous vacuoles. In this species the contractile vacuole is formed by fusion of many small vacuoles. (After Taylor, 1923: J. Exp. Zool. *37*:259; and 1920: Univ. Calif. Publ. Zool. *19*:403.)

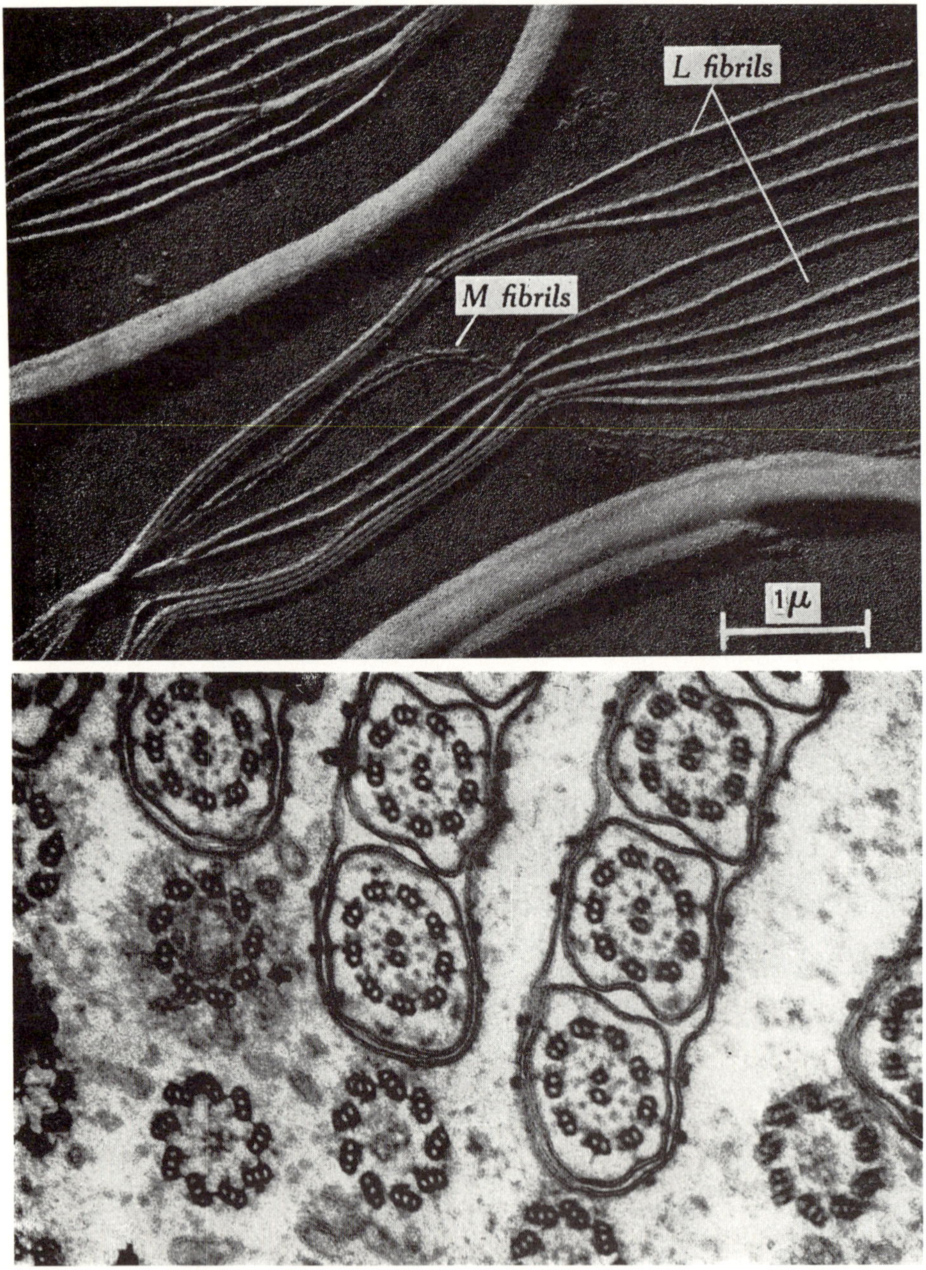


***Figure 24.6.*** Electron micrographs of flagella. *Upper,* Intact (large cable) and dismembered tails of cock spermatozoons, showing the eleven fibrils of which the flagellum is composed. M fibrils refers to the two, median, more delicate fibrils; L fibrils refers to the nine, larger, lateral fibrils. (From Grigg and Hodge, 1949: Austral. J. Sci. Res. *B2*:271.) *Lower,* Cross sections of flagella of *Pseudotrichonympha* (a flagellate from a termite). Note that the median fibrils are single while the lateral fibrils are double. Sections through basal granules are seen below the flagella. (From Gibbons and Grimstone, 1960: J. Biophys. Biochem. Cytol. *7*:697.)

cate than the former (Fig. 24.6). Attempts have been made to interpret ciliary movement in terms of principles of muscular contraction. Experiments show that glycerin-extracted cilia and flagella beat many times in a solution of ATP (Hoffmann-Berling, 1958a, b; Bishop, 1962; Gibbons and Rowe, 1965; Hammerle and Cameron, 1966). Some evidence has been presented that the two inner filaments of a cilium transmit excitation and the nine outer filaments are the seat of ATP splitting (Hayashi, 1961). Cilia have been isolated from ciliates and studied in suspension (Culbertson, 1966). The protein extracted from the cilia acts as an ATPase (Gibbons and Rowe, 1965).

In the flagellate *Noctiluca* the beat of the flagellum is accompanied by an action potential (Hisada, 1957; Kinosita *et al.*, 1964a, b; Eckert, 1965).

Akin to the cilia are the contractile myonemes seen, for example, as longitudinal fibers in the stalk of *Vorticella* (also found in *Spirostomum* and a few other ciliates). Studies on such myonemes indicate that they show many similarities to muscular movement (Hou and Brücke, 1930). Also, the energy source seems to be the same as that for muscular contraction (Hoffmann-Berling, 1964).

## MUSCULAR MOVEMENT

Contractility has reached its highest development in muscle cells in which, perhaps, it is most exaggerated and diagrammatic. Being adapted to their special function, the muscle cells are modified in many ways. In general, the units of structure are elongate and spindle shaped, and they consist either of single cells (e.g., visceral muscle cells) or of "composite cells" called *muscle fibers* (skeletal muscle fibers). In each case the muscle cells, or fibers, possess *contractile myofibrils* (fine elongate structures) and one or more nuclei and are enclosed in a membrane called the *sarcolemma*. Although some studies have been made upon single muscle fibers (Ramsey and Street, 1941; Nastuk, 1953; Street and Ramsey, 1965), most investigations of muscle contraction deal with the entire muscle or with muscle-nerve preparations, even though the major concern is the nature of cellular contractility. It will become clear that the behavior of muscle is explainable only in terms of the cells or fibers of which it is composed.

Three kinds of vertebrate muscle cells are recognized: *visceral* or smooth, *cardiac* or heart and *skeletal* or body muscles. Only the high points of the contractile properties of these various types of muscle cells can be discussed here.

## CONTRACTION OF VISCERAL MUSCLE CELLS

The contents of the gut and various other tubular internal organs are propelled by the contraction of the visceral muscle cells lining their walls.

Contraction of visceral muscle cells shows some interesting and unique features. In structure, visceral muscle consists of tiny, elongate spindle-shaped cells, each with a single nucleus and with interconnecting bridges (Thaenert, 1957; see Figs. 24.7 and 24.8). Minute longitudinal myofibrils (units of contraction) have been described in these cells and clearly seen in electron micrographs (Hanson and Lowy, 1957). The cells are arranged in the wall of the tubular organ in circular and longitudinal layers, which oppose one another in action, in such a way that contraction of the longitudinal layer shortens an organ and contraction of the circular layer lengthens it. The single cells are too small to be studied individually; therefore, for experiments on contraction, organs (e.g., stomach, intestine) or parts of them, containing visceral musculature have been used.

A striking property of the contraction of visceral muscle cells of the digestive tract is their rhythmic "beat," which occurs even when all nervous connections are severed if the preparation is aerated and bathed in a balanced salt solution. The tendency to contract is therefore inherent or *myogenic,* that is, the contractions are initiated in the muscle cells themselves. This rhythmic contraction leads to *peristalsis,* or waves of contraction, at intervals of about 1 minute. A peristaltic wave is highly coordinated; coordination is achieved by a slow spread of excitation (action potential). Preceding a contraction, an action potential can be measured at a "segment" (a small portion between two regions of contracted circular muscle) of the intestine, and after it has contracted, the electrical wave excites another segment, causing it to contract, until the contraction wave has passed the full length of the intestine. Long before this contraction wave has completed its course, another series of contraction waves has been initiated.

The resting potentials of visceral muscle cells are somewhat unstable and every once in a while a cell depolarizes and "fires," that is, sets off an action potential. This starts depolarization in other cells, serving as pacemaker (Fisher, 1944).

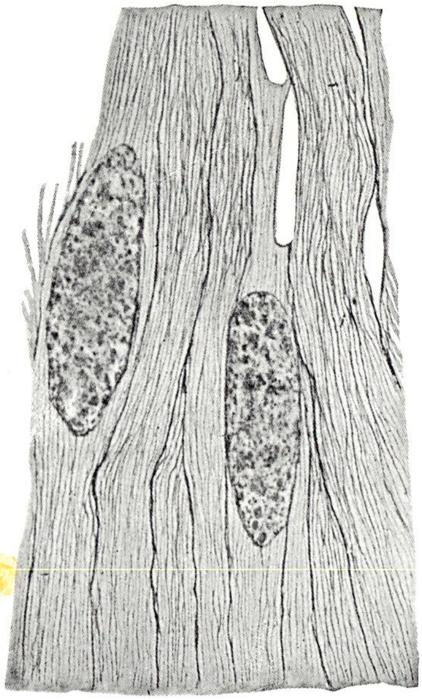

**Figure 24.8.** Smooth muscle cells of the intestine of *Triton vulgaris,* a salamander. Note the myofibrils (×1430). (After Levy, from Bloom and Fawcett, 1968: A Textbook of Histology. 9th Ed. W. B. Saunders Co., Philadelphia.)

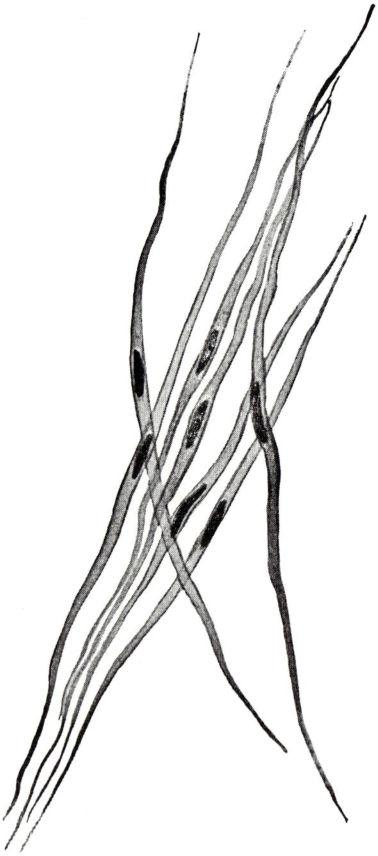

**Figure 24.7.** Isolated smooth muscle cells from the wall of the stomach of a cat (×220). (From Bloom and Fawcett, 1968: A Textbook of Histology. 9th Ed. W. B. Saunders Co., Philadelphia.)

Although contraction can be induced and maintained in visceral muscle cells without nervous connections, it is generally controlled by the nervous system through double innervation, that is, by excitatory and inhibitory (autonomic) nerve fibers. When denervated visceral muscle cells are stimulated with electricity, they remain contracted for many seconds before they relax. The time for a complete cycle of contraction and relaxation is called a *muscle twitch.* The twitch time for visceral muscle is generally long and the *refractory period* (that is, the period during which the muscle cells cannot be induced to contract) is relatively short. This means that a sustained state of contraction can be maintained even by a small number of periodic impulses from the nervous system (Fisher, 1944).

Visceral muscle cells maintain *tonus*

(the mild state of sustained contraction of a muscle brought about by contraction of a few of its units) over a wide range of levels of contraction. For example, the cells of the stomach distend as the organ receives food, yet the same tonus is exerted for each distention. As food is used up, the cells of the stomach contract and maintain a fairly constant tension on the food that remains. Visceral muscle cells are capable of contraction down to within a few per cent of the lengths of the maximal stretch, a contractility far greater than that of any other muscle cells.

Visceral muscle cells have relatively low sensitivity to stimuli for contraction. For example, the chronaxie for bladder musculature of a rat is about 50 to 100 ms., and for uterine muscle of a guinea pig it is 1500 to 2000 ms., as compared with the chronaxie for skeletal muscle of man, which is 0.15 to 0.5 ms.

The contraction of visceral muscle cells is influenced by the action of hormones. This field of research is very active, but it is beyond the realm of this discussion. It will suffice to mention that, in the vertebrate, epinephrine (adrenaline) inhibits the contraction of visceral muscle cells and acetylcholine enhances it, and female sex hormones and the pituitary gland make possible powerful contractions of the uterine musculature (Turner, 1966).

## CONTRACTION OF CARDIAC MUSCLE CELLS

An invertebrate heart may be a very simple organ—merely a contractile tube resembling a more active visceral muscle, as in the annelid worms. In fishes the heart is still a simple tube, but in all other vertebrates it is a fairly complex structure. The vertebrate heart muscle, viewed in sections under the light microscope, appears to be composed of an interlacing network of muscle fibers and was thought to be a syncytium, i.e., to form a single multinucleate unit. However, electron micrographs show that the network is composed of separate cells (fibers) which are in intimate contact rather than in a continuous syncytium (Sjöstrand *et al.,* 1958). The separate fibers can be demonstrated also by treatment with strong alkali. Each cell is indicated by a nucleus, and each mass of cytoplasm is separated from the next by membranes called intercalated discs.

The vertebrate heart, even when removed from the body, still continues to beat. The beat is, therefore, said to be *myogenic,* although a particular balance of ions is necessary for maintenance of this activity (see Chapters 8 and 13).

Even isolated tissue cultures of chick cardiac muscle cells (dispersed by action of proteolytic enzymes) are capable of contraction and are found to be of two kinds, "leaders" and "followers." The leaders initiate the beat, and the followers contract only if they are in contact with one another, so that a single rhythmic beat may follow (Harary, 1962; Harary and Sato, 1964; Smith and Berndt, 1964). The transmembrane potentials of such cells are like those of muscle cells generally, the resting potential being 59 mv. and the action potential 71 mv. (Lehmkuhl and Sperelakis, 1963).

Although the entire heart is capable of beating or contracting, a gradient of excitability is found in its parts, and the most irritable portion is called the *pacemaker.* In the frog heart the pacemaker is the sinus venosus (see Fig. 24.9); in the mammal heart the sinus venosus is represented by a vestige, the *sinoauricular node,* which is the pacemaker. From the pacemaker the wave of excitation (depolarization) sweeps over the heart, initiating the beat in the auricles, from which it spreads by way of the auriculoventricular node and the *bundle of His* to the ventricle (see Fig. 24.10). Modified muscular fibers (Purkinje fibers in the bundle of His) in the ventricle carry the action potential at the rate of about 5 meters per second to various portions of the heart, making possible a coordinated contraction (Ruch and Patton, 1965).

Initiation of the heart beat, as seen in the frog, is not the prerogative of the

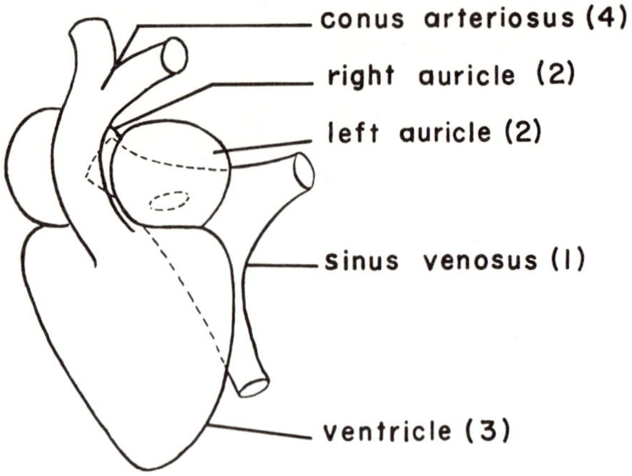


***Figure 24.9.*** Sequence (shown by numbers) of contraction of the parts of a frog's heart.

sinus venosus alone, since if the sinus is cut from the auricles, or the auricles from the ventricle, each piece, after it has recovered from the shock of the operation, will continue to beat. However, only the rate of beat of the portion containing the sinus venosus equals that of the intact heart; the isolated auricles and ventricles beat more slowly. The severed tip of the ventricle shows the least ability to initiate a contraction. Apparently the pacemaker is the most irritable portion of the heart and therefore the first to contract. If it is affected locally, by the application of a cold or hot rod, the rate of the heart beat is lowered or increased, respectively.

If a ligature is tied between the pacemaker and the rest of the heart, the ventricle no longer beats at its regular rate (see Fig. 24.11). If the ventricle, physiologically isolated in this manner, is stimulated by electrical means, a threshold stimulus evokes contraction to the full extent characteristic of the normal tonus level of the heart. Increasing the stimulus does not increase the degree of contraction. The cardiac muscle contraction therefore obeys the all-or-none rule, under the conditions of these experiments.

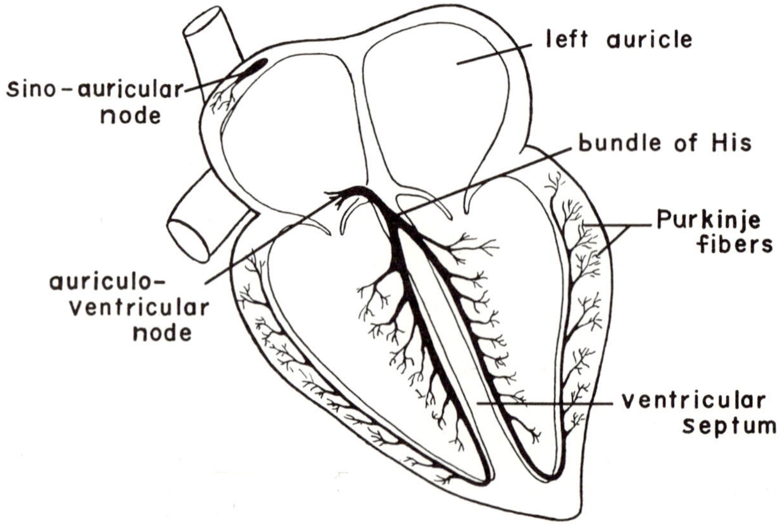


***Figure 24.10.*** Diagram of a mammalian heart, showing the bundle of His and the Purkinje fibers.

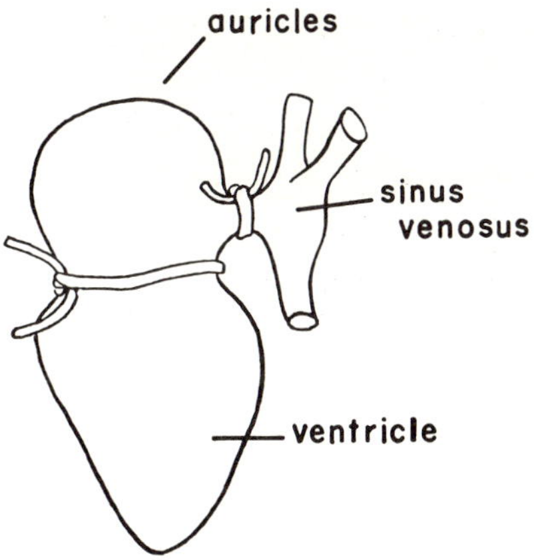


***Figure 24.11.*** Isolating the sinus venosus from the auricles and the auricles from the ventricle by means of ligatures.

The refractory period for cardiac contraction is fairly long, about 0.3 second in man (0.25 second absolute, 0.05 second relative—Guyton, 1966). It depends upon the size of the animal, being shorter for the rapidly beating heart of a small animal. It is shorter also in a warm blooded animal than in a cold blooded one of the same size. Because of the long refractory period, the heart muscle fibers cannot be tetanized on continued stimulation. Electrical stimulation is effective only if the heart muscle fibers are in the relative refractory phase (see Fig. 24.12), when an extra contraction may be induced. The extra contraction is followed by a compensatory phase (refractory time) during which contraction cannot be induced by further stimulation.

The long refractory period of the heart probably results from permeability changes and ionic fluxes at the cell membranes. The $Na^+$ influx is not cut off as rapidly as in nerve, nor does the $K^+$ efflux rise above the steady state condition. Repolarization is therefore slower (Katz, 1959).

The force of the isolated heart beat depends upon the stretch to which the fibers are subjected, e.g., upon the degree to which the heart is filled with blood. This relation is known as *Starling's law of the heart.* The heart muscle fibers maintain tonus at various levels of contraction; they can contract from any one of these levels. Such a response is adaptive and of importance in maintaining a good blood supply, since the heart may respond to varied demands upon it by a change in the strength or frequency of contraction. It is questionable whether such control operates in the animal in which neural controls are probably of primary importance (Ruch and Patton, 1965).

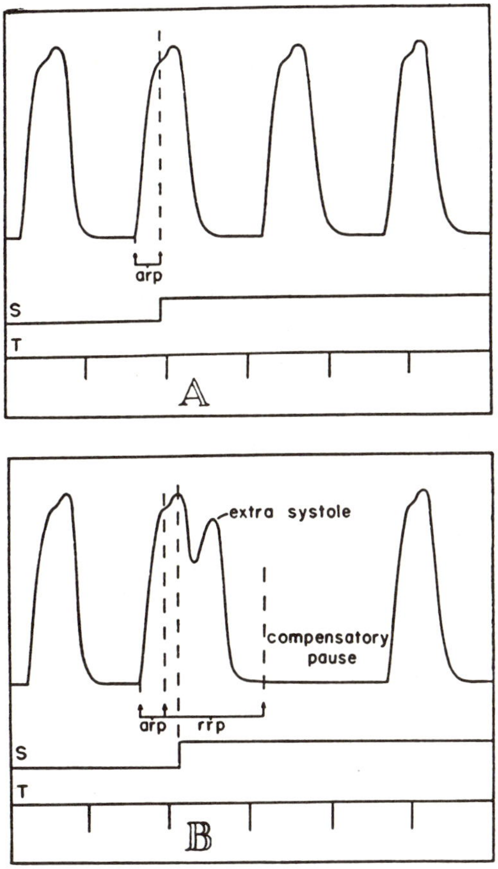


***Figure 24.12.*** *A,* Demonstration of the absolute refractory period (arp) of the heart. Each spike is a kymograph record of a heart beat, the peak being the height of ventricular systole, after which the heart begins to relax. When the heart is contracting, a single electric shock (*S*) has no effect. T is the time in seconds. *B,* Demonstration of the relative refractory period (rrp). When a single electric shock is sent into the ventricle at about the peak of ventricular contraction or soon afterwards, an extra systole appears, subnormal in strength and followed by a compensatory phase. T is the time in seconds. (From Amberson and Smith, 1948: Outline of Physiology. 2nd Ed. Williams & Wilkins, Baltimore.)

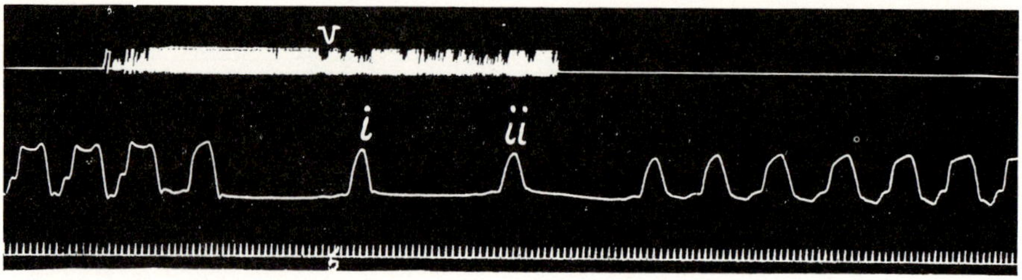


*Figure 24.13.* A kymograph record of the inhibition of heart beat by stimulation of the vagus. The signal magnet record at v shows the tetanic stimulation of the vagus. At i and ii the heart "escapes" from the inhibitory influence, indicating the strength of the tendency to beat. The lower markings are 0.2 second intervals. (Data from classroom experiment.)

The beat of the vertebrate heart, although self-initiated, is regulated by inhibitory and excitatory nerve fibers from the autonomic nervous system. Stimulation of the inhibitory nerve fibers, which come by way of the tenth cranial nerve (the vagus), slows or stops the heart beat (see Fig. 24.13). Loewi demonstrated that something diffusible appears in the heart after stimulation of the vagus and that the perfusate from such a heart, when circulated to another heart, will inhibit the beat of the second heart also (see Fig. 24.14). For studies in this field he received the Nobel prize in 1936. The diffusible substance, or "vagusstoff," was later identified as acetylcholine (equation 22.1), application of which slows or stops the heart beat. The mechanism by which this occurs has been considered in Chapter 23.

A heart with the beat slowed by acetylcholine, or by brief stimulation of the vagus nerve, usually beats more forcibly for a while and then resumes a normal rate of beating.

The beat is resumed because the acetylcholine is decomposed to choline and acetic acid. Decomposition is accelerated greatly by a highly active enzyme, choline esterase, which is widely distributed, especially in nervous and muscle cells. Sometimes, even if vagal stimulation is continued, the heart may occasionally "escape" from the inhibitory influence of the vagus and beat more forcibly than normally, thus indicating the strong tendency of the heart muscle cells to continue rhythmic contraction.

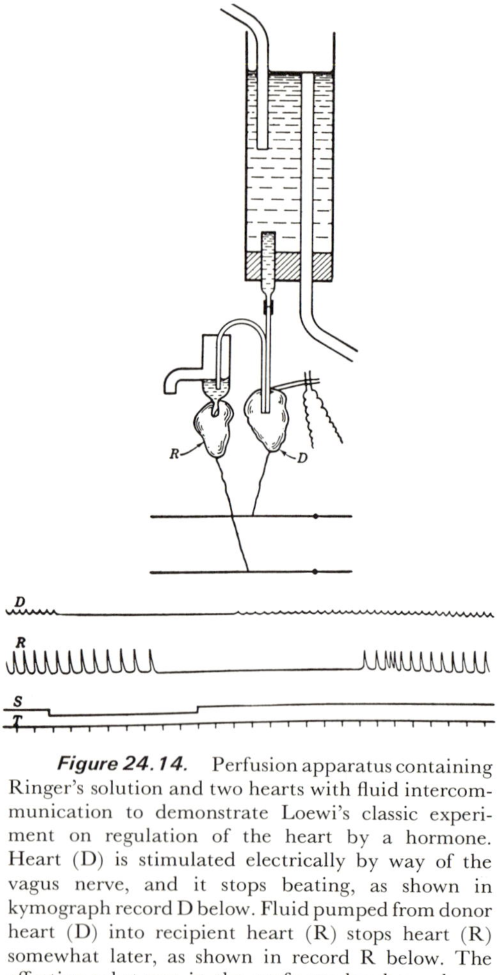


*Figure 24.14.* Perfusion apparatus containing Ringer's solution and two hearts with fluid intercommunication to demonstrate Loewi's classic experiment on regulation of the heart by a hormone. Heart (D) is stimulated electrically by way of the vagus nerve, and it stops beating, as shown in kymograph record D below. Fluid pumped from donor heart (D) into recipient heart (R) stops heart (R) somewhat later, as shown in record R below. The effective substance in the perfusate has been shown to be acetylcholine. S (below) indicates the time at which the electric stimulus is applied to the vagus. T represents the time in seconds for the 3 records above it. (From Scheer, 1953: General Physiology, John Wiley & Sons, New York.)

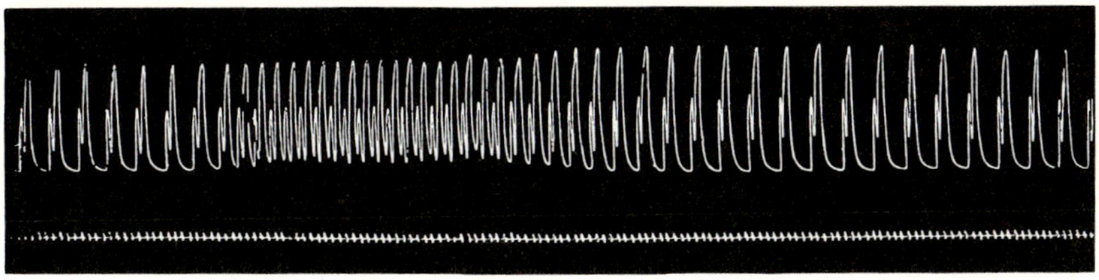

*Figure 24.15.* Kymograph record showing acceleration of the heart of a frog by stimulation of the sympathetic nerve. (From Zoethout and Tuttle, 1952: Textbook of Physiology. 11th Ed. C. V. Mosby Co., St. Louis.)

Stimulation of the excitatory nerve fibers, which come by way of the sympathetic nerve, accelerates the heart beat (Fig. 24.15). It is found that the perfusate of a heart excited in this manner contains sympathin, probably chiefly norepinephrine (differing from epinephrine by lack of a $CH_3$ radical). Application of epinephrine to the surface of the heart and its injection into the heart are familiar methods for stimulating heart muscle contractions (Vogt, 1934).

Contractions of some invertebrate hearts share many of the properties of contraction of the vertebrate heart. There are other invertebrate hearts, however (like that of the adult king crab *Limulus*), in which the heart contraction is *neurogenic;* that is, the heart will not beat when severed from nervous connections. It is also interesting to note that in the larval stage of this animal, the heart beat is myogenic, suggesting that the neurogenic heart beat is the result of a more specialized structure than a heart in which the contractions are myogenic. Studies on vertebrate and invertebrate hearts and their control are discussed in books on human and comparative physiology (Prosser and Brown, 1961; Ruch and Patton, 1965; Florey, 1966).

## CONTRACTION OF SKELETAL (BODY) MUSCLE FIBERS

Locomotion is made possible by skeletal muscle fibers, and the development of these highly differentiated structures indicates the premium placed upon speed in the animal world.

Individual muscle cells of the body musculature of lower invertebrates are relatively simple cells much like visceral muscle cells of vertebrates and show similar rhythmic pulsations (Pantin, 1952). It is only in some higher invertebrates (cephalopod mollusks and insects) and in the vertebrates that skeletal muscle cells become much elongated and show striking transverse striations. Such muscle cells probably represent highly specialized structures in which certain properties of contractility have been exaggerated while other properties, such as myogenicity, have been lost.

The skeletal muscles are made up of multinucleate muscle fibers containing contractile myofibrils, across which run transverse *striations,* enclosed in a *sarcolemma* (see Fig. 24.16). The contractile fibrils of a muscle fiber practically fill it, but a certain amount of cytoplasm (*sarcoplasm*) surrounds the fibrils, and in the cytoplasm are found the spindle-shaped nuclei, the entire fiber being a multinucleate unit. It is interesting that such multinucleate fibers have arisen not only in the vertebrates, but also in several invertebrate groups (Prosser and Brown, 1961). This suggests an evolutionary advantage of a unit of this type in which the contractile fibrils may become much longer than would be possible in a single cell. A long muscle fiber has an advantage in the propagation of excitation for contraction and in the mechanical work performed.

The skeletal muscle fibers of vertebrates are innervated only by excitatory nerve fibers (single innervation). An efferent neuron innervates a small number of muscle fibers, each branch of the distal

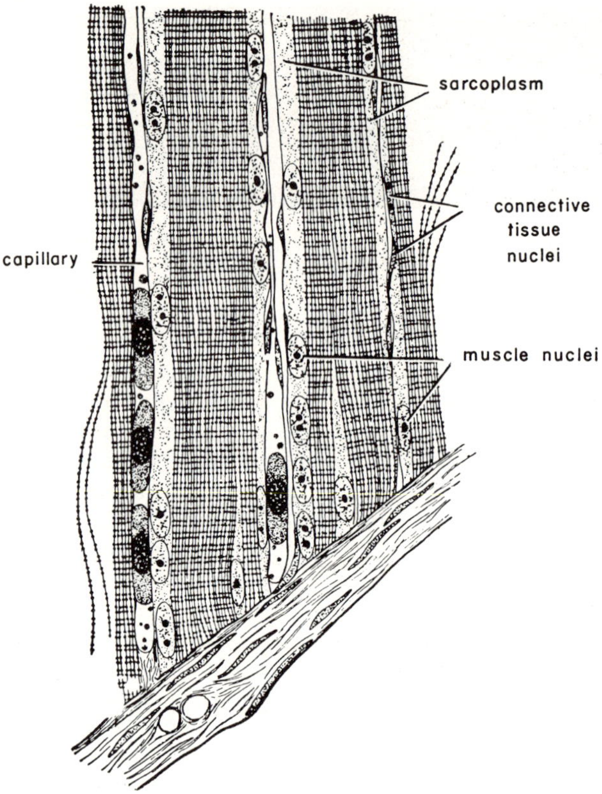


***Figure 24.16.*** Longitudinal section of striated muscle from the sucker *Catostomus.* (From Dahlgren and Kepner, 1925: Principles of Animal Histology. Macmillan Co., New York.)

portion of the axon attaching to a muscle fiber by a muscle end plate or *myoneural junction* (see Figs. 24.17 and 23.12). The motor neuron and the muscle fibers innervated by it respond as a unit and constitute the *motor unit.* A large number of motor units make possible a fine gradation of force exerted—from delicate movements to powerful sustained ones—depending upon the number of motor units called into action. When the nerve impulse traveling on the efferent neuron reaches the myoneural junction, it first induces a local excitatory state (local potential) on the muscle fiber near the myoneural junction. If this change is of sufficient magnitude (above threshold), an action potential (Chapter 22) sweeps the muscle fiber, following which the muscle fiber contracts (Fig. 24.18).

The muscle fibers may also be directly stimulated by heat, light, chemicals, pressure or electricity, and the excitation is always accompanied by action potentials. When a muscle fiber is electrically stimulated with a current below threshold, a local excitatory state develops around the cathode. This state of excitation grows with continued stimulation, but only when the current is above threshold and the time of application is sufficiently long will the excitation become propagated and induce contraction of the muscle fiber. Muscle fibers, like nerve fibers, maintain a resting potential of about 50 to 100 mv., the outside being positive to the inside. This sign of the potential reverses during excitation. Evidence indicates that during the development of the propagated action potentials the same type of ionic changes occur in muscle fibers as in nerve fibers. The action potential travels over the muscle fibers at the rate of 1 or 2 meters per second, which is much slower than the rate of the action potential in nerve fibers. Just how the electrical excitation induces muscular contraction is not known

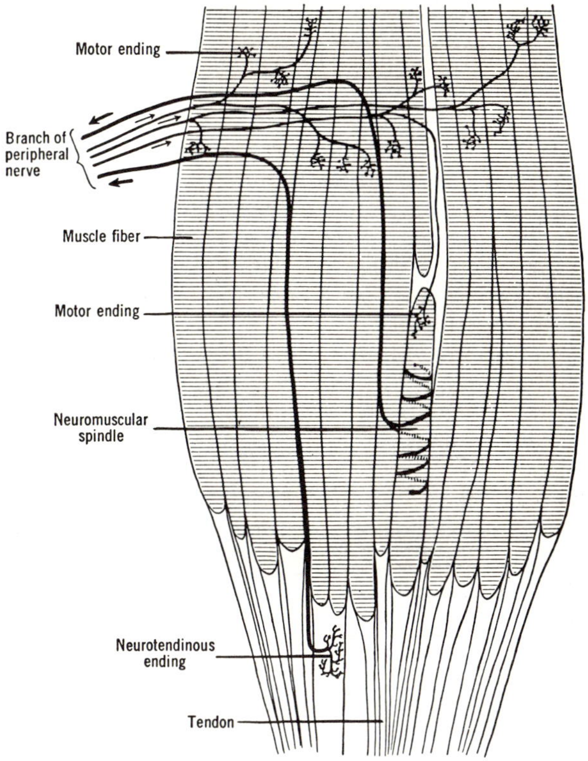


***Figure 24.17.*** Relations between neurons and muscle fibers. Efferent neurons end in myoneural junctions (muscle end-organs). Two afferent neurons are also shown, one from a muscle spindle and the other from a tendon receptor. (Adapted from Gardner, 1968: Fundamentals of Neurology. 5th Ed. W. B. Saunders Co., Philadelphia.)

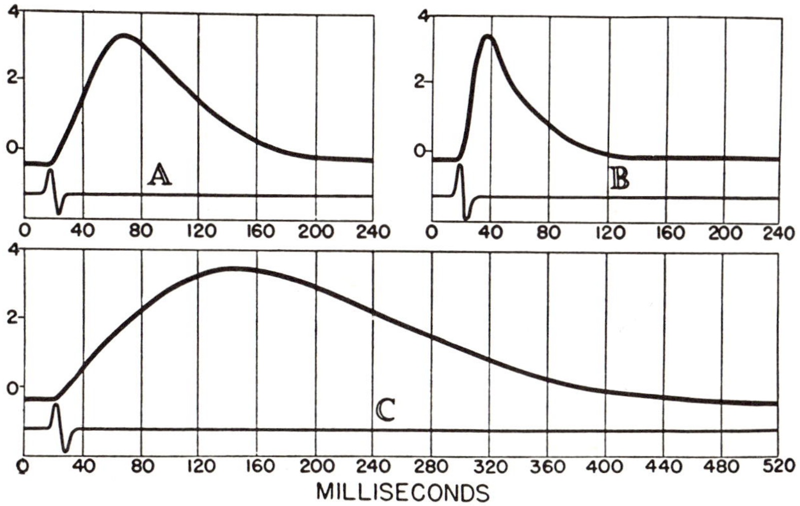


***Figure 24.18.*** Action potentials (lower record in each example) and muscle twitches in three muscles. *A,* The mainly white-fibered gastrocnemius; *B,* The wholly white-fibered internal rectus; *C,* the red-fibered soleus. (From Amberson and Smith, 1948: Outline of Physiology. 2nd Ed. Williams & Wilkins, Baltimore.)

with certainty. However, recent work implicates liberation of calcium ions from the *sarcoplasmic* (endoplasmic) *reticulum* in excitation-contraction coupling (Weber *et al.,* 1964; Peachey, 1965). This will be discussed in Chapter 25.

The chronaxie for muscle fibers is comparable to that for the motor neurons innervating it. The strength-duration curves resemble those of nerve fibers, only the absolute values being different. Muscle fibers are thus highly excitable systems and have many properties in common with nerve fibers.

If a muscle fiber is stimulated with a current above rheobase but of duration insufficient to excite, contraction does not occur. If the stimuli are frequently repeated, they may summate, each subliminal excitation producing a local potential which may be added to others until the local potential sets off a propagated action potential (spike) which, in turn, is followed by a contraction. When single muscle fibers are isolated by removal or destruction of surrounding muscle fibers, and then stimulated with a graded series of shocks, the individual muscle fiber always contracts as a unit and to the same extent. When individual muscle fibers in an intact muscle are labeled by spraying with mercury and then excited with electrodes small enough to contact only single fibers, individual fibers are seen to contract completely or not at all. These experiments show that single muscle fibers in the intact muscle, like single neurons, obey the all-or-none law under the conditions of the experiment.

An interesting but little understood exception to the all-or-none principle occurs in some frog muscle fibers of the back, which are innervated by small nerves. Stimulation of these nerves causes individual muscle fibers to contract sometimes to a greater degree than at other times. Analysis has shown that a local excitatory state in such muscle fibers is accompanied by a local contraction, the degree of local contraction depending upon the magnitude of the local excitatory potential. The purpose served by local contractions in these muscle fibers is unknown (Ruch and Patton, 1965). On the other hand, in some invertebrates, such as the crustacea, local contractions are the main method for obtaining graded responses, since the number of motor units is very small (Wiersma, 1952).

When a muscle is stimulated with electricity, by varying the voltage, it is possible to find a minimal stimulus threshold below which no response is observed. As the voltage is increased above the threshold, the force of contraction of the muscle increases in degree, until a maximal contraction is observed (see Fig. 24.19; Hayashi, 1961; Ruch and Patton, 1965). The same graded response is found upon stimulation, by way of the nerve, of an intact muscle in the animal, and is the consequence of the response of successively larger numbers of motor units, as stimulation increases in strength. A muscle has many motor units, each of which has a characteristic threshold. In a population of fibers some have low, some high, and some intermediate thresholds, the sensitivity being distributed about a mean. When a muscle is stimulated, gradation is obtained by calling into action all those fibers for which the stimulus is above threshold (Hayashi, 1961; Ruch and Patton, 1965).

A single electric shock induces a single *twitch* of a muscle fiber. The twitch is a laboratory phenomenon primarily useful for analysis of the properties of muscular contraction. The record of a muscle twitch may be spread out by being recorded on a rapidly moving drum of a kymograph, thereby making possible a study of the time relations in the contraction and relaxation of the muscle fiber. The entire cycle for the "form" curve of the gastrocnemius muscle of a frog occupies

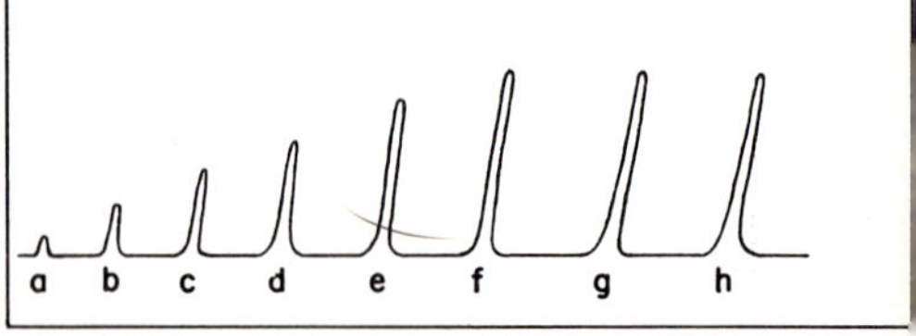


***Figure 24.19.*** Graded response of a skeletal muscle to increasing stimulation. A maximal response is reached at f. (Data from classroom experiment.)

about 0.1 second (Fig. 24.18). Immediately after stimulation, a latent period of about 0.0075 second elapses before the contraction begins. During this period the action potential, beginning at the point of stimulation, sweeps the muscle. Contraction occupies about 0.04 second, relaxation about 0.05 second. The exact times vary with the type of muscle fiber and the temperature. After stimulation muscle fibers become *refractory* for a period lasting a fraction of a millisecond to several milliseconds, during which they cannot be excited again. Since the twitch of a frog gastrocnemius muscle reaches its full contraction 0.04 second after excitation and relaxes only after another 0.05 second, it is evidently excitable long before it has relaxed. Consequently, it is possible for a skeletal muscle fiber to maintain continuous tension if stimulated frequently enough within a given time. A sustained contraction of this type is called *tetanus* (see Fig. 24.20).

If the stimuli are spaced so that they reach the muscle after it has begun to relax, incomplete tetanus is obtained (Fig. 24.20). The record shows that a relaxing muscle may be induced to contract from any degree of relaxation, as might be expected from the shortness of the refractory period.

Contractions of the muscle fibers in the living animal are normally initiated by continuous volleys of discharges along motor neurons. The impulses are spaced about a hundredth of a second apart, leading to tetanic contractions of some muscle fibers in various muscles of the body. By this means, muscular *tonus* is maintained. This tonus is responsible for maintenance of posture and for keeping the musculature in readiness for action. In a muscle stimulated to tetanic contraction, the degree of contraction increases during the course of the volley of impulses, and the final state of contraction achieved is greater than the initial state. This indicates that a muscle kept contracted does more work per stimulus than one just stimulated from a state of rest (Fig. 24.20), a behavior in apparent contradiction to the all-or-none principle. The fact is, individual skeletal muscle fibers will also contract to a greater extent if stimulated when already contracted than if stimulated from a state of rest, thus showing increased capacity for work under these conditions. The all-or-none law, then, holds under a given set of circumstances.

When a stimulated muscle is unable to move its load and does not shorten, the muscular contraction is said to be *iso-*

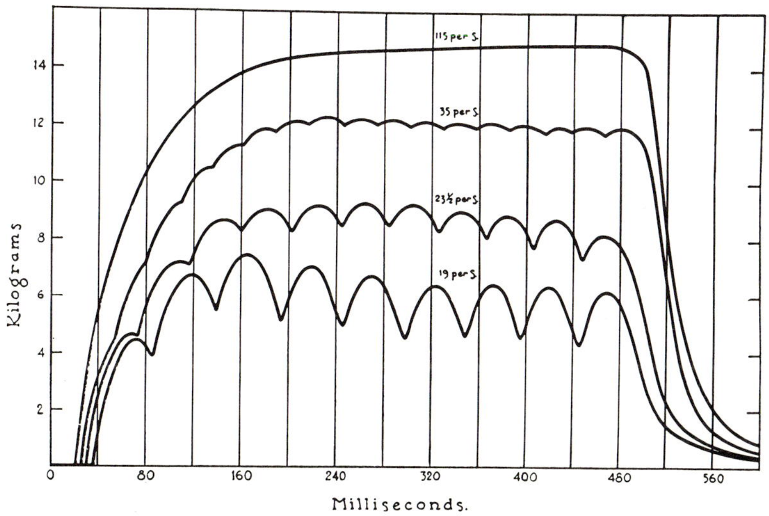


***Figure 24.20.*** Partial and complete tetanus caused by varying rhythms of stimuli applied to the gastrocnemius muscle. (From Amberson and Smith, 1948: Outline of Physiology. 2nd Ed. Williams & Wilkins, Baltimore.)

*metric* ("same length"). Although tension (force) is developed and the muscle tonus is increased, no mechanical work is done and all the expended energy appears as heat. For recording the maximum tension exerted by a particular muscle, use may be made of a lever and an attached spring (see Fig. 24.21). Muscles used for maintaining posture are contracting isometrically. On the other hand, if a muscle is able to move a load, that is, when the resistance offered by the load is less than the tension developed, the muscle shortens and performs mechanical work. The same tonus is maintained, irrespective of the degree of contraction; hence this contraction is said to be *isotonic* ("same tonus") (see Fig. 24.22). The muscles used in lifting a load are contracting isotonically. In lifting a load, an average of 20 to 25 per cent of the total chemical potential energy expenditure appears as mechanical work, the remainder being dissipated as heat. The muscle is therefore more efficient than the steam engine and comparable in efficiency to the gasoline engine (Ruch and Patton, 1965).

A muscle fiber is capable of maintaining tetanus for a considerable period, the limit being set by development of *fatigue* and resultant failure of contraction. Fatigue is also observed if individual twitches are induced continuously, by electricity, for example, although a frog leg muscle contracts about 12,000 times in $3\frac{1}{3}$ hours before fatiguing (Labhardt, 1927). As a muscle fatigues, its degree of contraction decreases gradually because some of the component fibers fail to contract; consequently, the capacity for work of the muscle as a whole decreases (see Fig. 24.23). The cause of muscular fatigue is not known with certainty. It is thought to be due to a number of changes occurring concurrently: first, the exhaustion of the substrate serving as the immediate source of energy and, second, the accumulation of waste products. Experiments have demonstrated that removal of wastes by continually washing a thin sheet of muscle in balanced salt solution delays fatigue. In any event, if stimuli are applied at such intervals that the muscle fibers have a chance to recover, they may be able to contract for hours without fatigue. Apparently, recovery is complete, or nearly so, in about 1 second. Intact muscle fibers, stimulated *in vivo* through a nerve, fatigue much more rapidly (e.g., frogs are fatigued after 29 to 82 jumps in 1 to 2 minutes (Galperin *et al.,* 1934). In this case the impulse passes by way of the myoneural junction and thence to the muscle fiber. The myoneural junction fatigues quite readily, even though the muscle fibers of a fatigued animal will continue to contract on direct stimulation. The nerve fiber fatigues even more slowly than the muscle fiber it innervates. The myoneural junction apparently acts as a safety valve,

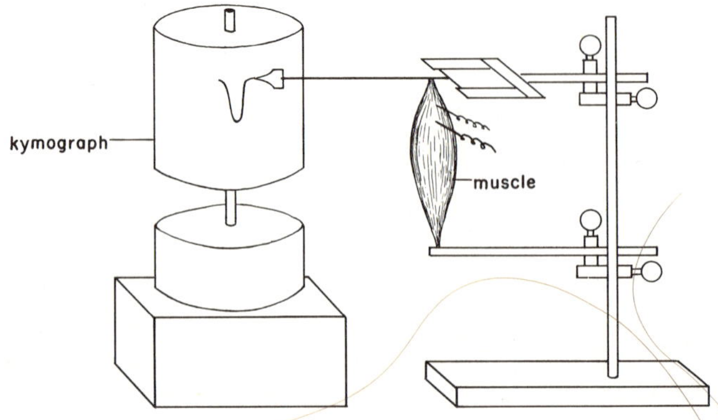


***Figure 24.21.*** Isometric muscle contraction. The muscle is attached to a lever which is soldered to a steel spring held tightly in the clamp. It therefore exerts tension with only a minimum of contraction which is magnified by the lever.

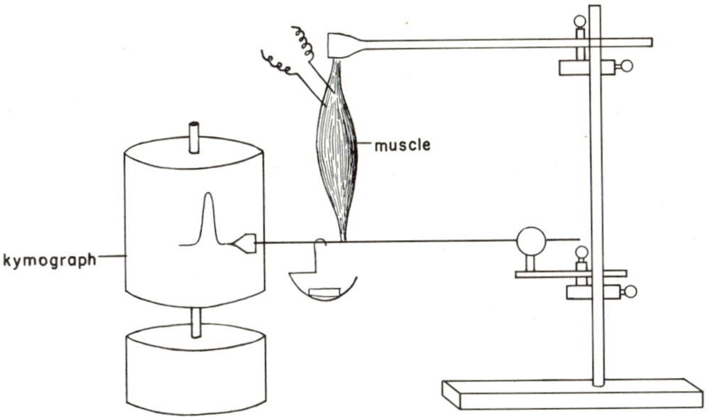


***Figure 24.22.*** Isotonic contraction. The muscle pulls an easily movable lever and shortens as it does so.

preventing injury to the muscle fiber by too prolonged a period of contraction.

Relaxation of a skeletal muscle fiber could conceivably be a passive process, since skeletal muscles occur in pairs which are opposite in action, e.g., an extensor and a flexor. Studies of an isolated muscle show that without any pull or friction upon it, a muscle after contraction relaxes and extends itself, but is not capable of doing work. Thus, if the muscle is placed in a bath of mercury it cannot become extended because of the weight of the mercury. Nonetheless, evidence indicates that relaxation of the muscle fiber is a spontaneous process (Ruch and Patton, 1965), and not merely the absence of the active state (see Chapter 25).

*Contracture* is the name given to a particular muscular contraction which is sustained and reversible, and which occurs with the liberation of heat but shows no propagated spike potentials. It is brought on by various stimuli, e.g., strong electric currents, heat, acids, narcotics, potassium, citrate, caffeine, acetylcholine, nicotine, veratrine and many other drugs. When veratrine, for example, is dropped on a single muscle fiber, it gives rise to several propagated spikes, after which a prolonged negativity (after-potential) develops, accompanied by a contracture. Spontaneous normal repolarization of the membrane of the muscle fiber, or anodal stimulation, brings the contracture to an end. Contracture precedes fatigue in a muscle which is continuously stimulated (Guyton, 1966).

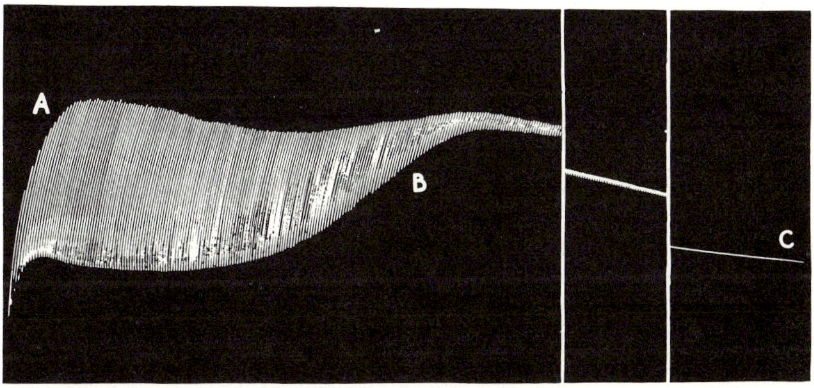


***Figure 24.23.*** Kymograph records of fatigue curve, showing staircase (A), contracture (B) and complete fatigue (C). (From Zoethout and Tuttle, 1952: Textbook of Physiology. 11th Ed. C. V. Mosby Co., St. Louis.)

If skeletal muscle fiber is stretched by weights, it may be considerably extended. After removal of the weights it returns to its original length, indicating that it is highly elastic. The elasticity is accounted for in part by the contractile fibrils and in part by the sarcolemma membrane which encloses each muscle fiber, since the membrane remains elastic, even when the contents of the muscle fiber have been expressed, although not as elastic as the whole fiber (Ramsey and Street, 1941).

## WORK AND HEAT RELATIONS IN MUSCLE CONTRACTION

A muscle can contract with shortening (isotonic contraction) or without shortening (isometric contraction). The muscle fibers can exert maximal tension (force) when they do not shorten; for example, a striated muscle can exert a tension of 3 kg. per square centimeter of cross section. Even though a muscle fiber in this state does no visible work in the physical sense of moving an object through a given distance, it needs energy to maintain the state of tension, and this energy appears as heat (Ruch and Patton, 1965).

When a muscle fiber shortens it exerts less tension than when it does not shorten. The tension also decreases as the speed of shortening increases. This decrease in tension is not due to changes in internal viscosity of the muscle cells (friction), since a muscle fiber shortening rapidly should liberate more heat than one shortening slowly over the same distance, yet this is not found to be the case. Therefore, the decrease in tension is presumably caused by other factors (A. F. Huxley, 1957, 1964; H. E. Huxley, 1960).

A shortening muscle fiber liberates heat in proportion to the distance of shortening, and not to the speed. But the "shortening heat" is independent of the load on the muscle fibers—a muscle fiber produces no more and no less "shortening heat" (i.e., at the time of shortening) when it lifts a large load than when it lifts a small one. Since the muscle fiber carrying a larger load does more work than the one carrying a small load, the total energy expended by the muscle fiber increases with the load. A muscle cell therefore shows some remarkable properties, and the chemical reactions which provide the energy for contraction must consequently be controlled not only by change in length of the muscle fiber, but also by the tension placed on it during this change (Hill, 1960).

Forced lengthening of a muscle fiber during contraction decreases the energy liberated. A. V. Hill (1960) has gathered evidence to indicate that the chemical processes which are induced in a muscle fiber by applying a stimulus are reversed by the application of external mechanical work (force working through a distance), although it is not possible to store energy in the muscle fiber in this way.

## *LITERATURE CITED AND GENERAL REFERENCES*


Allen, R. D., 1960: The consistency of amoeba cytoplasm and its bearing on the mechanism of ameboid movement. J. Biophys. Biochem. Cytol. *8*:379–397.

Allen, R. D., 1961a: Ameboid movement. *In* The Cell. Brachet and Mirsky, eds. Academic Press, New York, Vol. 2, pp. 135–216.

Allen, R. D., 1961b: A new theory of ameboid movement. Exp. Cell. Res. *8*(Suppl.):17–31.

Allen, R. D., 1964: Cytoplasmic streaming and locomotion in marine Foraminifera. *In* Primitive Motile Systems in Cell Biology. Allen and Kamiya, eds. Academic Press, New York, pp. 407–432.

Allen, R. D., and Kamiya, N. (eds.), 1964: Primitive Motile Systems in Cell Biology. Academic Press, New York.

Allen, R. D., Pitts, W. R., Jr., and Brault, J., 1963: Shuttle-streaming: synchronization with heat production in slime mold. Science *142*:1485–1487.

Allen, R. D., and Roslansky, J. D., 1958: An anterior-posterior gradient of refractive index in the ameba and its significance in ameboid movement. J. Biophys. Biochem. Cytol. *4*:517–524.

Ambrose, E. J., 1965. Cell movements. Endeavour *24*:27–32.

Bishop, D. W., 1962: Reactivation of extracted sperm cell models in relation to the mechanism of motility. *In* Spermatozoan Motility. American Association for the Advancement of Science, Washington, D. C., pp. 251–268.

Bonner, J. T., 1952: Morphogenesis. Princeton University Press, Princeton, N. J.

Bourne, C., ed., 1960: Structure and Function of Muscles. 2 vol. Academic Press, New York.

Culbertson, J. R., 1966: Physical and chemical properties of cilia isolated from *Tetrahymena pyriformis.* J. Protozool. *13*:397–406.
Davson, H., 1964: A Textbook of General Physiology. 3rd Ed. Little, Brown & Co., Boston, Chapter 4.
Eckert, R., 1965: Bioelectric control of bioluminescence in the dinoflagellate *Noctiluca:* I. Specific nature of triggering events; II. Asynchronous flash initiation by a propagated triggering potential. Science *147*:1140–1145.
Ewart, A. S., 1903: Protoplasmic Streaming in Plants. Clarendon Press, London.
Fawcett, D., 1961: Cilia and flagella. *In* The Cell. Brachet and Mirsky (eds.). Academic Press, New York, Vol. 2, pp. 218–270.
Fingerman, M., 1959: The physiology of chromatophores. Int. Rev. Cytol. *8*:175–210.
Fingerman, M., 1966: Chromatophores. Physiol. Rev. *45*:296–339.
Fisher, E., 1944: Vertebrate smooth muscle. Physiol. Rev. *24*:467–490.
Florey, E., 1966: An Introduction to General and Comparative Physiology. W. B. Saunders Co., Philadelphia.
Freed, J. J., 1965: Microtubules and saltatory movements of cytoplasmic elements in cultured cells. J. Cell. Biol. *27*:29A.
Galperin, L., Okun, M., Simonson, E., and Sirkina, G., 1934: Beiträge zur Physiologie der Ermüdung. Arbeitsphysiologie *8*:407–423.
Gibbons, I. R., and Rowe, A. J., 1965: Dynein: a protein with adenosine triphosphatase activity from cilia. Science *149*:424–426.
Goldacre, R. J., 1964: On the mechanism and control of ameboid movement. *In* Primitive Motile Systems in Cell Biology. Allen and Kamiya, eds. Academic Press, New York, pp. 237–255.
Gray, J., 1928. Ciliary Movement. Cambridge University Press, London.
Griffin, J. L., and Allen, R. D., 1960: The movement of particles attached to the surface of amebae in relation to current theories of ameboid movement. Exp. Cell Res. *20*:619–622.
Guyton, A. C., 1966: Textbook of Medical Physiology. 3rd Ed. W. B. Saunders Co., Philadelphia p. 64.
Hammerle, P., and Cameron, F., 1966: ATP-induced ciliary movements in glycerinated *Tetrahymena pyriformis.* J. Protozool. *13*:26 (Abstract).
Hanson, J., and Lowy, J., 1957: Structure of smooth muscles. Nature *180*:906–909.
Harary, I., May 1962: Heart cells *in vitro.* Sci. Am. *206*:141–152.
Harary, I., and Sato, E., 1964: Studies *in vitro* on single beating heart cells. V. Change in ATP-induced contractions of extracted models. Biochim. Biophys. Acta *82*:614–616.
Harvey, E. N., and Marsland, D. M., 1932: The tension at the surface of *Amoeba dubia* with direct observations on the movement of cytoplasmic particles at high centrifugal speeds. J. Cell. Comp. Physiol. *2*:75–87.
Hayashi, T., Sept. 1961: How cells move. Sci. Am. *205*:184–204.
Hill, A. V., 1960: Production and absorption of work by muscle. Science *131*:897–903.
Hill, A. V., and Solandt, D. Y., 1934: Myograms from the chromatophores of *Sepia.* J. Physiol. *83*:13–14.
Hisada, M., 1957: Membrane resting and action potentials from a protozoan, *Noctiluca scintillans.* J. Cell. Comp. Physiol. *50*(Suppl. 1):57–71.
Hoffmann-Berling, H., 1958a: Physiologie der Bewegungen und Teilungsbewegungen tierischer Zellen. Fortschr. der Zool. (N. F.) *11*:142–207.
Hoffmann-Berling, H., 1958b: Der Mechanismus eines neuen, von der Muskelkontraktion verschiedenen Kontraktionszyklus. Biochim. Biophys. Acta *27*:247–255.
Hoffmann-Berling, H., 1963: Mechanismus von Zubewegungen. Naturwiss. *50*:277–282.
Hoffmann-Berling, H., 1964: Relaxation of fibroblast cells. *In* Primitive Motile Systems in Cell Biology. Allen and Kamiya, eds. Academic Press, New York, pp. 365–375.
Holtzer, H., 1959: Some further uses of antibodies for analyzing the structure and development of muscles. Exp. Cell Res. *7*(Suppl.)324–343.
Hou, C. L., and Brücke, E., 1930: Reizvesuche an Vorticellen (Alle oder nichts Gesetz). Dekrement, der Erregnungsleitung in der Narkose, Chronaxie. Arch. Ges. Physiol. *226*:411–417.
Huxley, A. F., 1957: Molecular structure and theories of contraction. Progr. Biophys. *7*:255–318.
*Huxley, A. F., 1964: Muscle. Ann. Rev. Physiol. *27*:131–152.
Huxley, H. E., 1960: Muscle Cells. *In* The Cell. Brachet and Mirsky, eds. Academic Press, New York, Vol. 4, pp. 365–481.
*Huxley, H. E., Dec. 1965: The mechanism of muscular contraction. Sci. Am. *213*:18–27.
Jahn, T. L., and Rinaldi, R., 1959: Protoplasmic movement in the foraminiferan *Allogromia laticollaris;* and a theory of its mechanism. Biol. Bull. *117*:100–118.
Jarosch, R., 1964: Screw-mechanical basis of protoplasmic movement. *In* Primitive Motile Systems in Cell Biology. Allen and Kamiya, eds. Academic Press, New York, pp. 599–622.
Kamiya, N., 1959: Protoplasmic Streaming. Protoplasmatologia VIII.3a. Heilbrunn and Weber, eds. Springer-Verlag, Vienna.
Kamiya, N., 1960: Physics and chemistry of protoplasmic streaming. Ann. Rev. Plant Physiol. *11*: 323–340.
Kamiya, N., 1964: The motive force of endoplasmic streaming in the ameba. *In* Primitive Motile Systems in Cell Biology. Allen and Kamiya, eds. Academic Press, New York, pp. 257–277.
Kamiya, N., and Kuroda, K., 1958: Measurement of the motive force of the protoplasmic rotation in *Nitella.* Protoplasma *50*:144–148.
Karnovsky, M. L., 1962: Metabolic basis of phagocytic activity. Physiol. Rev. *42*:143–168.
Karpovich, P. V., 1965: Physiology of Muscular Activity. 6th Ed. W. B. Saunders Co., Philadelphia.
Katz, B., 1959: The nature of the nerve impulse. *In* Biophysical Science—A Study Program. Oncley, ed. John Wiley & Sons, New York, pp. 466–474.
Kelso, J. M., and Turner, J. S., 1955: Protoplasmic streaming in *Tradescantia.* I. The effects of indolacetic acid and other growth-promoting sub-

stances on streaming. Austral. J. Biol. Sci. *8*:19–35.

Kinosita, H., 1953: Studies on the mechanism of pigment migration within fish melanophores with special reference to their electric potentials. Annot. Zool. Japon. *26*:115–127.

Kinosita, H., 1963: Electrophoretic theory of pigment migration within the fish melanophore. Ann. N. Y. Acad. Sci. *100*:992–1003.

Kinosita, H., Dryl, S., and Naitoh, Y., 1964a: Relation between the magnitude of the membrane potential and ciliary activity in *Paramecium*. J. Faculty Sci. Univ. Tokyo Sect. IV *10*:303–309.

Kinosita, H., Dryl, S., and Naitoh, Y., 1964b: Changes in the membrane potential and the responses to stimuli in *Paramecium*. J. Faculty Sci. Univ. Tokyo Sec. IV *10*:291–301.

*Kinosita, H., and Murakami, A., 1967: Control of ciliary motion. Physiol. Rev. *47*:53–82.

Kishimoto, U., and Akabori, H., 1959: Protoplasmic streaming of an internodal cell of *Nitella flexilis*. J. Gen. Physiol. *42*:1167–1183.

Kuroda, K., 1964: Behavior of naked cytoplasmic drops isolated from plant cells. *In* Primitive Motile Systems in Cell Biology. Allen and Kamiya, eds. Academic Press, New York, pp. 31–41.

Labhardt, E., 1927: Untersuchungen der Muskelermüdung am ausgeschnitten Froschmuskel unter physiologischen Bodingungen. Zeits. Biol. *86*:27–38.

Lehmkuhl, D., and Sperelakis, N., 1963: Transmembrane potentials of trypsin-dispersed chick heart cell cultures *in vitro*. Am. J. Physiol. *205*:1213–1220.

Loewy, A. G., 1942: An actomyosin-like substance from the plasmodium of a myxomycete. J. Cell. Comp. Physiol. *40*:127–156.

Marsland, D., 1944: Mechanism of pigment displacement in unicellular chromatophores. Biol. Bull. *87*:252–261.

Marsland, D., 1964: Pressure-temperature studies on ameboid movement and related phenomena: an analysis of the effects of heavy water on the form, movement and gel structure in *Amoeba proteus*. *In* Primitive Motile Systems in Cell Biology. Allen and Kamiya, eds. Academic Press, New York, pp. 173–187.

Matthews, S. A., 1931: Observations on pigment migration within the fish melanophores. J. Exp. Zool. *58*:471–486.

Nakajima, H., 1964: The mechanochemical system behind streaming in *Physarum*. *In* Primitive Motile Systems in Cell Biology. Allen and Kamiya, eds. Academic Press, New York, pp. 111–123.

Nastuk, W. L., 1953: The electrical activity of the muscle cell membrane at the neuromuscular junction. J. Cell. Comp. Physiol. *42*:249–272.

Nicol, J. A. C., 1960: The Biology of Marine Animals. Interscience Publishers, New York.

Okajima, A., 1966: Ciliary activity and coordination in *Eupolotes eurystomus*. I. The effect of microdissection of neuromotor fibers. Comp. Biochem. Physiol. *19*:115–131.

Pantin, C. F. A., 1952: Croonian lecture: the elementary nervous system. Proc. Roy. Soc. London *B140*:147–168.

Paul, W. M., Daniel, E. E., Kay, C. M., and Monckton, G., (eds.), 1965: Muscle. Pergamon Press, Long Island City, N. Y.

Peachey, L. D., 1965: Transverse tubules in excitation-contraction coupling. Fed. Proc. *24*:1124–1134.

*Peachey, L. D., 1968: Muscular contraction. Ann. Rev. Physiol. *30*:401–440.

Pitelka, D. R., 1963: Electron-microscopic Structure of Protozoa. Macmillan Co., New York.

Prosser, C. L., and Brown, F. A., 1961: Comparative Animal Physiology. 2nd Ed. W. B. Saunders Co., Philadelphia, Chapters 14–17 and 19.

Ramsey, R. W., and Street, S. F., 1941: Muscle function as studied in single muscle fibers. Biol. Symp. *3*:9–34.

Rebhum, L. I., 1964: Saltatory particle movements in cells. *In* Primitive Motile Systems in Cell Biology. Allen and Kamiya, eds. Academic Press, New York, pp. 504–525.

Rinaldi, R. A., 1963: Velocity profile pictographs of ameboid movement. Cytologia *28*:417–427.

Rinaldi, R. A., 1964: Photographs and flow analysis of the hyaline cap in *Chaos chaos*. Protoplasma *58*:603–620.

Rinaldi, R. A., and Jahn, T. L., 1963: On the mechanism of ameboid movement. J. Protozool. *10*:344–357.

Rinaldi, R. A., and Jahn, T. L., 1964: Shadowgraphs of protoplasmic movement in *Allogromia laticollaris* and a correlation of this movement to striated muscle contraction. Protoplasma *58*:369–390.

Roth, L. E., 1958: Filamentous component of protozoan fibrillary systems. J. Ultrastruc. Res. *1*:223–234.

Ruch, T. C., and Patton, H. D. (eds.), 1965: Physiology and Biophysics. 19th Ed. W. B. Saunders Co., Philadelphia, Chapters 1–4.

Sandan, T., and Somura, T., 1959: Effect of ATP on the rate of the protoplasmic streaming in *Nitella*. Bot. Mag. (Tokyo) *72*:337–341.

Schaeffer, A. A., 1920: Ameboid Movement. Princeton University Press, Princeton, N. J.

Seravin, L. N., 1964: A critical survey of modern concepts of ameboid movement. Tsitologiya *6*:653–667.

Simard-Duquesne, N., and Couillard, P., 1962: Ameboid movement. I. Reactivation of glycerinated models of *Amoeba proteus* with ATP. Exp. Cell Res. *28*:85–91.

Sjöstrand, F. S., Andersson-Cedergran, E., and Dewey, M. M., 1958: Ultrastructure of intercalated discs of frog, mouse and guinea pig cardiac muscle. J. Ultrastruc. Res. *1*:271–287.

Slater, E. C., 1957: Sarcosomes (muscle mitochondria). Symp. Soc. Exp. Biol. *10*:110–133.

Sleigh, M. A., 1962: The Biology of Cilia and Flagella. Pergamon Press, Long Island City, N. Y.

Smirnova, N. A., 1955: Reversible changes in protoplasmic currents in *Elodea* after blocking and reduction of the sulfhydryl groups. Fiziologiya Rastenii, Akad. Nauk S. S. S. R. *2*:578–580.

Smith, T. E., Jr., and Berndt, W. O., 1964: The establishment of beating myocardial cells in long-term culture in fluid medium. Exp. Cell Res. *36*:179–191.

Steward, P. A., 1964: The organization of movement in slime mold plasmodia. *In* Primitive Motile Systems in Cell Biology. Allen and Kamiya, eds. Academic Press, New York, pp. 69–78.

Street, S. F., and Ramsey, R. W., 1965: Sarcolemma: transmitter of active tension in frog skeletal muscle. Science *149*:1379–1380.

Taylor, C. V., 1920: Demonstration of the function of the neuromotor apparatus of *Euplotes* by the method of microdissection. Univ. Calif. Publ. Zool. *19*:403–471.

Thaenert, J. C., 1957: Intercellular bridges as protoplasmic anastomoses between smooth muscle cells. J. Biophys. Biochem. Cytol. *6*:67–70.

Thimann, K. V., 1964: Introduction. *In* Primitive Motile Systems in Cell Biology. Allen and Kamiya, eds. Academic Press, New York, pp. 3–6.

Tiegs, O. W., 1953: Innervation of voluntary muscle. Physiol. Rev. *33*:90–144.

Turner, G. D., 1966: General Endocrinology. 4th Ed. W. B. Saunders Co., Philadelphia.

Tuttle, W. W., and Schottelius, B. A., 1965: Textbook of Physiology. 15th Ed. C. V. Mosby Co., St. Louis.

Vogt, M., 1934: The concentration of sympathin in different parts of the central nervous system under normal conditions and after administration of drugs. J. Physiol. *123*:451–481.

Weber, A., Herz, R., and Reiss, I., 1964: The regulation of myofibrillar activity by calcium. Proc. Roy. Soc. London *B160*:489–501.

Wiersma, C. A. G., 1952: Comparative physiology of invertebrate muscle. Ann. Rev. Physiol. *14*: 159–176.

Wilkie, R. D., 1959: Facts and theories about muscles. Progr. Biophys. *4*:288–324.

*Wilkie, R. D., 1966: Muscle. Ann. Rev. Physiol. *28*:17–38.

Wohlfarth-Botterman, K. E., 1961: Cytologische Studien VIII. Zum Mecanismus der Cytoplasmaströmung in dünnen Fäden. Protoplasma *54*: 1–26.

Wohlfarth-Botterman, K. E., 1964a: Differentiations of the ground cytoplasm and their significance for the generation of the motive force of ameboid movement. *In* Primitive Motile Systems in Cell Biology. Allen and Kamiya, eds. Academic Press, New York, pp. 79–109.

Wohlfarth-Bottermann, K. E., 1964b: Cell structures and their significance in ameboid movement. Int. Rev. Cytol. *16*:61–131.

Wolpert, L., 1965: Cytoplasmic streaming and ameboid movement. Symp. Soc. Gen. Microbiol. *15*:270–293.

Yagi, K., 1961: The mechanical and colloidal properties of *Amoeba* protoplasm and their relations to the mechanism of ameboid movement. Comp. Biochem. Physiol. *3*:73–91.

Yoneda, M., 1962: Force excited by a single cilium of *Mytilus edulis.* II. Free motion. J. Exp. Biol. *39*:307–317.

# CHAPTER 25 / THE CHEMICAL BASIS OF CONTRACTILITY

Movements such as are seen in cytoplasmic streaming, ameboid and ciliary movement or muscular contraction are, of course, possible only because of molecular and ionic movements that are taking place at all times inside the living cells. The nature of these molecular and ionic movements and the source of energy by which they occur are fundamental to an understanding of the mechanism of contractility.

Because energy conversions occur on such a large scale in muscle fibers, and because the quantities of enzymes and metabolic intermediates are large, muscle fibers, especially vertebrate skeletal muscle fibers, have been used extensively in experimental studies. Because they are highly differentiated, they are likely to show most clearly the basic features of a contractile system. Many of the enzyme systems have been isolated, and their reactions have been studied *in vitro* (see Chapter 16). However, the objective here is to consider only the broader aspects of muscle metabolism and the possible manner in which the muscle fiber uses the energy released to do mechanical work (Wilkie, 1966).

Studies on vertebrate cardiac and visceral and on invertebrate muscle fibers are also available in the literature and they indicate essential similarity in the chemical systems involved (Gergely, 1964; Ebashi *et al.*, 1966; Wilkie, 1966).

## EVIDENCE OF MOLECULAR ORIENTATION IN MUSCLE FIBERS

To perform work it is necessary that a muscle fiber exert force in a system with a favorable spatial relation. Muscle cells and individual muscle fibers are always elongate and they may even run the whole length of a muscle and join with the tendon. The membrane (sarcolemma) around each muscle fiber is electrically polarized, the inside being about a tenth of a volt negative to the outside, as already discussed (Chapters 22 and 23). A muscle fiber contains elongate *myofibrils* (Fig. 25.1), which lie in the cytoplasm (sarcoplasm), and around or between the myofibrils are found the nuclei. Each myofibril is about a micron in diameter and is presumably the seat of muscle contraction.

The striated appearance of a skeletal muscle fiber seen under the microscope is due to the striations on the myofibrils, the striations being similarly oriented across all the myofibrils of a muscle fiber. Under the polarizing microscope the dark bands are found to be doubly refracting, and the light bands are found to be singly refracting. This suggests molecular orientation (Barer, 1942).

Under high power of the light micro-

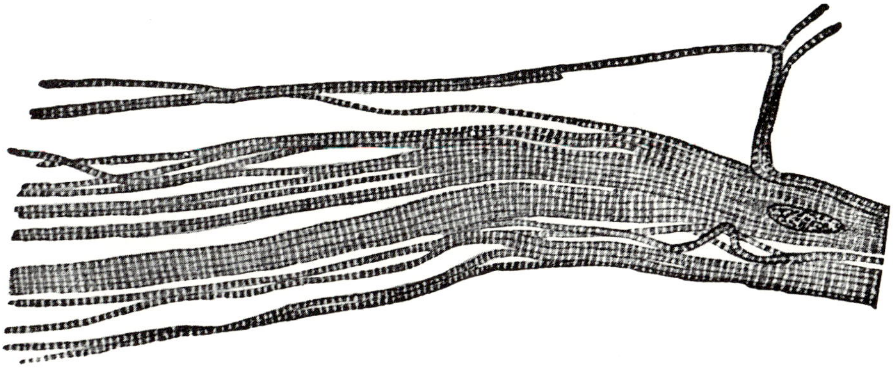

*Figure 25.1.* The separation of a muscle fiber into a mass of myofibrils after treatment with nitric acid. One nucleus is seen within a spindle-shaped mass of sarcoplasm (×530). (A. A. M.)

scope the striation pattern on an isolated myofibril is seen as a regular alternation of isotropic I-bands, the light bands, through which light passes equally in all directions, and anisotropic A-bands, the dark bands, possessing different refractive indices in different directions (see Fig. 25.3, top). When a vertebrate striated muscle fiber is almost fully relaxed, the length of one A-band of a myofibril is commonly about 1.5 microns and that of one I-band is about 0.8 micron. In the middle of the A-band is a portion called the H-zone which is lighter than the rest, and in the middle of the I-band is a darker Z-line which bisects the band. From one Z-line to the next is one repeating unit (*sarcomere*) of the myofibril (Figs. 25.2 and 25.3; H. E. Huxley, 1957). This also indicates a regular arrangement of units.

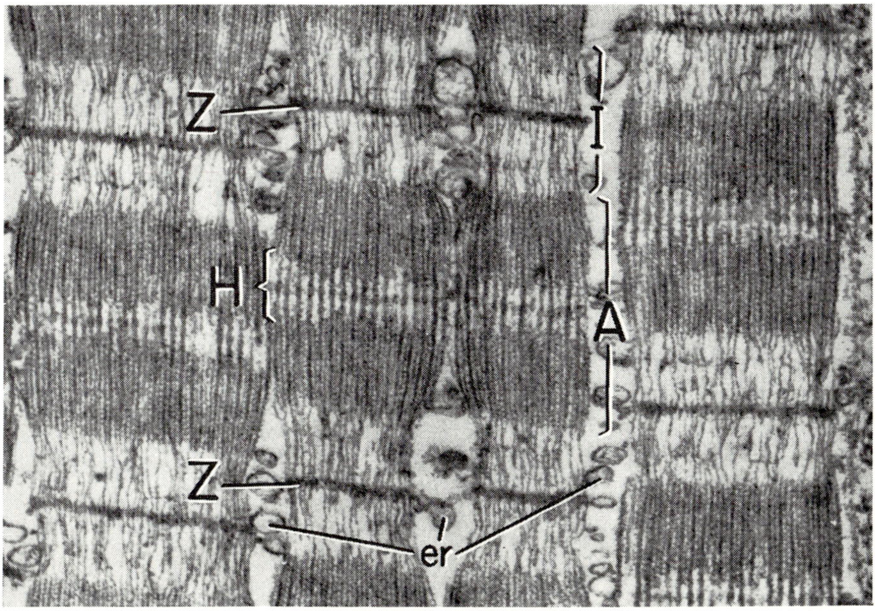


*Figure 25.2.* Electron micrograph of four myofibrils, showing the alternating sarcomeres with the Z-lines and the H-, A- and I-bands. The sarcoplasmic reticulum (er) is situated between the myofibrils. The finer structure of the myofibril represented by the thin (actin) and the thick (myosin) filaments is also clear (×60,000). (By courtesy of H. Huxley.)

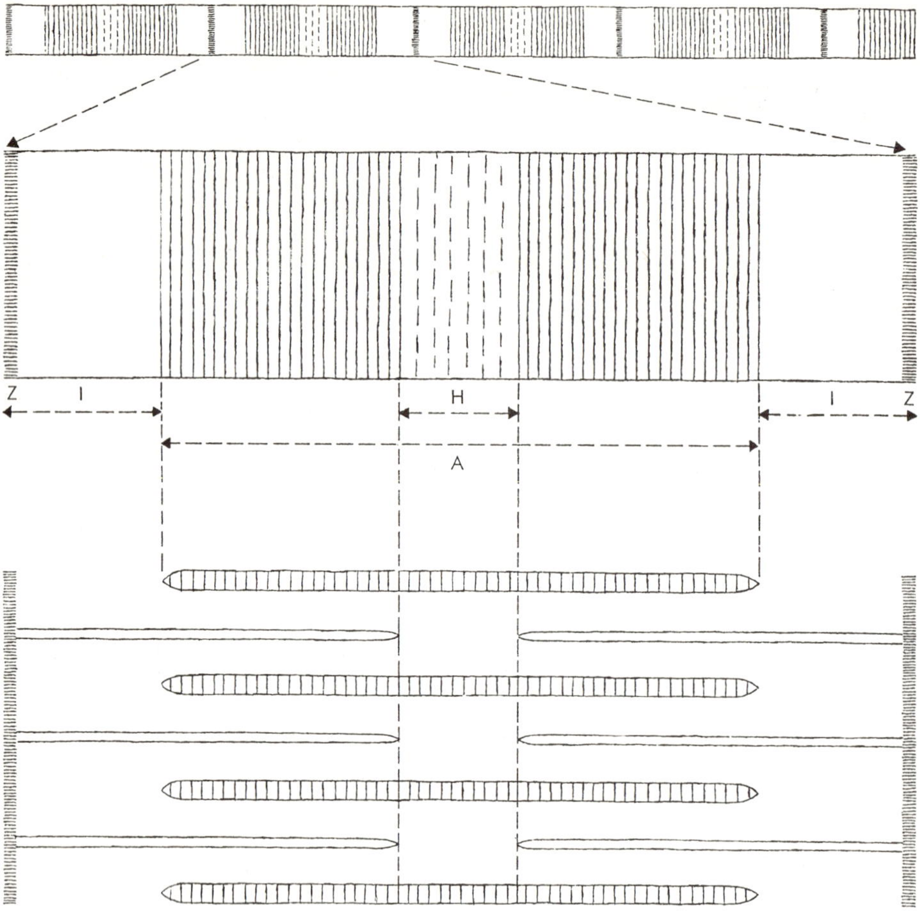


***Figure 25.3.*** Organization of a myofibril showing, at the top, the alternate light (isotropic) and dark (anisotropic) bands as seen with the light microscope. A single band is shown enlarged in the middle, the I-band at each end being half of a light band. The bottom figure shows the pattern of thick (myosin) and thin (actin) filaments as visualized with the electron microscope. The light bands are regions where thin filaments alone are present; the dark bands are regions where the two types of filaments overlap. The myofibril varies in diameter from 1 to 2$\mu$ and in length from a few millimeters to many centimeters depending on the muscle from which it is taken. The I-band is about 0.8 $\mu$ wide; the A-band is about 1.5 $\mu$ wide. The thin filament is about 50 Å in diameter and the thick filament is about 100 Å in diameter. (From The Contraction of Muscle, by H. E. Huxley. Copyright © 1958 by Scientific American, Inc. All rights reserved.)

Evidence of molecular orientation in muscle fibers comes also from x-ray diffraction studies. Briefly, such studies are based on the fact that x-ray wavelengths are diffracted by layers of molecules (or ions). If a beam of x-rays is passed through a crystal, the diffracted beam forms a circle of discontinuous images about the central beam (Fig. 3.7). If a beam of x-rays is passed through a powdered salt crystal, the diffracted beam does not form discontinuous images but rather a series of continuous rings about the central beam, indicating less perfect orientation than that found in true crystals because of fragmentation. When x-rays are passed through muscle fibers, the diffracted beam forms a series of continuous rings surrounding the central beam, indicating imperfect orientation. Each ring is the result of diffraction of the x-rays by a different set of oriented molecules. By measuring the angles between the central beam of x-rays and the rings it is possible to deduce the distance between the molecular units in the muscle fiber and the dimensions and arrangements of the structural units (Astbury, 1950). The x-ray diffraction method has the advantage in such studies because it may be used on living muscle fibers when at rest or contracted. The data so gathered give the true distances between the ordered molecular units, unmodified by fixatives and other agents (Elliott *et al.*, 1965). Without going into details, it is possible to state that the contractile units in the fibrils are oriented along the length of the fibrils with a regularity suggestive of a crystal. Since the indicated size of the contractile unit is that of a large protein molecule in every case, the x-ray studies make it seem likely that the regularly arranged constituents are protein molecules.

When the myofibril is examined under the electron microscope it is found to be made up of two kinds of smaller *filaments,* one twice as thick as the other. The thicker filaments are about 100 Å in diameter and 1.5 microns long, and the thinner ones are about 50 Å in diameter and 2 microns long. Each filament is seen arrayed with the other filaments of the same kind, and the two arrays overlap for part of their length. It is apparently this overlapping that gives rise to the cross bands of the myofibril: the dense *A-band* consists of overlapping thick and thin filaments; the lighter *I-band* consists of thin filaments alone; *the H-zone* consists of thick filaments alone. Halfway along their length, the thin filaments pass through a narrow zone of dense material; this constitutes the *Z-line* (A. F. Huxley, 1957; H. E. Huxley, 1957, 1965). In the A-band the two kinds of filaments are arranged in such a manner that each thin filament lies in a spatial symmetry among three thick ones (Fig. 25.4 and

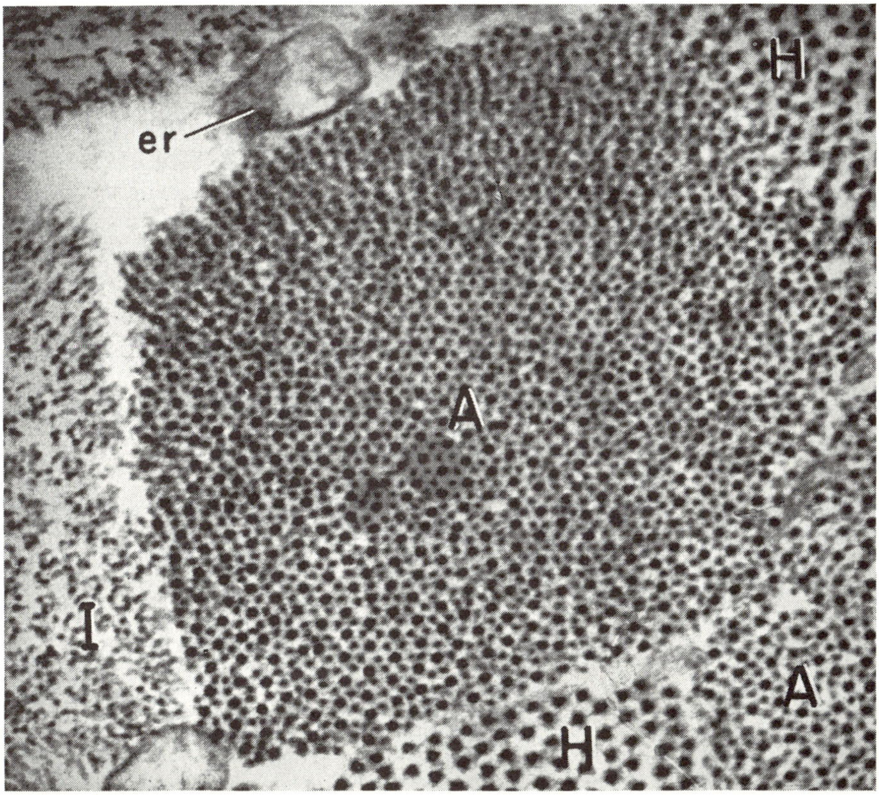


***Figure 25.4.*** Electron micrograph of cross sections passing through different levels of the sarcomere. The A-bands show the thick and thin filaments; the I-band only thin filaments; the H-band only thick ones. er, Sarcoplasmic reticulum (×120,000). (By courtesy of H. Huxley.)

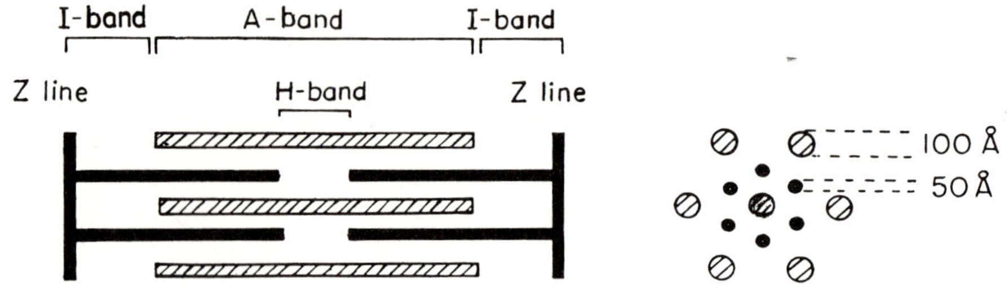


*Figure 25.5.* Diagram to show disposition of thick and thin filaments in longitudinal view (left) and cross section (right). A single sarcomere is shown on the left and a section through the A-band is shown on the right. (By courtesy of H. Huxley.)

25.5). The two kinds of filaments are linked together by *cross-bridges* which project outward from the thick filaments at a fairly regular interval of about 60 to 70 Å, each bridge being 60° further around the axis of the filament, so that the bridges form a helical pattern that is repeated every six bridges, or about every 400 Å along the filament (see Fig. 25.6).

The structures seen with the electron microscope are, of course, being observed on fixed and dried material. The regularity of filaments and their cross linkages is, however, also indicated in x-ray diffraction studies although the evidence is indirect and difficult to present.

Thick and thin filaments have been found in the myofibrils of many types of skeletal muscle fibers of both vertebrates and invertebrates, but they are apparently absent in visceral muscle fibers of the gut and arteries of vertebrates. The two major structural proteins of skeletal muscle fibers, myosin and actin, can be extracted from such visceral muscles, but it would seem that the molecular organization of these proteins in the visceral muscle cells differs in some fundamental manner from their organization in the skeletal muscle fibers (Huxley, 1963, 1965).

Perhaps the most convincing evidence of the reality of the filaments and their cross-bridges in myofibrils is obtained when these are separated intact by treatment of a live muscle in a Waring Blendor. Surprisingly, this drastic treatment separates a muscle fiber into its filaments and no further. In such preparations the

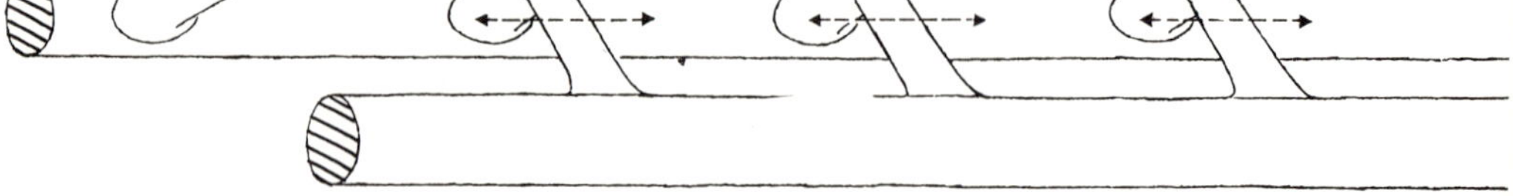

*Figure 25.6.* Arrangement of cross-bridges suggests that they enable the thick filaments to pull the thin filaments by a kind of ratchet action. In this schematic drawing one thin filament lies among three thick ones. Each bridge is a part of a thick filament, but it is able to hook onto a thin filament at an active site (dot). Presumably the bridges are able to bend back and forth (arrows). A single bridge might thus hook onto an active site, pull the thin filament a short distance, then release it and hook onto the next active site. (From The Contraction of Muscle, by H. E. Huxley. Copyright © 1958 by Scientific American, Inc. All rights reserved.)

cross-bridges bind the thick and thin filaments together strongly enough to resist separation in many cases, and some electron micrographs even show the two types of filaments connected by cross-bridges. Also, thin filaments are often seen attached to the Z-line. If the fresh muscle tissue or glycerol-extracted muscle tissue is treated with the appropriate salt solutions, these cross-bridges are broken. The separated thick and thin filaments can then be made visible in electron micrographs with negative staining and measured. They are found to have the same dimensions as measured in thin sections (Huxley, 1963, 1965).

## THE STRUCTURAL PROTEINS OF MUSCLE CELLS

About 20 per cent of the weight of the muscle fibers is represented by proteins. If a mass of muscle fibers is minced, wrapped in cheesecloth and squeezed in a press under many atmospheres of pressure, a fluid exudes which is called the *press juice.* Press juice contains the fluid from inside the cells as well as some intercellular fluid. If the muscle residue left after squeezing out the press juice is washed in very dilute salt solution, water-soluble albumins and readily soluble globulins are removed. The remaining material that did not dissolve contains, among other substances, two major proteins, namely *myosin* and *actin,* and one minor protein present in small quantities called *tropomyosin.* If this remaining mixture is extracted with a solution of 0.3 molar potassium chloride, buffered with 0.15 molar potassium phosphate to give the solution a pH of 6.5, and if the extraction is not too prolonged, myosin, being more soluble than actin, dissolves into the solution. It can be purified by diluting the solution slightly, whereupon its complex with actin, namely *actomyosin,* precipitates out.

Myosin is found to constitute 50 to 55 per cent of the myofibrillar protein in mammalian skeletal muscle fibers. Myosin is an $\alpha$ helix type of protein of high molecular weight (about 350,000). As seen in electron micrographs myosin molecules appear to be long and thin (2000 Å by 50 Å; see Fig. 25.7*C–F*). Myosin is composed of many dicarboxylic acids; therefore it has many ionizable residues, and, near neutrality, it has one charge for almost every angstrom of its length. This very high charge density explains why the molecule maintains its fibrous form, since one part of the molecule repels another part of the same molecule as well as another molecule of its own kind. Because of its needle-like, linear form, myosin shows beautiful streaming birefringence (see Chapter 2; Szent-Györgyi, 1960; Gergely, 1964).

Myosin is presumably present in the muscle fiber as the magnesium salt, magnesium myosinate. Myosin binds monovalent ions, not distinguishing between sodium and potassium ions, as well as divalent ions. Calculations also indicate that practically all the ATP present in muscle fibers is probably bound to myosin (Szent-Györgyi, 1958). Apparently ATP is strongly bound to the magnesium of the myosinate since the ultraviolet absorption spectrum of the myosin-ATP complex is different from that obtained by summing up the absorption of myosin and ATP measured separately. One mole of ATP is bound per 18,000 grams of myosin. In balanced salt solutions like those present in the cell, myosin is dispersed in solution by ATP. Since ATP is ionized with an overall negative charge, the myosin becomes negatively charged when ATP is added to it. When purified myosin, dissolved in 0.3 molar potassium chloride, is added to distilled water, it precipitates in the form of a thread. This thread binds ATP in the presence of salts (Szent-Györgyi, 1960; Kamin and Bell, 1966).

Myosin has been fragmented into two fractions (meromyosins), a heavy one and a light one, which together make up the myosin molecule. These fractions are of particular interest because electron microscope photographs indicate them to be of the right order of magnitude to account for some of the angular deviations of the

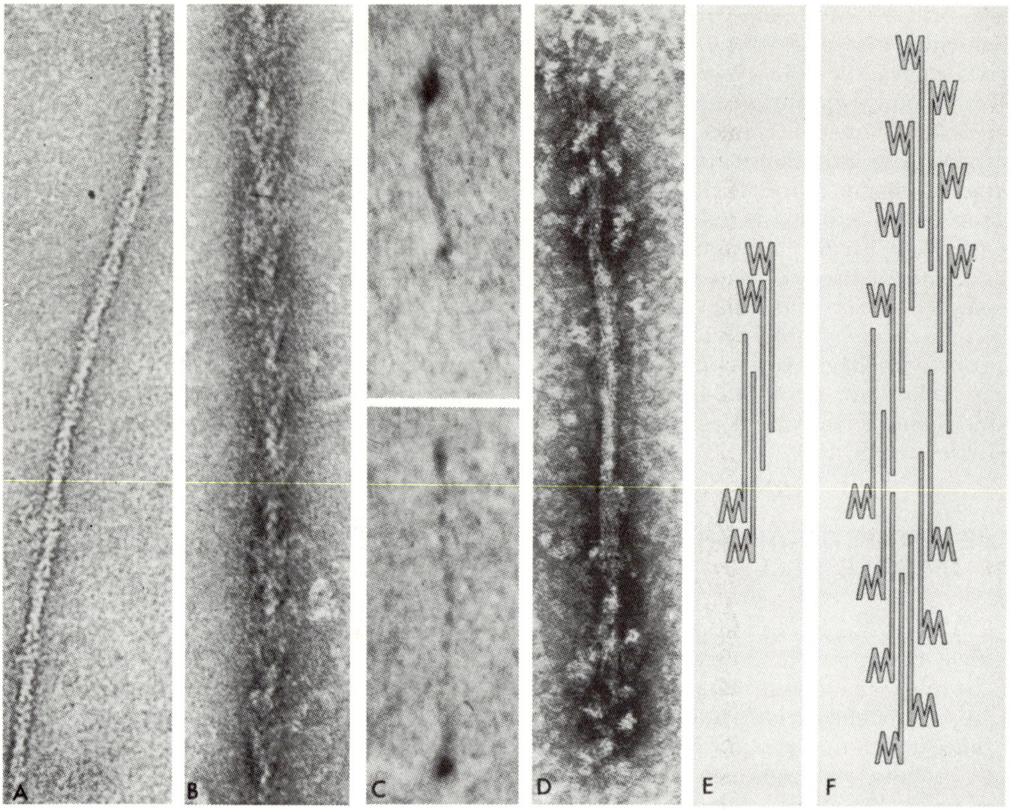


***Figure 25.7.*** Actin and myosin. *A,* Actin showing characteristic structure visible in an electron micrograph at enlargement of 100,000 diameters. The filament has the appearance of two coils of globular units wound in a double helix. *B,* Arrowheads pointing in one direction along each filament of actin labeled with heavy meromyosin, implying that actin has an a polarity of its own. *C,* Myosin molecules as seen in electron micrographs prepared by the shadow casting method. The wide "head" has the enzymatic property and combines with actin; the straight tail can aggregate with other myosin molecules ($\times$300,000). *D,* Aggregation of several molecules of myosin, negatively stained and magnified 175,000 diameters. *E* and *F,* Diagrams of the myosin aggregate. The head of the myosin molecule is indicated by the zigzag line, the tail by a straight line. Note that tails join in the center. (From The Mechanism of Muscular Contraction, by H. E. Huxley. 

x-ray diffraction rings (Fig. 3.7). The light and heavy fractions of myosin are each made up of much smaller fragments (protomyosins) held together by secondary links (Szent-Györgyi, 1958). From measurements on electron micrographs of the two fractions of myosin (light and heavy meromyosin) it appears that they are linked end to end in the myosin molecule.

When the meromyosins are shadow cast, heavy meromyosin appears to have a globular head and a short tail, and light meromyosin is just a straight rod. Thus the myosin molecule is asymmetrical, with a head 40 Å in diameter accounting for one-sixth of the length and the remainder, a tail 20 Å in diameter. Apparently the ATPase activity of intact myosin and the affinity of myosin for actin resides only in its heavy meromyosin head, whereas the affinity for other myosin molecules resides in its light meromyosin tail.

It is of interest that whether the filaments from the myofibrils are obtained by homogenizing a muscle in a Waring Blendor or reconstituted from solution (precipitated in dilute potassium chloride solution), they have the same linear form and head shape. The myosin molecules thus display a marked tendency to form long threads, like the filaments in muscle by attaching to one another.

Actin, the other major structural protein of muscle fibers, is less soluble in solutions of potassium chloride than is myosin and dissolves only when the concentration of potassium chloride is raised to about 0.6 molar. Actin constitutes 20 to 25 per cent of the protein in skeletal myofibrillar material of mammals. Actin binds four equivalents of calcium per molecule, and apparently it exists normally as calcium actinate. In the absence of salts, actin exists in globular form, with a molecular weight of about 70,000, but upon addition of ATP it forms dimers, and in the presence of 0.1 M potassium chloride and ATP, it polymerizes to form long strands. Polymerization is accompanied by splitting of ATP; the ADP formed becomes firmly bound to the actin and inorganic phosphate is given off. Each molecule of actin appears to bind one ADP. The actin appears to be asymmetrical and to be made up of two coils of globular units wound in a double helix with crossover points 360° apart (see Fig. 25.7*A*). This means that when myosin combines with actin it will repeat the arrangement (Huxley, 1963). The actin helix has a period of 400 Å according to both x-ray diffraction and electron microscope studies. It is not clear what role, if any, the globular-to-fibrous transformation that actin undergoes may play in muscular contraction (Huxley, 1960).

In polymerized form, actin shows streaming birefringence like myosin, another sign of the linear entities of which it is composed. Like logs in a stream of water, the long molecules are lined up in the solution, and differentially affect light passed at right angles to the long axes as compared with light passed parallel to the long axes of the molecules.

When actin and myosin are mixed in the presence of salts normally found in the muscle fibers, a striking change occurs. The viscosity of the mixture becomes much greater than the sum of the viscosities of the two constituent solutions. Electron micrographs show that a network is formed by electrostatic association between the two proteins. This complex is called *actomyosin.* Actin and myosin have been found to combine in stoichiometric ratios.

It has been possible, through studies with the electron microscope, to localize the proteins in the filaments of the myofibrils. Thus, when myosin of a muscle fiber is extracted with an appropriate potassium chloride solution, the thicker filaments of the myofibrils show only as ghosts, which indicates that they had contained the myosin. When muscle fibers are treated to remove actin, the electron micrographs show absence of the thin filaments in the fibrils, indicating that these consist mostly of actin (they also have some tropomyosin). Corroborating data come from analyses of the filaments using immunochemical techniques; fluorescent antibodies are produced against either myosin or actin (Marshall *et al.,* 1959).

*Tropomyosin* makes up about 2.5 per cent of the muscle fiber protein. Although it seems to be present in all muscle fibers, its function is still unknown. To obtain it, the muscle fibers are extracted with 1 M solution of potassium chloride (or weak acid) after preliminary treatment with alcohol (or ether). It has a relatively low molecular weight (about 60,000) and crystallizes readily. In water it aggregates to form strands 200 to 300 Å thick. Its localization in the muscle fiber is uncertain; it may be in the Z-line, since this remains after extraction of myosin, or it may be in the bridges between the actin and myosin filaments (Huxley, 1960).

Still another kind of muscle protein, *paramyosin,* is found in the catch muscle fibers of mollusks and annelids. It is extracted with weak acid (or strong salt solutions) after the muscle fibers are first treated with ethanol (or ether). It has a molecular weight of about 137,000 and a length of 1400 Å. The molecule shows pronounced asymmetry. Catch muscle fibers hold tension for prolonged periods with exceedingly low metabolic activity and without the characteristic tetanus of vertebrate striated muscle fibers, the "catch" mechanism being the setting of structural bonds of the paramyosin at a given length (Szent-Györgyi, 1960; Hanson and Lowy, 1960; Lowy *et al.,* 1964; Ruegg, 1964).

## ACTOMYOSIN

Actomyosin dissolved in 0.6 molar potassium chloride solution shows brilliant streaming birefringence. The electron microscope discloses actomyosin particles to be long threads of about 150 Å diameter and ten or more times the length of a myosin molecule (Fig. 25.7 *B*). If the salt solution is diluted to 0.1 molar potassium chloride, actomyosin is precipitated in the form of loose flakes. On addition of ATP the precipitate becomes more tightly set ("super precipitate"). At higher concentrations of potassium chloride, however, the actomyosin seems to be dispersed into solution by ATP, and the viscosity of the mixture drops to the sum of the viscosities of actin and myosin taken separately. Salts present in very high concentrations (0.6 molar potassium chloride) in solution also disperse (dissolve) the actomyosin molecules, even in the absence of ATP.

When actomyosin dissolved in 0.6 molar potassium chloride is squirted into 0.01 M potassium chloride (see Fig. 25.8), a gelled thread of actomyosin is obtained. When such a thread is placed in ATP solution, it contracts (Weber and Porzehl, 1954). Actomyosin is also found to possess an enzymatic function (adenosine triphosphatase), splitting ATP to ADP and inorganic phosphate. These observations

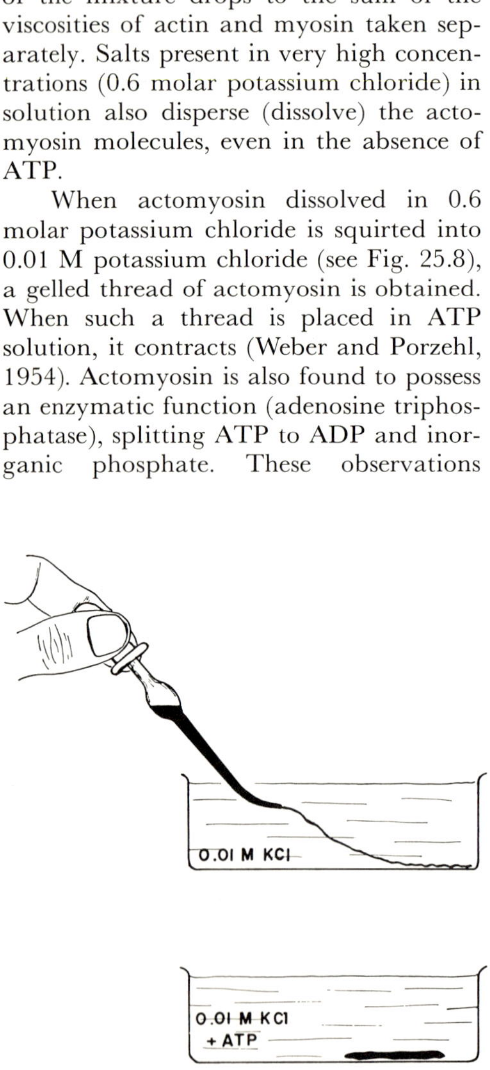


***Figure 25.8.*** Formation of an actomyosin thread by adding a jet of actomyosin to a dilute solution of KCl. Contraction of the thread is shown after addition of ATP (Harris, 1961).

created a sensation when they were first announced, because actomyosin thus appeared to be a model, or analogue, of muscular contraction. However, important differences were soon found between a muscle fibril and an actomyosin thread. The latter loses about 40 per cent of its water on contraction, whereas a muscle fibril does not. Furthermore, study of the actomyosin thread under the electron microscope shows that the molecules are in disarray, resembling the brush network in a protein gel rather than the linearly arranged and regularly spaced units disclosed by x-ray analysis of a muscle fiber. Moreover, if the thread represents a model of a muscle fibril, it should perform work like a myofibril, yet when the actomyosin thread is loaded, and then ATP is added, the thread stretches instead of contracting.

Later it was possible to construct an actomyosin model which contained linearly oriented actomyosin molecules. Because protein molecules become oriented at a surface, a film was made by dropping a solution of actomyosin on the surface of 0.06 molar potassium chloride solution in a trough. Since the monomolecular film formed is too weak to support weight, it is thickened by sweeping the trough, thus folding the surface sheet into an accordion-pleated thread which holds together because of cohesive forces (Hayashi, 1952). Such an actomyosin thread has tensile strength. Loaded with a weight, it contracts and performs work, on addition of ATP to the bathing fluid. It becomes relaxed when washed with appropriate salt solutions, and it can again be contracted by the addition of ATP (see Fig. 25.9). Furthermore, water is not lost during contraction. It was reasoned that if the energy for contraction of an actomyosin thread comes from ATP, the ATP should be used up when work is done by the model contracting thread. Quantitative determinations of ATP showed that this, in fact, is the case, and actomyosin therefore serves as an adenosine triphosphatase, liberating the energy of ATP to make possible the contraction of the actomyosin thread. Still the model, although it has many properties in common with a muscle fibril,

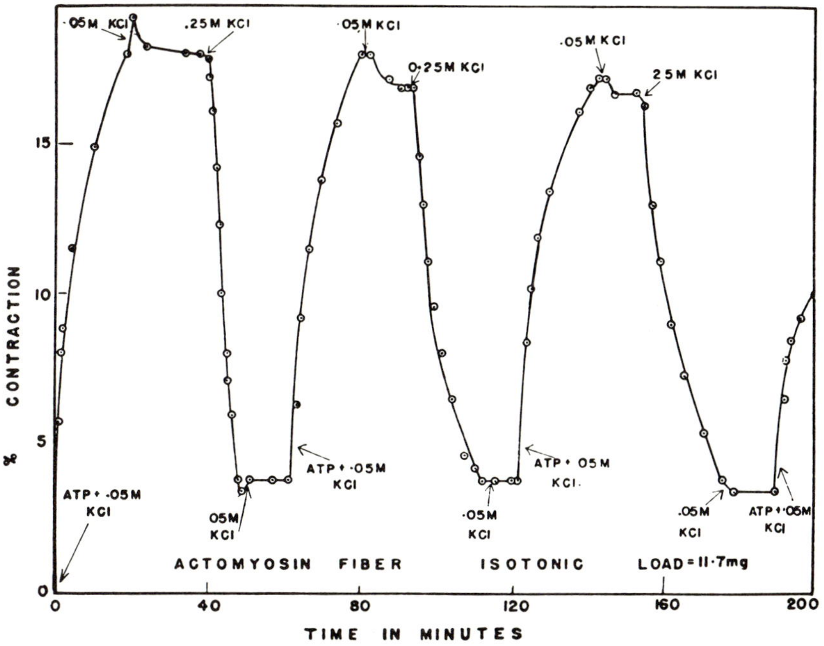


***Figure 25.9.*** Graph showing contraction of the actomyosin thread upon addition of ATP in 0.05 molar KCl, and showing relaxation upon washing in 0.25 molar KCl. (From Hayashi and Rosenbluth, 1952: J. Cell. & Comp. Physiol. *40*:501.)

is not a complete analogue to a muscle fibril, because the time required for its contraction is longer than that for the control muscle fibrils (Hayashi and Rosenbluth, 1952).

In muscle, myosin and actin exist as filaments, with connections by cross-bridges. They are not only aligned but also polarized. A model must therefore duplicate these arrangements. Although no model has yet been made that shows all the properties of muscular contraction, the elements of the filament systems have been reassembled. Assembly of actin and myosin filaments occurs spontaneously from solutions of actin and myosin, indicating that it is the result of properties of the respective actin and myosin molecules.

It will be recalled that actin has the form of a double helix, being made up of a double coil of globular units with a crossover distance of 360°. When actin is labeled with heavy meromyosin, which has a head and tail end, it combines to form a strand with distinct polarity and with an arrowhead appearance. Seen by negative staining, the arrows all point in the same direction (Fig. 25.7*B*). The nature of the reaction is unknown, but it appears that all the actin molecules of a given strand react in the same way with the meromyosin molecule.

To be effective in its function of contraction, the filament of the myofibril must have polarity and the polarity should be opposite on the two sides of the Z-line. To determine whether such polarity existed Huxley (1963) made use of the occasional Z-lines with attached thick filaments which appear in muscle homogenates. To a homogenate with such filaments attached to a Z-line he added heavy meromyosin and then studied the resulting structure under the electron microscope. He found that the arrow-like structures made by the attachment of the heavy meromyosin to the actin were in opposite directions on the two sides of the Z-line, pointing away from it. The polarity of the filaments is less readily seen when entire myosin molecules are used instead of the heavy meromyosin fraction, because the light meromyosin fraction attracts other

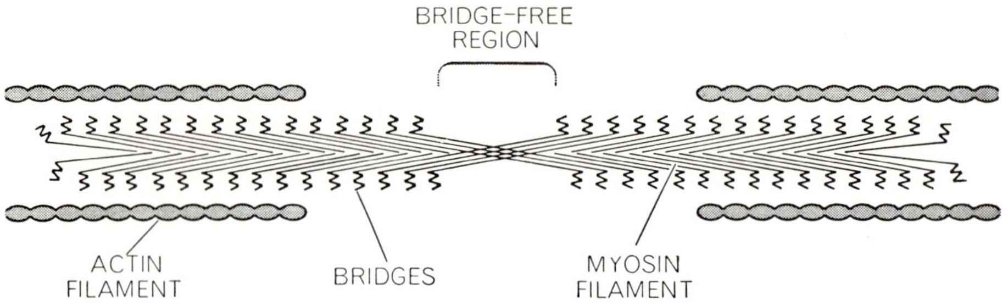


*Figure 25.10.* Diagram of the actomyosin molecule showing the arrangement of the actin and myosin molecules which result in polarity. The polarity probably resides in the actin molecules which orient the "heads" of the myosin molecules with respect to themselves as shown by the "arrowheads" in Figure 25.7*B*. The "heads" form the bridges; note that bridges are lacking in the center of the filament and oriented in opposite directions at the two ends of the filament. (From The Mechanism of Muscular Contraction, by H. E. Huxley. Copyright © 1965 by Scientific American, Inc. All rights reserved.)

myosin molecules to form the large myosin filament. Presumably the same orientation exists. In any case, the myosin molecules are arranged in such a way that all the globular portions are at the two ends, the middle portion being free of "heads." Such a strand is interpreted to consist of myosin molecules attached to one another by the light meromyosin portions and attached to the actin filaments in opposite polarities at the two ends, as shown in Figure 25.10.

## SOURCE OF ENERGY FOR CONTRACTION

Whereas the fibrils of the muscle fiber are the seat of muscular contraction, the sarcoplasm is the seat of the biochemical reactions from which high energy compounds are made available to the muscle fibrils. Mitochondria, with all the essential oxidation-reduction and phosphorylation enzyme systems of the Krebs cycle, have been identified in muscle fibers. They are arranged in close proximity to the region requiring energy liberation. They are also more complex internally than mitochondria of less active cells, containing more cristae, presumably in connection with their greater activity in oxidation-reduction than in less active cells. These mitochondria are especially prominent in size and number in the flight muscle fibers of birds and insects.

The press juice of minced muscle fibers is largely extracted from the sarcoplasm and contains most of the high energy phosphate compounds, as well as salts and the globular proteins which constitute almost all of the enzymes. Many of the glycolytic enzymes characteristic of the Krebs cycle may be identified in the press juice.

Glycogen constitutes about 1 per cent of the wet weight of whole muscle fiber, creatine phosphate (CP) about 0.5 per cent, and adenosine triphosphate (ATP) about 0.025 per cent, indicating a considerable store of ready nutrient. Experiments show that during muscular contraction, glycogen disappears and a corresponding quantity of lactic acid appears when the rate of activity is too rapid to permit complete oxidation of the pyruvic acid formed during glycolysis. In vertebrates the excess lactic acid is carried by the blood to the liver where about four-fifths of it is resynthesized into glycogen, while one-fifth of the lactic acid is completely oxidized to carbon dioxide and water in the process. New supplies of glucose are continually delivered to muscle cells by the blood (Gergely, 1964).

The diagram of the anaerobic breakdown of glucose (Fig. 16.4) reveals the mechanism whereby pyruvic acid produced in glycolysis is converted into lactic acid. In the absence of oxygen, lactic acid accumulates, since the reactions cannot go any further. When oxygen is available, lactic acid is oxidized in the reactions of the Krebs cycle, as shown in the diagram of the metabolic mill (Fig. 16.8). The glycolytic reactions occur whether oxygen is present or absent. If oxygen is present the glycolytic reactions are then followed by the reactions of the Krebs cycle* (Fig. 16.7).

According to measurements of the heat produced during muscular contraction, energy is liberated in two stages. A small amount of heat is given off upon contraction, and a much greater amount of heat is liberated for many seconds, or even for several minutes, after contraction (see Fig. 25.11). Studies indicate that the initial heat is evolved during breakdown of high energy phosphates and that the large amount of heat measured later is evolved during glycolysis and the complete breakdown of some of the lactic acid to carbon dioxide and water. Actually, the sequence of events is much more complex, the initial and delayed heat each being made up of several components. (For example, tension or shortening heat, maintenance heat and relaxation heat all contribute to the initial heat. The delayed heat phase has anaerobic and aerobic components.) Hill and Meyerhof received the Nobel Prize in 1922 for their pioneer studies on muscle metabolism.

If a muscle fiber is stimulated in the presence of adequate oxygen, it fatigues slowly. In the absence of oxygen, the muscle fatigues much more rapidly because it must then depend upon glycolysis for its energy, and anaerobic processes always yield considerably less energy than aerobic processes. Since glycolysis precedes aerobic oxidations and continues in their absence, is it, then, the immediate source of energy for contraction of muscle fibers? To test this, Lundsgaard (1930) used iodoacetic acid to poison the enzymes catalyzing the glycolytic reactions, and he found that contraction still continued but that the muscle fatigued even more rapidly than an unpoisoned anaerobic control. Clearly, some immediate

*Muscle cells make use of fat as well as glycogen, red muscles more so than white. During migration, butterflies, fishes and birds make extensive use of fat for muscular contraction and show a corresponding low respiratory quotient (Drummond and Black, 1960).

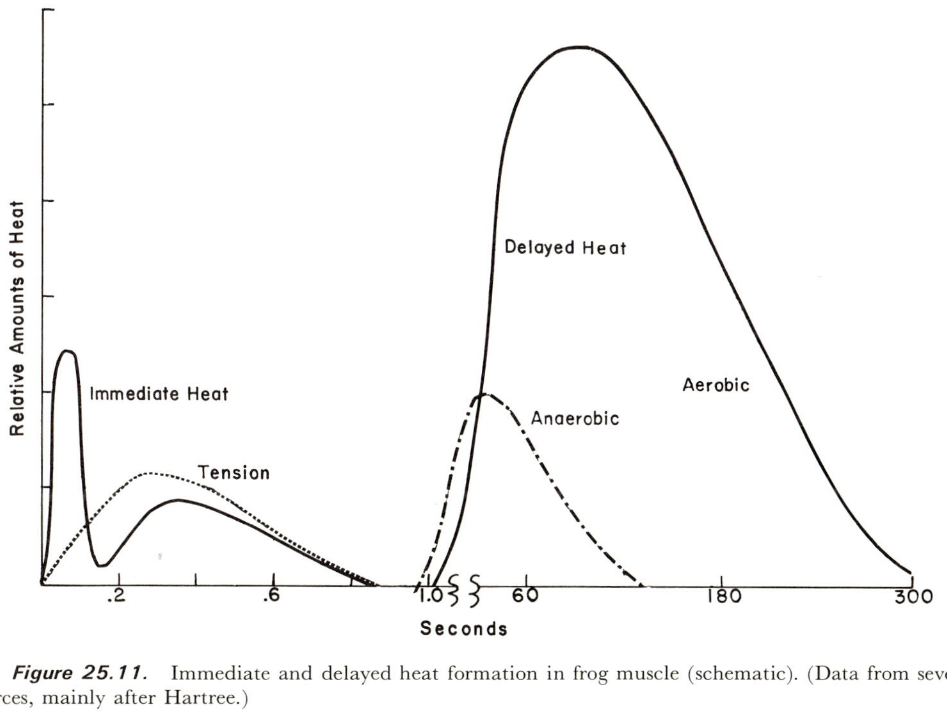


***Figure 25.11.*** Immediate and delayed heat formation in frog muscle (schematic). (Data from several sources, mainly after Hartree.)

source of energy for muscular contraction, other than glycolysis, had to be sought.

Analyses showed that the compounds of high phosphate transfer potential decreased in amount as contractions continued in the poisoned anaerobic preparations. Presumably, these compounds were being used in muscular contractions, but in the absence of both glycolysis and aerobic reactions these high energy phosphate bonds could not be replenished, and the muscle fibers fatigued very rapidly.

If ATP is the immediate source of energy for contractility it should provoke contraction when applied directly to a muscle fiber (Bozler, 1953). As expected, it has been demonstrated that when a solution of ATP is injected into an artery, tetanus is induced in the muscle fiber supplied with blood by this artery. If a muscle is cut up into small bundles and placed at low temperature in glycerin, all the soluble constituents leach into the glycerin, but the structural proteins presumably remain intact and arranged as they were in the live muscle. Muscle strands leached in this manner are called *muscle models* and have been extensively used in research. Contraction with a tension equal to that exerted by living muscle fibers may then be induced by placing the dead leached muscle fibers in either fresh muscle press juice or a balanced salt solution containing ATP (Filo *et al.*, 1965). Press juice in which the ATP has been decomposed does not induce contraction (Szent-Györgyi, 1953).

That ATP is involved in some way in muscular contraction is incontestable. The decomposition of ATP to ADP, liberating the energy of one high energy phosphate bond, appears as the likely immediate source of energy for muscular contraction. ADP is "recharged" to ATP by glycolysis during which glucose is decomposed to pyruvate (or lactate in absence of oxygen), and by oxidation of pyruvate to carbon dioxide and water in the reactions of the Krebs cycle. ATP can be immediately replenished by the reaction between ADP and creatin phosphate which occurs as a reserve of high energy phosphate bonds in the muscle cells.

Working with other motile systems such as fibroblasts, histiocytes, amebas, flagella, contractile myonemes (*Vorticella*) and mitotic spindles extracted with glycerol, Hoffman-Berling has shown that they, too, contract when ATP is added. Contraction does not occur if the adenosine triphosphatase present on the fibers is first poisoned with Salyrgan, which is specific to this type of enzyme. This suggests that all of these systems, like muscle contraction, depend upon ATP as a source of energy for contraction (Jewell, 1959). Proteins that contract on addition of ATP have been extracted from embryonic fibroblasts and from the flagella of sperm (Hayashi, 1961; Bishop, 1962) as well as from cilia (Gibbons and Rowe, 1965) and ameba (Simard-Duquesne and Couillard, 1962).

## THE SLIDING FILAMENT THEORY OF MUSCULAR CONTRACTION

Another approach to the study of muscular contraction has been used by H. E. Huxley, J. Hanson and A. F. Huxley, who investigated the structural changes occurring during a cycle of contraction, as seen in a series of electron micrographs of ultrathin sections (about 150 Å) of muscle fibers (A. F. Huxley, 1957; H. E. Huxley, 1957, 1965). Such an electron micrograph is shown in Figure 25.2 and an interpretation of what is seen is given diagrammatically in Figures 25.3 and 25.5.

During muscular contraction, as seen with electron micrographs, the A-band remains constant, but the length of the I-band varies with the state of contraction. Since the length of the A-band is apparently equal to the length of the thick filaments, it can be assumed that the length of these filaments is constant. The length of the H-zone in the middle of the A-band increases and decreases proportionally with the length of the I-band, so that the distance from the end of one H-zone to the beginning of the next H-zone remains approximately the same. Since this distance is equal to the length of the thin filaments, they also appear to remain approximately constant in length. Consequently, it is concluded that when a muscle fiber changes length the two sets of filaments slide past one another (Huxley and Hanson, 1959).

When a muscle fiber is contracted to its fullest extent, the filaments now meet and their ends may even crumple (see Fig. 25.12). Under such conditions, new bands appear on the electron micrographs, which are explained as a result of

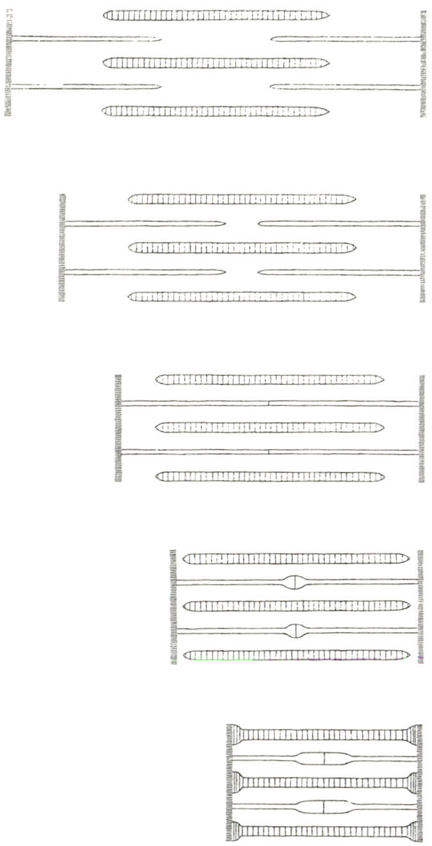


***Figure 25.12.*** Muscular contraction as envisaged from the contractile filament theory. Change in length of the muscle fiber changes the arrangement of the filaments. In *A* the muscle fiber is stretched; in *B* it is at its resting length; in *C, D* and *E* it is contracted. In *C* the thin filaments meet; in *D* and *E* they crumple up. In *E* the thick filaments also meet adjacent thick filaments (not shown) and crumple. The crumpling gives rise to new band patterns. The central thick filaments are presumably myosin; the thin filaments, actin. (See text.) (From The Contraction of Muscle, by H. E. Huxley. Copyright © 1958 by Scientific American, Inc. All rights reserved.)

the shortening process, not as a cause of contraction (H. E. Huxley, 1957).

As was pointed out earlier, actin and myosin must be combined to form actomyosin before contraction will occur in the presence of ATP. However, since the actin and myosin filaments in the myofibrils are too far apart for any "action at a distance," the motive power for muscular contraction is postulated to be the formation of chemical couplings between the myosin and actin filaments through the bridges between the two (Fig. 25.6). The bridges are permanent parts of the myosin filaments jutting outward toward the actin filaments and the number of such bridges appears to be the same as the number of molecules of myosin. A chemical reaction is postulated to occur between the myosin and actin filaments, as a result of which the actin filament slides along the myosin filaments for perhaps 100 Å, and then returns to its original configuration, ready for another pull, completing a cycle (Huxley, 1965). To account for the rate of contraction and of energy liberation in the rabbit psoas muscle fiber, for example, it has been calculated that it is necessary for each bridge to go through 50 to 100 cycles per second. Each such reaction is thought to be at the expense of one ATP molecule. When removal of phosphate groups from ATP stops (presumably as a result of the action of an inhibitor or relaxing factor, the myosin and actin dissociate from each other and the fiber returns to its relaxed state. There is evidence to show that in the presence of the relaxing factor, ATP can break the combination between actin and myosin (Weber, 1959).

The sliding filament theory effectively explains a number of the features of muscular contraction. For example, it is possible to envisage why a rapidly contracting muscle fiber exerts less tension than one contracting slowly, i.e., because it takes a certain amount of time to form the bridges between actin and myosin; hence, if the contraction is too fast, fewer bridges can form, and fewer bridges means less tension exerted. Again, it is possible to see why a muscle fiber carrying a larger load uses more energy (heat plus work) than one under a lesser load. When the load is larger, more bridges form and more ATP molecules are used up. It is also possible to conceive how a muscle fiber can continue to exert tension, since at any site the reaction can occur more than once, the bonds between actin and myosin breaking and re-forming, permitting a muscle fiber to continue to do work. Bridges have not yet been identified in electron micrographs of contractile systems other than striated muscle (Hayashi, 1961).

## EXCITATION OF MUSCULAR CONTRACTION

Although evidence for activation of the muscle fiber by acetylcholine secreted by the terminations of the efferent neurons at the myoneural junction is convincing (see Chapter 23), there still remains the problem of how the individual units of the filaments are activated. It has been shown that if minute microelectrodes (about 2 microns in diameter) are brought alongside the isotropic and anisotropic bands of a single isolated frog muscle fibril and the flow of current is allowed to depolarize it, contraction occurs when the stimulating electrode is against the Z-line, but not if it is placed midway between the two Z-lines, even though the current is increased severalfold (Huxley and Taylor, 1958). The Z-lines, as was pointed out previously, appear to line up and perhaps connect transversely through a whole series of muscle fibers. It is tempting to suggest that the Z-lines are involved in excitation coupling, but the mechanism by which this is accomplished is as yet not quite clear.

Electron micrographs of the myoneural junction show a tubular system extending from the membrane of the muscle fiber inward as a sort of microsynaptic system to the fibrils. This system has been called the *triad* or *T-system* (Fig. 25.13). It consists of a central tubular extension inward of the sarcolemma (plasma membrane of the muscle cell) which contacts the sarcomeres (the myofibril units) at the Z-line (or at the boundary of the A- and I-bands in some muscles), and of two lateral tubular extensions of the endoplasmic (sarcoplasmic) reticulum. That the tubules are truly extensions of the cell membrane has been demonstrated by placing muscle fibers in a suspension of labeled ferritin (an electron-dense, iron-containing compound) or in thorium dioxide (Philpott and Goldstein, 1967), or in a solution of fluorescent protein (Endo, 1960). These tracers are shown to appear in sarcolemmal tubules of the triad next to the sarcomeres (Fawcett, 1961; Peachey, 1965; Smith, 1966). The lateral vesicles of the triad (extensions of the sarcoplasmic reticulum) accompany the sarcolemmal tubules to the sarcomere.

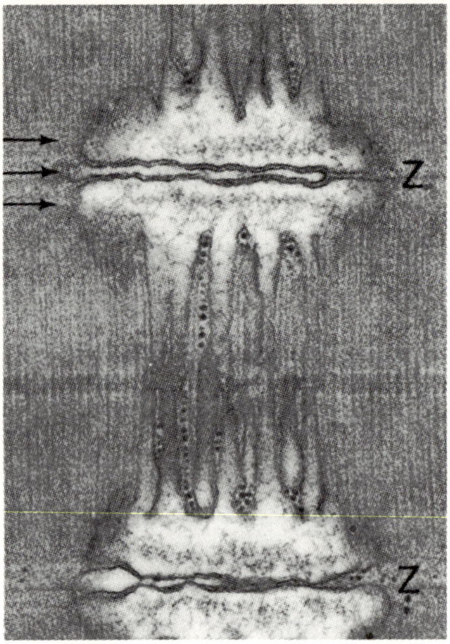


***Figure 25.13.*** Diagram of triads by which the impulse is spread in excitation of muscular contraction. A triad consists of the incursion of the sarcolemma (follow middle arrow) and the tubules of the sarcoplasmic reticulum which impinge upon the tubule from the sarcolemma. The impulse comes to the incursive tubule of the sarcolemma and a current is incited which leads to the release of calcium ions from the tubules of the sarcoplasmic reticulum at the Z-line, leading to muscular contraction. (From DeRobertis *et al.,* 1965: Cell Biology. 4th Ed. W. B. Saunders Co., Philadelphia.) Courtesy of K. R. Porter, Dept. of Biol. Univ. of Harvard, Cambridge, Mass.

Such a T-system is quite prominent in fibers of active muscle, but less so in cardiac and slow muscles (Franzini-Armstrong and Porter, 1964; Pease *et al.,* 1965; Smith, 1966). Because the T-system functions in both excitation and relaxation of the muscle (see later), it is more than a telegraphic connection signaling contraction. In invertebrate muscles studied the triad is replaced by a *dyad* of sarcolemmal and sarcoplasmic tubules (Hoyle, 1965).

The detailed manner in which the triad functions is not fully understood. Excitation of the sarcolemma induces a spike potential but the potential does not appear to extend to the limits of the sarcolemmal tubules at the sarcomeres. However, the current flow initiated by excitation of the sarcolemma initiates ionic flow in the tubules (Freygang, 1965). This in turn in some way results in liberation of calcium ions from the sarcoplasmic cisternal vesicles of the triad (Ebashi, 1965; Sandow *et al.,* 1965; Winegrad, 1965a). That calcium is liberated from the triads has been demonstrated by treating the muscle preparation during stimulation with oxalate and then studying it by electron microscopy. Vesicles so treated show deposits of electron-dense calcium oxalate which is not present in unstimulated controls (Costantin and Podolsky, 1965). Calcium release from the triads is also shown by autoradiography experiments with $^{45}Ca$. Muscle stimulated to contract in the presence of $^{45}Ca$ shows the presence of silver grains which appear to be in the filaments over the triad; muscles treated with relaxant (calcium chelators) show much less (Winegrad, 1965b).

It is believed that the T-system would help explain spread of excitation since no ordinary diffusion of a substance across a membrane would be fast enough to account for the rapidity of a contractile response (Bennett, 1960).

It is inviting to assume that in the living muscle fibers ATP must play the same role as in contraction of muscle fiber models and actomyosin threads. That ATP is present in muscle fibers in abundance is shown by biochemical extraction. Furthermore, a muscle fiber which has exhausted its ATP goes into a state of rigor (stiffness) which is reversed by addition of ATP. In fact, one of the major functions of ATP is believed to be the maintenance of suppleness and plasticity of muscle fibers (Weber, 1959).

An attempt has been made to localize ATP in the bands of the sarcomeres in muscle fibers by autoradiography, using tritiated adenine in the ATP. The resolution of the photographs is not quite adequate to make localization certain, but concentration in the boundary between the A-bands and the I-bands is suggested (Huxley and Hanson, 1959; Hill, 1959).

## RELAXATION OF MUSCLE CELLS

Since ATP is used in muscular contraction, either it is present in an inactive form to be released for action only when the impulse reaches the muscle fiber, or the enzymatic breakdown of ATP is in some way normally inhibited by some substance until the action potential reaches the muscle fibers, at which time the inhibitory influence vanishes and the ATP in the region of the active enzyme is decomposed. An inhibitor called the *relaxing factor* or the *Marsh-Bendall (MB) factor* has been demonstrated in muscle contraction. Marsh (1952) and Bendall (1954) independently showed that ATP is decomposed in muscle homogenate preparations more readily when the particulates which include vesicular elements of the sarcoplasmic reticulum (microsomes) have been removed than in the preparations in which this has not been done. This suggests the presence in the latter case of an inhibitor of the ATPase. The mitochondrial fraction has little relaxing effect; instead, the relaxing effect is concentrated almost entirely in the microsomal fragments of the sarcoplasmic reticulum. As a consequence of these experiments and other studies, particularly electron microscopic investigations, it became evident that the sarcoplasmic vesicles in muscle fibers, and the microsomes derived from them in homogenates, are the probable seat of the inhibitor (Marsh-Bendall factor).

Marsh and Bendall and others attempted to separate a soluble inhibitory substance from the particular fraction but failed. Meanwhile, it was demonstrated that the equivalent of relaxation in muscle fibers can be brought about in muscle homogenates or model systems by such chelating agents as ethylenediamine tetraacetic acid (EDTA) which removes calcium ions. It was also demonstrated that the microsomes in muscle homogenates and the sarcoplasmic reticulum in muscle fibers perform similar functions in removing the calcium ions (Hasselbach, 1965; Lorand *et al.,* 1965).

Most of the studies have been carried out with muscle homogenates or model systems containing actomyosin, ATP and salts of potassium, calcium and magnesium, because it is easier to control the conditions of such experiments, but corroborative experiments were then performed on muscle fibers. It has been demonstrated that both calcium and magnesium ions are essential to muscular contraction. If both ions are reduced to below $10^{-7}$ M in concentration, contraction and actomyosin-ATPase activity cease. Raising the calcium concentration alone, in presence of ATP, has no restorative effect; raising the concentration of magnesium ions alone is also without effect. In the presence of minimal magnesium ion concentration raising the calcium ion concentration to $3 \times 10^{-6}$ M restores ATPase activity and permits contraction. Maintaining calcium ion concentration at $3 \times 10^{-6}$ M and raising the magnesium ion concentration to $10^{-2}$ M once again inhibits the actomyosin-ATPase activity. Counteracting this inhibition with more calcium depends in part on the ATP concentration, which is optimal for these conditions at about $4 \times 10^{-4}$ M (Hasselbach, 1965; Weber *et al.,* 1964).

The inhibitory action of high concentrations of magnesium ions on actomyosin-ATPase activity is of interest to the antagonistic relation between calcium and magnesium ions. In presence of excess magnesium ions, actomyosin appears to dissociate into actin and myosin, losing thereby its characteristic strong ATPase activity. Calcium ion presumably competes with magnesium for a site upon the enzyme and maintains the integrity of the actomyosin and its enzymatic activity as a strong ATPase, if magnesium is also present in adequate concentration. Magnesium was found necessary for actomyosin-ATPase activity. Evidence has accumulated that it complexes with ATP, forming $Mg\text{-}ATP^{2-}$, and that $MgATP^{2-}$, serving as the substrate for actomyosin-ATPase, is decomposed during muscular activity (Hasselbach, 1965).

Some confusion is introduced into experimental results by the myosin-ATPase activity. This is at a lower level than actomyosin-ATPase activity and is activated by calcium and depressed by excess magnesium.

The manner in which microsomes in muscle analogues and sarcoplasmic reticulum in muscle fibers remove calcium is postulated to be by the operation of a calcium pump. The mechanism of active transport of calcium across the membranes of the sarcoplasmic reticulum against a concentration gradient is similar to that for other substances already discussed in Chapter 14. Presumably a carrier, X, attaches to the calcium ion at the outer boundary of the membrane of the sarcoplasmic vesicle and, at the expense of ATP, carries it to the other side and into the cavity of the vesicle (Philpott and Goldstein, 1967).

Quantitative estimates indicate that at highest efficiency two calcium ions are carried per ATP molecule decomposed (Hasselbach, 1964). Such ionic movements must occur rapidly indeed, since liberation of calcium-inciting contraction followed by reabsorption of the ions permitting relaxation must be accomplished within the short period of a muscle twitch.

While in all of our thinking ATP is closely linked with muscular contraction, the evidence for its action in the live muscle fiber can only be inferred. In live muscle fibers it is difficult to demonstrate any changes in ATP concentration of sufficient magnitude, even by spectrophotometric means. Changes in concentration of creatine phosphate, however, can be demonstrated, and the decrease in creatin phosphate is proportional to the number of contractions of the muscle. Creatine phosphate, it will be remembered, is always ready to donate its high energy phosphate bond to ADP, resulting in the formation of ATP and creatin (see Chapter 16), since its equilibrium in the reaction with ADP lies in the direction of ATP production (in the presence of the enzyme creatine phosphokinase). When creatine phosphokinase is inactivated by a specific poison, fluorodinitrobenzene, however, ATP is decomposed to ADP and inorganic phosphate in proportion to the number of contractions induced. This considerably strengthens the hypothesis that ATP furnishes the energy for muscular contraction (Wilkie, 1966).

Here in the muscle fiber is a machine which elegantly converts chemical energy to mechanical energy. Mechanical work is done directly at the expense of chemical energy in ATP, which is converted to ADP and inorganic phosphate. In man-made machines the chemical energy must first be converted to electrical or heat energy. Engineers have yet to find a way to imitate the direct conversion of chemical to mechanical energy which occurs in the muscle cell. The biochemical mechanism for liberation of the chemical potential energy in the living cell is well known. However, one large and critical portion of the puzzle yet remains to be found: How is the final chemical to mechanical energy conversion completed inside the living muscle fiber? (Peachey, 1967.)

## LITERATURE CITED AND GENERAL REFERENCES


Astbury, W. T., 1950: Muscular contraction and relaxation: X-ray studies of muscle. Proc. Roy. Soc. (London) *B137*:58–63.

Barer, R., 1942: The structure of the striated muscle fiber. Biol. Rev. *23*:159–200.

Bendall, J. R., 1954: Relaxing effect of myokinase on muscle fibres; its identity with "Marsh" factor. Proc. Roy. Soc. (London) *B142*:409–426.

Bennett, H., 1960: The structure of striated muscle as seen by the electron microscope. *In* The Structure and Function of Muscle. Bourne, ed. Academic Press, New York, Vol. 1, pp. 137–181.

Betz, E. H., Firket, H., and Regnik, M., 1966: Some aspects of muscle regeneration. Int. Rev. Cytol. *19*:203–227.

Bishop, D. W., 1962: Reactivation of extracted sperm cell models in relation to the mechanism of motility. *In* Spermatozoan Motility. American Association for the Advancement of Science, Washington, D. C., pp. 251–268.

Bozler, E., 1953: The role of phosphocreatine and ATP in muscular contraction. J. Gen. Physiol. *37*:63–70.

Bourne, G. H. (ed.), 1960: Structure and Function of Muscle. 3 vol. Academic Press, New York.

Bratton, C. B., Hopkins, A. L., and Weinberg, J. W., 1965: Nuclear magnetic resonance studies of living muscle. Science *147*:738–739.

Carsten, M. E., and Mommaerts, W. F. H. M., 1965: The accumulation of calcium ions by sarcotubular vesicles. J. Gen. Physiol. *48*:183–197.

Costantin, L. L., and Podolsky, R. J., 1965: Calcium localization of the activation of striated muscle fibers. Fed. Proc. *29*:1141–1145.

Costantin, L. L., Franzini-Armstrong, C., and Podolsky, R. J., 1965: Localization of calcium accumulating structures in striated muscle fibers. Science *147*: 158–160.

Davies, R. E., Cain, D., and Delluva, A. M. M., 1959: The energy supply for muscle contraction. Ann. N. Y. Acad. Sci. *81*:468–476.

Davson, H., 1964: Textbook of General Physiology. 3rd Ed. Little, Brown & Co., Boston, Section 5.

Drummond, G. I., and Black, E. C., 1960: Comparative physiology: fuel of muscle metabolism. Ann. Rev. Physiol. *22*:169–190.

Ebashi, S., 1965: The sarcoplasmic reticulum and excitation-contraction coupling. *In* Molecular Biology of Muscular Contraction. Ebashi, S., Oosawa, F., Sekine, T., and Tonomura, Y., eds. Igaku Shoin Ltd., Tokyo (American Elsevier Publishing Co., New York), pp. 197–206.

Ebashi, S., Oosawa, F., Sekine, T., and Tonomura, Y. (eds.), 1966: Molecular Biology of Muscular Contraction. Igaku Shoin Ltd., Tokyo (American Elsevier Publishing Co., New York).

Elliott, G. F., Lowy, J., and Millman, B. M., 1965: Diffraction from living striated muscle during contraction. Nature *206*:1357–1358.

Endo, M., 1966: Entry of fluorescent dye into the sarcotubular system of the frog muscle. J. Physiol. *185*:224–238.

Excitation-Contraction Coupling in Striated Muscle. 1966. Fed. Proc. Vol. 24.

Fawcett, D. W., 1961: The sarcoplasmic reticulum of skeletal and cardiac muscle. Circulation *24*: 336–348.

Filo, R. S., Bohr, D. F., and Ruegg, J. C., 1965: Glycerinated skeletal and smooth muscle: calcium and magnesium dependence. Science *147*:1582–1583.

Franzini-Armstrong, C., and Porter, K. R., 1964: Sarcolemmal invaginations constituting the T system in fish muscle fibers. J. Cell Biol. *22*: 675–696.

Freygang, W. H., Jr., 1965: Tubular ionic movements. Fed. Proc. *24*:1135–1140.

Gergely, J. (ed.), 1964: Biochemistry of Muscle Contraction. Little, Brown & Co., Boston.

Gibbons, I. R., and Rowe, A. J., 1965: Dynein: a protein with adenosine triphosphatose activity from cilia. Science *149*:424–425.

Hanson, J., and Huxley, J., 1955: The structural basis of contraction in striated muscle. *In* Fibrous Proteins and Their Biological Significance. Symp. Soc. Exper. Biol. *9*:228–264.

Hanson, J., and Lowy, J., 1960: Structure and function of the contractile apparatus in the smooth muscles of invertebrate animals. *In* Structure and Function of Muscle. Bourne, ed. Academic Press, New York, pp. 265–330.

Harris, R. J. C. (ed.), 1961: Cell Movement and Cell Contact. Academic Press, New York.

Hasselbach, W., 1964: ATP-driven active transport of calcium in the membranes of the sarcoplasmic reticulum. Proc. Roy. Soc. (London) *B160*:501–504.

*Hasselbach, W., 1965: Relaxing factor and the relaxation of muscle. Progr. Biophys. *14*:167–222.

Hayashi, T., 1952: Contractile properties of compressed monolayers of actomyosin. J. Gen. Physiol. *36*:139–152.

*Hayashi, T., Sept. 1961: How cells move. Sci. Am. *205*:184–204.

Hayashi, T., and Rosenbluth, R., 1952: Muscle contraction-elongation cycle of loaded surface-spread actomyosin fibers. J. Cell. Comp. Physiol. *40*:495–506.

Hayashi, T., and Rosenbluth, R., 1953: Muscle mechanochemical properties of fibers of surface-spread actomyosin. Proc. Natl. Acad. Sci. U. S. A. *39*:1285–1290.

Heilbrunn, L. V., and Wiercinski, F. J., 1947: The action of various cations on muscle protoplasm. J. Cell. Comp. Physiol. *29*:15–32.

Hill, A. V., 1960: Production and absorption of work by muscle. Science *131*:897–903.

Hill, D. K., 1959: Autoradiographic localization of adenine nucleotide in frog's striated muscle. J. Physiol. *145*:132–174.

Hodge, A. J., 1959: Fibrous proteins of muscle. *In* Biophysical Science—A Study Program. Oncley, ed. John Wiley & Sons, New York, pp. 409–425.

Hoffmann-Berling, H., 1960: Other mechanisms producing movements. Comp. Biochem. *2*:341–370.

Hoyle, G., 1965: Nature of the excitatory sarcoplasmic reticular junction. Science *149*:70–72.

*Huxley, A. F., 1957: Muscle structure and theories of contraction. Progr. Biophys. *7*:255–319.

Huxley, A. F., and Peachey, L. D., 1961: The maximum length for contraction in vertebrate striated muscle. J. Physiol. *156*:150–165.

Huxley, A. F., and Taylor, R. E., 1958: Local activation of muscle fibers. J. Physiol. *144*:426–441.

Huxley, H. E., 1957: The double array of filaments in cross-striated muscle. J. Biophys. Biochem. Cytol. *3*:631–647.

*Huxley, H. E., 1960: Muscle cells. *In* The Cell. Brachet and Mirsky, eds. Academic Press, New York, Vol. 4, pp. 365–481.

Huxley, H. E., 1963: Electron microscope studies of natural and synthetic protein filaments from striated muscle. J. Molec. Biol. *7*:281–308.

*Huxley, H. E., Dec. 1965: The mechanism of muscular contraction. Sci. Am. *213*:18–27.

*Huxley, H. E., and Hanson, J., 1959: The structural basis of the contraction mechanism in striated muscle. Ann. N. Y. Acad. Sci. *81*:403–408.

Huxley, H. E., and Niedergerke, R., 1954: Structural changes in muscle during contraction. Interference microscopy of living muscle fibers. Nature *173*:971–973.

Jewell, B. R., 1959: The nature of the phasic and the tonic responses of the anterior byssal retractor muscle of *Mytilus*. J. Physiol. *149*:154–177.

Johnson, W. H., and Kahn, J. S., 1959: Paramyosin and contraction of catch muscles. Science *130*: 160–161.

Kamin, B., and Bell, A. L., 1966: Synthetic myosin filaments. Science *151*:323–324.

Konigsberg, I. R., 1964: The embryological origin of muscle. Sci. Am. *211*:61–66.

Lorand, L., Schuel, H., Demorsky, R. A., and Meisler, J., 1965: Distribution and some properties of muscle-relaxing particles. *In* Molecular Biology of Muscular Contraction. Ebashi, S., Oosasa, F., Sekine, T., and Tonomura, Y., eds. Igaku Shoin Ltd., Tokyo (American Elsevier Publishing Co., New York).

Lowy, J., Millman, B. M., and Hanson, J., 1964: Structure and function in smooth tonic muscle of lamellibranch molluscs. *In* A Discussion of the Physical and Chemical Basis of Chemical Contraction. Proc. Roy. Soc. (London) *B160*:525–536.

Lundsgaard, E., 1930: Studies on muscle contraction without heat production. Biochem. Zeitschr. *217*: 162–177.

Marsh, B. B., 1952: The effects of ATP on the fiber volume of a muscle homogenate. Biophys. Acta *9*:247–260.

Marshall, J. M., Jr., Holtzer, H., Finck, H., and Pepe, F., 1959: The distribution of protein antigens in striated myofibrils. Exp. Cell Res. *7*(Suppl.): 219–233.

*Mommaerts, W. F. H. M., Brady, A. J., and Abbott, B. C., 1961: Major Problems in muscle physiology. Ann. Rev. Physiol. *23*:529–567.

*Morales, M. F., 1959: Mechanism of muscle contraction. *In* Biophysical Science—A Study Program. Oncley, ed. John Wiley & Sons, New York, pp. 426–432.

Needham, D. M., 1960: Biochemistry of muscular action. *In* Structure and Function of Muscle. Bourne, ed. Academic Press, New York, Vol. 2, pp. 55–104.

Peachey, L. D., 1965: Transverse tubules in excitation-contraction coupling. Fed. Proc. *24*:1124–1134.

Peachey, L. D., 1967: Muscle. Ann. Rev. Physiol. *30*:401–440.

Pease, D. C., Jenden, D. J., and Howell, J. N., 1965: Calcium uptake in glycerol-extracted rabbit psoas muscle fibers. II. Electron microscope localization of uptake sites. J. Cell. Comp. Physiol. *65*:141–154.

Philpott, C. W., and Goldstein, M. A., 1967: Sarcoplasmic reticulum of striated muscle: localization of potential calcium binding sites. Science *155*:1019–1021.

*Porter, K. R., and Franzini-Armstrong, C., March 1965: The sarcoplasmic reticulum. Sci. Am. *212*: 73–80.

Porzehl, H., 1951: Muskelkontraktion und Modellkontraktion. Z. Naturforsch. *6b*:355–361.

Ruch, T. C., and Patton, H. D., 1965: Physiology and Biophysics. 19th Ed. W. B. Saunders Co., Philadelphia, Chapters 1–4.

Ruegg, J. C., 1964: Tropomyosin-paramyosin system and 'prolonged contraction' in a molluscan smooth muscle. Proc. Roy. Soc. (London) *B160*:536–542.

Sandow, A., Taylor, S. R., and Preiser, H., 1965: Role of the action potential in excitation-contraction coupling. Fed. Proc. *24*:1116–1123.

The Sarcoplasmic Reticulum. 1961. J. Biophys. Biochem. Cytol. Vol. 10 (Suppl.).

Simard-Duquesne, N., and Couillard, P., 1962: Ameboid movement. I. Reactivation of glycerinated models of *Amoeba proteus* with ATP. Exp. Cell Res. *28*:85–91.

*Smith, D. S., 1966: The organization and function of the sarcoplasmic reticulum and T-system of muscle cells. Progr. Biophys. *16*:107–142.

Szent-Györgyi, A., 1953: Chemical Physiology of Contraction in Body and Heart Muscle. Academic Press, New York.

Szent-Györgyi, A. G., 1955: Structural and functional aspects of myosin. Advances Enzymol. *16*:313–360.

Szent-Györgyi, A., 1958: Muscle research. Science *128*:699–702.

Szent-Györgyi, G., 1960: Proteins of the myofibril. *In* The Structure and Function of Muscle. Bourne, ed. Academic Press, New York, Vol. 2, pp. 1–43.

Tso, P. O. P., Eggman, L., and Vinograd, J., 1957: Physical and chemical studies of myxomyosin, an adenosine-triphosphate (ATP)-sensitive protein in cytoplasm. Biochim. Biophys. Acta *25*: 532–542.

Weber, A., Herz, R., and Reiss, I., 1964: The regulation of myofibrillar activity by calcium. Proc. Roy. Soc. (London) *B160*:489–501.

Weber, H. H., 1958: The Motility of Muscle and Cells. Harvard University Press, Cambridge, Mass.

Weber, H. H., 1959: The relaxation of the contracted actomyosin system. Ann. N. Y. Acad. Sci. *81*:409–421.

Weber, H. H., and Porzehl, H., 1954: The transference of the muscle energy in the contraction cycle. Progr. Biophys. *4*:60–111.

Weinstein, H. J., 1954: An electron microscope study of cardiac muscle. Exp. Cell Res. *7*:130–146.

Wilkie, D. R., 1954: Facts and theories about muscle. Progr. Biophys. *4*:288–324.

*Wilkie, D. R., 1966: Muscle. Ann. Rev. Physiol. *28*:17–38.

Winegrad, S., 1965a: Role of intracellular calcium movements in excitation-contraction coupling in striated muscle. Fed. Proc. *24*:1146–1152.

Winegrad, S., 1965b: Autoradiographic studies of intracellular calcium in frog skeletal muscle. J. Gen. Physiol. *48*:455–479.

# SECTION VI

# Growth and Cell Division

In preceding chapters an actively functioning cell has always been assumed without a question as to its origin or its fate. In turn, the cell's structure and chemical organization, its environment, its manner of achieving a steady state relationship with its environment, its release of energy for cell activities and its use of the energy in maintaining within itself a state of excitability have been considered. In this section the temporal parameter of cell physiology is explored, with the fate of the cell as the main theme. What is the mechanism of cell growth and cell division, and how does the cell derive the energy for these processes?

Either the cell continues to function until it ages and dies or it grows and divides. In dividing it reproduces its kind, for all cells arise only from preexisting cells. It is almost impossible to study cell growth apart from cell division, since growth precedes cell division, and it must be recognized that the two are always intimately related. To the degree that separation of the two subjects is possible, cell growth is considered in Chapter 26 and the cell division in Chapter 27.

Multicellular organisms grow by an increase in number and by differentiation of cells. Because the problem of differentiation of cells is usually considered in developmental biology it is omitted from discussion here (Berrill, 1961; Bell, 1965).

# CHAPTER 26 / CELL GROWTH

The cell grows from within (by intussusception). Growth occurs only if temperature, pH, oxygen and other conditions are favorable, and if the required nutrients and adequate water are available. In the absence of water, the cell may pass into such dormant resting stages as cysts or spores, and remain thus for many years (see Chapters 3, 7 and 10). If nutrient is lacking, or if certain required nutrients are limiting factors, growth may stop or become abnormal. Growth is also easily inhibited by poisons, radiations and, in fact, by all agents that injure the cell or interfere with normal cell functions.

## THE VIRUS AS A MODEL REPLICATING SYSTEM

Under favorable conditions a virus particle will have produced between 100 and 300 replicas of itself in half an hour after entering a host cell, each replica consisting of a DNA molecule (chromosome) and a covering of protein. As soon as the virus enters the host cell it usually represses the synthesis of host nucleic acids, starts replication of its own DNA and directs transcription of its genes to form messenger RNA. The proteins synthesized under direction of the virus mRNA are synthesized on the host ribosomes, using the soluble RNA of the host cell. Among the proteins synthesized by the virus are the packaging proteins for the viral coating and enzymes such as the RNA polymerase necessary for mRNA synthesis and the enzymes facilitating entry of the virus into the host cell. The substances repressing the host DNA synthesis may be proteins also. Although the virus has given much insight into our knowledge of DNA and RNA structure and function which has helped in the solution of the genetic code (see Chapters 3 and 6), it offers less help in understanding many of the problems of cell growth.

## LIFE CYCLE OF A CELL CULTURE

One bold viewpoint holds that unicellular organisms are potentially immortal and that in multicellular organisms only cells of the "germ plasm" which are passed during reproduction, from generation to generation, are potentially immortal, but the specialized cells of the body (soma) must die (Weismann, 1913).

It has been thought in the past that a stock of protozoans always grows old and dies after prolonged subculture. Thus individuals of the protozoan *Paramecium bursaria,* for example, isolated from cultures which had been kept in the laboratory for many years, divide more and more slowly with time, often contain nuclei in the process of degenerative fragmentation and ultimately die out. Similar life cycles had been observed in other species of protozoa, and attempts were made to determine the span between youth, old age and death for various kinds of protozoans. It was also demonstrated that, after conjugation, some individuals in each of many kinds of protozoans were rejuvenated and more viable; presumably as a result of a favorable genotype (Calkins, 1933). It was then concluded that unicellular organisms age and ultimately die unless periodically rejuvenated by some type of nuclear reorganization. But it was soon realized that this conclusion is not valid, because it is based on observations of protozoans grown in a medium which was not controlled.

To control conditions of growth, it is necessary to grow the protozoans in the absence of any other microorganisms (axenic culture) and preferably on a synthetic nutrient known to be fully adequate (Table 26.1). *Tetrahymena pyriformis* (*geleii*) grown under such conditions has been kept in axenic culture for years by Lwoff, Kidder and Dewey, growing at a rate even in excess of the wild state and without loss of vigor (Hall, 1967). It is of interest that conjugation has never been observed in any of these cultures, yet wild stocks of *Tetrahymena,* isolated from ponds, undergo conjugation each time the proper mating types are mixed (Elliott and Hayes, 1953). Similar experiments with a variety of other kinds of microorganisms in repeated transfer or in steady state growth devices show that many microorganisms may be grown at a constant growth rate for thousands of generations (Brock, 1966). To that extent it may be stated that microorganisms are probably potentially immortal and that growth of protoplasm and cell division follow without abatement if nutrient is available and other conditions are favorable.

Yet a growth cycle ending in death will occur in a pure culture of microorganisms if they are grown in a limited or unbalanced environment, or under other unfavorable conditions. For example, if a single vigorous protozoan is inoculated into a limited volume of a known culture medium, it divides regularly until lack of food limits its growth. A plot of the number of divisions against time gives a characteristic curve (see Fig. 26.1, A). After the initial rise the curve flattens out and forms a plateau (the stationary phase) as food becomes a limiting factor. The curve declines as the individuals begin to die because of inanition. If an identical medium is initially inoculated with a starving individual, a lag precedes the rising straight

**Table 26.1.** Nutrient Requirements of *Tetrahymena pyriformis*

| L-AMINO ACIDS | VITAMINS, ETC. | SALTS |
|---|---|---|
| Arginine | Thiamin* | $MgSO_4$ |
| Histidine | Riboflavin | $Fe(NH_4)_2(SO_4)_2$ |
| Isoleucine | Nicotinic acid | $MnCl_2$ |
| Leucine | Pyridoxin* | $ZnCl_2$ |
| Lysine | Folic acid | $CaCl_2$ |
| Methionine | Pantothenic acid | $CuCl_2$ |
| Phenylalanine | — | $FeCl_3$ |
| Serine* | Purine (guanine) | $KH_2PO_4$ |
| Threonine | Pyrimidine (uracil, cytidine) | $K_2HPO_4$ |
| Tryptophan | | |
| Valine | $\alpha$ lipoic acid | |

* The asterisk indicates that mutants exist not requiring these nutrients. Culture media usually contain in addition to the nutrients listed some carbohydrate or acetate, which supposedly have sparing action for amino acids, but neither these nor sterols nor unsaturated fatty acids are required. For other species of *Tetrahymena* the requirements may be different.

Data from Kidder, Dewey, and Parks, 1951: Physiol. Zool. *24*:70; and Holz, G. G., 1966.

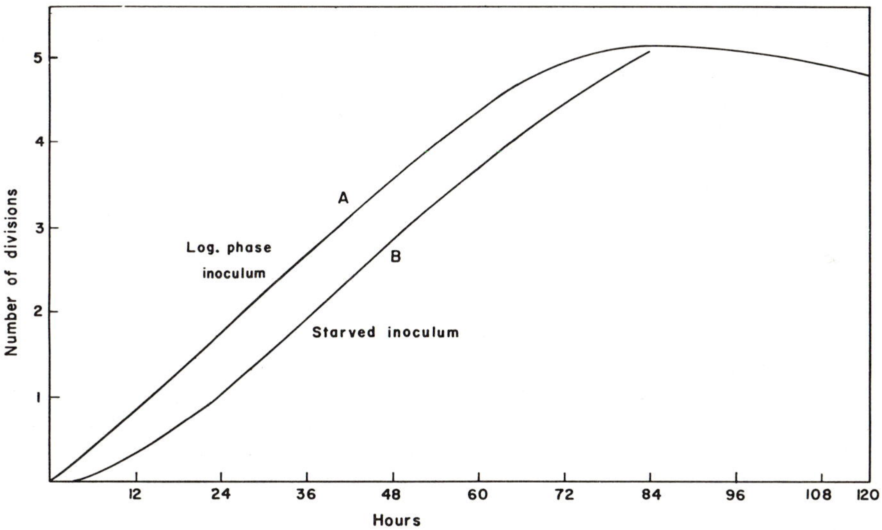


*Figure 26.1.* Population curve for a protozoan culture (*Paramecium caudatum*). If paramecia dividing at the maximal rate (logarithmic phase) are used in inoculum, division continues at a maximal rate. If starved animals are used as inoculum, a lag precedes division at the maximal rate.

line (Fig. 26.1, B), representing the period of recovery from the effects of previous starvation. If individuals are removed from a culture in which they are about to die for lack of food, and are inoculated into fresh medium identical to the first one, after a lag phase they resume growth and continue to multiply at the same rate as in the preceding experiments.

Even if food is available, microorganisms may die when other conditions become unfavorable. For example, yeast grown anaerobically produces alcohol and carbon dioxide; alcohol, as it accumulates, eventually becomes toxic to the yeast. Similarly, lactic acid bacteria produce lactic acid in such concentrations that the low pH of the medium kills off all other microorganisms and finally the lactic acid bacteria themselves (Brock, 1966). Man has made use of some of these special byproducts of microorganisms, e.g., yeast to bake and to make alcoholic beverages and lactic acid bacteria to preserve ensilage and cabbage.

If microorganisms did not die for lack of food or because of other unfavorable conditions such as extremes of pH, temperature, radiations, poisons and so forth, or because of the presence of enemies, they would soon exhaust the nutrients of the earth, since their growth is exponental:

$$N = 2^n \qquad (26.1)$$

In this equation N is the number of individuals, and n is the number of divisions.

It will be seen that such a growth curve resembles a compound interest curve, because each increment resulting from a cell division doubles the population, which will double again at the next division. Such an exponential increase in population is often said to be explosive.

In population growth studies, it is found that equation 26.1 is more convenient in its logarithmic form:

$$\text{Log } N = n \log 2 \qquad (26.2)$$

$$\text{or } n = \frac{\log N}{\log 2} \qquad (26.3)$$

Equation 26.3 is used in plotting the data in Figure 26.1. The equation may be put into a more generalized form:

$$N = 2^{t/\tau} \qquad (26.4)$$

where t is the time and $\tau$ the generation time in the same units.

In multicellular organisms the cells divide actively in the early embryo, but the rate of cell division rapidly declines. When maturity is reached, cell division occurs only in restricted "immature" tissues, or under special conditions.

Although mature and aging cells are no longer growing and dividing, replacement apparently continues at all times. Schoenheimer (1942) and his associates, using radioactive tracers, have claimed that proteins in the cells of even such relatively inert tissues as tendons are continually being replaced. This conclusion, however, has been questioned. In fact, some studies suggested that it is probably not biochemical replacement in old cells that accounts for the incorporation of radioactive materials, but rather that these materials are incorporated into new cells which are continually being formed to replace dying and disintegrating, aging cells. Apparently such a replacement of cells takes place at a much more rapid rate than had been presumed (Leblond and Walker, 1956). Nevertheless, some evidence indicates that animal cells in tissue culture kept for a long time from dividing do indeed incorporate new material (Mazia, 1963).

Cells of aging multicellular individuals usually become somewhat dehydrated, respire more slowly (Zeuthen, 1947, 1953), accumulate calcium, various pigments, e.g., pigments in lipid called lipofuscin granules, and other inert matter (Shock, 1962; Strehler, 1962, 1963), and show a decrease in enzymatic activity. The inert matter may be insoluble excretory residues of metabolism which are incapable of passing through a cell membrane. Dehydration of the cytoplasm, resulting perhaps from insolubilization of proteins and loss of their water-binding capacity, is noted in cells of many but not all species.

It would be of interest in this respect to follow the molecular changes of various cellular constituents with aging of the cell. Such a comparative study has been made with collagen, which forms about 40 per cent of the protein in the human body, being the major constituent in tendon and connective tissue. Experiments indicate that on heating, the tension exerted by collagen fibers from old individuals is greater than that of young individuals and the change is progressive with age. Although the molecular basis of the change in collagen has not been fully established, it appears that cross-linking by ester linkages increases with age, conferring on collagen this particular property (Verzár, 1963).

Why do cells of multicellular organisms age? It could be because the individual cells making up the body have completed their life cycle, or because some unfavorable conditions have arisen. Again, perhaps the body no longer is able to provide adequately for the constituent cells, like an exhausted culture medium for unicellular organisms. Tests of some of these possibilities are considered in the following paragraphs.

## MEASURING GROWTH RATES OF CELLS

Growth of individual cells as expressed by increase in mass can be measured using the diver balance, provided the cells are large enough. An *Amoeba* serves such a project well (Fig. 27.10). It is doubtful, however, whether it will ever be possible to study growth in a minute cell such as a bacterium in this manner.

Inasmuch as in a given population cells are generally of the same size, the number of cells at any one time can be determined by direct counts on an aliquot of the culture. Protozoans can be counted directly in aliquots of the medium, even under low magnification. Bacteria, which are smaller, must be counted under greater magnification than protozoans in a chamber containing a grid and so arranged as to hold a fixed volume of fluid, e.g., the Petroff-Hauser counting chamber. Cells in the blood, eggs of marine animals, protozoans and so forth may also be counted in electronic counters (e.g., the Coulter counter) in which each cell passes through a chamber in which the resistance between two oppositely stationed electrodes changes as the cell passes through and is recorded as a pulse. A sample can thus be placed in the counter and the total number of cells recorded. It is even possible to count cells of different sizes and to record them separately.

A much simpler method which combines the direct count as a calibration with an optical density reading for total numbers has been widely employed for study of growth of populations of small cells such as bacteria. The optical density of a dense population is first measured and the optical densities of known dilutions of the sample are determined. For a standard curve counts are made of the number of cells in several of the samples for which the optical density is known.

The optical density method is satisfactory, provided growth is normal and cells divide regularly. When abnormal growth occurs, for example, when instead of dividing the cells form filaments, then the optical density may increase linearly even though the genome may not have replicated. The optical density by itself does not, therefore, give sufficient information about growth.

Neither counts of cells nor optical density measurements tell how many of the cells recorded are capable of division. The best way to find this is to test a sample of a bacterial culture, for example, on a nutrient agar plate for a colony count. Since a colony is visible only when a relatively large number of cells is present, the presence of a colony indicates an initial viable, healthy cell which has undergone many rounds of replication. With protozoans it is possible to determine periodically the number of cells in a small test tube (or other container) and so to follow the division rate of a single cell lineage (Giese, 1946).

Because, under strictly controlled environmental conditions, oxygen consumption is proportional to the number of cells, provided an adequate oxygen supply is available, it is possible to determine the growth of cells by the increase in oxygen consumption (Fig. 26.2). The change in total concentration of certain respiratory enzymes in growing cells under an unlimited oxygen supply may also be

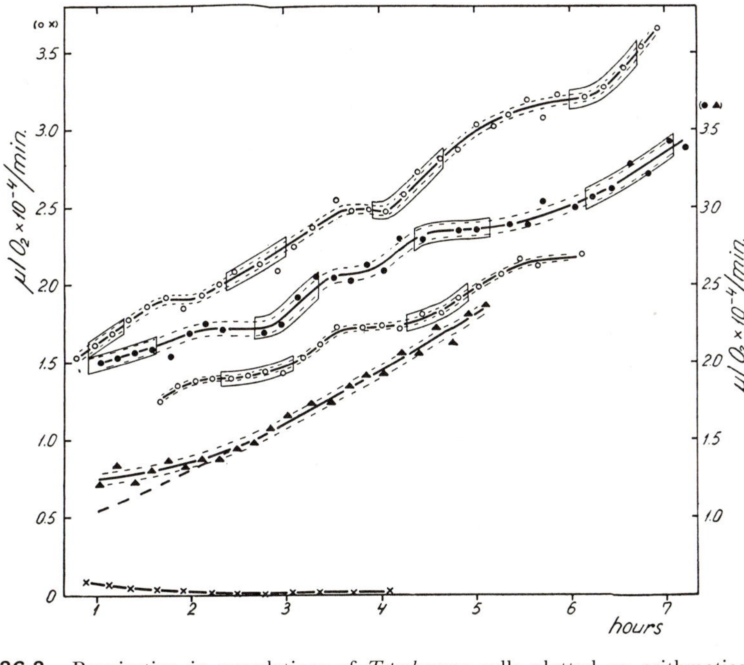


***Figure 26.2.*** Respiration in populations of *Tetrahymena* cells plotted on arithmetical scales against time in hours. Open and full circles: three experiments with synchronized populations. Upper curve: 10 cells initially present, about 100 cells after the experiment. Second curve from above: 11 cells initially present. Third curve from above: 9 cells initially present, 70 in the end. In the three curves, frames indicate division periods observed in the divers. The fourth curve (triangles) is a control run with 20 cells from an unsynchronized population. A control run with empty diver is represented by crosses. (Experiments by H. Thormar; from Zeuthen and Scherbaum, 1954: Colston Papers *7*:151.)

used as a measure of the rate of growth of cells, but again precautions must be taken to control the environmental factors.

Growth of the amount of certain cell constituents may be followed in living cells by the rate of incorporation of tracers, e.g., $^{35}S$ incorporated into protein. Since protein increases continuously in a healthy, growing culture, the label concentration in the culture may be a satisfactory measure of the overall rate of growth. Tritiated thymidine and $^{14}C$-thymidine have been used in a similar manner to label the DNA. It must be remembered that replication is semiconservative. Therefore, the ratio of the unlabeled to the labeled chromosomes will become smaller the larger the number of cell generations and replications taking place in labeled phymidine.

## GROWTH OF CELLS IN TISSUE CULTURE

Cells in culture, when the cells are transferred periodically to fresh, controlled medium, pose a unique opportunity to

**Table 26.2.** A Serum-Free Medium That Supports Quantitative Growth of Single S3-HeLa Cells

| COMPONENTS | GRAMS/LITER | COMPONENTS | GRAMS/LITER |
|---|---|---|---|
| *1. Amino Acids* | | | |
| L-Arginine HCl | 0.015 | L-Valine | 0.010 |
| L-Histidine HCl | 0.015 | L-Glutamic acid | 0.030 |
| L-Lysine HCl | 0.032 | L-Aspartic acid | 0.012 |
| L-Tryptophan | 0.008 | L-Proline | 0.010 |
| β-Phenyl-L-alanine | 0.010 | Glycine | 0.040 |
| L-Methionine | 0.010 | Glutamine | 0.080 |
| L-Threonine | 0.015 | L-Tyrosine | 0.016 |
| L-Leucine | 0.010 | L-Cystine | 0.003 |
| L-Isoleucine | 0.005 | | |
| *2. Vitamins and Growth Factors* | | | |
| Hypoxanthine | 0.010 | Choline | 0.0012 |
| Thiamine HCl | 0.0020 | Calcium pantothenate | 0.0012 |
| Riboflavin | 0.00020 | Niacinamide | 0.0012 |
| Pyridoxine HCl | 0.00020 | Inositol | 0.0004 |
| Folic acid | 0.00004 | Vitamin $B_{12}$ | 0.0010 |
| Biotin | 0.00004 | | |
| *3. Salts and Other Small Molecules* | | | |
| NaCl | 7.40 | $CaCl_2 \cdot 2H_2O$ | 0.016 |
| KCl | 0.285 | $NaHCO_3$ | 1.20 |
| $Na_2HPO_4 \cdot 7H_2O$ | 0.29 | Glucose | 1.10 |
| $KH_2PO_4$ | 0.083 | Phenol red | 0.0012 |
| $MgSO_4 \cdot 7H_2O$ | 0.154 | | |
| *4. Proteins* | | | |
| Normal serum albumin | 2.0 | Fetuin | 2.0 |

From Puck, T. T., 1959–1961: Quantitation of growth of mammalian cells. Harvey Lect. 55:1–12. A completely synthetic medium for mammalian cells has been developed for growth of NCTC strain 2071, for example. The medium contains 69 crystalloidal compounds. Growth is somewhat slower than when protein, especially blood serum, is added (McQuilkin *et al.*, 1957).

study cell growth and aging processes in a variety of cells.

Cells from various plant or animal tissues have been successfully cultured in the absence of other living things. In a typical culture medium for cells of animal tissues, plasma serves as a substrate for attachment of the cells, and an extract from an embryo usually serves as nutrient, although synthetic media are now widely used—see Table 26.2 (White, 1963; Waymouth, 1965). The cells must be periodically washed free of metabolic wastes. If these conditions are met, the tissue cells are capable of living, growing and dividing at a constant rate year after year and may be alive and fully vigorous many years after the animal from which the tissue cells were taken would have normally died. Thus Carrel isolated a piece of chick heart and kept it alive in culture from 1913 to 1946, when the experiment was terminated by choice. Plant cells, too, have been grown in tissue culture and in completely synthetic media, and the requirements for different types of plant tissues have been determined (White, 1963).

When cells of human tissues, kept in tissue culture, are dissociated, clones can be grown from single cells in colonies much like microorganisms (Puck, 1961). Such cultures are invaluable for studying many problems. They show that cells of man, e.g., epithelial cells and fibroblasts (cells of connective tissue), like those of the chick, grow rapidly and multiply indefinitely if supplied with the proper nutrients and conditions and washed free of wastes (Paul, 1960; Eagle, 1963; Evans *et al.*, 1964; Willmer, 1965). Protein-free culture media have been developed for this purpose (Eagle, 1960; Willmer, 1965).

## GROWTH OF CELLS IN TUMORS

Under certain conditions a cell previously in stasis (not growing or dividing) begins to grow and divide actively, forming a *tumor*, or a cancerous growth (Congdon and Mari-Chavez, 1963; Gause, 1966; Todaro *et al.*, 1966). A *cancer* is the result of uncontrolled proliferation of undifferentiated cells when the cells grow at an exponential rate comparable to that of microorganisms, as shown in Figures 26.3 and 26.4. The cancer may remain localized in a given area, or it may *metastasize* (spread by way of the lymph or blood) to all parts

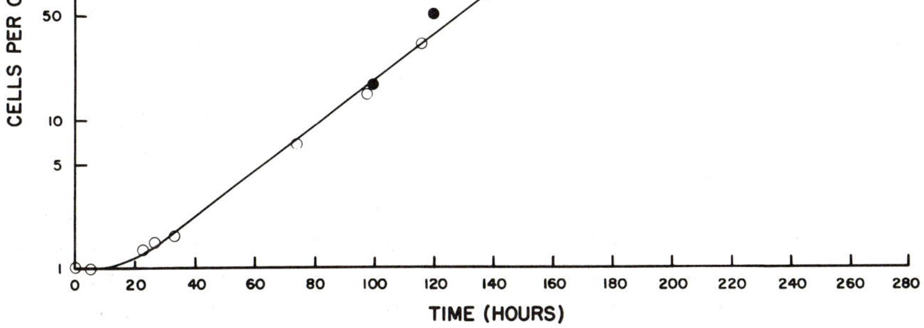


***Figure 26.3.*** Growth curve of HeLa cell colonies plotted on a logarithmic scale, using points from two sets of experiments (dots and circles). The cells require about 24 hours to become established (lag phase). Then they multiply at a steady rate (logarithmic phase) which can be maintained almost indefinitely if food supply of the cells is continually renewed. Note the resemblance to Figure 26.1. The HeLa cell is from a tumor. (From Puck, 1956: J. Exp. Med. *103*:273.)

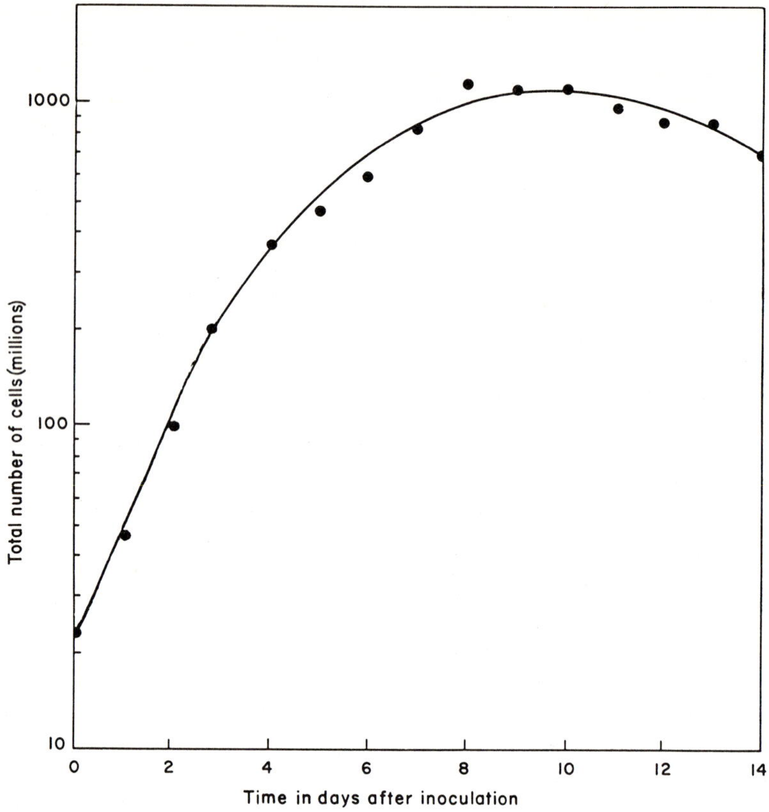


***Figure 26.4.*** Population growth curve for cells of the Ehrlich ascites tumor grown in the mouse. This tumor is made up of individual cells growing in the body (ascitic) fluid. The initial injection on "0" days consisted of $22 \times 10^6$ cells. Note the initial logarithmic rise for almost 4 days, the later decline in rapid growth, and then the decrease in number of cells as some of them die, the whole curve resembling that in Figure 26.1. The mean survival time for mice with the tumor so inoculated is 14 days. Each point is the average of six determinations (By courtesy of A. Furst and M. Dollinger, Stanford University.)

of the body. The undifferentiated cells grow and disorganize the tissues among which they become located. The cancer cells do not appear to be markedly different in nucleolar (Busch *et al.,* 1963) or nuclear fine structure (Bernhard and Granboulan, 1963; Kollner, 1963) or in macromolecular syntheses (Seed, 1966) and biochemistry (Busch, 1962). To prevent or cure cancer it will be necessary to gain a fundamental knowledge of the nature of cell growth and of the mechanism by which the body normally holds the growth and division of cells under control.

At present cancer is treated by various techniques which aim only to suppress its rate of growth once it has started. Most of these techniques originated from study of cell activities. These techniques include ionizing radiations, to suppress division of cells in certain types of cancer (Alexander, 1965); radioactive salts selectively incorporated into a tissue containing cancerous cells, to suppress cell division; and substances which block the synthesis of metabolites necessary for the growth and division of cancerous cells (Pullman and Pullman, 1955).

On a cellular basis interpretation of cancer induction is still lacking, in spite of many attempts. However, one interesting suggestion is that cancer results from injury of the cellular respiratory equipment, with a consequent increase in glycolysis (anaerobic metabolism). Glycolysis is considered a lower order of energy-liberating mechanism, incapable of maintaining structural integrity and organization in cells of higher forms. Many poisons and other agents that induce cancer have such effects upon the respiratory mechanisms of cells (Warburg, 1956). The question arises whether such changes are a consequence, rather than the cause, of cancer. When cells from a tumor are grown in tissue culture, they often behave and respire like normal cells under similar conditions. However, when placed inside an organism, tumor cells multiply, whereas normal cells do not. Tumor cells synthesize protoplasm so actively that they soon exhaust the supplies of high energy phosphates required for the normal pathways of aerobic respiration. It is then that they obtain a major part of the required energy from reactions proceeding along other pathways. They also become more susceptible to injury or death from hydrogen peroxide, much like anaerobic cells (Warburg *et al.,* 1957).

It has been suggested that a mutation could so change a normal cell as to remove it from the control of hormones and other influences which normally repress multiplication of cells of tissues in the body. For example, bone marrow cells from a normal individual and bone marrow cells from a leukemic individual, both grown in tissue culture, are inhibited by antagonists to folinic acid, one of the B vitamins required for cell division (see Table 3.9). Within 15 minutes, division dramatically stops. Normal cells remain inhibited from dividing as long as they are exposed to the antagonist. However, the leukemic cells begin to divide actively after 24 hours, though still exposed to the antagonist. During this time they appear to have inactivated the antagonist because the filtrate from this culture no longer inhibits either normal cells or leukemic cells, whereas the filtrate from the experiment with the normal cell culture is still inhibitory. (The inactivated material has been separated, crystallized and reactivated.) Apparently, the leukemic cells differ from the normal ones in being able to develop an enzyme system capable of handling the folinic acid antagonist. Since enzymes are considered to be produced by genes, it would seem likely that the leukemic cells differ genetically from the normal ones (Jacobson, 1958).

Another interpretation is also possible. It will be recalled that lysogenic viruses "hide" in bacterial cells, remaining inactive except under certain circumstances. But they do produce a change in the host cell since the bacterium is then not infected by the active or lytic variety of the same species of phage (see Chapter 6). It is conceivable that cells become leukemic as a result of invasion by viruses which change the properties of the cells, enabling them to escape the various controls that the body normally exerts upon the bone marrow cells which restrict multiplication to a level adequate for replacement of lost cells. Such a change has been observed in tissue culture cells from each of a number of rodents (hamster, rat, mouse) when infected by one of several viruses of the papova viruses (polyoma virus and simian virus 40, or SM40). Since the cells form colonies in culture medium, the physical and biochemical characteristics of the normal and virus-infected cell lines can be readily compared. Two striking changes were noticed in the virus-infected cells. Normal cells in a tissue culture on making contact with other cells stop moving and orient with respect to one another. Virus-infected cells, on the other hand, move right on, crawling over one another, and they continue to divide as long as nutrients are available (Fig. 26.5). This indicates not only a change in surface properties but also in the signal for replication. Some change has apparently been produced in the genome of the tissue culture cells by virus infection (transduction? see Chapter 6) (Dulbecco, 1967).

A virus-infected cell offers a favorable opportunity to study the gene changes which convert a normal cell into a cancer cell, because the number of genes present in a virus of the type discussed here is small, perhaps nine or ten in all. An attempt has been made to identify all of the virus genes and then to identify which of these genes may cause changes resulting

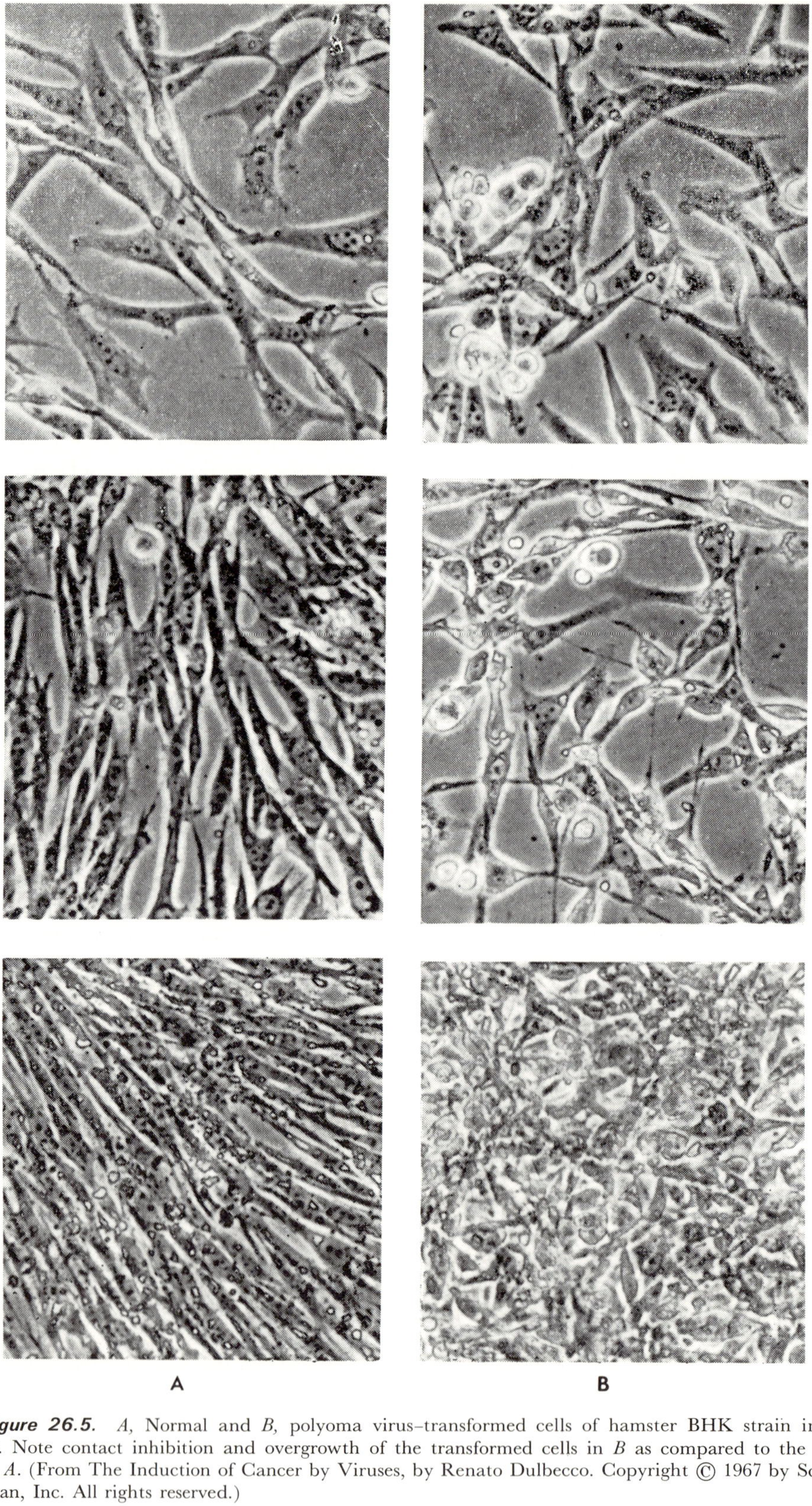

A B

**Figure 26.5.** *A,* Normal and *B,* polyoma virus-transformed cells of hamster BHK strain in tissue culture. Note contact inhibition and overgrowth of the transformed cells in *B* as compared to the normal cells in *A*. (From The Induction of Cancer by Viruses, by Renato Dulbecco. Copyright © 1967 by Scientific American, Inc. All rights reserved.)

in cancer induction. In the polyoma virus there is enough DNA (about 5000 base pairs) to code for about eight or nine proteins. Probably one or two code for the protein in the viral coat. Six functions, each arising in the host cell after infection and each coded by one gene in the virus, have now been studied in the infected cells. These functions involve changes in biochemical or immunological properties and probably depend upon proteins produced by these genes. None of these functions have been specifically related to the two cancerous properties which were observed in virus-infected tissue culture cells—contact inhibition and uncontrolled multiplication (Dulbecco, 1967).

It is recognized that mutation, without virus infection, may occur in a cell of an animal (or plant). Such a mutation may result in a change to a cancerous condition, releasing the cell from regulation which normally controls cell division. However, it is difficult to identify such a genetic change because there are so many genes in the genome of the animal cell that analysis is made immensely more complicated.

Reversal of the tumorous condition of cells in plants and animals has raised the question whether the change in some cases might not be an altered expression of a gene rather than a change in genome. Thus the crown gall tumor can be reversed by a series of transfers to healthy plants. Reversion of tumors has also been observed in both newt and mouse (Braun, 1965).

A variety of evidence for viral and for genetic origins of cancer has accumulated, none yet permitting a successful analysis of the basic change involved. Cancer remains an important problem to the cell physiologist (Hieger, 1961; Sanford, 1965).

## REGULATION OF FORM AND SIZE IN CELLS

The ability of a cell to maintain its organization and characteristic size at a given temperature and in the presence of adequate nutrients is perhaps best shown by the regeneration of protozoans after removal of parts of a cell (see Fig. 26.6).

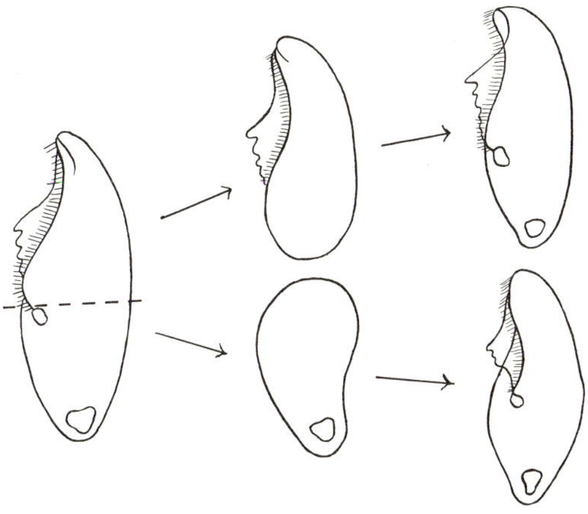

***Figure 26.6.*** Regeneration of missing parts in two pieces of a bisected *Blepharisma.* Note especially the regeneration of the hypostome (mouth region) in the posterior segment. Regeneration occurs within about 6 hours at room temperature. (From a classroom experiment. For details, see Moore, 1924: J. Exp. Zool. *39*:249.)

In fact, a *Blepharisma* from which the entire hypostome (mouth area) has been removed regenerates it sufficiently well to feed within 6 hours (Giese and McCaw, 1963). If large portions of the cell are removed, but macronuclear fragments and micronuclei remain, regeneration will result in a miniature edition of the species. After feeding, however, the regenerated organism rapidly regains its original size. Redifferentiation is usually restricted to portions of the cell adjoining those removed by microsurgery; however, even then, materials required for regeneration must be obtained from existent structures. In some species of ciliates, amazingly enough, even a small injury results in dedifferentiation of the entire animal and redifferentiation of all the organelles (such as membranelles, motile organs and mouth parts). *Stentor* mangled into about 40 adherent pieces survives and regenerates its normal form (Tartar, 1960). As a result of this morphogenetic activity, organelles which have been removed or damaged are regenerated, and the characteristic form and size of the species are reconstituted (Tartar, 1967).

When pieces are removed from eggs of some chordates and echinoderms, the eggs produce complete embryos of reduced size. This shows that the egg maintains a certain organization, just as does the unicellular organism. When pieces are removed from the eggs of annelids or mollusks, the eggs develop into partial embryos only. In such cases the pattern is supposedly already laid down earlier in these eggs and each part is already destined to form a particular organ.

It is probable that most vertebrate cells can recover from minor injuries, although experimental studies on this subject have been limited. The neuron of a vertebrate, for example, regenerates damaged axons or dendrites provided the cell body is intact. Like any other cell, the neuron needs a nucleus for normal functioning.

The nature of the process of cell regeneration and redifferentiation and also the nature of the cellular interactions leading to differentiation in development and morphogenesis are not known. In the slime mold *Dictyostelium* it has been shown that apparently some chemical attraction and cellular interactions result in the aggregation of the ameboid cells to form the plasmodium, and also result in the differentiation of the fruiting stalks. (Bonner, 1952).

Some ciliates, such as *Stylonychia* and *Blepharisma,* both of which are ordinarily bacterial feeders, may be fed successively a series of protozoans of increasing size, ultimately including their own kind (cannibalism). As a result they become considerably enlarged, having many times the bulk of the normal bacteria-fed individuals. The mouth parts, membranes, motile organs, macronuclei and other structures become proportionally enlarged, and the large size is maintained so long as the abnormal diet is continued. In all cases, however, it has been noticed that when the giants are again put on a bacterial diet, division is more rapid than it is in controls fed continuously on bacteria, so that the giants soon return to the characteristic species size (Giese 1938; Giese and Alden, 1938).

Conversely, ciliates may be starved for long periods, during which they undergo reduction in size. *Tetrahymena,* for example, will become a minute particle with little resemblance to its former self. Yet on addition of nutrients it quickly enlarges to the normal size, and when it has grown double this size it divides, thus maintaining the normal size of the species.

That size of cells is regulated by division is most clearly demonstrated with *Amoeba proteus* (Prescott, 1956a, b). During growth of an ameba the mass increases until it just doubles. Then no further increase occurs, and after a lapse of time it divides, each daughter cell receiving half the mass of the mother cell. If a part of an ameba is cut off during the period of growth, division will not occur at the time it does in control amebas, but only after the removed mass of protoplasm has been replaced by growth. An ameba may be prevented from dividing, presumably indefinitely, by the periodic removal of a piece of protoplasm, although there is some evidence to the contrary (Hirshfield *et al.,* 1960).

Heredity plays an important role in the maintenance of size and shape of cells. This is evident since clones of a given species of microorganisms under a given set of environmental conditions tend to be quite constant in size and form for a long series of generations—a constant nuclear to cytoplasmic ratio is maintained. However, overall size was found to vary with a change in each environmental factor studied (Adolph, 1931).

Experiments have shown that increasing the number of chromosomes by polyploidy is apparently often attended with an increase in cell size (Fankhauser, 1952). If the number of chromosomes is increased, each chromosome maintains its characteristic size; therefore the amount of chromatin present in the polyploid nucleus is correspondingly increased. The nucleus increases proportionally in size, and the cytoplasm compensates. A polyploid salamander, for example, has giant cells. The same is true of the cells in certain mosses which are usually haploid but may be experimentally produced in diploid, triploid or tetraploid form. Polyploidy wherever found is always correlated with increased size of the polyploid cell (White, 1961).

In plant cells, growth might appear to be limited by the rigidity of the cell walls. However, the cell wall is known to be rendered softer and plastic by the plant growth hormone, auxin (indole acetic acid), this being, presumably, what occurs to cells in the region of growth. For example, when plants illuminated on one side only turn toward the light, auxin was shown to accumulate on the side away from the light, inducing extra elongation of young cells in the region of growth on that side. The curvature is thus a result of increase in cell size and not from cell multiplication. Even in the dark application of auxin to one side of a plant results in curvature from increase in cell size on that side. Similar softening of cell walls occurs whenever plant cells increase in volume after cell division, but here the action of auxin is less evident than in tropistic responses. Again, during embryogeny plant cells lack a central sap vacuole. Differentiation usually involves the development of a large central sap vacuole, along with the resultant enlargement of the cell. During such increases in cell size, auxins render the cell walls plastic. It is conceivable that regulation of growth in plant cells may be controlled in part by the auxin supply (Hillman, 1962).

## UNBALANCED GROWTH

When a mutant form of *E. coli*, which requires for its growth the pyrimidine base thymine, is grown in a medium deficient in thymine but adequate in all the other nutrients, growth (as measured by turbidity) apparently occurs at essentially the normal rate. However, when this culture has been grown in the deficient medium for a period equivalent to that required for one generation and then is plated out, even on a balanced medium containing all necessary nutrients, including thymine, it is observed that about 90 per cent of the individuals fail to produce colonies. Injury therefore appears to occur when replication is attempted. If so grown for two generations, about 99 per cent fail to form colonies. Tests show that protein and RNA increase in the cells, but there is only a slight increase in DNA (Cohen and Barner, 1954). Such disproportionate growth of different cell constituents is spoken of as *unbalanced growth.* A similar phenomenon has been observed in yeast (Ridgeway and Douglas, 1958) and *Neurospora* (Strauss, 1958). However, it does not appear to be general, since some cells recover even after becoming much enlarged (Mazia, 1960).

Recent studies on repair mechanisms in cells (see Chapter 9) have indicated that during the course of replication some breaks occur in the double-stranded DNA as it opens before the growing point. Such breaks occur, perhaps at a relatively high frequency, and are apparently readily repaired provided the nucleoside triphosphates and the necessary enzymes are present. For example, if thymine is withdrawn from the medium of a thymineless mutant of *E. coli* and DNA replication is attempted, the DNA molecule breaks and cannot be repaired because of the lack of the nucleoside containing thymine. Consequently the cell cannot replicate, the chromosome fragments, and the cell dies (Hanawalt, 1966). A similar course of events may occur in other cases as well.

An interesting experiment is one in which a mutant requiring some nutrient common to synthesis of both RNA and DNA is maintained alive on a medium deficient in the nutrient (Strauss, 1958). In this case, no growth, or very little, can occur. Presumably it is only when growth is unbalanced, that is, when most constituents can increase in quantity but one does not, that injury occurs.

It might be expected that any factor (other than limitation of nutrients) which leads to unbalanced growth might injure the cell. Experiments with ultraviolet radiations (to which synthesis of DNA is one of the most susceptible processes in the cell—Kelner, 1953) show that the irradiated cells in which DNA synthesis is just inhibited may continue to grow (as measured by optical density) and individual cells become giants accumulating RNA and proteins. Plated out, however, they also fail to produce colonies—

i.e., they are unable to divide (Cohen and Barner, 1954). It is argued that the unbalanced growth of various structures in the absence of DNA synthesis has deranged the cell so that it is no longer able to divide and reproduce, even when later the DNA synthesis has apparently recovered from the radiation injury. If, however, the cells are irradiated sufficiently to stop DNA synthesis and then kept in non-nutrient solution so that they cannot grow, DNA synthesis recovers slowly, and addition of required nutrients will restore balanced growth. These are the very conditions under which dark repair can best occur (Setlow, 1964).

## REGULATION OF GROWTH

*Balanced growth* is a steady state condition in which every component of the cell doubles in a cell generation (time between divisions). The generation time depends upon the species and in part upon the conditions of the environment. Thus, *E. coli,* on a medium containing water, glucose, ammonium chloride, potassium basic phosphate, sodium sulfate and trace metals, kept at 37° C., divides every 45 minutes. If supplied with amino acids and some other organic materials at 37° C., it divides every 20 minutes. The cells have thus reached a new steady state in which duplication of cell constituents occurs much quicker. Division is considerably retarded at 15° C., at which temperature a cell gives rise to only about 128 cells after 24 hours, whereas at 37° C. the number would be about $2 \times 10^9$.

Cell growth remains a challenging problem for experimental study. The question at hand is to determine how a cell achieves balanced growth regulation. Since growth is a result of metabolic activity along various fronts, the questions really asked are: how is metabolic activity regulated? and how are the various types of metabolic activity coordinated? The induction of enzyme synthesis and the repression of enzyme synthesis have already been discussed (see Chapter 21), and it will be recalled that the end product inhibition is a means of regulating many of the steps in synthesis, producing immediate changes in rate of reactions and thus preventing accumulation of cell products. In Figure 21.3 the interplay between the synthesis and end product inhibition of various substances needed for the synthesis of DNA and RNA is considered. The net effect is a proportionality between synthesis of the two important nucleic acids. A similar interplay in other synthetic pathways perhaps leads to production of useful amounts of proteins and other substances needed by the cell. Thus a balance may be achieved in the amounts synthesized of the various requirements for cell growth. It is likely that steady state necessary for balanced growth is thus achieved.

## LITERATURE CITED AND GENERAL REFERENCES

Adolph, E. F., 1931: The Regulation of Size as Illustrated in Unicellular Organisms. Charles C Thomas, Springfield, Ill.

Alexander, P., 1965: Atomic Radiation and Life. 2nd Ed. Penguin Books, Baltimore, Chapter 8.

Bell, E. (ed.), 1965: Molecular and Cellular Aspects of Development. Harper & Row, New York.

Bernhard, W., and Granboulan, N., 1963: The fine structure of the cancer cell nucleus. Exp. Cell Res. *9*(Suppl.):19–53.

Bernstein, E., and Zeuthen, E., 1966: Relation of RNA synthesis to temperature as inducer of synchronous division. Comp. Rend. Lab. Carlsberg *35*:501–517.

Berrill, N. T., 1961: Growth, Development and Pattern. W. H. Freeman & Co., San Francisco.

Bonner, J. T., 1952: Morphogenesis. Princeton University Press, Princeton, N. J.

Boyland, E., (ed.), 1964: Mechanisms of carcinogenesis: chemical, physical and viral. Brit. Med. Bull. *20*:87–164.

Braun, A. C., Nov. 1965: The reversal of tumor growth. Sci. Am. *213*:75–83.

Brock, T. D., 1966: Principles of Microbial Ecology. Prentice-Hall, Englewood Cliffs, N. J.

Busch, H., 1962: Biochemistry of the Cancer Cell. Academic Press, New York.

Busch, H., Byvoet, P., and Smetana, K., 1963: The nucleolus of the cancer cell: a review. Cancer Res. *23*:313–338 (401 references).

Calkins, G. N., 1933: The Biology of the Protozoa. Lea & Febiger, Philadelphia.

Cohen, S. S., and Barner, H. D., 1954: Studies on unbalanced growth in *Escherichia coli.* Proc. Natl. Acad. Sci. U. S. A. *40*:885–893.

Congdon, C. C., and Mori-Chavez, P., 1963: Control of Cell Division and the Induction of Cancer. Monograph No. 14, U.S. Dept. Health, Education and Welfare, Bethesda, Md.

Daudel, P., and Daudel, R., 1966: Chemical Carcinogenesis and Molecular Biology. Interscience Publishers, New York.

Dulbecco, R., April 1967: The induction of cancer by viruses. Sci. Am. *216*:28–37.

Eagle, H., 1960: The sustained growth of human and animal cells in a protein-free environment. Proc. Natl. Acad. Sci. U. S. A. *46*:427–432.

Eagle, H., 1963: Population density and nutrition of cultured mammalian cells. *In* The General Physiology of Cell Specialization. Mazia and Tyler, eds. McGraw-Hill Book Co., New York, pp. 151–170.

Elliott, A. M., and Hayes, R. E., 1953: Mating types in *Tetrahymena.* Biol. Bull. *105*:269–284.

*Emmelot, P., and Muhlbock, D. (eds.), 1964: Cellular Control Mechanisms and Cancer. American Elsevier Publishing Co., New York.

Evans, V. J., Bryant, J. C., Kerr, H. A., and Schilling, E. L., 1964: Chemically defined media for cultivation of longer-term cell strains from four mammalian species. Exp. Cell Res. *36*:439–474.

Fankhauser, G., 1952: Nucleo-cytoplasmic relations in amphibian development. Int. Rev. Cytol. *1*: 165–193.

Frei, E., III, and Freireich, E. J., May 1964: Leukemia. Sci. Am. *210*:88–96.

Gause, G. F., 1966: Microbial Models of Cancer Cells. W. B. Saunders Co., Philadelphia.

Giese, A. C., 1938: Cannibalism and gigantism in *Blepharisma.* Trans. Am. Micros. Soc. *57*:245–255.

Giese, A. C., 1946: A simple method for division rate determination in *Paramecium.* Physiol. Zool. *18*:158–161.

Giese, A. C., and Alden, R. H., 1938: Cannibalism and giant formation in *Stylonychia.* J. Exp. Zool. *78*:117–134.

Giese, A. C., and McCaw, B. K., 1963: Regeneration rate of *Blepharisma* with special reference to the effect of temperature. J. Protozool. *10*:173–182.

*Hall, R. P., 1967: Nutrition and growth of protozoa. *In* Research in Protozoology. Chen, ed. Vol. 1. Pergamon Press, New York.

Hanawalt, P. C., 1966: The U. V. sensitivity of bacteria: its relation to the DNA replication cycle. Photochem. Photobiol. *5*:1–12.

Hieger, I., 1961: Carcinogenesis. Academic Press, New York.

Hillman, W. S., 1962: The Physiology of Flowering. Holt, Rinehart & Winston, New York.

Hirshfield, H. I., Tulchin, H., and Fong, B. A., 1960: Regeneration and cell division in two protozoan species. Ann. N. Y. Acad. Sci. *90*:523–528.

Holz, G. G., Jr., 1966: Is *Tetrahymena* a plant? J. Protozool. *13*:2–4.

Jacobson, W., 1958: The toxic action of drugs on the bone marrow. *In* A Symposium on the Evaluation of Drug Toxicity. Walpole and Spinks, eds. Little, Brown & Co., Boston, pp. 76–101.

Jennings, H. S., 1945: *Paramecium bursaria.* J. Exp. Zool. *99*:15–31.

Kelner, A., 1953: Growth, respiration and nucleic acid synthesis in ultraviolet-irradiated and in photoreactivated *Escherichia coli.* J. Bact. *65*:252–262.

Kollner, P. C., 1963: The nucleus of the cancer cell. Exp. Cell Res. *9*(Suppl. 90):3–14.

Leblond, C. P., and Walker, B. E., 1956: Renewal of cell populations. Physiol. Rev. *36*:255–276.

Leopold, A., 1955: Auxins and Plant Growth. University of California Press, Berkeley, Calif.

*Mazia, D., 1960: Mitosis and the physiology of cell division. *In* The Cell. Brachet and Mirsky, eds. Academic Press, New York, Vol. 3, pp. 77–412.

Mazia, D., 1963: Synthetic activities leading to mitosis. J. Cell. Comp. Physiol. *62*:123–140.

McQuilkin, W. T., Evans, V. J., and Earle, W. R., 1957: The adaptation of additional lines of NCTC clone 929 (strain L) cells to chemically defined protein-free medium NCTC 109. J. Natl. Cancer Inst. *19*:885–907.

Oberling, C., and Bernhard, W., 1961: Morphology of the cancer cells. *In* The Cell. Brachet and Mirsky, eds. Academic Press, New York, Vol. 5, pp. 405–496.

Paul, J., 1960: Cell and Tissue Culture. Williams & Wilkins, Baltimore.

Paul, J., 1964: Cell Biology, a Current Summary. Stanford University Press, Stanford, Calif.

Prescott, D. M., 1956a: Relation between cell growth and cell division. II. The effect of cell size on cell growth rate and generation time in *Amoeba proteus.* Exp. Cell Res. *11*:86–94.

Prescott, D. M., 1956b: Relation between cell growth and cell division. III. Changes in nuclear volume and growth rate and prevention of cell division in *Amoeba proteus* resulting from cytoplasmic amputations. Exp. Cell Res. *11*:94–98.

*Puck, T. T., 1961. Quantitation of growth of mammalian cells. Harvey Lect. *55*:1–12.

Pullman, A., and Pullman, B., 1955: Electronic structure and carcinogenic activity of aromatic molecules. Advances Cancer Res. *3*:117–169.

Ridgeway, G. J., and Douglas, H. C., 1958: Unbalanced growth of yeast due to inositol deficiency. J. Bact. *76*:163–166.

Sanford, K. N., 1965: Malignant transformation of cells *in vitro.* Int. Rev. Cytol. *18*:249–311.

Schoenheimer, R., 1942: The Dynamic State of Body Constituents. Harvard University Press, Cambridge, Mass.

Seed, J., 1966: The synthesis of DNA, RNA and protein in normal and tumor strain cells. I. Fresh embryo human cells. J. Cell Biol. *28*:233–248.

*Setlow, R. E., 1964: Physical change and mutagenesis. J. Cell. Comp. Physiol. *64*(Suppl. 1):51–68.

Shock, N. W., Jan. 1962: The physiology of aging. Sci. Am. *206*:100–110.

Strauss, B. S., 1958: Cell death and 'unbalanced growth' in *Neurospora.* J. Gen. Microbiol. *18*:658–669.

Strehler, B. L., 1962: Time, Cells and Aging. Academic Press, New York.

*Strehler, B. L., 1963: The senescence of differentiated cells; some chemical bases of cellular aging. *In* The General Physiology of Cell Specialization.

Mazia and Tyler, eds. McGraw-Hill Book Co., New York, pp. 116–148.
Tartar, V., 1960: Reconstitution of minced *Stentor coeruleus*. J. Exp. Zool. *144*:187–207.
Tartar, V., 1967: Morphogenesis in protozoa. *In* Research in Protozoology. Chen, ed. Vol. 2. Pergamon Press, New York, pp. 1–116.
Thompson, D. W., 1942: Growth and Form. Macmillan Co., New York.
Todaro, G. T., Lazar, G. K., and Green, H., 1966: The initiation of cell division in a contact-inhibited mammalian cell line. J. Cell. Comp. Physiol. *66*:325–334.
Verzár, A., April 1963: The aging of collagen. Sci. Am. *208*:104–114.
Warburg, O., 1956: Origin of cancer cells. Science *123*:309–314.
Warburg, O., Gawehn, K., and Geissler, A. W., 1957: Effect of hydrogen peroxide on cancer cells and on embryonic cells. Z. Naturforsch. *12*:393–395; Chem. Absts. *52*:1431 (1958).
Waymouth, C., 1965: Construction and use of synthetic media. *In* Cells and Tissues in Culture. Willmer, ed. Academic Press, New York, pp. 99–142.
Weismann, A., 1913: Vortrage über Deszendenztheorie. Jena.
White, M. J. D., 1961: The Chromosomes. Ed. 5. John Wiley & Sons, New York.
*White, P. R., 1963: The Cultivation of Animal and Plant Cells. 2nd Ed. Ronald Press Co., New York.
Willmer, E. N. (ed.), 1965: Cells and Tissues in Culture. Academic Press, New York.
Yasumura, Y., Tashjian, A. H., and Sato, G. H., 1966: Establishment of four functional clonal strains of animal cells in culture. Science *154*: 1186–1189.
Zarrow, M. X. (ed.), 1961: Growth in Living Systems. Basic Books, New York.
Zeuthen, E., 1947: Body size and metabolic rate in the animal kingdom. Compt. Rend. Lab. Carlsberg Series Chim. *26*:17–161.
Zeuthen, E., 1953: Oxygen uptake as related to body size in organisms. Quart. Rev. Biol. *28*:1–12.

# CHAPTER 27 / CELL DIVISION

## CELL DIVISION CYCLE

Cells that are growing are destined to divide. Cell division takes place at different intervals which depend upon the species and the environmental conditions. As already mentioned, *Escherichia coli,* on a medium containing water, glucose, ammonium chloride, potassium basic phosphate, sodium sulfate and trace metals, kept at 37° C., divides every 45 minutes. If supplied with amino acids and some other organic materials at 37° C., it divides every 20 minutes. Eukaryotic cells divide at a slower rate than bacteria. It is true that eggs of some marine animals from warm climates divide about every 30 minutes, and those from cooler climates divide at about one hour intervals. However, such cells have reserves and need only convert these to cell constituents before dividing. Protozoan cells divide somewhat more slowly; *Tetrahymena pyriformis,* under favorable conditions, divides every 3 to 4 hours, and *Paramecium aurelia,* every 6 hours. Cells of higher plants and animals divide about once or twice a day. Thus the generation time (time between divisions) for cells of the broad bean *Vicia faba* is about 14.3 hours; for the cells of *Tradescantia paludosa,* about 17 hours; for Chinese hamster fibroblasts, 11 hours; and for mouse fibroblasts, 22 hours.

Cell division or *mitosis,* which consists of prophase, metaphase, anaphase and telophase, occupies only a small part of the generation time, as may be seen in Figure 27.1. The major portion of the division interval is interphase (Prescott, 1964; Kihlman, 1966). Interphase consists of three main stages: $G_1$ (Gap 1) from telophase to the beginning of S (synthesis); S, during which DNA synthesis takes place; and $G_2$ (Gap 2), from the end of DNA synthesis to the beginning of mitosis (division). The amount of time for DNA synthesis varies, as seen in Figure 27.1.

It is evident that in a multicellular plant or animal cell division goes on until the adult stage is reached. Thereafter, cell division occurs only in germinal tissues such as the cambium in the stem and the root of a plant and the germinal layer of the epidermis, the gametogenic cells in the gonads and blood cell-forming tissues of the animal. Divisions also occur during repair of wounds in both plants and animals.

## SYNCHRONIZED CELL DIVISION

For many of the studies of cell division it has been found desirable to have all of a population of cells in a culture divide at one time. Such *synchronized* cell division makes possible more effective analysis of the various components of the process.

Eggs of marine animals have been

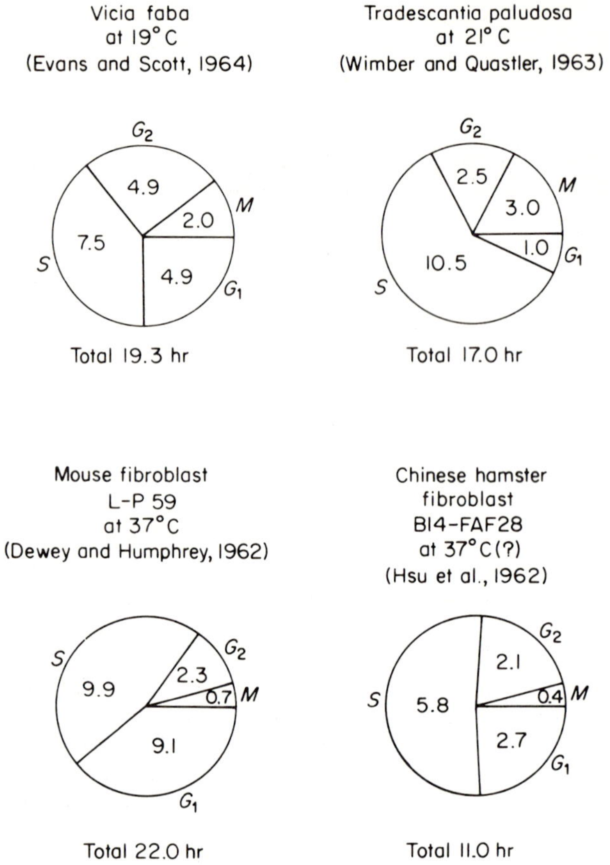


***Figure 27.1.*** Duration of phases in the division cycle of a cell: $G_1$ (Gap 1), $S_1$ (synthetic phase), $G_2$ (Gap 2) and M (mitosis). $G_1$ follows mitosis and lasts until the beginning of DNA synthesis ($S_1$); $S_1$ lasts until replication of DNA is over, when $G_2$ begins and lasts until the beginning of mitosis. Data for root tip cells of *Vicia faba* and *Tradescantia paludosa* and in tissue culture cells of mouse fibroblast and Chinese hamster fibroblast. (From Kihlman, 1966: Actions of Chemicals on Dividing Cells. Prentice-Hall, Englewood Cliffs, N. J., p. 60.)

favorite objects for such studies since practically all the eggs divide at the same time in a suspension of healthy and normal cells (Fig. 27.2). This synchrony in dividing marine eggs implies that some event has occurred to achieve the synchrony, e.g., the almost simultaneous entry of sperm in all eggs starts a train of events that takes about the same amount of time in each of the cells. Cleavage in the embryo continues in synchrony for several generations unless cells are separated from one another (Ducoff & Ehret, 1959). When the eggs have been affected by unfavorable conditions, e.g., temperature, radiations, pH, division may be delayed, and when it starts, some eggs are found to cleave before others. As seen in Figure 27.2, the susceptibility of the cells to the unfavorable conditions varies about a mean for the population and a characteristic distribution curve is obtained for onset of division in the population.

In a culture of bacteria, protozoa or tissue culture cells the number of cells in division at any one time is limited, being from 5 to 10 per cent. The ratio of the number of cells in division to the total number of cells is called the *mitotic index.* The mitotic index appears to be a function of the *generation time* for a given species under the conditions provided and the actual length of time a cell of this species remains in mitosis (Zeuthen and Scherbaum, 1954). The mitotic index may be increased by selecting a single cell from such an asynchronous culture and using it as the progenitor of a culture. For a few generations thereafter, synchrony is obtained, but it also gradually disappears, as shown in Figure 27.3. A variation of this technique consists of filtering bacteria of a given size from a mixed population of cells, thus obtaining cells in essentially the same stage of growth which will divide at approximately the same time (Maruyama, 1964).

The failure to achieve synchrony of division in a culture of cells is probably a result, in part, of differential exposure to

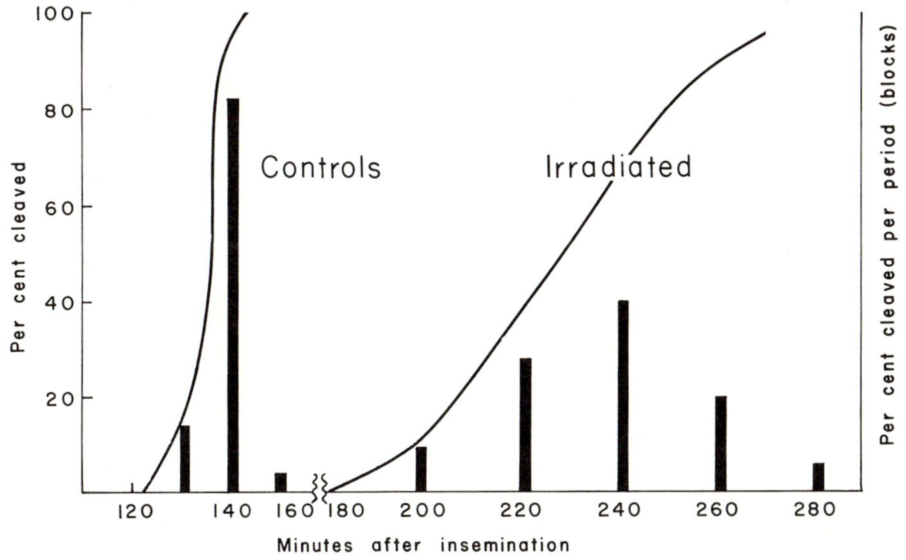


***Figure 27.2.*** High degree of synchrony in division of normal sea urchin eggs (inseminated simultaneously) as shown in the curve for "controls." Note that the histogram plotting the number of eggs cleaving in a 10 minute period indicates that over 80 per cent of the eggs cleave simultaneously. The same sample of eggs fertilized with ultraviolet radiation-treated sperm (50 ergs/mm.$^2$ of wavelength 2654 Å) show much less synchronous division. The division figures were recorded photographically over a period of about 100 minutes. (Data from P. Wells, 1952: PhD. Thesis, Stanford University.)

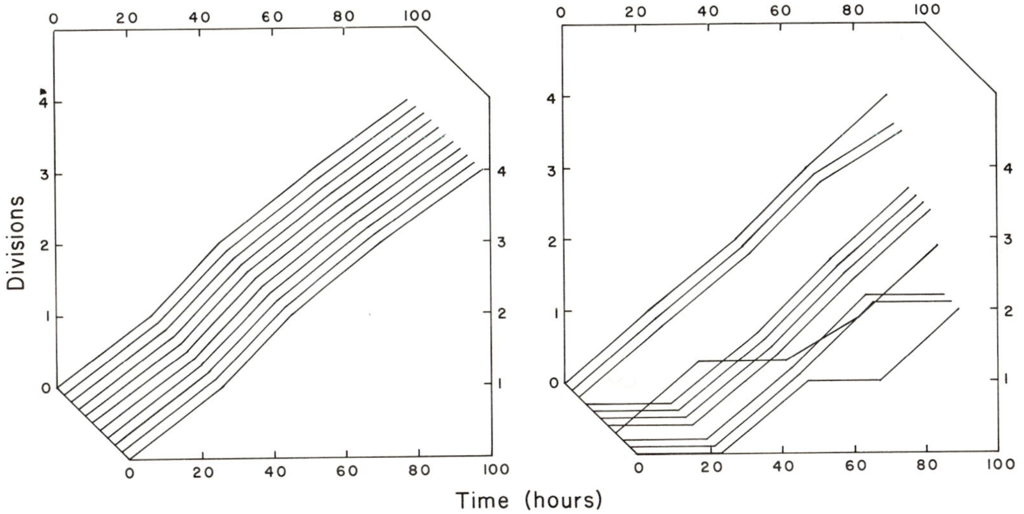


***Figure 27.3.*** Synchrony and asynchrony in division of the ciliate *Blepharisma intermedium. A,* Plot of the division of 11 isolated individuals which are descended from a single individual four generations back. Perfect synchrony is maintained for the four divisions that were traced. *B,* The relatively asynchronous division of 11 other individuals of the same species and stock taken at random from a large stock culture, many generations removed from a common ancestor. The time and division axes are staggered to make more evident sychrony and asynchrony. (Data by courtesy of B. McCaw.)

various environmental conditions. In a syncytium (multinucleate cell) in which mixing of all materials occurs and the conditions of the environment are similar, the division of nuclei is usually perfectly synchronized. Also, in the syncytial insect egg nuclear division continues synchronously for as many as eleven generations. The same is true for the nuclei in the syncytial endosperm of plant embryos. It is thought likely that nuclei in division may secrete division-stimulating materials. To explain the low mitotic index in most cell cultures it has been suggested by some that the division-stimulating material from the few cells in division does not affect many other cells, either because they are distant or because the membranes of such cells are relatively impervious to the hypothetical division-controlling substances (Zeuthen and Scherbaum, 1954).

This concept of control of synchrony by secretion of materials from nuclei has some experimental backing. For example, grafts can be made between two multinucleate amebas (*Chaos chaos*), each with nuclei synchronously dividing but out of phase with one another. After one cycle the division of all nuclei from both amebas now present in one cell is synchronous, the larger piece imposing its time upon the smaller. There are many examples of synchronized cell division which can be found in cells of higher plants (Erikson, 1964) and animals (Agrell, 1964). Cells in synchronous division are found, for example, in spermatogenic cells lining the lumen of a germinative tubule, which are connected to one another by cytoplasmic bridges (Fawcett *et al.,* 1959). Such bridges insure entry of the division-stimulating material from one cell to another.

It is important at this time to emphasize that a suspension of cells, whether they are cells in tissue culture, protozoans, yeast or bacteria, are fundamentally different from a suspension of marine eggs. Marine eggs have undergone a period of growth in the ovary and they have self-contained supplies which will serve for many cell divisions before intake of nutrients need take place. Cells other than eggs, however, must incorporate nutrients and grow before they can divide. Perhaps not all cells are able to incorporate the same nutrients at the same rate. Studies on *Tetrahymena* (ciliate protozoan) show that the mass of the cell increases linearly as measured by oxygen consumption, by increase in volume and by $^{14}$C-methionine incorporation, during the entire *synthetic period,* (S), from the onset of furrowing in one division until about 10 to 20 minutes before the next division. The division cycle has a generation time of 2.25 to 2.5 hours at 28 to 29° C. Furthermore, $^{3}$H-histidine incorporation (protein synthesis) and $^{3}$H-uridine incorporation (RNA synthesis) continue during division with little change (Prescott, 1964). During the 10 to 20 minute *predivision period,* which precedes furrowing, there is no increase in mass while the cell gets ready to divide. Such a period has also been found in other cells; e.g., in *Amoeba* it occupies about one-sixth of the generation time (Fig. 27.4).

Studies on *Paramecium aurelia* show that the mass, as measured by increase in dry weight, changes little for a period after division, and then it increases almost exponentially (Kimball *et al.,* 1959). From the data recorded this would appear to be exceptional compared to most cells studied (Prescott, 1964). On the other hand, Lovlie (1963) in a careful study on single *Tetrahymena* (using the diver technique) has shown that the type of curve obtained for increase in mass with time is related to the conditions of growth. He found three types of curves for *Tetrahymena,* exponential, linear and linear with a plateau, depending on whether the growth was balanced (doubling during the interdivision period) or unbalanced.

Since it is not possible to achieve synchrony of cell division in cell suspensions other than marine eggs, attempts have been made to *shock* cells into synchrony. Two methods have been used: chemical shock and physical shock. Chemical shock consists of withholding or limiting the supply of some nutrient necessary for division and then supplying it to the culture at one time in a large quantity, inducing in this manner a high level of simultaneous

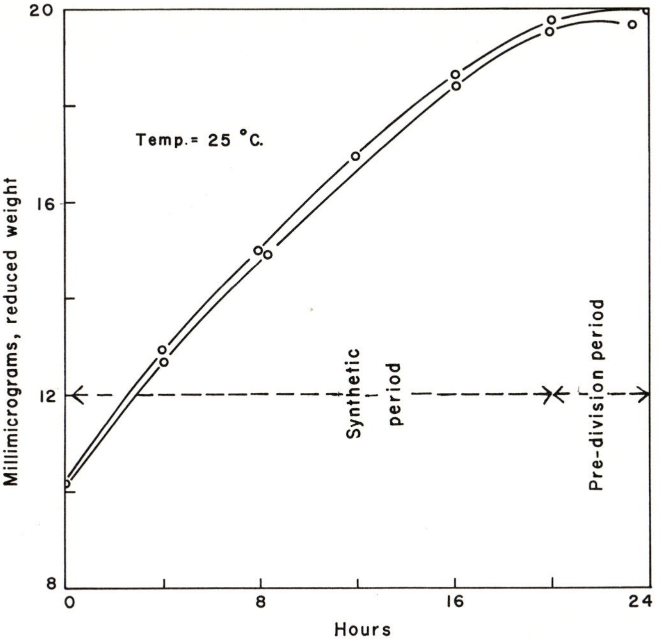


***Figure 27.4.*** Growth in mass of *Amoeba proteus.* The pair of lines represents the growth of a pair of sister cells from the time of division of their mother cell to the time they themselves divide. "Reduced weight" is defined as the total weight of the object (in this case the cell) minus the weight of water which is displaced by the object. It is the value obtained with the Cartesian diver balance for weighing single cells. (See Zeuthen, 1948: Compt. Rend. Lab. Carlsberg Series Chimique *26*:243). (Data from Prescott, D. M., after Mazia, 1956: Am. Scientist *44*:9.) Part of the increase represents ingested *Tetrahymena.*

metabolic activity. The physical shock is one that is unfavorable for the act of cell division yet favorable to other metabolic activities preceding division, thus allowing the cells in the predivision stages of the division cycle to catch up with those in the later stages of the division cycle.

Only a few experiments using shocks for synchronization of cell divisions will be considered here. Cell division and DNA synthesis in *Escherichia coli* T-15, a thymine-requiring mutant (thymineless), are blocked immediately upon transfer of the organism to a thymine-free medium. However, RNA and protein synthesis continue, apparently at the original rate. When thymine is added 30 minutes later, DNA synthesis is resumed and nearly all the cells are found to divide simutaneously after a lag of 35 to 40 minutes (Barner and Cohen, 1956). Similarly, in *Lactobacillus acidophilus,* synchrony can be induced by the addition of thymidine to a thymidine-starved culture (Burns, 1964). Yeast cells starved in succinate buffer until some of the reserves are gone will divide synchronously after return to complete nutrient medium, including carbohydrate and nitrogen sources (Sylvan *et al.,* 1959). Similar results were obtained with a variety of cells needing some particular metabolite. For example, synchronous division in the cells of the epidermis of the insect *Rhodnius* follows its periodic ingestion of blood. In *Chlorella* and various algae, all photosynthetic, the daily periodic lighting regimen makes possible synchronization of cell division, presumably by periodic accumulation of food reserves during the period of illumination (Tamiya, 1964). Lighting, too, may be considered a physical shock, much like temperature; in fact, all the arguments used for temperature apply to light in light-sensitive cells. Environmental changes produce oscillations in growth of many types of cells (Scherbaum and Loefer, 1964).

Use of temperature as a physical shock to obtain synchronized division stems from the notion that the processes that occur during the division cycle are differentially sensitive to temperature. If some reaction in the predivision period is more sensitive to heat than are the reactions in the synthetic period, then a high temperature might prevent division without stopping syntheses. Thus, cells lagging behind in preparations for division might be given a chance to catch up with the others. As expected, a single temperature shock synchronizes only a small fraction of the cell population because it allows only a small proportion of the cells to accumulate in predivision stages. On the other hand, a series of temperature shocks which block division—each alternating with an exposure to a near optimal temperature for a period insufficient to allow the cells in the predivision stages to divide—synchronizes most of the cell population, because of the gradual accumulation of a large proportion of the cells in predivision stages before cessation of heat shocks. Zeuthen and Scherbaum (1954)

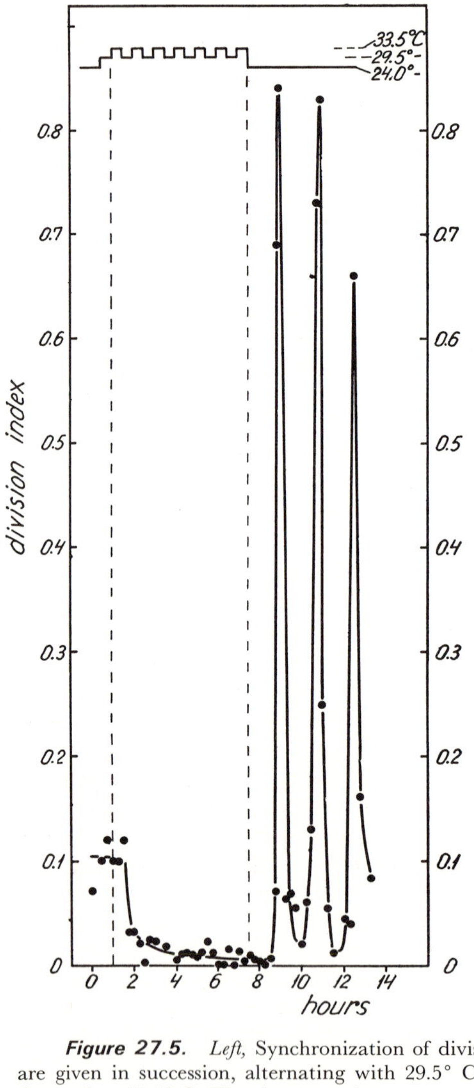


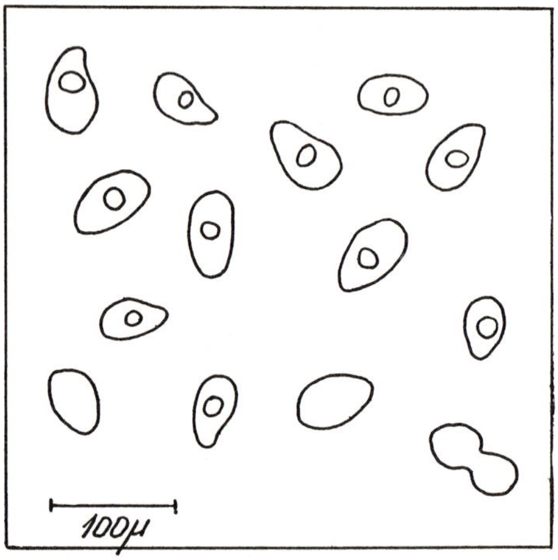


(*a*) *normal*

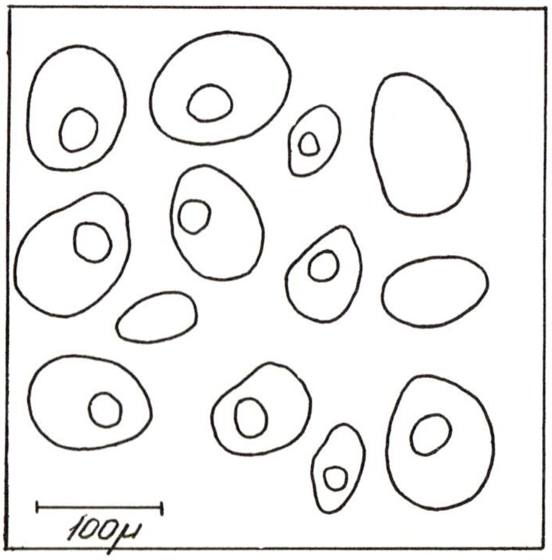


(*b*) *heat-treated*

***Figure 27.5.*** *Left,* Synchronization of division by heat-shocking *Tetrahymena.* Seven shocks at 33.5° C. are given in succession, alternating with 29.5° C. At the conclusion of the heat shocks, the temperature is dropped to 24.0° C. The three post heat treatment divisions are synchronized to the extent indicated by the division index. *Right* (*a*), Normal *Tetrahymena* drawn with camera lucida and (*b*) *Tetrahymena* after heat treatment. The circles inside the cell are the macronuclei. (From Zeuthen and Scherbaum, 1954: Colston Papers 7:141.)

found that alternate exposure of suspensions of the ciliate *Tetrahymena*, for half-hour periods at 28 to 29° C. (optimal) and 34° C. (inhibitory) for seven cycles resulted in 85 per cent synchronization of the cells in division when the cultures were subsequently kept at 28° C. Synchrony persisted for several cell generations and then division became random (Fig. 27.5). Thermal shocks have also been effective for synchronizing division of many other kinds of cells, including bacteria (Fig. 27.6; Zeuthen, 1958).

Cold shocks have also been used to induce synchronization of cell division in a number of protozoans and bacteria. Nutritional deficiency, coupled with temperature shocks, has also been very effective (Scherbaum, 1960a, b). X-rays have also been shown to induce synchrony in division, again presumably by holding back the cells in the radiation-sensitive predivision stages and causing accumulation of the cells which will ultimately divide at nearly the same time (Spoerl and Looney, 1959).

Much effort has gone into production of synchronized cultures of cells because in such cultures division and growth are essentially uncoupled. Therefore, growth can be studied in the cells prior to division, and the process of division and its accompaniments can then be studied in the cells accumulated in the predivision stages. The cells shocked into synchrony have accumulated sufficient reserves for a series of divisions, which follow one another more rapidly than the usual generation time for the species until the accumulated reserves have been partitioned among the descendants. This, of course, is also true of a suspension of marine eggs. Heat-shocked cells thus grow until they are considerably larger than controls (Fig. 27.5*b*). In a heat-shocked culture of *Tetrahymena*, the generation time for the first few synchronized divisions is about 60 per cent of that required for a culture that was not subjected to heat shock.

Division of heat-shocked *Tetrahymena* continues even in the presence of inhibitors of synthesis of DNA and steroids.

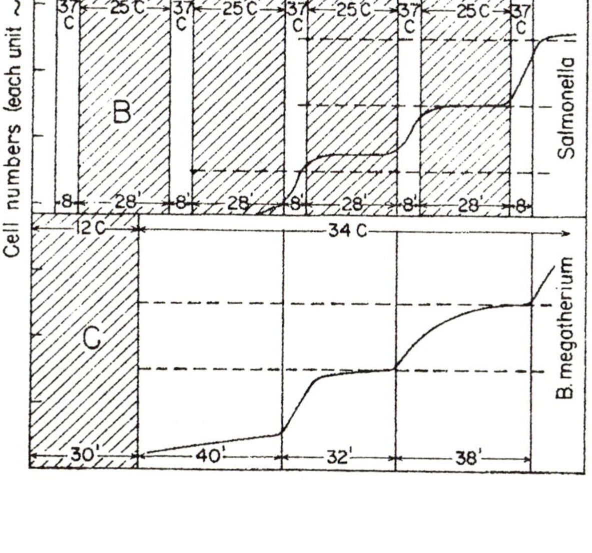


***Figure 27.6.*** Three different examples of synchronization in Bacteria. *A, Pneumococcus* (Hotchkiss, 1954). *B, Salmonella typhimurium* (Maaløe and Lark, 1954). *C, Bacillus megatherium* Szybalski and Hunter-Szybalska, 1955).

However, certain long chain unsaturated fatty acids must be synthesized, as must certain specific proteins (Holz *et al.,* 1963). The specific protein synthesis continues even if the cells are starved (Crockett *et al.,* 1965). The specific protein is presumably "division protein" thought to be required for structuring the cell for division (Frankel, 1965). It has been extracted and characterized (Watanabe, 1965). Interestingly, DNA synthesis, as measured by $^3$H-thymidine incorporation, is not synchronized with cell division in heat-shocked cultures (Cerroni and Zeuthen, 1962). It would be of interest to determine the DNA polymerase activity at various times during the cell division cycle (Mazia, 1963).

Synchronized cultures of various cells in suspensions are now being widely used in biochemical and cytochemical research (Zeuthen, 1958). It is important to note, however, that changes in size and composition of the cells after synchronization must be taken into consideration. For example, resistance of such cells to ultraviolet radiations is markedly altered (Iverson and Giese, 1957).

Synchronization of cell division in the photosynthetic dinoflagellate *Gonyaulax,* in the normal diurnal rhythm of day and night, occurs in 85 per cent of all cells destined to divide during a 5 hour period spanning the end of darkness and the beginning of the light period (Sweeney and Hastings, 1958). Although it might appear that cell division here, as in *Chlorella,* is related to periodic changes in nutritional conditions, keeping *Gonyaulax* in continuous dim light only sufficient to maintain nutritional balance in the cells does not stop synchronous division for at least 14 days. The synchronous division here is also relatively independent of small variations in light intensity and tempera-

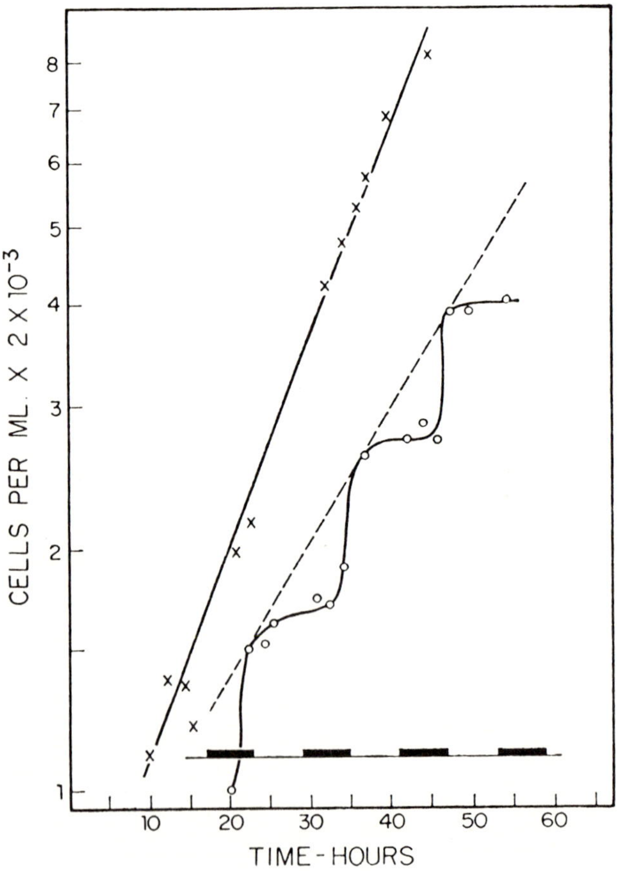


***Figure 27.7.*** Increase in cell population of the photosynthetic dinoflagellate *Gonyaulax polyhedra* in continuous light (X) and in alternating light and dark periods of 12 hours (open circles). Light intensity, 900 foot-candles; temperature, 21.5° C. Average generation time in continuous light, 1 day; in alternating light and darkness, 1.5 days. Note the synchronization achieved by the alternating light-darkness regime. (From Sweeney and Hastings, 1958: J. Protozool. *5*:217.)

ture. In high light intensity, however, synchrony is lost in 4 to 6 days (Fig. 27.7). This is an interesting case worthy of detailed study. At present it is interpreted as an example of inherent biological rhythm (see Chapter 9; Hastings and Sweeney, 1964).

Bruce (1965) lists the generation time of cells from a large number of species of plants and animals. In many cases the division of cells appears to be circadian, almost every 24 hours, but it has not yet been determined whether the day-night cycle will synchronize these to a 24 hour rhythm.

## SOURCE OF ENERGY FOR CELL DIVISION

Cell division, as has been pointed out many times throughout the text, is a fundamental activity of the living cell. During cell division the cell does work at the expense of energy derived from nutrients, as witness the case just cited of synchronized cell division in yeast obtained whenever the cells are supplied with adequate food (Sylvan *et al.,* 1959). It has also been shown that addition of glucose to isolated epidermal tissue culture results in synchronous division of many cells. Presumably, cell division had previously been blocked by lack of nutrient since such cells store little glycogen (Bullough, 1952). Even if glucose is supplied, division fails if oxygen is not available (Rasmussen, 1963).

Glucose may be replaced by lactate, glutamate, fumarate or citrate. All the experiments suggest that operation of the Krebs cycle supplies the energy for division of the cells. As expected, mitosis is inhibited by Krebs' cycle poisons such as malonate, cyanide and carbon monoxide, and by phlorhizin, a phosphorylation inhibitor (Bullough, 1952). The latter finding suggests that high energy phosphate bonds are an energy source or are involved in building this source. Marine eggs and protozoans require oxygen for division, but frog eggs and many embryonic tissue cells do not, presumably supplying their energy for cell division by glycolysis. Cells which require oxygen for division have a lower rate of glycolysis compared to their rate of respiration. Cells which divide in absence of oxygen have rates of glycolysis much higher than that of respiration, sometimes severalfold greater. As might be expected, in cells which can supply their energy needs for division by glycolysis cell division is very sensitive to glycolytic inhibitors such as iodoacetic acid (Bullough, 1952).

In order to have an effect on cell division the energy sources must be supplied early in the cell division cycle during what Bullough calls the *antephase.* Once a critical concentration of the energy-rich substances has been built up and the prophase begins, he found that nothing short of killing the cell will stop it from dividing. This has been noticed not only for cells deprived of oxygen or metabolically poisoned, but also for those damaged by radiations or subjected to other injuries; such cells divide and soon thereafter cytolyze. Initiation of mitosis appears to begin a series of concatenated and irreversible reactions which stop only when the cells have divided (Bullough, 1952).

By employing carbon monoxide, working under green light, to block cytochrome oxidase in sea urchin eggs and releasing the block by shifting from green light (which was not thought to be absorbed by the enzyme–carbon monoxide complex) to white light, Swann (1953) obtained the following results. If the inhibitor is applied before sea urchin eggs enter mitosis, the first cleavage is delayed by a time about equal to the time of application of carbon monoxide (Fig. 27.8). If the inhibitor is applied after the cells have entered mitosis, they complete mitosis and cleave with little or no delay, but the second cleavage is delayed for a period which is roughly equal to the period of inhibition. It was suggested that these results can best be explained by a hypothetical *energy reservoir* which is continually being filled and in which energy is stored for a cell division during the preceding mitosis and cleavage. Once the energy has siphoned out it carries the cell through mitosis and cleavage; if at this time the cell is poisoned for a period by monoxide and energy is no longer being accumu-

Diagrams to illustrate the working of the reservoir hypothesis under normal conditions, carbon monoxide and ether inhibitions. Times slightly modified for simplicity—critical point for siphoning out of reservoir, 40 minutes after fertilisation; cleavage 60 minutes after fertilisation.
1. Normal conditions.
2 and 3. Carbon monoxide.
Action: complete block to filling of reservoir.
2. Inhibition: from 20–50 minutes after fertilisation.
Result: 1st cleavage delayed 30 minutes.
3. Inhibition: from 50–80 minutes.
Result: 1st cleavage not delayed.
2nd cleavage delayed 30 minutes.
4. 1 per cent ether.
Action: (a) no effect on filling of reservoir.
(b) block to building up of mitotic figure, either by preventing siphoning out or by preventing effect of reservoir contents.
Inhibition: from 20–50 minutes.
Result: 1st cleavage delayed 10 minutes.
5. 2 per cent ether.
Action: (a) halving of rate of filling of reservoir.
(b) block to building up of mitotic figure, as with 1 per cent ether.
Inhibition: from 20–70 minutes.
Result: 1st cleavage delayed 30 minutes.

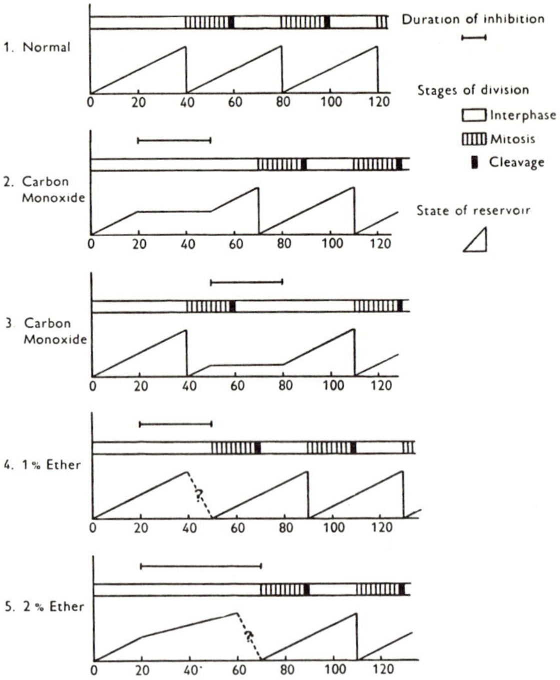


***Figure 27.8.*** Interpretation of the action of poisons on cell division based on the energy reservoir hypothesis. (From Swann, 1954: Exp. Cell Res. *7*:513.)

lated in the reservoir, cleavage will nevertheless continue to completion. However, the next cell division is delayed for an interval equal to the period of application of the poison. Lack of oxygen and various other poisons that affect the aerobic enzymes would also affect cell division in a similar manner (Swann, 1953).

Poisons which do not affect energy-liberating systems act in a different manner. For example, ether (1 per cent) is almost without effect if applied to a suspension of sea urchin eggs before the mitotic spindle has formed, but if applied after that time, it blocks development of the spindle fibers, maintaining the proteins in a solvated state. If the treated eggs are then washed free of the ether, development proceeds at once, the only delay being the period during which the eggs were subjected to the ether since in this case the energy reservoir, if it exists, is unaffected by ether (Swann, 1954).

The need for a postulated *energy reservoir* is difficult to accept on the basis of data collected both on animal (Epel, 1963) and plant cells (Amoore, 1963). Epel found that cleavage rate in the eggs of the purple sea urchin was decreased when the concentration of carbon monoxide was at a level which inhibited respiration. At this time the ATP level was also decreased. When the ATP level had dropped to 50 per cent of normal, mitosis was completely inhibited. By varying the carbon monoxide concentration, and thus obtaining varying degrees of inhibition of ATP production, Epel demonstrated that the mitotic rate paralleled the ATP level in cells. He also found that division could be blocked at any stage of mitosis if the inhibitor was applied at the appropriate time. When phosphorylation is uncoupled by dinitrophenol (DNP), ATP fails to accumulate and cell division is also blocked. Such a concentration of DNP, which uncouples phosphorylation, causes an increased rate of respiration, but the respiration is now useless and idling (Epel, 1963). This points again at ATP as the likely source of energy for cell division.

The discrepancy between the results

of Epel and those of Swann (1954) perhaps rests upon a difference in experimental technique. Swann did his "dark" experiments in green light, which is theoretically not effective in reversal of carbon monoxide poisoning; he used white light for full reversal of carbon monoxide poisoning. Epel, on the other hand, put his carbon monoxide experiments in full darkness because he found that in green light the rate of respiration and ATP production were reduced to only 75 per cent of the level obtained by the same concentration of carbon monoxide in full darkness.

Heat shock, as shown earlier, does not stop growth although it prevents division, and, consequently, the cells may reach a size four times that of the controls. Returning the cells to the optimal temperature after one or more heat shocks permits cleavage in less than the generation time. Presumably the mitogenic and growth channels are separate at their definitive ends, though both are using the same sources of energy and materials. When the mitogenic channel is blocked by physical or chemical means, the energy and material which might normally go through this channel pass instead into synthetic reactions leading to extra cell growth. Consequently, after removal of the block, some time must elapse before the specific molecules necessary for cell division are synthesized to the necessary level (Zeuthen, 1964).

In *Tetrahymena* if heat shocks as used by Zeuthen (1964) continue well beyond seven cycles, the cells may ultimately divide, even during a heat shock. This is taken to mean that accumulation of products of growth in a cell, and especially of the material needed for cell division has reached a point at which division is initiated. Division can then no longer be prevented by thermal shocks that previously blocked it.

Perhaps an even more clear-cut separation of growth processes from cell division is shown in experiments with *Tetrahymena* cells which are heat shocked in nutrient medium and then transferred to balanced salt medium with no nutrients (Hamburger and Zeuthen, 1957). Such cells subsequently divide at least twice in absence of nutrients. Cleavage of such cells under these conditions can be "set back" or delayed by various metabolic poisons or physical shocks.

Experiments with dinitrophenol also distinguish two phases of metabolism in *Tetrahymena.* One, constituting about 30 per cent of total metabolism, relies on endogenous reserves and can support cell division; the other phase (about 70 per cent of total metabolism) depends upon exogenous supples and is coupled with growth (Scherbaum, 1960a, b).

It has become increasingly evident that specific protein (*division protein*) probably plays an important role in cell division in *Tetrahymena* (Zeuthen, 1961). It is thought that a critical concentration of this particular protein is needed for cell division. It has also been shown that ATP, GTP and RNA accumulate in the period just preceding cytoplasmic division and that the changes are reversed at mitosis. (Zeuthen, 1961). However, such an accumulation of free energy sources is interpretable on the basis of protein synthesis, because protein synthesis decreases for a period beginning just about two-thirds of the time to the next cell division. At this time the protein formed on the ribosomes is less readily released than it was previously (Plesner, 1961). The call for ATP is then less than its concentration in the cell, along with that of GTP, and RNA therefore increases. This change in concentration thus appears to be incidental to the slowing or cessation of protein synthesis rather than due to accumulation of an energy reservoir. The high energy phosphates then play a part in reversing the binding of the protein to the ribosomes (Plesner, 1964).

## BLOCKING CELL DIVISION

***By Radiations.*** Blocking of cell division by x-radiation was observed soon after the discovery of x-rays, and this led to the use of ionizing radiations in the treatment of cancer (see Chapter 9). Almost without exception dividing cells are most sensitive to x-rays during a short period just prior to the onset of mitosis (Gray,

1959). Cells already well along in the division cycle complete the division, but those in earlier stages regress. The type of cell seems of little consequence provided the cell is growing and dividing either in culture or in the living organism (Puck, 1957, 1960). (The cells in a mature tissue are apparently radioresistant, i.e., not killed, when they have ceased dividing.) As a consequence, as already mentioned, x-rays produce a degree of synchronous division of cells in tissue culture (Spoerl and Looney, 1959).

The locus of action of the x-rays is not fully known. The rate of incorporation of precursors into DNA is decreased or inhibited in cells of tissues subjected to x-rays. Since in most cases assays were made many hours or even days after the irradiation, at which time the cell population may have declined as a result of x-raying, the results from this type of experiment have been questioned. However, by using cell suspensions, it has been shown that application of a certain x-ray dose, at a time just before synthesis of DNA has begun, is much more effective in stopping DNA synthesis than if the radiations are applied after synthesis has started. There thus appears to be a system connected to, but not identical with, DNA synthesis that is even more sensitive to x-rays than DNA synthesis itself. Larger doses affect both this process and synthesis of DNA as well (Stocken, 1959).

It is interesting that synthesis of protein and RNA continue in a *radiation-sterilized* cell (a cell that will not divide again). The result is the production of giant cells. Unbalanced growth of this type is followed by death, but the undivided cell may live for a considerable time. During this time, various materials leak from these cells which are found useful in experiments with tissue culture cells. For example, Puck has used mats of similar nondividing (sterilized) cells on which to place a single normal unirradiated human tissue cell. The cell lives using leaking material as nutrient and gives rise to a colony, the cells of which form a *clone;* that is, they are all the descendants of the single cell placed there (Puck, 1960).

Ultraviolet radiations also retard or stop cell division, but, because the radiations are absorbed so superficially, they are not suitable for cancer therapy. The action spectrum for cell division-inhibition in a suspension of cells resembles the absorption spectrum of nucleic acids or nucleoproteins. Also, it has been shown that DNA synthesis is retarded by small doses of ultraviolet radiations (Kelner, 1953). As in the case of x-rays, exceedingly small doses may inhibit division of cells without having any detectable effect on DNA synthesis, indicating the existence of a process even more sensitive than DNA synthesis but probably closely related to it. The action spectrum for the effect implicates nucleic acid (Deering and Setlow, 1957).

In general, synthesis of DNA appears more sensitive to ultraviolet radiations than is RNA synthesis or protein synthesis. Since the blockage of DNA synthesis leads to unbalanced growth and larger doses of ultraviolet radiation inhibit RNA and protein syntheses, the interrelations between all these syntheses are complex.

***By Poisons.*** A great number of poisons block cell division (Figs. 27.8 and 27.9), but only those which help identify metabolic reactions or events in the complex which makes up cell division can be mentioned here. The action of such poisons has been studied extensively because of its theoretical interest and its possible use in control of cancer (Gelfant, 1963a, b; Cromwell, 1965; Kihlman, 1966).

Many mitotic poisons interfere with interphase growth. For example, *metabolic poisons,* such as the inhibitors of respiration and glycolysis or those which uncouple phosphorylations from energy-yielding reactions, prevent the accumulation of high energy compounds and prevent release of energy for synthesis or division. Another group of mitotic poisons—the *antimetabolites*—interfere with syntheses. Still others, such as the *antibiotic chloramphenicol* (isolated from *Streptomyces*), which resembles phenylalanine and presumably competes with the latter on the surface of the protein-synthesizing enzymes, block protein synthesis. If applied

A

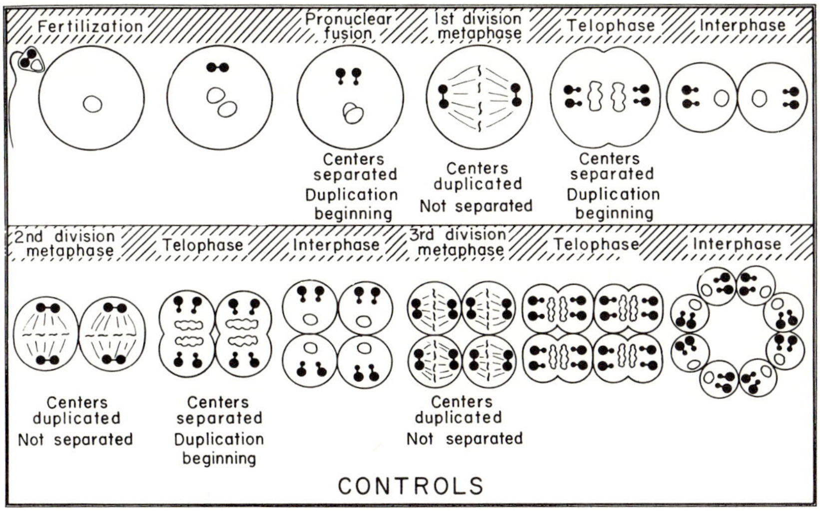


B

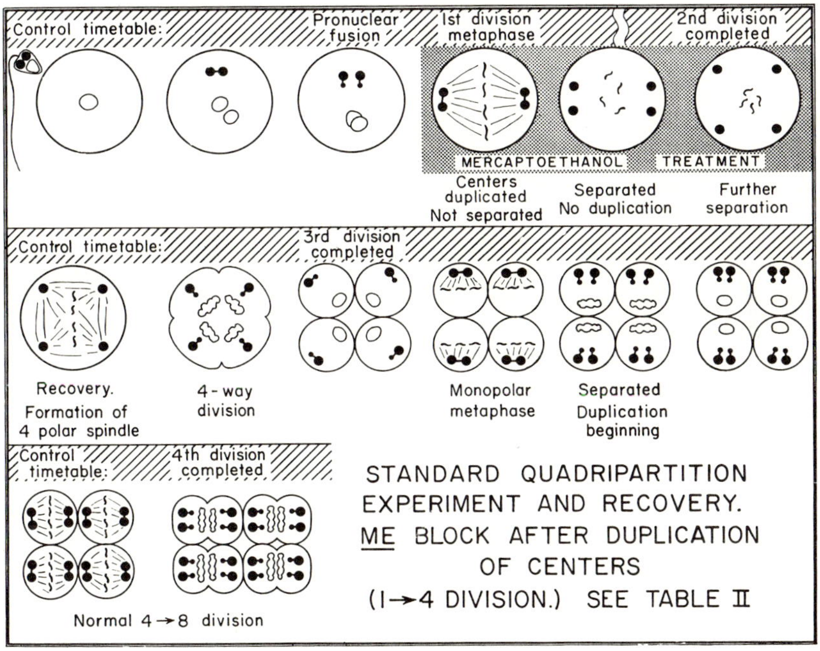


***Figure 27.9.*** *A,* Diagrammatic interpretation of the reproductive cycle of the cell centers. All nuclear figures shown are purely symbolic and are introduced merely to identify mitotic stages. Functional units of the centers are represented by solid dots. Small dots indicate newly formed unit. Bars connecting the units indicate that they have not yet split from each other. *B,* An interpretation of the basic quadripartition experiment, in which it is assumed that the mercaptoethanol block is imposed after the duplication of the centers but before they have split. During blockage, the centers split and separate, forming four poles, each of which is composed of half the normal number of units. The following mitosis is monopolar. (From Mazia, Harris, and Bibring, 1960: J. Biophys. Biochem. Cytol. *7*:17.)

in sufficient concentration early in the division cycle, chloramphenicol also blocks cell division. *Tetracyclines* (isolated from *Streptomyces*) and *puromycin* interfere with protein synthesis, although probably in some manner other than that of chloramphenicol (Abraham, 1959).

A few mitotic poisons block nucleic acid synthesis. These compounds resemble some normal metabolite, and thus compete with the normal metabolite for the surface of the particular cell enzyme (competitive inhibition). For example, *azaguanine,* which resembles the purine guanine, acts in this manner, and many other purine and pyrimidine analogues have been tested. Other such substances that have been used extensively are *5-fluorouracil* and *5-fluorodeoxyuridine,* which at relatively low concentrations interfere with DNA synthesis by inhibiting thymidine synthetase. At higher concentrations these substances also inhibit RNA synthesis and protein synthesis (Kihlman, 1966). However, whereas such agents may act in this manner, it is possible that they also act in other ways (Bass, 1959). Unfortunately, many such analogues have too great an effect on cell division in germinative areas (e.g., on blood-forming cells of man) to be useful for treatment of cancer. *Mitomycin* C interferes with DNA synthesis without markedly affecting RNA synthesis or protein synthesis. *Actinomycin* D at concentrations which have no effect on DNA synthesis blocks DNA-dependent RNA synthesis, probably mainly messenger RNA production (Kihlman, 1966). Both antibiotics retard cell division and have been used extensively in recent years as selective inhibitors of the synthesis of the two nucleic acids. Such poisons generally prevent cells from entering mitosis.

Antagonists to folic acid (a B vitamin, see Table 3.9) such as amethopterin and aminopterin affect a specific step in the separation of the chromosome between metaphase and anaphase. They have therefore been classed as *specific metaphase inhibitors* (Jacobson and Webb, 1952). These experiments demonstrate that one of the B vitamins controls a specific step in the mitoses of the cells (chick fibroblasts) upon which they were tested. However, according to Taylor (1963) cells are arrested by such poison in either the $G_1$ or the early S phase and do not progress unless thymidylic acid becomes available.

*Alkylating agents* (e.g., those which replace an H atom of a compound with an alkyl group such as the methyl or ethyl) do not affect growth but prevent cell division. The nitrogen mustards (used as mutagens and in the treatment of cancer) and methane sulfonates (e.g., Myleran used in treatment of chronic myeloid leukemia) are such alkylating agents. The exact manner in which inhibition of cell division occurs is unknown, but some of these alkylating agents cause chromosome breakage or failure of normal chromosome movement similar to the effects following x-radiation (Alexander, 1960; Mazia and Gontcharoff, 1964). This could presumably occur as a result of translocation of parts of two chromosomes, resulting in two kinetochores (chromosome constrictions for spindle fiber attachment) at two places on a single chromosome which is torn in half during anaphase. Abnormal disjunction could occur from failure of attachment of kinetochores (Mazia, 1956a, b). These agents also appear to affect the formation of spindle fibers, generally reducing the viscosity of the cytoplasm. Esterification of carboxyl groups of proteins and combinations with nucleic acids have also been demonstrated, but the precise nature of the reactions is unknown (Bass, 1959).

The antimitotic agents such as *colchicine, podophyllin* and the *urethans* inhibit the formation, or the breakdown of the mitotic apparatus. Colchicine apparently inhibits cell division by disorganizing the mitotic spindle, without stopping growth or duplication of various organelles in the cell. Chromosomes duplicate themselves, but the spindle fibers that form are disoriented, resulting in polyploidy which, in some species, persists even after the poison is removed. It does not interfere with the making or breaking of disulfide bonds between proteins forming the spindle fibers, but seems to interfere with the secondary bonding in the formation of an oriented and symmetrical mitotic spindle (Mazia, 1956a, b).

The *sulfhydryl reactants* (e.g., chloracetophenone), on the other hand, block cell division by interfering with the disulfide and sulfhydryl cycle which is necessary for the formation and dissolution of the spindle fibers. All the other processes continue up to the formation of the spindle (Hughes, 1950).

Another interesting poison which has given considerable insight into mitosis is *mercaptoethanol* which, among other things, prevents the duplication of *mitotic centers* (centrioles) of the cell (Fig. 27.9). Centrioles always appear in animal cells in pairs. Each member of a pair must divide before they can act as division centers for the cell. The precise way in which mercaptoethanol affects the pattern of cell division depends upon just when it is applied during the division cycle. Exposure of the sea urchin eggs to this poison delays cell division to the extent that it delays duplication of the mitotic centers, which divide only after the cells are removed from the poison. In addition to this delay, cell division is made abnormal because, although the division of centrioles is inhibited by mercaptoethanol, the separation of the centrioles from one another is not affected. Along with all the other processes preparing for cell division, the members of the pair of centrioles must separate and take their positions at the poles of the cell. If this poison is applied to an egg cell just before its first division when it already has two pairs of centrioles and the egg is kept in mercaptoethanol beyond the time for the first division, it will then prepare for the second division except that the required duplication of the centrioles will not occur. The two pairs of centrioles separate and single centrioles now occupy four coordinate radii of the egg. Released from mercaptoethanol after being washed in sea water, the centrioles first divide and then the egg divides into four cells at once (Fig. 27.9). The data are of interest because they show that the poison specifically affects only the duplication of centrioles, not their separation or movement in the cell (Mazia *et al.,* 1960).

Some antibiotics, such as *penicillin,* prevent cell division of bacteria by interfering with the incorporation of amino acids into the cell walls and causing some bacteria to develop as protoplasts—without cell walls. Such protoplasts cannot divide—apparently the cell wall is necessary for bacterial cell division. Since animal cells do not form cell walls they are unaffected by this action of penicillin (Abraham, 1959).

A search is being made for naturally occurring substances with dual actions, on the one hand promoting growth, and on the other, retarding growth. Such substances are induced in animals after injection with active cancerous cells (Szent-Györgyi, 1965). Chemical analysis suggests keto-aldehydes as the compounds produced in response to such injections. Such compounds are being tested as possible regulators of cell division (Szent-Györgyi *et al.,* 1967).

## THE INITIATION OF CELL DIVISION

The cell, as we have seen, grows to a certain stage and then divides. In some unknown way each unit (organelle?) in the cell presents a template upon which duplication occurs. When cell division is completed each daughter cell contains in itself a duplicate of all the varied structures present in the mother cell.

Cell division has been studied in a multitude of cells, and most cells, under normal conditions, are known to divide by mitosis. Chromosomes condense and appear on an equatorial plane. The daughter chromosomes, produced by splitting, separate from one another, and a duplicate set of chromosomes is moved toward each of the two poles. Finally the cytoplasm separates into two portions. The complete process may take no more than half an hour in some marine eggs. Mitosis is described in detail in textbooks of cytology (DeRobertis *et al.,* 1965). The interest here centers on the mechanism initiating cell division.

The trigger which sets off the cell division has been of interest for a long time and numerous suggestions have been made about the nature of this process. One possibility that suggests itself is the doubling in mass of the cell, since this might be the last growth requirement for a series of processes, each resulting in doubling of cell organelles. Division then occurs because the ratio of cytoplasm to nucleus is upset. Some evidence points this way. As previously indicated, when an ameba has doubled its mass it divides, but not immediately (see Fig. 27.4). The division takes place after a lapse of time (rest period) during which something occurs which triggers the division of the cell (Mazia 1956a, b; Prescott, 1956). This time lapse before division is about 4 hours, or about one-sixth of the generation time of 24 hours for *Amoeba proteus.* Also, when an ameba about to divide is shaken, it divides unequally, but each daughter ameba grows until it equals the mass of the mother ameba before it divides again (see Fig. 27.10). Moreover, amputation of a part of a growing ameba is followed by replacement of the mass removed before division occurs at all. In all these cases division takes place 4 hours after the ameba has doubled its mass. Growth before division in all cells is not alike however. *Tetrahymena* shows an increase in mass followed by a predivison "rest period" similar to *Amoeba* (Zeuthen and Scherbaum, 1954), but in *Escherichia coli* B, synthesis proceeds in exponential fashion (Scherbaum, 1960a).

Another suggestion for the trigger in cell division is the upsetting of the surface to volume ratio. However, if a piece of cytoplasm is removed at a "critical" time just before an ameba has doubled its mass, division will still occur, even though smaller offspring than normal are produced. This experiment might be taken to indicate that, by this time, division has been determined and follows even if the ameba is not as large as a typical mother cell about to divide, and that neither the ratio of nucleus to cytoplasm nor the ratio of surface to volume can in itself be the probable trigger for cell division. When cells are starved they may undergo division producing smaller daughters. However, such alterations in cell size are also produced by a number of other changes in environment (Prescott, 1964).

Doubling the DNA content has been suggested as still another possibility. By the use of labeled precursors, it has been shown in most cells, however, that DNA is doubled in the interphase (during the S period, see Fig. 27.1) long before division actually occurs, in fact even before the formation of the extra pairs of chromosomes. Doubling of DNA content is there-

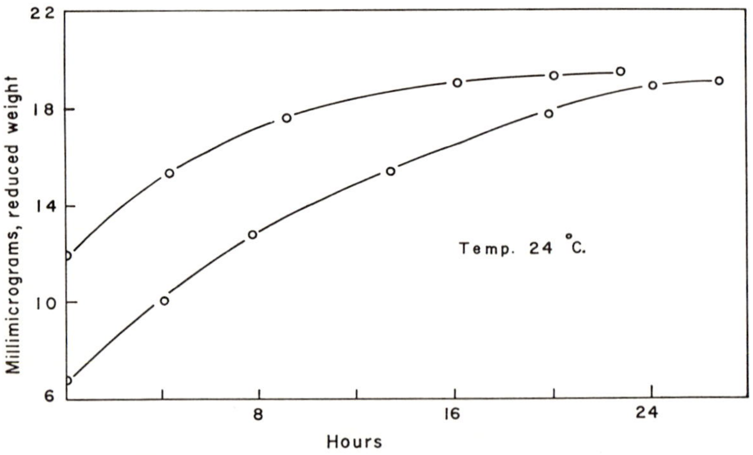


***Figure 27.10.*** Relation between the weight of *Amoeba proteus* at "birth" and the weight at maturity. Curves represent the growth of sister cells following unequal division of their mother cell. The last point on each curve gives the weight at the time of division. (Data from Prescott, after Mazia, 1956: Am. Scientist *44*:11.)

fore not the likely trigger. As the nucleolus re-forms in the "resting" nucleus, intense protein synthesis appears to occur just outside the nuclear membrane. The interphase is apparently not a period of rest for the cell but rather a period during which the necessary cell constituents are manufactured by it (see Fig. 6.18).

The possibility that the nucleolus has a vital role in initiating cell division is suggested by experiments in which microbeam irradiation of a nucleolus (neuroplast of a grasshopper) at certain critical times (telophase to middle prophase) permanently stops mitosis in the cell, while comparable irradiation of the surrounding areas of the cell does not (Gaulden and Perry, 1958).

It was also thought possible that the chromosomes or other organelles produce a specific substance which induces the division of a cell. One team of workers found that kinetin (6-furfurylaminopurine) serves this function in some plant cells in which division is otherwise infrequent (Miller *et al.,* 1955). No convincing evidence has accumulated to show this to be a general phenomenon. In fact, it has been shown that nuclear structures need not be present for the cytoplasm of the sea urchin egg to divide, since non-nucleated egg fragments can be excited to divide by parthenogenetic agents (Harvey, 1936). However, such cells develop abnormally and eventually die since they have no nuclei.

According to Heilbrunn, cell division is initiated by a coagulative process similar to the formation of a clot in blood, and any stimulus which induces clot formation is likely to initiate formation of spindles and cell division. It has been shown that agents which inhibit clot formation, e.g., heparin, interfere with the division of normal fertilized eggs. Agents which release calcium, which is also required for clotting blood, facilitate cell division, whereas agents which bind calcium, such as oxalate or citrate, inhibit cell division (Heilbrunn, 1955).

Whatever the immediate cause of cell division, it is clear that there are apparently many factors that contribute toward a favorable state for division of the cell and that preparations must be completed along many parallel lines before cell division can occur.

## THE NATURE AND FORMATION OF THE MITOTIC APPARATUS

That the division spindles are gels has long been known. This was demonstrated by micromanipulative studies and by experiments in which intact mitotic figures were isolated from dividing eggs by the use of mild detergents which solubilize the rest of the cell (see Fig. 27.11; Mazia, 1954). The mitotic apparatus has also been isolated from dividing eggs by changes in pH (Kane, 1962) and from dividing eggs from which the membrane and the hyaline outer cytoplasm has been removed by immersion in Versene (ethylenediamine tetraacetic acid), dextrose and dithioglycol solutions (Mazia *et al.,* 1961). The mitotic spindle is made up of about 3 to 5 per cent ribonucleic acid (RNA), and the remainder appears to be largely one type of protein. Electrophoretic diagrams indicate two peaks, one of which is the protein, and the other its conjugate with RNA; both act as antigens (Zimmerman, 1963). Electron microscope studies of fixed cells demonstrate that the spindle gel consists of definitely organized and oriented microtubules of protein, the centers of orientation being the centrioles (see Chapter 5; Mazia, 1961). When three pairs of centrioles are present, as in polyspermic eggs, the orientation is toward three poles. Presumably, something in or from the centrioles leads to the development of such oriented tubules, although no evidence for its origin or transport has yet been obtained.

On the basis of chemical evidence, it was suggested that the linkages between protein molecules in the microtubules of the mitotic spindle are disulfide (S—S) bonds (Mazia, 1956a, b). In the preparatory phase the *intramolecular* disulfide bonds of protein molecules are presumably reduced by glutathione. This is considered to take place before the metaphase, at a

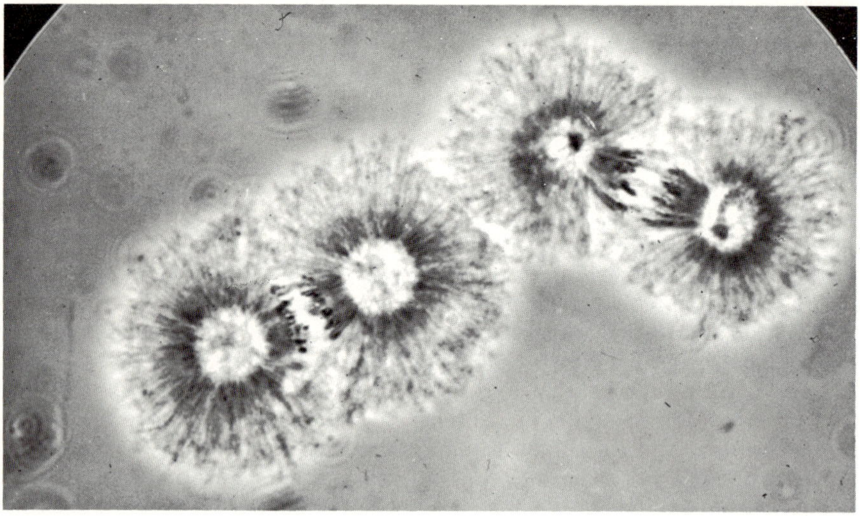

***Figure 27.11.*** Isolated mitotic apparatus at anaphase (developing eggs of the sea urchin *Strongylocentrotus purpuratus*). Note mitotic centers, astral rays, chromosomal fibers connecting chromosomes to centers, and interzonal regions of spindles between separated chromosomes. (From Mazia, 1956: Am. Scientist *44*:21.)

time when the glutathione content of the egg is known to be declining (Rapkine, 1931). When the mitotic apparatus is fully formed and growing, the disulfide bonds are supposedly restored, but now as *intermolecular* disulfide bonds, binding the protein molecules into tubules (microtubules, previously called fibers; see Chapter 5 for a discussion of this type of organelle) and liberating the glutathione. The concentration of the glutathione in the egg cell was found to increase at this time (Rapkine, 1931). Postulation of intermolecular disulfide bonds rests on the assumption that the probability of forming a bond between sulfur atoms on two protein molecules would be greater than the probability of re-forming the original intramolecular disulfide bonds. In essence, then, the sulfhydryl cycle has converted intramolecular disulfide bonds into intermolecular disulfide bonds, the glutathione acting essentially like a catalyst, being recovered unaltered at the end of the cycle (Mazia, 1956a, b). Although the cyclic variation in SH bonds present in dividing cells has been confirmed independently (Dan, 1966), there is little evidence to back the specific hypothesis just outlined.

If orientation and organization of the spindle tubules are a prime necessity for cell division, then any agent which interferes with the spindle should interfere with cell division, even if this agent permits gelation. This is exactly what happens when a cell is poisoned with colchicine. A gel appears, but the secondary bonds between protein molecules fail to develop, the gel is disoriented and separation of the doubled chromosomes fails to occur ("flabby anaphase"). Because of this, as we have already seen, polyploid cells are produced by the use of colchicine. When such cells are washed free of colchicine, they later divide, but the polyploid condition is maintained.

When Hoffmann-Berling placed cells about to divide (i.e., showing spindles) in glycerin solutions, the division process ceased. When they were later placed in a balanced salt solution, the spindles remained, but division still failed to occur. It is perhaps possible that some material necessary for cell division had leached into the glycerin solution. When ATP was applied to such mitotic "models" (glycerin-extracted dividing cells) the spindle tubules contracted and the chromosomes moved toward the poles (Hoffmann-

Berling, 1960). Furthermore, an enzyme for decomposing ATP is present in the spindle (Mazia, 1961). On the other hand, ATP produced no movement when applied to mitotic spindles isolated from sea urchin eggs by mild detergents and subsequently alcohol hardened (Mazia, 1956a, b). Hoffmann-Berling points out that hardening the spindles denatures the proteins of the tubules, making them incapable of contraction in response to ATP, even though hardening permits their separation and purification from other cellular constituents.

It is an oversimplification to consider only the contraction of the spindle tubules. Careful analysis of the movements during mitosis indicates that only the early part of mitosis (to anaphase) occurs by contraction. This is indicated by the broadening of the spindle. Evidence suggests that in anaphase the two sets of chromosomes in the cell are actually pushed apart by the tubules between the poles. For example, when the spindles of grasshopper neuroblasts are immersed in hypertonic solutions which reduce the volume of the cell and thus crowd the mitotic spindle, the chromosomes are crumpled and spiraled during the anaphase and pushed against the ends of the cell (Hoffmann-Berling, 1960).

The active elongation of the spindle poses the same problems as elongation and relaxation of muscle fibers, both active processes. It has been shown that low concentrations of ATP applied to the mitotic models cause contraction; higher concentrations, elongation (Hoffmann-Berling, 1960). On the other hand, the spindle fibers appear to consume themselves as contraction occurs, unlike a muscle which maintains its mass in spite of shortening. A single contraction of the spindle tubules, however achieved, suffices to separate the duplicated chromosomes from one another. It would almost appear as if the chromosomes, as they move toward the poles, release some material which dissolves the spindle tubules. The fact that glycerin-extracted models of dividing cells contract when placed in ATP, and that the chromosomes on the metaphase plate separate, could conceivably be explained as the result of contraction of the entire cell in such a way as to actually simulate mitosis rather than by contraction of the spindle tubules (Mazia, 1956a, b). The problem of chromosome movement should not be considered solved, despite the appeal of the contractile tubule hypothesis (Dan, 1966).

## FURROW FORMATION AND CLEAVAGE (CYTOKINESIS)

The formation of a furrow in a dividing egg or other animal cell appears to be accompanied by increased viscosity in the area of the furrow. High pressures, which liquefy protein gels, also prevent the formation of cleavage furrows and cause such dividing cells to fuse again. The furrow disappears with even a slight pressure applied to both sides with a microneedle, but cell division resumes when the pressure is released (Guttes and Guttes, 1960). An egg ruptured at the time of furrow formation disintegrates, but a part of the cortex and the furrow remain intact, indicating that the furrow is apparently a rigid part of the cell. Furthermore, it has been noticed that as the cortex forms in the cleaving eggs, a slight separation between the cortical layers of the two blastomeres (the two-celled stage) is maintained for a time, indicating that some change has occurred which prevents coalescence of the two layers (Marsland, 1957).

An attempt has been made to explain the formation of a furrow in a dividing cell by means of the *contractile ring theory* (or *cortical gel contraction theory*). According to this theory the gelated ring in the cortex of the dividing cell is considered to contract like the nonmotile portion of an ameba, and it might be expected, therefore, to decrease the surface area. However, measurement in some cells showed that before division the surface increased by about 26 per cent (Fig. 27.12; Swann and Mitchison, 1958).

The increase in surface area accompanying cell division does not invalidate contraction theories of cell division. It is

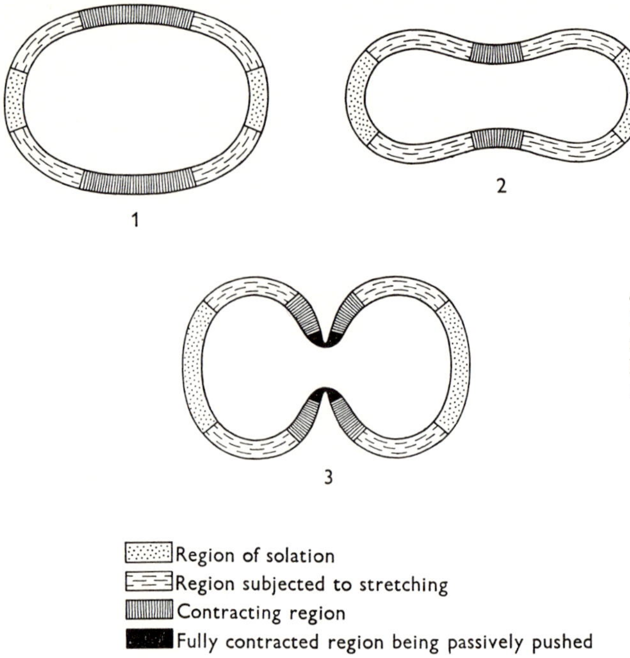


***Figure 27.12.*** Diagram of the *cortical gel contraction theory* of Marsland and Landau. *1,* Early telophase; *2,* early furrow; *3,* late furrow. (After Marsland and Landau, 1954: from Swann and Mitchison, 1958: Biol. Rev. *33*:109.

essential that the cell elongate before it cleaves, and at this time the surface area increases. But elongation is part and parcel of any contractile system, as already seen in muscle fibers (Chapter 25). When relaxation of muscle cells is better understood it may also be possible to understand cell division better. In modified form the contractile furrow theory, illustrated in Figure 27.12, allows for a passive increase in surface at the poles as a result of active contraction of the furrow region (Marsland and Landau, 1954).

The *expanding surface theory* (Mitchison, 1952), on the other hand, suggests that a nuclear substance is liberated (probably from chromosomes) which causes expansion of the cellular membrane at the poles (Fig. 27.13). As the polar area expands, the equator contracts, leading to division. This theory is supported by two lines of evidence. First, movement of material on the surface of cells, consistent with such a suggestion, has been observed (Dan and Dan, 1947). Second, expansion at the poles should weaken the membrane there. The fact that cytolysis of an egg occurs more readily during division than during interphase suggests some change in the cell membrane at the poles (Mitchison, 1952). The equatorial part persists after cytolysis, apparently being more rigid. However, even cells without nuclei or chromosomes cleave (Harvey, 1936; Hiramoto, 1956, 1965; Nachtwey, 1965), indicating that the excitant may have origins other than the nucleus. Another form of this theory postulates expansion of surface area as a result of growth (Selman and Waddington, 1955). It is interesting that the resting potential of a cleaving starfish egg does not change (Ashman *et al.,* 1964).

The *spindle elongation theory* (Dan and Dan, 1947; Dan, 1958) assigns a decisive role to the spindle and asters in cell division. The evidence comes from observation of movement, with respect to one another, of kaolin particles attached to the surface of an egg membrane. The basic observation is that the elongation of the cell at anaphase is accompanied by a shrinkage at the equator, the two kaolin particles on either side of the equator coming closer together while there is a corresponding stretching at the poles. Since the spindle and asters appear to be rigid structures, as tested with a micromanipulator, the driving force is believed to be the elonga-

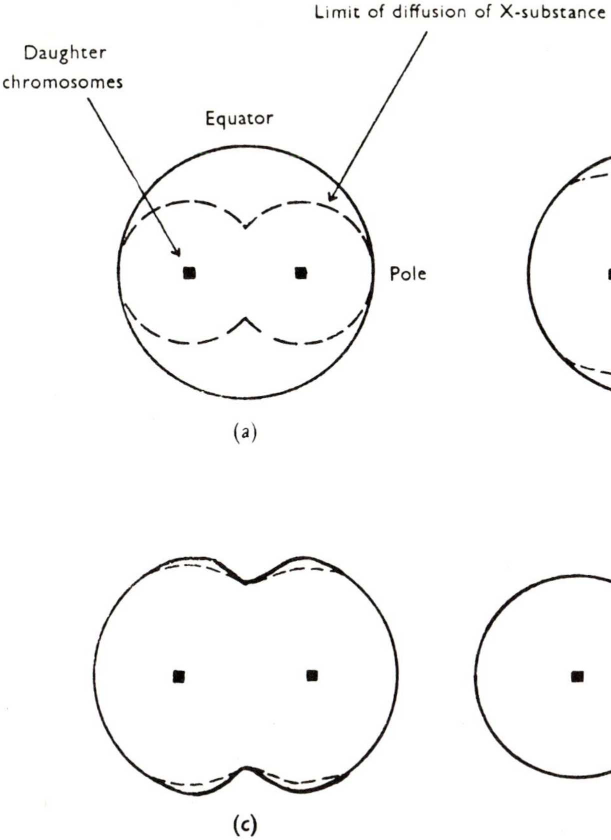


***Figure 27.13.*** Diagram to illustrate the *expanding membrane theory* of cleavage of Mitchison and Swann. The X-substance is the hypothetical substance which presumably comes from the chromosomes, and which leads to an increase in polar surface area upon contact with the poles. (From Mitchison, 1952: Symp. Soc. Exp. Biol. *6*:105.)

tion of the spindle tubules which pushes the centers apart, bringing the astral rays attached to the equator close together (Fig. 27.14).

Because a nucleus-free half of a sea urchin egg will cleave when artificially stimulated to divide (Fig. 6.4; Harvey, 1936), the *astral relaxation theory,* based only on changes produced by the asters, has been proposed to account for cleavage (Wolpert, 1960, 1963). Surface tension and elasticity measurements show that an egg surface is under uniform tension before cleavage begins. When the asters reach the poles of the egg it is thought that they produce a change (the nature of which is unknown), and the surface tension at the poles is lowered. This permits the

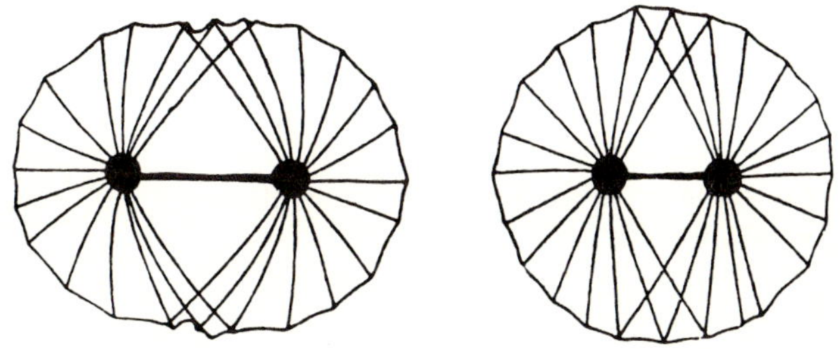

***Figure 27.14.*** Model illustrating Dan's *spindle elongation theory.* Note especially the effect of the spindle elongation upon the astral ray attachments at the surface of the cell. Spindle elongation results in a pull at the furrow region and furrow formation. (From Dan, 1948: Physiol. Zool. *21*:191.)

furrow region, which maintains its surface tension, to contract while the polar regions expand.

Experiments of Hiramoto (1956, 1965), however, in which the mitotic apparatus was either sucked out or displaced by injection of mineral oil into the egg, indicate that even in the absence of spindle tubule attachments (asters), cleavage furrows are formed in the same place as they are when mitotic apparatus is intact and in place. Also, a piece of cortex of an amphibian egg develops a furrow even in the absence of a nucleus and a spindle (Dan and Kojima, 1963). Stiffness of the cortex of the sea urchin egg increases before cleavage furrows form and varies cyclically during the division interval (Hiramoto, 1963). The contracting power of the equatorial ring of the sea urchin egg reaches a maximum at the middle of the cleavage cycle (Dan, 1963).

Work with eggs under pressure and subjected to various inhibitory reagents suggests that the ATP system accounts for the energy of whatever movements occur during cleavage, since addition of ATP often relieves the effect of an inhibitor (Zimmerman *et al.,* 1957). The energy required for division of an egg has been calculated to be about three times that normally available from respiration. Since the rate of respiration rises only a few per cent during cleavage, it would appear that the energy for cleavage must come from a specific source. It is suggested that compounds with high energy phosphate bonds, formed by the cell in preparation for each successive cleavage, might well serve this purpose (Zeuthen 1949; Scholander, *et al.,* 1958).

It is apparent that all of these theories have much in common and that they are developed primarily from data on the marine egg cell. They do not apply to such phenomena as multiple fission, which occurs in some plant and animal cells, nor do they take into consideration the division of plant cells with cell walls—where the laying down of a new wall is of prime importance for cleavage (cytokinesis; see Chapter 2). In bacteria and yeast, as was seen earlier, removal of the cell wall prevents cell division even though the cells continue to increase in size (Robinow, 1960). We are still without a generalized theory of cytokinesis.

## LITERATURE CITED AND GENERAL REFERENCES


Abraham, E. P., 1959: Antibiotics in microbiology. Endeavour *18*:212–220.

Agrell, I., 1964: Natural division synchrony and mitotic gradients in metazoan tissues. *In* Synchrony in Cell Division and Growth. Zeuthen, ed. Interscience Publishers, New York, pp. 39–67.

Alexander, P., Jan. 1960: Radiation-imitating chemicals. Sci. Am. *202*:99–108.

Alexander, P., 1965: Atomic Radiations and Life. Penguin Books, Baltimore, Chapter 4.

Amoore, J. E., 1963: Non-identical mechanisms of mitotic arrest by respiratory inhibitors in pea root-tips and sea urchin eggs. J. Cell Biol. *18*:555–567.

Anderson, N. G., 1956: Cell division. Quart. Rev. Biol. *31*:166–169; 243–269.

Ashman, R. F., Kanno, Y., and Loewenstein, W. R., 1964: Intercellular electrical coupling at a forming membrane junction in a dividing cell. Science *145*:604–605.

Barner, H. D., and Cohen, S. S., 1956: Synchronization of division of a thymineless mutant of *Escherichia coli.* J. Bacteriol. *68*:80–88.

Bass, A. D., 1959: Chemical influences on cell division and development. Ann. Rev. Physiol. *22*:49–68.

Bruce, V. G., 1965: Cell division rhythms and the circadian clock. *In* Circadian Clocks. Aschoff, ed. North-Holland Publishing Co., Amsterdam, pp. 123–138.

Bullough, W. S., 1952: Energy relations of mitotic activity. Biol. Rev. *27*:133–168.

Burns, V. W., 1964: Synchronization of division in bacteria by nutritional change. *In* Synchrony of Cell Division and Growth. Zeuthen, ed. Interscience Publishers, New York, pp. 433–439.

Cameron, I., 1965: Macromolecular events leading to cell division in *Tetrahymena pyriformis* after removal and replacement of required pyrimidines. J. Cell Biol. *25*:9–18.

*Cameron, I. L., and Padilla, G. M. (eds.), 1966: Cell Synchrony: Studies in Biological Regulation. Academic Press, New York.

Cerroni, R. E., and Zeuthen, E., 1962: Asynchrony of nuclear incorporation of tritiated thymidine into *Tetrahymena* cells synchronized for division. Compt. Rend. Lab. Carlsberg *32*:499–512.

Crockett, R. L., Dunham, P. B., and Rasmussen, L., 1965: Protein metabolism in *Tetrahymena pyriformis* cells dividing synchronously under starvation conditions. Compt. Rend. Lab. Carlsberg *34*:451–486.

Cromwell, N. H., 1965: Chemical carcinogens, carcinogenesis and carcinostasis. Am. Sci. *55*: 213–236.

Dan, K., 1958: On 'the stationary surface ring' in heart-shaped cleavage. J. Exp. Biol. *35*:400–406.

Dan, K., 1963: Force of cleavage of the dividing sea urchin egg. *In* Cell Growth and Cell Division. Harris, ed. Academic Press, New York, pp. 261–276.

Dan, K., 1966: Behavior of sulfhydryl groups in synchronous division. *In* Cell Synchrony: Studies in Biological Regulation. Cameron and Padilla, eds. Academic Press, New York, pp. 307–327.

Dan, K., and Dan, J. C., 1947: Behavior of the cell surface during cleavage. VII. On the division mechanism of cells with excentric nuclei. Biol. Bull. *93*:139–162.

Dan, K., and Kojima, M. K., 1963: A study of the mechanism of cleavage of the amphibian egg. J. Exp. Biol. *40*:7–14.

Deering, R. A., 1958: Studies on division inhibition and filament formation of *Escherichia coli* by ultraviolet light. J. Bacteriol. *76*:123–130.

Deering, R. A., and Setlow, R. B., 1957: Inhibition of cell division of *E. coli* by low doses of UV light. Science *126*:397–398.

DeRobertis, E. D. P., Nowinski, W. W., and Saez, F. A., 1965: Cell Biology. 4th Ed. W. B. Saunders Co., Philadelphia.

Ducoff, H. S., and Ehret, C. F. (eds.), 1959: Mitogenesis. University of Chicago Press, Chicago.

Epel, D., 1963: The effects of carbon monoxide inhibition on ATP level and the rate of mitosis in the sea urchin egg. J. Cell Biol. *17*:315–319.

Erikson, R. O., 1964: Synchronous cell and nuclear division in tissues of the higher plants. *In* Synchrony in Cell Division and Growth. Zeuthen, ed. Interscience Publishers, New York, pp. 11–37.

Fawcett, D. W., Ito, I., and Slautterback, D., 1959: The occurrence of intercellular bridges in groups of cells exhibiting synchronous division. J. Biochem. Biophys. Cytol. *5*:453–460.

Frankel, P., 1965: The effect of nucleic acid antagonists on cell division and oral organelle development in *Tetrahymena pyriformis.* J. Exp. Zool. *157*: 113–147.

Gaulden, M. E., and Perry, R. P., 1958: Influence of nucleolus on mitosis as revealed by ultraviolet microbeam irradiation. Proc. Natl. Acad. Sci. U. S. A. *44*:554–559.

Gelfant, S., 1963a: Inhibition of cell division. Int. Rev. Cytol. *14*:1–39.

Gelfant, S., 1963b: A new theory of the mechanism of cell division. *In* Cell Growth and Cell Division. Harris, ed. Symposium of the International Society for Cell Biology. Vol. 2. Academic Press, New York, pp. 229–259.

Gray, L. H., 1959: Cellular radiobiology. Radiat. Res. *1*(Suppl.):73–101.

Gross, P., and Mazia, D. (eds.), 1960: Second Conference on the Mechanism of Cell Division. Ann. N. Y. Acad. Sci. *90*:345–613.

Guttes, E., and Guttes, S., 1960: Suppression of cytoplasmic division in *Amoeba proteus.* Nature *187*:520–521.

Hamburger, K., and Zeuthen, E., 1957: Synchronous divisions in *Tetrahymena pyriformis* as studied in an inorganic medium. The effect of 2, 4-dinitrophenol. Exp. Cell. Res. *13*:443–453.

Harvey, E. B., 1936: Parthenogenic merogony or cleavage without nuclei in *Arbacia punctulata.* Biol. Bull. *71*:101–121.

Hastings, J. W., and Sweeney, B. M., 1964: Phased cell division in marine dinoflagellates. *In* Synchrony in Cell Division and Growth. Zeuthen, ed. Interscience Publishers, New York, pp. 307–349.

Heilbrunn, L. V., 1952: The physiology of cell division. *In* Modern Trends in Physiology and Biochemistry. Barron, ed. Academic Press, New York, pp. 123–134.

Heilbrunn, L. V., 1955: Dynamics of Living Protoplasm. Academic Press, New York.

Hiramoto, Y., 1956: Cell division without mitotic apparatus in sea urchin eggs. Exp. Cell Res. *11*: 630–636.

Hiramoto, Y., 1963: Mechanical properties of sea urchin eggs. II. Changes in mechanical properties from fertilization to cleavage. Exp. Cell Res. *32*:76–89.

Hiramoto, Y., 1965: Further studies on cell division without mitotic apparatus in sea urchin eggs. J. Cell Biol. *25*:161–167.

Hoffmann-Berling, H., 1960: Other mechanisms producing movements. Comp. Biochem. *2*:341–370.

Holz, G. G., Rasmussen, L., and Zeuthen, E., 1963: Normal versus synchronized division in *Tetrahymena pyriformis.* Comp. Rend. Lab. Carlsberg *33*:289–300.

Hughes, A. F. W., 1950. The effect of inhibitory substances on cell division. A study of living cells in tissue cultures. Quart. J. Micr. Sci. *91*:251–278.

Iverson, R. M., and Giese, A. C., 1957: Nucleic acid content and ultraviolet susceptibility of *Tetrahymena pyriformis.* Exp. Cell Res. *13*:213–223.

Jacobson, W., and Webb, M., 1952: Nucleoproteins and cell division. Endeavour *11*:200–207.

Kane, R. E., 1962: The mitotic apparatus—isolation by controlled pH. J. Cell Biol. *12*:47–55.

Kelner, A., 1953: Growth, respiration and nucleic acid synthesis in ultraviolet-irradiated and in photoreactivated *Escherichia coli.* J. Bact. *65*:252–262.

*Kihlman, B., 1966: Actions of Chemicals on Dividing Cells. Prentice-Hall, Englewood Cliffs, N. J.

Kimball, R. F., Caspersson, T. O., Svensson, G., and Carlson, L., 1959: Quantitative cytochemical studies on *Paramecium aurelia.* I. Growth in the dry weight measured by the scanning interference microscope and x-ray absorption methods. Exp. Cell Res. *17*:160–172.

Lovlie, A., 1963: Growth in mass and respiration rate during the cell cycle of *Tetrahymena pyriformis.* Compt. Rend. Lab. Carlsberg *33*:377–415.

Marsland, D., 1957: Temperature-pressure studies on the role of sol-gel reactions in cell division. *In* Influence of Temperature on Biological Systems. Johnson, ed. American Physiological Society, Washington, D. C., pp. 111–126.

Marsland, D. A., and Landau, J. V., 1954: The mechanism of cytokinesis—temperature pressure studies on the critical gel system in various marine eggs. J. Exp. Zool. *125*:507–539.

Maruyama, Y., 1964: Synchrony of bacterial populations as established by filtration techniques. *In* Synchrony of Cell Division and Growth. Zeuthen, ed. Interscience Publishers, New York, pp. 421–432.

Mazia, D., 1954: The organization of the mitotic apparatus. Symp. Soc. Exp. Biol. *9*:335–357.

Mazia, D., 1956a: Materials for the biophysical and biochemical study of cell division. Advances Biol. Med. Phys. *4*:69–118.

Mazia, D., 1956b: The life history of the cell. Am. Sci. *44*:1–32.

Mazia, D., Sept. 1961: How cells divide. Sci. Am. *205:* 100–120.

*Mazia, D., 1961: Mitosis and the physiology of cell division. *In* The Cell. Brachet and Mirsky, eds. Academic Press, New York, Vol. 3, pp. 77–412.

Mazia, D., 1963: Synthetic activities leading to mitosis. J. Cell. Comp. Physiol. *62*:123–140.

Mazia, D., Chaffee, R. R., and Iverson, R. M., 1961: Adenosine triphosphatase in the mitotic apparatus. Proc. Natl. Acad. Sci. U.S.A. *47*:788–790.

Mazia, D., and Gontcharoff, M., 1964: The mitotic behavior of chromosomes in Echinoderm eggs following incorporation of bromodeoxyuridine. Exp. Cell Res. *35*:14–25.

Mazia, D., Harris, P. J., and Bibring, T., 1960: The multiplicity of mitotic centers and the time course of their duplication and separation. J. Biochem. Biophys. Cytol. *7*:1–20.

Miller, C. O., Skoog, F., Von Saltza, M. H., and Strong, F. M., 1955: Kinetin, a cell division factor from deoxyribonucleic acid. J. Am. Chem. Soc. *77*:1392.

Mitchison, J. M., 1952: Cell membranes and cell division. Symp. Soc. Exp. Biol. *6*:105–127.

Nachtwey, D. S., 1965: Division of synchronized *Tetrahymena pyriformis* after emacronucleation. Compt. Rend. Lab. Carlsberg *35*:23–25.

Plesner, P., 1961: Changes in ribosome structure and function during synchronized cell division. Cold Spring Harbor Symp. Quant. Biol. *26*:159–162.

Plesner, P., 1964: Nucleotide metabolism during synchronized cell division in *Tetrahymena pyriformis.* Compt. Rend. Lab. Carlsberg *34*:1–76.

Prescott, D., 1956: Relation between cell growth and cell division. III. Changes in nuclear volume and growth rate and prevention of division in *Amoeba proteus* resulting from cytoplasmic amputation. Exp. Cell Res. *11*:94–98.

Prescott, D. M., 1964: The normal cell cycle. *In* Synchrony in Cell Division and Growth. Zeuthen, ed. Interscience Publishers, New York, pp. 71–97.

Puck, T. T., Aug. 1957: Single human cells in vitro. Sci. Am. *197*:91–100.

Puck, T. T., 1960: *In vitro* studies on the radiation biology of mammalian cells. Progr. Biophys. *10*: 237–258.

Rapkine, L., 1931: Chemical processes of cell division. Ann. Physiol. Physiochim. Biol. *7*:382–417.

Rasmussen, L., 1963: Delayed divisions in *Tetrahymena* as induced by short-time exposures to anaerobiosis. Compt. Rend. Lab. Carlsberg *33*: 53–71.

*Roberts, H. S., 1961: Mechanism of cytokinesis: a critical review. Quart. Rev. Biol. *36*:155–177.

Robinow, C. F., 1960: Outline of the visible organization of bacteria. *In* The Cell. Brachet and Mirsky, eds. Academic Press, New York, Fol. 4, pp. 45–108.

*Scherbaum, O. H., 1960a: Possible sites of metabolic control during the induction of synchronous cell division. *In* Second Conference on Mechanism of Cell Division. Gross and Mazia, eds. Ann. N. Y. Acad. Sci. *90*:565–579.

Scherbaum, O., 1960b: Synchronous division of microorganisms. Ann. Rev. Microbiol. *14*:283–310.

Scherbaum, O. H., and Loefer, J. B., 1964: Environmentally induced growth oscillations in protozoa. Biochem. Physiol. Protozoa *3*:9–59.

Scholander, P. F., Leivestad, H., and Sundnes, G., 1958: Cycling in the oxygen consumption of cleaving eggs. Exp. Cell Res. *15*:501–511.

Selman, G. G., and Waddington, C. H., 1955: The mechanism of cell division in the cleavage of the newt's egg. J. Exp. Biol. *32*:700–733.

Spoerl, E., and Looney, D., 1959: Synchronized budding of yeast cells following x-irradiation. Exp. Cell Res. *17*:320–327.

Stern, H., 1956: The physiology of cell division. Ann. Rev. Plant Physiol. *7*:91–114.

Stocken, L. A., 1959: Some observations on the biochemical effects of x-radiations. Proc. Int. Congr. Rad. Res. Burlington, Vt., Rad. Res. *1*(Suppl.):53–72.

Swann, M. M., 1953: The mechanism of cell division. A study with carbon monoxide on the sea-urchin egg. Quart. J. Micr. Sci. *94*:369–379.

Swann, M. M., 1954: The control of cell division. *In* Seventh Colston Symposium: Recent Developments in Cell Physiology. Kitching, ed. Academic Press, New York, pp. 185–196.

Swann, M. M., and Mitchison, J. M., 1958: The mechanism of cleavage in animal cells. Biol. Rev. *33*:103–135.

Sweeney, B. M., and Hastings, J. W., 1958: Rhythmic cell division in populations of *Gonyaulax polyedra.* J. Protozool. *5*:217–224.

Sylvan, B., Tobias, C. A., Malmgren, H., Otteson, R., and Thorell, B., 1959: Cyclic variations in the peptidase and catheptic activities of yeast cultures synthesized with respect to cell multiplication. Exp. Cell Res. *16*:75–87.

Szent-Györgyi, A., 1965: Cell division and cancer. Science *149*:34–37.

Szent-Györgyi, A., Egyud, L. G., and McLaughlin, J. A., 1967: Keto-aldehydes and cell division. Science *155*:539–541.

Tamiya, H., 1964: Growth and cell division of *Chlorella. In* Synchrony in Cell Division and Growth. Zeuthen, ed. Interscience Publishers, New York, pp. 247–305.

Taylor, E. W., 1959: Dynamics of spindle formation and its inhibition by chemicals. J. Biophys. Biochem. Cytol. *6*:193–196.

Taylor, J. H., 1963: DNA synthesis in relation to chromosome reproduction and the reunion of breaks. J. Cell. Comp. Physiol. *62*(Suppl. 1):73–86.

Watanabe, Y., 1965: Isolation of characterization of the division protein in *Tetrahymena pyriformis.* Exp. Cell Res. *39*:443–452.

*Wilson, G. B., 1967: Cell Division and the Mitotic Cycle. Reinhold Publishing Corp., New York.

Wolpert, L., 1960: The mechanics and mechanism of cleavage. Int. Rev. Cytol. *10*:163–216.

Wolpert, L., 1963: Some problems of cleavage in relation to the cell membrane. *In* Cell Growth and Cell Division. Harris, ed. Academic Press, New York, pp. 277–298.

Zeuthen, E., 1949: Oxygen consumption during mitosis; experiments on fertilized eggs of marine animals. Am. Nat. *83*:303–322.

Zeuthen, E., 1958: Artificial and induced periodicity in living cells. Advances Biol. Med. Phys. *4*:37–73.

Zeuthen, E., 1961: Cell division and protein synthesis. *In* Biological Structure and Function. Goodwin and Linberg, eds. Academic Press, New York, Vol. 2, pp. 537–548.

Zeuthen, E., 1964a: The temperature-induced division synchrony in *Tetrahymena. In* Synchrony in Cell Division and Growth. Zeuthen, ed. Interscience Publishers, New York, pp. 99–158.

Zeuthen, E. (ed.), 1964b: Synchrony in Cell Division and Growth. Interscience Publishers, New York.

Zeuthen, E., and Scherbaum, O., 1954: Synchronous division in mass culture of the ciliate protozoan *Tetrahymena pyriformis,* as induced by temperature changes. *In* Recent Developments in Cell Physiology. Kitching, ed. Academic Press, New York, pp. 141–157.

Zimmerman, A. M., 1963: Chemical aspects of the isolated mitotic apparatus. *In* The Cell in Mitosis. Levine, ed. Academic Press, New York, pp. 159–179.

Zimmerman, A. M., Landau, J. V., and Marsland, D., 1957: Cell division: pressure-temperature analysis of effects of sulfhydryl reagents on cortical plasma gel structure and furrowing strength of dividing eggs (*Arbacia* and *Chaetopterus*). J. Cell. Comp. Physiol. *49*:395–435.

SECTION
VII

# Trends in Cell Physiology

# CHAPTER 28 / TRENDS IN CELL PHYSIOLOGY: AN OUTLINE

When the subject of cell physiology is presented, one cannot help inquiring into its origin and development, yet no comprehensive historical account is available. Perhaps such an account is lacking because the subject is little more than a hundred years old, and it is difficult to gain perspective in so short a time. The summary that follows here is an attempt to outline the origins of some of the major trends in cell physiology, rather than to give a chronological history of it, and it was written with the feeling that these developments may be seen in quite a different light in another few years.

Three phases in the history of cell physiology might be recognized: the first, an exploratory period which began with the acceptance of the cell doctrine and continued through the beginning of the twentieth century; the second, an analytical period which has occupied the time since; and the third, a new period of development, just beginning, which might be best characterized as molecular biology in the broadest sense—interpretation of all cell activities at the molecular level. In the limited space available here it is impossible to name all the investigators of merit in the field of cell physiology. Only a few are chosen as examples, with the understanding that those omitted are not necessarily of lesser merit.

It is perhaps self-evident that cell physiology, as a separate field of study, could develop only after the cell was recognized as a unit of structure in plant and animal organisms, but many of the sources of cell physiology are of much older origin.

The *cell doctrine* was fairly widely accepted by 1839, although it would be difficult to give the credit for its inception to any one or two individuals. Even before that time, some workers had pointed out that the cell had a sort of double existence as part of an organism and as an independent unit, and Schleiden wrote, "It is, however, apparent that the vital process of the individual cell must form the very first absolutely indispensable basis of vegetable physiology and comparative physiology" (Conklin, 1940a). It has been pointed out that Schleiden and Schwann contributed to the development of the cell doctrine primarily by publicizing the idea and bringing to a culmination the work of a large number of other investigators who already clearly understood the cell to be a unit of structure (Conklin, 1940b; Baker, 1948, 1949, 1952, 1953, 1955).

Before 1839 biological research had progressed independently in a variety of directions in zoology, botany and physiology. The cell doctrine was the first of the great unifying principles that caused botanists, zoologists and physiologists to realize that organisms—whether animals or plants—and their functions were not quite so unrelated as had been previously supposed, and that there is unity in life, a concept which was to grow as other principles of biology were developed.

In the period following 1839 in connection with cell physiology two other new fields of biological study arose, histology and cytology, initially concerned with the structure of tissues and of cells, respectively, the latter centering on the structure and behavior of the nucleus. Cell physiologists continued studies on cytoplasmic structure and used both cells and tissues for experiments but primarily as a means of interpreting the way in which the cell performs its functions.

The vast fund of information amassed during the early period indicates that cell physiology was preoccupied with the study of various types of cells and cell organelles, and with the development of experimental techniques to study the properties and behavior of cells under a variety of conditions and to determine the types of substances that enter and leave the cell (Singer, 1950). Various kinds of movements made by cytoplasm, such as streaming or movement by means of cilia, flagella, pseudopodia or massed cells in various types of muscles, had to be studied and characterized. The nature of the excitability process in diverse organisms and the role of the individual cells in these responses of organisms had to be ascertained. The nature of cell division and cell growth had to be investigated and the information codified.

All these objectives and the investigations performed indicate that the point of view introduced by Descartes (1596–1650) and developed by others—namely, that the phenomena of life are explainable in terms of the *organism as a machine* expending energy, rather than in terms of supernatural forces—had been accepted by this time (Nordenskjold, 1936). The cell was spoken of as a unit of function, but few did more than assume this fact.

Perhaps the earliest studies of the functions of cells were those by Purkinje (1787–1869), who introduced the term protoplasm, on the physical properties of protoplasm, and those by others on cell division in connection with the chromosome cycle. In 1858 Gerlach introduced the staining technique which greatly stimulated cytological studies, culminating in the interpretation of the chromosomal behavior in *mitosis* and *meiosis* (Strasburger, 1875; Farmer and Moore, 1905), covering a span of about 65 years following the acceptance of the cell doctrine. Toward the end of the nineteenth century the botanists Pfeffer (1877) and Overton (1902) used plant cells in studies on permeability. Some embryologists began manipulating single blastomeres of larger embryos (1895). Soon (1914–1937) micromanipulators were devised, enabling workers (Kite, Barber, Seifritz, Taylor, Chambers, Peterfi, DeFonbrunne) to handle single small cells and even to dissect them. These workers were thereby able to demonstrate that many of the cell organelles seen in stained preparations were not mere artifacts produced by fixation and staining.

In parallel with these events, *organic chemistry* was developed and was to contribute tools of vast importance to cell physiology. Wöhler synthesized urea in 1828, beginning the study of organic syntheses and organic chemistry which, at its inception at least, was concerned with biological compounds and was therefore closely allied to biology. The great work of Liebig (1803–1873) demonstrated some of the food transformations and contributed methods of organic analysis of importance to cell physiology. Various organic acids occurring in plant and animal cells were identified and synthesized. Proteins were recognized and characterized as early as 1838 (Mulder, Boussingault). Various *amino acids* were isolated and then synthesized, and by 1842 a difference between plant and animal proteins was recognized. It was further determined that the animal cell remakes its proteins from amino acids found in the digest of plant proteins (Johnson). In 1902 the *peptide bond* was discovered independently by Fischer and Hofmeister, and it was recognized that the protein molecule could be made from a relatively small number of amino acids. At about this time Ostwald (1853–1932) studied catalysts, and the concept of catalytic action was accepted into biological thinking (Fischer, Nobel Prize, 1902) to explain the many oxidations and syntheses found occurring in animal and plant cells (Betrand) (Dawes, 1952).

Miescher's researches on nucleic acid, from 1868 to 1892, laid the foundation for the study of these cellular constituents, the importance of which has been fully realized and exploited only in recent years (Dawes, 1952).

In the meantime, techniques for handling cells were continually improved and many pieces of physical equipment for measuring and recording changes in organs and cells were adopted. In this respect, the physiologist Ludwig (1816–1895), one of the great teachers of his time, is outstanding. Many pieces of equipment used even today in elementary physiology courses are traceable to his adaptations and inventions. Du Bois-Reymond (1818–1896) devised methods to measure potentials and action currents in animal cells. Contributions were made and ways were found to measure muscle tension and electrical changes on nerve and muscle and in individual cells. The discovery of x-rays (1895), radioactivity (1898) and other phenomena in physics stimulated researches in cell physiology but the main impact of these discoveries was to be felt only later.

A great impetus to many fields of biological research was the foundation of a number of marine laboratories, where investigators from all parts of the world could study together and exchange ideas. Four such laboratories of especial note are the first marine laboratory, established at Concarneau, France, in 1859; the international biological station at Naples, Italy, established by Anton Dohrn about 14 years later; the Marine Biological Laboratory at Woods Hole, Massachusetts, which succeeded the small laboratory set up by Louis Agassiz on the island of Penikese in 1873; and the Plymouth Laboratory of the Marine Biological Association, founded in England in 1888.

The main theoretical efforts of the period were attempts to formulate analogues to visualize or duplicate the structure and the operation of the cell, but the information available at the time was inadequate for more than suggestive models (Barnes, 1937).

Various conceptions were proposed to account for the structure of cytoplasm. Flemming (1882) thought cytoplasm was fibrillar, Altmann (1893), that it was granular; and Bütschli (1894), that is was alveolar; E. B. Wilson (1925) thought it was a colloid. The last of these conceptions makes room for all the others, since it can account for granules, fibrils and emulsified materials, all of which occur in the cell. The Brownian movement (discovered in 1827) in portions of it suggested a liquid. In 1899 Hardy and Fischer independently showed that some of the structures seen in fixed preparations of cytoplasm could be duplicated by action of fixatives upon albumin.

The *cell membrane* was conceived as a lipoidal film by Overton (1902). Michaelis (1929) later made collodion analogues of the cell membrane to study the dynamics of movements of substances through it. Using iron wire, Lillie (1929) made a model that shows many of the characteristics of the nerve impulse. From such a model Lillie predicted saltatory conduction in nerve fibers, which was later established in living myelinated nerve fibers. A model to show the origin of potentials on the nerve fibers was made of a collodion sac filled with a salt solution. On the basis of such findings and others Bernstein (1902) developed the membrane theory of resting and action potentials. A variety of models for muscular contraction, ciliary movement, ameboid movement, cell division and even for contraction of the heart appeared during this time. Some of the models are ingenious and elaborate and are still useful pedagogically (Bayliss, 1931; Barnes, 1937).

Many attempts were made to search for some particles in the cell which were as fundamental to an explanation of the properties of life as are the atoms to an explanation of the properties of matter in the physical world. Perhaps one of the most notable of these was the *biogen hypothesis* of Verworn (1863–1921). The biogen was conceived of as a hypothetical albuminoid molecule whose chief characteristic consisted of high energy content and chemical lability, "a constant and simultaneous alternation of disintegration and reconstruction." This property, according to Verworn, is the true essence of life, there being no difference between animate and inanimate matter except that brought about by the very extraordinary metabolic possibilities of the biogen molecule. His book *Allgemeine Physiologie* and his monograph on *Irritability* included many important contributions from his laboratory. If, in retrospect, the general theories developed at this time are often inadequate, in terms of available information they were worthy accomplishments.

During the next period, the analytical period of development of cell physiology, the newer and more precise methods of the chemist and physicist were applied to biological problems. This approach was the basis of the famous textbook by Höber, called the *Physical Chemistry of Cells and Tissues,* the first German edition of which appeared in 1902. An edition of it in English with the collaboration of a number of eminent American physiologists appeared in 1945. Perhaps a quotation from the preface to the last edition will serve to illustrate the tenets of general physiology at this time: ". . . not physiology originated essentially to fill human needs and help suffering individuals, but physiology as a branch of physical science dealing with life as a physical, though exceedingly complex, system that may be subjected to scientific analysis like any other natural object." Bayliss' equally famous *General Physiology** was modeled along somewhat similar lines, but in addition treated the effects of light on organisms, as well as studies on many aspects of what is essentially mammalian physiology. Both these books had a considerable amount of influence on training and thinking in the field of cell physiology. Another influential book in cell physiology has been Heilbrunn's *An Outline of General Physiology,* which first appeared in 1937.

At the time of World War I, the concept of molecular orientation at surfaces was developed by Langmuir, Hardy and Jarkins. This concept was to have far-reaching effects in research of cell activities and organization of cell membranes. By *x-ray diffraction,* it became possible to determine the size of regularly arranged units in such structures as cell membranes, intact tendon and muscle cells. With the *polarizing microscope,* the probable orientation of molecules in cell structures was demonstrated. The application of the *phase microscope,* invented in 1934 by Zernicke, in making observations needs no further emphasis here. With it many structures of the living cell have been clearly identified, and cyclic changes have been readily followed for the first time. After the invention of the *electron microscope* in 1938, the size and shape of many submicroscopic structures in the cell and many other molecules or molecular aggregates of biological interest could be determined.

*Electronic methods* adapted to the investigation of biological problems also have been of tremendous importance. These studies were begun in 1922 by Gasser and Erlanger who made use of the oscilloscope to investigate features of the nervous impulse not previously susceptible to analysis by slower methods.

In recent years, following the pioneering work of Schoenheimer (1942), radioactive tracers have facilitated investigation of many fundamental problems in the metabolism of cells and in the dynamic aspects of bodily and cellular constituents. The tracer technique has led to many other important discoveries in cell physiology, for example, by use of radioactive sodium, Brooks demonstrated that it enters the cell, exchanging places with its nonradioactive isotope. Again Calvin, Benson and Bassham have used tracers to determine the pathway of synthesis of

*A new edition of this book has recently appeared in two volumes. See Literature Cited at the end of the chapter.

carbon compounds in photosynthesis (Arnon, 1961), for which Calvin received the Nobel Prize in 1961.

*Biochemistry,* also, has led to the elucidation of the degradation of nutrients and the dynamics of energy liberation in cells. The field of *enzyme chemistry* has had a spectacular development and the Nobel Prize in 1955 went to Theorell for his studies on enzymes of respiration. Syntheses as well have been studied in cells, and many enzymes and enzyme reactions in cell activities have been studied by a large number of workers. The enzymes and coenzymes concerned with these activities have been isolated and purified, following pioneer experiments by Fischer. Protein chemistry has also shown great progress; e.g., Pauling's contribution to the knowledge of the structure of the protein molecule won him the Nobel Prize in 1954. Sanger's analysis of the sequence of amino acids in a protein, for which he received the Nobel Prize in 1958, opened the way for "fingerprinting" proteins with an opportunity to analyze how genes work. The study of nucleic acids has been intensely stimulated by the greater insight achieved by the Watson-Crick double-helix model of DNA for which they and Wilkins received the Nobel Prize in 1962. The synthesis of DNA and RNA *in vitro* by Kornberg and Ochoa, respectively, for which they received the Nobel Prize in 1959, makes possible an attack on the most basic of biological problems—the nature of inheritance—at the molecular level.

Perhaps the evolution in thinking and conceptions in cellular problems during the development of cell physiology can be best brought out by some examples. Sixty years ago mitochondria would have been described as cell organelles which stain specifically with Janus green B (Michaelis, 1900). They were known to vary in size, shape, position and number, depending on cellular activity and treatment of the preparation. They were thought to play some part in cell respiration, but the issue was controversial (Kingsbury, 1912). Some workers even thought that mitochondria were symbiotic microorganisms. At present, mitochondria, which may be separated from other constituents of cells by the fractional centrifugation of ruptured cells (Hoogeboom, Schneider and Palade, 1948), are known to be cell organelles bounded by selectively permeable membranes. Washed free of contaminants, mitochondria have been analyzed and found to consist primarily of proteins, lipids and ribonucleic acid, the proteins being present in greatest proportion. In fragmented mitochondria many of the enzymes concerned with aerobic oxidations in the cell have been identified, as have also many of the enzymes concerned with phosphorylations and transfer of high energy phosphate bonds (Lehninger and Kennedy, 1948). It is thought that in the mitochondria occur all the reactions of the Krebs tricarboxylic acid cycle, as a result of which most of the energy required for the work of the cell is liberated. Palade (1940) demonstrated by electron microscopy that the structure of a mitochondrion consists of a matrix in which are found cristae or lamellae which give a large surface for enzyme activity, and the positions of some of the enzymes in a mitochondrion have been identified.

In like fashion, the vague conception of protoplasm as a colloid has given place now to the more critical physical-chemical appraisal of the types of forces and bonds that give cytoplasm its rigidity and at the same time its fluidity (Frey-Wyssling, 1948; Bernal, 1935). In fact, the word protoplasm has lost its meaning now as the organization of the cell into compartmented organelles and their functions have become clearer.

The early workers considered cell membranes to be lipoidal films (Overton, 1902), and cells to be simple osmometers. At the present time, data are available from many lines of inquiry to show that cell membranes contain proteins, lipids (Parpart, 1940) and other materials. Furthermore, the arrangement of these substances in the membrane has been ascertained to some extent (Schmitt, Davson, Danielli, Wolpers 1930–1940). Studies on permeability have demonstrated that although some materials pass through the cell membrane in response to a concentration gradient, many ions and molecules move against concentration gradients, and are taken up actively with expenditure of energy by the cell (Blinks, Steinbach, Rothstein, Ussing, 1935–1950). For example, potassium is often accumulated inside cells in greater amounts than are present outside the cells, and many wastes are carried into the kidney tubules against a concentration gradient (Richards, Homer Smith, Marshall, Wearn, 1914–1938; see Smith, 1943). The cell membrane is now viewed, therefore, as dynamic and changeable. Some enzymes have been located on the cell surface, which also shows that it is apparently an active structure rather than just a selective sieve.

A great change has also taken place in the conceptions of biological oxidation-reduction reactions and of energy liberation in the cell. The early work of Lavoisier (1780) and the later studies of Liebig (1803–1873) had demonstrated that animals liberate energy from foodstuffs. About 1903, Wieland and Thunberg emphasized the need for activation of hydrogen, and Wieland (Nobel Prize, 1927) made models, with activated platinum as a catalyst, for the dehydrogenation of organic molecules. He found that vigorous oxidations occur only if a substance which can carry hydrogen to oxygen is added.

Warburg (1908) emphasized that activation of oxygen precedes oxidations in cells. He made models of cellular oxidation, using as a catalyst animal blood charcoal, which contains iron. Many organic materials, shaken with such charcoal in the presence of oxygen, were oxidized. Warburg (Nobel Prize, 1931) insisted that the iron activated the oxygen, making possible its reaction with substrate molecules, and he postulated an iron-containing, oxygen-activating enzyme, which was later isolated by others and is now known as cytochrome oxidase.

Michaelis (1935) next elucidated the stepwise nature of electron transfer in biological oxidation-reduction reactions. Since that time biochemists and cell physiologists have succeeded in identifying the steps in the decomposition of cell nutrients, especially in yeast and muscle cells. The enzymes concerned in most of the steps have been identified, isolated and in some cases purified. The two Nobel Prize winners in medicine and physiology in 1953 were Krebs and Lipmann, who did much to elucidate the cyclic changes in metabolism, but many others contributed to the discoveries which enabled them to relate all the reactions into the cyclic oxidations. The function of the B vitamins as prosthetic groups in the enzymes has in many cases been clarified, and attempts are being made to put the enzymes together on particles in order to simulate their action in the sequences operative in the cell (Green, 1950).

Fundamental researches on muscular contraction by Hill and Meyerhof, especially on the heat production and the lactic acid cycle in the anaerobic metabolism of muscle cells, led to the awarding of the Nobel Prize to these men in 1922. Aerobic muscle metabolism and the enzymes involved have also been studied (Krebs, Warburg, Meyerhof and Embden in the twenties and thirties). In the last two decades the structural proteins of muscle have been isolated, and two of these, actin and myosin, have been found to combine in various stoichiometric ratios (Weber, Edsall, Szent-Györgyi and Mommaerts in the forties and fifties). Compressed films of actomyosin with oriented molecules, which contract to lift a load when placed in salt solutions containing adenosine triphosphate and relax when placed in other salt solutions, show many, though not all, of the properties of the contractile fibrils of muscle (Hayashi, 1952). H. E. and A. F. Huxley and J. Hanson (1954), on the basis of electron micrographic studies of ultrathin sections of muscle fibers and of x-ray diffraction studies, have demonstrated that myosin and actin occur in the muscle in the form of separate filaments. They postulate contraction to be a stero-spatial sliding of the two kinds of filaments as they react in the presence of ATP. The concept is in accord with most of our present knowledge of muscular contraction.

Studies on action potentials and injury potentials were made by Nernst (1899),

Bernstein (1913) and Höber (1910). The work of Lucas (1906) and Adrian (1912) developed many of the basic concepts of the excitability process, and that of Sherrington (1892) and Pavlov (1910; Nobel Prize, 1904) formulated reflexes as functions of the nervous system. With advances in knowledge of electronic devices, fundamental researches on the action potential and nerve impulse at the cellular level were made possible by the use of the cathode ray oscilloscope. The work of many distinguished investigators (among them Osterhout, Blinks, Bronk, Curtis, Cole, Gerard, Grundfest, Hodgkin, Huxley, Katz, Kuffler, Nachmansohn, Rushton and Wiersma) has led to a better understanding of resting and action potentials of cells, especially the large plant cells and the squid giant axon, on which it is easier to make electrical measurements. Gasser and Erlanger received the Nobel Prize in 1944 for their pioneering studies on the nerve impulse for which they adapted the cathode ray oscilloscope. Hodgkin, Huxley and Eccles received the Nobel Prize in 1963 for experimental support of the basic theory of resting and action potentials. Wald, Hartline and Granit received the Nobel prize in 1967 for their elucidation of visual responses.

The development of biochemical genetics, pioneered by Beadle and Tatum (1941), for which they received the Nobel Prize jointly with Lederberg in 1958, has led to a better understanding of biosyntheses. Selected mutants, each lacking capacity to produce some required nutrient, have made possible precise determinations of many biosynthetic pathways, adding a powerful tool to other biochemical techniques for studies of metabolic deficiencies. We now know the steps by which cells make many amino acids and many vitamins, and we know some of the steps in the formation of purines and pyrimidines. All these advances, as well as the elucidation of the synthesis of nucleotides, nucleic acids (Kornberg and Ochoa, Nobel Prize, 1959) and proteins (Ingram, Magasanik, Dintzis, Zamecnik, Spiegelman, Tissieres, Brenner, Meselson and others) and the spectacular deciphering of the genetic code (Nirenberg, Matthei, Gamow, Khorana, Ochoa, Holley, Watson, Crick and others, 1954–1965), have helped to define a new field of study in cell physiology. The field, currently called *molecular biology,* has continued to focus on the problems of biochemical genetics, but problems in all fields of cell physiology are being attacked at the molecular level.

Startling advances in understanding photosynthesis have come as a result of the application of comparative biochemical approaches (van Niel, 1931), and of radioactive tracers (Calvin, Bassham, Benson and others). Recently, most of the enzymes (Racker, 1955) and many of the conditions required for both the photochemical and the thermochemical reactions in photosynthesis have been analyzed, and the biochemical pathway has been indicated. For his analysis of the dark reactions in photosynthesis, Calvin received the Nobel Prize in 1961. It has been shown that the chloroplast alone is capable of photophosphorylation and photoreduction (Arnon, 1955). The cytochromes appear to be involved in photophosphorylation and their activity is currently being studied with rapid spectrophotometric methods (Duysens, Kok, San Pietro, Jagendorff and others).

Newer techniques have made possible studies on cell division and of the complex series of reactions involved therein. Here might be mentioned the isolation and analysis of the division spindle and the study of the trigger mechanism that initiates cell division (Mazia, 1954). Development of a synthetic medium for culture of plant cells (White, 1953) and the development of a similar one for animal cells offer further opportunities for an analysis of cell division.

All over the world in both the western (Hughes, 1959) and eastern countries (Christman, 1952; Suskind, 1960; Setlow and Setlow, 1961) attempts are made to construct for each cellular structure a molecular concept from the chemical constituents of that structure, and to analyze the operation of each cell organelle in terms of physics and chemistry by any tool available. A few books representing steps in this direction are: *Modern Trends in Physiology and Biochemistry* (1952), honoring Michaelis, edited by Barron and containing articles by various leaders in their fields, and the book by Johnson, Eyring and Polissar called *The Kinetic Basis of Molecular Biology* (1954). Several other excellent books written along these lines have also appeared (Setlow and Pollard, 1962; Clayton, 1965; Lehninger, 1965; Watson, 1965).

Perhaps at this time we must stress that the role of cells as parts of an organism should not be forgotten and that the fundamental aspects of organismal development are cell problems also (Picken, 1960; Bell, 1965).

Cell physiology during its development has contributed to the thought of the times and has reinforced the mechanistic concept of life, since the behavior of cells has proved to be explainable in terms of the principles of chemistry and physics. For example, utilization of energy and matter in the cell appears to be governed by the principles of conservation of energy and matter that govern the physical world.

A second contribution of cell physiology is the observation that basic cell activities, and constituents, are remarkably similar in cells of all organisms, no matter how diverse the organisms may be. Many of the cell organelles are common to all eukaryotic cells and are alike in chemistry and function; for example, the plasma membrane, the nucleus and its constituents, the mitochondria and the microsomes are among the oldest structural homologies of the living world. Many of the chemical constituents of cells are universal. This is true not only of the proteins, lipids and carbohydrates, but also of substances present in only catalytic quantities. For example, vitamin $B_1$ has been found in all organisms assayed and it forms part of the enzyme system involved in conversions of that universal metabolite, pyruvic acid. Again, the flavoproteins, which carry hydrogen from a dehydrogenase to some hydrogen acceptor, are omnipresent, and according to Barron (1952), represent the most archaic enzymes—presumably having been present in the primeval anaerobic cells. For their synthesis, flavoproteins require vitamin $B_2$, which the animal gets from plants, or from microorganisms in its gut. Szent-Györgyi (Nobel Prize, 1937), speaking of the role of vitamins and other chemicals universally required in cells but in many cases obtained by an animal cell from plant food, says: "This simple fact involves a point of greatest philosophical importance. If I look upon the cell as a mechanism and upon the molecule as a wheel of this mechanism, then by saying that I take my vitamins from the plant, I say that there are two mechanisms, the plant cell and my cells, whose parts, the single wheels, are interchangeable. Two mechanisms, whose parts are interchangeable, cannot be very different. This is the first scientific evidence for the great, fundamental chemical unity of living Nature. There is no real difference between cabbages and kings, we are all recent leaves on the old tree of life" (Szent-Györgyi, 1940).

It goes without saying that advances in knowledge of cell physiology have not always followed along a perfectly smooth and continuous curve. Decisive advances have been the result of the operation of several factors: (1) choice of some exceedingly favorable experimental cell material in which one kind of phenomenon can be studied with a minimum of other biological complications, e.g., the use of *Neurospora* in biosynthetic studies; (2) advances in chemistry and physics or other related fields of study, resulting in parallel developments in cell studies; (3) the development of new experimental tools which make possible new kinds of study and measurement, e.g., radioactive tracer techniques which have played so vital a role in investigations of metabolism, in studies of photosynthesis and also in an analysis of the function of the nucleus; (4) the imagination and thorough scholarship which has enabled an investigator to develop a generalization based on previously uncorrelated information, e.g., the Krebs cycle and the metabolic mill, which together correlate a vast mass of biochemical information and make possible a logical attack on unsolved problems.

The great number of original publications, reviews and symposia on cell physiology are all evidences of an actively growing field.

## LITERATURE CITED AND GENERAL REFERENCES


Annual Reviews, 1965: The Excitement and Fascination of Science. Annual Reviews, Palo Alto, Calif. (essays by eminent physiologists).

Arnon, D. I., 1961: Changing concepts of photosynthesis. Bull. Torrey Bot. Club *88*:215–259.

Baker, J. R., 1948: The cell theory: a restatement, history and critique. Quart. J. Micr. Sci. *89*:103–125.

Baker, J. R., 1949: The cell theory: a restatement, history and critique. Part II. Quart. J. Micr. Sci. *90*:87–108.

Baker, J. R., 1951: Remarks on the discovery of cell division. Isis (Cambridge) *42*:285–287.

Baker, J. R., 1952: The cell theory: a restatement, history and critique. Part III. The cell as a morphological unit. Quart. J. Micr. Sci. *93*:157–190.

Baker, J. R., 1953: The cell theory: a restatement, history and critique. Part IV. The multiplication of cells. Quart. J. Micr. Sci. *94*:407–440.

Baker, J. R., 1955: The cell theory: a restatement, history and critique. Part V. The multiplication of nuclei. Quart. J. Micr. Sci. *96*:449–481.

Barnes, T. C., 1937: Textbook of General Physiology. Blakiston, New York.

Bayliss, L. E., 1959-1960: Principles of General Physiology. I. The Physico-Chemical Background. II. General Physiology. Longmans Green & Co., New York.

Bayliss, W. M., 1931: Principles of General Physiology. 4th Ed. Longmans Green & Co., New York.

Bell, E., (ed.) 1965: Molecular and Cellular Aspects of Development. Harper & Row, New York.

Christman, R. (ed.), 1952: Soviet Science. American Association for the Advancement of Science, Washington, D. C.

Clayton, R. K., 1965: Molecular Physics in Photosynthesis. Blaisdell Publishing Co., Waltham, Mass.

Commoner, B., 1961: In defense of biology. Science *133*:1745–1748.

Conklin, E. G., 1940a: Predecessors of Schleiden and Schwann. Biol. Symp. *1*:58–66.

Conklin, E. G., 1940b: Cell and protoplasm concepts: historical account. *In* The Cell and Protoplasm. Taylor, ed. The Science Press, Washington, D. C.

Dawes, B., 1952: A Hundred Years of Biology. Macmillan Co., New York.

Franklin, K. J., 1949: A Short History of Physiology. Staples Press, London.

Hardin, G., 1956: Meaninglessness of the word protoplasm. Sci. Monthly *82*:112–120.

Höber, R., 1945: Physical Chemistry of Cells and Tissues. Blakiston, New York.

Hughes, A., 1952: Some historical features in cell biology. Int. Rev. Cytol. *1*:1–8.

Hughes, A., 1959: A History of Cytology. Abelard-Schuman, New York.

Lehninger, A. L., 1965: Bioenergetics. W. A. Benjamin, New York.

Nobel Prize Lectures: Physiology and Medicine. 1901–1962. 3 vol. American Elsevier Publishing Co., New York.

Nordenskjold, E., 1936: The History of Biology. Tudor Publishing Co., New York.

Picken, L., 1960: The Organization of Cells and Other Organisms. Clarendon Press, Oxford.

Sarton, G., 1952: A Guide to the History of Science. Ronald Press Co., New York.

Setlow, R. B., and Pollard, E. C., 1962: Molecular Biophysics. Addison-Wesley Publishing Co., Reading, Mass.

Setlow, R. B., and Setlow, J. E., 1961: Soviet cellular biophysics 1950–60. Quart. Rev. Biol. *36*:1–49.

Singer, C., 1950: A History of Biology. Henry Schuman Publishers, New York.

Singer, C., 1959: A Short History of Scientific Ideas. Clarendon Press, Oxford.

Smith, H. W., 1943: Lectures on the Kidney. University of Kansas, Lawrence, Kans., Renal physiology between two wars, pp. 65–82.

Sturtevant, A. H., 1965: A History of Genetics. Harper & Row, New York.

Suskind, S. R., 1960: Microbial genetics in the USSR. Quart. Rev. Biol. *35*:41–79.

Szent-Györgyi, A., 1940: Vitamins. *In* The Cell and Protoplasm. Taylor, ed. The Science Press, Washington, D. C.

Veith, I., 1954: Perspectives in Physiology. American Physiological Society, Washington, D. C.

Watson, J. D., 1965: Molecular Biology of the Gene. W. A. Benjamin, Inc., New York.

Wright, S., 1959: Genetics and the hierachy of biological sciences. Science *130*:959–965.

Zuckerman, S., 1960: Classics in Biology. Philosophical Library, New York.

# Index

In this index *italic* page numbers indicate illustrations; those followed by (t) indicate tables.